Third Edition

Clinical Exercise Physiology

Jonathan K. Ehrman, PhD, FACSM
Henry Ford Hospital

Paul M. Gordon, PhD, MPH, FACSM
Baylor University

Paul S. Visich, PhD, MPH
University of New England

Steven J. Keteyian, PhD, FACSM
Henry Ford Hospital

Editors

Human Kinetics

Library of Congress Cataloging-In-Publication Data

Clinical exercise physiology / [edited by] Jonathan K. Ehrman, Paul M. Gordon, Paul S. Visich, Steven J. Keteyian. -- 3rd edition.
p. ; cm.
Includes bibliographical references and index.
I. Ehrman, Jonathan K., 1962- editor of compilation. II. Gordon, Paul M., 1960- editor of compilation. III. Visich, Paul S., 1955- editor of compilation. IV. Keteyian, Steven J., editor of compilation.
[DNLM: 1. Exercise Therapy--Case Reports. 2. Exercise--physiology--Case Reports. 3. Exercise Movement Techniques--Case Reports. WB 541]
RM725
615.8'2--dc23

2013000868

ISBN-10: 1-4504-1280-7 (print)
ISBN-13: 978-1-4504-1280-3 (print)

The web addresses cited in this text were current as of April 30, 2013, unless otherwise noted.

Acquisitions Editor: Amy N. Tocco; **Developmental Editor:** Kevin Matz; **Assistant Editor:** Susan D. Huls; **Copyeditor:** Joyce Sexton; **Indexer:** Andrea Hepner; **Permissions Manager:** Dalene Reeder; **Graphic Designer:** Joe Buck; **Graphic Artist:** Denise Lowry; **Cover Designer:** Keri Evans; **Photo Asset Manager:** Laura Fitch; **Photo Production Manager:** Jason Allen; **Art Manager:** Kelly Hendren; **Associate Art Manager:** Alan L. Wilborn; **Illustrations:** © Human Kinetics; **Printer:** Sheridan Books

Printed in the United States of America 10 9 8 7 6 5 4 3 2

The paper in this book is certified under a sustainable forestry program.

Human Kinetics
Website: www.HumanKinetics.com

United States: Human Kinetics, P.O. Box 5076, Champaign, IL 61825-5076
800-747-4457
e-mail: humank@hkusa.com

Canada: Human Kinetics, 475 Devonshire Road Unit 100, Windsor, ON N8Y 2L5
800-465-7301 (in Canada only)
e-mail: info@hkcanada.com

Europe: Human Kinetics, 107 Bradford Road, Stanningley, Leeds LS28 6AT, United Kingdom
+44 (0) 113 255 5665
e-mail: hk@hkeurope.com

Australia: Human Kinetics, 57A Price Avenue, Lower Mitcham, South Australia 5062
08 8372 0999
e-mail: info@hkaustralia.com

New Zealand: Human Kinetics, P.O. Box 80, Torrens Park, South Australia 5062
0800 222 062
e-mail: info@hknewzealand.com

E5501

Contents

Contributors

Helmut Albrecht, MD
School of Medicine
University of South Carolina
Columbia, SC

Ann L. Albright, PhD, RD
Division of Diabetes Translation
Centers for Disease Control
Atlanta, GA

Edward Archer, PhD, MS
Arnold School of Public Health
University of South Carolina
Columbia, SC

Krista A. Barbour, PhD
Scientific Program Director
Sprout Pharmaceuticals, Inc
Raleigh, NC

Tracy Baynard, PhD
Department of Kinesiology and Nutrition
College of Applied Health Sciences
University of Illinois at Chicago
Chicago, IL

Julie Biller, MD
Children's Hospital of Wisconsin
Milwaukee, WI

James A. Blumenthal, PhD
Department of Psychiatry and Behavioral Sciences
Duke University Medical Center
Durham, NC

Brian W Carlin, MD
Drexel University College of Medicine
Philadelphia, PA
Allegheny General Hospital
Pittsburgh, PA

Ivan P. Casserly, MB, BCh
Cardiac and Vascular Center
University of Colorado Hospital
Aurora, CO

Linda H. Chung, PhD
Department of Kinesiology
University of Massachusetts
Amherst, MA

A.S. Contractor, MD
Department of Preventive Cardiology and Rehabilitation
Asian Heart Institute
Bombay, India

Michael J. Danduran, MS
Physical Therapy Department
Marquette University
Milwaukee, WI

Paul G. Davis, PhD
Department of Kinesiology
University of North Carolina at Greensboro
Greensboro, NC

J. Larry Durstine, PhD
Department of Exercise Science
University of South Carolina
Columbia, SC

Jonathan K. Ehrman, PhD
Division of Cardiovascular Medicine
Henry Ford Hospital
Detroit, MI

Bo Fernhall, PhD
Department of Kinesiology and Nutrition
College of Applied Health Sciences
University of Illinois at Chicago
Chicago, IL

Stephen F. Figoni, PhD
PM&RS (117) VA West Los Angeles Healthcare Center
Los Angeles, CA

Jerome L. Fleg, MD
National Heart, Lung, and Blood Institute
Bethesda, MD

Daniel E. Forman, MD
Division of Cardiovascular Medicine
Brigham and Women's Hospital
Boston, MA

David R. Gater, Jr., MD, PhD
Chief, Spinal Cord Injury
Hunter Holmes McGuire Veterans Affairs Medical Center
Professor, Physical Medicine and Rehabilitation
Virginia Commonwealth University
Richmond, VA

Mike Germain, MD
Professor of Medicine
Tufts University School of Medicine
Medical Director Transplant
Baystate Medical Center
Springfield, MA

Benjamin Gordon, MS
Department of Exercise Science
University of South Carolina
Columbia, SC

Neil F. Gordon, MD, PhD
Medical Director, INTERVENT USA, Inc.
Savannah, GA

Paul M. Gordon, PhD, MPH
Professor and Chair
Department of Health, Human Performance, and Recreation
Baylor University
Waco, TX

Terri L. Gordon, MPH
INTERVENT USA, Inc.
Savannah, GA

Peter W. Grandjean, PhD
Health, Human Performance, and Recreation
Baylor University
Waco, TX

Gregory A. Hand, PhD, MPH
Arnold School of Public Health
University of South Carolina
Columbia, SC

Sam Headley, PhD
Department of Exercise Science and Sport Studies
Springfield College
Springfield, MA

Gregory W. Heath, DHSc
Department of Health and Human Performance
University of Tennessee at Chattanooga and Department of Medicine
Chattanooga, TN

Benson M. Hoffman, PhD
Department of Psychiatry and Behavioral Sciences
Duke University Medical Center
Durham, NC

Lee W. Jones, PhD
Duke Cancer Institute
Durham, NC

N. Brian Jones, PhD
Health and Sport Sciences
University of Louisville
Louisville, KY

Jane Kent-Braun, PhD
Department of Exercise Science
University of Massachusetts
Amherst, MA

Dennis Kerrigan, PhD
Division of Cardiovascular Medicine
Henry Ford Hospital
Detroit, MI

Steven J. Keteyian, PhD
Division of Cardiovascular Medicine
Henry Ford Hospital
Detroit, MI

Andrew B. Lemmey, PhD
School of Sport, Health, and Exercise Sciences
Bangor University
Bangor, Gwynedd, UK

Ryan J. Mays, PhD
School of Medicine, Division of Cardiology
University of Colorado
Aurora, CO

Timothy J. Michael, PhD
Department of Exercise Science
Western Michigan University
Kalamazoo, MI

David C. Murdy, MD
Internal Medicine, Dean Medical Center
Clinical Instructor, University of Wisconsin
Janesville, WI

David L. Nichols, PhD
Department of Kinesiology and Institute for Women's Health
Texas Woman's University
Denton, TX

Quinn Pack, MD
Division of Cardiovascular Diseases and Internal Medicine
Mayo Clinic
Rochester, MN

Mark A. Patterson, MEd
Cardiovascular Service
Kaiser Permanente
Denver, CO

Andjelka Pavlovic, MS
Department of Kinesiology and Institute for Women's Health
Texas Woman's University
Denton, TX

Jan Perkins, PhD
School of Rehabilitation and Medical Sciences
Central Michigan University
Mt. Pleasant, MI

Mark D. Peterson, PhD
Laboratory for Physical Activity and Exercise Intervention Research
University of Michigan
Ann Arbor, MI

Amy E. Rauworth, MS
National Center on Physical Activity and Disability (www.ncpad.org)
Department of Disability and Human Development
University of Illinois at Chicago
Chicago, IL

Judith G. Regensteiner, PhD
School of Medicine, Division of Cardiology
University of Colorado
Aurora, CO

James H. Rimmer, PhD
National Center on Physical Activity and Disability (www.ncpad.org)
Department of Disability and Human Development
Professor, University of Illinois at Chicago Chicago, IL

William Saltarelli, PhD
School of Health Sciences
Central Michigan University
Mt. Pleasant, MI

John R. Schairer, DO
Division of Cardiovascular Medicine
Henry Ford Hospital
Detroit, MI

David D. Spragg, MD
Johns Hopkins School of Medicine
Baltimore, MD

Ray W. Squires, PhD
Division of Cardiovascular Diseases
Mayo Clinic
Rochester, MN

Kerry J. Stewart, EdD
Division of Cardiology
Johns Hopkins School of Medicine
Baltimore, MD

Ann M. Swank, PhD
Department of Health and Sport Sciences
Exercise Physiology Laboratory
University of Louisville
Louisville, KY

Paul S. Visich, PhD, MPH
Chair of Exercise and Sport Performance Department
Westbrook College of Health Professions
University of New England
Biddeford, ME

Christopher J. Womack, PhD
Associate Professor
James Madison University
Harrisonburg, VA

J. Tim Zipple, DScPT, OCS
School of Rehabilitation and Medical Sciences
Central Michigan University
Mt. Pleasant, MI

Preface

Since the first edition of *Clinical Exercise Physiology* was published in 2003, much has occurred in both the field of practice and the science behind it. This text has quickly become the primary textbook for both upper level undergraduate and graduate students preparing to become clinical exercise physiologists. This text also is an excellent resource for those preparing to sit for the American College of Sports Medicine Registered Clinical Exercise Physiologist (RCEP) or clinical exercise specialist examinations. In 2003 clinical exercise physiology as a profession was still just taking shape. Now in 2013 it has blossomed into a profession that works within the healthcare system to deliver evidence-based care to a broad range of patients with chronic diseases. Much important work remains to be performed with respect to the profession of clinical exercise physiology. However, the two original purposes for developing this book, which were to disseminate the research associated with clinical exercise physiology and to provide a comprehensive resource for people working in the field, continue to be accomplished in this third edition.

Evidence of the expanding role of clinical exercise physiology in healthcare continues. This includes the conduct of an increasing amount of clinical exercise physiology–related research and the incorporation of the results from such research into evidence-based guidelines for the treatment of patients with a variety of diseases. Since the second edition of this text the number of PubMed.gov referenced articles that include the keywords 'clinical exercise physiologist' has more than doubled. The Clinical Exercise Physiology Association (CEPA), established in 2008 to serve practitioners in the field through advocacy and education, published the inaugural edition of a profession-specific journal (*i.e., Journal of Clinical Exercise Physiology*). The CEPA also offers continuing education and credits which are vitally important to maintaining clinical exercise physiology related certifications such as the American College of Sport's Medicine's (ACSM) Registered Clinical Exercise Physiologist. Additionally, organizations such as the ACSM, the American Council on Exercise (ACE), and the Canadian Society of Exercise Physiologists (CSEP) continue to offer other relevant certifications and registry examinations, helping to provide uniformity to the level of preparedness of those working in the field or aspiring to do so. Also in 2004 clinical exercise physiology was formally recognized by the Commission on the Accreditation of Allied Health Education Programs (CAAHEP) as a field of study that required a formal and uniform graduate-level curriculum. This has led to the development, albeit slow, of accredited university programs for those who wish to study clinical exercise physiology. This strong foundation serves to demonstrate the permanency needed for the advancement of the field.

Although the day-to-day duties of people working today as clinical exercise physiologists mostly involve patients with cardiovascular disease, more than ever before clinical exercise physiologists are helping to care for patients with cancer, musculoskeletal disorders, and metabolic diseases such as chronic kidney disease and diabetes. Therefore, the timing is right for this revised version of what we believe has become a staple in the preparation of students interested in clinical exercise physiology. This third edition of *Clinical Exercise Physiology* is fully revised and, we believe, better organized than its predecessors were. The initial part of the book presents five foundational chapters, including an excellent review of the history of clinical exercise physiology, a description of the essentials of the physical examination, and a review of the general properties of drugs and pharmacotherapy. The remaining chapters, the core of the book, cover specific diseases and conditions in populations. These chapters are organized into seven parts, each relating to a physiologic system or population: endocrinology, cardiovascular, respiratory, oncology and immunology, orthopedic and musculoskeletal, neuromuscular, and special populations.

Based on reviews and feedback from students and professors, we made the general construction of the chapters in parts II through VIII of this third edition of *Clinical Exercise Physiology* the same. Each of the chapters in these sections begins with an introduction to the specific disease that includes the definition and scope of the disease and a discussion of the relevant pathophysiology. A focus on the medical and clinical considerations follows, including signs and symptoms, diagnosis, exercise testing, and evidence-based treatment. Each chapter concludes with an overview of the exercise prescription for the disorder being discussed, with special emphasis placed on any unique disease-specific issues that might alter the exercise prescription.

Each chapter also contains several practical application boxes that provide additional information summarizing unique chapter-specific information. In each of the disease-specific chapters two of these practical application boxes focus on the exercise prescription and on practical information to consider when interacting with the patient (i.e., Patient-Client Interaction). A third practical application box reviews the relevant exercise-training literature and discusses the physiological adaptations to exercise training and the ways that exercise can influence primary and secondary disease prevention. Each clinical chapter (parts II thru VII) also contains a case study involving an actual patient, progressing from initial presentation and diagnosis to therapy and exercise treatment. Each case study concludes with several questions aimed at facilitating group discussion in the classroom or for the individual learner to consider when preparing for the RCEP examination.

To keep abreast of trends in the field, the chapters on pharmacology, metabolic syndrome, and graded exercise testing and prescription have been reconstructed. Additionally, all of the remaining chapters have undergone a thorough revision to ensure that the material is consistent with current science and evidence-based practice guidelines. Finally, a new chapter has been added to address individuals with an intellectual disability.

Few, if any, upper-level undergraduate courses or graduate-level clinical exercise physiology programs currently provide students with the breadth of information required to sit for the ACSM RCEP examination. Those who plan to study to take this or any similar certification examination should understand that no single text provides in-depth coverage of all the clinical populations that benefit from physical activity and exercise. But this text may be as close as one can come. In all, *Clinical Exercise Physiology* addresses 27 different diseases and populations. And each chapter has been compared against the latest ACSM Guidelines for Exercise Testing and Prescription text (9th edition) for consistency.

Besides serving as a textbook for students studying in the field, *Clinical Exercise Physiology* is an excellent resource guide that the professional will want to have in her office. The features of consistent organization, case studies, discussion questions, up-to-date references, and feature boxes are designed to provide information required for effective study. In fact, the content was developed based on the KSs (i.e., knowledge and skills) of the ACSM RCEP examination. We believe that this text serves as a valuable textbook for the student and as a useful desk reference for the practicing clinical exercise physiologist.

INSTRUCTOR RESOURCES

The text also features a test package, presentation package, and image bank for instructors—some of which are new to this edition. These instructor ancillaries are available online at **www.HumanKinetics.com/ClinicalExercisePhysiology**.

- **Test package.** Created with Respondus 2.0, the test package includes over 450 true-or-false and multiple-choice questions. With Respondus, instructors can create versions of their own tests by selecting from the question pool; select their own test forms and save them for later editing or printing; and export the tests into a word-processing program. Additionally, instructors can access the test package in RTF format or in a format that can be imported into learning management systems.
- **Presentation Package.** The presentation package includes approximately 800 slides of text, artwork, and tables from the book that instructors can use for class discussion and presentation. The slides in the presentation package can be used directly within PowerPoint or printed to make transparencies or handouts for distribution to students. Instructors can easily add, modify, and rearrange the order of the slides.
- **Image Bank.** The image bank provides almost all the figures, photos, and tables from the print book, organized by chapter. Instructors can use these loose images to create their own presentations, handouts, or other class materials.

Acknowledgments

I am very appreciative of the contributors to this third edition of Clinical Exercise Physiology who agreed to devote their time and knowledge to the development of this text. Working with each has been a pleasure. Thank you to those who have purchased the first two editions of this text and the university faculty who have adopted it for their courses. It is my hope that this text provides the information needed to keep abreast of the ever growing field of clinical exercise physiology. I also wish to recognize the colleagues I work with at Henry Ford Health System. It is your knowledge and passion that keeps me enjoying what I do on a daily basis. To my wife and children, thank you for all of your enduring support. Finally, I dedicate this book in loving memory to my father, George A. Ehrman, Jr., a devoted father and grandfather.

Jonathan K. Ehrman

As we continue to see the value of exercise with various chronic diseases, I am excited to see how this textbook has become accepted in our field. Only through the expertise of many contributors has this book been possible. I thank each of them. As a professor for the past 15 years I am excited that this comprehensive book in clinical exercise physiology can be offered to colleagues and, most important, to students in the field of exercise physiology who are interested in improving the health of others. I continue to thank my wonderful wife, Diane, who is always supportive of my endeavors; my two sons, Matt and Tim; and my parents, Frank and Mary, who have always encouraged their children to give their best effort in whatever they choose to do (success = drive + intelligence).

Paul S. Visich

Many thanks to everyone at Human Kinetics who helped us prepare and bring life to this much revised third edition of a textbook that we believe will continue to do much to advance and support the rapidly expanding practice and profession of clinical exercise physiology. Thanks as well to my coeditors and the many contributing authors for the quality writing put forth on behalf of this book. My sincere thanks to all past and current staff of Henry Ford Health System's Preventive Cardiology Unit for their tireless efforts to provide outstanding patient care, to contribute to the profession through original research, and to train others entering the field. As always, many blessings to Lynette, Stephanie, Courtland, Jacob, and Aram for their support of my professional interests. Finally, in loving memory of Albert Z. and Virginia Keteyian.

Steven J. Keteyian

The clinical exercise physiology specialty is now taking shape, and knowledge in the area is rapidly expanding. We are blessed to have a distinguished international group of authors who have worked diligently to impart state-of-the-art knowledge about their respective subspecialties. Many thanks go to them for their tireless efforts as well as to the reviewers and staff who often go unnoticed behind the scenes to assure an exceptional final product.

To my wonderful wife, Ina, and my children, Joshua, Natalie, and Liam—thanks for putting up with me and providing much love, support, and understanding. Finally, in loving memory of my father, Edwin W. Gordon III, who always believed in me.

Paul M. Gordon

PART I

Introduction to Clinical Exercise Physiology

As mentioned in the preface, although the day-to-day duties of people working in our field most often involve patients with cardiovascular disease, more than ever before clinical exercise physiologists now contribute to the evidence-based care provided to patients with cancer, musculoskeletal disorders, and metabolic diseases such as chronic kidney disease and diabetes. As a result, the chapters in this first part of *Clinical Exercise Physiology* not only review the rich history and expanding scope of our profession and the foundational knowledge that we use for safe and effective exercise testing and exercise prescription, but the chapters in this section also address other key areas that the practicing clinical exercise physiologist must also demonstrate proficiency in if they are to help care for patients across a broad range of chronic diseases. These include behavioral approaches to the maintenance of regular exercise training, essential principles in pharmacology, and general interview and examination skills. A thorough reading of these chapters will provide the necessary background for the important work that lies ahead.

Chapter 1 is a general introduction to the field of clinical exercise physiology. With each passing year clinical exercise physiology as a profession deepens its roots, along with a variety of other allied health professions, into the delivery of evidence-based healthcare. Since the second edition of this text there has been a continuation of this natural process. Chapter 1 reviews the beginnings, current practices, and future directions for clinical exercise physiology, particularly as it relates to disease prevention and management. The chapter also reviews the various professional organizations and certifications that are continuing to influence the profession.

Chapter 2 centers on behavioral approaches. Although most of the material in this textbook and in the classroom pertains to the knowledge and physical skills needed to write safe exercise prescriptions, interpret exercise responses, and lead or supervise exercise, the fact remains that at the end of the day, after we have written our prescriptions and measured and tested our patients, we are in the business of changing human behavior. This challenge differs little from the one that confronts the practicing physician or nurse. They, too, play an important role and should be relied on to help educate and motivate patients to take an active part in improving their own health. To that end, this chapter reviews and applies the various behavioral approaches known to help patients adopt long-term habits aimed at improving their health. The information presented here can be applied across demographic categories and disease conditions.

Chapter 3 focuses on pharmacotherapy. Those who want to help care for and understand the underlying pathophysiology of a particular condition should take the time to understand how and why a particular medication is being used to help treat it. Adopting this approach will help the clinical exercise physiologist understand the underlying disease process and improve his or her ability to develop a proper exercise prescription and conduct safe exercise evaluations and training sessions. Learning the essential drug properties in general and pharmacotherapy taught in this chapter, along with the key factors that influence drug compliance, will go a long way toward preparing a person to work effectively in the field.

Chapter 4 discusses general interview and examination skills, so that, on any given day, the clinical exercise physiologist is better prepared to determine whether a patient can safely exercise. Rarely is a chronic disorder stagnant. Instead, diseases are dynamic. As a result, the clinical exercise physiologist must regularly inquire about or recognize signs, symptoms, and other evidence of possible disease progression. This approach applies regardless of whether a patient is being considered for surgery or about to undergo a graded exercise stress test. Safety remains paramount. Although this chapter will help the clinical exercise physiologist learn about what questions to ask, what signs to look for, and how to interpret the evaluations performed by others, there is no substitute for taking additional time to practice newly acquired observational, history-taking, and examination skills on patients.

Chapter 5 looks at graded exercise testing and prescription. Like the other chapters in part I, this chapter reviews and provides foundational information needed to work through the subsequent chapters of the book. Fortunately, graded exercise testing and prescription are topics that most have likely had prior coursework in, freeing them up to begin to think about conducting an exercise test or writing an exercise prescription in a manner that integrates exercise physiology knowledge with the unique clinical issues germane to the chronic condition of the current patient. This approach will serve students well when they are someday working an 8 h shift in an exercise-testing lab, allowing them to integrate information about patients presenting with a variety of clinical conditions, using an approach that considers all aspects of a patient's condition and a physician-directed care plan.

Chapter 1

Introduction

Jonathan K. Ehrman, PhD
Paul M. Gordon, PhD, MPH
Paul S. Visich, PhD, MPH
Steven J. Keteyian, PhD

Previous editions of this textbook published in 2003 and 2009 provided a contemporary review of clinical exercise physiology as a profession. This introduction briefly reviews current issues relevant to this topic and, where appropriate, makes comparisons to past experiences.

THE PAST, PRESENT, AND FUTURE OF CLINICAL EXERCISE PHYSIOLOGY

Clinical exercise physiology is a subspecialty of exercise physiology, which also includes applied exercise physiology (13). The discipline of clinical exercise physiology investigates the relationship of exercise and chronic disease; the mechanisms and adaptations by which exercise influences the disease process; and the role and importance of exercise testing and training in the prevention, evaluation, and treatment of these diseases. This knowledge is used to develop and implement exercise training programs for those who will benefit. Therefore someone who practices clinical exercise physiology (i.e., the clinical exercise physiologist) must be knowledgeable about the broad range of exercise responses, including those that occur both within a disease class and across different chronic diseases. Although the response of various organ systems is usually the focus, important behavioral, psychosocial, and spiritual issues are usually present as well. A capable clinical exercise physiologist must have a wide knowledge base in these areas:

- Anatomy
- Physiology (organ systems and exercise)
- Chemistry (organic and biochemistry)
- Biology (cellular and molecular)
- Psychology (behavioral medicine, counseling)

In addition, a clinical exercise physiologist must complete an undergraduate or graduate degree in a related major. Although those desiring to work in the field are not currently required to perform the job responsibilities of a clinical exercise physiologist, it is strongly recommended that they perform a clinical internship of at least 600 h (diagnostic and functional exercise testing; exercise assessment, prescription, leadership, and supervision; counseling; and education) and pass a certification examination accredited according to standards set by an accrediting agency independent and outside of the organization offering the certification. See practical application 1.1 for more information.

The Past

The formal use of exercise in the assessment and treatment of chronic disease has existed for more than 40 yr, but exercise physiology can trace its roots back to the late 1800s (Fernand LaGrange's textbook *Physiology of Bodily Exercise*, published in 1889) and the early 1900s (e.g., the Harvard Fatigue Laboratory). Although at that time much of the focus was on the physiological response to exercise in healthy and athletic populations, in the late 1930s Sid Robinson and colleagues at Harvard and Indiana University studied the effects of the aging process on exercise performance. Outside of the United States, other countries have also contributed greatly to the knowledge base of exercise physiology, notably the Scandinavian and other European countries in the 1950s and 1960s.

The development of the modern-day clinical exercise physiologist dates back to the 1960s, around the time when the term *aerobics* was popularized by Dr. Ken Cooper—a time when regular exercise was beginning to be considered important to maintaining optimal health (14) and prolonged bed rest was found to be associated with marked loss of exercise tolerance (19). In addition, pioneers such as Herman Hellerstein demonstrated that bed rest was detrimental to people with heart disease. These findings gave way to the development of inpatient and, subsequent rapid proliferation of, outpatient cardiac rehabilitation programs throughout the 1970s and into the 1990s.

The Present

Although published in 1996, *Physical Activity and Health: A Report of the Surgeon General* remains a landmark report for the field of clinical exercise physiology (20). Using an evidence-based approach, it identified numerous chronic diseases and conditions whose risk increases among people lacking in exercise or physical activity (see "Diseases and Conditions Related to a Lack of Exercise"). More recently, in 2007, the American College of Sports Medicine (ACSM) and American Heart Association (AHA) jointly published physical activity guidelines for both healthy and older adults (17, 18). These were updates to the 1995 joint Centers for Disease Control and ACSM physical activity guidelines. Each guideline suggests a recommended physical activity dose for reducing the risk of a number of chronic health diseases and conditions.

It is now accepted that exercise limits disability and improves outcome for many diseases and conditions, including cardiovascular, skeletal muscle, and pulmonary diseases. For these conditions, exercise training is often a part of a comprehensive treatment plan. Additionally, some population groups are at greater risk than others for developing a chronic disease or disability because of physical inactivity; these include women, children, and people of certain races and ethnicities. This risk also increases in all persons as they age. More than ever before, clinical exercise physiology is at the forefront of advances in clinical care and public policy directed at improving health and lowering future disease risk through regular exercise training and increased physical activity.

Because of the Surgeon General's report, published practice guidelines, position statements, and ongoing research, cardiac rehabilitation programs have evolved into sophisticated exercise training and behavioral management programs administered by multidisciplinary teams that include clinical exercise physiologists. Additionally, pulmonary rehabilitation programs, bariatric surgery and weight management programs, exercise oncology programs, diabetes management programs, and other services have developed similarly, although less rapidly, due to the research and clinical contributions of clinical exercise physiologists.

Diseases and Conditions Related to a Lack of Exercise

- Cancer (breast, colon, prostate)
- Cardiovascular disease (coronary artery, peripheral vascular, cerebral vascular)
- Falling
- Frailty
- Health-related quality of life
- Non-insulin-dependent diabetes
- Mood and mental health
- Obesity
- Osteoarthritis
- Osteoporosis
- Overall mortality
- Premature mortality
- Sarcopenia-related functional loss

Adapted from the Surgeon General's Report on Physical Activity and Health, 1996.

Practical Application 1.1

SCOPE OF PRACTICE

The following are definitions of the (clinical) exercise physiologist or specialist from several organizations. Careful reading of these definitions reveals that there is still no clear consensus in place regarding the title of the person who works with patients in an exercise or rehabilitative setting. Titles include exercise physiologist, exercise specialist, clinical exercise specialist, certified exercise physiologist, and registered clinical exercise physiologist.

U.S. Department of Labor, Bureau of Labor Statistics

The following is the description of an exercise physiologist according to the U.S. government (21). This classification was added on March 11, 2010.

29-1128 Exercise Physiologists

Assess, plan, or implement fitness programs that include exercise or physical activities such as those designed to improve cardiorespiratory function, body composition, muscular strength, muscular endurance, or flexibility. Excludes "Physical Therapists" (29-1123), "Athletic Trainers" (29-9091), and "Fitness Trainers and Aerobic Instructors" (39-9031).

Examples

Kinesiotherapist, Clinical Exercise Physiologist, Applied Exercise Physiologist

Broad Occupation: 29-1120, Therapists

Minor Group: 29-1000, Health Diagnosing and Treating Practitioners

Major Group: 29-0000, Healthcare Practitioners and Technical Occupations

Clinical Exercise Physiology Association (CEPA)

What is a clinical exercise physiologist (CEP)? A CEP is a certified health professional who uses scientific rationale to design, implement, and supervise exercise programming for those with chronic diseases, conditions, or physical shortcomings. Clinical exercise physiologists also assess the results of outcomes related to exercise services provided to those individuals. Clinical exercise physiology services focus on the improvement of physical capabilities for the purpose of (1) chronic disease management, (2) reducing risks for early development or recurrence of chronic diseases, (3) creating lifestyle habits that promote enhancement of health, (4) facilitating the elimination of barriers to habitual lifestyle changes through goal setting and prioritizing, (5) improving the ease of daily living activities, and (6) increasing the likelihood of long-term physical, social, and economic independence (12).

American College of Sports Medicine (ACSM)

The ACSM Certified Clinical Exercise Specialist℠ (CES) is a professional with a minimum of a Bachelor's degree in exercise science. The CES works with patients and clients challenged with cardiovascular, pulmonary, and metabolic diseases and disorders, as well as with apparently healthy populations in cooperation with other healthcare professionals to enhance quality of life, manage health risk, and promote lasting health behavior change.

The CES conducts preparticipation health screening, maximal and submaximal graded exercise tests, and performs strength, flexibility, and body composition tests. The CES develops and administers programs designed to enhance aerobic endurance, cardiovascular function, muscular strength and endurance, balance, and range of motion. The CES educates their clients about testing, exercise program components, and clinical and lifestyle self- care for control of chronic disease and health conditions (5).

The ACSM Registered Clinical Exercise Physiologist (RCEP) is an allied health professional, with a minimum of a master's degree in exercise science, who works in the application of physical activity and behavioral interventions for clinical diseases and health conditions in which such interventions have been shown to provide therapeutic or functional benefit. Services provided by an RCEP include, but are not limited to, services for individuals with cardiovascular, pulmonary, metabolic, orthopedic, musculoskeletal, neuromuscular, neoplastic, immunologic,

(continued)

and hematologic disease. The RCEP provides primary and secondary prevention and rehabilitative strategies designed to improve physical fitness and health in populations ranging across the life span.

The RCEP performs exercise screening, exercise and physical fitness testing, exercise prescription, exercise and physical activity counseling, exercise supervision, exercise and health education and promotion, and measurement and evaluation of exercise- and physical activity-related outcome measures. The RCEP works individually or as part of an interdisciplinary team in a clinical, community, or public health setting. The practice and supervision of the RCEP is guided by published professional guidelines, standards, and applicable state and federal laws and regulations (6).

American Council on Exercise (ACE)

As defined by ACE, the ACE-certified Advanced Health & Fitness Specialist designation demonstrates that the professional has the knowledge to provide in-depth preventive and postrehabilitative fitness programming for individuals who are at risk for or are recovering from a variety of cardiovascular, pulmonary, metabolic, and musculoskeletal issues (3). Requirements include either an ACE Personal Trainer or Lifestyle & Weight Management Coach certification; or a National Commission for Certifying Agencies (NCCA)-accredited personal trainer or fitness certification; or a bachelor's degree in an exercise science or related field.

Canadian Society of Exercise Physiology (CSEP)

A CSEP-certified exercise physiologist (CEP) performs assessments; prescribes conditioning exercise; and provides exercise supervision, counseling, and healthy lifestyle education. The CEP serves apparently healthy individuals and populations with medical conditions, functional limitations, or disabilities associated with musculoskeletal, cardiopulmonary, metabolic, neuromuscular, and aging conditions (11). The CSEP specifies duties that the CEP is sanctioned or not sanctioned to perform, as follows.

A CSEP clinical exercise physiologist is sanctioned by CSEP to do the following:

1. Administer appropriate assessment protocols for the evaluation of physical fitness to individuals who have been screened, have signed an informed consent form, or who have been cleared for unrestricted or restricted activity by a licensed health care professional.
2. Provide physical activity clearance following further queries to positive responses to questions 4, 5, and/or 7 on the PAR-Q. For example, an individual could be cleared for physical activity or exercise by a CSEP CEP if (1) with question 4, it was determined that the dizziness was associated with overbreathing during heavy exercise or sudden postural changes; (2) with question 5, it was determined that the joint problem was an old knee, ankle, shoulder, or other old joint constraint; and (3) with question 7, it was determined that the individual had a "cold" or a relative contraindication such as, but not limited to, controlled diabetes or stable medicated blood pressure.
3. Provide physical activity clearance to clients who are screened out by PAR-Q question 1, 6, or both. In these instances, until additional information is gathered, the CSEP CEP can recommend tailored low-intensity, progressive physical activity (such as walking).
4. Seek medical clearance for clients of any age who are screened out by PAR-Q question 2, 3, or both, which deal with potential heart problems, before providing physical activity recommendations.
5. Provide physical activity clearance and recommend tailored progressive physical activity for clients over age 69 who do not respond positively to PAR-Q question 2, 3, or both, which deal with potential heart problems.
6. Provide physical activity clearance to clients over age 69 and recommend tailored progressive physical activity.
7. Provide physical activity clearance to youths under age 15 who have consent of their parent or guardian.
8. Interpret the results of an individual's fitness assessment to determine the individual's health-related fitness level, performance-related (function, work or sport) fitness level, or both.
9. Use the outcomes from objective assessments to guide decisions regarding physical activity and exercise (prescription, demonstration, supervision and monitoring, fitness and healthy lifestyle counseling) and act as a personal trainer.

10. Suggest healthy dietary practices in concert with physical activity and exercise programs for healthy weight management.
11. Suggest dietary practices for health-related nutrition and performance-related nutrition.
12. Use a heart rhythm tracing to observe heart response during a fitness assessment and a structured exercise session.
13. Evaluate and treat both asymptomatic and symptomatic populations with medical conditions, functional limitations, and disabilities through the application of exercise and physical activity for the purpose of improving health and function.
14. Perform evaluations, prescribe conditioning exercise, and provide exercise supervision, health education, and outcome evaluation.
15. Work with apparently healthy asymptomatic and symptomatic populations such as older adults, children and youth, obstetric populations, and society as a whole in health enhancement and the prevention of impairment and disability.
16. Provide appropriate exercise therapy to clients including, but not limited to, those with musculoskeletal, cardiorespiratory, and metabolic conditions.
17. Accept referrals from licensed health care professionals trained to diagnose and treat musculoskeletal conditions or medical conditions.

A CSEP CEP is not sanctioned by CSEP to do the following:

1. Administer assessment protocols and prescribe exercise or therapy to acutely injured and diseased individuals who are not within the boundaries of the scope of practice just outlined.
2. Diagnose pathology based on any assessment performed.

Although some may consider that differences in these definitions for the clinical exercise professional are simply a matter of semantics, such differences often lead to confusion among the public about which title or type of clinical exercise professional they should look for when referred to or considering participation in a clinical exercise program. Additionally, there may be confusion among other health care professionals about the job titles and the duties of those who hold them. For instance, in many institutions, someone who performs the technical duties in a noninvasive cardiology laboratory is typically titled a cardiovascular technician (an occupation defined by the Department of Labor). But since the duties performed by a cardiovascular technician can overlap with those of a clinical exercise physiologist, it is not uncommon for cardiovascular technicians to be hired into exercise physiologist types of positions or for exercise physiologists to be hired into cardiovascular technician types of positions.

The Future

Expansion of employment opportunities for those with clinical exercise physiology credentials continues to take shape. Traditionally, most clinical exercise physiologists worked in the cardiac rehabilitation setting. More recently opportunities have expanded to include other diseases and conditions either within a cardiac rehabilitation program or in independent disease management programs in which exercise plays an important role. These conditions and diseases primarily include obesity, peripheral arterial disease, diabetes, and cancer, although there are many others that positively respond to exercise. Emphasis has also recently emerged on the impact of physical activity and exercise on those who are mentally or physically disabled. The National Center on Physical Activity and Disability has taken a leading role in this promising focus; and the ACSM recently developed the Inclusive Fitness Trainer certification, aimed at identifying people with the minimal skills to work with these populations on exercise programming and implementation. Additionally, clinical exercise professionals continue to gain employment in clinical research trials at sponsor and site investigation locations, the latter of which may be increasing given the greater emphasis on exercise-related research. As in research projects involving exercise, noninvasive exercise testing laboratories, including those specializing in cardiopulmonary exercise testing, often regard clinical exercise professionals as an asset because their academic training is well suited for this type of employment. Finally, the emerging importance of physical activity and health and an increase in both professional and public awareness of the skills of clinical exercise physiologists have led to increasing employment in nonclinical settings. These include, but are not limited to, personal training, corporate fitness programs, medically affiliated fitness centers, schools and communities, professional and amateur sport consulting, and weight management programs.

A recent survey by the CEPA reported a total of approximately 4,450 individuals in the United States who consider themselves a clinical exercise physiologist (12). Of these, about 3,600 are ACSM-certified clinical exercise specialists and 850 are ACSM-registered clinical exercise physiologists.

The profession of clinical exercise physiology continues to develop a unique body of knowledge that includes exercise prescription development and implementation of both primary and secondary prevention services. This information has been published through the years in a growing number of biomedical journals that contribute to the clinical exercise physiology body of knowledge (see "Selected Biomedical Journals"). These now include many important medical journals such as the *New England Journal of Medicine (NEJM), Journal of the American Medical Association (JAMA),* and *Circulation.* "Selected Biomedical Journals" lists other journals that regularly publish manuscripts written either by or for the clinical exercise physiologist.

This body of knowledge has led to the emergence of evidence-based recommendations both for the general public and for patients with a chronic disease. These pronouncements and scientific statements are presented in various chapters through this text. In addition, the clinical exercise physiologist is uniquely trained to identify individual lifestyle-related issues that promote poor health and to design and implement a behavior-based treatment plan aimed at modifying lifestyle factors. Thus, the profession of clinical exercise physiology and the clinical exercise physiologist fill a void in health care and are becoming increasingly important, particularly in the United States, as the average age of the population rapidly increases. And as the baby boom generation (those born between 1946 and 1964) begins to reach retirement age there will be an increasing number of individuals who are living with chronic diseases both in the United States and in other developed countries.

In summary, the involvement of clinical exercise physiologists in a variety of settings has grown dramatically over the last 30 yr. Historically, most exercise physiologists were engaged in human performance–related research or academic instruction. But many now provide professional advice and services in clinical, preventive, and recreational fitness programs located in health clubs, corporate facilities, and hospital-based complexes. Clinical exercise physiologists provide a number of important services, including fitness assessments or screenings, exercise testing and outcome assessments, exercise prescriptions or recommendations, exercise leadership, and exercise supervision. To date in the United States, only the state of Louisiana has

Selected Biomedical Journals

ACSM's Health and Fitness Journal
American Journal of Cardiology
American Journal of Clinical Nutrition
American Journal of Physiology
American Journal of Sports Medicine
Annals of Internal Medicine
Archives of Internal Medicine
British Journal of Sports Medicine
Canadian Journal of Applied Physiology
Circulation
Clinical Journal of Sport Medicine
Diabetes
Diabetes Care
European Journal of Applied Physiology
European Journal of Sport Science
Exercise and Sport Science Reviews
International Journal of Obesity
Journal of Aging and Physical Activity
Journal of Applied Physiology
Journal of Cardiopulmonary Rehabilitation and Prevention
Journal of Clinical Exercise Physiology
Journal of Obesity
Journal of Sport Science and Medicine
Journal of the American Medical Association (JAMA)
Medicine and Science in Sports and Exercise
New England Journal of Medicine (NEJM)
Pediatric Exercise Science
Pediatric Obesity
Sports Medicine
The Physician and Sportsmedicine
Research Quarterly for Exercise and Sport

opted to license clinical exercise physiologists as a means to govern the delivery of specified and defined services to consumers. However, efforts are under way in other states such as Massachusetts, Minnesota, California, and Kentucky. Practical application 1.2 presents information about licensure of the exercise physiologist.

Practical Application 1.2

THE CHALLENGE OF CEP LICENSURE IN THE UNITED STATES

The profession of clinical exercise physiology is largely unregulated in the United States. At the time of this writing, efforts to enact legislation were ongoing in several U.S. states. The process toward licensure has been long and tedious. With its beginnings during Socratic debates at the annual ACSM meetings in the 1990s, the move toward licensure is slow. To date, only the U.S. state of Louisiana has passed and enacted (in 1995) a licensure process for the clinical exercise physiologist. Some have been critical of the Louisiana enactment, suggesting that it too narrowly limits the scope of practice of the CEP. Others believe it is sufficient and should be a model of future legislative attempts. Additionally, other professional groups, such as physical therapists (PTs), are critical of the movement to license CEPs. Some have argued that the PT professional scope of practice sufficiently covers the suggested scope of practice of the CEP and that licensing CEPs would add to an already "too large" allied health licensing process. Moreover, some have stated that CEPs are not sufficiently trained in the neuromuscular and orthopedic areas.

Evidence justifying licensure must first be gathered and published and must then be appreciated by reasonable segments of both the public and state legislators. Recently a point/counterpoint presentation of the pros and cons of licensure was published in the *Journal of Clinical Exercise Physiology* (9, 15). Several state exercise physiology societies have been involved in this process also. For instance, in the summer of 2011 in Massachusetts, testimony was provided by five clinical exercise physiologists associated with the Massachusetts Association of Clinical Exercise Physiologists (MACEP) to the Committee for Consumer Protection and Professional Licensure (16). Some professions are opposed to this attempt at CEP licensure (7). This is an example of the process and opposition that every licensure attempt in the United States will likely face. Although opposition is expected and will occur, one must appreciate the fact that CEPs cannot only deliver relevant services, and often do so in a cost-effective manner, but that they also provide unique services (e.g., prescription of exercise in patients with known disease, evaluation of program effectiveness) outside the scope of other health care providers.

In large part, efforts toward licensure will likely focus on whether such legislation is necessary to protect the public from harm due to services provided by exercise physiologists. However, few data are available to suggest that a significant risk of harm exists. This is possibly related to the close medical supervision and evidence-based exercise interventions used by practicing CEPs. (22). But one might also argue that the diligence that has operated to date to keep patient safety at a high level does not suggest that any loosening of the standards for CEPs should occur. In fact, the aging population and the incidence of chronic disease risk factors and outright chronic disease at younger ages might suggest a need for tightening these standards as would occur with licensure.

This lack of applicable state licensing requirements for exercise physiologists has, in some respects, muddied the waters for those who attempt to assess and define the practice roles for such professionals in the delivery of services to consumers. This confusion may exist particularly from the perspective of the legal system, where questions arise relative to a variety of concerns related to service delivery. Important legal questions include the following:

- What services may exercise physiologists lawfully provide given the absence or lack of licensure?
- What practices performed by exercise physiologists may be prohibited as a matter of law because of state statutes regarding unauthorized practice of medicine?
- What practices performed by exercise physiologists may be prohibited because they are in the scope of practice of other licensed health care professionals (e.g., physical therapy, nursing, dietetics) and thus prohibited by current law?
- What potential liabilities may exercise physiologists face when their delivery of service results in harm, injury, or death attributable to alleged negligence or malpractice?
- What recognition may be given to exercise physiologists and their opinions in a variety of legal settings (such as evaluating disability or working capacity in matters involving insurance, personal injury, or workers' compensation)? Can they serve as reputable expert witnesses?

PROFESSIONAL ORGANIZATIONS AND CERTIFICATIONS THROUGHOUT THE WORLD

Literally hundreds of exercise- and fitness-related certifications are available, but only a few exist for the purpose of certifying exercise professionals to work with people who have chronic diseases. The primary professional organizations in the United States that provide clinically oriented exercise physiology certifications are the American Council on Exercise (ACE [2]), the American College of Sports Medicine (ACSM [4]), and the Canadian Society of Exercise Physiology (CSEP [10]). In addition, Exercise and Sports Science Australia (ESSA) accredits exercise physiologists in Australia. In the United Kingdom, the British Association of Sport and Exercise Sciences (BASES) assesses individuals to earn accreditation in both supervised exercise and as a certified exercise practitioner (22). To date there do not appear to be any organized efforts for certification or accreditation of clinical exercise physiologists in mainland Europe, Asia, Africa, or South America. However, ACE, ACSM, and CSEP are working to provide their certifications to exercise physiologists throughout the world, including these continents. Table 1.1 outlines aspects of each of these organizations' certifications, which are also reviewed in the next several paragraphs.

Clinical Exercise Physiology Association (CEPA): www.cepa-acsm.org

Established in 2008 as a member of the American College of Sports Medicine's Affiliate Societies, CEPA is an autonomous professional member organization with the sole purpose of advancing "the scientific and practical application of clinical exercise physiology for the betterment of health, fitness, and quality of life among patients at high risk or living with a chronic disease." Although no clinical exercise certifications are offered through CEPA per se, the focus of this organization is to serve the profession of clinical exercise physiology and practicing clinical exercise physiologists through advocacy, education, and career development.

American Council on Exercise (ACE): www.acefitness.org

The ACE (2) continues its long history of certifying exercise and fitness professionals, predominantly in the area of personal training. Nevertheless, with the foreseeable growth in the aging population and the rise in the prevalence of chronic diseases, particularly obesity, ACE has developed four certifying levels for the general public. Besides the personalized trainer certification, other available certifications include advanced health and fitness specialist, group fitness instructor, and lifestyle and weight management coach.

Advanced Health & Fitness Specialist The ACE clinical advanced health and fitness specialist is most closely aligned with the clinical certifications offered by other credentialing bodies. This certification is designed to test individual competencies to work with chronic diseased populations including those with cardiovascular, pulmonary, metabolic, and musculoskeletal disorders. There is also an emphasis on knowledge of psychological aspects and the physiologic and biomechanical effects of many different diseases, disorders and ailments, assessed through a written examination. To sit for the examination, one must be at least 18 yr old, have proof of current cardiopulmonary resuscitation (CPR) and automated external defibrillator (AED) certification, a bachelor's degree in an exercise-related field, and a current ACE personal trainer certification or NCCA-accredited university degree. In addition, 300 h of practical experience designing and implementing exercise programs for apparently healthy or high-risk people is required. The ACE website states that currently there are almost 50,000 certified individuals with over 55,000 certifications. ACE has international appeal and collaborates with several international groups dedicated to health and fitness, including the European Health and Fitness Association (EHFA) and the International Health, Racquet and Sportsclub Association (IHRSA).

American College of Sports Medicine (ACSM): www.acsm.org

The American College of Sports Medicine has been internationally acknowledged as a leader in exercise and sports medicine for almost 60 yr. Besides clinical exercise physiologists, the ACSM comprises professionals with expertise in a variety of sport medicine fields, including physicians, general exercise physiologists and those from academia, physical therapists, athletic trainers, dietitians, nurses, and other allied health professionals.

Clinical Exercise Specialist and Registered Exercise Physiologist Since 1974, ACSM has been recognized throughout the industry for its certification

Table 1.1 Comparison of Qualifications for Clinical Exercise Physiologists

	Advanced health and fitness specialist (ACE)*	Clinical exercise specialist (ACSM)*	Registered clinical exercise physiologist (ACSM)*	Certified exercise physiologist (CSEP)*	Accredited exercise physiologist (ESSA)
Eligibility	Bachelor's degree in an exercise science or related field; or current ACE Personal Trainer certification or the Lifestyle & Weight Management Coach certification or another National Commission for Certifying Agencies (NCCA)–accredited personal trainer certification Current CPR/AED certification 300 h of practical experience	Bachelor's degree in kinesiology, exercise science, or other exercise-type degree 500 h practical experience in a clinical exercise program or 400 h for students graduating from a CoAES-accredited program Current BCLS or CPR for the Professional Rescuer certification	Master's or doctorate degree in clinical exercise physiology More than 600 h clinical experience, preferably across all domains Current CPR or BCLS certification	4 yr degree in exercise sciences or a related area (note that there are two eligibility categories—see www.csep.ca for details) At least 200 h of related experience	Undergraduate degree in field of exercise, sport science, or exercise physiology 140 h practical experience related to the field Member of ESSA as an Exercise Scientist/ Full member
Examination and requirements	125 scored and 25 nonscored multiple-choice questions Specific exam content available on ACE website Exam study materials available for purchase Content domains: program implementation and management (40%), assessment (26%), program design (26%), professional responsibility (8%)	Computer-based exam at Pearson VUE–authorized testing centers 100 scored and 35 nonscored questions in 3.5 h Approximate weightings and content areas: health appraisal and fitness exercise testing (26%); exercise prescription (training) and programming (19%); ECG and diagnostic techniques (17%); exercise physiology and related exercise science (10%); pathophysiology and risk factors (10%); human behavior (5%); safety, injury prevention, and emergency procedures (5%); nutrition and weight management, patient management and medications, medical and surgical management, and program administration (2% each)	Computer-based exam at Pearson VUE–authorized testing centers 125 scored and 15 nonscored questions in 3.5 h Approximate weightings and content areas: health appraisal, fitness and clinical exercise testing (25%); exercise prescription and programming (21%); exercise physiology and related exercise science (19%); medical and surgical management (13%); pathophysiology and risk factors (9%); human behavior (5%); safety, injury prevention, and emergency procedures (4%); program administration, quality assurance, and outcome assessment (4%)	130 multiple-choice questions Written examination requiring 75% passing score One or two examinations per Canadian province held each year	No written examination Completion of all knowledge criteria through formal university study 500 h work or practicum experience in apparently healthy clients (140 h); cardiopulmonary and metabolic clients (140 h); musculoskeletal, neurological, neuromuscular clients (140 h); other clinical health delivery activities (80 h) Note that the cardiopulmonary and musculoskeletal hours must be supervised by an accredited exercise physiologist.

(continued)

Table 1.1 *(continued)*

	Advanced health and fitness specialist (ACE)*	Clinical exercise specialist (ACSM)*	Registered clinical exercise physiologist (ACSM)*	Certified exercise physiologist (CSEP)*	Accredited exercise physiologist (ESSA)
Practical examination	No practical testing	No formal exam; assessment by the computer-based exam	No formal exam; assessment by the computer-based exam	"Objective standard practical examination"	See above
Certification maintenance	Every 2 yr Requires 20 h continuing education credits (CECs); CPR and first aid certification not required, but CECs can be received for those certifications	Every 3 yr Requires 60 CECs and BCLS or CPR certification $45 recertification fee	Every 3 yr Requires 60 CECs and CPR certification $45 recertification fee	Every 2 yr Requires CPR certification, 12 fitness appraisals, and 30 professional development credits (PDCs)	20 points of continuing professional development (CPD) and current CPR certification required to maintain the accreditation
Pass rate	67% in 2007-2008 (note that this is cumulative of all ACE examinations)	50% first-time candidates	63% first-time candidates	Not available	Not applicable

*Information gathered from the ACE (2), ACSM (3), ASEP (4), and CSEP (5)websites and promotional materials.

program. Although ACSM does provide certifications for the exercise professional interested in preventive exercise programming (e.g., certified personal trainer, health fitness specialist), the ACSM rehabilitative track is most appropriate for the clinical exercise physiologist. For example, the ACSM clinical exercise specialist certification is designed for the professional who plans to administer and supervise an exercise program for patients with cardiovascular, pulmonary, or metabolic disease.

In response to the ever-growing body of evidence outlining the benefits of exercise for both disease prevention and rehabilitation across a variety of diseases and populations (e.g., cardiovascular, pulmonary, metabolic, musculoskeletal, neuromuscular, immunology, oncology, geriatric, obstetric, and pediatric), in 1998 the ACSM developed a registry for clinical exercise physiologists. The registry is designed to credential the clinical exercise physiologist who cares for patients with a chronic disease. Practical hours should be accumulated from experiences across various diseases and conditions to help prepare candidates for both examination and employment. Goals of the RCEP certification are (1) to provide the consumer and employer with the assurance that successful candidates are adequately prepared to work with patients with a variety of chronic diseases and special populations, and (2) to improve the visibility and acceptance of the clinical exercise physiologist among the public and other health professionals.

The ACSM has an emphasis on international outreach not only for membership but also for its professional certifications. The ACSM states on its website that there are over 25,000 certified professionals (both clinical and health fitness tracks) in 44 countries. The organization also maintains an international subcommittee on the Certification and Registry Board, which handles all issues related to certification.

Canadian Society of Exercise Physiology (CSEP): www.csep.ca

According to its website, the CSEP (10) is the principal body for physical activity, health and fitness research, and personal training in Canada. The CSEP is considered the gold standard for health and fitness professionals who are dedicated to getting Canadian citizens physically active. The CSEP directs a part of its effort toward providing a certification for those who are exercise physiologists.

Certified Exercise Physiologist The CSEP-certified exercise physiologist (CSEP-CEP) title was developed and approved as a certification in 2006. According to the CSEP website, the CSEP-CEP "performs assessments, prescribes conditioning exercise, as well as exercise supervision, counseling and healthy lifestyle education in apparently healthy individuals and/or populations with medical conditions, functional limitations or disabilities associated with musculoskeletal, cardiopulmonary, metabolic, neuromuscular, and ageing conditions." Certification requires a university degree in physical activity, exercise science, kinesiology, human kinetics, or a related field and passing written and practical examinations. The CSEP offers regular workshops to assist candidates in preparing for the national examination process, although attendance is not a requirement to sit for the examination.

Exercise and Sports Science Australia (ESSA): www.essa.org.au

ESSA (formerly the Australian Association for Exercise and Sports Science) website states that it is a professional organization committed to establishing, promoting, and defending the career paths of exercise and sport science practitioners. Its vision is to enhance the health and performance of all Australians by supporting exercise and sport science professionals.

Accredited Exercise Physiologist ESSA does not offer a certification examination for its clinical exercise professionals. The accredited exercise physiologist (AEP) accreditation was designed to fit the needs of individuals who are working in a clinical setting. In Australia this accreditation is required for people who seek to become a Medicare (Australian), Department of Veterans Affairs and Healthfund or WorkCover provider. ESSA approves courses and other studies that are delivered by AEPs to maintain continuing professional development credits. ESSA also approves published journal articles, scientific conference presentations, and grant funding submissions.

PROFESSIONALIZATION OF CLINICAL EXERCISE PHYSIOLOGY

The biomedical literature continues to expand on our understanding that physical activity plays a direct role in preventing the development of many chronic diseases, as well as playing an important role in inhibiting disease progression. In addition, there are several emerging areas of research, including inactivity physiology and its relationship to metabolic health risk; the relationship of exercise and physical activity to cognitive health and progression of neurodegenerative disease; and the importance of exercise in the treatment of cancer and peripheral artery disease. Given these and other findings on the importance of exercise, in combination with the increasing number of people 65 yr of age and older in the world population, employment opportunities for the clinical exercise physiologist continue to expand. Possibly the most important task of the future clinical exercise physiologist will be to review and disseminate the vast amounts of exercise-related research and recommendations to the masses. The trend of late is increasingly away from the "one-size-fits-all" exercise prescription and toward recommendations specific to a given population. The complexity of today's health information often makes it difficult for the general public and patients to comprehend. The clinical exercise physiologist should prepare to be the intermediary between knowledge and understanding of the exercise prescription.

In the United States, beginning in April 2004 under the auspices of the Commission on Accreditation of Allied Health Education Programs (CAAHEP), the Committee on Accreditation for the Exercise Sciences (CoAES) accomplished its goal of establishing guidelines and standards for postsecondary academic institutions for personal fitness trainer and exercise physiology/science academic programs at both the bachelor's and master's degree levels. To date, six academic institutions have attained the CoAES exercise physiology program accreditation. It is anticipated that over the next several years, more academic programs throughout the United States will apply for and receive accreditation at the undergraduate (exercise science) and graduate (applied exercise physiology and clinical exercise physiology) levels. No other independent, third-party organization currently accredits academic exercise science or exercise physiology programs in North America.

Students interested in becoming a clinical exercise physiologist should seek the guidance of professionals in the field to select a program that also provides a well-rounded set of practical, laboratory, and research experiences aimed at best preparing them to successfully work in the field. Enrolling in a CoAES-accredited program would help to ensure the quality of education with respect to these program offerings. Currently, although most clinical exercise physiologists are employed in cardiopulmonary rehabilitation programs, more programs designed to help care for patients with other chronic diseases and disabilities are being launched each year. These other programs are usually targeted at patients with chronic kidney disease, cancer, chronic fatigue, arthritis, and metabolic syndrome, among others.

Those interested in cardiac or pulmonary rehabilitation should become familiar with the American Association of Cardiovascular and Pulmonary Rehabilitation (AACVPR). The AACVPR publishes a directory of programs in the United States and in other countries, which is an ideal resource for identifying potential employers or clinical internship sites. The AACVPR also offers a career hotline to members and nonmembers that lists open positions and provides a site for posting resumes (1). The ACSM's monthly journal *Medicine and Science in Sports and Exercise (MSSE)* lists exercise physiology–related employment and educational positions. In addition, posted positions can be accessed through the ACSM website (4). The website www.exercisejobs.com lists positions nationwide and provides the opportunity to post a resume online. In Australia, the website www.simplyhired.com.au provides a listing of potential places of employment for the exercise physiologist.

Another area within the clinical exercise profession that is experiencing growth involves programs that use exercise in primary prevention. Many hospitals and corporations recognize that regular exercise training can reduce future medical expenses and increase productivity, and many of these programs are implemented in a medical fitness center setting. In fact, most medically based fitness facilities now exclusively hire people with exercise-related degrees as a means to ensure safety for an increasingly diverse clientele that includes both healthy people and individuals with clinically manifest disease. To learn more about this exercise physiology specialty, visit the website of the Medical Fitness Association (MFA) at www.medicalfitness.org.

Also important in the area of primary prevention is the "Exercise is Medicine" (EIM) initiative, launched as a joint effort by the ACSM and the American Medical Association. The EIM has become a global entity and lists initiatives currently in Latin America, the middle-East, and Australia in addition to the ongoing work in the United States. According to the EIM website (www.exerciseismedicine.org), the EIM sees itself as an ongoing initiative that:

1. Creates broad awareness that exercise is indeed medicine.
2. Makes "level of physical activity" a standard vital sign question in each patient visit.
3. Helps physicians and other health care providers to become consistently effective in counseling and referring patients as to their physical activity needs.
4. Leads to policy changes in public and private sectors that support physical activity counseling and referrals in clinical settings.
5. Produces an expectation among the public and patients that their health care providers should and will ask about and prescribe exercise.
6. Appropriately encourages physicians and other health care providers to be physically active themselves.

The EIM initiative recently developed a credential aimed at exercise professionals. The goal of the credential is to accredit individuals who will work closely with the medical community. This three-level credential is based on three components:

1. Professional preparation necessary to safely and effectively prescribe exercise to a patient population
2. Development of the skills needed to work within the health care system
3. Development of the skills needed to support sustained behavior change

The level of certification one can achieve is based on previous levels of professional certification, university degree, and amount of clinical exercise experience.

Those seeking employment or attempting to maintain professional certification would also be wise to join one or more professional organizations. Besides the national-level organizations previously mentioned, many clinical

CoAES Sponsors

American Association of Cardiovascular and Pulmonary Rehabilitation	www.aacvpr.org
American College of Sports Medicine	www.acsm.org
American Council on Exercise	www.acefitness.org
American Kinesiotherapy Association	www.akta.org
Cooper Institute	www.cooperinst.org
National Academy of Sports Medicine	www.nasm.org
National Strength and Conditioning Association	www.nsca.com

Reprinted from www.coaes.org/sponsors.

exercise physiologists also join a regional chapter of the ACSM or a state chapter of the AACVPR, CEPA, or CSEP. These state or regional chapters all hold local educational programs that provide professionals an avenue through which to gain continuing education credits and an opportunity to network with those who make decisions to hire clinical exercise physiologists.

Several other professional organizations publish or present exercise-related research or information that focuses on specific diseases or disorders. These include the American Heart Association (AHA), the American College of Cardiology (ACC), the American Diabetes Association (ADA), the American Academy of Orthopedic Surgeons, the American Cancer Society, the Multiple Sclerosis Society, and the Arthritis Foundation. The National Center on Physical Activity and Disability (NCPAD) rounds out this listing of such organizations with its focus on various disabilities and conditions. Table 1.2 lists the websites for these and other organizations, most of which serve as excellent information resources. Finally, clinical exercise physiologists should stay abreast of current research in general exercise physiology, as well as research specific to their disease area of interest. The body of scientific information specific to clinical exercise physiology continues to grow rapidly. Regular journal reading is an excellent way to keep current in the field.

Table 1.2 Selected Internet Sites of Chronic Disease and Disability Organizations and Institutes

Category	Organization or institute	Internet site
Endocrinology and metabolic disorders	American Diabetes Association	www.diabetes.org
	National Institute of Diabetes and Digestive and Kidney Diseases	www.niddk.nih.gov
	National Kidney Foundation	www.kidney.org
	North American Association for the Study of Obesity	www.naaso.org
Cardiovascular diseases	American Heart Association	www.americanheart.org
	American College of Cardiology	www.acc.org
	Heart Failure Society of America	www.hfsa.org
	National Heart, Lung, and Blood Institute	www.nhlbi.nih.gov
	Vascular Disease Foundation	www.vdf.org
Respiratory diseases	American Lung Association	www.lungusa.org
	National Heart, Lung, and Blood Institute	www.nhlbi.nih.gov
Oncology and immune diseases	American Cancer Society	www.cancer.org
	National Cancer Institute	www.nci.nih.gov
Bone and joint diseases and disorders	American Academy of Orthopedic Surgeons	www.aaos.org
	American College of Rheumatology and Association of Rheumatology Health Professionals	www.rheumatology.org
	Arthritis Foundation	www.arthritis.org
	International Osteoporosis Foundation	www.osteofound.org
	National Institute of Arthritis and Musculoskeletal and Skin Diseases	www.niams.nih.gov
	Spondylitis Association of America	www.spondylitis.org
Neuromuscular disorders	National Multiple Sclerosis Society	www.nmss.org
	National Institute of Neurological Disorders and Stroke	www.ninds.nih.gov
	National Spinal Cord Injury Association	www.spinalcord.org
Special populations	American Geriatrics Society	www.americangeriatrics.org
	National Institute on Aging	www.nia.nih.gov
	National Institute of Child Health and Human Development	www.nichd.nih.gov
	National Women's Health Information Center	www.4woman.gov
Disabilities	National Center on Physical Activity and Disability	www.ncpad.org

After a person has begun practicing as a clinical exercise physiologist, he should consider using several safeguards. As only one state, Louisiana, now requires licensing of exercise physiologists, almost anyone in the other 49 states can claim to be a clinical exercise physiologist. Licensure may help to solve this issue, but as previously mentioned, this is a slow, state-by-state process. A major issue related to licensure and professionalization is the risk that practicing professionals assume when working with clients, as discussed in practical application 1.2. Professionals must protect themselves against potential litigation related to their practice. Practical application 1.3 describes an interesting case that highlights both the potential risk and the important role that a clinical exercise physiologist can play in litigation. Clinical exercise physiologists should consider malpractice insurance as a safeguard in the event of such incidents. In a hospital setting the institution often covers malpractice insurance; but if working in fitness, small clinical, ambulatory care, or personal training settings, exercise professionals should inquire about and obtain, if necessary, their own insurance coverage. Organizations such as the ACSM offer discounted group rates for malpractice insurance for the exercise professional.

Limiting legal risk is important to the practicing exercise physiologist. Within the confines of the civil justice system, under which personal injury and wrongful death cases are determined, clinical exercise physiologists could be held accountable to a variety of professional standards and guidelines. These standards and guidelines, sometimes referred to as practice guidelines, can apply to some practices carried out by clinical exercise physiologists and others who provide services within health and fitness facilities. A variety of professional organizations have developed and published guidelines or standards dealing with the provision of service by fitness professionals, including clinical exercise physiologists. Practically, professionals use standards and guidelines statements to identify probable and evidence-based benchmarks of expected service delivery owed to patients and clients, which helps ensure appropriate and uniform delivery of service.

In the course of litigation, these standards statements are used by expert witnesses to establish the standard of care to which providers will be held accountable in the event of patient injury or death. In years past, expert witnesses often used their subjective and personal opinions to establish the standard of care and then provided an opinion as to whether that standard was violated in the actual delivery of service. Today, most professional organizations use an evidence-based approach to establish standards of care for the following purposes:

- To achieve a consensus of standards of care among professionals that practitioners could aspire to meet in their delivery of services in accordance with known and established benchmarks of expected behavior
- To reduce cases of negligence and findings of negligence or malpractice
- To minimize the significance of individually and subjectively expressed opinions in court proceedings
- To reduce potential inconsistent verdicts and judgments that arose from the expression of individual opinions

Practice guidelines are a reference that is readily available when one wishes to compare the actual delivery of service to what the profession as a whole has determined to be established benchmarks. Consequently, clinical exercise physiologists should review relevant published practice guidelines and then consider in what situations the standards might act as a shield to protect against negligent actions. Others may engage in the same process to determine whether such standards should be used as a sword to attack the care provided to clients in particular cases. Clinical exercise physiologists can use the following strategies to limit exposure and risk.

- In programs that include patients with known disease, and in which diagnosis, prescription, or treatment is the probable or actual reason for performing the procedures, a physician should be involved in a significant way. This involvement must be meaningful and real if it is to be legally effective. Examples include referring patients, receiving and acting on results provided by rehabilitation staff, and discussing findings with the rehabilitation or therapy staff.
- In programs that have patients who demonstrate a high risk of suffering an adverse event or injury during exercise, the physician should be within immediate proximity to the participant during initial testing. The physician need not be watching the patient or the electrocardiogram monitor, but she should be controlling the staff during the procedure. Thereafter, the physician's proximity of supervision in further exercise testing should be dictated by the medical interpretation of each patient's initial test result.
- Evaluate and screen participants before recommending or prescribing activity, especially in rehabilitative or preventive settings.
- Secure a medically mandated consent-type document when involved in procedures such as a graded exercise test.
- Develop and assist individuals with implementing a safe and evidence-based exercise prescription that

is aimed at a patient's goals and addresses disease-related disability.

- Recognize and refer people who have conditions that need evaluation before commencing or continuing with various activity programs.
- Provide feedback to referring professionals relative to the progress of participants in activity.
- Provide appropriate, timely, and effective emergency care, as needed.

Practical Application 1.3

LEGAL CASE

To appreciate what can be at stake for health professionals who become involved in courtroom litigation, consider this case of an exercise physiologist working in the clinical setting. She wrote of her litigation experiences in the first person. Excerpts follow.

"The hospital where I work is a tertiary care facility. . . . The staff members in Cardiac Rehabilitation are required to have a master's degree in exercise physiology, American Heart Association basic cardiac life support certification, and American College of Sports Medicine certification as an exercise specialist. . . ."

"In May of 1991, Cardiac Rehabilitation was consulted to see a 78 yr old woman 4 d post–aortic valve surgery. Her medical history included severe aortic stenosis, coronary artery disease, left ventricular dysfunction, carotid artery disease, atrial fibrillation, hypertension, and noninsulin dependent diabetes. . . . She was in atrial fibrillation but had stable hemodynamics, and was ambulating without problems in her room."

"The patient was first seen by Cardiac Rehabilitation 7 d postsurgery. Chart review . . . revealed no contraindications to activity. . . . Initial assessment revealed normal supine, sitting, standing, and ambulating hemodynamics. She walked independently in her room, and approximately 100 ft (30 m) in 5 min . . . using the handrails for support. . . . General instructions for independent walking later that day were given. . . ."

"On postop day 8, there was documentation in the chart regarding the patient's need for an assist device while ambulating. . . . The patient ambulated approximately 200 ft (60 m) with a quad cane. . . . She appeared stable. . . ."

"On postop day 9, I was informed by the patient's nurse, as well as by documentation in the chart, that the patient had fallen in her room that morning, apparently hitting her head on the floor. . . . A computed tomography scan of the head was negative."

"On postop day 10, I observed the patient sitting with her daughter in the hall. . . . Her mental status had noticeably changed. She did not recognize her daughter, where she was, or why she was in the hospital. . . . No further formal exercise was performed. . . . The patient was transferred to intensive care and later died of a massive brain hemorrhage."

"Two years after the incident, a hospital lawyer contacted me and informed me that the family had brought suit against the hospital contending that the patient fell because she was not steady enough to walk independently. . . . I was called to testify and explain my participation in her care. . . ."

"I met with the hospital's defense lawyer on two separate occasions. The first meeting, the lawyer inquired about my education, certification, years of employment, and what specifically my job entailed. . . . Two weeks before trial, the hospital's defense lawyer again questioned me. . . . He also played the role of the plaintiff's lawyer and reworded similar questions in a slightly different manner. The defense lawyer encouraged me to maintain consistency in my answers, and remain calm and assured. . . . To prove our case, we had to establish, by a preponderance of the evidence (i.e., more likely than not) that the patient was indeed ambulating independently prior to her fall. My documentation was crucial. . . ."

"Answering the plaintiffs' lawyer's questions during cross-examination was like a mental chess game. . . . My answers were crucial, as was my composure and the belief that all my training and certification *did* qualify me as a professional health care provider. . . . Although the only hard evidence I had to work with in the courtroom was my chart notes, the judge did allow me to supplement them with oral testimony. . . ."

"Whether you document in the patient's chart or on a separate summary sheet that is filed in your office, your documentation is a permanent record that provides powerful legal evidence. . . ."

"Later I was told that my testimony helped the hospital prove the patient's fall was not a result of negligence. The court ruled in favor of our hospital."

CONCLUSION

More than ever before, clinical exercise physiology is becoming an allied health profession recognized both by other practicing clinicians and by the general public. Every person who considers himself or herself a clinical exercise physiologist should become involved in local or national organizations dedicated to promoting this profession. Given the growing evidence for and acceptance of the role of physical activity and exercise in the prevention and treatment of chronic disease, the timing has never been better for permanently solidifying the role of the clinical exercise physiology profession in the health care of the population. A primary purpose of this text is to provide aspiring and practicing clinical exercise physiologists with up-to-date and practical information to prepare for or to maintain certification and to continue to provide contemporary and evidence-based patient care. To that end, this third edition of *Clinical Exercise Physiology* provides a comprehensive and practical review dealing with the most common chronic diseases. Each chapter serves as a guide for important issues regarding client–clinician interaction through the development of a comprehensive exercise prescription.

Chapter 2

Behavioral Approaches to Physical Activity Promotion

Gregory W. Heath, DHSc, MPH

The clinical exercise physiologist can use a number of behavioral strategies in assessing and counseling individual patients or clients about their physical activity behavior change. The behavioral strategies discussed in this chapter are intended to be used in the context of supportive social and physical environments. A nonsupportive environment is considered a barrier to regular physical activity participation; consequently, if this barrier is not altered, change is unlikely to occur. Thus, one of the clinician's goals is to identify environmental barriers with the client, include steps on how to overcome these barriers, and build supportive social and physical environments as part of the counseling strategy. Although no guarantees can be made, the literature suggests that if clinicians take a behavior-based approach to physical activity counseling, within the context of a supportive environment, they may indeed experience greater success in getting their clients moving. Therefore, this chapter also presents information about the role of social and contextual settings in promoting health- and fitness-related levels of physical activity. The most important task of the clinical exercise physiologist is to guide a client into a lifelong pattern of regular, safe, and effective physical activity.

BENEFITS OF PHYSICAL ACTIVITY

The exercise physiologist, who understands the physiological basis for activity as well as the impact of pathology on human performance, is well positioned for such counseling. However, to be an effective counselor, the clinical exercise physiologist also needs to understand human behavior in the context of the individual client's social and physical milieu. This chapter seeks to underscore some of the important theories and models of behavior that have been shown to be important adjuncts for clinicians seeking to help people make positive changes in their physical activity behavior.

The historical literature has established evidence that persons who engage in regular physical activity have an increased physical working capacity (59); decreased body fat (51); increased lean body tissue (51); increased bone density (52); and lower rates of coronary heart disease (CHD) (39), diabetes mellitus, hypertension (55), and cancer (27). Increased physical activity is also associated with greater longevity (40). Regular physical activity and exercise can also assist persons in improving mood and motivational climate (38), enhancing their quality of life, improving their capacity for work and recreation, and altering their rate of decline in functional status (50). A recent international review of the health benefits of physical activity reinforces these findings (28).

When one is promoting planned exercise and physical activity, one must pay attention to specifically designed outcomes. Notably, one needs to account for the health and fitness outcomes of a well-designed exercise prescription. Finally, a number of physiological, anatomical, and behavioral characteristics should also be considered to ensure a safe, effective, and enjoyable exercise experience for the participant.

Health Benefits

Physical activity has been defined as any bodily movement produced by skeletal muscles that results in caloric expenditure (13). Because caloric expenditure uses energy and because energy use enhances weight loss or weight maintenance, caloric expenditure is important in the prevention and management of obesity, CHD, and diabetes mellitus. Healthy People 2020 Physical Activity Objective PA-2.1 (58) highlights the need for *every adult to engage in moderate aerobic physical activity for at least 150 min per week or 75 min of vigorous aerobic physical activity per week or an equivalent combination.* Current research suggests that engaging regularly in moderate aerobic physical activity for at least 150 min per week or 75 min of vigorous physical activity per week will help ensure that the calories expended confer specific health benefits. For example, daily physical activity equivalent to a sustained walk for 30 min per day for 7 d would result in an energy expenditure of about 1,050 kcal per week. Epidemiologic studies suggest that a weekly expenditure of 1,000 kcal could have significant individual and public health benefits for CHD prevention, especially among those who are initially inactive. More recently, the American College of Sports Medicine and the American Heart Association concluded that the scientific evidence clearly demonstrates that regular, moderate-intensity physical activity provides substantial health benefits (18). In addition, following an extensive review of the physiological, epidemiological, and clinical evidence, the Scientific Committee for the National Physical Activity Guidelines formulated this guideline:

> *Every U.S. adult should accumulate a minimum of 150 min or more of moderate-intensity aerobic physical activity or 75 min of vigorous aerobic physical activity per week* (18).

This guideline emphasizes the benefits of moderate and vigorous aerobic physical activity that can be accumulated in bouts of 10 min of exercise or more. Intermittent activity has been shown to confer substantial benefits. Therefore, the recommended minutes of activity can be accumulated in shorter bouts of 10 min spaced throughout the day. Although the accumulation of 150 min of moderate-intensity or 75 min of vigorous-intensity aerobic physical activity per week has been shown to confer important health benefits, these guidelines are not intended to represent the optimal amount of physical activity for health but instead a minimum standard or base on which to build to obtain more specific outcomes related to physical activity and exercise. Specifically, selected fitness-related outcomes may be a desired result for the physical activity participant, who may seek the additional benefits of improved cardiorespiratory fitness, muscle endurance, muscle strength, flexibility, and body composition. Indeed, Healthy People 2020 actually took these guidelines even further by proposing additional objectives. Objective PA-2.2 is to increase the proportion of adults who engage in aerobic physical activity of at least moderate intensity for more than 300 min/week, or more than 150 min/week of vigorous intensity, or an equivalent combination; objective PA-2.3 is to increase the proportion of adults who perform muscle-strengthening activities on 2 or more days of the week (58).

Fitness Benefits

Regular vigorous physical activity helps achieve and maintain higher levels of cardiorespiratory fitness than moderate physical activity. There are five components of health-related fitness: cardiorespiratory fitness, muscle strength, muscle endurance, flexibility, and enhanced body composition (see "Examples of Health and Fitness Benefits of Physical Activity").

Cardiorespiratory fitness or aerobic capacity refers to the body's ability to perform high-intensity activity for a prolonged period of time without undue physical stress

Examples of Health and Fitness Benefits of Physical Activity

Health benefits

- Reduction in premature mortality
- Reduction in cardiovascular disease risk
- Reduction in colon cancer
- Reduction in type 2 diabetes mellitus
- Improved mental health

Fitness benefits

- Cardiorespiratory fitness
- Muscle strength and endurance
- Enhanced body composition
- Flexibility

or fatigue. Having higher levels of cardiorespiratory fitness enables people to carry out their daily occupational tasks and leisure pursuits more easily and with greater efficiency. Vigorous physical activities such as the following help to achieve and maintain cardiorespiratory fitness and can also contribute substantially to caloric expenditure:

Very brisk walking
Jogging, running
Lap swimming
Cycling
Fast dancing
Skating
Rope jumping
Soccer
Basketball
Volleyball

These activities may also provide additional protection against CHD over moderate forms of physical activity. People can achieve higher levels of cardiorespiratory fitness by increasing the frequency, duration, or intensity of an activity beyond the minimum recommendation of 20 min per occasion, on three occasions per week, at more than 45% of aerobic capacity (3).

Muscular strength and endurance are the ability of skeletal muscles to perform work that is hard or prolonged or both (3). Regular use of skeletal muscles helps to improve and maintain strength and endurance, which greatly affects the ability to perform the tasks of daily living without undue physical stress and fatigue. Examples of tasks of daily living include home maintenance and household activities such as sweeping, gardening, and raking. Engaging in regular physical activity such as weight training or the regular lifting and carrying of heavy objects appears to maintain essential muscle strength and endurance for the efficient and effective completion of most activities of daily living throughout the life cycle (3). The prevalence of such physical activity behavior is still quite low, with recent prevalence estimates indicating that only 19.6% of the adult population engages in strength training at least twice per week (56).

Musculoskeletal flexibility refers to the range of motion in a joint or sequence of joints. Joint movement throughout the full range of motion helps to improve and maintain flexibility (3). Those with greater total body flexibility may have a lower risk of back injury (10). Older adults with better joint flexibility may be able to drive an automobile more safely (3, 61). Engaging regularly in stretching exercises and a variety of physical activities that require one to stoop, bend, crouch, and reach may help to maintain a level of flexibility that is compatible with quality activities of daily living (3).

Excess body weight occurs when too few calories are expended and too many consumed for individual metabolic requirements (41). The maintenance of an acceptable ratio of fat to lean body weight is another desired component of health-related fitness. The results of weight loss programs focused on dietary restrictions alone have not been encouraging. Physical activity burns calories, increases the proportion of lean to fat body mass, and raises the metabolic rate (62). Therefore, a combination of caloric control and increased physical activity is important for attaining a healthy body weight. The 2010 United States Dietary Guidelines (57) have highlighted the importance of increasing the duration of moderate-intensity physical activity to 60 to 90 min per day as the necessary dose to prevent weight gain and regain, respectively.

PARTICIPATION IN REGULAR PHYSICAL ACTIVITY

In designing any exercise prescription, the professional needs to consider various physiological, behavioral, psychosocial, and environmental (physical and social) variables that are related to participation in physical activity (47). Two commonly identified determinants of physical activity participation are **self-efficacy** and **social support**.

Self-efficacy, a construct from social cognitive theory, is most characterized by the person's confidence to exercise under a number of circumstances and appears to be positively associated with greater participation in physical activity. Social support from family and friends has consistently been shown to be associated with greater levels of physical activity participation. Incorporating some mechanism of social support within the exercise prescription appears to be an important strategy for enhancing compliance with a physical activity plan (23). Common barriers to participation in physical activity are time constraints and injury. The professional can take these barriers into account by encouraging participants to include physical activity as part of their lifestyle, thus not only engaging in planned exercise but also incorporating transportation, occupational, and household physical activity into their daily routine.

Participants can also be counseled to help prevent injury. People are more likely to adhere to a program of low- to moderate-intensity physical activities than one comprising high-intensity activities during the early phases of an exercise program. Moreover, moderate activity is less likely to cause injury or undue discomfort (43).

A number of physical and social **environmental factors** can also affect physical activity behavior (48). Family and friends can be role models, can provide encouragement, or can be companions during physical activity. The physical environment often presents important barriers to participation in physical activity, including a lack of bicycle trails and walking paths away from vehicular traffic, inclement weather, and unsafe neighborhoods (49). Sedentary behaviors such as excessive television viewing or computer use may also deter persons from being physically active (49).

Risk Assessment

An exercise prescription may be fulfilled in at least three different ways: (1) on a program-based level that consists primarily of supervised exercise training (24); (2) through exercise counseling and exercise prescription followed by a self-monitored exercise program (26); and (3) through community-based exercise programming that is self-directed and self-monitored (63).

Within supervised exercise programs and programs offering exercise counseling and prescription, participants should complete a brief medical history and risk factor questionnaire and a preprogram evaluation (3). More information on the medical history and risk factor questionnaire and preprogram evaluation is presented in chapter 4.

When one is developing a community-based, self-directed program, medical clearance is left to the judgment of the individual participant. An active physical activity promotion campaign in the community seeks to educate the population regarding precautions and recommendations for moderate and vigorous physical activity (14). These messages should provide information that participants must know before beginning a regular program of moderate to vigorous physical activity. This information should encompass the following:

1. Awareness of preexisting medical problems (e.g., CHD, arthritis, osteoporosis, or diabetes mellitus)
2. Consultation before starting a program, with a physician or other appropriate health professional, if any of the previously mentioned problems are suspected
3. Appropriate mode of activity and tips on different types of activities
4. Principles of training intensity and general guidelines as to rating of perceived exertion and training heart rate
5. Progression of activity and principles of starting slowly and gradually increasing activity time and intensity
6. Principles of monitoring symptoms of excessive fatigue
7. Making exercise fun and enjoyable

Theories and Models of Physical Activity Promotion

Historically, the most common approach to exercise prescription taken by health professionals has been direct information. In the past, the counseling sequence often consisted of the following:

1. Exercise assessment, usually cardiorespiratory fitness measures
2. Formulation of the exercise prescription
3. Counseling the patient regarding
   - mode (usually large-muscle activity),
   - frequency (three to five sessions per week),
   - duration (20-30 min per session), and
   - intensity (assigned target heart rate based on the exercise assessment) (25)
4. Review of the exercise prescription by the health professional and participant
5. Follow-up
   - Visits (reassessments and revising of the exercise prescription)
   - Phone contact

Most of the research evaluating this traditional approach to exercise prescription has not been too favorable in terms of its results with respect to long-term compliance and benefits (42). That is, most people who begin an exercise program drop out during the first 6 mo. Why has the traditional information-sharing approach been used? Because it's easiest for the clinician, requires less time, and is prescriptive. However, it is not interactive with the client. More recently, contemporary theories and models of human behavior have been examined and developed for use in exercise counseling and interventions (1, 2, 4-6, 32, 33, 44, 45). These theories, referred to as cognitive–behavioral techniques, represent the most salient theories and models that have been used to promote the initiation of and adherence to physical activity. These approaches vary in their applicability to physical activity promotion. Some models and theories were designed primarily as guides to understanding behavior, not as guides for designing intervention protocols. Others were specifically constructed with a view toward developing cognitive–behavioral techniques for physical activity behavior initiation and maintenance. Consequently, the clinical exercise physiologist may find that the majority of the theories summarized in table 2.1 will assist in understanding physical activity behavior change. Nevertheless, other theories have evolved

Table 2.1 Summary of Theories and Models Used in Physical Activity Promotion

Theory or model	Level	Key concepts
Health belief model (45)	Individual	Perceived susceptibility Perceived severity Perceived benefits Perceived barriers Cues to action Self-efficacy
Relapse prevention (32, 33)	Individual	Skills training Cognitive reframing Lifestyle rebalancing
Theory of planned behavior (1, 2)	Individual	Attitude toward behavior Outcome expectations Value of outcome expectations Subjective norm Beliefs of others Motive to comply with others Perceived behavioral control
Social cognitive theory (4, 6)	Interpersonal	Reciprocal determinism Behavioral capability Self-efficacy Outcome expectations Observational learning Reinforcement
Social support (5)	Interpersonal	Instrumental support Informational support Emotional support Appraisal support
Ecological perspective (29)	Environmental	Multiple levels of influence: • Intrapersonal • Interpersonal • Institutional • Community • Public policy
Transtheoretical model (11, 30, 31, 44)	Individual	Precontemplation Contemplation Preparation Action Maintenance

Adapted from K. Glanz and B.K. Rimer, 1995, *Theory-at-a-glance: A guide for health promotion practice.* (U.S. Department of Health and Human Services).

sufficiently to provide specific intervention techniques to assist in behavior change.

The Patient-Centered Assessment & Counseling for Exercise & Nutrition (PACE and PACE+) materials were developed for use by the primary care provider in the clinical setting targeting apparently clinically healthy adults (30). The materials have been evaluated for both acceptability and effectiveness in a number of different clinical settings (11, 30, 31). Sample materials taken from PACE have been included in practical application 2.1. These materials are intended to provide a quick look at the steps in assessing and counseling an individual for physical activity. The materials incorporate many of the principles from a number of theoretical constructs reviewed in table 2.1. For further explanation of PACE materials, visit the PACE website at www.paceproject.org. Wankel and colleagues (60) demonstrated the effectiveness of cognitive–behavioral techniques for enhancing physical activity promotion efforts in showing that the use of increased social support and decisional strategies improved adherence to exercise classes among participants. Martin and colleagues (34) demonstrated through a series of studies the positive effects of personalized praise and feedback and the use of flexible goal setting among participants on exercise class adherence. Participants in the enhanced intervention group demonstrated an 80% attendance rate during the intervention compared with the control group's 50% attendance rate (34).

McAuley and coworkers (35) successfully emphasized strategies to increase self-efficacy and thereby increase physical activity levels among adult participants in a community-based physical activity promotion program. These successful strategies included social modeling and social persuasion to improve compliance and exercise adherence. Promoting physical activity through home-based strategies holds much promise and might prove to be cost-effective (54). Through tailored mail and telephone interventions, significant levels of social support and reinforcement have been shown to enhance participants' self-efficacy in complying with exercise prescription, thus significantly improving levels of physical activity (15).

Finally, it has been demonstrated that **lifestyle-based physical activity** promotion increases the levels of moderate physical activity among adults. Lifestyle-based physical activity focuses on home- or community-based participation in many forms of activity that include much of a person's daily routine (e.g., transport, home repair and maintenance, yard maintenance) (16). This approach evolved from the idea that physical activity health benefits may accrue from an accumulation of physical activity minutes over the course of the day (59). Because lack of time is a common barrier to regular physical activity, some researchers recommend promoting lifestyle changes whereby people can enjoy physical activity throughout the day as part of their lifestyle. Taking the stairs at work, taking a walk during lunch, and walking or biking for transportation are all effective forms of lifestyle physical activity. Assessing the common barriers (table 2.2) to physical activity among participants can be helpful for developing individual awareness and targeting strategies to overcome the barriers.

Ecological Perspective

A criticism of most theories and models of behavior change is that they emphasize individual behavior change and pay little attention to sociocultural and physical environmental influences on behavior (7). Recently, interest has developed in ecological approaches to increasing

Table 2.2 Barriers to Being Active Quiz: What Keeps You From Being More Active?

Instructions: Listed here are reasons that people give to indicate why they do not get as much physical activity as they should. Please read each statement and circle the number that represents how likely you are to say it.

How likely are you to say?	Very likely	Somewhat likely	Somewhat unlikely	Very unlikely
1. My day is so busy now that I just don't think I can make the time to include physical activity in my regular schedule.	3	2	1	0
2. None of my family members or friends like to do anything active, so I don't have a chance to exercise.	3	2	1	0
3. I'm just too tired after work to get any exercise.	3	2	1	0
4. I've been thinking about getting more exercise, but I just can't seem to get started.	3	2	1	0
5. I'm getting older so exercise can be risky.	3	2	1	0

How likely are you to say?	Very likely	Somewhat likely	Somewhat unlikely	Very unlikely
6. I don't get enough exercise because I have never learned the skills for any sport.	3	2	1	0
7. I don't have access to jogging trails, swimming pools, bike paths, and so forth.	3	2	1	0
8. Physical activity takes too much time away from other commitments—time, work, family, and so on.	3	2	1	0
9. I'm embarrassed about how I will look when I exercise with others.	3	2	1	0
10. I don't get enough sleep as it is. I just couldn't get up early or stay up late to get some exercise.	3	2	1	0
11. It's easier for me to find excuses not to exercise than to go out to do something.	3	2	1	0
12. I know of too many people who have hurt themselves by overdoing it with exercise.	3	2	1	0
13. I really can't see learning a new sport at my age.	3	2	1	0
14. It's just too expensive. You have to take a class or join a club or buy the right equipment.	3	2	1	0
15. My free times during the day are too short to include exercise.	3	2	1	0
16. My usual social activities with family or friends do not include physical activity.	3	2	1	0
17. I'm too tired during the week and I need the weekend to catch up on my rest.	3	2	1	0
18. I want to get more exercise, but I just can't seem to make myself stick to anything.	3	2	1	0
19. I'm afraid I might injure myself or have a heart attack.	3	2	1	0
20. I'm not good enough at any physical activity to make it fun.	3	2	1	0
21. If we had exercise facilities and showers at work, then I would be more likely to exercise.	3	2	1	0

Follow these instructions to score yourself:

- Enter the numbers you circled in the spaces provided, putting the number for statement 1 in space 1, for statement 2 in space 2, and so on.
- Add the three scores on each row. Your barriers to physical activity fall into one or more of seven categories: lack of time, social influences, lack of energy, lack of willpower, fear of injury, lack of skill, and lack of resources. A score of 5 or above in any category shows that the barrier is an important one for you to overcome.

1 =	8 =	15 =	Row sum =	Lack of time
2 =	9 =	16 =	Row sum =	Social influence
3 =	10 =	17 =	Row sum =	Lack of energy
4 =	11 =	18 =	Row sum =	Lack of willpower
5 =	12 =	19 =	Row sum =	Fear of injury
6 =	13 =	20 =	Row sum =	Lack of skill
7 =	14 =	21 =	Row sum =	Lack of resources

From Centers for Disease Control and Prevention. Available: www.cdc.gov

Practical Application 2.1

THE PACE+ MODEL

Telling patients what to do doesn't work, especially over the long term. An effective behavioral model helps to facilitate long-term changes by telling them how to change. PACE (Patient-Centered Assessment & Counseling for Exercise & Nutrition) is a comprehensive approach to physical activity and nutrition counseling that uses materials developed by a team of researchers at San Diego State University. The curriculum draws heavily on the "stages of change" model, which suggests that individuals change their habits in stages. Taking into account each person's readiness to make changes, PACE provides tailored recommendations for patients in each stage. PACE offers three different counseling protocols. Empirically derived behavioral strategies are applied in each protocol. The development of the PACE model began in 1990, with funding originally from the Centers for Disease Control and Prevention, the Association of Teachers of Preventive Medicine, and San Diego State University. The original PACE materials, first released in 1994, dealt only with physical activity. The program was originally developed to overcome barriers to physician counseling for physical activity—especially lack of time for counseling, lack of standardized counseling protocols, and lack of training in behavioral counseling. Counseling was designed to be delivered in 2 to 5 min during a general patient checkup.

The PACE+ materials were thoroughly tested and found to be acceptable and usable by health care providers and patients across the United States. Physicians also found PACE to be practical, improving their confidence in counseling patients about physical activity (12). In a controlled efficacy study of 212 sedentary adults, patients who received PACE counseling increased their minutes of weekly walking by 38.1 compared with 7.5 among the control group. Additionally, 52% of the patients who received PACE counseling adopted some physical activity compared with 12% of the control group (12).

Since these earlier studies, the current PACE+ materials have been revised to include the recommendations from the Surgeon General's report *Physical Activity and Health* (59), as well as the *Physical Activity Guidelines* (42). The current materials also address nutrition behaviors such as decreasing dietary fat consumption; increasing fruit, vegetable, and fiber consumption; and balancing caloric intake and expenditure for weight control (12).

---

participation in physical activity (46). These approaches place the creation of supportive environments on par with the development of personal skills and the reorientation of health services. Creation of supportive physical environments is as important as intrapersonal factors when behavior change is the defined outcome. Stokols (53) illustrated this concept of a health-promoting environment by describing how physical activity could be promoted through the establishment of environmental supports such as bike paths, parks, and incentives to encourage walking or bicycling to work. An underlying theme of ecological perspectives is that the most effective interventions occur on multiple levels. Interventions that simultaneously influence multiple levels and multiple settings (e.g., schools, worksites) may be expected to lead to greater and longer-lasting changes and maintenance of existing health-promoting habits.

In addition, investigators have recently demonstrated that behavioral interventions primarily work by means of mediating variables of intrapersonal and environmental factors (46). Mediating variables are those that facilitate and shape behaviors—we all have a set of intrapersonal factors (e.g., personality type, motivation, genetic predispositions) and environmental factors (e.g., social networks like family, cultural influences, and the built and physical environments). However, few researchers have attempted to delineate the role of these mediating factors in facilitating health behavior change. Sallis and colleagues (46) recently discussed how difficult it is to assess the effectiveness of environmental and policy interventions because of the relatively few evaluation studies available. However, based on the experience of the New South Wales (Australia) Physical Activity Task Force, a model has been proposed to help with understanding the steps necessary to implement these interventions (37). Figure 2.1 presents an adaptation of this model as prepared by Sallis and colleagues (46) and outlines the necessary interaction between advocacy, coordination, or planning, agencies, policies, and environments to make such interventions a reality. Another pragmatic model that appears to have relevance for the promotion of physical activity has been proposed by McLeroy and colleagues (36). This model specifies five levels of determinants for health behavior:

1. Intrapersonal factors, including psychological and biological variables, as well as developmental history
2. Interpersonal processes and primary social groups, including family, friends, and coworkers
3. Institutional factors, including organizations such as companies, schools, health agencies, or health care facilities
4. Community factors, which include relationships among organizations, institutions, and social networks in a defined area
5. Public policy, which consists of laws and policies at the local, state, national, and supranational levels

Important in implementing this concept of behavioral determinants is realizing the key role of behavioral settings, which are the physical and social situations in which behavior occurs. Simply stated, human behavior such as physical activity is shaped by its surroundings—if you're in a supportive social environment with access to space and facilities, you are more likely to be active. It is important to acknowledge the determinant role of selected behavioral settings: Some are designed to encourage healthy behavior (e.g., sport fields, gymnasiums, health clubs, and bicycle paths), whereas others encourage unhealthy (or less healthy) behaviors (e.g., fast food restaurants, vending machines with high-fat and high-sugar foods, movie theaters). We need to understand the environment in which our client lives. These structures (e.g., fields, gymnasiums, community centers) are part of each of our living environments; people who disregard them are less likely to be active. For physical activity providers, a potentially important adjunct in assessing and prescribing physical activity interventions for participants is understanding the physical and social contexts in which their patients live. This information can be obtained from various sources and can be at the level of the individual or at the more general community level. When individual physical activity behavior information is coupled with sociodemographic, physical, and social context information, physical activity

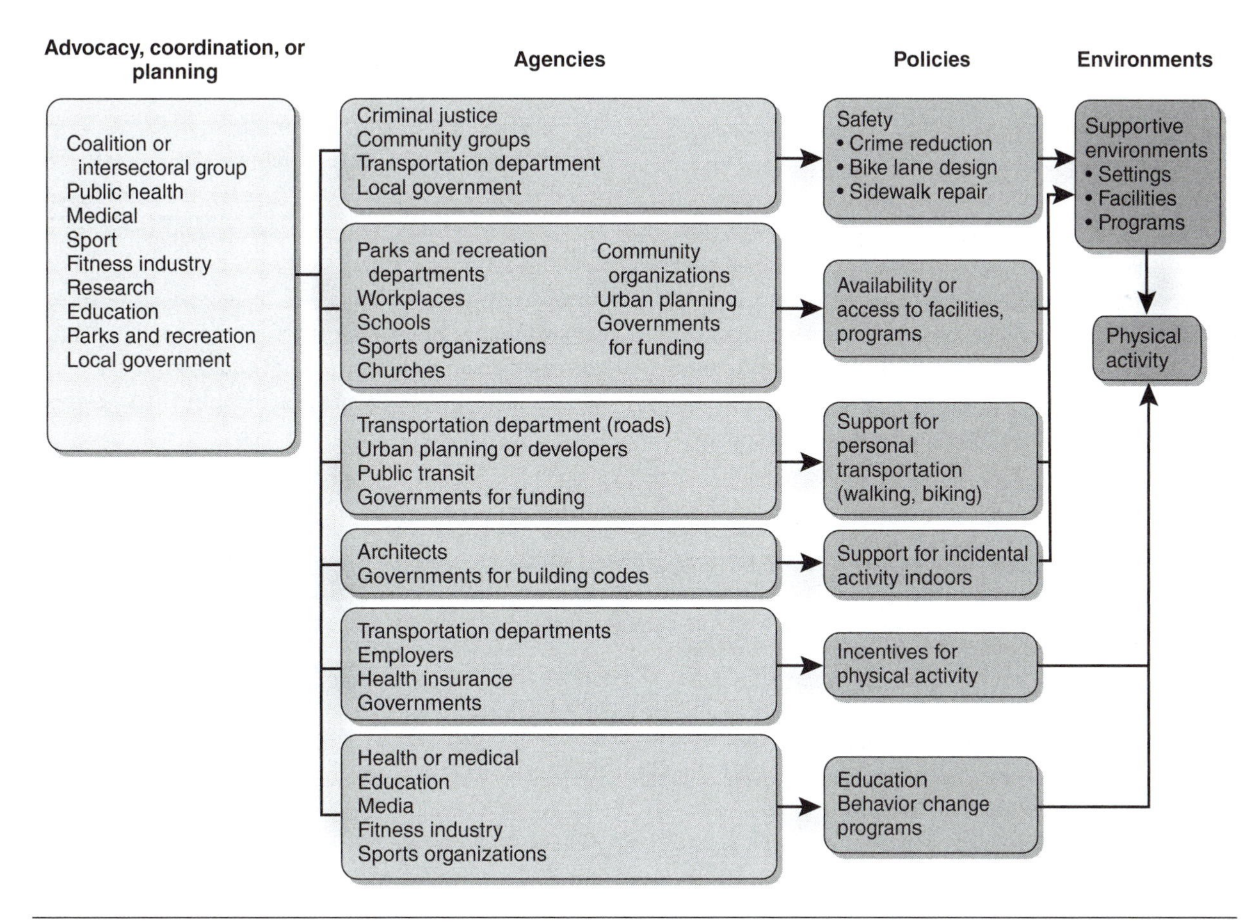

Figure 2.1 Conceptualization of the development of policy and environmental interventions to promote physical activity.

Reprinted from *American Journal of Preventive Medicine*, Vol. 15, J.F. Sallis et al., "Environmental and policy interventions to promote physical activity," pgs. 379-397, Copyright 1998, with permission from Elsevier.

interventions can be further tailored to maximize the participant's physical activity behavior change and maintenance plan. Exercise physiologists cannot alter the client's physical environment; however, they should be able to address environmental barriers and provide insights into how to overcome these barriers. In the long run, we all should be a part of changing our environments for the better.

An example of such tailoring for physical activity promotion that alters physical activity behaviors is the work of Linenger and colleagues (29), an effort to increase physical activity levels among naval personnel through a multifactorial environmental and policy approach to physical activity promotion. These investigators compared an "enhanced base" to a "control base." The enhancements involved increasing the number of bike trails on base, acquiring new exercise equipment for the local gym, opening a women's fitness center, instituting activity clubs, and providing released time for physical activity and exercise (29). The changes were positive for those living on the enhanced base—that is, they increased their physical activity levels.

Another example, this time emphasizing an incentive-based approach to promoting physical activity, is the work of Epstein and Wing (17). Although this work was undertaken quite some time ago, the lessons are very relevant in today's inactive culture. In this study, contracts and the use of a lottery (a popular enterprise today!) were used to boost exercise attendance with the consequence of increasing participants' overall physical activity levels. Compared with results for a "usual care" comparison group, adherence and activity levels were significantly improved and sustained (37). However, caution in using an incentive-based approach has been urged by some researchers who believe that over the long term, participants never internalize the health behavior—meaning that they are likely to stray back to sedentary habits once the incentive is removed or loses its appeal. Nevertheless, incentives have been proven to be effective in the short term. Additional community-based environmental efforts to influence physical activity behavior have included the use of signs in public settings to increase use of stairs and walkways (8, 9). These latter studies are examples of single intervention efforts that can be carried out in concert with systematic exercise prescription efforts among individuals. Thus, the increase in stair usage as a result of a promotional campaign can help individuals meet their prescribed energy expenditure requirements. Table 2.3 outlines some of the common barriers to people becoming more physically active. Also listed are some suggested solutions, although these can vary from client to client.

Useful resources for the clinical exercise physiologist are the very recent evidence-based recommendations for physical activity promotion in communities (www.thecommunityguide.org/pa/default.htm). The evidence base for these recommendations provides insights into how exercise practitioners can integrate their clinical efforts to assess and counsel patients into supportive and reinforcing environments. The recommendations are summarized with respect to informational, behavioral–social, and environmental approaches to promoting physical activity (table 2.4) (20, 22). A more recent review further establishes these physical activity intervention domains and expands the number of evidence-based approaches to promoting physical activity (21).

Table 2.3 Tips on Overcoming Potential Barriers to Regular Physical Activity

Barriers	Suggestions for overcoming physical activity barriers
Lack of time	Identify available time slots. Monitor your daily activities for 1 wk. Identify at least three 30 min time slots you could use for physical activity.
	Add physical activity to your daily routine. For example, walk or ride your bike to work or shopping, organize school activities around physical activity, walk the dog, exercise while you watch TV, park farther from your destination.
	Make time for physical activity. For example, walk, jog, or swim during your lunch hour, or take fitness breaks instead of coffee breaks.
	Select activities that require minimal time, such as walking, jogging, or stair climbing.
Social influence	Explain your interest in physical activity to friends and family. Ask them to support your efforts.
	Invite friends and family members to exercise with you. Plan social activities that involve exercise.
	Develop new friendships with physically active people. Join a group, such as the YMCA or a hiking club.

Barriers	Suggestions for overcoming physical activity barriers
Lack of energy	Schedule physical activity for times in the day or week when you feel energetic. Convince yourself that if you give it a chance, physical activity will increase your energy level; then try it.
Lack of motivation	Plan ahead. Make physical activity a regular part of your daily or weekly schedule and write it on your calendar. Invite a friend to exercise with you on a regular basis. Then both of you write it on your calendars. Join an exercise group or class.
Fear of injury	Learn how to warm up and cool down to prevent injury. Learn how to exercise appropriately considering your age, fitness level, skill level, and health status. Choose activities that involve minimum risk.
Lack of skill	Select activities that require no new skills, such as walking, climbing stairs, or jogging. Exercise with friends who are at the same skill level as you are. Find a friend who is willing to teach you some new skills. Take a class to develop new skills.
Lack of resources	Select activities that require minimal facilities or equipment, such as walking, jogging, jumping rope, or calisthenics. Identify inexpensive, convenient resources available in your community (community education programs, park and recreation programs, worksite programs, and so on).
Weather conditions	Develop a set of regular activities that are always available regardless of weather (e.g., indoor cycling, aerobic dance, indoor swimming, calisthenics, stair climbing, rope skipping, mall walking, dancing, gymnasium games). Look at outdoor activities that depend on weather conditions (e.g., cross-country skiing, outdoor swimming, outdoor tennis) as bonuses—extra activities possible when weather and circumstances permit.
Travel	Put a jump rope in your suitcase and jump rope. Walk the halls and climb the stairs in hotels. Stay in places with swimming pools or exercise facilities. Join the YMCA or YWCA (ask about reciprocal membership agreement). Visit the local shopping mall and walk for 30 min or more. Take a portable audio player and listen to your favorite upbeat music as you exercise.
Family obligations	Trade babysitting time with a friend, neighbor, or family member who also has small children. Exercise with the kids—go for a walk together, play tag or other running games, or get aerobic dance or exercise music for kids (several are on the market) and exercise together. You can spend time together and still get your exercise. Hire a babysitter and look at the cost as a worthwhile investment in your physical and mental health. Jump rope, do calisthenics, ride a stationary bicycle, or use other home gymnasium equipment while the kids are busy playing or sleeping. Try to exercise when the kids are not around (e.g., during school hours or their nap time). Encourage exercise facilities to provide child care services.
Retirement years	Look at your retirement as an opportunity to become more active instead of less. Spend more time gardening, walking the dog, and playing with your grandchildren. Children with short legs and grandparents with slower gaits are often great walking partners. Learn a new skill that you've always been interested in, such as ballroom dancing, square dancing, or swimming. Now that you have the time, make regular physical activity a part of every day. Go for a walk every morning or every evening before dinner. Treat yourself to an exercise bicycle and ride every day while reading a favorite book or magazine.

Content in the "Personal Barriers" taken from *Promoting physical activity: A guide for community action,* 1999 (USDHHS).

Table 2.4 Summary of Recommended Physical Activity Interventions—Guide to Community Preventive Services

Intervention	Recommendation
INFORMATIONAL APPROACHES TO INCREASING PHYSICAL ACTIVITY	
Community-wide campaigns	Recommended (strong evidence)
Point-of-decision prompts	Recommended (sufficient evidence)
Mass media campaigns	Insufficient evidence
BEHAVIORAL AND SOCIAL APPROACHES TO INCREASING PHYSICAL ACTIVITY	
Individually adapted health behavior change	Recommended (strong evidence)
Health education with TV and video turnoff	Insufficient evidence
College-age physical and health education	Insufficient evidence
Family-based social support	Insufficient evidence
School-based physical education	Recommended (strong evidence)
Community social support	Recommended (strong evidence)
ENVIRONMENTAL AND POLICY APPROACHES TO INCREASING PHYSICAL ACTIVITY	
Creation of or enhanced access to places for physical activity	Recommended (strong evidence)
Community-scale urban design and land use	Recommended (sufficient evidence)
Street-scale urban design and land use	Recommended (sufficient evidence)
Transport policy and practices	Insufficient evidence

Reprinted from Guide to Community Preventive Services. Available: www.thecommunityguide.org/pa/pa.pdf

CONCLUSION

Within the past decade, physical activity has emerged as a key factor in the prevention and management of chronic conditions. Although the role of exercise in health promotion has been appreciated and applied for decades, recent findings regarding the mode, frequency, duration, and intensity of physical activity have modified exercise prescription practices. Included in these modifications has been the delineation between health and fitness outcomes relative to the physical activity prescription. Most importantly, new approaches to physical activity prescription and promotion that emphasize a behavioral approach with documented improvements in compliance have now become available to health professionals. Behavioral science has contributed greatly to the understanding of health behaviors such as physical activity. Behavioral theories and models of health behavior have been reexamined in light of physical activity and exercise. Although more research is needed to further develop successful, well-defined applications that are easily adaptable for intervention purposes, behavioral principles and guidelines have evolved that are designed to help the health professional understand health behavior change and guide people into lifelong patterns of increased physical activity and improved exercise compliance.

New frontiers in the application of exercise prescription to specific populations, as well as efforts to define the specific dose (frequency, intensity, duration) of physical activity for specific health and fitness outcomes, are now being explored. As this information becomes available, it must be introduced to the participant via the most effective behavioral paradigms, such as the models discussed in this chapter. Moreover, positive changes in the participant's physical and social environments must occur to enhance compliance with exercise prescriptions. In turn, increased levels of physical activity among all people will improve health and function.

Key Terms

cardiorespiratory fitness (p. 20)
environmental factors (p. 22)
excess body weight (p. 21)
lifestyle-based physical activity (p. 24)
muscular strength and endurance (p. 21)
musculoskeletal flexibility (p. 21)
physical activity (p. 20)
self-efficacy (p. 21)
social support (p. 21)

CASE STUDY

MEDICAL HISTORY

Mrs. KY is a 45 yr old Caucasian female, married with two teenage sons. She is employed as a senior manager at a large bank and reports experiencing an "above average" level of tension and stress. She presents at the referral of her primary care physician, who has observed that the client has elevated blood pressure and cholesterol levels that may be attributed to her stressful and highly sedentary job. In addition, the client admits that she would like to lose 30 to 40 lb (14 to 18 kg) and improve her fitness so that she can ride her bike with her husband on a community rail trail recently installed by her neighborhood.

DIAGNOSIS

Mrs. KY is a sedentary but otherwise healthy middle-aged female with significant risk factors for cardiovascular disease including obesity, dyslipidemia, and psychosocial stress. At the request of her physician she seeks to start exercising as part of a disease prevention program.

EXERCISE TEST RESULTS

The client is 5 feet 6 in. tall (168 cm) and weighs 196 lb (89 kg), with a body mass index of 31.7. She has a resting heart rate of 85 beats $\cdot$ min^{-1} and a resting blood pressure of 136/89 mmHg. Her total cholesterol is 198 mg $\cdot$ dl^{-1} untreated, and her high-density lipoproteins are 34 mg $\cdot$ dl^{-1}. Her graded treadmill stress test revealed that she has a $\dot{V}O_2$max of 20.5 ml $\cdot$ kg^{-1} $\cdot$ min^{-1}, which is normal for an unfit woman of her age range. Her electrocardiogram was also unremarkable at rest, as well as during and following her test. In addition, she reported smoking from ages 17 to 40. She also complains of occasional joint stiffness in her hands and ankles. The client describes herself as nonathletic and admits to never participating in an organized sport or exercise setting. She is aware of the benefits of exercise but did not feel an incentive to begin a formal program until her doctor's recommendation. The client jokes that although her workday is highly organized and structured, the rest of her life is chaotic and that it is due only to the support of her husband and kids that anything gets done at home. She laments that her eating habits are atrocious and that she is so tired when she gets home from work that she has only enough energy to make dinner before crashing in front of the television. She presents to you to start a workout program that will help achieve her goals.

EXERCISE PRESCRIPTION

The exercise plan including the traditional components—frequency, intensity, duration, and modality (discussed in detail in later chapters)—may be tailored to address specific risk factors. The subject's medical history, however, clearly indicates that this person had not prioritized exercise participation until she received her doctor's recommendation. Moreover, she presents with numerous potential barriers to engaging in a physically active lifestyle, as well as behaviors that contribute to her overweight status. The clinician should assist the participant in establishing awareness and developing strategies to address those barriers. Furthermore, the clinician should consider tailoring strategies for motivating the participant toward the adoption of a healthy, physically active lifestyle.

(continued)

DISCUSSION QUESTIONS

1. Applying the transtheoretical model, at what stage of exercise adoption is this client?
2. Based on your response to question 1, what types of interventions are most appropriate for this stage of change and why?
3. If you used the health belief model, what factors would you emphasize to achieve optimal exercise adherence?
4. How would Bandura's social cognitive theory be relevant to fostering exercise adherence for this client?

Chapter 3

General Principles of Pharmacology

Steven J. Keteyian, PhD

The material in this chapter assumes that most clinical exercise physiologists currently practicing in the field or students interested in becoming a clinical exercise physiologist likely did not or are yet to complete a separate semester-long course on pharmacology or drug therapeutics and exercise. Instead, most people now involved in or interested in the field received their pharmacology or medication learning through several lectures given as part of a broader exercise testing or prescription course. They have gathered or are gathering additional knowledge during a clinical internship experience, self-study, or on-the-job training and experience.

Obviously, clinical exercise physiologists are not involved in the prescription of medications, but we do work in environments where many, if not all, of the patients we help care for are prescribed medications. That said, a robust presentation on the topic of exercise and medications is beyond the scope of this book; therefore, this chapter instead focuses on essential principles and information pertaining to drug therapy in general. The disease-specific chapters in this textbook provide more specific information about which drugs are used to treat patients with specific clinical conditions, as well as how these agents might interact with exercise testing or training responses. In fact, most of these chapters include a table devoted to the pharmacology pertinent to the disease under discussion.

GENERAL PROPERTIES OF DRUGS

Drugs have two names: generic and brand. The generic name identifies the drug no matter who makes or manufactures it. The brand name refers to the name given by the company that makes or manufactures the drug. For example, carvedilol is the generic name for a drug that is manufactured by a drug company and sold under the brand name of Coreg. When used in a living organism, a drug is a chemical compound that yields a biologic response. By themselves, drugs do not confer new functions on tissues or an organ; instead they only modify existing functions. This means that a drug attenuates, accentuates, or replaces a response or function (10). For example, in a 50 yr old patient, a β-adrenergic blocking agent (i.e., a so-called β-blocker) attenuates the rate of the increase in heart rate during exercise, whereas glipizide is used to treat patients with type 2 diabetes because it binds to the plasma membrane of beta cells in the pancreas and accentuates or enhances insulin release from secretory granules. Finally, a 20 yr old patient with insulin-dependent diabetes is prescribed insulin to replace what is no longer produced by his own pancreas.

Another general principle about the effect of drugs is that none exert just a single effect. Instead, almost all drugs result in more than one response; some of these might be well tolerated by a patient whereas others might represent an undesirable response (i.e., side effect).

Routes of Administration

Before discussing how drugs are "moved" through the body and the effect of the drug on the body, it is important to address the various pathways through which a drug can enter the body. Deciding which route of administration works best for each patient is necessary to ensure that the drug is suitable, well tolerated, effective, and safe. Keep in mind that drugs are usually given for either local or systemic effects, with any significant absorption or loss of drug from the intended local site considered a negative for locally administered agents, whereas absorption into circulation is a must for agents targeting a more generalized or systemic effect.

With respect to drug administration, the two main routes are **enteral** and **parenteral**. Certain agents can also be effectively and safely delivered as an inhaled gas or fine mist through the pulmonary system or topically through the skin, eyes, ears, or mucosal membranes of the nasal passages.

Enteral

Although rectal absorption is an effective method for drug delivery in certain instances (e.g., patient vomits when given an oral agent), the majority (up to 80%) of all medications prescribed today are delivered enterally through oral ingestion; such a route is safe, economical, and convenient. Once placed in the mouth, drugs can be absorbed through the oral mucosa, in the stomach (gastric absorption), or in the small intestine. Although the mouth has a good blood supply and is lined with epithelial cells that make up the oral mucosa, few agents can be sufficiently disintegrated and dissolved to the extent that that the desired systemic response is carried out. One commonly prescribed agent that is rapidly broken down and absorbed through the oral mucosa is sublingual nitroglycerin. When it is placed under the tongue and the patient refrains from swallowing the saliva that contains the dissolved drug, this agent can act within minutes on angina pectoris that is due to myocardial ischemia in patients with coronary artery disease. Although the stomach has the potential to be an excellent site for the absorption of orally ingested drugs, the small intestine is by far the most responsible for this task. The small intestine has the large surface area (i.e., villi), the vascular blood supply, and the pH to facilitate absorption.

Parenteral

Parental means delivered by injection. Parenteral routes are many, but the three predominant routes are subcutaneous, intravenous, and intramuscular. Obviously, with the intravenous route, 100% of the drug is delivered into circulation, allowing for a higher concentration of the drug to be most quickly delivered to target tissues. A few other parenteral routes are intrathecal (directly into the cerebrospinal fluid), epidural, and intra-articular (directly into a joint).

Phases of Drug Effect

To produce the desired effect, the molecules in an administered drug must move from the point at which they enter the body to the target tissue. The magnitude of the effect then rendered is heavily influenced by the concentration of these molecules at the target site. For all of this to occur, we usually look at the administered drug passing through three distinct phases, the pharmaceutical phase, the **pharmacokinetic phase**, and the **pharmacodynamic phase** (8, 10, 14).

Pharmaceutical Phase

The pharmaceutical phase pertains to how a drug progresses from the state (e.g., solid) in which it was administered through the stages in which it is disintegrated and then dissolved in solution (i.e., dissolution). Obviously, the pharmaceutical phase applies only to drugs taken orally, because drugs taken subcutaneously, intramuscularly, or intravenously are already in solution. A drug taken orally in liquid form (vs. solid form) is more quickly dissolved in gastric fluid and ready for absorption through the small intestine of the gastrointestinal (GI) tract. Some common factors can accentuate or interfere with the dissolution process for certain drugs that are taken in solid form; these include the enteric coating of a tablet, crushing of the tablet, and the presence of food in the GI tract.

Pharmacokinetic Phase (The Effect of the Body on the Drug)

The movement of the drug's molecules through the body comprises the pharmacokinetic phase, which encompasses four discrete subphases: absorption, distribution, metabolism, and excretion. The absorption of dissolved drugs consumed orally from the GI tract to the blood occurs via either passive absorption (i.e., diffusion, no energy required), active absorption (i.e., involves a carrier such as a protein, requires energy), or pinocytosis. The latter refers to cells engulfing a drug to move it across a cell membrane. Factors that affect absorption include the pH of the drug, local blood flow to the GI tract (which can be negatively influenced during exercises that involve the skeletal muscles of the limbs), hunger, and food content in the GI tract. Certainly there are drugs given via eye drops, nasal sprays, respiratory inhalants, or transdermally or intramuscularly that are absorbed without the involvement of the GI system. Due to the number of surrounding blood vessels, drugs administered intramuscularly are taken up more readily than those given subcutaneously.

Bioavailability refers to the percentage of a drug that makes it all the way into the bloodstream. The bioavailability of drugs given intravenously is obviously 100%, whereas in agents taken orally the bioavailability is almost always less than 100%. Agents administered parenterally (e.g., subcutaneously, intramuscularly, intravenously) avoid first being transported through the liver via portal circulation, which occurs when agents are taken orally and absorbed through the GI tract. Therefore, more of a drug administered parentally is available to the tissues during its first pass through systemic circulation. For example, when taken orally, 90% of the drug nitroglycerin is metabolized before it even enters systemic circulation, because it is first transported from the GI tract through the liver.

The process of moving an absorbed drug to the target tissues is influenced not only by the concentration of the drug in the body and the flow of blood into the target tissue but also by the percentage bound to protein. Most drugs are bound to protein (usually albumin) at varying levels (high, moderately high, moderate, low) and while in this bound state are inactive with respect to causing a pharmacologic response. The portion of the drug not bound to protein, or active and able to exert its effect, is replenished from the bound portion as free drug decreases in circulation.

The biotransformation or metabolism of a drug occurs mainly in the liver, within which most drugs are both rendered inactive by the enzymes in this organ and readied for excretion. In some instances, the enzymatic actions that occur on a drug in the liver yield an active metabolite that accentuates the pharmacologic response.

With respect to metabolism, the **half-life (t-1/2)** of a drug refers to the time it takes for one half of the drug concentration to be eliminated. For example, 500 mg of drug A is absorbed, and it has a known half-life of 3 h. That means it takes 3 h to eliminate 250 mg, another 3 h to eliminate another 125 mg, and so on. Knowledge of a drug's half-life helps us compute how long it will take a drug to reach a stable concentration in the blood. Drugs with a half-life less than 8 h or more than 24 h are considered to have short or long half-lives, respectively.

The fourth and final subcomponent within the pharmacokinetic phase pertains to elimination, the main avenue of which is through the kidneys as urine; however, other routes such as bile, feces, saliva, and sweat contribute to elimination as well. All drugs that are free in plasma (not bound to protein) are filtered through the kidneys. Factors that can affect the ability of the kidneys to filter and therefore eliminate a drug include any disease that decreases glomerular filtration rate, any disorder that hampers renal tubular secretion, or an overall decrease in blood flow to the kidneys themselves.

Pharmacodynamic Phase (The Effect of the Drug on the Body)

The third phase pertains to the effect of drug molecules in the body and is referred to as the science of pharmacodynamics. Such effects are usually categorized as primary or secondary, with the former referring to the planned or therapeutic effect and the latter pertaining to either a desired or an unwanted effect (side effect). Again using β-blockers as an example, in patients recovering from a myocardial infarction, these agents are planned and prescribed because they improve one's likelihood of survival significantly over the next 5 yr. Coincidently, these same agents also lessen the frequency and severity of migraine headaches, which could easily be a welcomed or desired secondary effect for the patient who suffers from migraines and is also recovering from a myocardial infarction.

The association between the amount or dose of a drug and the body's response is referred to as dose–response relationship. Although no two patients yield the exact same response to the same dose of a drug, due to a variety of factors discussed in more detail later in this chapter, in general, less of a drug yields a lesser response and more of a drug (up to a point) yields a greater response. The amount of any drug that is associated with its greatest response (i.e., beyond which no further increase in drug yields an observed increase in response) is called the drug's maximum drug effect or maximal efficacy. For example, two agents might be used to treat the pruritis (itching) and the disrupted skin associated with eczema. The first agent might be an over-the-counter product with a maximal drug effect that helps relieve the itching but has little effect on the disrupted skin. The maximal drug effect of the second medication not only relieves itching but also heals the broken or disrupted skin.

Mechanism of Action

The means by which a drug produces an alteration in function at the target cells is referred to as its *mechanism of action,* with most drugs either (a) working through a protein receptor found on the cell membrane, (b) influencing the effect of an enzyme, or (c) involving some other nonspecific interaction. The drug–protein receptor mechanism is based on the premise that certain drugs have a selective, high affinity for an active or particular portion of the cell. This portion of the cell is referred to as a receptor and once stimulated produces a biologic effect. Simply speaking, the receptor represents the "keyhole" that a unique key (i.e., the drug) can fit into and yield a specific response (i.e., open the door). Those keys that best fit the keyhole yield the greatest response, whereas any other keys that also fit the keyhole, but just not quite as well as the best-fitting key, yield a lesser response.

With respect to the drug–receptor interaction, a drug that combines with the receptor and leads to physiologic changes or responses is referred to as an **agonist**. Conversely, an agent or drug that interferes with or counteracts the actions of an agonist is called an **antagonist**. A *competitive antagonist* is an agent that inhibits or interferes with an agonist by binding to the exact same receptor that the agonist is targeting, thus preventing its action. β-blockers slow heart rate and lessen inotropicity by blocking the β-1 receptors found on cardiac cells—cells that when normally stimulated by the agonist norepinephrine contribute to increases in both heart rate and force of contraction.

As already mentioned, a second type of mechanism of action involves the interaction between an enzyme and a drug. Since enzymes work as catalysts that control biochemical reactions, inhibition of the enzymes via their binding with certain drugs can inhibit subsequent catalyzed chemical reactions. For example, among patients with an elevated blood concentration of low-density lipoprotein (LDL), a class of drugs commonly referred to as statins are used because they inhibit the normal action of a rate controlling enzyme in the liver. This enzyme normally facilitates the production of cholesterol by the liver, so the inhibition of this process leads to reductions in both total cholesterol and LDL cholesterol levels.

It is also important to know that many agonists and antagonists lack specific effects, in that the response produced by the receptor that they influence varies based on the location of the receptor in the body. For example, in the salivary glands, stimulation of the cholinergic receptors with the agonist acetylcholine leads to an increase in saliva production, the correct response for someone preparing to chew food. However, stimulation of another cholinergic receptor located elsewhere in the body, again by acetylcholine, results in different responses such as pupillary constriction in the eye and a slowing of heart rate.

Before leaving our discussion about the pharmacodynamic phase of drugs, there is one more term you should become familiar with, **therapeutic index** (TI). The therapeutic index helps describe the relative safety of a drug, representing the ratio between toxic $dose_{50}$ (TD_{50}) and effective $dose_{50}$ (ED_{50}), where TD_{50} equals the dose of drug that is toxic in 50% of humans and ED_{50} is the dose of the drug needed to be therapeutically effective in 50% of a like population of humans. The closer the TI for a drug is to 1, the greater the danger of toxicity; the higher the TI, the wider the margin of safety and the less likely the agent will result in harmful or toxic effects. For the higher-TI drugs, plasma drug levels do not need to be monitored as frequently. Figure 3.1 shows how the initial dose and subsequent doses maintain drug concentration within the therapeutic range.

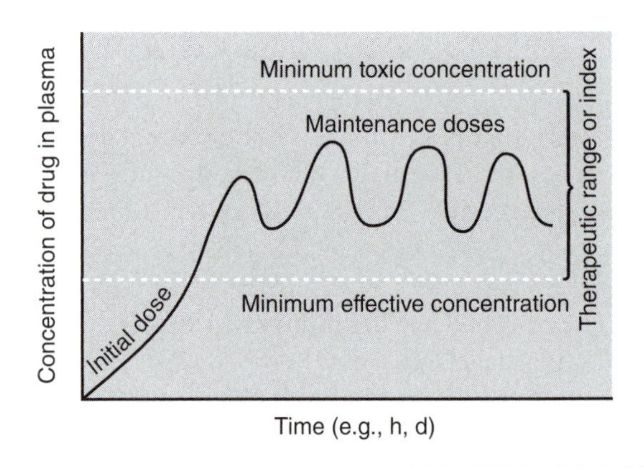

Figure 3.1 The relationship between the regular administration of a drug over time and the maintenance of a plasma concentration that yields the desired therapeutic effect.

PHARMACOTHERAPY

The science of prescribing or individualizing drug therapy based on patient-specific characteristics is referred to as *therapeutics* or *pharmacotherapy*. Just as with developing a safe exercise prescription, the prescription of a medication by a licensed professional involves choosing the correct mode of administration (e.g., oral, subcutaneous), intensity (i.e., dose), frequency of administration per day or per week, and duration or time (e.g., 3 d, 10 d, permanently). To accomplish this, "orders" are written to communicate to the patient, the person filling the prescription, and other health care professionals what the prescribed therapy plan is. Table 3.1 shows a brief list of the acceptable and common terms and abbreviations used in health care. As mentioned earlier, since clinical exercise physiologists work in environments or settings where many, if not all, of the patients are under the care of a physician and taking prescribed medications, it is important that you familiarize yourself with this language in order to effectively perform your work duties and contribute to patient care.

The majority of the remainder of this chapter provides other important medication-related information that the clinical exercise physiologist needs to be familiar with. Although exercise physiologists play no role in determining what medications are prescribed, how often they are to be taken, or in what form they are to be taken, we may be called upon to administer a medication, such as sublingual nitroglycerin, if the agent is part of a program-related "standing order" approved by a physician. Additionally, we often care for patients over many visits and as a result are in a unique position to help ensure that patients are taking their drug as prescribed (see practical application 3.1) and are aware of any factors that might influence an agent's efficacy.

Table 3.1 Acceptable and Common Terms and Abbreviations Pertinent to Drug Measurements, Delivery, and Times of Administration

DRUG MEASUREMENTS AND DRUG FORMS		ROUTES OF DRUG ADMINISTRATION		TIMES OF ADMINISTRATION	
Abbreviation	**Meaning**	**Abbreviation**	**Meaning**	**Abbreviation**	**Meaning**
Cap	capsule	A.D., ad	right ear	AC, ac	before meals
dr	dram	A.S., as	left ear	ad lib	as desired
elix	elixir	A.U., au	both ears	B.i.d., b.i.d.	twice daily
g, gm, G, GM	gram	ID	intradermal	$\bar{c}$ (c with line above)	with
gr	grain	IM	intramuscular	Hs	hour of sleep
gtt	drops	IV	intravenous	NPO	nothing by mouth
kg	kilogram	IVPB	intravenous piggyback	PC, pc	after meals
l, L	liter	KVO	keep vein open	PRN, p.r.n.	whenever necessary, as needed
m^2	square meter	L	left	Q	every
mcg, μg	microgram	NGT	nasogastric tube	Qam	every morning
mEq	milliequivalent	O.D., od	right eye	Qh	every hour
mg	milligram	O.S., os	left eye	q2h	every 2 hours
mL, ml	milliliter	O.U., ou	both eyes	q4h	every 4 hours
oz	ounce	PO, po, os	by mouth	q6h	every 6 hours
pt	pint	R	right	q8h	every 8 hours
qt	quart	SC, subc, sc, SQ, subQ	subcutaneous	$\bar{s}$ (s with line above)	without
SR	sustained release	SL, sl, subl	sublingual	Stat	immediately
supp	suppository	TKO	to keep open	T.i.d., t.i.d.	three times a day
susp	suspension				
T.O.	telephone order				
T, tbsp	tablespoon				
t, tsp	teaspoon				
V.O.	verbal order				

Practical Application 3.1

CLIENT–CLINICIAN INTERACTION

Over the decades, if not over a full century, the use of medicine to treat illness has progressed from an effort that was somewhat simple and ineffective to one that is highly complex and quite effective. Added concerns about the polypharmacy experienced by many patients today, complex physician–nurse workflows, and lack of patient knowledge about their medications (4) all contribute to increasing the potential risk of harm to patients. The formal process that compares a patient's current medications to those in the patient record or medication orders is referred to as **medical reconciliation** (4, 6).

Making these comparisons at each encounter that occurs between a patient and a health care provider is solely intended to avoid errors or conflicts and prevent unwanted omissions. Clinical exercise physiologists who help care for a patient in any setting (hospital, ambulatory clinic, physician's office, medical fitness center) are also responsible for reconciling a patient's medications, with the goal of ensuring that all medications are being continued, discontinued, or modified as prescribed.

The reason for the importance of medical reconciliation is that in the United States alone, inaccurate medications lists in an ambulatory clinic cause a larger number of fatal adverse drug effects (1 of 131 outpatient deaths) than in a hospital setting (1 of 854 inpatient deaths) (9). Many of these are due to omitting a medication, taking a drug at the wrong dose or frequency, or taking the wrong drug. The process of medical reconciliation works, with several studies showing that emergency room visits and adverse drug events are reduced as a result.

Although reconciling with patients the medications they are taking with what is prescribed might seem tedious or unnecessary, it helps to reduce errors and improve documentation (6, 11, 13, 15). To begin the medical reconciliation process, you need a current medication list that should probably include all drugs being taken at home, regardless of whether they are prescribed or over-the-counter medications (including herbal and nutritional supplements). Also, gather information about dose, route, and frequency; time of last dose; who prescribed or recommended the agent (self, pharmacist, or physician); and how the patient is complying. Don't hesitate to ask patients to bring in all the medications they are currently taking to help ensure that their list reflects actual actions.

There is no perfect list, but given most of this information, clinicians can go on to the next step in the reconciliation process, which is comparing the information they are provided by the patient against documents in the patient's medical records. Forms or source documents to seek out to obtain the correct information on prescriptions might include hospital discharge summaries, nursing notes, pharmacy records, and notes from the last clinic encounter with a physician or physician extender (nurse practitioner, physician assistant).

In this process, not only are you looking to be sure that patients are taking medications as prescribed in their medical records; you should also be taking the initiative and looking beyond the obvious. For example, it may be that a physician has switched a patient's blood pressure–lowering medication because another drug is less expensive, but the patient forgot or was confused and did not stop taking the drug that was being replaced. Any "red flag" concerns like this should be identified for the patient and discussed with her doctor, then followed up with the patient to correct any errors. Again, all health care professionals are obligated to look for and assist with correcting any potential errors in a patient's drug regimen.

Factors That Modify Drug Response or Activity

As one might assume, a host of factors can influence an individual's response to a given drug. Some of these factors are easily recognized, well understood, and predictable, whereas others are less obvious. Several of the more common influencing factors or characteristics are discussed next.

Age

Younger children and older adults respond more than do postpubescent youth through age 65 yr, and as a result the doses may have to be modified. As well, older adults may be taking multiple agents, and various drug interactions can affect a drug's efficacy or lead to unwanted side effects.

Body Mass

Adjusting the dosage of a drug to match a patient's body mass or body surface area helps ensure that the concentration of drug that reaches specific body tissues is not higher or lower than desired. For adults between the ages of 18 and 65 yr, most drug calculations assume a 154 pound (70 kg) person.

Sex

Women are smaller than men, and the dose may have to be adjusted for this difference in body mass. Other sex-specific reasons that might lead a physician to adjust the dose in women include the states of pregnancy and lactation and differences in the amount of body fat and water in women, either of which might influence the action of a particular drug compared to that in men.

Other Factors

Many other factors can also affect a drug's actions; these include time of administration, severity of coexisting diseases (e.g., diseases of the kidney), genetics, patient mood or mental state, and surrounding environment (e.g., ambient temperature, altitude). Concerning time of day, this is tied to the absence or presence of food, which must be considered if the absorption of a drug in the GI tract is or is not hampered by the presence of food. Finally, as vague as the concept might seem, keep in mind that a patient's mood or attitude regarding the effectiveness of a drug can influence its effect. Clearly, the placebo effect is but one example of how mental state can affect drug response; conversely, distrust or skepticism might attenuate a drug's effect.

Compliance

What happens when a patient does not go to the pharmacy and get her prescription filled, or stops taking her medication too soon, or does not take the medications as prescribed? Are these likely occurrences, and are there factors or characteristics we can identify beforehand that would allow us to identify a patient at risk for noncompliance? These are important questions that can not only affect the health of the patient but also are associated with great public health and economic burdens.

The obvious main concern is that noncompliance may not allow a patient to get better or will worsen a patient's disease state. This can translate to increased morbidity, treatment failure, more frequent physician or emergency room visits, and even premature death. Noncompliance, which includes both not taking a medication as prescribed once the prescription is filled and not filling the prescription, costs more than $100 billion annually. Noncompliance also contributes to 11% of all hospitalizations and 125,000 unnecessary deaths and may be higher than 50% for some medications (3, 16). The economic impact of noncompliance includes increased absenteeism from work, lost productivity at work, and lost revenues to pharmacies and pharmaceutical manufacturers.

Persistence is the opposite of noncompliance and often decreases over time, with the greatest declines observed within the first 6 mo after starting a drug. Persistence for some drugs may fall as low as 35% at 5 yr. Factors that negatively influence persistence (or contribute to noncompliance) include prior compliance behaviors, one's ability to integrate the complexity of a drug regimen into daily life, health beliefs, health literacy, and social support. Patients at very high risk for noncompliance include those who are asymptomatic (e.g., hypertension), are suffering from a chronic versus an acute condition, are cognitively impaired (i.e., dementia), are taking frequent doses per day (vs. once a day), are taking multiple medications, and are experiencing drug-related side effects (2, 7).

The factor of health literacy is an increasingly important issue that pertains to how well an individual is able to obtain, process, and understand basic health information and to do so in a manner that allows him to make appropriate health decisions and comply with health-related directions (e.g., restrict or increase physical activity, take medications) (12). Although health literacy is related to functional literacy (e.g., ability to read, write, solve problems), it is also related to age, physical disabilities, psychological or emotional state, culture, and past experiences. Health literacy affects an individual's ability to navigate the health care system, communicate with providers, engage in self-care, and adopt health-promoting behaviors. These individuals are more likely to have chronic diseases, miss appointments, experience preventable hospital admissions, and suffer poorer overall care. As a clinician you can help patients who may have limited health literacy if you take the time needed during each conversation and keep technical information simple by using "living room" language. Additionally, associate difficult concepts with a context that is familiar to the patient; use multiple teaching modes (e.g., pictures and demonstrations); assess comprehension by having the patient "teach back"; and be patient, respectful, and sensitive.

In general, documenting noncompliance is not an easy task, and it is important to appreciate that compliance in general is essentially independent of age, sex, race, income, education level, and intelligence. Strategies shown to improve compliance include effective counseling and communication between patients and clinicians and between patients and pharmacists, special labels and written materials, self-monitoring, and follow-up. The RIM (Recognize, Identify, Manage) technique is a structured tool to facilitate patient compliance.

> R = Recognize: Using both subjective (i.e., ask the patient) and objective (pill counts) evidence to document that compliance is, in fact, an issue.
>
> I = Identify: Determine the causes of noncompliance using supportive, probing questions and empathetic statements. Responses pertaining to noncompliance may range from lack of financial resources and side effects to doubts about a drug's effectiveness.
>
> M = Manage: Using counseling, contracting, and multiple options for teaching (devices, telephones, calendars, daily pill boxes) and communication, to establish effective short-term and intermediate-term plans aimed at improving compliance and monitor, follow up, and adjust strategies as indicated.

Almost all medications taken orally by patients require some period of time before they exert their desired action in the body. Because of this delay, medications that might influence the exercise response may need to be timed such that they appropriately affect or do not affect the exercise response. For example, in patients with heart disease or heart failure, the β-adrenergic blocking agent carvedilol, as well as other agents within the same class such as metoprolol or atenolol, should be taken at least 2.5 h before exercise training or before exercise testing is performed for the purpose of developing or revising an exercise prescription. This will help ensure that the medication is sufficiently absorbed and exerting its heart rate–controlling effect. Additionally, some medications are best taken before sleep because any associated mild side effects are better tolerated when the person is less active, inactive, or asleep.

Finally, some patients will approach you and ask what they should do if they have missed a dose of their medication: They are unsure whether they should take the missed dose late or simply wait and take the next scheduled dose as planned. Ideally patients should have been informed about how to handle this scenario beforehand, but that does not always occur. One option is for patients to check the literature that came with the prescription, which often includes instructions about what to do if a dose is missed. Another option is to recommend that patients speak with a pharmacist (or their doctor). If neither of these options is available, a general recommendation is for people to take the medication as soon as they remember that they missed a dose. However, if the amount of time before the next planned dose is less than 50% of what it would be on the normal schedule, the patient should probably not take the missed dose and should instead wait until the next scheduled time. What patients must *not* do is take a double dose; they must avoid taking the dose they missed and the next scheduled dose at the same time.

Table 3.2 provides a list of common medications that may affect exercise capacity or heart rate or blood pressure response to exercise (1). The list is not meant to be comprehensive but should help the clinical exercise physiologist appreciate the wide range of drugs that can affect exercise-related responses, as well as reinforce the helpful approach of learning about medication effects by class or type of drug rather than individual agents.

Table 3.2 Common Medications and a Summary of Their Effect on Heart Rate and Blood Pressure During Exercise and on Exercise Capacity

Organ system, drug class or disease state	Heart rate response	Blood pressure response	Exercise capacity	Comments
ANTIARRHYTHMIC AGENTS				
Amiodarone (Cordarone) Sotalol (Betapace)	↓	↔	↔	
β-ADRENERGIC BLOCKING AGENTS				
Atenolol (Tenormin) Metoprolol (Lopressor SR, Toprol XL) Carvedilol (Coreg)	↓	↓	↑ in patients with angina ↓ or ↔ in patients without angina	Carvedilol has both α- and β-adrenergic blocking properties

Organ system, drug class or disease state	Heart rate response	Blood pressure response	Exercise capacity	Comments
BRONCHODILATORS				
Albuterol (Proventil, Ventolin) Ipratropium (Atrovent)	↑ ↔	↑ ↔	↔	
CALCIUM CHANNEL BLOCKERS				
Diltiazem (Cardizem CD) Verapamil (Calan, Isoptin) Amlodipine (Norvasc) Nifedipine (Adalat, Plendil, Procardia XL)	↓ ↔	↓	↑ in patients with angina ↔ in patients without angina	Negative chronotropic effect more pronounced with diltiazem and verapamil
CARDIAC GLYCOSIDES				
Digoxin (Lanoxin)	↓ ↔	↔	Possible ↑ in patients with atrial fibrillation or heart failure	Negative chronotropic effect more likely in patients with atrial fibrillation or heart failure
Nicotine	↑ ↔	↑	Possible ↓ in patients with angina	
NITRATES				
Isosorbide mononitrate (ISMO, Imdur, Monoket) Isosorbide dinitrate (Isordil)	↑ ↔	↔	↑ in patients with angina ↔ in patients without angina	
PSYCHOTROPICS				
Fluoxetine (Prozac) Sertraline (Zoloft) Paroxetine (Paxil) Citalopram (Celexa) Venlafaxine (Effexor) Bupropion (Wellbutrin)	↑ ↔	↓ ↔	↔	
THYROID AGENTS				
Levothyroxine (Synthroid)	↑ ↔	↑ ↔	↔ but may lead to a ↓ if angina is worsened	
VASODILATORS				
Angiotensin-converting enzyme inhibitors Captopril (Capoten) Enalapril (Vasotec) Lisinopril (Prinivil, Zestril) α-Adrenergic blocking agents Cardura (Doxazosin) Flomax (Tamsulosin) Minipress (Prazosin) Terazosin (Hytrin)	↔ ↔	↔ ↓	↑ in patients with chronic heart failure ↔	

↑ = increase; ↓ = decrease; ↔ = no change.

CONCLUSION

The widespread use of pharmacological agents in our society requires the clinical exercise physiologist to be familiar with many of the essential elements associated with how drugs are administered and handled in the body, as well as factors that affect a drug's action and patient compliance. Exercise physiologists are likely to encounter patients who are taking various pharmacological agents; they should do the extra work needed to learn about agents they are unfamiliar with and assist whenever and wherever possible with ensuring that patients comply with their prescribed medication regimens.

Key Terms

agonist (p. 36)
antagonist (p. 36)
enteral (p. 34)
half-life (t-1/2) (p. 35)
medical reconciliation (p. 38)
parenteral (p. 34)
pharmacodynamic phase (p. 34)
pharmacokinetic phase (p. 34)
therapeutic index (p. 36)

CASE STUDY

MEDICAL HISTORY

Mr. MT is a 46 yr old white male with a family history of cardiovascular disease, high blood pressure (150/94 mmHg), and hypercholesterolemia (total cholesterol = 284 mg · dl^{-1}, high-density lipoprotein cholesterol = 35 mg · dl^{-1}). Because of his extremely high risk of cardiac disease, his primary care physician has started him on the following medications:

Metoprolol, 100 mg per day (a β-blocker)
Atorvastatin, 10 mg per day (a cholesterol-lowering agent)
Niacin, 1,000 mg per day (an agent that will increase high-density lipoprotein cholesterol)
Aspirin, 81 mg per day (anti-platelet agent)

DIAGNOSIS AND PLAN

Mr. MT is referred to you for an exercise prescription as part of a comprehensive therapeutic lifestyle modification program to deal with his cardiovascular risk factors.

EXERCISE TEST RESULTS

No previous exercise test results are available.

EXERCISE PRESCRIPTION

The goal is to determine an appropriate exercise regimen given this patient's current medical therapy. Given the lack of a preliminary exercise test and the potential influence of his medical therapy (β-blocker) on heart rate responses, rating of perceived exertion (RPE) is used to determine the appropriate exercise intensity. The following exercise prescription is developed:

Frequency = 5 d per week
Intensity = 98 to 118 beats · min^{-1}
RPE = 11 to 15
Duration = 30 min
Mode = aerobic, such as walking, swimming, and cycling

At a scheduled follow-up visit, the patient complains of increased sluggishness throughout the day and increased fatigue when performing the prescribed exercise.

DISCUSSION QUESTIONS

1. If the patient also complained of angina before starting metoprolol, how would you explain the effect this drug has on such symptoms?
2. Would regular exercise lower resting blood pressure or total cholesterol for this person?
3. Given a change (increase or decrease) in his β-blocker dosage, how might you update the exercise prescription? How would completing an exercise test help with such an update?

Chapter 4

General Interview and Examination Skills

Quinn R. Pack, MD

Clinical exercise physiologists work in settings that require them to assess patients with various health problems. This chapter focuses on helping clinical exercise physiologists better understand the elements of the clinical evaluation conducted by a physician or physician extender, as well as the measurements that they may need to obtain to determine whether the patient can exercise safely.

The clinical evaluation of any patient usually involves two steps. First, a general interview is conducted to obtain historical and current information. A physical assessment or examination follows, the extent of which may vary based on who is conducting the examination and the nature of the patient's symptoms or illness. After these are completed, a brief numerical list is generated to summarize the assessment, relative to both prior and current findings and diagnoses. Finally, a numerical plan is generated to indicate the one, two, or three key actions that are to be taken in the care of the patient. This chapter describes in detail the general interview and physical examination components of the clinical evaluation.

GENERAL INTERVIEW

The general interview is a key step in establishing the patient database, which is the working body of knowledge that the patient and the clinical exercise physiologist will share throughout the course of treatment. This database is primarily built from information obtained from the patient's hospital or clinical records. But as the clinical exercise physiologist, you will need to interview the patient to obtain information that is missing, as well as update data to address any changes in the patient's clinical status since his last clinic visit of record. With experience you will learn which information is incomplete or necessitates questioning of the patient. A list of the relevant components, some of which you will need to enter into the patient file or database, is shown in "Essentials of Clinical Evaluation for the New Patient Referred."

Reason for Referral

The reason for referral for exercise therapy is generally self-explanatory and may include one or more of the following: to improve exercise tolerance, improve muscle strength, increase range of motion, or provide relevant intervention and behavioral strategies to reduce future risk. But the clinical exercise physiologist may need to reconcile the physician's reason for referral and the patient's understanding of the need for therapy. Differences can exist between the two. For example, some patients who undergo coronary artery angioplasty with stent deployment may return to work in just a few days without physical limitations. Unfortunately, sometimes these patients perceive that they are "cured" and surmise that they do not need to engage in rehabilitation or make lifestyle

Acknowledgment: Much of the writing in this chapter was adapted from the first and second editions of *Clinical Exercise Physiology*. Thus I wish to gratefully acknowledge the previous efforts of and thank Peter A. McCullough, MD, MPH, FACC.

Essentials of Clinical Evaluation for the New Patient Referred

- General interview
  - Reasons for referral
  - Demographics (age, gender, ethnicity)
  - History of present illness (HPI)
  - Current medications
  - Allergies
  - Past medical history
  - Family history
  - Social history
- Physical examination
- Laboratory data and diagnostic tests
- Assessment
- Plan

Adapted, by permission, from P.A. McCullough, 1999, *Clinical Exercise Physiology* 1(1): 33-41.

adjustments. When such a patient is referred for cardiac rehabilitation, she must understand that coronary artery disease is a dynamic disorder that is influenced, just like her original problem, by lifestyle habits and medications. The clinical exercise physiologist plays an important role in enforcing long-term compliance to physical activity, hypertension and diabetes management, proper nutrition, and medical compliance—all key components of secondary prevention.

Demographics

Patient demographics such as age, sex, and ethnicity are the basic building blocks of clinical knowledge. A great deal of medical literature describes the relationship between this type of demographic information and health problems. For example, nearly 40% of adult Americans with at least one type of heart disease are age 65 yr or older (15). Age is also the most important factor in the development of osteoarthritis, with 68% of those over the age of 65 having some clinical or radiographic evidence of the disorder. Finally, age is an independent predictor of survival in virtually every cardiopulmonary condition, including acute myocardial infarction, stroke, peripheral arterial disease, and chronic obstructive pulmonary disease (13). Because these and other age-related diseases can influence a patient's ability to exercise, age becomes a key piece of information to consider when developing an exercise prescription.

Sex also relates to outcomes such as behavioral compliance or disease management in patients with chronic diseases. For example, rheumatoid arthritis (RA) is more common in women than men (3:1 ratio) and seems to have an earlier onset in women. Also, although the onset of cardiovascular disease is, in general, 10 yr later in women than in men, the morbidity and mortality after revascularization procedures (i.e., coronary bypass or angioplasty) are higher in women (11). Finally, keep in mind that exercise capacity, as measured by peak oxygen consumption, decreases progressively in men and women from the third through the eighth decade. Ades and colleagues (1) showed that the rate of decline in men with age (-0.242 ml · kg^{-1} · min^{-1} per year) is greater than the rate of decline for women (-0.116 ml · kg^{-1} · min^{-1} per year). These and other sex-based difference remain an area of intense investigation as clinical scientists strive to determine which biological or socioeconomic factors account for the poorer outcomes sometimes observed in women. Additionally, a few data describe the positive or negative effects of exercise as an intervention for many chronic diseases in women, especially those with cardiopulmonary disease (3). When an exercise prescription is developed for women, unique compliance- and disease-related barriers and confounders need to be solved to improve exercise-related outcomes.

A great deal of information is available about differences in health status between various ethnic groups. Most of these differences are attributable to socioeconomic status and access to care, but a few ethnic-related differences are worth mentioning. For example, obesity, hypertension, renal insufficiency, and left ventricular hypertrophy are all more common in African American patients with cardiovascular disease than in their age- and sex-matched Caucasian counterparts. Likewise, diabetes and insulin resistance are more common in Hispanic Americans and some tribes of Native Americans.

The clinical exercise physiologist should consider these issues when developing, implementing, and evaluating an exercise treatment plan. This information may influence the clinician's decision about which risk factors to address first. Finally, program expectations and outcomes may be influenced as well.

History of Present Illness

The purpose of this element is to record and convey the primary information related to the condition that led to the patient's referral to a clinical exercise physiologist.

The history usually begins with the "chief complaint," which is typically communicated as one sentence that sums up the patient's comments (see practical application 4.1). The body of the history of present illness is a paragraph that summarizes the manifestations of the illness as they pertain to pain, mobility, nervous system dysfunction, or alterations in various other organ system functions (e.g., circulatory, pulmonary, musculoskeletal, skin, and gastrointestinal). Important elements of the illness are reviewed such as the date of onset, chronicity of symptoms, types of symptoms, exacerbating or alleviating factors, major interventions, and current disease status. Traditionally, this is a paragraph that describes events in the patient's own words. A practical approach for the clinical exercise physiologist is to incorporate reported symptoms with information from the patient's medical record.

It is important to fully describe any reported symptom by asking about the characteristics of that symptom. Ask about the symptom onset, provocation, palliation, quality, region, radiation, severity, timing, and associated signs and symptoms. An easy way to remember this is the mnemonic OPQRST&A. (See Clinical Exercise Pearl sidebar.) For example, knee pain could be described as follows: started 3 years ago (O); worse at the end of the day and better with ibuprofen (P); sharp, knife-like (Q); right knee, front of knee, radiation to feet (R); 8 out of 10 pain with walking but 4 out of 10 pain at rest (S); worse after recent fall 3 days ago (T); and swollen, hot to the touch, with a "popping sound" with motion (A).

The importance of taking a thorough history cannot be overstated. In many instances the history alone is sufficient to determine the diagnosis and often guides subsequent testing and treatment (9). In an age of advanced diagnostic testing, *listening to and carefully observing* the patient is usually the most important clinical tool for making a diagnosis and communicating your findings. As you ask questions and thoughtfully listen, you will better understand your patients and be better equipped to help them.

For patients with cardiovascular disease, the features of chest pain should be described (table 4.1). Such a description can help in the future application of diagnostic testing when it comes to assessing pretest probability of underlying obstructive coronary artery disease (6). Standard classifications should be used, if possible, such as the Canadian Cardiovascular Society functional class for angina or the New York Heart Association functional class (table 4.2) (2). For patients with pain attributable to muscular, orthopedic, or abdominal problems, the important elements of the illness such as chronicity, type of symptoms, and exacerbating or alleviating factors need to be addressed.

Clinical Exercise Pearl

OPQRST and A: A Useful Mnemonic to Describe the Characteristics of Any Symptom

O = Onset
P = Provocation and palliation
Q = Quality
R = Region and radiation
S = Severity
T = Timing
A = Associated signs and symptoms

Practical Application 4.1

CLIENT–CLINICIAN INTERACTION

During any history or physical examination, exercise physiologists should remember two important tenets:

1. All interactions are confidential, and any information obtained should remain private and protected. A breach of patient confidentiality represents serious misconduct and in some cases could result in termination of your job.
2. When examining patients, maintaining their modesty is always very important; never become casual about this. Remember that some patients have modesty standards that are very different from those you yourself might hold.

Table 4.1 Features of Chest Pain

Type of discomfort	Quality	Radiation	Exacerbating and alleviating factors
Typical*	Heaviness, pressure, squeezing, generalized left to midchest	To neck, jaw, back, left arm, less commonly the right arm	Worsened with exertion or relieved with rest or nitroglycerin
Atypical	Sharp, stabbing, pricking, tingling	None	None clearly present; can happen any time
Noncardiac	Discomfort clearly attributable to another cause	Not applicable	Not applicable

*Note the difference between typical stable and typical unstable angina. The difference between these two types of angina pertains to no change in (or stable pattern for) intensity, duration, or frequency of pain in the past 60 d, as well as no change in precipitating factors.

Reprinted, by permission, from P.A. McCullough, 1999, *Clinical Exercise Physiology* 1(1): 33-41.

Table 4.2 Classification of Cardiovascular Disability

Class I	Class II	Class III	Class IV
NEW YORK HEART ASSOCIATION FUNCTIONAL CLASSIFICATION			
Patients with cardiac disease but no marked physical limitations (e.g., undue fatigue, palpitation, dyspnea, or anginal pain)	Patients with cardiac disease resulting in slight limitation (fatigue, dyspnea, angina) of physical activity	Patients with cardiac disease with marked limitation of physical activity	Patients with cardiac disease who are unable to carry out physical activity without discomfort
	Comfortable at rest	Symptoms such as fatigue, palpitations, and dyspnea occur with less than ordinary physical activity	Symptoms may even be present at rest
			Physical activity worsens symptoms
CANADIAN CARDIOVASCULAR SOCIETY FUNCTIONAL CLASSIFICATION			
Routine daily activity does not cause angina	Slight limitation of ordinary physical activities such as climbing stairs rapidly or exertion after meals, in cold or windy conditions, under emotional stress, or within a few hours after awakening	Marked limitation of routine daily activities such as walking one or two blocks on level ground or climbing more than one flight of stairs in normal conditions	Inability to carry out physical activity without discomfort
Angina occurs with rapid pace or prolonged exertion	No symptoms when walking more than two blocks on level ground or climbing more than one flight of ordinary stairs at a normal pace and in normal conditions		Angina may be present at rest

Adapted, by permission, from American Alliance of Health, Physical Education, Recreation and Dance, 2004, *Physical education for lifelong fitness: The physical best teacher's guide* (Champaign, IL: Human Kinetics), 127.

Medications and Allergies

A current medication list is an essential part of the clinical evaluation, especially for the practicing clinical exercise physiologist, because certain medications can alter physiological responses during exercise. In fact, asking about current medications is an excellent segue into obtaining relevant past medical history. Compare the medications that patients state they are taking against what you think they should be taking given the medical information they report. For example, medical therapy for patients with chronic heart failure usually includes a β-adrenergic blocking agent (i.e., β-blocker) and a vasodilator such as an angiotensin-converting enzyme inhibitor. If during your first evaluation of a new patient with chronic heart failure you learn that the patient is not taking one of these agents, you should ask if his doctor has prescribed it in the past. In doing so, you may learn whether the patient has been found to be intolerant to an agent or you may simply refresh his memory.

When describing the current medical regimen, be sure to include frequency, dose of administration, and time taken during the day. The latter may be especially important if a medication affects heart rate response at rest or during exercise, because you must allow sufficient time (usually 2 to 3 h minimum) between when the medication is taken and when the patient begins to exercise. Specifically, the medication must have time to be absorbed and exert its therapeutic effect.

A medical allergy history, with a comment on the type of reaction, is also a necessary part of your patient database. If a patient is unaware of any drug allergies, note this as "no known allergies (nka)" in your database. Note as well medicines that the patient does not tolerate (e.g., "Patient is intolerant to nitrates, which cause severe headache").

Medical History

This section should contain a concise, relevant list of past medical problems with attention to dates. Be sure that your list is complete because orthopedic, muscular, neurological, gastrointestinal, immunological, respiratory, and cardiovascular problems all have the potential to influence the exercise response and the type, progression, duration, and intensity of exercise. For example, for patients with coronary heart disease, the record must include the severity of coronary lesions, types of conduits used if bypass was performed, target vessels, and most current assessment of left ventricular function (i.e., ejection fraction). For patients with cerebral and peripheral disease, the same degree of detail is needed with respect to the arterial beds treated.

Among patients with intrinsic lung disease, attempt to clarify asthma versus chronic obstructive pulmonary disease attributable to cigarette smoking. Such information may help explain why certain medications are used when the patient is symptomatic and why others are part of a patient's long-term, chronic medical regimen. Inquire about and investigate other organ systems as well. For example, if a diagnostic exercise test is ordered for a patient with intermittent claudication, using a dual-action stationary cycle may be better than using a treadmill. Additionally, knowing about any previous low back pain is important when you are developing an exercise prescription.

Family History

This element should be restricted to known, relevant, heritable disorders in first-degree family members (parents, siblings, and offspring). Relevant, heritable disorders include certain cancers (e.g., breast), adult-onset diabetes mellitus, familial hypercholesterolemia, sudden death, and premature coronary artery disease defined as new-onset disease before the age of 55 in men or 65 in women. While assessing family history, you may wish to discuss with the patient that first-degree family members should be screened for pertinent risk factors and possible early disease detection, if indicated.

Social History

The social history section brings together information about the patient's lifestyle and living patterns. Always inquire about tobacco use. If patients smoke, strongly encourage them to quit and refer them for treatment. Ask routinely about alcohol, marijuana, cocaine, and other substance use. Ask about marital or significant partner status, occupation, transportation, housing, and routine and leisure activities. Because long-term compliance to healthy behaviors is influenced, in part, by conflicts with transportation, work hours, and childcare and family responsibilities, inquire about and discuss these issues. Conclude by making reasonable suggestions to improve long-term compliance.

Inquire about prior exercise habits. Be specific and try to quantify the amount, intensity, and type of activities patients engage in, as well as barriers to regular exercise. Attempt to estimate your patient's aerobic capacity (table 4.3) (12). Inquire about preferred forms of exercise and exercise dislikes. Build on your patient's prior successes. If there have been problems in the past, help patients understand why they had problems so that they will be more successful in the future when trying to

Table 4.3 Estimating Aerobic Capacity by History

Can your patient do the following without symptoms?	Estimated metabolic equivalents (METs)
Get dressed without stopping	2 METs
Do general housework such as making the bed, washing the car, hanging laundry	3 METs
Push a lawnmower or grocery cart	4 METs
General ballroom dancing	5 METs
Carry a 15 lb (7 kg) load of laundry up two flights of stairs	6 METs
Moderate jogging	7 METs
Running, basketball, swimming	9 METs or more

establish daily habits of exercise. Try to identify any possible roadblocks to improving their exercise habits, such as a busy job, lack of access to equipment, safety concerns, and family responsibilities. Work with your patients to help them overcome such obstacles.

Ask about nutrition patterns. Find out what kinds of food patients eat on a typical day. Do they eat fast food? Do they eat breakfast? How many servings of fruit and vegetables do they eat each day? How many desserts do they have each day? Do they try to avoid salt? Have they ever been on a diet? Was it successful? Finding out this kind of information is essential for properly encouraging and planning good nutrition.

Ask about sleep and snoring. Do patients get the recommended 8 h of sleep per night? Do they awake refreshed? Are they tired and fatigued throughout the day? Do they rely on caffeinated beverages? These types of questions may help you identify patients with insufficient sleep or perhaps even **obstructive sleep apnea**. Increasingly, abnormal sleep patterns are being identified as significant contributors to heart disease and stroke (10).

In summary, the social history is usually a rich source of information about a patient's risk factors. Unfortunately, doctors rarely take the time to ask about and address such issues. However, assisting patients with modifying lifestyle behaviors is within the training and expertise of clinical exercise physiologists. You are a professional who is able to help others identify and overcome common and unique barriers that hamper improvement in long-term health and well-being.

PHYSICAL EXAMINATION

Physicians and physician extenders are taught to take a complete head-to-toe approach to the physical examination. For every part of the body examined, an orderly process of inspection, palpation, and, if applicable, **auscultation** and percussion is followed. The clinical exercise physiologist, however, can take a more focused approach and concentrate on abnormal findings, based on patient complaints or symptoms and information from prior examinations performed by others.

Specifically, you must develop the skills needed to determine whether, on any given day, it is safe to allow exercise by a patient who presents with signs or symptoms that may or may not be related to a current illness. For example, consider the patient with a history of dilated cardiomyopathy who complains of being short of breath and having had to sleep on three pillows the previous two nights, just so she could breathe more comfortably. This complaint should raise a red flag for you, because it may indicate that the patient is experiencing pulmonary edema attributable to heart failure. Besides asking other questions, you or another clinician should assess body weight, peripheral edema, and lung sounds before allowing the person to exercise. A telephone conversation with the patient's doctor concerning your findings, if meaningful, might also be warranted.

At no time should your examination be represented as one performed in lieu of the evaluation conducted by a professional licensed to do so. Still, you are responsible for ensuring patient safety. Therefore, the information and data that you gather are important. They should be communicated to the referring physician and become part of the patient's permanent medical record. Identifying the relevant aspects of an earlier examination provides a reference point for comparison, particularly if complications occur. Important red flags that should be identified and evaluated by the clinical exercise physiologist and, if needed, a physician, are shown in "Red Flag Indicators of a Change in Clinical Status."

Red Flag Indicators of a Change in Clinical Status

The following are red flag indicators that, if detected by the clinical exercise physiologist, should be discussed with a physician before a patient is allowed to exercise:

- New-onset or definite change in pattern of shortness of breath or chest pain
- Complaint of recent syncope (loss of consciousness) or near syncope
- Neurologic symptoms suggestive of transient ischemic attack (vision or speech disturbance)
- Recent fall
- Lower leg pain at rest (also called critical leg ischemia)
- Severe headache
- Pain in a bone area (i.e., in patients with history of cancer)
- Unexplained tachycardia (>100 beats · min^{-1}) or bradycardia (<40 beat · min^{-1})
- Systolic blood pressure >200 mmHg or <86 mmHg; diastolic blood pressure >110 mmHg
- Pulmonary rales or active wheezing

General State

The initial general survey is an important marker of overall illness or wellness. Take a careful look at patients. Are they comfortable, well nourished, and well groomed? Or do they look ill, fatigued, or worn down? Are they cheerful, engaged, and motivated? Or do they look distressed, anxious, frail, or malnourished? In time, this so-called "eyeball" test will become a valuable tool in your physical examination skills. It will alert you to a change in your patient's clinical status. In fact, you may sometimes find that your intuition tells you something is wrong even though the rest of the examination appears normal. Although this skill can take years to develop, if you get a sense that something could be amiss, take the time to investigate.

Blood Pressure, Heart Rate, and Respiratory Rate

Determinations of blood pressure, heart rate, and respiratory rate are the foundation of any physical evaluation and are sometimes made before and after an exercise session. These vital signs should be repeated in the event of any clinical change or symptom such as chest pain or dizziness. You should become expert at taking an accurate blood pressure reading. Ideally, patients should have rested for 5 min and should be seated, with arm and back supported, bare skin exposed, and arm at the level of the heart. Additionally, they should have an empty bladder and not have used caffeine or tobacco products within 30 min.

Hypertension is a blood pressure consistently greater than 140/90 mmHg. Hypertension is an important risk factor for cardiovascular disease and is discussed fully in chapter 8. **Hypotension** (i.e., blood pressure <90/60 mmHg) is uncommon and may signify an important clinical change. When hypotension is accompanied by symptoms of light-headedness or dizziness it often requires immediate medical attention. The one exception to this rule is patients with chronic heart failure who are treated with blood pressure–lowering medications to the point that a blood pressure of 90/60 is "normal" and well tolerated without symptoms.

Tachycardia (heart rate >100 beats · min^{-1}) after 15 min of sitting is always abnormal and can be found in many medical conditions, including very low left ventricular systolic function, severely impaired pulmonary function, hyperthyroidism, anemia, volume depletion, pregnancy, infection, or fever or as a medication side effect. **Bradycardia** (heart rate <60 beats · min^{-1}) can be either abnormal or normal, depending on the patient condition and medication profile. However, a heart rate below 40 beats · min^{-1} usually represents an underlying problem and requires further evaluation.

Tachypnea is a respiratory rate greater than 20 breaths · min^{-1}. **Bradypnea** is a respiratory rate less than 8 breaths · min^{-1}. Labored, tachypneic breathing is worrisome when it occurs at rest or with minimal exertion. Measurement of the oxygen saturation using pulse oximetry is important in the case of abnormal breathing rates or bluish skin discoloration. **Hypoxia** is a blood oxygen saturation below 95%. In patients with chronic obstructive pulmonary disease (COPD), an oxygen saturation greater than 88% is acceptable. The presence of tachypnea, bradypnea, or hypoxia usually requires medical attention.

Body Fatness

Height and weight should be measured in each patient to obtain a sense of body fatness. The simplest measure of body fatness is the **body mass index (BMI)**. The BMI is mass in kilograms (weight) divided by the height in meters squared ($kg \cdot m^{-2}$). A BMI less than 24.9 is normal, between 25 and 29.9 is overweight, and 30 or more is obese. You should record in your database the BMI of every patient whom you are evaluating for the first time. Practically, you accomplish this by weighing the patient in kilograms and then asking for self-reported height in feet and inches. An easy-to-use conversion nomogram, such as the one in figure 4.1, should be accessible for the quick calculation of BMI. Obesity (BMI >30) is a major comorbidity that doubles a patient's risk for heart disease and worsens bone and joint problems.

Body fatness can be further assessed by measuring the patient's waist circumference with a measuring tape. In general, a waist circumference greater than 40 in. (100 cm) in a male or more than 35 in. (90 cm) in a female is indicative of central adiposity. Central (android) obesity is an important risk factor for heart disease, hypertension, diabetes, and dyslipidemia. Although gynecoid obesity (around the legs) is a significant comorbidity, it does

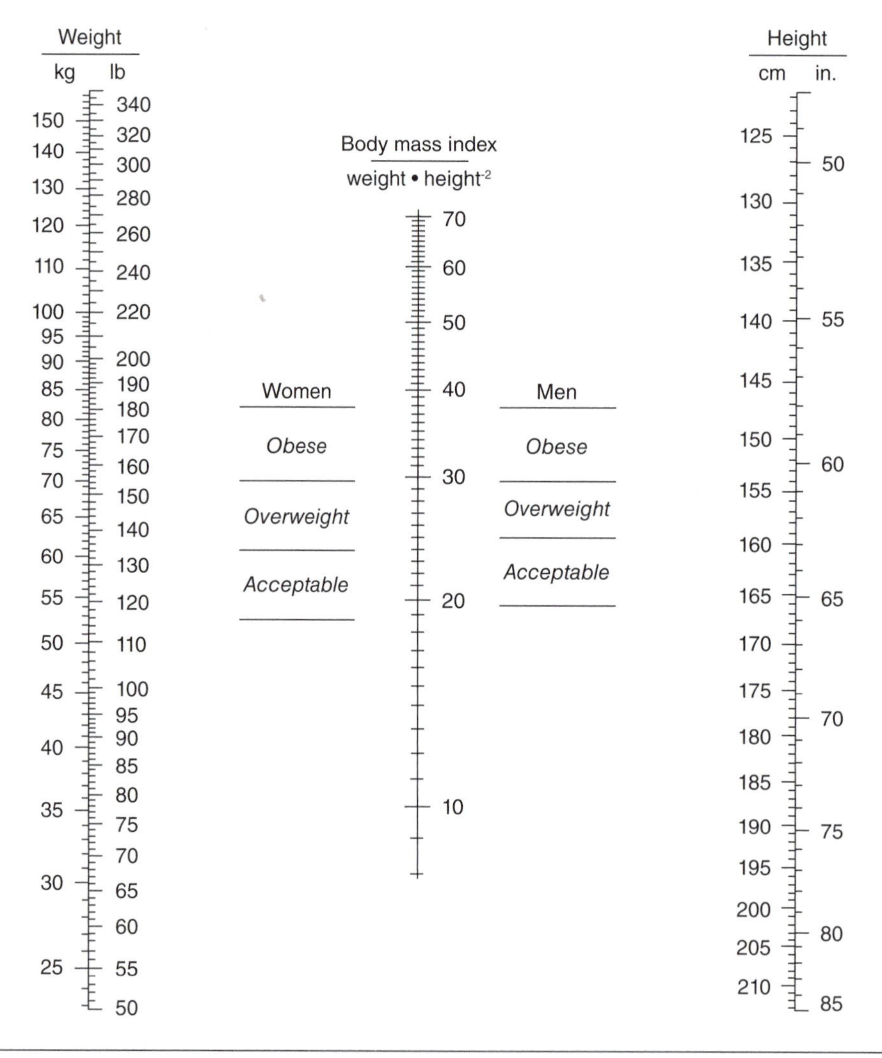

Figure 4.1 Body mass index (BMI) conversion chart that is used with weight and height. To use this nomogram, place a straightedge between the body weight (without clothes) in kilograms or pounds located on the left-hand line and the height (without shoes) in centimeters or inches located on the right-hand line. The point where the straightedge intersects the middle line indicates the BMI.

From Center for Disease Control.

not carry the same disease risk as does android obesity (5). Waist circumference is the best measure of android obesity and should be used routinely.

Pulmonary System

The thorax should be inspected and deformities of the chest wall and thoracic spine noted. Common abnormalities to look for include kyphosis (i.e., curvature of the upper spine resulting in rounding of the back and appearing as slouching posture), barrel chest (i.e., large torso suggesting upper body strength, or emphysema among patients that smoke cigarettes), and pectus excavatum (i.e., so-called sunken or hollowed chest due to a congenital deformity of the ribs and sternum in the anterior chest wall). Thoracic surgical incisions and implantable pacemaker or defibrillator sites should be inspected and palpated. Redness, warmness, or tenderness of any incision is always abnormal and often signifies a wound infection, which should prompt a physician evaluation. With the patient sitting, the lungs should be auscultated with the diaphragm of the stethoscope in both anterior and posterior positions, and breath sounds should be characterized as normal, decreased, absent, coarse, wheezing, or crackling (i.e., rales). Decreased or absent breath sounds should prompt **percussion** of the chest wall for dullness. Dullness signifies a pleural effusion, which is an abnormal collection of fluid in the pleural space that does not readily transmit sound. This finding on physical examination would prompt withholding exercise and notifying the patient's physician. Coarse breath sounds can signify pulmonary congestion or chronic bronchitis. Crackles or rales can be caused by atelectasis (inadequate alveolar expansion after thoracic surgery), pulmonary edema attributable to congestive heart failure, or intrinsic lung disease such as pulmonary fibrosis.

Cardiovascular System

With the patient supine, the cardiac examination should start with inspection of the anterior chest wall. A cardiac pulse that can be visualized on the chest wall is often abnormal and represents a left ventricular hyperdynamic state. Standing at the patient's right side and while the patient lies comfortably, palpate the heart with the right hand. Make an effort to characterize where you feel or palpate the cardiac **point of maximal cardiac impulse (PMI)**, or cardiac apex, as shown in figure 4.2. A normal PMI is the size of a dime and is located in the fourth or fifth intercostal space at the midclavicular line. The PMI should be characterized as normal, diffuse (enlarged), hyperdynamic, or sustained. The location of the PMI should be identified as normal or laterally displaced. A diffuse and laterally displaced PMI indicates left ventricular systolic impairment, often with enlargement of that chamber. If you feel two cardiac impulses, this finding often indicates a right and left ventricular heave suggestive of biventricular impairment in patients with heart failure.

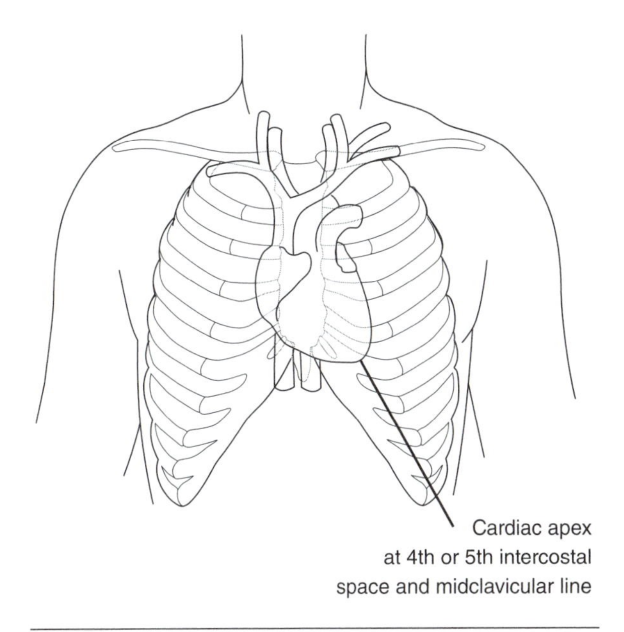

Figure 4.2 Surface topography of the heart.

An essential component in the examination of the heart is auscultation. An introduction to the assessment of heart sounds is reviewed in practical application 4.2. Practice listening to heart and lung sounds in apparently healthy people and patients you help care for who are known to have cardiac and pulmonary disorders as described in their medical records.

The cardiovascular physical examination should also include some evaluation of the status of peripheral vascular circulation. Extremities that are well perfused with blood are warm and dry. Poorly perfused extremities are often cold and clammy. Measuring and grading the characteristics of the arterial pulse in a region assesses adequacy of blood flow in arteries. Arterial pulses are graded (7) as follows: 3 = bounding, 2 = normal, 1 = reduced or diminished, and 0 = absent or nonpalpable. Using a stethoscope, one can also listen for **bruits**, which are high-velocity swooshing sounds created as blood becomes turbulent when it flows past a narrowing artery or through a tortuous artery. Volume should be assessed, and bruits should be characterized as soft or loud. Bruits detected in the carotid arteries that were not previously mentioned in the medical record should be brought to

Practical Application 4.2

AUSCULTATION OF THE HUMAN HEART

Auscultation of, or listening to the sounds made by, the heart is but one part of a comprehensive cardiac examination. Begin a habit of auscultating the heart in a systematic fashion that is the same for every patient whom you evaluate. Establishing a uniform approach will more quickly familiarize you with normal heart sounds and help you identify abnormal sounds. To aid your concentration, try to auscultate the heart in a quiet room. Approach the patient from the right side and do all that you can to minimize anxiety. Attempt to warm your stethoscope before using it and communicate with the patient as you progress through this part of the physical examination. Remember that maintaining patient modesty is always a priority.

Auscultation is usually done with the patient lying on his back, before an exercise test; however, auscultation can be performed with the patient in the sitting position. As mentioned earlier, when you auscultate the heart for cardiac sounds, you should do so systematically. Begin by placing the diaphragm of the stethoscope firmly on the chest wall in the lower-left parasternal region. First, characterize the rhythm as regular, occasionally irregular, or irregular. An irregular rhythm is usually attributable to atrial fibrillation. The diaphragm of the stethoscope is best used to hear high-pitched sounds, whereas the bell portion is used for low-pitched sounds.

Next, move the stethoscope to the point on the chest where the first heart sound (S_1, the sounds of mitral and tricuspid valves closing) is best characterized. For most people, this location is found at the apex of the heart at the left midclavicular line and near the fourth and fifth intercostal spaces. You will hear two sounds. The first sound (S_1) is the louder and more distinct sound of the two. Then, move your stethoscope upward and to the right side of the sternal border at the second intercostal space. This location is generally the best place to characterize the second heart sound (S_2, the sounds of aortic and pulmonic valves closing) because it is louder and more pronounced here.

Soft heart sounds occur with low cardiac output states, obesity, and significant pulmonary disease (e.g., diseased or hyperinflated lung tissue between the chest wall and the heart). Loud heart tones occur in thin people and in hyperdynamic states such as pregnancy. The second heart sound normally splits with inspiration as right ventricular ejection is delayed with the increased volume it receives from the augmented venous return of inspiration. This delays or splits the pulmonic component of S_2 (sometimes referred to as P_2) from the aortic component of S_2 (referred to as A_2, figure 4.3).

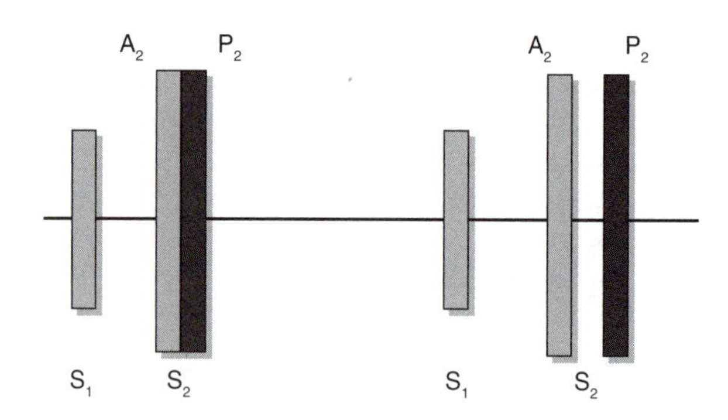

Figure 4.3 Normal physiologic splitting of the second heart sound because of augmented venous return to the right heart during inspiration.

Reprinted, by permission, from P.A. McCullough, 1999, *Clinical Exercise Physiology* 1(1): 33-41.

With practice and as you begin to care for patients with various cardiac problems, you will become exposed to and appreciate third (S_3) and fourth (S_4) heart sounds. S_3 and S_4 are low-pitched sounds that are heard during diastole and best appreciated with use of the bell portion of the stethoscope. The presence of either of these two heart sounds is most often associated with a heart problem and, if not previously noted in the patient's medical record, should be brought to the attention of a physician. S_3 is best heard at the apex and occurs right after S_2. S_4 is also well heard at the apex and occurs just before S_1. An S_3 commonly indicates severe left ventricular systolic impairment with volume overload and dilation. An S_4 commonly indicates chronic stiffness or poor compliance of the left ventricle, usually attributable to long-term hypertension. As you can appreciate, learning and identifying heart sounds require a great deal of practice.

The clinical exercise physiologist should be able to appreciate systole and diastole and, with advanced training, listen for murmurs in the mitral, tricuspid, pulmonic, and aortic areas (figure 4.4). Murmurs are characterized by the timing in the cardiac cycle (systolic, diastolic, or both), location where best heard, radiation, duration (short or long), intensity, pitch (low, high), quality (musical, rumbling, blowing), and change with respiration (8). A central concept to keep in mind while listening with the stethoscope is that the sounds heard are attributable to changes in blood velocity and the movement of cardiac valve leaflets, both of which are driven by pressure gradients and result in flow. Systolic murmurs are more common and are often characterized as ejection type (e.g., diamond shaped or holosystolic). Diastolic murmurs are distinctly less common and are always abnormal.

Exercise physiologists working in the clinical exercise testing setting would be well served to acquire the basic skills needed to auscultate the heart. For further information and to hear audio clips of murmurs go to www.blaufuss.org or Heart Songs 3 at www.cardiosource.org.

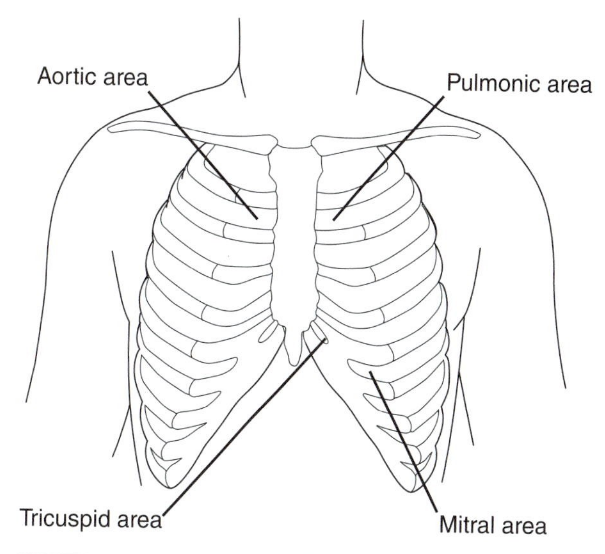

Figure 4.4 Auscultatory areas of the heart.

Reprinted, by permission, from P.A. McCullough, 1999, *Clinical Exercise Physiology* 1(1): 33-41.

the attention of a physician, because they may indicate severe carotid atherosclerosis. Bruits in the common femoral arteries are suggestive of peripheral arterial disease but by themselves do not call for immediate physician evaluation.

Peripheral **edema** (e.g., swelling of the lower legs, ankles, or feet) is a cardinal sign of congestive chronic heart failure. Because of elevated left ventricular end-diastolic pressure and consequently the backward cascade of increased pressure to the left atrium, pulmonary capillaries, pulmonary artery, and right-sided cardiac chambers, increased hydrostatic forces move extracellular fluid from within the blood vessels into the tissue spaces of the lower extremities. Edema is graded on a 1 to 3 scale, with 1 being mild, 2 being moderate, and 3 being severe. Additionally, "pitting edema" can be present, which is easily identified by pressing a thumb into an edematous area (e.g., distal anterior tibia) and observing that an indentation remains. A patient with 3+ pitting edema of the lower legs and ankles obviously has a great deal of fluid that has left the vascular compartment and moved into the surrounding tissue.

Not all edema, however, results from congestive heart failure. Minor edema can be a side effect of many medications such as slow-channel calcium entry blockers (e.g., nifedipine). In addition, chronic venous incompetence associated with prior vascular surgery, obesity, or lymphatic obstruction can cause edema in the setting of normal cardiac hemodynamics and heart function. A practical point for the clinical exercise physiologist caring for patients with cardiovascular disease is that an increase in edema or body mass (>1.5 kg) over just a 2 or 3 d period is often the first sign of congestive heart failure and warrants a call to a physician.

Musculoskeletal System

Approximately one person in seven suffers from some sort of musculoskeletal disorder. The history of present illness or past medical history should note the major areas of discomfort and self-reported limitation of motion. In addition, prior major orthopedic surgeries, such as a hip or knee joint replacement, should be noted.

The approach to the musculoskeletal physical examination should be grounded in observation. For example, observe the patient as she gets up from a chair and walks into a rehabilitation area, gets onto an examination table, or handles personal belongings. Observe gait and characterize it as normal (narrow based, steady, deliberate), antalgic (limping because of pain), slow, hemiplegic (attributable to weakness or paralysis), shuffling (parkinsonian), wide based (cerebellar ataxia or loss of position information), foot drop, or slapping (sensory ataxia or loss of position information) (figure 4.5). An antalgic gait is a limp, which reflects unilateral pain and compensation for that pain. A slow gait is often a tipoff for back disease, hip arthritis, or underlying neurological problems. Hemiplegic, shuffling, wide-based, foot-drop, and slapping gaits all represent compensation for underlying neurologic disease such as a spinal cord injury, cerebellar dysfunction (e.g., attributable to alcohol), or midbrain dysfunction (e.g., Parkinson's disease). These "neurological gaits" are all unsteady and leave the patient prone to falling. For safety reasons, the clinical exercise physiologist must pay special attention to gait and modify the exercise prescription as necessary.

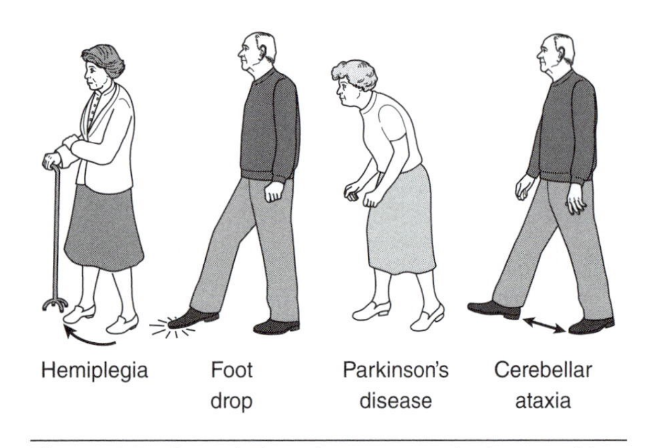

Figure 4.5 Description of common gait abnormalities.

Reprinted from *Physical Diagnosis: History and Examination*, M.H. Swartz, pg. 452, © 1999, with permission from Elsevier Science.

The core of the musculoskeletal physical examination is an assessment of range of motion of the movable joints. Important terminology for describing limb motion is given in table 4.4. As needed, palpation of the major joints (elbows, wrists, hips, knees, and ankles) can be performed to note thickening of the joint capsules, swelling or effusion, and tenderness of ligaments or tendons. Redness, warmth, swelling, and fever are all signs of active inflammation and require evaluation by a physician before you proceed with exercise testing or therapy. If these signs are found in conjunction with a prosthetic joint, such as a total knee replacement, they may indicate an infection and require immediate contact with the patient's surgeon.

In addition to joint health, muscle strength can be examined and graded on a scale of 0 to 5, with 0 indicating flaccid paralysis and 5 indicating sufficient power to overcome the resistance of the examiner. Muscle stiffness and soreness (not related to exercise) should be noted because they are often the sign of a chronic underlying inflammatory condition that requires medical evaluation.

Low Back Pain

Because low back pain (see chapter 24) is one of the most common physical complaints in human medicine,

Table 4.4 Definitions for Selected and Common Joint Movements

Motion	Definition
Flexion	Motion in sagittal plane which decreases the angle at a joint (away from zero position)
Extension	Motion in sagittal plane which increases the angle at a joint (toward zero position)
Dorsiflexion	Movement decreases the 90 degree angle between the dorsum (superior surface) of the foot and the leg (toes are brought closer to the shin)
Plantar (or palmar) flexion	Movement increases the 90 degree angle between the surface of the foot and the leg (e.g., standing on tip toes)
Adduction	Movements in the frontal plane that involves moving the body part toward, and perhaps beyond, the central line or midline
Abduction	Movements in the frontal plane that involves moving the body part away from the central line or midline
Plantar inversion/ supination	Movement of turning the sole (plantar surface) of the foot inwards
Plantar eversion/ pronation	Movement of turning the sole (plantar surface) of the foot outward
Pronation	Movement or rotation of the palm over or to face downwards
Supination	Movement or rotation of the palm upwards (e.g., "face-up")
Internal (medial) rotation	Movement of the hip or shoulder joints in the transverse plane; limb surface rotates inward (toward the body) or downward
External (lateral) rotation	Movement of the hip or shoulder joints in the transverse plane; limb surface rotates outward (away from the body) or upwards

it is worth mentioning that the etiology of this problem can range from a mild muscle strain to a life-threatening ruptured abdominal aortic aneurysm. The clinical exercise physiologist must have a rational approach to this problem and tailor aspects of care based on etiology, severity, and prognosis. In the young, stable patient with no evidence of neurological compromise (e.g., radiating pain, numbness), a physician evaluation is likely not necessary before exercise testing or therapy. In the geriatric patient (i.e., >65 yr old), however, new-onset low back pain that has not been evaluated by a physician deserves referral to the primary care physician or a back specialist before exercise testing or training because it could be something serious, such as a spontaneous compression fracture attributable to osteoporosis.

Arthritis

A final musculoskeletal condition worth mentioning, and one that is especially common among the elderly, is arthritis (see chapter 22). The two most common types of arthritis are **rheumatoid arthritis (RA)** and **osteoarthritis (OA)**. Rheumatoid arthritis is a chronic inflammatory condition manifested by functional disability and early morning stiffness (>1 h) followed by improvement through the rest of the day. Most damage experienced in RA occurs in the first 5 yr, and approximately 40% of patients who suffer from RA become completely disabled in 15 yr. Life expectancy falls by 3 to 18 yr.

Rheumatoid arthritis begins insidiously with fatigue, anorexia, generalized weakness, and vague musculoskeletal symptoms until the appearance of synovitis (inflammation of joint lining) becomes apparent. It has a predilection for the small joints in the hand, especially the metacarpophalangeal joints (knuckles). Rheumatoid arthritis results in pain, inflammation, thickening of the joint capsule, lateral deviation of the fingers, and significant disability. It can involve large joints and the spine, especially the cervical vertebrae. Rheumatoid arthritis is usually apparent to the clinical exercise physiologist through the history and the presence of disease-modifying antiarthritic drugs (DMARDs).

Patients with RA can generally exercise safely unless neck pain or lancinating pains in the shoulders or arms are reported. These symptoms are indicative of cervical spine involvement and require a physician's attention. In some cases, bracing or surgery is needed to prevent cervical spine subluxation and paralysis.

Osteoarthritis is the most common skeletal disorder in adults. It can result from longstanding wear and tear on large and small joints including shoulders, elbows, wrists, hands, hips, knees, and back. The correlation between the pathological severity of OA and patient symptoms is poor. Patients with OA may show signs of synovitis and secondary muscle spasm.

Exercise rehabilitation may improve the symptoms of OA in one location (e.g., the back) only to worsen pain in another location (e.g., hips and knees). The clinical exercise physiologist should take a pragmatic approach and work in a coordinated fashion with physical and occupational therapy colleagues to find activities that minimize joint loading at involved sites.

Nervous System

Like the examination of the musculoskeletal system, a neurological examination performed by the clinical exercise physiologist mainly involves observation. In general, the clinician should make a comment in the patient file regarding level of understanding, orientation, and cognition. Obvious disabilities of speech, balance, and muscle tremor and disabilities of the eyes, ears, mouth, face, and swallowing should be cross-referenced for confirmation with the patient's medical record.

The practicing clinical exercise physiologist rarely performs a detailed neurological examination, so this chapter does not describe the conduct for such an examination. Nevertheless, a notation regarding gross **hemiparesis** or complete paralysis of a limb should be made. As mentioned previously (figure 4.5), "neurological gaits" should be identified as well. For obvious reasons, patients with any of these problems may require an appropriately modified exercise prescription.

Although the clinical exercise physiologist may work with the spinal cord–injured patient (see chapter 25) or patients with multiple sclerosis (see chapter 26), the most common neurological problem that most exercise physiologists encounter is with patients who have suffered a previous cerebrovascular event or stroke. Stroke is defined as an ischemic insult to the brain resulting in neurological deficits that last for more than 24 h (see chapter 28). The etiology of stroke includes local arterial thrombosis, cardiac and carotid thromboembolism, and intracranial hemorrhage. Risk factors for stroke include the following:

- History of stroke
- Atrial fibrillation
- Left ventricular dysfunction
- Aneurysms
- Carotid artery stenoses
- Uncontrolled hypertension

If the clinical exercise physiologist determines that any of these risk factors are new, he should notify the physician before the patient starts exercise. Common abnormalities after a stroke are a loss in or diminished limb function, drooping of one side of the face, drooling, and garbled speech (dysarthria).

Skin

A full examination of the skin is not appropriate for the exercise physiologist, but examination of the hands, arms, and surgical incisions for coronary bypass or pacemaker implantation should be routine. If the skin is hot, warm, swollen, or draining fluids, this should be reported and may represent a new infection. For patients with diabetes, addressing the skin can include asking them to take off their shoes so you can inspect for blisters, cuts, scrapes, ulcers, and discoloration. Patients with diabetes may have a condition called **peripheral neuropathy**, in which the nerves of the legs are damaged and stop functioning properly, with the result that people can no longer feel their toes and feet. They lose the ability to sense pain. Consequently, diabetic patients sometimes develop wounds and infections in their skin that they are completely unaware of. This makes it important to be sure to examine the feet of all diabetic patients.

When examining the skin of the arms and hands, pay careful attention to discoloration or **clubbing**, which may be seen in patients with chronic lung disease. Clubbing is a condition where a buildup of tissue in the fingers causes the finger nails to become rounded, like an upside-down spoon; seen often with diseases (e.g., pulmonary diseases) that result in less oxygen in the blood. Blueness of the hands, lips, and nose suggests hypoxia. These findings should be evaluated by a physician.

Significant bruising of the skin may be the result of medications such as aspirin, clopidogrel, or warfarin or could result from frequent falls. Pale skin may signify **anemia**, in which a patient has a low blood count. Anemia is defined as a hemoglobin less than 12 $g \cdot dl^{-1}$ in women and less than 13 $g \cdot dl^{-1}$ in men and may result from recent operations, procedures, or bleeding. In addition, some patients have anemia due to prolonged chronic illness. In general, anemia requires further physician evaluation.

Laboratory Data and Diagnostic Tests

This section of the clinical evaluation summarizes the relevant testing that has previously been performed. In general, test results or reports from a patient's resting electrocardiogram, chest X-ray, echocardiogram, cardiac catheterization, and coronary artery bypass grafting surgery are relevant and should be included. Consultation results from physical therapy, orthopedics, or physical medicine and rehabilitation should be noted. Furthermore, laboratory data such as hemoglobin, electrolytes, creatinine, cholesterol, and blood glucose levels should be noted. Detailed discussion of these tests is beyond the scope of this chapter, but they are relevant and can affect the exercise prescription.

The most important diagnostic test to incorporate into your evaluation is the most recent stress test. What was the level of exercise attained? What was the peak heart rate? Were there any symptoms? Was there any evidence of ischemia or arrhythmia and if so, at what heart rate did these abnormalities first appear? These kinds of results help you to build an appropriate exercise prescription and should be included in this section of the general evaluation. Exercise testing and prescription are discussed fully in chapter 5.

CONCLUSION

This chapter provides a platform for the clinical evaluation of the new patient by the clinical exercise physiologist. We have given special emphasis to the day-to-day interview and examination skills that a practicing clinical exercise physiologist may need to help decide whether a patient can exercise safely. Also, we have emphasized that the information gathered through a clinical evaluation, when viewed in conjunction with existing medical record information and physical examination findings reported by others, provides a point of reference should complications occur during exercise treatment. If the evaluation is broadened to include a comprehensive assessment of future health risk and consideration of disease, the clinical evaluation also becomes a useful decision-making tool and guide for patient education.

Key Terms

anemia (p. 58)
auscultation (p. 50)
body mass index (BMI) (p. 52)
bradycardia (p. 51)
bradypnea (p. 51)
bruits (p. 53)
clubbing (p. 58)
edema (p. 55)
hemiparesis (p. 57)
hypertension (p. 51)
hypotension (p. 51)
hypoxia (p. 51)
obstructive sleep apnea (p. 50)
osteoarthritis (OA) (p. 57)
percussion (p. 53)
peripheral neuropathy (p. 58)
point of maximal cardiac impulse (PMI) (p. 53)
rheumatoid arthritis (RA) (p. 57)
tachycardia (p. 51)
tachypnea (p. 51)

CASE STUDY

MEDICAL HISTORY

Ms. WY is a 63 yr old woman a history of diabetes, hypertension, and tobacco use. She experienced chest pain 2 wk ago while shoveling snow. The pain was midchest, caused nausea, and was worse with exertion. It improved when she rested. She was hospitalized and diagnosed with an ST-segment elevation myocardial infarction (STEMI), and a stent was placed in the left anterior descending coronary artery. She now has shortness of breath on moderate exertion such as carrying a full laundry basket up one flight of stairs, but is able to go shopping and do general housework.

Medications: Simvastatin 40 mg daily at bedtime, aspirin 81 mg daily, metoprolol 100 mg twice daily, metformin 500 mg three times per day with meals, clopidogrel 75 mg daily, lisinopril 10 mg daily, nitroglycerin 0.4 mg sublingual as needed.

Allergies: Penicillin causes a rash.

Medical history: Diabetes for 6 yr, hypertension for 10 yr.

Family history: Mother had diabetes. Father died of a heart attack at age 62.

Social history: Current smoker (one pack per day), has one alcoholic drink per week, employed as a secretary, married, two adult children, college graduate.

Lifestyle history: Sedentary with very rare physical exertion. Diet history is notable for skipping breakfast and eating fast food four or five times per week, with one or two servings of fruits and vegetables each day. She averages 6 h of sleep at night and snores heavily.

PHYSICAL EXAMINATION

General appearance: Pleasant, alert, interactive, but looks tired. Smells of smoke.

Vital signs: Blood pressure = 152/87 mmHg, pulse = 48 beats · min^{-1}, respiration = 14 breaths · min^{-1}, body mass = 185 lb (84 kg), height = 5 ft 4 in. (163 cm), waist circumference = 38 in. Oxygen saturation is 98%.

Cardiovascular: No carotid bruits, no jugular venous distension, normal point of maximal impulse, regular rate and rhythm, normal S_1 and S_2, 2/6 short early peaking systolic ejection murmur best heard in the aortic area without radiation, no extra heart sounds, no edema, normal peripheral vascular exam.

Pulmonary: Breathing comfortably. Mild expiratory wheeze in bilateral lung fields.

Skin: No blisters, wounds, or rashes, but has decreased sensation in her feet.

Musculoskeletal: Normal gait, joints are normal, good range of motion throughout.

Laboratory data and testing: Normal ECG. Echocardiogram shows an ejection fraction of 40%. Blood glucose is elevated, but the rest of the labs are normal. No prior stress testing done.

(continued)

DIAGNOSIS

(a) Recent STEMI, (b) diabetes, (c) hypertension, (d) obesity, (e) tobacco abuse

Plan: (a) Complete a graded exercise test and begin cardiorespiratory training program with associated weight training program. (b) Initiate counseling about low-fat, low-calorie diet and refer to registered dietitian for a weight loss plan. (c) Strongly encourage smoking cessation and refer patient for specialized treatment.

DISCUSSION QUESTIONS

1. Classify Ms. WY's chest pain at the time of her hospitalization as typical, atypical, or noncardiac. What features of the chest pain are missing? Why might these be important features?
2. What is her functional class? What is her estimated aerobic capacity? Explain your answer.
3. Calculate her BMI. Does she have gynecoid or android features? How does her obesity contribute to her medical problems?
4. Are any red flags present on the clinical evaluation that require contacting, at least by telephone, her physician or another supervising physician before exercise testing?
5. What prevention issues need to be addressed in the near future?

Chapter 5

Graded Exercise Testing and Exercise Prescription

Steven J. Keteyian, PhD

The material presented in this chapter represents the minimum foundational knowledge that the clinical exercise physiologist needs to acquire to be sufficiently prepared to (a) help conduct a graded exercise test (GXT) and (b) prescribe exercise in apparently healthy people and those with clinically manifest disease. The reader is expected to use this material and adapt it based on the population-specific exercise testing and prescription information provided in each of the chapters that follow.

EXERCISE TESTING

The earliest form of a GXT was first used in approximately 1846, when Edward Smith began to evaluate the responses of different physiological parameters (heart rate, respiratory rate) during exertion (16). In 1929 Master and Oppenheimer began to use the GXT to evaluate a patient's cardiac capacity (43); however, they did not yet recognize the value of incorporating an electrocardiogram (ECG) during or following exercise in the detection of **myocardial ischemia**. In fact, Master's original two-step test, so called because patients repeatedly walked up and over a box-like device with two steps that stood a total of 18 in. (45 cm) high, counted the number of ascents completed during 90 s as a measure of exercise capacity. It was not until several years later that Master obtained an ECG before and immediately after the test to assess myocardial ischemia. The Master two-step remained the clinical standard for assessing exercise-induced ischemia until the early 1970s.

Graded exercise testing has evolved greatly over the past 40 yr. In addition to serving as a tool to help detect the presence of ischemic heart disease using various exercise-based modalities (e.g., treadmill, stationary bike), such testing yields data that are now also used to assess prognosis. Additionally, in combination with various cardiac imaging modalities, some in conjunction with exercise and others using drugs to induce myocardial stress, these tests have an increased ability to detect coronary artery disease (CAD). All of this notwithstanding, even with the addition of **radionuclide agents** or **stress echocardiogram** for imaging, the standard GXT with a 12-lead ECG remains the first choice to evaluate myocardial ischemia in most individuals who have a normal resting ECG and the ability to exert themselves physically.

Indications

Although the overall risk of death during a GXT is small (0 to 5 sudden cardiac deaths per 10,000 tests) (38, 50), before a GXT is performed it is important to understand or appreciate (a) which patients the GXT is useful in; (b) the reason for completing the test; and (c) that although the risks for experiencing an event are rare, this does not mean there is no risk whatsoever. Knowing this information will help ensure that the benefits and data derived from the test outweigh any potential risks (31, 61). If a question arises concerning the safety or rationale for completing the GXT, the supervising or referring physician should be contacted for clarity.

With respect to indications for GXT, the American College of Cardiology and the American Heart Association provide guidelines for exercise testing in various clinical situations (30, 50). In general, a GXT is ordered by the physician for two main reasons: first, to evaluate chest pain as a means to assist in the diagnosis of CAD, and second to identify a patient's future risk or prognosis. In addition, there are several other conditions in which a GXT can assist the physician, several of which are discussed a little later in this section. In many cases, a GXT is the test of choice because of its noninvasive nature and relatively low cost.

The diagnostic value of determining the presence of CAD is greatest when one is evaluating individuals with an intermediate probability of CAD, which is based on the person's age, sex, the presence of symptoms (typical or atypical), and the actual prevalence of CAD in other persons of the same age and gender. In a man 60 yr of age who complains of typical angina (i.e., substernal chest discomfort brought on with exertion and relieved by rest), the pretest probability for CAD is quite high because of the patient's symptoms and the higher prevalence of CAD in 60 yr old men. Therefore, a test that shows ECG changes (e.g., ST-segment depression) indicative of exercise-induced myocardial ischemia really provides little new information. Conversely, the true prevalence of CAD in 30 yr old women with atypical angina is very low, so any ECG changes suggestive of ischemia in these patients would cause the physician to seriously consider a false-positive finding. There appears to be little value for improving patient outcomes when exercise testing is performed on individuals who are asymptomatic.

To determine a patient's prognosis related to CAD or help stratify the person's future risk for experiencing a clinical event, a GXT can again prove helpful. In addition to true angina and the presence of ST-segment depression evident on the ECG during exercise or in recovery, other factors that influence prognosis include the magnitude of ST depression (1 mm vs. 3 mm, where more ST depression represents greater risk), the number of ECG leads showing significant ST-segment depression, time of onset for ST depression during exercise and time to resolution during recovery, and estimated **functional capacity** (FC) as measured by exercise duration or **metabolic equivalents (METs)**. For example, a patient who complains of anginal pain and demonstrates 2.5 mm of ST-segment depression in four ECG leads just 3 min into an exercise test at a workload or FC that approximates just 4 or 5 METs has a much poorer prognosis than the patient who is symptom free and completes 10 METs of work with no ECG evidence of ST-segment depression. Based on the severity of disease, a GXT can be a useful tool to stratify patients. In people found to be at high risk (i.e., abnormal GXT response at low workloads, <5 METs), further medical intervention may be required. But if an abnormal ECG response does not occur until the person reaches 10 METs, the prognosis is much more favorable and additional medical or surgical intervention may not be needed. Gathering together all this information allows one to estimate prognosis using the Duke nomogram to determine the 5 yr survival and average annual mortality rate (42). The take-home point here is that much information can be gathered from a GXT, beyond simply the presence of ST depression or anginal symptoms—information that can be used to guide a patient's future care (medical management vs. additional testing vs. surgical intervention).

The important association between FC and prognosis can be expanded if the GXT is performed in conjunction with the direct measurement of respiratory gas exchange. A major advantage of measuring gas exchange is that it is the most accurate measurement of FC. The American Thoracic Society/American College of Chest Physicians statement on cardiopulmonary gas exchange methods provides a complete review of this topic (7). Besides providing a more accurate measurement of exercise capacity, gas exchange data may be particularly useful in defining prognosis (and thus help guide the timing for cardiac transplantation) in patients with heart failure (41, 49, 51, 57). In addition to peak oxygen uptake ($\dot{V}O_2$), the slope of the change in minute ventilation ($\dot{V}_E$) to change in carbon dioxide ($\dot{V}CO_2$) production (i.e., $\dot{V}_E$-$\dot{V}CO_2$ slope) during an exercise test has been shown to be related to prognosis, especially in patients with heart failure (9). Other measurements that can be determined through the measurement of gas exchange include the ventilatory-derived anaerobic threshold, oxygen pulse, slope of the change in work rate to change in $\dot{V}O_2$, oxygen uptake efficiency slope, partial pressure of end-tidal CO_2, breathing reserve, and the respiratory exchange ratio (9). Gas exchange measurements are also particularly useful in identifying whether the cause of unexplained dyspnea has a cardiac or pulmonary etiology.

A regular GXT can also be performed for reasons other than diagnosis or prognosis. These include quantification of change in FC due to an exercise training program or a medical or surgical intervention; assessment of syncope or near-syncope, exercise-induced asthma, exercise-induced arrhythmias, and pacemaker or heart rate response to exercise; assessment of blood pressure response to exercise; assessment of FC for the purpose of guiding return to work; and preoperative clearance.

Contraindications

Although the risk of death or a major complication during a GXT is very small, such a test is simply not recommended in some situations because the risks of safely completing the test outweigh the value of the potential derived information (30). Current guidelines separate absolute versus relative contraindications for test termination (see "Contraindications to Graded Exercise Testing"). Just as the phrase suggests, absolute contraindications are conditions in which a GXT should not be performed if the conditions listed are observed; the risk for potential serious medical consequences are unsatisfactorily high. Relative contraindications suggest the presence of a potential medical issue and indicate that concerns regarding the worsening of the patient's condition should be considered before the GXT is conducted (30).

In addition to contraindications for GXT, there are conditions in which GXT is not recommended for diagnostic purposes because abnormalities in the resting ECG (e.g., left bundle branch block, ventricular pacemaker) render the exercise ECG unable to accurately identify the presence of CAD (i.e., ST-T changes).

To increase the sensitivity of the standard GXT to detect exertional ischemia, a GXT with nuclear perfusion imaging or echocardiographic imaging is recommended for people who have these abnormalities in their resting ECG. These tests are clearly more costly than the ECG-only GXT, so ideally they should be used primarily for the detection of exertional ischemia when the ECG abnormalities are present at rest. Note that if the primary reason for ordering the GXT is a reason other than evaluation of exertional myocardial ischemia, such as evaluation of FC or the efficacy of medical therapy to control arrhythmia or hypertension, then the GXT with ECG alone (without imaging) remains appropriate.

Contraindications to Graded Exercise Testing

Absolute Contraindications

1. Significant change on the resting ECG suggesting ischemia, acute myocardial infarction (within past 2 d), or other acute cardiac event
2. Unstable angina not controlled by medical therapy
3. Uncontrolled cardiac arrhythmia causing symptoms or hemodynamic compromise
4. Symptomatic severe aortic stenosis
5. Uncontrolled symptomatic heart failure
6. Acute pulmonary embolus or pulmonary infarction
7. Acute myocarditis or pericarditis
8. Suspected or known dissecting aortic aneurysm
9. Acute infections (influenza, rhinovirus)

Relative Contraindications

1. Left main coronary stenosis
2. Moderate stenotic valvular heart disease
3. Severe arterial hypertension at rest (systolic blood pressure >200 mmHg or diastolic blood pressure >110 mmHg)
4. Tachycardic or bradycardic rhythm
5. Hypertrophic cardiomyopathy and other forms of outflow tract obstruction
6. Mental or physical impairment that leads to inability to exercise adequately or is exacerbated with exercise
7. High-degree atrioventricular block
8. Uncontrolled metabolic disease (diabetes, thyrotoxicosis) or electrolyte abnormality (hypokalemia)
9. Chronic infectious disease (mononucleosis, hepatitis)

Performing a GXT on a person with a relative contraindication may be acceptable if the benefits outweigh the risks.

Adapted, by permission, from ACSM, 2013, *ACSM's guidelines for exercise testing and prescription,* 9th ed. (Philadelphia, PA: Lippincott, Williams, and Wilkins) 53.

Procedures for Preparing, Conducting, and Interpreting a Graded Exercise Test

With an introduction to graded exercise testing and a review of who should and should not undergo such testing behind us, we can turn our attention to the procedural and interpretive elements associated with such testing. In general, these elements can be organized or grouped into seven topics: pretest considerations; appearance and quantification of symptoms; assessment and interpretation of heart rate (HR) response during exercise and recovery; assessment and interpretation of blood pressure (BP) response during exercise and recovery; ECG findings; assessment of FC; and interpretation of findings and generation of final summary report.

Pretest Considerations

Preparing a person to undergo any exercise test (i.e., regular, with imaging, or with measurement of expired gases) involves several common pretest considerations. These are personnel, informed consent, general interview and examination, pretest instructions and subject preparation for resting ECG, and selection of exercise protocol and testing modality.

Personnel Before 1980, GXTs conducted in the clinical setting were primarily (90%) supervised by a cardiologist (63). Since that time, GXTs have been performed by many types of health care professionals (clinical exercise physiologists and exercise specialists, physicians trained in internal medicine or family medicine, physician assistants, nurses, and physical therapists). The use of health care professionals other than cardiologists is likely due to a better understanding of the risks associated with a GXT, cost-containment initiatives, time constraints on physicians, and more sophisticated ECG analysis (computerized exercise ST-segment interpretation). The American Heart Association guidelines for exercise testing laboratories state that the health care professionals just mentioned, when appropriately trained and possessing specific performance skills (e.g., American College of Sports Medicine certification), can safely supervise clinical GXTs (50). In addition, these health care professionals should be certified in basic life support and probably in advanced cardiac life support as well.

Since diagnostic and most other forms of GXT usually necessitate that both healthy people and patients with a clinically manifest disease put forth a near-maximal or maximal effort, the competency, knowledge, skills, and ability of physician and nonphysician professionals to safely supervise and conduct all aspects of the test are essential (50, 58). The incidence of death or a major event requiring hospitalization during a GXT is rare (50), approximately 1 to 10 events per 10,000 tests; and the rate for such events is the same regardless of whether the tests are supervised by physicians or nonphysicians. Interestingly, with respect to the preliminary interpretation of the ECG and other test findings made by nonphysicians, the level of agreement with a cardiologist appears to be quite good (50). Key needed skills for test supervision include

- knowledge of absolute and relative contraindications for testing;
- knowledge of specificity, sensitivity, and predictive value of positive and negative tests;
- understanding of the causes of false-positive and false-negative tests;
- demonstrated ability to select the correct test protocol and modality;
- knowledge of normal and abnormal cardiorespiratory and metabolic responses in normal, healthy individuals and in patients with a clinical condition;
- ability to assess and respond appropriately to important clinical signs and symptoms;
- knowledge of and ability to demonstrate proper test termination skills using established indications;
- knowledge of and ability to correctly respond to normal and abnormal ECG findings at rest and during exercise; and
- appropriate skills for the measurement of BP at rest and during exercise.

Informed Consent Obtaining adequate informed consent from individuals before exercise testing is an important ethical and legal step. The form that you ask the patient to read and sign should lay the foundation for a "meeting of the minds" between the testing staff and the patient relative to the reason for the test, an understanding of test procedures, and safety issues or potential risks. Enough information must be exchanged that the patient knows, understands, and verbalizes back to you (a) test purpose, (b) test procedures, and (c) risks. In addition to asking the patient to read the informed consent form, it is appropriate for you to orally summarize the content of the form. The patient should be given several defined opportunities to ask questions. Since most consent forms contain information explaining the risks associated with the test, the likelihood of their occurrence, and the fact that emergency equipment is available should it be required, you should also explain that appropriately trained staff are present in the rare event that an emergency occurs. If the client agrees to participate, he must sign the form; clients under 18 yr of age must have a parent or guardian sign the consent form as well.

General Interview and Examination In hospital-based or outpatient ambulatory care settings, the clinical exercise physiologist will be involved only with supervising or assisting with GXTs that were ordered by a licensed physician or physician extender (e.g., nurse practitioner, physician assistant). This means that sometime in the usually not-too-distant past, perhaps yesterday or within the past month or two, a health care provider familiar with the indications and contraindications for GXT considered the patient's clinical status and ordered the test. However, although the patient has been referred, it is still the responsibility of the clinical exercise physiologist supervising the test to be sure that there have been no changes in the patient's clinical status since she was last seen by her doctor; a change might preclude taking the test today for safety reasons.

The complete elements of the interview and examination performed by a physician are detailed in chapter 4. Fortunately, for the purpose of deciding that in fact it is safe to conduct the test today, the questions that must be asked and the examination that must be performed can be much more focused. The test supervisor should have already reviewed the patient's medical record for past medical and surgical history; this knowledge will help determine whether any new contraindications to the GXT have arisen since the patient was last seen by the physician. It will also aid in choosing an appropriate testing protocol and will help identify areas that may require more supervision during the test (e.g., history of hypertension, previous knee injury). A review of the medical record should pay attention to height and weight; cardiovascular risk factors and any recent signs or symptoms of cardiovascular disease; previous cardiovascular disease; other chronic diseases that may influence the outcome (e.g., arthritis, gout); and information about current medications (including dose and frequency), drug allergies, recent hospitalizations or illnesses, and exercise and work history. In addition to collecting information about the patient's medical history and status through both the review of the patient's medical record and your interview of the patient, you may need to perform a brief physical examination, one that focuses on just peripheral edema and possibly heart and lung sounds. Specific physical assessment skills are presented in chapter 4.

Obviously, any "red flags" (e.g., marked recent weight gain or worsening edema over the past 3 d, worsening or unstable symptoms, recent fall or injury) detected by the clinical exercise physiologist supervising the test should be brought to the attention of the physician overseeing testing that day or the referring physician. A decision can then be made about proceeding with the test, having the patient first be seen by a doctor, or rescheduling the test.

Patient Pretest Instruction and Preparation

The patient must receive clear instructions on how to prepare for the GXT along with an appropriate explanation of the test. A lack of proper instruction can delay the test, increase patient anxiety, and potentially increase the health risk to the patient. Instructions for patient preparation should include information on clothing; footwear; prior food consumption; avoidance of alcohol, cigarettes, caffeine, and over-the-counter medicines that could influence the GXT results; and whether the patient should follow his prescribed drug regimen. A prescribed medication may need to be discontinued because certain agents can inhibit the ability to observe ischemic responses during exercise (e.g., β-adrenergic blocking agents attenuate HR and BP, which can inhibit the occurrence of ST-segment changes associated with ischemia). However, withholding prescribed medication before testing is not a requirement. Note that if a physician wants the patient to discontinue medications before GXT, the patient should receive those instructions from the physician. The process of weaning a person off various medications can vary, and a rebound effect can be a concern.

It is important to recognize that the patient may be anxious before completing the GXT, and the test supervisor should answer all questions in a professional and caring manner. If the test supervisor is prepared and confident, patients will be more at ease and potentially more willing to give their best effort.

The clarity of the ECG tracing during exercise is of utmost importance, especially for diagnostic purposes. Therefore, properly preparing and connecting the patient for an ECG at rest is mandatory. For males, hair from all electrode positions should be removed with a razor or battery-operated shaver. Conductance through the skin surface is critical for achieving a clear ECG recording, and shaving away hair allows the electrodes to lie flat on the skin. Removal of body oils and lotions with an alcohol wipe and skin abrasion is also required for optimal conductance. Silver chloride electrodes are ideal because they offer the lowest offset voltage and are the most dependable in minimizing **motion artifact** (50). After electrodes are in place and cables are attached, each electrode placement site should be assessed for excessive resistance or insufficient skin preparation. On the newer ECG stress systems, a function built into the equipment can help determine whether the electrode sites have excessive resistance that would require reprepping of a specific site. Another good test is to lightly tap on the electrode and observe the ECG; if the ECG concomitantly shows a great deal of motion artifact, check the electrode and lead wire for potential interference. Adequate electrode preparation can decrease the likelihood of motion artifact, whereas artifact due to muscle contraction cannot be easily reduced unless upper body skeletal muscle activity is decreased. Because the skin moves while the patient walks on the treadmill, keeping electrode and wire movement to a minimum is important, especially in people who are obese.

After the patient has rested for a several minutes, a 12-lead ECG and BP should be recorded in both the supine and standing positions; with respect to the latter, at least 2 min should elapse after the patient stands upright and measurements are again taken. Evaluating both conditions allows the test supervisor to assess the effects of body position on hemodynamics. Also, the resting ECG from today should be compared, if available, to a resting ECG taken on a prior day. If any significant differences are found between the two ECGs that contraindicate starting the GXT (see section on contraindications), the health care provider supervising the test should contact the supervising or referring physician with the updated information so that a decision can be made about completing the GXT.

Graded Exercise Testing Protocols and Testing Modalities Numerous protocols are used to perform GXT; a few common examples are shown in table 5.1. The supervisor of the GXT must use a protocol that enables the client to achieve maximal effort without premature fatigue. The protocol should allow the patient to exercise for a minimum of 8 min and preferably not more than 12 min. This will provide enough time for significant physiologic adaptations to exercise to occur but will reduce the likelihood of ending the GXT due to skeletal muscle fatigue. The GXT supervisor is responsible for choosing the most appropriate protocol. Before the test, when the patient's medical record is reviewed and the patient is interviewed, evaluation of current exercise and activity habits (e.g., is the patient able to climb a flight of stairs?) can help with the selection of the appropriate exercise protocol. Ideally, all clinical testing laboratories have specific protocols for testing patients with a variety of functional capacities and should avoid changing protocols, if possible, for any repeat testing performed on the same patient over time (unless the exercise time is <8 min or >12 min). Such consistency between tests over time makes comparisons of results possible from one test to the next, enabling the clinician, for example, to identify the workload achieved on each test that corresponds to the onset of any chest pain or ST changes.

For diagnostic purposes, the standard Bruce treadmill protocol (i.e., starting at 1.7 mph [2.7 kph] and a 10% grade) is the most widely used test because of its longstanding history of use in clinical laboratories, use for training other professionals, frequent citation in the medical literature, and normative data for many populations. In people who are not frail, do not have an extremely low FC (e.g., difficulty walking a flight of stairs), and are free from orthopedic problems that are exacerbated by walking, the Bruce protocol is acceptable for diagnostic purposes. That said, the Bruce protocol does have some limitations; these include large increments in workload (2 to 3 METs per stage) that can make it difficult to determine a person's true FC; 3 min stages that for some are too fast for walking and too slow for running; and a large vertical component, which can lead to premature leg fatigue. When trying to determine whether the Bruce protocol is appropriate, the GXT supervisor should ask the client whether she can comfortably walk a flight of stairs without stopping. If no, another treadmill protocol may need to be considered, one that employs smaller MET increments between stages (e.g., modified Bruce protocol). Alternately, the Balke-Ware and Naughton protocols use 1 and 2 min per stage, respectively, with increments of 1 MET or less per stage. These protocols are commonly used with the elderly and people who are very deconditioned due to a chronic medical problem such as stage C heart failure or moderate chronic obstructive pulmonary disease. One of the more common problems with these protocols is that the supervisor underestimates the client's FC and the person exercises for longer than 12 min.

Ramping protocols are also an option within the more automated ECG treadmill or bicycle ergometer systems that are available. The advantage of ramping tests is that there are no large incremental changes, in that ramp rates can be adjusted and the client's FC can be more accurately determined. Equipment manufacturers now often include a ramping option that allows conversion of a standardized protocol (e.g., Bruce) to a ramping format in which treadmill speed and grade can change in small time increments (i.e., 6 to 15 s). When the Bruce protocol was compared using the standard 3 min stage and ramping in 15 s intervals, similar hemodynamic changes were observed, but the ramping test produced a significantly greater duration in time and a higher peak MET level. In addition, subjects perceived the ramping protocol to be significantly easier than the standard Bruce protocol (73).

Table 5.1 Commonly Used Treadmill and Bicycle Protocols

Protocol	Stage	Time (min)	Speed in mph (kph)	Grade (%)	Estimated $\dot{V}O_2$ (ml · kg^{-1} · min^{-1})	METs
Bruce[a]	1	3	1.7 (2.7)	0.0	8.1	2.3
	2	3	1.7 (2.7)	5.0	12.2	3.5
	3	3	1.7 (2.7)	10.0	16.3	4.6
	4	3	2.5 (4.0)	12.0	24.7	7.0
	5	3	3.4 (5.5)	14.0	35.6	10.2
	6	3	4.2 (6.8)	16.0	47.2	13.5
	7	3	5.0 (8.0)	18.0	52.0	14.9
	8	3	5.5 (8.9)	20.0	59.5	17.0
Naughton	1	2	1.0 (1.6)	0.0	8.9	2.5
	2	2	2.0 (3.2)	0.0	14.2	4.1
	3	2	2.0 (3.2)	3.5	15.9	4.5
	4	2	2.0 (3.2)	7.0	17.6	5.0
	5	2	2.0 (3.2)	10.5	19.3	5.5
	6	2	2.0 (3.2)	14.0	21.0	6.0
	7	2	2.0 (3.2)	17.5	22.7	6.5
Balke-Ware	1	1	3.3 (5.3)	2.0	15.5	4.4
	2	1	3.3 (5.3)	3.0	17.1	4.9
	3	1	3.3 (5.3)	4.0	18.7	5.3
	4	1	3.3 (5.3)	5.0	20.3	5.8
	5	1	3.3 (5.3)	6.0	21.9	6.3
	6	1	3.3 (5.3)	7.0	23.5	6.7
	7	1	3.3 (5.3)	8.0	25.1	7.2
Standard Balke	1	2	3.0 (4.8)	2.5	15.2	4.3
	2	2	3.0 (4.8)	5.0	18.8	5.4
	3	2	3.0 (4.8)	7.5	22.4	6.4
	4	2	3.0 (4.8)	10.0	26.0	7.4
	5	2	3.0 (4.8)	12.5	29.6	8.5
	6	2	3.0 (4.8)	15.0	33.2	9.5
	7	2	3.0 (4.8)	17.5	36.9	10.5
Protocol	**Stage**	**Time (min)**	**RPM**	**kgm · min^{-1}; watts**	**Estimated $\dot{V}O_2$ (ml · kg^{-1} · min^{-1})**	**METs**
Standard Bicycle Test	1	2 or 3	50.0	150; 25	10.9	3.1
	2	2 or 3	50.0	300; 50	14.7	4.2
	3	2 or 3	50.0	450; 75	18.6	5.3
	4	2 or 3	50.0	600; 100	22.4	6.4
	5	2 or 3	50.0	750; 125	26.3	7.5
	6	2 or 3	50.0	900; 150	30.1	8.6

Note. When trying to determine the appropriate protocol, consider the client's age, current functional capacity, activity level, and medical history and disease status. MET = metabolic equivalent.

[a]$\dot{V}O_2$ is calculated while the client is walking through stage 6. If the client is running at stage 6, the $\dot{V}O_2$ would be 42.4 ml · kg^{-1} · min^{-1} and 12.1 METs. Also, the conventional or standard Bruce protocol begins at stage 3 (1.7 mph [2.7 kph]) and a 10% grade); modified versions start at either stage 1 (1.7 mph [2.7 kph] at 0% grade) or stage 2 (1.7 mph [2.7 kph] and 5% grade).

With respect to deciding which mode or type of equipment to use for the GXT, in the United States if a treadmill cannot be used, the bicycle ergometer is the usual second choice. But comparison between the treadmill and the bicycle ergometer demonstrated that people achieve a 5% to 20% lower FC on the bicycle ergometer, based on leg strength and conditioning level (2). Bicycle ergometer testing is helpful for diagnostic reasons and to assess FC in patients who have difficulty weight bearing or have an abnormal gait that can affect their ability to perform the test. In electronically braked bicycles, ramping protocols are commonly used in which a slow increase in workload occurs each minute. A typical bicycle ramping protocol consists of an increase between 15 and 25 W · min^{-1}. In more standardized bicycle ergometers, a work rate increase of 150 kpm · min^{-1} or 25 W every 2 to 3 min stage is commonly used for the general population. In frail clients or in individuals with a low FC, smaller work rate increments should be considered. In contrast, in heavier clients, the standard incremental increase of 25 W may be insufficient to allow them to reach maximal effort within 12 min. With respect to cadence, 50 to 60 rev · min^{-1} is suggested with standard bicycle ergometers. In the electronically braked bicycles, cadence is not a major concern because resistance automatically changes, based on cadence, to maintain a specific work rate (e.g., 100 W). Among heavier and more active clients who need to accomplish a high workload to reach maximum effort, increasing work rate by 300 kgm · min^{-1} or 50 W every 2 or 3 min may be more appropriate. After the person being tested achieves 75% of his predicted maximal HR or a rating of perceived exertion (RPE) of 3 (0 to 10 scale) or 13 (6 to 20 scale), the workload would increase by 150 kgm · min^{-1} or 25 W every stage so that the ending work rate and peak FC can be more precisely determined.

In years past, arm ergometry testing was used for diagnostic purposes in clients with severe orthopedic problems, peripheral vascular disease, lower extremity amputation, and neurological conditions that inhibited the ability to exercise on a treadmill or bicycle ergometer. In a clinical setting, however, the arm ergometer is now routinely replaced by pharmacological stress testing, which means that drugs are administered as a method to increase myocardial stress (in place of exercise). Clients for whom an arm ergometry GXT may still be beneficial include those with symptoms of myocardial ischemia during dynamic upper body activity and those who have suffered a myocardial infarction and plan to return to an occupation that requires upper body activity. Because of a smaller muscular mass in the arms versus the legs, a typical arm protocol involves 10 to 15 W increments for every 2 to 3 min stage at a cranking rate of 50 to 60 rev · min^{-1}.

After the testing staff prepares the patient, chooses the appropriate protocol, and reviews the resting ECG for any abnormalities that may inhibit the evaluation, the GXT can start after the patient receives instruction on how to use the mode identified for testing (treadmill, bicycle ergometer). In patients unfamiliar or uncomfortable with walking on a treadmill, the test supervisor should take the needed time to make sure that they are comfortable walking with little or no handrail support for assistance. Clients who are not comfortable may stop prematurely simply because of apprehension. In addition to the supervisor of the GXT (physician or allied health professional), a technician is needed to monitor the patient's BP and potential signs and symptoms of exercise intolerance. The test supervisor is responsible for discontinuing the test at the appropriate time.

Appearance and Quantification of Symptoms

Symptoms such as angina or claudication should be noted and quantified when they first occur and then periodically throughout the test and at peak exercise. The time when symptoms resolve in recovery should also be noted. Heart rate, RPE, and BP should be recorded at the end of each stage. Approximately 30 to 45 s should be given to measure BP manually. Automated BP units are commonly used for resting measurements, but their accuracy under exercise conditions continues to be a question.

As the exercise test progresses from one stage to the next, testing staff must observe the client and communicate regularly. Before the test begins, the patient must understand that she is responsible for communicating the onset of any discomfort. At minimum, at the end of each stage of testing, staff should ask clients how they are feeling and if they are having any discomfort. Testing staff should record any such discomforts, including the absence of any stated discomfort, because doing so implies that the question was asked. During the GXT, the ECG must be continually monitored. A 12-lead ECG should be reviewed at the end of each stage, because it is possible that ECG changes can take place in any of the lead combinations. If gas exchange is analyzed during the GXT, communication becomes more difficult because of the use of a mouthpiece. Establishing specific hand signals or use of handheld posters is important so that symptoms (e.g., chest pain, shortness of breath) along with RPE can be accurately determined. A client can be apprehensive during her first experience with the measurement of gas exchange because breathing through a mouthpiece is uncommon and speaking abilities are inhibited. The test supervisor must properly instruct the client on the use of the gas exchange apparatus and allow the client to adjust to the gas exchange apparatus while resting, so that she is comfortable before starting the test.

The use of RPE with exercise testing is now common and provides a monitor of how hard the patient perceives the overall work and how much longer he will be able to continue. The client's RPE is generally recorded during the last 15 s of each stage. Two RPE scales are commonly used: the 6- to 20-point category scale and the 0- to 10-point category–ratio scale. To improve the accuracy of the RPE scales, the following instructions should be given to the patient:

During the exercise test we want you to pay close attention and rate how hard you feel the exercise is overall. Your response should rate your total amount of exertion and fatigue, combining all sensations and feelings of physical stress, effort, and fatigue. Don't focus on or concern yourself with any one factor such as leg pain, leg fatigue, or shortness of breath, but try to concentrate on and rate your total, inner feeling of overall exertion.

Other scales that are helpful in evaluating clients' specific symptoms include angina, **dyspnea**, and peripheral vascular disease scales for claudication (table 5.2). The angina scale is beneficial because it evaluates the intensity of chest discomfort and helps determine whether the test should be terminated based on standard criteria (see "Indications for Termination of Graded Exercise Test"). Another important point is to document the onset of angina with the corresponding MET level and **rate–pressure product (RPP)**. The progression of angina can be evaluated with the angina scale, which can be beneficial when a comparison is made to previous test results (i.e., whether the intensity of angina is occurring at a lower or higher workload). The dyspnea scale, which concerns rating the intensity and progression of dyspnea, is commonly used when testing patients with pulmonary disease and serves a similar role as the scale used for patients who have angina. The peripheral vascular disease scale is beneficial in evaluating clients with documented or questionable **claudication** based on their medical history.

The information gathered during the GXT using the RPE scale and the other three scales described in table 5.2 can also contribute to the safe prescription of exercise. The RPE during testing that is associated with a subsequent prescribed training HR range can help the clinician determine whether the client or patient can achieve and sustain that level of intensity. Be aware, however, that differences may exist when mode of testing differs from

Table 5.2 Angina, Dyspnea, and Peripheral Vascular Disease Scales

ANGINA SCALE	
1+	Light, barely noticeable
2+	Moderate, bothersome
3+	Moderately severe, very uncomfortable
4+	Most severe or intense pain ever experienced
DYSPNEA SCALE	
1+	Mild, noticeable to patient but not observer
2+	Mild, some difficulty, noticeable to observer
3+	Moderate difficulty, but patient can continue
4+	Severe difficulty, patient cannot continue
PERIPHERAL VASCULAR DISEASE SCALE FOR ASSESSMENT OF INTERMITTENT CLAUDICATION	
1+	Definite discomfort or pain but only of initial or modest levels (established but minimal)
2+	Moderate discomfort or pain from which the patient's attention can be diverted by a number of common stimuli (e.g., conversation, interesting TV show)
3+	Intense pain from which the patient's attention cannot be diverted except by catastrophic events (e.g., fire, explosion)
4+	Excruciating and unbearable pain

mode of training. Also, research shows that patients with cardiovascular disease sometimes perceive exercise to be more difficult during exercise training compared with exercise testing (14). Besides using RPE, comparing prescribed training HR ranges to the onset of angina, dyspnea, and claudication symptoms can be beneficial in determining a safe and feasible training range. Finally, although these scales can be helpful, they need to be interpreted and considered in conjunction with other medical findings.

As clients approach their maximum effort, they may become anxious and want to stop the test, so the supervisor and technician must be prepared to quickly record peak values (ECG, BP, RPE). A submaximal GXT results when the patient achieves a certain predetermined MET level or the test is terminated because the patient achieved a percentage of maximal HR (i.e., 85% of predicted maximum HR). Symptom-limited or maximal GXTs are commonly used for diagnostic purposes and assessment of FC. A symptom-limited test refers to a test that is terminated because of the onset of symptoms that put the client at increased risk for further medical problems. A maximal test terminates when an individual reaches his maximal level of exertion, without being limited by any abnormal signs or symptoms. When a client gives a maximal effort, the test is normally terminated because of volitional or voluntary fatigue. Several criteria are often used to determine whether a person has reached maximal effort:

1. A plateau in $\dot{V}O_2$ (<2.1 $ml \cdot kg^{-1} \cdot min^{-1}$ increase)
2. A respiratory exchange ratio value greater than 1.1
3. Venous blood lactate exceeding 8 to 10 mM (24)

These criteria rely on having a metabolic cart to measure oxygen consumption or a lactate analyzer, which is not feasible at many clinical facilities. Therefore, other criteria are commonly used in conjunction with voluntary fatigue; these include an RPE greater than 17, a plateau in HR despite an increasing workload, attainment of 85% of predicted maximal HR in patients taking a β-adrenergic blocking agent, or attainment of a rate–pressure product ≥24,000 $mmHg \cdot min^{-1}$.

The supervisor of the GXT must be aware of and understand potential complications that may occur and know when to terminate the GXT. This knowledge is especially crucial if a physician is not present in the room where the test is being conducted. Knowledge of the normal and abnormal physiological responses that take place with GXT is imperative. Reasons for test termination are presented as absolute or relative criteria (see "Indications for Termination of Graded Exercise Test"). Apart from a patient's request to stop the test or technical difficulties that arise during testing, absolute indications for stopping comprise high-risk criteria that have the potential to result in serious complications if the test is continued. Therefore, the test should be terminated immediately. If the physician is not present, she should be contacted and informed of the outcome as soon as possible. With respect to relative indications for stopping, these findings might represent reasons for test termination, but generally the test is not stopped unless other abnormal signs or symptoms occur simultaneously. The reason for performing the GXT may also influence relative reasons for test termination. For example, when evaluating a patient for suspected CAD, if the test supervisor observes left bundle branch block with exertion, this precludes interpretation of ST changes and is a reason for test termination. Each facility must have policies and procedures that specify absolute and relative indications for test termination. Having these policies and procedures eliminates any confusion about why a test was terminated and helps protect the facility from potential negligence if complications arise. If a question ever arises about whether a test should be stopped based on what is observed, the tester should always err on the conservative side and consider the safety of the patient as the highest priority (i.e., the test can always be repeated if necessary).

Resting, Exercise, and Recovery ECG Abnormalities

This section discusses analysis of the 12-lead ECG, which is based on recognizing normal and abnormal findings at rest, during exercise, and in recovery. Reading this section should not replace an in-depth study of ECG analysis; the material is intended as a review of GXT-specific issues. When reviewing the resting ECG before the GXT, the clinician should also assess the client's medical history to see whether any discrepancies are present (e.g., the client states no previous myocardial infarction, but the current ECG shows significant Q waves throughout the anterior leads). Also important is a comparison of the client's current resting 12-lead ECG with a previous ECG, especially when an abnormality is detected (e.g., the client's current resting ECG rhythm shows atrial fibrillation; is this arrhythmia new or old?). When a clinically significant difference is detected on the resting ECG and appears to be a new finding, the supervising or referring physician should be informed before the test is undertaken. When determining the safety of performing the GXT, refer to "Contraindications to Graded Exercise Testing."

The most common reason to complete a GXT with a 12-lead ECG is to assess potential CAD. However, in certain patients, abnormalities in their ECG at rest preclude use of the ECG during exercise to accurately detect or determine exercise-induced myocardial ischemia. In these patients, an exercise- or pharmacology-induced stress test is completed in conjunction with cardiac imaging by echocardiography or radionuclide testing. The following are common abnormalities in the resting ECG that limit the sensitivity of the exercise ECG to detect ischemia:

- Left bundle branch block (figure 5.1)
- Right bundle branch block (ST changes in anterior leads, V1-V3, cannot be interpreted, but remaining leads are interpretable) (figure 5.2)
- Pre-excitation on syndrome (Wolff-Parkinson-White [WPW])
- Nonspecific ST-T wave changes with >1 mm depression
- Abnormalities due to digoxin therapy or left ventricular hypertrophy
- Electronically paced ventricular rhythm

During the GXT, one person should continually observe the ECG monitor to identify the onset and nature of any ECG change. The evaluation of the ST segment is of great importance because of its ability to suggest the onset of ischemia. In addition, the onset of arrhythmias should be identified, especially those that are related to indications for test termination (see "Indications for Termination of Graded Exercise Test"). Although the onset and progression of ST depression are subtle in most cases, arrhythmia can occur suddenly and can be brief, intermittent, or sustained.

All 12 ECG leads should be monitored; however, V5 is most likely to demonstrate ST-segment depression, whereas the inferior leads (II, III, and aVF) are associated with a relatively higher incidence of false-positive findings (i.e., ST depression occurs, but it is not truly related to ischemia). V5 represents the most diagnostic lead because when true ischemia occurs with exertion, it is most commonly observed in this lead, which reflects a large area of the left ventricle. When the test supervisor recognizes ST-segment changes, he should note when (time and work rate) it begins, the morphology, and the magnitude, as well as document any symptoms associated with the ST changes (e.g., chest pain, shortness of breath, dizziness).

• ST-Segment Depression: Of the potential ST-segment changes, ST-segment depression is the most frequent response and is suggestive of subendocardial ischemia. One or more millimeters of horizontal or downsloping ST-segment depression that occurs 0.08 s

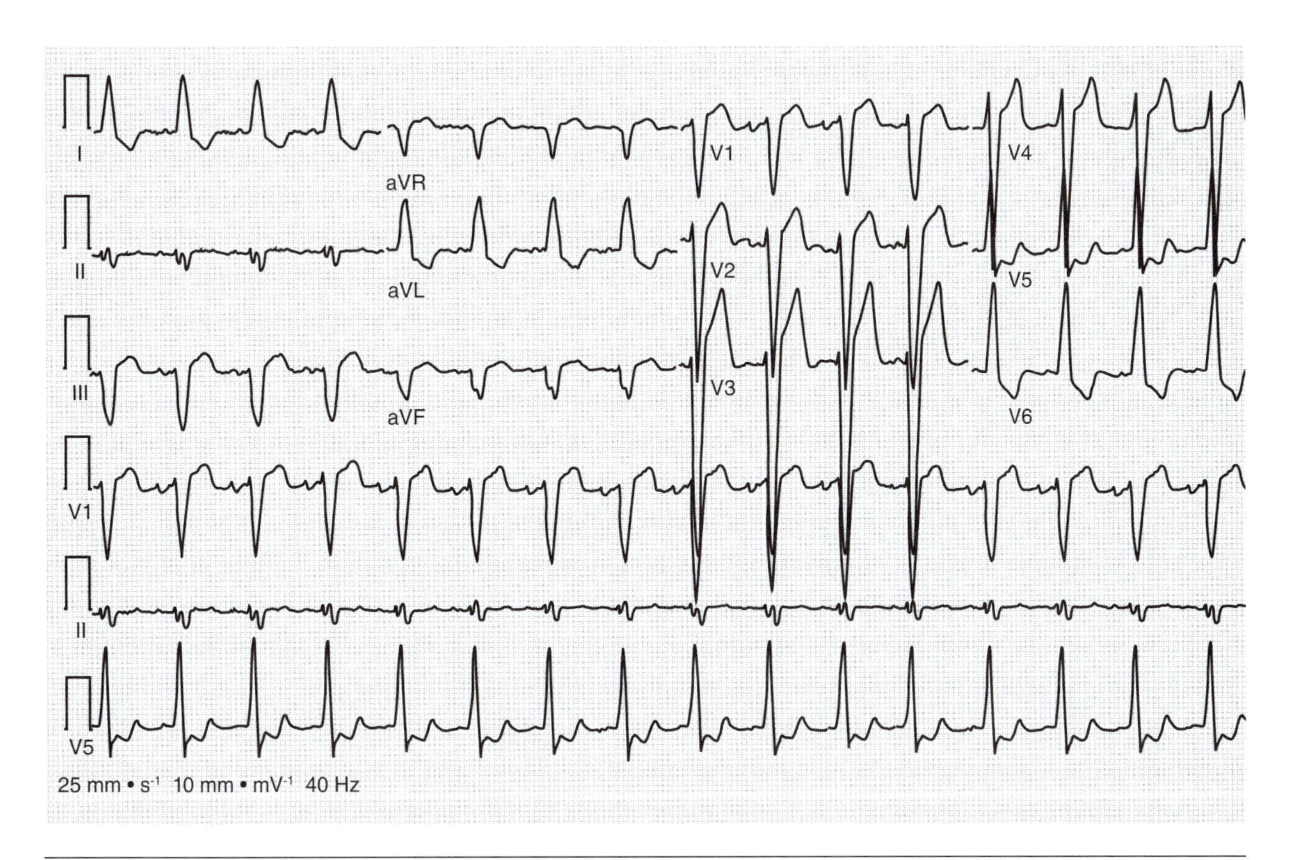

Figure 5.1 Electrocardiogram showing left bundle branch block.

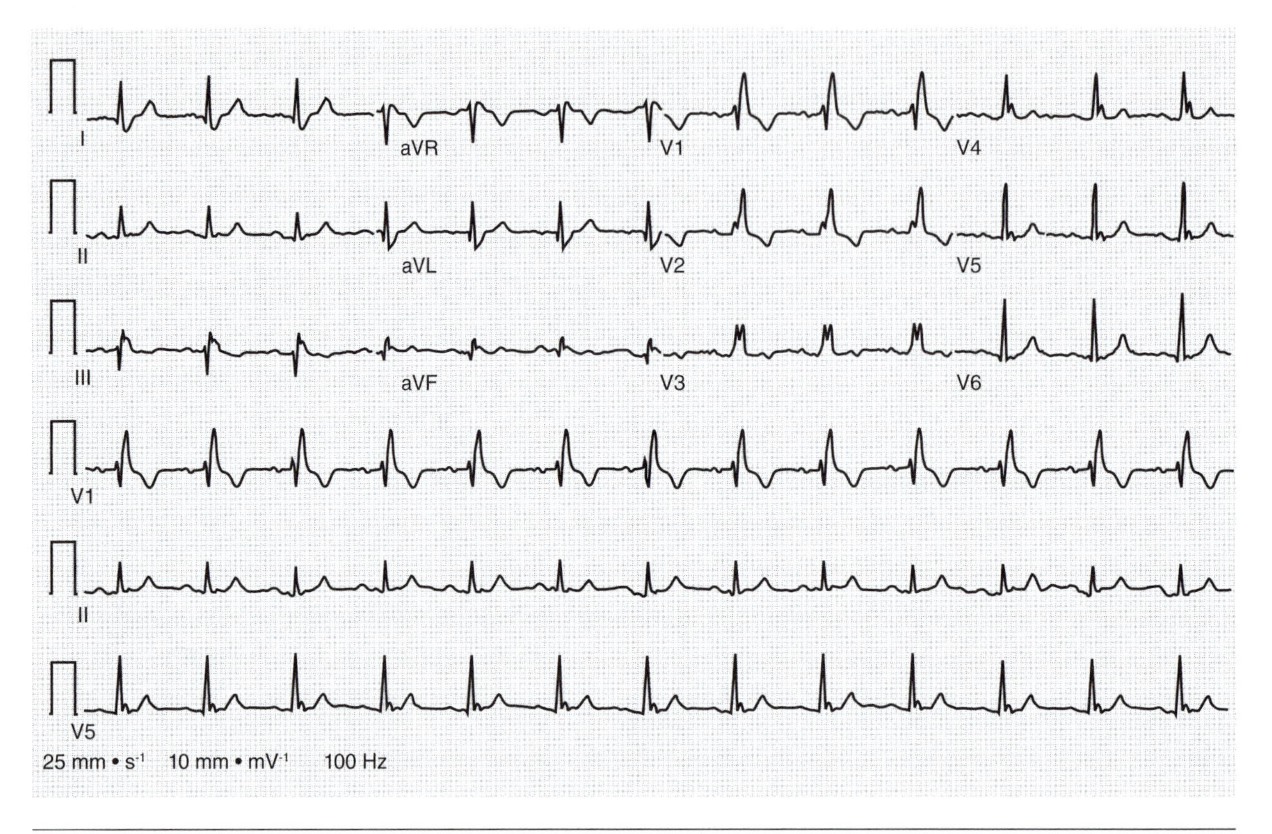

Figure 5.2 Electrocardiogram showing right bundle branch block.

(i.e., 80 ms) past the J point is recognized as a positive test for myocardial ischemia (figure 5.3). When ST-segment changes of this type occur along with typical angina symptoms, the likelihood of CAD is extremely high. In addition, the earlier the onset, the greater the amount of ST depression, the more leads with ST depression, and the more time it takes for the ST depression to resolve in recovery, the more likely it is that significant CAD is present. In some cases ST-segment depression is observed only in recovery, yet it should be treated as an abnormal response. Additionally, J-point depression with upsloping ST segments that are more than 1.5 mm depressed at 0.08 s past the J point are suggestive of exercise-induced ischemia. If the rate of upsloping ST depression is slow or gradual, the probability of CAD is increased.

- ST-Segment Elevation: ST-segment or J-point elevation observed on a resting ECG is often attributable to early repolarization and is not necessarily abnormal in healthy people, but this should be documented on the GXT report. With exertion, this type of ST elevation normally returns to the isoelectric line. New ST-segment elevation with exertion (assuming that the resting ECG is normal) is a somewhat rare finding and may suggest transmural ischemia or a coronary artery spasm. This type of ST-segment elevation can be associated with serious arrhythmias and is an absolute reason for stopping the test (see "Indications for Termination of Graded Exercise Test"). When Q waves are present on the resting ECG from a previous infarction, ST elevation with exertion may simply reflect a left ventricular aneurysm or wall motion abnormality. ST-segment elevation can localize the ischemic area and the arteries involved, whereas this is not always the case with ST-segment depression (52, 66).

- T-Wave Changes: When exercise is started in healthy individuals, T-wave amplitude gradually decreases initially. Later, at maximal exercise, T-wave amplitude increases. An increase in serum potassium following exercise is postulated to be responsible for increased T-wave amplitude. In the past, T-wave inversion with exertion was thought to reflect an ischemic response. But it is now believed that flattening or inversion of T waves may not be associated with ischemia. T-wave inversion is common in the presence of left ventricular hypertrophy. It presently remains undecided whether or not inverted T waves present on the ECG at rest that normalize with exertion reflect myocardial ischemia. Normalization of T waves is also present during ischemic responses associated with coronary spasms; however, this finding has the greatest significance under resting conditions. Overall, T-wave changes with exertion are not specific to

Indications for Termination of Graded Exercise Test

Absolute Indications

1. Decrease in systolic BP of >10 mmHg below baseline BP despite an increase in workload and when accompanied by other evidence of ischemia
2. Moderate to severe angina (2^+ to 3 or more on a 4-point scale)
3. Increasing nervous system symptoms (e.g., ataxia, dizziness, or near-syncope)
4. Signs of poor perfusion (cyanosis or pallor)
5. Technical difficulties monitoring the ECG or systolic BP
6. Subject's desire to stop
7. Sustained ventricular tachycardia
8. ST elevation (>1.0 mm) in leads without diagnostic Q waves (other than V1 or aVR)

Relative Indications

1. Decrease in systolic BP of >10 mmHg below baseline BP despite an increase in workload and in the absence of other evidence of ischemia
2. ST or QRS changes such as excessive ST depression (>2 mm horizontal or downsloping ST-segment depression) or marked axis shift
3. Arrhythmias other than sustained ventricular tachycardia, including multifocal premature ventricular contractions (PVCs), triplets of PVCs, supraventricular tachycardia, heart block, or bradyarrhythmias
4. Fatigue, shortness of breath, wheezing, leg cramps, or claudication
5. Development of bundle branch block or intraventricular conduction delay that cannot be distinguished from ventricular tachycardia
6. Increasing chest pain
7. Hypertensive response
8. A systolic BP of >250 mmHg or a diastolic BP of >115 mmHg

Adapted, by permission, from American College of Sports Medicine, 2013, *ACSM's guidelines for exercise testing and prescription,* 9th ed. (Philadelphia, PA: Lippincott, Williams, and Wilkins), 131.

exercise-induced ischemia but should be correlated with ST-segment changes and other signs and symptoms. Note that T-wave changes frequently follow ST changes and can be difficult to isolate with exertion.

- Arrhythmia: In relation to ST changes with exertion, arrhythmias are of equal clinical importance and potentially more life threatening, based on the suddenness in which arrhythmias may appear. Although a GXT is most commonly used to diagnose potential CAD, this test can also be used to evaluate symptoms (e.g., near-syncope) attributable to arrhythmia. In addition, GXT may be used to evaluate the effectiveness of medical therapy in controlling an arrhythmia. The supervisor of the GXT must have a strong knowledge of arrhythmia detection and be able to respond appropriately and quickly. When arrhythmias appear during the GXT, the onset of the arrhythmia and any signs or symptoms associated with the arrhythmia should be documented, along with any other ECG changes (e.g., ST depression). The health care provider supervising the GXT should be knowledgeable about and should be able to recognize three major types of ECG rhythm or conduction abnormalities during exercise, which include

1. supraventricular arrhythmias that compromise cardiac function,
2. ventricular arrhythmias that have the potential to progress to a life-threatening arrhythmia, and
3. the onset of high-grade conduction abnormalities.

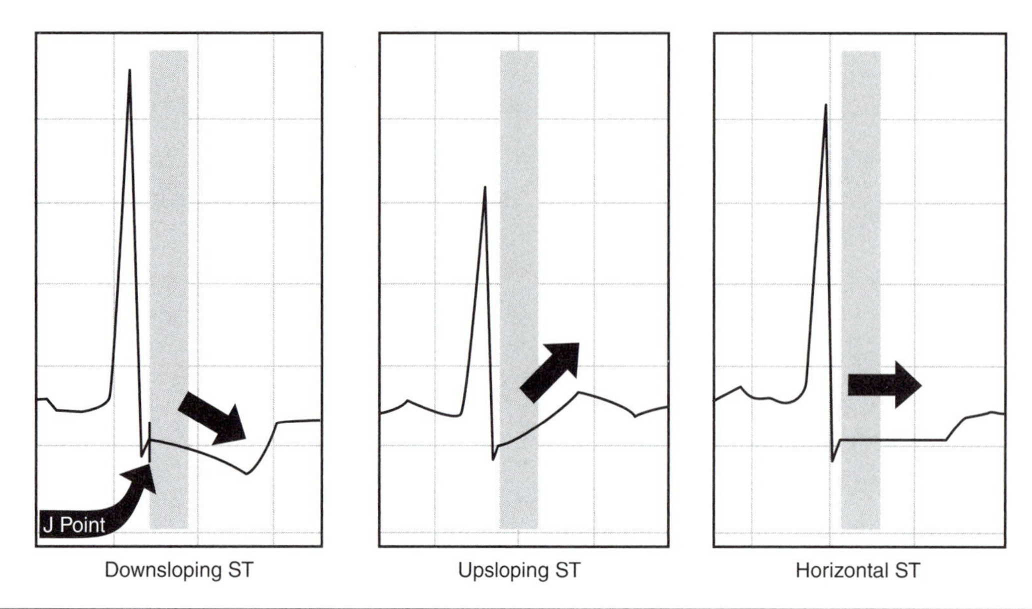

Figure 5.3 Location of J point and examples of the three types of ST-segment depression (downsloping, upsloping, and horizontal). See text for criteria for determining a positive test for myocardial ischemia based on the type of ST-segment depression.

A thorough review of these abnormalities is beyond the scope of this chapter. In general, supraventricular (i.e., above the ventricles) arrhythmias such as isolated premature atrial and junctional contractions and short runs of supraventricular tachycardia are considered benign and provide little diagnostic or clinical value. Supraventricular tachycardia may be related simply to an increased adrenergic response that occurs with physical exertion, although sustained supraventricular tachycardia is generally considered an absolute indication for test termination. If the patient also experiences associated symptoms (e.g., dizziness, syncope), this may be related to the fast rate of supraventricular tachycardia. The supervising physician should be informed immediately if she is not in the testing room. Also, exercise-induced atrial flutter or fibrillation is a reason for terminating the GXT.

In general, ventricular arrhythmias are considered to be of greater importance than supraventricular arrhythmia because they can more easily progress to more life-threatening arrhythmia. Occasional (fewer than six per minute) ventricular premature beats, however, are benign and do not warrant stopping the test. Estimates are that 20% to 30% of the healthy population and 50% to 60% of patients with cardiovascular disease experience ventricular premature beats under resting conditions. Ventricular premature beats at rest that disappear with exertion are considered benign, but when they increase in frequency with exertion, there is more reason to suspect an underlying cardiac problem. Ventricular couplets or pairs should be noted but generally are not considered an absolute reason for test termination. Ventricular premature beats that occur as a triplet (ventricular tachycardia) at a rate between 100 and 150 beats $\cdot$ min^{-1} are considered a relative indication for stopping the test. Ventricular tachycardia that reaches 150 to 250 beats $\cdot$ min^{-1} is an absolute indication for test termination because of the increased likelihood of progressing to a faster rate and eventually ventricular flutter and ventricular fibrillation. The test supervisor must recognize ventricular tachycardia and record the onset. Generally, ventricular tachycardia is easily recognized by the wide QRS complex; however, the onset of left bundle branch block during exercise can resemble ventricular tachycardia. If an individual converts to a bundle branch block, P waves still precede the QRS complex, whereas this is not true in ventricular tachycardia. Because HR increases during exercise, however, identifying P waves is often difficult. If the test supervisor has difficulty trying to distinguish between exercise-induced ventricular tachycardia and left bundle branch block, the test should be terminated.

Assessment of Functional Capacity

As mentioned earlier in this chapter, FC as measured by estimated METs, exercise duration, or peak $\dot{V}O_2$ using gas exchange is a strong predictor of prognosis. Clinically, such information is used often in the management of patients with heart failure, and exercise MET level is a key parameter in the Duke nomogram for determining 5 yr survival and 1 yr mortality.

Additionally, the assessment of FC during testing can help the clinician guide decision making relative to return to work after an illness or injury. For example, for the patient recovering from a myocardial infarction that occurred 4 wk ago who now asks to return to an occupation that requires periodic bouts of exertion equivalent to 6.5 METs (e.g., moderately paced shoveling of dirt), FC measured during a GXT can help determine if the patient is able to achieve a peak MET level that is at least 20% more than the task he wishes to resume—in this example, almost 8 METs. If such a peak work rate cannot yet be achieved, return to work with restrictions or further time in rehabilitation may be indicated.

Finally, assessment of FC during a GXT can help quantify, for clinical reasons or research purposes, the impact of either an exercise training program or a medical device on FC, be it change in exercise duration, peak $\dot{V}O_2$, or work rate (e.g., watts). Because of the inconvenience and the slight increase in expense of directly measuring a patient's FC using respiratory gas exchange, some clinicians and researchers use equations to predict peak $\dot{V}O_2$. The standard American College of Sports Medicine (ACSM) metabolic equations (2) are commonly used; but they rely on achieving a steady state, which does not occur at maximal effort, and therefore can lead to an overestimation of peak FC and potential misclassification of a patient's risk or inaccurate assessment of the response to a therapy. Other metabolic prediction equations exist for predicting FC, such as those for the Bruce GXT protocol with and without handrail support and a ramping bicycle ergometer protocol (table 5.3).

Interpretation of Findings and Generation of Final Summary Report

The recovery period, which follows the exercise portion of the GXT, allows the clinical exercise physiologist to monitor a patient's physiologic and clinical responses after exercise, as well as provide an opportunity to briefly speak with the patient about one or two important recommendations aimed at pertinent lifestyle-related risk factors (practical application 5.1). Once the GXT is done, the clinical exercise physiologist who helped supervise the test also often helps prepare a final summary report that is later read and signed off on by a physician. The key elements to be included and succinctly discussed and interpreted in this report are (1) angina status, (2) ECG findings pertaining to ischemia, (3) ECG findings pertaining to arrhythmia, (4) functional capacity, (5) HR response, and (6) BP response. When applicable, each of these elements should be interpreted relative to diagnosis, prognosis, or future risk for developing a disorder.

The statement pertaining to angina status should specify whether or not angina was present and, if so, when it began (at what HR or MET level). Also include whether angina was the reason for stopping the test, what intensity or grade it reached (see angina scale in table 5.2), and what intervention was required during recovery to relieve the discomfort (rest, medications). A statement addressing ECG evidence of exercise-induced myocardial ischemia should be included in the summary report and should be based on ST-segment information gathered at rest, during exercise, and in recovery. If the ECG at rest was normal, then simply state, based on the presence or absence of ST depression during exercise, that the test either does or does not meet the criteria to be considered positive for exercise-induced myocardial ischemia.

Information concerning FC, stated as estimated METs or measured peak $\dot{V}O_2$, should be provided in the final report both as an absolute number and in qualified terms (e.g., above average, poor, superior) relative to data from other people of similar age and gender. The ACSM provides a table for making such comparisons (2).

Another test-related variable that must be included in the final report pertains to HR and BP response.

Table 5.3 Estimating Functional Capacity at Maximal Effort

GRADED EXERCISE TREADMILL TEST USING THE STANDARD BRUCE PROTOCOL	
Without holding the handrail	$\dot{V}O_2$max (ml · kg^{-1} · min^{-1}) = 14.8 – 1.379 (time in min) + 0.451 (time^2) – 0.012 (time^3)
With holding the handrail	$\dot{V}O_2$max (ml · kg^{-1} · min^{-1}) = 2.282 (time in min) + 8.545
CYCLE ERGOMETRY (BASED ON THE FINAL POWER COMPLETED IN A 15 W/MIN RAMP PROTOCOL)	
Males	$\dot{V}O_2$max (m · min^{-1}) = 10.51 (power in W) + 6.35 (body mass in kg) – 10.49 (age in yr) + 519.3
Females	$\dot{V}O_2$max (ml · min^{-1}) = 9.39 (power in W) + 7.7 (body mass in kg) – 5.88 (age in yr) + 136.7

Adapted from ACSM 2013.

Practical Application 5.1

CLIENT–CLINICIAN INTERACTION

When exercise is stopped during a stress test, there is usually a 6 to 8 min recovery period during which HR, BP, and the ECG are monitored to ensure that the patient's cardiovascular system has returned to a level that allows for safe discharge from the laboratory. Although some other duties and tasks must be performed during this brief recovery period, this time also provides an excellent opportunity for the testing staff to discuss with the patient any interest she may have or methods she might be employing relative to modifying risk factors in an attempt to decrease future risk for a lifestyle-related health problem. This can include beginning an exercise program.

Obviously, any clinical concerns or findings that arose during the test need to be brought to the patient's attention; and without going into specific findings or interpreting the results, staff should ask the patient to be sure to follow up concerning the results with the clinician who ordered the test. During recovery it is also appropriate to possibly talk with patients about one of their major risk factors. You can simply ask them if they would mind talking about their exercise habits, weight management practices, or smoking cessation. If they agree to do so, you'll have a few minutes to learn about their goals and about steps they are currently taking or have tried in the past, and then bring in your expertise relative to effective behavior methods that may assist them. Despite all the important work that clinical exercise physiologists perform with respect to supervising or conducting exercise tests, prescribing or leading exercise, or measuring changes in fitness, always remember that in the end, a big portion of our time is spent counseling others about how to make positive changes in unhealthy behaviors. After an exercise test, the patient represents a "captive audience," so seize this time to help others using your expertise.

Quantifying whether HR was normal (exceeded 80% of age predicted) or consistent with a chronotropic incompetence (less than 80% of age predicted) response is essential. The prognostic value of a poor HR response is as great as that of an exercise-induced myocardial perfusion deficit. In a similar fashion, the failure of the HR to slow promptly after exercise provides independent information related to prognosis (39). The failure of the HR to decrease by at least 12 beats during the first minute or 22 beats by the end of the second minute of recovery is independently associated with an increased risk for mortality over the next 3 to 5 yr period (39). However, HR recovery cannot discriminate those with coronary disease and instead should be used to supplement other predictors of mortality, such as the Duke nomogram (62).

Systolic BP normally increases in a negatively accelerated manner during incremental exercise. The magnitude of the increase approximates 10 mmHg per stage of work. An absolute peak systolic pressure of >250 mmHg is considered an indication to stop the test. An increase in systolic to >210 mmHg in men or >190 mmHg in women, or a relative increase of >140 mmHg above resting levels, is considered a hypertensive response and is predictive of future resting hypertension (4). Patients with limitations of cardiac output show either an inappropriately slow increase in BP or possibly even a slight decrease in systolic BP during the exercise test. A decrease of systolic BP to below the resting value or by >10 mmHg after a preliminary increase, particularly in the presence of other indices of ischemia, is abnormal.

There is typically no change or a slight decrease in diastolic BP during an exercise test. An increase by >10 mmHg is generally considered to be an abnormal finding and may be consistent with exertional ischemia, as is an increase to >115 mmHg (4).

During the postexercise period, systolic BP normally decreases promptly (8, 44). Several recent investigators have demonstrated that a delay in the recovery of systolic BP is related both to ischemic abnormalities and to a poor prognosis (8, 44). As a general principle, the 3 min postexercise systolic BP should be <90% of the systolic BP at peak exercise. If peak exercise BP cannot be measured accurately, the 3 min postexercise systolic BP should be less than the systolic BP measured at 1 min after exercise.

EXERCISE PRESCRIPTION

The optimal adaptation of a client or patient to an exercise training program requires the development of, and adherence to, an evidence-based exercise prescription. This section of the chapter discusses the key underlying principles and considerations associated with developing such a prescription. In each of the remaining chapters in this book, these same underlying principles and tenets may be modified so as to take into account and accommodate the pathology and type of care discussed in the chapter.

An exercise prescription is a specific guide provided to an individual for the performance of an exercise training program. Despite the use of the word *prescription*, the development of an exercise prescription does not necessarily require approval by a physician. In some situations, however, it may, especially when the prescription is developed for a patient with a clinically manifest disease or disorder. For example, a physician's approval and signature may be required to secure reimbursement from Medicare for a patient in cardiac rehabilitation or if the clinical exercise professional feels that this is warranted to limit personal liability. Individuals practicing clinical exercise physiology who are charged with developing an exercise prescription find that doing so is both an art and a science; they must possess both the requisite knowledge and skills and must be able to articulate and put into action a prescription that is safe and practical (see practical application 5.2).

The primary purpose of the exercise prescription is to provide a valid and safe guide for optimal health and improvements in physical fitness. The specificity of the exercise prescription should be made relevant to the nature of clinical population. The ACSM has published several position statements on particular populations, including healthy people and those with coronary artery disease, osteoporosis, hypertension, and diabetes, as well as older people (2-6). The exercise prescription can also be specific to the five health-related components of physical fitness:

1. Cardiorespiratory endurance (aerobic fitness): ability of the cardiorespiratory system to transport oxygen to active skeletal muscles during prolonged submaximal exercise and the ability of the skeletal muscles to use oxygen through aerobic metabolic pathways.
2. Skeletal muscle strength: peak ability to produce force. Force may be developed by isometric, dynamic, or isokinetic contraction.

Practical Application 5.2

THE ART OF EXERCISE PRESCRIPTION

The art of prescribing exercise involves the successful integration of exercise science with behavioral techniques in a manner that results in long-term program compliance and attainment of the individual's goals (2). Unlike the disciplines of chemistry or physics, physiology and psychology are not always exact. We cannot always precisely predict physiological or psychological responses because numerous factors and confounders have influence over the outcome. These include, but are not limited to, age, physical and environmental conditions, sex, previous experiences, genetics, and nutrition. When developing an exercise prescription, you should follow the basic guidelines provided in this chapter. In doing so, you can help elicit the desired response both during a single exercise training session and over the course of an extended training period. Keep in mind, however, that not all people respond as expected, especially those with a chronic disease. For example, people with coronary artery disease may require modification of exercise intensity because of myocardial ischemia. Additionally, those currently undergoing treatment for cancer often fatigue easily and may better tolerate training sessions of shorter duration and lesser intensity. The *ACSM* (2) lists several reasons for altering an exercise prescription in selected individuals:

Variance in objective (physiological) and subjective (perceptual) responses to an exercise training bout

Variance in the amount and rate of exercise training responses

Differences in goals between individuals

Variance in behavioral changes relative to the exercise prescription

Each of these reasons should be considered for both the initial development and subsequent review of the exercise prescription. A modified exercise prescription should not be considered adequate unless it is evaluated for effectiveness over time. As a rule, a person's exercise prescription should be reevaluated weekly until its parameters appear to be safe as well as adequate to improve health-related behaviors and selected physiological indexes.

3. Skeletal muscle endurance: ability to produce a submaximal force for an extended period.
4. Flexibility: the ability of a joint to move through its full, capable range of motion.
5. Body composition: the relative percentage of fat and nonfat mass composing total body weight. Chapter 7 provides details regarding body composition assessment and exercise prescription for fat or weight loss.

Each of these components of physical fitness is related to at least one aspect of health, and each component of health-related physical fitness is positively influenced by exercise training, likely reducing the risk of a primary or secondary chronic disease (11, 53, 70). Thus the importance of regular exercise and physical activity for overall health is well established. A summary of the general benefits of exercise training is presented in table 5.4 (70). Several principles or tenets must be considered in the development of an exercise prescription:

1. Specificity of training
2. Progressive overload
3. Reversibility

The clinical exercise physiologist must also consider several aspects of a person's psychosocial conditions, such as those that are relevant to beginning and adhering to an exercise training program (see chapter 2).

Exercise Training Sequence

A comprehensive training program should include flexibility, resistance, and cardiorespiratory (aerobic) exercises. The order of the exercise training routine is important for both safety and effectiveness. Scientific data on this topic, however, are lacking. Generally, it is recommended that flexibility training take place following a warm-up period or following an aerobic or resistance training routine to reduce the risk of muscular injury and soreness. In a clinical population, if an aerobic bout and a resistance training bout take place on the same day, the best approach is to first perform the activity that is the primary focus of that day's training.

Goal Setting

A comprehensive exercise prescription should consider the goals that are specific for each person. Common goals include the following:

Appearance

Improved quality of life

Weight management

Preparing for competition

Improving general health to reduce risk for primary or secondary occurrence of disease

Reducing the burden of a chronic disease or condition (early fatigue, depression, loss of personal control, economic impact)

People with specific diseases often have goals that relate directly to reversing or reducing the progression of their disease and its side effects or the side effects of the therapies used to treat the disorder. Because of these and other goals, a clinical exercise physiologist must have a comprehensive understanding of how to alter the general exercise prescription so as to provide the patient with the best chance of success in achieving a desired goal. Also, the exercise physiologist should help assess whether goals are realistic and discuss them with patients when they are not.

Principles of Exercise Prescription

To gain the optimal benefits of exercise training, regardless of the area of emphasis (i.e., cardiorespiratory,

Table 5.4 Physiological, Health, and Psychological Benefits of Exercise Training (70)

Improved cardiorespiratory and musculoskeletal fitness	Improved glucose metabolism
Improved metabolic, endocrine, and immune function	Improved mood (depression and anxiety)
Reduced all-cause mortality	Reduced risk of obesity
Reduced risk for cardiovascular disease	Overall improved health-related quality of life
Reduced risk for certain cancers (colon, breast)	Reduced risk of falling
Reduced risk of osteoporosis and osteoarthritis	Improved sleep patterns
Reduced risk of non-insulin-dependent diabetes mellitus	Improved health behavior

Adapted from the Surgeon General's Report on Physical Activity and Health.

strength, muscular endurance, body composition, range of motion), several principles must be followed; these are discussed next.

Specificity of Training

Long-term changes or adaptations in physiological function occur in response to a chronic or repeated series of stimuli. The principle of specificity of exercise training states that these physiologic changes and adaptations are specific to the cardiorespiratory, neurologic, and muscular responses that are called upon to perform the exercise activity. Specifically, the neuromuscular firing patterns or the cardiorespiratory responses (or both) that are needed to perform an activity are the ones that undergo the greatest degree of adaptation. If a 1,600 m college runner wants to do all he can to improve his race performance, he should spend the majority of his training time running for mid- and long durations; this type of training engages the cardiorespiratory processes involved with the transport and use of oxygen with the firing of the neurons used during running at a high velocity. In contrast, an offensive lineman in football should spend very little of his practice time in distance running and instead engage in explosive activities and blocking techniques.

The classic example of specificity of training would be seen if we measured a person's peak $\dot{V}O_2$ on the treadmill one day and then on a stationary bike the next day. We know that for this person, $\dot{V}O_2$ peak will be 5% to 15% higher on a treadmill versus a cycle ergometer (34, 45, 72, 74). The majority of this difference is related to both the weight independence associated with sitting on a bike and the smaller total muscle mass used during cycling (vs. running). However, if we repeated this same experiment using highly trained cyclists, we would observe that they achieve a strikingly similar peak $\dot{V}O_2$ regardless of whether testing is on the bike or the treadmill. This observation among competitive cyclists is attributable to the fact that they spend virtually all of their training time cycling and that their physiologic adaptations are in specific response to such. Some modest crossover adaptations likely do occur from one mode to another and are partly the result of a combination of central cardiac improvements (e.g., increased stroke volume and cardiac output) and involvement of the skeletal muscles used in the alternative exercise mode (e.g., training the leg muscles by cycling and then using many of those same leg muscles when running), which is partly the basis for the **cross-training** concept.

The following sidebar provides questions that can be asked of a person for whom an exercise professional is developing an exercise prescription. This approach will help ensure that the exercise prescription is specific to that person.

Progressive Overload

The progressive overload principle refers to the relationship between the magnitude of the dose of the exercise stimulus and the benefits gained. There appears to be a level of physical activity or exercise beyond which favorable physiological adaptations and health benefits are incurred; likewise, there appears to a much higher level or stimulus of exercise beyond which benefits plateau or possibly even diminish. For instance, in the Harvard Alumni study, a dose response relationship was observed between all-cause mortality and the number of kcals expended each week (53). Below an energy expenditure of 500 kcal/wk the mortality benefit was negligible, and beyond an expenditure of 3,500 kcal per week, the risk for death actually rose slightly.

Overload refers to the increase in total work performed above and beyond that normally performed on a day-to-day basis. For example, when a person performs walking as part of an exercise training regimen as a means to improve fitness, the pace and duration should be above those typically experienced on a daily basis. Progressive overload is the gradual increase in the amount of work performed in response to the continual adaptation of the body to the work.

Applying this principle would relate to walking more often, farther, or at a faster pace. Overload is often applied using the FITT principle. FITT is an acronym that stands for frequency, intensity, time (i.e., duration), and type of exercise.

Before we discuss frequency, intensity, and duration separately, it is important to understand how these three parameters can be combined as a means to quantify the total volume of exercise that an individual is engaging in. Although kilocalories per week can be used, another common unit used to express total exercise volume is MET-min per week or MET-h per week, where one MET approximates a resting $\dot{V}O_2$ of 3.5 $ml \cdot kg^{-1} \cdot min^{-1}$. For example, during a week's period of time if a 50 yr old person square-dances three times for 40 min at a moderate pace (~4 METs), she engages in 480 MET-min or 8 MET-h per week. Although it is not easy to use MET-min or MET-h per week to prescribe exercise to individuals or patients, this unit of measure does come in handy for comparisons of estimates of exercise volume across different research studies or for conveying broad recommendations for public health. Concerning the latter, current recommendations are that all adults engage in at least 8 MET-h of exercise per week (5, 71).

Questions to Ask a Person When Developing an Exercise Prescription

Specificity

What are your specific goals when performing exercise (health, fitness, performance)?

Do you want to exercise more?

Do you want to be able to do more activities of daily living?

Do you want to perform something that you currently cannot? If yes, describe.

Mode

What types of exercise or activity do you like the best?

Do you already have any exercise equipment in your home?

What types of exercise do you like the least?

Frequency

Do you know how many days per week of exercise or physical activity are required for you to reach your goals?

How many days during a week do you have 30 to 60 min of continuous free time?

Intensity

Do your goals include optimal improvement of your fitness level?

Or are your goals primarily related to your health?

Do you have any musculoskeletal problems that would limit your intensity level?

Time

How much time per day do you have to perform an exercise routine?

What is the best time for you to exercise?

Can you get up early or take 30 to 40 min at lunchtime for exercise?

Frequency Frequency is the number of times an exercise routine or physical activity is performed (per week or per day).

Intensity The intensity of exercise or physical activity refers to either the objectively measured work or the subjectively determined level of effort performed by an individual. Typical objective measures of work that are important to the clinical exercise professional include oxygen uptake ($\dot{V}O_2$, or metabolic equivalent), caloric expenditure (kilocalories [kcal] or joules [J]), and power output (kilograms per minute [$kg \cdot min^{-1}$] or watts [W]). The anaerobic or lactate threshold may also be used to determine exercise intensity; but it often is impractical to use as a guide during exercise training. The subjective level of effort can be evaluated through use of either a verbal statement from a person performing exercise (e.g., "I'm tired" or "This is easy") or a standardized scale (e.g., Borg rating of perceived exertion). The patient must be taught the proper use of this assessment tool to obtain accurate indications of perceived effort (47).

Duration Duration (or Time in the FITT acronym) refers to the amount of time that is spent performing exercise or physical activity. During exercise training, the duration is typically accumulated without interruption (i.e., continuously) or with very short rest periods to gain fitness benefits. In 1995, the Centers for Disease Control and Prevention (CDC) and ACSM advocated that every U.S. adult *accumulate* 30 min or more of physical activity each day of the week, suggesting that people can gain health-related benefits from three 10 min bouts or two 15 min bouts that occur over a day (54). Compared to the continuous method, the discontinuous model appears to yield similar gains in cardiorespiratory fitness and improvement in BP, with its effect on body composition, lipoproteins, and mental health inconclusive (48). The discontinuous approach is appealing from an adherence perspective and to people with very low fitness levels.

Reversibility

The reversibility principle refers to the loss of exercise training adaptations because of inactivity. Positive adaptations accrue at a rate specific to the overloaded physiological processes. Typically, most untrained people can expect a 10% to 30% improvement in $\dot{V}O_2$peak and work capacity following an 8 to 12 wk period of training (5). Alterations in other physiological variables, such as body weight and BP, may take a variable amount of time. As a rule, less fit people can expect to achieve gains at a faster rate and to a greater relative degree than more fit people. Maintaining all these improvements is an important issue. If a minimal training volume is not maintained, training effects will begin to reverse or erode. This reversal of fitness is often called deconditioning or detraining.

Cardiorespiratory Endurance

Cardiovascular conditioning to improve cardiorespiratory or aerobic endurance requires that individuals perform those types of training that use large muscle groups and are continuous and repetitive or rhythmic (5). Satisfactory modes of exercise that involve the legs include walking, running, cycling, skating, stair stepping, rope skipping, and group aerobics (e.g., dance, step, tae bo, spinning, water). Exercises using strictly the arms is more limited but includes upper body crank ergometry, dual-action stationary cycling using only the arms, and wheelchair ambulation. Several popular modes of exercise use both the arms and legs: rowing, swimming, and exercise on some types of stationary equipment (e.g., dual-action [arm and leg] cycles, cross-country skiing, elliptical trainers, seated dual-action steppers).

Specificity of Training

General benefits gained from aerobic exercise training appear to be independent of any specific type or mode of training. For instance, the Harvard Alumni study reported that men who were physically active and had a high weekly caloric expenditure, regardless of mode, had a lower incidence of all-cause mortality than those who were less active (53). Further interpretation of the Harvard Alumni database demonstrates that all-cause mortality rate is improved in those who perform higher-intensity exercises compared to those who perform less vigorous activity (40). These reports suggest that the specific type of physical activity is less important for general mortality benefits than the amount and intensity of the activity.

Progressive Overload

To derive benefits from cardiorespiratory training, people must follow the principle of progressive overload, which involves the appropriate application of frequency, intensity, and duration (or time).

Frequency Recommendations for the frequency of aerobic activity vary between sources. The ACSM recommends a frequency of more than 5 d/wk for cardiorespiratory training (5). An earlier and slightly different recommendation was published in the joint recommendation paper from the CDC and the ACSM (54), which recommends "physical activity on most, preferably all, days of the week." The difference in these recommendations might lie in the fact that the 2011 ACSM position stand recommends exercise to improve cardiorespiratory fitness level, whereas the CDC-ACSM recommendation focuses more on improving overall health. Cardiorespiratory fitness and health benefits related to the performance of regular exercise training or physical activities, respectively, are not mutually exclusive, with the difference in these recommendations being the desired effects of exercise or physical activity. A second difference in these recommendations is that the CDC-ACSM joint statement suggests a reduced intensity of activity.

Research suggests that an increased time commitment to exercise train more than 3 d/wk may not be an efficient use of time for the nonathlete with little time to spare (10, 36, 55). In fact, in a study that held total exercise volume constant, no difference in $\dot{V}O_2$peak was reported for those who exercised 3 versus 5 d/wk (63). Still, exercising more than 5 d/wk may play a positive role by increasing total caloric expenditure, optimizing health improvements, and reducing all-cause mortality rates.

Intensity and Duration Intensity and duration of training are often interdependent with respect to the overall cardiorespiratory training load and adaptations. Generally, the higher the intensity of an exercise training bout, the shorter the duration, and vice versa. The selection of the intensity and duration should consider several factors, including the current cardiovascular conditioning level of the individual; the existence of underlying chronic diseases such as cancer, CAD, or obesity; the risk of an adverse event; and the individual's goals.

To achieve an adequate training response, most people must exercise at an intensity between 50% and 85% of their $\dot{V}O_2$peak in order for the cardiorespiratory system to be sufficiently stimulated to adapt and for aerobic capacity to increase (37). In clinical patients suffering from marked deconditioning (e.g., those with heart failure or those being treated for cancer; the elderly; obese individuals), improvements in $\dot{V}O_2$peak may be observed at intensities as low as 40% of their maximal ability (2). Generally, the lower a person's initial fitness level, the lower the required intensity level to produce adaptations. The upper level for active, healthy individuals who want to improve $\dot{V}O_2$peak can be set as high as 85% of peak $\dot{V}O_2$ (5). This upper level is generally regarded as the threshold between optimal gains in fitness and increased risk of orthopedic injury or adverse cardiovascular event. Training at too high an intensity may also be difficult for healthy individuals and patients with a chronic disease who have a lactate threshold that is less than 85% of their $\dot{V}O_2$peak. Practical application 5.3 provides additional information about determining appropriate training HR and practical application 5.4 highlights a growing area of research interest that pertains to the potential role of higher intensity interval training in patients with a chronic disease.

The ACSM's position stand (5) recommends a minimum of 30 min of exercise per day, or 20 min per day (4 d/wk) if a more vigorous exercise intensity is employed.

Practical Application 5.3

DETERMINING THE APPROPRIATE HEART RATE RANGE

It is impractical and nearly impossible for an individual to guide his exercise intensity using $\dot{V}O_2$. But because HR has a near-linear relationship with $\dot{V}O_2$ at levels between 50% and 90% of $\dot{V}O_2$ peak in both healthy people and most patients with a chronic disease, HR becomes an excellent surrogate to guide exercise intensity. Several methods exist to determine an exercise training HR range. The following outlines how to develop an exercise prescription based on HR.

When the true maximal HR (HRmax) is unknown, in apparently healthy people it can be approximated using either of these two equations:

Equation 1: 220 – Age = Estimated HRmax

Equation 2: 208 – 0.7 (Age) = Estimated HRmax (69)

Example: to predict HRmax in a 32 yr old person:

Using equation 1: 220 – 32 = 188

Using equation 2: = 208 – 0.7(32) = 198

A disadvantage of estimating HRmax is that the standard deviation (SD) can be as high as ±10 to 12 beats $\cdot$ min^{-1}, and thus this method has a high variability from person to person. Another disadvantage is the inaccuracy of equations 1 and 2 in people taking a β-adrenergic blocking agent. Equation 3 is an equation for use in patients with stable CAD taking a β-blocker (12).

Equation 3: 164 – 0.7 (Age) = Estimated HRmax

Given the drawbacks associated with estimating HRmax, using the HRmax from a GXT test is always preferred as long as the patient or client

- attained a true peak exercise capacity on the exercise stress evaluation (e.g., did not stop because of intermittent claudication, arrhythmias, poor effort, severe dyspnea) and
- took her prescribed chronotropic medication (e.g., β-blocker) before the exercise stress and no change has occurred in the type or dose of this medication.

Once a maximal HR has been estimated using equation 1, 2, or 3 or measured during a GXT, two methods can be used to determine a client's or patient's training HR range (THRR); the HR reserve (HRR) method and %HRmax method.

HRR method (a.k.a. Karvonen method):

Equation 4: Step I. HRmax – HRrest = HRR

Step II. (HRR × Desired $\dot{V}O_2$ percentages) + HRrest = THRR

Example: HRmax (estimated or maximal) = 170, HRrest = 68, and desired $\dot{V}O_2$ percentages are 50% and 80%.

Step I. 170 – 68 = 102 (HRR)

Step II. 102 × .5 + 68 = 119 (lower end)

102 × .8 + 68 = 150 (upper end)

Note: $\dot{V}O_2$ reserve is defined as the difference between resting $\dot{V}O_2$ and peak exercise $\dot{V}O_2$ ($\dot{V}O_2$max – $\dot{V}O_2$rest). The %HRR method has a closer 1:1 relationship with %$\dot{V}O_2$ reserve than with % peak $\dot{V}O_2$ (13, 67, 68). This means that a 50% HRR ≈ 50% $\dot{V}O_2$reserve and 75% HRR ≈ 75% $\dot{V}O_2$ reserve. As recommended by the ACSM (5), the range used for intensity with the HRR method should remain at 50% to 80% of peak $\dot{V}O_2$ reserve for most patient populations.

%HRmax method:

Equation 5: HRmax × Desired HR percentages = THRR

Example: HRmax (estimated or measured) = 170, and desired HR percentage range is 60% to 90%.

170 × .6 = 102 (lower end)

170 × .9 = 153 (upper end)

Note: Because the relationship between $\dot{V}O_2$ and HR is not a straight line, the % peak exercise HR does not match up to the exact same %$\dot{V}O_2$peak. At lower intensity levels, the % peak HR is approximately 10% higher than the %$\dot{V}O_2$peak (i.e., 60% of peak HR is equivalent to about 50% of $\dot{V}O_2$peak). This difference in percentages is reduced at higher intensity levels (i.e., 90% of peak HR is roughly equivalent to 85% of $\dot{V}O_2$peak).

Both regimens will improve fitness (likely more so with the more intense approach), and both will reduce the primary and secondary risks for developing a chronic disease.

Reversibility

Several studies have investigated the physiological effects of detraining before or after a period of conditioning. The classic study of Saltin and colleagues (60) reported on the effects of bed rest over a 3 wk period in five subjects. All subjects had reductions in peak cardiac output resulting from reduced stroke volume, and this result was related to reductions in total heart volume. This deleterious trend was reversed when exercise training was implemented. Another study of seven subjects who exercise trained for 10 to 12 mo followed by 3 mo of detraining evaluated the time-course effect of inactivity (18). $\dot{V}O_2$peak decreased by 7% from 0 to 12 d of detraining and by another 7%

Practical Application 5.4

RESEARCH FOCUS: HIGH-INTENSITY INTERVAL TRAINING

Over the past decade, several clinical researchers have used intermittent higher-intensity aerobic interval training (vs. continuous moderate-intensity exercise) to further improve exercise capacity. This method of training is quite common among competitive athletes, yields higher-intensity training levels (i.e., training up to 95% of peak HR vs. up to 75% of peak HR with continuous exercise), and results in the completion of more total work during a single training session. Among patients with stable chronic heart failure, Wisloff and colleagues (75) showed a 6 $ml \cdot kg^{-1} \cdot min^{-1}$ (+46%) increase in peak $\dot{V}O_2$ after 12 wk of intermittent higher-intensity aerobic interval training. This was a level of improvement that far exceeded the 1 to 2 $ml \cdot kg^{-1} \cdot min^{-1}$ increase in peak $\dot{V}O_2$ typically observed in patients with heart failure who engage in a continuous steady-state type of exercise training.

It is important to point out, however, that despite what appears to be a greater impact of interval training on exercise capacity, adequately powered studies evaluating the safety of such a method in patients with cardiovascular disease have not yet taken place. Furthermore, any additional incremental effect that interval training might have on further lowering a patient's future risk for a subsequent or second clinical event (beyond that already known to occur in patients with cardiovascular disease who engage in continuous training) has not yet been demonstrated.

from 21 to 56 d. The early reduction in $\dot{V}O_2$peak was the result of a near-equal percent reduction in stroke volume. The reduction during days 21 to 56 appeared to more associated with a decline in arteriovenous oxygen difference. Note that the absolute reduction in $\dot{V}O_2$ was greater in the complete bed rest study than in the study that allowed limited daily activity (18, 60). This finding suggests that a minimal amount of activity can be effective at maintaining fitness levels or can attenuate the effects of deconditioning.

Some other take-home points concerning the reversibility of acquired cardiorespiratory training adaptations are as follows:

- Cardiorespiratory conditioning can be maintained with a reduced level of exercise training.
- The more sedentary a person becomes, the greater the loss of fitness.
- All people, no matter what their conditioning level or disease status, are prone to deconditioning.
- A sedentary lifestyle results in a loss in fitness that is in addition to the loss of fitness that occurs with aging.

Skeletal Muscle Strength and Endurance

Resistance training improves muscular strength and power and reduces levels of muscular fatigue. The definition of muscular strength is the maximum ability of a muscle to develop force or tension. The definition of muscular power is the maximal ability to apply a force or tension at a given velocity. See "Benefits of Resistance Training."

In general, the focus of a resistance training program should be on the primary muscle groups. Proper lift technique is important to reduce the risk of injury and increase the effectiveness of an exercise. General recommendations include the following:

- Lift throughout the range of motion unless otherwise specified.
- Breath out (exhale) during the lifting phase and breathe in (inhale) during the recovery phase.
- Do not arch the back.
- Do not recover the weight passively by allowing weights to crash down before beginning the next lift (i.e., always control the recovery phase of the lift).

In certain clinical populations, it may be prudent for the professional to follow these recommendations:

- Initially monitor BP before and after a resistance training session and periodically during a session.
- Try to involve the same clinical exercise professional who assisted with a patient's initial orientation and evaluation in regular reevaluations of lifting technique.
- Regularly assess for signs and symptoms of exercise intolerance that may occur during resistance training.
- Instruct participants to train with a partner.

Benefits of Resistance Training

Improved muscular strength and power

Improved muscular endurance

Modest improvements in cardiorespiratory fitness

Reduced effort for activities of daily living as well as leisure and vocational activities

Improved flexibility

Reduced skeletal muscle fatigue

Elevated density and improved integrity of skeletal muscle connective tissue

Improved bone mineral density and content

Reduced risk of falling

Improved body composition

Possible reduction of blood pressure

Improved glucose tolerance

Possible improvement in blood lipid profile

Resistance exercises should be sequenced so that large muscle groups are worked first and smaller groups thereafter (32). If smaller muscle groups are trained first (i.e., those associated with fine movement), the large muscle groups may become fatigued earlier when they are used.

Resistance training, like other types of training, can be general or specialized to result in specific adaptations. Most of the specialized training routines have little relation to improvement of skeletal muscle fitness for healthy individuals or patients with a chronic disease. These populations will improve sufficiently with a standard resistance training program as suggested by the ACSM (5). Circuit programs (aerobic and resistance training) are a popular way of trying to incorporate a cardiovascular stimulus during a resistance training program (27, 46). But although impressive strength gains have been reported, only modest cardiorespiratory benefits result, if any (28, 35).

Specificity of Training

Resistance training for improvement in specific muscles should follow a lifting routine that closely mimics the activity or sport-specific muscular movements in which gains in muscular fitness are desired. Because the components of skeletal muscle fitness are related, any type of resistance training program will provide some benefits in each area of muscular fitness (i.e., strength, power, endurance). The resistance training program for general health should emphasize dynamic exercises involving concentric (shortening) and eccentric (lengthening) muscle actions that recruit multiple muscle groups (multijoint) and target the major muscle groups of the chest, shoulders, back, hips, legs, trunk, and arms. Single-joint exercises involving the abdominal muscles, lumbar extensors, calf muscles, hamstrings, quadriceps, biceps, and triceps should also be included.

Progressive Overload

For the client or patient seeking general or overall muscular fitness, the ACSM recommends performing resistance exercise training for 8 to 12 repetitions per set to produce maximum improvement in both skeletal muscle strength and endurance (5). In general, the greater the overload, the greater the improvement. Excessive and prolonged overload, however, can lead to increased risk for overtraining effects (e.g., worsening performance) or skeletal muscle injury.

Frequency Most studies report optimal gains when subjects perform resistance training from 1 to 3 d per week (19, 25, 33). The ACSM recommends performing a general or circuit resistance training program at least 2 or 3 d per week (5). There is little evidence that substantial additional gains can be realized from performing resistance exercise on more than 3 d per week.

Intensity and Duration The intensity of resistance training is also important in determining the load or overload placed on the skeletal muscle system to produce adaptation. For maximal strength and endurance improvement, the resistance should be at 60% to 80% of a person's one-repetition maximum (1RM). An alternative is to use 8- to 12RM, such that the person is at or near maximal exertion at the end of these repetitions (although this level of resistance may be too strenuous for some clinical conditions). The RM can be determined using either a direct or an indirect method (15, 39, 65).

The total training load placed on the skeletal muscle system is a combination of the number of repetitions performed per set and the number of sets per resistance exercise. The ACSM recommends between one and three sets per exercise (5). Several well-designed studies indicate little benefit of performing resistance training for more than one set per resistance exercise (22). If more than one set is performed per exercise, it may be prudent to keep the between-set period to a minimum to reduce the total exercise time. Generally, a 2 min rest is sufficient between sets.

Reversibility

A majority of the adaptations in neuromuscular function and skeletal muscle strength and endurance can be expected within 12 wk (56). The anticipated mean improvement in strength is about 25% to 30% (23). As with aerobic training, a reduction in total resistance training volume without complete cessation of training allows for maintenance of much of the gained resistance training effects (32). Complete loss of training-related muscular adaptations occurs after several weeks to months of inactivity.

Flexibility Training

Flexibility is the ability to move a joint throughout a full capable range of motion (ROM). Proper flexibility is associated with good postural stability and balance, especially when exercises aimed at improving flexibility are performed in conjunction with a resistance training program. According to a recent ACSM statement (5), no consistent link exists between regular flexibility exercise and reduction of incidence of musculotendinous injuries or prevention of low back pain.

Several devices can be used to assess ROM. These include a goniometer, which is a protractor-type device; the Leighton flexometer, which is strapped to a limb and reveals the ROM in degrees as the limb moves around its joint; and the sit-and-reach box, which assesses the ability to forward flex the torso while in a seated position and measures torso forward flexion as attributed to lower back, hamstring, and calf flexibility. An excellent review of the methodology of the sit-and-reach test is provided by Adams (1). In patients with certain diseases, improved flexibility may decrease as the course of the disease progresses (e.g., multiple sclerosis, osteoporosis, obesity). These and other populations benefit from regular ROM assessment and an exercise training program designed to enhance flexibility. A well-rounded flexibility program focuses on the neck, shoulders, upper trunk, lower trunk and back, hips, knees, and ankles.

The following are brief descriptions of the three primary stretching modes of flexibility training.

1. Static: A stretch of the muscles surrounding a joint that is held without movement for a period of time (e.g., 10-30 s) and may be repeated several times. Within 3 to 10 wk, static stretching yields improvements in joint ROM of 5° to 20°.
2. Ballistic: A method of rapidly moving a muscle to stretch and relax quickly for several repetitions, often used for sports that involve ballistic movements such as basketball.
3. Proprioceptive neuromuscular facilitation (PNF): A method whereby a muscle is isometrically contracted, relaxed, and subsequently stretched. The theory is that the contraction activates the muscle spindle receptors or Golgi tendon organs, which results in a reflex relaxation (i.e., inhibition of contraction) of either the agonist or the antagonist muscle. There are two types of PNF stretching (17, 20):
    - Contract-relax occurs when a muscle is contracted at 20% to 75% of maximum for 3 to 6 s and then relaxed and passively stretched. Enhanced relaxation is theoretically produced through the muscle spindle reflex.
    - Contract-relax with agonist contraction occurs initially in the same manner as contract-relax, but during the static stretch the opposing muscle is contracted. This action theoretically induces more relaxation in the stretched muscle through a reflex of the Golgi tendon organs.

Static and ballistic types of stretching are simple to perform and require only basic instruction. Proprioceptive neuromuscular facilitation stretching is somewhat complex, requires a partner, and may require close supervision by a clinical exercise professional. Static stretching is typically believed to be the safest method for enhancing the ROM of a joint. Both ballistic and PNF stretching may increase the risk for experiencing delayed-onset muscle soreness and muscle fiber injury. Generally, PNF is the most effective of the three methods of stretching at improving joint ROM (21, 59).

Specificity of Training

The flexibility of a joint or muscle–tendon group depends on the joint structure, the surrounding muscles and tendons, and the use of that joint for activities. Improved flexibility and ROM of a specific joint are developed through a flexibility training program that is specific to that joint. A joint used during an activity, especially if it requires a good ROM, typically demonstrates good flexibility.

Progressive Overload

As ROM increases, people should enhance the stretch to a comfortable level. This practice will produce optimal increases in ROM.

Frequency An effective stretching routine should be performed a minimum 2 to 3 d per week (5). As stated previously, however, daily stretching is advised for optimal improvement in ROM.

Intensity and Duration Static and PNF stretches should be held for 10 to 30 s. For PNF, this should follow a 6 s contraction period. Each stretch should also be performed for three to five repetitions and to a point of only mild discomfort or a feeling of stretch.

Reversibility

Few data exist on the rate of loss of ROM, and many factors are involved, including injury, specific individual physiology, degree of overall inactivity, and posture. The reintroduction of a flexibility training routine should result in rapid improvements in ROM.

CONCLUSION

The GXT is a useful and often the first diagnostic tool used to assess the presence of significant CAD with or without nuclear perfusion or echocardiography imaging. In past years, a cardiologist normally supervised the GXT. Today, other health care providers such as clinical exercise physiologists are supervising these tests; reasons include a better understanding as to which patients are at increased risk for a complication during testing, improved ECG technology, cost-containment initiatives within health care organizations, greater constraints on the cardiologists' time, and improved knowledge and training of other health care providers. Data from the test can be used not only to help diagnose the presence of CAD but also to determine prognosis and help design an exercise training program.

Any type of exercise training routine, whether it is cardiorespiratory conditioning, resistance training, or ROM training, should follow the FITT principle to ensure an optimal rate of improvement and safety during training. When an exercise physiologist is working with specific clinical populations, modifications of these general principles may be necessary. The chapters in this textbook addressing specific patient populations and diseases contain information pertinent to adapting the general exercise testing and training principles to these individuals.

Key Terms

claudication (p. 69)
cross-training (p. 79)
dyspnea (p. 69)
functional capacity (p. 62)
metabolic equivalents (METs) (p. 62)
motion artifact (p. 65)
myocardial ischemia (p. 61)
radionuclide agents (p. 61)
ramping protocols (p. 66)
rate–pressure product (RPP) (p. 69)
stress echocardiogram (p. 61)

CASE STUDY

MEDICAL HISTORY

Ms. WB is a former a high school and college cross country runner. She is Caucasian, 40 yr old, and the mother of two children, ages 8 and 6. Her children are in school, and she works full-time. She wishes to begin a regular exercise training routine. She has an interest in improving her health-related physical fitness. She does not smoke and has no known serious medical condition. Her body mass index is 31.4, and she has a goal for weight loss of 50 lb, or 23 kg (she is currently 64 in. tall [163 cm] and 180 lb [82 kg]). She has not run regularly in 10 yr and walks only occasionally.

DIAGNOSIS

Ms. WB is evaluated by her primary care provider and cleared for participation in an exercise program. She is apparently healthy and decides to take advantage of her local hospital's performance assessment program to establish baseline fitness levels and acquire specific exercise training recommendations.

EXERCISE TEST RESULTS

A symptom-limited exercise text is performed and the following reported:

Protocol: ramp running

Rest HR: 76 beats · min^{-1}

Rest BP: 136/88 mmHg

Rest ECG: normal sinus rhythm

Peak HR: 187 beats · min^{-1}

Peak BP: 210/90 mmHg

Exercise ECG: Rare premature atrial contractions and premature ventricular contractions; 0.5 mm J-point depression with quickly upsloping ST segments at peak exercise. Negative for exercise-induced myocardial ischemia.

$\dot{V}O_2$peak: 32.7 ml · kg^{-1} · min^{-1}

$\dot{V}O_2$ at anaerobic threshold: 25.6 ml · kg^{-1} · min^{-1}

OTHER PROCEDURES

Body composition demonstrating a body fat percentage of 32%

Bench press 1RM: 40 lb (18 kg)

(continued)

Sit and reach: lowest quintile (poor)
Leg press 1 RM: 150 lb (68 kg)

DISCUSSION QUESTIONS

1. What would be an appropriate comprehensive exercise prescription for Ms. WB?
2. Ms. SJ is 56 yr old and healthy. She wants to begin her own exercise program. What are your general recommendations and concerns?
3. Mr. BW performed a stress test on a treadmill with the following results: completed stage IV of the Bruce protocol (4.2 mph, 16% grade), HRmax = 190 per minute, maximum BP = 200/70 mmHg, $\dot{V}O_2peak$ = 43.2 $ml \cdot kg^{-1} \cdot min^{-1}$, no symptoms, stopped because of leg fatigue. Resting HR is 71 $beats \cdot min^{-1}$. What specific intensity recommendations can you make for him with regard to beginning a jogging program? Use each of the HR based methods discussed in this chapter.

PART II

Endocrinology and Metabolic Disorders

Issues related to abnormal metabolic function of the human body are complex and extremely important with respect to the many potential detrimental health effects that may occur. Some of these conditions result in their own specific effects or hazards to health, such as diabetes, high blood pressure, and renal failure. But these and the other disorders addressed in Part II (metabolic syndrome, obesity, dyslipidemia) increase the risk for many other chronic diseases including cardiovascular disease and cancer.

Diabetes has risen to epidemic proportions in the United States with about 1 in every 11 people afflicted. And this effect is not isolated; it is seen throughout the world. Behaviors such as poor eating habits and lack of exercise are components which influence the risk and development of type 2 diabetes. Chapter 6 presents information about the rise of diabetes prevalence and the role that the clinical exercise physiologist can play in the prevention and treatment of diabetes with exercise. Since we continue to better understand the connection and role of exercise in the prevention and treatment of diabetes this chapter is very important for any practicing clinical exercise physiologist.

Paralleling the rise in the incidence of diabetes is the rate of rise of those considered obese, with a body mass index (BMI) of at least 30 $kg \cdot m^{-2}$. The link of obesity to diabetes and heart disease is strong. Obesity is also strongly linked to increased risks of cancer, arthritis, disability, hypertension, and many other chronic diseases. As shown in chapter 7, weight-loss strategies can be effective, and exercise physiologists must play an active role in implementing these plans, particularly for weight-loss maintenance. The clinical exercise physiologist will be looked to more and more for the development and implementation of sustainable exercise programs particularly for those actively involved in a weight-loss program.

Although hypertension awareness, diagnosis, and treatment have been enhanced tremendously since the 1960s, hypertension remains a leading contributor to as many as 10% of all deaths in the United States. Exercise training can undoubtedly enhance both the prevention and treatment of hypertension. Chapter 8 provides specific exercise recommendations for both at-risk and hypertensive disease populations.

Although the role that dyslipidemia plays in the development of atherosclerosis is debatable, the association is irrefutable. Therefore, therapies such as statins and nutritional counseling have been developed to treat abnormal blood lipid values. Exercise can play a role in this treatment regimen with respect to controlling weight, which can positively affect lipids, and raising HDL levels. Chapter 9 presents this information in detail.

Chapter 10 delves into how conditions such as diabetes, obesity, hypertension, and dyslipidemia, each of which is the topic of a subsequent chapter, tend to cluster to make up what is termed *metabolic syndrome*. Although specific definitions of metabolic syndrome are subject to debate, it is clear that when these conditions are combined they generate greatly increased risk for cardiac and other diseases.

Chronic and deteriorating kidney function is staged across five levels and chapter 11 thoroughly discusses those patients with stage 5 chronic kidney disease or so called, end-stage renal disease. Among the nearly 600,000 patients in the United States suffering from end-stage kidney disease many also suffer from other debilitating disorders including heart failure, diabetes, and hypertension. These patients also experience marked fatigue and exercise intolerance and are often treated with hemodialysis. As discussed in chapter 11, exercise is an important treatment modality for not only the comorbid conditions associated with chronic renal failure but it also helps combat the exercise intolerance and depression that often accompanies the condition. Additionally, because hemodialysis requires up to 4 hours several times per week, exercise is increasingly being incorporated into the hemodialysis setting.

Chapter 6

Diabetes

Ann L. Albright, PhD

Exercise has long been recognized as an important component of diabetes care (67) and now is considered an important component in the prevention or delay of type 2 diabetes (33, 102). However, only 39% of adults with diabetes are physically active (defined as engaging in moderate or vigorous activity for at least 30 min, three times per week) compared to 58% of other adults (79). It is important that exercise professionals be equipped to assist people with diabetes and those at risk for diabetes in adopting and maintaining more physically active lifestyles.

DEFINITION

Diabetes mellitus (diabetes) is a group of metabolic diseases characterized by an inability to produce sufficient insulin or use it properly, resulting in hyperglycemia (7). Insulin, a hormone produced by the beta cells of the **pancreas**, is needed by muscle, fat, and the liver to use glucose. The hyperglycemia resulting from diabetes places people with the disease at risk for developing microvascular diseases including retinopathy and nephropathy, macrovascular disease, and various neuropathies (both autonomic and peripheral).

SCOPE

Approximately 26 million people in the United States have diabetes. About one-quarter of these are undiagnosed, in large part because symptoms of the most common form of diabetes, type 2, may develop gradually and years can pass before severe symptoms appear, if ever (23). Even before symptom development, however, these individuals are at increased risk for developing complications (13, 44, 66, 99, 104).

Diabetes continues to become increasingly common in the United States. From 1980 through 2009, the number of Americans with diagnosed diabetes more than tripled (from 5.6 million to 19.7 million) (22). Currently, 8.3% of the United States population has diabetes, and 1.9 million new cases of diabetes among people 20 yr and older were diagnosed in 2010 (23). Estimates are that the number of Americans with diagnosed diabetes will double in the next 15 to 20 yr. Diabetes is a worldwide problem. The Centers for Disease Control and Prevention considers diabetes to be at epidemic proportions in the United States. The reasons for the epidemic are likely threefold:

1. An increasingly sedentary lifestyle and poor eating practices, resulting in a rise in overweight and obesity
2. The increase in high-risk ethnic populations (discussed next) in the United States
3. Aging of the population

Diabetes is currently the seventh leading cause of death in the United States (23). African Americans, Hispanics, American Indians, Alaskan Natives, Native Hawaiians, other Pacific Islanders, and some Asians have higher rates of diabetes than the non-Hispanic white population (23). Death rates of people with diabetes are twice those of people without diabetes of the same age. As serious as these mortality statistics are, they underestimate the effect of diabetes. Because people with diabetes usually die from the complications of the disease, diabetes is underreported as the cause or underlying cause of death. In a large multicenter research study, diabetes was reported to have been listed on 39% of death certificates and as the underlying cause of death for 10% of decedents with diabetes (76). The economic effect of diabetes is staggering. The estimated direct costs (for medical treatment and services) and indirect costs (in time lost from work) are estimated to be $174 billion per year (in 2007 dollars) (8). A large portion of the economic burden of diabetes is attributable to long-term complications and hospitalizations.

PATHOPHYSIOLOGY

The various forms of diabetes affect the options for treatment. All forms share the risk for developing complications. This section reviews the types of diabetes and associated complications.

Diabetes Categories

The American Diabetes Association recognizes four categories of diabetes, as listed in "Etiologic Classification of Diabetes Mellitus."

Etiologic Classification of Diabetes Mellitus

Type 1 Diabetes

- Beta cell destruction, usually leading to absolute insulin deficiency
- Immune mediated
- Idiopathic

Type 2 Diabetes

May range from predominant insulin resistance with relative insulin deficiency to predominant secretory defect with insulin resistance

Other Specific Types

- Genetic defects of beta cell function
- Genetic defects in insulin action
- Diseases of the exocrine pancreas
- Endocrinopathies
- Drug or chemical induced
- Infections
- Uncommon forms of immune-mediated diabetes
- Other genetic syndromes sometimes associated with diabetes

Gestational Diabetes Mellitus

- Any degree of glucose intolerance with onset or first recognition during pregnancy

Note: Patients with any form of diabetes may require insulin treatment at some stage of their disease. Such use of insulin does not, of itself, classify the patient.

Adapted from American Diabetes Association 2011.

Type 1 Diabetes

Type 1 diabetes comprises two subgroups: immune mediated and **idiopathic**. Type 1 immune-mediated diabetes was formerly known as juvenile-onset or insulin-dependent diabetes and accounts for approximately 5% to 10% of those with diabetes. This form of diabetes is considered an **autoimmune** disease in which the immune system attacks the body's own beta cells, resulting in absolute deficiency of insulin. Consequently, insulin must be supplied by regular injections or an insulin pump. Type 1 immune-mediated diabetes usually occurs in childhood and adolescence but can occur at any age. Symptoms appear to develop quickly in children and adolescents, who may present with ketoacidosis, and more slowly in adults (7). Type 1 idiopathic diabetes is type 1 diabetes with no known etiologies and is present in only a small number of people. Most who fall into this category are of African or Asian ancestry. This form of type 1 diabetes is strongly inherited and lacks evidence of beta cell autoimmunity. The requirements for insulin therapy in patients with type 1 idiopathic diabetes are sporadic (7).

Type 2 Diabetes

Type 2 diabetes was formerly called adult-onset or non-insulin-dependent diabetes. This is the most common form of the disease and affects approximately 90% to 95% of all those with diabetes (7). The onset of type 2 diabetes usually occurs after age 40, although it is seen at increasing frequency in adolescents (28, 93).

The pathophysiology of type 2 diabetes is complex and multifactorial. Insulin resistance of the peripheral tissues and defective insulin secretion are common features. With insulin resistance, the body cannot effectively use insulin in the muscle or liver even though sufficient insulin is being produced early in the course of the disease (82, 89). Type 2 diabetes is progressive. Over time, the pancreas cannot increase insulin secretion enough to compensate for the insulin resistance, and hyperglycemia occurs (figure 6.1). The treatment

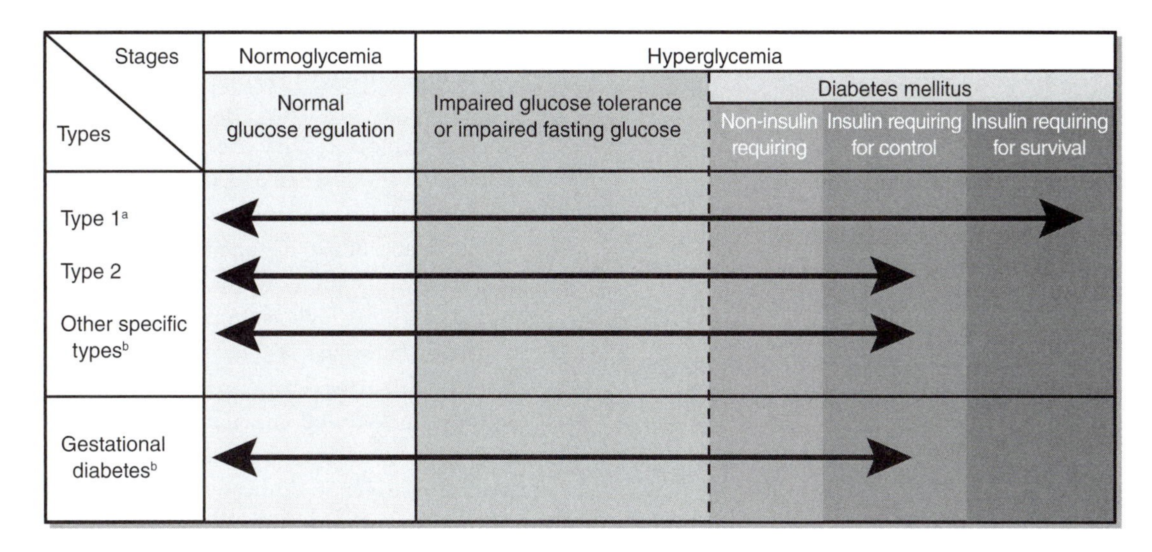

Figure 6.1 Disorders of glycemia.

options are medical nutrition therapy (MNT) and exercise and, if medication is needed, oral agents or insulin. Bariatric surgery may also be added to the treatment plan for those who are obese (body mass index [BMI] >35 kg · m^{-2}), especially if the diabetes or comorbidities are unmanageable with lifestyle and medication (12). Ketoacidosis rarely occurs.

A clear genetic influence is present for type 2 diabetes. The risk for type 2 diabetes among offspring with a single diabetic parent is 3.5-fold higher, and the risk for those with two diabetic parents is 6-fold higher, than that for offspring without diabetic parents (77). Along with genetic influences, other risk factors are present. Obesity contributes significantly to insulin resistance, and most people (80%) with type 2 diabetes are overweight or obese at disease onset (23, 58). An abdominal distribution of body fat (e.g., male belt size >40 in. [>102 cm] and female >35 in. [>89 cm]) is associated with type 2 diabetes (58, 84). The risk of developing type 2 diabetes also increases with age, lack of physical activity, history of **gestational diabetes**, and presence of hypertension or dyslipidemia (12, 23). The combination of hypertension, dyslipidemia, obesity, and diabetes is often termed *metabolic syndrome* or *cardiometabolic risk*. See chapter 10 for more information on metabolic syndrome.

Other Specific Types

The third category of diabetes, termed *other specific types*, accounts for only 1% to 2% of all diagnosed cases of diabetes (7). In these cases, certain diseases, injuries, infections, medications, or genetic syndromes cause the diabetes. This form may or may not require insulin treatment.

Gestational Diabetes

Gestational diabetes occurs during 2% to 10% of pregnancies (23). It is usually diagnosed during pregnancy by an oral glucose tolerance test, often performed routinely during the second trimester. Risk factors for developing this form of diabetes include family history of gestational diabetes, previous delivery of a large birth weight (>9 lb [4 kg]) baby, and obesity. Although glucose tolerance usually returns to normal after delivery, women who have had gestational diabetes have a 35% to 60% chance of developing type 2 diabetes in the next 10 to 20 yr and should receive long-term follow-up and proven interventions to prevent type 2 diabetes (23, 33).

Complications of Diabetes

Complications associated with diabetes are categorized as acute and chronic. This section reviews those complications.

Acute Complications

The acute complications of diabetes are **hyperglycemia** (high blood sugar) and **hypoglycemia** (low blood sugar). Each of these acute complications must be quickly identified to ensure proper treatment and reduce the risk of serious consequences.

Hyperglycemia The manifestations of hyperglycemia are as follows:

1. Diabetes out of control
2. Diabetic ketoacidosis
3. Hyperosmolar nonketotic syndrome

Diabetes out of control is a term used to describe blood glucose levels that are frequently above the patient's **glycemic goals** (see following discussion and table 6.1). High blood glucose levels cause the kidneys to excrete glucose and water, which causes increased urine production and dehydration. Symptoms of high blood glucose levels and dehydration are

- headache,
- weakness, and
- fatigue.

The best treatment for a patient with diabetes out of control includes drinking plenty of non-carbohydrate-containing beverages, regular self-monitoring of blood glucose, and, when instructed by a health care professional, an increase in diabetes medication. Frequent high blood glucose levels damage target organs or tissues, which increases the risk of chronic complications.

Diabetic ketoacidosis occurs in patients whose diabetes is in poor control and in whom the amount of **effective insulin** is very low or absent. This result is much more likely to occur in those with type 1 diabetes. **Ketones** form because without insulin, the body cannot use glucose effectively and a high amount of fat metabolism occurs to provide necessary energy. A byproduct of fat metabolism in the absence of adequate carbohydrate is ketone body formation by the liver, causing an increased risk of coma and death. Ketone levels in the blood are approximately 0.1 mmol $\cdot$ L^{-1} in a person without diabetes and can be as high as 25 mmol $\cdot$ L^{-1} in a person with diabetic ketoacidosis. This level of ketosis can be evaluated with a urine dipstick test. Other symptoms of ketoacidosis include abdominal pain, nausea, vomiting, rapid or deep breathing, and sweet- or fruity-smelling breath. Exercise is contraindicated in anyone experiencing diabetic ketoacidosis.

Hyperglycemic hyperosmolar nonketotic syndrome occurs in patients with type 2 diabetes when hyperglycemia is profound and prolonged. This circumstance is most likely to happen during periods of illness or stress, in the elderly, or in people who are undiagnosed (12). The syndrome results in severe dehydration attributable to rising blood glucose levels, causing excessive urination. Extreme dehydration eventually leads to decreased mentation and possible coma. Exercise is contraindicated during periods of hyperglycemic hyperosmolar nonketotic syndrome.

Table 6.1 Suggested Treatment Goals for Blood Glucose, Blood Pressure, and Lipids in Nonpregnant Adults

GLYCEMIC CONTROL	
A1c	<7.0%*
Preprandial capillary plasma glucose	70-130 mg $\cdot$ dl^{-1} (3.9-7.2 mmol $\cdot$ L^{-1})
Peak postprandial capillary plasma glucose^	<180 mg $\cdot$ dl^{-1} (<10.0 mmol $\cdot$ L^{-1})
Blood pressure	<130/80 mmHg
LIPIDS**	
LDL	<100 mg $\cdot$ dl^{-1} (<1.6 mmol $\cdot$ L^{-1})
Triglycerides	<150 mg $\cdot$ dl^{-1} (<1.7 mmo $\cdot$ L^{-1})
HDL	>40 mg $\cdot$ dl^{-1} (>1.0 mmol $\cdot$ L^{-1})**

Key concepts in setting glycemic goals:

- A1c is the primary target for glycemic control.
- Goals should be individualized based on duration of diabetes, age and life expectancy, comorbid conditions, known CVD or advanced microvascular complications, hypoglycemia unawareness, individual patient considerations.
- More stringent glycemic goals (closer to normal) may further reduce microvascular complications.
- Postprandial glucose may be targeted if A1c goals are not met despite reaching preprandial glucose goals.

*Referenced to a nondiabetes range of 4.0% to 6.0% using a Diabetes Control and Complications Trial (DCCT)-based assay.

^Postprandial glucose measurements should be made 1 to 2 h after the beginning of the meal, generally peak levels in patients with diabetes.

**For women, it has been suggested that the HDL goal be increased by 10 mg $\cdot$ dl^{-1} (14).

Adapted from American Diabetes Association 2011.

Hypoglycemia Hypoglycemia (also called insulin shock and insulin reaction) is a potential side effect of diabetes treatment and usually occurs when blood glucose levels drop below 60 to 70 mg · dl^{-1}. Hypoglycemia may occur in the presence of the following factors:

- Too much insulin or selected antidiabetic oral agent(s)
- Too little carbohydrate intake
- Missed meals
- Excessive or poorly planned exercise

Hypoglycemia can occur either during exercise or several hours later. Postexercise, late-onset hypoglycemia generally occurs following moderate- to high-intensity exercise that lasts longer than 30 min. This kind of hypoglycemia results from increased insulin sensitivity, ongoing glucose use, and physiological replacement of glycogen stores through gluconeogenesis (27). Patients should be instructed to monitor blood glucose before and periodically after exercise to assess glucose response. This approach is also recommended in clinical exercise programs, such as cardiac rehabilitation, especially in patients new to exercise. Recommendations for pre- and postexercise blood glucose assessment are provided later in this chapter.

The two categories of symptoms of hypoglycemia are autonomic and neuroglycopenic. As blood glucose decreases, **glucagon**, epinephrine, growth hormone, and cortisol are released to help increase circulating glucose. Autonomic symptoms such as shakiness, weakness, sweating, nervousness, anxiety, tingling of the mouth and fingers, and hunger result from epinephrine release. As the blood glucose delivery to the brain decreases, neuroglycopenic symptoms such as headache, visual disturbances, mental dullness, confusion, amnesia, seizures, or coma may occur. Some people with diabetes lose their ability to sense hypoglycemic symptoms (termed hypoglycemia unawareness). Instituting tight control of blood glucose may lower the threshold so that symptoms do not occur until blood glucose drops quite low. Intensity of control may need to be slightly reduced to alleviate hypoglycemia unawareness. Hypoglycemic unawareness may also result from autonomic neuropathy. In contrast, patients who have been in poor control may sense low blood glucose symptoms at levels much higher than 60 to 70 mg · dl^{-1}. Treatment of hypoglycemia consists of testing blood glucose to confirm hypoglycemia and, if the person is conscious, consumption of approximately 15 g of a carbohydrate (e.g., glucose, sucrose, or lactose) product that does not contain fat. Commercial products (glucose tablets) are available that allow a person to eat a precise amount of carbohydrate. Other sources include 1 C (240 ml) of nonfat milk; 1/2 C (120 ml) of orange juice; one-half can (180 ml) of regular soda; six or seven Life Savers; 2 tbsp (30 ml) of raisins; or 1 tbsp (15 ml) of sugar, honey, or corn syrup. The person with diabetes should wait about 15 or 20 min to allow the symptoms to resolve. If necessary, he should consume another 15 g of carbohydrate (28). If the patient becomes unconscious because of hypoglycemia, an injection of glucagon should be administered. If glucagon is not available, 911 should be called immediately.

Chronic Complications

Diabetes is the leading cause of adult-onset blindness, nontraumatic lower limb amputation, and end-stage renal failure (23). In addition, those with diabetes are at two to four times the normal risk of heart disease and stroke (23). The hyperglycemia of diabetes is considered of primary importance in development of the chronic complications, along with hypertension and hyperlipidemia. Intensive blood glucose control can reduce the risk of developing microvascular diabetic complications in patients with either type 1 or 2 diabetes (31, 103). Current evidence from the ACCORD trial does not support intensive glycemic control (A1c <6%) to reduce the rate of cardiovascular events in middle-aged and older people with type 2 diabetes and either established cardiovascular disease or additional cardiovascular risk factors (2). In this trial the mortality rate was higher in the intensive therapy group. The clinical exercise professional who is involved in the exercise training of people with diabetes must obtain information about the presence and stage of complications. The clinician should then use this information when developing an exercise prescription and behavior modification plan designed to help those with diabetes reduce their risk of developing or amplifying the complications of the disease. Cardiac rehabilitation programs may be suitable for patients, including those at high risk, who wish to incorporate an exercise program into their lifestyle. It is helpful to divide the chronic complications into three categories:

1. Macrovascular (large vessel or atherosclerotic) disease, which includes coronary artery disease with or without angina, myocardial infarction, cerebrovascular accident, and peripheral arterial disease
2. Microvascular (small vessel) disease, which includes diabetic retinopathy (eye disease) and diabetic nephropathy (kidney disease)
3. Neuropathy that involves both the peripheral and autonomic nervous systems

Macrovascular Disease Diabetes is a risk factor for macrovascular disease. The vessels to the heart, brain, and lower extremities can be affected. Blockage of the blood vessels in the legs results in peripheral artery disease, intermittent claudication (see chapter 15), and exercise intolerance (6). Reduction and control of vascular risk factors are especially important in those with diabetes. The methods used for this purpose are similar to those used for coronary heart disease. The chapter on myocardial infarction (chapter 12) in this text reviews the vascular risk factor control methods in detail. The symptoms of peripheral arterial disease can be improved with exercise training, as reviewed in chapter 15.

Microvascular Disease **Microvascular disease** causes retinopathy and nephropathy, which result in abnormal function and damage to the small vessels of the eyes and kidneys, respectively. The ultimate result of retinopathy can be blindness, whereas end-stage renal failure is the most serious complication of nephropathy (see chapter 11). Prevention or appropriate management requires periodic (often yearly) dilated eye examinations and renal function tests, along with optimal blood glucose and blood pressure control. The exercise professional must give careful attention to the stage of complications when prescribing exercise for those with microvascular involvement; this topic is discussed in detail in the exercise prescription section of this chapter.

Peripheral and Autonomic Neuropathy Both peripheral and autonomic neuropathy have implications for exercise. **Peripheral neuropathy** typically affects the legs before the hands. Patients initially experience sensory symptoms (paresthesia, burning sensations, and hyperesthesia) and loss of tendon reflexes. As the complication progresses, the feet become numb and patients are at high risk for foot injuries because they have difficulty realizing when they are injured. Muscle weakness and atrophy can also occur. Foot deformities can result, causing areas to receive increased pressure from shoe wear or foot strike, placing them at risk for injury. The large number of lower limb amputations from diabetes results from loss of sensation that places the patient at risk for injury and from diminished circulation attributable to peripheral artery disease. This circumstance impairs healing and can lead to severe reductions in blood flow, potential gangrene, and amputation. Persons with diabetes must be given instruction on how to examine their feet and practice good foot care. Foot care is especially important when someone with peripheral neuropathy begins an exercise program, because increased walking and cycle pedaling increase the risk of foot injury. Diabetic autonomic neuropathy may occur in any system of the body (e.g., cardiovascular, respiratory, neuroendocrine, gastrointestinal). Many of these systems are integral to the ability to perform exercise. **Cardiovascular autonomic neuropathy** is manifested by high resting heart rate, attenuated exercise heart rate response, abnormal blood pressure, and redistribution of blood flow response during exercise. This combination can severely limit exercise capacity and physical functioning.

CLINICAL CONSIDERATIONS

In the clinical setting, laboratory tests and examinations are used to diagnose diabetes or to facilitate ongoing monitoring. The following sections review these purposes.

Signs and Symptoms

The signs and symptoms of diabetes include excessive thirst (**polydipsia**), frequent urination (**polyuria**), unexplained weight loss, infections and cuts that are slow to heal, blurry vision, and fatigue. Many who develop type 1 diabetes have some or all of these symptoms, but those with type 2 diabetes may remain asymptomatic. About one-fourth of those with diabetes do not know that they have the disease (23).

History and Physical Examination

Patients who present for a clinic visit should have a thorough medical history review (see chapter 4). Those presenting with the risk factors or signs and symptoms for diabetes who have not been previously diagnosed as having diabetes should be evaluated appropriately. The evaluation includes performing the associated diagnostic testing as presented in the next section. The physical examination should focus on potential indicators of diabetes complications. These may include elevated resting pulse rate, loss of sensation or reflexes especially in the lower extremities, foot sores or ulcers that heal poorly, excessive bruising, and retinal vascular abnormalities. Exercise testing may be appropriate before beginning an exercise program (see the "Exercise Testing" section that follows) (3, 4).

When reviewing the medical history of a patient with diabetes for exercise training clearance, the exercise professional should consider the following:

1. The presence or absence of acute and chronic complications and, if chronic complications exist, the stage of complications

2. Laboratory values for hemoglobin A1c, plasma glucose, lipids, and **proteinuria**
3. Blood pressure
4. Self-monitoring blood glucose results
5. Body weight and BMI
6. Medication use and timing
7. Exercise history
8. Nutrition plan, particularly timing, amount, and type of most recent food intake
9. Other non-diabetes-related health issues

Diagnostic Testing

The American Diabetes Association recommends that diagnostic testing be considered in all adults who are overweight (BMI ≥25 kg · m^{-2}; at-risk BMI may be lower in some ethnic groups) and have additional risk factors:

1. Physical inactivity
2. First-degree relative with diabetes
3. High-risk race or ethnicity (e.g., African American, Latino, Native American, Pacific Islander)
4. Having delivered a baby weighing more than 9 lb (4 kg) or having been diagnosed with gestational diabetes
5. Hypertension (≥140/90 mmHg or on therapy for hypertension)
6. High-density cholesterol ≤35 mg · dl^{-1} or triglyceride of ≥250 mg · dl^{-1}
7. A1c ≥5.7, an impaired fasting glucose or glucose tolerance test
8. Polycystic ovarian syndrome
9. Other clinical conditions associated with insulin resistance (e.g., severe obesity, acanthosis nigricans)
10. History of cardiovascular disease

In the absence of these criteria, testing should begin at age 45 yr. If results are normal, testing should be repeated at least at 3 yr intervals, with consideration of more frequent testing depending on risk status (7).

Four criteria are used to diagnose diabetes (see "Criteria for the Diagnosis of Diabetes"). In the absence of unequivocal hyperglycemia, these criteria should be confirmed by repeat testing. Those found to meet the criteria for the diagnosis of diabetes should be told they have diabetes, not "borderline" diabetes. The latter explanation may give the patient the impression that the disease is not serious. The method of therapy used to treat diabetes should not be interpreted as an indication of the seriousness of the disease. Regardless of treatment, diabetes is a serious disease that requires diligent self-care and appropriate medical intervention. A fasting blood glucose ranging from 100 to 125 mg · dl^{-1} is considered a risk factor for developing type 2 diabetes and is termed impaired fasting glucose. When an oral glucose tolerance test is used, a 2 h postload glucose between 140 mg · dl^{-1} and 199 mg · dl^{-1} is termed impaired glucose tolerance. An A1c of 5.7% to 6.4% is now considered increased risk for developing diabetes. Impaired fasting glucose, impaired glucose tolerance, and at-risk A1c are often called prediabetes.

Criteria for the Diagnosis of Diabetes

1. A1c ≥6.5%. The test should be performed in a laboratory using a method that is National Glycohemoglobin Standardization Program (NGSP) certified and standardized to the Diabetes Control and Complications Trial (DCCT) assay.*
2. Fasting plasma glucose ≥126 mg · dl^{-1} (7.0 mmol · L^{-1}). Fasting is defined as no caloric intake for at least 8 h.*
3. Two-hour plasma glucose ≥200 mg · dl^{-1} (11.1 mmol · L^{-1}) during an oral glucose tolerance test. The test should be performed as described by the World Health Organization, using a glucose load containing the equivalent of 75 g of anhydrous glucose dissolved in water.
4. Classic symptoms of hyperglycemia or hyperglycemic crisis plus a random plasma glucose ≥200 mg · dl^{-1} (11.1 mmol · L^{-1}). The classic symptoms of diabetes include polyuria, polydipsia, and unexplained weight loss.

*In the absence of unequivocal hyperglycemia, results should be confirmed with repeat testing.

This group of patients should receive instruction and encouragement to lower their risk of developing type 2 diabetes, including beginning an exercise training program. The Centers for Disease Control (CDC)-led National Diabetes Prevention Program, authorized in the Patient Protection and Affordable Care Act, is under way to implement and increase access to cost-effective versions of the lifestyle intervention proven to prevent or delay type 2 diabetes in the National Institutes of Health (NIH)-led Diabetes Prevention Program research trial (1, 22, 32-34).

Exercise Testing

Most people with diabetes can benefit from participating in regular exercise. Participation in exercise is not without risk, however, and each individual should be assessed for safety (3, 4). Priority must be given to minimizing the potential adverse effects of exercise through appropriate screening, program design, monitoring, and patient education (3, 4, 6). Exercise testing may be viewed as a barrier or as unnecessary for some patients. Discretion must be used to determine the need for exercise testing. For participation in low-intensity exercise, health care professionals should use clinical judgment in deciding whether to recommend preexercise testing (27, 95, 96). Conducting exercise testing before walking is considered unnecessary (27). For exercise more vigorous than brisk walking or exceeding the demands of everyday living, sedentary and older individuals with diabetes will likely benefit from being assessed for conditions that might be associated with cardiovascular disease (CVD), that contraindicate certain activities, or that predispose to injuries. The assessment may include a graded exercise test, but should include a medical evaluation and screening for blood glucose control, physical limitations, medications, and macrovascular and microvascular complications (discussed on p. 96). In general, electrocardiogram (ECG) stress testing may be indicated for individuals meeting one or more of these criteria (27):

1. Age >40 yr, with or without CVD risk factors other than diabetes
2. Age >30 yr and any of the following
   - Type 1 or type 2 diabetes of >10 yr
   - Hypertension
   - Cigarette smoking
   - Dyslipidemia
   - Proliferative or preproliferative retinopathy
   - Nephropathy including microalbuminuria
3. Any of the following, regardless of age:
   - Known or suspected coronary artery disease, cerebrovascular disease, or peripheral artery disease
   - Autonomic neuropathy
   - Advanced nephropathy with renal failure

There is no evidence to determine if stress testing is necessary or useful before participation in anaerobic or resistance training. Coronary ischemia is less likely to occur during resistance compared to aerobic training at the same heart rate response. There is doubt that resistance exercise induces ischemia (41, 46). Contraindications for exercise testing are listed in chapter 5. The clinical exercise professional must be prepared to provide input to the physician to assist in the decision-making process to determine need for exercise testing. Table 6.2 summarizes exercise testing specifics. Practical application 6.1 provides important information about the client-clinician interaction.

Treatment

There is currently no cure for diabetes. The disease must be managed with a program of exercise, **medical nutrition therapy**, self-monitoring of blood glucose, diabetes self-management education, and, when needed, medication (always needed in type 1) or significant weight loss from bariatric surgery or a complete meal replacement diet. When medication or bariatric surgery is used, it should be added to lifestyle improvements, not replace them (12). The patient and his health care team must work together to develop a program to achieve individual treatment goals. Few diseases require the same level of ongoing daily patient involvement as does diabetes. Because so much patient involvement is required, patients must receive information and training on disease management. Other members of the health care team may include the patient's primary care physician or an endocrinologist, a nurse practitioner, a physician assistant, a diabetes educator, a registered dietitian, a clinical exercise professional, a behavioral or psychosocial counselor, and a pharmacist. In many instances, these health care professionals work together in a diabetes education program. The American Diabetes Association, along with many other contributing organizations, has developed standards for diabetes education programs (9). The patient must understand and be involved in developing appropriate treatment goals, which take into consideration the patient's desires, abilities, willingness, cultural background, and comorbidities. Suggested treatment goals for blood glucose, blood pressure, and

Table 6.2 Exercise Testing Review

Test type	Mode	Protocol specifics	Clinical measures	Clinical implications	Special considerations
Cardiovascular	Treadmill Ergometer (leg or arm)	Low level for many (≤2 METs per stage or 20 W/min increases in work rate)	Peak $\dot{V}O_2$ or estimated METs Heart rate and blood pressure responses 12-lead ECG	Watch for ischemia and arrhythmias because these are often undiagnosed and patients are at high risk for heart disease.	Chest pain due to myocardial ischemia may not be perceived in those with neuropathy (also may blunt peak HR achieved). Patients with peripheral vascular disease probably should use the cycle ergometer mode. Consider testing blood glucose before exercise test to reduce the risk of hypoglycemia.
Resistance	Machine weights Isokinetic dynamometer	1RM or indirect 1RM method	Strength and power		1RM may not be recommended in those with severe disease and those who are sedentary. Those with retinopathy most typically should not perform resistance training.
Range of motion	Sit-and-reach Goniometry		Major muscle groups range of motion		Patients should not hold breath; any exercise may result in excessive blood pressure response.

Data from Lohman 1992.

Practical Application 6.1

CLIENT–CLINICIAN INTERACTION

The interaction between the client and the clinician at the time of exercise evaluation, and especially during ongoing exercise training visits, is important. Living with diabetes poses many challenges and fears for patients and their families. The exercise professional must be aware of the psychosocial components of living with a chronic disease and must be able to apply strategies to help the patient maintain participation in exercise.

The clinician should consider the following guidelines. Treat the patient as an individual who is much more than her diagnosis. Be cautious about referring to the patient as a diabetic, because this terminology labels the patient by the disease. Remember that the person usually needs to apply a great deal of effort and discipline to live with diabetes. Acknowledge that diabetes is challenging, and listen to the patient's particular challenges. In general, do not use terms like *noncompliance* when discussing an exercise program. Inherent in the definition of noncompliance is the concept that a person is not following rules or regulations enforced by someone else. This concept is incongruent with self-management and patient empowerment, which consider the patient the key member of the health care team. The health care professional should not make decisions for the patient. Instead, the clinical exercise professional should equip patients with information so that they can make their own decisions.

Several strategies are helpful for exercise maintenance. Ask the patient to consider the following questions: How easily can I engage in my activity of choice where I live? How suitable is the activity in terms of my physical attributes and lifestyle (74)? Have the patient identify exercise benefits that she finds personally motivating. Be sure that exercise goals are not too vague, ambitious, or distant. Establish a routine to help exercise become more habitual. Have the patient identify any social support systems she may have. Provide positive feedback to the patient.

Basic Guidelines for Diabetes Care

Physical and Emotional Assessment

Blood pressure, weight or BMI—every visit

- For adults: Blood pressure target goal <130/80 mmHg, BMI <25 kg · m^{-2}.
- For children: Blood pressure target goal <90th percentile adjusted for age, height, and gender; BMI for age <85th percentile.

Foot exam (for adults): Thorough visual inspection every diabetes care visit; pedal pulses, neurological exam yearly

Dilated eye exam (by trained expert)

- Type 1: Five years postdiagnosis, then every year.
- Type 2: Shortly after diagnosis, then every year. *Note:* Internal quality assurance data may be used to support less frequent testing.

Depression: Probe for emotional and physical factors linked to depression yearly; treat aggressively with counseling, medication, referral, or some combination of these.

Dental—exam at least twice yearly; assess oral symptoms that require urgent referral

Lab Exam

A1c (HbA1c)

- Quarterly, if treatment changes or if not meeting goals; one or two times a year if stable.
- Target goal <7.0% or <1% above lab norms.
- For children: Modify as necessary to prevent significant hypoglycemia.

Microalbuminuria (albumin/creatinine ratio)

- Type 1: Begin with puberty once the duration of diabetes is more than 5 yr unless proteinuria has been documented.
- Type 2: Begin at diagnosis, then every year unless proteinuria has been documented.

Glomerular filtration rate (GFR): Estimate whenever chemistries are checked.

Blood lipids (for adults): On initial visit, then yearly for adults. Target goals (mg · dl^{-1}): LDL <100 (<70 for high CVD risk); triglycerides <150; HDL >40 for men; HDL >50 for women.

Self-Management Training

Management principles and prevention of complications

- Initially and ongoing: Focus on helping the patient achieve the American Association of Diabetes Educators (AADE) seven self-care behaviors: healthy eating, being active, monitoring, taking medications, problem solving, healthy coping, and reducing risks. Screen for problems with and barriers to self-care; assist patient to identify achievable self-care goals.
- For children: As appropriate for developmental stage.

Self-monitoring of glucose

- Type 1: Typically test four times a day.
- Type 2 and others: As needed to meet treatment goals.

Medical nutrition therapy (by trained expert)

- Initially: Assess needs and condition; assist patient in setting nutrition goals.
- Ongoing: Assess progress toward goals; identify problem areas.

Physical activity—initially and ongoing: Assess and prescribe physical activity based on patient's needs and condition; goal of at least 150 min/week of moderate-intensity exercise.

Weight management—initially and ongoing: Must be individualized for patient.

Interventions

Preconception, pregnancy, and postpartum counseling and management

- Consult with high-risk, multidisciplinary perinatal/neonatal programs and providers where available (e.g., California Diabetes and Pregnancy Program "Sweet Success").
- For adolescents: Age-appropriate counseling advisable, beginning with puberty.

Aspirin therapy (for adults): 75 to 162 mg · d^{-1} as primary and secondary prevention of cardiovascular disease unless contraindicated.

Smoking cessation: Ask all patients if they use tobacco; advise them to quit; refer them to California Smokers' Helpline, 1-800-NO-BUTTS (662-8887).

Immunizations: Influenza and pneumococcal, per CDC recommendations.

Developed by the Diabetes Coalition of California and the California Diabetes Program, revised August 2009. For further information: www.caldiabetes.org.

plasma lipids from the American Diabetes Association are provided in table 6.1. **Evidence-based** care guidelines include regular hemoglobin A1c testing, dilated eye exam, foot exam, blood pressure monitoring, lipid panel, renal function tests, smoking cessation counseling, flu or pneumococcal immunizations, and diabetes education (see "Basic Guidelines for Diabetic Care") that should be followed to help ensure appropriate care. The patient should be educated about the purpose and importance of the medical tests and feedback on the results.

Medical nutrition therapy, often the most challenging aspect of therapy, is essential to the management of diabetes. Nutrition recommendations were developed by the American Diabetes Association (10). These guidelines promote individually developed dietary plans based on metabolic, nutrition, and lifestyle requirements in place of a calculated caloric prescription. This approach is appropriate because a single diet cannot adequately treat all types of diabetes or individuals. Consideration must be given to each macronutrient (i.e., protein, fat, carbohydrate) in the development of a nutrition plan for the person with diabetes. Protein intake should be approximately 10% to 20% of daily caloric intake because no evidence indicates that lower or higher intake is of value. Based on the risk of atherosclerosis, fat intake should be limited, with less than 10% from saturated fats and up to 10% from polyunsaturated fats. Cholesterol intake should be limited to 300 mg daily. Carbohydrate and monounsaturated fat make up the remaining calories and need to be individualized based on glucose, lipid, and weight goals. The most common nutritional assumption about diabetes is that sugars should be avoided and replaced with starches. Little evidence supports this assumption (15, 68). Priority should first be given to the total amount of carbohydrate consumed rather than the source, because all carbohydrates can raise blood glucose. Nutritional value must also be considered.

Self-monitoring of blood glucose is also an important part of managing diabetes. No standard frequency for self-monitoring has been established, but it should be performed frequently enough to help the patient meet treatment goals. Increased frequency of testing is often required when people begin an exercise program to assess blood glucose before and after exercise and to allow safe exercise participation. Patients must be given guidance about how to use the information to make exercise, food, and medication adjustments. Those who require glucose-lowering medication must understand how their medications work with food and exercise to ensure the greatest success and safety. Clinical exercise professionals must understand diabetes medications so that they can safely prescribe exercise and provide guidance on exercise training to patients with diabetes. Refer to table 6.3 for specific information on glucose-lowering medications and to chapter 3 for information on other pharmacology-related issues. Consult the chapters that address blood pressure and lipid medications, since these considerations are often part of diabetes management.

EXERCISE PRESCRIPTION

Exercise is a vital component of diabetes management. Exercise is considered a method of treatment for type 2 diabetes because it can improve insulin resistance. Although exercise alone is not considered a method of treating type 1 diabetes because of the absolute requirement for insulin, it is still an important part of a healthy lifestyle for people in this group.

Table 6.3 Pharmacology

ORAL GLUCOSE-LOWERING MEDICATIONS			
Medication name and class	**Primary effects**	**Exercise effects**	**Special considerations**
Tolbutamide (Orinase) Tolazamide (Tolinase) Chlorpropamide (Diabenese) Class is sulfonylureas (1st generation)	Increases insulin production in the pancreas	Risk of hypoglycemia	Chlorpropamide remains active for up to 60 h. Use extreme caution with elderly patients or patients with hepatic or renal dysfunction. Use of these agents is not recommended unless the patient has a well-established history of taking them. Second-generation sulfonylureas provide more predictable results with fewer side effects and more convenient dosing.
Glyburide (Micronase, Diabeta, Glynase) Glipizide (Glucotrol, Glucotrol XL) Glimepiride (Amaryl) Class is sulfonylureas (2nd generation)	Increases insulin production in the pancreas	Risk of hypoglycemia	Clearance may be diminished in patients with hepatic or renal impairment. Glipizide is preferred with renal impairment. Doses >15 mg should be divided. Glimepiride is indicated for use with insulin. Shown to have some insulin-sensitizing effect.
Repaglinide (Prandin)	Increases insulin release from pancreas	Risk of hypoglycemia	Use with caution in patients with hepatic or renal impairment. Patients should be instructed to take medication no more than 30 min before a meal. If meals are skipped or added, the medication should be skipped or added as well.
Nateglinide (Starlix) Class is meglitinide	Increases insulin release from pancreas	Risk of hypoglycemia	Use with caution with moderate to severe hepatic disease. Has only a 2 h duration of action. If meals are skipped or added, the medication should be skipped or added as well.
Metformin (Fortamet, Glumetza, Glucophage) Class is biguanide	Primarily decreases hepatic glucose production Minor increase in muscle glucose uptake, which may improve insulin resistance	Risk of hypoglycemia after prolonged or strenuous exercise	Due to increased risk of lactic acidosis, should not use if frequent alcohol use, liver or kidney disease, or CHF is suspected. Contraindicated if serum creatinine is >1.5 mg $\cdot$ dl^{-1} in men or >1.4 mg $\cdot$ dl^{-1} in women. Do not use if creatinine clearance is abnormal.
Pioglitazone (Actos) Rosiglitazone (Avandia, Avandamet, Avandaryl) *Note:* On May 18, 2011, FDA released an updated risk evaluation and mitigation strategy further restricting the use of rosiglitazone. Class is thiazolidinedione	Decreases insulin resistance, increasing glucose uptake, fat redistribution Minor decrease in hepatic glucose output Preserves beta cell function Decreases vascular inflammation		Should not be used in patients with CHF or hepatic disease. Can cause mild to moderate edema.
Acarbose (Precose) Miglitol (Glyset) Class is α-glucosidase inhibitor	Slows absorption of starch, disaccharides, and polysaccharides from GI tract		Gas and bloating, sometimes diarrhea. Should not be used if GI disorders are present. Avoid if serum creatinine is >2.0 mg $\cdot$ dl^{-1}.

Medication name and class	Primary effects	Exercise effects	Special considerations
INCRETINS AND AMYLINS			
Exenatide (Byetta) Class is glucagon-like peptide-1 receptor agonist	Decreases postmeal glucagon production Delays gastric emptying Increases satiety, leading to decreased caloric intake	Risk of hypoglycemia	Not for use in patients with type 1 diabetes, severe renal disease, or severe GI disease. Consider lowering dose of sulfonylurea to avoid hypoglycemia when starting. May reduce the rate of absorption of oral medication.
Pramlintide (Symlin) Class is amylin analogue	Decreases postmeal glucagon production Delays gastric emptying Increases satiety, leading to decreased caloric intake Degree of response dependent on plasma glucose levels	Risk of hypoglycemia	Indicated for insulin-treated type 2 diabetes or for type 1 diabetes. Contraindicated in patients with hypoglycemia unawareness, gastroparesis, or poor adherence. Should never be mixed with insulin and should be injected separately. Reduce insulin dose by 50% when starting. Requires patient testing of blood sugars before and after meals, frequent physician follow-up, and thorough understanding of how to adjust doses of insulin and pramlintide. May reduce the rate of absorption of orally administered medication. Medications requiring threshold concentrations should be taken 1 h before injection.
Sitagliptin (Januvia) Saxagliptin (Onglyza) Vildagliptin (Galvus) Class is DPP-4 inhibitor	Inhibits the DPP-4 enzyme that degrades GLP-1 and GIP, resulting in two- to threefold increased levels of these incretins Increases insulin secretion in presence of elevated plasma glucose Reduces postmeal glucagon secretion	Low risk of hypoglycemia	Observe patients carefully for signs and symptoms of pancreatitis. Rare reports of hypersensitivity reactions. Not for use in type 1 diabetes. Assessment of renal function is recommended before initiation and periodically thereafter.
INSULIN, RAPID ACTING			
Lispro (Humalog)	Onset <15 min Peak 1 to 2 h Effective duration 2 to 4 h	Risk of hypoglycemia	Should be taken just before or just after eating.
Aspart (Novolog)	Onset <15 min Peak 1 to 3 h Effective duration 3 to 5 h	Risk of hypoglycemia	Should be taken just before or just after eating.
Glulisine (Apidra)	Onset 0.5 to 1 h Peak 1 to 2 h Effective duration 3 to 5 h	Risk of hypoglycemia	Should be taken just before or just after eating.
INSULIN, SHORT ACTING			
Regular (Novolin R, Humulin R)	Onset 0.5 to 1 h Peak 2 to 4 h Effective duration 3 to 5 h	Risk of hypoglycemia	Best if taken 30 min before a meal.

(continued)

Table 6.3 *(continued)*

Medication name and class	Primary effects	Exercise effects	Special considerations
INSULIN, INTERMEDIATE ACTING			
NPH (Novolin N, Humulin N)	Onset 2 to 4 h Peak 4 to 10 h Effective duration 10 to 16 h	Risk of hypoglycemia	Bedtime dosing minimizes nocturnal hypoglycemia.
Lente (Novolin, Humulin L)	Onset 3 to 4 h Peak 4 to 12 h Effective duration 12 to 18 h	Risk of hypoglycemia	
INSULIN, LONG ACTING			
Glargine (Lantus)	Onset 4 to 6 h Peak: none Duration 24 h	Risk of hypoglycemia	Cannot be mixed in same syringe with other insulins, and do not use same syringe used with other insulins.
Detemir (Levemir)	Onset 3 to 4 h Peak 50% in 3 to 4 h, lasting up to 14 h Effective duration 5.7 to 23.2 h	Risk of hypoglycemia	Cannot be mixed in same syringe with other insulins, and do not use same syringe used with other insulins.
Ultralente	Onset 6 to 10 h No Peak Minimal effective duration 18 to 20 h	Risk of hypoglycemia	

Combination oral medications and combination insulins are available.

Adapted from Diabetes Medications Supplement. National Diabetes Education Program.

Special Exercise Considerations

When developing an exercise prescription for persons with diabetes, the exercise professional should consider the topic of fitness versus the health benefits of exercise (88). Methods to enhance maximal oxygen uptake are often extrapolated to the exercise prescription for disease prevention and management (6). Changes in health status, however, do not necessarily parallel increases in maximal oxygen uptake. In fact, evidence strongly suggests that regular participation in light- to moderate-intensity exercise may help prevent diseases such as coronary artery disease, hypertension, and type 2 diabetes but does not necessarily have an optimal effect on maximal oxygen uptake (33, 38). Therefore, when frequency and duration are sufficient, exercise performed at an intensity below the threshold for an increase in maximal oxygen uptake can be beneficial to health. Exercise must be prescribed with careful consideration given to risks and benefits. The consequences of disuse combined with the complications of diabetes are likely to lead to more disability than the complications alone. Exercises that can be readily maintained at a constant intensity, and in which there is little interindividual variation in energy expenditure, are preferred for those with complications in whom more precise control of intensity is needed (6).

Macrovascular Disease

Macrovascular disease is a complication that often affects patients with diabetes. The primary macrovascular diseases are coronary artery disease and peripheral artery disease. Chapters 13 and 15 review specifics regarding preexercise evaluation and exercise prescription for coronary and peripheral artery disease. These approaches should be incorporated for patients with diabetes and macrovascular disease.

Practical Application 6.2

PREVENTION AND TREATMENT OF ABNORMAL BLOOD GLUCOSE BEFORE AND AFTER EXERCISE

Preexercise Hypoglycemia

Blood glucose levels should be monitored before an exercise session to determine whether the person can safely begin exercising, especially someone using insulin or selected glucose-lowering oral agents. Consideration must be given to how long and intense the exercise session will be. The following general guidelines can be used to determine whether additional carbohydrate intake is necessary (26).

If diabetes is managed by diet or oral glucose-lowering medications with little to no risk of hypoglycemia, most patients will not need to consume supplemental carbohydrate for exercise lasting less than 60 min. If blood glucose is less than 100 mg · dl^{-1} and the exercise will be of low intensity and short duration (e.g., bike riding or walking for <30 min), 5 to 10 g carbohydrate should be consumed. If blood glucose is greater than 100 mg · dl^{-1}, no extra carbohydrate is likely needed. If blood glucose is less than 100 mg · dl^{-1} and exercise is of moderate intensity and moderate duration (e.g., jogging for 30-60 min), 25 to 45 g carbohydrate should be consumed. If blood glucose is 100 to 180 mg · dl^{-1}, then 15 to 30 g of carbohydrate is needed. If blood glucose is less than 100 mg · dl^{-1} and exercise is of moderate intensity and long duration (e.g., 1 h of bicycling), then 45 g of carbohydrate should be consumed. If blood glucose is 100 to 180 mg · dl^{-1}, 30 to 45 g of carbohydrate is needed. Remember that these guidelines may need to be modified. In addition, someone trying to lose weight might benefit from a medication adjustment rather than increased food intake.

Preexercise Hyperglycemia

If the preexercise blood glucose is greater than 300 mg · dl^{-1}, urine should be checked for ketones. If ketones are present (moderate to high), exercise should be postponed until glucose control is improved. If it is determined safe to exercise a patient with a blood glucose greater than 300 without ketones, be sure that the patient is hydrated. These blood glucose values are guidelines, and actions should be verified with the patient's physician. Patients who use medication as part of diabetes treatment should be assessed to determine whether the timing and dosage of medication will allow exercise to have a positive effect on blood glucose. For example, a patient who uses insulin and had blood glucose of 270 mg · dl^{-1}, had no ketones, and took regular insulin within 30 min will see a reduction in blood glucose from both the insulin and exercise. If this patient has not just administered fast-acting insulin and the previous insulin injection has run its duration, the patient is under-insulinized, and additional insulin is needed to help reduce the blood glucose before she exercises. In this case, exercise would likely increase her blood glucose level. In all cases, adding medication must be cleared by a clinician with prescriptive authority. Those with type 2 diabetes who are appropriately managed by diet and exercise alone usually experience a reduction in blood glucose with low to moderate exercise. Timing of exercise after meals can help many patients with type 2 diabetes reduce **postprandial** hyperglycemia. Blood glucose should be monitored after an exercise session to determine the patient's response to exercise.

Postexercise Hypoglycemia

Patients are more likely to experience hypoglycemia (usually <70 mg · dl^{-1}) after exercise than during exercise because of the replacement of muscle glycogen, which uses blood glucose (41, 100). Periodic monitoring of blood glucose is necessary in the hours following exercise to determine whether blood glucose is dropping. More frequent monitoring is especially important when initiating exercise. If the patient is hypoglycemic, he needs to take appropriate steps to treat this medical emergency as previously discussed.

Postexercise Hyperglycemia

In poorly controlled diabetes, insulin levels are often too low, resulting in an increase in counterregulatory hormones with exercise. This circumstance causes glucose production by the liver, enhanced free fatty acid release by adipose tissue, and reduced muscle uptake of glucose. The result, more likely in type 1 than type 2 diabetes, can be an increased blood glucose level during and after exercise. High-intensity exercise can also result in hyperglycemia. In this case, the intensity and duration of exercise should be reduced as needed.

Peripheral Neuropathy

The major consideration in patients with peripheral neuropathy is the loss of protective sensation in the feet and legs that can lead to musculoskeletal injury and infection. Those without acute foot wounds or injuries can undertake moderate weight-bearing exercise, but anyone with a foot injury, open sore, or foot ulcer should be restricted to non-weight-bearing activities (e.g., chair exercises, arm exercises) (27). Proper footwear and examination of the feet are especially important for these patients. The clinical exercise professional should reinforce instruction given to the patient on self-examination of the feet and ways to recognize related injuries and should encourage patients to have their feet examined regularly.

Autonomic Neuropathy

Cardiovascular autonomic neuropathy can affect the patient with diabetes. This disease is manifested by abnormal heart rate, abnormal blood pressure, and redistribution of blood flow. Patients with cardiovascular autonomic neuropathy have a higher resting heart rate and lower maximal exercise heart rate than those without this condition (57, 105). Thus, estimating peak heart rate in this population may lead to an overestimation of the training heart rate range if heart rate–based methods are used (see chapter 5). Exercise intensity may be accurately prescribed using the heart rate reserve method with maximal heart rate directly measured (27). Rating of perceived exertion can also be used to guide intensity. Early warning signs of ischemia may be absent in these patients. The risk of exercise hypotension and sudden death increases (40, 56, 105). An active cool-down reduces the possibility of a postexercise hypotensive response. Moderate-intensity aerobic training can improve autonomic function in those with and without cardiovascular autonomic neuropathy (27). These patients should have an exercise stress test and physician approval before starting an exercise program. Because of difficulty with thermoregulation, they should be advised to stay hydrated and not to exercise in hot or cold environments (11).

Retinopathy

Those with proliferative or severe nonproliferative diabetic retinopathy or macular degeneration should be carefully screened before beginning an exercise program. Activities that increase intraoccular pressure (e.g., high-intensity aerobic and resistance training with large increases in systolic blood pressure), head-down activities, and jumping and jarring activities are not advised with severe nonproliferative or proliferative disease (27).

Nephropathy

Elevated blood pressure is related to the onset and progression of diabetic nephropathy. Placing limits on low- to moderate-intensity activity is not necessary, but strenuous exercises should likely be discouraged in those with diabetic nephropathy because of the elevation in blood pressure (11, 27). Patients on renal dialysis or patients who have received a kidney transplant can also benefit from exercise (5, 54, 65). See chapter 11 for details about renal failure.

Exercise Recommendations

Endurance, resistance, and range of motion exercise training are all appropriate modes for most patients with diabetes. Patients who are trying to lose weight (especially those with type 2 diabetes) should expend a minimum cumulative total of 2,000 kcal per week in aerobic activity and participate in a well-rounded resistance training program (3). Patient interests, goals of therapy, type of diabetes, medication use (if applicable), and presence and severity of complications must be carefully evaluated in developing the exercise prescription. The following exercise prescription recommendations are guidelines, and individual patient circumstances must always determine the specific prescription. Table 6.4 presents a summary of the exercise prescription recommendations.

Cardiovascular Exercise

The value of cardiovascular exercise for persons with diabetes is strong. Given the high risk for developing atherosclerotic disease in those with diabetes, the ameliorating effects of cardiovascular exercise may help to reduce this risk. In most patients, cardiovascular exercise should be performed more often than resistance or flexibility exercise.

Mode Personal interests and the desired goals of the exercise program should drive the type of physical activity that is selected. Caloric expenditure is often a key goal for those with diabetes. Walking is the most commonly performed mode of activity. Walking is a convenient, low-impact activity that can be used safely and effectively to maximize caloric expenditure. Non-weight-bearing modes should be used if necessary (as described earlier). For a given level of energy expenditure, the health-related benefits of exercise appear to be independent of the mode.

Intensity Programs of moderate intensity are preferable for most people with diabetes because the cardiovascular risk and chance for musculoskeletal injury are

Table 6.4 Exercise Prescription Review

Training method	Mode	Intensity	Frequency	Duration	Progression	Goals	Special considerations
Aerobic	Walking, cycling, swimming	40% to 60% of $\dot{V}O_2$reserve	Three to seven times per week No more than two consecutive days between bouts of activity	Minimum of 150 min/wk of moderate activity (30 min, 5 d/wk) or 60 to 75 min of vigorous activity (20 min, 3 d/wk) and up to ≥ 300 min/wk	Rate of progression depends on many factors including baseline fitness, age, weight, health status, and individual goals. Gradual progression of both intensity and volume is recommended.	Patient dependent Energy expenditure of 1,000 to 2,000 kcal/wk	*Note.* All special considerations listed in the chapter apply to all these training methods. Avoid exercise during peak insulin action time. Search for vascular and neurological complications, including silent ischemia. Warm-up and cool-down are important. Promote patient education. Assess for proper footwear and inspect feet daily. Avoid extreme environmental temperatures. Avoid exercise when blood glucose control is poor. Adequate hydration should be maintained. Instruct patient on blood glucose monitoring and on following guidelines to prevent hyper- and hypoglycemic events.
Resistance	Free weights, machines, elastic bands 5 to 10 exercises involving the major muscle groups	Moderate at 50% of 1RM or vigorous at 75% to 80% of 1RM	Two to three times per week, but never on consecutive days	8 to 15 repetitions per set, one to three sets per type of specific exercise	As tolerated. Increase resistance first, followed by a greater number of sets, and then increased training frequency.		
Range of motion	Static stretching		Along with resistance training	10 to 30 s per exercise of each muscle group	As tolerated. May increase range of stretch as long as patient does not complain of pain (acute or chronic).		

lower and the likelihood of maintaining the exercise program is greater. Exercise intensity appears to predict improvements in overall blood glucose control to a greater extent than exercise volume (18). With attention to safety, it may be beneficial to those already exercising at a moderate intensity to consider some vigorous physical activity (27). Exercise should generally be prescribed at an intensity of 40% to 60% of $\dot{V}O_2$reserve, or a rating of perceived exertion of 11 to 13 (28). Chapter 5 provides specifics for determining and calculating proper exercise intensity.

Frequency The frequency of exercise should be 3 to 7 d per week. Exercise duration, intensity, weight loss goals, and personal interests determine the specific frequency. Additionally, the blood glucose improvements with exercise in those with diabetes are seen for 2 to 72 h (17, 27). These data indicate that exercise done on 3 nonconsecutive days each week with no more than 2 consecutive days between bouts is necessary and that 5 d/wk is recommended. Those who take insulin and have difficulty balancing caloric needs with insulin dosage may prefer to exercise daily. This schedule may result in less daily adjustment of insulin dosage and caloric intake than a schedule in which exercise is performed every other day or sporadically, and it will reduce the likelihood of a hypoglycemic or hyperglycemic response. In addition, patients who are trying to lose weight will maximize caloric expenditure by participating in daily physical activity (11).

Duration and Rate of Progression Exercise duration for those with diabetes should be a minimum of 150 min per week of moderate activity (30 min, 5 d/wk) or 60 to 75 min of vigorous activity (20 min, 3 d/wk). Bouts of exercise should be a minimum of 10 min. No specific studies in diabetes have compared rates of progression in intensity or volume. Gradual progression of both intensity and volume is recommended to reduce the risk of injury (27).

Timing Exercise should be performed at the time of day most convenient for the participant. Because of the risk of hypoglycemia, those taking insulin should give careful consideration to the time of day they perform exercise. They should not exercise when insulin action is peaking. Because exercise acts like insulin in that it promotes peripheral glucose uptake, the combination of exercise and peak insulin action increases the risk of hypoglycemia. Because of this effect and the need to replace muscle glycogen, hypoglycemia is more likely to occur after exercise than during exercise (27). Exercising late in the evening for those on insulin and some oral medications is not recommended because of the possible occurrence of hypoglycemia during sleep.

Resistance Exercise

Evidence supports the inclusion of resistance training in a patient's program. Resistance training programs can improve cardiovascular function, glucose tolerance, strength, and body composition in people with diabetes (20, 49, 97). All patients should be screened for contraindications before they begin resistance training. Proper instruction and monitoring are also needed. A recommended resistance training program consists of 5 to 10 exercises involving major muscle groups performed with one to three sets of 8 to 15 repetitions to near fatigue. Resistance training exercises should be done 2 to 3 d per week on nonconsecutive days (27). Progression of intensity, frequency, and duration should occur slowly. Increases in resistance should be made first and only once the target number of repetitions per set can be exceeded, followed by a greater number of sets and then increased training frequency (27). Modifications such as lowering the intensity of lifting and eliminating the amount of sustained gripping or isometric contractions should be considered to ensure safety. Chapter 5 provides specific information about resistance training that should be considered for these patients.

Range of Motion Exercise

Range of motion exercises can be included as part of an exercise program but should not be substituted for aerobic or resistance exercise. Flexibility exercises combined with resistance training has been shown to increase range of motion in people with type 2 diabetes (27).

EXERCISE TRAINING

When beginning a regular exercise training program, those with diabetes can anticipate significant and meaningful improvements.

Cardiovascular and Resistance Exercise

Benefits for persons with diabetes are seen with both acute and chronic exercise. Acute bouts of exercise can improve blood glucose, particularly in those with type 2 diabetes (17, 19, 27). The response of blood glucose to exercise is related to preexercise blood glucose level as well as to the duration and intensity of exercise. Several studies on type 2 diabetes have demonstrated a reduction in blood glucose levels that is sustained into the postexercise period following mild to moderate exercise (17, 19, 27, 78). The reduction in blood glucose is attributed to an attenuation of hepatic glucose production along with a normal increase of muscle glucose use (63, 78). The effect of acute exercise on blood glucose levels in those with type 1 diabetes and in lean patients with type 2 diabetes is more variable and unpredictable. A rise in blood glucose with exercise can be seen in patients who are extremely insulin deficient (usually type 1) and with short-term, high-intensity exercise (63, 73). Also, an elevation in blood glucose has been shown in patients with type 2 diabetes, but this result occurred with short-term, high-intensity exercise (78).

Most of the benefits of exercise for those with diabetes come from regular, long-term exercise. These benefits can include improvements in metabolic control (glucose control and insulin resistance), hypertension, lipids, body composition and weight loss or maintenance, and psychological well-being.

Like acute exercise, exercise training can improve blood glucose. Exercise training (both aerobic and resistance) improves glucose control as measured by hemoglobin A1c or glucose tolerance, primarily in those with type 2 diabetes (27, 113). Improved glucose tolerance was seen in as little as 7 consecutive days of training in subjects with early type 2 diabetes (39, 113). Improve-

Clinical Exercise Pearls: Exercise Training Considerations for Diabetes

- Exercise testing or training promotes glucose absorption. People who use insulin or selected oral glucose-lowering medications may run the risk of hypoglycemia.
- The heart rate response of those with longstanding diabetes may be impaired. They may have a higher than typical testing heart rate and a lower peak heart rate on an exercise test.
- Do not exercise a patient who reports a blood glucose >300, especially if ketones are present. If it is determined safe to exercise a patient with a blood glucose >300 without ketones, be sure that the patient is hydrated.

ments in blood glucose deteriorate within 72 h of the last bout of exercise, emphasizing the need for consistent exercise (17, 27).

Following exercise training, insulin-mediated glucose disposal is improved. Insulin sensitivity of both skeletal muscle and adipose tissue can improve with or without a change in body composition (37, 75). Exercise may improve insulin sensitivity through several mechanisms, including changes in body composition, muscle mass, fat oxidation, capillary density, and glucose transporters in muscle (GLUT 4) (24, 48, 52, 59, 81, 107). The effect of exercise on insulin action is lost within a few days, again emphasizing the importance of consistent exercise participation.

Hypertension affects approximately 67% of adults with diabetes (23). Most of the data supporting a positive effect of aerobic and resistance (slightly greater effect with aerobic) exercise on blood pressure come primarily from studies in subjects without diabetes (29, 60, 61). Most observational studies show that both forms of exercise lower blood pressure in those with diabetes (21, 25, 106). Some randomized controlled trials have shown some reduction in blood pressure (particularly systolic), but others have shown no change in blood pressure with training (55, 62, 69, 87, 94).

Information about the effect of exercise on lipids in diabetes shows mixed results. Improvements have been demonstrated in total cholesterol, low-density lipoprotein cholesterol, and high-density lipoprotein cholesterol following aerobic training in patients with type 2 diabetes (50, 55, 59, 92). Several studies have shown no effect of training on cholesterol profile or triglycerides (14, 69, 94, 101). Lipid profiles may benefit more from both exercise training and weight loss. The Look AHEAD (Action for Health in Diabetes) trial found that lifestyle participants had greater decreases in triglycerides and high-density lipoprotein cholesterol than controls and that both the intensive lifestyle group and the controls had decreases in low-density lipoprotein cholesterol (70).

Weight loss is often a therapeutic goal for those with type 2 diabetes, because most are overweight or obese. Moderate weight loss improves glucose control and decreases insulin resistance (17, 108, 112). Medical nutrition therapy and exercise combined are more effective than either alone in achieving moderate weight loss (6, 91, 111). Visceral or abdominal body fat is negatively associated with insulin sensitivity in that increased abdominal body fat decreases peripheral insulin sensitivity. This body fat is a significant source of free fatty acids and may be preferentially oxidized over glucose, contributing to hyperglycemia (16). Exercise results in preferential mobilization of visceral body fat, likely contributing to the metabolic improvements (45, 80). Exercise is one of the strongest predictors of success of long-term weight control (64). This feature of exercise is extremely important because weight lost is often regained. The amount of exercise required to maintain large weight loss is probably much larger than the amount needed for improved blood glucose control and cardiovascular health (35, 53).

Epidemiological evidence supports the role of exercise in the prevention or delay of type 2 diabetes (83). Early studies showing an increase in type 2 diabetes in societies that had abandoned traditional active lifestyles suggest a relationship between physical activity and diabetes (109). Several cross-sectional studies have shown that blood glucose and insulin values after an oral glucose tolerance test were significantly higher in less active, compared with more active, individuals (36, 42, 85, 90). Prospective studies of several groups have also demonstrated that a sedentary lifestyle may play a role in the development of type 2 diabetes (43, 51, 71, 72, 86). Some of the early data in support of exercise in the prevention of type 2 diabetes come from a 6 yr clinical trial in which subjects with impaired glucose tolerance were randomized into one of four groups: exercise only, diet only, diet plus exercise, or control. The exercise group was encouraged to increase daily physical activity to a level that was comparable to a brisk 20 min walk. The incidence of diabetes in the exercise intervention groups was significantly lower than in the control group (83). A randomized, multicenter clinical trial of type 2 diabetes prevention in those with impaired glucose tolerance at numerous sites around the United States showed a 58% reduction in the incidence of type 2 diabetes with lifestyle intervention that had goals of 7% weight loss and 150 min of physical activity per week (32, 33). Table 6.5 provides a brief review of the effects of exercise training for individuals with diabetes.

Psychological Benefits

Psychological benefits of regular exercise have been demonstrated in those without and with diabetes, including reduced stress, reduction in depression, and improved self-esteem (30, 98, 110). However, a meta-analysis

Table 6.5 Exercise Training Review

Cardiorespiratory endurance	Skeletal muscle strength	Skeletal muscle endurance	Flexibility	Body composition
Associated with a lower risk of all-cause mortality and cardiovascular mortality. Prevention or delay of type 2 diabetes. Improves insulin action and fat oxidation and storage in muscle, and can improve blood glucose control. Exercise intensity affects blood glucose control to a greater extent than exercise volume. May result in a small reduction in LDL cholesterol. May result in a small reduction in systolic blood pressure, but reductions in diastolic blood pressure are less common.	Improves insulin action and fat oxidation and storage in muscle, and can improve blood glucose control.	Improves insulin action and fat oxidation and storage in muscle, and can improve blood glucose control.	Flexibility training combined with resistance training can increase range of motion. Flexibility training has not been shown to reduce risk of injury.	Helps produce and maintain weight loss. Combined weight loss and exercise may be more effective with respect to lipids than aerobic training alone.

Practical Application 6.3

RESEARCH FOCUS

The Look AHEAD (Action for Health in Diabetes) study compared the effects of an intensive lifestyle intervention to those of diabetes support and education on the incidence of major cardiovascular events in overweight or obese people with type 2 diabetes. The four year results from this randomized multicenter clinical trial demonstrated that the intensive lifestyle intervention can achieve weight loss and improvements in fitness, glycemic control, and cardiovascular risk factors in those with type 2 diabetes (70). In October 2012, the Look AHEAD trial was stopped because the intensive lifestyle intervention resulting in weight loss did not achieve the primary study goal of reducing cardiovascular events such as heart attack and stroke in people with type 2 diabetes (www.niddk.nih.gov/news). Other health benefits were seen including decreasing sleep apnea, reducing the need for diabetes medications, helping maintain physical mobility, and improving quality of life. A structured lifestyle change program has been shown to prevent type 2 diabetes in those at risk. Once type 2 diabetes develops, weight loss and exercise provide benefits but do not seem to reduce cardiovascular events more than diabetes support and education.

Reference: Look AHEAD Research Group. Long-term effects of a lifestyle intervention on weight and cardiovascular risk factors in individuals with type 2 diabetes mellitus. *Arch Intern Med* 2010;170:1566-1575.

Practical Application 6.4

AMPUTATION

Diabetes is the leading cause of nontraumatic lower limb amputations. Amputations occur because of impaired sensation and circulation in the extremity, resulting in wounds that are not able to heal properly. Because the energy cost of walking increases markedly for someone with an amputation and a prosthetic limb, walking exercises are more difficult. In addition, prolonged walking may cause trauma and ulceration of the stump. Various upper body exercises, including chair exercises, weights, and arm ergometry, or non-weight-bearing exercise such as swimming, may be better choices. The clinical exercise staff should closely monitor the patient, and the patient should have regular visits with her health care professional.

found that those who exercise to prevent disease have significantly improved psychological well-being, while it deteriorated in those who exercised for management of diseases (e.g., cardiovascular disease, end-stage renal disease, cancer) (47). These findings suggest that benefits may vary and that people with fewer existing health concerns benefit the most. Mechanisms for the impact of exercise include psychological factors such as increased self-efficacy and changes in self-concept, as well as physiological factors like increased norepinephrine transmission and endorphins (30, 98).

CONCLUSION

Living with a chronic illness poses special issues. Diabetes management requires ongoing dedication; in addition, the patient must cope with complications if they develop and, at the very least, deal with the threat of their development. Exercise training should be an essential component of the treatment plan for patients with diabetes because it can improve blood glucose control, lipid levels, blood pressure, and body weight; reduces stress; and has the potential to reduce the burden of this metabolic disease.

Key Terms

autoimmune (p. 92)
cardiovascular autonomic neuropathy (p. 96)
diabetic ketoacidosis (p. 112)
effective insulin (p. 94)
evidence-based (p. 101)
gestational diabetes (p. 93)
glucagon (p. 95)
glycemic goals (p. 94)
hyperglycemia (p. 93)
hypoglycemia (p. 93)
idiopathic (p. 92)
ketones (p. 94)
medical nutrition therapy (p. 98)
microvascular disease (p. 96)
pancreas (p. 91)
peripheral neuropathy (p. 96)
polydipsia (p. 96)
polyuria (p. 96)
postprandial (p. 105)
proteinuria (p. 97)

CASE STUDY

MEDICAL HISTORY

Medication: Glucophage taken two times per day, captopril (for blood pressure control and protection of kidneys), and Lipitor for control of hyperlipidemia. Laboratory values: Last HbA1c = 8.8% (normal 4-6%); cholesterol 200 mg · dl^{-1}; low-density lipoprotein cholesterol 130 mg · dl^{-1}; high-density lipoprotein cholesterol 35 mg · dl^{-1}; triglycerides 160 mg · dl^{-1}; microproteinuria. Physical exam: Blood pressure 130/80 mmHg; resting heart rate 70 beats · min^{-1}; height 5 ft 11 in. (190 cm); weight 230 lb (104 kg) with 27% body fat

(continued)

(skinfold). Complications history: Acute periodic episodes of diabetes out of control but has never experienced hyperosmolar nonketotic syndrome or **diabetic ketoacidosis**. Two-vessel bypass surgery 5 yr ago, moderate peripheral neuropathy, and early diabetic nephropathy.

DIAGNOSIS

Mr. SR is 63 yr old and was diagnosed with type 2 diabetes 5 yr ago.

EXERCISE TESTING RESULTS

No abnormal electrocardiogram changes; maximum blood pressure 180/83 mmHg; maximum heart rate 150 beats · min^{-1}; $\dot{V}O_2max$ 25.5 ml · kg^{-1} · min^{-1}; random blood glucose before test 180 mg · dl^{-1}.

EXERCISE PRESCRIPTION

The goals of the exercise program, mutually agreed on by the patient and the clinical exercise professional, are to lose weight and improve body composition, improve blood glucose levels, and reduce risk for another cardiac event. When asked about interests and hobbies, the patient indicates that he enjoys traveling, wine tasting, playing with his dog, and classic movies. Participation in a supervised exercise program and frequent contact with an exercise professional are advised. A warm-up and cool-down of static stretches and low-intensity aerobic activity are prescribed.

Mode: Stationary cycling, water exercise, or moderate walking, if tolerated. Attention should be given to the patient's ability to safely perform weight-bearing activities because of the peripheral neuropathy. Walking his dog and walking during traveling are discussed. The patient must take care of his feet and do these activities as safely as possible.

Frequency: 3 to 5 d per week with a goal of increasing to daily.

Intensity: This patient is taught to monitor heart rate and to use the rating of perceived exertion (RPE) scale. The intensity is prescribed at 40% to 60% of heart rate reserve, or 102 to 118 beats · min^{-1}. An RPE rating of 11 to 13 on a 6- to 20-point Borg scale is advised.

Duration: An initial duration of 15 to 30 min is suggested and should eventually be increased to 60 min per session to facilitate weight loss.

Rate of progression: Attention is first given to frequency of exercise. After he has reached 5 d per week or more, duration will be increased.

Other information: Mr. SR is instructed to increase his blood glucose–monitoring frequency to assess the effect of exercise on his blood glucose control.

DISCUSSION QUESTIONS

1. Is this patient likely to have problems with hypoglycemia during or following exercise? Why or why not?
2. What are potential risks of exercise for this patient? What cautions should he be given?
3. What strategies can be suggested to help the patient stay motivated and maintain a regular exercise program?
4. What general nutrition suggestions might be helpful for this patient?
5. What other health care team members should this patient work with as he begins his exercise program? Why?
6. What else should be added to his program to help him attain his goals?

Chapter 7

Obesity

David C. Murdy, MD
Jonathan K. Ehrman, PhD

The National Center for Chronic Disease Prevention and Health Promotion, part of the Centers for Disease Control and Prevention (CDC), developed an "At A Glance" report in 2010 titled "Obesity: Halting the Epidemic by Making Health Easier" (21). In this document the U.S. government considers obesity as an epidemic associated with, if not the principal cause of, much disease, disability, and even some discrimination. Many still see obesity as a social or moral issue and stigmatize its management, adding to the considerable barriers to understanding and treating this expanding epidemic. Short-term intervention has limited effectiveness, and long-term success is rare without ongoing medical efforts or weight loss surgery. Clinicians should approach obesity as an illness in itself and manage its comorbidities as well. Few medical treatments have such far-ranging positive effects as assisting an overweight or obese patient achieve and sustain medically significant weight loss.

Exercise is an essential component in the management of obesity, along with diet and lifestyle change. These three elements of therapeutic lifestyle change are critical first steps in the prevention and treatment of many medical conditions but are often overlooked because of the complexity of their practical application. Exercise combined with calorie reduction, lifestyle change, and in some cases weight loss medication and surgery is best provided by medical clinicians who work in a team environment with a long term horizon. And the clinical exercise professional should be a key component of this group. Using this approach, it is possible to reduce the burden of obesity in a cost- and care-efficient manner, and its comorbidities can be controlled or eliminated. Today, clinical exercise physiologists are much more involved in the treatment of obesity than ever before. Considering that the level of obesity in the United States and around the world will not decrease significantly in the short term, the use of these professionals in this type of programming continues to look promising.

DEFINITION

Because many assessments are used to determine threshold values for overweight and obesity, the definitions vary (67). Overweight can be most simply defined as weight that exceeds the threshold of a criterion standard or reference value (67). The World Health Organization (WHO) defines **overweight** and **obesity** as abnormal or excessive fat accumulation that may impair health (136).

A common current approach in the clinical setting for determining overweight and obesity is to use **body mass index (BMI)**. BMI is recommended by the National Institutes of Health to classify overweight and obesity and to estimate relative risk of associated disease (88). Body mass index indicates overweight for height but does not discriminate between fat mass and lean tissue. This is demonstrated in figure 7.1. The BMI does, however, significantly correlate with total body fat (79). Therefore, BMI is an acceptable measure of overweight and obesity in the clinical setting. Body mass index is calculated as weight in kilograms divided by height in meters squared.

Body mass index can also be determined using pounds and inches. The following are the BMI formulas:

In most cases excess weight is due to being overfat.

$$\text{BMI} = \frac{\text{Weight (kg)}}{(\text{Height [m]})^2}$$

or

$$\text{BMI} = \frac{\text{Weight (lb)} \times 703}{(\text{Height [in]})^2}$$

Reference values used to define overweight and obesity for BMI are derived from population data and may be specific to certain characteristics, including sex. Distributions of these data are determined, and criteria are then based on threshold values associated with increased risk or morbidity and mortality. Threshold determinations may vary internationally by race. For instance, in Japan an overweight BMI ranges from 23.0 to 24.9 kg · m^{-2} and obesity is any BMI ≥25.0 kg · m^{-2} (105), while in the United States, overweight is defined as a BMI ranging from 25.0 to 29.9 and obesity is ≥30 (116).

Despite its ease of use, there may be important inaccuracies with respect to determining if a given BMI is "unhealthy" or if a person with a BMI categorized as overweight (or obese) actually has an elevated body fat percentage. For instance, some studies have reported that a BMI range of 25.0 to 29.9 may be no different from a BMI <25.0 or even superior with respect to all-cause death (38, 93). Others report that all-cause mortality is lowest for the BMI range of 20.0 to 24.9 (29).

Many methods of assessing fatness (e.g., skinfolds, bioelectrical impedance, underwater weighing, and dual-energy X-ray absorptiometry [DXA]) are available. But using these techniques requires technical skill, patient cooperation, time, and space, all of which may preclude their use. It is for this reason that BMI is a popular choice.

The classification of overweight and obesity by BMI for U.S. adults is based on the 1998 "Clinical Guidelines on the Identification, Evaluation, and Treatment of Overweight and Obesity in Adults" and is shown in table 7.1 (88). However, as recently as 1987, the U.S. government used BMI values of 27.8 (males) and 27.3 (females) kg · m^{-2} to define overweight in the population (86). And other countries have developed their own normative data for BMI. For instance, Japan's normal BMI range is 18.5 to 22.9 kg · m^{-2}, and obesity begins at a BMI value of 25.0 kg · m^{-2} (105). These differences are based on the specific relationship of BMI in a given country to morbidity and mortality.

Medically significant obesity applies to people who have gained 30 lb (14 kg) or more as adults and as a result have increased their waist circumference >6 in. (15 cm) (12). The National Institutes of Health (NIH) guidelines define this population as adults with a BMI >30 kg · m^{-2} or a BMI >25 kg · m^{-2} with a family history of obesity or an obesity-related comorbidity such as diabetes, hypertension, heart disease, or stroke (88).

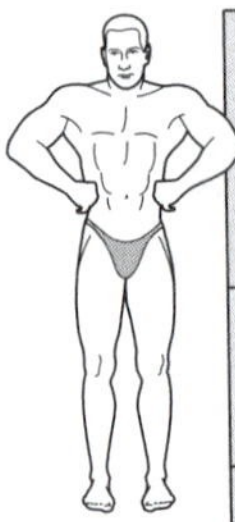

75 in 191 cm	**Height**	75 in 191 cm
250 lb 114 kg	**Weight**	250 lb 114 kg
31.0	**BMI (kg·m^{-2})**	31.0
48 lb 22 kg	**Above ideal weight**	48 lb 22 kg
24%	**Above ideal weight (%)**	24%
15%	**Body fat (%)**	45%
32 in 81 cm	**Waist**	52 in 132 cm
38 lb 17 kg	**Fat weight**	113 lb 51 kg

Figure 7.1 The two men in this example are the same height and weight and thus have the same BMI and weight relative to ideal weight. But it is clear from their physical appearance that the person on the left has a lower body fat percentage than the man on the right. The body fat, waist, and fat weight determination clarify this difference.

Although BMI is recommended to evaluate obesity, patients may request that their healthy weight range for their height be determined. *Dietary Guidelines for Americans, Seventh Edition*, recommends healthy weights for Americans (116). These guidelines are similar to the often cited Metropolitan Life Insurance ideal weight tables and provide ranges of weight that vary by frame size (80). Frame size can be specifically determined using elbow widths, but usually individuals subjectively self-select into a frame category (i.e., small, medium, large).

SCOPE

Table 7.2 shows the prevalence of normal weight, overweight, and obesity in the population using the 2008 report from the U.S. Department of Health and Human Services (87). Of note is the prevalence of the extreme obesity category (i.e., BMI ≥40 kg · m^{-2}) from 1988 to 1994 (2.9%), from 1999 to 2000 (4.7%), and from 2003

Table 7.1 Body Weight–Related Classifications

	Underweight	Normal	Overweight	Mildly obese (class I)	Moderately obese (class II)	Morbidly obese (class III)
BMI*	<18.5	18.5-24	25-29	30-34	35-39	≥40
% over ideal weight**	NA	0-10%	10-20%	20-40%	40-100%	>100%
% fat+	<20	20-25	26-31	32-37	38-45	>45
Waist circumference ++	NA	NA	High	Very high	Very high	Extremely high

*BMI: body mass index (kg · m^{-2}).

**Weight: percent over standard height–weight tables.

+Fat (%): calculated body fat expressed as percent of total weight.

++Waist circumference: risk additive to BMI of overweight- or obesity-related comorbidities for values ≥40 in. for men and ≥35 in. for women.

Adapted from National Institutes of Health 1998.

Table 7.2 Current (2003-2006) U.S. Population and Age-Adjusted Body Weight Demographics (87)

Population	Overweight and obese	Normal weight or underweight	Overweight	Obese
Males	73%	29%	39%	34%
Females	61%	39%	26%	35%
All	67%	34%	33%	34%
White males	72%	26%	39%	33%
Black males	72%	27%	36%	36%
Hispanic males	77%	22%	47%	30%
White females	57%	37%	24%	33%
Black females	81%	18%	27%	54%
Hispanic females	74%	25%	31%	43%

to 2004 (4.8%) (60). Extreme obesity appears to be increasing in younger people and is greater among black women than in men, as well as among blacks than in non-Hispanic whites. Black women have the highest rates of extreme obesity in terms of ethnic and gender makeup (60). In the same period, the prevalence of overweight and obesity also rose in every age, gender, and ethnic category. And throughout the United States, each state demonstrated an increase in the prevalence of obesity with a continual rise throughout the 1990s and into the early 21st century (figure 7.2).

Although obesity often begins in childhood and early adolescence, 70% of all obesity begins in adulthood. Approximately 17% to 19% of children in the United States are obese, and many will carry their obesity into adulthood (87, 91). The risks of childhood obesity persisting into adult obesity depend on the severity of obesity, age of onset, childhood BMI value, and parental obesity (55, 131). Weight gain tends to be more rapid in early adulthood and can be as much as 10 to 16 kg over a 15 yr period (0.75 to 1.0 kg annually) (70). This weight gain is attributed to an environment that discourages physical activity and promotes excessive food intake, including large portion sizes and high fat content.

Weight that is 20% above ideal weight (or ~BMI = 30 kg · m^{-2}) carries increased health risk for obesity-related comorbidities (5). The patterns of fat distribution also affect risk (89). Central or android obesity (upper body, with a waist size of ≥35 in. [88 cm] for women or ≥40 in. [102 cm] for men) carries a higher risk for diabetes and coronary heart disease than lower body obesity (42, 102, 103). The enzyme lipoprotein lipase, which regulates the

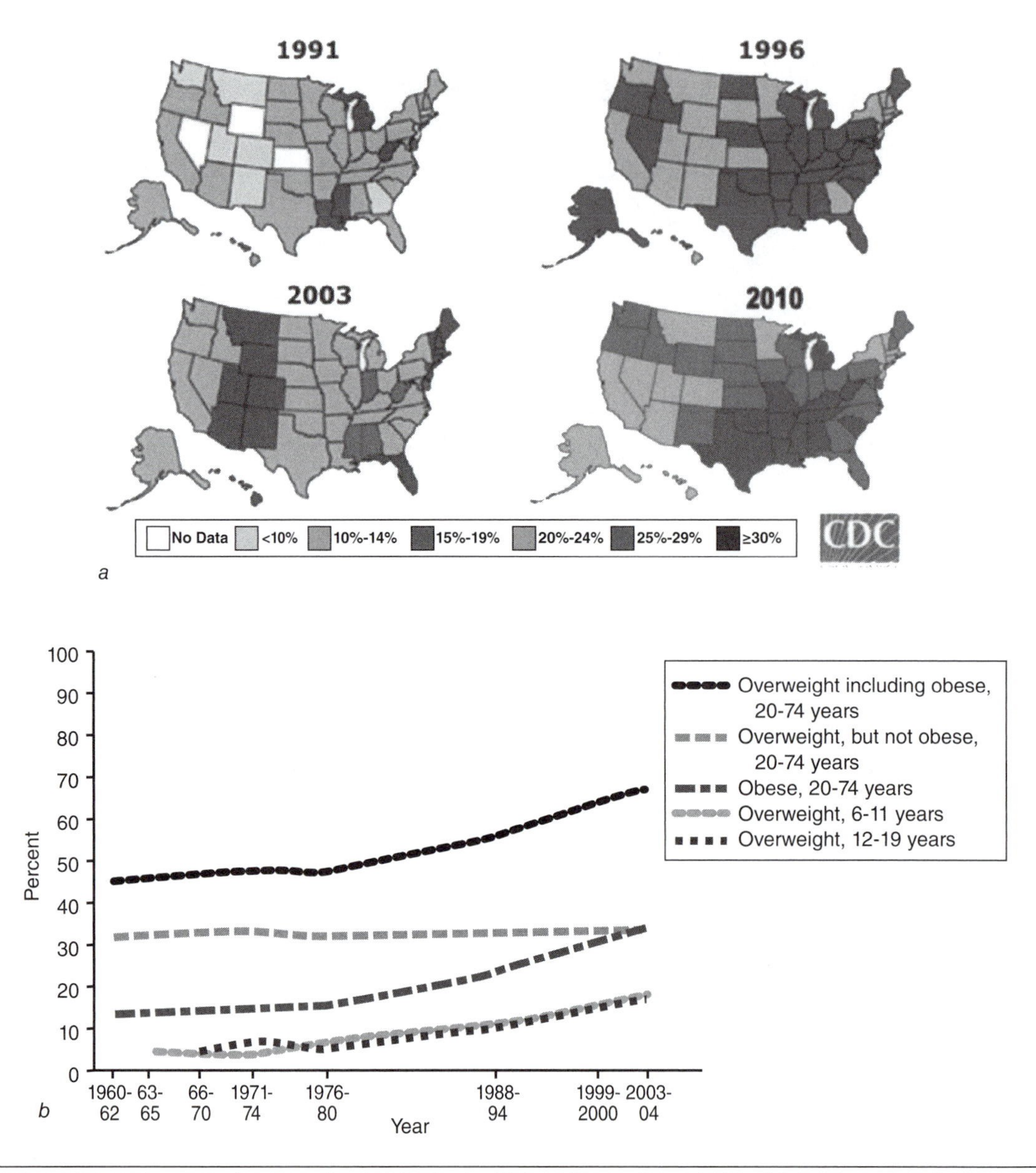

Figure 7.2 *(a)* State-by-state increase in obesity in the United States from 1990; *(b)* rise in overweight and obesity since the early 1960s in children and adults, respectively.

Reprinted from CDC.

storage of fats as triglyceride, is more active in abdominal obesity and therefore increases fat storage. Upper body obesity is measured by the waist-to-hip ratio (W/H). Increased health risk is present in women when the W/H ratio exceeds 0.8 and in men when the W/H ratio exceeds 1.0 (108). A more important factor in fat distribution is distinguishing abdominal visceral fat from subcutaneous fat. Visceral fat lies deep within the body cavities and is associated with a higher risk than subcutaneous fat and waist circumference because of the metabolic characteristics, which include insulin resistance, glucose intolerance, and clustering of metabolic risk factors (76). Visceral fat is best measured by magnetic resonance imaging (MRI) or computed tomography (CT), which are expensive and unavailable to most clinical exercise practitioners.

A more practical measurement of visceral fat uses sagittal diameters (106). This technique requires the patient to lie on his back. The clinician obtains the sagittal diameter by measuring the distance from the examination table to a horizontal level placed over the abdomen at the site of the iliac crest. This promising technique is currently the best practical predictor of visceral fat. The assessment of visceral fat provides additional information regarding health risk and should be used in counseling patients about realistic weight loss. Realistic weight loss should be based on total weight loss and the resulting redistribution of fat.

For those who are overweight or obese, estimated direct medical costs (including preventive, diagnostic, and treatment services) in 2008 accounted for 9.1% of total U.S. medical expenditures or about $147 billion (36). Andreyeva and colleagues suggested that each increment of 5 in BMI (e.g., 30 to 35, 35 to 40) results in a 25% to 100% increase in the lifetime health care expenditures for an individual (8). Per capita medical spending is as much as $1,429 higher each year for an obese person compared to a person of normal weight (36). As noted in table 7.3, Medicaid and Medicare pay up to half of obesity-related medical care expenditures, which exemplifies the growing burden of overweight and obesity on federal and state spending. Indirect costs relate to morbidity and mortality costs and are difficult to capture. Morbidity costs result from decreased productivity and absenteeism. Mortality costs are the value of future income lost by premature death.

Obesity appears to reduce life expectancy, and this effect is most powerful in those who develop obesity earlier in life (43). Flegal and colleagues (using National Health and Nutrition Examination Survey [NHANES] III data) reported that obesity, but not overweight, is related to excess premature death (39). They estimated 112,000 excess deaths per year because of obesity and reported that as obesity levels became more severe (class I to II to III), the mortality rate increased. They also reported no effect on longevity in the cohort defined as overweight (BMI = 25-29.9). This finding is similar to the findings of NHANES I and II. In addition, several studies have found a protective effect on mortality for overweight individuals when other conditions are present such as heart failure and revascularization per percutaneous intervention (26, 56, 59). It is important to understand that there are limitations to this "obesity paradox" data. These include the cross-sectional nature of the observations, limited data on change in weight over the follow-up period, and assessment of variables that may affect weight including pulmonary disease, human immunodeficiency virus (HIV) infection, alcohol and drug use, smoking, and cancer (2). The obesity paradox may, in part, be explained by cardiorespiratory fitness levels in overweight and obese individuals. Higher fitness values based on estimated metabolic equivalents (METs) during an exercise test are associated with a lower all-cause mortality rate (77). Recent findings from the legendary Harvard Alumni Health Study reported a doubling of heart disease risk in those who were obese in early childhood (51). But, importantly, this risk was reduced when weight loss occurred in midlife or later. This study is important because it is one of the few to provide these data longitudinally with prospective measured BMI in both early and later life.

Table 7.3 Estimated U.S. Medical Spending, in Billions of Dollars, Attributable to Overweight and Obesity

Insurance category	OVERWEIGHT AND OBESITY	
	MEPS	NHEA
Private	$49.4	$74.6
Medicaid	$8.1	$27.6
Medicare	$19.7	$34.3
Total	$85.7	$146.6

MEPS = Medical Expenditure Panel Survey

NHEA = National Health Expenditure Accounts

Americans may spend as much as $60 billion each year in efforts to control their weight. These attempts include participation in commercial and clinical weight loss programs and purchases of food and nonfood supplements. In recent years several sets of professional guidelines have been written that have focused on obesity, including *Dietary Guidelines for Americans* (116); *Physical Activity and Health: A Report of the Surgeon General* (118); NIH's "*Clinical Guidelines on the Identification, Evaluation, and Treatment of Overweight and Obesity in Adults*" (89); the American College of Sports Medicine position stand (31); the American Heart Association scientific statements (98, 100); and "*Healthy People 2020: National Health Promotion and Disease Prevention Objectives*" (117).

PATHOPHYSIOLOGY

Obesity results from longstanding positive energy balance. The simplicity of this restatement of the first law of thermodynamics can blind us to the complex physiology of food intake and calorie expenditures and tempt us to tell patients only to "eat less and run more." Such an approach is oversimplistic and ineffective. Positive energy balance has a myriad of contributors in current societies, ranging from increasing availability of lower-cost foods that some refer to as a toxic food environment to decreasing physical activity at work and in leisure. Average daily calorie intake has increased by over 200 kcal over the last several decades (43) as food costs have fallen dramatically and as more calories are consumed outside the home. At the same time, physical activity has fallen because of advances in equipment and ergonomics in the workplace.

Beyond daily behavioral and dietary influences, recent research has identified an ever-increasing number of genetic and physiologic factors pointing to a large array of neurological and peripheral endocrine messengers that influence food intake and nutrient utilization and that regulate weight in a way that often frustrates patients' weight loss efforts (58). Advanced neuroimaging research has recently shown similar patterns of brain activation between addictive-like eating and substance abuse (49).

A revolution in the understanding of obesity began with the identification of **leptin**. Leptin is secreted by fat cells and was found to regulate body weight in mice. Administration of leptin to mice, genetically altered so that they were deficient in leptin, reduced their extreme obesity and poor growth, resulting in marked reduction in food intake and weight, as well as return of normal growth and metabolic function. This research proved that a molecular defect could be the basis of some forms of obesity and provide a new approach for the treatment of obesity (11). Humans produce leptin from their fat cells in proportion to their weight and particularly their girth. Humans are not leptin deficient, however, and exogenous leptin administration has limited benefit for weight reduction except in the rarest of cases of human obesity because of a genetic absence of this fat cell hormone. Nevertheless, leptin research changed the perception of many clinicians and researchers. Instead of shunning obesity or considering it an untreatable moral failing, they now see it as a complex behavioral and neuroendocrine disorder that may be unlocked with additional study (17).

Genetic causes of obesity are a feature of rare disorders such as Prader-Willi syndrome (20) but are thought to be a factor in at least half of human obesity based on studies of twins raised separately or children who were adopted (112). Genetics plays a significant part in explaining responses to overfeeding or weight loss in metabolic lab settings, which points to variation in inheritable control of food intake, fat storage, and energy expenditure (15). Patients do not conveniently follow equations of calorie intake or energy utilization.

Body fat distribution is also genetically determined and gender specific, and predictable changes throughout life may confound patients' efforts at weight control. Increasing age reduces growth hormone and gonadal hormone secretion, which may predispose to greater visceral fat storage with its links to metabolic and cardiovascular abnormalities that in turn are linked to hypertension, coronary artery disease, and diabetes mellitus (14).

Energy expenditure has complex genetic and environmental variation that can promote weight gain in some and frustrate weight loss in others. The **set-point theory** of weight regulation can be seen in metabolic lab studies of overfeeding compared with calorie restriction. Weight loss decreased total and resting energy expenditure, which slowed further weight loss. Weight gain through overfeeding was associated with an increase in energy expenditure, which slowed further weight gain. These studies show that after weight loss, individuals require 15% fewer calories to maintain their lower weight. This tendency causes weight loss patients to return to higher weights unless they restrict calories over the long term or expend calories through greater physical activity (68). Quantifying the effect of energy restriction shows that dynamic models are required to accurately predict weight loss and that both weight loss and prevention of weight regain are multifaceted and complicated (57).

Variation in energy regulation is seen in genetic studies of the Pima Indians, who are at significantly increased risk of obesity and its comorbid conditions (98). Efforts to augment energy expenditure through medications have been complicated by adverse effects of increased blood pressure and heart rate. This area will remain a target for ongoing pharmaceutical research because resting metabolic activity accounts for about 70% of energy expenditure. The metabolic cost of food digestion accounts for about 10% of energy expenditure. The remainder is physical activity, which is influenced by sedentary pursuits such as television and computer viewing compared with intentional exercise or dedicated increases in lifestyle activities.

Of the millions of calories eaten or expended over a lifetime, a mere fraction of 1% can result in clinically significant obesity. The neurological and peripheral control system for this process is biased toward preserving weight in times of famine and receives signals from fat cells and the gastrointestinal tract. An interesting observation in patients following gastric bypass weight loss surgery is the dramatic reduction in **ghrelin**, a potent appetite-increasing gut hormone, which may explain, in part, the success of this surgical procedure (25). Ghrelin, produced during stomach distention, triggers the central nervous system appetite stimulant **neuropeptide Y (NPY)**, growth hormone, and norepinephrine. Neuropeptide Y and norepinephrine predominantly stimulate carbohydrate intake. Weight loss can increase serum ghrelin levels, which may explain part of the process that makes sustaining weight loss difficult.

Cholecystokinin, serotonin, and peptide YY are among a group of central nervous system and gut peptides involved in satiety and reduced food intake (69). Serotonin has been targeted in the treatment of depression with selective serotonin reuptake inhibitors such as fluoxetine, but these medications have had limited effect in weight loss treatment. Nutrient selection during weight loss treatment may be a factor because appetite-reducing peptide YY is increased during high-protein diets. This may partially explain the appetite-reducing effect of high-protein diets. Table 7.4 lists the monoamines and peptides that affect appetite.

Table 7.4 Monoamines and Peptides That Affect Appetite

Stimulatory	Inhibitory
Norepinephrine	Leptin
Neuropeptide Y	Cholecystokinin
Opioids	Serotonin
Melanin-concentrating hormone	Corticotropin-releasing hormone (CRH) or urocortin

The pathophysiology of obesity has been known for millennia. The degree of obesity predicts the morbidity and mortality for a long list of common afflictions in humans (19). Obesity and extreme obesity are associated with an increased rate of death from all causes, particularly from cardiovascular disease. In the United States, estimates of annual excess deaths due to obesity range from 112,000 to 365,000 (1, 39, 83). Obesity and sedentary lifestyle is considered the second leading cause of preventable death in America and may overtake tobacco abuse within the next decade. During the last 40 yr, other cardiovascular risk factors have decreased: High total cholesterol levels ($\geq$240 mg · dl^{-2}) decreased from 34% to 17%; high blood pressure (BP $\geq$140/$\geq$90 mmHg) decreased from 31% to 15%; and smoking dropped from 39% to 26%. Weight-related cardiometabolic risks have increased at an accelerated rate; diagnosed diabetes mellitus rose from 1.8% to 6.0%, and metabolic syndrome increased from 27% to 56% for men (52). The effects of obesity are so dramatic that projections of life expectancy, which have risen with declining cardiovascular risks over 40 yr, are predicted to decline as the full effect of increasing obesity is felt in society (92). For obese individuals, life expectancy decreased by about 7 yr compared with normal-weight individuals in the Framingham Study (97). A J-shaped relationship exists between BMI and all-cause mortality, with the lowest mortality for a BMI of 20 to 24.9 increasing to 2.5 times greater mortality risk in those with a BMI of 40 to 49.9 (29).

Type 2 diabetes mellitus is strongly associated with obesity; up to 80% of cases are weight related. Compared with people with a BMI of 22, those with a BMI of 35 have a 61-fold increased risk of developing diabetes (24). The Diabetes Prevention Project (DPP) showed that even a modest 7% weight loss reduced the progression of prediabetes to diabetes by 58% over 4 yr, exceeding medication treatment (i.e., metformin) by 25% (66). Obesity, particularly abdominal (or visceral) obesity, is also among the current diagnostic criteria for **metabolic syndrome**, as explained in chapter 10 (53).

Intensive lifestyle intervention has been studied in the NIH-sponsored, multicenter Look AHEAD trial (72). This decade-long study evaluated the benefit of modest weight loss through low-calorie diets, meal replacement programs, and weight loss medications coupled with thorough behavioral education, support, and exercise over 5 yr of initial treatment and a planned 6 yr of follow-up in over 5,000 individuals. The 1 yr results of the Look AHEAD trial showed an average 8.6% weight loss in the group treated with intensive lifestyle interventions (ILI) compared with a 0.7% loss in the diabetes support and education group (DSE). For the year, ILI participants attended an average of 35 treatment sessions and reported exercising a mean of 137 min/wk and consuming a total of 361 meal replacement products. Greater self-reported physical activity was the strongest correlate of weight loss, followed by treatment attendance and consumption of meal replacements (126). At year 4, the ILI group maintained a weight loss of 4.7% compared to 1.1% in the DSE group. The ILI group's success was related to greater attendance at education sessions and more favorable diet and activity patterns (124). See Practical Application 6.3 for more about the Look AHEAD study.

Hypertension often increases in obese people and may be present in up to half of those seeking weight loss treatment. In the Framingham Heart Study, excess body weight explained over 25% of the incidence of hypertension (135). A 10 kg weight loss can reduce blood pressure by 10 mmHg, which is comparable to the effect required to achieve Food and Drug Administration (FDA) approval for a new antihypertensive drug.

Cardiovascular morbidity and mortality are greater in obese individuals, and the role of adipose tissue, particularly visceral fat mass, is becoming more clearly established as a key coronary disease risk factor. Women in the Nurses' Health Study showed a fourfold increased risk of death from cardiovascular disease if their BMI was over 32 compared with those with a BMI under 19 (132). The INTERHEART study showed that central obesity, particularly elevated waist circumference, explained 20% of the risk of first myocardial infarction (138). Obesity may also increase cardiovascular disease through elevated low-density lipoprotein (LDL) cholesterol levels and reduced high-density lipoprotein (HDL) cholesterol levels. Heart failure, mostly those with preserved ejection fraction (PEF), is increased twofold in obese people. This consequence is often largely reversible with significant weight loss. Atrial fibrillation and flutter are increased about 50% in obese individuals. Stroke risk is also doubled for obese women, based on the findings of the Nurses' Health Study. In addition, venous thromboembolic disease, including deep venous thrombosis and pulmonary embolus, is increased in obese individuals, particularly those who are sedentary (128).

Respiratory illness, with adverse effects on exercise capacity, is also common in obese people (101). Obesity is the greatest predictor of obstructive sleep apnea, a syndrome of interrupted sleep with snoring and apneic periods that lowers oxygen saturation levels to those associated with potentially lethal ventricular arrhythmias. Dyspnea and asthma are increased with obesity through adverse effects on respiratory mechanics and through gastroesophageal reflux disease (GERD)-associated bronchospasm.

Obesity is associated with increased gastroesophageal disease and hepatobiliary illness. GERD is increased through reflux of stomach acid into the esophagus because of increased intra-abdominal pressure, which may cause esophagitis and lead to esophageal cancer. Gallstones, cholecystitis, and biliary dyskinesia (biliary colic without cholelithiasis) are increased in obese individuals because of increased biliary excretion of cholesterol (111). Active weight loss may increase the risk of cholelithiasis and cholecystitis because cholesterol is removed from reducing fat stores and secreted; it may then crystallize in the gallbladder. Weight loss diets with daily modest amounts of fat may empty the gallbladder and reduce this risk. Nonalcoholic fatty liver disease is linearly related to obesity and can lead to cirrhosis and liver failure but is often reversible with weight loss (9).

Osteoarthritis, particular weight-bearing joint disease, is increased with increasing obesity. Elevated weight multiplies the effects on the hips, knees, and ankles and increases the risk of foot pain and plantar fasciitis, which may further increase weight through secondary reductions in physical activity. Significant weight loss is associated with a 50% reduction in arthritic joint pain and may postpone or obviate the need for joint replacement in some individuals.

Depression and eating disorders, particularly binge-eating disorder, are increased in obese women, especially those seeking professional help with weight loss. Evaluation and management of these issues are important to eliminate them as barriers to weight loss. Psychosocial function is decreased by obesity, and outright prejudice is common toward obese people, particularly women, who may be significantly less likely to marry or to complete their education and thus are likely to face increased poverty and lower annual incomes (50).

CLINICAL CONSIDERATIONS

Although medically significant obesity exists in most patients who seek medical care, only a minority present requesting medical help with weight reduction. Increasing the awareness of those not aware of the relationship between their weight and medical problems is an effective way of helping them become determined to address their weight. Compassion and understanding coupled with flexible and practical weight loss recommendations are essential to building rapport with obese patients. Most individuals seeking professional help with weight loss have a BMI of 38 or more, and invariably they have attempted to lose weight several times in the past. Proper assessment of the barriers to and benefits of weight loss should be the initial step in the clinical evaluation of the obese patient. Comprehensive assessment of obesity can lead to appropriate treatment and effective long-term control of obesity and its related comorbidities, as shown in "Health Consequences of Obesity."

Health Consequences of Obesity

Greatly increased risk (relative risk >3)

- Diabetes
- Hypertension
- Dyslipidemia
- Breathlessness
- Sleep apnea
- Gallbladder disease

Moderately increased risk (relative risk about 2-3)

- Coronary heart disease or heart failure
- Osteoarthritis of the knees
- Hyperuricemia and gout
- Complications of pregnancy

Increased risk (relative risk about 1-2)

- Cancer
- Impaired fertility or polycystic ovary syndrome
- Low back pain
- Increased risk of anesthesia
- Fetal defects arising from maternal obesity

Based on Haslam, Sattar, and Lean 2006 (58).

Signs and Symptoms

Although obesity would seem the most obvious medical condition, its key determinants, weight and height, are often omitted in medical records or ignored in patient problem lists. Ideally, patients should be measured at each visit in lightweight clothing without shoes. Comparable measurements would ideally be at the same time of day (intraday variation of up to 2% is common) with an empty bladder. Appropriate scales that can measure extreme weights (up to 600 lb [270 kg]) are essential or wheelchair scales should be used, because many patients who weigh more than 600 lb cannot stand in order to be weighed. A wall-mounted statiometer can be used for heights, although patients typically report accurate height (but not weight). Waist circumference should be measured at the umbilicus while the patient is standing, because measurement at that location is the best predictor of central, or visceral obesity, and its significance is not related to a patient's height. Patients should not suspend their breathing during waist measurements, and a measuring tape should not be overtightened.

Body fat measurements, typically by bioimpedance but also through underwater weighing, skinfold measurements, or DXA scans, can be performed; but they add little to therapeutic management and have significant methodological errors and costs. The U.S. Surgeon General recommends that BMI be added to blood pressure and pulse as a routinely recorded vital sign because further evaluation and management of obesity and its related comorbidities depend on it.

History and Physical Examination

Most clinicians and patients are aware that a blood pressure of 150/90 mmHg or a cholesterol of 240 mg · dl^{-1} or more is associated with increased cardiovascular risk, and immediate aggressive therapy would be started. Few are aware, however, that a BMI of 33 conveys the same risk of cardiovascular death (16). Additionally, only 25% of physician visits with patients who are obese address issues of weight reduction (110).

Therefore, many overweight and obese patients who begin to work with an exercise professional have not been approached about their weight by a medical professional. The exercise professional should address this issue by simply asking the patient about the subject. One can assess the patient's readiness to address her weight with the simple question "Have you been trying to lose weight?" Potential responses are the following:

No, and I do not intend to in the next 6 mo (precontemplation).

No, but I intend to in the next 6 mo (contemplation).

No, but I intend to in the next 30 d (preparation).

Yes, but for less than 6 mo (action).

Yes, for more than 6 mo (maintenance).

The response of the exercise professional should be used to assess the patient's readiness to lose weight. For instance, patients who are not intending to lose weight (precontemplators) might simply be educated about the health risks of being overweight and the benefits of losing weight. Those who are preparing to lose weight soon might be best served by being directed into a clinically based weight loss program.

The evaluation of the obese person, whether he is seeking medical help for weight control or for routine care of unrelated medical concerns, should include relevant factors of history and physical exam as well as laboratory testing that can better characterize the person's risks from obesity, its often silent comorbid conditions, and barriers to effective treatment. The medical approach to obesity should consider, in compassionate cooperation with the patient, the risks of obesity, appropriate treatment, and the most effective form of treatment.

Medical causes of obesity, including illnesses, medications, and lifestyle changes known to increase weight, must be identified. Examples include diabetes mellitus; polycystic ovary syndrome; Cushing's syndrome (although rare); many medications (some neuroleptics, psychotropics, anticonvulsants, antidiabetics, β-blockers, ACE inhibitors, and hormone therapies); and behavioral changes such as smoking cessation, job changes, injuries, sleep habits, and life stressors. History of obesity should be obtained whenever possible, including the age of onset of obesity, weight gain since adolescence, changes in weight distribution, peak and lowest maintained adult weight, any history of weight loss (intentional or unintentional), and methods used to lose weight. Assessment of readiness for weight loss, social support, weight control skills, and history of any past or present eating disorders (binge eating, bulimia nervosa, or, rarely, anorexia nervosa) should be considered. Family history of obesity predisposes toward weight problems, and social history can identify key issues such as divorce, job changes, and upcoming weddings or class reunions that can affect weight loss efforts. An early family history of cardiovascular disease may motivate overweight patients to control their cardiovascular risk factors. A thorough review of systems can help identify medical conditions such as joint pain or asthma that may be overlooked barriers to weight control.

An exercise history should identify current intentional exercise or opportunities for increased lifestyle activities, evaluate any exercise barriers and patient-specific benefits to help in the assessment of any contraindications to moderate or vigorous exercise, and develop a comprehensive exercise plan. Exercise testing may be appropriate for the assessment of cardiovascular disease or to explain dyspnea on exertion, which is common in obese patients.

Diagnostic Testing

Laboratory testing should screen for diabetes (fasting blood sugar), elevated cholesterol and triglycerides as well as LDL and HDL cholesterol levels (lipid profile), and clinical or subclinical hypothyroidism (thyroid-stimulating hormone [TSH] and free T4). Comprehensive chemistry profiles can assess for nonalcoholic fatty liver and renal disease. Electrocardiograms are rarely necessary except to evaluate specific cardiovascular problems such as elevated blood pressure or palpitations.

Exercise Testing

Routine exercise testing in the overweight and obese population is not indicated. And it has been suggested that pharmacological stress testing using dipyridamole is preferred for evaluating obese individuals (98). When performed, the purpose is primarily to assess for the presence of coronary artery disease (40). Exercise testing may also be performed to determine functional capacity, develop an exercise prescription based on heart rate, and assess for risk related to bariatric surgery (78). Several studies demonstrate that obese and morbidly obese patients can exercise to maximal exertion on a variety of treadmill protocols (40, 45).

Although walking is the preferred mode of exercise for testing, it is not always practical in those who are obese. Patients in this group, especially those with BMI values greater than 40 $kg \cdot m^{-2}$, often have concomitant gait abnormalities and joint-specific pain during weight-bearing exercise. Seated devices such as upper body ergometers, stationary cycles, or seated stepping machines offer excellent alternatives that allow patients to achieve maximal exercise effort in a non-weight-bearing mode. Despite this, McCullough and colleagues reported that in a group of 43 consecutive patients referred for bariatric surgery (mean BMI = $48 \pm 5\ kg \cdot m^{-2}$), only one could not perform a walking protocol (78).

Testing should be performed with the normal routine (see chapter 5). Prediction equations for METs from the work rate achieved on an exercise device are typically inaccurate in people who are obese. Assessment of cardiopulmonary gas exchange provides an accurate measurement of exercise ability. Gallagher and colleagues reported a peak oxygen consumption (peak $\dot{V}O_2$) level of $17.8 \pm 3.6\ ml \cdot kg^{-1} \cdot min^{-1}$ (equivalent to 5.1 METs) in a morbidly obese group of patients who achieved peak respiratory exchange ratio (RER) values greater than 1.10 (48). No complications related to exercise testing were reported in this cohort, suggesting that exercise testing is safe in this extremely obese population. See table 7.5 for a review of exercise testing methods.

Treatment

Although all weight loss treatment should include specific diet, behavioral, and exercise prescriptions, and in some cases pharmacotherapy or weight loss surgery, the specifics of treatment must be matched to the patient's circumstances, including her BMI and related considerations. Patients at lower BMI and without comorbid conditions should be offered less intensive treatment than patients with extreme obesity (BMI ≥40) who have significant obesity-related comorbidities such as diabetes, hypertension, and obstructive sleep apnea. Table 7.6 illustrates the strategy used for weight management as discussed in this chapter.

Treatment goals need to consider both the medical benefits of modest (10%) weight loss and the patient's expectations. The NIH has recommended a 10% weight loss within 4 to 6 mo and weight loss maintenance as an initial weight loss goal because this amount is associated with several health-related benefits. See "Health Benefits of a 10% Weight Loss." Improvements in obesity-related functional limits and medical comorbidities should be identified (89). Patients, on the other hand, commonly want to lose 35% of their weight to attain their dream weight, seek to lose 25% to reach a satisfactory weight, and would consider a weight loss of 17% disappointing (46). Few commonly prescribed weight loss programs achieve average weight losses that match patients' expectations. Physicians' expectations are not much lower; they consider a weight loss level of 13% disappointing. Patient expectations have been remarkably resistant to change, which may contribute to treatment dissatisfaction, treatment discontinuation, and treatment recidivism.

Diet Therapy

To lower weight, energy balance must be negative. Calorie reduction is the essential first step. For normal adults, 10 kcal per pound (22 kcal per kilogram) per day is required to maintain weight. Average calorie needs can vary by as much as 20% between individuals. Hence, some

Table 7.5 Exercise Testing Review

Test type	Mode	Protocol specifics	Clinical measures	Clinical implications	Special considerations
Cardiovascular	• Treadmill • Cycle ergometer	Typically low level because of deconditioned state (e.g., ≤2 MET increments)	Standard ECG, blood pressure, rating of perceived exertion	• Typically necessary to perform only if at risk for ischemic heart disease • Also may be useful for setting an exercise prescription based on heart rate	• Must use equipment that can handle high body weight • Non-weight-bearing equipment useful in presence of joint problems
Strength	Resistance machines	• Standard • Assess major muscle groups (e.g., chest, arms, legs)	Strength of various muscle groups	Serves as a baseline for developing a resistance training program for use in a weight loss or weight maintenance program	May consider performing a 2RM to 10RM assessment if patient is limited
Range of motion	• Sit-and-reach • Goniometry	Standard	Range of motion of major joints	• Used to identify joints with limited range on which to focus • Also used to assess progress	• Patient may have difficulty with standard sit-and-reach because of requirement to sit on floor • Excess fat tissue may limit range in affected joints

Health Benefits of a 10% Weight Loss

Blood Pressure

Decline of about 10 mmHg in systolic and diastolic blood pressure in patients with hypertension (equivalent to that with most BP medications)

Diabetes

Decline of up to 50% in fasting glucose for newly diagnosed patients

Prediabetes

>30% decline in fasting or 2 h post-glucose insulin level >30% increase in insulin sensitivity

40% to 60% decline in the incidence of diabetes

Lipids

10% decline in total cholesterol

15% decline in LDL cholesterol

30% decline in triglycerides

8% increase in HDL cholesterol

Mortality

>20% decline in all-cause mortality

>30% decline in deaths related to diabetes

>40% decline in deaths related to obesity

Data from Haslam et al. 2006.

Table 7.6 A Systematic Approach to Management Based on BMI and Other Risk Factors

BMI	Suggested weight loss	Deciding factors for treatment level
< 18.5	• None • Consider weight gain	• Keep at same or move to less intensive strategy: • No cardiovascular risk factors • Lower body obesity/overweight • No previous weight-loss attempts • < 25 lb to lose or goal of less than 5% to 10% weight loss
18.5 – 26.9	• Exercise • Diet modification • Counseling	
27 - 29.9	• Weight loss program • Behavioral health services • Self-help materials or program	
30 - 34.9	• Weight loss program • Pharmacotherapy • Meal-replacement	• Consider move to more intensive strategy: • Any cardiovascular risk factors present • Metabolic syndrome present • Abdominal obesity/overweight • Previous weight-loss attempt failure at current level • > 25 lb to lose or goal of more than 10% weight loss
35 - 39.9	• Very low energy diet • Residential programs • Pharmacotherapy	
≥ 40	• Very low energy diet • Suggest surgery	

Adapted from: WHO Obesity: Preventing and Managing the Global Epidemic, Report of WHO Consultation on Obesity, World Health Organization Tech Rep Ser 2000;894i.

individuals will require 12 kcal per pound (26 kcal per kilogram) and others only 8 kcal per pound (18 kcal per kilogram) per day to maintain weight. The lowest calorie level for weight maintenance is about 1,200 kcal daily, even for those at bed rest. The minimum calorie intake to assure adequate intake of essential micro- and macronutrients is 500 kcal per day, commonly provided under medical supervision in very low-calorie diets. Exercise and activity levels affect maintenance calorie levels by 25% or more, depending on the degree of physical activity, and physical activity needs to be considered in determining maintenance calorie levels.

Hypocaloric diets for weight loss typically set intake at 500 to 750 kcal less than predicted maintenance requirements. Typically, a deficit of 3,500 kcal is needed to lose 1 lb (7,700 kcal to lose 1 kg), and such diets can average about a 1 lb (0.45 kg) per week weight loss. If losses are slower, more aggressive diet therapy should be considered. Many popular variations in hypocaloric diet composition have been developed. Because of the high calorie content of fat (9 kcal per gram) compared with carbohydrates or proteins (4-5 kcal per gram) and the heart health benefits for cholesterol lowering, most national guidelines recommend low-fat diets.

Recently, higher-protein and lower-carbohydrate diets have become favored by patients because of greater weight losses and better satiety. Higher-protein diets tend to promote satiety; and lower carbohydrate levels can promote greater fat utilization (ketosis), which can boost the rate of weight loss. Overall reviews of popular commercial diet programs have shown about 5% weight loss at 1 yr and a 3% weight loss for standard hypocaloric diets at 2 yr (114). Although published data are supportive of higher-protein and lower-carbohydrate diets, no specific low-calorie diet composition is clearly superior to any other, and all fall short of patient and provider expectations. Systematic reviews of weight loss trials show moderate weight loss at 6 mo and favor intensive counseling, pharmacotherapy, and meal replacements as best-practices strategies (47, 115).

Recent research has focused on more structured lower-calorie diets using meal replacement supplements. These diets are combined with exercise and lifestyle efforts to increase weight loss and achieve greater long-term results (47). Better portion control is achieved in these programs, which often use higher protein content to promote satiety and reduce snacking. Most use 1,000 kcal diets initially and combine weight loss medications to optimize results. Wadden and colleagues showed 18% weight loss results at 6 mo and 17% at 12 mo when a structured meal replacement diet was followed by a low-calorie diet in conjunction with behavioral and lifestyle changes along with the weight loss medication sibutramine (123). Of note, sibutramine has been removed from the market in the United States by the FDA and in Europe by the European Medicines Agency as related to an increased risk of heart attack and stroke.

Because most patients who present for medical help to lose weight have BMIs of 38 or more, even structured meal replacement programs may not be sufficient. When rapid weight loss is critical, medically supervised very low-calorie diets (<800 kcal per day; a.k.a. complete meal replacement [CMR]), very low-calorie diet [VLCD], or protein-sparing modified fast) can be used. These diets routinely consist of highly engineered powdered supplements rich in protein. Weight losses can begin at 1 lb (0.45 kg) per day but average 3 to 5 lb (1.4 to 2.3 kg) weekly for a typical 16 wk period (85). First developed in the 1920s, these diets resurfaced in the 1960s in surgical research centers. Popularized in the 1970s, such "last-chance" diets fell from favor when excess protein losses led to deaths from ventricular arrhythmias. Reformulated in the 1980s with better-absorbed nutrients, these diets, often called liquid protein diets, peaked in popularity when celebrities used them for rapid success, although most people regained significant weight after resuming their previous eating and activity patterns. In studies comparing CMR diets to low-calorie diets, outcomes were similar at 1 yr (122). Now, most very low-calorie diets are combined with long-term combination treatment. Among completers, weight loss ranges from 10% to 35%, averaging over 15% at 2 and 4 yr (7).

The CMR plan, using high biological grade protein-based supplement, is often criticized for supposedly increasing the risk of putting excess weight back on or for hampering future success with weight loss attempts. Li and colleagues (71) present data on over 480 individuals who used a 700 to 800 kcal/d CMR diet from two to four times. Initial-attempt weight loss was 21.3 kg for women and 28.8 kg for men. There was no difference in total weight loss at CMR diet restart attempt 2, 3, or 4 as compared with the initial attempt when the rate of weight lost per week was assessed. The authors concluded that their data refute the notion that repeated CMR dieting reduces the effect of subsequent weight loss diet attempts. This finding supports a modern approach to clinical weight management suggesting that people desiring to lose large amounts of weight and maintain this weight loss should consider the periodic or continuous use of high biological grade protein-based supplement.

Practical application 7.1 presents a study that assessed the use of CMR diet plans in the primary care setting.

Practical Application 7.1

RESEARCH FOCUS

Despite the success of CMR diets, there remains relatively little traction for this type of plan in the medical community, particularly for the extremely obese individual. To address this issue, Ryan and colleagues designed a randomized controlled trial (LOSS) that provided CMR primarily in the primary care setting for those with an average BMI of >45 $kg \cdot m^{-2}$. The CMR diet was implemented for 12 wk. At 2 yr follow-up, the control group did not demonstrate any weight change. Completers in the CMR group maintained a 9.6% weight loss. Maximal weight loss was 15.5% at 38 wk. In a subanalysis using the LOSS study data, Johnson and colleagues reported that many cardiometabolic risk factors, including fasting plasma glucose, triglycerides, HDL and LDL cholesterol, and C-reactive protein (CRP), were improved incrementally as weight loss increased. Reductions in glucose began after only a 5% weight loss and exceeded a 25% reduction when weight loss exceeded 20%. These findings suggest that CMR plans can be easily implemented in new settings, including primary care, and are effective for long-term (2 yr) weight loss in morbidly obese individuals. Many clinical exercise physiologists are active in providing exercise and behavior change counseling in these types of programs. Given the current state of health care spending and the push for reform, providing this type of effective weight loss programming in the primary care setting can be an important low-cost method to combat obesity and its comorbid conditions.

Johnson WD. Incremental weight loss improves cardiometabolic risk in extremely obese adults. *Am J Med* 2011;124:931-938.

Ryan DH, et al. Nonsurgical weight loss for extreme obesity in primary care settings: Results of the Louisiana Obese Subjects Study (LOSS). *Arch Intern Med* 2010;170:146-154.

Behavioral Therapy

Providing any level of diet advice without behavior change is typically futile. Most medically significant weight loss efforts require frequent contact and support in order for people to adopt the healthier weight behaviors necessary for weight loss and maintenance of weight loss. Regular accountability, problem solving, and skill building are necessary over a 20 wk period to establish long-term success. Such behavioral change can be supported by individual or group therapy and augmented by phone and Internet follow-up. Weight loss efforts typically move from precontemplation to contemplation to determination and then to action phases. Maintenance efforts must follow action steps in weight loss; otherwise, relapse is common. Motivation and realistic goal setting must be supported in a compassionate environment focused on measurable progress (82).

Record keeping and review predicts success because most people make better choices when they are made aware of the significance of those choices. Stimulus control helps patients identify stress and emotional eating cues and make other choices. Unhealthy eating behaviors like eating while driving or eating in front of a television or computer screen should be discouraged. Increased intentional exercise or lifestyle activity should be planned and monitored. Cognitive restructuring is used to detect black-and-white thinking and to help patients avoid an all-or-nothing pattern. Addressing emotional issues such as depression and shame with supportive therapy is essential, and referral for significant mental health or eating disorders may be necessary. Nutrition education and planning for maintenance, including relapse, can help reduce recidivism and the need for retreatment (18).

Exercise Therapy

Certainly exercise and physical activity are important in order for people to avoid becoming overweight or obese. But for the treatment of overweight and obesity, exercise alone has not shown long-term weight loss success (135). Exercise in conjunction with diet therapy or other treatment modalities, however, is effective in slightly accelerating weight loss (135). In the National Weight Control Registry (NWCR), over 90% of the successful subjects combined exercise with diet therapy (65). And evidence from the NWCR suggests that regular exercise of 60 to 90 min on most days of the week, expending 2,500 to 2,800 kcal per week, may be required to maintain large amounts of weight loss for the long term (65, 116). Regardless of the timing of exercise and its effectiveness in causing or maintaining weight loss, all overweight and obese patients will likely demonstrate improvements in cardiovascular function and physical fitness as a result. Additionally, exercise can improve self-esteem, which may improve adherence to weight loss-based treatments. The exercise prescription for overweight and obese persons is reviewed later in this chapter.

Pharmacotherapy

Medications have been used to reduce weight for over 100 yr. The first medication advertised for weight loss was thyroid extract. Now known to cause more muscle loss than fat losses, thyroid replacement, necessary for hypothyroidism, is no longer used for weight loss. Weight loss drugs have faced significant prejudice. Many people expect long-term benefits from short-term treatment, something not expected with antihypertensives, for example, and are reluctant to use them at all. Although not the "cure," some weight loss medications are an important tool for achieving and maintaining medically significant weight loss. Several weight loss medications have been removed from the market (e.g., Meridia). Until recently, orlistat, which is now marketed over the counter, was the only FDA-approved weight loss drug for long-term use. Current weight loss drugs are appropriately recommended for individuals with a BMI ≥30 or with a BMI ≥27 if they have obesity-related comorbidities. See table 7.7 for a brief review of weight loss medications.

Phentermine (Adipex-P, Ionamin) is the most commonly prescribed weight loss drug in the United States. Approved in 1958 and currently recommended for short-term use, it acts as an appetite suppressant but can cause dry mouth, palpitations, and anxiety. It was part of the famous phen-fen combination therapy effort pioneered in the mid-1990s by a study published by Weintraub until the fenfluramines (Pondimin and Redux) were withdrawn from the market after fears of cardiac valve abnormalities were discovered. Cleared by the FDA, phentermine remains the least expensive common weight loss medication.

Orlistat (Zenical, Alli), approved in 1999, is an intestinal lipase inhibitor that causes a 30% malabsorption of dietary fat. It is not absorbed into the bloodstream, and patients who eat high-fat meals may have oily diarrhea. Patients who follow a typical fat diet (30% fat diet) do not have significant gastrointestinal symptoms from fat malabsorption. Orlistat produces an energy deficit of about 180 kcal daily (27). Orlistat has been approved by the FDA for over-the-counter sale.

Table 7.7 Pharmacology Review

Medication name and class	Primary effects	Exercise effects	Special considerations
Phentermine (Adipex-P, Ionamin)	Appetite suppression	None	May acutely increase heart rate and blood pressure
Lorcaserin hydrochloride (Belviq)	Satiety enhancement	None known	Average weight loss 3 to 3.7% of body weight
Phentermine and topiramate (Qsymia)	Appetite suppression	None known	May acutely increase heart rate and blood pressure
Orlistat (Xenical, Alli)	Intestinal lipase inhibitor	None	Can cause intestinal discomfort, flatulence, and oily stools

In 2012, the FDA approved two new oral therapies for chronic weight loss and weight loss maintenance. Qsymia (phentermine/topiramate ER) is capable of between 12% and 14% weight loss and Belvig is capable of 4% weigh loss at 1 yr in completers. Additional combination medications are pending FDA approval.

Surgical Therapy

The fastest-growing area of obesity treatment is surgical procedures to restrict the stomach or to cause malabsorption of food or both. Frustration with less invasive medical therapies now leads over 250,000 people to have one of these types of procedures each year. Surgery for weight loss has been performed for over 50 yr, and newer laparoscopic techniques can cause patients to lose a third of their weight (>50% of their excess weight) within 18 mo. Such weight loss results in marked reduction in obesity-related medical conditions and improved life expectancy in these seriously obese patients (22). Recent research has confirmed reduced mortality with weight loss surgery (1).

Although not without significant risks (up to 1% for death and 15% for morbidity), surgery can lead to long-term weight loss. Additionally, patients undergoing bariatric surgery must commit to a lifelong program of restricted diet, lifestyle changes, vitamin supplementation, and follow-up testing to ensure safety. Weight regain after 2 yr can cancel out some of the initial benefits such that, in the Swedish Obesity Study, longer-term weight loss at 10 yr averaged between 15% and 24% of initial weight depending on the procedure chosen (107).

Surgery is typically restricted to those with a BMI ≥40 or those with a BMI ≥35 if they have obesity-related comorbid conditions. Many insurers require a 6 to 12 mo trial of comprehensive medical therapy before surgery, and many surgeons require that patients attempt to lose 5% to 10% of initial weight before surgery to reduce complications, aid exposure interoperatively, and allow a laparoscopic approach (109).

The initial intestinal bypass procedures (jejunoileal bypass) of the 1950s were abandoned because of excessive nutritional deficiencies and were replaced by gastric banding procedures (gastric stapling) and vertical banded gastroplasty from the 1960s through the early 1990s (84). Roux-en-Y gastric bypass, combining the restrictive effect of gastric stapling with the malabsorptive effect of intestinal bypass, became the gold standard in the mid-1990s (75). Now done laparoscopically, this procedure is still recommended for those with BMIs ≥50. Laparoscopic adjustable gastric banding has been adopted as a lower-risk procedure that has fewer postoperative complications and almost no mortality for those at lower weights. This procedure places a restriction on the stomach that is adjusted to promote early satiety and lower calorie intake. Outside the United States this device has become the procedure of choice (30). The procedure is not without long-term risks (e.g., obstruction requiring band removal in up to 50% of procedures), but mean excess weight loss has been shown to be 43% or approximately 21% of the baseline weight (61).

Comprehensive Long-Term Therapy

The imperative for health providers interested in obesity, and for those who face the many comorbid conditions directly related to obesity, is to match patients to treatment options that will achieve meaningful medical and personal benefits and not merely tell patients to "eat less and run more." The NIH-sponsored Look AHEAD trial was designed around intensive biweekly comprehensive treatment over 5 yr in hopes of reducing obesity-related morbidity and mortality from diabetes in obese subjects (72). This level of treatment also requires combinations of diet, exercise, behavioral, and pharmacologic therapy, and in many cases weight loss surgery for those who fail with less invasive approaches. Effective January 1, 2012, the Centers for Medicare and Medicaid Service added a yearlong weight management services intensive counseling benefit to Medicare (119) for those with a BMI ≥ 30 $kg \cdot m^{-2}$.

Medical and surgical therapy for obesity are associated with significant benefits including remission of metabolic syndrome and diabetes (1, 74). The maintenance of weight loss also requires diligent follow-up to offset the metabolic penalty of the reduced obesity (68). The Study to Prevent Regain (STOP) looked at face-to-face and Internet options for long-term weight loss management and found that both have a role in preventing the regain of weight so commonly seen with termination of treatment (134). The chronic disease model applies to obesity as much as it does to the comorbidities of obesity such as diabetes mellitus. Comprehensive approaches that depend on long-term lifestyle training will be those that can tame obesity and reduce its effect on mortality in this century. Finally, long-term weight loss maintenance is also important because there is evidence that weight loss attempts after weight regain may not result in a similar degree of success (120), although this has been refuted in the CMR literature (71).

EXERCISE PRESCRIPTION

The American College of Sports Medicine's *Guidelines for Exercise Testing and Prescription* recommends that an exercise program focus on physical activities and intentional exercise for 60 to 90 min/wk to promote and maintain weight loss (4). These recommendations are beyond the general recommendation of 1,000 to 2,000 kcal expenditure per week (30 min on most days) for general health benefits. Exercising for 250 to 300 min per week (or ~60 min per day) is equivalent to about 2,000 kcal energy expenditure per week. But this figure is less than the 2,500 to 2,800 kcal per week expenditure recommended by the National Weight Control Registry mentioned previously in this chapter (133). Therefore, the following exercise prescription recommendations are based on a weekly caloric expenditure of 2,000 to 2,800 kcal per week. This goal range is appropriate for all obese individuals, although some obesity class II patients and most class III patients will have to progress gradually to these higher levels of daily energy expenditure. Counseling about physical activity provided by a clinical exercise physiologist will help people develop realistic goals, establish appropriate exercise progression schedules, and gain control of their exercise programs (96). See table 7.8 for a summary of the exercise prescription for overweight and obese individuals.

Cardiovascular Exercise

Initially, exercise and physical activity should focus on cardiovascular (i.e., aerobic) modes. The primary reason for this approach is to focus on the greatest amount of energy expenditure possible in a given period of time. To achieve the target of 2,000 to 2,800 kcal per week expenditure, exercise must be predominantly aerobic. Although resistance training may provide added benefits, the caloric expenditure of resistance training is less than that of aerobic exercise because (1) it is performed discontinuously, (2) a single training session incorporates less exercise time than an aerobic session does, and (3) resistance training should be performed on only 2 or 3 d/wk. Resistance training should be considered only after a regular aerobic program meeting the weekly caloric expenditure goal has been in place for a minimum of 1 mo.

Exercise mode selection is important for enhancing adherence and reducing the risk of injury. Some people have preexisting musculoskeletal issues that could prevent certain modes of cardiovascular exercise. These issues often relate to pain in the lower back, hip, knee, and ankle joints that may be chronic. However, these problems may improve as weight is lost. The clinical exercise physiologist should assess any painful conditions and make recommendations to avoid this type of pain.

In general, aerobic exercise should be categorized as either weight bearing or non-weight bearing. When possible, walking is the best form of exercise for several reasons. Walking has few disadvantages; all patients have experience with the activity and a goal to remain functional and independent. Walking is an excellent, low-intensity activity with little risk of injury. It is available to most patients and does not require special facilities. Neighborhoods, parks, walking trails, shopping malls, fitness centers, and so on offer walking opportunities. A minimum amount of attention is necessary, so socializing is easy and convenient. If a patient wishes to walk on a treadmill, care should be taken to assess the weight limits of the treadmill. Many are rated to handle only 350 lb (160 kg). Issues relate to the ability of the walking board and motor to handle the weight of an individual. If a motor seems to "bog down" when an individual is walking on it, then the weight of the person is likely too much. Treadmills especially designed for obese individuals up to 500 lb (227 kg) are

Table 7.8 Exercise Prescription Review

Training method	Mode	Intensity	Frequency	Duration	Progression	Goals	Special considerations and comments
Aerobic	• Walking • Consider any non-weight-bearing mode where appropriate	• Initially at moderate intensity of 40% to 59% of heart rate reserve • Increase to 60% to 80%	Progress to daily	Progress to 60 to 90 min	• Initial bouts are typically 10 to 20 min if person has no recent exercise history • Advance duration and frequency initially and intensity later • Consider beginning at lower end of intensity for very deconditioned people	• Achieve regular exercise pattern • Achieve 60 to 90 min per day by the time weight maintenance program begins • 2,000 to 2,800 kcal per wk expended	• Non-weight-bearing modes should be considered if joint pain or injury exists • Watch for indications of hyperthermia • Provide guidelines on water consumption during exercise
Resistance	• Machines • Free weights • Elastic bands • Calisthenics	• 10- to 15RM (i.e., 10-15 repetitions per set) • RPE of 11-15 (6-20 scale)	2 or 3 d per week	30 min involving two sets per major muscle group with minimum 1 min rest between sets	As tolerated to maintain 10 to 15 reps per set at a RPE of 11 to 15	• Regular resistance training • Improved skeletal muscle strength and endurance • Maintenance of lean mass during rapid weight loss phase	• Because of range of motion limitations, some equipment may be difficult to use (e.g., machines) • Because of high incidence of hypertension, consider reducing breath hold or Valsalva maneuver
Range of motion	Static and proprioceptive or passive stretching	Within comfortable ranges	Daily	10 to 30 s per major joint	Increased range of motion as tolerated	Enhanced range of motion	Keep in mind that certain stretching techniques may be difficult for some obese or overweight patients (e.g., because of poor balance, coordination, inability to sit on floor)

available. Jogging should usually be avoided, especially in patients with no previous jogging history or people who have a preexisting musculoskeletal issue that may be aggravated by jogging. Some class I patients may be appropriate candidates for jogging.

Non-weight-bearing exercise options include stationary cycling, recumbent cycling, seated stepping, upper body ergometry, seated aerobics, and water activities. These activities are useful at any time but are particularly useful for those with joint injury or pain. The clinical exercise physiologist should adapt these modes of exercise by providing larger seats and stable equipment. People who are obese often have difficulty getting on or off these types of equipment or moving through the range of motion required by a given piece of equipment. For some individuals, seated aerobics may be an excellent option to reduce the typical orthopedic limitations that some people experience, including back, hip, knee, and ankle pain. Another advantage is that seated or chair aerobics can be performed in the comfort of a person's home. In extremely obese individuals, it is important to use a chair that is rated to handle a very heavy body weight. In some chair aerobic exercise routines, the force on the chair from the movement and body weight can be quite large. Water provides an alternative to walking or aerobic dance activities performed on land. The buoyancy of water takes much of the body weight off the joints. Additionally, patients who experience heat intolerance with other activities are often more comfortable performing water-based exercise. Most patients are not efficient swimmers, so swimming laps should be avoided. An experienced exercise leader can make a workout session fun and effective. For example, the resistance of the water can be used creatively to increase intensity. Many patients do not consider water activities because of the effort necessary to get into and out of the pool and because of their concern about their appearance in a bathing suit. The exercise physiologist should work to overcome these issues by using zero-entry pools and locations where the public does not have a direct view of the aquatic facility.

Frequency

Behavioral changes in activity must be consistent and long lasting if the patient is to lose weight and maintain weight loss over the long term. Daily exercise and physical activity at the recommended levels of duration and intensity are required to achieve and sustain long-term, significant weight loss. All people who are obese can exercise daily, typically from the beginning of a program. Key factors are to minimize the duration and intensity initially to avoid excessive fatigue or muscle soreness that may sabotage the patient's willingness to exercise the next day. Altering the exercise mode may also help reduce the risk of injury.

Intensity

The intensity of exercise must be adjusted so that the patient can endure up to 1 h of activity each day. For those who have never exercised previously, intensity in the range of 50% to 60% of peak $\dot{V}O_2$ (50-60% of heart rate reserve) is typically low enough for sustained exercise. However, an intensity closer to 40% may be required by some individuals, particularly those who have not exercised recently. As an individual progresses, a goal of 60% to 80% of heart rate reserve is adequate. People without significant comorbid conditions can perform at these intensities in either a supervised or a nonsupervised setting. Many individuals who are obese are hypertensive and may be taking a β-blocker. This possibility must be considered when intensity is prescribed using heart rate. Typical rating of perceived exertion values of 11 to 15 (6 to 20 scale) may be substituted when assessing heart rate is not convenient.

Duration

For those with little or no previous recent exercise history, beginning with 20 to 30 min each day is appropriate. Breaking this exercise time into two or three sessions per day of shorter duration (5 to 15 min) may be required for highly deconditioned people. Progression of approximately 5 min every 1 to 2 wk, until the person can perform at least 60 min of exercise, is usually appropriate. This progression scheme is intended to increase compliance to the duration of each session as well as to daily exercise. An accumulation of time over several sessions in a day is as beneficial as one continuous work bout with respect to total caloric expenditure. Besides performing this intentional exercise duration, all obese people should be continuously encouraged to maximize daily physical activity by considering all options. For instance, they could park at the far end of the parking lot when visiting a store or get off one or two stops early when taking public transportation. The duration of daily physical activity should typically not be restricted unless the individual appears to be suffering from effects of excessive activity. But the contribution of incidental exercise to total caloric expenditure is significant and may be as beneficial as that of the planned exercise (6).

Intensity and duration must be manipulated so that the intensity is low enough to allow suitable duration to expend the recommended caloric energy. For many obese patients, the intensity will not be great enough to improve cardiovascular fitness. The initial focus, however, should be on weight loss and therefore caloric expenditure. As the exercise progresses and the individual is able to better tolerate the exercise routine, higher-intensity activities should be encouraged. Patients should be encouraged to increase the duration from 20 min per day to 60 min or more per day on every day of the week so that they expend >2,000 kcal per week. An exercise program for obese patients should include both the supervised and nonsupervised phases with adaptations in modes, intensity, duration, and frequency to provide adequate calorie expenditure while preventing soreness and injury. Patients with existing comorbidities should preferably participate in a supervised exercise program 3 to 5 d/wk with a prescribed intensity and duration to treat their comorbidities. Patients should be physically active a minimum of 60 min each day, including the days of supervised exercise; therefore, they may have to supplement with walking to accumulate 60 min. The remaining days of the week (2-4 d) can be nonsupervised with self-reported exercise to accumulate 60 min of physical activity each day.

Resistance Exercise

If resistance training is incorporated, careful attention must be given to beginning this type of program. Strength equipment may not be an option for some morbidly obese individuals. In general, the exercise prescription for obese people should include resistance intensity in the range of 60% to 80% of an individual's one-repetition maximum (1RM,) performed for 8 to 15 repetitions for two sets each, with 2 to 3 min of rest after each bout. This plan will allow the person to perform 6 to 10 exercises in a 20 to 30 min session. Resistance exercises can be performed maximally on 2 or 3 d/wk. These exercises should focus on the major muscle groups of the chest, shoulders, upper and lower back, abdomen, hips, and legs. The primary acute benefit of the prescribed resistance program is to improve muscle endurance; the secondary benefit is to increase muscle strength. For obese individuals, the long-term benefit may be related to a higher resting metabolic rate (RMR) and protection of lean mass loss during rapid weight loss attempts.

Practical application 7.2 provides guidance for interactions with those who are overweight or obese.

Range of Motion

Obese patients may have a reduced range of motion as a result of increased fat mass surrounding joints of the body (46), in conjunction with a lack of stretching. As a result, these patients often respond slowly to changes in body position and have poor balance. Persons who are obese are also at a greater risk of low back pain and

Practical Application 7.2

CLIENT–CLINICIAN INTERACTION

Interactions with those who are overweight or obese should be sensitive to the language used. Many of these individuals do not want the terms "obese" or "fat" used when discussing their health condition. Additionally, research demonstrates that people who are overweight or obese typically have lower self-esteem than others, are less likely to be married or in a relationship, are less likely to be employed, or if employed tend to make less money than similar-aged individuals. Also, many may have psychological issues that hamper their ability to function from day to day. These might include depression and anxiety. When you are attempting to help someone lose weight, it is important to be sensitive to these issues. And if there are behavioral issues that are beyond the scope of a clinical exercise physiologist, the individual should be referred to a behavioral health specialist for assistance. This is all-important in weight loss and management efforts.

Regarding an exercise program, the exercise physiologist must be able to discuss realistic expectations regarding exercise for weight loss and to design a program that begins at a person's level of readiness and comfort, provides proper progression, and provides an avenue for regular follow-up for program adjustments to enhance adherence. The clinical exercise physiologist must be comfortable with obese clientele, especially when discussing exercise options that may not be appropriate. For instance, some patients may wish to jog, and the exercise physiologist must be prepared to discuss the issues that may confront a morbidly obese individual when jogging. Others may be reluctant to consider using a swimming pool because of issues of appearance or ease of entering and exiting the pool. Time may not be available to build rapport, so the exercise physiologist must be able to discuss these issues almost immediately with any patient.

joint-related osteoarthritis because of their condition (35, 121). Therefore, range of motion may improve spontaneously with weight loss (62). Still, to the degree possible, patients should perform a brief flexibility routine focused on the legs, lower back, and arm and chest regions. Normal flexibility routines (see chapter 5) are recommended as tolerated.

EXERCISE TRAINING

Low fitness levels add to the risk of mortality in overweight and obese people (129). In fact, low fitness is similar in risk to diabetes, hypertension, elevated cholesterol, and smoking. Ample evidence indicates that regular aerobic exercise training improves physical functioning, independent of weight loss, in those who are obese or overweight (23). Although their data are not specific to the obese and overweight individual, Blair and colleagues have reported that men who increase their cardiovascular fitness level by one quartile have significant reductions in their long-term morbidity and mortality profile. Gulati and colleagues report similar benefits for women (13, 54).

Exercise training is likely most important in the weight maintenance process of weight control (81). Data from the National Weight Control Registry project indicate that regular exercise training expending more than 2,000 kcal per week is a strong predictor of long-term weight loss maintenance (65). Walking has been shown to be effective for long-term (2 yr) weight loss maintenance in women (41).

Table 7.9 briefly reviews the anticipated responses to exercise training in obese individuals.

Cardiovascular Exercise

The focus of exercise training in the overweight and obese person should initially be on caloric expenditure, which is best achieved with aerobic-based training. Although dietary changes appear to be more effective than structured exercise alone at reducing body weight, data suggest that exercise alone can result in weight loss similar to caloric restriction (104). The amount of exercise required, however, is likely beyond the capacity of most obese individuals with no recent exercise history (e.g., 700 kcal per day for 3 mo). However, a recent study of obese cardiac rehabilitation patients used a high caloric expenditure exercise routine, compared to standard cardiac rehabilitation, to assess the effects on weight loss (3). Energy expenditure in the groups was 3,000 to 3,500 versus 700 to 800 kcal/wk. The high-expenditure group lost significantly more weight (8.2 ± 4 vs. 3.7 ± 5 kg) over 5 mo of intervention. Benefits in fat mass loss, waist reduction, and a lower prevalence of metabolic syndrome were also noted. Importantly, this group of patients with a low initial fitness level and little exercise experience was apparently able to tolerate and adhere to this type of exercise regimen.

Some data support a synergistic effect of combined caloric restriction and exercise for weight loss in women (33). This finding, however, has not been universally replicated (125). Jakicic and colleagues reported that

Table 7.9 Exercise Training Review

Cardiorespiratory endurance	Skeletal muscle strength	Skeletal muscle endurance	Flexibility	Body composition
Improved peak $\dot{V}O_2$ similar to that of normal-weight individuals (about 15-30% on average) In some cases, extremely deconditioned individuals can improve peak exercise capacity upwards of 50% to 100%	• Evidence that aerobic and resistance training maintains or enhances skeletal muscle strength • Does not affect the amount or rate of weight loss	Likely improvement similar to that expected in normal-weight individuals	• No randomized studies available that assess flexibility training in the obese • Expect normal increases in flexibility with ROM training, but may be limited by excessive fat tissue	• Standard exercise training alone does not result in a significant reduction in weight • Up to 700 kcal per day expenditure from exercise will reduce weight • Evidence that aerobic and resistance training may preserve lean mass during weight loss attempts

exercise intensity (moderate or vigorous) and duration (moderate or high) adjusted to expend between 1,000 and 2,000 kcal per week, combined with calorie intake restriction, did not have an effect on the amount of weight lost over 12 mo in a group of obese women (63). Others have demonstrated that diet combined with increased daily lifestyle physical activity may be as effective as a program of diet and intentional exercise (7).

The expectations for cardiovascular exercise training with regard to hemodynamic and other physiologic system adaptations are similar to those for people who are not obese or overweight. Increases in relative peak $\dot{V}O_2$ of 28% have been reported following a cardiovascular training program (130). Blood pressure declines acutely following exercise, and accumulation of exercise over time may chronically reduce blood pressure (94). These blood pressure effects appear to be independent of weight loss (34). An important adaptation to exercise in the overweight or obese population is an increase in insulin sensitivity and improved glucose metabolism (32, 99). But reports give conflicting evidence about whether exercise alone, in the absence of weight loss, results in enhanced insulin sensitivity in women (99). The Diabetes Prevention Project demonstrated that lifestyle intervention of weight loss and exercise training was superior to medical therapy for preventing or delaying the development of diabetes in obese individuals with prediabetes (66). Favorable changes in lipoprotein profiles, vascular function, inflammatory biomarkers, and the risk of blood clot development may also be positive effects of exercise training, independent of weight loss, in obese people (28, 90, 127). But the long-term effects of these exercise-related responses on the development of disease have not been studied.

Resistance Exercise

Resistance training is suggested for those losing weight, particularly large amounts, to offset potential losses in lean mass. Recently Weiss and colleagues reported on a group of overweight individuals who lost approximately 10% of their initial body weight through either caloric restriction or cardiorespiratory training (6 d/wk for 60 min, about 320 kcal expended per day) (130). Thigh skeletal muscle mass and knee extensor strength declined in both groups following the intervention, but to a significantly greater degree in the caloric restriction group, suggesting a beneficial effect of exercise training to ameliorate loss of lean mass during weight loss. The use of resistance training to stimulate increases in lean mass may also be beneficial for reducing lean mass losses (10). No evidence suggests that resistance training during caloric restriction for weight loss results in additive weight loss (31). A study that compared aerobic training and aerobic training plus resistance training in patients with type 2 diabetes showed that aerobic training plus resistance training provided no additional weight loss benefit (73).

Range of Motion Exercise

Little research has dealt with weight loss and changes in range of motion. In a nonrandomized study, 10 obese adults had lost 27.1 ± 5.1% of their weight (34 ± 9.4 kg) by 7 mo and an additional 6.5 ± 4.2% (8.2 ± 6 kg) at 13 mo following bariatric surgery (62). Assessment of range of motion showed significant increases for the hip during the swing phase of walking, for maximal knee flexion, and for ankle joint function toward plantar flexion. This was also associated with a greater swing time during self-paced walking with increased stride length and speed. These findings suggest that weight loss in the morbidly obese (BMI = 43 ± 6.5 kg · m^{-2}) can positively affect range of motion and that this may have positive affects on ambulation.

Clinical Exercise Pearls

- Exercise alone is not effective for significant weight loss. Dietary changes are a must for weight loss.
- Exercise is the best predictor of long-term weight loss maintenance. People must approach 60+ min daily for best maintenance results.
- Obesity is a risk factor for cardiovascular disease and many other chronic diseases. This must be considered in those beginning an exercise program.
- Consider aggressive techniques for weight loss in the extremely obese. These include complete meal replacement and surgery and may also include prolonged bouts of exercise when appropriate.

CONCLUSION

Exercise and behavioral modification are cornerstones of sound weight management programs. Clinical exercise physiologists, especially those who have a strong background in behavioral or lifestyle counseling, are playing an increasing role in the primary prevention, treatment phase, and secondary prevention of overweight and obesity. Patients benefit from seeking out exercise physiologists who are providing services in weight management programs. Given the number of overweight individuals in the United States and worldwide, it is important to continue to advance clinical exercise physiologists to work with overweight patients.

Key Terms

body mass index (BMI) (p. 113)
ghrelin (p. 118)
leptin (p. 118)
metabolic syndrome (p. 119)
neuropeptide Y (NPY) (p. 118)
obesity (p. 113)
overweight (p. 113)
set-point theory (p. 118)

CASE STUDY

MEDICAL HISTORY AND PHYSICAL EXAMINATION

Ms. KB (age 47) came to an orientation session for a clinical weight management program weighing 368 lb (167 kg). She is 65 in. (165 cm) tall, and her waist measurement was 48 in. (122 cm). She is married and works full-time in a sedentary job. She has a goal to weigh under 200 lb (91 kg). She indicated that she would be disappointed if she achieved a weight loss of "only" 100 lb (46 kg). She has the following comorbidities: diabetes, high blood pressure, and lower body joint pain. Medications include glyburide, metformin, hydrochlorothiazide, and protonix. Her baseline HR was 68 beats · min^{-1}, and BP was 130/82 mmHg. She reports previous success in weight loss with a variety of methods but has gained the weight back over time. She is currently at her highest adult weight. She has difficulty exercising, primarily because of lack of time and joint pain, but is able to walk occasionally for 30 min. The physical exam was unremarkable.

DIAGNOSIS

Morbid obesity with secondary comorbid conditions of diabetes, high blood pressure and possible arthritis.

EXERCISE TEST RESULTS

A modified Bruce protocol test was performed on the patient within the past several months because of a complaint of atypical chest pain to her primary care physician. Results indicated a peak work capacity estimated at 10 METs (12 min test), a normal HR, excessive BP increase (peak BP = 246/116 mmHg), and no indication of myocardial ischemia. She complained of lower back and left knee pain when the treadmill was elevated.

EXERCISE PRESCRIPTION

The patient is not interested in exercising regularly because of a lack of time and the pain associated with walking. Time was spent discussing the benefits of exercise, and she was provided with a basic plan.

DISCUSSION

The patient self-selected a diet based on reducing caloric intake by 500 kcal per day using foods purchased at grocery stores. The diet is based on the exchange system and focuses on low-fat eating and portion control. The patient did not do well on this initial diet plan and was not exercising over the first 2 mo.

DISCUSSION QUESTIONS

1. Considering the diet and exercise prescription information, how might this be modified so the patient is successful?

2. Discuss Ms. KB's goal and the weight goal she would consider disappointing.
3. Given the patient's exercise limitations, how might you go about developing and implementing an exercise plan for her?

   The patient switched to a complete meal replacement plan and followed the advice of the exercise physiologist with respect to exercise. She lost 188 lb (85 kg) over the next 9 mo.
4. Given the previous information, discuss why this patient was more successful after the changes in diet and exercise plan.
5. What do you think might have occurred with respect to Ms. KB's comorbid medical conditions as a result of weight loss?
6. What behavioral issues and plan would be prudent to discuss and develop for this patient? Consider this in light of her apathy toward exercise and the potential eating habits of someone who has weighed 368 lb.
7. Discuss the patient's exercise routine and its importance to weight management.
8. What other plans might be put in place to help this patient achieve long-term weight loss maintenance?

Chapter 8

Hypertension

Aashish S. Contractor, MD, MEd
Terri L. Gordon, MPH
Neil F. Gordon, MD, PhD, MPH

Hypertension is recognized as the most common modifiable cardiovascular disease risk factor. Because it increases with aging and given the emergence of the baby boomer population to elderly status (65 yr or older), public health officials acknowledge that managing blood pressure will become a major priority for preventing future cardiovascular disease and other comorbidities. A complete evaluation of all risk factors, including accurate and repetitive blood pressure measurements in patients, is essential. An appropriate treatment course can vary depending on whether the patient has existing coronary artery disease or not. Lifestyle modifications may be the only treatment necessary in some patients. Nevertheless, a clinical team approach may be warranted to ensure that all risk factors are controlled.

DEFINITION

Hypertension is defined as a transitory or sustained elevation of systemic arterial blood pressure (BP) to a level likely to induce cardiovascular damage or result in other adverse consequences (52). The seventh report of the Joint National Committee on Prevention, Detection, Evaluation, and Treatment of High Blood Pressure (JNC 7) defines hypertension as having a resting **systolic** BP of 140 mmHg or greater, or a resting **diastolic** BP of 90 mmHg or greater, or taking antihypertensive medication, or any combination of these (19). The report also defines an additional class of untreated patients with systolic BP ranging from 120 to 139 mmHg or diastolic BP ranging from 80 to 89 mmHg (or both) as prehypertensive, or at heightened risk of developing hypertension in the future (19). The classification of BP for adults (age 18 or older) is based on the average of two or more properly measured, seated BP readings on each of two or more office visits and is shown in table 8.1.

Hypertension increases the risk for cardiovascular disease, stroke, heart failure, peripheral arterial disease, and chronic kidney disease. Indeed, BP readings as low as 115/75 mmHg are associated with a higher than desirable risk for ischemic heart disease and stroke, and the risk for cardiovascular disease doubles for every increment increase in systolic BP of 20 mmHg or diastolic BP of 10 mmHg (11, 19, 36).

In approximately 90% of cases, the **etiology** of hypertension is unknown, and it is called essential, **idiopathic**, or primary hypertension. Secondary hypertension is **systemic** hypertension with a known cause (36, 55). Although essential hypertension and secondary hypertension are the major classifications of hypertension, several other descriptive terms are used to define various types of hypertension. Isolated systolic hypertension is defined as systolic BP of 140 mmHg or more and diastolic BP of less than 90 mmHg. Malignant hypertension is the syndrome of markedly elevated BP associated with

Table 8.1 Classification and Management of Blood Pressure for Adults

BP classification	SBP (mmHg)*	DPB (mmHg)*	Lifestyle modification	INITIAL DRUG THERAPY: Without compelling indication*	INITIAL DRUG THERAPY: With compelling indications[†]
Normal	<120	and <80	Encourage	No antihypertensive drug indicated	Drugs for compelling indications[‡]
Prehypertension	120-139	or 80-89	Yes		
Stage 1 hypertension	140-159	or 90-99	Yes	Antihypertensive drugs indicated	Drugs for compelling indications[‡] Other antihypertensive drugs as needed
Stage 2 hypertension	≥160	or ≥100	Yes	Antihypertensive drugs indicated Two-drug combination for most[†]	

Note. DBP = diastolic blood pressure; SBP = systolic blood pressure.

*Treatment determined by highest BP category.

[†]Initial combined therapy should be used cautiously in those at risk for orthostatic hypotension.

[‡]Compelling indications include heart failure, postmyocardial infarction, high coronary artery disease risk, diabetes, chronic kidney disease, and recurrent stroke prevention. Treat patients with chronic kidney disease or diabetes to BP goal of <130/80 mmHg.

Data from Chobanian et al. 2003.

papilledema. In cases of malignant hypertension, the systolic BP and diastolic BP are usually greater than 200 mmHg and 140 mmHg, respectively. White-coat hypertension is the situation in which a person's BP is elevated when measured by a physician or other health care personnel but is normal when measured outside a health care setting.

SCOPE

For statistical purposes, the American Heart Association defines hypertension as having a resting systolic BP of 140 mmHg or greater, having a resting diastolic BP of 90 mmHg or greater, taking antihypertensive medication, or being told by a physician or health professional on at least two occasions that one has high BP. Using this definition, hypertension affects about 76 million Americans in the age group of 20 and older (7).

Over one-third of adult Americans have hypertension, and another 30% or so have prehypertension (7, 26). The prevalence of hypertension increases substantially with age; and estimates are that among adults older than age 50, the lifetime risk of developing hypertension approaches 90% (12, 54). According to the American Heart Association's 2011 statistical update, the prevalence of hypertension among the various ethnic populations and sexes living in the United States is as follows (7):

- For non-Hispanic whites 20 yr or older, 33.9% of men and 31.3% of women
- For non-Hispanic blacks 20 yr or older, 43.0% of men and 45.7% of women
- For Mexican Americans 20 yr or older, 27.8% of men and 28.9% of women

On average, blacks have higher BP than nonblacks, as well as increased risk of BP-related complications, particularly stroke and kidney failure (7, 19). Awareness, treatment, and control of hypertension have increased substantially in recent decades (table 8.2), although there is still considerable room for improvement. In this respect, data from the National Health and Nutrition Examination Survey (NHANES) 2005 to 2008 showed that of those with hypertension who were 20 or older, 79.6% were aware of their condition, 70.9% were under current treatment, 47.8% had their hypertension under control, and 52.2% did not have it controlled (7).

From 1997 to 2007, the death rate from hypertension increased 9.0% and the actual number of deaths rose 35.6%. For 2007, the estimated direct and indirect cost of hypertension was $43.5 billion (7).

Table 8.2 Trends in the Awareness, Treatment, and Control of High Blood Pressure in Adults, United States, 1976-2008

	NATIONAL HEALTH AND NUTRITION EXAMINATION SURVEY, %				
	1976-1980	1988-1991	1991-1994	1999-2000	2005-2008
Awareness	51	73	68	70	79.6
Treatment	31	55	54	59	70.9
Control*	10	29	27	34	47.8

*Systolic blood pressure below 140 mmHg and diastolic blood pressure below 90 mmHg and on antihypertensive medication.

Based on Chobanian et al. 2003; V.L. Roger et al., 2011, "Heart disease and stroke statistics - 2011 update: A report from the American Heart Association," *Circulation* 123: e18-e209.

PATHOPHYSIOLOGY

A variety of systems are involved in the regulation of BP, including renal, hormonal, vascular, peripheral, and central adrenergic systems. Blood pressure is the product of cardiac output (CO) and total peripheral resistance (TPR): BP = CO × TPR. The pathogenic mechanisms leading to hypertension must lead to increased TPR, to increased CO, or to both. Hypertension is frequently associated with a normal CO and elevated TPR (36).

Essential hypertension tends to cluster in families and represents a collection of genetically based diseases and syndromes with a number of underlying inherited biochemical abnormalities. Factors considered important in the genesis of essential hypertension include genetic factors, salt sensitivity, inappropriate renin secretion by the kidneys, and the environment. Environmental factors that have been implicated in the development of hypertension include obesity, physical inactivity, and excessive alcohol and salt intake.

Although secondary hypertension forms a small percentage of cases of hypertension, recognizing these cases is important because they can often be improved or cured by surgery or specific medical therapy. Nearly all the secondary forms of hypertension are renal or **endocrine** hypertension. Renal hypertension is usually attributable to a derangement in the renal handling of sodium and fluids, leading to volume expansion or an alteration in renal secretion of vasoactive materials that results in a systemic or local change in arteriolar tone. Endocrine hypertension is usually attributable to an abnormality of the adrenal glands.

Untreated hypertension leads to premature death; the most common cause is heart disease, followed by stroke and renal failure. Hypertension damages the **endothelium**, which predisposes the individual to **atherosclerosis** and other **vascular pathologies**. Increased **afterload** on the heart caused by hypertension may lead to left **ventricular hypertrophy** and is an important cause of **heart failure**. Hypertension-induced vascular damage can lead to stroke and transient ischemic attacks as well as end-stage renal disease. A meta-analysis of nine studies, involving 420,000 people, revealed that prolonged increases in usual diastolic BP of 5 and 10 mmHg were associated with at least 34% and 56% increases in stroke risk and with at least 21% and 37% increases in coronary heart disease (CHD) risk, respectively (38). Although both systolic and diastolic BP are important, in persons older than age 50 the systolic BP is a more important cardiovascular disease risk factor than diastolic BP (12, 19).

CLINICAL CONSIDERATIONS

The clinical evaluation of a person with hypertension should be aimed at assessing secondary forms of hypertension, assessing factors that may influence therapy, determining whether target organ damage is present, and identifying other risk factors for cardiovascular disease. Establishing an accurate pretreatment baseline BP is also vital.

Signs and Symptoms

Hypertension is often referred to as the silent killer because most patients do not have specific symptoms related to their high BP. Headache is popularly considered a symptom of hypertension, although it occurs only in severe hypertension; most commonly, such headaches are localized to the **occipital** region and are present on awakening in the morning. Dizziness, palpitations, and easy fatigability are other complaints that have been linked to elevated BP. Some symptoms, such as **epistaxis**, **hematuria**, and blurring of vision, may be attributable to underlying vascular disease.

History and Physical Exam

A thorough medical history should assess the duration and severity of hypertension, symptoms, and signs, if any. The history should include questions concerning the individual's risk factors for CHD and stroke, as well as symptoms and signs of CHD, heart failure, renal disease, and endocrine disorders. Information should be obtained about the past and present use of medications and about lifestyle habits. A comprehensive medical history will help in the treatment of primary hypertension and in the diagnosis and treatment of causes of secondary hypertension. Common medications for hypertension are provided in table 8.3.

Diagnostic Testing

An accurate BP reading is the most important part of the diagnostic evaluation. A person is classified as hypertensive based on the average of two or more BP readings, at each of two or more visits after an initial screening visit.

The auscultatory method of BP measurement with a properly calibrated and validated instrument should be used. Persons should be seated quietly for at least 5

Table 8.3 Pharmacology of Most Commonly Used Antihypertensive Agents

Medication name and class	Primary effects	Exercise effects	Special considerations
ACE (angiotensin-converting enzyme) inhibitors and angiotensin II receptor blockers (ARBs)	Reduce BP; compelling indications when used for antihypertensive therapy include heart failure, postmyocardial infarction, high risk for CHD, diabetes, chronic kidney disease, and recurrent stroke prevention	Decrease BP; may increase exercise capacity in patients with heart failure	Typically have minimal side effects
β-blockers	Reduce BP; compelling indications when used for antihypertensive therapy include heart failure, postmyocardial infarction, high risk for CHD, and diabetes	Decrease BP; decrease heart rate; may decrease exercise capacity (especially nonselective agents) in patients without angina; may increase exercise capacity in patients with angina	May blunt exercise training-induced lowering of triglycerides and increase in HDL cholesterol; may adversely affect thermoregulatory function; may increase predisposition to hypoglycemia in patients taking insulin or insulin secretagogues
Calcium channel blockers	Reduce BP; compelling indications when used for antihypertensive therapy include high risk for CHD and diabetes	Decrease BP; may increase (dihydropyridines) or decrease (non-dihydropyridines) heart rate; may increase exercise capacity in patients with angina	May predispose to postexertion hypotension (an adequate cool-down may be especially important)
Diuretics	Reduce BP; compelling indications when used for antihypertensive therapy include heart failure, high risk for CHD, diabetes, and recurrent stroke prevention	Decrease BP; may increase exercise capacity in patients with heart failure	May result in serum potassium derangements and thereby accentuate the risk for exercise-induced dysrhythmias; may adversely affect thermoregulatory function
Other vasodilators, including antiadrenergic agents without selective blockade, α-adrenergic blockers, and nonadrenergic agents	Reduce BP	Decrease BP; may increase or decrease (antiadrenergic agents without selective blockade) heart rate	May predispose to postexertion hypotension (an adequate cool-down may be especially important)

min in a chair (rather than on an exam table), with both feet placed on the floor and the arm supported at heart level. Caffeine, exercise, and smoking should be avoided for at least 30 min before measurement. Measurement of BP in the standing position is indicated periodically, especially in those at risk for postural hypotension. An appropriate-sized cuff (cuff bladder encircling at least 80% of the arm) should be used to ensure accuracy. At least two measurements should be made (19). Systolic BP is the point at which the first of two or more Korotkoff sounds is heard (phase 1), and diastolic BP is the point before the disappearance of Korotkoff sounds (phase 5).

Measurement of BP at home by the patient (or family) or by automated ambulatory monitoring helps verify the diagnosis of hypertension and response to treatment. Advantages of self-measurement of BP include distinguishing sustained hypertension from white-coat hypertension. Table 8.4 provides follow-up recommendations based on the initial set of BP measurements.

Laboratory tests for hypertension should include urinalysis, **hematocrit**, blood chemistry (including sodium, potassium, creatinine, and lipid profile), and an electrocardiogram (10). These routine tests will help determine the presence of target organ damage and other risk factors as well as helping to guide certain therapeutic decisions. From an exercise prescription point of view, the electrocardiogram is an important test because it may reveal the presence of arrhythmias and baseline ST-segment changes. Other tests that can be of value, depending on indications, include **creatinine clearance**; microscopic urinalysis; chest X-ray; echocardiogram; hemoglobin A1c or fasting plasma glucose or both; and serum calcium, phosphate, and uric acid.

Exercise Testing

From an exercise prescription perspective, recommendations for individuals with hypertension vary depending on their actual BP and the presence of other cardiovascular disease risk factors, target organ disease, or clinical cardiovascular disease. On the basis of published guidelines from the American College of Sports Medicine (ACSM) and other expert groups, the following would appear to be prudent recommendations (2, 3, 5, 28, 30):

- Hypertensive individuals with an additional CHD risk factor, males older than 45 yr of age, and females older than 55 yr of age should perform an exercise test with electrocardiogram monitoring before starting a vigorous-intensity (i.e., ≥60% $\dot{V}O_2$ reserve [$\dot{V}O_2R$]) exercise program.
- Irrespective of the intensity of exercise training, the following groups of hypertensive individuals should perform an exercise test with electrocardiogram monitoring before commencing an exercise program: individuals with known cardiovascular, pulmonary, or metabolic disease; individuals with target organ disease (e.g., left ventricular hypertrophy and retinopathy); and individuals with one or more major symptoms or signs suggestive of cardiovascular, pulmonary, or metabolic disease.

Contraindications

The American College of Cardiology, American Heart Association, and ACSM guidelines on exercise testing state that severe arterial hypertension, defined as systolic

Table 8.4 Recommended Length of Follow-Up Based on BP Reading

Initial blood pressure (mmHg)*	Recommended BP recheck†
Normal	Recheck in 2 yr
Prehypertension	Recheck in 1 yr**
Stage 1 hypertension	Confirm within 2 mo**
Stage 2 hypertension	Evaluate or refer to source of care within 1 mo. For those with higher pressures (e.g., >180/110 mmHg), evaluate and treat immediately or within 1 wk depending on clinical situation and complications.

*If systolic and diastolic categories are different, follow recommendations for shorter-time follow-up (e.g., 160/86 mmHg should be evaluated or referred to source of care within 1 mo).

†Modify the scheduling of follow-up according to reliable information about past BP measurements, other cardiovascular risk factors, or target organ disease.

**Provide advice about lifestyle modifications.

Based on Chobanian et al., 2003.

BP greater than 200 mmHg or diastolic BP greater than 110 mmHg at rest, is a relative contraindication to exercise testing (2, 5).

Additional Recommendations and Anticipated Responses

Standard exercise testing methods and protocols may be used for persons with hypertension (5, 30). Before the hypertensive person undergoes graded exercise testing, obtaining a detailed health history and baseline BP in both the supine and standing positions is important. Certain medications, especially β-blockers, affect BP at rest and during exercise and may affect the heart rate response to exercise. When testing is conducted for diagnostic purposes, BP medication may be withheld before testing with physician approval. In contrast, the person should be taking his usual antihypertensive medications when exercise testing is performed for the purpose of exercise prescription.

Abnormal BP Response

Blood pressure is the product of CO and TPR. Normally during exercise, the TPR decreases, but the increase in CO more than compensates for the decrease in TPR and systolic BP increases. Diastolic BP usually remains the same or may decrease slightly because of the decrease in TPR. But hypertensive patients often experience an increase in diastolic BP both during and after exercise. They are often unable to reduce TPR to the same extent as normotensive people (those with normal BP). Impaired endothelial function in the early stage and, later, a reduced lumen-to-wall thickness could be responsible for the increased resistance during exercise (29).

Indications for Terminating a Graded Exercise Test

A significant decrease in systolic BP (>10 mmHg) from baseline systolic BP despite an increase in workload is an indication for terminating an exercise test. An excessive increase in BP, defined as a systolic BP greater than 250 mmHg or diastolic BP greater than 115 mmHg, is also an indication for terminating an exercise test (5).

Predictive Value of BP Response

An exaggerated BP response to graded exercise testing in normotensive people has been associated with an accentuated future risk of developing hypertension and cardiovascular disease. An exaggerated response can be arbitrarily defined as a level of BP higher than that expected for the individual being tested. Data from the Framingham Heart Study (50) showed that an exaggerated diastolic BP response to exercise is predictive of risk for new-onset hypertension in normotensive men and women and that an elevated recovery systolic BP is predictive of the future development of hypertension in men. The Framingham Heart Study is one of the few studies on the predictive value of exercise BP that included large numbers of men and women. After multivariate adjustment, an exaggerated diastolic BP response during stage 2 of the Bruce treadmill protocol was observed to have the strongest association with new-onset hypertension in both men (odds ratio = 4.16) and women (odds ratio = 2.17). The study defined an exaggerated exercise BP response as either a systolic or diastolic BP above the 95th percentile of sex-specific, age-predicted values during stage 2 of the Bruce protocol. More recent data from the Framingham Heart Study also suggest that the diastolic BP response to exercise may be a better predictor of new-onset cardiovascular disease than the systolic BP response to exercise (41).

Studies at the Cooper Clinic (42) and Mayo Clinic (1) revealed that an exaggerated exercise BP response had an odds ratio of 2:4 for predicting future hypertension. The subjects were mostly men, and the average follow-up time was about 8 yr. The Mayo Clinic study also found that exercise hypertension was a significant predictor for total cardiovascular events but not for death or any individual cardiovascular event.

In a review article, Benbassat and Froom (15) found that the prevalence of hypertension on follow-up among normotensive subjects with a hypertensive response to exercise testing was 2.06 to 3.39 times higher than that among subjects with a normotensive response. But the predictive value was limited because 38.1% to 89.3% of those with a hypertensive response to exercise did not have hypertension on follow-up, and a normotensive response only marginally reduced the risk of future hypertension.

After 17 yr of follow-up in 4,907 men, Filipovsky and colleagues (27) found that the exercise-induced increase of systolic BP was a risk factor for death from cardiovascular as well as noncardiovascular causes independent of resting BP. Similar findings have been shown in other studies (44). Some studies, however, showed no additional prognostic information regarding total mortality rates and cardiovascular events from exercise BP readings. According to the ACSM, mass exercise testing is not advocated to identify those at high risk for developing hypertension in the future because of an exaggerated exercise BP response (3).

Table 8.5 provides a review of exercise testing considerations.

Table 8.5 Exercise Testing Review

Test type	Mode	Protocol specifics	Clinical measures	Clinical implications	Special considerations
Cardiovascular	Aerobic	Cycle (ramp protocol 17 W · min^{-1}; staged protocol 25-50 W/3 min stage); treadmill (1-2 METs/3 min stage)	12-lead ECG; heart rate; systolic and diastolic BP; rate–pressure product; RPE; respired gas analysis	Evaluate for untoward symptoms, abnormal BP responses, arrhythmias, and myocardial ischemia	Use standard endpoints for test termination, including systolic BP above 250 mmHg or diastolic BP above 115 mmHg; cardioactive medications should be taken at usual time relative to exercise session
Strength	Free weights or resistance machines	Determine 1RM or maximal voluntary contraction; 1RM may be estimated from a higher RM (e.g., 5RM)	1RM	Evaluate for untoward symptoms and abnormal BP responses; if ECG is monitored, evaluate for arrhythmias and myocardial ischemia	Observe for exaggerated pressor response (i.e., systolic BP above 250 mmHg or diastolic BP above 115 mmHg)
Range of Motion	As for otherwise healthy persons	As for otherwise healthy persons	As for otherwise healthy persons	As for otherwise healthy persons	As for otherwise healthy persons

Treatment

The goal of prevention and management of hypertension is to reduce morbidity and mortality rates by the least intrusive means possible. This goal may be achieved through lifestyle modification alone or in combination with pharmacological treatment. People with an untreated systolic BP of 120 to 139 mmHg or diastolic BP of 80 to 89 mmHg or both should be considered prehypertensive and also require lifestyle intervention.

The BP goal is <130/80 mmHg in patients with hypertension and diabetes or chronic kidney disease and <140/90 mmHg in other patients with hypertension (19, 51). Because most patients with hypertension, especially those older than age 50, will reach the diastolic BP goal after systolic BP is at the goal level, JNC 7 recommends that the primary focus be on achieving the systolic BP goal. Figure 8.1 depicts the algorithm recommended by JNC 7 for the treatment of hypertension (19).

Lifestyle modification (table 8.6) helps in controlling BP as well as other risk factors for cardiovascular disease (19, 28, 32). For example, a review of randomized controlled trials of over 6 mo duration analyzing the effect of weight reduction in reducing BP showed a decrease of 5.2/5.2 mmHg and 2.8/2.3 mmHg in hypertensive and normotensive participants, respectively (25). Weight reduction enhances the effects of antihypertensive medications and positively affects other cardiovascular risk factors, such as diabetes and **dyslipidemia**.

Similarly, a more recent meta-analysis of lifestyle interventions in BP management showed statistically significant reductions in systolic BP/diastolic BP of 5.0/3.7 mmHg when people followed an improved diet (24). Most of these diets included a target of weight reduction. The Dietary Approaches to Stop Hypertension (DASH) eating plan is a diet rich in fruits, vegetables, and low-fat dairy products with reduced content of dietary sodium, cholesterol, and saturated and total fat. The plan has been shown to reduce BP by 8 to 14 mmHg (6, 49).

Patients should be questioned in detail about their current alcohol consumption, because excessive alcohol consumption is a risk factor for high BP. The JNC 7 report recommends that men who drink alcohol should be counseled to limit their daily intake to no more than 1 oz (30 ml) of ethanol and that women and lighter-weight individuals should be told not to exceed 0.5 oz (15 ml). These quantities are equivalent to two drinks per day in most men and one drink per day in women. A drink is 12

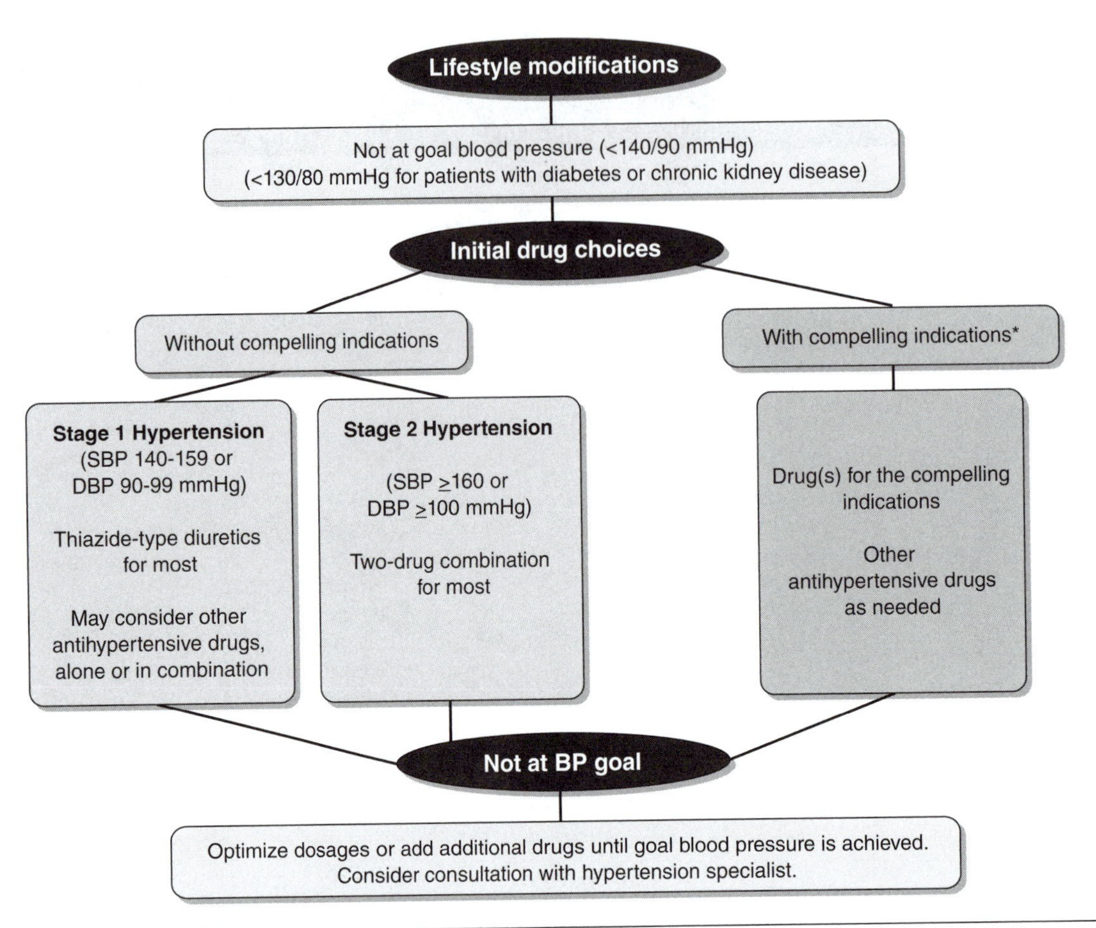

Figure 8.1 The algorithm recommended by JNC 7 for the treatment of hypertension. DBP = diastolic blood pressure; SBP = systolic blood pressure. *Compelling indications include heart failure, postmyocardial infarction, high coronary heart disease risk, diabetes, chronic kidney disease, and recurrent stroke prevention. Treat patients with chronic kidney disease or diabetes to BP goal of <130/80 mmHg.

Reprinted from A.V. Chobanian et al., 2003, "The Seventh Report of the Joint National Committee on prevention, detection, evaluation, and treatment of high blood pressure: The JNC 7 Report," *Journal of American Medical Association* 289: 2560-2572.

oz (360 ml) of beer, 5 oz (150 ml) of wine, or 1.5 oz (45 ml) of 80-proof liquor (57).

Epidemiological data demonstrate a positive association between sodium intake and level of BP. Patients with essential hypertension may be classified as salt sensitive and salt resistant, based on the absolute changes in BP that originate from dietary salt intake (8, 48). African Americans, older people, and patients with hypertension or diabetes are more sensitive to changes in dietary sodium than are others in the general population. A review of randomized controlled trials of 6 mo or longer duration in adults over the age of 45 found a small but statistically significant effect of lowered BP through salt reduction. Salt reduction resulted in pooled net systolic/diastolic BP changes of 2.9/2.1 mmHg in hypertensive individuals and 1.3/0.8 mmHg in normotensive individuals (25). The level of BP reduction was related to the level of salt reduction. A moderate sodium restriction to no more than 100 mmol per day (2,400 mg of sodium) is recommended in the JNC 7 report, whereas 1,500 mg per day is the upper level of sodium intake more recently recommended by the American Heart Association (8, 19). The average American sodium consumption is more than 4,100 mg a day in men and 2,750 mg per day in women (19).

The decision to initiate pharmacological therapy should be guided by the degree of BP elevation, the presence of target organ damage, and the presence of clinical cardiovascular disease or other cardiovascular risk factors (table 8.1 and figure 8.1). The presence of cardiovascular risk factors is assessed during the initial evaluation of the patient with hypertension. Their presence independently modifies the risk for future cardiovascular disease. After the clinician has determined the person's BP and the presence of risk factors, target organ damage, and clinical cardiovascular disease, the person's risk group can be

Table 8.6 Lifestyle Modifications for Hypertension*

Modification	Recommendation	Approximate systolic BP reduction (range)†
Weight reduction	Maintain normal body weight (body mass index 18.5-24.9 kg · m^{-1}).	5-20 mmHg/10 kg weight loss
Adoption of DASH eating plan	Consume a diet rich in fruits, vegetables, and low-fat dairy products with reduced content of saturated and total fat.	8-14 mmHg
Dietary sodium reduction	Reduce dietary sodium intake to no more than 100 mmol/d (2.4 g sodium or 6 g sodium chloride). Note that 1,500 mg per day is now the upper level of sodium intake recommended by the American Heart Association.	2-8 mmHg
Physical activity	Engage in regular aerobic physical activity such as brisk walking (at least 30 min/d, most days of the week).	4-9 mmHg
Moderation of alcohol consumption	Limit consumption to no more than two drinks (e.g., 24 oz [720 ml] beer, 10 oz [300 ml] wine, or 3 oz [90 ml] 80-proof whiskey) per day in most men and to no more than one drink per day in women and lighter-weight persons.	2-4 mmHg

DASH: Dietary Approaches to Stop Hypertension.

*For overall cardiovascular risk reduction, stop smoking.

†The effects of implementing these modifications are dose and time dependent and could be greater for some individuals.

From Chobanian et al. 2003.

determined. This classification into a risk group helps guide therapeutic decisions.

Most patients with hypertension who require drug therapy in addition to lifestyle modification need two or more antihypertensive medications to achieve goal BP. If BP is >20/10 mmHg above the goal, consideration should be given to starting antihypertensive therapy with two agents, one of which should usually be a thiazide-type diuretic. Thiazide-type diuretics should be used in drug treatment for most patients with uncomplicated hypertension, either alone or combined with drugs from other classes. Certain high-risk conditions are compelling indications for the initial use of other antihypertensive drug classes. Compelling indications include heart failure (diuretics, β-blockers, ACE inhibitors or ARBs, and aldosterone antagonists), postmyocardial infarction (β-blockers, ACE inhibitors or ARBs, and aldosterone antagonists), patients at high risk for CHD (diuretics, β-blockers, ACE inhibitors or ARBs, and calcium channel blockers), diabetes (diuretics, β-blockers, ACE inhibitors or ARBs, and calcium channel blockers), chronic kidney disease (ACE inhibitors or ARBs), and recurrent stroke prevention (diuretics and ACE inhibitors or ARBs). A complete listing of hypertensive medications is presented in chapter 3.

After initiation of drug therapy, most patients should return for follow-up and adjustment of medications at approximately monthly intervals until the BP goal is reached. More frequent follow-up may be needed for patients with stage 2 hypertension or with complicating comorbid conditions. Serum potassium and creatinine should be monitored at least once or twice per year. Follow-up visits can usually be at 3 to 6 mo intervals after BP is at goal and stable (19).

Resistant hypertension, defined as BP that remains above goal in spite of the concurrent use of three antihypertensive agents of different classes, is a relatively common clinical problem, involving an estimated 20% to 30% of hypertensive patients. Resistant hypertension is almost always multifactorial in etiology; and successful treatment requires appropriate lifestyle intervention, diagnosis and appropriate treatment of secondary causes of hypertension, and use of effective multidrug regimens (10).

Practical application 8.1 provides advice on interacting with patients with hypertension.

EXERCISE PRESCRIPTION

Exercise training has been recommended as one of the important lifestyle modifications for the prevention and management of hypertension (3, 5, 28, 30). When compared with active and fit individuals, those who are sedentary have a 20% to 50% increased risk of developing hypertension. Aerobic exercise training by individuals

Practical Application 8.1

CLIENT–CLINICIAN INTERACTION

Patients with hypertension are at heightened risk for atherosclerotic cardiovascular disease. Atherosclerotic cardiovascular disease is by far the leading cause of death in hypertensive patients. Therefore, besides assisting the patient with an appropriate exercise prescription, the clinician should attempt to educate the patient about atherosclerotic cardiovascular disease and its risk factors. In view of this, education also should be provided about factors that may help minimize the risk for exercise-related cardiac complications. These points include the importance of an adequate warm-up and cool-down and the warning symptoms and signs of an impending cardiac event. Drug therapy is often needed to optimize hypertension management and to facilitate cardiovascular disease risk reduction. Consequently, hypertensive patients are often receiving treatment with one or more medications. The clinician should educate the patient about the effect, if any, of specific medications on exercise performance and training. The clinician also should emphasize to the patient the importance of taking medications as prescribed and not discontinuing drug therapy without notifying her personal physician. If the clinician believes that the patient may be experiencing medication-related adverse effects, the clinician should refer the patient to her personal physician. Clinicians are likely to interact with patients on many occasions throughout the course of a year. Therefore, they will probably have an opportunity to measure a patient's BP on many occasions. When discussing the patient's BP recordings, the clinician must strike a balance between not alarming the patient about minor day-to-day fluctuations in BP and expressing appropriate concern about excessive elevations in systolic or diastolic BP. However, a clinician who believes that the patient's BP is not under adequate control should consult with the patient's physician.

who are at high risk for developing hypertension may also reduce the increase in BP that occurs with age (3, 5).

Although regular aerobic exercise (also referred to in this chapter as cardiovascular or endurance exercise) has been shown to reduce BP, the precise mechanisms responsible for this remain largely unknown. Some evidence shows that exercise training is associated with a decrease in plasma norepinephrine levels, which may be responsible for the decrease in BP (3). The kidneys play an important role in BP regulation. In this respect, exercise training may decrease BP by improving renal function in patients with essential hypertension. Another postulated mechanism is that regular physical activity causes favorable changes in arterial structure, which would presumably reduce peripheral vascular resistance (53). Hyperinsulinemia has been postulated to raise BP through renal sodium retention, sympathetic nervous activation, and induction of vascular smooth muscle hypertrophy (37, 47). Hypertension and hyperinsulinemia, along with insulin resistance, abdominal obesity, increased triglycerides, and decreased high-density lipoprotein cholesterol, often cluster together to form what has been called the metabolic syndrome. Even a single bout of exercise has a well-known insulin-like effect and dramatically increases skeletal muscle glucose transport. Exercise training increases insulin sensitivity, which can decrease serum insulin and BP (14, 28).

ACSM's recommendations (3, 5, 30) for aerobic exercise programming for patients with hypertension include the following:

- Frequency: Aerobic exercise on most, preferably all, days of the week
- Intensity: Moderate-intensity aerobic exercise, that is, 40% to <60% $\dot{V}O_2R$ or heart rate reserve (HRR) (typically, this intensity corresponds to a rating of perceived exertion of 11-14)
- Time: 30 to 60 min of continuous or intermittent (minimum of 10 min bouts) aerobic activity
- Type: Primarily aerobic exercise supplemented by resistance exercise

The frequency, duration, and intensity of aerobic exercise should be modulated to achieve a weekly energy expenditure of 700 (initial goal) to 2,000 (long-term goal) kcal per week. Patients should also be advised to increase the amount of leisure-time activity (5, 28). The clinician should be aware that β-blockers attenuate the heart rate response to exercise; β-blockers, calcium channel blockers, and vasodilators may cause postexertion hypotension (an adequate cool-down may be especially important for patients taking these medications); and certain diuretics may cause a decrease in serum potassium levels, thereby predisposing the patient to arrhythmias (28, 30).

Moderate-intensity resistance training is an important component of a well-rounded exercise program for the prevention, treatment, and control of hypertension. A recent meta-analysis that pooled data from studies published between 1996 and 2003 showed that the overall effect of resistance training was a decrease of 3.2 mmHg in systolic BP and a decrease of 3.5 mmHg in diastolic BP (17).

When performing resistance training, hypertensive patients generally should adhere to the American Heart Association's guidelines (9), which have been endorsed by the ACSM. These include the following:

- Frequency: 2 or 3 d per week
- Sets: at least one
- Repetitions (reps): 8 to 12 per set for healthy adults or 10 to 15 per set at a lower level of resistance for older (more than 50 to 60 yr of age), more frail persons or cardiac patients
- Stations or devices: 8 to 10 exercises that condition the major muscle groups

People with more marked elevations in BP (resting systolic BP ≥180 mmHg, diastolic BP ≥110 mmHg, or both) should add exercise training to their treatment only after starting appropriate drug therapy. Individuals should not be allowed to exercise on a given day if their resting systolic BP is more than 200 mmHg or diastolic BP is more than 110 mmHg. Although BP termination criteria for exercise testing are generally established at >250/115 mmHg, lower BP thresholds for termination of an exercise training session may be prudent (i.e., >220/105 mmHg) (3, 5, 30). Diuretics and β-blockers may impair thermoregulation during exercise in hot or humid environments. Those taking these medications should be well informed about signs and symptoms of heat intolerance and should know how to make prudent modifications in their exercise routine to prevent heat illness.

Strength or resistance training is not recommended as the only form of exercise training for people with hypertension. With the exception of circuit weight training, resistance training has not consistently been shown to lower BP. Resistance training is recommended as a component of a well-rounded fitness program but not when done independently. BP values of >180/110 mmHg and >160/100 mmHg are considered absolute and relative contraindications to resistance training, respectively (9).

Table 8.7 provides a summary of exercise prescription considerations and goals.

Table 8.7 Summary of Exercise Prescription

Training method	Mode	Frequency	Intensity	Duration	Progression	Goals	Special considerations
Cardiovascular	Aerobic exercise; large muscle group activities such as walking, jogging, cycling	Most, preferably all, days of the week	40-<60% $\dot{V}O_2R$	30-60 min	Progress gradually, avoiding large increases in any of the components of the exercise prescription; increase exercise duration over first 4 to 6 wk and then increase frequency, intensity, and duration (or some combination of these) to achieve recommended quantity and quality of exercise over next 4 to 8 mo	Help control BP at rest and during exercise; improve CAD risk factors; increase $\dot{V}O_2R$ and ventilatory threshold; increase caloric expenditure	Do not perform cardiovascular exercise if resting systolic BP exceeds 200 mmHg or diastolic BP exceeds 110 mmHg
Resistance	Circuit training; 8 to 10 different exercises for major muscle groups	2 or 3 d/wk	60% to 80% 1RM (lower level of resistance for older, more frail persons or cardiac patients, e.g., <40% 1RM)	At least one set of 8 to 12 reps for healthy adults or 10 to 15 reps for older, more frail persons, or cardiac patients	To increase strength, apply the progressive overload principle by using a greater resistance, performing more reps, or exercising more frequently; once satisfied with strength, adopt a maintenance program	Increase muscle strength and endurance; decrease BP at any given resistance (e.g., during lifting and carrying heavy objects)	Do not perform resistance exercise if resting systolic BP exceeds 180 mmHg or diastolic BP exceeds 110 mmHg

(continued)

Table 8.7 *(continued)*

Training method	Mode	Frequency	Intensity	Duration	Progression	Goals	Special considerations
Range of motion	Static stretches; major muscle–tendon groups	2 or 3 d/wk	Perform each stretch to the point of mild tightness without discomfort	At least four repetitions per muscle group; hold each static stretch for 15 to 60 s	Progress as tolerated to at least four repetitions per muscle group	Decrease risk of injury; enhance ability to perform activities of daily living	Stretching is most effective when the muscles are warm

EXERCISE TRAINING

Exercise training recommendations for individuals with hypertension should take into account their medical history, current BP levels, and presence of cardiovascular disease and its risk factors. Comorbid conditions such as diabetes, CHD, and heart failure should be adequately controlled before the start of exercise training. The program should include cardiovascular endurance training, resistance training, and flexibility exercises. See "Client–Clinician Interaction" for additional information relating to the exercise prescription for patients with hypertension.

Cardiovascular Exercise

Reviews of studies on the BP-lowering effects of aerobic exercise training in hypertensive people have shown somewhat differing results, perhaps because of the different study inclusion criteria and study designs. All meta-analyses, however, concluded that BP decreases significantly in response to aerobic exercise training.

Recently, Dickinson and colleagues conducted a meta-analysis of lifestyle intervention trials from 1998 to 2006 (24). Only randomized controlled trials with at least 8 wk of follow-up were included. A total of 105 studies that randomized 6,805 participants were identified. These trials showed that BP reductions occurred because of a host of lifestyle interventions. Robust statistically significant effects were found for improved diet, aerobic exercise, alcohol and sodium restriction, and fish oil supplements. With aerobic exercise, a mean reduction in systolic BP of 4.6 mmHg and a mean reduction in diastolic BP of 2.4 mmHg occurred.

The ACSM, as reflected in its position stand on exercise and hypertension, found that endurance training significantly reduced systolic BP and diastolic BP, and those reductions were greater in the hypertensive as compared with the normotensive population (3). There are, however, some concerns that long-term BP-lowering effects of exercise can be attenuated, mostly because of high levels of exercise dropout.

Research shows that 24 h ambulatory BP monitoring may be more predictive of target organ damage than casual resting measures, but nighttime BP appears to be less influenced by aerobic exercise training than daytime BP. Previous trials have led to conflicting results; and even in trials showing a significant reduction in ambulatory BP, the magnitude of reduction was generally less than that observed in casual BP (3).

Recently, it has been hypothesized that improvements in BP and plasma lipids induced by exercise training in hypertensive individuals may be genotype dependent. Hagberg and colleagues (34) found evidence to support the possibility that ACE, apoE, LPL PvuII, and Hind III genotypes may identify hypertensive individuals likely to reduce BP the most with exercise training. More recent research has identified additional genotypes that may be useful in identifying individuals who may be especially responsive to the antihypertensive effects of aerobic exercise (13).

The prevalence of modifiable CHD risk factors is higher in hypertensive individuals. Aerobic exercise training has been shown to have a beneficial effect on obesity, lipid profiles, and insulin sensitivity. Dengel and colleagues (23) showed that a 6 mo intervention of aerobic exercise and weight loss substantially reduced BP, improved insulin sensitivity by 39%, and resulted in a 50% reduction in the number of metabolic abnormalities associated with the metabolic syndrome in obese, hypertensive, middle-aged men. Brown and colleagues (18) found that an aerobic exercise program of 7 d duration improved insulin sensitivity in African American hypertensive women independent of changes in fitness levels, body composition, or body weight. Subjects performed aerobic exercise on the treadmill or stationary bicycle for 50 min on 7 consecutive days. The exercise intensity corresponded to 65% of their maximum heart rate reserve.

Aerobic exercise training does not, however, necessarily have to be performed for extended periods or at high intensities to achieve significant reductions in systolic and diastolic BP. A recent report by Guidry and coworkers showed that even short bouts of exercise at low intensity (15 min in duration at 40% $\dot{V}O_2max$) had effects on BP reduction comparable to those with longer and more intense bouts of exercise (20 min at 60% $\dot{V}O_2max$) (33). Although some studies have yielded seemingly conflicting results, moderate-intensity aerobic exercise appears to lower BP as much as, if not more than, high-intensity aerobic exercise (3, 5, 20, 28, 30, 45).

Blair and colleagues (16) found that hypertensive men who were more fit had lower death rates than less fit men. Between 1970 and 1981, these authors tested 1,832 men who reported a history of hypertension but were otherwise healthy. Mortality surveillance was conducted on the group through 1985. The inverse relation between fitness and all-cause mortality held even after the investigators adjusted for the influence of age, serum cholesterol, resting systolic BP, body mass index, current smoking, and length of follow-up.

Most studies show a beneficial effect of exercise on CHD risk factors even if the exercise is not enough to increase fitness levels or decrease body weight (28).

On the basis of existing research, the ACSM, in its position statement on physical activity, physical fitness, and hypertension, states that moderate-intensity cardiovascular exercise training appears effective in lowering BP acutely and chronically (3). Although vigorous exercise is not necessarily contraindicated, endurance exercise corresponding to 40 to <60% of $\dot{V}O_2R$ is generally recommended for those with hypertension to maximize the benefits and minimize possible adverse effects of more vigorous exercise (3, 5, 30). This intensity range corresponds to approximately 11 to 14 on the Borg rating of perceived exertion (RPE) 6 to 20 scale. The mode (large-muscle activities), frequency (most, preferably all, days of the week), and duration (30-60 min of continuous or intermittent aerobic activity) are similar to those recommended for healthy adults (4, 5).

The ACSM now views exercise and physical activity for health and fitness in the context of an exercise dose continuum. That is, there is a dose response to exercise by which benefits are derived through varying quantities of physical activity ranging from approximately 700 to 2,000 or more kilocalories of effort per week (5, 28).

Resistance Exercise

In the past, hypertensive patients have been discouraged from participating in resistance training because of fear of overloading an already compromised myocardium (31). These fears were increased by a study in which BPs in excess of 400/200 mmHg were recorded in weightlifters during high-intensity resistance exercise (31). Similarly, there have been concerns that resistance training could give rise to increased arterial stiffness with a subsequent deleterious effect on TPR. Studies by Cortez-Cooper and colleagues (22) and Miyachi and colleagues (43) that used resistance training protocols consisting of high-intensity supersets showed a significant increase in the carotid augmentation index along with a decrease in the central arterial compliance in men and women. In contrast, a study by Rakobowchuk and coworkers that used a circuit training protocol showed no changes in arterial compliance after 3 mo of resistance training in young men. The authors hypothesized that circuit weight training would be more appropriate for a hypertensive population (46).

The rationale for resistance training as an adjunct to aerobic exercise for controlling BP stems from multiple studies. Kelley and Kelley (39), in a meta-analysis of progressive resistance exercise, found a 2% and 4% reduction in systolic BP and diastolic BP, respectively, and concluded that progressive resistance exercise training is efficacious in reducing resting systolic BP and diastolic BP. Harris and Holly (35) evaluated a circuit weight training program in male subjects with BP between 140/90 and 160/99 mmHg. Subjects exercised at approximately 79% of their maximum heart rate 3 d per week for 9 wk. They improved their muscular strength and cardiovascular endurance and lowered their diastolic BP from 96 to 91 mmHg. Resting heart rate and systolic BP did not change. In a more recent meta-analysis, Cornelissen and Fagard pooled data from studies published between 1996 and 2003 that included nine randomized controlled trials involving 341 participants (21). The overall effect of resistance training was a decrease of 3.2 mmHg in systolic BP and a decrease of 3.5 mmHg in diastolic BP. At present, insufficient data are available on the effects of resistance training on ambulatory BP.

Resistance training for hypertensive people should ideally involve lower resistance with higher repetitions. The recommendations are to do at least one set of 8 to 10 different exercises that condition the major muscle groups 2 to 3 d per week. Each set should consist of 8 to 12 repetitions for healthy adults or 10 to 15 repetitions for older, more frail persons or cardiac patients (5, 9, 28, 30).

Circuit weight training, defined as lifting a weight equal to 40% to 60% of one-repetition maximum (1RM) for 10 to 20 reps in a 30 to 60 s period, has also been found to be beneficial for hypertensive people. After a rest of 15 to 45 s, the person moves to the next exercise (40). See "Guidelines for a Circuit Weight Training Program."

Blood pressure should be monitored frequently before, during, and after resistance training during the initial few weeks of participation. The improvement in strength from a resistance training program will help hypertensive persons better perform both occupational and leisure tasks and may enhance their quality of life (40).

Range of Motion

Flexibility exercises should be included in the exercise routine. They should include a variety of upper and lower body range of motion activities, which should be performed on at least 2 or 3 d of the week. At least four repetitions per muscle group should be done, and each static stretch should be held for 10 to 30 s (4, 5). The goal of these exercises is to reduce the risk of musculoskeletal injury and improve the individual's flexibility.

CONCLUSION

Hypertension affects approximately 76 million Americans and is one of the leading causes of death. In approximately 90% of cases, the etiology of hypertension is unknown. Hypertension is one of the major risk factors for CHD and is often found clustered with other CHD risk factors. Lifestyle modification helps control BP and reduces other CHD risk factors. Exercise training helps prevent the development of hypertension and reduces the BP of those with hypertension. Chronic aerobic exercise training has been shown to reduce both systolic and diastolic BP by about 5 to 10 mmHg. Studies have shown that aerobic exercise training at somewhat lower intensity (40-<60% of $\dot{V}O_2R$) appears to decrease BP as much as, or more than, exercise at higher intensity. The mode (large muscle activities), frequency (most, preferably all, days of the week), and duration (30-60

Guidelines for a Circuit Weight Training Program

Select 8 to 12 exercises to create a well-balanced program.

Establish a conservative 1RM in each exercise or a 10RM on three to four key exercises.

Use 40% to 60% of 1RM.

Do 10 to 15 reps in 30 to 60 s.

Do one set of each exercise in a circuit pattern, alternating between upper and lower body and moving from large muscle group exercises to small muscle group exercises.

Begin with a 45 s rest between sets and gradually reduce to a 15 to 30 s rest.

Train 2 or 3 d a week.

Machine weights are preferable for hypertensive patients.

Emphasize full range of motion and proper posture during all exercises.

Warm up with stretching and 10 to 15 min of moderate aerobic exercise (11 to 13 on the RPE scale).

Perceived exertion during the circuit should be 12 to 14.

Avoid straining or heavy lifting; emphasize aerobic circuit training.

Use proper breathing technique to avoid the Valsalva maneuver.

Maintain a firm handgrip, but not too tight, to avoid a pressor response that may cause excessive rises in BP.

When you can comfortably complete two or three circuits with a given load, increase the load by 2.5 to 10 lb (1.1 to 4.5 kg).

Be sure to adhere to the medical regimen prescribed by your physician.

Exercise caution if you have diabetic retinopathy or any other condition resulting in raised intraophthalmic pressures.

Reprinted, by permission, from T. LaFontaine, 1997, "Resistance training for patients with hypertension," *Strength and Conditioning* 19: 5-9.

Practical Application 8.2

RESEARCH FOCUS: EFFECTS OF COMPREHENSIVE LIFESTYLE MODIFICATION ON BLOOD PRESSURE CONTROL

Design and Results

A total of 810 men and women who were not taking antihypertensive medication and had a systolic BP of 120 to 159 mmHg and diastolic BP of 80 to 95 mmHg were randomized to one of three interventions for 6 mo: (1) "established" = behavioral intervention that implemented established recommendations (including >180 min/wk of moderate-intensity physical activity); (2) "established plus DASH," which also implemented the Dietary Approaches to Stop Hypertension (DASH) diet; and (3) an "advice only" comparison group (56). Compared with the baseline hypertension prevalence of 38%, the prevalence at 6 mo was 26% in the advice-only group, 17% in the established group ($p = 0.01$ vs. advice only), and 12% in the established plus DASH group ($p < 0.001$ vs. advice only; $p = 0.12$ vs. established); the prevalence in the established plus DASH group corresponded to a 53% risk reduction compared with the advice-only group. The established and established plus DASH interventions significantly increased the percentage of individuals who achieved a systolic BP <120 mmHg and diastolic blood BP <80 mmHg.

Conclusions and Implications

It is feasible for individuals with above-optimal BP, including stage 1 hypertension, to make multiple lifestyle changes (including exercise training) that lower BP and reduce cardiovascular risk. These findings extend those of the original DASH study in that physical activity was combined with dietary modification and, whereas participants in the original DASH study were provided with prepared meals, participants in the PREMIER trial purchased their own food.

Reference: Writing Group of the PREMIER Collaborative Research Group. Effects of comprehensive lifestyle modification on blood pressure control: main results of the PREMIER clinical trial. *JAMA* 2003;289:2083-2093.

Clinical Exercise Pearls

- Chronic aerobic exercise training has been shown to reduce both systolic and diastolic BP by about 5 to 10 mmHg.
- Although resistance training has a favorable effect on resting BP, the magnitude of reduction is less than that reported for aerobic exercise.
- Hypertensive individuals with a resting systolic BP ≥180 mmHg, diastolic BP ≥110 mmHg, or both should add exercise training to their treatment only after starting appropriate drug therapy.
- Hypertensive individuals with an additional CHD risk factor, males older than 45 yr of age, and females older than 55 yr of age should perform an exercise test with electrocardiogram monitoring before starting a vigorous-intensity (i.e., ≥60% $\dot{V}O_2R$) exercise program.
- Irrespective of the intensity of exercise training, the following groups of hypertensive individuals should perform an exercise test with electrocardiogram monitoring before commencing an exercise program: individuals with known cardiovascular, pulmonary, or metabolic disease; individuals with target organ disease (e.g., left ventricular hypertrophy and retinopathy); and individuals with one or more major symptom or sign suggestive of cardiovascular, pulmonary, or metabolic disease.

(continued)

Clinical Exercise Pearls *(continued)*

- A systolic BP greater than 200 mmHg or diastolic BP greater than 110 mmHg at rest is a relative contraindication to cardiovascular exercise testing.
- Although BP termination criteria for exercise testing are generally established at >250/115 mmHg, lower BP thresholds for termination of an exercise training session may be prudent (i.e., >220/105 mmHg).
- Studies have shown that aerobic exercise training at somewhat lower intensity (40-<60% of $\dot{V}O_2R$) appears to decrease BP as much as, or more than, exercise at higher intensity.
- For cardiovascular exercise training, the mode (large-muscle aerobic activities), frequency (most, preferably all, days of the week), and duration (30-60 min of continuous or intermittent aerobic activity) are similar to those recommended for healthy adults.
- Resistance training recommendations for hypertensive individuals are to do at least one set of 8 to 10 different exercises that condition the major muscle groups, 2 to 3 d per week; each set should consist of 8 to 12 repetitions for otherwise healthy adults or 10 to 15 repetitions for older, more frail persons or cardiac patients.
- Be aware that β-blockers attenuate the heart rate response to exercise; α-blockers, calcium channel blockers, and vasodilators may cause postexertion hypotension; certain diuretics may cause a decrease in serum potassium levels, thereby predisposing to arrhythmias; and diuretics and β-blockers may impair thermoregulation during exercise in hot or humid environments.

min of continuous or intermittent aerobic activity) are similar to those recommended for healthy adults. Although limited data suggest that resistance training has a favorable effect on resting BP, the magnitude of the acute and chronic BP reductions are less than those reported for aerobic exercise. Resistance training for hypertensive people should involve lower resistance and more repetitions.

Key Terms

afterload (p. 139)
atherosclerosis (p. 139)
creatinine clearance (p. 141)
diastolic (p. 137)
dyslipidemia (p. 143)
endocrine (p. 139)
endothelium (p. 139)
epistaxis (p. 139)
etiology (p. 137)
heart failure (p. 139)
hematocrit (p. 141)
hematuria (p. 139)
idiopathic (p. 137)
occipital (p. 139)
papilledema (p. 138)
systemic (p. 137)
systolic (p. 137)
vascular pathologies (p. 139)
ventricular hypertrophy (p. 139)

CASE STUDY

MEDICAL HISTORY

Mr. AB, a 60 yr old Caucasian male, comes to your fitness center for an exercise program. He is apparently healthy but has a 15 yr history of hypertension. His goal is to improve his health and fitness and reduce his risk for heart disease. In the past he did heavy resistance training and powerlifting. At present he has a body mass index of 31 $kg \cdot m^{-2}$. His medications include aspirin (325 mg daily) and atenolol (50 mg daily). He has a family history of premature CHD; his low-density lipoprotein (LDL) cholesterol is elevated (165 mg · dl^{-1}); his resting BP is 148/88 mmHg; and his resting heart rate is 60 beats · min^{-1}. He drinks three or four cans of beer with dinner each evening.

DIAGNOSIS

Mr. AB has a lengthy history of hypertension. Additional risk factors include obesity, age, family history, dyslipidemia, sedentary lifestyle, and elevated alcohol consumption.

EXERCISE TEST RESULTS

During a recent maximal exercise test, Mr. AB exercised for 6 min using the Bruce treadmill protocol and achieved a maximum heart rate of 130 beats · min^{-1} and BP of 240/90 mmHg. He did not develop any significant arrhythmias or ST-segment changes.

EXERCISE PRESCRIPTION

Mr. AB achieved 81% of his age-predicted maximal heart rate (HRmax). The test was negative for underlying cardiovascular ischemia. The initial prescription should be set at a modest intensity (40% $\dot{V}O_2R$), and emphasis should be placed on building exercise duration to increase energy expenditure. Non-weight-bearing activities should be emphasized initially to minimize risk of musculoskeletal injury.

DISCUSSION QUESTIONS

1. Is it necessary for Mr. AB to exercise at high intensity to help manage his hypertension?
2. How will atenolol (a β-blocker) affect the prescription of a target heart rate range for Mr. AB?
3. What is an appropriate aerobic exercise prescription for Mr. AB?
4. What advice would you give Mr. AB about going back to heavy resistance training and powerlifting?
5. What additional lifestyle modifications should be discussed with Mr. AB to optimize his BP control?

Chapter 9

Hyperlipidemia and Dyslipidemia

Peter W. Grandjean, PhD
Benjamin Gordon, MS
Paul G. Davis, PhD
J. Larry Durstine, PhD

The term "lipids" refers to a variety of fats that are necessary for normal body functioning. However, abnormal blood lipid levels, particularly of cholesterol and triglyceride, are modifiable cardiovascular disease (CVD) risk factors. Over the last several years, empirically based lifestyle intervention recommendations have been developed and refined to address blood lipid management within the U.S. population (5, 57, 124). Regular physical activity and exercise contribute to managing healthy blood lipid and **lipoprotein** levels and to ameliorating **dyslipidemias** (49, 96, 97, 99, 100). Yet regularly practiced physical activity is just one of the healthy lifestyle choices that can have a tangible effect on blood lipid management. This chapter presents insights concerning healthy lifestyle choices to better manage blood lipids and lipoproteins. Diet and nutrient composition, body weight management, physical activity, and exercise programming are discussed as important interventions for blood lipid and lipoprotein control (5, 57, 124). The independent and combined effects of lifestyle factors and lipid-lowering medications are also addressed, as these medications are prescribed when dyslipidemias persist after the adoption of healthy lifestyles.

DEFINITION

Hyperlipidemia and dyslipidemia are terms that are frequently used interchangeably. However, these terms are not completely synonymous. Hyperlipidemia is a general term used to refer to chronic elevations in the fasting blood concentrations of triglyceride, cholesterol, or specific subfractions of each, whereas dyslipidemia is a combination of genetic, environmental, and pathological factors that can work together to abnormally alter blood lipid and lipoprotein concentrations (40, 84, 162). Most severe forms of dyslipidemia are linked to genetic defects in cholesterol or triglyceride metabolism (149). Secondary dyslipidemia usually results from metabolic dysfunction that arises from the accumulation of ectopic fat (169), insulin resistance (72), diabetes mellitus (122, 162), hypothyroidism (147), and renal insufficiency and nephrotic kidney disease (80).

Lipids are substances such as fatty acids, cholesterol, and triglycerides. Lipids serve a variety of crucial physiological roles. These roles include energy storage, provision of body insulation, maintenance of bile acids, steroid hormone production, structure of cell membranes, and metabolic regulation. Because lipids are not soluble in body fluids, they must combine with proteins called **apolipoproteins** (apo) to form lipoproteins (table 9.1) (83). Thus, lipoproteins are spherical macromolecules generally composed of apolipoproteins, phospholipids, triglycerides, and free and esterified cholesterol.

Scientists and clinicians use a variety of techniques to classify and study different lipoproteins (69). In fact, the measurement technique dictates how lipoproteins are identified and described. However, the clinician should remember that the lipid and apolipoprotein composition constantly changes in the circulation as these macromolecules interact with each other and with the surrounding body tissues. We'll incorporate a traditional measurement technique, ultracentrifugation, to discuss four general classes of lipoproteins based on their gravitational density (table 9.2) (69).

- **Chylomicrons** originate from intestinal absorption of dietary or exogenous triglyceride. These triglyceride-rich lipoproteins (TRL) are distinguished from liver-derived TRLs because they contain a single copy of apolipoprotein B-48 (117).
- Very low-density lipoproteins (VLDL or pre-β-lipoprotein) are TRLs formed by the liver with a single copy of apolipoprotein B-100. VLDLs are precursors to intermediate- and low-density lipoproteins and are the primary transport mechanism for endogenous triglyceride to body tissues (83).
- Low-density lipoproteins (LDL or β-lipoprotein) are formed in the circulation from lipid and protein exchanges between VLDL and other lipoproteins and tissues. Therefore, LDL exists as a heterogeneous group of lipoproteins with a variety of sizes and lipid compositions. Low-density lipoproteins are the principal means by which cholesterol is transported throughout the body. Lipoprotein(a) is a subfraction of LDL with high atherogenic potential (83, 88).
- High-density lipoproteins (HDL or α-lipoprotein) arise in the liver and intestine and are characterized by their apolipoprotein A, C, and E content (83). High-density lipoproteins are known to be antiatherogenic because they possess antioxidant effects and serve to transport cholesterol from the body's tissues to the liver in a process known as reverse cholesterol transport (168). Like LDL, HDL is a heterogeneous class of lipoproteins with a variety of sizes and densities due to the continuous exchange of apolipoproteins and lipids while in circulation.

Table 9.1 Major Human Apolipoproteins

Apolipoprotein	Major function	CAD risk factor
A-I	LCAT activator	Inversely related with CAD risk
A-II	LCAT inhibitor or activator of heparin releasable hepatic triglyceride hydrolase or both	Not associated with CAD risk
B-48	Required for synthesis of chylomicron	Directly associated with CAD risk
B-100	LDL receptor binding site	Directly associated with CAD risk
(a)	Similar characteristics between apo(a) and plasminogen, thus may have a prothrombolytic role by interfering with function of plasminogen, possible acute phase reactant to tissue damage	Directly associated with CAD risk
C-I	LCAT activator	Not associated with CAD risk
C-II	LPL activator	Not associated with CAD risk
C-III	LPL inhibitor, several forms depending on content of sialic acids	Not associated with CAD risk
D	Core lipid transfer protein, possibly identical to the cholesteryl ester transfer protein	Not associated with CAD risk
E	Remnant receptor binding, present in excess in the β-VLDL of patients with type III hyperlipoproteinemia and exclusively with HDL cholesterol	Not associated with CAD risk

VLDL = very low-density lipoprotein; IDL = intermediate-density lipoprotein; LDL = low-density lipoprotein; HDL = high-density lipoprotein; CAD = coronary artery disease; LCAT = lecithin-cholesterol acyltransferase; LPL = lipoprotein lipase.

Based on J.L. Durstine and P.D. Thompson, 2001, "Exercise in the treatment of lipid disorders," *Cardiology Clinics: Exercise in Secondary Prevention and Cardiac Rehabilitation,* 19(3): 471-488

Table 9.2 Characteristics of Plasma Lipids and Lipoproteins

Lipid/ lipoprotein	Source	COMPOSITION						Apolipoprotein
				PERCENTAGE OF TOTAL LIPID				
		Protein %	Total lipid %	TG	Chol	Phosp	Free chol	
Chylomicron	Intestine	1-2	98-99	88	8	3	1	Major: A-IV, B-48, B-100, H Minor: A-I, A-II, C-I, C-II, C-III, E
VLDL	Major: liver Minor: intestine	7-10	90-93	56	20	15	8	Major: B-100, C-III, E, G Minor: A-I, A-II, B-48, C-II, D
IDL	Major: VLDL Minor: chylomicron	11	89	29	26	34	9	Major: B-100 Minor: B-48
LDL	Major: VLDL Minor: chylomicron	21	79	13	28	48	10	Major: B-100 Minor: C-I, C-II, (a)
HDL_2	Major: HDL_3	33	67	16	43	31	10	Major: A-1, A-II, D, E, F Minor: A-IV, C-I, C-II, C-III
HDL_3	Major: liver and intestine Minor: VLDL and chylomicron remnants	57	43	13	46	29	6	Major: A-1, A-II, D, E, F Minor: A-IV, C-I, C-II, C-III
Chol	Liver and diet		100			70-75	25-30	
TG	Diet and liver		100	100				

VLDL = very low-density lipoprotein, IDL = intermediate-density lipoprotein, LDL = low-density lipoprotein, HDL = high-density lipoprotein; chol = cholesterol; TG = triglyceride; phosp = phospolipid.

Based on J.L. Durstine, and P.D. Thompson, 2001, "Exercise in the treatment of lipid disorders," *Cardiology Clinics: Exercise in Secondary Prevention and Cardiac Rehabilitation,* 19(3): 471-488; H.B. Brewer et al., 1988, Apolipoproteins and lipoproteins in human plasma: An overview," *Clinical Chemistry* 34(8)(Suppl B): B4-B8.

The following are general terms for dyslipidemias that are used in the clinical setting:

- Hyperlipidemia is a general term for elevated blood triglyceride and cholesterol.
- **Hypertriglyceridemia** denotes only elevated triglyceride concentration. Fasting blood triglyceride concentrations over 150 mg · dl^{-1} are classified as hypertriglyceridemic regardless of the triglyceride origins (9).
- **Postprandial lipemia** (PPL) is characterized by exaggerated levels of triglycerides in the blood and failure to return to baseline levels within 8 to 10 h after consumption of dietary fat (103).
- Elevated cholesterol is recognized as total blood cholesterol levels >200 mg · dl^{-1}. Hypercholesterolemia implies elevated blood cholesterol concentration >240 mg · dl^{-1} (124).
- Elevated LDL cholesterol (LDLc) is classified according to CVD risk stratification. LDLc is elevated when concentrations exceed 160 mg · dl^{-1} in low-risk individuals with less than one CVD risk factor, 130 mg · dl^{-1} in those at moderate risk, and 100 mg · dl^{-1} in high-risk individuals. Optimal LDLc is recognized at levels <70 mg · dl^{-1} (table 9.3) (124).
- Low HDL cholesterol (HDLc) is defined as concentrations lower than 50 mg · dl^{-1} in women and 40 mg · dl^{-1} in men (124).
- **Hyperlipoproteinemia** or **dyslipoproteinemia** refers to elevated lipoprotein concentrations. Hyperlipoproteinemia is associated with genetic abnormalities or may be secondarily related to an underlying disease such as diabetes mellitus, renal insufficiency, hypothyroidism, biliary obstruction, dysproteinemia, or nephrotic kidney disease (21, 46).

Table 9.3 LDL Cholesterol Goals and Cut Points for Therapeutic Lifestyle Changes (TLC) and Drug Therapy in Different Risk Categories

Risk category	LDL goal	LDL level at which to initiate therapeutic lifestyle changes	LDL level at which to consider drug therapy
CHD or CHD risk equivalents (10 yr risk >20%)	<100 mg · dl^{-1}	<100 mg · dl^{-1}	<130 mg · dl^{-1} (100-129 mg · dl^{-1}: drug optional)*
Two or more risk factors (10 yr risk >20%)	<130 mg · dl^{-1}	<130 mg · dl^{-1}	10 yr risk 10% to 20%: <130 mg · dl^{-1} 10 yr risk <10: <160 mg · dl^{-1}
Zero to one risk factor[H]	<160 mg · dl^{-1}	<160 mg · dl^{-1}	<190 mg · dl^{-1} (160-189 mg · dl^{-1}: LDL-lowering drug optional)

*Some authorities recommend use of LDL-lowering drugs in this category if an LDL cholesterol <100 mg · dl^{-1} cannot be achieved by therapeutic lifestyle changes. Others prefer use of drugs that primarily modify triglycerides and HDL, for example, nicotinic acid or fibrate. Clinical judgment also may call for deferring drug therapy in this subcategory.

[H]Almost all people with zero to one risk factor have a 10 yr risk <10%; thus 10 yr risk assessment in people with zero to one risk factor is not necessary.

Reprinted from National Cholesterol Education Program 2001.

SCOPE

A wealth of data support the relationship between dyslipidemias and greater risk for CVD. Arterial atherosclerotic lesions, characterized first by fatty streaks and later by fibrous plaque, begin in childhood and increase in prevalence and size with age. Development of these lesions, leading to coronary artery disease (CAD) and cardiovascular complications, is significantly related to concentrations of total blood cholesterol, LDLc, and triglycerides and inversely with levels of HDLc. As such, relationships between blood lipid concentrations and CVD morbidity and mortality have been regularly demonstrated across the life span (15, 128). Among a variety of adult populations, the relative risks for coronary heart disease (CHD) mortality and a subsequent CHD-related event are reduced 24.5% and 29.5% with every 1 mmol · L^{-1} (~39 mg · dl^{-1}) decrease in total cholesterol. For every 1 mmol · L^{-1} reduction in LDLc, the relative risks for CHD mortality and coronary-related events are decreased 28% and 26.6%, respectively (68). Similar estimates of relative risk reduction for all-cause and CVD mortality have been presented in recent meta-analyses of lipid-lowering trials (22, 64). This apparent dose-response relationship between LDLc lowering and reduced CHD risk is also observable at lower levels of acuity. Briel and colleagues (22) demonstrated approximately 7% reductions in CHD events and deaths for every 10 mg · dl^{-1} (0.26 mmol · L^{-1}) reduction in LDLc. Genzer and Marz (64) estimated from their meta-analysis that LDLc lowering in adults accounted for 75% of the variance in risk reduction for cardiovascular endpoints. At present, HDLc has not been demonstrated to have the same predictive power as total and LDLc (22, 126).

These relationships between blood lipid concentrations and CVD appear to be consistent in men and women and in those with and without documented CVD. In older individuals, the relationship between cholesterol and CVD becomes nuanced. In men 65 to 80 yr of age, the relative risk for CHD incidence increases 28% and CVD mortality by 22% for each 1 mmol · L^{-1} increase in total cholesterol. After 80 yr of age, the positive association no longer exists. In fact, an inverse relationship between total cholesterol and all-cause mortality is found. In older women, the relationship between total cholesterol and CHD incidence and mortality is not as strong as in men (4).

The association between triglyceride concentrations and the incidence of CVD is a topic of ongoing debate in the literature. At least part of the debate is based on the inverse and biologically plausible relationships between triglyceride and HDLc and non-HDLc concentrations (44, 132). Because of these relationships, triglycerides, HDLc, and non-HDLc would account for much of the same variance when predicting CVD. Therefore, some would suggest that triglycerides do not predict CVD outcomes independently after HDLc, non-HDLc, or both have been accounted for. This argument is countered by Austin (9), who calculated that the relative risk for incident CVD was increased 76% in women and 32% in men for every 1 mmol · L^{-1} (~89 mg · dl^{-1}) increase in triglyceride. These risk estimates were reduced to 37% and 14% in women and men, respectively after adjusting for HDLc; however, the association between triglyceride and CVD incidents remained significant (9, 85). Nonfasting, or postprandial, triglyceride concentrations predict CVD risk in men and women independent of other established risk factors and at least as well as fasting triglyceride levels (102). Meta-analyses of large population studies report a strong and consistent relationship between nonfasting triglyceride levels and a variety of cardiovascular events, such as myocardial infarct, ischemic stroke, coronary revascularization, and cardiovascular mortality (44, 85, 153). The relationship between nonfasting triglyceride concentrations and CVD outcomes remains after adjusting for age, body weight, blood pressure, smoking, HDLc, LDL particle size, and hormone therapy use in women.

Based on results from these meta-analyses, triglyceride concentrations appear to provide predictive information beyond that obtained for cholesterol and LDLc especially in women. On the other hand, results from a recent meta-analysis that included 68 long-term prospective studies and over 302,000 participants indicate that triglyceride and non-HDLc account for the same variation in predicting future CVD outcomes (44). Here, the clinical exercise physiologist is reminded to keep current with the available literature and to be mindful that traditional and emerging risk factors will vary in their predictive efficacy among different subgroups within our population. Race, gender, genetics, disease states, and physical characteristics, among other factors, will influence the strength of association between blood lipids and lipoproteins, the atherosclerotic process, and CVD outcomes (25, 91, 149, 162, 169).

Current Statistics

Recent statistics from the Centers for Disease Control and Prevention's Behavioral Risk Factor Surveillance System indicate that 67.5% to 85% of American adults had their cholesterol checked within the last 5 yr (111). According to the latest prevalence data from the American Heart Association (AHA), Centers for Disease Control and Prevention (CDC), and National Institutes of Health (NIH), 44.4% of American adults 20 yr of age or older—41.8% of men and 46.3% of women—have total serum cholesterol ≥200 mg · dl^{-1} (≥5.18 mmol · L^{-1}) (148). Just over half of Mexican American men (50.1%), 41.2% of white males, and 37% of black males have high total cholesterol (e.g., total cholesterol ≥200 mg · dl^{-1}). Among females, the greatest prevalence of high serum cholesterol is found in white women (47%) versus 46.5% in Mexican American and 41.2% in black women. One in six, or roughly 16.3%, of American adults has hypercholesterolemia, defined as total serum cholesterol >240 mg · dl^{-1} (154).

LDLc remains the primary target for lipid-lowering interventions (124). Approximately 71.3 million adults, or 31.9% of the U.S. adult population, have LDLc ≥130 mg · dl^{-1} (148). Elevated LDLc is present in 32.5% of men and 31% of women and follows the same pattern as total serum cholesterol across racial demographics. Prevalence estimates are slightly lower (25.3%) when the National Cholesterol Education Program Adult Treatment Panel III (NCEP-ATP III) cut points according to CHD risk are used to identify individuals with high LDLc (table 9.3) (87). Low serum HDLc, generally defined as <40 mg · dl^{-1}, is present in 41.8 million U.S. adults or 18.9% of the population. Low HDLc is three times more prevalent in men than in women. Among both men and women, low HDLc is observed more often in whites and Mexican Americans than in blacks (148). Elevated triglyceride levels (e.g., serum triglyceride >150 mg · dl^{-1}) are observed in approximately 30% of the U.S. adult population and, as with low HDLc, are more prevalent in whites and Mexican Americans than in black adults (124).

Trends

The percentage of U.S. adults 20 to 74 yr of age with hypercholesterolemia has decreased from 33% to 16.3% since 1960 (148). The reduction in those with hypercholesterolemia represents a Healthy People 2010 objective that has been attained. Tracking over the same time, the population-wide average total serum cholesterol has decreased from 222 mg · dl^{-1} to 200 mg · dl^{-1} (29, 148, 154). These population statistics are generally good news from a public health perspective since the AHA estimates that the incidence of CAD can be reduced by 30% for every 10% decrease in the total serum cholesterol among U.S. adults (148).

Average LDLc values have also declined from 129 to 123 mg · dl^{-1}(29). Much of the reduction in our population-wide total and LDLc values has been attributed to dietary changes (27) and the increased use of lipid-lowering drugs (87, 89, 104). Indeed, the use of lipid-lowering medication has increased dramatically over the last few decades, and these drugs had a greater prescription volume than any other drug class in 2010 (89). Statins are recommended by the NCEP-ATP III, National Heart Lung and Blood Institute, AHA, the American College of Cardiology, and others for primary and secondary prevention. As such, statins are by far the most frequently prescribed of the lipid-lowering agents, and approximately 75% of those taking statins do so for primary prevention (89, 104).

Although the trends for total and LDLc are promising, continued and more aggressive lipid-lowering interventions appear to be warranted. Approximately two-thirds of apparently healthy adults at high risk for CAD are not taking needed medications for dyslipidemia (87, 93, 104). Only 23% to 26% of those at high CVD risk meet the target LDLc goal of <100 mg · dl^{-1} (93). This problem is greater among younger adults compared with those aged 65 yr and older (87). Moreover, the average HDLc and triglyceride values have not changed appreciably over the last five decades (29). This result is likely due to the increased incidence of overweight and obesity among our adult population over the same time period (29, 56). In fact, Burke and colleagues (25) recently reported that fasting glucose and triglyceride concentrations were greater (8 to 17 mg · dl^{-1} and 18 to 55 mg · dl^{-1}), HDLc was lower (4 to 14 mg · dl^{-1}), and concentrations of small LDL particles were greater (109 to 173 mmol · L^{-1}) in obese adults versus those of normal weight across ethnicities.

PATHOPHYSIOLOGY

Over the past several years, convincing evidence has accumulated to suggest that atherosclerotic disease processes originate in early childhood and through the interactions of genes and modifiable and nonmodifiable environmental exposures. These inheritable or environmental exposures influence a complex network of lipid and lipoprotein pathways that provide transportation for delivery, metabolism, or elimination.

Lipoprotein Metabolic Pathways

Blood lipoproteins provide a system of transportation for the movement of lipids between the intestine, liver, and extrahepatic tissue (figure 9.1). Several reviews deal with the complex interactions that take place within the circulation (39, 65, 155). Four important enzymes facilitate these interactions: **lipoprotein lipase (LPL)**, **hepatic lipase (HL)**, **lecithin-cholesterol acyltransferase (LCAT)**, and **cholesterol ester transfer protein (CETP)**. Although scientists and clinicians recognize the dynamic exchange of lipid and lipoproteins among the various lipoprotein classes and their subfractions, the transport of cholesterol and triglyceride is generally described in terms of two general processes (21, 39, 46, 168). The LDL receptor pathway consists of a sequence of chemical steps designed for the delivery of lipids to extrahepatic tissue (39, 103). A second pathway, reverse cholesterol transport, involves a sequence of chemical reactions necessary for returning lipids from peripheral

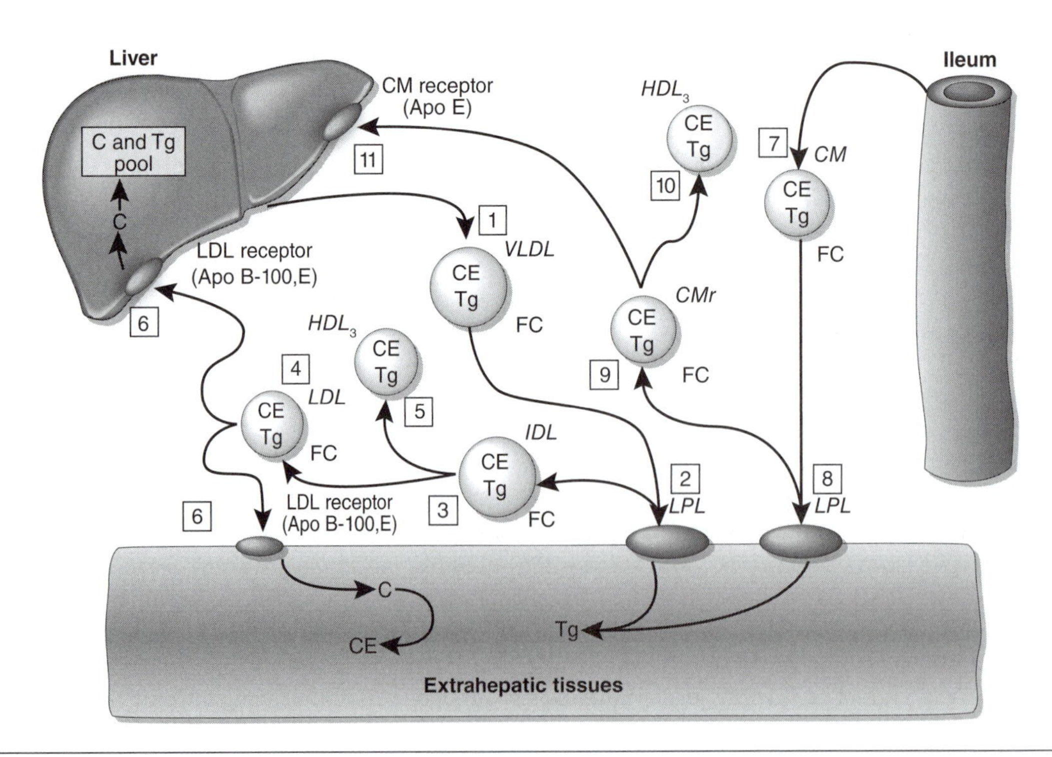

Figure 9.1 Schematic diagram of the transport of lipids between the intestine, liver, and extrahepatic tissues.

tissue to the liver for metabolism and excretion (48, 168). These two intravascular transport processes are further described next.

LDL Receptor Pathway

Dietary or exogenous fat is absorbed by the small intestine as fatty acids and free cholesterol. During absorption, these lipids combine with apolipoproteins B-48, A-I, A-II, A-IV, and E and are internalized into the core of the chylomicron and enter the circulatory system via the lymphatic system and the thoracic duct. Blood chylomicrons then react with LPL (apo C-II facilitates this reaction), hydrolyzing the triglyceride core and releasing fatty acids. During this process, chylomicron surface remnants move to nascent HDL particles while apo C and E are acquired by the chylomicron remnants. Apo E and apo B-48 receptors on the surface of hepatic cells bind to the chylomicron remnants and are removed from circulation. A similar endogenous pathway exists for VLDL synthesized by the liver. In this pathway, LPL hydrolyzes the VLDL triglyceride core, releasing fatty acids that are taken up by extrahepatic tissue. The remnant intermediate-density particles interact with LPL and HL to form LDL, with the remaining VLDL remnants being removed from the blood by hepatic apo E receptors. Low-density lipoprotein particles formed in this process are the primary cholesterol carriers and deliver LDLc to extrahepatic tissue where LDL receptors located on the cell's surface mediate its uptake. Once the LDLc is recognized by the LDL-apo B-100/apo E receptor, LDL is moved inside the cell and used for cellular metabolic needs. This process also initiates within the cell a negative feedback response causing a reduction in cellular cholesterol synthesis while promoting cholesterol storage. As these processes occur, cellular LDL receptor synthesis is suppressed and further cellular LDL uptake is halted.

Reverse Cholesterol Transport

Cholesterol is moved to the liver for catabolism by a process termed reverse cholesterol transport. This elimination of cholesterol from the peripheral tissue involves several processes. The most notable pathway uses nascent HDL particles secreted by the liver and enriched with free cholesterol and phospholipid derived from LPL-mediated chylomicron and VLDL catabolism. In this particular process, free cholesterol is esterified by the action of LCAT with apo A-I as a cofactor. The cholesteryl ester is moved into the HDL_3 core, causing a chemical gradient with a constant cholesterol supply for the LCAT reaction. As the HDL_3 core expands, the HDL_3 is transformed into HDL_2. While this process is happening, two other separate sets of metabolic reactions occur. In one series, CETP facilitates an exchange for the newly formed HDL_2 cholesteryl ester with triglyceride obtained from TRL remnants. The remaining lipid-depleted TRL remnants move to the liver for metabolism and removal. A second set of reactions mediated by hepatic HL removes triglyceride from the HDL_2 particle that previously gained triglyceride by CETP action. Smaller, more dense HDL_3 particles return to the circulation once HL-mediated triglyceride removal is complete. In addition, two other cholesterol removal pathways exist. One pathway is the direct HDL_2c withdrawal from circulating blood by liver cells through the action of phospholipase and HL. A final pathway is the hepatic apo E receptor–mediated removal of an HDL_2 cholesteryl ester. Here cholesteryl ester-rich HDL_2 particles containing apo E are withdrawn from circulating blood by hepatic LDL receptor-mediated endocytosis.

Postprandial Lipemia

Normally, blood triglycerides will not appreciably increase or will quickly and modestly rise after a meal and then steadily decline (103). Generally, the time needed after a meal for blood triglyceride levels to return to fasting levels is 6 to 8 h. Exaggerated or prolonged lipemia is associated with increased CAD risk (102). Elevated PPL may initiate a number of harmful events associated with endothelial dysfunction and arterial plaque buildup (42). Possibly the most significant finding is the formation of the highly atherosclerotic small dense LDL particles and a reduction in the concentration of HDLc. Prolonged PPL also promotes inflammation, oxidative stress, and thrombosis or blood clot formation (173). Genetic factors affect the magnitude of the postprandial responses (135), but postprandial lipid responses are also affected by exercise (119, 137) and nutrient composition (108).

Metabolic Dyslipidemia

The metabolic syndrome is a well-recognized but contentious concept in the scientific literature and in clinical practice (63, 144, 158). There are tremendous efforts to understand the complex pathophysiology that arises from ectopic fat accumulation, insulin resistance, and the dysfunction in glucose and lipid metabolism that results (50, 145, 169). An accompanying complication is hypertriglyceridemic dyslipidemia, which occurs mostly because of overproduction of TRLs. In recent years scientists have focused on the mechanisms regulating the overproduction of these atherogenic apo B-containing lipoproteins, and evidence from animal models and human studies has identified hepatic VLDL overproduction as a critical underlying factor in the development of hypertriglyceridemia and metabolic dyslipidemia (10, 11).

Elevation of bloodborne fatty acids observed with metabolic dyslipidemia affects the vascular endothelium by reducing nitric oxide production, inducing adhesion characteristics, facilitating oxidative damage and inflammation, and resulting in diminished vascular compliance and reactivity (31, 173). Oversupply of these fatty acids to the liver facilitates triglyceride synthesis, reduces apolipoprotein B-100 degradation, and enhances the production of triglyceride-rich VLDL. Elevated plasma triglyceride concentrations and TRLs may be further exacerbated by impaired LPL activity (121). The increased actions of HL and CETP, observed with insulin resistance and an overabundant amount of liver fat, potentiate the transfer of triglyceride from VLDL to HDL and result in the formation of small dense LDL. These smaller, dense LDL are less likely to be taken up by the normal LDL receptor pathway, more readily penetrate the vascular endothelium, and are more susceptible to oxidative damage (10). Once in the subendothelial tissue, the modified LDL induces vascular inflammation and contributes to atherosclerotic plaque accumulation and lesion instability (107). At the same time, the activities of HL and CETP and dysfunctional apolipoprotein metabolism contribute to greater hepatic HDL clearance and a reduction in HDLc and HDL particle number (31). Effects on HDL number and composition further impair reverse cholesterol transport and attenuate HDL antioxidant potential. These conditions, in turn, contribute to the aforementioned processes that exacerbate dyslipidemia, vascular dysfunction, and atherosclerotic plaque burden.

CLINICAL CONSIDERATIONS

Elevated LDL proportionally increases one's risk for CHD. Although national guidelines for treating dyslipidemias have existed for several decades and some improvement has been seen, many individuals still do not reach recommended goals. The Adult Treatment Panel III (ATP III) introduced several years ago calls for more aggressive approaches to the clinical management of dyslipidemias. The result substantially increases the number of individuals in need of lifestyle and drug treatment programs.

Signs and Symptoms

Given that most physical manifestations of dyslipidemic diseases are absent, these conditions are considered silent diseases. Because signs and symptoms are not physically evident, dyslipidemic diagnosis is based almost solely on the lipid profile itself; thus the primary means for dyslipidemia diagnosis are beyond the scope of practice for the typical clinical exercise physiologist. Most importantly, each dyslipidemic condition requires laboratory testing to determine the specific abnormal blood lipid or lipoprotein level, allowing for classification and diagnosis. Remember that each type of dyslipidemia carries a different level of CHD risk, making the evaluation of individual lipid and lipoprotein levels crucial.

Hypercholesterolemia

Typical hypercholesterolemia has no outward physical symptoms; however, more severe forms of hypercholesterolemia are accompanied by physical signs. For example, familial hypercholesterolemia, a rare genetic condition of hypercholesterolemia, is typified by a hyper-β-lipoproteinemia resulting in drastically reduced LDL clearance rates. Because clearance rates are reduced, blood total cholesterol levels can reach 500 mg · dl^{-1} or more. As a result of these extreme blood cholesterol levels, **xanthomas** (a deposition of yellowish cholesterol-rich material in tendons or in subcutaneous tissue in which lipids accumulate) and **atheromas** (lipid-filled large foam cells) may arise. Another more common genetic form of hypercholesterolemia is familial polygenic hypercholesterolemia, the most common cause of elevated serum cholesterol concentrations. In this case LDLc elevations are moderate (140 to 300 mg · dl^{-1}) with serum triglyceride concentrations within the normal range (149). Another prominent genetic form of hypercholesterolemia is familial combined hypercholesterolemia, which is typified by elevated total cholesterol and triglyceride concentrations (24). All forms of hypercholesterolemia increase the risk for premature atherosclerotic disease; however, familial hypercholesterolemia carries the highest risk (66).

Hypertriglyceridemia

In addition to familial combined hypercholesterolemia, there are other genetic conditions that can result in elevated triglycerides. Three primary isoforms of apo E exist—E2, E3, and E4. One particular apo E genotype, the apo E2, promotes hypertriglyceridemia and is associated with increased CHD risk. Individuals with the apo E2 genotype are ineffective in binding to the apo E receptor. Because binding is required for triglyceride hydrolysis and removal from blood, both blood chylomicrons and VLDL–triglyceride levels increase (41). Deficient levels of blood apo C-II are also associated with elevated blood triglyceride levels and with limited clearance of chylomicrons and VLDL (20, 60). Familial LPL deficiency, usually found in childhood, is characterized by severe hypertriglyceridemia with episodes of abdominal pain, recurrent acute pancreatitis, and development of cutaneous xanthomata (12). Blood chylomicron clearance is impaired, causing blood triglycerides to accumulate and causing the plasma to appear milky. These physical signs usually resolve with restriction of total dietary fat to 20 g per day or less (60). Other genetic defects involving single nucleotide polymorphisms and haplotypes directly influence lipoprotein–triglyceride transport in the blood (135). Familial hypertriglyceridemia is caused by a genetic defect and is passed on in an autosomal-dominant fashion. In this condition, VLDL are overproduced, causing elevated blood VLDL–triglyceride concentrations. Familial hypertriglyceridemia does not usually become noticeable until puberty or early adulthood. Obesity and high blood glucose and insulin concentrations often accompany this genetic form of dyslipidemia (60).

Physical Examination and Diagnostic Testing

Common dyslipidemias are diagnosed by obtaining a blood sample and measuring the blood for what is known in the clinical setting as a "blood lipid profile." Dyslipidemias traditionally were classified according to the Frederickson classification system, in which the pattern of lipoproteins was described using electrophoresis or ultracentrifugation (61). The current classification for dyslipidemias was adopted in 1988 as part of the first statement paper from the NCEP-ATP III. The NCEP-ATP III guidelines for U.S. adults established priorities and rationale for treating and managing blood lipids for primary and secondary CHD prevention (124). The current NCEP-ATP III guidelines include precise clinical cut points that are tailored to the individual's CHD risk (table 9.3). Clinical screening for dyslipidemia starts with analysis of the blood lipid profile measured every 5 yr and can begin at age 20. LDLc levels are measured directly or most likely estimated through the use of the Friedewald formula (LDLc = total serum cholesterol – HDLc – [triglyceride/5]) (62). However, the Friedewald formula is not used when an individual's triglyceride levels exceed 400 $mg \cdot dl^{-1}$. The basic lipid profile of total, LDLc, HDLc, and triglyceride estimates may not provide the clinician with enough information to properly diagnose some dyslipidemias. In these cases, the clinician may order more sophisticated analytic procedures to determine lipoprotein particle numbers and sizes or apolipoprotein concentrations (69).

Some clinicians include the measurement of postprandial lipids in patients who exhibit refractory hypertriglyceridemia (i.e., hypertriglyceridemia that remains after treatment). Assessment of triglyceride and other lipid parameters, such as the number and size of TRLs, apolipoprotein responses, and calculated non-HDLc during the postprandial period, may help the clinician develop a specific therapeutic strategy for ameliorating the dyslipidemia (129, 130). Although the NCEP-ATP III does not recognize hypertriglyceridemia as a primary target of lipid lowering, exaggerated triglycerides can prohibit an accurate assessment of the primary lipid target (LDLc), contribute to vascular endothelial dysfunction, and increase CHD risk (102). A strong argument can be made that PPL is a serious but treatable condition since most people live in a postprandial state throughout the day and those with PPL may not fully clear blood lipids from a previous meal before eating again (102, 120). Typically a patient performs 8 to 12 h overnight fasting before consuming a test meal with known amounts of fats, protein, and other macronutrients. Before ingestion of the test meal, blood samples are taken, and after the meal is ingested, blood samples are taken at timed intervals for up to 8 h. The changes in blood triglyceride are plotted and a removal curve is developed. The more quickly blood triglycerides are removed after the fat test meal, the less the CHD risk (130).

Exercise Testing

Secondary forms of dyslipidemia are void of signs and symptoms for other comorbidities (e.g., CAD or renal insufficiency), and in these cases exercise testing should follow protocols used in populations at risk for CAD. Exercise testing protocols for individuals at risk for CAD have been developed by the AHA and the American College of Sports Medicine (ACSM) (see chapter 5). The possibility of dyslipidemic patients having latent CVD is higher than in the healthy population. In light of this possibility, the clinical exercise physiologist should be aware of and understand the contraindications for exercise and exercise testing in this particular group.

Practical application 9.1 provides information about interacting with dyslipidemic patients.

Cardiovascular Testing

The purpose of cardiovascular testing in an individual with dyslipidemia is to help diagnose CAD, determine the functional capacity of the individual, and help determine an appropriate exercise intensity range. Various appropriate cardiovascular tests exist for the dyslipidemic population, and as previously stated, cardiovascular testing should follow normal protocols unless comorbidities are present (see chapter 5). When comorbidities are pres-

Practical Application 9.1

CLIENT–CLINICIAN INTERACTION

When interacting with dyslipidemic patients, the clinical exercise physiologist must clearly outline the dangers and implications of having altered levels of cholesterol. This information will help lend seriousness to the patient's condition and communicate the importance of ameliorating the condition. The clinical exercise physiologist should then give options on improving the client's condition, starting with high-volume aerobic exercise. When discussing aerobic exercise, the clinical exercise physiologist must be simple and direct, clearly presenting the beneficial effects of high-volume aerobic exercise. The clinical exercise physiologist needs to highlight the improvements seen in HDLc and triglycerides with exercise training. In addition, emphasis should be placed on the beneficial effects of exercise on other CVD risks. This is particularly important in patients who have multiple risk factors. With regard to aerobic training, the patient needs to understand that the benefits are increased with greater exercise energy expenditure.

Along with aerobic exercise, a calorie-restricting diet can be an effective tool in decreasing total cholesterol, LDLc, and triglycerides while increasing HDLc. Therefore, the clinical exercise physiologist should ultimately encourage the client to combine high volumes of aerobic exercise of moderate to vigorous intensity with a calorie-restricting diet to result in high weekly caloric deficits. The collaborative team approach, invoking the knowledge and expertise of the clinical exercise physiologist and the clinical nutritionist and the advocacy and guidance of a physician is strongly recommended.

ent, testing protocols should allow for the needs of the individual and his particular condition(s) (table 9.4). Also when exercise testing and training are being considered for patients with a medical condition, a referral and a medical evaluation performed by a physician are required before testing (159).

Musculoskeletal Testing

The purpose of musculoskeletal testing in a patient with dyslipidemia is to determine the musculoskeletal strength, endurance, and performance. Various appropriate muscular strength and endurance tests are

Table 9.4 Dyslipidemia: Exercise Testing

Test type	Mode	Protocol specifics	Clinical measures	Clinical implications	Special considerations
Cardiovascular	Cycle ergometer Treadmill	Ramp protocol 17 W/min; staged protocol 25 to 50 W/3 min stage 1-2 METs/3 min stage	12-lead ECG, HR, BP, rate–pressure product, RPE	Determine fitness level and whether CAD condition exists	Evaluation of risk stratification should be made before test to ensure safety and supervision level necessary for given test.
Strength	Free weights Machine weights	1RM of all major muscle groups Estimation of 1RM of all major muscle groups	1RM	Determine musculoskeletal fitness	Evaluation of patient's medical history needs to be performed to ensure orthopedic safety and stability.
Range of motion	Static stretching	Sit-and-reach (hip flexion) Thomas test (hip flexor tightness) Ober's test (iliotibial band tightness)	Various ROM	Determine specific joint ROM	Evaluation of patient's medical history needs to be performed to ensure orthopedic safety and stability.

available for use in the dyslipidemic client. Most often one-repetition maximum tests are used; however, if comorbidities are present, other tests may be employed (138). Other less hazardous tests include the estimation of one-repetition maximum (see chapter 5). Special attention should always be given to any possible orthopedic issues the patient may have.

Flexibility Testing

The purpose of flexibility testing in a patient with dyslipidemia is to determine the range of motion in all joints and, most practically, to assist in the assessment of functional abilities as they relate to exercise prescription and activities of daily living. Patients with dyslipidemia can perform testing protocols for the healthy population (see chapter 5).

Contraindications

Contraindications for general exercise testing have been identified and described by the ACSM (3). An in-depth review of each patient's medical history is performed to find possible contraindications and to stratify the patient's level of risk. As stated before, exercise is not contraindicated for those with a dyslipidemic blood profile. Rather, attention is given to possible existing comorbidities. When diseases are suspected or present, the protocols for testing a patient are likely to change. The clinical exercise physiologist should pay careful attention to the facility's procedures for determining eligibility for exercise testing, including obtaining a physician's release and having the appropriate personnel on hand for exercise and functional tests. The clinical exercise physiologist should be cognizant and observant of signs and symptoms that may suggest underlying or latent CVD while testing dyslipidemic patients.

Anticipated Exercise Responses

Considering the dyslipidemic patient as an otherwise healthy individual, the anticipated responses to an exercise test are no different than what is expected in healthy individuals. However, some special circumstances may exist. Prescribed medications can alter anticipated exercise responses. In some cases lipid-lowering medications such as statins and fibric acid, when used individually or in combination, have the potential for causing muscle damage, releasing proteins such as myoglobin into the blood (e.g., rhabdomyolysis), and causing reduced exercise performance. Besides reducing exercise performance, these proteins are harmful to the kidneys. Elevated levels may lead to kidney failure and, in extreme cases, death (52). Also, chronically high cholesterol levels can result in the formation of tendon xanthomas. Individuals with this condition may experience biomechanical problems that contribute to reduced exercise performance. Exercise testing is a crucial step in evaluating the health and fitness status of patients with dyslipidemia. This information is extremely valuable for creating an exercise prescription. Table 9.4 summarizes considerations regarding exercise testing in patients with dyslipidemia.

Treatment

The primary goal in the clinical management of dyslipidemias is to reduce the global risk for CVD. Attention focuses on reducing the severity and number of traditional CVD risk factors. The current goals and recommendations for managing dyslipidemia are described in consensus statements for dyslipidemia, high blood pressure, obesity, diabetes, and physical activity (32, 47, 81, 124, 167). Aggressive treatment, often including the use of lipid-lowering medication, is of primary importance for preventing future CVD outcomes in dyslipidemic patients with documented or known disease (5, 57, 80, 122, 163, 165). The AHA describes a collaborative approach to managing dyslipidemia in a 2005 monograph (57). The AHA advocates teamwork by physicians and allied health professionals (e.g., nutritionists, nurses, clinical exercise physiologists) to help patients normalize their blood lipids and improve their health.

The fundamental interventions for those with dyslipidemia are to engage in regularly practiced physical activity, consume a heart-healthy diet, lose weight, and prevent weight regain after weight loss. All of the characteristics of secondary dyslipidemia are mitigated when these lifestyle behaviors are consistently practiced (57). In addition, efforts should be made to quit smoking, improve stress management, and practice individually appropriate behavioral techniques for long-term adherence to healthy lifestyle changes (5, 57). A clinician's guide that explains how to prescribe exercise for most patients and delineates useful strategies for promoting healthy lifestyle changes has recently been published (26). Strategies for improving patient adherence to lifestyle behavior changes are also available to the physician and members of the collaborative health care team (54, 57).

Diet

Adopting a healthy diet is crucial for treating dyslipidemia and managing healthy blood lipid concentrations (5, 57, 74, 112, 146, 185). Evidenced-based dietary recommendations from the AHA, NCEP-ATP III, and the Dietary Approaches to Stop Hypertension (DASH) diet are widely used in clinical practice (32, 106, 124). A Mediterranean-style diet is consistent with the

NCEP-ATP III recommendations for a dietary fat intake of 25% to 35% of total calories (124). A Mediterranean-style diet, composed of whole grain foods, fruit, vegetables, fish, fiber, and nuts, has demonstrated efficacy in mitigating CVD risk and reducing future CVD events (43, 151). The benefits of a Mediterranean-style diet are due, in part, to the fact that it is low in saturated fats, trans fats, cholesterol, sodium, red meat, simple sugars, and refined grains (14, 57, 165). In addition, the Mediterranean diet provides nutrients such as omega-3 fatty acids, polyunsaturated fatty acids, soluble fiber, and carotenoids, all of which are associated with cardiometabolic health (14). Although red wine is often associated with the Mediterranean diet, the health benefits of moderate alcohol consumption, such as an increase in HDLc and enhanced vascular endothelial function, appear to be unrelated to the type of alcoholic beverage that is consumed (45). Therefore, patients who are responsible drinkers should be encouraged to continue moderate alcohol use.

Adopting a Mediterranean diet will help those with dyslipidemia balance macronutrient intake and avoid the temptation to try one of the ever-present fad diets. A fad diet may generally be characterized as one that is based on omitting an important food group and promising unsupported benefits. Current fad diets include those that recommend high or low fat intake, high carbohydrate consumption, or high protein intake. With the exception of Mediterranean fare, diets with fat intake exceeding 35% are likely to include too much saturated fat, contributing to elevated LDLc, insulin resistance, and weight gain (74, 106). Low-fat, high-carbohydrate diets promote features of atherogenic dyslipidemia, such as elevated triglycerides and lower HDLc. High-protein diets can increase blood phosphorus levels and may lead to hypercalcuria, acidosis, and insulin resistance, especially in individuals with poor renal function (74, 106, 185).

Functional foods are foods containing biologically active substances that impart medicinal or health benefits beyond their basic nutritional components. These foods appear promising in the prevention and treatment of CVD (156). What makes functional foods attractive to the clinician and patient is the fact that they do not seem to have major side effects and therefore are safely incorporated into the diet (57). In addition, functional foods can impart health benefits when substituted into or added to the current diet without major alterations in eating habits or changes in nutrient intake. For individuals who are taking "lipid-lowering" medications, the effects of drug therapy can be augmented when combined with functional foods because these foods seem to exert their effects differently from the medications (124). Functional foods that may be effective for blood lipid management include soluble fiber, plant stanols and sterols, psyllium, flaxseed, a variety of nuts, omega-3 fatty acids, garlic, flavonoids found in dark chocolate, and soy protein. Recent literature reviews on the topic provide a more thorough understanding of the proposed mechanisms of action and lipid-lowering effects (82, 150, 156). The AHA's recommendations for managing dyslipidemias delineate a practical approach to including functional foods in an individualized plan (57).

Dietary habits are hard to break, and the clinician must think beyond itemized recommendations for dietary nutrient composition. Patients should be provided with resources to make incremental and lasting behavioral changes in meal planning, food selection, and food preparation. Contingencies for dietary challenges, such as food choices and portion control when dining out, and habits associated with altered eating patterns need to be addressed (5, 54, 57).

Physical Activity and Exercise

The amount of physical activity recommended for improving and maintaining health in most adults is very attainable. The health benefits of regularly practiced physical activity, such as improved triglyceride and HDLc concentrations, blood pressure, blood glucose control, and cardiovascular function, will be enjoyed regardless of weight change or in the absence of noticeable improvements in fitness (5, 30, 81, 90, 105). The following section on exercise prescription discusses the amount and type of physical activity appropriate for those with dyslipidemia.

Weight Loss

Weight loss is a top priority for overweight and obese patients with dyslipidemia. The current recommendations are to achieve a 7% to 10% reduction in body weight over a 6 to 12 month period through caloric restriction and increased physical activity. Most scientists and practitioners agree that this target is best attained through a modest reduction in daily caloric intake (i.e., lowering total calories by 500 to 1,000 kcal) (47, 125). Achieving a modest weight loss of 7% to 10% will positively affect all lipid characteristics (125, 139, 184). Daily physical activity adds to the caloric deficit and imparts health benefits that may not be achieved through hypocaloric diets alone (47, 81). Weight loss achieved through healthy lifestyle behavior may be the most effective means of preventing and treating dyslipidemia (164, 176, 179, 181). Additional health benefits can be realized with greater weight loss and long-term maintenance of the lower body weight (47, 115, 139, 143). Therefore, individuals who respond well to the initial weight loss goal should be encouraged to continue behaviors conducive to further weight loss or maintenance of their new lower body weight.

Pharmacology: Lipid-Lowering Medications

The primary target for lipid-lowering therapy is LDLc (table 9.5). Statin therapy is by far the primary drug used to lower LDLc (89). Among the statins, rosuvastatin exhibits the greatest effects (104). In standard dosages, statins are capable of reducing LDLc by as much as 40% but are limited in their efficacy for lowering triglycerides, increasing HDLc, and modifying other aspects of atherogenic dyslipidemia like small dense LDL particles (75). In addition to lipid-lowering effects, statin treatment may reduce serum uric acid levels (6), improve renal function (7), and lower CVD events in patients with metabolic syndrome and diabetes mellitus (13, 34, 142). Statins inhibit cholesterol production in the liver and increase LDL receptor numbers in hepatic and other tissues. Downstream effects are stabilization of atherosclerotic lesions and decreasing endothelial dysfunction. The most common side effects reported with statin therapy are elevated liver enzymes, myopathy, myalgia, and dyspepsia (160). Statins should be avoided or discontinued in those with liver disease or elevated liver enzymes and during pregnancy (53, 104, 134).

Elevated fasting and postprandial triglycerides, low HDLc, and small dense LDL particles contribute to residual CVD risk after LDLc goals have been achieved (16, 75). This atherogenic dyslipidemia is approximated in the clinic through calculation of non-HDLc and measurement of triglyceride concentrations (44, 102). Therefore, non-HDLc, which is an aggregate measure of VLDL, chylomicrons, and all TRLs, is considered the next target for lipid-lowering interventions after LDLc goals are met (124). Targeting triglyceride concentrations is also of importance for avoiding acute pancreatitis in those with severe hypertriglyceridemia (74, 102). Non-HDLc goals are generally 30 $mg \cdot dl^{-1}$ less than those for LDLc, and triglyceride concentrations should be reduced

Table 9.5 Pharmacology

Medication name and class	Primary effects	Exercise effects	Special considerations
HMG-CoA reductase inhibitor	Blocks cholesterol synthesis; increases tissue LDL receptors	No direct effect	Benefits: convenient dosage schedule; useful in those with multiple risk factors; low drug interaction potential; low level of system toxicity Side effects: increased liver enzymes; myopathy; myalgia; dyspepsia Contraindications: chronic liver disease; pregnancy; in some instances, use with erythromycin, antifungal agents, cyclosporine
Fibric acid derivatives	Decreased triglyceride synthesis and VLDL production; increased lipoprotein lipase activity	No direct effect	Benefits: used with well-controlled diet to reduce the number of small dense atherogenic LDL particles; is generally better tolerated than niacin; also decreases fibrinogen levels Side effects: GI distress; gallstones; myopathy; increased excretion of uric acid Contraindications: renal disease; liver disease
Nicotinic acid	Decreased VLDL production; decreased liver clearance of HDL particles	No direct effect	Benefits: low cost and availability (OTC); favorably improves all atherogenic components of the lipid profile; often used with bile acid binding resins Side effects: flushing; GI distress; hyperglycemia; hyperuricemia Contraindications: severe hypotension; liver disease; diabetes; gout; hyperuricemia; active peptic ulcer
Bile acid sequestrants	Binds bile acids in the intestine	No direct effect	Benefits: low system toxicity; use permitted in children and during pregnancy Side effects: GI distress; constipation; interferes with absorption of warfarin, digoxin, and thyroxine Contraindications: bile duct obstruction; dysbetalipoproteinemia; triglyceride >200 $mg \cdot dl^{-1}$

to <150 m · dl^{-1} or 1.7 mmol · L^{-1} (74). HDLc, essentially a marker of reverse cholesterol transport, becomes a focus of lipid management after LDLc, non-HDLc, and triglyceride values have been addressed. There are no clinical targets for raising HDLc (124).

Fibrates and niacin are effective for lowering triglycerides, increasing HDLc, and increasing LDL particle size (74, 112, 165, 172). As such, these agents are chosen to address atherogenic dyslipidemia—either as first-line therapy when LDLc is within normal limits or in combination with statins. The combination of simvastatin and fenofibrate appears to have an additive and powerful effect on ameliorating all atherogenic characteristics of the lipid and lipoprotein profile (8, 170). Likewise, extended-release niacin improves LDLc and triglyceride concentrations and is very effective at increasing HDLc (136). Niacin also has the added benefit of lowering lipoprotein(a) and fibrinogen levels (76) and, in combination with simvastatin, shows promise in decreasing CVD events (23).

Fibrates reduce fatty acid uptake by the liver, thereby slowing hepatic triglyceride synthesis and VLDL production. Fibrates may also lower serum triglyceride and increase HDLc by stimulating the activity of LPL (134, 136). Common side effects of fibrates include constipation and those reported for statins. Fibrates are contraindicated for people with hepatic or renal dysfunction (73). Niacin primarily reduces serum triglyceride and LDLc by suppressing hepatic synthesis of VLDL. A longstanding limitation for niacin has been that it is not well tolerated (18), although extended-release forms have improved tolerance. Niacin therapy often results in flushing, skin rashes, gastrointestinal problems, and pruritus (53, 136). Niacin is not recommended for individuals with hypotension, liver dysfunction, or peptic ulcers or for those with diabetes mellitus, as it often causes increases in blood glucose concentrations (73).

Bile acid sequestrants are considered for persons who remain resistant to the lipid-lowering effects of statins and for younger individuals facing long-term pharmacologic management of dyslipidemia (75). Ezetimibe, an intestinal cholesterol absorption inhibitor, is very effective at lowering LDLc. However, it is currently under review for its possible role in increasing cancer risk (140). Bile acid resins promote cholesterol elimination through the digestive tract by binding bile acids in the intestine. The use of bile acid resins is limited because they interfere with the absorption of several pharmacological agents, such as digoxin, thyroid hormones, and coumadin. In addition, bile acid sequestrants tend to elevate triglyceride concentrations, and constipation and gastrointestinal disturbances are commonly reported (53). Tota-Maharaj and colleagues (165) have published an excellent contemporary review of recent clinical trial data supporting the beneficial cardiometabolic effects of antihyperlipidemic medications.

Choosing to follow a Mediterranean-style diet, incorporating functional foods in the diet, exercising regularly, and losing a little weight are each effective means for improving blood lipid levels. When these strategies are combined, they may work in concert to improve blood lipids. If patients respond conservatively to each of these therapeutic lifestyle changes, they may expect to lower LDLc and triglyceride levels and improve HDLc to the same magnitude that would be expected with lipid-lowering medications (57). For patients who are not on lipid-lowering medication, the lifestyle strategies could be enough to keep them off these medicines. For patients already taking these drugs, lifestyle strategies may add to their medicine's effect. However, this does not always occur, and more research on combination therapy is clearly needed (1, 137). For all patients, these therapeutic lifestyle changes will provide health benefits above and beyond what can be attained with lipid-lowering medication alone.

EXERCISE PRESCRIPTION

Exercise can have a profound impact either directly or indirectly (i.e., weight changes) on blood lipid and lipoprotein levels. While various exercise modalities can provide health benefits that may ultimately improve lipid status, the majority of evidence suggests that cardiovascular or endurance-type activity is preferred, although a yet-defined combination program may be optimal. Nevertheless, the specific doses for optimizing improvements have not yet been completely defined. Here we review currently recommended programming.

Cardiovascular Exercise

The ranges of physical activity volumes recommended in the 2008 U.S. Department of Health and Human Services' *Physical Activity Guidelines for Americans* (167) and informed by position statements from the ACSM (33, 81, 127) are appropriate for individuals with dyslipidemia (table 9.6). These guidelines suggest ranges of 150 to 300 min of moderate-intensity or 75 to 150 min of vigorous-intensity physical activity per week, or some combination of the two. A dose–response relationship is recognized, with activity at the upper ends of these ranges likely to result in more positive health benefits. Indeed, some of the earliest research investigating the effect of exercise training on lipid and lipoprotein concentrations indicated that a threshold of 8 to 10 mi (12.9 to 16 km) of running per week was necessary to significantly change

HDLc concentration (178, 180). Because the lipid- and lipoprotein-related benefits of physical activity rely more on the total volume of activity performed than on the intensity of the activity (49, 105), walking an equivalent distance is thought to yield similar benefits. Assuming common walking and running speeds (e.g., 3 and 6 mph [4.8 and 9.7 kph], respectively), this 8 to 10 mi threshold falls within the lower end of the moderate-intensity and vigorous-intensity time ranges mentioned.

In support of a dose–response relationship, Kraus and colleagues (105) reported greater favorable changes in LDL particle size and HDLc and triglyceride concentrations in dyslipidemic men and women who averaged 20 mi (32 km) of running per week versus those who ran or walked an average of 12 mi (19 km) weekly. Therefore, while fulfilling the minimum recommendations in the *Physical Activity Guidelines for Americans* is likely beneficial, exercising at or slightly above the upper end of these guidelines may be necessary to optimize lipid and lipoprotein changes. In fact, in some cases, practice of physical activity at the lower end of the recommended levels may result in a "stabilization" of lipid and lipoprotein levels rather than an actual improvement. It should be noted, however, that a sedentary person's lipid and lipoprotein profile is likely to worsen over time, particularly if she is gaining weight, so such stabilization should be recognized as a positive effect if lipid and lipoprotein levels had previously been getting worse. In addition, as mentioned in the next section on exercise training, regularly practiced physical activity can directly affect the metabolism of some lipids and lipoproteins, while alteration of others may depend on changes in body composition. Therefore, energy expenditure at the upper end of the continuum, combined with prudent nutrition, is the best behavioral prescription for improving the overall blood lipid and lipoprotein profile.

While progressing toward 300 weekly minutes of moderate-intensity or 150 weekly minutes of vigorous-intensity physical activity is recommended, little research is available to suggest the number of days per week the activity should be performed. Studies employing single sessions of exercise typically show that lipid and lipoprotein concentrations change the most 1 or 2 days afterward and that they reapproach baseline by 3 days. Thus a sensible approach might be to allow no more than 1 or 2 inactive days between exercise sessions. This would require performing leisure-time physical activity on at least 3 days across the week. Because of the amount of time involved, most people require 5 or more days per week to complete the optimal amount of moderate-intensity physical activity. On the other hand, a person's daily dose of physical activity does not need to be completed all at once. Limited research indicates that breaking physical activity up into three sessions of 10 min or more within a day is at least as effective in altering blood lipid and lipoprotein concentrations as is continuous physical activity (2, 118, 123).

In summary, people should accumulate 150 to 300 min of moderate-intensity or 75 to 150 min of vigorous-intensity dynamic physical activity using large muscle groups, or some combination of the two, throughout each week. To optimize lipid and lipoprotein changes, clients are encouraged to progress to the upper end of the exercise dose range. Leisure-time physical activity should be practiced on at least 3 days throughout the week, and 5 or more days are required for most people to meet the dose recommendations. Moderate- to vigorous-intensity physical activity may be accumulated throughout the day and does not necessarily need to be accomplished within a single daily session.

Table 9.6 Dyslipidemia: Exercise Training Review

Cardiorespiratory endurance	Skeletal muscle strength	Skeletal muscle endurance	Flexibility	Body composition
Dynamic, large muscle group exercise (e.g., walking, jogging, cycling, elliptical training). 150 to 300 weekly minutes of moderate-intensity physical activity OR 75 to 150 weekly minutes of vigorous-intensity physical activity (or some combination of the two). Clients should be encouraged to progress toward the upper end of the dose range.	8 to 10 exercises involving all major muscle groups. One set of 8 to 15 repetitions to fatigue for each exercise. Lighter resistance may be recommended for patients with certain comorbidities (e.g., cardiovascular disease). Two or three sessions per week on nonconsecutive days.	See exercises for skeletal muscle strength.	Stretch all major muscle-tendon groups. Four or more static repetitions of 15 to 60 s per exercise. At least 2 or 3 d/wk.	See exercises for cardiorespiratory endurance.

Clinical Exercise Pearls

1. A dyslipidemic condition alone does not necessitate a change in exercise testing protocol or limiting intensity for exercise training. However, the dyslipidemic patient could have pre-existing comorbidities that would significantly change exercise testing or training protocols.
2. Dyslipidemic patients have an elevated risk of CVD. Therefore, the patients themselves and professionals providing exercise testing and training should be keenly aware of CVD signs and symptoms.
3. Xanthomas that are present in some dyslipidemic patients may be severe enough to cause biomechanical problems.
4. Physical activity and exercise benefits for dyslipidemias are best seen with training that results in large caloric expenditures.
5. Lipid and lipoprotein changes sometimes take several months to reach peak levels.
6. Flushing, sweating, and nausea are associated with niacin. The clinical exercise physiologist should remind the patient to take this medication before going to sleep at night in order to avoid these unwanted side effects during the day and especially with exercise. The clinical exercise physiologist should make the patient aware of these side effects in order to distinguish between the effect of the drug and what might be experienced during exercise (53).
7. Rhabdomyolysis, which is caused by muscle damage and is characterized by the presence of myoglobin in the blood, is a rare side effect of some dyslipidemia medications (statins and fibric acid). Although up to 5% of patients taking these medications may experience less severe muscle soreness, rhabdomyolysis is also characterized by reduced and dark urine output and a general feeling of weakness. Since kidney damage may result, a patient experiencing these symptoms should immediately contact his physician (160).
8. Though grapefruit and grapefruit juice are healthful, providing many important nutriments, grapefruit juice or grapefruit can interact with dozens of medications and have undesirable effects. The chemicals in grapefruit do not interact directly with the medication. Rather these chemicals bind to an intestinal enzyme, blocking its action and making the passage of the medication easier from gut to bloodstream. As a result, blood medication levels may rise faster and remain at higher than normal levels, and in some cases the abnormally high levels are dangerous. With regard to statins and grapefruit use, some statins are affected more than others. For example, with atorvastatin (Lipitor), simvastatin (Zocor), and lovastatin (Mevacor), blood levels are boosted more than when fluvastatin (Lescol), pravastatin (Pravachol), or rosuvastatin (Crestor) is the prescribed medication. As a recommendation, any grapefruit product or grapefruit juice consumption should be avoided when taking a statin. If you have any questions about grapefruit or grapefruit juice consumption and statin use, ask your physician.

Resistance Exercise

Resistance training alone may have a very limited effect on improving blood lipid and lipoprotein concentrations (49, 95). However, given that resistance exercise has numerous benefits not associated with lipoprotein metabolism, dyslipidemic patients should still follow the recommendations presented in chapter 5. Since cardiovascular exercise should be prioritized, more than one set per resistance training exercise is not recommended unless the client is clearly motivated to be compliant with both cardiovascular and resistance exercise.

Range of Motion Exercise

Although it provides no known benefit to lipid and lipoprotein profiles, range of motion exercise is important to overall fitness and should be practiced as discussed in chapter 5. Unless certain comorbidities exist, there are no special considerations for dyslipidemic patients.

EXERCISE TRAINING

Exercise has become a valuable therapeutic treatment for improving blood lipids. Several studies and decades of research provide accumulating evidence of the benefits of exercise training on blood lipids and lipoproteins. While many initial studies were either cross sectional or were conducted in individuals with normal lipid concentrations, more recent approaches have attempted to clarify the expected improvements among dyslipidemia patients.

Cardiovascular Exercise

Blood lipid profiles of physically active groups generally reflect a reduced risk for the development of CVD compared to that of their inactive counterparts (48, 58, 133). There is strong evidence for lower triglyceride and greater HDLc concentrations in physically active individuals. Triglyceride levels are almost always lower in endurance athletes, aerobically trained people, and physically active individuals when compared to sedentary controls. Significant triglyceride differences of up to 50% exist between these groups in over half of all related cross-sectional studies. Blood levels of HDLc are between 9% and 59% higher in those having physically demanding jobs and in individuals engaged in endurance exercise compared to their less active counterparts (48, 49). There is only limited evidence to suggest that people who are physically active exhibit lower concentrations of total cholesterol and LDLc than those who are less active.

In longitudinal studies, total cholesterol and LDLc infrequently change with exercise training in either men or women. When these lipid fractions are altered with exercise training, the reductions are minimal or moderate, averaging only 4% to 7% when compared to values in nonexercising control subjects (48, 49, 77). Based on the frequency of reported changes, HDLc and triglyceride are more responsive to regular exercise than are total cholesterol and LDLc. Significantly greater HDLc concentrations are reported after exercise training in over half of the manuscripts that have been reviewed, while reductions in triglyceride levels are found in a third of the related literature (48, 49). When HDLc is significantly elevated after exercise training, the increases are similar in men and women, ranging from 4% to 22%. Likewise, significant reductions in triglyceride concentrations range from 4% to 37% after aerobic exercise training in males; the magnitude of change is similar in women but is seen less frequently. Resistance training also has positive but more modest effects on blood lipid and lipoprotein concentrations than observed for aerobic exercise (table 9.7) (95).

Exercise and Dyslipidemia

A meta-analysis of lipid changes in normo- and hyperlipidemic groups suggested that exercise training may have only limited influence on lowering total and LDLc. The effects of exercise training on HDLc and triglyceride also favored more conservative estimates, with HDLc increases averaging 4% and triglyceride decreases averaging 6% to 19% (77).

LDLc is lowered by a modest 5.5 mg · dl^{-1} (95% CI: −9.9 to −1.2 mg · dl^{-1}) through regularly practiced walking programs lasting 8 wk or more (100). In men, the pooled

Table 9.7 Lipid and Lipoprotein Changes Associated with Exercise

Lipid/lipoprotein	Single exercise session	Exercise training
Triglyceride	Decreases of 7% to 69% Approximate mean change 20%	Decrease of 4% to 37% Approximate mean change 24%
Cholesterol	No change*	No change†
LDL cholesterol	No change	No change
Small dense LDL cholesterol particles	Literature unclear	Can increase LDL particle size, usually associated with triglyceride lowering
Lp(a)	No change	No change
HDL cholesterol	Increase of 4% to 18% Approximate mean change 10%	Increase of 4% to 18% Approximate mean change 8%

*No change unless the exercise session is prolonged (see text).

†No change if body weight and diet do not change (see text).

average reduction in total cholesterol and increase in HDLc were only 2%, whereas triglyceride was decreased by 9% (94). In both men and women, walking lowered non-HDLc by a pooled mean of 5.6 mg · dl^{-1} (95% CI: −8.8 to −2.4 mg · dl^{-1}), which equated to a modest 4% decrease (99). The pooled estimated changes in total cholesterol, LDLc, HDLc, and triglyceride concentrations indicate that those with documented CVD respond similarly to regular exercise training (96). There is solid evidence that regular moderate-intensity aerobic exercise can increase LDL and HDL particle sizes in men and women (78, 105). In women, however, the lipid responses appear to be more "favorable" among those who are at greater risk for heart disease than among their healthy counterparts (114).

Overweight and Obesity

Meta-analytic results of 13 studies and 31 groups of overweight and obese adults were strikingly similar to those reported for most adults regardless of weight status (77, 94, 99). Total cholesterol decreased by 2% with endurance exercise training of at least 8 wk. Triglycerides were lowered by 11%, and LDLc and HDLc were not significantly affected (98).

More rigorous physical activity programs are required for weight loss, and they result in greater health benefits (47); healthful changes blood lipid and lipoproteins, particularly total and LDLc, are greatest when physical activity is accompanied by weight loss (176, 177, 179). These findings are consistent with an initial meta-analysis of 95 studies examining the question of exercise-induced lipid changes with and without weight loss (166).

Immediate and Transient Effects of a Single Exercise Session

The characteristic antiatherogenic lipid profile of physically active and exercise-trained individuals was recognized long ago as primarily a transient response to the last session of physical activity or exercise and independent of chronic exercise training (131). Early support for this acute response hypothesis came from studies showing that blood triglyceride concentration was reduced for up to 2 d after a single session of aerobic exercise, and that this beneficial reduction was evident even in hyperlipidemic men (28, 28, 86, 131). Pioneering work in this field provided evidence that HDLc concentrations were higher and total cholesterol concentrations lower shortly after a single session of exhaustive exercise (51, 101, 161).

Total and LDLc responses to an exercise session are highly variable. Small postexercise reductions in total cholesterol (3% to 5%) have been reported for male and female hyper and normocholesterolemic subjects (19, 36, 37, 70, 92, 141). Lower serum LDLc has been reported in trained men immediately and up to 72 h after completion of intense endurance events (19, 55) and in women after exercise of relatively high intensity and volume (67, 141). In contrast, LDLc remained unaffected in normolipidemic obese women after 1 h of exercise (183), and LDLc concentration may increase or decrease 5% to 8% in hypercholesterolemic men after exercise (36, 37, 70).

Low-density lipoprotein density, a measure of the atherogenic potential of the LDL, was not altered in either normal or hypercholesterolemic women by a single session of aerobic exercise (35, 182). However, an increase in LDL particle size has been shown to follow completion of a marathon in men (109). The transient effects of exercise on LDL may not always benefit health, as it appears that circulating LDL is more susceptible to harmful oxidation after very intense, long-duration exercise such as a marathon (113).

A single aerobic exercise session of sufficient volume can raise serum HDLc. The postexercise increase in HDLc is strikingly similar to what is generally attributed to long-term endurance exercise training. However, HDLc levels peak 24 to 48 h after exercise and last up to 72 h before returning to preexercise levels (49). An exercise energy expenditure threshold of about 350 kcal is enough to elevate HDLc in deconditioned individuals (36), but a caloric expenditure threshold of 800 kcal or more may be needed in those who are well conditioned (55). High-density lipoprotein density is reportedly reduced for up to 2 d after aerobic exercise, a finding that provides further evidence that exercise acutely influences lipoprotein metabolism in the circulation (105).

A single session of endurance exercise lowers blood triglyceride concentrations (152, 171), and this exercise effect is observed in apparently healthy normo- and hyperlipidemic men (19, 36, 37, 55, 59, 70). Similar to the HDLc responses to a single exercise session, the exercise effect on serum triglyceride is influenced by the training status of the subjects and volume of exercise performed. Regardless of mode or intensity, the exercise effect is lost after about 48 to 72 h (49). Existing evidence also suggests a relationship between preexercise triglyceride concentration and the magnitude of the postexercise change. In other words, people with elevated preexercise serum triglyceride concentrations exhibit the greatest postexercise reductions, while those with relatively low preexercise triglyceride concentrations show only modest or no change after exercise (37, 38, 71, 92). Genetic variations are known to play a role in triglyceride metabolism (91, 135); however, there is a paucity of data on the transient lipid-altering effects of a single exercise session in individuals with genetically determined hypertriglyceridemia.

Postprandial lipemia is lower in the hours after completion of aerobic exercise of sufficient volume (116, 119, 137) but is increased when exercise is withdrawn for several days (79). Together, these data suggest that the beneficial influence of exercise on circulating lipids and lipoproteins is an acute phenomenon that is lost rather quickly after cessation of exercise, even in the most highly trained individuals. The message for the clinical exercise physiologist is that exercise must be repeated regularly to maintain the acute benefit.

Resistance Exercise

Resistance exercises appear to have a small beneficial influence on blood lipids that is somewhat similar to what is reported for endurance exercise. Pooled data from a recent meta-analysis on 29 studies that included 1,329 male and female adult participants provide evidence that regular resistance training modestly lowers total cholesterol, LDLc, non-HDLc, and triglyceride while increasing HDLc (95). The clinician should interpret these recent findings with caution, as relatively fewer calories are expended in resistance versus aerobic activity and therefore resistance training per se may be less effective than endurance activities for modifying blood lipid levels (157).

CONCLUSION

Exercise is an essential element of management of patients with dyslipidemias, with proven benefits that likely extend to a reduction in mortality and morbidity. However, a comprehensive lifestyle treatment approach should be of utmost priority to best attain targeted patient cholesterol goals. As such, in addition to exercise programming, nutritional choices and body composition improvements are needed and, when necessary, use of appropriate medications. A multidisciplinary team approach may prove most beneficial to helping patients achieve long-term success in meeting stringent guidelines.

Practical Application 9.2

RESEARCH FOCUS

Background

Modest weight loss is recommended by several health organizations for improving overall health in overweight and obese individuals (47). However, weight loss is often temporary in people who are initially successful. In fact, almost all obese individuals are unable to sustain permanent weight loss (110), and a third of those achieving significant short-term weight loss will have returned to their original body weight within a year of reaching their weight loss goal (174). The weight relapse is often so discouraging that many individuals quit the very interventions that led to lower body weight and improved health. Clinical exercise physiologists are among many health practitioners who advocate strongly for regularly practiced exercise to improve and maintain health. Unfortunately, there is currently very little evidence to determine whether physical activity attenuates or prevents the adverse health consequences of a positive energy balance and weight regain.

Research Methodology

Researchers from the University of Missouri employed a well-designed randomized control trial to help us understand the health benefits of regularly practiced exercise during weight regain. Sixty-seven overweight or obese participants (27 women and 40 men) lost 10% of their initial body weight over a 4 to 6 mo period through prescribed diet and exercise. After reaching their target weight loss, participants underwent partial weight regain (up to 50% of the initial weight lost) in a supervised program of increased energy intake. An exercise group continued regular exercise while a control group discontinued exercise altogether throughout a 4 to 6 mo weight regain period. Weight loss was induced by nutritionist-supervised caloric restriction (−600 kcal/d) and 2,000 kcal/wk of walking/jogging. The exercisers continued with their program of 2,000 kcal/wk in exercise energy expenditure during the weight regain period. Weight regain occurred with a caloric intake that caused steady positive energy balance in both exercisers and controls. The increased caloric intake over this period was administered through healthy snacks (e.g., protein bars, dried fruits, nuts). Most of the health parameters that were addressed in this investigation are commonly measured in the clinical setting. Measurements before the

(continued)

intervention, after weight loss, and again after weight regain included markers of cardiovascular and metabolic health; blood lipids; markers of glucose homeostasis, inflammation, and oxidative stress; blood pressure; and body fat distribution.

Results

With respect to blood lipids, exercise during weight regain helped to maintain weight loss-induced decreases in LDLc and oxidized LDL. The improvements in triglyceride and total and HDLc measured after weight loss were not countered by exercise as body weight was regained. The likely reason was that regular exercise did not prevent the restoration of abdominal visceral and subcutaneous fat.

Regular exercise during weight regain was successful at maintaining the improvement in cardiovascular fitness that was observed with increased physical activity during weight loss. This finding alone represents a significant protective effect against cardiovascular and metabolic disease risk (175). The exercisers were successful at maintaining improvements in clinical markers of glucose homeostasis and some humoral markers of inflammation.

Commentary and Significance

Exercise alone during weight regain has only a limited ability to maintain a healthy blood lipid profile. During weight relapse, approximately 200 min of moderate-intensity exercise resulting in 2,000 kcal/wk is effective at defending the LDLc reductions that were incurred by weight loss during weight relapse. However, this does not appear to be enough exercise to preserve the weight loss-induced improvements in other blood lipids during positive energy balance. In addition, one might argue that blood lipid profiles may have deteriorated more with additional weight regain, or if a high-fat nutrient composition typical for many adult Americans had been part of the increased caloric intake used in the weight regain protocol (17). We don't know if exercise of this volume would offer a protective effect from unhealthy dieting. The ethics of a research protocol to address this issue would certainly be called into question. Even so, this important innovative research provides evidence that regularly practiced endurance exercise may help to maintain many of the health gains attributed to weight loss during a subsequent period of weight regain. The clinical exercise physiologist is now armed with new evidence that regularly practiced exercise can preserve LDLc reductions and other health benefits of weight loss even when weight loss cannot be sustained.

Key Terms

apolipoproteins (p. 156)
atheromas (p. 162)
cholesterol ester transfer protein (CETP) (p. 160)
chylomicrons (p. 156)
dyslipidemias (p. 155)
dyslipoproteinemia (p. 157)
hepatic lipase (HL) (p. 160)
hyperlipidemia (p. 155)
hyperlipoproteinemia (p. 157)
hypertriglyceridemia (p. 157)
lecithin-cholesterol acyltransferase (LCAT) (p. 160)
lipoprotein (p. 155)
lipoprotein lipase (LPL) (p. 160)
postprandial lipemia (p. 157)
xanthomas (p. 162)

CASE STUDY

MEDICAL HISTORY

Mr. R is a 34 yr old executive manager for a very popular national chain restaurant. Between his family, work, and volunteer duties, he has rarely had the time for physical activity. His chronic knee problems, attributed to mild osteoarthritis, are most evident after he stands for long periods of time at work. Mr. R has a family history of CHD, as his father and brother each underwent cardiovascular revascularization surgeries in their late 40s.

As part of a recent medical checkup, Mr. R's physician strongly recommended that he get on a weight reduction program that included both diet and exercise. Mr. R was told that he would more than likely have to address his burgeoning weight problem and other health-related issues through expensive medication or even surgery if he did not do something about it right now. Mr. R's physician has sent him to see you and the hospital nutritionist about developing an exercise and diet plan for targeting his most pressing health concerns over the next 6 mo (24 wk). After receiving Mr. R's permission, you obtain his medical information, consult with the nutritionist, and perform several physical tests to develop your exercise plan. Information from the physical exam and your preliminary tests is as follows:

Anthropometric

Height: 67 in. (170 cm)

Weight: 113 kg

34% body fat

Waist circumference: 44 in. (112 cm)

Hemodynamics

Resting HR: 72 beats · min^{-1}

Resting BP: 138/84 mmHg

Fasting Blood Panel

Cholesterol: 262 mg · dl^{-1}

LDLc: 180 mg · dl^{-1}

HDLc: = 41 mg · dl^{-1}

TG: = 204 mg · dl^{-1}

Blood glucose: 111 mg · dl^{-1}

DIAGNOSIS

Mr. R has no known disease and exhibits no signs or symptoms of latent heart disease. He has multiple CVD risk factors. Together, these characteristics place him at moderate risk according to the current ACSM risk stratification criteria.

EXERCISE TEST RESULTS

A physician-monitored exercise test was ordered because of Mr. R's family history and his history of physical inactivity. Mr. R's treadmill test results indicated a normal blood pressure, ECG, and heart rate responses to exercise. His maximum heart rate was measured at 7:20 min of the Bruce protocol and was 190 beats · min^{-1}. His $\dot{V}O_2max$ was estimated to be 24.2 ml · kg^{-1} · min^{-1}.

EXERCISE PRESCRIPTION

Mode

Any mode that uses large muscle groups in dynamic activity.

Frequency

Begin with 3 d/wk of supervised exercise.

Intensity

Try 40% to 50% of HRR (HR range = 119 to 131 beats · min^{-1}) and rating of perceived exertion (RPE) of 4 to 6 on scale of 1 to 10.

Duration

Complete 30 min. Consider intermittent bouts using the same or different modes if Mr. R cannot complete 30 continuous minutes of exercise.

(continued)

Energy Expenditure

Target a total exercise energy expenditure of 200 kcal per session or 600 kcal per week.

DISCUSSION

The exercise physiologist should educate Mr. R regarding the benefits to be reasonably expected from regularly practiced exercise, the potential risks, and how to recognize when it may not be safe to continue exercising. Help Mr. R choose a mode that he prefers, as his choice may help with adherence. In addition, consider recommending to Mr. R that he alternate between modes (e.g., treadmill walking, elliptical walking, cycle ergometer). Mixing up the exercise type may help to make exercise more enjoyable and may help to prevent knee pain. Initially, it is more important to develop a regular habit of exercise than it is to attain the energy expenditure needed to address Mr. R's dyslipidemia. Work with Mr. R to include a basic resistance exercise routine consisting of one set of 8 to 12 repetitions for each major muscle group (8 to 10 exercises total). Advocate for Mr. R to become more physically active every day (81, 167).

DISCUSSION QUESTIONS

1. Identify and discuss the different ways to progress the exercise training plan to increase weekly caloric expenditure without increasing the session duration.
2. Explain to Mr. R why resistance training is an integral part of his exercise plan.
3. Describe your specific resistance training recommendations. Include the specific exercise, the number of sets and reps, and any other information that would be necessary for Mr. R to clearly understand his program. Provide a rationale for your response.
4. List the primary risks of exercise that you would be concerned about. Next to each of the risks you mention, list the precautions you will take in order to make Mr. R's exercise safe.
5. Identify the dietary components and caloric intake that a nutritionist may recommend to address Mr. R's dyslipidemia. Describe the benefits of each dietary recommendation that you identify.

Chapter 10

Metabolic Syndrome

Mark D. Peterson, PhD
Paul M. Gordon, PhD, MPH

The metabolic syndrome is a collection of interrelated **cardiometabolic risk factors** that are present in a given individual more frequently than may be expected by a chance combination. Patients with the metabolic syndrome are usually characterized as overweight or obese, and have significantly greater prospective risk for developing atherosclerotic cardiovascular disease (**ASCVD**) (114) and type 2 diabetes (169). Although the metabolic syndrome has received a great deal of research attention over the past decade, the accumulation and clustering of obesity-related cardiometabolic risk factors have been documented by physicians and investigators for much longer. Indeed, the metabolic syndrome was originally alluded to by a Swedish physician who in 1923 reported an association between hypertension, hyperglycemia, and hyperuricemia (99). Moreover, in 1947, Vague documented a robust link between "android" (i.e., central or abdominal) adiposity and metabolic abnormalities (162); and in 1965 and 1967, Avogaro and colleagues confirmed the co-occurrences of obesity, hyperinsulinemia, hypertriglyceridemia, and hypertension (15, 16). However, it wasn't until 1977 that the term "metabolic syndrome" was used, by Haller, who suggested that obesity, diabetes mellitus, hyperlipoproteinemia, hyperuricemia, and hepatic steatosis cluster to pose significant risk for developing or worsening arteriosclerosis (81). On the basis of those findings, and during his Banting lecture to the American Diabetes Association in 1988 (135), Dr. G.M. Reaven proposed that hyperinsulinemia and insulin resistance were the pathophysiologic links between these risk factors, in a condition for which he coined the term "Syndrome X." Since then, the metabolic syndrome has also been referred to as "dyslipidemic hypertension" (150), "the deadly quartet" (91), and "insulin resistance syndrome" (52), among other names. Although at present *metabolic syndrome* is a widely recognized public health term, the clinical diagnosis and classification of this condition have been equivocal. This is in large part due to historical inconsistencies regarding the operationalization and classification of the condition.

DEFINITION

An initial formal definition for the metabolic syndrome was proposed by a consultation group for the World Health Organization (WHO) in 1998; however, it was ultimately deemed impractical due to the requirement of direct measures of insulin resistance (13). Since this original definition, numerous other public health entities have proposed alternative criteria to provide a basis for clinical diagnostics and epidemiological surveillance. Specifically, the European Group for the Study of Insulin Resistance, the American Association of Clinical Endocrinologists, the American Heart Association (AHA), the International Diabetes Federation (IDF), the National Cholesterol Education Program Adult Treatment Panel III (ATP III), and the National Heart, Lung, and Blood Institute (NHLBI) have all contributed separate definitions (5, 6, 14, 46, 77). Whereas most of these demonstrated redundancy regarding the individual risk factors of the metabolic syndrome (i.e., dysglycemia, raised blood pressure, elevated triglyceride levels, low high-density lipoprotein cholesterol levels, and central obesity), the conflicting criteria for cut points across definitions have caused confusion in the clinical context for identifying patients at risk, as well as inconsistencies in population-based prevalence studies.

More recently, a new harmonized definition has been proposed by the IDF Task Force on Epidemiology and Prevention, AHA, NHLBI, the World Health Federation, the International Atherosclerosis Society, and the International Association for the Study of Obesity (12). The purpose of this harmonized definition was to standardize the criteria for metabolic syndrome by publishing an agreed-upon set of cut points for individual factors (table 10.1). Further, whereas several previous definitions incorporated obligatory factors (e.g., insulin resistance and abdominal obesity [13, 14]), the newest definition recommends that a diagnosis of metabolic syndrome reflect the presence of any three or more abnormal findings. Although this diagnostic scheme is superior to the previous discordant recommendations, there is still some disagreement pertaining to the definition of and cut points for elevated waist circumference. This is due to the significant variability in abdominal obesity phenotypes between sexes and among races and ethnic groups, as well as the subsequent association with other metabolic risk factors and predictive values for ASCVD and diabetes. Therefore, the harmonized definition defers to the IDF recommendations for cut points of abdominal obesity to be used as a single potential factor in the diagnosis of the metabolic syndrome (table 10.2) (12, 14). Table 10.2 also includes the waist circumference thresholds being recommended in several different populations and ethnic groups and by different organizations.

SCOPE

Before the harmonized definition was proposed, the reported prevalence of metabolic syndrome varied widely depending on the criteria used for diagnosing the condition. According to the original report by Ford and colleagues (56), during the late 1980s and early 1990s, approximately 47 million Americans were affected by the metabolic syndrome. By the year 2000, this estimate had increased to nearly 65 million persons (57). Since then, prevalence of the metabolic syndrome has continued to steadily rise to 34.1% based on National Health and Nutrition Examination Survey (NHANES) 2003-2006 data and ATP III criteria (adults ≥20 yr) (50), or to approximately 40% according to IDF criteria (55). This is not entirely surprising considering the high prevalence of obesity among U.S. adults (7, 54). Whereas the latest NHANES data actually suggest a tapering in the upward

Table 10.1 Harmonized Clinical Criteria for the Metabolic Syndrome

Measure	Categorical cut points
Elevated waist circumference*	Population- and country-specific definitions
Elevated triglycerides (drug treatment for elevated triglycerides is an alternate indicator)**	≥150 mg · dl^{-1} (1.7 mmol · L^{-1})
Reduced HDLc (drug treatment for reduced HDLc is an alternate indicator)**	<40 mg · dl^{-1} (1.0 mmol · L^{-1}) in males <50 mg · dl^{-1} (1.3 mmol · L^{-1}) in females
Elevated blood pressure (antihypertensive drug treatment in a patient with a history of hypertension is an alternate indicator)	Systolic ≥130 mmHg, diastolic ≥85 mmHg, or both
Elevated fasting glucose+ (drug treatment of elevated glucose is an alternate indicator)	≥100 mg · dl^{-1}

HDLc indicates high-density lipoprotein cholesterol.

*It is recommended that the IDF cut points be used for non-Europeans and either the IDF or AHA/NHLBI cut points be used for people of European origin until more data are available.

**The most commonly used drugs for elevated triglycerides and reduced HDLc are fibrates and nicotinic acid. A patient taking one of these drugs can be presumed to have high triglycerides and low HDLc. High-dose omega-3 fatty acids presumes high triglycerides.

+Most patients with type 2 diabetes mellitus have the metabolic syndrome by the proposed criteria.

Table 10.2 Current Recommended Waist Circumference Thresholds for Abdominal Obesity by Organization

		RECOMMENDED WAIST CIRCUMFERENCE THRESHOLD FOR ABDOMINAL OBESITY	
Population	Organization	Men	Women
Europid	IDF	≥94 cm	≥80 cm
Caucasian	WHO	≥94 cm (increased risk) ≥102 cm (higher risk)	≥80 cm (increased risk) ≥88 cm (higher risk)
United States	AHA/NHLBI (ATP III)	≥102 cm	≥88 cm
Canada	Health Canada	≥102 cm	≥88 cm
European	European Cardiovascular Societies	≥102 cm	≥88 cm
Asian (including Japanese)	IDF	≥90 cm	≥80 cm
Asian	WHO	≥90 cm	≥80 cm
Japanese	Japanese Obesity Society	≥85 cm	≥90 cm
China	Cooperative Task Force	≥85 cm	≥80 cm
Middle East, Mediterranean	IDF	≥94 cm	≥80 cm
Sub-Saharan Africa	IDF	≥94 cm	≥80 cm
Ethnic Central and South American	IDF	≥90 cm	≥80 cm

Recent AHA/NHLBI guidelines for metabolic syndrome recognize an increased risk for CVD and diabetes at waist circumference thresholds of ≥94 cm in men and ≥80 cm in women and identify these as optional cut points for individuals or populations with increased insulin resistance.

Reprinted, by permission, from K.G. Alberti et al., 2009, "Harmonizing the metabolic syndrome: a joint interim statement of the International Diabetes Federation Task Force on Epidemiology and Prevention; National Heart, Lung, and Blood Institute; American Heart Association; World Heart Federation; International Atherosclerosis Society; and International Association for the Study of Obesity," *Circulation* 120: 1640-1645.

trend for overweight and obesity prevalence (54, 118), current statistics are no less daunting—based on body mass index most Americans are overweight, and more than one-third are obese (54, 66). Interestingly, abdominal obesity has recently been identified as the most prevalent component of the metabolic syndrome and, depending on criteria, affects one-half to three-quarters of all U.S. adults (58). Further, evidence continues to mount in support of the robust association between excess abdominal, visceral, and hepatic adiposity and a broad spectrum of deleterious cardiometabolic outcomes (41), prompting the alarming public health message that obesity leads to more chronic disorders and poorer health-related quality of life than smoking (159). Since current statistics do not indicate a tapering of metabolic syndrome prevalence in the United States (58), further efforts should be supported that target the diagnosis, prevention, and treatment of both obesity and the metabolic syndrome as separate yet overlapping conditions.

Previous epidemiological studies have used the definitions set forth by either the WHO, ATP III, or IDF guidelines. Whereas the new harmonized definition (12) represents a significant departure from the WHO definition, the only modifications to previous ATP III guidelines are a lower cutoff for systolic blood pressure, ≥130 mm Hg, and an endorsement of IDF race- and ethnicity-specific cutoffs for abdominal obesity (table 10.2) (i.e., rather than abdominal obesity defined uniformly across all races and ethnicities as ≥102 cm for men and ≥88 cm for women). In a recent study performed to reestimate prevalence using the new harmonized definition, as well as IDF waist circumference criteria and NHANES 2003-2006 data, Ford and colleagues (58) demonstrated the age-adjusted frequency of the metabolic syndrome to be 38.5% among all adults, 41.9% among men, and 35% among women.

Many reports have suggested similar prevalences of the metabolic syndrome between men and women, but a significant disproportion by age and ethnicity. However, greater disparities were recently revealed when sex-specific prevalence was stratified for decade of life. Thus, it may well be that the cardiometabolic consequences of aging are different in men and women. Specifically, in their recent report, Ford and colleagues (58) found that prevalence of the metabolic syndrome was 19.8% and 15%

for men and women aged 20 to 29 yr, 33% and 17.5% for men and women aged 30 to 39 yr, 45.9% and 33.4% for men and women aged 40 to 49 yr, 49.5% and 46.2% for men and women aged 50 to 59 yr, 67.3% and 57.6% for men and women aged 60 to 69 yr, and 51.9% and 63.5% for men and women ≥70 yr. The prevalence of the metabolic syndrome among pregnant women has also increased significantly, from approximately 18% during 1988 to 1994 to nearly 27% during 1999 to 2004 (134).

The prevalence of metabolic syndrome varies significantly by race and ethnicity (8). The highest incidence rates of metabolic syndrome in the United States have historically been reported in Hispanic Americans, Native Americans, and African Americans, which also may contribute to the disproportionately higher risk of cardiovascular disease and diabetes among these groups (8). Although the specific mechanisms to fully explain these disparities are at present unclear, it has been suggested that differential susceptibility to abdominal obesity and insulin resistance may play a substantial role (8). Based on current estimates, prevalence is lower among African American men than among white or Mexican American men, and lower among white women than among African American or Mexican American women (58) (figure 10.1). Prevalence is also high among immigrant Asian Indians and is estimated to be between 27% and 38%, depending on the criteria used (111, 139).

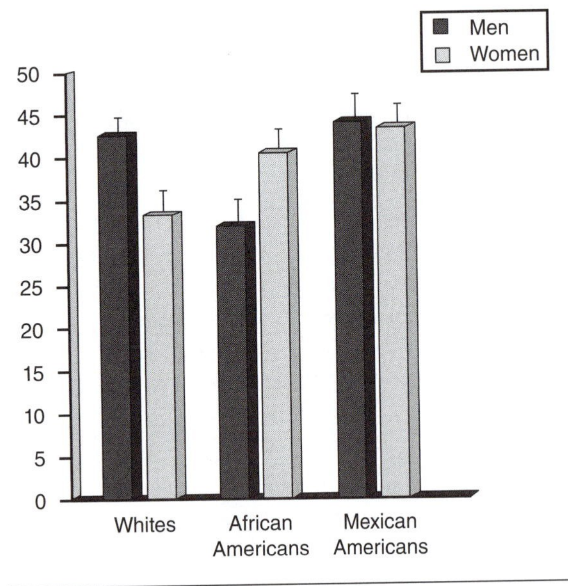

Figure 10.1 Ethnic diversity in relation to the metabolic syndrome.

Normal-Weight Obesity

Based on current IDF cutoffs, nearly 67% of men and 80% of women have abdominal obesity (i.e., as opposed to ~46% and 61% for men and women, respectively, using ATP III criteria) (58). As previously mentioned, there is a well-documented and strong association between obesity and the metabolic syndrome; however, there are exceptions in which these phenotypes do not overlap. One exception, the "metabolically healthy obese" (27, 82, 155) phenotype, is characterized by high body mass index (BMI) or waist circumference yet subclinical (i.e., normal) cardiometabolic health. Conversely, there are also cases of **"normal-weight obesity"** (108, 141), in which an individual is classified in the normal range for body mass or BMI, but has elevated adiposity or risk factors (or both) for metabolic syndrome, ASCVD, or diabetes.

Based on the new harmonized definition and NHANES 2003-2006 data, age-adjusted prevalence of metabolic syndrome among normal-weight (BMI <25) adults (13.3%) is lower than among overweight (BMI 25.0-29.9) (40.1%) and obese (BMI ≥30) (60.7%) adults (58). Despite its relative infrequent occurrence, recent data have confirmed that the normal-weight obesity phenotype may be present in as many as 30 million Americans and is strongly associated with cardiometabolic dysregulation, a high prevalence of metabolic syndrome, and increased risk for cardiovascular mortality (141). The vast majority of population-based studies use surrogate indicators of adiposity (e.g., BMI and waist circumference) (140), which can be somewhat problematic since such metrics do not discriminate between adipose tissue and muscle and lack sensitivity to identify nonobese individuals with excessive body fat (119).

Metabolic Syndrome and the Aging Adult

Incidence of normal-weight obesity is of particular concern for older adults with age-related losses of bone mass (i.e., osteopenia or osteoporosis) and muscle mass (i.e., sarcopenia) or frailty, as these conditions often coincide with subclinical chronic inflammation and oxidative stress (33), insulin resistance (157), myosteatosis (i.e., intra- and intermuscular adipose tissue infiltration), and increased overall fat mass (i.e., sarcopenic obesity) (40, 146). Therefore, under these conditions an individual may have a normal BMI and yet still have excessive body fat or increased risk of cardiometabolic decline or both. Indeed, prevalence of the metabolic syndrome increases with age and peaks among individuals aged 60 to 69 yr (58).

Pediatric Metabolic Syndrome

Evidence from NHANES III (38) has indicated that approximately two-thirds of adolescents aged 12 to 19 yr present with one or more metabolic abnormalities. Moreover, based on the NHANES 1999-2002 data and ATP III criteria, nearly 10% of this population had clustering of three or more metabolic abnormalities (36). Those data notwithstanding, the utility of identifying and diagnosing pediatric metabolic syndrome has been heavily debated due largely to the instability of dichotomized diagnoses in this population (158). According to the Princeton School District Study using ATP III criteria, over 1,000 adolescents diagnosed with the metabolic syndrome lost the diagnosis within 3 yr (70). Conversely, the occurrence of obesity and insulin resistance during childhood has been demonstrated to track into adulthood (79, 156), and appears to be an antecedent to adult-onset type 2 diabetes (117) and cardiovascular disease (113). In addition, several ASCVD risk factors among children, including obesity, glucose intolerance, and hypertension, have been associated with higher risk of premature death (61); and there is evidence to suggest that risks for ASCVD originate during childhood. Specifically, in the Bogalusa Heart Study, Berenson and colleagues (22) observed preclinical coronary atherosclerotic lesions in young autopsies, and the extent of the lesions was robustly associated with BMI, blood pressure, serum concentrations of total cholesterol, triglyceride, low-density lipoprotein (LDL), and high-density lipoprotein (HDL).

Since the risk for cardiometabolic diseases can originate during childhood, screening children to identify emergent risk is crucial for early intervention and public health preventive efforts. However, consistent guidelines for research and clinical practice are still lacking with regard to the structure and norms of cardiovascular disease (CVD) risk factors for children and adolescents (159). In 2009 the AHA released a scientific statement about the progress and challenges pertaining to metabolic syndrome in children and adolescents (158). In particular, barriers to a universally accepted definition include the fact that clinical endpoints emerge infrequently at an early age, and there is substantial variability in the established "normal values" across different ages, between the sexes, and across races. Moreover, there is a complete lack of cut points for abdominal obesity (thus forcing the use of sex- and age-adjusted BMI cut points), as well as for a "normal" range of insulin concentration or cutoffs for physiologic insulin resistance during puberty (158).

Although there has also been debate regarding the designation of "high BMI" (118), a recent expert committee (97) has recommended that children and adolescents aged 2 through 19 yr at or above the 95th percentile of BMI for age be labeled obese, and that children between the 85th and 95th percentiles be labeled overweight. The use of current adult criteria to stratify children and adolescent patients with metabolic abnormalities could significantly underestimate emergent cardiometabolic risk. Due to the progressive nature of cardiometabolic disease, dichotomizing individual risk factors may thus lead to losses of valuable information and possible misclassification (83). Rather than simple binary definitions for classification or diagnosis, the use of aggregate metabolic syndrome continuous scores (cMetS) (48, 90, 120) has been reported to be potentially superior in pediatric populations (47, 160), particularly for epidemiological studies (168). Using this method, each factor (i.e., BMI, blood pressure or mean arterial pressure, glucose or homeostasis model assessment of insulin resistance, HDLc and triglyceride levels) is weighted (e.g., through principal component analysis, summing of standardized residuals or z-scores, or a combination of the two), and a higher aggregate score is indicative of diminished metabolic health (see practical application 10.1). Since the discrete associations between individual risk factors and their interactions during atherosclerosis disease proliferation are at present undetermined, the proper weighting of each factor for a scoring system is not uniformly agreed upon. However, in a recent investigation to determine the validity of a cMetS among a nationally representative sample of 1,239 adolescents (2003-2004 and 2005-2006 NHANES), data revealed high construct and predictive validity for the metabolic syndrome (120). Indeed, if such a method is to be useful for widespread public health endeavors, future efforts are needed not only to determine the appropriate weighting for individual components of cardiometabolic health risk, but also to identify how these methods could be readily implemented in a clinical setting to improve diagnostic and treatment capabilities.

Economic Burden

The metabolic syndrome is frequently referred to as "the disease of the new millennium" (59).Although much debate currently surrounds the definition, criteria, and diagnostic utility for the metabolic syndrome, high prevalence rates combined with the morbidity and mortality associated with this condition imply significant economic consequences (107). There is general consensus that the increasing incidence of the metabolic syndrome represents a huge burden on current and future medical costs coincident with obesity, CVD, and diabetes (107). However, since the explicit associations between the metabolic syndrome, its component risk factors, long-term clinical outcomes, and future health care costs have not been entirely unraveled, it is difficult to pinpoint actual public health ramifications of the syndrome. This is due,

Practical Application 10.1

RESEARCH FOCUS: GENDER-SPECIFIC PREDICTORS OF CARDIOMETABOLIC RISK IN 6TH GRADERS

This study (127) was designed to determine the biological and behavioral variables associated with cardiometabolic disease risk among adolescents, as well as to assess the pattern of risk component clustering. Using a cross-sectional research design, a large cohort (*n* = 2,866) of 6th grade students was assessed for cardiometabolic profiles. Cardiometabolic risk components included waist circumference, fasting glucose, blood pressure, plasma triglyceride levels, and HDL cholesterol. Principal components analysis was used to determine the pattern of risk clustering and to derive a continuous aggregate score (MetScore). Individual risk components and MetScore were then analyzed for association with age, general adiposity (i.e., BMI), cardiorespiratory fitness, physical activity (PA), and several parental factors (e.g., age, adiposity, family history of CVD). The findings demonstrated that BMI was associated with multiple risk factors and overall MetScore among boys and girls. Moreover, cardiorespiratory fitness was a strong negative predictor, such that greater fitness conferred lower risk. In addition to BMI and cardiorespiratory fitness, various parental and familial factors were found to influence adolescent health. Specifically, maternal smoking was associated with multiple risk factors in girls and boys even after controlling for children's BMI. Moreover, paternal family history of early CVD and parental age were associated with increased blood pressure and MetScore for girls. Children's PA levels, maternal history of early CVD, and paternal BMI were also indicative for various risk components in girls or boys. These findings revealed numerous modifiable biological and behavioral factors (i.e., in both children and parents) that shared robust associations with children's cardiometabolic health risk. Factors associated with risk clustering might represent the ideal targets for prevention and treatment therapies to reduce the cardiometabolic health burden in children. This study serves to bolster the value of fitness, PA, and family-oriented healthy lifestyles for improving children's cardiometabolic health.

Reference: Peterson MD, Liu D, IglayReger HB, Saltarelli WA, Visich PS, Gordon PM. Principal Component Analysis Reveals Gender-Specific Predictors of Cardiometabolic Risk in 6th Graders. *Cardiovasc Diabetol.*, 2012;11:146.

in large part, to the lack of definitive data to confirm a greater health risk from the dichotomized metabolic syndrome diagnosis per se than that conferred by the sum of the component risk factors (90, 153). This issue aside, the cumulative annual costs of medical care for CVD (~$286 billion [139]) and diabetes (~$116 billion [10]) have reached a seemingly insurmountable magnitude, much of which may be attributable to the prevalence of and the cost incurred by obesity (~$147 - 190 billion per year [31]) and the metabolic syndrome.

In a 2007 study to quantify the long-term costs of the metabolic syndrome and constituent risk factors among nearly 4,000 elderly adults (≥65 yr), Curtis and colleagues (37) demonstrated that total Medicare costs were 20% higher among participants with metabolic syndrome over a 10 yr period as compared to healthy older adults. Findings further revealed that abdominal obesity (15%), low HDL cholesterol (16%), and high blood pressure (20%) were also independent predictors of long-term costs. More recently, Schultz and Edington (149) examined 4,188 employees of a large manufacturing corporation in an attempt to identify associations between metabolic syndrome prevalence and other health risks or conditions, as well as with total "economic costs" (i.e., health care costs, pharmaceutical costs, short-term disability absenteeism, and on-the-job productivity). Findings revealed that approximately 30% of all employees met criteria for metabolic syndrome (77) and that among those individuals, risk of additional health complications or conditions (e.g., allergies, asthma, arthritis, back pain, depression) were significantly greater, as were health care costs ($3,340 vs. $1,788), pharmacy costs ($570 vs. $270), and costs stemming from short-term disability ($106 vs. $59) (149). Even more recently, in a small cohort of Chevron Texaco Corporation employees, Birnbaum and colleagues (25) demonstrated that employees with metabolic syndrome had double the costs of those without any risk factors ($4,603 vs. $1,859).

PATHOPHYSIOLOGY

At present, the etiology of metabolic syndrome is widely debated; however, most in the field agree that this condition leads to ASCVD and diabetes (114, 169) and is characterized by a confluence of atherogenic dyslipidemia, hypertension, elevated glucose, chronic low-grade inflammation, and prothrombosis (75). In conjunction with several behavioral factors (i.e., sedentary behavior and atherogenic diet), genetic predisposition, and advancing age, the clustering of multiple risk components within the metabolic syndrome is widely thought to occur as a result of obesity (more specifically, abdominal obesity) and insulin resistance (75, 130, 158) (figure 10.2). However, although obesity and insulin resistance are considered two hallmarks of chronic health risk, not all obese or insulin-resistant individuals develop the metabolic syndrome (123). Indeed, the specific trajectory of cardiometabolic decline leading to or coinciding with excessive accumulation of adiposity, altered fat partitioning, and diminished insulin sensitivity is a multifactorial, complex issue to disentangle.

Insulin Resistance and the Metabolic Syndrome

Gradual decreases in cardiometabolic integrity start to occur long before an individual reaches obesity or is diagnosed as insulin resistant. For that matter, the metabolic syndrome is often referred to as a premorbid condition (153). In a healthy, insulin-sensitive person, glucose stimulates the release of insulin from pancreatic beta cells, which in turn reduces plasma glucose concentration through suppression of hepatic glycogenolysis and gluconeogenesis and simultaneous glucose uptake, utilization, and storage by the liver, muscle, and adipose tissue. Conversely, under conditions of insulin resistance, there is a chronic failure of insulin to maintain glucose homeostasis. The role of insulin resistance in the development of the metabolic syndrome has been controversial, as a direct causal role has not yet been identified (90). Part of this confusion has been driven by the consistently oversimplified definition of insulin resistance, which although useful in the clinical context, does not account for the fact that insulin regulates various other processes in addition to glucose metabolism (80) and cannot discriminate the origin of the "resistance" (i.e., defects in insulin signaling vs. those in the insulin receptor [24]). It is postulated that the insulin resistance associated with the metabolic syndrome is indeed "pathway specific" (80), and that multiple forms of molecular insulin resistance could contribute to abnormal glucose homeostasis (23). In any case, in the clinical setting, insulin resistance is characterized as the extent to which the liver manifests insulin resistance in proportion to the periphery (i.e., skeletal muscle insulin resistance) (158). Over time, resistance to insulin in the metabolic syndrome causes hyperinsulinemia and thus increased lipogenesis, **hypertriglyceridemia**, hypertension, and **steatosis** (23, 80, 158). Although a thorough discussion of insulin action and insulin signaling is beyond the scope

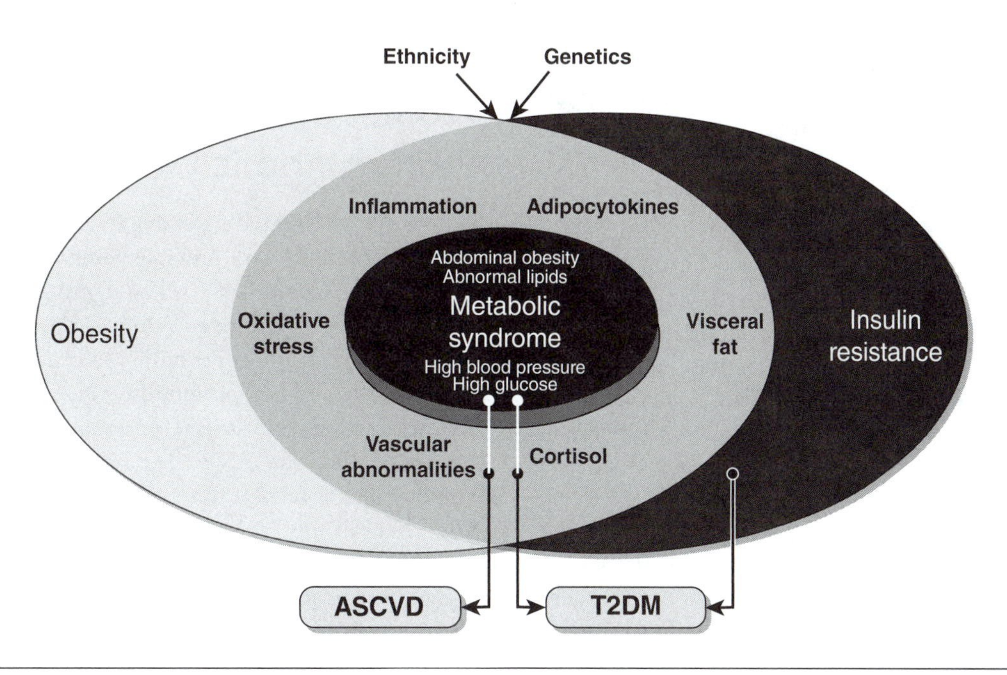

Figure 10.2 Schematic of the components of metabolic syndrome.

Reprinted, by permission from J. Steinberger et al., 2009, "AHA scientific statement: Progress and challenges in metabolic syndrome in children and adolescents," *Circulation* 119: 628-647.

of this chapter, research continues to emerge that confirms a pathophysiologic link not only between insulin resistance, the metabolic syndrome, and glucose intolerance, but also with ASCVD (80). Thus, if the metabolic syndrome is indeed a premorbid condition, as has been suggested (90), identifying and treating at-risk individuals before the emergence of categorical insulin resistance or hyperglycemia is an exceedingly important directive.

Obesity and Ectopic Adiposity

Obesity is an independent risk factor for insulin resistance, hyperglycemia, hypercholesterolemia, and hypertension. Left untreated, this combination of pathophysiologic factors precipitates increased risk for chronic disease and early all-cause mortality (41, 98). Despite the robust association between obesity and cardiometabolic health decline, BMI is suggested to account for only 60% of the variance in insulin resistance among adults (9). Rather, in conjunction with several abnormalities in adipose tissue metabolism, abnormal fat distribution and partitioning may actually be the pathophysiologic link between obesity per se and the numerous hormonal and metabolic derangements that compose the metabolic syndrome. Accumulation of fatty acids in nonadipose tissues (i.e., **ectopic adiposity**) is a dynamic, "lipotoxic" (147) process that occurs as a result of chronic disequilibrium between energy intake and energy expenditure—and is robustly associated with skeletal muscle insulin resistance (86, 95, 101). Specifically, visceral adipose tissue (VAT) is broadly recognized to have metabolic, endocrine, and immune system interactions; and increases in VAT precipitate heightened risk for metabolic and cardiovascular disorders. However, during conditions in which fat infiltrates the muscle (i.e., inter- and intramuscular adipose tissue [**IMAT**]), it appears to independently contribute to impaired glucose metabolism and decreased insulin sensitivity (101). Most often characterized with cross-sectional data from aging adults (i.e., muscle attenuation on computed tomography [71] or localized IMAT with magnetic resonance imaging [143]), this fat infiltration also occurs with certain disease processes (e.g., Duchenne muscular dystrophy and type 2 diabetes) (64, 103), spinal cord injury (73, 151), and obesity (74, 92, 154), as well as during extended periods of sedentary behavior (106) and with vitamin D insufficiency (67). IMAT is a dynamic tissue with both paracrine and endocrine properties (68) and is suggested to arise from satellite stem cells or distinct fibrocyte–adipocyte progenitor cells, which form adipocytes within skeletal muscle during conditions of metabolic dysregulation and hyperglycemia (11). Moreover, evidence reveals a robust link between IMAT and elevated levels of proinflammatory, adipocyte-derived hormones and cytokines (19, 173), which also lead to insulin resistance (94) and muscle dysfunction (165).

Mitochondrial Dysfunction

Concurrent with an increased storage of ectopic adiposity, reductions in mitochondrial size, density, and function (125, 129, 148) have been implicated in the etiology of insulin resistance, metabolic syndrome, and diabetes. More specifically, obese, sedentary, and insulin-resistant individuals have smaller and fewer mitochondria, with impaired function. Diminished mitochondrial density and function may lead to or coincide with decreased or incomplete lipid oxidation and subsequent accumulation of lipid metabolites (diacylglycerol, ceramides, and acyl coenzyme A [CoA]), impaired insulin signaling, metabolic inflexibility, and oxidative stress (32, 112, 122). Moreover, each of these outcomes is also thought to cause further impairment of mitochondrial function, and thus provokes a vicious and chronic circular chain of cause and consequence of cardiometabolic events. Impaired mitochondrial function may also lead to diminished adenosine triphosphate (ATP) production, energy deficit, and decreased functional capacity. Sedentary behavior is a robust predictor of mitochondrial dysfunction; and more importantly, evidence suggests improvements in ATP synthesis and fatty acid oxidation after exercise interventions, independent of weight loss.

Proinflammatory and Prothrombotic Characteristics

Adipose tissue is considered a dynamic organ with pleiotropic properties (166). Previous research among nondiabetic obese adults has revealed significantly elevated levels of adipocyte-derived hormones and cytokines, which are significant contributors to insulin resistance (94). Specifically, ectopic adiposity is known to play a role in secreting proinflammatory cytokines (e.g., tumor necrosis factor alpha [TNF-α] and interleukin-6 [IL-6]), adipocytokines (e.g., leptin, resistin, and adiponectin), and chemokines (e.g., monocyte chemoattractant protein-1 [MCP-1]). Produced by adipose tissue macrophages, the inflammatory cytokines IL-6 and TNF-α are positively associated with triglycerides and total cholesterol, are inversely associated with HDL cholesterol, can interfere with insulin signaling (152), and can lead to increased cellular oxidative stress (i.e., accumulation of reactive oxygen species [ROS]) (172). Moreover, IL-6 stimulates hepatic production of C-reactive protein (CRP) (18), an acute-phase protein and robust predictor of various features in the metabolic syndrome (62). Clinically, high-sensitivity C-reactive protein (hs-CRP) has become accepted as a useful biomarker for chronic, low-grade inflammation, and elevated hs-CRP is known to be robustly associated with ASCVD (138).

Markers of thrombosis (i.e., clotting factors) are also known to be increased in the metabolic syndrome, thus representing an additional link with ASCVD. Specifically, elevated levels of plasma plasminogen activator inhibitor-1 (PAI-1) (88) and fibrinogen (62) are linked to thrombosis and fibrosis, insulin resistance, and abdominal obesity. However, it is conceivable that the proinflammatory and prothrombotic states associated with the metabolic syndrome are elevated before presentation of standard syndrome risk components. Certainly, adipocyte proliferation and accumulation occur long before an individual meets the criteria for obesity or is dichotomized as at risk for cardiometabolic disease. Therefore, in conjunction with pronounced changes in the hormonal-metabolic milieu (72, 74), increases in adiposity could yield a general, inhospitable physiologic environment that contributes to diminished insulin sensitivity even in the absence of a clinical diagnosis of the metabolic syndrome. A recent study among prepubescent children demonstrated that obesity without established comorbidities of the metabolic syndrome was indeed associated with proinflammatory and prothrombotic states (110). These findings are significant because they support the need for aggressive preventive strategies for children with obesity, even in the absence of comorbidities.

Although various pathophysiologic underpinnings to the metabolic syndrome have been identified, present debate pertaining to the nature of these associations has raised questions about directional link, that is, whether steatosis and chronic inflammation cause mitochondrial dysfunction or insulin resistance (or both) or vice versa (87, 115, 125), as well as about primary versus secondary targets for intervention. Unraveling these associations has been considered a critical directive for future studies (104), and may be fundamental to optimizing therapeutic interventions to reverse the sequelae caused by—or coinciding with—obesity, insulin resistance, and the metabolic syndrome.

CLINICAL CONSIDERATIONS

The state of the literature pertaining to the clinical utility of the metabolic syndrome is in constant flux, and thus several points of clarification and caution are warranted. First, the ATP III and IDF and the new harmonized definitions do not exclude hyperglycemia in the diabetes range as a potential criterion for diagnosis of the metabolic syndrome (5, 12, 14). Therefore, most patients with type 2 diabetes also meet criteria for the metabolic syndrome. But this overlap between diabetes and the metabolic syndrome has not been without dispute (89, 153).

In fact, ever since the original definition was published (13), there have been many opponents to initiatives to impose a clinical definition, as well as to initiatives to derive threshold cut points for the syndromic clustering of the respective risk factors (89, 145). The most frequent argument against the use of a metabolic syndrome pertains to the clinical relevance of the condition (26, 42). According to a recent WHO Expert Consultation Group (153), the diagnostic criteria for the metabolic syndrome does not add predictive value beyond the sum of the individual risk factors in forecasting future CVD, diabetes, or disease progression. In this 2010 WHO report, various other rationales against the clinical and epidemiological use of the metabolic syndrome were presented. Most notably, criticisms from previous reports (89) were reiterated to conclude that while the metabolic syndrome may be a useful educational concept, "it has limited practical utility as a diagnostic or management tool (153: pp. 600 and 604)." Specific limitations that were proposed as rationale against widespread clinical adoption of the metabolic syndrome include the following (153):

- Lack of a single agreed-upon pathophysiologic mechanism
- Reliance of definitions on dichotomization of the diagnosis and individual risk factors
- The fact that the syndrome describes relative risk as opposed to absolute risk
- Differing predictive value of various risk factor combinations
- Inclusion of individuals with established diabetes and CVD
- The omission of important risk factors for predicting diabetes and CVD

Despite these longstanding criticisms regarding the clinical utility of the metabolic syndrome, and since the release of the 2010 WHO report, a large study was recently published demonstrating that among nearly 1 million patients (n = 951,083), the metabolic syndrome was associated with a twofold increase in cardiovascular outcomes and a 1.5-fold increase in all-cause mortality (114). Moreover, this investigation revealed that ASCVD risk was significantly higher among patients with metabolic syndrome even in the absence of diabetes. This is consistent with two previous meta-analyses (63, 65), and thus these studies collectively represent strong support for the metabolic syndrome as a robust predictor of cardiovascular events, even when adjusting for the individual risk components.

In conjunction with the elevated risk for cardiometabolic disease and mortality, secondary clinical outcomes that are associated with the metabolic syndrome include nonalcoholic fatty liver disease (NAFLD), cholesterol gallstones, polycystic ovary syndrome, and sleep apnea (75, 77). In addition to general increased risk of cardiovascular and metabolic complications in patients with the metabolic syndrome, a wide variety of pharmacological treatments might coincide with secondary comorbidities (table 10.3). Moreover, among older adults, various studies suggest that the metabolic syndrome is associated with increased risk for age-related dementia and overall cognitive decline (132, 133, 171). Most recently, Raffaitin and colleagues (132) demonstrated that among a prospective cohort of 4,323 women and 2,764 men (≥65 yr), presence of the metabolic syndrome at baseline was

Table 10.3 Pharmacology

Medication name and class	Primary effects	Exercise effects	Special considerations
Telmisartan, irbesartan, losartan, olmesartan, valsartan Class: angiotensin II receptor antagonists	Block the action of angiotensin, dilate blood vessels, and reduce blood pressure	Reduce blood pressure during exercise	Among individuals with impaired glucose tolerance, this class of drug may decrease the risk of developing type 2 diabetes.
Atorvastatin, fluvastatin, lovastatin, simvastatin, pravastatin, rosuvastatin Class: statins (HMG-CoA reductase inhibitors)	Inhibit the enzyme HMG-CoA reductase, which results in decreased hepatic cholesterol Improve lipid profile (LDLc) and may inhibit the progression of atherosclerosis	No effect	May improve (increase) HDLc and lower triglycerides. Many statins are metabolized by cytochrome p450 and therefore have higher risks of drug–drug interactions. Patients should avoid eating grapefruit (contains furanocoumarins), as it inhibits the metabolism of statins.
Bezafibrate, gemfibrozil, fenofibrate Class: fibrates (fibric acid sequestrants)	Activate PPAR-α signaling and lead to decreased hepatic cholesterol secretion and increased β-oxidation Improve LDLc, but also may improve HDLc and decrease triglycerides	Combined with statins, may increase risk of rhabdomyolysis (i.e., severe muscle damage)	
Cholestyramine, colesevelam, colestipol Class: bile acid sequestrants	Antilipemic agents	Increase risk of gastrointestinal distress and constipation, which may be exaggerated during exercise	May bind with fat-soluble vitamins (i.e., vitamins A, D, E, and K) in the gut and lead to insufficiency.
Niacin Class: nicotinic acid	Increases HDL	May help to decrease blood pressure with exercise	Can worsen glycemic control in patients with metabolic syndrome or diabetes.
Metformin Class: antidiabetic (biguanides)	Antihyperglycemic: Decreases hepatic glucose production and intestinal glucose absorption	No effect	
Orlistat Class: lipase inhibitor	Prevents absorption of fats from the diet and thus reduces caloric intake	Increases risk of gastrointestinal distress (e.g., fecal incontinence and loose, oily stools).	Absorption of fat-soluble vitamins and nutrients is inhibited.
Omega-3 fatty acids Class: nutritional supplement	Decrease triglycerides and improve insulin sensitivity	No effect	

predictive of "global" cognitive decline and in particular executive function, independent of previous CVD or depression.

Although children, adolescents, and adults with the metabolic syndrome have significantly elevated risk for chronic disease and secondary comorbidity, it stands to reason that not everyone diagnosed will eventually develop insulin resistance. However, since the metabolic syndrome is known to double the risk for ASCVD and to serve as a strong precursor to the development of diabetes, there are indeed profound long-term clinical implications regarding the diagnosis of the syndrome. Moreover, there is an urgent need to develop preventive and treatment strategies for the syndrome, as well as the underlying causes (76). Identifying individuals with emergent risk for and early-phase diagnosis of the metabolic syndrome is an exceedingly important clinical and public health directive, as this represents a critical opportunity to intervene.

Signs and Symptoms

Considering the pathophysiology and trajectory of the condition, much overlap exists between the signs and symptoms of the metabolic syndrome and those associated with diabetes (chapter 6) and obesity (chapter 7). In addition to the actual diagnostic risk components of the metabolic syndrome (i.e., elevated glucose, hypertension, elevated triglyceride levels, low HDL cholesterol levels, and abdominal obesity), many patients also present with **microalbuminuria**, **hyperuricemia**, fatty liver disease, high levels of PAI-1 and fibrinogen (i.e., prothrombotic state), elevated hsCRP (i.e., proinflammatory state), cholesterol gallstones, polycystic ovary syndrome, and disordered sleeping (e.g., sleep apnea). Among older adults with the metabolic syndrome, patients may report losses of memory, impaired cognition, general confusion, or more than one of these.

History and Physical Exam

Many patients diagnosed with the metabolic syndrome also have hyperglycemia (i.e., type 2 diabetes). Thus it is extremely important to take into consideration specific precautions related to the screening, diagnosis, and physical exam for patients with diabetes (chapter 6). In general, one should identify the core risk components of the metabolic syndrome during a routine physical examination. However, since many physicians do not specifically monitor abdominal obesity (i.e., height, weight, and BMI calculations are more common), an assessment for this should be a fundamental aspect of the patient history and physical exam. In fact, since there is ample variation in the overlap between categorical obesity according to BMI and the criteria for abdominal obesity, it is plausible that a physician may fail to diagnose or identify a patient with the metabolic syndrome. Because abdominal obesity is a fundamental risk component in the new harmonized definition for the metabolic syndrome (12), it is a parameter that must be accounted for. A simple clinical measure of waist circumference is highly predictive of this risk factor.

Moreover, with regard to individual parameters, estimates of relative risk for ASCVD incrementally and significantly increase as the number of syndrome components increases (167). Thus, individuals who present with one or two of the components are at significantly lower risk for cardiac events, as well as for mortality from coronary heart disease and CVD, than individuals with three or more risk components (figure 10.3) (105). Estimates of relative risk are also contingent upon the particular combination of components. Specifically, numerous combinations of the syndrome components warrant a diagnosis, and recent evidence reveals that ASCVD and mortality risk is highest when elevated blood pressure, abdominal obesity, and elevated glucose are simultaneously present (60, 139). A careful review of the patient's medical history will provide beneficial insights relevant to starting a comprehensive lifestyle approach for the patient. Toward that end, and in conjunction with the standard metabolic syndrome components and medical history, the exercise professional should also take into consideration patient exercise history.

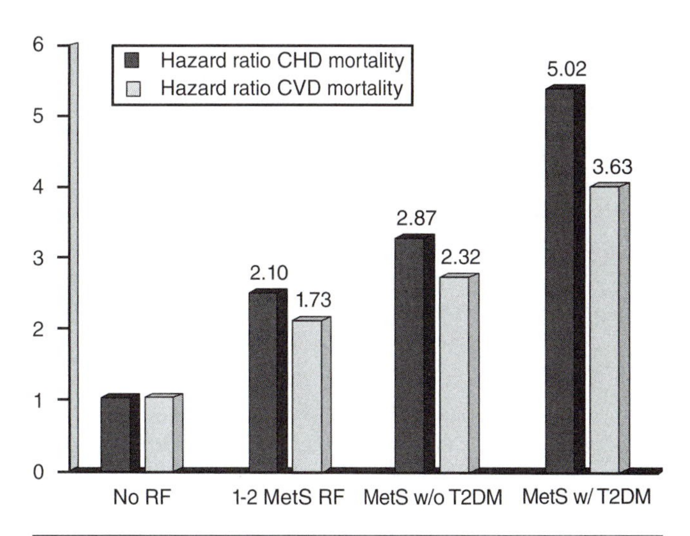

Figure 10.3 Coronary heart disease (CHD) and cardiovascular disease (CVD) mortality increases with number of cardiometabolic risk factors (RF).

Adapted from Malik et al. 2004.

Diagnostic Testing

In addition to the five risk components of the metabolic syndrome, and especially in the case of a positive diagnosis, laboratory screening for diabetes should take place (chapter 6). This is particularly necessary for patients over the age of 45 yr. Individuals may also need to be screened for prediabetes, which is characterized by impaired fasting glucose (fasting plasma glucose levels 100 mg · dl^{-1} [5.6 mmol · L^{-1}] to 125 mg · dl^{-1} [6.9 mmol · L^{-1}]) or impaired glucose tolerance (2 h values in the oral glucose tolerance test of 140 mg · dl^{-1} [7.8 mmol · L^{-1}] to 199 mg · dl^{-1} [11.0 mmol · L^{-1}]) (4). Moreover, depending on age, the number of risk components identified, and the specific combination of those components, additional testing may be warranted to screen for clinical inflammation and thrombosis, **hyperandrogenemia** (to rule out polycystic ovary syndrome), microalbuminuria, hyperuricemia, cholesterol gallstones, and sleep apnea.

Exercise Testing

For the purposes of exercise testing, and in accordance with the American College of Sports Medicine (ACSM), appropriate risk stratification for patients diagnosed with the metabolic syndrome should be based on the presence of dyslipidemia, hypertension, and hyperglycemia (1). Thus, patients with the metabolic syndrome are likely to be classified as either "moderate risk" (i.e., asymptomatic men and women who have two or more ASCVD risk factors) or "high risk" (i.e., individuals who have known cardiovascular, pulmonary, or metabolic disease, or one or more signs and symptoms of these diseases) (1). Stratifying by risk is useful to the exercise professional in that it identifies specific people at greater risk for untoward events during exercise testing or participation. Risk stratification may also result in the need for additional medical screening. Specifically, for individuals with diabetes, the ACSM recommends a clinical exercise stress test before engaging in exercise. However, considering the robust association between sedentary behavior and the metabolic syndrome, an assessment of aerobic fitness is applicable for all patients. Standardized treadmill protocols that use relatively small workload increases are recommended and are often well tolerated by those with obesity and the metabolic syndrome. However, for individuals with morbid obesity, walking may not be practical due to gait abnormalities. In such instances, testing for aerobic fitness may be performed using a seated recumbent cycle ergometer or upper body ergometer. In addition to aerobic fitness, patients diagnosed with the metabolic syndrome should be assessed for muscular fitness (i.e., muscle strength and local muscular endurance), as well as for flexibility and range of motion. Exercise testing should follow standard procedures as described in chapter 5. All exercise testing should be performed with caution and should be completed for the purpose of designing individualized PA and exercise prescriptions. See table 10.4 for information on exercise testing methods for individuals with the metabolic syndrome.

Treatment

Although the primary goals in the management of metabolic syndrome are to reduce the risk for clinical ASCVD and to prevent type 2 diabetes, clinical and public health efforts should lend equitable support to treating the established risk factors, as well as the underlying causes of the condition. All five risk components of the metabolic syndrome are modifiable, but abdominal obesity is often considered a primary driver of the other risk components (42). Thus, treatment strategies should address the contributing risk factors for abdominal obesity, which in turn may positively influence all other components. Indeed, interventions designed to promote weight loss (e.g., caloric restriction, increased PA, pharmacological agents, and even surgical procedures when necessary) should be a central feature of the treatment of the metabolic syndrome (see practical application 10.2). The NHLBI recommends a minimum weight loss of 10% (3); however, ample evidence also exists to demonstrate significant health benefit after as little as 2% to 3% weight reduction (43, 100). According to Grundy, a reasonable weight loss goal for the first year of intervention is 10 kg, which people may attain by creating a caloric deficit of about 400 calories per day through reduced calorie intake (e.g., 300 kcal/d) and increased energy expenditure (e.g., 100 kcal/d) (75).

In addition to managing abdominal obesity through improvements in diet quality and participation in PA, smoking is a lifestyle risk factor for ASCVD that should continue to be directly targeted by public health and clinical interventions.

EXERCISE PRESCRIPTION

Although numerous studies have addressed the value of PA and exercise for patients with the metabolic syndrome, no specific guidelines focusing directly on this topic have been published. However, the ACSM has recently published separate guidelines for weight loss (44) and diabetes (2); the following recommendations on PA and exercise for the metabolic syndrome are representative

Table 10.4 Exercise Testing Review

Test type	Mode	Protocol specifics	Clinical measures	Clinical implications	Special considerations
Cardiovascular	Step test, cycle ergometer, or graded treadmill	Standard tests (e.g., Harvard step test, YMCA bicycle test, Bruce treadmill protocol, or some combination of these)	$\dot{V}O_2$, RER, METs, RPE	$\dot{V}O_2$ is typically diminished among patients with metabolic syndrome. Normative data from healthy populations may be used for each test to provide stratification and percentile rank. Each test may also be used to track longitudinal progress of and adaptation to cardiovascular exercise interventions.	Patients with diabetes or hypertension or both should follow specific guidelines for each condition to avoid undue risk of cardiovascular events. Obese individuals may not be able to complete the step test or treadmill tests and thus may require use of the cycle ergometer assessment.
Strength	Maximal voluntary contraction (MVC), one-repetition maximum (1RM), or multiple RM (e.g., 5RM)	Dynamometer (e.g., Biodex isokinetic dynamometer), selectorized or pneumatic machines (e.g., Nautilus, Keiser), free weights (e.g., chest press, squat), or some combination	Isometric torque (N·m), maximum dynamic force (newtons), or maximal strength (kg or lb)	Lower extremity body mass–adjusted strength (e.g., 1RM leg press/body mass) and muscle quality (i.e., muscle cross-sectional area and strength) are valuable clinical measures that have direct relevance to functional mobility and activities of daily life. Results may be used to prescribe relative intensity for resistance exercise interventions.	Individuals with preexisting hypertension should avoid maximal strength (i.e., 1RM) testing. Rather, multiple RM testing (e.g., 5-10RM) is a safe and valid alternative technique.
Range of motion	Sit-and-reach, overhead squat test	Sit-and-reach or modified (i.e., unilateral) sit-and-reach Overhead squat test provides comprehensive qualitative assessment of functional mobility and hip, low back, knee, ankle, and shoulder ROM and stability.	Sit-and–reach (distance in cm) Overhead squat test is a qualitative assessment with the following requirements: (1) Upper torso is parallel with shin or toward vertical; (2) hips are lower than knees; (3) knees are aligned over feet; and (4) with arms outstretched, hands or bar/dowel are aligned directly over feet.	Results from the overhead squat test can indicate tightness or weakness in respective musculature through coordinated, full ROMs; excessive joint stiffness (e.g., limited ankle dorsiflexion); and strength or ROM asymmetries.	Sit-and-reach may not be possible for individuals with abdominal obesity.

of these collective suggestions. Chapter 5 presents additional detail on the fundamentals of exercise prescription. Since a subset of the metabolic syndrome population also has diabetes, specific suggestions pertaining to exercise prescription and contraindications in chapter 6 may be relevant. Table 10.5 identifies exercise prescription recommendations for the management of metabolic syndrome.

Practical Application 10.2

CLIENT–CLINICIAN INTERACTION

Most patients with the metabolic syndrome have been sedentary and are obese. As a result, establishing PA or structured exercise interventions can be a challenging endeavor for the exercise physiologist. Since the decline associated with cardiometabolic conditions is a gradual, often asymptomatic process, many people with the metabolic syndrome do not perceive significant deficits in physical function or general well-being and thus are not likely to understand the seriousness of the condition. Therefore, it is vital to educate patients regarding the accumulation and clustering of risk factors, as well as the role of appropriate strategies to help prevent further health declines. At present, weight loss is the inevitable clinical recommendation for treatment of obesity-related cardiometabolic decline; thus PA is generally regarded as an adjunct to dietary restriction. However, whereas weight loss is indeed effective for ameliorating health risk, very few studies have documented long-term sustainability or efficacy of weight loss; on the contrary, the vast majority of individuals (approximately 80%) who lose significant weight experience complete weight regain (17, 170). This underscores the importance of educating patients about the relative value of PA and exercise as a complement to healthy dietary practices for improving insulin sensitivity and cardiometabolic health. Although the expected weight loss attributed to PA without dietary manipulation is generally small, improvements in cardiovascular risks such as resting blood pressure, lipid profiles, and glucose tolerance may provide the motivation people need to adhere to a physically active lifestyle.

Cardiovascular Exercise

In 2009 and 2010, the ACSM released updated position stands providing recommendations on PA and exercise intervention strategies for weight loss and prevention of weight regain (44), as well as for type 2 diabetes (2). These recommendations center on PA and cardiovascular exercise for weight loss and aerobic fitness, encouraging people to accumulate 150 to 250 min of moderate-intensity PA per week with an energy equivalent of 1,200 to 2,000 kcal. Specifically, people should perform continuous or intermittent low-intensity (40-60% $\dot{V}O_2$ or heart rate reserve [HRR]) to moderate-intensity (50-75% $\dot{V}O_2$ or HRR) cardiovascular exercise with the explicit goals of calorie expenditure and improved aerobic fitness. Sessions should take place on at least 3 d/wk, but preferably 5 d or more, with no more than 2 consecutive days between bouts. Sessions should be 30 to 60 min in duration (minimum of 10 min for intermittent cardiovascular exercise) and intended to reach energy expenditure goals of a minimum of 100 to 400 kcal per session. Gradual progression in duration and intensity may be effective for chronic weight maintenance, additional weight loss, or further improvement in aerobic fitness capacity beyond the baseline requirements for health. However, caution is warranted, since progressing sedentary individuals too quickly may exacerbate untoward responses such as musculoskeletal injury. According to the new ACSM guidelines, PA for weight loss is dose dependent, such that <150 min per week leads to minimal weight loss, PA >150 min per week leads to modest weight loss (i.e., 2-3 kg), and PA >225-420 min per week results in 5-7.5 kg weight loss (44). Activities such as brisk walking, swimming, and cycling are usually well tolerated by those with the metabolic syndrome; however, unless specified by a primary care physician or cardiologist, individuals may also progress to other modalities such as jogging, running, hiking, rowing, and stair climbing.

Resistance Exercise

Recommendations pertaining to resistance exercise for the metabolic syndrome are at present nonexistent. Since resistance exercise is generally not considered effective for weight loss per se, it has received very little attention from the clinical and public health community regarding its relative value in the fight against obesity or cardiometabolic disease. Interestingly, during the past 5 to 10 yr, research has begun to shed light on the utility of resistance exercise in stimulating positive cardiovascular, endocrine, metabolic, neuromuscular, and morphological adaptations, independent of weight loss. These adaptations are briefly discussed in the next section on exercise training. With regard to exercise prescription, current minimum recommendations call for resistance exercise training to supplement cardiovascular exercise and to be performed on 2 (preferably 3) nonconsecutive days per week, using a single set of 5 to 10 exercises for the whole body and at a moderate level of intensity that allows 10 to 15 repetitions (2). As is generally accepted for

Table 10.5 Exercise Prescription Review

Training method	Mode	Intensity	Frequency	Duration	Progression	Goals	Special considerations
Cardiovascular	Walking, jogging, swimming, elliptical, stair climbing, hiking, rowing	Low (50% to 60% $\dot{V}O_2$ or HRR) to moderate (50% to 75% $\dot{V}O_2$ or HRR)	5 to 7 d/wk (minimum 3 d)	30 to 60 min (minimum 10 min for intermittent)	Short-duration, low-intensity PA may progress to longer-duration or moderate-intensity exercise bouts or both.	Weight loss and improved $\dot{V}O_2$max	For obese individuals, a mix of weight-bearing and non-weight-bearing modalities is better tolerated and may reduce risk of overuse injury or joint pain.
Resistance	Exercises for major muscle groups: full ROM body weight movements, selectorized or pneumatic machines, free weights	Low (12-15RM) to moderate (8-12RM)	2 to 4 d/wk	Low volume (one or two sets for all major muscle groups) to moderate volume (three or four sets)	After familiarization to the movements, progress from low relative intensities and volumes to moderate to vigorous relative intensities and higher volumes.	Local muscular endurance, hypertrophy, strength adaptation	Full-body program can be split into a combination upper–lower program to accommodate progression.
Range of motion	Dynamic ROM exercises and static stretching	NA	Dynamic ROM can be done prior to or in conjunction with resistance training; static stretching can be performed postexercise.	10 to 30 s for each major muscle group	NA	Preservation of or improved dynamic ROM and static posture	All stretches should be completed only to the point of mild discomfort, and normal breathing should take place at all times.

novice trainees, prescription of resistance exercise should include a familiarization period in which very low dosage training (i.e., minimal sets and intensity) takes place one or two times per week. Following the familiarization, one can expect adults with the metabolic syndrome to benefit from gradual increases in dosage to accommodate improvements in strength and muscle hypertrophy. In particular, although the established guidelines provide a basis for maintaining muscular fitness, there is now ample evidence to confirm the viability of progressive resistance exercise for improving strength and muscle mass among adults. Additional suggestions on progression in resistance exercise include (1) gradual increases in intensity from moderate (i.e., 50% of one-repetition maximum [1RM]) to vigorous (75-80% of 1RM); (2) gradual increases in the number of sets from a single set to as many as four sets per muscle group; (3) gradual decreases in the number of repetitions performed (i.e., to coincide with progressively heavier loading), from 10 to 15 repetitions per set to approximately 8 to 10 repetitions per set; and (4) progression in mode from primarily machine-based resistance exercise to machine plus free-weight resistance exercise.

Range of Motion Exercise

Range of motion and flexibility training may be included as an adjunct modality to supplement cardiovascular and resistance exercise for individuals with the metabolic syndrome. Current ACSM recommendations for stretching and range of motion exercises suggest that stretching be performed at least 2 or 3 d/wk following a warm-up or workout, when muscles are warm (1). Static, dynamic, or proprioceptive neuromuscular facilitation (PNF) activities are suggested for all major muscles and joints of the body; however, considering that most patients with the metabolic syndrome have been sedentary, static stretching may be preferable during the initial phases of exercise participation. For static stretching, the recommendation is that each stretch be repeated at least four times and held for 15 to 60 s to the point of mild discomfort.

EXERCISE TRAINING

At present there is a divergence between the basic research intended to uncover the biomolecular etiology of obesity and associated cardiometabolic fallout, and that at the clinical and public health level, aimed at preventing or reversing these phenotypes. Increased adiposity (29, 101) and insufficient PA (49, 109, 121, 136, 144) have both been implicated in the etiology of CVD, insulin resistance, metabolic syndrome, and diabetes. However, with regard to correcting the obesity-related milieu, research is yet to explicate an optimal PA strategy not only to confer benefit across standard markers of health (e.g., lipid profiles, fasted glucose), but also to reverse various etiologic characteristics of physiological decline. In the clinical context, since weight loss is known to elicit improvements in overall patient health (131, 161), blanket recommendations regarding healthy dietary practices and PA to promote weight reduction have become the inevitable norm. Although intuitive, this paradigm has been at the expense of understanding the mechanisms of change that accompany PA, independent of those conferred by weight loss. As pointed out previously, weight loss is effective for ameliorating health risk, but little research has documented its long-term sustainability or efficacy, and the vast majority of people who lose significant weight gain it back (17, 170). Further, as compared to interventions with PA, data suggest that purely diet-induced weight loss may have limited utility for establishing a sustainable, insulin-sensitive phenotype (35, 102, 142). This underscores the importance of identifying alternative choices for improving cardiometabolic health among obese individuals with the metabolic syndrome.

Indeed, emerging evidence suggests improved cardiometabolic profile and improved insulin sensitivity after cardiovascular exercise (116) and resistance exercise (163), independent of weight loss. These changes may in part be attributable to enhanced muscle function and cardiorespiratory fitness, insulin-stimulated glucose disposal, and fatty acid oxidation (84, 128). Our recent work has identified an inverse association between adiposity and muscle function even among nonobese individuals (126), underscoring the value of general PA and targeted exercise as primary elements in prescriptive metabolic syndrome treatment. Therefore comprehensive clinical and public health interventions to treat the metabolic syndrome must include simultaneous directives for improved dietary habits and gradual weight loss, increased daily PA, and a targeted exercise prescription. Table 10.6 provides a review of exercise training and benefits for metabolic syndrome.

Cardiovascular Exercise

Physical inactivity has been recognized as a primary contributing factor to progression of obesity and subsequent risk of metabolic syndrome, ASCVD, and diabetes (75, 78). Conversely, when combined with dietary manipulation or intervention, cardiovascular exercise or moderate-intensity PA has been widely regarded as the most acceptable means to modify excess adiposity and reduce cardiometabolic health risk (2, 44). The utility of cardiovascular exercise is suggested to be dose dependent, such that greater volumes of PA are associated with greater cardiometabolic health risk reduction (44). It is commonly thought that the health benefits of PA for individuals with metabolic syndrome are directly related to mobilization of free fatty acids and gross decreases in adiposity. Indeed, since extended-duration cardiovascular exercise is effective for large absolute energy expenditure, recommendations to prioritize this type of exercise are intuitive for the promotion of weight loss. However, recent research suggests that improvements in metabolic disturbances are possible with exercise interventions independent of overall weight loss or changes in body composition (116). This is an important message for clinical and public health constituents to accept and convey to patients, because it underscores the relative value of healthy lifestyles and fitness rather than focusing on body habitus or aesthetics.

The positive benefits of exercise on cardiometabolic health in the metabolic syndrome are multifaceted, and can be explained by three phenomena (137). During the acute phase, a single bout of exercise can significantly increase whole-body glucose disposal and thus temporarily attenuate hyperglycemia. Moreover, for several hours after a given bout of exercise, insulin sensitivity is increased. Lastly, repeated bouts of exercise lead to a chronic adaptive response, characterized by enhanced cardiorespiratory function and global improvements in insulin action (39, 116). Cardiovascular exercise is also

Table 10.6 Exercise Training Review

Cardiorespiratory endurance	Skeletal muscle strength	Skeletal muscle endurance	Flexibility	Body composition
Low-intensity, long-duration PA and structured cardiorespiratory exercise are beneficial for energy expenditure, weight management, and improving cardiometabolic risk factors. Moderate-intensity, medium- and short-duration structured cardiorespiratory exercise is beneficial for aerobic fitness ($\dot{V}O_2$max), enhancing body composition, and improving cardiometabolic risk factors.	Resistance exercise for strength is important for preservation of muscle function, lean body mass, and bone density, and can improve certain parameters of cardiometabolic health (e.g., insulin sensitivity, glucose tolerance, and lipid profiles).	Resistance exercise for local muscular endurance is beneficial for fatigue resistance during activities of daily living and bouts of physical activity, as well as for low back health and posture.	Flexibility and ROM exercises are vital for musculoskeletal and joint integrity and low back health.	Decrease in relative adiposity (i.e., % body fat) is vital for improved overall cardiometabolic health. Increase in absolute and relative lean body mass (LBM) is necessary for preservation of muscular function and basal metabolic rate.

effective for improving blood pressure and lipid profiles, decreasing visceral adiposity even in the absence of weight loss, enhancing fatty acid oxidation, increasing mitochondrial function and content, and attenuating the proinflammatory state (2, 44, 124).

Resistance Exercise

According to the ACSM position stand, resistance exercise does not promote clinically significant weight loss (evidence category A) (44). However, with regard to the restoration of cardiometabolic health among at-risk, obese individuals, mounting evidence indicates that resistance training may be a viable treatment option, comparable to aerobic exercise. In particular, among adults with existing risk factors for morbidity, several longitudinal studies examining the role of exercise in metabolic disturbance have reported significantly improved insulin sensitivity and glucose tolerance (45, 51, 96) with structured, progressive resistance exercise interventions. Resistance exercise has also been associated with improvements in various ASCVD risk factors in the absence of weight loss, including decreased LDL cholesterol (69, 85), decreased triglycerides (69), reductions in blood pressure (93), and increases in HDL cholesterol (85). Moreover, several studies have documented the superiority of resistance exercise over traditional aerobic exercise for glycemic control and insulin sensitivity among type 2 diabetic adults (30). Chronic resistance training has been traditionally regarded as an appropriate means to augment or preserve skeletal muscle tissue, and thus may serve as an essential stimulus to improve and sustain whole body glucose-disposal capacity (51). Despite the well-known independent value of resistance exercise for cardiometabolic health, strong evidence exists to confirm the superiority of combined aerobic and resistance exercise over either individual modality (34).

Among children and adolescents, research evidence suggests that muscular strength capacity is an important component of metabolic fitness and provides protection against insulin resistance (21). In this investigation, stronger children were 98% less likely to be insulin resistant compared with those who were in the lowest percentiles for strength, even after adjustment for central adiposity, body mass, and maturation (21). Part of the reason may be that the high-force muscle actions associated with strength capacity are generally attributable to a higher ratio of myosin heavy chain II isoforms (i.e., type II muscle fibers, or fast fibers) over myosin heavy chain I isoforms (i.e., type I muscle fibers, or slow fibers). Increases in strength capacity have been suggested to promote superior glucose metabolism among individuals with impaired baseline glucose tolerance following an exercise intervention (164). Moreover, research has documented the efficacy of isolated high-intensity resistance training (i.e., without dietary intervention or aerobic exercise) to safely reduce adiposity among overweight children (20).

Of particular relevance to the generalizability of structured exercise in obese individuals with the metabolic syndrome, is the sustainability of such treatments. Some evidence indicates that resistance exercise elicits an earlier adaptive response in muscular function and anthropometric characteristics than traditional aerobic exercise, and as such may be associated with superior adherence rates over the long term. It is conceivable that early positive reinforcement could enhance self-efficacy and physical self-perception, which may substantially influence the success of and commitment to sustainable health behavior change (28).

Range of Motion Exercise

Regular range of motion exercise may be particularly relevant to older adults as a way to maintain or improve balance and posture and may help to prevent slip-and-fall accidents. Although "stretching" is commonly recommended as a means to improve joint range of motion, very little if any research supports the utility of this modality in the treatment or prevention of cardiometabolic disease. Rather, and in conjunction with full range of motion resistance training, stretching exercises may allow patients to preserve functional capacity and musculoskeletal integrity during activities that require full range of motion.

CONCLUSION

The metabolic syndrome is a collection of interrelated cardiometabolic risk factors that are present in a given individual more frequently than may be expected by a chance combination. Patients with the metabolic syndrome are usually overweight or obese and have significantly greater risk for developing CVD, insulin resistance, and type 2 diabetes, as well as early mortality. Diagnosis of the metabolic syndrome requires presence of three or more of the following risk factors (table 10.1): (1) elevated waist circumference, (2) elevated triglycerides, (3) reduced HDL cholesterol, (4) elevated blood pressure, and (5) elevated fasting glucose. The current age-adjusted prevalence of the metabolic syndrome is approximately 40% among all adults, thus representing a tremendous financial burden on public health. Although no uniform criteria exist for diagnosing the metabolic syndrome among pediatric populations, overweight or obese children and adolescents have a significantly elevated risk for cardiometabolic health decline. Treatment for the metabolic syndrome must be multifaceted, and behavioral modification including diet manipulation and PA are fundamental. Indeed, when combined with dietary interventions, cardiovascular exercise has been widely regarded as the most acceptable means to induce weight loss and reduce cardiometabolic health risk. However, and as an important point of clarification, ample research exists to demonstrate that improvements in metabolic disturbances are possible with exercise interventions even in the absence of overall weight loss or changes in body composition. Moreover, with regard to the restoration of cardiometabolic health among at-risk obese individuals, mounting evidence indicates that resistance training may also be a viable treatment option, comparable to aerobic exercise. Thus, there is a critical need to inform the appropriate clinical and public health audiences in an effort to bolster tailored preventive and treatment efforts specific to individuals with the metabolic syndrome.

Clinical Exercise Pearls

- The metabolic syndrome is a collection of interrelated cardiometabolic risk factors that are present in a given individual more frequently than may be expected by a chance combination.
- Individuals with the metabolic syndrome are at heightened risk of type 2 diabetes, atherosclerotic CVD, and early mortality.
- Diagnosis of the metabolic syndrome requires presentation of three or more cardiometabolic risk factors, including (1) elevated waist circumference, (2) elevated triglycerides, (3) reduced HDL cholesterol, (4) elevated blood pressure, and (5) elevated fasting glucose.
- Exercise training promotes weight loss, weight management, or both, as well as improvements in cardiometabolic risk profile.
- Obese individuals with the metabolic syndrome benefit most from a combination of exercise and healthy dietary intervention.
- Combined aerobic and resistance exercise is superior for improvements in serum lipid profiles, insulin sensitivity, glucose tolerance, body composition, and physical function.
- For individuals who are diabetic, hypertensive, or both, cardiorespiratory and strength testing should be modified or closely monitored to avoid undue risk of cardiovascular events.

Key Terms

ASCVD (p. 177)
cardiometabolic risk factors (p. 177)
ectopic adiposity (p. 184)
hyperandrogenemia (p. 188)
hypertriglyceridemia (p. 183)
hyperuricemia (p. 187)
IMAT (p. 184)
microalbuminuria (p. 187)
normal-weight obesity (p. 180)
steatosis (p. 183)

CASE STUDY

MEDICAL HISTORY AND PHYSICAL EXAMINATION

Ms. AL is a 54 yr old white woman who has been referred by her primary care physician to an exercise program for ASCVD risk reduction. She has a positive family history of heart disease; her father had a nonfatal myocardial infarction at age 55. She is 5 ft 3 in. (160 cm) and weighs 175 lb (79.4 kg), for a BMI of 31 $kg \cdot m^{-2}$ (i.e., an "obese" classification). Her waist circumference is 36 in. (91.4 cm). She has been taking Lipitor (20 $mg \cdot d^{-1}$) for the last 6 mo to lower LDLc. She reports being completely sedentary at work and at home and no history of PA or exercise participation for at least 5 yr.

DIAGNOSIS

Aside from occasional knee pain when climbing stairs, her medical history is unremarkable. The results from her exercise test, body composition assessment, and fasting blood lipid profile, obtained before she started the exercise program, were as follows.

EXERCISE TESTING RESULTS

Resting heart rate was 84 $beats \cdot min^{-1}$, and resting blood pressure was 135/80. The subject performed a modified Bruce treadmill protocol and achieved a peak metabolic equivalent (MET) level of 8.0 at a heart rate of 162 $beats \cdot min^{-1}$ (~98% of predicted maximum heart rate), blood pressure of 194/88, and an RPE of 18. An occasional premature ventricular contraction (PVC) was noted at rest and during recovery, but this resolved with exercise. Body composition testing using the BodPod system revealed a body fat percentage of 40%.

Total cholesterol = 190 $mg \cdot dl^{-1}$

HDL = 45 $mg \cdot dl^{-1}$

LDL = 145 $mg \cdot dl^{-1}$

Triglycerides = 180 $mg \cdot dl^{-1}$

Fasting glucose = 130 $mg \cdot dl^{-1}$

INITIAL EXERCISE PRESCRIPTION

Frequency

- PA and cardiovascular exercise: 4 or 5 d/wk
- Resistance exercise: 2 d/wk

Intensity

- PA and cardiovascular exercise: 83 to 116 $beats \cdot min^{-1}$ (50-70% max heart rate)
- Resistance exercise: 12- to 15RM

Duration

- PA: 20 to 40 min per session
- Cardiovascular exercise: 10 to 20 min per session
- Resistance exercise: one or two sets per muscle group (full body), 15 to 25 min per session

(continued)

Mode

- PA: walking, cycling, yard work
- Cardiovascular exercise: brisk walking, swimming, rowing ergometer, elliptical trainer
- Resistance exercise: body weight movements, selectorized machines, pneumatic machines

EXERCISE PROGRESSION

The first 4 to 6 wk are intended to introduce exercise and accommodate familiarization. Thereafter, and as the patient begins to notice adaptation, gradual progression in exercise dosage is recommended to ensure continued success and further cardiometabolic risk reduction.

Frequency

- PA and cardiovascular exercise: 5 to 7 d/wk
- Resistance exercise: 2 to 4 d/wk

Intensity

- PA: 83 to 100 beats · min^{-1} (50-60% max heart rate)
- Cardiovascular exercise: 100 to 125 beats · min^{-1} (60-75% max heart rate)
- Resistance exercise: 8- to 12RM

Duration

- PA: 30 to 60 min per session
- Cardiovascular exercise: 20 to 30 min per session
- Resistance exercise: two or three sets per muscle group (full body or upper-lower split), 20 to 30 min per session

Mode

- PA: walking, cycling, yard work
- Cardiovascular exercise: brisk walking, jogging, swimming, rowing ergometer, elliptical trainer
- Resistance exercise: body weight movements, selectorized machines, pneumatic machines, free weights

Ms. AL should be educated about the signs and symptoms of exercise intolerance as well as the long-term significance if her health condition (i.e., metabolic syndrome) is not aggressively managed. Blood pressure should be monitored initially during resistance training. Proper breathing during performance of resistance training needs to be demonstrated to prevent Valsalva. Additionally, the patient should be taught how to self-monitor exercise intensity by checking heart rate and using the Borg rating of perceived exertion scale.

In addition to fitness improvements, decreases in abdominal obesity will significantly help with reduction of metabolic risk. Therefore, a decrease in total caloric consumption combined with PA should be used to achieve a reduction in total body weight of 7% to 10% from baseline during the first 12 mo. Daily PA should continue to supplement all structured exercise and will assist in weight reduction or maintenance and decreased cardiometabolic health risk. Examples of daily PA include walking the dog for 25 min, taking the stairs instead of the elevator (depending on knee pain), and parking farther from the door for work or when shopping.

DISCUSSION QUESTIONS

1. How can an exercise program help treat this woman's cardiometabolic health status?
2. If this woman experiences minimal losses of body weight after several weeks of exercise and dietary restriction, what cardiometabolic benefits might still be occurring?
3. Which of the following represents the optimal combination of behavior modification tips for this woman to experience successful improvements in health: modest dietary restriction; modest dietary restriction and aerobic physical activity; aerobic exercise and resistance exercise; reduced sedentary behavior and modest dietary restriction; or reduced sedentary behavior, aerobic physical activity, resistance exercise, and modest dietary restriction.

Chapter 11

End-Stage Renal Disease

Samuel Headley, PhD
Michael Germain, MD

Chronic kidney disease (CKD) results from structural renal damage and progressively diminished renal function. Chronic kidney disease is divided into five stages, depending on the extent of kidney damage and the **glomerular filtration rate** (GFR). After it begins, the disease can progress to **end-stage renal disease** (ESRD), requiring some form of **renal replacement therapy** (RRT) such as **dialysis** (either hemo- or peritoneal) or transplantation.

DEFINITION

As the name implies, ESRD represents the complete or almost complete failure of the kidneys to work. In this stage, sometimes referred to as stage 5 chronic kidney disease, GFR drops to only 10% to 15%, and the presence of disease-related symptoms is common. Unfortunately, many CKD patients do not progress to ESRD but instead die prematurely of cardiovascular disease (16). In recent years, the progression of CKD patients to ESRD has stabilized (28).

SCOPE

Based on current estimates, approximately 23 million Americans have CKD (23). In 2008, the most recent year for which we have statistics, the ESRD population in the United States rose to 547,982, with the total cost for the treatment of the disease estimated to be $39.5 billion (28).

In the United States before 1972, access to dialysis treatment was limited, and selection of patients for treatment was made by committees of medical professionals, clergy, and laypeople. Essentially, these committees decided who would receive the lifesaving therapy of dialysis. In 1972, Congress passed landmark legislation that extended Medicare coverage to patients with ESRD. This legislation hinged on the expectation of successful vocational rehabilitation of these patients (an expectation that has not been realized). Renal replacement therapy is expensive. The estimated cost of dialysis is $77,506 per patient per year; kidney transplant costs less over time ($26,688 per year) (28).

Although the overall outcomes and well-being of patients with renal failure have significantly improved because of advances in technology and pharmacology that improved these patients' potential for rehabilitation, it is generally acknowledged that rehabilitation has not been addressed nationally in this patient group in a sustained, consistent, and integrated fashion (30). Low levels of physical functioning contribute significantly to the low levels of rehabilitation, thus indicating the need for physical rehabilitation as a part of the routine medical therapy of these patients. An exercise intervention is critical since the sedentary behavior that is characteristic of the CKD population contributes to the excess morbidity and mortality observed in this population (16).

Acknowledgment: Much of the writing in this chapter was adapted from the first and second editions of *Clinical Exercise Physiology*. As a result, we wish to gratefully acknowledge the previous efforts of and thank Patricia Painter, PhD.

PATHOPHYSIOLOGY

Damage to the kidney can result from longstanding diabetes mellitus, hypertension, autoimmune diseases (e.g., lupus), glomerulonephritis, pyelonephritis, some inherited diseases (i.e., polycystic kidney disease, Alport's syndrome), and congenital abnormalities. The damaged kidney initially responds with higher filtration and excretion rates per nephron, which masks symptoms until only 10% to 15% of renal function remains. Progressive renal failure causes loss of both excretory and regulatory functions, resulting in uremic syndrome. Uremia is characterized by fatigue, nausea, malaise, anorexia, and subtle neurological symptoms. Patients present with these symptoms and often with peripheral edema, pulmonary edema, or congestive heart failure. Diagnosis is made from elevated serum creatinine (>8 $mg \cdot dl^{-1}$), blood urea nitrogen (>100 $mg \cdot dl^{-1}$), and reduced GFR (i.e., 5-8 $ml \cdot min^{-1} \cdot 1.73\ m^{-2}$).

The loss of the excretory function of the kidney results in the buildup of toxins in the blood, any of which can negatively affect enzyme activities and inhibit systems such as the sodium pump, resulting in altered active transport across cell membranes and altered membrane potentials. The loss of regulatory function of the kidneys results in the inability to regulate extracellular volume and electrolyte concentrations, which adversely affects cardiovascular and cellular functions. Most patients are volume overloaded; this results in hypertension and often congestive heart failure. Other malfunctions in regulation include impaired generation of ammonia and hydrogen ion excretion, resulting in metabolic acidosis and decreased production of erythropoietin, which is the primary cause of the anemia in ESRD (6, 29).

Normal substances may be excessively produced or inappropriately regulated in response to renal failure. Parathyroid hormone may be the most important of these. Parathyroid hormone is produced in excess secondary to hyperphosphatemia and reduced conversion of vitamin D to its most active forms (6, 29).

Several metabolic abnormalities are associated with uremia, including insulin resistance and hyperglycemia. Hyperlipidemia is characterized in patients treated with dialysis by hypertriglyceridemia with normal (or low) total cholesterol concentrations. Several interventions associated with the dialysis treatment or immunosuppression therapy (following transplant) can contribute to these metabolic abnormalities (6, 29).

CLINICAL CONSIDERATIONS

The staging criteria for CKD, which have been published elsewhere, range from stage 1 with a GFR of ≥90 $ml \cdot min^{-1} \cdot 1.73\ m^{-2}$ to stage 5 with a GFR of <15 $ml \cdot min^{-1} \cdot 1.73\ m^{-2}$ (1). Renal failure produces the specific signs and symptoms noted earlier. The diagnosis is made by the serum creatinine, blood urea nitrogen, and urinalysis. The management of renal failure is multitiered. First, the cause of the kidney disease needs to be determined, and either cure or control of the primary kidney disease is pursued. If this is not possible, then the clinician should strive to slow or arrest the progression of the decline in kidney function. Treatment is multipronged: control of blood pressure and diabetes, use of an angiotensin-converting enzyme inhibitor/angiotensin receptor blocker (ACEI/ARB), lowering urine protein, weight loss, smoking cessation, and exercise. Dietary adjustment for protein, sodium, and fluid intake plays an important role in the initial management of renal failure. If these treatments are not successful, RRT is required. Currently, the three options for RRT are hemodialysis, peritoneal dialysis, and transplantation. Although transplantation is the preferred method, patients need to be free of other life-threatening illnesses to be considered for transplantation. **Hemodialysis** is the most common therapy for renal failure, although it requires significant time throughout the week at a renal center. The third alternative for RRT, **peritoneal dialysis**, is the method least used in United States, but it is more frequently used in other countries. Home dialysis either as hemodialysis or peritoneal dialysis is the modality of choice. Recent RCTs have demonstrated a better outcome for patients treated with five-plus dialysis treatments a week either in-center or at home.

Signs and Symptoms

Deterioration in renal function results in an overall decline in physical well-being. Signs include anemia, fluid buildup in tissues, loss of bone minerals, and hypertension. Patients experience fatigue, shortness of breath, loss of appetite, restlessness, change in urination patterns, and overall malaise. Muscle mass, muscle endurance, and peak oxygen uptake decline as the disease progresses.

History and Physical Examination

The medical history usually focuses on renal-related issues, including cause of renal failure, current renal function (if any), and current treatment. The comorbidities are listed and the treatments for each. There is rarely any information on physical functioning or recommendations for activity. The dialysis chart may include additional information that can be helpful (for people on dialysis). The dialysis chart includes evaluations by the social worker and dietitian, both of whom more often address limitations in physical functioning (in terms of activities of daily living or need for assistance in the home). A regular assessment of quality of life (i.e., kidney disease quality of life) is now required in all dialysis patients. The dialysis chart also typically includes a problem list. Of additional interest may be the short- and long-term patient care plan, a multidisciplinary plan that is included in the dialysis chart. This information will give the clinical exercise physiologist a better view of the overall plan for the patient in terms of RRT and social considerations. The clinical exercise physiologist should pay attention to any cardiac history and the type and frequency of dialysis treatment in order to develop the best strategy for exercise that considers the treatment burden experienced by the patient.

Diagnostic Testing

Diagnosis of renal failure is typically made by determination of levels of serum creatinine and blood urea nitrogen through a blood test and a urinalysis. Renal biopsy can be done to determine the etiology of disease, and a renal ultrasound or computed axial tomography (CAT) scan can be performed to rule out obstruction or congenital abnormalities that may contribute to increased creatinine levels in the blood. Treatment of chronic renal failure consists of medical management until the creatinine clearance is less than 10 ml · min^{-1}, at which time RRT is usually required. Management is directed at minimizing the consequences of accumulated uremic toxins normally excreted by the kidneys. Dietary measures play a primary role in the initial management, with very low-protein diets being prescribed to decrease the symptoms of uremia and possibly to delay the progression of the disease. In addition to protein restriction, dietary sodium and fluid restrictions are critical (29) because the fluid regulation mechanisms of the kidney are deteriorating in function. Any excess fluid taken remains in the system and, with progressing deterioration in renal function, ultimately results in peripheral edema, congestive heart failure, and pulmonary congestion.

The decision to begin dialysis is determined by many factors, including cardiovascular status, electrolyte levels (specifically potassium), chronic fluid overload, severe and irreversible oliguria (i.e., urine output less than 0.5 cc/kg of body weight divided by height) or anuria (i.e., absence of urine output), significant uremic symptoms, and excessively abnormal laboratory values (usually creatinine >8-12 mg · dl^{-1}, blood urea nitrogen >100-120 mg · dl^{-1}) and creatinine clearance (<10 ml · min^{-1}). Renal replacement therapy does not correct all signs and symptoms of uremia and often presents the patient with other concerns and side effects to deal with. Table 11.1 lists laboratory values for healthy patients versus those undergoing dialysis.

Table 11.1 Normal Laboratory Values Compared With Acceptable Values for Dialysis Patients

Laboratory value	Normal range	Target range for dialysis patients[a]
Hemoglobin (g · dl^{-1})	12.0-16.0	Goal 10-12[b]
Sodium (mEq · L^{-1})	136.0-145.0	135.0-142.0
Potassium (mEq · L^{-1})	3.5-5.3	4.0-6.0
Chloride (mEq · L^{-1})	95.0-110.0	95.0-100.0
HCO_3 (mEq · L^{-1})	22.0-26.0	23-28
Albumin (g · dl^{-1})	3.7-5.2	Goal >4.0
Calcium (mg · dl^{-1})	9.0-10.6	Goal 8.5-9.5
Phosphorus (mg · dl^{-1})	2.5-4.7	Goal <5.5
BUN (mg · dl^{-1})	5.0-25.0	60.0-110.0
Creatinine (mg · dl^{-1})	0.5-1.4	3.0-25.0
pH	7.35-7.45	7.38-7.39
Creatinine clearance (ml · min^{-1})	85.0-150.0	0 (or residual of <10)
Glomerular filtration rate (ml · min^{-1})	90.0-125.0	0 (or residual of <10)

Note. BUN = blood urea nitrogen.

[a]Assuming well-dialyzed, stable patient.

[b]Hemoglobin levels depend on the level of erythropoietin treatment.

Hemodialysis

Hemodialysis is the most common form of RRT in the United States. Approximately 95% of all patients undergo hemodialysis in a center or at home. In other countries, some patients prefer more home-based treatments such as peritoneal dialysis (discussed later). Hemodialysis is a process of ultrafiltration (fluid removal) and clearance of toxic solutes from the blood. It necessitates vascular access by way of an arteriovenous connection (i.e., fistula) that uses either a prosthetic conduit or native vessels. Two needles are placed in the fistula; one directs blood out of the body to the artificial kidney (dialyzer), and the other directs blood back into the body. The dialyzer has a semipermeable membrane that separates the blood from a dialysis solution, which creates an osmotic and concentration gradient to clear substances from the blood. Factors such as the characteristics of the membrane, transmembrane pressures, blood flow, and dialysate flow rate determine removal of substances from the blood. Manipulation of the blood flow rate, dialysate flow rate, dialysate concentrations, and time of the treatment can be used to remove more or less substances and fluids (29).

The duration of the dialysis treatment is determined by the degree of residual renal function, body size, dietary intake, and clinical status. A typical dialysis prescription is 3 to 4 h three times per week. Complications of the dialysis treatment include hypotension, cramping, problems with bleeding, and fatigue. Significant fluid shifts can occur between treatments if the patient is not careful with dietary and fluid restrictions. Table 11.2 and "Long-Term Complications of Dialysis" list the complications.

Peritoneal Dialysis

Approximately 7% of dialysis patients in the United States are treated with peritoneal dialysis (28). Other countries tend to have a higher percentage of patients treated with this form of dialysis. This form of therapy is accomplished via introduction of a dialysis fluid into the peritoneal cavity through a permanent catheter placed in the lower abdominal wall. The peritoneal membranes are effective for ultrafiltration of fluids and clearance of toxic substances in the blood of uremic individuals. The dialysis fluid is formulated to provide gradients to remove fluid and substances. The fluid is introduced either by a machine (cycler), which cycles fluid in and out over an 8 to 12 h period at night, or manually with 2 to 2.5 L bags that are attached to tubing and emptied by gravity into and out of the peritoneum. The latter process, known as continuous ambulatory peritoneal dialysis, allows the patient to dialyze continuously throughout the day. Con-

Table 11.2 Complications Associated With Hemodialysis Treatment

Complication	Pathophysiology
Hypotension	Decreased plasma volume with slow refilling Impaired vasoactive or cardiac responses Vasodilation Autonomic dysfunction
Cramping	Contraction of intravascular volume Reduced muscle perfusion
Anaphylactic reactions	Reaction to dialysis membrane (usually at first use)
Pyrogen- or infection-induced fever	Bacterial contamination of water system Systemic infection (often at the access site)
Cardiopulmonary arrest	Dialysis line disconnection Air embolism Aberrant dialysate composition Anaphylactic membrane reaction Electrolyte abnormalities Intrinsic cardiac disease
Itching	Unknown etiology
Restless legs	Unknown etiology
Fatigue	Most prevalent symptom

Data from Johansen 1999.

Long-Term Complications of Dialysis

Metabolic abnormalities

- Metabolic acidosis
- Hyperlipidemia (type 4)
  - Increased triglycerides
  - Increased very low-density lipoprotein cholesterol
  - Decreased high-density lipoprotein cholesterol
  - Normal total cholesterol
- Hyperglycemia
- Other hormonal dysfunction

Malnutrition

Cardiovascular disease

- Hypertension
- Ischemic heart disease
- Congestive heart failure
- Pericarditis

Renal-related metabolic bone disease (secondary hyperparathyroidism, osteoporosis, osteomalacia, adynamic)

Peripheral neuropathy

Amyloidosis

Severe physical deconditioning

Frequent hospitalizations

Continuation of progressive complications in diabetic patients

Table 11.3 Complications Associated With Peritoneal Dialysis Treatment

Complication	Comments
Infections	Possible at exit site or along catheter—"tunnel infection" May be the result of a break in sterile procedures during exchange
Peritonitis	Most frequent complication of peritoneal dialysis
Hypotension	Excessive ultrafiltration and sodium removal
Hernia, leaks	Associated with the increased intra-abdominal pressure

tinuous ambulatory peritoneal dialysis requires exchange of fluid every 4 h using a sterile technique (29). Table 11.3 lists complications associated with peritoneal dialysis.

Patients may choose peritoneal dialysis so that they can experience more freedom and less dependency on a center for use of a machine. This method of treatment allows patients to travel and dialyze on their own schedules. Patients who have cardiac instability may also be placed on peritoneal dialysis because this method does not involve the major fluid shifts experienced with hemodialysis.

Complications of peritoneal dialysis include problems with the catheter or catheter site, infection, hernias, low back pain, and obesity. Hypertriglyceridemia is a problem caused by the exposure and absorption of glucose from the dialysate. Patients may absorb as many as 1,200 kcal from the dialysate per day, contributing to the development of obesity and hypertriglyceridemia (29).

Kidney Transplant

Transplantation of kidneys is the preferred treatment of ESRD. In 2008 in the United States, 17,413 transplants were performed (28). The source of the kidneys available for transplant can be a living relative, an unrelated individual, or a cadaver. Because of the shortage of organs available for transplantation and improvements in immunosuppression medications, living nonrelated transplants are becoming more frequent. Patients considered for transplant are typically healthier and younger than the general dialysis population, although there are no age limits to transplantation. Patients with severe cardiac, cerebrovascular, or pulmonary disease and neoplasia are not considered candidates. Table 11.4 lists long-term complications of transplantation.

Following transplantation, patients are placed on immunosuppression medication, which includes combinations of glucocorticosteroids (prednisone), cyclosporine derivative, and monoclonal antibody therapy. New immunosuppression medications are constantly being developed, allowing for minimization of side effects through alteration of therapies or combinations of therapies. Many centers now use a steroid-free protocol. Patients may experience rejection early (acute) or later (chronic), which is detected by elevation of creatinine. Rejection is treated immediately with increased dosing of immunosuppression (mostly prednisone), with a tapering back to a maintenance dose. Patients must remain on immunosuppression for the lifetime of the transplanted organ. Nationwide 1 yr graft survival is 85%, and patient survival is 90%. Five-year rates are 67% for graft survival and 85% for patient survival. Causes of graft loss include chronic rejection (25%), cardiovascular deaths (20.3%), infectious deaths (8.7%), acute rejection (10.2%), technical complications (4.7%), and other deaths (10.2%) (table 11.4). Short-term transplant survival has been improved with new immunosuppression medications, leaving the major challenges of long-term survival of graft and patients to be investigated. Loss of kidney results in the need to return to dialysis (13).

Complications of kidney transplantation are primarily related to immunosuppression therapy and include infection, hyperlipidemia, hypertension, obesity, steroid-induced diabetes, and osteonecrosis. The incidence of atherosclerotic cardiovascular disease is four to six times higher in kidney transplant recipients than in the general population, and cardiovascular risk factors are prevalent in most patients (43).

Exercise Testing

Most patients on dialysis are severely limited in exercise capacity (see practical application 11.1), primarily by fatigue. Peak oxygen uptake ($\dot{V}O_2$peak) is reported to be only 17 to 20 ml · kg^{-1} · min^{-1}; this represents 55% to 65% of

Table 11.4 Long-Term Complications Associated With Transplantation

Complication	Comments
Rejection	Can be acute or chronic; in most cases treated with increased immunosuppression dosages.
Cardiovascular disease	Most frequent cause of death posttransplant. All known risk factors are prevalent, and immunosuppression medications may exacerbate risk.
Infections	Immunosuppression may increase infection risk.
Musculoskeletal disorders	Glucocorticoid therapy (prednisone) reduces bone density, increases risk for aseptic necrosis of the hip, and causes muscle protein breakdown. Many centers now use a steroid-free protocol.
Obesity	Very prevalent, often associated with prednisone therapy, but more likely attributable to imbalance between calorie intake and expenditure (i.e., lifestyle issues).

Practical Application 11.1

RESEARCH FOCUS: EFFECTS OF MODALITY CHANGE AND TRANSPLANT ON PEAK OXYGEN UPTAKE IN PATIENTS WITH KIDNEY FAILURE

The uremia associated with ESRD is believed to contribute to the reduced $\dot{V}O_2$peak observed in patients with ESRD. There is some evidence to suggest that immediately following kidney transplantation, patients show an improvement in aerobic capacity (36). This question was further investigated by Painter and colleagues (37), who studied four groups of patients: conventional dialysis patients; some who changed from receiving conventional dialysis to receiving short, frequent dialysis sessions; transplant recipients; and a group described as having preemptive transplant. Patients were tested at baseline (i.e., before dialysis change or before receiving a transplant) and after 6 mo. A branching treadmill protocol was used, and medications that could affect the chronotropic response (i.e., β-blockers and calcium channel blockers) were withheld for at least 15 h before testing. Cardiac output was measured using an acetylene rebreathing method. Following the 6 mo period there was no change in the aerobic capacity of hemodialysis patients, but a significant increase in $\dot{V}O_2$peak was seen among transplant recipients. This increase was mediated by an increase in cardiac output due to a better chronotropic response. Despite the increase in $\dot{V}O_2$peak, the observed value was approximately 79% of that of age- and gender-matched controls. The same authors (37) also noted that there was no change in the peripheral oxygen extraction after transplant, suggesting that exercise training would be necessary to improve this parameter.

normal age-expected levels but can be as low as 39% (2, 18, 21, 32, 41). The degree to which exercise capacity is limited in these patients is difficult to determine because of the complex nature of uremia, which affects nearly every organ system of the body. Reduced exercise capacity is almost certainly a multifaceted problem that is influenced by anemia, muscle blood flow, muscle oxidative capacity, myocardial function, and the person's activity levels. Muscle function may be affected by nutritional status, dialysis adequacy, hyperparathyroidism, and other clinical variables (18). The estimation of $\dot{V}O_2$peak from submaximal responses is not currently recommended in CKD patients since prediction equations have not been validated in this population (21).

Most studies that have measured $\dot{V}O_2$peak have included only the healthiest patients; thus, the average CKD patient may have an even lower exercise capacity. One school of opinion holds that information obtained from graded exercise testing in this patient group is not diagnostically useful since most patients stop exercise because of leg fatigue and do not achieve age-predicted maximal heart rates. Furthermore, many patients have abnormal left ventricular function (11), and some have conditions that make interpretation of the stress electrocardiogram difficult; these include left ventricular hypertrophy (LVH) with strain patterns, electrolyte abnormalities, and digoxin effects on the electrocardiogram. Thus, stress testing may not be routinely performed before initiation of exercise training, and requiring stress testing may prevent some patients from becoming more physically active (40). However, some experts still recommend the use of incremental testing, particularly with gas exchange, since functional capacity is considered useful for prognostic purposes (21). Because exercise capacity is so low, most patients do not exercise train at levels that are much above the energy requirements of their daily activities. Therefore, risk associated with such training is minimal. Heart rate is not recommended for determining training intensity because of the effects of the varied medications used and the impact of uremia. Thus, exercise testing is not needed to develop a training heart rate prescription. Table 11.5 summarizes a few of the important issues associated with various fitness assessments. Table 11.6 lists the medications commonly used to treat patients with ESRD and ways in which these medications might affect the responses observed during exercise. Practical application 11.2 deals with client–clinician interaction for patients with ESRD.

It may be best to test physical performance in dialysis patients using tests such as stair climbing, 6 min walk test, sit-to-stand-to-sit test, or gait speed testing. These tests, which have been standardized and used in many studies of elderly people, have been shown to predict outcomes such as hospitalization, discharge to nursing home, and mortality rate (21, 40). A walking–stair-climbing test was validated in hemodialysis patients by Mercer and colleagues (25). However, according to Koufaki and Kouidi, more research needs to be done on the use of these physical performance tests in patients with ESRD since the minimum difference in scores that is meaningfully linked to important outcomes is not known (21). Additionally, self-reported physical functioning scales such as those on the SF-36 Health Status Questionnaire are highly predictive of outcomes in dialysis patients, specifically hospitalization and death (12). Exercise training improves scores on these self-reported scales in hemodialysis patients (35).

Exercise capacity is similarly low in peritoneal dialysis patients (3, 24, 39). Following successful renal transplant, exercise capacity increases significantly, to near sedentary normal predicted values (3, 8, 13, 23, 36, 39). The improvement in aerobic capacity after kidney transplantation peaks at 16 mo posttransplant, after which there is a decline that seems to be linked to poorly controlled blood pressures (42). Renal transplant recipients who were active and who participated in the 1996 U.S.

Table 11.5 Exercise Testing Review

Cardiorespiratory endurance	Skeletal muscle strength	Skeletal muscle endurance	Flexibility	Body composition
• Functional tests may be more appropriate for very deconditioned ESRD patients. • Abnormal cardiovascular responses are common.	• Isotonic RM testing is not recommended in ESRD. • Isokinetic testing and functional testing is recommended.	Due to the low weights used, this can be similar to testing in normal healthy individuals.	Standardize the timing of testing in relation to dialysis treatments or in relation to bag changes in peritoneal patients.	• Specific equations have not been validated in ESRD population. • Measurement must be taken in relation to dialysis treatment and may be affected by fluid status.

Table 11.6 Pharmacology

Medication class	Primary effects	Exercise effects	Special considerations
Diuretics (e.g., furosemide)	↓ blood pressure	↓ blood pressure	Hypotensive response postexercise
ACE I (e.g., lisinopril)	↓ blood pressure	↓ blood pressure	Hypotensive response postexercise
ARB (losartan)	↓ blood pressure	↓ blood pressure	Hypotensive response postexercise
Calcium channel blockers (diltiazem)	↓ blood pressure	↓ blood pressure ↓ heart rate	Hypotensive response postexercise
β-blockers (e.g., metoprolol)	↓ heart rate and blood pressure	↓ heart rate and blood pressure	Can blunt the exercise response
Statins (e.g., pravastatin, simvastatin)	Modification of lipids	None	Muscle myalgia
Oral hypoglycemic agents (e.g., glyburide)	↓ blood glucose	↓ blood glucose	Must monitor blood sugars to avoid hypoglycemia
Insulin (e.g., Humalog, Lantus)	↓ blood glucose	↓ blood glucose	Must monitor blood sugars to avoid hypoglycemia
Anticoagulant (e.g., aspirin)	↓ blood clotting	None	Increased tendency to bruise or bleed with any trauma
Phosphate binders (sevelamer)	↓ phosphate	None	None
Erythropoietin (Epogen)	↑ red blood cells	Small ↑ in aerobic capacity	None

↑ = increase; ↓ = decrease.

Transplant Games had exercise capacities that averaged 115% of normal age-predicted values (24, 38). Along with recent findings of improved survival, this underscores the need for transplant recipients to be physically active posttransplant (42, 43). Exercise testing with standard protocols is more appropriate for transplant recipients who are able to push themselves in their training programs above their daily levels of activity. Exercise heart rate responses are normalized after transplant. The major abnormality noted in transplant recipients is excessive blood pressure response to exercise.

EXERCISE PRESCRIPTION

The exercise prescription for patients on dialysis should include flexibility, resistance, and cardiovascular exercises. Weight management considerations may be needed for many transplant recipients. For the dialysis patient, the key to prescription is understanding the multiple barriers to exercise that may exist. These include general feelings of malaise, time requirements of treatment, fear, and adaptation of lifestyles to low levels of functioning. Thus, any prescription should start slowly and progress gradually to prevent discouragement and additional feelings of fatigue or muscle soreness (30). Practical application 11.3 reviews the literature about exercise training and ESRD.

Special Exercise Considerations

When patients are diagnosed with ESRD, many may not be given information on exercise and physical activity. The absence of such education or counseling may pose questions and plant doubt in the minds of patients and their families, who can be protective. Patients often do not know how much activity is too much activity, and because they do not feel well and are fatigued, they opt for no activity. The dialysis staff, who see the patients regularly for their treatments, may also unintentionally reinforce an inactive lifestyle. Patients interact primarily with their dialysis providers, and therefore the exercise professional must take the time to learn more about

Practical Application 11.2

CLIENT–CLINICIAN INTERACTION

Exercise professionals can best serve the needs of CKD patients if they understand the medical complications associated with the disease, the patients' treatment regimens, and the setting in which they receive treatment. Predialysis CKD patients tend to be deconditioned and are at higher than average risk of cardiovascular disease. Most hemodialysis patients receive treatments three times per week for 3 to 4 h, and following these treatments patients tend to feel fatigued for much of the day. They are transported either by friends and family or by medical transportation services, so they have minimal flexibility with their schedule.

The staffing at outpatient dialysis clinics typically consists of a charge nurse and patient care technicians who are not medically trained except for administration of dialysis. The schedules for dialysis are quite tight, and the opportunity for patient education is minimal in terms of both staff and time. These circumstances may represent a significant barrier to attending exercise class at a supervised program. Thus, to optimize adherence, every effort should be made to implement home exercise programs or programs at the dialysis clinic.

For a program to be successful, the exercise professional should first interact with and educate the dialysis staff, which requires the support of the administration in coordinating in-service training. The dialysis staff are close to the patients and can be influential in their participation (or nonparticipation) in exercise. This support is critical to the efforts of the patient and the exercise personnel. Educational and motivational programs at the dialysis clinic not only increase staff and patient awareness of the importance of exercise but also can help change the environment of the clinic from one of illness to one of wellness.

Following kidney transplant, some patients may be afraid to exert themselves vigorously. This concern may be due to the absence of information and recommendations from health care professionals regarding exercise, weakness despite a significant improvement in overall health, fears or concerns on the part of the patient's family, or lack of experience with exercise (because of health concerns before transplant). Exercise is not routinely addressed following kidney transplant—either at the time of transplant or in the routine follow-up care in the clinic. The exercise professional should attempt to educate the transplant team about the importance of exercise and, as much as possible, become part of the routine care team so that exercise counseling is incorporated into a patient's care plan following transplantation.

dialysis and kidney transplant to appreciate the burden that patients experience because of the need to undergo dialysis three times a week or receiving a transplanted kidney. This learning experience could entail watching a patient being put on the dialysis machine and visiting with a few patients and the dialysis staff during a patient's treatment. Patient support groups are also a good source of information. Patients often talk freely about their experiences. By listening carefully, the exercise professional will understand more about patient responses to major changes in lifestyle such as initiating exercise, and they may be able to devise more effective ways to motivate patients. The exercise professional should reach out to dialysis staff about how exercise can benefit their patients and assure them that the programs will be safe and will not interfere with the treatments (8). This education should also include ideas about how the dialysis staff can encourage patients to be physically active. Additional encouragement and reinforcement can greatly facilitate patient efforts in rehabilitation. Lack of support and understanding on the part of the dialysis staff can sabotage efforts to increase patients' activity.

Most exercise professionals practice in their own laboratory and depend on referrals of patients to their exercise facility. Many patients may be unable to participate in exercise at the designated facility but may benefit significantly from counseling on home-based, independent exercise. Thus, exercise professionals should reach out to other health care providers (other than cardiologists and pulmonologists), educate them on the exercise-related services that they can offer their patients, and discuss the benefits of exercise for their patient groups. Although the nephrology community is becoming more interested in improving physical functioning of patients, most nephrologists and kidney transplant staff are not familiar with how exercise may benefit their patients or how to evaluate physical functioning or prescribe exercise. A trained professional who knows about the problems associated with dialysis and transplant may be a valuable addition to the patient care team.

Practical Application 11.3

LITERATURE REVIEW

Dialysis patients are less active than healthy gender- and age-matched sedentary individuals (34). A number of barriers have been identified that contribute to this reduced physical activity level in ESRD patients, the chief of which seems to be "a lack of motivation" (34). ESRD patients have low $\dot{V}O_2$peak scores, poor performances on physical functional tests (e.g., 6 min walk test, gait speed test, and sit-to-stand tests), and low self-reported functioning (31). Dialysis patients with $\dot{V}O_2$peak values >17.5 ml · kg^{-1} · min^{-1} have been found to have better survival rates during a 3.5 yr follow-up period than patients with lower values (21). In a meta-analysis of the relevant literature, Cheema and Singh reviewed 29 trials including ESRD patients; 9 were uncontrolled trials, 7 were controlled, and 13 were randomized clinical trials (10). From this analysis, the average increase in $\dot{V}O_2$peak was 17% to 23%; but when resistance training was combined with aerobic training, the average increase went to 41% to 48% (10, 16, 33). The timing of training is complicated in hemodialysis patients by the need of these patients to be dialyzed for 3 to 4 h three times per week. Exercise programs can therefore be structured for off-dialysis days or during the dialysis treatment. The major problem with structured exercise programs on nondialysis days relates to compliance despite the potential for greater improvements in fitness compared to exercising during dialysis (5). There is currently some interest in promoting home exercise programs in ESRD patients (20, 33). **Intradialytic exercise** has many benefits, including good compliance, less boredom during the dialysis treatment, fewer hypotensive episodes, less cramping, improved blood pressure, more energy during the off-dialysis days, and better urea and phosphate removal (9, 14, 33). To be effective, exercise training should be performed early in the treatment (17, 22).

In general, exercise training studies lasting from 8 wk to 10 mo lead to approximately 10% to 40% improvements in $\dot{V}O_2$peak (3-8 ml · kg^{-1} · min^{-1} absolute values) (5). However, the sample sizes in these studies are typically small, and the groups consist of the healthier patients. Despite the improvement in $\dot{V}O_2$peak, CKD patients do not normalize their values with exercise training (16).

Besides the improvements in aerobic capacity, aerobic exercise training leads to improvements in cardiovascular risk factors including arterial stiffness, heart rate variability, high blood pressure, and inflammatory markers (5, 16). Researchers have investigated the impact of exercise training on morbidity and mortality in the ESRD population, but to date the evidence is not conclusive (5).

ESRD patients are weaker than their age- and gender-matched controls. Surprisingly, Diesel and colleagues found a significant relationship between muscle strength and $\dot{V}O_2$peak in a sample of ESRD patients. Researchers have found that a 12 wk moderate-intensity resistance training program increased muscular strength and the physical functioning (6 min walk test, sit-to-stand-to-sit test, gait speed test) of 10 ESRD patients (15). Other researchers performed intradialytic exercise training and found evidence of an anabolic effect in addition to improvements in physical function (16, 17).

One study showed significant improvements in gait speed and the sit-to-stand-to-sit test following exercise counseling interventions for independent home exercise and in-center cycling. This study also reported significant improvements in the physical scales of the SF-36 Health Status Questionnaire (35).

An exercise training study that included patients treated with peritoneal dialysis showed significant improvements in $\dot{V}O_2$peak (16.2%) and physical functioning dimensions of quality of life (24). Two studies reported significant improvements in exercise capacity in renal transplant recipients (19, 26).

Exercise Recommendations

As previously mentioned, the timing of exercise in relation to the dialysis treatment should be considered. Hemodialysis patients can exercise any time, although they may feel best on their nondialysis days and may be able to tolerate higher intensity or duration. Most feel extremely fatigued after their dialysis treatment, and problems with hypotension may develop following the treatment when vasodilation induced by physical activity occurs. Immediately before the dialysis treatment, some patients may have excessive fluid in their systems because they are unable to rid their bodies of fluid taken in between treatments. Thus, they may not tolerate as much exercise before dialysis because of increased volume overload on the left ventricle; increased blood

Clinical Exercise Pearls: Special Considerations for Clients with End-Stage Renal Disease and Chronic Kidney Disease

- End-stage renal disease patients are very deconditioned and complain of fatigue possibly due to anemia.
- Intradialytic exercise should be recommended whenever possible; if it is not possible, the exercise professional needs to schedule exercise in relation to dialysis.
- A high percentage of ESRD patients are diabetic and hypertensive.
- ESRD and CKD patients are at higher than normal risk of cardiovascular disease. Exercise training should be encouraged to decrease this risk.
- Renal disease may impose a ceiling on the improvement that can occur with training, so the exercise professional should not be discouraged if the magnitude of the change in fitness variables over time is small. The patient's quality of life improves markedly.

pressure at rest (and during exercise), which may increase ventricular preload and afterload; and, in extreme cases, pulmonary congestion. Exercise should be deferred if the patient is experiencing shortness of breath related to excess fluid status. No specific guidelines regarding the upper limit of fluid weight gain contraindicate exercise, although the guidelines for blood pressure established by American College of Sports Medicine (1) should be followed. Practical application 11.4 discusses exercise prescription for patients who are undergoing dialysis and those with transplants.

The ideal mode of exercise and the ideal time for the patient to exercise, in terms of adherence and convenience, may be recumbent stationary cycling during the hemodialysis treatment. This form of exercise does not interfere with the dialysis treatment and should be encouraged in most dialysis clinics. If this is not possible, a home exercise program may be the best approach for these patients.

Cycling during dialysis is best tolerated during the first 1 to 1.5 h of the treatment because after that time the patient has a greater risk of becoming hypotensive even while sitting in the chair, which makes cycling difficult (22). This response is caused by the continuous removal of fluid throughout the treatment, which decreases cardiac output, stroke volume, and mean arterial pressure at rest (28). Therefore, after 2 h of dialysis, cardiovascular decompensation may preclude exercise (28).

For patients treated with continuous ambulatory peritoneal dialysis, the exercise may be best tolerated at a time when the abdomen is drained of fluid; this allows for greater diaphragmatic excursion and less pressure against the catheter during exertion, reducing the risk of hernias or leaks around the catheter site (7). Patients may choose to exercise in the middle of a dialysis exchange—after draining of fluid and before introduction of the new dialysis fluid. This option requires capping off the catheter for exercise, a technique that must be discussed with the dialysis nurse.

Mode

There is no restriction on the type of activity that can be prescribed for dialysis or transplant patients. Range of motion and strengthening exercises are critical for most patients because they are stiff and weak after long periods of inactivity. Because many patients have weak muscles and joint discomfort, non-weight-bearing cardiovascular activity may be best tolerated. As with anyone else, if jarring activity causes joint discomfort, then a change in mode of exercise is indicated. The access site for the hemodialysis may be in the arm or upper leg. This circumstance should not inhibit activity at all, although many patients are told by the vascular surgeon to not use the arm with the fistula in it for 6 to 8 wk after it is implanted to ensure sufficient time for postoperative healing. The only precaution for the fistula is to avoid any activity that would close off the flow of blood (e.g., having weights lying directly over the top of the vessels). Although the patient should be protective of the access site, use of the extremity will increase flow through it and actually help develop muscles around the access site, which should make placement of needles easier.

Patients with a peritoneal catheter should avoid full sit-ups and activities that involve full flexion at the hip. They can accomplish abdominal strengthening by performing isometric contractions and "crunches." Swimming may be a challenge for those with peritoneal catheters because

Practical Application 11.4

EXERCISE TRAINING REVIEW FOR ESRD PATIENTS TREATED WITH DIALYSIS OR RECEIVING A TRANSPLANT

Training method	Mode	Intensity	Frequency	Duration	Progression	Goals
Cardiovascular (CV)	Intradialytic cycling, walking, swimming, low-level aerobics, stepping	RPE of 12 to 15 (on 6- to 20-point scale)	3 to 5 d/wk	Work up to 30 min of continuous exercise	Start with intervals of intermittent exercise and gradually increase the work intervals	Improve CV fitness Reduce CV risk
Resistance	Machine weights Resistance bands, isometric, very low-level hand and ankle weights, body weight resistance	10RM	2 to 3 d/wk	Repetitions: 10 to 15 per set	Start with one set of 10 repetitions with 1 to 2 lb (0.5 to 1 kg) weights; increase gradually	Increase muscular strength, decrease muscle wasting Improve physical function
Range of motion	Static stretching	To the point of mild discomfort	Daily	10 to 30 s	Start at 10 s and increase to 30 s	Decrease stiffness after dialysis

Special Considerations for Hemodialysis Patients

Patients have extremely low fitness levels.

Timing of exercise sessions should be coordinated with dialysis sessions; nondialysis days are usually preferred.

Patients will experience frequent hospitalizations and setbacks.

Gradual progression is critical.

Heart rate prescriptions are typically invalid—use of rating of perceived exertion (RPE) is recommended.

Maximal exercise testing may not be feasible in all patients but should be used whenever appropriate.

Performance-based testing should be used in patients who cannot tolerate graded exercise testing.

One-repetition maximum testing for strength is not recommended because of secondary hyperparathyroidism-related bone and joint problems.

Prevalence of orthopedic problems is significant.

Motivating patients is often a challenge.

Every attempt should be made to educate dialysis staff about the benefits of exercise so that they can help motivate patients to participate.

Special Considerations: Transplant

Patients are initially weak, so gradual progression is recommended.

Patients may experience a lot of orthopedic and musculoskeletal discomfort with strenuous exercise.

Weight management often becomes an issue following transplant.

Patients and their families are often fearful of "overexertion"; thus, gradual progression should be stressed.

Prednisone may delay adaptations to resistance training.

Exercise should be decreased in intensity and duration during episodes of rejection, not eliminated completely.

Patients may experience frequent hospitalizations during the first year after transplant. Because patients are immunosuppressed, every effort must be made to avoid infectious situations (e.g., strict sterilization procedures must be followed for exercise testing and training equipment).

of the possibility of infection. Patients must be advised to cover the catheter with protective tape and to clean around the catheter exit site after swimming. Freshwater lake swimming is not recommended, whereas swimming in chlorinated pools and in the ocean involves less risk of infection.

Although renal transplant recipients are often told not to participate in vigorous activities, the actual primary concern is to avoid any contact sport that may involve a direct hit to the area of the transplanted kidney (e.g., football). Vigorous noncontact sports and activities are generally well tolerated by transplant recipients who have exercise trained to build adequate muscle strength and cardiovascular endurance through a comprehensive conditioning program.

Frequency

Flexibility exercises should be encouraged daily. Hemodialysis patients feel especially stiff after their dialysis session because of the 3 to 4 h of sitting as well as removal of fluid and often cramping. Stretching during and after the dialysis treatment may relieve this stiffness. Resistance exercise should be performed 2 or 3 d per week. Cardiovascular exercise should be prescribed for at least 3 d per week, although a prescription of 4 to 5 d per week may be most beneficial.

Intensity

Cardiovascular exercise intensity should be prescribed through use of a rating of perceived exertion (RPE), because heart rates are highly variable in dialysis patients as a result of fluid shifts and vascular adaptations to fluid loss during the dialysis treatment. Many patients initially may tolerate only a few minutes of very low-level exercise, which means that warm-up and cool-down intensities are irrelevant. These individuals should just be encouraged to increase duration gradually at whatever level they can tolerate. After they achieve 20 min of continuous exercise, the warm-up, conditioning, and cool-down phases can be incorporated, with an RPE of 9 to 10 for warm-up and cool-down and of 12 to 15 for the conditioning time (on a 6- to 20-point scale).

Duration and Progression

The exercise professional must start patients slowly and progress them gradually. In practice, this means that the patient should determine the duration of activity that he can comfortably tolerate during the initial sessions. This duration will be the starting duration of activity. If the patient tolerates only 2 to 3 min of exercise, the prescription may be written for several intervals of 2 to 3 min each, with a gradual decrease in rest times as the patient is progressed to continuous activity. A progressive increase in duration of 2 to 3 min per session or per week is recommended, depending on individual tolerance. Extremely weak patients may need to start with a strengthening program of low weights and high repetitions as well as flexibility exercise before beginning any cardiovascular activity. To progress they should gradually work up to 20 to 30 (or more) min of continuous activity at an RPE of 12 to 15 (on a 6- to 20-point scale).

When a patient begins cycling during dialysis, the initial session is usually limited to 10 min, even if the patient is able to tolerate a longer duration. This precaution assures the dialysis staff and the patient that cycling does not have any adverse effects on the dialysis treatment. The patient can then increase duration in subsequent sessions according to tolerance as described previously. Rating of perceived exertion is also used for intensity prescription during the dialysis treatment, because removal of fluid from the beginning to the end of dialysis can cause resting and exercise heart rates (standard submaximal level) to vary by 15 to 20 beats (27).

EXERCISE TRAINING

The results of an exercise training program in renal patients are variable. Most often, there is an improvement in physical functioning (as measured by exercise testing or physical performance testing or self-reported functioning on questionnaires). Some dialysis patients (and possibly diabetic transplant recipients), particularly those who have a very low functional capacity and have multiple comorbidities, may not show dramatic improvements.

However, the natural course of the condition is a deterioration in physical function; thus, if the outcome of the exercise intervention is to maintain exercise capacity and physical function over time, that is a positive result. Most patients experience improvement in muscle strength and often an increase in lean muscle mass. This increase in lean mass may have implications for hemodialysis patients, since the amount of fluid removed during dialysis is gauged by body weight. Thus, if their body weight (lean mass) increases, their target weight for dialysis may need to be adjusted. Many dialysis patients also experience improvements in blood pressure, often requiring a reduction in antihypertensive agents. Diabetic patients may experience improved glucose control, thus requiring adjustments in insulin requirements.

Patients who undergo dialysis and transplant recipients respond to exercise training with a magnitude of change in strength and exercise capacity similar to that for normal healthy individuals. However, they may not achieve similar maximal levels of functioning. Likewise, the time course for improvement may be longer in patients with renal disease. For transplant recipients on prednisone, improvements in muscle strength may be slower, and absolute gains may take longer than in healthy individuals; however, these patients can achieve normal levels of muscle strength, counteracting the negative effects of prednisone on the muscles (i.e., sarcopenia).

Most patients experience significant improvements in energy level and ability to perform their activities of daily living, and may experience fewer symptoms associated with dialysis such as muscle stiffness, cramping, and hypotensive events during dialysis. If exercise is performed during the hemodialysis treatment, clearance of toxins may be improved (22). Overall, quality of life improves, particularly in the physical domains.

CONCLUSION

The exercise prescription for patients with renal failure depends on their treatment. The prescription must be individualized to the patient's limitations; this includes the type of exercise (cardiovascular, flexibility, resistance), frequency of exercise, timing of exercise in relation to treatment, duration, intensity (prescribed primarily based on RPE), and progression. The progression should be gradual in those who are extremely debilitated. The starting levels and progression must be according to tolerance because fluctuations in well-being, clinical status, and overall ability frequently change with changes in medical status. Hospitalization or a medical event (e.g., clotting of the fistula or placement of a new fistula) may set patients back in the progression of their program, requiring frequent evaluation of the prescription. The goal is for patients to become more active in general and, if possible, for them to work toward a regular program of 3 to 5 d per week of cardiovascular exercise, 30 min or more per session, at an intensity of 12 to 14 on the RPE scale. Resistance exercise two or three times per week is recommended.

Key Terms

chronic kidney disease (CKD) (p. 197)
dialysis (p. 197)
end-stage renal disease (ESRD) (p. 197)
glomerular filtration rate (p. 197)
hemodialysis (p. 198)
intradialytic exercise (p. 206)
peritoneal dialysis (p. 198)
renal replacement therapy (p. 197)

CASE STUDY

MEDICAL HISTORY

Mrs. HN is a 68 yr old Hispanic female with known ESRD. She has been on hemodialysis for 28 mo, and her treatment prescription is 3 per week for 3 h per treatment. She has a graft in her right upper arm as her access site. She presents with the complaints of lack of energy, weakness, and decreased endurance. Her nephrologist refers her to the staff clinical exercise physiologist.

DIAGNOSIS

The clinical exercise physiologist reviews Mrs. HN's chart to find that her ESRD is secondary to long-term non-insulin-dependent diabetes (15 yr). Mrs. HN has also developed severe peripheral neuropathy as a result of her diabetes.

EXERCISE TEST RESULTS

The clinical exercise physiologist then conducts a battery of physical function tests to assess Mrs. HN's physical ability. These tests consist of the sit-to-stand test, the 6 min walk, and the 20 ft (6 m) gait speed test at both a comfortable and a fast pace. The results are as follows: sit-to-stand test, 33.01 s, which is 28% of normal age-predicted values (39); 6 min walk, 350 ft (107 m); and 20 ft normal gait speed, 55.01 cm/s, which is 42% of normal age-predicted values (4). Her self-reported physical function scale on the SF-36 Health Status Questionnaire is 55 (average age value is 84). During her walking tests, Mrs. HN exhibits poor balance and endurance as a result of her peripheral neuropathy and general weakness, respectively. Physical activity questionnaires are administered to assess her current activity as well as degree of difficulty of those activities.

EXERCISE PRESCRIPTION

With the assessment complete, an exercise prescription is developed. Mrs. HN is first counseled on exercise as it relates to her diabetes and glycemic control. Written information is also provided. The exercise prescription proceeds only when it is certain that Mrs. HN fully understands the balance between exercise and glycemic control. Because of her poor balance and endurance, a stationary bicycle is the preferred mode for cardiovascular exercise. She is asked to begin with a frequency of 3 or 4 d per week on nondialysis days because she generally feels better on those days. The duration of the exercise is 10 min, with two bouts each exercise day, totaling 20 min of exercise each exercise day. The initial prescribed intensity should be light to moderate, or enjoyable. Mrs. HN is asked to progress gradually each week with a goal of 30 min of continuous cycling 3 to 4 d per week minimum. Her initial exercise prescription also includes various flexibility exercises for the upper and lower body as well as for the back. She is asked to do these exercises daily. Resistance exercises are also prescribed. Again, both upper and lower body exercises that use the major muscle groups are encouraged. These exercises are prescribed for 3 d per week on nonconsecutive days. She will initially perform the exercises without resistance weight and gradually progress to performing them with weight. Her initial prescription consists of one set of 10 repetitions of each exercise. The clinical exercise physiologist reviews Mrs. HN's progress weekly at her dialysis treatments. Progression and exercise participation are noted in the patient's chart. At the end of 8 wk of exercise, Mrs. HN's physical functioning is again assessed with the battery of physical function tests and the activity questionnaires.

DISCUSSION QUESTIONS

1. Why is performance-based testing preferred for this patient?
2. What causes Mrs. HN to have poor balance during walking? How does this condition affect her exercise prescription?
3. Why is no heart rate prescription given to Mrs. HN for her cycling program?
4. How would the cycling program differ if she chose to cycle during the dialysis treatment instead of at home?
5. Does anything in her history suggest that cycling during dialysis might be advantageous for Mrs. HN?

PART III

Cardiovascular Diseases

A reduction in cardiovascular disease (CVD) mortality has been observed over the last 20 years, but the incidence of CVD has not experienced a commensurate decrease. At the same time, many advances have been made in the technology of assessing and correcting significant disease; however, given a societal trend toward more people being overweight, obese, and physically inactive, we can expect many healthcare dollars to be directed toward CVD in the future. The clinical exercise physiologist should be familiar with the areas addressed in the chapters in part III because all these forms of CVD can be positively altered by an appropriate rehabilitation program that also addresses specific lifestyle interventions.

Chapter 12 centers on myocardial infarction. The prevalence of myocardial infarctions continues to be very high in our society. The value of secondary prevention in this population has been shown to reduce the risk of mortality. These patients can strongly benefit from positive lifestyle alterations by participating in a cardiac rehabilitation program. The clinical exercise physiologist plays an important role in assessing the patient's cardiovascular limitations through graded exercise testing and developing an appropriate exercise prescription. This chapter provides a clear understanding of the disease process and preventive measures that can be used to ward off further events.

Chapter 13 discusses revascularization of the heart. Revascularization procedures have become a common way to address significant coronary artery disease. Emergency percutaneous coronary intervention is becoming the standard way of decreasing the risk of myocardial damage. The challenges that the clinical exercise physiologist must address vary according to the clinical procedure completed. He or she must have a good understanding of the procedures and potential issues that may arise when performing a graded exercise test and prescribing exercise.

Chapter 14 deals with chronic heart failure. Given improved emergency and long-term care for patients with CVD and the increasing number of people entering the age group greater than 65 years, both the incidence and prevalence of chronic heart failure in our society is increasing. Current estimates suggest that these trends and the public health burden associated with heart failure (HF) will increase even further over the next 10 years. As a result, the clinical exercise physiologist will be required to have a strong knowledge base to work with patients with HF in both the rehabilitation and exercise testing settings.

Chapter 15 explores peripheral artery disease. Peripheral artery disease (PAD) is common in our society, especially in those with existing CVD, because the major risk factors for CVD are similar to those of PAD. Significant PAD can severely limit a person's exercise tolerance, depending on when the subject develops intermittent claudication. The degree of a person's PAD will have a large influence on testing procedures offered. The clinical exercise physiologist must learn specific strategies to develop an appropriate exercise program for reducing a patient's symptoms of PAD.

Chapter 16 considers pacemakers and implantable cardioverter defibrillators (ICDs). As our population becomes older, the need for the use of pacemakers and

ICDs becomes greater. Because of the complexity of these devices, the clinical exercise physiologist needs to be knowledgeable in how they work, how they can be influenced when a graded exercise test is performed, and what issues may arise under normal exercise training (e.g., what exercise heart rate should be avoided to prevent the risk of premature firing of an ICD).

Chapter 12

Acute Coronary Syndromes: Unstable Angina Pectoris and Acute Myocardial Infarction

Ray W. Squires, PhD

The purpose of this chapter is to provide information on the pathophysiology of the acute coronary syndromes unstable angina and acute myocardial infarction, the treatment options, and the role of cardiac rehabilitation, with emphasis on exercise training. The chapter covers the following specific topics:

1. The scope of cardiovascular disease
2. The disease process of atherosclerosis and thrombosis
3. Myocardial blood flow, ischemia, and angina pectoris
4. Acute coronary syndromes (definition, acute myocardial infarction, clinical assessment, diagnosis, classification by ECG, management strategies, potential complications, right ventricular infarction)
5. Factors associated with poor prognosis
6. Stress testing after acute myocardial infarction
7. Exercise training and cardiac rehabilitation for acute coronary syndromes

The chapter also provides an illustrative case study.

SCOPE

The burden of diseases of the cardiovascular system on our society is horrendous. Since 1900, cardiovascular diseases, such as coronary heart disease, stroke, heart failure, and hypertension, have been the leading cause of death in the United States every year with the exception of 1918, the year of the great influenza epidemic (4). Fifty percent of all cardiovascular deaths are a result of coronary heart disease (CHD). This is the most common cause of death for both men and women (61, 42). For 2007, CHD caused a total of 406,351 deaths and accounted for one of every six deaths. Acute coronary syndromes are almost always the result of coronary artery atherosclerosis and subsequent thrombosis. Each year, 785,000 Americans will have a new acute coronary syndrome, and 470,000 will suffer a recurrent coronary event. Silent (painless) myocardial infarction will occur in 195,000 persons per year. The numbers of Americans who underwent percutaneous coronary intervention (coronary angioplasty, stenting, or both) or coronary bypass graft surgery in 2007 were 622,000 and 232,000, respectively. Every 25 s, an American will have a coronary event, and each minute someone will die from one. Approximately 16,300,000 Americans have a history of myocardial infarction, and 9,000,000 experience angina pectoris (chest pain). Estimated total costs for diseases of the heart were $177.5 billion for 2007 (61).

PATHOPHYSIOLOGY

The pathology of acute coronary syndromes is complex; it involves the development of atherosclerotic lesions in the walls of the coronary arteries with subsequent thrombosis formation resulting in an abrupt decrease in vessel blood flow.

Atherosclerosis and Thrombosis

Atherosclerosis is a disease process that may result in blood flow-limiting lesions in the epicardial coronary, carotid, iliac, and femoral arteries, as well as the aorta. Some arteries are resistant to atherosclerosis (brachial, internal thoracic, intramyocardial) for unknown reasons (13). The processes of atherosclerosis and thrombosis are interrelated, and the term "atherothrombosis" has been adopted by some investigators to emphasize this point (13).

The Normal Artery

The channel for the flow of blood within the artery is the *lumen*. The inner, single-cell layer of the artery is the *endothelium*. The endothelium plays a critical role in maintaining vasomotion (the degree of vasoconstriction) and regulating hemostasis (balancing pro- and antithrombotic properties). When intact, the endothelium produces nitric oxide, a vasodilator, and substances such as plasminogen that inhibit thrombosis formation. Various receptors, such as those for low-density lipoprotein and growth factors, are located on the endothelial cells (63). Under normal circumstances, the endothelium protects against the development of atherothrombosis, but when damaged it plays a central role in the development of the disease (63).

Underneath the endothelial basement membrane is the *intima*, consisting of a thin layer of connective tissue with an occasional smooth muscle cell. The lesions of atherosclerosis form in the intima (63).

The *media* contains most of the smooth muscle cells of the arterial wall, in addition to elastic connective tissue, and is located underneath the intima between the internal and external elastic laminae. The smooth muscle cells maintain arterial tone (partial vasoconstriction). Smooth muscle cells have receptors for low-density lipoprotein, insulin, and growth factors. When appropriately stimulated, smooth muscle cells are capable as functioning as synthetic tissue, producing connective tissue (23).

The outermost layer of the arterial wall is the *adventitia*, consisting of connective tissue (collagen, elastin), fibroblasts (cells capable of synthesizing connective tissue), and a few smooth muscle cells. This tissue is highly vascularized (its blood supply is provided by small vessels called the vasa vasorum) and provides the media and intima with oxygen and nutrients (63).

Atherogenesis

Our understanding of the development and progression of atherosclerosis (atherogenesis) is incomplete. However, it is clear that *endothelial injury* resulting in *endothelial dysfunction* and a subsequent *inflammatory response* play critical roles (62). The disease process may begin in childhood and progress for decades before a clinical event occurs. The rate of progression of atherosclerosis may not be consistent over time and is impossible to predict.

Under normal conditions, the endothelium may experience periodic minimal amounts of injury. In these situations, the inherent repair processes of the endothelium are adequate to restore normal function. However, chronic, excessive injury to endothelial cells initiating the process of atherogenesis may result from multiple causes, such as the following (14, 17, 23, 24, 48, 52, 62, 65):

- Tobacco smoke and other chemical irritants from tobacco
- Low-density lipoprotein cholesterol (LDL-C)
- Hypertension
- Glycated substances resulting from hyperglycemia and diabetes mellitus
- Plasma homocysteine
- Infectious agents (e.g., *Chlamydia pneumoniae* and herpes viruses)

Endothelial dysfunction may result from these potentially injurious factors, leading to the following abnormalities characteristic of an inflammatory response:

- Increased adhesiveness resulting in platelet deposition, monocyte adhesion
- Increased permeability to lipoproteins and other substances in the blood
- Impaired vasodilation, increased vasospasm

Platelets adhere to the damaged endothelium (platelet aggregation), form small blood clots on the vessel wall (mural thrombi), and release growth factors and vasoconstrictor substances, such as thromboxane A2 (23, 63). These changes indicate a switch in endothelial function favoring a prothrombotic, vasoconstrictive state.

Monocytes, a type of white blood cell, also adhere to the injured endothelium and migrate into the intima. LDL-C enters the arterial wall and undergoes the process of oxidation. Monocytes accumulate LDL-C, augmenting the oxidation process, and are transformed into a distinctly different type of cell, the macrophage (62, 63).

Growth factors expressed by platelets, monocytes, and damaged endothelium result in growth and proliferation (increase in cell numbers and cell size) of certain types of cells (*mitogenic effect*), as well as the migration of cells into the area of injury (*chemotactic effect*) (62, 63). In response to the growth factors, smooth muscle cells and fibroblasts (undifferentiated connective tissue cells that can synthesize fibrous tissue) migrate from the media to the intima. Smooth muscle progenitor cells from bone marrow also migrate to the intima (11). Some of these cells, in addition to monocytes, accumulate cholesterol, forming *foam cells* that may release their cholesterol into the extracellular space—giving rise to *fatty streaks*, the earliest visually detectable (yellow macroscopic appearance) lesion of atherosclerosis (24, 64). Immune system cells, *T lymphocytes*, are also present in fatty streaks and are part of the inflammatory state in the arterial wall (81).

With continued migration, proliferation, and growth of tissue, the lesion progresses in complexity and size and becomes a *fibromuscular plaque* (64). The composition of the plaque now includes a fibrous cap, connective tissue extracellular matrix, lipids, inflammatory cells such as macrophages and T-lymphocytes, smooth muscle cells, thrombus, and calcium. The typical plaque is firm in texture and pale gray in color, and may contain a yellow cholesterol core.

As the intimal lesions of atherosclerosis progress and thicken the vessel wall, a compensatory outward expansion of the vessel occurs (to a point) and lumen size remains unchanged. This is called *arterial remodeling* and may be effective in compensating for plaques whose bulk may represent up to 40% of the vessel diameter (43). With continued progression in plaque bulk, the area of the lumen decreases, which may ultimately result in a reduction in blood flow.

The progression of the size and volume of atherosclerotic lesions is highly variable. Some lesions appear relatively stable over many years; other plaques may slowly progress in size, while still other areas of atherosclerosis may enlarge very rapidly (23). The slowly progressing plaques are thought to gradually internalize monocytes and lipids, while rapidly progressing lesions incorporate thrombus into the plaque (13). Local stressors (from turbulent blood flow or vasoconstriction, for example) or chemical factors (enzymes such as metalloproteinases that weaken the fibrous cap) within the lesion may result in *plaque rupture* or *fissuring* of the fibrous cap, exposing the internal contents of the plaque to the blood (12, 23). Various amounts of thrombus form in response to this prothrombotic environment and may be incorporated into the plaque. The scenario of plaque rupture, subsequent thrombus formation, and incorporation into the arterial wall may repeatedly occur, giving a layered appearance to the lesion and resulting in rapid progression in the size of the plaque. These lesions, which include organized thrombus, are called *advanced atherosclerotic plaques*.

Atherosclerosis affects arteries in an extremely diffuse manner, with occasional discrete, localized areas of more pronounced narrowing of the vessel lumen (44). Selective coronary angiography is the "gold standard" (best available test) for determination of the severity of coronary lesions. However, based on comparisons of angiographic and autopsy findings, with the exception of the situation of complete occlusion of the vessel in question (100% stenosis), the degree of stenosis is greatly underestimated by angiography because of the diffuse nature of the disease process (6). Obstructive coronary lesions (severe enough to reduce blood flow) occur most frequently in the first 4 to 5 cm of the epicardial coronary arteries, although more distal disease may be also seen. Obstructive lesions at the origin (ostial lesions) of the left main and main right coronary arteries may also occur. For reasons not fully understood, women generally lag 5 to 20 yr behind men in the extent and severity of coronary atherosclerosis (76).

Risk Factors for Atherosclerosis

Risk factors are associated with an increased likelihood that atherosclerosis will develop over time. Such factors have been identified on the basis of observational studies evaluating common characteristics of persons with the disease (38, 46). Possible mechanisms of atherogenic effect have been identified for some risk factors. The effects of reducing the severity of some risk factors, especially LDL-C, have been demonstrated to reduce progression of the disease. Predicting whether an individual patient will or will not develop atherosclerosis based on the presence and severity of risk factors is very imprecise, however. Less than half of future cardiovascular events can be predicted using conventional risk factors (40) such as the following:

- Tobacco use
- Dyslipidemia, especially elevated LDL-C and low levels of high-density lipoprotein cholesterol (HDL-C)
- Hypertension
- Sedentary lifestyle
- Obesity
- Diabetes mellitus
- Metabolic syndrome (a combination of conventional risk factors associated with obesity and insulin resistance)
- Family history of premature coronary disease (male first-degree relatives <55 yr of age, female first-degree relatives <65 yr of age)
- Male sex
- Obstructive sleep apnea
- Psychosocial factors, such as depression, anxiety, social isolation, coronary-prone personality (type A), lower socioeconomic status, and chronic life stressors

Less well-established risk factors, termed emerging risk factors, are being investigated, including these (40, 56, 60):

- Elevated plasma homocysteine, an intermediary in the metabolism of the essential amino acid methionine
- Fibrinogen, a protein factor in the blood coagulation cascade
- Lipoprotein (a)
- LDL particle concentration
- High-sensitive C-reactive protein, a marker for systemic inflammation

Myocardial Blood Flow, Metabolism, and Ischemia

Normal contraction and relaxation of cardiac myocytes requires the presence of adequate amounts of adenosine triphosphate (ATP; a high-energy phosphate molecule) in the myocardium. The heart is a highly aerobic organ with an extensive circulatory system and abundant mitochondria (26). Figure 12.1 illustrates the epicardial coronary arteries. The coronary arterial system includes epicardial arteries that bifurcate into intramyocardial and endomyocardial branches. At rest, coronary blood flow averages 60 to 90 $ml \cdot min^{-1} \cdot 100g^{-1}$ myocardium and may increase five- to sixfold during exercise (9). Under usual conditions, the heart regenerates ATP aerobically, and myocardial cells are not well adapted to anaerobic energy production. At rest, myocardial oxygen uptake is approximately 8 to 10 $ml\ O_2 \cdot 100g^{-1}$ tissue per minute. During intense exercise, the oxygen requirement may increase by 200% to 300% (25). The myocardium (unlike skeletal muscle) extracts nearly all of its oxygen from the capillary blood flow, and coronary blood flow must be closely regulated to the needs of the myocardium for oxygen (33). With an increase in myocardial work, oxygen demand increases, and coronary blood flow must also increase to provide the necessary amount of oxygen.

Blood flow through any regional circulation, including the coronary system, is determined by the blood pressure and the vascular resistance (33). During systole, intramyocardial pressure is increased (as is vascular resistance) and the intramural vessels are compressed. Therefore, most coronary blood flow occurs during diastole, when intramyocardial pressure is lower (lower vascular resistance).

Before a decrease in flow can be measured distal to an atherosclerotic, narrowed coronary artery segment, a substantial reduction in vessel luminal diameter must occur. When plaque bulk reduces the luminal cross-sectional area by 75% or more, flow is reduced under resting conditions (a hemodynamically significant lesion) (9). Beyond this amount of *critical stenosis*, further small decreases in cross-sectional area of the vessel result in large reductions in flow.

A reduction in the lumen diameter may be caused by several factors (22, 47):

- Significant atherosclerotic plaque
- Vasospasm without underlying plaque
- Vasospasm superimposed over plaque
- Thrombus associated with plaque rupture

Coronary vasospasm may result from several factors, such as endothelial dysfunction, sympathetic nervous system activation (for example, vasospasm resulting from exposure to very cold ambient temperatures), and bloodborne substances like epinephine (49).

Myocardial ischemia results when myocardial blood flow is inadequate to provide the required amounts of oxygen for ATP regeneration (oxygen supply < oxygen demand) (9, 26). Ischemia may result in progressive

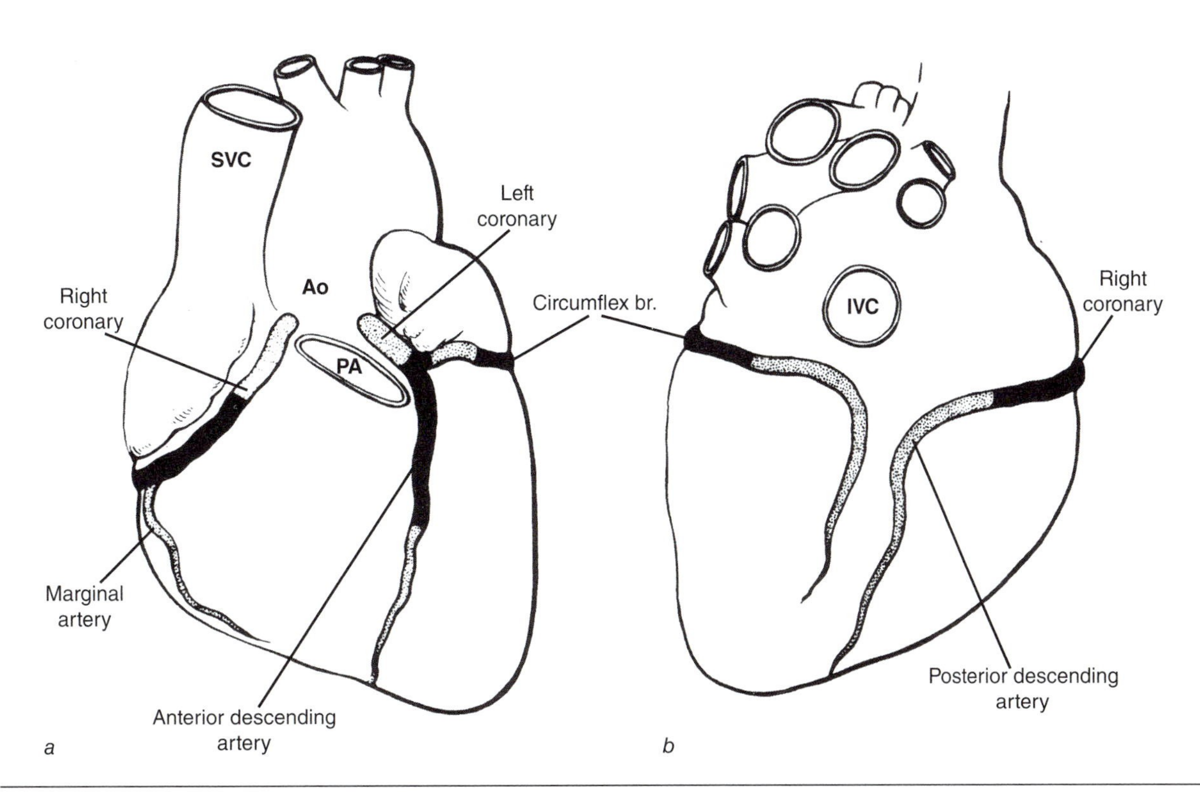

Figure 12.1 The epicardial coronary arteries. *(a)* Anterior view. *(b)* Posterior view. Black segments are prime sites for the development of obstructive atherosclerotic plaques. Ao = aorta; IVC = inferior vena cava; SVC = superior vena cava; PA = pulmonary artery.

Reprinted, by permission, from J.T. Lie, 1991, Pathology of coronary artery disease. In *Cardiology: Fundamentals and practice,* 2nd ed., edited by E.R. Giuliani et al. (Chicago: Mosby Yearbook).

abnormalities in cardiac function termed the *ischemic cascade* (54). The first abnormality is stiffening of the left ventricle, which impairs diastolic filling of the heart (*diastolic dysfunction*). Second, systolic emptying of the left ventricle becomes impaired (*systolic dysfunction*). Localized areas of the myocardium develop abnormal contraction patterns such as *hypokinesis* (reduced systolic contraction). **Left ventricular ejection fraction** may decrease. Third, *electrocardiographic abnormalities* associated with altered repolarization (ST-segment changes, T-wave inversion) or arrhythmias may occur. Finally, symptoms of *angina pectoris* may develop.

Angina pectoris is transient, referred cardiac pain resulting from myocardial ischemia (68). A minority of patients with substantial amounts of ischemia do not report pain (*silent ischemia*). The pain of angina may be located in the substernal region, jaw, neck, or arms, although pain may also occur in the epigastrium and interscapular regions. It is usually described as a feeling of pressure, heaviness, fullness, squeezing, burning, aching, or choking. The pain may vary in intensity and may radiate. The patient may experience dyspnea (*anginal equivalent*) if the ischemia results in increased left ventricular end-diastolic pressure and increased pulmonary vascular pressure. *Typical angina* is provoked by exertion, emotions, cold and heat exposure, meals, and sexual intercourse; it is relieved by rest or nitroglycerin or both. *Atypical angina* refers to similar symptoms but with features that set it apart from typical angina, such as no relationship with exertion. *Stable angina* is reproducible and predictable in onset, severity, and means of relief. *Unstable angina* is defined as new onset of typical angina; increasing frequency, intensity, or duration of previously stable angina; or angina that occurs at rest or in the first few days after acute myocardial infarction.

If an episode of ischemia is brief, the contractile abnormalities that have been described are quickly reversible. Brief, postischemic left ventricular dysfunction is called *stunned myocardium* (18). Chronic, substantial, nonlethal ischemia may result in prolonged but reversible left ventricular dysfunction called *hibernating myocardium.* Myocytes remain viable but exhibit depressed contractile function. Elimination of the chronic ischemia with revascularization results in a gradual return of normal contractile function, although resolution may require up to 1 yr (59). Prolonged, severe ischemia results in myocyte necrosis (irreversible damage, *myocardial infarction*).

CLINICAL CONSIDERATIONS

This section addresses clinical considerations for patients presenting with acute coronary syndromes. Information on clinical assessment, diagnosis, classification by ECG, management strategies, potential complications, and right ventricular infarction are addressed.

Definition of Acute Coronary Syndromes

Unstable angina pectoris, acute myocardial infarction, and some instances of sudden cardiac death constitute the *acute coronary syndromes* (ACS) (13). The underlying mechanism resulting in these syndromes is atherosclerotic plaque erosion, rupture, or other type of plaque disruption resulting in thrombus formation, and possibly vasoconstriction with subsequent vessel occlusion and acute myocardial ischemia.

The type of acute coronary syndrome that occurs is related to the duration of vessel occlusion. Unstable angina is probably the result of transient vessel occlusion (<10 min) followed by spontaneous thrombolysis (clot dissolution) and vasorelaxation. Vessel occlusion persisting for more than 60 min results in acute myocardial infarction. Ischemia resulting from atherothrombotic vessel occlusion may trigger ventricular tachycardia or ventricular fibrillation and sudden cardiac death (13).

Why do some plaques rupture and thrombose? Approximately two-thirds of patients with ACS have "high risk" or "vulnerable" atherosclerotic lesions with thin fibrous caps overlying a lipid-rich core with an abundance of macrophages (13). Inflammation mediated by various cytokines, including proteases and tumor necrosis factor, erodes the plaque from within. The physical forces assisting with plaque disruption include increased blood pressure or heart rate, local vasoconstriction, and nicotine or immune complexes. After plaque rupture, circulating blood platelets come into direct contact with the thrombogenic internal environment of the plaque, resulting in clot formation (21). Angiographic studies have demonstrated that most of these rupture-prone lesions are less than 50% occlusive before they become disrupted (66). This explains why many patients who experience an ACS do not have warning symptoms. However, angiographically severe coronary atherosclerosis does increase the likelihood of a coronary event by serving as a marker for the presence of extensive disease, including rupture-prone lesions. Autopsy studies have also demonstrated that many patients have disrupted plaques but no history of an ACS. Thus, not all plaque disruption results in clinical events. In approximately one-third of cases, an ACS results from only superficial erosion of a severely stenotic and fibrotic plaque with clot formation due to a hyperthrombotic state caused by factors such as smoking, hyperglycemia, or elevated LDL-C (13).

Acute Myocardial Infarction

Acute myocardial infarction is the necrosis (death) of cardiac myocytes resulting from prolonged ischemia attributable to complete vessel occlusion (58). The key event in distinguishing reversible from irreversible (infarction) ischemia is disruption of the myocyte membrane (a lethal event). The myocyte cannot recover if membrane disruption occurs and cytoplasmic contents spill into the circulation.

In some patients, a precipitating event or "trigger" for the myocardial infarction may be determined, such as physical exertion, emotional stress, or anger (27, 50). It may also occur in the setting of surgery associated with substantial loss of blood.

There is evidence for circadian variation, with slightly more myocardial infarctions occurring in the early morning hours than at other times, suggesting a role for sympathetic nervous system activation as a trigger (27, 58).

When a patient presents to a medical facility and acute coronary syndrome is suspected, time is of the essence. Prompt diagnosis and treatment is of paramount importance in terms of ensuring the best possible outcome for the patient. The following assessment is typically performed.

Signs and Symptoms

Symptoms of myocardial infarction include chest pain or other anginal sensations, gastrointestinal upset, dyspnea, sweating, anxiety, or syncope. Myocardial infarction may be painless (*silent myocardial infarction*) in approximately 25% of cases (86). Pain is often severe, but all intensities of discomfort may be experienced. Elderly patients report more dyspnea, while women are more likely to report atypical symptoms, such as shoulder, middle back, or epigastric pain; fatigue; and general weakness (10).

Physical Examination

Patients with ACS may demonstrate the following findings on physical examination (36):

- Systolic hypotension
- Diaphoresis
- Sinus tachycardia
- Tachypnea
- New murmur of mitral regurgitation
- Third, fourth heart sounds
- Pulmonary rates

Diagnostic Testing

Three primary pieces of information are valuable in the diagnostic setting.

1. Electrocardiogram. The ECG may show ST-segment elevation or nonspecific ST-T wave abnormalities (see "Classification of Myocardial Infarction").

2. Chest radiograph. This is useful for patients with evidence of hemodynamic instability or pulmonary edema.

3. Laboratory results. The biomarker cardiac troponin (cTn) is measured twice, 6 to 12 h apart. An ECG is performed on admission and repeated serially, as needed. A blood lipid profile should be obtained within 24 h of symptom onset. An echocardiogram may be performed to assess ventricular and valvular function.

Diagnosis of Acute Myocardial Infarction

The diagnosis of myocardial infarction is based on the presence of blood markers (biomarkers) of myocyte necrosis, preferably cTn, with at least one value above the 99th percentile upper reference limit and with at least one of the following:

- Symptoms of ischemia, especially chest pain persisting for ≥30 min
- New or presumed new electrocardiographic ST-segment or T-wave changes or new left bundle branch block
- Development of pathological Q-waves in the electrocardiogram
- Imaging evidence (nuclear perfusion scan, echocardiogram) of new loss of viable myocardium or new regional wall motion abnormality
- Identification of an intracoronary thrombus by angiography (86, 80a)

Differentiation of unstable angina from acute myocardial infarction is based on biomarker elevation. Cardiac Tn is the preferred biomarker for the detection of cardiomyocyte necrosis. It is highly sensitive and specific for cardiac necrosis. Elevation occurs 2 to 3 h after the onset of infarction and remains elevated for 1 or 2 wk (67). It should be noted that cTn is elevated after percutaneous intervention or cardiac surgery. The standard cardiac biomarker used in the past was the MB fraction of creatine kinase (CK-MB) (5). While not as specific and sensitive as cTn, it is still an acceptable diagnostic test if cTn is not available.

After myocardial infarction, myocardial cells do not regenerate, and healing occurs via scar formation. Depending on the extent of infarction, scar formation may take days to weeks for completion. The larger the size of the infarction, the larger the scar.

Classification of Myocardial Infarction

Infarctions are classified as ST-segment elevation (STEMI) or non-ST-segment elevation (NSTEMI) (86). Electrocardiographic criteria for STEMI and NSTEMI are as follows:

- STEMI: ST-segment elevation of at least 1 mV in two contiguous leads or new left bundle branch block
- NSTEMI: ST-segment depression or T-wave inversion persisting for at least 24 h

STEMI are the result of an occluded epicardial coronary artery with more extensive myocardial damage and a worse prognosis. NSTEMI have less myocardial damage due to spontaneous thrombolysis (clot dissolution). Figure 12.2 shows the evolution of the ECG after STEMI with the formation of a Q-wave indicating infarction of all or most of the thickness of the ventricular wall. The ST-segment elevation results from ischemic injury, and inverted T waves are due to ischemia around the outside borders of the infarct. Anatomic localization of myocardial infarctions is possible if Q-waves are formed, as shown in the following list.

Criteria for Anatomic Localization of Myocardial Infarction (MI) by Q-Wave Appearance

- Inferior wall MI (usually right coronary artery occlusion): Q-wave (>40 ms duration, amplitude >25% of the R-wave) in leads II, III, and aVf

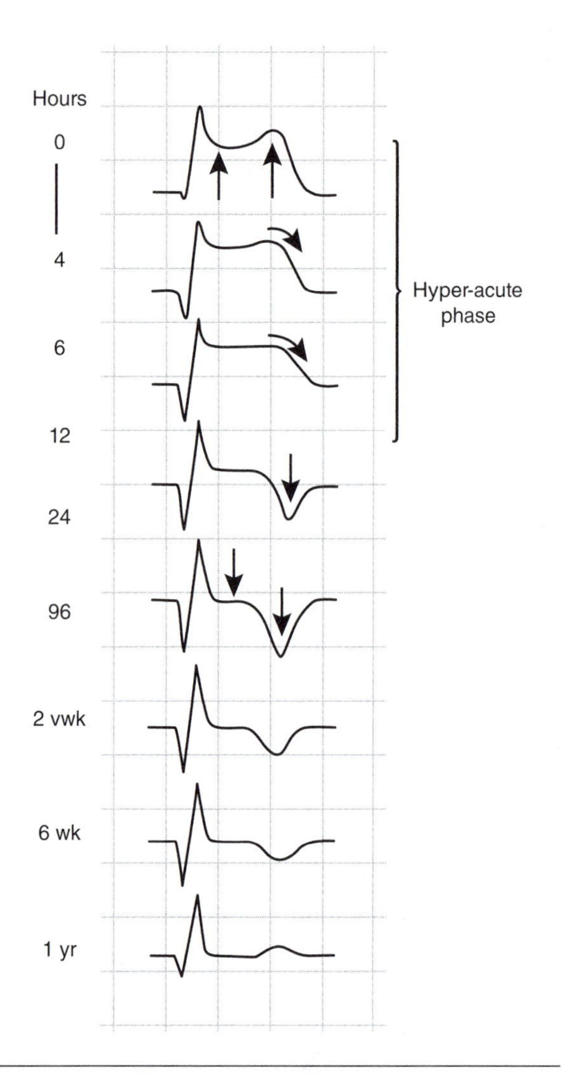

Figure 12.2 The evolution of the electrocardiogram in acute ST-segment elevation myocardial infarction resulting in Q-wave formation.

Reprinted, by permission, from G.T. Gau, 1991, Standard electrocardiography, vectorcardiography and signal-averaged electrocardiography. In *Cardiology: Fundamentals and practice*, 2nd ed., edited by E.R. Giuliani (St. Louis: Mosby Year Book).

- Anterior wall MI (left anterior descending coronary artery occlusion): Q-wave in leads V1 to V3 (anteroseptal), QS pattern in leads V1 to V3 (anteroseptal), Q-wave in leads V2 to V4 (anterior), QS pattern in leads V2 to V4 (anterior)
- Lateral wall MI (usually circumflex coronary artery occlusion): Q-wave in leads V4 to V6 or QS pattern in leads V4 to V6
- Posterior wall MI (usually right coronary artery occlusion): prominent R-wave in leads V1 to V2 with positive T-waves
- High lateral wall MI (usually circumflex coronary artery occlusion): Q-wave in leads I and aVL or QS pattern in leads I and aVL

From: Gau GT. Standard electrocardiography, vectorcardiography and signal-averaged electrocardiography. In: Giuliani ER, Fuster V, Gersh BJ, McGoon MD, McGoon DC, eds. Cardiology: Fundamentals and Practice, 2nd ed. St. Louis: Mosby Year Book, 1991.

Treatment of Acute Coronary Syndromes

The following are therapeutic approaches to treating patients with acute coronary syndromes (5, 87):

Anti-Ischemic Therapy

- Bed or chair rest with continuous ECG monitoring
- Supplemental oxygen when arterial saturations are less than 90%
- Nitroglycerin; sublingual, oral, or intravenous administration
- β-Blocker
- Nondihydropyridine calcium channel blocker in the absence of severe left ventricular dysfunction when β-blockers are contraindicated
- Angiotensin-converting enzyme inhibitor if left ventricular ejection fraction is less than or equal to 40% or with anterior wall myocardial infarction
- Angiotensin receptor blocker for patients with ejection fractions less than or equal to 40%, or anterior wall myocardial infarction, if intolerant of angiotensin-converting enzyme inhibitors
- Intra-aortic balloon pump counterpulsation

Oral Antiplatelet Therapy

- Aspirin
- Clopidogrel or prasugrel

Intravenous Antiplatelet Therapy (Glycoprotein IIb/IIIa Antagonists)

- Abciximab
- Eptifibatide
- Tirofiban

Anticoagulants

- Unfractionated heparin
- Low molecular weight heparin
- Bivalirudin (direct thrombin inhibitor)

Pain Relief (Despite Anti-Ischemic Therapy)

- Morphine

Reperfusion Therapy (36, 81)

Reperfusion of the infarct-related artery may be accomplished via thrombolytic therapy, percutaneous coronary intervention (PCI), or coronary artery bypass graft (CABG) surgery. Thrombolytic therapy has not been demonstrated to be beneficial in NSTEMI.

Non-ST-Segment Elevation Myocardial Infarction or Unstable Angina Pectoris

Two management options are available for NSTEMI or unstable angina (36):

1. Early invasive strategy: Depending on the results of coronary angiography, the patient may undergo PCI (by far the most common occurrence) or CABG if the anatomy is not favorable for PCI. Most patients undergo the early invasive strategy if they are hospitalized at a center capable of PCI.
2. Conservative treatment: Optimal medical management with graded exercise testing performed either before discharge or early after discharge. If the patient achieves 5+ metabolic equivalents (METs) without signs of ischemia, prognosis is favorable and angiography may be deferred.

ST-Segment Elevation Myocardial Infarction

The optimal management of patients with STEMI is prompt reperfusion of the infarct-related artery (81). The earlier the reperfusion, the better the outcome for the patient in terms of limiting infarct size, preserving left ventricular function, and improving survival.

Two reperfusion strategies are available: fibrinolysis and primary PCI (81). PCI is the best option for the patient if available. **Fibrinolysis** is performed using tissue plasminogen activators, such as tenecteplase or alteplase. It is most effective if given during the first 3 h after symptom onset. These are the benefits of fibrinolysis:

- It is readily available in rural and community hospitals.
- It doesn't require a cardiologist or a cardiac catheterization laboratory.
- It can usually be administered within minutes of the patient's reaching the hospital.

However, fibrinolytic therapy has substantial limitations (81):

- Restoration of full coronary blood flow occurs in only 60% to 70% of cases treated within 3 h of symptom onset.
- Contraindications to fibrinolytic agents, such as prior intracranial hemorrhage, prior ischemic stroke, or active bleeding, occur in 30% to 40% of patients.

Patients treated with fibrinolysis should be transferred immediately to a PCI-capable facility (facilitated PCI) unless this is impossible due to weather or other circumstances. PCI is beneficial after both successful or failed fibrinolysis.

Primary PCI has impressive benefits (81):

- It improves myocardial "salvage" (less necrotic myocardium).
- It leads to less reinfarction and vessel reocclusion.
- It identifies coronary anatomy more suitable for CABG, such as left main or three-vessel coronary disease.
- The invasive strategy helps to identify causes of ST-segment elevation other than atherosclerosis, such as pericarditis, myocarditis, apical ballooning syndrome (Takotsubo cardiomyopathy), cocaine use, and coronary vasospasm resulting from endothelial dysfunction (5).

The major limitation of primary PCI is delay in opening the infarct-related artery once symptoms begin (81). For every 30 min delay, there is an 8% increase in mortality at 1 yr. Delays in reperfusion may be patient or hospital dependent. Patient dependent delays involve failure to recognize and appreciate the importance of cardiac symptoms and delays in activating the emergency response system (911). Typical patient-related delays are 2 to 3 h. Hospital-dependent delays are related to system and process failures for the prompt delivery of reperfusion therapy (83).

The "open artery hypothesis" implies that even after 12 to 48 h have elapsed from symptom onset, PCI may limit infarct size and improve survival (81). Another term used in invasive cardiology is "no-reflow," which is angiographically observed slow flow in the infarct-related coronary artery after primary PCI as a result of swollen endothelial cells and subsequent red blood cell plugging of the microcirculation (53).

Complications of Acute Myocardial Infarction (53)

Many patients who suffer an acute myocardial infarction and receive prompt reperfusion therapy recover quickly and require only 2 or 3 d of hospitalization. However, complications do occur frequently and may be life threatening. This section discusses the most common complications after acute myocardial infarction.

Arrhythmia

Several types of supraventricular arrhythmias are common after myocardial infarction, including these:

- Sinus bradycardia due to excessive vagal tone or ischemia of the sinoatrial node
- Sinus tachycardia related to pain, fear, heart failure, or excessive sympathetic nervous system activation
- Premature atrial contractions—provide no prognostic information
- Atrial fibrillation—observed in up to 20% of patients, usually transient, more frequent in older patients, associated with increased mortality

Ventricular arrhythmias are also common after myocardial infarction:

- Ventricular fibrillation occurs in approximately 5% of hospitalized patients. β-blockers are effective in decreasing the incidence of this arrhythmia in the peri-infarct period.
- Ventricular tachycardia is observed in 10% to 40% of hospitalized patients; it is usually transient and benign in the early postinfarct period.
- Accelerated idioventricular rhythm is also observed in 10% to 40% of hospitalized patients. It is not associated with increased mortality.
- Premature ventricular contractions are very common during the peri-infarct period; they have no clear relationship to the risk of ventricular tachycardia or ventricular fibrillation.
- Asystole or electromechanical dissociation is rare, but portends an extremely poor prognosis.

Conduction Disturbances

The following conduction disturbances may occur in the setting of acute myocardial infarction:

- First-degree atrioventricular (AV) block occurs in 5% to 10% of hospitalized patients.
- Type I second-degree AV block (Wenckebach) occurs in 10% of hospitalized patients.
- Type II second-degree AV block is rare; usually requires pacemaker implantation.
- Third-degree AV block requires at least temporary pacing, may resolve spontaneously in inferior wall myocardial infarction.

Bundle Branch Block

This occurs in approximately 15% of hospitalized patients; right bundle branch block is more common than left bundle branch block (LBBB). LBBB is associated with an increased risk of third-degree heart block and increased mortality.

Cardiogenic Shock

Cardiogenic shock is the result of inadequate cardiac output with signs of persistent hypotension (systolic blood pressure <80 mmHg) and a cardiac index <2.0 $L \cdot min^{-1} \cdot m^{-2}$ (normal is approximately 3.0 $L \cdot min^{-1} \cdot m^{-2}$). It is usually the result of a large myocardial infarction. Treatment options are direct monitoring of pulmonary capillary wedge pressure with a Swan-Ganz catheter to determine the effects of positive inotropic agents such as dobutamine, insertion of an intra-aortic counterpulsation balloon pump to maintain blood pressure, and prompt percutaneous **coronary revascularization**. Mortality is high with this condition.

Infarct Extension and Expansion

Infarct extension is recurrent necrosis occurring 2 to 10 d after myocardial infarction in an area remote from the original infarction. Infarct expansion is thinning and dilatation of the infarcted myocardium without new necrosis. It occurs most commonly with anterior wall myocardial infarctions and may result in aneurysm formation, congestive heart failure, and serious ventricular arrhythmias.

Myocardial Rupture

Rupture of the ventricular free wall presents catastrophically with either sudden death due to **electromechanical dissociation (pulseless electrical activity)** or **cardiac tamponade** with cardiogenic shock. Fortunately it is rare; it occurs most commonly within 4 d of the myocardial infarction. Left ventricular rupture is much more common than right ventricular rupture. Predisposing factors include advanced age and female sex. The only available treatment is emergent surgery.

New Mitral Valve Regurgitation

This is most often a result of papillary muscle or chordae rupture caused by the infarction. It may also occur due to left ventricular dilatation attributable to heart failure. It typically occurs during the first few days after infarction and presents abruptly with hypotension and right ventricular failure. Treatment is prompt surgical intervention with repair or replacement of the mitral valve.

Pericardial Effusion and Pericarditis

Pericardial effusion is fluid accumulation in the pericardial space. In the setting of acute myocardial infarction, it is associated with pericarditis and occurs in approximately 10% of hospitalized patients. The usual treatment is high-dose aspirin.

Postinfarction Syndrome (Dressler's Syndrome)

This is pleuritic or pericardial chest pain associated with a friction rub heard on auscultation of the heart. A form of pericarditis, it usually occurs several weeks after myocardial infarction and is treated with high-dose aspirin.

Left Ventricular Mural Thrombus

Fifty percent of patients with anterior wall myocardial infarction develop a blood clot on the endocardial surface of the left ventricle. Anticoagulation with warfarin for a period of time is the usual treatment.

Right Ventricular Myocardial Infarction

This entity is most commonly associated with inferior wall myocardial infarction and is the result of occlusion of the proximal right coronary artery (32). Patients with right ventricular (RV) infarction present with hypotension and signs of right heart failure. This form of infarction is associated with high-degree AV block and increased in-hospital morbidity and mortality.

Therapy consists of prompt reperfusion of the right coronary artery and maintenance of adequate RV function. For patients who survive the period of hospitalization, the RV dysfunction resolves and there is no additional long-term increase in mortality compared with that in patients with inferior wall myocardial infarction without RV involvement.

Factors Associated With Poor Prognosis

The following characteristics (one or more) are associated with a poor prognosis (71):

- Left ventricular ejection fraction less than or equal to 35%, or congestive heart failure due primarily to diastolic dysfunction
- Extremely poor exercise capacity, <5 METs
- Evidence of extensive, severe myocardial ischemia during exercise or pharmacologic stress testing
- Having survived primary (not in the setting of an acute myocardial infarction) sudden cardiac death without treatment with an implantable cardioverter-defibrillator
- Severe nonrevascularized coronary artery disease (left main, severe proximal three-vessel disease)

Medications for Outpatients After Myocardial Infarction

Table 12.1 provides a summary of common medications prescribed for patients after myocardial infarction. Several classes of drugs are associated with improved survival, such as aspirin, β-blockers, statins, and angiotensin-converting enzyme inhibitors or angiotensin receptor blockers (for patients with depressed left ventricular ejection fraction or anterior wall myocardial infarction) (87). After PCI, dual antiplatelet therapy with aspirin plus clopidogrel or prasugrel is standard therapy.

Stress Testing After Acute Myocardial Infarction

A brief review of exercise testing is provided in table 12.2. Exercise testing is helpful after myocardial infarction for the following purposes (28):

- To evaluate symptoms, potential myocardial ischemia
- To determine the need for coronary angiography in patients treated initially with a noninvasive strategy
- To determine the effectiveness of medical therapy
- To assess future risk and prognosis
- To objectively determine the exercise capacity (this information is used for exercise prescription, entry

Table 12.1 Common Medications Used in Outpatients After Myocardial Infarction

Class and example	Primary effects	Exercise effects	Side effects
Antiplatelet (aspirin, clopidogrel)	Blocks platelet aggregation, improves survival	None	Increased bleeding
β-blocker (metoprolol)	Reduces heart rate, blood pressure, improves survival	Decreases heart rate, blood pressure	Fatigue, hypotension, bradycardia
ACEi (lisinopril)	Reduces blood pressure, improves survival	Reduces blood pressure	Cough, hypotension
ARB (losartan)	Reduces blood pressure, improves survival	Reduces blood pressure	Hypotension
Nondihydropyridine calcium channel blocker (verapamil)	Vasodilation, decreases blood pressure	Decreases heart rate, blood pressure	Peripheral edema, constipation
Statin (atorvastatin)	Decreases blood cholesterol, improves survival	None	Muscle pain, weakness
Nitrate (isosorbide mononitrate)	Coronary vasodilation	Raises ischemic threshold	Headache, hypotension

ACEi = angiotensin-converting enzyme inhibitor; ARB = angiotensin receptor blocker.

Table 12.2 Stress Testing After Myocardial Infarction

Exercise stress test type	Endpoint	Advantages
Predischarge exercise test	Submaximal effort	Determines need for coronary angiography in patients not treated initially with an invasive strategy
Standard exercise test	Symptom-limited maximal effort	Inexpensive; diagnosis of ischemia; assessment of symptoms, exercise capacity, and blood pressure
Cardiopulmonary exercise test	Symptom-limited maximal effort	Provides extensive information regarding the oxygen transport system; diagnosis of ischemia; assessment of symptoms, exercise capacity, and blood pressure
Nuclear exercise test	Symptom-limited maximal effort	Localizes and quantifies areas of ischemia; provides ejection fraction and infarct size
Echocardiographic exercise test	Symptom-limited maximal effort	Localizes and quantifies areas of ischemia; provides ejection fraction and infarct size
Pharmacologic stress test Nuclear or echocardiographic imaging	Test type dependent	Localizes and quantifies areas of ischemia; provides ejection fraction and infarct size in patients who cannot exercise adequately

into an outpatient cardiac rehabilitation program, and return to work and other activities)

Absolute contraindications to exercise testing are as follows (3, 28):

- Acute myocardial infarction, within 2 d, or other acute cardiac event
- Unstable angina not previously stabilized by medical therapy
- Uncontrolled cardiac arrhythmias causing symptoms or hemodynamic compromise
- Symptomatic, severe aortic stenosis
- Uncontrolled symptomatic heart failure
- Acute pulmonary embolus or pulmonary infarction
- Acute myocarditis or pericarditis
- Acute aortic dissection

- Acute systemic infection, accompanied by fever, body aches, or swollen lymph glands

Exercise test factors associated with an increased risk of a recurrent cardiac event and poor prognosis include the following (3, 28):

- Inability to exercise
- Substantial exercise-induced myocardial ischemia
- Exercise capacity <5 METs
- Failure of systolic blood pressure to increase at least 10 mmHg

The timing of the performance of a postmyocardial infarction graded exercise test varies, depending on the clinical situation and the preferences of the treating cardiologist. For patients treated medically, a predischarge or early postdischarge exercise test may be performed to determine the need for coronary angiography (3, 28). Generally, if testing occurs before 7 d postinfarction, a submaximal protocol is selected, although some studies have evaluated symptom-limited protocols as early as 4 d after infarction. Symptom-limited tests are generally performed 7 or more days after infarction. These tests may be performed at 14 to 21 d or 6 or more weeks after the event, depending on the practice patterns of the individual cardiologist. Exercise testing after myocardial infarction is safe if the previously mentioned contraindications are not present. Ideally, a symptom-limited graded exercise test should be performed before starting outpatient cardiac rehabilitation, although local practice patterns may preclude this.

Assessment of exercise capacity and the presence and extent of myocardial ischemia may be accomplished using several different techniques (3, 28):

- Standard exercise ECG, with or without expired air analysis (cardiopulmonary exercise testing)
- Nuclear imaging modalities to measure myocardial perfusion
- Echocardiographic imaging of ventricular systolic function and regional wall motion

In all testing modalities, the clinical interpretation of the test result requires integration of all of the available clinical data. These tests are not infallible with regard to detecting the presence or absence of myocardial ischemia. An excellent discussion of test interpretation is provided by Ellestad and colleagues (15).

Standard exercise electrocardiography, with ST-segment depression (or less commonly, ST-segment elevation) >1 mm at 0.08 s after the J-point required for the diagnosis of ischemia, provides a sensitivity of approximately 65% to 70% (28). Other factors, such as the time of onset of ST depression (early in exercise vs. near maximal exertion), the maximal amount of ST change, and the presence of typical angina, increase the accuracy of the assessment of ischemia. Limitations of the exercise ECG are its inability to diagnose ischemia in the setting of digoxin use or an abnormal rest ECG (particularly left bundle branch block), inability to localize the area of the myocardium that is ischemic by ST-segment depression, lower sensitivity than with imaging techniques (sensitivities of 85%+), and the lack of information provided regarding the extent of ischemia. The exercise ECG does provide evidence of the "ischemic threshold" (the heart rate and systolic blood pressure that corresponds to the first evidence of ischemia), which is valuable in prescribing physical activity for patients.

Direct measurement of oxygen uptake, carbon dioxide, minute ventilation, and associated variables during graded exercise (cardiopulmonary exercise testing) is particularly useful in determining prognosis in patients with chronic heart failure (7). The technique is also helpful for determining the cause of unexplained dyspnea with exertion. In addition, the direct measurement of maximal oxygen uptake is much more accurate in establishing aerobic exercise capacity than are estimations of exercise capacity based on achieved workload.

Myocardial perfusion imaging using radioisotopes (technetium sestamibi or technetium tetrofosmin) is based on the premise that myocardial uptake of these substances is proportional to myocardial blood flow (3, 28). Images are obtained at rest and after exercise with a single-photon emission computed tomography camera system. Reversible defects (better perfusion at rest than with exercise) represent ischemia. Fixed defects (present at rest and with exercise) represent infarct scar or, less commonly, stunned or hibernating myocardium. The images provide quantification of infarct size. In exercise echocardiography, quantitative echo images of left ventricular ejection fraction and end-systolic volume, as well as subjective echo images of regional wall motion and thickening, are obtained before and immediately after maximal exercise (3, 28). The nuclear and echocardiographic imaging techniques are capable of localizing areas of ischemia as well as quantifying the extent and severity of ischemia. Imaging techniques do not provide serial information regarding ischemia during graded exercise and thus do not provide information regarding the ischemic threshold.

Exercise for standard exercise electrocardiography (with or without cardiopulmonary measurements) and the imaging techniques is usually performed on a motorized treadmill, although cycle ergometry, arm-only, or arm-leg ergometry may be preferable in certain situations (3). Exercise testing protocols and procedures for patients with coronary artery disease are described in several excellent references and are not reviewed here (3, 15).

Pharmacologic stress using intravenous administration of a coronary vasodilator (adenosine or regadenoson) or a positive chronotropic and inotropic agent (dobutamine) may be used in conjunction with imaging techniques for detection of myocardial ischemia in patients who cannot exercise adequately (3, 28). Obviously, pharmacologic stress testing provides no information regarding exercise capacity and the hemodynamic responses to exercise. Adenosine and the newer agent regadenoson are the most common coronary vasodilators used in conjunction with nuclear perfusion imaging techniques. An abnormal flow reserve (less flow in a particular region of the myocardium relative to other regions) in the territory supplied by a stenotic coronary artery represents an ischemic response. Dobutamine is a synthetic sympathomimetic that increases both myocardial contractility and heart rate, thus elevating myocardial oxygen requirement in a manner analogous to exercise. It is most commonly used in conjunction with continuous echocardiographic assessment of ventricular function. If at the maximum dose of dobutamine the heart rate is below 85% of age predicted, atropine may be given to further increase the heart rate.

Treatment: Exercise Training and Cardiac Rehabilitation for Acute Coronary Syndromes

As part of comprehensive secondary prevention after the occurrence of an acute coronary syndrome, exercise training and the involvement of patients in a well-organized cardiac rehabilitation program are crucial. The benefits of such efforts are well established. The following section describes the basic elements of inpatient and outpatient cardiac rehabilitation with emphasis on exercise prescription.

Inpatient Cardiac Rehabilitation

Lengths of hospital stay after acute myocardial infarction have declined over the past two decades. Currently, patients are hospitalized for no more than 2 or 3 d unless complications arise. The vast majority of patients with unstable angina or acute myocardial infarction are treated invasively with PCI, as noted previously. The detailed, expansive inpatient rehabilitation protocols described by this author in 1987 seem archaic compared with today's practice patterns (73). There is little opportunity for formal exercise training during hospitalization.

In order to minimize the deleterious effects of bed rest, patients are mobilized as soon as they are stable. Exposure to the normal stress of gravity is emphasized in this stage of rehabilitation to prevent orthostatic intolerance (19). Patients sit, stand, perform active range of motion exercises for the major joints, and walk short distances as soon as possible to prevent further deconditioning. Various types of allied health professionals may be involved in inpatient cardiac rehabilitation, such as a registered nurse, physical therapist, occupational therapist, and exercise physiologist. Patients with neuromuscular diseases or other conditions that limit their ability to ambulate benefit from formal physical therapy treatments. After hospital dismissal, frail elderly and other debilitated patients may be referred to outpatient postacute or transitional care facilities for various time periods before returning home (69).

A critical aspect of inpatient cardiac rehabilitation is the introduction of concepts of secondary prevention of atherosclerosis to patients and family members (69). Basic information regarding the importance of cardioprotective medications, avoidance of tobacco, heart-healthy eating patterns, blood pressure and blood lipid goals, exercise (including any temporary restrictions imposed by the coronary event), potential depression after the cardiac event, and return to usual activities is provided. Patients should be referred to an outpatient cardiac rehabilitation program if one is available in the patient's home area. A home exercise prescription providing guidance as to recommended exercise types, intensity, duration, frequency, and progression for the first few weeks after hospital dismissal should be given to the patient. Typically, the prescription recommends walking or stationary cycling at a comfortable intensity, beginning with 10 to 20 min once or twice daily and progressing to 30 to 45 min daily.

Outpatient Cardiac Rehabilitation

This phase of rehabilitation can begin soon after hospital dismissal, often within 1 or 2 wk after the patient leaves the hospital (3, 69). During the interval between hospital dismissal and beginning outpatient rehabilitation, patients are encouraged to exercise independently, using the home exercise prescription provided in the hospital.

The initial assessment, performed by cardiac rehabilitation professionals, may involve the following (3, 69):

- Review of the medical records with emphasis on the recent coronary event.
- Physical examination, performed by a qualified medical professional, usually a physician, midlevel provider, or registered nurse. This ideally should include a review of the neuromuscular systems in terms of potential limitations to exercise.
- Graded exercise test, which ideally should be performed to assist with exercise prescription and risk and prognosis assessment. Depending on the local practice patterns, this may not be available for the patient when outpatient cardiac rehabilitation begins. Alternatively, a 6 min walk may be performed as a check of submaximal exercise response.
- Medication review with emphasis on compliance, side effects, and optimal cardioprotective medication use.
- Review of standard coronary risk factors with emphasis on avoidance of tobacco, proper nutrition habits, control of blood pressure and blood lipids, body weight optimization, and stress and depression management.
- Patient-selected and provider-selected goals to work on during the program.
- Anticipated return-to-work date if needed.
- Explanation of the short-term and long-term benefits of exercise training, optimal control of coronary risk factors, and the use of cardioprotective medications.
- Anticipated rehabilitation schedule, including number of rehabilitation sessions per week and number of weeks in the program.

Cardiac rehabilitation programs are ideal for helping patients achieve the secondary prevention goals of taking appropriate cardioprotective medications; avoiding tobacco; following heart-healthy eating patterns; performing regular exercise and general physical activity; achieving a desirable body weight; optimally controlling blood lipids, blood pressure, and blood glucose; and psychosocial health (69). Counseling is a major component of patient visits to cardiac rehabilitation. Case management, more recently renamed disease management, involves one or more cardiac rehabilitation staff members taking responsibility for oversight of an individual patient's secondary prevention program (2, 39). Disease management by cardiac rehabilitation professionals has been demonstrated to be highly effective in promoting patient adherence to secondary prevention measures. My own program includes disease management over years of follow-up after completion of the standard 36 cardiac rehabilitation sessions covered by insurance (74). A technique used by health care providers to assist patients in making healthy changes in behavior is motivational interviewing (51). The technique is not difficult to learn and involves developing a partnership with the patient that honors the individual's perspectives and talents. These are the four general principles of motivational interviewing:

1. Express empathy.
2. Develop discrepancy: The patient needs to perceive a discrepancy between present behavior and important goals.
3. Roll with resistance: Avoid arguing for change.
4. Support self-efficacy: An important motivator for change is the patient's belief in the possibility of change.

EXERCISE PRESCRIPTION

The prescription of exercise after an acute coronary syndrome should be individualized, based on factors such as these (3):

- Exercise capacity
- Ischemic or anginal threshold or both
- Cognitive or psychological impairment
- Vocational or avocational requirements
- Musculoskeletal limitations
- Obesity
- Prior physical activity history
- Patient goals

Table 12.3 gives an overview of the components of an exercise prescription for a coronary patient.

Frequency

Exercise sessions should occur on most days of the week, that is, 4 to 7 d each week. With the exception of extremely high-risk patients, one or more sessions per week should be performed independently, outside of the supervised environment of a cardiac rehabilitation program. Patients with very limited endurance may perform multiple short-duration (1-10 min) sessions daily.

Table 12.3 Outpatient Exercise Training Recommendations After an Acute Coronary Syndrome

Type	Modes	Intensity	Frequency and duration	Progression	Goals	Considerations
Aerobic	Walking, jogging, cycling, arm–leg ergometer, elliptical, arm ergometer, water-based exercise, stair climber, rower	RPE 11 to 16, 40% to 80% of exercise capacity	4 to 7 d/wk, 20 to 60 min	Begin with 5 to 10+ min at RPE 11 to 13; gradually increase 1 to 5 min per session and RPE as tolerated.	Improve aerobic activity and submaximal exercise endurance	Use intensity below ischemic threshold. Take medications on schedule before exercise. Use intermittent exercise for very deconditioned patients. Use interval training for selected patients.
Resistance	Free weights, weight machines, elastic bands, stability ball (include major muscle groups)	RPE 11 to 14, 30% to 80% of 1RM	2 or 3 d/wk, 8 to 10 exercises, 8 to 15 slow repetitions, one to four sets	Begin with one set of eight repetitions at RPE 11 or 12 for each exercise; gradually increase repetitions, sets, resistance, and RPE to 13 or 14 as tolerated.	Increase skeletal muscle strength and endurance	Avoid Valsalva maneuver or straining. Use circuit training for selected patients.
Flexibility	Static stretching, major muscle groups	Hold to point of mild discomfort	Daily, 5 to 15 min	Begin with 15 s per static stretch exercise; gradually increase to 30 to 60 s per exercise as tolerated.	Improve joint range of motion	Use full range of motion. Avoid breath holding.

RPE = ratings of perceived exertion (Borg scale 6-20); 1RM = one-repetition maximum.

Intensity

Exercise intensity may be prescribed using one or more of the following:

- Ratings of perceived exertion (RPE) of 11 to 16 on a scale of 6 to 20
- 40% to 80% of exercise capacity using the heart rate reserve method, or the percent peak oxygen uptake reserve technique if maximal exercise test data are available
- When exercise test or pharmacologic stress data are not available, an upper-limit heart rate of rest heart rate + 20 beats · min^{-1} and RPE 11 to 14, gradually titrating the heart rate to higher levels according to RPE, signs and symptoms, and normal physiologic responses
- Exercise intensity that is kept below the ischemic threshold if one has been demonstrated to exist

Patients should take their medications on schedule for performance of exercise training sessions. Changes in the doses of β-blockers or other drugs that affect the chronotropic response to exercise may occur. In these cases, it is unlikely that a new exercise test will be performed solely for the purpose of determining a target heart rate. Using RPE and signs and symptoms, as well as determining the new heart rate at previously performed work intensities, is recommended. It is common for cardiac rehabilitation professionals to diagnose hypotension, either orthostatic, at rest, or during or after exercise, due to excessive medication dosage.

Duration

Patients should perform warm-up and cool-down activities, including static stretching to improve range of motion and low-intensity aerobic exercise, for 5 to 20 min before and after the conditioning phase of exercise. The goal duration for conditioning aerobic exercise is

20 to 60 min per session. Patients may begin with an easily tolerable duration of 5 to 10+ min with a gradual increase of 1 to 5 min per session, as tolerated. The progression in duration should be individualized for each patient; some patients may be able to progress much more rapidly than others, based on fitness, exercise habits immediately prior to the coronary event, symptoms, and musculoskeletal limitations. Frail and otherwise extremely deconditioned patients may require intermittent exercise, alternating short periods of exercise with rest breaks, although most patients can perform continuous exercise.

Type

Aerobic forms of exercise should include rhythmic large muscle group activities. Ideally, both upper and lower extremity exercises should be incorporated into the program, such as these:

- Treadmill or track for walking (jogging or inclined walking for more fit patients)
- Upright or recumbent cycle
- Combination upper and lower extremity ergometer
- Elliptical
- Stair climber
- Arm ergometer
- Rower

Independent exercise often includes walking outdoors, with or without hills, or walking indoors at shopping malls or schools that are open to the public for this purpose.

Resistance Training

The vast majority of patients with treated acute coronary syndromes should perform strength training as a component of their exercise program. Patients with severe ischemia or hemodynamic instability should be adequately stabilized before beginning strengthening exercises. The purposes of strength training for patients include the following (3):

- To increase muscular strength and endurance
- To decrease cardiac demands of muscular work (reduced rate–pressure product with lifting and carrying activities of daily life)
- To improve self-confidence
- To maintain independence (enable patient to perform household and personal care duties)
- To prevent diseases such as osteoporosis, type 2 diabetes, obesity
- To slow the age-related declines in skeletal muscle mass and strength

Although the most recent guidelines for strength training in patients with cardiovascular diseases, published in 2004, recommend waiting 5 wk from the date of myocardial infarction before beginning strength training, in my own program patients begin such training within 2 wk of the event. Several hundred patients have followed this practice, and there have been no complications associated with strength training.

Equipment for strength training may include free weights, elastic bands, and weight machines including pulley devices, stability ball, or weighted wands. Patients should perform movements in a slow, controlled manner while maintaining a regular breathing pattern and avoiding straining and musculoskeletal pain. RPE ratings initially should range from 11 to 14 on the 6 to 20 scale. Initial resistances should allow 12 to 15 repetitions (approximately 30% to 60% of one-repetition maximum [1RM]). The resistance may be gradually increased, and most patients may progress to 8 to 12 repetitions with a resistance of 60% to 80% of 1RM, with one to four sets of repetitions. Exercises should be performed for the major muscle groups, usually 8 to 10 different exercises. Frequency should be two or three sessions per week on nonconsecutive days.

Aerobic Interval Training

Over the past 30 yr, there has been a gradual easing of restrictions regarding exercise training for patients with cardiovascular diseases. For example, outpatient exercise training after myocardial infarction now begins within a few days of hospital dismissal (rather than several weeks) with more aggressive aerobic exercise and strength training (forbidden in the past). Patients with chronic heart failure are now encouraged to exercise rather than to rest. The latest exercise taboo to be lifted is high-intensity interval training for cardiac patients. Investigators have documented superior improvements in $\dot{V}O_2$peak, measures of endothelial function, and left ventricular systolic and diastolic function in coronary patients with high-intensity interval training (up to 95% of maximal heart rate for short periods of time interspersed with lower-intensity exercise) compared with conventional continuous aerobic training (35, 84). In my own program, we have used interval training for several hundred cardiac patients over the past 4 yr and have not encountered safety issues or patient reluctance to try it. We start with approximately 2 wk of standard continuous aerobic training, increasing the duration to at least 20 min before beginning interval training. We start with 30 s intervals three to five times during the exercise session interspersed with lower-intensity exercise. Over a period of days to weeks, the length of the high-intensity intervals is increased to 60 to 120 s or longer, as tolerated. Interval training occurs 2 or 3 d/wk.

Balance Exercises

Many participants in cardiac rehabilitation programs are older than 65 yr of age. With increasing age, the neuromuscular reflexes involved with proprioception and balance become less effective. Falls are a common cause of morbidity and mortality in older individuals. A balance abnormality was found in over half of a consecutive sample of 284 participants in my program (31). Women and patients over 65 yr of age were more likely to exhibit poor balance. As part of the baseline assessment of patients beginning outpatient cardiac rehabilitation, a balance assessment using simple techniques such as the single-leg stand and the tandem gait (walking heel to toe in a straight line) should be performed. For patients with poor balance, specific exercises may be prescribed (3).

Lifestyle or General Physical Activity

In addition to formal exercise sessions, patients should be encouraged to gradually return to general activities of life, such as walking for transportation, household tasks, shopping, gardening, nonsedentary hobbies, and sport activities (3). Measuring the amount of lifestyle activity is relatively easy using a pedometer (steps per day) or various smartphone applications. The more time spent sitting, such as time watching television or on the computer, the greater the risk of cardiovascular and all-cause mortality as well as nonfatal cardiovascular events (57, 75). Greater energy expenditure resulting from increased amounts of general physical activity facilitates weight loss.

EXERCISE TRAINING FOR OVERWEIGHT CORONARY PATIENTS

The vast majority of coronary patients who enter outpatient cardiac rehabilitation are either overweight or obese. Standard exercise protocols used in most programs, that is, 30 to 45 min of moderate-intensity aerobic exercise three times per week (approximately 800 kcal/wk) for 3 mo, are generally not adequate to result in substantial body fat loss. Investigators used a high-calorie exercise protocol (approximately 3,000 kcal/wk) that required 45 to 60 min of walking more than 4 d/wk (1). Compared with standard cardiac rehabilitation, the high-calorie exercise protocol resulted in twice the weight loss (8.2 kg vs. 3.7 kg). In my own program, we encourage patients who desire to lose considerable weight to consult with our registered dietitian, gradually increase exercise duration to 60 min 5 or 6 d/wk, incorporate high-intensity interval training 2 or 3 d/wk, and perform moderate strength training using free weights and weight machines 2 or 3 d/wk. This high-volume exercise protocol requires very motivated patients.

Safety of Exercise Training for Coronary Patients

Supervised exercise training has been demonstrated to be relatively safe for patients with cardiovascular diseases. Franklin and colleagues reported the average incidence of cardiac arrest, nonfatal myocardial infarction, and death as 1 for every 117,000, 220,000, and 750,000 patient-hours of participation, respectively (20). A large trial of over 1,000 higher-risk patients with chronic heart failure (more than half with coronary disease) studied the safety of supervised and independent moderate-intensity aerobic exercise training. Over approximately 30 mo, there were 37 events associated with exercise that resulted in hospitalization. In over several million patient-hours of exercise, only five deaths occurred (55). The benefits of exercise training far outweigh the risks for coronary patients without absolute contraindications to exercise.

Types of Outpatient Exercise Programs

In an ideal world, patients with a recent acute coronary syndrome would exercise with medical supervision in a cardiac rehabilitation program for at least several weeks. Unfortunately, some geographical areas do not have sufficient programming to ensure easy accessibility for all patients. Patients may present multiple barriers for attendance at the standard three supervised exercise sessions per week of a typical cardiac rehabilitation program, such as return to full-time employment within a few days of hospital dismissal, difficulty in getting reliable transportation to and from the rehabilitation center, and excessive out-of-pocket costs (e.g., coinsurance, parking fees). High-risk and debilitated patients are best served with frequent supervised rehabilitation sessions. Non-high-risk patients may benefit from flexibility in program design. Medicare reimburses for up to 36 rehabilitation sessions over a 36 wk period, as does most private insurance. Patients may attend one to three sessions weekly with additional independent exercise as prescribed by cardiac rehabilitation staff. Some patients are not able to attend a formal program and require an entirely independent program. Recently published performance measures for the referral to and delivery of cardiac rehabilitation and secondary prevention services should assist patients in obtaining appropriate posthospital coronary disease prevention care (78).

Maintenance of the Exercise Habit

Most patients, after completion of a standard cardiac rehabilitation program of 36 sessions, need to continue exercise training independently (outside of a cardiac rehabilitation program). Compliance with independent exercise may be improved with follow-up visits with health care providers. In my own program, patients are encouraged to return for regular follow-up visits with the cardiac rehabilitation staff on a continuing basis (74). Some cardiac rehabilitation programs have sufficient capacity to offer patients a "maintenance" exercise program (usually patient funded) for a period of months up to lifelong participation. Some programs have offerings for spouses to aid in retention of patients in maintenance programs.

Benefits of Cardiac Rehabilitation and Secondary Prevention Programs

The following list provides a summary of the important benefits available to patients with coronary disease who participate in cardiac rehabilitation. As a result of training, $\dot{V}O_2$peak may improve 10% to 20% or more (72). In general, the magnitude of the relative improvement is inversely proportional to the baseline exercise capacity. The rate of improvement is greatest during the first 3 mo of training, but increases in aerobic capacity may continue for 6 or more mo. High-intensity interval training results in greater improvements in $\dot{V}O_2$peak than traditional continuous-intensity training (84).

Strength training results in significant gains in skeletal muscle strength. Symptoms related to coronary disease usually improve with cardiac rehabilitation. Left ventricular function may improve. Improvement in coronary risk factors is observed. Atherogenesis is slowed; endothelial function is improved; and myocardial ischemia may be lessened. Psychosocial function may improve (29, 41, 45, 72). A meta-analysis of several randomized trials demonstrated a 31% reduction in mortality for patients who participated in cardiac rehabilitation (37). In my own program, located at an academic medical center, we have demonstrated a 56% and a 45% reduction in all-cause mortality for participants with acute myocardial infarction or percutaneous coronary intervention, respectively, compared with nonparticipants (30, 84). In addition, in a large sample of Medicare beneficiaries, a dose-response relationship between the number of cardiac rehabilitation sessions attended and various cardiovascular outcomes was determined; that is, the more sessions attended, the better the outcomes (34).

Summary of Benefits of Cardiac Rehabilitation Programs

- Improvement in aerobic capacity
- Increased submaximal exercise endurance
- Increase in muscular strength
- Reduction in symptoms: angina pectoris, dyspnea on exertion, fatigue, claudication
- Vascular regeneration via bone marrow–derived endothelial progenitor cells
- Decrease in myocardial ischemia, potential increase in myocardial perfusion
- Improved endothelial function
- Improved left ventricular systolic and diastolic function in chronic heart failure
- Potential retardation of coronary disease progression, actual regression of plaque
- Incorporation of heart-healthy dietary practices
- Improved blood lipid profile
- Improvement in indices of obesity
- Reduced blood pressure
- Improved glucose intolerance, insulin resistance
- Decrease in inflammatory markers (hsCRP)
- Improved autonomic tone: less sympathetic activity, more parasympathetic activity
- Reduction in ventricular arrhythmias
- Improved blood platelet function, blood rheology

Improved psychosocial function (less depression, anxiety, somatization, hostility)

Repeated surveillance of blood pressure, symptoms, arrhythmias, and so on, leading to earlier treatment

Improved patient compliance with taking cardioprotective medications

Reduced health care costs

Reduced mortality

Note: hsCRP = high-sensitive C-reactive protein.

Data from 29, 41, 45, 72.

Underutilization of Cardiac Rehabilitation Services

In spite of the impressive benefits of exercise-based cardiac rehabilitation summarized in the preceding section, most patients after an acute coronary syndrome do not attend. Among Medicare beneficiaries, only 19% of eligible patients participate in outpatient cardiac rehabilitation (77). Older adults, nonwhites, patients with comorbidities, patients with low socioeconomic status, people who are unemployed, single parents, and women are less likely to participate (8, 16, 79). Suggested steps to improve the enrollment rate in cardiac rehabilitation programs include (a) automation of the referral process, (b) designation of referral and enrollment in cardiac rehabilitation as a quality indicator in cardiovascular care, (c) design and implementation of programs using the telephone or Internet for patients who live in areas without programs, (d) determination of how best to include the underserved populations just mentioned, and (e) multimedia education programs on the benefits of cardiac rehabilitation directed at both patients and health care providers (70).

Clinical Exercise Pearls

1. There is not a clear relationship between left ventricular ejection fraction and aerobic capacity in postmyocardial infarction patients.
2. Cardiac medications that lower blood pressure at rest, such as β-blockers, angiotensin-converting enzyme inhibitors, or angiotensin receptor blockers, may result in orthostatic or exercise hypotension. Reducing the dose of medication often eliminates this side effect.
3. For patients with exercise-induced angina pectoris, preexercise training treatment with sublingual nitroglycerin often reduces anginal pain and improves exercise performance.
4. Patients who exercise first thing in the morning should be encouraged to take their morning cardiovascular medications 20 min before they begin to exercise.

CONCLUSION

Acute coronary syndromes are the result of atherosclerotic plaque development with subsequent plaque disruption and thrombus formation leading to myocardial ischemia and potential necrosis. Unstable angina pectoris is the result of transient coronary artery occlusion with spontaneous clot dissolution and no demonstrable myocardial necrosis. Vessel occlusion persisting for more than 1 h results in myocardial necrosis, which is the hallmark of acute myocardial infarction. Myocardial infarctions are categorized, based on electrocardiographic findings, as either ST-segment elevation or non-ST-segment elevation. Preferred treatment is prompt reperfusion of the occluded vessel, and this results in less myocardial damage. Several classes of cardioprotective medications are given to survivors of acute coronary syndromes. In addition to these medications, comprehensive cardiac rehabilitation including exercise training forms the basis for secondary prevention of future cardiac events for these patients. Cardiac rehabilitation results in impressive benefits for patients, including reduced mortality.

Key Terms

cardiac tamponade (p. 224)
coronary revascularization (p. 224)
electromechanical dissociation (pulseless electrical activity) (p. 224)
fibrinolysis (p. 223)
infarct-related artery (p. 236)
left ventricular ejection fraction (p. 219)

CASE STUDY

MEDICAL HISTORY

A 31 yr old man, a two pack per day smoker, presented to the emergency department complaining of sudden-onset, 10/10 crushing substernal chest pain associated with shortness of breath and diaphoresis that began 30 min prior to arrival. He had no previous history of cardiovascular disease.

Physical exam: Height 192 cm, weight 153.8 kg, BMI 41.6 kg · m^{-2} (class III obesity); blood pressure 178/101 mmHg, pulse 75 beats · min^{-1}, respiratory rate 20 breaths · min^{-1}, temperature 36.7 ∞C; normal heart sounds, lungs clear to auscultation, normal peripheral pulses

Electrocardiogram: Normal sinus rhythm, rate 72 beats · min^{-1}; ST-segment elevation V1-V4

Chest radiograph: Normal

Blood work: Elevated cTn and CK-MB; glucose normal; blood lipids: total cholesterol 200 mg · dl^{-1}, HDL-C 45 mg · dl^{-1}, LDL-C 134 mg · dl^{-1}, triglycerides 104 mg · dl^{-1}

Chest CT: No evidence of aortic dissection

Echocardiogram: Left ventricular ejection fraction 45%, hypokinesis of the anteroseptal wall, borderline left ventricular enlargement, normal diastolic function, normal cardiac valves

DIAGNOSIS

Acute anterior wall STEMI.

TREATMENT

1. Emergent cardiac catheterization (began 46 min after initial presentation to the hospital): total occlusion of the proximal left anterior descending (LAD) coronary artery, mild diffuse narrowing of the circumflex and right coronary arteries. The LAD occlusion was treated with thrombectomy, percutaneous coronary angioplasty, and placement of a drug-eluding stent, with good result.
2. Medical treatment: Aspirin, heparin, intravenous nitroglycerin, clopidogrel, abciximab, atorvastatin, carvedilol, lisinopril, nicotine patch.
3. Nicotine dependence consultation.
4. Referral to outpatient cardiac rehabilitation.

Complications: None

Hospital dismissal: Occurred approximately 48 h after admission; appointments made to see primary physician and to begin outpatient cardiac rehabilitation program in 5 d; follow-up blood work, ECG, and cardiology appointment in 8 wk

Dismissal medications: Carvedilol, aspirin, clopidogrel, lisinopril, nicotine patch, simvastatin

OUTPATIENT EXERCISE TEST RESULTS (PERFORMED 8 D AFTER HOSPITAL DISMISSAL)

8.0 min duration (treadmill), normal blood pressure and electrocardiographic responses, peak heart rate 164 beats · min^{-1}, $\dot{V}O_2$peak = 3.04 L · min^{-1} (20.8 ml Σ kg^{-1} Σ min^{-1}; 47% of expected), no angina reported

EXERCISE PRESCRIPTION

Outpatient cardiac rehabilitation program: Began 5 d after hospital dismissal; initial evaluations included assessments of neuromuscular function (normal), depression (normal), and potential sleep apnea (abnormal

(continued)

Berlin questionnaire, abnormal overnight oximetry resulted in referral for a sleep consultation); meeting with registered dietitian.

Exercise prescription: Goals included weight loss, increased aerobic and muscular fitness; *frequency* of three supervised sessions per week in cardiac rehabilitation facility plus three or four independent sessions per week (total frequency of six or seven sessions each week); average *intensity* using target heart rate 130 to 140 beats · min^{-1} (60-70% heart rate reserve), RPE 12 to 14; high-intensity intervals (to begin 2 wk after starting cardiac rehabilitation) using RPE 17, heart rate <160 beats · min^{-1}; *duration* initially set at 10 to 15 min (not including warm-up, cool-down, strength training), gradually increasing to 45 to 60 min per session; *types* of exercise to include outdoor or treadmill walking, elliptical trainer; *resistance training* (to begin 2 wk after starting cardiac rehabilitation) using free weights and weight machines, initially one set of 8 to 15 slow repetitions, exercises for the major muscle groups, two or three sessions per week; *lifestyle or general physical activity* including walking short distances during the day, yard work, physical activity at work (return to work as restaurant equipment installer 4 wk after myocardial infarction).

OUTCOMES FROM 11 WK OF CARDIAC REHABILITATION PROGRAM

Medication compliance: Excellent (self-report)

Exercise training: Frequency six or seven sessions per week (two or three supervised sessions per week); aerobic exercise duration 50 to 60 min per session; aerobic interval training two or three sessions per week (initially treadmill grade walking for 30 s progressing to slow jogging on treadmill for up to 4 min, three to five high-intensity intervals per session); resistance training two or three sessions per week, two or three sets of 8 to 15 repetitions, major muscle groups.

Body weight: Stopped eating fast food and desserts, had more fruit, vegetables, poultry, fish; weight loss = 20.8 kg (BMI still excessive, 33.8 kg · m^{-2}).

Repeat cardiopulmonary exercise test: 10.5 min duration (treadmill), normal blood pressure and electrocardiographic responses, peak heart rate 162 beats · min^{-1}, $\dot{V}O_2$peak = 4.16 L · min^{-1} (31.2 ml Σ kg^{-1} Σ min^{-1}; 71% of expected), 37% increase in $\dot{V}O_2$peak (L · min^{-1}), no angina reported.

Tobacco use: One or two cigarettes daily.

Blood pressure: 126/80 mmHg.

Blood lipids: Total cholesterol 126 mg · dl^{-1}, HDL-C 33 mg · dl^{-1}, LDL-C 80 mg · dl^{-1}, triglycerides 65 mg · dl^{-1}.

Recommendations: Stop smoking; continue medications, exercise program, and heart-healthy eating habits; continue efforts for weight loss; return appointment with cardiac rehabilitation team in 3 mo.

DISCUSSION

This patient was extremely young and was fortunate to realize the importance of promptly seeking medical attention with the onset of symptoms. His hospital course included standard diagnostic testing for acute myocardial infarction with subsequent timely coronary angiography with PCI of the **infarct-related artery**. Fortunately he did not develop serious complications and was discharged from the hospital after only 2 d. He enrolled in outpatient cardiac rehabilitation 1 wk after his MI. Of concern, he was markedly obese and may have had obstructive sleep apnea based on his abnormal Berlin questionnaire; he was referred for further evaluation in the Sleep Clinic. His course during rehabilitation was very favorable, and he did not experience postinfarction angina. He was successful in exercise and weight loss and demonstrated a remarkable increase in aerobic capacity. His challenge will be to continue taking his medications, stop smoking completely, eat a healthy diet, maintain his exercise program, and control his other coronary risk factors.

DISCUSSION QUESTIONS

1. What were this patient's coronary risk factors?
2. What clinical tests were used to make the diagnosis of an acute ST-segment elevation anterior wall myocardial infarction?
3. What classes of cardiovascular medications were prescribed for this patient?
4. Compare the two cardiopulmonary exercise tests for this patient. What were the similarities? What were the differences?
5. Review the patient's exercise prescription and program. Were the established goals achieved? Would you alter the exercise program in any fashion?
6. How effective was the patient in controlling his coronary risk factors by the time he completed 11 wk in the outpatient cardiac rehabilitation program?
7. What is his most important remaining coronary risk factor?
8. What component of the recommendations given to him at the completion of 11 wk of cardiac rehabilitation will potentially help him in sustaining his efforts at secondary prevention?

Chapter 13

Revascularization of the Heart

Mark A. Patterson, MEd

Heart disease remains the leading cause of death according to the Centers for Disease Control and Prevention. Considering this reality is of upmost importance to have an understanding of the procedures used to potentially alleviate symptoms, reduce risk of significant heart injury, or even save the lives of those afflicted with heart disease. Equally important is knowing how to utilize exercise as a therapy to maintain or improve physical functioning of the heart and body to maintain lower risk of further adverse cardiac events and improve a person's quality of life.

DEFINITION

The term revascularization refers to a surgical procedure to help provide new or additional blood supply to a body part or organ. Several organs, such as the heart, lungs, kidneys, liver, and muscles (in situations such as gangrene), can benefit from this procedure. Typically, diagnostic tests may involve using magnetic resonance imaging (MRI), CT scans, or X ray fluoroscopy to identify the need for revascularization or help in guidance of the procedure.

Although there are multiple parts of the body and organs for which revascularization may be indicated, the focus of this chapter will be on procedures concerning the heart.

You will see in this chapter terms such as percutaneous transluminal coronary angioplasty (PTCA) and percutaneous coronary intervention (PCI) with or without stenting and coronary artery bypass surgery (CABS) that are used to describe techniques to establish proper blood flow back to the heart.

SCOPE

When a person has coronary artery disease, clinical procedures may be elected to restore myocardial blood flow with the specific intent of symptom relief and improved morbidity and mortality. The two most commonly used techniques are coronary artery bypass surgery (CABS) and percutaneous transluminal coronary angioplasty (PTCA) or percutaneous coronary intervention (PCI) with or without stenting. The National Center for Biotechnology Information estimated rates of CABS and PCI in the adult U.S. population in 2008. More than 240,000 CABS and 817,000 PCI were performed.

Over the past couple of decades, patterns of referral to cardiac rehabilitation for patients having coronary artery revascularization have evolved beyond those who have undergone surgical revascularization with CABS (68). Today, because of advances in coronary invasive technology, cardiac rehabilitation programs are seeing a growing number of people who have experienced percutaneous interventions, including percutaneous transluminal coronary **angioplasty** alone and, more frequently, in combination with **stent** therapy, which involves the placement of a mesh tube along the artery wall to prevent reocclusion (2, 43). Even though minor convalescent differences exist among the different percutaneous interventional procedures, the standards of practice and expected outcomes for cardiac rehabilitation are similar (26, 68); thus, this chapter focuses on patients who have undergone CABS or PTCA with or without stent therapy. An important note concerns the emerging information regarding mechanical revascularization versus aggressive medical management. In a 2008 paper, Dr. Barry Franklin, PhD, outlined several articles that indicated questionable mortality benefit for those who underwent CABS or PCI versus medical management. It was still advantageous for those with evolving myocardial infarction or unstable angina to undergo revascularization, but for those with stable angina the benefit is not as clear (25). Regardless of whether patients with coronary artery disease are revascularized or not, multiple studies have demonstrated that for each 1 MET (metabolic equivalent) improvement in exercise capacity there appears to be between an 8% and 17% reduction in mortality (10).

PATHOPHYSIOLOGY

Coronary artery disease (CAD) involves a buildup of lipids, macrophages, platelets, calcium, and fibrous connective tissue within the coronary arteries. This results in the formation of a plaque that progressively narrows the lumen. This may eventually cause a limitation or obstruction of normal blood flow. While symptoms or changes on an electrocardiogram (ECG) during a stress test may not occur until a coronary artery has a 75% or greater stenosis, lesions that compromise 50% or more of the lumen in a major coronary artery might be considered clinically significant. Multiple factors such as location of the lesion, stability of the plaque, symptoms, short- and long-term prognosis, and quality of life may all influence the decision on whether a particular patient is a candidate for revascularization procedures.

Coronary Artery Bypass Surgery

Coronary artery bypass surgery involves revascularization using a venous graft from an arm or leg or an arterial graft (both ends free or from a regional intact native vessel, e.g., internal mammary, **gastroepiploic artery**) to provide blood flow to the myocardium beyond the site of the occluded or nearly occluded area in a coronary artery. Although CABS has traditionally involved a **sternotomy** and the use of a **heart and lung bypass**, technical advances now permit a growing number of procedures to be performed:

1. Procedures performed through small port incisions (or "minisurgery") using microscopic procedures
2. Procedures performed with robotic technology
3. Surgery performed on the beating heart without the use of cardiopulmonary bypass

Subsequent to these technical advances, postoperative morbidity has significantly decreased. The postsurgical hospital stay for CABS patients without complications is now less than 5 d. As a result of the evolution of revascularization procedures (particularly percutaneous intervention), the role for CABS has changed; it is now reserved for the following patients:

1. Patients who are post-PTCA or stenting (or both) with restenosis
2. Patients who are no longer candidates for angioplasty but still have target vessels offering preservation of left ventricular systolic function
3. Those with multivessel disease not amenable to angioplasty or stenting
4. Those with technically difficult vessel lesions, for example, a lesion on the curve of a vessel or in a distal location not readily amenable to angioplasty or stenting

The number of surgical revascularization procedures has declined, but these procedures still play an essential role in higher-risk occlusive disease. Successful CABS results in myriad improved exercise responses; and when combined with medical therapy, CABS may more effectively relieve significant residual exercise-induced symptomatic or silent myocardial ischemia. There is also some evidence that in persons with diabetes, the need for future revascularization procedures is less with CABS than with PTCA or stenting (14, 35). Thus, the symptom relief, improved functional capacity, and improved quality of life may be the most practical and important patient benefits of CABS.

Percutaneous Interventions

Coronary angioplasty, which is also called percutaneous transluminal coronary angioplasty (PTCA), is less invasive than CABS. Several techniques have been developed for use in restoring adequate blood flow in diseased coronary arteries. Often the procedure is combined with stent therapy to reduce the likelihood that the artery will reocclude.

Percutaneous Transluminal Coronary Angioplasty

PTCA is a well-established, safe, and effective revascularization procedure for patients with symptoms attributable to CAD. The procedure may use one or more techniques alone or in combination to open the vessel:

1. Balloon dilation is most commonly used in conjunction with stent placement (figure 13.1).
2. Rotational atherectomy, a rotational device used for removing plaque, may be applied to central bulky lesions in a minority of cases.
3. Directional atherectomy and laser may be used to debulk large lesions, but the risk of vessel wall perforation or dissection may be greater. The use of these devices is limited to a few centers in contemporary practice.

The complications of angioplasty are acute vessel closure (rebound vasoconstriction) or chronic restenosis, thrombotic distal **embolism,** myocardial infarction (MI), arrhythmias, dissection of the coronary artery, and bleeding.

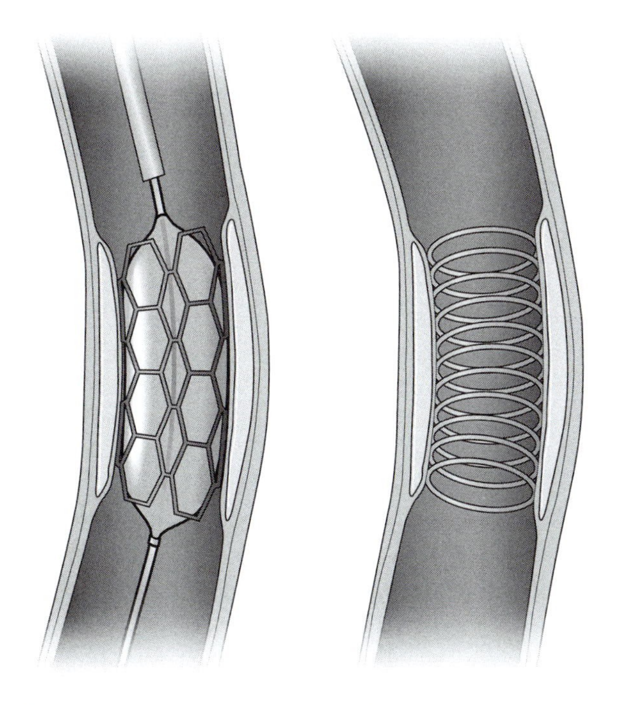

Figure 13.1 Percutaneous transluminal coronary angioplasty balloon catheter and two types of stents: latticed steel (left) and the coiled stent (right).

Stent Therapy

To reduce the risk of acute closure and restenosis of coronary arteries after PTCA (37), several models of intracoronary stents have been advocated. Stent therapy is frequently used in conjunction with one of the previously described techniques to preserve the patency of the vessel. Stents are stainless steel mesh tube bridges that are advanced on the end of a balloon catheter, passed across the culprit lesion, and expanded. The stent, serving as a permanent intravascular prosthesis, compresses the lesion, resulting in an open vessel. This is the final treatment following balloon angioplasty or debulking in over 95% of cases. After removal of the balloon catheter, the stent remains permanently in the coronary artery and is eventually covered with endothelium, becoming part of the luminal wall structure.

A quantum leap in the last 3 yr has been the availability of drug-eluting stents. The metal scaffolding prevents acute closure and also provides a vehicle for local drug delivery. This innovation plus the use of aspirin and thienopyridines (clopidigrel and prasugrel) has reduced the problem of in-stent restenosis to a great extent (1, 55, 66, 72).

Many stent procedures require a one-night hospital stay or are a same-day procedure. The loss in functional capacity following PCI is less than that following a bypass procedure. Subsequently, PCI patients begin cardiac rehabilitation at higher functional capacity, quality of life, and **self-efficacy** (30, 47).

CLINICAL CONSIDERATIONS

The success rate of a revascularization procedure may be predicted, in part, by the patient's age, existing comorbidities, and severity and location of the lesion.

Coronary Artery Bypass Surgery

Elective CABS improves the likelihood of long-term survival in patients who have the following:

- Significant left main CAD
- Three-vessel disease
- Two-vessel disease with a proximal left anterior descending stenosis
- Two-vessel disease and impaired left ventricular function (2)

In patients experiencing failed angioplasty with persistent pain or hemodynamic instability, acute MI with persistent or recurrent ischemia refractory to medical therapy, **cardiogenic shock**, or failed PTCA with an area of myocardium still at risk, revascularization by CABS offers effective relief of angina pectoris and improves the quality of life (2). The occlusion rate of grafts is approximately 10%, 20%, and 40% after 1, 5, and 11 yr, respectively (10). The CABS patient's postoperative education should include wound care, appropriate management of recurring symptoms, and risk factor modification.

Percutaneous Transluminal Coronary Angioplasty

In select cases of unstable angina, PTCA has an acute success rate of over 84%. Following successful PTCA, restenosis occurs in approximately 25% of patients, almost always within the first 6 mo (2, 4). Following the development of drug-eluting stents, restenosis rates have dropped dramatically in many cases, to below 5% after 2 yr (27). Table 13.1 lists several predictors of restenosis (56).

Other very important potential predictors include comorbidities such as diabetes mellitus and whether patients are on optimal medical therapy that includes use of aspirin and thienopyridines such as Plavix and Effient (1, 27).

Table 13.1 Predictors of Restenosis After PTCA-Stent

PTCA	Stent
Degree of residual stenosis after PTCA	Lesion eccentricity
Diameter of the parent vessel	Diameter of the parent vessel
Number of diseased vessels	Type of vessel stented (artery vs. vein)
Degree of reduction of the stenosis	Location of stent in vessel
Presence or type of coronary dissection	Presence of multiple stents
Presence of documented variant angina	Recurrence of unstable angina

Note. All predictors are positively associated with risk for the revascularized vessel to reocclude. PTCA = percutaneous transluminal coronary angioplasty.

Patients who have had PTCA in the setting of unstable angina should have close surveillance following hospital discharge and should be advised to seek prompt medical attention in the event of a recurrence of the symptoms that were occurring before their PTCA (17).

Stent Therapy

Improved technology currently confers procedural success rates in excess of 95% in most centers. Acute closure and restenosis remain as limitations to short- and long-term success, respectively, although the incidence of both these complications has decreased dramatically in recent years. The incidence of thrombosis and acute closure is in the range of 1% to 2% with use of thienopyridenes (clopidogrel or ticlopidine); therefore, chronic **anticoagulation therapy** is no longer required. Restenosis rate ranges from 25% to 40% with bare metal stents and less than 10% with drug-eluting stents (4, 48). These results vary depending on comorbidities and efficacy of aggressive medical management.

Table 13.2 provides a summary of medications commonly prescribed.

Exercise Testing

The graded exercise test (GXT) is commonly used for continued diagnosis of possible ischemic myocardium, prognostication, and the establishment of functional status for exercise prescription purposes. Although an integral component for exercise prescription, the timing of the GXT is somewhat controversial. Standard administration procedures and contraindications to testing, discussed in chapter 5, should be followed. Practical application 13.1 outlines exercise testing for revascularized patient.

Coronary Artery Bypass Surgery

Requiring all patients to have an exercise test after successful bypass surgery for the purpose of beginning a supervised and monitored exercise program is of questionable clinical benefit and an unnecessary financial burden (51). The patient's coronary anatomy is known, and unless surgical complications or postsurgical symptoms are present, the chance of detecting unknown ischemia is extremely low. In addition, because of the acute convalescent period, the patient may not be able to give a physiological maximal effort, sacrificing test sensitivity.

A more opportune time for testing the patient is after incisional pain has resolved, blood volumes and hemoglobin concentrations have normalized, and skeletal muscular strength and endurance have improved from participating in low-level exercise and activity. At least 3 to 4 wk postsurgery, the patient will be able to give a near-maximal physiological effort, providing test results with greater diagnostic accuracy for assessing functional capacity, determining return-to-work status, or recommending the resumption of physically vigorous recreational activities. For patients whose surgical revas-

Table 13.2 Pharmacology

Medication name and class	Primary effects	Exercise effects	Special considerations
β-blockers: atenolol, metoprolol, propranolol, carvedilol	Used for hypertension, angina, arrhythmias; increase atrioventricular (AV) block to slow ventricular response	Lower heart rate and blood pressure at submaximal and maximal exercise, increase exercise tolerance in patients with angina and heart failure	Traditional heart rate prescriptions are not valid; use of RPE may be more desirable.
Nitrates: nitroglycerin, isosorbide (can come in pills, sprays, and patches)	Used for angina; vasodilator	May decrease HR with exercise; usually no effect	Hypotension; may increase exercise tolerance in patients with angina.
Calcium channel blockers	Used for angina, hypertension; increase AV block to slow ventricular response	Decrease in or no effect on submaximal and maximal heart rates, decrease in blood pressure at rest and exercise	Increase exercise tolerance in patients with angina.
Digitalis: digoxin	Increases contractility; used primarily in chronic heart failure	Decrease in resting heart rate in people with atrial fibrillation and heart failure; not significantly altered in patients with sinus rhythm	Exercise tolerance improvement in patients with atrial fibrillation and heart failure only; can cause ST depression on ECG leading to false-positive stress test results.
Diuretics: hydrochlorothiazide, furosemide, triamterene, spironolactone	Used for edema, chronic heart failure, certain kidney disorders	No effect on resting or exercise heart rates; decrease resting blood pressure but not with exercise	No effects on exercise response except for possibly in patients with heart failure; watch for increased ventricular ectopy due to hypokalemia and hypomagnesemia.
ACE inhibitors: captopril, enalapril, lisinopril	Used for hypertension, coronary artery disease, chronic heart failure, diabetes, chronic kidney disease	No effects on exercise heart rate; may lower exercise blood pressure	Some improvement in exercise tolerance in patients with heart failure.
Antiarrhythmic agents: procainamide, lidocaine, flecainide	Specific for individual drug, but may be used for suppression of arrhythmias such as atrial fibrillation	May increase HR at rest; may decrease BP at rest; typically no effects on exercise HR or blood pressures	Watch for QRS widening with exercise; some agents may cause false-negative stress tests (e.g., quinidine).
Antilipemics and statins: fenofibrate, atorvastatin, lovastatin, simvastatin, ezetimibe	Used for elevated blood cholesterol, triglycerides, and metabolic syndrome	No effects on heart rate and blood pressure at rest or with exercise	Myalgia may result; sometimes difficult to discern whether muscle pain is from exercise or statins.
Blood modifiers (anticoagulant or antiplatelet): clopidogrel, prasurgrel, cilostazol, pentoxifylline	Prevent blood clots, heart attack, stroke, and intermittent claudication	No effects on heart rate and blood pressure at rest and with exercise	Cilostazol and pentoxifylline may increase ability to walk in people with claudication from PAD.

cularization was not successful or who are experiencing symptoms suggestive of ischemia, a clinical exercise test before starting an exercise program is recommended. All testing procedures should follow professional guidelines (8, 9, 11) as noted in chapter 5.

Percutaneous Transluminal Coronary Angioplasty

Debate exists regarding the proper timing of stress testing in PTCA patients. Several reports of acute thrombotic occlusion associated with exercise testing shortly after successful PTCA have been reported, although these have not been borne out as relevant in clinical practice. Although no chest pain is reported during the test, ischemia within 1 h after testing has been reported (18, 60). The mechanisms for the apparently abnormal test responses are unclear but are possibly related to the following:

1. Higher levels of **platelet aggregation** during exercise testing

Practical Application 13.1

EXERCISE TESTING REVIEW

Test type	Mode	Protocol specifics	Clinical measures	Clinical implications	Special considerations
Cardiovascular	Treadmill Cycle (if treadmill not possible)	Bruce, Ellestad (for younger or more physically fit individuals) Naughton or Balke-Ware (for older, deconditioned, or symptomatic patients) Ramping protocols appear more tolerable for many patients Pharmacologic testing for those unable to exercise	Heart rate and rhythm Blood pressure 12-lead electrocardiogram Symptoms Rating of perceived exertion Nuclear or echocardiographic imaging as prescribed Gas exchange analysis as prescribed	Rhythm disturbances Hemodynamics Myocardial ischemia Ischemic threshold Perception of work difficulty Ischemic myocardium LV function Dyspnea with exertion	CABS: Chest and leg wounds (4-12 wk for complete healing) PTCA-stent: Reocclusion—recurrence of previous symptoms
Strength	Isometric Isotonic Isokinetic	Peak force RM procedures as described in the text 3- to 10RM dependent on patient conditioning, experience, and clinical condition Peak torque	Maximal strength of the muscle or muscle group tested Blood pressure Heart rate Symptoms	Functional fitness	CABS: Incisional healing No Valsalva
Range of motion	Trunk flexion Shoulder flexion, extension, abduction	Sit-and-reach Goniometer	Posterior leg and lower back flexibility Shoulder flexibility	Functional fitness Maintenance of ability to perform ADLs after sternotomy	Orthopedic complications that may preclude testing

Testing procedures outlined in *ACSM's Guidelines for Exercise Testing and Prescription, Eighth Edition.*

Valve Surgery

- For severe regurgitant (leaking) or stenotic (narrowing) valve disease, valve replacement may be essential for symptom relief and improved exercise tolerance. Regarding the open-heart surgical process, precautions similar to those for the bypass surgery patient should be followed, with attention to a few special considerations.
- Special considerations: Symptom resolution may not occur immediately following surgery. Symptoms may gradually resolve because of heart remodeling after valve replacement, for example, for aortic stenosis. Avoid strength training for severe stenotic or regurgitant valvular disease. Isometric exercises are not recommended. Common exercise-induced symptoms include shortness of breath, fatigue, and dizziness or light-headedness.
- Exercise prescription: Follow procedures similar to those for revascularization surgery.

2. An increase in thromboxane A2
3. Platelet activation and hyperreactivity increase during exercise
4. Increased arterial wall stress associated with increased coronary blood flow
5. The higher blood pressure that occurs during exercise, which may traumatize an already disrupted **intima** (18, 42)

On the other hand, exercise testing of patients with PTCA has been accepted standard practice, particularly for those with incomplete revascularization. A large body of evidence supports the use of early postprocedure exercise testing (1-2 d) to evaluate the functional status of the PTCA patient (5, 6, 15, 17, 64).

Stent Therapy

Controversy with regard to safety of early testing after stent placement has essentially been laid to rest. Most authorities now accept that performing a stress test after coronary stenting is safe. Recently, there has been increasing support for exercise testing before the start of an exercise program (13, 17, 20, 39, 40). The accuracy of these tests, however, can still be debated. In particular, there have been reports of false-positive stress results early on after coronary stenting. As with CABS and PTCA, the need to test all patients after stent therapy before starting cardiac rehabilitation is debatable. One scenario for which it might be useful is for patients who either cannot participate in a supervised program or choose to exercise independently and need some reassurance and initial guidance. In general, for the successfully revascularized patient, an exercise test may be redundant and may not provide any further useful clinical information before the patient starts a supervised exercise program. The primary concerns with PTCA and stent are **reocclusion** and **restenosis**. Subsequent restenosis may not be detected immediately following the procedure. The patient can exercise in a supervised exercise program and be tested at a later date if symptoms recur or for assessment of functional capacity before return to work. Additionally, a poor response to exercise training, demonstrating no improvement in functional capacity, may be indicative of restenosis (5, 6, 46).

Practical application 13.2 provides advice on helping revascularized patients adhere to a long-term exercise program.

EXERCISE PRESCRIPTION AND TRAINING

Over the past decade, the average length of hospital stay for cardiovascular patients has decreased dramatically. Currently, the hospital stay for uncomplicated cases of CABS is usually 2 to 5 d. For PTCA-stents, the stay is 1 to 2 d; or they are done on an outpatient basis, with the patient managed in an acute recovery suite and discharged on the same day. Although cardiac rehabilitation begins as soon as possible during hospital admission, the shorter length of hospital stay has changed the inpatient program to basic range of motion exercises and ambulation; and the educational focus is on discharge planning—teaching about medications, home activities, and follow-up appointments. Educational topics previously covered in the inpatient setting are now the responsibility of the outpatient program. Moreover, cardiac rehabilitation professionals must make every effort to enroll patients in an outpatient program. Table 13.3 reviews the exercise prescription for the revascularized patient.

Within the first few days of bed rest, many body composition changes (loss of lean body mass) occur (65), supporting the need for early exercise intervention during

Practical Application 13.2

CLIENT–CLINICIAN INTERACTION

Medical advances have come a long way to assist the clinical exercise professional in helping individuals return to exercise and more active lifestyles sooner. Despite all these advances, most revascularized patients eventually return to a more sedentary lifestyle. The field of behavior change has started to become an increasingly important part of helping these individuals maintain their exercise programs and more active lifestyles.

One of the most important tools you can use to increase exercise adherence is your ears. Take the time to listen to your patients. Understand what it is they have gone through. Listen to their fears and concerns when you are testing or designing exercise programs. Understand their physical, emotional, and environmental barriers. If you take the time to consider their sources of support and their resources for activity and exercise, and then remember to reassess their goals, to be flexible, and to follow up in a timely fashion, they will be more likely to succeed.

hospital admission. Patients who perform typical ward activities and moderate, supervised ambulation do not suffer the magnitude of loss in lean body tissue seen in those who remain inactive. Early standing and low-level activities, including range of motion and slow ambulation, may be all that are required to deter postsurgical lean body mass loss while the patient is in the hospital (65).

After hospital discharge, many positive physiological adaptations occur in revascularized patients who participate in a supervised exercise program (68):

- Improved cardiac performance at rest and during exercise
- Improved exercise capacity (aerobic and strength)

Table 13.3 Exercise Prescription Review

Training	Mode	Frequency	Intensity	Duration	Progression	Goals	Special considerations
Aerobic	Treadmill Walking Cycle Combined arm and leg exercise Rowing Stepper Swimming Elliptical Combination of above or others to ensure adequate utilization of major muscle groups and distribution between upper and lower extremities	Daily is ideal; at least 4 d/wk	Asymptomatic: 40% to 85% of HRmax RPE: 11 to 16 Symptomatic: below ischemic or anginal threshold RPE: 11 to 16	At least 30 min May be intermittent (e.g., 3 × 10) or continuous depending on patient tolerance	If appropriate, should start with some ambulation in hospital; first couple of weeks posthospitalization, progress to 5 to 10 min of very light intensity multiple times a day. After 4 wk posthospitalization, increase intensity to moderate levels, increase time to 15 to 30 min one or two times a day; at 6 wk or more posthospitalization, should be working toward 30 or more min at moderate or better intensities.	Adults should do at least 150 min (2 h 30 min) a week of moderate-intensity, or 75 min (1 h 15 min) a week of vigorous-intensity aerobic physical activity, or an equivalent combination of moderate- and vigorous-intensity aerobic activity, and more extensive health benefits. Adults should increase their aerobic physical activity to 300 min (5 h) a week of moderate-intensity, or 150 min a week of vigorous-intensity aerobic physical activity, or an equivalent combination of moderate- and vigorous-intensity activity. Additional health benefits are gained by engaging in physical activity beyond this amount.	Initially, need to be concerned with incisional discomfort in chest, arm, and leg of surgical patient. May need to restrict upper extremity exercises until soreness resolves. Also, those with PCI may have some groin soreness at the catheter insertion site that may restrict certain physical movements.

Training	Mode	Frequency	Intensity	Duration	Progression	Goals	Special considerations
Resistance	Elastic bands Hand weights Free weights Multistation machines Equipment selection based on patient progress (a rational progression is to use the equipment in the order listed)	Two or three times per week	Select a weight such that the last repetitions feel somewhat or moderately hard without inducing significant straining (bearing down and breath holding)	12 to 15 repetitions, slowly increasing weight and intensity so that 8 to 10 repetitions provide the appropriate response	Generally, in the first 4 wk posthospitalization, CABS patients are asked to use little to no resistance and primarily perform ROM exercises and some strengthening exercises that do not produce significant strain on the incision site. PCI patients can generally start a bit earlier, but also use light weights that can be completed for 12 to 15 repetitions without producing a Valsalva effect. After 4 wk posthospitalization, CABS patients can start to increase the amount of weight—can start with 12 to 15 repetitions without Valsalva and should not have any clicking or grinding of the sternotomy. PCI patients should look to increase efforts to moderate levels at this point, and ultimately all revascularization patients should progress to multiple muscle group exercises, using enough resistance to produce muscular fatigue in 10 to 12 repetitions. Days per week and sets of each exercise should be individualized according to patient needs and goals.	Adults should do muscle strengthening activities that are moderate or high in intensity and involve all major muscle groups on 2 or more days a week.	For surgical patients, the initial upper extremity exercises may be range of motion without resistance—progressing initially with elastic bands or 1 to 3 lb (0.5 to 1.5 kg) increments. Slightly higher weight may be employed for muscle groups and movements that do not put sternal healing at risk. Further progression depends on sternal healing and stability. Exercises should be selected that employ muscle groups involved in lifting and carrying.

(continued)

Table 13.3 *(continued)*

Training	Mode	Frequency	Intensity	Duration	Progression	Goals	Special considerations
Warm-up and cool-down exercises	Range of motion and flexibility exercises	Daily	Static stretching	5 to 10 min	In hospital and immediately posthospitalization, CABS patients should be doing daily ROM exercises to the point of mild stretch on the sternum, but should not feel any pain. As they continue to heal, the exercises should continue to progress to greater ranges of motion to tolerance. PCI patients should resume or start a daily stretching routine without restrictions.	Daily, 5 to 10 min or as needed for activities participated in or for increases in range of motion.	Exercises should emphasize major muscle groups, especially lower back and posterior leg muscles.

- Greater total work performed
- Improved angina-free exercise tolerance, much of which is attributable to peripheral muscular adaptations (34)
- Improved neurohumoral tone (44)

Patients in such a program gain in several ways:

- They more often achieve full working status.
- They have fewer hospital readmissions.
- They are less likely to smoke at 6 mo following completion of exercise therapy (9).

When we compare the physiological and psychosocial outcomes between CABS, PTCA–stent, and MI patients at the beginning and end of 12 wk of cardiac rehabilitation, some group trends are apparent. CABS patients may begin with lower functional capacities and lower ratings of quality of life and self-efficacy attributable to the surgical recuperative process (69), but they show greater improvement during the program and obtain similar or greater values compared to other cardiac patients at program completion, regardless of age (21, 30, 47). This result may reflect lower rates of **ischemia** than in MI patients, greater confidence in their own ability, and the potential psychological feeling that "something was done" about their heart disease and that they are "cured." Regardless of age, the CABS patients demonstrate functional improvement but may require a longer training period to obtain the same magnitude of effect (59). The PTCA–stent groups have not suffered the loss in functional capacity because of the more prolonged recuperative process following an MI or bypass surgery and have greater functional capacity when starting cardiac rehabilitation (30, 47).

Exercise prescription guidelines for **revascularization** patients have been published by the American College of Sports Medicine (11), American Association of Cardiovascular and Pulmonary Rehabilitation (9), and American Heart Association (8).

Special Exercise Considerations

Although revascularized patients are just as knowledgeable about risk factors as post-MI patients, they are less compelled to make changes (31). Post-MI patients initiate considerably greater lifestyle changes than revascularized patients do (54). Patients undergoing revascularization may be less motivated to adhere to risk factor behavior change because of a perception of that they are less sick or have been cured, which has a negative effect on compliance with risk factor modification (32, 38, 53).

Revascularized patients may encounter, or anticipate, restrictions differently than other cardiac patients do.

Most PTCA–stent patients are capable of resuming normal activities of daily living following hospital discharge, but patients frequently perceive considerable restrictions after the procedure with respect to all activities of daily living—leisure activities, sexual activity, and early return to work (23, 57, 62).

Depression remains prevalent in patients with coronary heart disease after major cardiac events (CABS and PTCA included). Cardiac rehabilitation does reduce the prevalence and severity of depression. Therefore, cardiac patients should be routinely screened and offered the benefits of comprehensive cardiac rehabilitation including psychosocial support and pastoral care (52). There has also been some research on preprocedure counseling ("prehabilitation") and extended telephone counseling after the procedure and rehabilitation sessions have ended; these studies have shown promise in reduction in depression surrounding revascularization procedures (29, 63).

Spouses may be more likely to seek information about the patient's psychological reactions and recovery, whereas patients are more likely to seek information about their physical condition and recovery (53). Patients tend to be more positive than spouses, who tend to be more fearful of the future (33). Also, patients and spouses differ in their views on the causes of CAD and about the responsibility for lifestyle changes and the management of health and stress (58). Therefore, assessing both the patient's and the spouse's educational needs is important. "Health Care Considerations for PTCA Patients" lists important concerns in the rehabilitation of the post-PTCA patient, taking into account that exercise may be contraindicated for patients who continue to be symptomatic postevent or postprocedure, particularly at low workloads (<5.0 METs).

Another consideration is the potential uncovering of claudication in peripheral artery disease (PAD) with ambulation after revascularization. It is well established that persons with CAD are at higher risk than others for PAD (and vice versa). This may be more of an issue with patients who have been very sedentary and then postprocedure start a new exercise regimen and develop symptoms of claudication.

Coronary Artery Bypass Surgery

Primary concerns for the CABS patient when entering outpatient cardiac rehabilitation are the state of incisional healing and sternal stability, hypovolemia, and low hemoglobin concentrations. During the initial patient interview, the rehabilitation professional needs to ensure that the surgical wound has no signs of infection, significant draining, or instability. Questions should focus on the following:

- Excessive or unusual soreness and stiffness
- Cracking, grinding, or motion in the sternal region
- Whether the patient is sleeping at night
- How the patient's chest and leg incisions are responding to current activities of daily living since discharge

Also, knowing how patients performed during the inpatient program may help determine how soon they can begin the outpatient program and at what level they can begin exercising. For example, was the patient out of bed, upright, and walking soon after surgery without problems? If not, was the patient's lack of activity attributable to extreme physical discomfort, clinical or orthopedic difficulties, or lack of motivation?

Historically, surgical patients did not begin cardiac rehabilitation for 4 to 6 wk postsurgery or longer and avoided upper extremity exercise for even longer periods. Today, standard practice is for patients to begin the outpatient program soon after discharge, often within a week of surgery. For the uncomplicated revascularized patient, light upper extremity exercises are now

Healthcare Checklist for PCI/Stent Patients

- Control of hypertension, obesity, and smoking
- Progressive exercise and weight reduction
- Awareness of other cardiac risk factors
- Identification of stressful factors
- Counseling services for weight reduction, stress management, and smoking cessation
- Maintaining close contact between health professionals
- Organizing and maintaining long-term follow-up records
- Reinforcing the noncurative nature of PCI/stent as cardiac treatment modality
- Encouraging revascularization patients, and PCI/stent patients to take a proactive approach to improve health outcomes.

prescribed, including range of motion exercises, light hand weights progressing to light resistive machinery, and gradually progressive upper extremity ergometry beginning at zero resistance.

Percutaneous Transluminal Coronary Angioplasty

The primary concern for the PTCA or stent patient is restenosis. At the patient's initial orientation session, questioning should address the presence of signs or symptoms indicative of angina or the person's particular anginal equivalent. Education should include information about symptoms, including anginal equivalents; management of angina (e.g., how to use nitroglycerin, going to the emergency department); precipitating factors (exertion or anxiety related); and care of the catheter insertion site. Patients with PTCAs and stents may begin the outpatient program as soon as they are discharged from the hospital or immediately following the procedure if it has been performed on an outpatient basis (67).

Exercise training may alleviate the progression of coronary artery stenosis after PTCA by inhibiting smooth muscle cell proliferation, lowering serum lipids, improving insulin resistance and glucose intolerance, and causing hemostatic changes (45, 70). Aerobic training for 30 to 40 min, four to six times per week for 12 wk, improves treadmill time and myocardial perfusion and reduces the restenosis rate at 3 mo following PTCA (45, 47). Other benefits include an improved sense of well-being, relief of depression, stress reduction, and sleep promotion.

As a result of angioplasty with improved techniques of revascularization, more patients with low-risk profiles are being referred to cardiac rehabilitation (i.e., patients with a greater exercise capacity, no evidence of ischemia, normal left ventricular function, and no arrhythmias). Specific examples include patients who are younger, have single-vessel disease, and did not experience an MI before their PTCA. Regarding exercise prescription, these individuals may be treated similarly to apparently healthy individuals with the addition of education concerning the recognition of anginal equivalents, self-monitoring, self-care, and risk factor modification. Optimized medical therapy along with appropriately prescribed exercise training can be an alternative approach to interventional strategies in selected patients who are asymptomatic (36). When PCI is the therapy of choice, it should be combined with daily physical exercise and increased physical activity to optimize success (36). Cardiac rehabilitation results in early and sustained improvement in quality of life and is highly cost-effective (73). In addition, angioplasty patients commonly experience restenosis. Supervised exercise training and education improve recognition of signs and symptoms associated with closure. Most important, angioplasty patients need instruction concerning appropriate exercise training, dietary modifications, medications, and general risk factor reduction to slow or reverse the coronary disease process.

Because the PTCA patient remains on complete bed rest while the sheath is in situ for approximately 18 to 24 h, the immobilization often causes back pain. Appropriate flexibility exercises that enhance range of motion often help to resolve low back pain.

Stent Therapy

Because placement of stents uses the same catheter procedure as in the PTCA, the same considerations exist. But the risk for thrombosis is greater following stent therapy. Consequently, patients are often placed on anticoagulant therapy for preventive purposes. Although no specific contraindications preclude exercise following recent stent placement, proceeding with similar caution is prudent.

Cardiovascular Exercise

The multiple improvements in patients' tolerance to acute bouts of exercise after revascularization include the following (30, 47):

- Improved myocardial blood flow
- Increased functional capacity
- Improved cycle ergometer or treadmill performance
- Variable improvements in left ventricular function
- Increased maximal heart rate
- Increased rate–pressure product
- Reduction in ST-segment depression
- Relief or improvement in anginal symptoms with exercise
- Improved heart rate recovery
- Reduction in exertional hypotension

The initial exercise prescription is based on information gained from the patient's orientation interview for the outpatient cardiac rehabilitation program. Patients are questioned concerning the presence of signs or symptoms, their activity while in the hospital, and their activity level since their return home from the hospital. Of equal importance are their level and consistency of conditioning before their cardiac event. Depending on how long they were in the hospital and the amount of deconditioning, better-conditioned, higher-functioning patients may be able to return to higher levels of exercise volume and intensity more quickly than most patients. Initially, patients are closely observed and monitored to establish appropriate exercise intensities and durations that are within their tolerance. A starting program may include treadmill walking (5-10 min), cycle ergometry (5-10 min), combined arm and leg ergometry (5-10 min), and upper body ergometry (5 min). Initial intensities may approximate 2 to 3 METs (multiple of resting oxygen uptake of 3.5 $ml \cdot kg^{-1} \cdot min^{-1}$), but starting MET levels may be a bit higher depending on exercise history and prior conditioning. The patient's heart rate, blood pressure, rating of perceived exertion, and signs and symptoms are monitored and recorded. Programs are gradually titrated during the initial sessions to a rating of perceived exertion of 11 to 14 in the absence of any abnormal signs or symptoms.

In general, exercise intensity is progressed by 0.5 to 1.0 MET increments (i.e., 0.5 mph [0.8 kph] or 2.0% grade on the treadmill or 12.5-25 W on the cycle). The rate of progression is based on the patient's symptoms, signs of overexertion, rating of perceived exertion, indications of any exercise-induced abnormalities, and prudent clinical judgment on the part of the cardiac rehabilitation staff. Patients with greater exercise capacities (PTCA–stent patients with no MI) are started according to their exercise capacities and progressed more rapidly. The selection of exercise modality depends on the person's program objectives. For example, those who are employed in a labor-type occupation or perform many upper extremity activities at home spend a greater portion of their exercise time doing upper extremity exercises. If specific limitations preclude certain exercise modalities, program modifications are made that allow more time on tolerable equipment to obtain the greatest cardiovascular and muscular advantage. For patients who have an exercise test, standard recommended procedures for exercise prescription are followed (8, 9, 11).

Resistance Training

Muscular strength and endurance exercise training should be incorporated equally with cardiovascular endurance and flexibility exercise training during the early outpatient recovery period. Following revascularization, low-risk patients can perform muscular strength and endurance exercise training safely and effectively (41). Depending on the patient's clinical and physical status, successful approaches for upper and lower extremity strength enhancement include 10 to 12 repetitions with a variety of types of equipment that may include elastic bands, Velcro-strapped wrist and ankle weights, hand weights, and various multistation machines. Usual guidelines include maintenance of regular breathing patterns (avoiding the **Valsalva maneuver**), selection of weights so that the last repetition of a set is moderately or somewhat hard, and progression when the perception of difficulty decreases. CABS patients may start range of motion exercises with light weights of 1 to 3 lb (0.5-1.5 kg) within 4 wk of surgery as long as sternal stability is ensured and excessive incisional discomfort is not present. PTCA-stent patients may start resistance training immediately. Exercises should be selected that will strengthen muscle groups used during normal activities of daily living for lifting and carrying and occupational or recreational tasks (2).

Weights are selected that allow the completion of 12 to 15 repetitions initially; the patient then progresses to higher weights and 10 to 12 repetitions with the last three repetitions feeling moderately hard. Those who cannot securely hold hand weights should use wrist weights with Velcro straps. Typically, exercises are selected that use upper and lower extremity muscle groups involved in routine lifting and carrying and other activities of daily living. Patients are progressed from stretch bands and light hand weights to resistance machines, again using resistances that result in a perception of difficulty of moderately hard for two or three sets of 10 to 12 repetitions. Figures 13.2-13.4 outline possible exercises for a rehabilitative exercise progression for revascularized patients.

Progression of exercise is based on patients with open-heart procedures; patients undergoing PTCA and stenting can perform all exercises listed for the beginning of the program with proper progression based on prior exercise history and clinical judgment of the cardiac rehabilitation team. The list that follows is based partly on information from Adams and colleagues (7).

Early-Phase Rehabilitative Exercise (2 to 4 wk Postdischarge)

Traditionally patients at this phase have been told to avoid strength training and do only range of motion or very light if any strength training. You may consider having

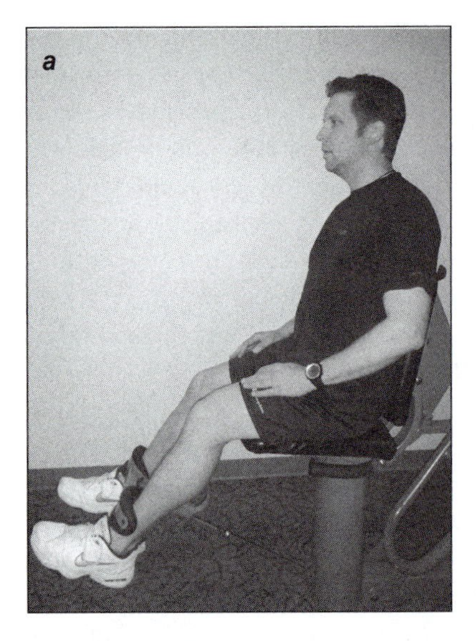

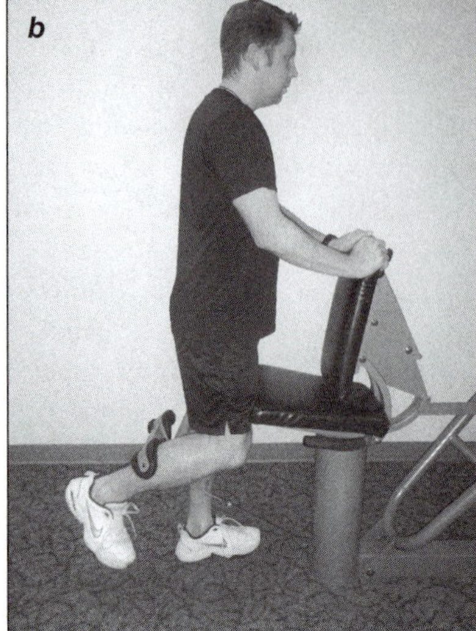

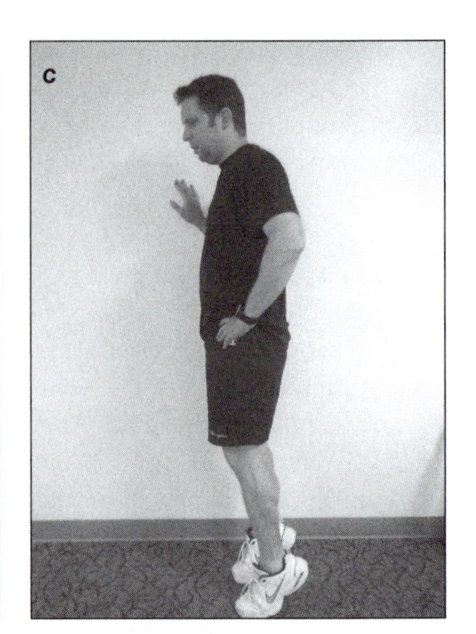

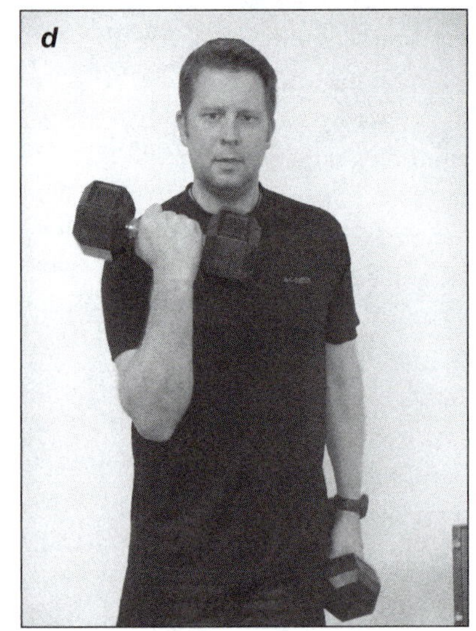

Figure 13.2 Early-phase rehabilitative exercises: (*a*) seated leg extension, (*b*) standing leg curl, (*c*) standing calf raise, (*d*) dumbbell curl, and (*e*) tricep push-down.

Photos courtesy of Mark A. Patterson.

patients start these exercises earlier in their recovery if sternal healing is going well and they are ambulatory:

- Seated leg extension
- Seated or standing leg curl (supine late phase only)
- Standing calf raise (without significant weight on shoulders)
- Dumbbell curl
- Tricep push-down

Midphase Rehabilitative Exercise (4 to 6 wk Postdischarge)

- Dumbbell bent-over row
- Seated row
- Lateral dumbbell raise
- Shoulder press
- Tricep kickback

Late-Phase Rehabilitative Exercise (6 wk or More Postdischarge)

- Lat pull-down
- Dumbbell bench press
- Dumbbell fly
- Front raise

The potential benefits of resistance training in the revascularized population include improving muscular strength and endurance and possibly attenuating the heart rate and blood pressure response to any given workload (lower workload on the heart). General resis-

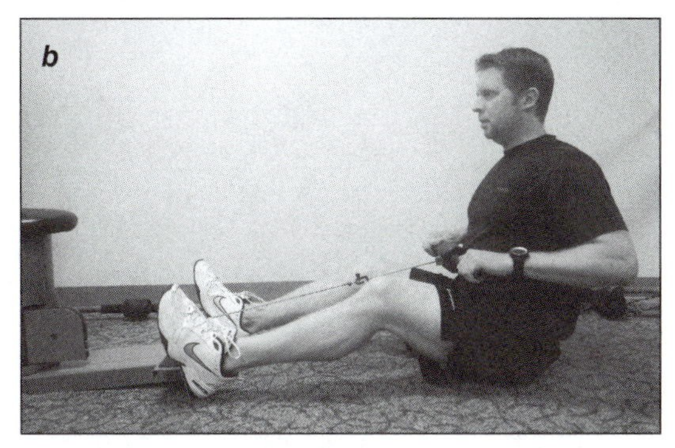

Figure 13.3 Midphase rehabilitative exercises: (*a*) dumbbell bent-over row, (*b*) seated row, (*c*) lateral dumbbell raise, (*d*) shoulder press, and (*e*) tricep kickback.

Photos courtesy of Mark A. Patterson.

tance training guidelines for cardiac rehabilitation are presented in "Patient's Guide for Resistance Training." Risk stratifying patients to determine eligibility is important.

Range of Motion

Each exercise session begins with a series of range of motion and flexibility exercises designed to maintain or improve the range of motion around joints and maintain or improve flexibility of major muscle groups (figure 13.5). The exercises begin in the standing position (or seated in a chair, if the person has difficulty standing) with the neck, progressing downward to the shoulders and trunk and eventually to the lower extremities. The final stretches for the improvement of posterior leg muscles and lower back flexibility may be performed on the floor, or in a chair for those with difficulty getting to the floor.

The following are examples of stretches for the revascularized patient. Patients who underwent an open-heart procedure should perform stretches with caution in the early stages after hospital discharge. Patients should do the stretches daily, and ideally multiple times a day, but should take each stretch only to the point where they may feel some mild tugging on the sternotomy site; the stretch should not be painful.

Head to shoulder

Arm circles

Lateral arm over head

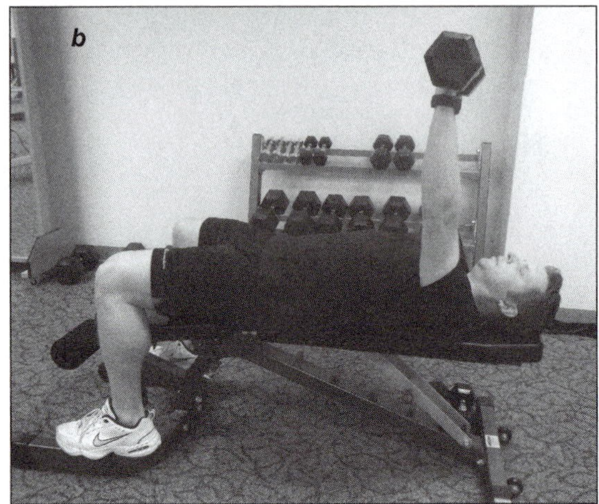

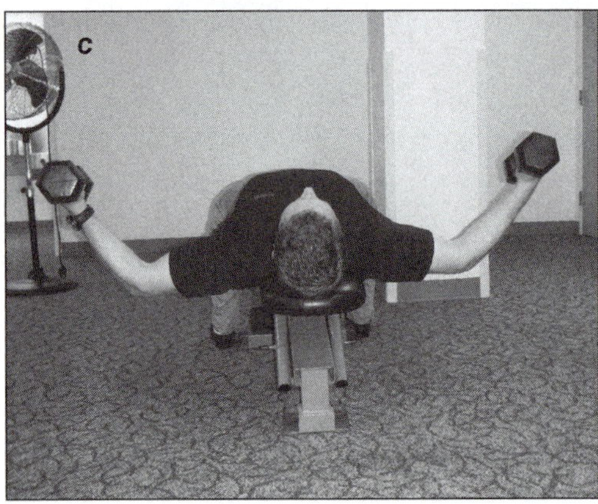

Figure 13.4 Late-phase rehabilitative exercises: (*a*) lat pull-down, (*b*) dumbbell bench press, (*c*) dumbbell fly, and (*d*) front raise.

Photos courtesy of Mark A. Patterson.

Shoulder shrug

Seated hamstring

Calf stretching

Table 13.4 outlines exercise prescription for revascularized patients.

CONCLUSION

Advances in coronary revascularization procedures and an aging population have led to a greater number of patients presenting for rehabilitation following CABS, PCTA, and stenting. In addition to exercise programming, risk factor modification is essential for prevention of recurrent events (19, 49). Furthermore, barring no new symptoms, the GXT, although a good prognostic tool, may better serve its purpose of assessing functional status if it is postponed until later in the rehabilitation program. Provided that no untoward events occur over the course of rehabilitation, CABS and PTCA patients usually outperform their MI counterparts, achieving greater fitness improvements at a faster rate.

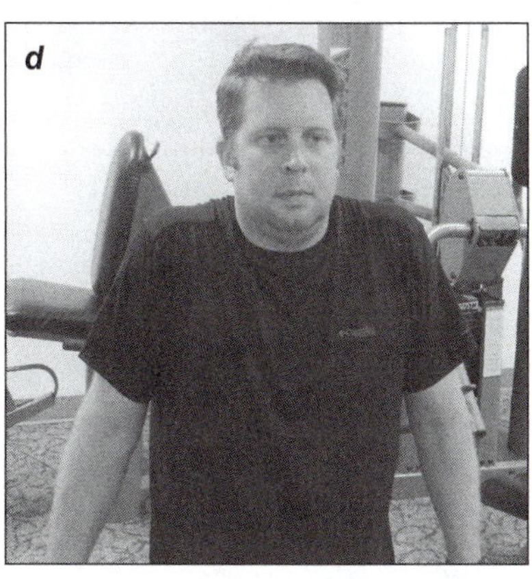

Figure 13.5 Stretching exercises: (*a*) head to shoulder, (*b*) arm circles, (*c*) lateral arm over head, (*d*) shoulder shrug, (*e*) seated hamstring, and (*f*) calf stretching.

Photos courtesy of Mark A. Patterson.

Table 13.4 Exercise Training Review

Cardiorespiratory endurance	Skeletal muscle strength	Skeletal muscle endurance	Flexibility and balance	Body composition
Increase $\dot{V}O_2max$, improve RPEs, HR, and BP at absolute submaximal levels; reduce possibility of multiple risk factors for further progression of CAD; also very important in overall CAD and all-cause mortality risk.	Maintain or increase muscular strength, improve ability to perform activities of daily living; help in return to work after hospitalization, improved self-esteem; key component in decreasing risk of injuries and fall risk in the elderly	Maintain or increase muscular endurance, improve ability to perform activities of daily living, assist in reduction of fatigue with repetitive activities	Maintain or increase flexibility; very important in proper recovery from open-heart procedures; help in return to work after hospitalization; key component in decreasing risk of injuries and fall risk in the elderly	If loss of weight and body fat is a goal, then program design should help to reduce body composition, BMI, and waist circumference

Patient's Guide for Resistance Training

Choose a weight initially that you can comfortably lift for 12 to 15 repetitions; as you progress over time, increase the weight accordingly so that your muscles will get significantly tired by 10 to 12 repetitions but you are not struggling to complete them.

Avoid tight gripping during pushing, pulling, and lifting exercises.

Do not hold your breath during the activity. Exhale during the exertion phase and avoid straining.

Perform two or three sets of each exercise and train three times per week.

Rest 30 to 45 s between sets.

Increase weight modestly (1-2 lb, or 0.5 to 1.0 kg) after you can easily perform 12 to 15 repetitions of a given weight.

Practical Application 13.3

RESEARCH FOCUS: EARLIER PROGRESSION AND FEWER RESTRICTIONS CONCERNING STRENGTH TRAINING

Guidelines for starting and progressing strength training after revascularization may occasionally be considered "overly restrictive," as many tasks of daily living demand lifting or pulling objects that exceed the suggested strength training weight limits. Clinicians may use clinical judgment to start strength training at earlier stages while taking into consideration the patient's prior conditioning and experience with strength training exercises, and with the knowledge that no absolute contraindications to strength training are present. The 2006 paper by Adams and colleagues (7) suggests earlier and relatively more aggressive use of strength training even for CABS patients when exercises are selected individually with use of their safety and efficacy of resistance exercise tool. This method puts various strength training exercises into categories by their potential to cause harm to the patient (such as damage to the sternum of the CABS patient). The investigators used a dynamometer to measure push and pull forces in many commonly performed activities. Patients with open-heart procedures were surveyed as to the advice given to them by physicians and surgeons about limits for lifting and strength training. Finally the researchers assessed the potential damage to surgical sites from resistance exercises by using kinesiologic analysis (anatomy, physiology, and movement patterns of each exercise). Figure 13.2 and its suggestions are based partly on the information provided from this study. The results suggested that the usual advice from physicians and surgeons, and even the recommendations from the American Association of Cardiovascular and Pulmonary Rehabilitation (AACVPR) and American College of Sports Medicine (ACSM), are vague and possibly overrestrictive—that they do not take into account which muscle groups patients could work safely without risking damage to sternal surgical sites. The advice to lift no more than 5 lb would actual preclude patients from opening a standard door to their physician's office. The study and the tool devised by Adams and colleagues suggest the use of a risk–benefit ratio to decide which exercises can be started sooner rather than later to help enhance the rehabilitation process and reduce the total decline in muscular strength and function after hospitalization.

Reference: Adams J, Cline MJ, Hubbard M, et al. A new paradigm for post-cardiac event resistance exercise guidelines. *Am J Cardiol* 2006;97:281-286.

Clinical Exercise Pearls

- CABS patients are very protective of their sternotomy sites and are sometimes reluctant to do range of motion and stretching exercises. Be very persistent in ensuring that they are working on these exercises to restore normal function to upper body movements as soon as possible to limit future issues with upper body strength and flexibility.
- Revascularized patients sometimes feel "cured" of their disease; the rehabilitation team needs to reinforce proper lifestyle and medical management to ensure continued lower risk of future cardiac events and hospitalizations.
- Many revascularized patients are on β-blockade medications, which makes exercise prescription by heart rate difficult if not impossible. It is better to use ratings of perceived exertion, spend time helping them "feel" how hard to be exercising when they are on their own, and help them get some practical experience in understanding the sensation of proper exercise intensity.
- While many revascularized patients struggle with depression, some are on the other end of the spectrum and feel better than they have in years; these patients may be too aggressive in progressing exercise.

Key Terms

angioplasty (p. 239)
anticoagulation therapy (p. 242)
cardiogenic shock (p. 242)
embolism (p. 241)
gastroepiploic artery (p. 240)
heart and lung bypass (p. 240)
intima (p. 245)
ischemia (p. 248)
platelet aggregation (p. 243)
reocclusion (p. 245)
restenosis (p. 245)
revascularization (p. 248)
self-efficacy (p. 241)
stent (p. 239)
sternotomy (p. 240)
Valsalva maneuver (p. 251)

CASE STUDY

MEDICAL HISTORY

Ms. RW is a 58 yr old white woman with no prior history of heart disease. She has hypertension under good control with medications and was diagnosed with diabetes 15 yr ago. Her last fasting blood sugar was 234, and her HbA1c was 8.7. She is a former smoker (quit 10 yr ago) and leads a rather sedentary lifestyle as a computer analyst for a large local corporation. Over the past 3 mo she has started to notice increased shortness of breath when climbing two flights of stairs at work; at the top of the stairs, she feels some moderate chest pressure that resolves in a couple of minutes after she sits down at her desk. Her primary care physician sends her for a routine exercise stress test.

EXERCISE TEST RESULTS

Exercise Stress Test

Resting ECG: appears normal

Heart rate: 65 beats · min^{-1}

Blood pressure: 138/92mmHg

Heart and lung sounds: within normal limits

(continued)

She exercises on a standard Bruce protocol. At 4:30 there is some horizontal ST depression, about 1 mm in inferior and lateral leads; by peak exercise (5:20) it is about 2 mm downsloping in the same leads, and she is developing the chest pressure she described in her symptom history. ECG changes resolve by 10 min of recovery, and symptoms resolve in about 5 min of recovery.

DIAGNOSIS

Principal diagnosis: Severe two-vessel CAD. Stenting was performed to 95% proximal left anterior descending coronary artery (LAD) lesion; Ms. RW's 75% distal left circumflex lesion was not a candidate for revascularization at the time of the procedure.
Medications: Metoprolol 50 mg twice daily, lisinopril 10 mg once daily, aspirin, Plavix, and simvastatin.

CARDIAC REHABILITATION

A referral for cardiac rehabilitation was placed by her cardiologist and was to start as soon as she was able to set it up after hospitalization.

EXERCISE PRESCRIPTION

Resting heart rate: 54 beats · min^{-1}

Resting blood pressure: 112/64mmHg

Initial exercise program:

Treadmill walking = 2.0 mph (3.2 kph), 0% grade for 10 min

Combined arm and leg ergometry = 100 W for 10 min

Upright stationary leg ergometry = 30 to 50 W for 10 min

Short circuit of resistance machines = one set of six exercises for 10 repetitions

Patient completed 6 wk in the program at the following workloads:

Treadmill walking = 2.7 mph (4.0 kph), 3% grade for 10 min

Combined arm and leg ergometry = 100 to 125 W for 10 min

Upright stationary leg ergometery = 100 to 125 W for 10 min

Rowing = 50 to 75 W for 10 min

Short circuit of resistance machines = two or three sets of eight exercises for 10 repetitions

Exercise heart rate: 100 to 110 beats · min^{-1}

Exercise rating of perceived exertion: 12 to 14

The remainder of program was uneventful. The patient completed a total of 12 wk from the start of the program and returned to her home exercise program and activities of daily living.

DISCUSSION QUESTIONS

1. What changes may have to be made to this patient's exercise regimen in view of the residual 75% blockage in her left circumflex coronary artery?
2. If she hits a plateau in her ability to increase her intensity of cardiovascular exercise that is not due to further complications with her heart, what issues may be limiting her ability to increase exercise intensity?

Chapter 14

Chronic Heart Failure

Steven J. Keteyian, PhD

With advancing age we are subject to a variety of disorders that disable us, impair our function and quality of life, and contribute to loss of exercise capacity. Perhaps chief among these is heart failure, the "final common pathway" for a multitude of cardiovascular disorders. Both the incidence and prevalence of heart failure are increasing, due in part to (a) the aging of our population (increasing percentage of people who reach greater than 65 yr) and (b) improved survival of middle-aged patients with cardiovascular disease due to advancements in pharmacotherapy, mechanical interventions (e.g., coronary artery stents, implantable cardioverter-defibrillators), and surgical techniques.

DEFINITION

Patients with **heart failure** (HF) suffer from a condition in which the heart is unable to pump blood at a rate sufficient to match the metabolic requirements of the body's organs and tissues.

The inability of the left ventricle (LV) to pump blood adequately can be due to a failure of either systolic or diastolic function. Patients with HF due to **systolic dysfunction**, or the inability of cardiac myofibrils to contract or shorten against a load, have a reduced **ejection fraction** (HFREF; heart failure due to reduced ejection fraction).

Alternatively, about one-half of patients with HF have normal or near-normal systolic contractile but still suffer from HF (HFPEF; heart failure with preserved ejection fraction). In these patients, the disorder is not associated with an inability of the heart to contract; instead it is attributable to an abnormal increase in resistance to filling of the LV—referred to as **diastolic dysfunction**. Think of diastolic dysfunction as involving a stiff or less compliant chamber that is partially unable to expand as blood flows in during diastole.

SCOPE

The public health burden associated with HF is immense and increasing. More than 5 million people are afflicted with the syndrome, and in people >45 yr of age approximately 670,000 new cases are identified each year. Because of the aging U.S. population and increased survival of patients with cardiovascular disorders, the prevalence of HF will increase over the next 10 yr.

Heart failure is a leading reason for hospitalizations in people 65 yr of age and older and directly or indirectly contributes to nearly 280,000 deaths annually. The 5 yr mortality rate for a person newly diagnosed with HF is ~50%. The economic burden imposed by HF in the United States is enormous, standing at almost $32 billion annually.

PATHOPHYSIOLOGY

Heart failure remains a final common denominator for many cardiovascular disorders. Although LV diastolic dysfunction can be an important cause of HF, most exercise research to date has focused on patients with HFREF. Thus this chapter primarily addresses this disorder. Note that in patients with HFREF, some degree of diastolic dysfunction is usually present as well.

Figure 14.1 depicts the complex abnormalities and changes that occur following loss of systolic function.

When cardiac cells (myocytes) die because of infarction, chronic alcohol use, longstanding hypertension, disorders of the cardiac valves, viral infections, or still yet unknown factors, diminished LV systolic function results. Among all cases of HFREF, approximately 60% are caused by **ischemic heart disease** (i.e., coronary atherosclerosis). For this reason, these patients are commonly referred to as having an ischemic versus a **nonischemic cardiomyopathy**, where a nonischemic cardiomyopathy refers to some other disease process having affected the heart muscle (e.g., viral cardiomyopathy or alcoholic cardiomyopathy).

As figure 14.1 also shows, a variety of physiological adaptations and compensatory changes occur in response to LV systolic dysfunction. Most of the current medical therapies used to treat patients with chronic HFREF are aimed at modifying one or more of these abnormalities.

Key characteristics particular to the pathophysiology of HFREF, HFPEF, or both include the following:

1. An ejection fraction that is reduced (systolic) or unchanged or slightly increased (diastolic) at rest
2. An increase in LV mass, with end-diastolic and end-systolic volumes that are increased (systolic failure) or decreased (diastolic failure)
3. Edema or fluid retention because of elevation of diastolic filling pressures or activation of the renin-angiotensin-aldosterone system, causing sodium retention
4. More commonly with systolic (vs. diastolic) HF, an imbalance of the autonomic nervous system such that parasympathetic activity is inhibited and sympathetic activity is increased

Additionally, abnormalities of other hormones and chemicals such as an increased release of brain naturetic peptide (BNP), diminished production of nitrous oxide (endothelium-derived relaxing factor), increased endothelin-1, and increased cytokines (e.g., tumor necrosis factor-alpha) all contribute to adverse cardiac and vascular remodeling or function and changes within and around the skeletal muscles.

Substantial clinical evidence now indicates that many of those factors contribute to the remodeling of the LV, reshaping it from a more elliptical form to a spherical form. This change in shape itself contributes to a further loss in LV systolic function. Currently, several of the treatment strategies used in patients with HFREF interrupt this process, referred to as reverse remodeling.

In patients with HFPEF, LV end-diastolic and -systolic volumes are generally reduced, and ejection fraction may be normal or increased. Table 14.1 describes normal LV

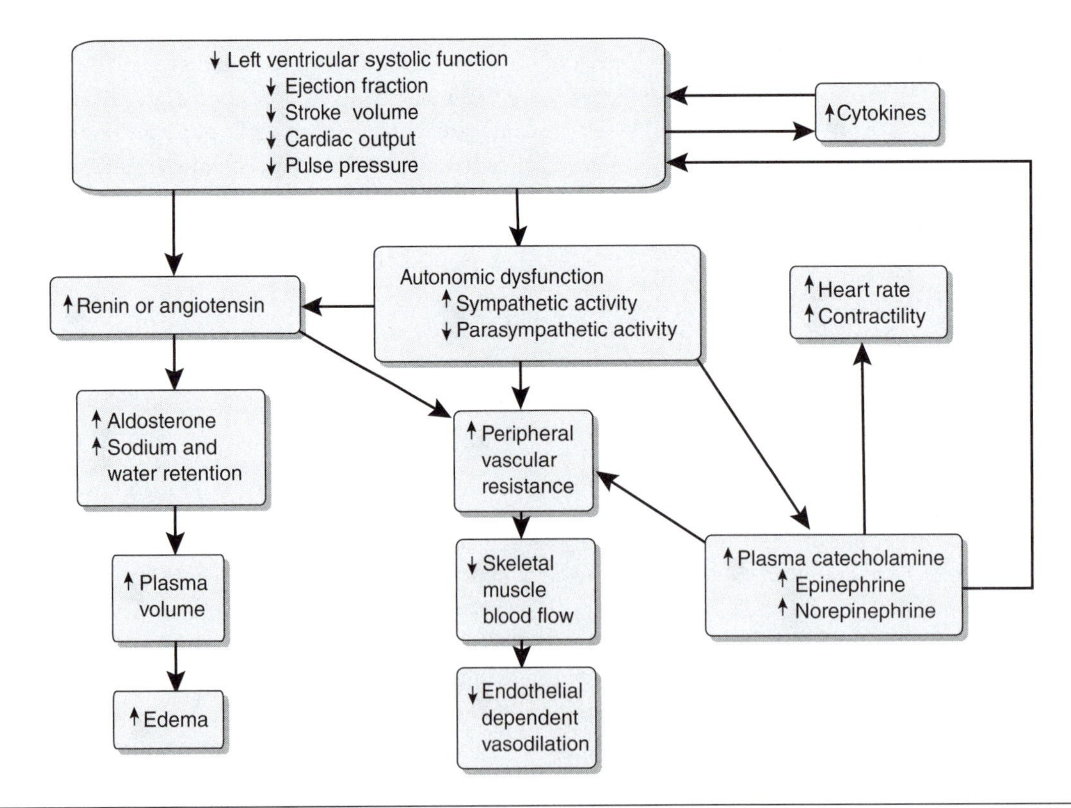

Figure 14.1 Schematic representation of some of the main physiological and pathophysiological adaptations that occur at rest in patients with heart failure attributable to left ventricular systolic dysfunction.

Table 14.1 Comparison of Typical Left Ventricle Characteristics

	End-diastolic volume (ml)	End-systolic volume (ml)	Stroke volume (ml)[a]	Ejection fraction (%)[b]
Normal	120	55	65	55
HFREF[c]	160	110	50	30
HFPEF[d]	85	35	50	60

[a]Stroke volume = end-diastolic volume – end-systolic volume.
[b]Ejection fraction = stroke volume/end-diastolic volume.
[c]HFREF = heart failure due to reduced ejection fraction.
[d]HFPEF = heart failure with preserved ejection fraction.
Courtesy of Henry Ford Health System.

characteristics and those associated with LV systolic and diastolic dysfunction.

MEDICAL AND CLINICAL CONSIDERATIONS

Before a patient with HF can be cleared for exercise rehabilitation, his past and current medical history should be reviewed and functional status evaluated. Exercise testing provides important information about the patient's functional status, but signs, symptoms, and medications must also be considered.

Signs and Symptoms

Clinically, patients with HF present with several key characteristics or findings, two of which are

1. exercise intolerance, as manifested by fatigue or shortness of breath on exertion; and
2. fluid retention, as evidenced by peripheral edema, recent weight gain, or both.

These signs and symptoms are often associated with complaints of difficulty sleeping flat or awakening suddenly during the night to "catch my breath." Sudden awakening caused by labored breathing is referred to as **paroxysmal nocturnal dyspnea**.

Labored or difficult breathing during exertion is called dyspnea on exertion (DOE), and difficulty breathing while lying supine or flat is referred to as **orthopnea**. Clinically, the severity of orthopnea is rated based on the number of pillows that are needed under a patient's head to prop her up sufficiently to relieve dyspnea. For example, three-pillow orthopnea means that a patient needs three pillows under the head and shoulders to breathe comfortably while recumbent.

History and Physical Examination

The signs and symptoms just discussed represent findings that the clinical exercise physiologist may need to evaluate to ensure that on any given day, the patient can safely exercise. For example, a patient's complaint of increased DOE or recent weight gain may or may not be clinically meaningful, but such signs warrant further inquiries by the clinical exercise physiologist to ascertain whether they are associated with new or increased peripheral (e.g., ankle) edema or fluid accumulation in the lungs. The severity of ankle edema is typically evaluated on a scale of 1 to 3, as discussed in chapter 4. Lung sounds, called rales, are associated with pulmonary congestion and are best heard using the diaphragm portion of a stethoscope. Rales appear as a crackling noise during inspiration.

In patients with HFREF, an abnormal third heart sound (S_3) can often be heard when the bell portion of the stethoscope is lightly placed on the chest wall over the apex of the heart. This S_3 sound occurs soon after S_2 and is most likely attributable to vibrations caused by the inability of the LV wall and chamber to accept incoming blood during the early, rapid stage of diastolic filling. Listening to the audio materials available in most medical libraries is an effective way to begin learning both normal and abnormal breath and heart sounds (see also www.blaufuss.org).

Diagnostic Testing

The diagnosis of HF attributable to LV systolic dysfunction, although based on the presence of signs and symptoms, also requires a reduced ejection fraction. Most often, ejection fraction is measured using an **echocardiogram** machine, although a radionuclide test or cardiac catheterization can assess this parameter as well. Normally, LV ejection fraction is greater than 55%, which means that slightly more than one-half of the blood in the LV at the end of diastole is ejected into the systemic circulation during systole. In patients with HFREF, LV ejection fraction is typically reduced below 45%. Severe LV dysfunction may be associated with an ejection fraction of 30% or lower. The decrease in ejection fraction is qualitatively proportional to the amount of myocardium that is no longer functional. Chronic HFREF is also usually associated with an enlarged LV.

Additionally, in patients with recent-onset HFREF caused by a large anteroseptal myocardial infarction, a marked increase in serum troponin is typically observed in the blood (see chapter 12). In these patients' electrocardiograms (ECGs), changes such as Q waves in leads V1 through V4 are usually evident. In most patients, a cardiac catheterization is performed to determine whether, and to what extent, ischemic heart disease contributed to the problem. Ischemic heart disease is the underlying cause for HF in about 60% of cases.

The diagnosis of HFPEF is somewhat less exact. Although an echocardiogram can be used to evaluate unique characteristics during diastole (i.e., left-sided filling pressures), quite often the diagnosis is made when the clinical syndrome of congestive HF (i.e., fatigue, dyspnea, and peripheral and pulmonary edema) requires hospitalization in the presence of a somewhat normal ejection fraction.

Although much of the diagnosis of HF relies on the use of echocardiography and patient symptoms, some laboratory tests, such as those measuring BNP levels in the plasma, may be used to help support or lend weight to the suspected diagnosis of HF. Synthesized in excess and released by the myocardium when the ventricles themselves are stretched by mechanical and pressure overload, BNP (>100 $pg \cdot ml^{-1}$) can be a sensitive index of decompensated HF even when other clinical signs are nebulous. Serum BNP can, for example, distinguish dyspnea from HF versus pulmonary disease, a particularly useful tool in patients whose physical examination is nonspecific or in people with poor echocardiographic images.

Despite their utility, neither echocardiography, BNP, nor other assessments of cardiac function quantify the full scope of HF as a disease that has both a cardiac and a systemic impact. But assessment of exercise capacity as a means to evaluate integrated physiologic function stands out as an important exception, in terms of maximal oxygen uptake measured during a cardiopulmonary exercise test (along with other parameters indicative of ventilatory and circulatory efficiency).

The routine evaluation of patients with HF now includes a graded exercise test with measured gas exchange to determine ventilatory efficiency and peak oxygen uptake ($\dot{V}O_2$peak) (figure 14.2). The 3 yr risk of death for patients who achieve a $\dot{V}O_2$peak greater than 17 $ml \cdot kg^{-1} \cdot min^{-1}$ is about 20%, clearly better than the 3 yr risk (approximately 50%) in patients who achieve a peak value less than 14 $ml \cdot kg^{-1} \cdot min^{-1}$ (figure 14.3) (62). Among patients with HFREF who are treated with β-adrenergic blocking agents, a peak $\dot{V}O_2$ value below 10 to 12 $ml \cdot kg^{-1} \cdot min^{-1}$ or below 55% of age-predicted $\dot{V}O_2$ are each independent, strong markers that can identify a marked increase in future 3 yr risk. Ventilatory efficiency, computed as the slope of the relationship of minute ventilation to carbon dioxide production ($\dot{V}_E$-$\dot{V}CO_2$) during exercise, is also an excellent predictor of future risk of death, with a value >30 indicating moderate increased risk and a value >45 indicating very high risk (3). Clearly, both measured $\dot{V}O_2$peak and $\dot{V}_E$-$\dot{V}CO_2$ can help determine which

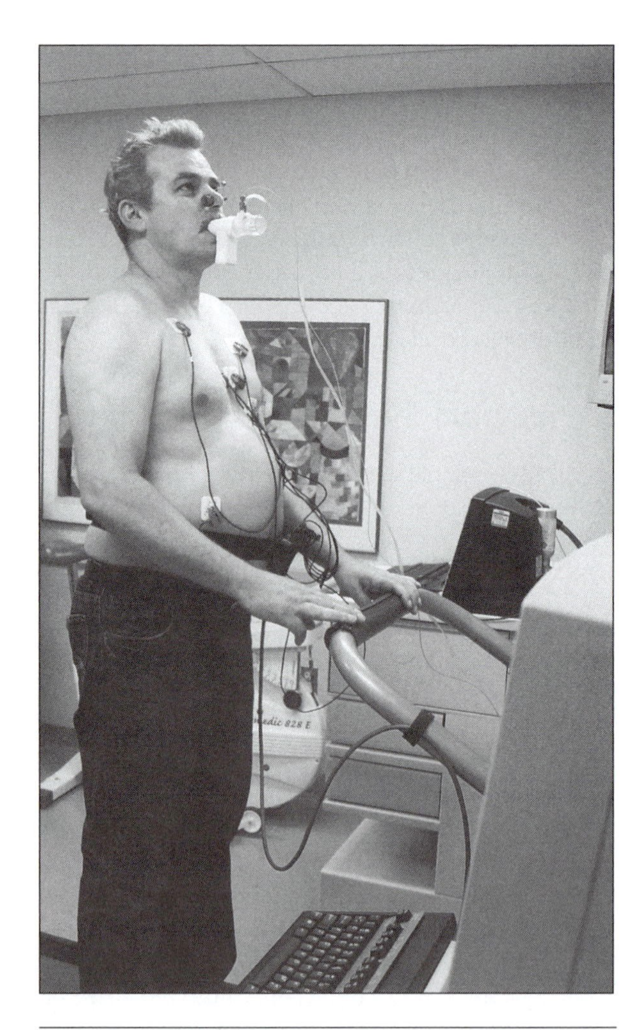

Figure 14.2 Example of a patient ready to undergo a cardiopulmonary exercise test, often used to estimate prognosis in patients with chronic heart failure.

Courtesy of Henry Ford Health System.

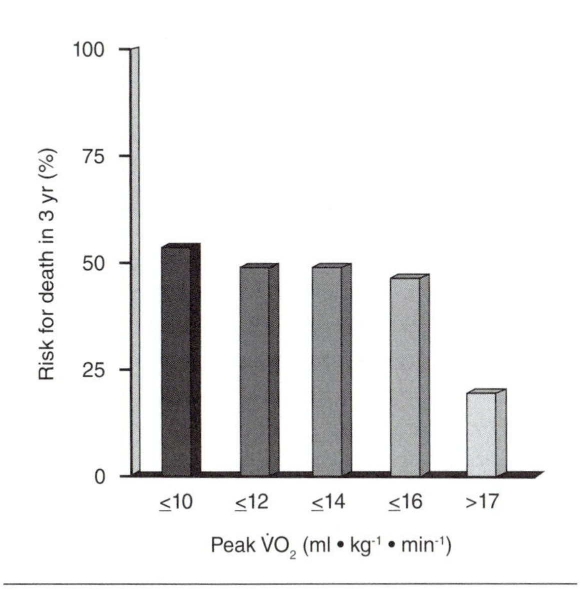

Figure 14.3 Three-year risk for death based on achieved peak $\dot{V}O_2$.

patients require aggressive medical therapy or possible cardiac transplant.

Although clinicians often use New York Heart Association (NYHA) functional class (see chapter 4 and table 14.2) to characterize a patient's clinical status, use of that system has limitations in that it does not fully reflect the breadth of the disorder. Table 14.2 shows the various stages in the development of HF as designated by the American College of Cardiology and American Heart Association (37). This staging system covers all patients with HF, regardless of whether symptoms are present. Note that those patients experiencing symptoms (New York Heart Association classes II-IV) fall within stage C and stage D alone.

Exercise Testing

The use of exercise testing in patients with HFREF provides an enormous amount of useful information. As mentioned previously, information is gathered not only on severity of illness and 3 yr survival but also on response to medications and response to an exercise training program. Information to guide exercise training intensity is obtained as well.

Medical Evaluation

The methods for exercise testing patients with HFREF differ little from the testing in patients with other types of heart disease. Although most exercise tests conducted in these patients use a steady-state (2-3 min per stage) protocol like the modified Bruce or Naughton (24), a ramp protocol can also be performed with a stationary cycle. With this method, external work rate is increased 10 to 15 W every minute (61). A ramp protocol generally results in less variable data during submaximal exercise because of the more gradual increments in work rate that it provides. $\dot{V}O_2$peak is approximately 10% to 15% lower when measured with cycle ergometry than with a treadmill (61, 64).

Because an accurate measure of functional capacity is needed in these patients, the use of prediction equations to estimate $\dot{V}O_2$ is discouraged because they tend to overpredict functional capacity (24). Measured exercise capacity using a cardiopulmonary cart (figure 14.2) is preferred, and such equipment is usually available in either the cardiac noninvasive or the pulmonary function laboratory of most hospitals. In addition to measuring $\dot{V}_E$-$\dot{V}CO_2$ and $\dot{V}O_2$peak, determining **ventilatory-derived lactate threshold (V-LT)** might also be helpful. An

Table 14.2 Stages in the Development of Heart Failure: ACC/AHA Guidelines (2009)

Stage	Description	Examples	New York Heart Association functional class
A (patient at risk)	High risk for heart failure; no anatomic or functional abnormalities; no signs or symptoms	Hypertension, coronary artery disease, diabetes, alcohol abuse, family history	
B (patient at risk)	Structural abnormalities associated with heart failure but no symptoms	Left ventricular hypertrophy, prior myocardial infarction, asymptomatic valvular disease, low ejection fraction	
C (heart failure present)	Current or prior signs and structural abnormalities	Left ventricular systolic dysfunction with or without dyspnea on exertion or fatigue, reduced exercise tolerance	II or III
D (heart failure present)	Advanced structural heart failure with symptoms at rest despite maximal medical therapy	Frequent hospitalizations, awaiting transplant, intravenous support	III or IV

adequate discussion of this parameter is provided elsewhere (24), but one approach often used when determining V-LT is the V-slope method (5).

Contraindications

Thirty years ago, standard teaching in most medical schools was that moderate or vigorous physical activity should be avoided or withheld in patients with HF. The increased hemodynamic stress that exercise places on an already weakened heart was thought to worsen heart function further. As a result, most guidelines listed HF as a contraindication to exercise testing. Today, however, patients with stable HF routinely undergo symptom-limited maximum cardiopulmonary exercise testing to evaluate cardiorespiratory function, and such testing has been shown to be safe (47).

All other contraindications to exercise testing still apply to patients with HF. Since arrhythmias are common in patients with HF, the person supervising the exercise test should review prior examination and testing reports to determine if arrhythmias have been previously reported in the person being tested. Note that the use of exercise testing to assess myocardial ischemia can be problematic because many patients with HF present with ECG findings at rest that invalidate or reduce the sensitivity of the test (e.g., left bundle branch block, left ventricular hypertrophy, and nonspecific ST-wave changes attributable to digoxin therapy).

Recommendations and Anticipated Responses

Compared with healthy normal people, patients with HFREF exhibit differences in their cardiorespiratory and peripheral responses at rest and during exercise (table 14.3) (49). Resting stroke volume and cardiac output are both lower in patients with HFREF versus controls (stroke volume: approximately 50 vs. approximately 75 ml · beat^{-1}, respectively; cardiac index: <2.5 vs. >2.5 L · min^{-1} · m^{-2}, respectively). Resting heart rate (HR) is increased (HF = 75-105 beats · min^{-1} vs. controls = 60-80 beats · min^{-1}) and systolic blood pressure may be reduced, attributable to both the underlying LV systolic dysfunction and the use of afterload-reducing agents such as angiotensin-converting enzyme inhibitors.

During submaximal exercise, patients with HF exhibit a higher HR, a lower stroke volume response, and attenuated increases in cardiac output and $\dot{V}O_2$ compared with persons without HF (11, 73, 79). To compensate, the extraction of oxygen (i.e., arterial–mixed venous oxygen difference) in exercising muscles is higher in patients with

Table 14.3 Resting and Exercise Characteristics of Patients With HFREF

	Resting	Submaximal exercise	Peak exercise
CARDIORESPIRATORY			
Ejection fraction	↓	↓	↓
Cardiac output	↓	↓	↓
Stroke volume	↓	↓	↓
Heart rate	↑	↑	↓
Oxygen consumption	↔	↓↔	↓
Arterial–mixed venous oxygen difference	↑↔	↑	↑↔
$\dot{V}O_2$ at ventilatory-derived lactate threshold	NA	↓	NA
PERIPHERAL			
Arterial blood lactate	↔	↑	↓
Total systemic vascular resistance	↑	↑	↑
Blood flow in active muscle	↔	↓	↓
Skeletal muscle mitochondrial density	↓	NA	NA
Skeletal muscle oxidative enzymes	↓	NA	NA
Reliance on anaerobic metabolism	NA	↑	↔

Note. ↑ = increased response compared with healthy subjects; ↓ = decreased response compared with healthy subjects; ↔ = similar responses compared with healthy subjects; NA = not applicable.

HF than in persons without HF. Also during exercise, plasma norepinephrine, an endogenous **catecholamine**, is released at increased levels by the sympathetic postganglionic fibers. A disproportionate increase occurs as well in plasma norepinephrine levels, and the magnitude of the increase is generally related to the severity of the illness (53).

Increasing blood flow to metabolically active skeletal muscles is part of a complex interplay between blood pressure and vasoconstriction–vasodilation response to exercise. Patients with HFREF have a reduced exercise-induced vasodilation, which is attributable to both increased plasma norepinephrine (53) and impaired endothelial function. The latter is the result of both the lesser release into the blood of local chemicals such as endothelium-derived relaxing factors (i.e., nitrous oxide; 15, 57) and the abnormal accumulation of inducible nitric oxide synthase within skeletal muscle cells that inhibit mitochondrial processes (1).

Compared with healthy normal persons, at peak exercise, patients with HFREF exhibit a lower power output (30%-40% decrease), lower cardiac output (40% decrease), lower stroke volume (50% decrease), and lower HR (20% decrease) (11, 12, 36, 73). Concerning exercise capacity as measured by peak $\dot{V}O_2$, this too is decreased compared to that in normal persons—30% to 35% or more in patients with HFREF or HFPEF (55). Depending on severity of illness, $\dot{V}O_2$peak typically ranges from 8 to 21 $ml \cdot kg^{-1} \cdot min^{-1}$. For patients not taking a β-adrenergic blocking agent, peak HR typically does not exceed 150 $beats \cdot min^{-1}$. In patients with cardiovascular risk factors or known coronary heart disease and taking a β-adrenergic blocking agent, achieving a peak HR that is ≤62% of age predicted (50) is referred to as **chronotropic incompetence**, a characteristic that occurs in 20% to 25% of patients with HF. In 1999 Robbins and coworkers (66) showed that, like $\dot{V}O_2$peak, chronotropic incompetence during exercise is a powerful and independent predictor of mortality in patients with HF.

Despite the diminished peak exercise capacity of these patients, the degree of exercise intolerance is not related to the magnitude of LV systolic dysfunction. This finding suggests that several other factors besides the impaired LV ejection fraction contribute to the exercise intolerance that these patients experience. Although this may seem a bit perplexing, it is in fact the case that abnormalities develop in tissues other than the damaged heart. Two such noncardiac mechanisms that limit exercise capacity are as follows:

1. An inability to dilate peripheral vasculature sufficiently as a means to increase blood flow to the metabolically active muscles (77)
2. Histological and biochemical abnormalities within the skeletal muscle itself (18, 59, 72).

In patients with HFREF, there is clear evidence of endothelial dysfunction, along with a decrease in the percentage of myosin heavy chain type 1 isoforms, diminished oxidative enzymes, and decreased capillary density. As a result, and in comparison with normal people, these patients rely more on anaerobic pathways to produce energy earlier during exercise.

Finally, participation in traditional aerobic-based exercise, including symptom-limited exercise testing, appears to be safe and well tolerated in patients with a left ventricular assist device (LVAD; 14, 38-40). The LVAD can increase cardiac output during exercise, and contemporary models are able to increase flow up to approximately 10 $L \cdot min^{-1}$ (38).

Treatment

Over the past 20 yr, great strides have been made in the medicines and devices used to treat patients with HF. These advances have led to fewer HF-related deaths, fewer symptoms, and increased exercise tolerance. In patients who are refractory to optimal medical therapy and who demonstrate a deteriorating clinical state consistent with a poor 1 yr survival rate, LVAD or cardiac transplant is a possible consideration.

This section summarizes the guidelines for the treatment of patients with HFREF (37). The current medical therapy for these patients includes angiotensin-converting enzyme (ACE) inhibitors, β-adrenergic receptor blockers, diuretics, and, possibly digoxin or an aldosterone antagonist. The possible influence of several of these agents on exercise responses is summarized in table 14.4.

ACE inhibitors reduce 5 yr mortality rate by approximately 30% among patients with HFREF. They also improve exercise tolerance, alter the rate of HF progression, influence structural remodeling, and decrease future hospitalizations. β-blockers also reduce morbidity and mortality rates by approximately 30%, as well as improving resting ejection fraction by approximately 7 percentage points. Congestion and fluid overload remain important complications in many patients with HF. To address this, diuretic therapy is commonly used in many patients; and in some, fluid intake may also be restricted to 2,000 ml or less per day. And the role that sodium restriction plays in minimizing fluid congestion cannot be overemphasized. Sodium restriction can help reduce the need for diuretics, and a dietary plan that aggressively restricts sodium intake may actually allow for the discontinuation of diuretic therapy.

Table 14.4 Types of Medications Commonly Used to Treat Heart Failure, Primary Clinical Effects, and Effects on the Exercise Response

Medication name and class	Primary effects	Exercise effects	Special considerations
Diuretics (e.g., furosemide)	↓ volume overload	↓ blood pressure	Attenuate blood pressure response during exercise and cause possible hypotensive response after exercise
Angiotensin-converting enzyme inhibitors (e.g., lisinopril)	↓ afterload and blood pressure at rest; improved survival	↓ blood pressure	Attenuate blood pressure response during exercise and cause possible hypotensive response after exercise
Angiotensin receptor blockers (e.g., losartan)	↓ afterload and blood pressure at rest	↓ blood pressure	Cause possible hypotensive response after exercise
β-adrenergic blockers (e.g., carvedilol)	↓ heart rate and blood pressure at rest; improved survival	↓ heart rate and possibly blood pressure	Improve exercise capacity in patients with angina
Digitalis	↓ hospitalization in patients with symptomatic heart failure; antiarrhythmic	↓ heart rate at rest and slight decrease, if any, during exercise	Can result in nonspecific ST-T wave changes in resting ECG (so-called dig effect), which can reduce the sensitivity of any ST-segment changes observed during an ECG stress test

↓ = decreased response compared with healthy subjects.

Digoxin may also be useful in patients with HFREF who are in sinus rhythm and can help reduce future hospitalizations. The use of digoxin in patients with HF and atrial fibrillation is also common. Other agents that may be used to treat patients with HFREF include antiplatelet and anticoagulation therapies, angiotensin receptor blockers, and an aldosterone antagonist.

For patients with NYHA class II through IV HF who also have electrocardiographic evidence of dyssynchronous contractions of the left and right ventricles (based on a QRS duration greater than 120 ms), a special type of pacemaker therapy (called cardiac resynchronization therapy) is commonly used. As the name implies, cardiac resynchronization therapy involves implanting pacemaker leads in both the right and left ventricles and pacing the ventricles to reestablish the correct firing pattern of one ventricle relative to the other. Resynchronization therapy often improves cardiac function and exercise tolerance and lowers future hospitalizations and deaths (37). Additionally, the benefits of implantable defibrillators, either by themselves or combined with cardiac resynchronization pacemakers, are well documented for patients who have dilated hearts with low ejection fractions, particularly as a means to reduce mortality that results from the common occurrence of ventricular arrhythmias.

In some instances, and despite aggressive attempts to optimize medical therapy, a patient's clinical condition continues to deteriorate such that special medications, an LVAD, or cardiac transplantation is required. Without these efforts, many such patients die in months, if not weeks.

LVADs now represent a standard therapy option either as a bridge to transplant or as destination therapy for patients who do not qualify for transplant, extending survival to 48% at 2 yr (68). LVADs provide circulatory or hemodynamic support to underperfused organs and reverse the pathophysiological sequelae of HF. In brief, the LVAD pump is implanted intra-abdominally or in a preperitoneal pocket external to the abdominal viscera. The LV is cannulated at the apex of the heart for inflow to the pump, which then sends blood into the ascending aorta distal to the aortic valve. The technology of LVADs has evolved dramatically over the past 20 yr, such that current-generation devices have smaller internal components, due to the use of continuous versus pulsatile-flow pumps, and a longer life span (29). Functional capacity and health-related quality of life are improved following LVAD implant, and in some patients the increases are comparable to the changes observed after heart transplantation (23, 66). In short, patients receiving an LVAD feel better and may be more willing to participate in regular exercise.

Cardiac transplantation is a surgical procedure that represents standard therapy for patients with end-stage

HF refractory to maximal medical therapy. Practical application 14.1 provides an overview of this procedure.

An important part of the care for any chronic disease remains secondary prevention. For the exercise professional working with patients with HF, this care includes counseling to help manage behavioral habits known to exacerbate the condition. For example, in patients with an ischemic cardiomyopathy, every attempt should be made to help stabilize existing coronary atherosclerosis through aggressive risk factor management and to prevent further loss of cardiac cells (myocytes) attributable to reinfarction.

For the patient with HF attributable to other causes such as alcohol abuse, the exercise physiologist should support healthy behaviors and be alert for signs of relapse that may require referral to a specialist. For all patients with HF, observing a low-sodium diet is an important step toward preventing congestion and fluid overload, thus reducing chances of subsequent hospitalizations. Consistent with this practice, compliance with all prescribed medications, especially diuretics aimed at removing excess fluid, is an important variable that an exercise professional can assess during clinical appointments or before exercise class. See practical application 14.2.

Practical Application 14.1

CARDIAC TRANSPLANTATION

Cardiac transplantation is an effective therapeutic alternative for persons with end-stage HF. Each year, approximately 3,000 transplants are performed worldwide, with 1 and 3 yr survival rates approximating 85% and 78%, respectively (71). In most patients undergoing cardiac transplant, the atria of the recipient's heart are attached at the level of the atria of the donor's heart.

After surgery, many patients with cardiac transplant continue to experience exercise intolerance. In fact, $\dot{V}O_2$peak is typically between 14 and 22 ml · kg^{-1} · min^{-1} among untrained patients. For this and other reasons, these patients are commonly enrolled in a home-based or supervised cardiac rehabilitation program as soon as 4 wk after surgery. The expected increase in $\dot{V}O_2$peak ranges between 15% and 25% (45, 56).

There is an increasing probability of developing accelerated atherosclerosis in the coronary arteries of the donor heart. For this and other reasons, the traditional risk factors for ischemic heart disease such as hypertension, obesity, diabetes, and hyperlipidemia are aggressively treated. Additionally, immune system–mediated rejection of the donor heart remains a constant concern for heart transplant recipients, and therefore most patients receive a variety of immunosuppressive medications. Common agents include cyclosporine, tacrolimus, mycophenolate mofetil, and prednisone.

Patients with cardiac transplant represent a unique physiology in that the donated heart they received is decentralized. This circumstance means that except for the parasympathetic, postganglionic nerve fibers that are left intact, other cardiac autonomic fibers are severed. Regeneration of these fibers can occur after 1 yr in some patients, but it is likely best to assume that decentralization, for the most part, is permanent.

Because of the decentralized myocardium, the cardiovascular response of cardiac transplant patients to a single bout of acute exercise differs from the response of normally innervated people. At rest, HR may be elevated to between 90 and 100 beats · min^{-1} because parasympathetic (i.e., vagal nerve) influence is no longer present on the sinoatrial node. When exercise begins, HR changes little because (a) no parasympathetic input is present to be withdrawn and (b) no sympathetic fibers are present to stimulate the heart directly. During later exercise, HR slowly increases because of an increase in norepinephrine in the blood. At peak exercise, HR is lower, and both cardiac output and stroke volume are approximately 25% lower than in age-matched controls.

Because of the absence of parasympathetic input, the decline of HR in recovery takes longer than normal. This effect is most observable during the first 2 min of recovery, during which HR may stay at or near the value achieved at peak exercise. Systolic blood pressure recovers in a generally normal fashion after exercise (20), which is why this measure can be used to assess adequacy of recovery.

The number of people needing a cardiac transplant each year far exceeds the available number of donors. Consequently, many patients die while awaiting a heart, or an increasing number receive an LVAD. The magnitude of this donor shortage will likely only increase in the future as the number of patients with end-stage HF continues to rise.

Practical Application 14.2

CLIENT–CLINICIAN INTERACTION

One of the important responsibilities for the exercise professional who finds himself working with a patient with HF is to ensure that on any given day, the patient is free of any signs or symptoms that might indicate the need to withhold exercise.

To accomplish this, you must not only persuade the patient to verbalize any problems that she might be having but also take the initiative to ask the patient key questions. Following are examples of these types of questions:

- How did you sleep over the weekend? Have you had any more bouts of waking up during the night short of breath? Are you still sleeping using just one pillow?
- How has your body weight been over the past three days, stable or increasing?
- It's been hot the past couple of days. Have you had any increased difficulty breathing?
- Do you ever get that increased swelling in your ankles anymore?

Each question enables you to assess change in HF-related symptoms, and with time and experience, you will develop a sense about when and how to assess these patients.

Also, patients with HF may not tolerate the first few days of their exercise rehabilitation well. Therefore, encouragement and guidance on your part are important. Explain that, over time, the patient can expect to be able to perform routine activities of daily living more comfortably. For some patients, achieving only this goal may be quite fulfilling. Other patients may wish to improve their exercise capacity to the point that they can resume activities they previously had to avoid. Although such a goal may be realistic, be sure to emphasize that the prudent approach is to advance their training volume in a progressive manner. Trying to do too much too soon may lead to disappointment if their functional improvement does not keep pace with their self-assigned interests.

Finally, when working with patients with HF, emphasize the importance of regular attendance at their exercise program. If they need to miss exercise for personal or medical reasons, let them know that you look forward to seeing them back when they feel better. Because you often see the same patient several times each week over several weeks, you are in a unique position to provide ongoing support, motivational counseling, and monitoring of medical compliance and symptom status over time. The issue of monitoring and intervening upon any lapse in prescribed medications is also important in that most of the medications a patient with HF takes have been prescribed because they have been proven to improve survival, lessen symptoms, prevent rehospitalizations, and improve quality of life. All these endpoints are key markers of a successful disease management plan.

EXERCISE PRESCRIPTION

For patients with HFREF, continuously monitoring exercise via ECG can be considered (2), with the idea that it may identify abnormal findings. In the HF-ACTION trial (Heart Failure: A Controlled Trial Investigating Outcomes of Exercise Training), such monitoring was not uniformly performed at all participating sites and there were very few exercise-related clinical events. After demonstrating for 1 to 3 wk that they can tolerate supervised exercise three times per week, patients can begin a one or two times per week home exercise program. As patients continue to improve and demonstrate no complications to exercise, they can transfer to an all home-based exercise program as needed.

Special Exercise Considerations

Despite the increased attention given to using moderate exercise training in the treatment plan of patients with HFREF, few HF-specific guidelines are needed relative for prescribing exercise in these patients. Prudent eligibility criteria for exercise training in patients with HF might be as follows:

- Ejection fraction less than 40%
- NYHA class II or III
- Receiving standard drug therapy of ACE inhibitors, β-adrenergic blockade, or diuretics for at least 6 wk
- Absence of any other cardiac or noncardiac problems that would limit participation in exercise

The ejection fraction cutoff of less than 40% was chosen for consistency with the majority of the exercise studies conducted in these patients to date. By no means is the intent to imply that patients with HF and an ejection fraction greater than 40% do not benefit from an exercise training regimen. On the contrary, patients with HFPEF seem to improve as well (54, 70). Likewise, besides enrolling stable NYHA class II and III patients in an exercise program, the exercise professional may be able to include carefully selected and motivated class IV patients who are free of pulmonary congestion. Our group previously reported a case detailing physiological outcomes in an ambulatory class IV patient who underwent 24 wk of exercise training while receiving continuous inotropic therapy (41). Initially, such patients may require more supervision during exercise. With respect to patients who received an LVAD or cardiac transplantation, it is possible that those <3 mo after surgery may still be extremely deconditioned and have unique medical and surgical concerns requiring additional attention from the rehabilitation staff.

Perhaps the most important exercise consideration for patients with HF is compliance. In the HF-ACTION trial, only ~40% of patients randomized to the exercise arm of the trial were exercising at or above the prescribed number of minutes per week (63). Additionally, using a program outcomes database, our group conducted an internal retrospective review of the records of all patients enrolled in the Henry Ford Hospital cardiac rehabilitation program over a 34 mo period. We observed that among patients who suffered a myocardial infarction, nearly 75% of both men and women completed the program. However, among patients with HF and no myocardial infarction, the percentages were lower; 71% of men and only 53% of women completed the program.

One reason related to the lower compliance in HF patients might be the fact that they often have comorbidities or other illnesses that interrupt or prevent regular program attendance. For example, arrhythmias, pneumonia, fluctuations in edema, adjustments in medications, and hospitalizations can all affect regular participation. Patients should be told to expect interruptions in their exercise therapy caused by both HF- and non-HF-related issues. Then, when they feel better, they should make every attempt to return to class.

Exercise Recommendations

Training for cardiorespiratory endurance is an obvious strategy for patients with HF, but muscular strength, muscular endurance, and flexibility training can also improve functional capacity and foster independence.

Cardiorespiratory Training

Unique issues pertinent to prescribing exercise in patients with HFREF are described next, including modality, intensity, duration, and frequency of exercise. Note that to date there are no large prospective exercise testing or training trials on the optimal exercise prescription methods in patients with LVAD.

- Mode: To improve cardiorespiratory fitness, the exercise professional should select activities that engage large muscle groups such as stationary cycling or walking. Benefits in exercise capacity or tolerance gained from either of these modalities transfer fairly well to routine activities of daily living. Consistent with this, upper body ergometry activities can improve function in the upper limbs as well.

With respect to comparing one modality to another, a review of the exercise training trials involving patients with HFREF indicates that similar increases in exercise capacity occur with walking and cycle exercise (6, 7, 10, 16, 17, 22, 31-33, 43, 48, 51, 78) (table 14.5).

- Frequency: Similar to people without HF, those with HF need to engage in a regular exercise regimen of four or five times per week. As already mentioned, patients should be counseled to remain regular with their exercise habits and, if their program is interrupted because of personal or medical reasons, to plan on restarting the program as soon as possible. After being in an exercise program for several weeks, patients often comment that they notice less DOE and less fatigue during routine daily activities. Part of exercise prescription involves educating patients about the detraining effect, which means informing them that many of the benefits they notice will fade if they stop exercising.

- Intensity: Exercise intensity can be established using variables derived from an exercise test; and in patients with HF, different ranges of exercise intensity have been used to improve cardiorespiratory fitness. If intensity of effort is set between 50% and 80% of heart rate reserve, the gains in cardiorespiratory fitness (i.e., $\dot{V}O_2$peak) typically range between 15% and 30%. Several centers have also used intermittent aerobic interval treadmill training (vs. continuous moderate-intensity training) to improve exercise capacity and have shown promising results. This method involves training intensities as high as 90% to 95% of peak HR and yields an even greater increase in $\dot{V}O_2$peak (78). However, the safety and clinical efficacy of this training method have not been fully described and therefore require additional investigation.

Table 14.5 Reported Changes in Peak $\dot{V}O_2$ in Aerobically Trained Subjects and Control Subjects From Four Single-Site Clinical Exercise Trials Involving Patients With Heart Failure

Authors	Training intensity	Training mode	Peak $\dot{V}O_2$ (ml · kg^{-1} · min^{-1})
Erbs et al. (21)	50% to 60% of peak $\dot{V}O_2$	Bike ergometer, walking	T: +2.0 C: −0.7
Feiereisen et al. (22)	60% to 75% of peak $\dot{V}O_2$	Bike ergometer	T: +1.6 C: −0.6
Passino et al. (64)	65% of peak $\dot{V}O_2$	Bike ergometer	T: +2.0 C: −1.0
Wisloff et al. (78)	70% to 75% of peak heart rate	Treadmill walking	T: +1.9 C: +0.2

Note. In all four trials, more than 70% of subjects were taking a β-adrenergic blocking agent and an angiotensin-converting enzyme inhibitor. T = change after training in treatment group; C = change after control period in control group.

Using the HR reserve method, the first few exercise sessions should be guided at an exercise intensity of about 60% and then progress to 70% to 80% of heart rate reserve, as tolerated. The exercise professional can also have patients titrate their training intensity using the rating of perceived exertion scale set at 11 to 14. One reason to start these patients at this lower intensity level is to allow them an opportunity to adjust to the exercise. Fatigue later in the day is not uncommon, and restricting exercise intensity at first may help minimize this effect.

For most patients with an LVAD, it is prudent to set the initial exercise training workload at or below 2 METs (metabolic equivalents), with increases guided to a perceived exertion of 11 to 13. Factors that can influence the overall functional capacity of a patient and initial workloads chosen for rehabilitation include time from LVAD implantation, current activity habits, comorbidities, and age.

- Duration: Most exercise trials involving patients with HF increased training duration to up to 30 to 60 min of continuous exercise. On occasion, it may be necessary to start an individual patient with two or three bouts of exercise that are each 4 to 8 min in duration.

Resistance Training

Patients with stable stage B or C (table 14.2) HF who have demonstrated that they can tolerate aerobic exercise training, which should be evident within 3 wk, also represent likely candidates for participation in a resistance training program. To date, studies show that mild to moderate resistance training is safe and improves muscle strength 20% to 45% in stable patients with HF. In 2007, Feiereisen and coworkers (22) compared standard moderate-intensity aerobic endurance-type training, strength training, and combined strength and aerobic training before and after 40 exercise sessions and observed, regardless of training type, improvements for all three groups in peak oxygen uptake, thigh muscle volume, and knee extensor endurance. None of the three regimens appeared to be superior to the other two. Given that loss of both muscle strength and endurance is common in these patients (60, 75), it is appropriate to assume that improvements in these measures through resistance training are helpful.

Because no specific guidelines address resistance training in patients with HF, resistance training recommendations in these patients are often drawn from a scientific advisory statement published in 2007 (76). Generally, exercise or lift intensity should start at 40% (for upper body) to 50% (for lower body) of one-repetition maximum (1RM) and then be progressively increased over several weeks to 60% to 70% of 1RM, with the patient starting with one set of 10 to 12 repetitions and progressing to one set of 12 to 15 repetitions. The specific exercises should address the individual needs of the patient but will likely include all the major muscle groups. As patients improve, load should be increased 5% to 10%.

For patients with an LVAD, the safety and efficacy of resistance training is an area void of definitive research. Similar to the situation with other patients, who have undergone a thoracotomy, rehabilitation staff should wait >12 wk after device implant before beginning patients on a light resistance program. Other considerations include avoiding any exercises that may dramatically increase intra-abdominal pressure (e.g., sit-ups) or have the potential for physical trauma (i.e., contact sports); the latter can cause a fracture of the driveline or create damage to the LVAD itself.

Summaries of the exercise prescriptions used for patients with HF and for patients with cardiac transplant are presented in practical applications 14.3 and 14.4.

Practical Application 14.3

SUMMARY OF EXERCISE RECOMMENDATIONS FOR PATIENTS WITH HEART FAILURE

Before participation in a regular exercise program, a limited evaluation should be performed to identify acute signs or symptoms that may have developed between when the patient was referred for exercise by his doctor and program participation. Elements to screen for might include an increase in body mass greater than 1.5 kg during the previous 3 to 5 d, complaints of increased difficulty sleeping while lying flat, sudden awakening during the night because of labored breathing, or increased swelling in the ankles or legs. Additionally, a graded exercise test is usually warranted both to identify any important exercise-induced arrhythmias and to develop a target HR range that allows determination of a safe exercise intensity.

Exercise Program

As tolerated, type of exercise, frequency, duration, and intensity of activity should be progressively adjusted to attain an exercise energy expenditure of 700 to 1,000 kcal/wk. The table provides a summary of the aerobic, resistance training, and range of motion recommendations.

Training method	Mode	Intensity	Frequency	Duration	Progression	Goals
Aerobic	Cycle or treadmill; arm ergometer as needed	60% to 75% of heart rate reserve or 11 to 14 on rating of perceived exertion scale	Four or five times per week	40 min or more per session; use interval training method as needed	10 min or more per session up to 40 min or more per session	Improve submaximal and peak endurance
Resistance	Fixed machines, free weights, and bands; six to eight regional exercises	Begin with 40% of 1RM for upper body and 50% of 1RM for lower body; progress both over time to 70% of 1RM	One or two times per week	One set of 12 to 15 repetitions for each of the involved muscle groups	Increase 5% to 10% as tolerated	Initial focus should be on increasing leg muscle strength and endurance
Range of motion	Static stretching		Before and after each aerobic or resistance training workout	5 min before and 5 to 10 min after each workout, with 10 to 30 s devoted to the major muscle groups and joints		

Special Considerations

- Many patients with HF are inactive and possess a low tolerance for activity. Exercise should be progressively increased in an individualized manner for every patient.
- For the first week or so of exercise training, patients may be a bit tired later in the day. To compensate for this, they should temporarily limit the amount of other home-based activities.

Practical Application 14.4

SUMMARY OF EXERCISE RECOMMENDATIONS FOR PATIENTS WITH CARDIAC TRANSPLANT

Because patients with cardiac transplant present with a decentralized heart, use of a graded exercise test within 12 mo after surgery may not be helpful if the test is being performed for the purpose of developing an exercise training HR range. Such a test, however, will help screen for exercise-induced arrhythmias, quantify exercise tolerance, and serve as a marker to assess exercise outcomes at a future date.

Exercise Program

The exercise professional should modulate type, frequency, duration, and intensity of activity so that the patient attains an exercise energy expenditure of 700 to 1,000 kcal/wk, as tolerated. The table provides a summary of the aerobic, resistance training, and range of motion recommendations.

Training method	Mode	Intensity	Frequency	Duration	Progression	Goals
Aerobic	Cycle or treadmill; arm ergometer as needed	Use 11 to 14 on rating of perceived exertion scale	Four or five times per week	40 min or more per session	10 min or more per session up to 40 min or more per session	Improve submaximal and peak endurance
Resistance	Fixed machines, free weights, and bands; six to eight regional exercises	Begin with 40% of 1RM for upper body and 50% of 1RM for lower body; progress both over time to 70% of 1RM	One or two times per week	One set of 12 to 15 repetitions for each of the involved muscle groups	Increase 5% to 10% as tolerated	Initial focus should be on increasing leg muscle strength and endurance; additional emphasis should then be placed on balanced program aimed at preventing corticosteroid-induced sarcopenia and decrease in bone mineral content
Range of motion	Static stretching		Before and after each aerobic or resistance training workout	5 min before and 5 to 10 min after each workout, with 10 to 30 s devoted to the major muscle groups and joints		

Special Considerations

- Loss of muscle mass and bone mineral content occurs in patients on long-term corticosteroids such as prednisone. As a result, resistance training programs play an important role in partially restoring and maintaining bone health and muscle strength (8). But high-intensity resistance training should be avoided because of increased risk of fracture in bone that has compromised bone mineral content.

- Although isolated cases of chest pain or angina associated with accelerated graft atherosclerosis have been reported, decentralization of the myocardium essentially eliminates angina symptoms in most patients.
- A regular exercise program performed within the first few months after surgery may result in an exercise HR during training that is equal to or exceeds the peak HR achieved during an exercise test taken before a patient started training. This response is not uncommon and supports using rating of perceived exertion as the method to guide intensity of effort.
- Marked increases in body fat leading to obesity sometimes occur in cardiac transplant patients. This increase is likely caused by both long-term corticosteroid use and restoration of appetite following illness. Exercise modalities chosen for training should take into account any possible limitation caused by excessive body mass.
- Because of the sternotomy, postoperative range of motion in the thorax and upper limbs may be limited for several weeks.

EXERCISE TRAINING

Improvements with exercise training in patients with HFREF can be seen not only in clinical and cardiorespiratory measures (practical application 14.5) but also in the skeletal muscle and other organ systems. Figure 14.4 provides a summary of these adaptations. Conversely, relatively little information is available regarding the physiologic effects and safety of exercise training in patients with HFPEF. Thus the material in this section focuses predominantly on exercise training responses in patients with HFREF.

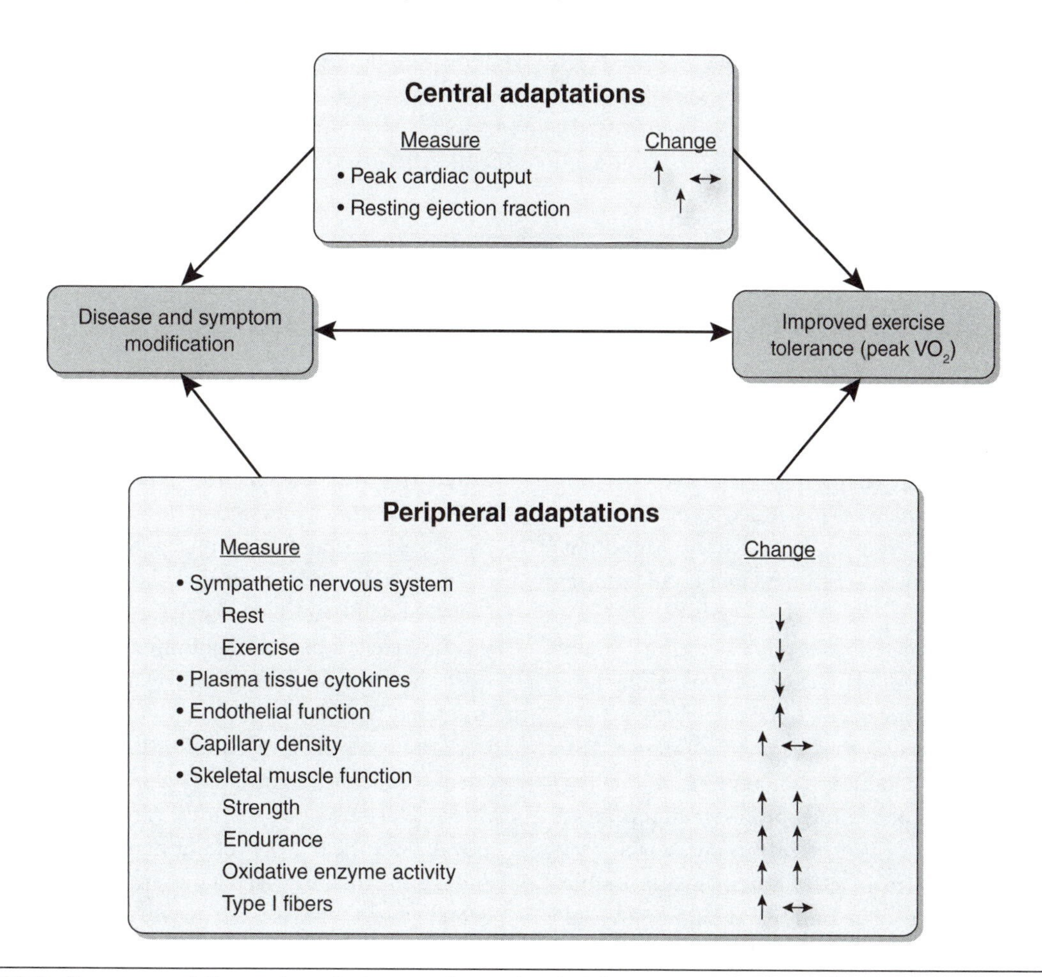

Figure 14.4 Summary of the various factors and exercise training responses that influence both disease-related symptoms and exercise tolerance

Practical Application 14.5

LITERATURE REVIEW

Although exercise is an important component in the treatment of HF today, not until the late 1980s and early 1990s did sufficient research show that patients could safely derive benefit (42, 49). Since that time, dozens of single-site randomized controlled trials have shown that exercise training safely results in a 15% to 30% increase in $\dot{V}O_2$peak (42, 46). Among these trials, in those in which at least 70% of patients were taking both an ACE inhibitor and a β-adrenergic blocking agent, the median increase was ~2 ml · kg^{-1} · min^{-1} in exercise-trained patients and unchanged in controls. In the HF-ACTION trial, the median increase in $\dot{V}O_2$peak was only 0.6 ml · kg^{-1} · min^{-1} after 3 m of training, with a 0.2 ml · kg^{-1} · min^{-1} increase observed in the usual-care group (63). One reason for the smaller than expected increase in $\dot{V}O_2$peak among patients in the exercise arm of the trial might be the aforementioned challenges with exercise compliance. Although the majority of this chapter focuses on patients with HFREF, patients with HFPEF also suffer from exercise intolerance; three single-site trials showed that exercise training improves $\dot{V}O_2$peak, as well as quality of life, depression, and physical function in these patients (25, 54, 70).

Presently, evidence-based guidelines for patients with HFREF include regular exercise training as a standard recommendation (37). In addition to partially reversing many of the physiologic abnormalities that accompany HF (42), exercise training has a favorable impact on many important clinical outcomes. The HF-ACTION trial compared the effects of exercise training plus usual care with usual care alone and showed an exercise training-related 11% reduction in the adjusted risk (compared to usual care alone) for all-cause mortality or hospitalization and a 15% reduction in the adjusted risk for cardiovascular mortality or HF hospitalization (63). A meta-analysis by Davies and colleagues in 2010 showed no effect on mortality alone due to exercise in patients with HFREF but did show a 28% decrease in HF hospitalization (13). Finally, Hammill and coworkers (34) reported that, among elderly Medicare beneficiaries with a diagnosis of HF, mortality and myocardial infarction were reduced by approximately 19% and 18%, respectively, over 4 yr in patients who attended 36 cardiac rehabilitation sessions versus those who attended 12 sessions.

Myocardial Function

Exercise training results in no change or a modest increase in peak cardiac output, with any such changes likely predominantly mediated through an increase in both peak HR (vs. stroke volume) (19). Given the current medication regimen recommended for patients with HFREF, the exact nature of adaptations in cardiac output, HR, and stroke volume can be problematic and requires further investigation.

Despite initial concerns that exercise training might further worsen myocardial function, several studies have shown that this does not occur. On the contrary, a meta-analysis by Haykowsky and colleagues (35) showed that aerobic-type training induced mild reverse remodeling, evidenced by a 2.6 percentage point weighted mean increase in resting ejection fraction and a weighted mean reduction of 12 ml in left ventricular end-diastolic volume. Using higher-intensity interval training, Wisloff and colleagues (78) showed a 45 ml reduction in left ventricular end-diastolic volume and a 10 percentage point increase in ejection fraction.

Skeletal Muscle

In response to exercise training, improvements occur in the skeletal muscle. These include improved ability to dilate the small blood vessels that nourish the metabolically active muscles, which leads to an approximate 25% to 30% increase in local blood flow (32, 33). The mechanisms responsible for the increase include an exercise training-induced decrease in plasma norepinephrine (10, 43) and other vasoconstrictor agents (9), as well as increased levels of endothelium-derived relaxing factor (30) and flow-mediated improvement in endothelial function (21, 58).

Although only a few studies have investigated the effects of exercise training on intrinsic characteristics of the skeletal muscle, a benefit appears to occur at this level as well. Specifically, the volume density of mitochondria and the enzymes involved with aerobic metabolism (e.g., cytochrome C oxidase) are improved up to 40% (33). In addition, a mild shift in fiber type occurs, involving a small increase in myosin heavy chain type I fibers (31, 44). Finally, Barnard and colleagues (4) showed a 31%

increase in single-repetition leg extension strength following 8 wk of combined aerobic and resistance training, compared with no change in a group of HF patients who performed aerobic training only.

Other Organ Systems

As mentioned in the section on physiology and pathophysiology in this chapter, patients with HFREF also have abnormalities of the autonomic nervous system, such as an increase in sympathetic activity and a decrease in parasympathetic activity. This imbalance likely makes the heart more prone to arrhythmia and less responsive to stressful situations.

Several studies show that exercise training partially restores autonomic, immune, and hormonal function in patients with HF (9, 10, 52, 65, 74). Down regulation of the sympathetic nervous system occurs (69), and parasympathetic activity increases. The levels of circulating norepinephrine in the blood both at rest and during exercise decrease with training (10, 43). Finally, Gielen and associates showed the anti-inflammatory effects associated with exercise training (26, 27).

CONCLUSION

Heart failure, today's fastest-growing cardiac-related diagnosis, represents an immense health burden. Current research establishes that among eligible patients with stable HF, regular exercise training improves exercise tolerance and quality of life and moderately reduces the risk for all-cause death or hospitalization and cardiovascular death or HF hospitalization. Health care practitioners involved in the care of such patients should include regular exercise training when developing their treatment strategies (37).

Key Terms

catecholamine (p. 265)
chronotropic incompetence (p. 265)
diastolic dysfunction (p. 259)
echocardiogram (p. 261)
ejection fraction (p. 259)
heart failure (p. 259)
ischemic heart disease (p. 260)
nonischemic cardiomyopathy (p. 260)
orthopnea (p. 261)
paroxysmal nocturnal dyspnea (p. 261)
systolic dysfunction (p. 259)
ventilatory-derived lactate threshold (V-LT) (p. 263)

CASE STUDY

MEDICAL HISTORY

Mr. WT is a 55 yr old African American male who was initially hospitalized in 1996. The patient stated that he had no history of diabetes, angina, or myocardial infarction. He did have a positive history for cocaine use, as recently as 1 wk before hospitalization. He had a strong family history of ischemic heart disease and denied alcohol and tobacco use. He worked as a computer programmer and was sedentary during leisure.

DIAGNOSIS

At the time of admission, Mr. WT complained of being unable to lie flat without being short of breath and said that the condition had worsened over the previous several days. Cardiac catheterization revealed three-vessel coronary artery disease, which included a 20% left main lesion, a 95% stenosis in the left anterior descending artery, total occlusion of the circumflex, and a 70% narrowing in the proximal right coronary artery. An echocardiogram performed during the same admission demonstrated a left ventricular ejection fraction of 30%. Diagnosis was ischemic cardiomyopathy. The patient was discharged from the hospital and scheduled to see a cardiologist in the outpatient clinic later in the week. Medications at the time of hospital discharge were lisinopril, simvastatin, lopressor, digoxin, and furosemide.

(continued)

The patient was seen in the outpatient clinic 6 d after discharge from the hospital. Blood lipids were obtained, revealing total cholesterol of 227 mg · dl^{-1}, high-density lipoprotein cholesterol of 27 mg · dl^{-1}, and low-density lipoprotein cholesterol of 172 mg · dl^{-1}. A dobutamine echocardiogram showed improved contractility. Based on this test and the cardiac catheterization results, coronary artery bypass surgery was recommended. The patient refused surgery, opting for medical management, lifestyle changes, and cardiac rehabilitation.

EXERCISE TEST RESULTS

Before the patient enrolled in cardiac rehabilitation, a graded exercise test was completed. Resting blood pressure was 138/96 mmHg. Resting ECG showed sinus rhythm with a rate of 72 beats · min^{-1}. Heart rate during seated rest was 74 beats · min^{-1}. Occasional ventricular premature beats were noted at rest, along with left ventricular hypertrophy. An old anterior-lateral infarction pattern was present, age indeterminate.

The patient exercised for 6 min on a treadmill to a peak $\dot{V}O_2$ of 17.2 mL · kg^{-1} · min^{-1}. Chest pain was denied and exercise was stopped because of fatigue and dyspnea. Peak HR was 123 beats · min^{-1}, peak blood pressure was 158/98 mmHg, and 1.0 mm additional ST-segment depression was observed in lead V6. Isolated ventricular premature beats were again observed.

EXERCISE PRESCRIPTION

Mr. WT's goal was to make important and aggressive changes in his lifestyle, including increasing his activity levels from being inactive to exercising 4 or 5 d/wk. During rehabilitation he was able to tolerate 30 min of exercise without complication. The training HR range in this patient that corresponds to 60% to 70% of heart rate reserve is 103 to 108 beats · min^{-1}.

DISCUSSION QUESTIONS

1. Given Mr. WT's history of being inactive during leisure, what is the likelihood that he will be following a regular exercise program 3 yr in the future? What steps can you take now to ensure long-term compliance?
2. Which symptoms did Mr. WT complain of at the time of hospitalization that were consistent with the diagnosis of cardiomyopathy or heart failure?
3. What target levels would you recommend that Mr. WT achieve for his blood lipid levels?
4. Based on data from Mr. WT's exercise test, did he show chronotropic incompetence? Is this a common or an uncommon finding in patients with HF?
5. Given the $\dot{V}O_2$peak measured during his exercise test, what magnitude of improvement, if any, would you expect after 12 wk of exercise training? How would this compare with the improvement that would occur in a sedentary, apparently healthy person who also undergoes 12 wk of training?

Chapter 15

Peripheral Artery Disease

Ryan J. Mays, PhD
Ivan P. Casserly, MB, BCh
Judith G. Regensteiner, PhD

Peripheral artery disease (PAD) occurs as the result of the development of atherosclerotic plaque in the intimal layer of the major arteries of the legs. Peripheral artery disease reduces patient-reported outcomes as well as functional capacity, increases the risk of cardiovascular events, and is associated with an increased risk of mortality (35, 79, 139, 168). This chapter focuses on the diagnosis, assessment, and treatment of PAD.

DEFINITION

Peripheral artery disease refers to the blockage of the leg arteries by plaque, leading to gradual narrowing of the arteries in the lower extremities. The resulting **stenoses** or **occlusions** (or both) lead to decreased blood flow to the muscles of the leg.

SCOPE

An estimated 8 to 12 million adults have PAD in the United States alone, with worldwide prevalence estimated at 3% to 10% (27, 79, 161). Approximately 35% to 40% of PAD patients have **intermittent claudication** (IC), which is characterized by pain, cramping, or aching in the calves, thighs, or buttocks, while 1% to 2% have the most severe manifestation of the disease, **critical limb ischemia** (CLI) (126, 172). The majority of patients with PAD are asymptomatic or have atypical symptoms. According to the Trans-Atlantic Inter-Society Consensus II international guidelines, the prevalence of PAD with IC is approximately 2% to 7% in patients 50 to 70 yr of age (126). Patients with PAD have a high burden of cardiac and cerebrovascular disease (e.g., myocardial infarction, stroke), with an annual event rate of approximately 5% to 7% (126). Additionally, PAD constitutes a major burden to national health care expenditures, regardless of the stage of disease. In a review of Medicare costs in 2001, the treatment for PAD was estimated at $4.37 billion annually, with 88% of the expenditures for inpatient care and 6.8% of the elderly Medicare population being treated for PAD (80). In addition to the economic burden of PAD, the disease can lead to other indirect costs such as absence from work and loss of productivity.

Peripheral artery disease has an overall prevalence rate slightly higher in men than in women (126). However, recently it has been suggested that the greater prevalence with increasing age may be more significant in women. Thus in the older patient groups, PAD may be more common in women than men (173). Race and ethnicity play a role in the prevalence of PAD (4, 126). For instance, non-Hispanic blacks have a 2.9% higher prevalence of PAD than non-Hispanic whites (28). European and Asian population prevalence rates of PAD are generally lower than in other ethnicities; and in Native Americans, PAD is estimated to have a prevalence of ~6.0%, with similar prevalence rates among tribal groups (47, 93, 101).

Peripheral artery disease is often accompanied by multiple comorbidities in addition to cardiac and cerebrovascular disease. For instance, the prevalence of PAD is estimated to be 29% in patients with diabetes, which is one of the most common risk factors (smoking being another) (67, 105, 126, 177). Other concurrent health problems such as congestive heart failure (CHF) and severe pulmonary disease may prevent sufficient physical activity to produce limb symptoms, so that PAD may be underdiagnosed in these patient populations. Thus, the exercise physiologist must be aware of the PAD patient's comorbid conditions when conducting stress tests and prescribing exercise.

PATHOPHYSIOLOGY

The process of developing clinically significant PAD likely begins with endothelial damage in the arteries of the periphery (73). The following risk factors, which are similar to those for coronary artery and cerebrovascular disease, can lead to increased levels of oxygen-derived free radicals and ultimately endothelial injury (14): 1) smoking, 2) diabetes, 3) hypertension, 4) hypercholesterolemia, 5) high levels of triglycerides, 6) high leukocyte count, 7) high levels of homocysteine and fibrinogen, 8) increased blood viscosity, and 9) elevated C-reactive protein levels. Reducing these risk factors can aid in normalizing vascular superoxide anion production and improve endothelium-dependent vascular relaxation (127). Following the damage to the endothelium, a series of further events occurs; these are described elsewhere and are beyond the scope of this chapter (14). It is critical, however, to understand that the events following endothelial damage (e.g., accumulation of macrophages and foam cells within a fibrous cap) result in blockage and subsequent abnormal blood flow in the peripheral arteries. The abnormal blood flow in the lower extremities is predicated on the percent blockage of the artery due to the stenoses, as well as the location of the blockage (e.g., whether there are multiple blockages within a vessel) (25, 78). The blockages impair adequate blood flow to meet metabolic demands of active muscles with even low-intensity exercise. For patients with IC, blood flow is typically adequate at rest; but for patients with CLI, even perfusion pressure to the tissues at rest is impaired due to the increased severity of disease, thus resulting in rest pain (132, 183).

CLINICAL CONSIDERATIONS

A patient presenting with signs and symptoms suggestive of PAD should first be assessed for the presence of PAD, most often via measurement of the **ankle–brachial index** (ABI). If PAD is present, the patient should be treated as if heart disease is also present in terms of risk factor modification, since PAD is a heart disease equivalent. Supervised exercise training is a first-line medical therapy if IC is present. If symptoms are more severe, non-invasive and invasive imaging studies are typically performed in the event that surgical or peripheral interventions are planned. Thus, evaluating the severity of the PAD patient's clinical condition is essential for planning and conducting appropriate medical treatment as well as providing an optimal individualized exercise program.

Signs and Symptoms

As mentioned previously, the majority of patients with PAD have atypical symptoms or an asymptomatic presentation of the disease (111). In clinical practice, a vascular specialist uses these terms to describe symptoms that are felt to be consistent with PAD but do not meet the classic definition of IC. In essence, the primary components of such a symptom complex would include symptoms that are largely exertional, are somewhat relieved with rest, and are primarily located in muscle groups of the lower extremity.

For patients who have IC, a wide variety of terms are often used to describe leg pain symptoms, including cramping, aching, tightening, and fatigue. Regardless of the anatomic location of PAD, the calf is the most commonly reported location for leg pain. Intermittent claudication of the thigh and buttock is predicated on compromise of flow to the profunda femoris and internal iliac artery, respectively. Hence, when present, thigh or buttock IC (or both) may provide some insight into the anatomic location of PAD underlying the clinical presentation, although specific diagnostic tools should be used to assess disease localization when needed, for example for interventional therapy.

Critical limb ischemia is defined as the presence of chronic ischemic rest pain, foot ulcers (non-healing wounds), or gangrene attributable to objectively proven arterial occlusive disease (48, 172). The burden of atherosclerosis in patients with CLI is such that perfusion to the lower extremity is compromised even at rest. Typically, the disease is present at multiple levels of the tibial vessels in the leg and the small vessels of the foot (150, 162). The diagnosis of CLI carries a grave prognosis, in terms of both the risk for limb loss and overall cardiovascular morbidity and mortality. Patients with ischemic rest pain usually report severe pain located in the forefoot that is precipitated by assuming the recumbent position. Hence, many patients report significant difficulty with sleep that is interrupted by episodes of severe foot pain, which they relieve by hanging their lower extremity over the edge of the bed. The use of narcotic analgesia is commonly required to provide adequate relief from ischemic rest pain. Tissue loss in patients with CLI is most commonly located in the foot, and specifically in the forefoot (125). Atypical locations, such as the heel and over the medial and lateral malleoli, may also be seen in patients with arterial disease that compromises flow to these territories. Arterial wounds are classically described as being painful and having a pale base with an irregular margin. Superimposed soft tissue infection and underlying bony infections (e.g., osteomyelitis) are complications that may occur, particularly in wounds that are not aggressively managed.

A variety of symptom severity scales have been devised to provide consistency for the grading of PAD in clinical practice and during clinical investigation. The two widely accepted scales are the Fontaine stages and Rutherford categories (151).

The Fontaine stages:

I: Asymptomatic

IIa: Mild claudication

IIb: Moderate to severe claudication

III: Ischemic rest pain

IV: Ulceration or gangrene

The Rutherford categories:

0: Asymptomatic

1: Mild claudication

2: Moderate claudication

3: Severe claudication

4: Ischemic rest pain

5: Minor tissue loss

6: Major tissue loss

History and Physical Examination

The key components of the clinical history and physical examination that should be documented in a patient with PAD are summarized in tables 15.1 and 15.2. The primary element of the clinical history should be the characterization of the patient's leg symptoms and an assessment (both qualitative and quantitative) of the impact of these symptoms on the patient's functional capacity. A careful history should be able to assess with a reasonable degree of certainty that a patient's symptoms are due to ischemic or non-ischemic causes. The

Table 15.1 Components of Medical History Important in Assessment of a Patient With Peripheral Artery Disease

Medical history	Comments
Assessment of leg pain site, quality of pain or discomfort, precipitating and relieving factors, effect of leg exertion and rest, temporal pattern of pain	Assess whether leg discomfort represents true ischemia of the leg versus other nonischemic causes of leg pain.
Inquiry about rest pain in foot or tissue loss	Determine if patient has any evidence of CLI.
Perceived impact of leg symptoms on patient's quality of life	Affects treatment decisions.
Patient expectations about ambulation	Important component in decision making.
Walking distance, claudication onset distance,† peak walking distance‡	These provide more objective measure of severity of PAD and degree of functional impairment.
Medical comorbidities	Assess whether comorbidities may be contributing limitations to exercise (e.g., CHF, COPD, prior stroke). Hence, information on comorbidities has an important role in making treatment decisions.
Atherosclerotic risk factors	Important targets for medical treatment.
Medications	Ensure that patient is on optimal medical therapy (e.g., antiplatelet, statin).
Social history	Helps assess lifestyle of patient and need for ambulation to meet patient's needs (e.g., role of ambulation in current employment).

CLI, critical limb ischemia; PAD, peripheral artery disease; CHF, congestive heart failure; COPD, chronic obstructive pulmonary disease.

†Typical distance walked at the initial onset of any claudication pain.

‡Typical distance at which claudication becomes so severe that the patient is forced to stop (90).

remainder of the history centers on the documentation of atherosclerotic risk factors, a survey of conditions that may contribute to functional limitation, and a review of evidence for atherosclerotic involvement of other vascular beds beyond the lower extremities.

The physical examination should be centered on the objective assessment of the lower extremity arterial circulation. Palpation of lower extremity pulses and documentation of the presence and quality of the Doppler signal from the tibial pulses are mandatory. The ABI provides an objective measure of the severity of PAD. Ankle–brachial index assessment performance among medical practitioners is poor, emphasizing the need for appropriate education of those assigned to perform this critical part of the lower extremity examination (118, 181). In patients with CLI, the ABI has a number of limitations and should rarely be used to make clinical decisions in this patient subset. Beyond the assessment of the lower extremity circulation, the physical examination should carefully document any evidence of atherosclerotic involvement of other vascular territories (e.g., abdominal aortic aneurysm) and any other physical manifestations of pathologies that might influence the functional performance of the patient.

As to laboratory testing in patients with PAD, routine testing should include a lipid panel and assessment of glucose control in patients with diabetes. When imaging using iodinated contrast agents is considered, such as computed tomography (CT) **angiography**, a renal profile is mandatory to assess the risk of contrast-induced nephropathy.

Diagnostic Testing

Diagnostic testing in patients with PAD falls into two broad categories: pressure measurement(s) (i.e., hemodynamic assessment) and imaging studies. Hemodynamic testing provides important functional information about disease severity (149), whereas imaging studies provide anatomic detail about the vascular obstruction. These

Table 15.2 Summary of Primary Components of Physical Exam in Patients With Peripheral Artery Disease

Physical examination	Comments
Signs of limb ischemia	Examples: hair loss, dystrophic nails
Signs of CLI	Examples: dependent rubor, nonhealing wounds, gangrene, associated soft tissue infection
Pulse exam	
Palpation of femoral pulse	Palpable femoral pulse suggests infrainguinal disease. Diminished femoral pulse suggests suprainguinal disease.
Palpation of popliteal pulse	Rule out presence of popliteal aneurysm—increased incidence in patients with PAD.
Palpation of radial pulse	Asymmetry of radial pulses suggests unilateral subclavian disease. Bilateral diminished radial pulse suggests bilateral subclavian disease and that ABI will likely underestimate severity of PAD.
Doppler of DP and PT arteries	Objective assessment of presence of tibial flow with much better inter- and intraobserver variability compared with palpation. Quality of Doppler signal also important—biphasic signal suggests superior flow compared with monophasic signal.
ABI measurement	Objective assessment of severity of PAD: ≥1.30 noncompressible/uninterpretable; 1.0 to 1.29 normal; 0.90 to 0.99 equivocal; 0.40 to 0.89 mild to moderate PAD; <0.40 severe PAD (81).
Pulmonary	Assess for presence of significant pulmonary disease.
Cardiac	Assess for evidence of structural heart disease (e.g., valvular heart disease) or CHF.
Abdomen	Palpation for presence of AAA, auscultation for renal or mesenteric bruits.
Neck	Auscultation for carotid bruits.
Other	Assess for any other issues that might contribute to functional limitation (e.g., osteoarthritis of knee, peripheral neuropathy).

CLI = critical limb ischemia; PAD = peripheral artery disease; ABI = ankle–brachial index; DP = dorsalis pedis; PT = posterior tibial; CHF = congestive heart failure; AAA = abdominal aortic aneurysm.

tests are complementary in allowing optimal decision making in patients with PAD.

Hemodynamic Testing

The ABI is the simplest hemodynamic assessment in a patient with PAD. As such, it is the most widely used screening tool for PAD.

The ABI measurement is executed as follows:

1. Brachial pressures are acquired through the placement of a blood pressure cuff around the arm and a Doppler probe (5 MHz) over the brachial artery. The blood pressure cuff is inflated until the Doppler signal disappears. The cuff is then slowly deflated, and the pressure at which the Doppler signal returns is recorded for each arm.

2. Using a similar technique, the ankle pressure is recorded from the dorsalis pedis (DP) and posterior tibial (PT) arteries of both lower extremities. It is important to apply an appropriately sized blood pressure cuff (i.e., typically a 10 cm cuff) over the lower half of the calf (i.e., away from the bulk of the calf muscles). Use of an inappropriately small-sized cuff or placement over the bulk of the calf muscles will result in a falsely elevated ankle pressure.

In calculation of the ABI, the brachial pressure is used as a surrogate for the intra-aortic pressure, which is a reasonable assumption in the absence of significant subclavian artery disease. Therefore, the higher of the two brachial pressures is used for the ABI calculation of both lower extremities. By convention, the higher of the two ankle pressures (i.e., DP or PT) is used for the ankle pressure component of the calculation. It is important to understand that the effect of using the higher of the two ankle pressures is to decrease the sensitivity and increase the specificity of the ABI measurement for making the diagnosis of PAD (160). This may seem counterintuitive for a screening test, but it remains the standard method for ABI calculation in clinical practice (152). The standardized interpretation of ABI values is provided in table 15.2 (81).

Additional hemodynamic testing beyond the ABI measurement requires specific equipment and is generally performed in vascular laboratories. These tests include the following:

1. Toe pressure: The toe pressure is typically acquired using a small pneumatic cuff that is applied to the great toe of each foot. The absolute toe pressure is normally 30 mm lower than the ankle pressure, and a toe–brachial index of ≥0.70 is regarded as normal. In clinical practice, the toe pressure is rarely measured in patients with IC. It may be used to confirm the diagnosis of PAD in a patient with IC with a non-compressible vessel and therefore uninterpretable ABI measurement. In contrast, it is widely used in patients with CLI to confirm the diagnosis (i.e., toe pressure of <50 mmHg). In addition, the absolute toe pressure is helpful in predicting the likelihood of limb salvage and the need for revascularization.

2. Segmental limb pressures (SLPs): SLPs are used to help provide hemodynamic evidence of the anatomic location of PAD. They are acquired via placement of blood pressure cuffs at various locations along the length of the leg (i.e., upper thigh, lower thigh, upper calf, ankle). The SLP at each level is recorded by inflation of the cuff of interest to above systolic pressures and recording the pressure at which the Doppler signal returns to the DP and PT arteries at the ankle. A drop of ≥20 mmHg between adjacent SLPs suggests significant disease in the intervening arterial segment.

3. Pulse volume recordings (PVRs): PVRs are recorded using the same cuffs that are used to perform SLPs. For the purpose of recording PVRs, each of the cuffs is attached to a plethysmograph. Each cuff is sequentially inflated to ~60 mmHg, and the volume change over time at that cuff level is recorded. Since arterial inflow is the only significant volume changing over time in the lower extremity, the waveform generated resembles an arterial waveform. The amplitude of the waveform provides important information about the adequacy of arterial inflow at that level, and comparison with the PVR above and below that level is also helpful in assessing the location of disease in that arterial segment. Table 15.3 provides a summary of presumed anatomic location of disease based on a drop of ≥20 mmHg in pressure between SLPs or change in amplitude of PVRs between adjacent pressure cuffs.

4. Transcutaneous oxygen pressure ($TcPO_2$) and skin perfusion pressure: These measurements are performed in a small number of laboratories and are used to provide functional information about tissue perfusion. As such, they are used only in patients with CLI. Clinically, these measurements are especially useful in patients who have had prior digital or transmetatarsal amputation, where the toe pressure measurement is not possible. A cutoff of 30 mmHg for both $TcPO_2$ and skin perfusion pressure is believed to be consistent with the diagnosis of CLI. Figure 15.1 depicts a full hemodynamic evaluation of the lower extremities.

Imaging Studies

The evaluation of patients with PAD using imaging studies has evolved significantly over the last decade. This is based in large part on the evolution of noninvasive imaging studies, such as CT and magnetic resonance (MR) angiography, that offer anatomic detail rivaling that with

Table 15.3 Summary of Presumed Anatomic Location of Disease Using Segmental Limb Pressures and Pulse Volume Recordings

Location of drop in SLP or change in PVR†	Presumed location of diseased arterial segment
Brachial–upper thigh	Aortoiliac, CFA
Upper thigh–lower thigh	SFA
Lower thigh–upper calf	Distal SFA, popliteal artery
Upper calf–ankle	Tibial arteries
Ankle–toe	Small vessels of the foot

SLP = segmental limb pressure; PVR = pulse volume recordings; CFA = common femoral artery; SFA = superficial femoral artery.

†Based on a drop of ≥20 mmHg in pressure between sequential limb pressures or change in amplitude of pulse volume recordings between adjacent pressure cuffs

invasive angiography (10, 20, 43, 98, 116, 148). As a result, in contemporary practice, the majority of invasive angiographic studies are performed as a preamble to planned interventional procedures based on the diagnostic findings of non-invasive studies (98).

Computed tomography and MR angiography are performed following the administration of iodinated and gadolinium-based contrast agents through a peripheral intravenous cannula. Each of these imaging techniques has specific advantages and disadvantages (table 15.4). Most vascular specialists currently use CT angiography as the first-line option for non-invasive imaging. The emergence of nephrogenic systemic fibrosis as a potential complication of exposure to gadolinium-based contrast

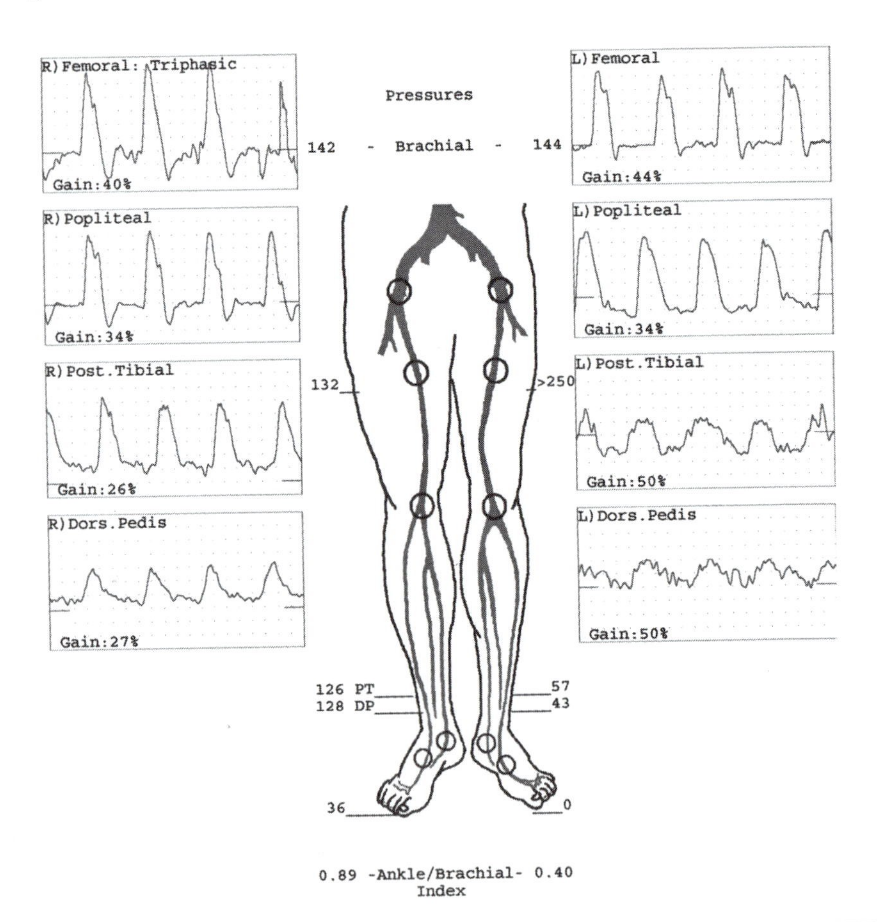

Figure 15.1 Hemodynamic assessment of lower limbs.

agents during MR angiography has resulted in the use of similar thresholds of renal insufficiency as a contraindication to imaging by MR and CT angiography in most radiology groups, removing the major clinical advantage previously held by MR angiography in this patient population (134, 176).

Duplex ultrasound is unique among imaging studies in that it offers both functional and anatomic detail about the peripheral arteries (1). B-mode imaging provides some basic information regarding vessel size and the presence of plaque. Doppler examination allows an assessment of flow within the arterial segment of interest and provides information regarding the functional significance of a stenosis (figure 15.2). The clues to the presence of functionally significant stenosis are as follows:

1. Alteration of the normal triphasic Doppler waveform. The earliest alteration in the waveform is a loss of the transient reverse flow component in systole, resulting in a biphasic waveform. This is followed by loss of the late diastolic forward flow component, resulting in a monophasic waveform. These changes are qualitative and need to be interpreted in the context of the assessment of flow velocities.

2. Peak systolic velocity (PSV). A predictable increase in the PSV occurs at the site of an arterial obstruction. A PSV of ≥200 cm/s is generally regarded as indicating a stenosis of ≥50% in the vessel segment. However, there is significant variation in the PSV along the length of the lower extremity. Hence a comparison of the PSV at the site of a stenosis with the PSV in the segment of the vessel proximal to the site of stenosis appears to be a more specific measure of the functional significance of a stenosis. A peak velocity ratio of ≥2.0 is generally regarded as indicating the presence of a significant stenosis.

Invasive angiography remains an important diagnostic imaging tool for a subset of patients. These include patients who have significant renal insufficiency that precludes the use of a bolus of iodinated contrast or gadolinium-based contrast for CT or MR angiography, respectively. In such patients, targeted invasive angiography may be used to answer a specific clinical question. In addition, in patients with CLI, invasive angiography remains the gold standard for the assessment of the infrapopliteal and small vessel anatomy of the foot (21, 162). This information is critical in clinical decision making in this patient group. Magnetic

Table 15.4 Summary of Advantages and Disadvantages of Noninvasive Imaging Using Computed Tomography or Magnetic Resonance Angiography

Imaging study	Advantages	Disadvantages
CT angiography	-Widespread availability -Simplicity of imaging protocols -Higher spatial resolution -Rapid scanning times -Large gantry size -Ability to visualize calcium -Ability to visualize lumen within stents -Lower cost	-Artifact in presence of severe calcium -Exposure to ionizing radiation -Use of nephrotoxic contrast
MR angiography	-Absence of ionizing radiation -No interference from calcium -Ability to obtain some information with specialized protocols without use of contrast agents	-Higher cost -More complicated imaging protocols -Contraindicated in patients with ferromagnetic materials (e.g., pacemakers, orthopedic hardware, shrapnel) -Gantry size small, limiting studies for patients with claustrophobia or marked obesity -Scanning times prolonged, requiring greater degree of patient cooperation -Venous contamination in legs -Difficulty visualizing stents; artifact from previously placed stents

CT = computed tomography; MR = magnetic resonance.

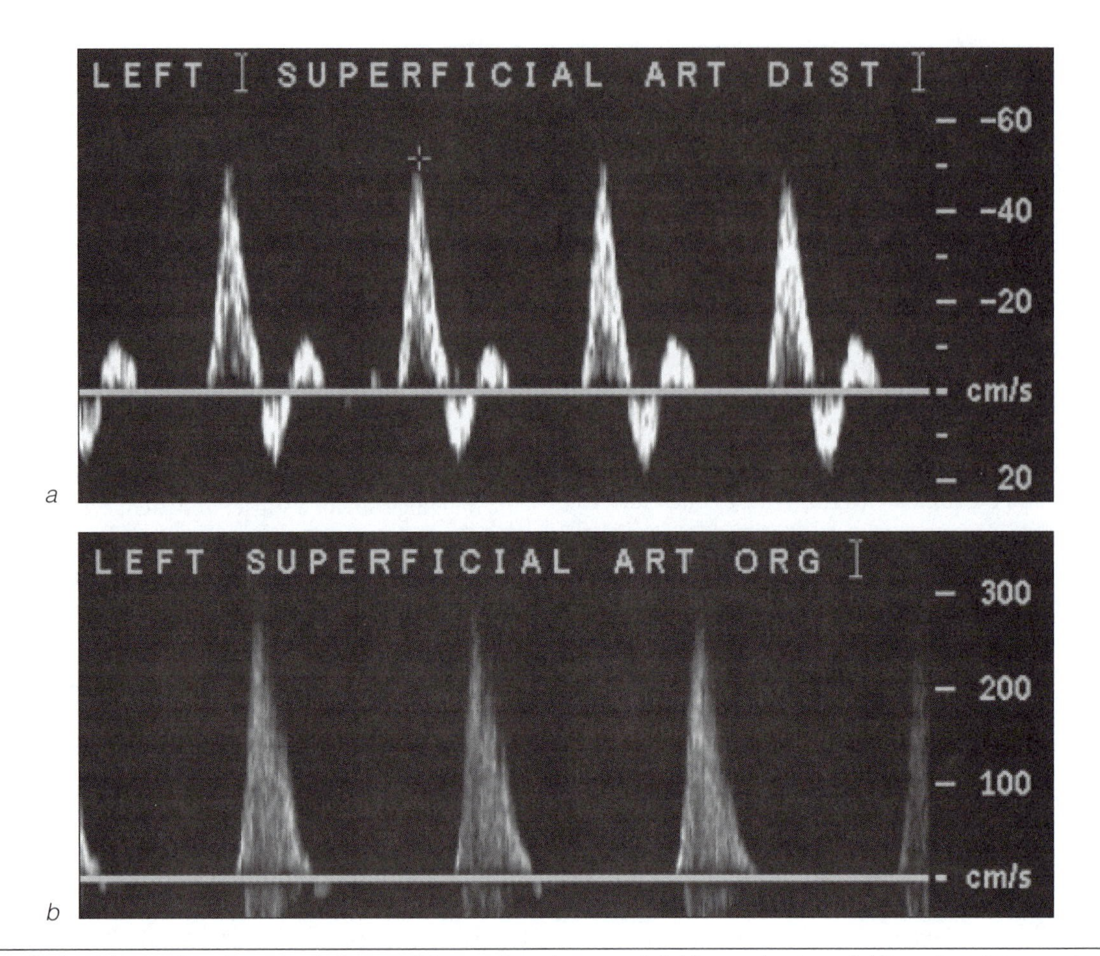

Figure 15.2 *(a)* Doppler ultrasound of normal flow pattern in superficial femoral artery. *(b)* Doppler ultrasound of flow in patient with severe stenosis in superficial femoral artery - high peak systolic velocity and loss of normal triphasic waveform (to monophasic waveform).

Photos courtesy of Ryan J. Mays, Ivan P. Casserly, and Judith G. Regensteiner.

resonance angiography suffers from significant issues with venous contamination in the assessment of tibial anatomy, and the resolution of CT angiography is not sufficient to provide this anatomical information in most individuals.

Exercise Testing

Exercise testing is useful for determining the degree of functional limitation in patients with PAD. Exercise testing using the treadmill is also a valuable tool for the exercise physiologist in designing a walking program for PAD patients. Assessment of a PAD patient's walking ability is critical in determining the clinical, functional, and quality of life limitations typically associated with the disease. The next sections summarize recommended exercise testing modalities and appropriate procedures for assessing the most important health-related fitness components in PAD (i.e., cardiovascular fitness, functional ability, muscular strength, and endurance).

Cardiovascular and Functional Testing

The hemodynamic assessments outlined are predicated on the detection of PAD based on the presence of a pressure gradient across a diseased segment of the artery at rest. Occasionally, however, moderate arterial obstructions (i.e., causing ~50-70% stenosis) are not associated with a resting gradient, but the increased flow across the lesion due to the increased cardiac output during exercise may provoke an exercise-induced gradient. It has traditionally been taught that this is most likely associated with moderate lesions in the aortoiliac segment but can be observed with moderate disease in any arterial segment. As a result, an exercise study is crucial in identifying this subset of patients, who typically provide a good clinical history that is consistent with IC and have a normal resting ABI (166). Exercise testing is also helpful in evaluating patients with atypical symptoms where a significant drop in perfusion pressures with exercise that is associated with the onset of symptoms, supports the conclusion that PAD is the probable underlying etiology. Thus, the presence of PAD is established by measuring an ABI at rest. If functional assessment is indicated, the patient then walks on a treadmill using a standardized protocol, and the ABI measurement may be repeated immediately following the walking exercise to determine the effects of exercise on disease severity. The study is considered abnormal if the ankle pressure drops by ≥20% from the baseline measurement and takes more than 3 min to normalize.

The most objective modality for testing the patient's walking ability and subsequently determining the effectiveness of any treatment to improve walking in PAD is a treadmill walking test. This type of test has been very well validated for use in patients with IC. Treadmill testing can be used to establish **claudication onset time or distance** (point of initial onset of any claudication pain) as well as **peak walking time or distance** (point at which claudication pain becomes so severe that the patient is forced to stop) (52). Exercise stress testing can also be helpful for providing information as to whether heart disease is present (154).

A familiarization session with the treadmill should be performed by PAD patients regardless of experience with treadmill walking. During the actual test, the endpoint is determined by the patient's perception of pain using the Claudication Symptom Rating Scale (ranges from 1 to 5 with 1 = no pain, 2 = onset claudication, 3 = mild, 4 = moderate, and 5 = severe pain). Patients decide when they need to stop. During testing, it is recommended that patients not use handrail support, as IC distances are affected (59). However, because patients may have balance issues in addition to a lack of familiarity with treadmill exercise, they may use handrails but only if needed. Several different treadmill protocols have been used in PAD populations to test for leg pain and evaluate other potential limiting factors in response to exercise (60). A typical exercise protocol for PAD patients uses either a graded treadmill test or a constant-load test. The graded exercise test keeps speed constant at 2 mph throughout. Grade begins at 0% and increases 2% every 2 min thereafter (58). The constant-workload or single-stage treadmill test holds speed and grade constant throughout. Gardner and colleagues (58) reported that the severity of PAD is better assessed with a graded protocol, as claudication onset distances and maximal walking distances demonstrated better reliability than constant-load tests. However, both types of test have been widely used.

Although optimal, walking exercise is not appropriate for all patients and may be unavailable in some clinic settings. Patients with abdominal aortic aneurysms, uncontrolled hypertension, or other exercise-limiting comorbidities, such as CHF or chronic obstructive pulmonary disease (COPD), may need aerobic capacity evaluation using a different modality (126). Wolosker and colleagues (178) determined that 16% of patients with PAD were unable to complete a treadmill test because of various limiting factors due to comorbid conditions. Thus, other modalities have been used to evaluate exercise capacity in PAD patients, including arm and leg ergometry, stair stepping, and active pedal plantar flexion (51, 57, 182, 185). Gardner and colleagues (57) demonstrated similar peak oxygen consumption ($\dot{V}O_2$peak) values between testing protocols when patients with IC performed level walking, graded walking, or progressive stair climbing. Thus, while graded exercise tests using the treadmill are preferred, other cardiovascular fitness assessments using different modalities are available if needed.

In addition to walking impairment, PAD patients have more overall functional impairment than persons without PAD (111, 113). Therefore, the ability to assess functional impairment is of key importance in assessing the PAD patient. Recently it was reported that PAD patients with declining functional performance are at increased risk for mobility loss and mortality with increasing age and that women may experience a faster functional decline than men (109, 112). Functional impairment can be subjectively assessed through validated questionnaires and objective measurements such as walking, balance, and other tests of general function. These performance test outcomes are more strongly associated with the volume of physical activity during daily life than are treadmill walking measures and thus may provide complementary information to augment that obtained with a peak aerobic exercise test (61, 107). The 6 min walk test in PAD patients has been examined extensively (96, 108, 110). The patient walks a defined course for 6 min, and the distance achieved is recorded. Additionally, incremental and constant-speed shuttle walking tests have been developed and demonstrate valid and reliable measures of a PAD patient's functional ability (29, 185). Patients typically walk a defined distance (e.g., 10 m) back and forth between two destinations (marked by cones or tape); a timer signals to increase the speed or holds the speed constant, depending on the protocol. Finally, the Short Physical Performance Battery has more recently been developed for functional performance testing. It combines several functional tests that provide a comprehensive objective measure of leg function, balance, leg strength, and ability to walk (64). This is important, as poor lower limb functioning is strongly related to disability, mortality rates, or both in the aging patient, regardless of disease. Functional performance tests may reflect usual walking and overall limb function during activities of daily life while providing clinicians with objective information for evaluating the progress of exercise training without expensive laboratory equipment.

Muscular Strength and Endurance Testing

Because of the advanced cardiovascular disease in PAD patients, cardiorespiratory fitness and functional testing remain the most important evaluation methods. However, recently, McDermott and colleagues (114) examined the relation between upper and lower limb strength and ABI of patients with and without PAD. Testing included handgrip, knee extension, and active pedal plantar flexion isometric strength as well as leg power during knee extension exercise. Results indicated that impaired strength may be limited to the lower extremities and that patients with PAD had lower plantar flexion strength and lower knee extension power than patients without PAD (114). Several other studies have evaluated strength and endurance in PAD patients with tests such as a one-repetition maximum using the leg press (120, 174), high-repetition muscular endurance testing (129), and isokinetic testing (143). Table 15.5 reviews testing options for PAD patients to evaluate health-related fitness components.

Treatment

A full discussion of the treatment of PAD is beyond the scope of this chapter. In general, the therapies offered can be broadly divided into two categories: optimal medical treatment and revascularization. Medical treatment focuses on therapies that improve the systemic cardiovascular complications associated with the diagnosis of PAD and agents that specifically may improve IC symptoms (8, 12, 63, 66, 69, 122). Aggressive modification of atherosclerotic risk factors is mandatory for all PAD patients, with the targets for therapy as outlined in table 15.6.

It is generally agreed that all patients with PAD should be treated with an antiplatelet agent, such as aspirin or clopidogrel, as well as other medications that treat risk factors for cardiovascular disease (e.g., statins) (146). Patients with PAD should have their risk factors treated with the same intensity as patients with other cardiovascular diseases. In contrast to the clear importance of drugs for optimally treating cardiovascular risk factors in PAD, pharmacological treatment of IC has been largely ineffective. Currently, despite numerous studies of proposed agents using many different mechanisms to treat IC, only two drugs have been approved by the Food and Drug Administration for these patients: pentoxifylline and cilostazol. Pentoxifylline improves the hemorrhagic profile of patients by improving red cell deformability, lowering fibrinogen levels, and decreasing platelet aggregation (133, 141). However, because of more recent evidence suggesting lack of efficacy in terms of walking distance, it is rarely used in clinical practice if at all (33, 38, 99, 133). Cilostazol, a phosphodiesterase inhibitor, is currently the only effective approved medication used to improve IC distance (70). One study in patients with IC who were randomized to receive cilostazol, pentoxifylline, or placebo for 6 mo showed that

Table 15.5 Exercise Testing Review

Cardiorespiratory endurance	Skeletal muscle strength	Skeletal muscle endurance	Flexibility	Body composition
1. Graded exercise testing *Primary modality* TM *Secondary modalities* CE AE Active plantar pedal flexion 2. Functional testing 6 min walk test SPPB Shuttle walking tests	1. Isometric testing Knee extension Plantar flexion Handgrip dynamometry 2. Maximal strength testing 1RM of upper and lower body muscle groups Leg press and plantar flexion tests 3. Isokinetic testing	Upper and lower body muscle group testing using plated or machine and pulley systems per standard guidelines outlined in chapter 5	Refer to chapter 5 for standard testing guidelines.	Refer to chapter 5 for standard testing guidelines.

TM = treadmill; CE = cycle ergometer; AE = arm ergometer; SPPB = Short Physical Performance Battery; 1RM = one-repetition maximum; PAD = peripheral artery disease.

Practical Application 15.1

CLIENT–CLINICIAN INTERACTION

Many patients with PAD have exercised very little in their lifetime and may not have the knowledge to start a safe and effective exercise program. The exercise physiologist needs to provide the patient detailed information about how to exercise and to reassure the patient that walking will not cause harm. More specifically, neither high-intensity shorter bouts or low-intensity longer bouts of exercise cause lasting harm to muscle beyond the possibility of overuse, which is commonly experienced by all populations, especially with aging (54, 124). It is also important for patients to understand that the arterial and venous systems are separate, as deep vein thrombosis and PAD are distinct problems and present different treatment options and risks. Patients need to understand that routine exercise is one of the most beneficial therapies for PAD, in addition to reducing risk factors of their underlying atherosclerosis. Smoking cessation should be strongly encouraged by the exercise physiologist, as the relative risk of IC was demonstrated to be 3.7 in current smokers versus 3.0 in ex-smokers who had discontinued smoking for less than 5 yr (49, 126). Finally, adequate foot care must be stressed to PAD patients, particularly those who have diabetes. Some PAD patients with IC may be hesitant to walk on a treadmill for fear of damaging their legs and feet. The risk is that ulceration from trauma, specifically from poorly fitting footwear or improperly cut toenails, may result in a nonhealing wound that could lead to amputation (167).

---

cilostazol improved maximal walking distance by 54% compared to a 30% increase in the pentoxifylline group and 34% in the placebo group (31). Additionally, a meta-analysis of six randomized controlled trials indicated that cilostazol improved walking ability and patient-reported outcomes in patients with IC compared to placebo (142). Many other pharmacological agents have been studied in PAD research trials in an attempt to establish an effective

Table 15.6 Major Components of Medical Management in Patients With Peripheral Artery Disease

Treatment	Comments
GENERAL MEDICAL TREATMENT	
Antiplatelet agents	Aspirin 81 to 325 mg daily po, clopidogrel 75 mg daily po in patients with contraindication or intolerance of aspirin
Statins	Statins indicated in patients with diagnosis of PAD independent of LDL cholesterol level; may have favorable impact on leg symptoms independent of lipid-lowering properties
MODIFICATION OF ATHEROSCLEROTIC RISK FACTORS	
Hypertension	Target <130/85; ACE inhibitors should be used as first-line therapy
Hyperlipidemia	Target LDL cholesterol <70 $mg \cdot dl^{-1}$
Smoking	Smoking cessation strongly recommended
Diabetes	HbA1c <7%
IMPROVING CLAUDICATION DISTANCE	
Cilostazol	Dose 100 mg twice daily po
ACE inhibitors	May have favorable impact on leg symptoms
Exercise	See section "Exercise Prescription" for PAD patients

po = orally; PAD = peripheral artery disease; LDL = low-density lipoprotein; ACE = angiotensin-converting enzyme; HbA1c = glycated hemoglobin.

treatment for IC; however, most have not shown significant benefits or are associated with significant side effects and intolerability (table 15.7).

The evolution of endovascular techniques to revascularize lower extremity arterial disease has resulted in a paradigm shift in the treatment of patients with IC (37, 68, 91). Previously, the primary option for revascularization was surgical bypass, which was associated with significant morbidity and mortality (2, 3, 39, 88, 131). As a result, the risk–benefit ratio of revascularization restricted its use to a small subset of patients with PAD. These included low risk surgical patients with severe IC symptoms and patients with CLI. In contrast, **endovascular revascularization** is characterized by very low rates of morbidity and mortality. With modern technology and techniques, acute technical success rates approach 100% for most procedures (147, 175). As a result, endovascular revascularization may be offered to a much broader spectrum of patients, although

Table 15.7 Pharmacological Treatment of Peripheral Artery Disease

Medication name, class	Primary effects	Exercise effects	Special considerations
Pentoxifylline, hemorrheologic agents	Platelet aggregation inhibitor; improves RBC deformity, lowers fibrinogen levels (133)	Equivocal findings, thus lack of efficacy for improving walking outcomes (31, 33, 38, 99, 133, 141)	First drug approved by FDA for treating IC (early 1980s); no longer in general use for treating IC
Cilostazol, phosphodiesterase inhibitors	Antiplatelet and vasodilatory effects; decreases vascular smooth muscle cell production and increases limb blood flow (83, 89, 170)	Effective for improving PWT and patient-reported outcomes in patients with IC (31, 141, 142)	Contraindicated in patients with heart failure
Naftidrofuryl, peripheral vasodilators and cerebral activators	Platelet aggregation inhibitor and peripheral vasodilatory effects (42)	Modest improvements in treadmill walking performance (62); mixed results for quality of life outcomes (40, 165)	Recommended as a second-line alternative therapy (126); approved in Europe but unavailable in the United States
Buflomedil, peripheral vasodilators and cerebral activators	Platelet aggregation and leukocyte adhesion inhibitor; increases tissue tolerance to ischemia (42)	Recent findings conclude lack of evidence for treating IC (102)	Safety concerns cited (32)
AT-1015, serotonin antagonists	Blocks serotonin-induced platelet aggregation and vasoconstriction (87)	Data indicate no improvement in exercise tolerance or patient-reported outcomes in individuals with IC (72)	Other serotonin receptor antagonists (ketanserin) shown to be effective for treating IC (23), but safety concerns cited (135)
Prostaglandins	Increase cAMP levels, subsequently increasing peripheral vasodilation and platelet inhibition (42)	Following 8 wk of treatment, WIQ scores improved significantly (106); however, insufficient evidence to provide recommendation for use in clinical settings (81, 126)	Discontinuation rate 20%; may be due to adverse effects such as headache, flushing, and gastrointestinal distress (11, 41, 42)
Propionyl-L-carnitine	Improves efficiency of oxidative phosphorylation and lessens symptoms of IC (17, 19)	Earlier studies showed improvement in walking outcomes, functional status, and patient-reported outcomes (18, 74); recent determination that the drug in combination with monitored home exercise not superior to exercise alone (71)	No major safety concerns for treating IC reported in several studies (18, 74)
Statins	May improve plaque stabilization, endothelial function, platelet activity, and inflammation, thus contributing to reduction of IC symptoms (42)	Improve maximum walking distances and community-based physical activity in patients with IC (117, 119)	Side effects may include myalgia with risk of rhabdomyolysis, increased liver enzymes, and polyneuritis (32, 65)

RBC = red blood cell; FDA = Food and Drug Administration; IC = intermittent claudication; PWT = peak walking time; cAMP = cyclic adenosine monophosphate; WIQ = Walking Impairment Questionnaire.

exercise training remains the gold standard treatment for improving peak walking time of patients with IC (123). The major limitation of endovascular compared to surgical revascularization is lower long-term **patency** rates. In the aortoiliac arterial segments, this difference in long-term patency is modest, whereas in the femoropopliteal segment, the difference is significant. Since restenosis generally results in a recurrence of IC, the lower patency rates associated with endovascular treatments are a concern in the treatment of patients with IC and warrant careful consideration before this treatment strategy is offered. In general, a more aggressive approach is adopted in younger, active patients without comorbidities that may contribute to functional limitation and with more favorable disease anatomies (e.g., aortoiliac vs. femoropopliteal, stenosis vs. occlusion, focal vs. diffuse disease, noncalcified vs. calcified plaque). In contrast, since CLI is associated with a significant risk of limb loss and restenosis is often clinically silent, an aggressive approach to endovascular revascularization is warranted in most patients with CLI. Table 15.8 summarizes the endovascular and surgical treatment options available for revascularization for disease in the major arterial territories of the lower extremity. Figure 15.3 depicts angiographic images of the lower extremity arterial system of a patient with CLI before and after angioplasty of the peroneal artery.

EXERCISE PRESCRIPTION

Walking exercise is particularly beneficial for helping PAD patients with IC to improve their walking ability, although all types of exercise are beneficial for general cardiovascular health. An exercise program causes physiological adaptations leading to higher $\dot{V}O_2$, increased walking time, and enhanced quality of life. Peripheral artery disease patients with IC are often not able to walk very well when they begin an exercise program, and their exercise prescription is set according to their individual limitations. Patients should begin with a warm-up and end with a cool-down. Bouts of exercise are intermittent, with the length of time for each bout limited by the onset of moderate IC (3 or 4 on the Claudication Symptom Rating Scale). Each bout is followed by a rest period to allow the leg pain to subside. These exercise and rest sessions are repeated for 35 to 50 min, at least three times a week, for a total duration of 3 to 6 mo, although some programs have been carried out for up to a year (56). Initially, the prescription for exercise is to walk at the speed and grade that elicits IC. Patients walk until they reach a 3 or 4 (moderate pain) on the Claudication Symptom Rating Scale. They then sit and rest until the pain abates. For the first few sessions, the goal is for the patient to walk and rest for a total of 35 min (with the time from all bouts added together). Walking bouts typically become longer with further training until the walking goal of 50 min per session has been reached. For patients who have received endovascular therapy or bypass surgery and may have other factors limiting exercise (e.g., orthopedic limitations or muscular fatigue), rating of perceived exertion (RPE) using the Borg 15-category scale or other walking perceived exertion scales (e.g., OMNI Walk/Run Scale) can be used to regulate exercise intensity at a moderately hard level (15, 171), as has been done in the past for exercise training in asymptomatic PAD patients (108). Exercise intensity based on age-predicted exercise heart rate responses rather than IC symptoms is not recommended for PAD patients.

Patients may exercise in a supervised setting (56, 75, 77, 140) or an unsupervised home or community exercise

Table 15.8 Endovascular and Surgical Options for the Revascularization of Various Arterial Segments of the Lower Extremity

Arterial segment	Endovascular	Surgical
Aortoiliac	Stenting*	Aortobifemoral bypass
CFA	PTA, atherectomy†	CFA endarterectomy
Femoropopliteal	PTA, atherectomy, stenting‡	Femoropopliteal bypass
Tibial++	PTA, atherectomy, stenting§	Femorotibial bypass

CFA = common femoral artery; PTA = percutaneous transluminal angioplasty.

*Typical stent types in use include balloon expandable, self-expanding, covered self-expandable, and covered balloon expandable.

†Surgical treatment of CFA disease is preferred over endovascular approach. Stents should be avoided in CFA.

‡Typical stent types include self-expanding and covered self-expanding. Stents should be avoided in popliteal artery.

§Typical stent types include bare metal and drug-eluting balloon, expandable and self-expanding.

++Tibial revascularization typically restricted for treatment of patients with critical limb ischemia.

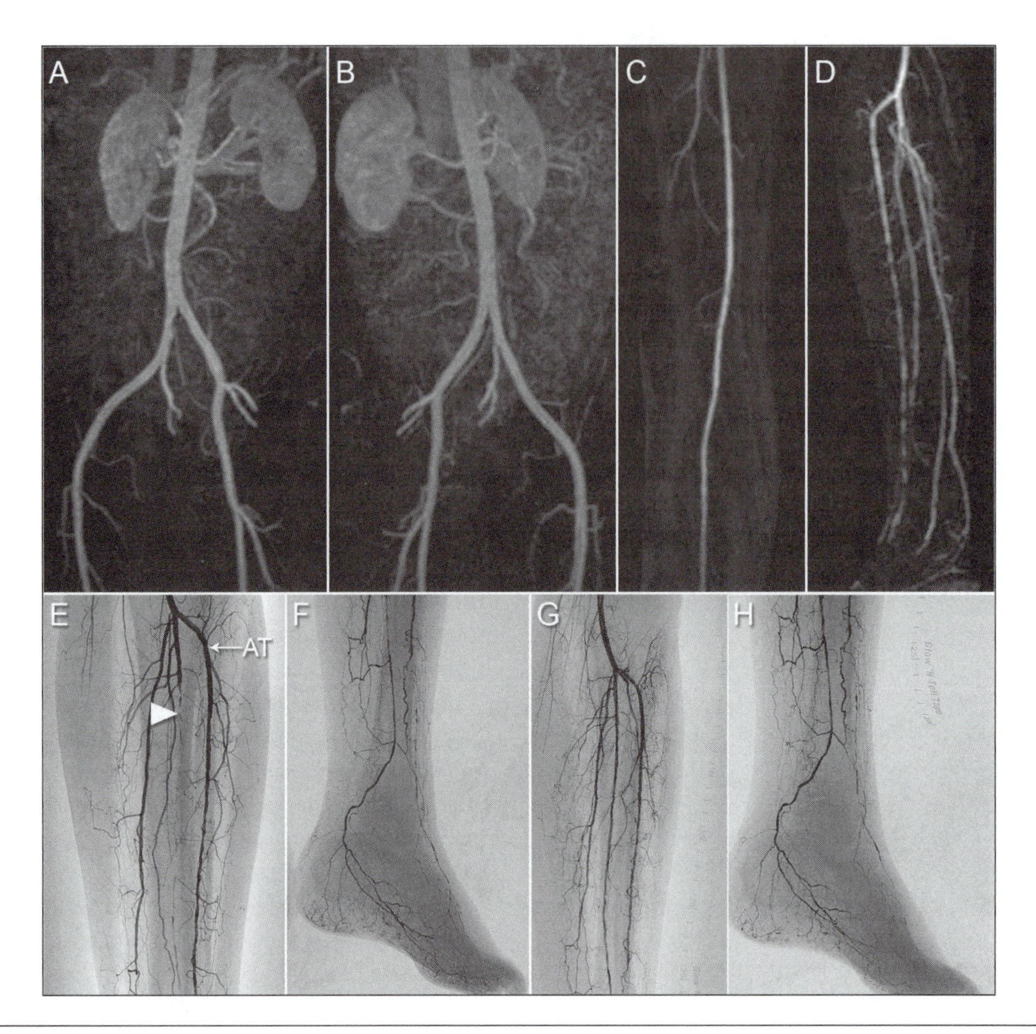

Figure 15.3 Anatomic evaluation of 68 year old diabetic male with a non-healing wound on the great toe of his right foot. (A-D) Magnetic resonance angiogram of pelvic region and right lower extremity demonstrating widely patent iliac system, right common femoral artery, superficial femoral artery, and popliteal artery. (E-F) Routine angiography of infrapopliteal vessels delineating anatomy of tibial pedal vessels, and showing that the peroneal artery (arrowhead) should be the target for revascularization. (G-H) Final angiogram of the tibial and pedal vessels following angioplasty of the peroneal artery. AT = anterior tibial.

Photos courtesy of Ryan J. Mays, Ivan P. Casserly, and Judith G. Regensteiner.

program (9, 13, 85, 94). Supervised exercise programs have shown a greater walking distance improvement, approximately 150 m, over unsupervised exercise training (13). However, supervised exercise training is often not used due to barriers including time constraints and habitual sedentary behavior, as well as lack of insurance coverage and proximity to clinics and health care facilities and transportation (7, 137, 141, 153).

Unsupervised exercise programs to date have primarily involved instructing patients to exercise at home or in the community with relatively little follow-up (13, 121, 140, 157, 180). As a result, home-based programs have not, for the most part, been successful. Factors resulting in the failure of many of these programs may include lack of motivation of patients and reliance on patient self-monitoring with no feedback (22, 34, 36, 86, 121, 130, 140). Kakkos and colleagues (86) randomized patients with IC into supervised and unsupervised exercise training programs and examined claudication onset time and peak walking time (baseline, 6 wk, and 6 mo). Patients in the home exercise program were "advised to exercise daily by walking," without further advice. The supervised exercise training group improved both claudication onset time and peak walking time, versus no improvement in either outcome for patients in the home training group. Other studies have demonstrated similarly poor results after patients were instructed to exercise at home or in the community with no feedback (22, 34, 75). A number of community-based exercise programs have provided some elements of coaching to improve health outcomes of PAD patients (e.g., counseling and therapy models, telephone contact). These programs have demonstrated inconsistent results in improving walking ability (26, 121, 130, 140, 157, 180). Briefly, Patterson and colleagues (130) compared a supervised exercise program and an unsupervised, home-based exercise program, with both groups receiving weekly lectures. Both programs improved peak walking time from baseline to 12 wk (207%, $p < 0.001$ vs. 70%, $p < 0.001$). In contrast, Regensteiner and associates (140) randomly assigned PAD patients to a supervised, hospital-based program or an unsupervised, home-based program for a 3 mo exercise treatment. Patients in the home-based program were given detailed instructions during the initial visit to the laboratory followed by a weekly phone call only. In both studies, the supervised exercise program improved treadmill exercise performance more than an unsupervised, home-based program. Despite the difficulties, home exercise programs are more conducive to a patient's schedule and are less costly than hospital-based programs. Thus, developing such programs with new elements that may improve compliance is likely to be highly beneficial.

Strength training alone, treadmill walking alone, and combinations of the two were examined in patients with IC to determine whether one modality of training was superior (77). Patients with symptomatic PAD showed improvements in peak walking times with treadmill exercise or with weight training for skeletal muscles of the legs. The treadmill group, however, showed improvement in $\dot{V}O_2$peak, but no changes were observed in $\dot{V}O_2$peak or claudication onset time for those participating in weight training alone. More recent studies have demonstrated improvements in walking and functional outcomes following muscular strength and endurance training in patients with IC (120, 129, 144, 174). However, while strength training is important, it should be only complementary to the PAD patient's exercise program, as walking is still preferred because of the beneficial effects on the cardiovascular system (81). Table 15.9 presents a brief review of the optimal training modalities for improving various health and fitness components of PAD patients.

EXERCISE TRAINING

In most studies, exercise training has resulted in excellent improvements in walking distances in PAD patients with IC, as well as a reduction in adverse cardiovascular event risk in a limited number of studies (81). A recent study determined that a low-intensity exercise training program is similar to high-intensity exercise rehabilitation in improving initial claudication onset distances and peak walking distances (54). However, achieving moderate exercise levels (guided by pain or general exercise discomfort) is still more often recommended (81). As patients improve their functional status and are able to exercise more, they may experience some adverse cardiac symptoms, at which time they should be re-evaluated by their primary care physician, who may order an exercise stress test (128).

Both walking distance on the treadmill and health status assessed by questionnaires are improved by supervised exercise training (56, 97, 138, 145, 169). Thus, supervised exercise training is considered the gold standard therapy for patients with PAD and IC (81, 126). Additionally, supervised exercise training programs improve peak walking time in PAD patients who do not experience classic IC (108, 115). Overall, exercise training improves the ability to walk for longer periods, and the consistency of these findings suggests that exercise training programs have a clinically important impact on functional capacity in PAD patients.

Table 15.9 Exercise Prescription Review

Mode	Intensity	Frequency	Duration	Progression	Goals	Special considerations
Walking exercise	For patients with IC, walking to moderate pain in 3 to 5 min followed by rest and resumption of walking (81)	At least three times per week (56), progressing to five times per week as tolerated	35 to 50 min (81)	Increase of a few min each session up to 50 min goal; increasing speed to 3.0 mph should also be a goal, as the average PAD patient's walking speed is 1.5 to 2.0 mph (126).	Improve: • PWT • COT • Patient-reported outcomes • $\dot{V}O_2$peak • Functional performance	For asymptomatic PAD, IC may not be a rate-limiting factor to exercise; thus RPE may be used to guide exercise intensity (15, 108).
Resistance	Refer to chapter 5 for standard resistance training guidelines.					
Range of motion	Refer to chapter 5 for standard flexibility training guidelines.					

IC = intermittent claudication; PAD = peripheral artery disease; PWT = peak walking time; COT = claudication onset time; $\dot{V}O_2$peak = peak oxygen consumption; RPE = rating of perceived exertion.

To date, the mechanisms by which exercise training may improve walking in patients with IC have not been completely elucidated. From a physiological perspective, some studies have shown that exercise rehabilitation for IC does not increase blood flow (44, 84, 95, 103, 104, 164, 184), but others have reported that exercise improved leg blood flow (5, 45, 75, 100). A review of studies using exercise training to treat patients with IC reported improvements in walking tolerance, but blood flow improvements alone could not account for the total improvement in pain-free walking distances (56). Following a 6 mo exercise training program, patients

Practical Application 15.2

RESEARCH FOCUS

Gardner and colleagues (55) conducted a randomized, controlled intervention trial assessing exercise performance outcomes of PAD patients at baseline and following 12 wk of supervised exercise training in a hospital setting (n = 33), home-based exercise training (n = 29) and no exercise training in a control group (n = 30). Patients were included if they had a history of any type of leg pain upon exertion, limitation in ambulation caused by leg pain during a graded treadmill exercise test, and an ABI of ≤0.90 at rest or an ABI ≤0.73 after exercise. Eligible patients performed a maximal treadmill walking test to determine peak walking time, claudication onset time, and $\dot{V}O_2$peak. Additionally, functional status and patient-reported outcomes were assessed pre- and postintervention using the Walking Impairment Questionnaire (WIQ) and physical component of the Medical Outcomes Study Short Form 36-item questionnaire (SF-36). Exercise compliance was monitored in patients randomized to the supervised exercise group and the home-based exercise group using an activity monitor. Patients were also asked to complete diaries or log books outlining the exercise sessions. In the supervised exercise program, patients walked for 15 min for the first 2 wk and increased duration each week until they had accomplished 40 min of walking during the final 2 wk of the program. Patients walked at a grade equal to 40% of the final workload from the baseline maximal treadmill test to the point of near-maximal IC pain; they then stopped to relieve the leg pain and began walking again when pain had attenuated. This process was repeated until the goal time was reached. The exercise program for the home-based group consisted of intermittent walking to near-maximal IC pain 3 d/wk at a self-selected pace. Walking duration began at 20 min for the first 2 wk and progressively increased 5 min biweekly until a total of 45 min of walking was accomplished. Patients randomized to the control group were encouraged to increase their walking activity but did not receive specific recommendations about an exercise program during the study.

Results indicated differences in pre- and post-intervention change scores for claudication onset time and peak walking time for both the supervised exercise group (+165 and +215 s) and the home-based exercise group (+134 and +124 s) versus the control group (–16 and –10 s, $p < 0.05$), with no differences between the two exercise groups. $\dot{V}O_2$peak change scores were significantly lower for the control group (13.7 to 12.8 ml · kg^{-1} · min^{-1}) compared to both exercise groups ($p < 0.05$); however, there were no differences in pre- and postintervention $\dot{V}O_2$peak within each individual exercise group (supervised: 11.4 to 11.7 ml · kg^{-1} · min^{-1}; home based: 11.8 to 12.4 ml · kg^{-1} · min^{-1}). Patients in the supervised and home-based exercise groups demonstrated significant improvements in the WIQ distance (+13% and +10%), speed (+9% and +11%), and stair climbing (+12% and +10%) scores, but were not significantly higher compared to the control group WIQ change scores (distance: +1%; speed: +4%; stair climbing: +3%). Finally, patient-reported outcomes change scores assessed by the physical component of the SF-36 were higher in the supervised exercise group compared to the control group only (+9% vs. –1% $p < 0.05$).

Supervised walking exercise is considered the gold standard physical activity option for treating PAD patients, and the study by Gardner and colleagues (55) adds to the existing body of literature for this recommendation. However, in contrast to other investigations of home-based exercise, this study also demonstrates that a home-based exercise program may be beneficial for improving the health of PAD patients. Because compliance was similar for the two exercise groups (supervised: 84.8%; home based: 82.5%), home-based exercise may be a feasible alternative to exercise programs at hospitals and clinics and should be examined in more detail.

Clinical Exercise Pearls

- Supervised exercise training has been given a class 1A recommendation within the American College of Cardiology/American Heart Association PAD guidelines. This indicates that exercise training is a powerful treatment option for PAD patients.
- Intermittent claudication is best relieved by sitting rather than standing. Sitting imposes less physical demand on patients and allows them to resume exercise sooner.
- Peripheral artery disease patients who have received endovascular therapy or surgical bypass may not experience leg pain during physical activity after the procedure. The use of RPE scales rather than pain scales may be an option to aid in regulating exercise intensity.

showed a 115% increase in claudication onset distance, and this improvement was independently related to the 27% increase in blood flow (53). Other studies suggest that the benefits of exercise conditioning for patients with IC appear more likely attributable to an improvement in calf muscle oxidative metabolism rather than to changes in skeletal muscle blood flow (76). Proposed mechanisms for the improvement in walking distance following exercise training include the following:

1. Improved biomechanics of walking resulting in decreased metabolic demands (155, 159, 179)
2. Increased angiogenesis and collateral circulation resulting in increased peripheral blood flow (5, 6, 45, 92, 156, 163)
3. A reduction in blood viscosity (46, 50)
4. An increase in blood cell filterability and a decrease in red blood cell aggregation (50)
5. Attenuation of atherosclerosis (75)
6. Increased extraction of oxygen and metabolic substrates resulting from improvements in skeletal muscle oxidative metabolism (30, 82, 164)
7. Increased pain tolerance (75, 184)
8. Improved endothelial function (16, 158)
9. An improvement in carnitine metabolism (75, 76)

CONCLUSION

This chapter provides a general review of PAD, focusing on symptomatic PAD and the benefits of exercise training. Optimal medical therapy especially with the inclusion of exercise is an important treatment option for PAD patients due to its low risk and high benefits. It is important for the exercise physiologist to be aware of concomitant diseases, as they may directly affect the type of exercise prescribed. Risk factor modification, such as smoking cessation, is critical for reducing cardiovascular morbidity and pre-mature mortality for PAD patients. Exercise training can improve the patient-reported outcomes, functional capacity, and the metabolic risk profile of these patients while also potentially reducing health care costs.

The current state of the health care system limits the treatment options for PAD patients. It is clear that supervised exercise training improves patients' health and is also cost-effective. However, until insurance reimbursement is available, other options such as home- and community-based exercise programs should be emphasized as a main treatment option for PAD patients. Unfortunately, most home and community exercise recommendations for PAD patients consist of advice given by their physicians to "go home and exercise" (24, 136). More research is needed to improve the effectiveness of these programs, particularly from the standpoint of increasing ease of adoption and long-term adherence.

Key Terms

angiography (p. 280)
ankle-brachial index (p. 278)
claudication onset time or distance (p. 285)
critical limb ischemia (p. 277)
endovascular revascularization (p. 288)
intermittent claudication (p. 277)
occlusions (p. 277)
patency (p. 289)
peak walking time or distance (p. 285)
stenoses (p. 277)

CASE STUDY

MEDICAL HISTORY

Mr. JH is a pleasant 55 yr old Caucasian male who came to the vascular clinic complaining of left leg calf pain. He reports no pain or symptoms in his right leg. He is currently able to walk 300 ft (one block) before the onset of moderate left leg pain, with initial onset at about 100 ft. Mr. JH also reported no pain at rest, only during exertion.

His history includes a positive stress test 4 mo earlier, during which he developed ST-segment depressions inferiorly with chest pain at peak exercise. He subsequently underwent cardiac catheterization with successful coronary artery stenting. He denies chest pain at rest and upon exertion. He has type 2 diabetes, diagnosed 4 mo ago, presenting with a HbA1c of 13.5%, now controlled with medications (6.0%). His other comorbid conditions controlled by medication include hyperlipidemia and hypertension (controlled: 118/70 mmHg). His medications include aspirin, 325 mg oral tablet daily; Toprol XL, 50 mg oral tablet daily; lisinopril, 20 mg oral tablet daily; Lantus, 2-unit injection at bedtime; metformin HCL, 1,000 mg oral tablet (0.5 tablet AM, 0.5 tablet PM); and Zocor, 40 mg oral tablet at bedtime.

Mr. JH has a family history of diabetes (mother) and coronary artery disease (father had myocardial infarction in his early 40s). He reports smoking 1.5 packs a day for 35 yr. He continues to smoke but reports having reduced smoking to 0.5 pack day. His resting supine pulse, respiratory rate, and temperature are normal (60 beats · min^{-1}; 16 breaths · min^{-1}; 36.9 °C). His femoral, popliteal, and pedal pulses were all palpable for both extremities, the left leg presenting with a much weaker pulse. He had no peripheral edema or evidence of gangrene or non-healing wounds on the forefoot or between the toes, but his foot was cold to touch. His right foot was warm and otherwise normal.

Vascular studies were completed to examine the peripheral vasculature. Mr. JH's ABIs were 0.80 for the right leg and 0.60 for the left leg. Baseline CT angiography indicated stenosis of the distal superficial femoral artery in his left leg. The peak systolic velocity was 300 cm/s proximal to the stenosis in the distal superficial femoral artery and 40 cm/s distal to the occlusion in the popliteal artery.

DIAGNOSIS

The patient's self-reported symptoms and the results of the non-invasive and angiographic studies confirm the diagnosis of PAD. It was determined that medical therapy would be the best course of action for treating Mr. JH given the location and severity of the disease. Mr. JH completed the WIQ and SF-36 questionnaires and then completed a graded exercise treadmill test in the clinic.

EXERCISE PRESCRIPTION

Mr. JH was referred to a supervised exercise program at the hospital for 3 d/wk for 3 mo before a return to clinic for follow-up. He was instructed to exercise to the point of moderate leg pain, ideally reaching this level in 3 to 5 min, and then rest until pain subsided, at which point he would resume walking. He repeated this until he reached a target window of 35 to 50 min. Bouts were increased by several minutes per week, as tolerated, up to 50 min of walk and rest time.

DISCUSSION

The patient improved his walking ability, increasing the distance before onset of pain as well as his total time walking. He initially was able to walk only one block before having to stop due to moderate leg pain but is now able to walk almost two full blocks before having to stop. Additionally, Mr. JH reported he was able to walk at a faster pace before onset of leg pain. He denied experiencing any chest pain throughout his exercise training program. He did report stiffness at the beginning of the exercise sessions and asked about starting the sessions at a slower pace. He would typically stand for rest sessions when walking on the treadmill and admitted that this also limited the number of days per week he attended the supervised sessions (showing up only two times per week for several weeks). Table 15.10 provides baseline and follow-up parameters.

Table 15.10 Case Study: Baseline and Follow-Up Parameters

Parameters	Pre-training	Post-training (3 mo)
ABI	Left leg: 0.60 Right leg: 0.80	Left leg: 0.62 Right leg: 0.84
COT (min)	1.1	2.2
PWT (min)	3.2	6.2
WIQ* (%)	66.7	77.9
SF-36** (%)	59.6	64.1
HDL cholesterol (mg · dl^{-1})	28	33
LDL cholesterol (mg · dl^{-1})	86	80
Triglycerides (mg · dl^{-1})	215	175
HbA1c (%)	6.0	5.8

ABI = ankle–brachial index; COT = claudication onset time; PWT = peak walking time; WIQ = Walking Impairment Questionnaire; SF-36 = Short Form 36-item questionnaire; HDL = high-density lipoprotein; LDL = low-density lipoprotein; HbA1c = glycated hemoglobin.

*Average of distance, speed, and stair climbing scores; **average of physical and mental component summary scores.

DISCUSSION QUESTIONS

1. Would you recommend that Mr. JH start with a warm-up and stretching routine before the exercise session? Why or why not?
2. What recommendations could be made to improve exercise compliance?
3. What recommendations, other than for the exercise prescription, do you have for Mr. JH to help improve his health?

Chapter 16

Cardiac Electrical Pathophysiology

Kerry J. Stewart, EdD
David D. Spragg, MD

The last several years have seen considerable advances in implantable cardiac devices that are used for regulating the heart rate, synchronizing the chambers of the heart in patients with heart failure, and shocking the heart back in the case of life-threatening arrhythmias like ventricular defibrillation and tachycardia. Many patients with an implantable cardiac device can resume their normal daily activities, including regular exercise. The purpose of this chapter is to describe some of the major indications for these devices, guidelines for physical activity, and strategies and precautions for living with a cardiac device to increase physical function and improve quality of life.

DEFINITION

The heart of the average person beats about 100,000 times per day. Each contraction results from an electric impulse that is initiated in the sinoatrial (SA) node, passes through the atrioventricular (AV) node, and is spread through the ventricles. An artificial pacemaker maintains a normal heart rate when the intrinsic electrical circuitry of the heart fails. The most common indications for **pacemaker** implantation are a heart rate that is too slow because of SA node dysfunction (i.e., a failure of impulse formation) or because of conduction block in the AV node (i.e., a failure of impulse propagation). When either of these two conditions results in a heart rate that is slow enough to cause symptoms (symptomatic bradycardia), permanent pacemaker implantation is indicated. Recently, pacing systems have been developed to normalize conduction and “resynchronize” the ventricles in patients with heart failure and myocardial conduction slowing, which is usually seen clinically as left bundle branch block (26). This mode of pacing, known as biventricular pacing, uses an additional pacing lead that is programmed to restore cardiac synchrony and mechanical activation that can lead to improvement in hemodynamics, ventricular remodeling, mitral regurgitation, exercise capacity, quality of life, and reduced mortality. This mode of pacing is discussed later in the chapter.

SCOPE

The need for pacing can occur at any age. Though some infants require a pacemaker from birth, about 85% of those who need a pacemaker are over the age of 65 yr, with an equal distribution among men and women. About 115,000 pacemakers are implanted in the United States each year. This number is likely to increase because of the growing number of elderly people in the population. In the 1950s, external pacemakers were used to treat symptomatic bradycardia. The 1960s produced AV pacemakers that provided a more physiological method of pacing. Today, with microcircuitry, pacemakers improve quality of life by optimizing the hemodynamic state at rest and can produce an appropriate heart response to meet the physiological demands of exercise. Because of newer technology, many patients can maintain or even begin an exercise program after pacemaker implantation. Therefore, clinical exercise physiologists should understand how pacemakers work to emulate normal cardiac rate, conduction, and rhythm in response to physiological and metabolic needs. They should also know about pacemaker programmed settings, how these settings can affect exercise capacity and the **exercise prescription**, and how to determine whether the patient's response to exercise is appropriate. The clinical exercise physiologist should communicate observations of the patient's responses to exercise to the pacemaker physician, who can reprogram the pacemaker to optimal settings.

PATHOPHYSIOLOGY

Rhythm disorders that involve the SA node are classified under the broad term **sick sinus syndrome**. This condition includes the inability to generate a heartbeat or increase the heart rate in response to the body's changing circulation demands. SA node dysfunction can cause fatigue, light-headedness, **exercise intolerance**, and syncope. A potentially more serious condition is failure of conduction through the AV node (heart block). When impulses are blocked at the AV node, ventricular activation is dependent on subsidiary **cardiac activation** arising from tissue below the level of block (typically from low in the AV node or from ventricular tissue). These escape rhythms can be unreliable, and heart block can lead to presyncope, syncope, or even death. Heart block may also cause a loss of *AV synchrony*, a term that refers to the sequence and timing of the atria and ventricles. Normally, the ventricles contract a fraction of a second after they have been filled with blood following an atrial contraction. Asynchrony may not allow the ventricles to fill with enough blood before contracting.

Depending on the patient's specific condition, the artificial pacemaker may replace SA node signals that are too slow, or that are delayed or blocked along the pathway between the upper and lower heart; maintain a normal timing sequence between the upper and lower heart chambers; and ensure that the ventricles contract at a sufficient rate. The 2008 Guideline Update for Implantation of Cardiac Pacemakers and Antiarrhythmia Devices issued by the American College of Cardiology, the American Heart Association, and the Heart Rhythm Society provides a review of the scientific literature and recommendations regarding which bradyarrhythmias and tachyarrhythmias may optimally be treated with a pacemaker (9). Table 16.1 shows pacemaker features that are used as therapy for different medical conditions (17).

Pacing System

A **pacing system** consists of separate but integrated components that stimulate the heart to contract with precisely timed electrical impulses. The pacemaker (also known as the pulse generator) is a small metal case that contains the circuitry controlling the electrical impulses, along with a battery. Modern pacemakers use lithium batteries that can last for many years, depending on the extent to which the pacemaker is used. Some patients depend on the pacemaker to provide cardiac rhythm and

Table 16.1 Pacemaker Therapies for Different Medical Conditions

Pacemaker features	Treatment conditions
Rate responsive for chronotropic incompetence	Feature is used by patients who need to sustain a heart rate that matches their metabolic needs to their daily lifestyle or condition.
Mode switch for managing atrial arrhythmias in patients with bradyarrhythmia	Many patients with sinus node dysfunction and atrioventricular (AV) block experience atrial fibrillation. Mode switch therapy reduces symptoms of atrial fibrillation during dual-chamber pacing.
Rate drop response for neurocardiogenic syncope	Patients with carotid sinus syndrome and vasovagal syncope are being treated for their symptoms by preventing their heart rate from falling below a prescribed level.
Ablation and pacing for atrial fibrillation	Patients with drug-refractory atrial fibrillation have been shown to benefit from ablation of the AV junction and implantation of a pacing system to maintain an appropriate heart rate.
Variable AV timing for patients with intermittent or intact AV conduction	Intrinsic AV activation is generally preferred to a ventricular-paced contraction because it provides improved hemodynamics and extended pacemaker longevity.

Based on http://www.medtronic.com/patients/bradycardia/index.htm.

conduction at all times. In others, the pacemaker acts as a backup that fires as needed when the sinus node fails to produce an appropriate rate or when the conduction system fails to transmit the impulses. The pacing leads are insulated wires connected to a pacemaker. The leads carry the electrical impulse to the heart and carry the response of the heart back to the pacemaker. These leads are extremely flexible to accommodate both the moving heart and the body. Depending on the type of pacemaker implanted, one, two, or three leads may be present. Each lead has at least one electrode that can deliver energy from the pacemaker to the heart and sense information about the electrical activity of the heart.

Pacemaker Implantation

This surgery, performed by surgeons or cardiologists, typically takes an hour or less and typically requires a single day of postoperative monitoring in the hospital. In some centers the implant may be done as an outpatient procedure in appropriate patients. Most patients receive a local anesthetic and remain awake during the surgery. The pulse generator is usually implanted below the collarbone just beneath the skin. The leads are threaded into the heart through a vein located near the collarbone. The tip of each lead is then positioned inside the heart. In rare cases, the pulse generator is positioned in the abdomen and the pacemaker leads are attached to the outside of the heart. After implantation, the pacemaker can be adjusted with an external programming device. The device works by sending radio frequency signals to the pacemaker via a transmitter placed on the chest.

Temporary External Pacemakers

An external pacemaker pulse generator is a temporary device that is commonly used in emergency and critical care settings, after open heart surgery, or until a permanent pacemaker can be implanted. This device is used outside the body as a temporary substitute for the heart's intrinsic pacing. Many of these devices have adjustments for impulse strength, duration, R-wave sensitivity, and other pacing variables.

Permanent Pacemaker Types

The two basic types of permanent pacemakers are single chamber and dual chamber. Both monitor the heart and send out pacing signals as needed to meet physiological demands.

1. A **single-chamber pacemaker** usually has one lead to carry signals to and from either the right atrium or, more commonly, the right ventricle. This type of pacemaker can be used for a patient whose SA node sends out signals too slowly but whose conduction pathway to the lower heart is intact. A single-chamber pacemaker is also used if there is a slow ventricular rate in the setting of permanent atrial fibrillation. In this case, the tip of the lead is usually placed in the right ventricle.

2. A **dual-chamber pacemaker** has two leads. The tip of one lead is positioned in the right atrium, and the tip of the other lead is located in the right ventricle. This type of pacemaker can monitor and deliver impulses to either or both of these heart chambers. A dual-chamber pacemaker is used when the conduction pathway to the lower chamber is partly or completely blocked. When pacing does occur, the contraction of the atria is followed closely by a contraction in the ventricles, resulting in timing that mimics the heart's natural way of working. Pacemakers are categorized by a standardized coding system developed by the Heart Rhythm Society (formerly the North American Society of Pacing and Electrophysiology) and the British Pacing and Electrophysiology Group.

Figure 16.1 shows the coding system for pacemaker functions. The letters refer to the chamber paced, the chamber sensed, what the pacemaker does when it senses an event, and other programmable features. This example is a ventricular demand pacemaker that is also rate responsive. The first V indicates that the ventricle is paced. The second V indicates that the pacemaker is programmed to sense for an impulse in the ventricle. The I indicates that when the pacemaker senses the patient's native ventricular impulse, the pacemaker is inhibited. The R indicates that the pacemaker is rate responsive or rate adaptive. A sensor in the pacemaker senses physical activity and adjusts the patient's pacing rate according to the level of activity.

Figure 16.2 shows the code for a dual-chamber pacemaker that is also rate responsive. The D stands for dual. Because of two leads, a dual-chamber pacemaker can pace both the atria and the ventricles. Likewise, the pacemaker can sense in the atria and ventricles. The third D indicates that the pacemaker can be either inhibited or triggered by the patient's own cardiac activity. The pacemaker will watch for atrial activity, and if it detects none, it will pace the atrium. After an appropriate AV time interval, the pacemaker will watch for a ventricular depolarization. If this is sensed, the pacemaker will be inhibited. If no ventricular activity is present, the pacemaker will pace the ventricle.

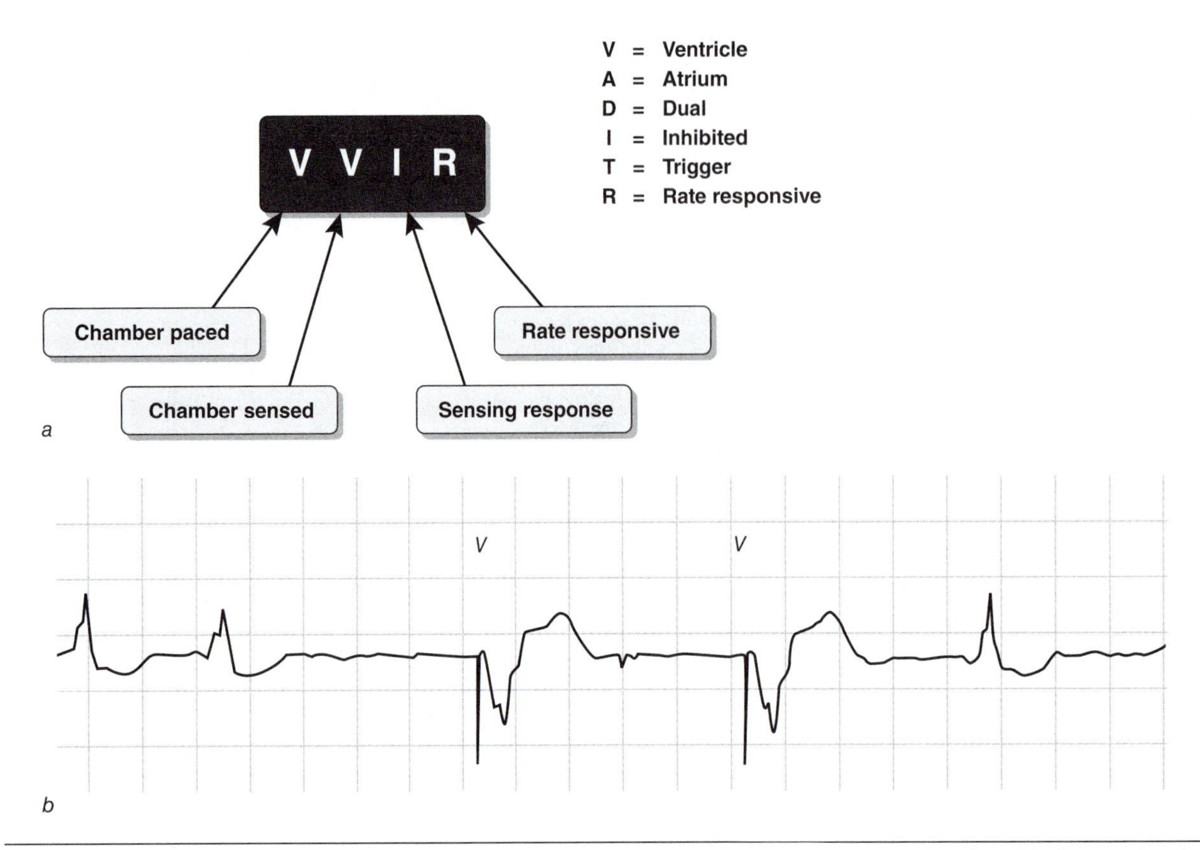

Figure 16.1 *(a)* Coding for a ventricular demand pacemaker that is also rate responsive. The first V indicates that the ventricle is paced. The second V indicates that the pacemaker senses in the ventricle. The I indicates that the pacemaker will be inhibited. The R indicates that the pacemaker is rate responsive. *(b)* VVI operation during atrial fibrillation. Ventricular pacing (V) occurs at the programmed lower rate limit of 60 beats · min^{-1} when the intrinsic ventricular activity falls below that level. Intrinsic ventricular activity at a faster rate inhibits ventricular pacing.

CLINICAL CONSIDERATIONS

The physiology of exercise for patients with pacemakers is generally the same as for other patients. The difference is in how their physiology interacts with the device. For patients who cannot provide an appropriate cardiac output response to exercise, pacemakers attempt to increase the cardiac output to meet changing physiological demands. The increase in oxygen uptake from rest to maximal exercise follows this formula: Oxygen uptake is equal to cardiac output multiplied by arteriovenous oxygen difference. From rest to maximal exercise, oxygen uptake can increase 700% to 1,200%, arteriovenous oxygen difference by 200% to 400%, and cardiac output by 200% to 400%. Cardiac output is equal to heart rate multiplied by stroke volume. With exercise, stroke volume can increase by 15% to 20%, whereas heart rate can increase by 200% to 300%. Thus, heart rate is the most important component for increasing cardiac output and is most closely related to metabolic demands. Though AV synchrony contributes to cardiac output, this factor is more important at rest and less important with exercise.

Physiological Pacing

The term **physiological pacing** refers to the maintenance of the normal sequence and timing of the contractions of the upper and lower chambers of the heart. AV synchrony provides higher cardiac output without increasing myocardial oxygen uptake. Dual-chamber pacemakers attempt to provide this physiologically beneficial function. The pacemaker senses the patient's sinus node and, in complete heart block, sends an impulse to the ventricle following an appropriate AV timing interval. Though the specific change in cardiac output depends on many factors, the optimal AV delay to produce the maximum cardiac output in normal people is about 150 ms from the beginning of atrial depolarization. The efficiency of cardiac work decreases with a shorter or longer AV interval. In normal subjects, the AV interval shortens with increased heart rate. The pacemaker can also set the AV interval based on heart rate. A dual-chamber pacer can also initiate an atrial impulse in sick sinus syndrome. AV synchrony augments ventricular filling and cardiac output, improves venous return, and assists in valve clo-

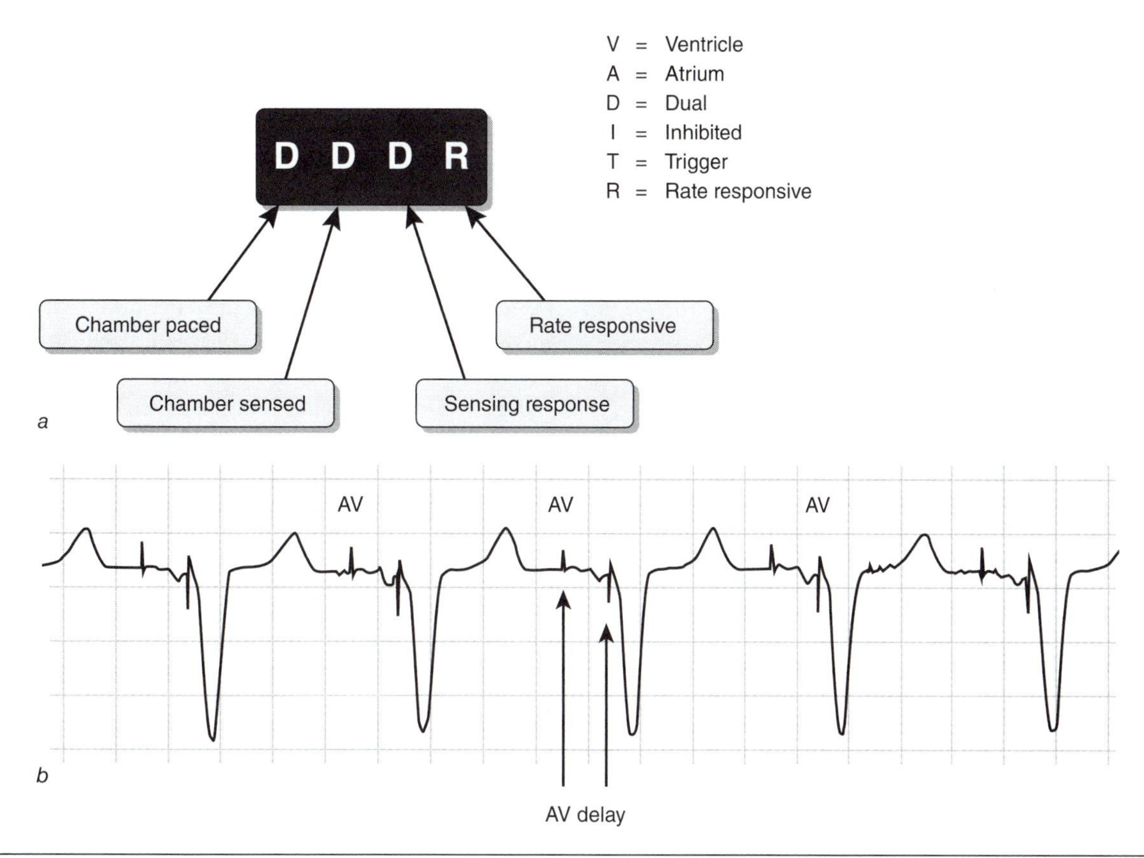

Figure 16.2 *(a)* Coding for a ventricular demand pacemaker that is also rate responsive. The first D stands for dual, indicating pacing in both the atria and the ventricles. The second D indicates sensing capability in both the atria and ventricles. The third D indicates inhibited or triggered, and the R stands for rate responsive. *(b)* DDD operation. Atrial pacing (A) occurs at the programmed lower rate limit of 75 beats · min^{-1}. Because of complete heart block, the pacemaker tracks the atrial rate to pace the ventricle (V) at the same rate after a programmed AV delay.

sure. The loss of atrial function increases atrial pressure and pulmonary congestion. The benefit of AV synchrony is independent of any measure of left ventricular function. The maintenance of normal AV synchrony allows for improved hemodynamic responses with a more normal increase in cardiac output (13). Because of higher cardiac output at any given level of work with synchronous pacing, the arteriovenous oxygen difference is narrower and the serum lactate is lower. Thus, synchronous pacing results (figure 16.3) in less anaerobic metabolism at the same level of work (15). AV synchrony and stroke volume provide their most important contributions to cardiac

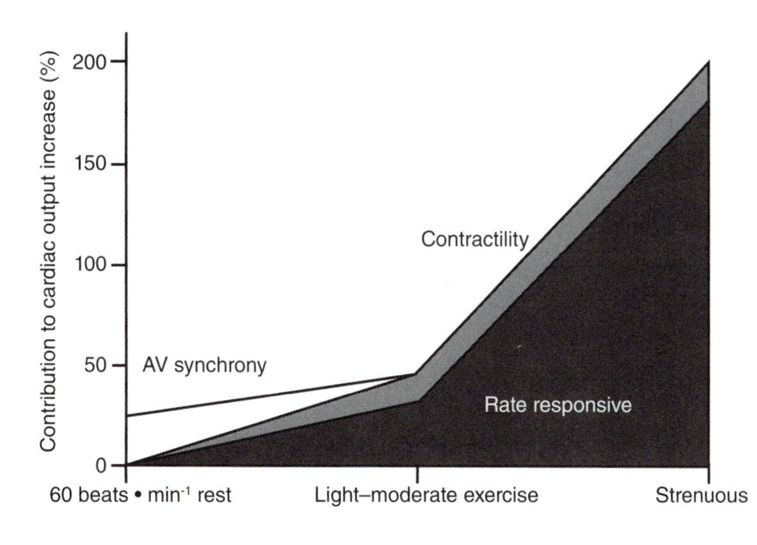

Figure 16.3 The relative contributions of atrioventricular (AV) synchrony, stroke volume (contractility), and heart rate to cardiac output at rest and exercise. Heart rate is the most important contributor to cardiac output during exercise.

output at rest, whereas an increase in heart rate is the predominant factor contributing to cardiac output during exercise (figure 16.4).

Mode Switching and Maximal Tracking Rates in Dual-Chamber Pacemakers

Many patients with sinus node dysfunction and AV block develop atrial arrhythmias. The most common arrhythmia is atrial fibrillation. To prevent the dual-chamber pacemaker from tracking or matching every atrial impulse with a ventricular pacing pulse, **mode switching** controls the ventricular rate. Mode switching temporarily reverts to a nontracking mode so that irregular or excessive atrial activity does not drive the ventricles to an extremely high rate. The mode-switching feature is programmable and, depending on the specific pacemaker model, can be adjusted for optimal performance in any given patient.

A different but related concept is the pacemaker maximal tracking rate. This rate refers to the maximal atrial rate that will trigger ventricular pacing. As the atrial rate begins to exceed the maximal tracking rate, many pacemakers allow individual atrial impulses to be ignored (resulting in single "dropped" ventricular beats). As the atrial rate continues to surpass the maximal tracking rate, the pacemaker reverts to 2:1 atrioventricular conduction, in which every other atrial beat results in ventricular pacing. As the atrial rate slows to below the maximal tracking rate, 1:1 atrioventricular activation resumes.

Rate-Responsive Pacing

The development of **rate-responsive pacing** (also called rate adaptive or rate modulated) has dramatically changed the application of pacing with regard to physical activity. The rate-responsive function is used when the native sinus node cannot increase heart rate to meet metabolic demands. Increasing heart rate in response to exercise is probably the single most important factor for increasing cardiac output and oxygen uptake. A sudden increase in exercise requires the heart rate to adjust quickly to the workload. Rate-responsive pacemakers can sense the body's physical need for increased cardiac output and produce an appropriate cardiac rate in patients with chronotropic incompetence. The highest rate at which the pacemaker will pace the ventricle in response to a sensor-driven rate is known as the maximal sensor rate. When rate-modulated pacing is compared with non-rate-modulated pacing (figure 16.5), exercise capacity is extremely limited without an appropriate increase in heart rate (6).

Several physiological and metabolic changes occur during exercise as the demand for energy increases:

- Movement that produces vibration and acceleration
- Respiration
- Heat that raises body temperature
- Electricity activity that produces electrocardiographic and electromyographic changes
- Carbon dioxide

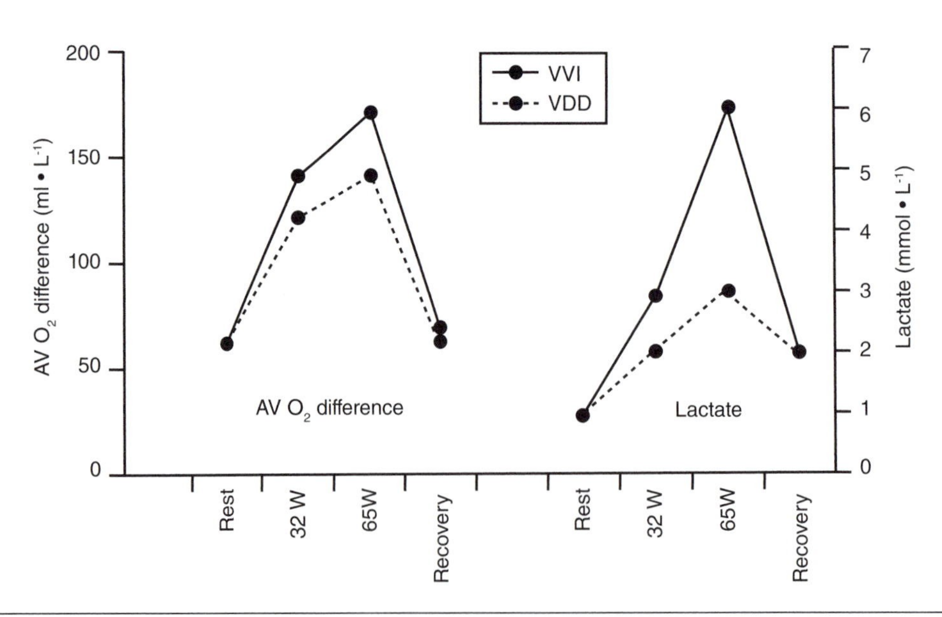

Figure 16.4 Synchronous pacing (VDD) results in less anaerobic metabolism at a given level of work compared with nonsynchronous pacing (VVI).

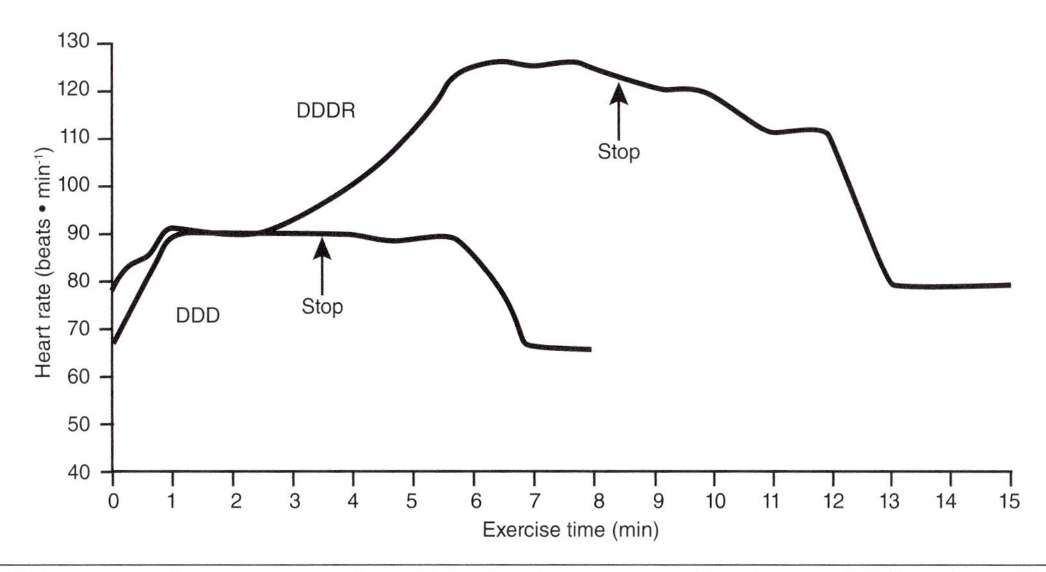

Figure 16.5 A comparison of rate-modulated (DDDR) and non-rate-modulated pacing (DDD) during treadmill exercise. Without an appropriate increase in heart rate, exercise capacity is extremely limited in a patient with chronotropic incompetence.

Adapted from Bodenhamer and Grantham 1993.

- Lactic acid, which reduces blood pH
- Intracardiac pressure

Various sensors have been developed to detect these changes and, based on computer algorithms, generate the electrical impulses that are used to pace the heart. The development of optimal sensors and algorithms for rate-modulated pacing systems must meet several requirements:

- The sensor should rapidly detect acceleration and deceleration of physiological changes.
- The response should be proportionate to the exercise workload and metabolic demands.
- The response should be sensitive to both exercise and nonexercise requirements such as posture, anxiety and stress, vagal maneuvers, circadian variations, and fever.
- The response should be specific and not be falsely triggered.

The most common rate-responsive pacemakers detect motion in response to physical activity. Vibration sensors use a piezoelectric crystal located in the pulse generator to detect forces generated during movement. These forces are transmitted to the sensor through connective tissue, fat, and muscle. Acceleration sensors detect body movement in anterior and posterior directions (8). The circuitry is also located in the pulse generator. Because the sensor is not in direct contact with the pacemaker case, no reaction to vibration or pressure occurs.

Producing an appropriate heart rate in response to certain work tasks poses a technological challenge. The simple task of walking up and down stairs can produce different heart rate responses, based on the type of sensor used to sense motion. Compared with a normal heart rate response, an accelerometer produces similar results going up stairs but overestimates the metabolic demand going down stairs. The vibration sensors produce a heart rate that is too low for stepping up and a rate that is too high for stepping down. This result occurs because the vibration of walking down stairs is greater than the vibration of walking up stairs, though the metabolic demand is greater for walking up stairs (4). In contrast to single sensors, dual-sensor rate response provided by activity and minute ventilation may help overcome these types of problems, as observed in appropriate heart rate responses while patients are ascending and descending stairs (2). Advances in pacemaker technology may allow the use of combined or blended sensors and advanced algorithms to improve rate performance over a single-sensor system (16). Blended sensors are designed to measure patient workload through respiration and motion, providing optimal rate response during changing levels of activity. Nevertheless, definitive clinical benefits from blended sensors have yet to be fully established (25).

Exercise Testing

Patients with pacemakers capable of rate modulation should undergo **exercise testing** to ensure appropriate rate responses (12). Exercise capacity and quality of life are improved by appropriately programmed rate-responsive pacemakers compared with fixed-rate units (28). These devices can be programmed to match the needs of the patient more closely. The primary pacemaker settings can be adjusted to optimize responses to physical activity:

- Sensitivity of the sensor
- Responsiveness to a physiological change
- Rate at which the cardiac rate changes
- Minimum rate at rest and maximal rate at peak activity

Exercise testing is used to guide the adjustment of these settings to improve exercise capacity and reduce symptoms. Exercise testing helps establish upper rate limits and adjust the sensitivity and responsiveness of the sensor. Exercise testing is also used to determine the anginal threshold, if any. Pacing the heart rate beyond the point at which ischemia would occur would not be prudent. Several approaches to exercise testing can be used. These include using informal or formal protocols with or without real-time electrocardiogram (ECG) monitoring and with or without determination of optimal rate-responsive parameters. The patient's health status and lifestyle, the type of pacemaker, and the facilities and experience of personnel also help determine the specific approach to exercise testing. For many patients, informal exercise testing is a reasonable and a less expensive alternative to formal treadmill testing (12). Empiric adjustment of the rate-response parameters is common.

With informal testing, the patient walks at a self-determined casual pace and at a brisk pace, usually for about 3 min each. The sensor-driven cardiac rate can be determined via examination of the ECG. Because pacemakers are capable of storing an electronic record of pacemaker activity, the physician, using a special computer, can also interrogate the pacemaker to examine a histogram display of the heart rate response during the walk. The optimal pacemaker rate is determined empirically. For casual exercise, the target is often 10 to 20 beats · min^{-1} above the lower rate limit. For brisk exercise, the target can be 20 to 50 beats · min^{-1} above the lower rate limit. This approach to exercise testing is best suited for less active patients who are unlikely to reach their upper rate limit. By examining a display of the sensed atrial rate as measured by an event counter in the pacemaker, or by measuring the heart rate by ECG and asking the patient about symptoms, the physician makes a clinical judgment about whether the patient is chronotropically competent. If not, the pacemaker will need to be programmed to elicit an appropriate response. Formal exercise testing allows a chronotropic evaluation that seeks to match the pacemaker-augmented response of the chronotropically incompetent patient to the metabolic requirements of the body (24). The 6 min walk can also be used as an alternative test and has good validity for predicting maximal oxygen uptake (19). Formal exercise testing is typically best for active patients likely to reach the programmed maximum sensor rate. Programming the upper rate of rate-adaptive pacing improves exercise performance and exertional symptoms during both low and high exercise workloads compared with a standard nominal value of 120 beats · min^{-1} (7).

With formal exercise testing, the protocol selected requires careful consideration. Many protocols are used for exercise testing. Nevertheless, many of the traditional protocols such as the Bruce and Naughton protocols are designed to test for coronary artery disease. Their usefulness in defining optimal programming for rate-responsive pacemakers may be limited. A widely used protocol for assessing patients with pacemakers is the chronotropic assessment exercise protocol (29, 30) shown in table 16.2. The advantage of this protocol is that the workload gradually increases to mimic the range of activities of daily living. This protocol allows a more complete assessment of how the pacemaker responds at the lower metabolic equivalent (MET) ranges where patients typically spend most of their time. The chronotropic assessment exercise protocol has five stages at a lesser exercise intensity than the Bruce protocol produces in the second stage (31). Because the Bruce protocol increases by 2 to 3 METs during each 3 min stage, assessing the patient's work capacity and the ability of the pacemaker sensor to provide an adequate hemodynamic response would be difficult. Table 16.3 provides a brief review of exercise testing.

Table 16.2 Chronotropic Assessment Exercise Protocol

Stage	Speed	Grade	Cumulative time	Metabolic equivalents
1	1.4	2	2	2.0
2	1.5	3	4	2.8
3	2.0	4	6	3.6
4	2.5	5	8	4.6
5	3.0	6	10	5.8
6	3.5	8	12	7.5
7	4.0	10	14	9.6
8	5.0	10	16	12.1
9	6.0	10	18	14.3
10	7.0	10	20	16.5
11	7.0	15	22	19.0

Table 16.3 Exercise Testing Review

Mode	Protocol specifics	Clinical measures	Clinical implications	Special considerations	Cardiovascular
Walking	Informal or formal	Heart rate response, blood pressure response, symptoms, time, METs, or maximal oxygen uptake	Used to guide adjustments of the device to optimize heart rate response to improve exercise capacity and reduce symptoms	Informal or formal testing depending on the patient's lifestyle and health status; formal testing best for active patients likely to reach higher heart rates	Testing focused on conduction system of the heart but can also be used for diagnosis of coronary artery disease; cardiac imaging may be needed because the ECG may not allow for assessment of ischemic changes.

EXERCISE PRESCRIPTION AND TRAINING

Dual-chamber pacemakers are in greater use today than in the past. Clinical exercise physiologists need to be familiar with the normal behavior of these devices during exercise. Figure 16.6 shows DDDR operation. The rate at which the sensor-driven heart rate increases follows algorithms that are programmed into the pacemaker. Among the key parameters are the slope of the heart rate increase and decline and the sensitivity of the sensor. With increased physical activity, the pacemaker will follow the sinus rate up to a maximal tracking rate. The activity sensor can be programmed to allow a further increase in the paced rate to the maximal sensor rate in response to physical activity. If the patient continues to exercise, the pacemaker may reach its maximal tracking rate or maximum sensor rate. When this occurs, the pacemaker will not further increase the heart rate. If the patient's native sinus rate continues to increase beyond this point, the pacemaker will switch to an AV block mode because the sinus rate now exceeds the rate at which the pacemaker will permit tracked ventricular pacing. The pacemaker first switches to a Wenckebach-type block to cause a gradual slowing of the ventricles, and 2:1 AV block ensues if the sinus rate continues to rise. This feature protects against nonexercise sinus tachycardia that might otherwise force the pacemaker to produce ventricular tachycardia.

At higher levels of exercise, the metabolic demands will be high, but 2:1 block may occur and slow the ventricular rate. In this situation, the development of 2:1 block is a normal feature of the pacemaker but can cause symptoms because of the sudden drop in heart rate. If this occurs, the patient is exercising too hard or the maximal

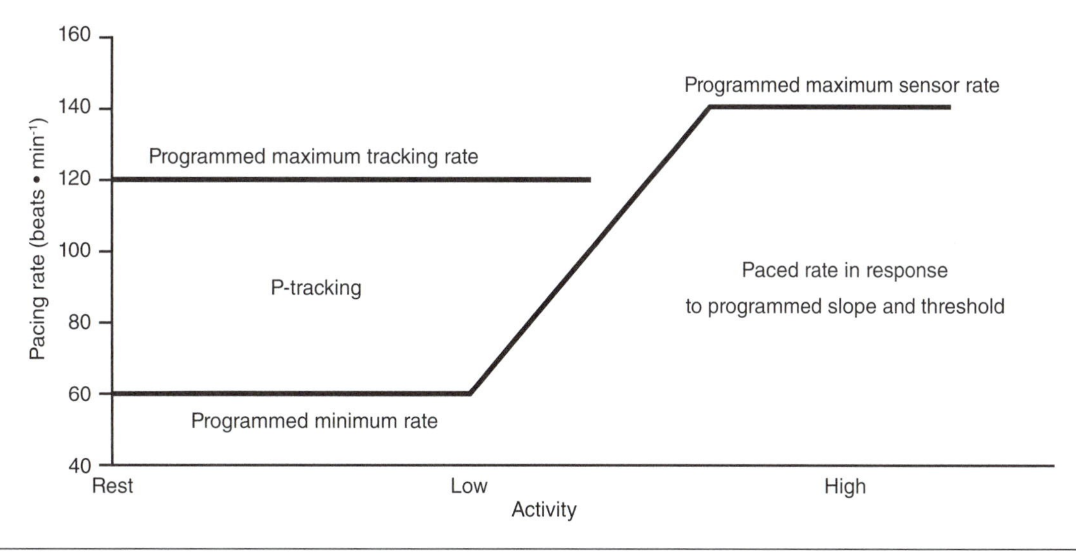

Figure 16.6 DDDR pacing and physical activity. The pacemaker follows the sinus rate to a maximal tracking rate. In response to physical activity, the sensor-driven response can drive the rate to the maximal sensor rate. The paced rate increases in response to programmed slope and threshold settings.

tracking rate is set too low. If the pacemaker is also rate adaptive, the sensor rate may be too low.

The exercise physiologist should record and communicate episodes of abrupt decreases in heart rate to the patient's pacemaker physician so that programmed settings can be evaluated for possible change. Several of the exercise modalities that are commonly prescribed in cardiac rehabilitation and adult fitness programs may pose a particular challenge in some patients with activity sensors (3). In the example shown in figure 16.7, because most of the increase in work is accounted for by raising the slope rather than speed on a treadmill, little change in the generated forces would be detected by the vibration sensor. Thus, the heart rate determined by the vibration sensor is too slow for the metabolic demand of the increased work. In this case, the accelerometer sensor is better able to provide a heart rate that more closely matches an appropriate rate response.

The clinical exercise physiologist should also be aware of how a vibration sensor responds to outdoor and stationary cycling (figure 16.8). The response of this type of sensor is particularly relevant to cardiac rehabilitation because stationary cycling and seated steppers are prevalent modes of exercise in many programs. This sensor response may explain why some patients complain of unusual fatigue and shortness of breath during stationary cycling but not other types of exercise such as treadmill walking. Patients with an artificial pacemaker require long-term surveillance by their physicians to ensure optimal adjustment of the programming for their individual needs, to maximize the life expectancy of the pacemaker through adjustment of pacemaker output settings, and to identify and treat complications. Pacemaker follow-up relies on clinical, electrocardiographic, and device assessment. Other tests may include exercise testing, Holter monitoring, and echocardiography. The device assessment requires a specialized programmer to verify pacemaker functions. In some cases, remote interrogation of the pacemaker over telephone lines is done periodically to provide useful information about selected functions of the pacemaker when a more complete test is not deemed necessary. The clinical exercise physiologist can play an important role in the overall evaluation of the patient by providing feedback to the physician about heart rate, blood pressure, and symptomatic responses to exercise. The case study at the end of the chapter illustrates the role of the clinical exercise physiologist in the management of the patient with a pacemaker.

Exercise prescription requires special attention to the type of pacemaker that is implanted. With fixed-rate pacemakers, cardiac output and arterial pressure are increased by stroke volume (1). Target heart rate cannot be used to guide exercise intensity. Instead, the patient follows ratings of perceived exertion (RPE). It is also important to monitor blood pressure to ensure an appropriate intensity. When the sinus node is normal, it is desirable to have the pacemaker "track" native sinus activity by pacing the ventricle after an appropriate AV delay (11). In all cases, the target heart rate must be lower than the anginal threshold in a patient with

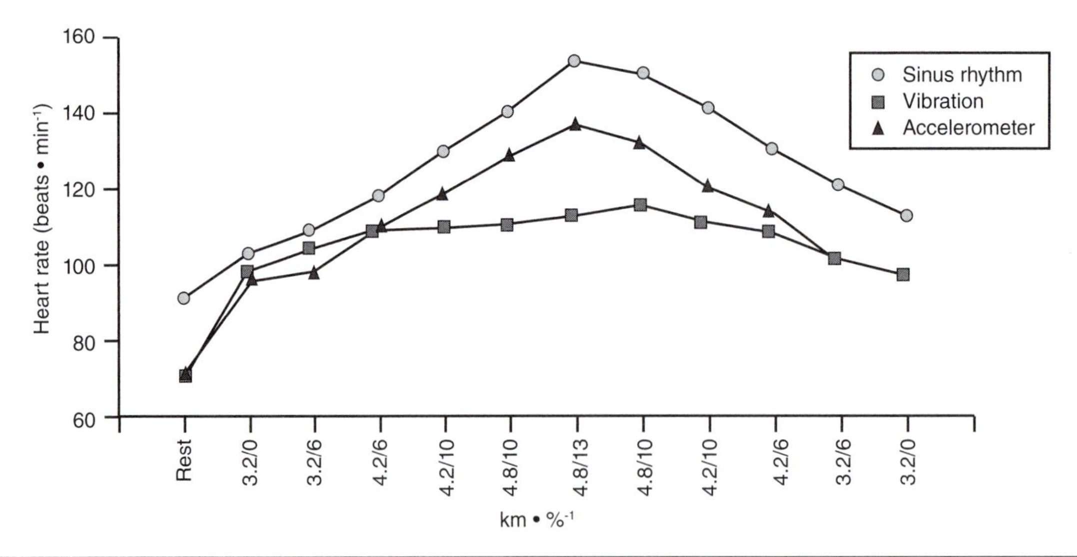

Figure 16.7 Generated forces. The circles show a normal sinus response at rest and with increased treadmill work. Work is increased primarily by raising the slope rather than speed. At rest, sinus rate is about 20 beats above pacemaker rate. This difference is maintained throughout the test. The accelerometer sensor (triangles) is able to produce a heart rate that better matches with the workload compared with the vibration sensor (squares).

Adapted from Alt and Matula 1992.

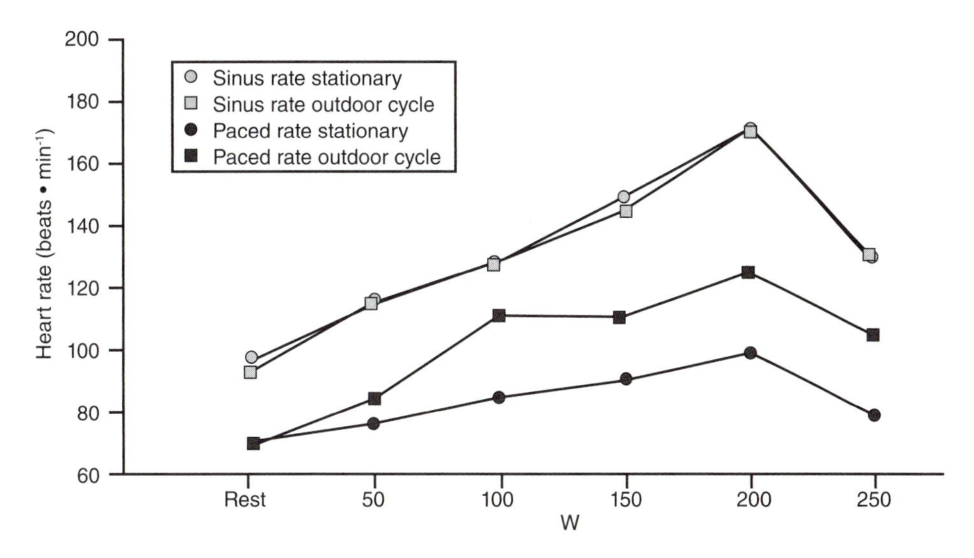

Figure 16.8 Vibration sensor. The upper lines represent a normal sinus rate with outdoor cycling (open squares) and stationary cycling (open circles) at increasing workloads and recovery. The lower lines represent the sensor. The difference of 20 beats at rest is maintained throughout the test during outdoor cycling (filled squares). During stationary cycling (filled circles), the paced cardiac rate is considerably slower than both the sinus rate and the rate during outdoor cycling. This occurs because stationary cycling produces less body motion and vibration.

Adapted from Alt and Matula 1992.

ischemia (5). Tailoring the exercise prescription and modifying the response rate of the pacemaker based on cardiopulmonary stress testing that determines the anaerobic threshold have been shown to provide functional advantages for patients in cardiac rehabilitation (10). Because rate-responsive pacemakers mediate the heart rate response to exercise, the type of sensor must be taken into account when exercise is prescribed (24). Sensors that detect movement may respond slowly to stationary cycling and increased treadmill slope. Again, the RPE and MET equivalents are extremely useful in establishing the exercise prescription. With modern pacemakers, pacing occurs only when needed. In many patients, such as those with normal sinus function with intermittent heart block, the exercise prescription can be written in the same way as for most other patients. Regarding heart rate monitoring, one study found that the use of dry-electrode heart rate monitors that transmit a signal to a monitor, such as those worn on the wrist, had no adverse effect on pacemaker function (15).

Exercise Recommendations

Patients with pacemakers can derive benefits from an exercise program similar to those gained by other people. The area of greatest consideration in the prescription of exercise is the issue of exercise intensity. Because of the variety of pacemakers, sensors used to detect activity, and mode of exercise prescribed, the appropriate heart rate response can vary considerably. Therefore, when patients with pacemakers start exercising, they should be monitored to make sure that the pacemaker is responding appropriately. See practical application 16.1 for a summary of the relevant exercise prescription.

Mode

Generally all forms of exercise are acceptable in patients with pacemakers, except activities that can cause direct contact with the pacemaker. Therefore, contact sports such as football, soccer, and hockey are generally not recommended. All forms of aerobic exercise are generally acceptable, and these most likely carry the greatest benefit with regard to improving overall health and decreasing risk factors for cardiovascular disease. When a patient with a pacemaker performs any form of aerobic exercise, rate-responsive pacemakers should be evaluated to see that they are increasing the heart rate appropriately relative to the intensity of exercise. Pacemakers that rely on vibration or accelerometer sensors to detect body motion during exercise may not produce an adequate response for activities such as stationary cycling and increased treadmill slope. Unusual shortness of breath and fatigue may indicate a lack of rate-responsive pacing and need to be monitored during different forms of activity. Weight training may also be acceptable, though weights or bars must not come in contact with the pacemaker.

Practical Application 16.1

SUMMARY OF EXERCISE PRESCRIPTION

The exercise prescription for those who use pacemakers is generally the same as for others, and many of the same cautions apply. The prescription must consider comorbidities such as angina and chronic heart failure, for example. The area of greatest consideration when one is prescribing exercise is the issue of exercise intensity, which is determined by the underlying reason for the pacemaker and the type of pacemaker implanted. For patients with an internal cardioverter defibrillator, the exercise heart rate should be kept at least 20 beats below the firing threshold. Activities that might result in contact with an implanted device should be avoided. Table 16.4 provides a brief review of exercise prescription guidelines.

Table 16.4 Exercise Prescription Review

Training method	Mode	Intensity	Frequency	Duration	Progression	Goals	Special considerations
Cardiovascular	All forms are acceptable; avoid contact sports that may affect area of device implantation.	Goal is to follow usual guidelines for comorbid conditions such as angina and heart failure.	Follow usual guidelines for comorbid conditions if present; other usual guidelines apply.	Follow usual guidelines for comorbid conditions if present; other usual guidelines apply.	Follow usual guidelines for comorbid conditions if present; other usual guidelines apply.	Increased cardiovascular endurance capacity	Patient may not be able to achieve target heart rate, depending on indication and type of device; keep heart rate 20 beats below firing rate if there is an internal defibrillator.
Resistance	Goal is to follow usual guidelines for comorbid conditions such as angina and heart failure.	Goal is to follow usual guidelines for comorbid conditions such as angina and heart failure.	Goal is to follow usual guidelines for comorbid conditions such as angina and heart failure.	Goal is to follow usual guidelines for comorbid conditions such as angina and heart failure.	Goal is to follow usual guidelines for comorbid conditions such as angina and heart failure.	Increased muscle strength and endurance; improved body composition; improved metabolism	Follow usual care guidelines for comorbid conditions.

Frequency

The frequency of activity is based on the goals of the program. If someone is interested in improving his health and is exercising at an intensity less than 60% of maximal aerobic capacity, daily activity is recommended. If the person is able and willing to exercise at a higher intensity (60-85% of maximal aerobic capacity), activity on 3 to 5 d a week is recommended.

Intensity

The intensity should be in the recommended range of 40% to 85% of maximal aerobic capacity but is primarily dependent on comorbid conditions (e.g., angina, chronic heart failure). Also, upper limits of the pacemaker (tracking and sensing) can influence the upper limit of exercise intensity. Because heart rate does not increase in a patient with a fixed-rate pacemaker, RPE and MET equivalents need to be used to evaluate exercise intensity. Additionally, blood pressure should be monitored to show appropriate increases with increasing workload. A patient with a fixed-rate pacemaker should not exceed exercise intensity above the point where blood pressure begins to plateau with increasing workload. With dual-chamber and rate-responsive pacemakers, heart rate can be used to determine exercise intensity and should be used along with RPE and METs. Knowledge of maximal tracking or sensing rates determines the upper intensity level. Patients should be monitored closely, and activities should be chosen based on the ability of the pacemaker to adjust heart rate with increasing metabolic demands.

Duration

The duration of activity is similar to that specified in the general recommended guidelines for promoting health and fitness (20-60 min). Duration depends on goals. Ideally, the duration should be adjusted so that the individual achieves an energy expenditure of at least 1,000 kcal per week.

Special Considerations

If a patient goes into second-degree type 1 block (Wenckebach) while exercising, the patient's native sinus rate likely exceeded the pacemaker's maximal tracking or sensor rate. If this occurs, the intensity of exercise should be reduced. If the exercise professional notices a decrease in heart rate well below the patient's tolerable limits, this information should be forwarded to the pacemaker physician so that programmed settings can be evaluated. Practical application 16.2 provides more information about living with a pacemaker.

Practical Application 16.2

PATIENT EDUCATION ABOUT LIVING WITH A PACEMAKER

The clinical exercise physiologist is often a primary source of patient education. Pacemaker patients frequently ask about what they should be aware of in their day-to-day lives. Advances in pacemaker technology have resulted in continuing improvements. Pacemakers are smaller and better shielded from external interference and magnetic fields than ever before. Recent research has shown that with appropriate protocols, magnetic resonance imaging (MRI) can be performed safely in patients with certain pacemakers and internal cardioverter defibrillator systems (18). Nevertheless, the clinical exercise physiologist should communicate some basic precautions to the patient. This section presents some common questions and issues.

Sports and Recreational Activities

Many active patients, after appropriate medical clearance, can travel, drive, bathe, shower, swim, resume sexual activities, return to work, walk, hike, garden, golf, fish, and participate in other similar activities. But contact sports that include jarring, banging, or falling such as football, baseball, and soccer should be avoided. Also, patients should avoid hunting if a rifle butt is braced against the implant site.

Work Activities

Most office equipment is unlikely to generate the type of electromagnet interference that can affect a pacemaker, but equipment with a large magnet should not be carried if it is held near the pacemaker. Patients who work with heavy industrial or electrical equipment need to consult with their physicians about resuming work because this equipment may produce high levels of electromagnetic interference that could affect pacemaker function.

Home Activities

People with pacemakers can participate in most activities of daily living and can be reassured that most home electrical devices will not interfere with pacemaker operation. But some precautions are recommended. Cellular phones should be kept at least 6 in. (15 cm) away from the pacemaker site, and phones transmitting above 3 W should be kept at least 12 in. (30 cm) away. The patient should hold the cell phone to the ear opposite the pacemaker site and should not carry a phone in a pocket or on a belt within 6 in. (15 cm) of the pacemaker. Ordinary cordless, desk, and wall telephones are considered safe. The patient should not lift or move large speakers because their large magnets may interfere with the pacemaker. General household electrical appliances like televisions and blenders, and outdoor tools such as electric hedge clippers, leaf blowers, and lawn mowers, do not usually interfere with pacemakers. But the patient should avoid using tools such as chain saws that require the body to come into close contact with electric spark–generating components. In addition, caution is advised when people are working near the coil, distributor, or spark plug cables of a running engine. The safe approach is to turn off the engine before making any adjustments.

(continued)

Practical Application 16.2 *(continued)*

Travel

Most people with pacemakers can travel but should tell airport security personnel that they have a pacemaker or other implanted medical device before going through security systems. Though airport security systems do not affect the pacemaker, the pacemaker's metal case could trigger the metal detection alarm. Home, retail, or library security systems are unlikely to be set off by the pacemaker.

Automatic Internal Cardioverter Defibrillators

Subsets of cardiac patients are at high risk for potential lethal ventricular tachycardias. This group includes patients who have survived a previous sudden death event and are at high risk for recurrent cardiac arrest, as well as patients who have never had sudden death but are at high risk for cardiac arrest because of prior myocardial injury. Though medications that stabilize heart rhythm are available, they are not entirely effective and often produce serious side effects. Increasingly, **automatic internal cardioverter defibrillators** (ICDs) are being used to control life-threatening ventricular arrhythmias. An ICD is a battery-driven implanted device, similar to a pacemaker, that is programmed to detect and then stop a life-threatening ventricular arrhythmia by delivering an electrical shock directly to the heart. Some models provide tiered therapy by including the capability of providing antitachycardia pacing, cardioversion, and defibrillation as needed. Modern ICDs are implanted beneath the skin and muscle of the chest or abdomen, and electrodes that sense the heart rhythm and deliver the shock are inserted into the heart through veins. Nevertheless, the site and placement of the electrode wires vary, depending on the patient and model of ICD used. In some cases, electrode patches are sewn to the surface of the heart. Other patients may receive electrodes that are placed under the skin of the chest near the heart. Rapid technological advances have produced devices that serve as both a pacemaker and an ICD and can be programmed to the patient's individual needs. Microchips inside the device record rhythms and shocks to be used to determine optimal therapy.

Practical application 16.3 provides considerations for dealing with patients with cardiac devices.

Special Considerations

In many cases, the failure of the heart's intrinsic pacing and conduction system is associated with comorbid conditions such as myocardial infarction and chronic heart failure. Many patients with artificial pacemakers or ICDs are elderly and have limited exercise capacity. The exercise prescription must consider not only the indications for the type of pacemaker implanted but also the limitations to exercise associated with comorbidities. Besides monitoring the patient for appropriate heart rate responses, the exercise physiologist must pay close attention to signs and symptoms that might occur with increased heart rate such as exercise-induced angina, failure of blood pressure to increase or decrease, and marked shortness of breath.

Exercise Recommendations

The American College of Cardiology/American Heart Association/Heart Rhythm Society Guidelines for Implantation of Cardiac Pacemakers and Antiarrhythmia Devices (9) provide recommendations for ICD therapy.

Practical Application 16.3

CLIENT–CLINICIAN INTERACTION

In a supervised exercise program, the clinician needs to observe whether the client can achieve the desired level of exercise without undue fatigue. Because the increase in heart rate during exercise is the largest contributor to cardiac output, limited exercise capacity may be a result of inappropriate heart rate response. Depending on the type of pacemaker, adjustment of the settings may allow the heart rate to respond more appropriately to the exercise demand. Carefully observing the client's exercise performance, asking the client about her fatigue level, and reporting those findings to the client's physician are key responsibilities of the exercise physiologist.

These guidelines emphasize the need for the physician to establish limitations on the patient's specific physical activities. The guidelines also refer to policies on driving, advising the patient with an ICD to avoid operating a motor vehicle for a minimum of 3 mo and preferably 6 mo after the last symptomatic arrhythmic event to determine if there is any recurrent ventricular fibrillation or tachycardia. After appropriate evaluation and observation, many patients with an ICD can participate in exercise programs. In most cases, the guidelines for exercise prescription are similar to those for any other patient with cardiovascular disease and should consider the patient's underlying diagnoses, medications, and symptoms. An exercise stress test is essential for establishing an appropriate exercise prescription. The prescribed target heart rate should be at least 20 beats below the heart rate cutoff point at which the device will shock. The exercise prescription must also take into account the existence of an **angina threshold** or exercise-induced hypotension, because many of these patients have severe coronary artery disease and poor left ventricular function. Furthermore, many patients with ICDs take β-blockers to limit heart rate to control symptoms and to prevent firing of the device. The benefits of pacing are

- alleviation or prevention of symptoms,
- restoration or preservation of cardiovascular function,
- restoration of functional capacity,
- improved quality of life,
- enhanced survival, and
- participation in exercise training programs with many forms of physical activity.

Cardiac Resynchronization Therapy

Cardiac resynchronization therapy (CRT) or biventricular pacing is an adjunctive therapy for patients with advanced heart failure (20, 21). Many of these patients have left bundle branch block or an intraventricular conduction delay, resulting in left ventricular dyssynchrony and a high mortality rate. The efficacy of CRT is based on the reduction in the conduction delay between the two ventricles. CRT is designed to keep the right and left ventricles pumping together by regulating how the electrical impulses are sent through the leads. This therapy contributes to the optimization of the ejection fraction, decrement in mitral regurgitation, and left ventricular remodeling, thus resulting in symptom improvement and enhanced quality of life. Several studies have shown the benefit of CRT in a subgroup of patients with heart failure with conduction delays (8, 14, 17, 22). Improvements have been found in the mean distance walked in 6 min, quality of life, New York Heart Association (NYHA) functional class, peak oxygen uptake, total exercise time, number of hospitalizations, LV function, and the LV end-diastolic diameter. Data suggest that improvements in functional capacity with CRT can be maintained long-term (23). In addition, ventricular–arterial coupling, mechanical efficiency, and chronotropic responses are improved after 6 mo of CRT. These findings may explain the improved functional status and exercise tolerance in patients treated with CRT (27). Exercise training after CRT helps to improve exercise tolerance, hemodynamic measures, and quality of life (18). Note that patients with these devices have serious heart disease, so the usual precautions regarding exercise participation should be applied.

Practical Application 16.4

RESEARCH FOCUS: LITERATURE REVIEW

Though it has been shown that cardiac resynchronization therapy has beneficial effects on clinical outcomes and cardiac remodeling, little is known about longer-term effects on myocardial function and exercise tolerance. Steendijk and colleagues (27) studied CRT in 22 patients with chronic heart failure. After 6 mo of the device therapy, marked improvements occurred in NYHA class, quality of life scores, 6 min walk distance, left ventricular ejection fraction, and stroke work at rest and with increased heart rates; there was also evidence of reverse remodeling. These results demonstrate that hemodynamic improvements with CRT can be maintained with chronic therapy and that they were associated with improved functional status and exercise tolerance. In another study, patients with CRT were randomized to 3 mo of exercise training or a control group (18). Exercise training resulted in further improvements in exercise capacity, hemodynamic measures, and quality of life in addition to the improvements seen after CRT. These studies suggest that exercise training allows maximal benefit to be attained after CRT.

Clinical Exercise Pearls

- Providing an adequate cardiac output is a key contributor to exercise capacity and endurance.
- Cardiac output is largely dependent on an appropriate increase in heart rate and, to a lesser degree, stroke volume.
- Implanted devices are used to maximize the cardiac output by regulating the heart rate in patients with abnormalities in the conduction system of the heart and by improving stroke volume through enhancing the efficiency of pumping action of the heart in patients with left ventricular dysfunction.

CONCLUSION

Because of the increased prevalence of pacemakers and ICDs, clinical exercise physiologists and cardiac rehabilitation professionals need to know how these devices function and what their limitations are. The two types of pacemakers are single-chamber units and dual-chamber units. Knowledge of the universal coding system is required to understand appropriate pacemaker function (figures 16.1 and 16.2). Initial pacemakers operated at fixed rates and were primarily used for patients who were symptomatic because of bradycardia or high-degree AV block. Because of the inability to increase heart rate with exertion in fixed-rate pacemakers, rate-responsive pacemakers have been developed, so that cardiac output can appropriately increase under physical activity. Motion sensors (vibration and accelerometers) are used in rate-responsive pacemakers. Each has advantages and disadvantages. In addition, dual-sensor (activity and ventilation) pacemakers have been developed to enhance normal heart rate response with exertion. In patients with rate-responsive pacemakers, exercise testing should be used to ensure an appropriate increase in heart rate and to allow for adjustment if the unit is not properly functioning. Exercise testing allows optimal programming of the pacemaker to provide maximal hemodynamic benefit and quality of life.

When prescribing exercise training with rate-responsive pacemakers, the clinical exercise physiologist must make sure that heart rate increases appropriately with exertion. In activities that do not involve a great deal of change in body movement (cycling, uphill walking), vibration sensors or accelerometers are not able to detect the real difference in activity level. Therefore, if possible, the physician should determine the type of activity that a person is planning to do before implanting a pacemaker. In addition, when prescribing different modes of activity, the clinical exercise physiologist must consider the type of pacemaker. Ideally, patients with rate-responsive pacemakers should be monitored to determine whether the physiological response is acceptable with exertion (i.e., heart rate, ECG, blood pressure, RPE, and METs). In addition, patients should avoid contact sports that carry a risk of direct contact with the pacemaker. Overall, not much in the way of limitation applies to prescribing exercise in pacemaker patients, other than making sure that an appropriate physiological response (increase in heart rate or blood pressure) occurs with increasing levels of physical exertion.

The use of ICDs is increasing to control life-threatening ventricular arrhythmias. Before patients with ICDs start an exercise program, an exercise test is recommended to determine the safety of exercise and rule out any other underlying diagnoses. When prescribing exercise, the major concern with patients with ICDs is to avoid reaching the threshold heart rate that will cause the device to shock. Training heart rate should stay 20 beats $\cdot$ min^{-1} below the preset heart rate that produces a shock. Otherwise, no specific limitations govern the prescription of exercise in this select population. The use of CRT to restore the coordinated pumping action of the ventricles is becoming more widespread in patients with chronic heart failure and delayed conduction. This type of device allows patients to be more active and have a better quality of life. Further research is needed to examine the long-term benefit of this therapy and to identify patients who are most likely to benefit from it.

Key Terms

angina threshold (p. 311)
automatic internal cardioverter defibrillators (p. 310)
cardiac activation (p. 298)
cardiac resynchronization therapy (CRT) (p. 311)
dual-chamber pacemaker (p. 299)
exercise intolerance (p. 298)
exercise prescription (p. 298)
exercise testing (p. 303)
maximum tracking rate (p. 313)
mode switching (p. 302)
pacemaker (p. 297)

(continued)

Key Terms *(continued)*

pacing system (p. 298)
physiological pacing (p. 300)
rate-responsive pacing (p. 302)
sick sinus syndrome (p. 298)
single-chamber pacemaker (p. 299)

CASE STUDY

MEDICAL HISTORY

Mrs. JD is a 64 yr old African American referred to cardiac rehabilitation 6 wk following implantation of a DDDR pacemaker with a vibration sensor. The indication for the pacemaker was marked sinus bradycardia and chronotropic incompetence. Her primary complaints were episodes of shortness of breath, undue fatigue, light-headedness, and weakness at rest and during exertion. She had no other significant cardiac history except for mild hypertension. Mrs. JD also has mild arthritis in her hands and knees. Her body mass index is 29. Currently, her symptoms at rest are resolved, but she complains of early exertional fatigue and shortness of breath while doing housework and taking short walks in her neighborhood. Her resting heart rate is paced at 60 beats $\cdot$ min^{-1}.

EXERCISE TESTING RESULTS

Because of limited exercise tolerance, Mrs. JD was referred to her cardiologist for a pacemaker evaluation. Her physician administered an office-based walking protocol during which Mrs. JD complained of shortness of breath after 3 min of "brisk" walking. Her heart rate reached a peak of 102 beats $\cdot$ min^{-1}. The diagnosis was that her increase in heart rate was inadequate for the amount of work being performed. As a result, the pacemaker was programmed to a maximum tracking rate of 110 beats $\cdot$ min^{-1} and a maximum sensor rate of 130 beats $\cdot$ min^{-1}, and the threshold of the sensor was lowered to be more sensitive to body movements.

EXERCISE PRESCRIPTION

The pacemaker settings relevant to her exercise prescription were a lower rate limit of 60 beats $\cdot$ min^{-1}, a programmed **maximum tracking rate** to 110 beats $\cdot$ min^{-1}, and a programmed sensor rate to 110 beats $\cdot$ min^{-1}. Mrs. JD was given a standard exercise prescription consisting primarily of stationary cycling for 15 min and walking on a treadmill for 15 min. Exercises with handheld weights and flexibility exercises were prescribed for warm-up. Because she did not have an exercise stress test before starting the cardiac exercise program, her exercise intensity for aerobic exercise was set at 12 to 14 on the Borg RPE scale.

RESPONSE TO INITIAL EXERCISE TRAINING SESSION

On the stationary cycle, Mrs. JD complained of shortness of breath and early fatigue, stopping at 4 min. Her heart rate peaked at 75 beats $\cdot$ min^{-1}. On the treadmill, she was able to walk for 10 min at 1.5 mph (2.4 kph), reaching a peak heart rate of 98 beats $\cdot$ min^{-1}. Her main complaint was shortness of breath.

DISCUSSION

In cardiac rehabilitation, the exercise prescription was changed to two bouts of 15 min each of treadmill walking with 2 min rest between bouts. The warm-up was unchanged. Mrs. JD tolerated 2.0 mph (3.2 kph) at 0% grade on the treadmill as prescribed, reporting only mild leg fatigue and shortness of breath. Her peak heart rate reached 122 beats $\cdot$ min^{-1}. In many instances, the pacemaker default factory settings are used. In this case, observation of the patient in cardiac rehabilitation resulted in a referral to her cardiologist, who performed an exercise test; this showed that the pacemaker needed to be reprogrammed to allow for a higher heart rate response, which in turn resulted a greater increase in cardiac output during exercise training. As a result, the patient was better able to meet her exercise training targets in cardiac rehabilitation with a marked reduction in symptoms.

(continued)

DISCUSSION QUESTIONS

1. Why was the maximum tracking rate left unchanged when the pacemaker settings were adjusted?
2. Why was stationary cycling dropped as an exercise modality for this patient?
3. What would be the initial choice for progressing the exercise intensity on the treadmill as the patient improves her fitness—an increase in speed, an increase in grade, or both?

PART IV

Diseases of the Respiratory System

Pulmonary diseases and conditions can play havoc with a person's ability to be physically active. The clinical exercise physiologist can play a role in diagnosis, functional assessment, and the recommendation of exercise training. Often these patients present in a rehabilitative setting in a pulmonary rehabilitation program. But given the level of association with other diseases (e.g., heart disease, cancer), these patients may also seek exercise therapy in cardiac rehabilitation programs and in general fitness center facilities. This section provides excellent chapters on three of the most common pulmonary conditions.

Chapter 17 examines chronic obstructive pulmonary disease, also known as COPD. Patients with COPD are most likely to seek out or be referred to pulmonary rehabilitation. Most are current or previous smokers, and they are often debilitated to the point of being unable to perform much physical activity without shortness of breath. The clinical exercise physiologist must learn the skills to educate and motivate these patients and implement appropriate exercise programming. This chapter is an excellent resource pertaining to these skills.

Chapter 18 explores asthma. Increasing numbers of people in the United States and the world are being diagnosed with asthma. This trend is likely a result of better recognition and diagnostic testing, although worsening environmental factors that trigger asthma attacks may also be contributing. Exercise-induced bronchospasm, a form of asthma, is also increasing in prevalence. People with asthma can benefit from exercise training and often are able to perform exercise independent of clinical supervision. As this chapter demonstrates, however, the clinical exercise physiologist can be a great asset to these patients in developing and implementing an exercise program to limit potential bouts of asthma and in helping with methods to treat an attack that is associated with exercise.

Chapter 19 deals with cystic fibrosis (CF). Patients with CF have multiple medical issues that affect not only the lungs but also the gastrointestinal tract, sinuses, and sweat glands. These parts of the body are primarily affected by inflammation and excess mucus production because of CF. The lungs are particularly affected, so a daily process of breaking up and expelling the mucus is a way of life for some people. Exercise is important for these individuals if they wish to maintain independence and functionality, but they must deal with potential respiratory difficulties and impaired thermoregulation. The clinical exercise physiologist can play an important role in helping these people deal with the effects of CF so that they can live more active and productive lives.

Chapter 17

Chronic Obstructive Pulmonary Disease

Ann M. Swank, PhD
N. Brian Jones, PhD

Chronic obstructive pulmonary disease includes the clinical conditions of chronic bronchitis and emphysema. This chapter discusses the underlying pathology of these conditions as well as the procedures and benefits of exercise testing and training for individuals with these conditions.

DEFINITION

Chronic obstructive pulmonary disease (**COPD**) is defined by the American Thoracic Society as a disease characterized by the presence of airflow obstruction that is attributable to either chronic bronchitis or emphysema (13). Chronic bronchitis is a clinical diagnosis for patients who have chronic cough and sputum production. The American Thoracic Society defines **chronic bronchitis** as the presence of a productive cough most days during 3 consecutive months in each of 2 successive years (13, 135). The cough is a result of hypersecretion of mucus, which in turn results from an enlargement of the mucus-secreting glands. In contrast to the clinical diagnosis for chronic bronchitis, **emphysema** is a pathological or anatomical diagnosis marked by abnormal permanent enlargement of the respiratory bronchioles and the alveoli, the air spaces distal to the terminal bronchioles, and is accompanied by destruction of the lung parenchyma without obvious fibrosis (135, 137, 149). Most patients with COPD have both chronic bronchitis and emphysema, and the relative extent of each varies among patients (figures 17.1 and 17.2). The World Health Organization's International Classification of Disease (ICD) codes used by nosologists to classify COPD are 490, 491, 492, 494, 495, and 496.

Patients with COPD experience acute **exacerbations**, or periods of worsening symptoms. The pathogenesis of an exacerbation is not well understood, and it may be difficult to define clinically. These exacerbations can lead to respiratory failure and are a major cause of hospitalizations in the United States. The major risk factors for the development of COPD include cigarette smoking, exposure to passive smoke, and air pollution. In addition, strong evidence implicates several rare genetic syndromes, including **α_1-antitrypsin deficiency**, and occupational exposure (15).

At times, asthma has been subsumed under the rubric of COPD. Asthma is characterized by inflammation and hyperresponsiveness of the tracheobronchial tree to a variety of stimuli. Although asthma patients experience exacerbations or attacks, these exacerbations are interspersed with symptom-free periods when airway narrowing is completely or almost completely reversed. In contrast, most patients with chronic bronchitis do not

Acknowledgment: Much of the writing for this chapter was adapted from the first and second editions of *Clinical Exercise Physiology*. As a result, we wish to gratefully acknowledge the previous efforts of and thank Michael J. Berry, PhD, and C. Mark Woodard, MS.

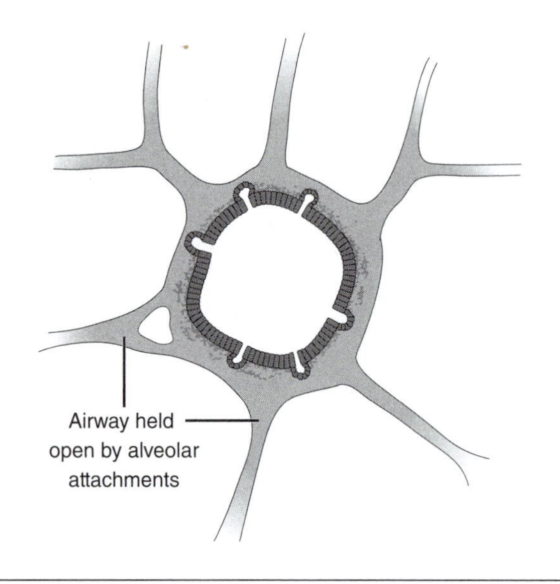

Figure 17.1 A normal airway that has little inflammation or mucus plugging and is being held open by parenchymal lung tissue.

Adapted from P.J. Barnes, 2000, "Chronic obstructive pulmonary disease," *The New England Journal of Medicine* 343: 269-280.

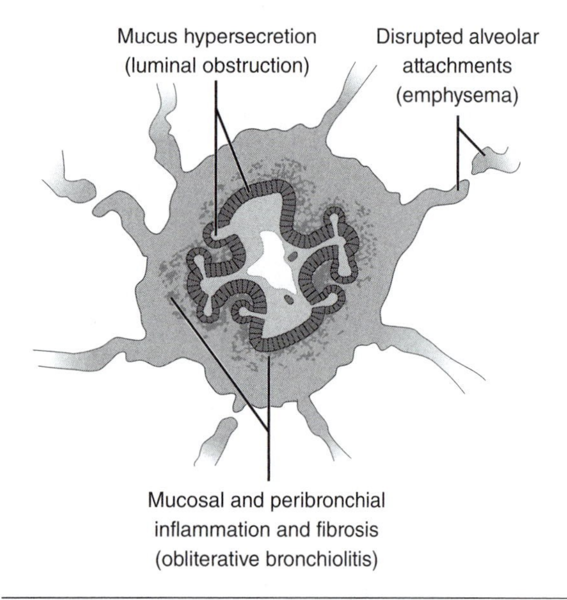

Figure 17.2 An obstructed airway that has significant inflammation and mucus plugging. Also shown is a loss of alveolar attachments, making airway collapse more likely.

Adapted from P.J. Barnes, 2000, "Chronic obstructive pulmonary disease," *The New England Journal of Medicine* 343: 269-280.

exhibit significant reversibility of airway narrowing, and they present with residual symptoms between exacerbations. Although patients with COPD and asthma share similar clinical characteristics, the pathology of the two syndromes differs considerably, suggesting that they are different diseases (91). Accumulating evidence suggests that COPD is a disease of multiple organ systems in which ventilatory impairments and muscle dysfunction contribute to the exercise intolerance seen in these patients (14). Because of the differences between COPD and asthma, they should be considered separately. Asthma is considered in the next chapter.

SCOPE

COPD is the fourth leading cause of death in the United States and is projected to be the third leading cause of death for both males and females by 2020. Estimates are that 16 million people in the United States may be currently diagnosed with COPD and an additional 14 million may be in the early stages without symptoms. Additionally, among the top five causes of death since 2000, COPD was the only one that showed an increase (103). Besides being the fourth leading cause of death in the United States, COPD is a major cause of morbidity and disability and a major health care cost, estimated at $24 billion in 2000. In 2000, COPD was given as the diagnosis at the time of discharge for 726,000 hospitalizations. The average length of stay for COPD hospitalizations was 6.3 d, and the mean cost of each of these hospitalizations was $10,684 (5, 103). In 1995, 16,087,000 physician office visits were attributable to COPD. Respiratory tract infections in COPD patients have been shown to have a major effect on use of health care resources (60). Between 1990 and 1992, 800,000 Americans reported that emphysema caused them to limit their activity (109). Serres and colleagues (125) reported that COPD patients have lower levels of physical activity than age-matched controls.

PATHOPHYSIOLOGY

The anatomical, physiological, and pathological abnormalities associated with COPD often result in debilitation for the COPD patient. Because of the direct insult that cigarette smoke, the primary risk factor for the development of COPD, has on the lungs, it has long been thought that the lungs were the primary organs affected by COPD. Recent research suggests, however, that the disease process itself or certain aspects of the disease process adversely affect not only the lungs but also skeletal muscle (14).

Cigarette smoking affects the large airways (**bronchi**), the small airways (**bronchioles**), and the pulmonary **parenchyma**. The pathological conditions that develop are a result of the effect that cigarette smoke has on each of these structures. Additionally, the degree of airway reactivity of the individual patient has an effect. Within the large airways, cigarette smoke causes the bronchial mucus glands to become enlarged and the gland ducts to become dilated. Excessive cough and sputum production are a result of these factors and are the characteristic symptoms of chronic bronchitis (148).

These alterations in the large airways have little effect on airflow or **spirometry**. The airflow obstruction that is characteristic of COPD does not occur until additional damage is incurred by the small airways and the lung parenchyma. The changes in the smaller airways include mucus plugging, inflammation, and an increase in the smooth muscle (figure 17.2). These changes decrease the cross-sectional area of the airways and can have a profound effect on airflow and the COPD patient's spirometry (148). The spirometry of a COPD patient is characterized by reductions in expiratory flow rates including the **forced expiratory volume in 1 s (FEV_1)**, the FEV_1/**forced vital capacity (FVC)** ratio, and the midexpiratory flow rate.

Destructive changes occur to the alveolar walls with emphysema. The net effect of these changes is twofold. First, destruction of the alveolar walls results in a loss of the tethering or supportive effect that the alveoli have on the smaller airways. This tethering effect helps keep the airways open during expiration. Without this alveolar support, the smaller airways are likely to collapse during expiration, thus adding further to the airway obstruction (figure 17.2). The second effect of destruction of the alveolar walls is to diminish the elastic recoil of the lungs, which in turn decreases the force that moves air out of the lungs. The combination of these two effects reduces airflow and increases the amount of work the respiratory muscles must perform to meet the ventilatory demands of the body.

The combined effects of airway obstruction and the reduced expiratory driving force increase the time needed for expiration. If inspiration occurs before the increased expiratory time requirement can be met, then the normal end-expiratory lung volume will not be reached, resulting in increased functional residual capacity and **hyperinflation** of the lungs. Furthermore, the diaphragm will assume a shorter, more flattened position. Because the diaphragm is a skeletal muscle, it operates according to the length–tension relationship (98), whereby the tension developed by skeletal muscle is a function of its resting length. As a muscle is shortened or lengthened beyond its optimal length, the potential for tension development decreases (57). Because the diaphragm is shortened with hyperinflation, it has less force-generating potential (114, 134). Some evidence suggests that the diaphragm adapts to these chronic changes by shortening the optimal length of its fibers (131). As such, each fiber would have the potential to generate its maximal force at its new length (52). Despite these adaptive changes, evidence still suggests that COPD patients have decreased capacity to generate diaphragmatic pressure.

For individuals with normal lung function, the end-expiratory lung volume decreases by approximately 200 to 400 ml with moderate exercise (66, 128). In contrast, patients with COPD demonstrate an increase in the end-expiratory lung volume with exercise, leading to dynamic hyperinflation of the lungs (141). This dynamic hyperinflation leads to further diaphragm weakness and may contribute to **dyspnea** (shortness of air) and reduced exercise tolerance. Because the diaphragm is a skeletal muscle, it has been hypothesized that positive adaptations may result from training of this particular muscle (84, 121) (see section on exercise training).

In addition to the damaging effect on the lungs, COPD contributes to skeletal muscle dysfunction (33, 139). The skeletal muscle dysfunction may contribute to the exercise intolerance seen in COPD patients. Several studies showed that COPD patients have diminished peripheral muscle strength (22, 36, 43, 48, 59, 64). Patients with COPD have been found to have a 20% to 30% reduction in quadriceps strength compared with age-matched controls (43, 59, 64). The decrease in strength is accompanied by a reduction in muscle cross-sectional area (22, 48). Other studies have reported reduced muscle mass in patients with COPD (47, 123, 166). An analysis of muscle biopsies from patients with moderate COPD showed decreases in type I fibers and an increase in the proportion of hybrid fibers compared with controls (58). Studies with advanced COPD patients have also reported a reduction in the proportion of type I fibers compared with control subjects (71, 160). These cumulative findings are consistent with the reported reduced oxidative enzyme activities in these patients (94, 95). Additionally, the reduction in type I fibers has been shown to be accompanied by a corresponding increase in type IIb fibers (160). Both chronic **hypoxemia** (low oxygen content in the blood) (67) and a lack of physical activity (160) may contribute to the changes in fiber types. Chronic steroid use has been suggested as a contributor to muscle weakness (42, 43). Other possible contributors to the skeletal muscle abnormalities seen in COPD patients include chronic **hypercapnia** (high levels of carbon dioxide in the blood), inflammation, nutritional depletion, and comorbid conditions that may impair skeletal muscle function (6, 14). Practical application 17.1 discusses malnutrition in COPD.

One hallmark of patients with COPD is a reduction in airflow, which is most prominent during maximal efforts. This reduction is typically quantified with use of

Practical Application 17.1

MALNUTRITION IN COPD

Malnutrition is a problem for as many as 25% of all COPD patients (30, 76). In addition, 50% of patients hospitalized for treatment of COPD demonstrate protein as well as calorie malnutrition (70). In a review of 90 COPD patients, researchers found that patients who required hospitalization and mechanical ventilation demonstrated the most severe nutritional decrements (53). Whereas the causes of malnutrition in COPD patients have not been clearly defined (162), the results of malnutrition have. Weight loss in COPD patients has been shown to be a predictor of mortality rate (164). Additionally, prolonged malnutrition results in deleterious changes to the diaphragm muscle such that its ability to generate force is decreased (74, 86). This fact, coupled with the fact that individuals with COPD exhibit dysfunction in other skeletal muscles, suggests the need for nutritional support in these patients. A study demonstrated that when individuals with COPD are given sufficient calories in excess of their needs, they gain weight and achieve significant improvement in ventilatory and peripheral muscle strength (163). Given the need for nutritional intervention and the positive outcomes that can result, the clinical exercise physiologist should consult with a nutritionist when working with underweight COPD patients.

the results of pulmonary function tests; the FEV_1 is one of the standards used to assess disease severity and monitor disease history. Additionally, the FEV_1 has been shown to be a strong predictor of mortality rate from COPD (147). In healthy nonsmokers, the FEV_1 declines by 20 to 30 ml per year (75, 136, 145). In male and female smokers, the rate of decline in FEV_1 increases (82, 145). The rate of decline is both age and sex dependent, and the greatest rate of decline occurs in men between the ages of 50 and 70. People who have characteristics compatible with emphysema have a rate of decline in the FEV_1 of 70 ml per year (30).

In those who quit smoking, the decline of FEV_1 is less pronounced than the decline observed in those who continue to smoke (82). In fact, ex-smokers show rates of decline of FEV_1 similar to those of nonsmokers; and in younger ex-smokers, the FEV_1 has been shown to increase following smoking cessation. Although complete cessation of smoking has beneficial effects on the decline of the FEV_1, the effect on FEV_1 decline of attempting to quit smoking and relapsing is equivocal. Sherrill and colleagues (129) reported that the rate of decline of the FEV_1 is greater in ex-smokers who resume smoking compared with those who continue to smoke. Murray and colleagues (102) reported that attempts to quit smoking in patients with mild COPD slow the rate of FEV_1 decline compared with those who continue to smoke.

CLINICAL CONSIDERATIONS

This section discusses signs and symptoms, history and physical examination, diagnostic exercise testing, and treatment specifically related to COPD.

Signs and Symptoms

The acute respiratory illness associated with COPD is characterized by increased cough, purulent sputum production, wheezing, dyspnea, and occasional fever. With progression of the disease, the interval between these illnesses decreases (13). In the early disease stages, slowed expiration and wheezing are noted during the physical examination. Additionally, breath sounds are decreased, heart sounds may become distant, and coarse crackles may be heard at the base of the lungs (13).

History and Physical Exam

The diagnosis of COPD is made based on patient history, a physical examination, and the results of laboratory and radiographic studies (**roentgenogram**). The diagnosis is suspected in patients who have a history of smoking and who present with an acute respiratory illness or respiratory symptoms such as a productive cough. A smoking history of 70 or more **pack-years** has been reported to be suggestive of COPD (18).

Diagnostic Testing

Results from pulmonary function tests are necessary for establishing a diagnosis of COPD and for determining the severity of the disease, but they cannot be used to distinguish between chronic bronchitis and emphysema. The FEV_1, FVC, FEV_1/FVC, and single-breath diffusing capacity are the primary pulmonary function tests recommended to aid in the diagnosis of COPD (12). In patients with COPD, the results from all these tests are less than what would be predicted for a person of the same sex and similar age and stature. Other recommended tests include lung volume measurements and determination of arterial blood gas levels. Lung volume measurements often reveal an increase in total lung capacity, functional residual capacity, and residual volume. Arterial blood gases may reveal hypoxemia in the absence of hypercapnia in the early stages of the disease, with a worsening of hypoxemia and hypercapnia presenting in the later stages of the disease (13).

Because emphysema is defined in anatomic terms, the chest roentgenogram can sometimes be used to differentiate between emphysema and chronic bronchitis. Patients with chronic bronchitis often have a normal chest roentgenogram; the roentgenogram of patients with advanced emphysema may reveal large lung volumes, hyperinflation, a flattened diaphragm, and vascular attenuation (13). Computed tomography has greater sensitivity and specificity than the chest roentgenogram and can be used for both qualitative and quantitative assessment of emphysema. Because the additional information gained from computed tomography rarely alters therapy, it is infrequently used in the routine care of COPD patients. Although computed tomography is not recommended for routine use with COPD patients, it is the best way of recognizing emphysema and probably has a significant role in recognizing localized emphysema that is amenable to surgical treatment (149).

The American Thoracic Society proposes staging patients with COPD into distinct categories based on the degree of airflow obstruction. This organization suggests using the FEV_1 as the staging criterion. Patients with an FEV_1 greater than or equal to 50% of predicted are categorized as having stage 1, mild disease. Those with an FEV_1 between 35% and 49% of predicted are categorized as having stage 2, moderate disease. Finally, those with an FEV_1 less than 35% of predicted are categorized as having stage 3, or severe disease (12, 13). Most COPD patients are categorized as having mild disease. In addition to these standards, the Global Initiative for Chronic Obstructive Lung Disease (GOLD) also proposes using the FEV_1 to define and categorize patients with COPD. According to the GOLD recommendations, an FEV_1 greater than or equal to 80% of normal indicates stage I, mild disease; FEV_1 between 50% and 79% indicates stage II, moderate COPD; FEV_1 between 30% and 49% indicates stage III, severe COPD; and FEV_1 <30% of normal indicates stage IV and very severe COPD.

Exercise Testing

Exercise testing is an integral component in the evaluation of patients with COPD. In patients with mild or moderate disease, symptoms generally do not present until increased demand is placed on the respiratory system, for example with exercise. In patients with severe disease, the functional capacity is reduced to such a level that even simple activities of daily living may impose a challenge to the respiratory system. Most patients with moderate to severe COPD have reduced exercise capacity because of reduced ventilatory capacity in the face of an increased ventilatory demand (46, 72). Because exercise places increased demand on the respiratory system, exercise testing provides an objective evaluation of the functional capacity of the COPD patient. Additionally, exercise testing can be used to detect COPD and cardiovascular disease, follow the course of the disease, detect exercise hypoxemia, determine the need for supplemental oxygen during exercise training, evaluate the response to treatment, and prescribe exercise (142).

Table 17.1 outlines procedures and guidelines for exercise testing of the patient with COPD (10, 11). During the exercise test, the minimum monitoring should include measurement of blood pressure, a 12-lead electrocardiogram (ECG), analysis of arterial oxygen saturation, and measurement of dyspnea. These variables should be measured before the start of the test, continuously throughout the test, at the termination of the test, and during recovery as needed. Blood pressure should be measured with the patient's arm relaxed and the manometer mounted at eye level. Automatic monitors for blood pressure measurement during exercise are available, and their use has been found acceptable (61). Placement for the 12-lead ECG is typically the Mason-Likar. The gold standard for the measurement of arterial oxygen saturation is co-oximetry using arterial blood. The use of a pulse oximeter is acceptable as long as the device has been validated during exercise. At oxygen saturations greater than 90%, these devices have a high degree of reliability; however, as oxygen saturation drops below 90% their reliability decreases (104). Because of the problems with precisely defining the degree of hypoxemia with these instruments, these devices should be used to determine whether desaturation is occurring and then to correct it with supplemental oxygen. The final variable that should be monitored during the exercise test in patients with COPD is dyspnea. A number of scales that have been validated and proved reliable are available for use (4, 11, 92, 93). One particular dyspnea scale of interest to the exercise specialist is the Borg scale. This scale has been adapted for use during exercise testing and training with COPD patients (25).

Table 17.1 Exercise Testing Recommendations, Procedures, and Guidelines for the COPD Patient

Test type	Mode	Protocol specifics	Clinical measures	Clinical implications	Special considerations
Cardiovascular	• Treadmill or cycle ergometer (preferred) • 6 min walk	• Duration of 8 to 12 min, small incremental increases in workload individualized to the patient • Treadmill: 1 to 2 METs per stage • Cycle ergometer: ramped protocol 10, 15, or 20 W/min, stage protocol 25 to 30 W per 3 min stage	• HR, 12-lead ECG (Mason-Likar placement) • BP • RPE, rating of perceived dyspnea • Oxygen saturation (pulse oximetry/ arterial PaO_2) • Ventilation measures and gas exchange • Blood lactate • Distance	• Serious dysrhythmias, >2 mm ST-segment depression or elevation, ischemic threshold, T-wave inversion with significant ST change • SBP >250 mmHg or DBP >115 mmHg • Maximum ventilations, $\dot{V}O_2$peak, lactate/ ventilatory threshold • Note rest stop distance or time, dyspnea index, vitals	No arm ergometer testing. Clients with COPD often have coexisting CAD. Breathing pattern may help analysis. Lactic acidosis may contribute to exercise limitation in some patients. Exercise testing in mid- to late afternoon is desirable. Useful for measurement of improvement throughout conditioning program.
Musculoskeletal	• Isokinetic, isotonic, or both • Sit-to-stand, stair climbing, lifting	Time to 10 reps	• Peak torque • Maximum number of reps • 1RM		• Clients may become more dyspneic when lifting objects (teach appropriate breathing strategies for lifting). • Specific evaluation and training may be needed.
Flexibility	• Sit-and-reach • Gait analysis • Balance	Hip, hamstring, and lower back flexibility			Body mechanics, coordination, and work efficiency are often impaired.

If the equipment is available, gas exchange and ventilatory measurements should be obtained. These measures provide valuable information that can be used to prescribe exercise intensity more accurately, to evaluate the effectiveness of an exercise intervention, and to provide information regarding the extent of the lung disease (96, 124). Of special concern with measuring gas exchange and ventilatory parameters for patients with COPD is the use of supplemental oxygen. These patients should be tested on an elevated fraction of inspired oxygen such that it equates with the flow rate established for the use of supplemental oxygen. Most commercially available gas exchange measurement systems have established procedures that allow for use of elevated fractions of inspired oxygen during exercise testing and conversions of oxygen flow rates to inspired oxygen fractions.

Exercise testing has been shown to be extremely safe, even in high-risk populations, with 1 or fewer deaths per 10,000 tests (55). To minimize risk to patients, recommended guidelines from the American College of Sports Medicine (11), the American Association of Cardiovascular and Pulmonary Rehabilitation (7, 8), and the American Heart Association (54) should be closely followed. In general, the procedures for testing patients with COPD follow those for testing other at-risk populations. Before conducting the exercise test, the clinician should review information from a medical exam and history to identify contraindications to testing as specified in the guidelines listed in table 17.1. An additional concern for patients with COPD is accompanying pulmonary hypertension. Some experts advise caution with these individuals because of the risk of serious cardiac arrhythmias or even sudden death during testing (8).

The exercise mode, the test protocol, and the monitoring equipment are all fundamental considerations in exercise testing. The exercise mode should be one that will increase total body oxygen demands by requiring the use of a large muscle group. The most common exercise testing modalities for the patient with COPD are the treadmill and the cycle ergometer. One exercise mode that should not be used routinely to test COPD patients is arm ergometry because patients with severe COPD often use the accessory muscles of inspiration for breathing at rest. Any additional burden placed on these muscles could result in significant symptoms and distress for the patient (35).

The testing protocol should start at a work rate that the patient can easily accomplish, use increments in the work rate that are progressively difficult, and last a total duration of 8 to 12 min. The initial stages should be of an intensity that allows the patient adequate time to warm up and become accustomed to the exercise. The work rate increments should be small and based on characteristics of the patient (e.g., sex, size, severity of disease, and previous level of physical activity). The increment of work rate affects the exercise response of patients with COPD (41). For example, the peak work rate achieved for a given level of oxygen consumption is greater when the work rate is increased quickly. Unfortunately, because of severe deconditioning and extreme dyspnea that some patients with COPD experience when performing even mild physical activity, having a test that lasts the minimum recommended duration may not be possible.

The responses of the patient with COPD to an exercise test vary depending on the severity of the disease. In patients with mild disease, the results of the exercise test may be consistent with those of normal individuals or may demonstrate abnormalities indicative of cardiovascular disease or deconditioning (49). Exercise responses for patients with moderate and severe COPD compared with age-matched healthy controls are shown in table 17.2. Peak oxygen uptake and peak work rates are usually reduced in patients with moderate or severe COPD (34, 94, 107, 130, 157). Concomitant with the lower peak oxygen consumption is a lower peak heart rate and a greater heart rate reserve (predicted peak heart rate minus measured peak heart rate). Oxygen pulse is also low in patients with COPD because they terminate exercise at a low work rate. In normal and deconditioned individuals and in cardiac patients, the ventilatory reserve (maximal voluntary ventilation minus the peak minute ventilation) is high at peak exercise. In contrast, the patient with COPD has a low ventilatory reserve and, in some cases, peak minute ventilation is equal to or even greater than the maximal voluntary ventilation (155). Additionally, the peak minute ventilation is lower than predicted (34). The partial pressure of oxygen in the arterial blood and the percentage saturation of hemoglobin in the arterial blood are often low at maximal exercise in the patient with moderate or severe COPD (13). Patients with COPD develop a significant anaerobiosis and demonstrate a lactate threshold, although these occur at relatively low work rates (34, 143). Because of their ventilatory impairment, patients with moderate and severe COPD do not show a disproportionate increase in minute ventilation with the development of anaerobiosis (143). Thus, the detection of a ventilatory threshold may not be possible in these patients.

Table 17.2 Exercise Test Responses in Patients With COPD Compared With Normal Healthy Subjects

Parameter	Finding
Peak work rate	Decreased
Peak oxygen consumption	Decreased
Peak heart rate	Decreased
Peak ventilation	Decreased
Heart rate reserve	Increased
Ventilatory reserve	Decreased
Arterial partial pressure of oxygen	Decreased
Arterial oxygen saturation	Decreased
Lactate threshold	Occurs at a lower work rate
Ventilatory threshold	Absent

Exercise testing is an important tool in assessing the patient with COPD. But the test and equipment used must be designed to meet the needs of the patient and the clinician administering the test. Additionally, because of the abnormal responses of the COPD patient, one must exercise care when interpreting the results of these tests.

Treatment

After a diagnosis of COPD has been made, a multifaceted approach to the treatment and management of the patient should be adopted. Comprehensive treatment should include smoking cessation, oxygen therapy, pharmacological therapy, and pulmonary rehabilitation (including exercise training) (31). Because smoking is a major cause of COPD, smoking cessation is a major therapy in the treatment of COPD patients. Smoking cessation is one of two interventions that have been shown to improve patient survival (13, 133).

The second therapy that has been shown to improve survival in patients with COPD is long-term oxygen therapy. The British Medical Research Council study (100) and the National Heart, Lung and Blood Institute's Nocturnal Oxygen Therapy Trial (105) showed that patients who received long-term oxygen therapy experienced a significant reduction in mortality rates. Other benefits of long-term oxygen therapy include a reduction in **polycythemia** (85), decreased pulmonary artery pressure (1-3), and improved neuropsychiatric function (65). More recently, domiciliary oxygen therapy has been demonstrated to prolong survival for patients with COPD but only for those with resting **hypoxia** (133). Other investigators are exploring the role of stem cells for the treatment of COPD, and this work is in the preliminary stages (63).

Acute administration of supplemental oxygen has been shown to preserve exercise tolerance in hypoxemic patients. Whether patients with COPD undergo an acute bout of exercise with administration of supplemental oxygen (27, 38, 40, 101, 113) or are trained with supplemental oxygen (44, 113, 122, 153, 167), the benefits are significant. The goal of oxygen therapy is to reverse hypoxemia and prevent tissue hypoxia (13). For COPD patients to realize benefits from supplemental oxygen therapy during training, they must demonstrate hypoxemia during training. If oxygen is to be prescribed for COPD patients, the goal is to maintain the partial pressure of oxygen in arterial blood above 60 mmHg or the percentage saturation above 90. Therefore, the delivery method and the dosage of oxygen, or the flow rate, need to be considered. COPD patients who need supplemental oxygen during exercise often use a liquid oxygen supply. These systems, although more expensive, are lightweight and easily refilled from larger stationary sources. Oxygen concentrators cannot be used during exercise because of their weight and need for an electrical supply.

Pharmacological therapy (table 17.3) in patients with COPD is aimed at inducing bronchodilation, decreasing the inflammatory reaction, and managing and preventing

Table 17.3 Pharmacology for COPD

Medications, name and class	Primary effects	Exercise effects	Special considerations
Albuterol, bitolterol, epinephrine, formoterol, isoetharine, isoproterenol, levalbuterol, metaproterenol, pirbuterol, salmeterol, terbutaline Sympathomimetic short- and long-acting bronchodilators	Bronchodilators (act primarily on smooth muscle)	May increase or have no effect on HR, ECG; may increase, decrease, or have no effect on BP; increase exercise capacity by limiting bronchospasm	
Ipratropium Tiotropium Anticholinergic bronchodilators	Bronchodilators (attenuate vagal tone)	May increase or have no effect on HR, ECG; no change in BP	
Aminophylline, dyphylline, oxtriphylline, theophylline Methylxanthine bronchodilators	Bronchodilators	May increase or have no effect on HR, ECG; no change in BP	May produce PVCs
Beclomethasone, budesonide, dexamethasone, flunisolide, fluticasone, triamcinolone Corticosteroid anti-inflammatories	Anti-inflammatory antiasthmatics	No effect on HR, BP, ECG, or exercise capacity	
Cromolyn, nedocromil Prophylatics for asthma	Asthma prophylactics	No effect on HR, BP, ECG, or exercise capacity	
Montelukast, zafirlukast, zileuton Anti-leukotriene-based antiasthmatics	Antiasthmatics		
Acetylcysteine, guaifenesin, potassium iodide Expectorants, mucolytics, or both	Reduce mucus production or act as expectorants		

respiratory infections (13, 151). **Bronchodilator** therapy includes the use of α_2-agonists, anticholinergic agents, and theophylline. The use of α_2-agonists may result in tremors, anxiety, palpitations, and arrhythmias. Because of these problems, careful dosing and monitoring of patients with known cardiovascular disease are necessary (13). After a patient develops persistent symptoms, anticholinergic agents such as ipratropium bromide may be prescribed because their effect is more intense and of longer duration. Additionally, they may have less potentially deleterious side effects than α_2-agonists.

Theophylline, one of the methylxanthines, is a third agent that may be used to induce bronchodilation. Besides its bronchodilator effects, theophylline increases cardiac output, decreases pulmonary vascular resistance, and may have anti-inflammatory effects (144, 168). Despite the beneficial effects of theophylline, its popularity has declined because of its toxicity (69, 127) and potential for adverse interaction with other drugs (120). Inhaled corticosteroids are indicated if a significant bronchodilator response occurs or if the patient has more severe disease with frequent exacerbations. Oral corticosteroids may be necessary for acute exacerbations (31, 151).

Pulmonary rehabilitation is an evidence-based, multidisciplinary, and comprehensive intervention for patients with chronic respiratory diseases who are symptomatic and often have decreased daily life activities. Integrated into the individualized treatment of the patient, pulmonary rehabilitation is designed to reduce symptoms, optimize functional status, increase participation, and reduce health care costs through stabilizing or reversing systemic manifestations of the disease (16). These services typically include patient assessment, patient education, exercise training, psychosocial intervention, and patient follow-up (7, 8). The various components of each of these services are listed in table 17.4. Practical application 17.2 deals with client–clinician interaction for patients with COPD. The goals of pulmonary rehabilitation are to decrease airflow limitations, improve exercise capacity or physical function, prevent and treat secondary medical complications, decrease respiratory symptoms, and improve the patient's quality of life (9, 16). As a result of participating in a comprehensive pulmonary rehabilitation program, patients have demonstrated improvements in quality of life (16, 56, 78), sense of well-being (16, 17), self-efficacy

Practical Application 17.2

CLIENT–CLINICIAN INTERACTION

Often, the first interaction between the clinical exercise physiologist and the patient with COPD occurs when the patient has been referred for pulmonary rehabilitation. Unfortunately, patients with COPD are often referred to pulmonary rehabilitation only after they have experienced an exacerbation of their disease or when their dyspnea has become so oppressive that they are severely disabled. As a result, these patients are often anxious, scared, frustrated, and depressed. The clinical exercise physiologist must be aware of these problems when working with the COPD patient and be able to present a positive, yet realistic, picture of the benefits of exercise training for the patient with COPD.

Dyspnea, the primary symptom of COPD, often results in a vicious cycle of fear and anxiety followed by inactivity resulting in deconditioning, which results in further dyspnea. Unless this cycle can be broken, patients with COPD are destined to become dependent on others to meet their most basic needs. Patients with COPD need to be made aware that they can learn to live with their dyspnea and that exercise can help reduce the intensity and distress associated with dyspnea. They should be taught strategies that will help them manage their dyspnea on a daily basis. These strategies include monitoring the effects of various medications on dyspnea and avoiding or minimizing factors that can result in dyspnea.

The clinical exercise physiologist must also be aware that patients with COPD often experience exacerbations or periods of worsening symptoms. As a result, these patients may not be able to exercise at their prescribed intensity or may miss exercise sessions completely. The patient should be encouraged to continue exercising, even if at a much lower intensity. If even extremely mild exercise is not possible, the patient should be encouraged to resume exercising after he has recovered from the exacerbation.

The successful clinical exercise physiologist can effectively interact with each person on a one-to-one basis. Such a professional is sensitive to the particular needs of patients and is able to tailor each program to meet individual needs. As a result of effective clinician–client interaction, patients will be able to take control of their disease with less fear and anxiety.

(16, 119), and functional capacity (16, 123). Additionally, functional status has been shown to be a strong predictor of survival in patients with advanced lung disease following pulmonary rehabilitation (28). Determining which of the specific components of pulmonary rehabilitation is responsible for these improvements is difficult, because they are all integrally related.

The American Thoracic Society recommends that patients with COPD be referred to a pulmonary rehabilitation program after they have been placed on optimal medical therapy and still demonstrate the following (12):

1. Severe symptoms
2. Several emergency room or hospital admissions within the previous year
3. Diminished functional status that limits their activities of daily living
4. Impairments in quality of life

Although patients who are referred to pulmonary rehabilitation typically have severe disease, research indicates that patients with mild and moderate disease benefit from participation in the exercise component of a pulmonary rehabilitation program similarly to those with severe disease (24, 156). Participation in a pulmonary rehabilitation program that includes at least 4 wk of exercise training can result in improvements that are clinically significant for patients with COPD (79), including increased quality of life, specifically in the relief of dyspnea, and improvement in their perceptions of how well they can cope with their disease.

EXERCISE PRESCRIPTION

Table 17.5 summarizes recommendations for the exercise prescription for the patient with COPD. Because heart rate is not a reliable indicator of exercise tolerance for

Table 17.4 Various Components of the Services Offered in Pulmonary Rehabilitation

Patient assessment	Patient training and education	Exercise training	Psychosocial interventions	Patient follow-up
Medical history	Anatomy and physiology	Mode, duration, frequency, and intensity	Identification of support systems	Outcome measurements of physical function and health-related quality of life
Pulmonary function tests	Pathophysiology of lung disease	Upper and lower extremity endurance training	Treatment of depression	Support groups
Symptom assessment	Description and interpretation of assessment tests	Upper and lower extremity strength training	Treatment of anxiety	Maintenance programs
Physical function assessment	Breathing retraining	Inspiratory muscle training	Anger management	
Nutritional assessment	Bronchial hygiene	Flexibility and posture	Sexuality issues	
Activities of daily living assessment	Medication information	Orthopedic limitations	Adaptive coping styles	
Educational assessment	Symptom management	Home exercise plans	Adherence to lifestyle modifications	
Psychosocial assessment	Activities of daily living and energy conservation Nutrition Psychosocial issues Smoking cessation		Relapse prevention	

Adapted from American Association of Cardiovascular and Pulmonary Rehabilitation 1998.

patients with COPD, intensity is monitored by dyspnea or ratings of perceived exertion, with a suggested 4 to 6 rating on the Borg dyspnea 10-point scale. At least 3 to 5 d/wk of aerobic training is recommended, with walking as preferred mode because of its importance in activities of daily living. Stationary cycling is a recommended alternative. Other recommendations include 2 or 3 d/wk of resistance training, daily flexibility training, and at least 4 or 5 d/wk of inspiratory muscle training (11).

EXERCISE TRAINING

Dyspnea and reduced exercise capacity are two of the most common complaints of COPD patients. In addition to the diaphragm muscle, the accessory muscles of inspiration (scalene, sternocleidomastoid, and serratus anterior) are activated during exercise. Even at low work rates, unsupported arm exercise results in greater levels of dyspnea compared with lower extremity exercise in patients with COPD (35). Arm exercise requires the use of the accessory muscles of inspiration, thereby decreasing their participation in ventilation and increasing the work of the diaphragm. This observation may explain, in part, why patients with COPD complain of dyspnea when performing activities of daily living with their upper extremities (146). Thus, strategies aimed at improving the function of the accessory muscles of inspiration, such as resistance training, could benefit COPD patients.

Skeletal muscle dysfunction contributes to the reduced exercise tolerance seen in COPD patients (33,

Table 17.5 Exercise Prescription Recommendations for the Patient With COPD

Training method	Mode	Intensity	Frequency	Duration	Progression	Goals	Special considerations and comments
Cardiovascular	Large-muscle activities (walking, cycling, swimming)	RPE 4 to 6 on 10-point scale (comfortable pace and endurance) Monitor dyspnea	One or two sessions, 3 to 5 d/wk	30 min sessions	Emphasize progression of duration more than intensity 2 to 3 mo to ensure compliance	Increase $\dot{V}O_2$peak, work Increase lactate threshold and ventilatory threshold Become less sensitive to dyspnea Develop more efficient breathing patterns Facilitate improvement in ADLs	Exercise compliance should be considered in the determination of exercise intensity. Shorter intermittent sessions may be necessary initially.
Resistance	Free weights, isokinetic or isotonic machines	Low resistance, high reps (fatigue by 8 to 15 reps)	2 or 3 d/wk		Resistance should be increased as strength increases 2 to 3 mo to goal	Increase maximal number of reps Increase isokinetic torque and work Increase lean body mass	Respiratory muscle weakness is common in pulmonary patients. Upper body exercise contributes to dyspnea. Inspiratory muscles may require training.
Range of motion	Stretching Tai chi		3 d/wk			Improve gait, balance, breathing efficiency	
Neuromuscular	Walking Balance exercises Breathing exercises		Daily			Improve gait, balance, breathing efficiency	

138). Impaired muscle strength has been found to be a significant contributor to symptom intensity during exercise in these patients (64, 139). Additionally in COPD, quadriceps muscle strength has been shown to be positively correlated with both the 6 min walk distance and maximal oxygen consumption (59, 64). These observations suggest that resistance training may also prove beneficial for the rehabilitation of patients with COPD.

In general, four exercise training strategies are recommended for improving respiratory and skeletal muscle dysfunctions. These include lower extremity aerobic exercise training, ventilatory muscle training, upper extremity resistance training, and whole-body resistance training. A brief discussion with review of the supporting literature for these four exercise training strategies follows.

Table 17.6 and practical application 17.3 review the literature about lower extremity exercise and COPD. Cumulative results indicate strong evidence for the use of lower extremity exercise as a therapeutic intervention for patients with COPD.

Ventilatory muscle training is recommended for COPD patients to increase ventilatory muscle strength and endurance. The ultimate goal is to improve exercise capacity, relieve the symptoms of dyspnea, and improve health-related quality of life. Three strategies have been used to train the ventilatory muscles: (1) voluntary isocapnic **hyperpnea**, (2) **inspiratory resistive loading**, and (3) **inspiratory threshold loading**.

With voluntary isocapnic hyperpnea, the patient is instructed to breathe at as high a level of minute ventilation as possible for 10 to 15 min. With this technique, the patient is hyperventilating, and therefore a rebreathing circuit must be used to maintain **isocapnia**. The rebreathing circuit is complex and not portable, and the patient requires constant monitoring to ensure isocapnia with use of this device. Because of these problems, this type of training has not been used or studied extensively.

During inspiratory resistive loading, the patient breathes through inspiratory orifices of smaller and smaller diameter while attempting to maintain a normal breathing pattern. A potential problem with the use of this device is that the patient may slow her breathing frequency in an attempt to decrease the sensation of effort. Because of the change in the breathing pattern, the load on the inspiratory muscles is reduced such that a training response may not occur.

With inspiratory threshold loading, the patient breathes through a device that permits air to flow through it only after a critical inspiratory pressure has been reached. These devices are small, do not require supervision, and avoid the problems associated with changing breathing patterns.

Practical Application 17.3

LITERATURE REVIEW

The American College of Chest Physicians and the American Association of Cardiovascular and Pulmonary Rehabilitation have released evidence-based guidelines for pulmonary rehabilitation (9). This document contains recommendations for pulmonary rehabilitation and reviews the supporting scientific evidence. Lower extremity exercise training received a grade of A. This grade reflects the fact that strong scientific evidence supports the use of lower extremity exercise training in COPD patients. This evidence is from the results of well-designed, well-conducted, controlled (both randomized and nonrandomized) trials with statistically significant results that support the use of lower extremity exercise training. The results of controlled randomized clinical trials that have included lower extremity exercise training as part of an intervention are shown in table 17.6. Shown are the effects of lower extremity exercise training on submaximal and maximal exercise capacity, peak oxygen consumption, and quality of life. In nearly all the studies, exercise capacity, as evaluated from time on the treadmill or a timed distance walk, was found to improve after lower extremity exercise training. Whether improvements in exercise capacity translate into improvements in domains such as physical function and activities of daily living has yet to be determined. According to a recent discussion at a workshop convened by the National Institutes of Health to investigate the efficacy of pulmonary rehabilitation, limiting the evaluation of interventions to outcomes such as timed walks or physiological measures is a myopic approach that provides incomplete measures of medical outcomes. The conclusion from this group was that the success of therapeutic interventions should be based on a variety of medical outcomes such as health-related quality of life; respiratory symptoms; frequency of exacerbations; activities of daily living; cost–benefit relationships; use of health care resources; and mental, social, and emotional function.

Table 17.6 Results of Controlled Randomized Clinical Trials on the Efficacy of Lower Extremity Exercise Training

Authors	Exercise capacity	Increased peak oxygen consumption	Improved quality of life
Lan et al. (81)	Yes	Yes	Yes
Berry et al. (23)	Yes	No	Not measured
Pereira et al. (112)	Yes	Not measured	Yes
Breyer et al. (29)	Yes	Not measured	Yes
Cambach (32)	Yes	Not measured	Yes
Goldstein et al. (56)	Yes	Not measured	Yes
Rejbi et al. (117)	Yes	Yes	Not measured
Lake et al. (80)	Yes	Not measured	Yes
Larson et al. (83)	Yes	Yes	Yes
Vogiatzis et al. (152)	Yes	Not measured	Yes
Reardon et al. (116)	Yes	No	Not measured
Ries et al. (119)	Yes	Yes	No
Strijbos et al. (140)	Yes	Not measured	Yes
Toshima et al. (150)	Yes	Not measured	No
Weiner et al. (158)	Yes	Not measured	Not measured
Wijkstra et al. (161)	No	Not measured	Yes

Results from studies on the efficacy of ventilatory muscle training for patients with COPD patients are equivocal. Table 17.7 shows the results of randomized controlled clinical trials that examined the effects of inspiratory resistive loading and inspiratory threshold loading. Of the 18 studies presented, inspiratory muscle strength was found to increase in 11 of them and inspiratory muscle endurance was found to increase in seven. These results suggest that inspiratory muscle training does not add significantly to a program of general exercise conditioning in patients with COPD. In the evidence-based guidelines for pulmonary rehabilitation (9), inspiratory muscle training received a grade of B. This grade reflected the fact that the scientific evidence from both observational and controlled clinical trials yielded inconsistent results. Because of this grade, it was recommended that inspiratory muscle training not be considered an essential component of pulmonary rehabilitation. But in patients who have decreased respiratory muscle strength and breathlessness and who remain symptomatic despite optimal therapy, inspiratory muscle training may be considered an adjunctive exercise therapy (37, 108).

Specific recommendations regarding the intensity, frequency, or duration of training for inspiratory muscle training have not been developed. Most studies reporting improvements in inspiratory muscle function have had patients perform inspiratory muscle training at a minimum of 30% of their maximal inspiratory pressure. The duration of the training has been at least 15 min and the frequency at least three times per week. These appear to be the minimal requisites for an exercise prescription if inspiratory muscle strength and endurance are to be improved.

Based on the results of preliminary studies that have evaluated upper body, lower body, and whole-body strength training in COPD patients, resistance training appears to offer distinct advantages over other forms of exercise training. As such, resistance training should be included in a comprehensive exercise rehabilitation program. Table 17.8 presents research regarding whole-body resistance training and COPD. Although studies are still limited in number, a reasonable recommendation on exercise dosage for demonstrating improvement in outcomes includes training 2 or 3 d/wk with 8 to 10 repetitions and loads of 50% to 85% of one-repetition maximum (138, 139).

Upper extremity resistance training has been proposed as a training modality to help reduce dyspnea

Table 17.7 Results of Randomized Clinical Trials on the Efficacy of Ventilatory Muscle Training

References	Type of training	Outcomes (compared with a control group)
Weiner and Weiner (159)	Resistive loading	Improved maximal inspiratory pressure and peak inspiratory flow
Berry et al. (23)	Threshold loading coupled with general exercise conditioning	No change in maximal inspiratory pressure No change in 12 min walk distance No change in dyspnea ratings
Battaglia et al. (20)	Resistive loading	Increased mean inspiratory pressure Increased mean expiratory pressure Decreased dyspnea ratings
Shahin et al. (126)	Resistive loading	Increased maximal inspiratory pressure Decreased dyspnea ratings Increased 6 min walk distance
Dekhuijzen et al. (45)	Resistive loading coupled with standard pulmonary rehabilitation	Improved inspiratory muscle strength No improvement in maximal work capacity Increased 12 min walk distance
Magadle et al. (90)	Resistive loading coupled with standard pulmonary rehabilitation	Improved score on St. George Respiratory Questionnaire No change in dyspnea ratings Increased 6 min walk distance
Hill et al. (68)	Resistive high-intensity loading	Improved maximal inspiratory muscle pressure Increased 6 min walk distance Decreased dyspnea and fatigue ratings
Guyatt et al. (62)	Resistive loading	No improvement in inspiratory muscle strength or endurance No improvement in 6 min walk distance No improvement in health-related quality of life
Koppers et al. (77)	Resistive loading	Improved inspiratory muscle endurance Increased endurance exercise capacity Decreased dyspnea ratings Increased quality of life ratings
Beckerman et al. (21)	Resistive loading	Improved inspiratory muscle strength Increased 6 min walk distance Decreased dyspnea ratings Improvement in health-related quality of life
Lisboa et al. (87)	Threshold loading	Improved inspiratory muscle strength and endurance Decreased dyspnea Increased 6 min walk distance
Lisboa et al. (88)	Threshold loading	Decreased dyspnea Improved 6 min walk distance
McKeon et al. (99)	Resistive loading	No change in inspiratory muscle strength Increased inspiratory muscle endurance No increase in maximal cycle exercise, 12 min walk distance, or treadmill walking
Noseda et al. (106)	Resistive loading	Increased inspiratory muscle endurance No change in maximal or constant-load cycle exercise

References	Type of training	Outcomes (compared with a control group)
Pardy et al. (111)	Resistive loading	Improved 12 min walk distance Improved submaximal exercise endurance
Preusser et al. (115)	Threshold loading	Improved inspiratory muscle strength and endurance Improved 12 min walk distance
Wanke et al. (154)	Threshold loading coupled with general exercise conditioning	Improved inspiratory muscle strength and endurance Improved maximal exercise capacity
Weiner et al. (159), Wuyam et al. (166)	Threshold loading coupled with general exercise conditioning	Improved inspiratory muscle strength and endurance Improved 12 min walk distance Improved submaximal exercise time

Table 17.8 Review of Studies of Whole-Body Resistance Training for Individuals With COPD

Authors	*n*	Age/ FEV_1 % predicted	Training intensity: % 1RM	Strength gains	Outcomes
Barnard et al. (19)	45	65.5/42.2%	60-80%	8-20%	No difference between groups
Clark et al. (36)	43	49/77%	70%	7.6 kg increase in max lifts	Increased quality of life
Kaelin et al. (73)	50	68/39%	RPE 4 to 7	NA	Increased physical function
Kongsgaard et al. (76)	18	65-80/46%	"Heavy RT"	14-18%	Increased health
Mador et al. (89)	32	74/44%	60%	17.5-26%	Increased health-related quality of life
Ortega et al. (109)	72	64/41%	70-85%	Significant increase in lifts	Decreased fatigue and anxiety
Panton et al. (110)	17	62/40%	32-64%	36%	Decreased effort with three of eight ADLs
Simpson et al. (132)	34	71.5/38 %	50-80%	73% max cycle ergometer test	Decreased shortness of breath with ADLs
Spruit et al. (138)	48	64/40%	70%	20-40%	Increased health-related quality of life
Wright et al. (165)	28	55.7/42	Maximal	18.7%	Increased quality of life

Note. RPE = rating of perceived exertion.

for patients with COPD. Ventilatory muscle fatigue and dyspnea occur when patients with COPD use their upper extremities to perform activities of daily living. Fatigue results because of the additional work that the accessory muscles of inspiration must perform in helping to support the arms during such activities (35). A summary of upper extremity resistance training is shown in table 17.9. Cumulative results show that patients with COPD can tolerate and benefit from a training program consisting of upper extremity strength exercise. But research has not conclusively demonstrated that upper extremity resistance training alone improves activities of daily living or physical function, whereas whole-body resistance training appears to be effective (table 17.8). Preliminary results, however, support the recommendation that upper extremity resistance training be included in a comprehensive rehabilitation program (9).

These preliminary studies do not provide clear recommendations on the specific upper body exercises that would benefit this population or on the resistance or number of repetitions that will provide the optimal benefits. The exercises should probably involve the

Table 17.9 Results of Trials Examining the Efficacy of Upper Extremity Resistance Training

References	Type of training	Outcomes
Couser et al. (39)	Arm ergometry at 60% of maximal workload. Unsupported arm exercise consisted of bilateral shoulder abduction and extension for 2 min. Weight was added as tolerated. Both groups performed leg cycle ergometry.	Following training, minute ventilation and oxygen consumption were lower during arm elevation. Respiratory muscle strength did not change following training. No differences were reported between the two groups.
Martinez et al. (97)	Arm ergometry at a workload that engendered an RPE of 12 to 14 and an RPD of 3. Unsupported arm exercise consisted of five shoulder and upper arm exercises for up to 3.5 min. Weight was added as tolerated. Both groups performed leg cycle ergometry at a workload that resulted in an RPE of 12 to 14, an RPD of 3.	No difference in improvements in 12 min walk, cycle ergometer test, or respiratory muscle function. Task-specific unsupported arm exercise tests improved in the group that performed unsupported arm exercise. Oxygen consumption in unsupported arm exercise tests decreased in the group that performed unsupported arm exercise.
Ries et al. (118)	Gravity resistance exercises that included five low-resistance, high-repetition exercises to improve arm and shoulder endurance. Proprioceptive neuromuscular facilitation that included lower-frequency progressive resistance training with weights to improve arm and shoulder strength and endurance. Both groups participated in standard pulmonary rehabilitation that included walking.	Compared with a control group, both training groups improved on training test specific to the exercise modality. Patients reported subjective improvement in ability to perform activities of daily living using the upper extremities. No change in performance of cycle ergometry tests, simulated activities of daily living tests, or ventilatory muscle endurance tests.

Note. RPE = rating of perceived exertion; RPD = rating of perceived dyspnea.

accessory muscles of inspiration. With respect to the amount of resistance used and number of sets and repetitions to be completed, the American College of Sports Medicine guidelines (11) and the recommendations of Evans (50) and Storer (139) should be followed.

Table 17.10 summarizes the benefits associated with exercise training on fitness components for the patient with COPD. COPD is a common condition that affects a large number of older people. The disease process spans several decades and eventually results in significant morbidity and mortality rates. Research suggests that exercise can be used as an effective therapeutic intervention in these patients. The review presented here supports the notion that participation in an exercise program will decrease dyspnea and increase exercise capacity, addressing two of the most common complaints of COPD patients.

Table 17.10 Benefits Associated With Exercise Training on Fitness Components

Cardiorespiratory endurance	Skeletal muscle strength	Skeletal muscle endurance	Flexibility	Body composition
Cardiovascular reconditioning	Improved muscle strength	Improved muscle endurance	Improved range of motion	Improved body composition
Desensitization to dyspnea	Better balance			Enhanced body image
Improved ventilatory efficiency				

CONCLUSION

This chapter presents background information regarding etiology, clinical history, and signs and symptoms of COPD, as well as exercise testing and prescription strategies for the clinical exercise physiologist and other exercise professionals. Exercise testing and training any patient with chronic disease involves individualization of treatment based on all patient information available. Thus, it is crucial that the exercise professional be familiar with each patient's history before developing an exercise program and that the program be individualized for each patient.

Key Terms

α_1-antitrypsin deficiency (p. 317)
bronchi (p. 318)
bronchioles (p. 318)
bronchodilator (p. 325)
chronic bronchitis (p. 317)
chronic obstructive pulmonary disease (COPD) (p. 317)
dyspnea (p. 319)
emphysema (p. 317)
exacerbations (p. 317)
forced expiratory volume in 1 s (FEV_1) (p. 319)
forced vital capacity (FVC) (p. 319)
hypercapnia (p. 319)
hyperinflation (p. 319)
hyperpnea (p. 328)
hypoxemia (p. 319)
hypoxia (p. 324)
inspiratory resistive loading (p. 328)
inspiratory threshold loading (p. 328)
isocapnia (p. 328)
pack-years (p. 320)
parenchyma (p. 318)
polycythemia (p. 324)
pulmonary rehabilitation (p. 325)
roentgenogram (p. 320)
spirometry (p. 319)
ventilatory muscle training (p. 328)

CASE STUDY

MEDICAL HISTORY

Mr. DM is a 69 yr old white male who complains of shortness of breath on exertion and occasionally at rest. The patient does not report any symptoms suggestive of myocardial ischemia. Additional findings from the medical history include treatment for hypertension and prostate cancer diagnosed within the past 5 yr. The patient quit smoking cigarettes approximately 3 yr ago and reports a 102 pack-year smoking history (average of two packs per day for 51 yr). The patient was admitted to a local hospital for an exacerbation of respiratory symptoms approximately 4 mo before enrolling in the exercise program. The remainder of the medical history is unremarkable. The patient did not use supplemental oxygen at the time of entrance into the exercise program; oxygen saturation at rest by pulse oximetry is 95%. The patient's score on the dyspnea subscale of the Chronic Respiratory Disease Questionnaire, a measure of health-related quality of life, is 5 (on a 1-7 scale), corresponding to "some shortness of breath" during performance of activities of daily living. The patient also reports a sedentary lifestyle, rarely walking outside the home, and not participating in any sport or recreational activities.

Mr. DM reported the following scores for performance of selected activities of daily living and their relation to shortness of air: grooming (2.4), shopping (2.7), driving (1.0), light housework (3.1), and walking (2.8). The scale parameters are as follows:

1 = able to perform without shortness of air

2 = able to perform with slight to moderate shortness of air

3 = able to perform with severe shortness of air

4 = unable to do because of shortness of air

(continued)

Results of the preexercise medical exam revealed the following:

- Height and weight of 70 in. (178 cm) and 222 lb (101 kg) (body mass index = 31.8)
- Resting heart rate of 85 and blood pressure of 144/98
- Enlarged anteroposterior chest diameter and decreased breath sounds, prolonged expiration, and wheezes
- Regular pulse with no murmurs, gallops, or bruits noted
- Normal hearing and vision, absence of edema in lower extremities, and good mobility

On enrollment into the exercise program, the patient reported the following medications:

- Atrovent inhaler, eight puffs twice a day (anticholinergic bronchodilator)
- Doxapram HCL, 50 mg three times a day (respiratory stimulant)
- Furosemide, 40 mg four times a day (diuretic)
- Hytrin, 2 mg four times a day (antihypertension drug 1-selective adrenoceptor-blocking agent)
- Prednisone, 5 mg four times a day (corticosteroid)
- Proventil, 0.5% twice a day (β_2-adrenergic bronchodilator)
- Serevent inhaler, two puffs twice a day (β_2-adrenergic bronchodilator)
- Theo-Dur, 300 mg twice a day (methylxanthine derivative)
- Ventolin inhaler, two puffs twice a day (β_2-adrenergic bronchodilator)

Pulmonary function testing revealed a forced vital capacity of 5.31 L (127% of predicted), an FEV_1 of 1.60 L (49% of predicted), an FEV_1/FVC ratio of 30%, and a maximal voluntary ventilation of 77 L (60% of predicted). After administration of 200 mg of albuterol by metered dose inhaler, the FVC improved by 80 ml and the FEV_1 improved by 20 ml. Blood gas analysis was not performed.

DIAGNOSIS

- Stage 2 (moderate) COPD with dyspnea on exertion
- Obesity
- Hypertension
- Physical deconditioning

EXERCISE TEST RESULTS

The patient performed a graded exercise test on the treadmill with continuous 12-lead ECG monitoring and blood pressure assessments. Ratings of perceived dyspnea were assessed with the Borg scale (1 to 10); oxygen saturation was assessed by pulse oximetry; and respired gas analysis was performed with a metabolic cart. Resting data included heart rate of 88, blood pressure of 144/100, and oxygen saturation of 94%. The resting ECG was essentially normal. Mild nonspecific T-wave flattening was noted in the lateral chest leads. The patient was able to complete only the first stage of the graded exercise test using a modified Naughton protocol, walking for 2 min at 1.5 mph (2.4 kph) and 1.0% grade. Heart rate was 125 beats · min^{-1} (83% of age-predicted maximum), and blood pressure was 194/100 at maximal exercise. The patient reported a dyspnea rating of 5, corresponding to "strong shortness of breath" on the Borg scale. No ECG changes consistent with ischemia were noted, and the patient did not report chest tightness, pain, or pressure. Rare premature ventricular contractions (PVCs) were observed during exercise. Oxygen saturation at maximal exercise decreased to 85%, and the peak oxygen consumption was 14.7 ml · kg^{-1} · min^{-1}. The test was terminated because of shortness of breath and oxygen desaturation.

The patient also performed a 6 min walk for distance and a hands-over-head task for time before beginning the exercise program. The distance covered during the 6 min walking trial was 948 feet (289 m); oxygen saturation decreased to 85%, and the patient reported a shortness of breath rating on the Borg dyspnea scale of 7 (severe shortness of breath or very hard breathing). The hands-over-head task is designed to assess upper body strength and susceptibility to dyspnea when the patient uses the upper extremities. The task involves removing and replacing 10 lb (4.5 kg) weights along a row of six pegs positioned at shoulder height. The patient completed this task in 57.4 s with a dyspnea rating of 3 (moderate shortness of breath) and an oxygen saturation of 86%. The average time for subjects in this particular rehabilitation program to complete this task is 50.3 s.

EXERCISE PRESCRIPTION

The primary consideration in prescribing exercise for this patient with COPD is his ability to maintain adequate oxygen saturation. The oxygen saturation values from the graded exercise test, the 6 min walk, and hands-overhead tasks indicate that the patient should be prescribed supplemental oxygen for use when exercising. In this case, the clinical exercise specialist can serve as patient advocate by providing the primary care physician or pulmonologist with documentation that supports the need for supplemental oxygen. During exercise, oxygen flow rate should be adjusted to maintain a minimum oxygen saturation of 90% or greater. The lack of ECG changes suggestive of myocardial ischemia during the treadmill test does not preclude the presence of coronary artery disease. Coronary artery disease is common in patients with COPD, and the diagnostic sensitivity of treadmill testing improves if the patient can attain a maximal or near-maximal level of exertion. In this case, the patient achieved a heart rate corresponding to approximately 83% of predicted. Signs and symptoms of myocardial ischemia should be carefully monitored during exercise training in this population. The exercise prescription for this patient includes the following components:

- Aerobic training through walking to improve functional capacity, perception of dyspnea, and ability to perform activities of daily living
- Upper body strength training exercises with dumbbells (biceps curl, triceps extension, shoulder flexion, shoulder abduction, and shoulder shrugs) to increase muscular strength and lean body mass
- Stretching exercises three times weekly, performed after walking to improve joint range of motion and mobility
- Frequency of three times weekly in a supervised setting to maximize training effects, minimize fatigue and risk of injury, and maximize compliance
- Intensity of aerobic training exercise at a dyspnea rating of 3 to 5 on Borg dyspnea scale and intensity of strength training at two sets of each exercise with a weight that allows 12 to 15 repetitions of the movement to maximize training effects, minimize risk of untoward cardiovascular or pulmonary events, and maximize compliance
- Duration of 30 min per session (interval training may be required, especially early in the training program) to maximize training effects, minimize fatigue and risk of injury, and maximize compliance

DISCUSSION QUESTIONS

1. How would the results of this patient's graded exercise test be expected to differ from those of a healthy age-matched nonsmoker?
2. What improvements can be expected in the graded exercise test and the other outcome measures because of this patient's participation in a program of exercise rehabilitation using the exercise prescription outlined?
3. How are the results of this patient's pulmonary function tests different from those of a healthy age-matched nonsmoker?
4. What physiological factors would account for these pulmonary function test differences?
5. Would involvement in an exercise program result in improvements in these pulmonary function tests? Why or why not?

Chapter 18

Asthma

Brian W. Carlin, MD

Asthma is a very common disease that affects people worldwide. Most people who have asthma can and should remain active throughout their life. Various strategies, both pharmacologic and nonpharmacologic, can be used to ensure that each person is able to be as active as possible. This chapter discusses the scope, pathophysiology, associated signs and symptoms, and treatment options available, with particular emphasis on exercise training and prescription.

DEFINITION

Asthma represents a continuum of a disease process characterized by inflammation of the airway wall. An operational definition of asthma is a chronic inflammatory disorder of the airways with **airway hyperresponsiveness** that leads to recurrent episodes of wheezing, breathlessness, chest tightness, and coughing occurring particularly at night or in the early morning. These episodes are usually associated with widespread but variable airflow obstruction that is often reversible either spontaneously or with treatment (30, 58).

SCOPE

Asthma is a worldwide problem with an estimated 300 million affected. More than 22 million Americans are affected (32, 57, 59). Studies have shown that the prevalence of asthma in the world has been increasing over the last several decades by 5% per year (17, 34, 53, 72). Most childhood asthma begins in infancy (before the age of 3), and viral infections are proposed to be a critical component in its development (81). The incidence is higher in some patient populations, with up to 23% of inner-city African Americans noted to have asthma compared with 5% of Caucasians (47). Asthma can also be found in elite-sport athletes (35).

Despite the availability of good medical therapy, the morbidity and mortality rates associated with asthma, both in the United States and worldwide, have increased, particularly in the African American population. The reasons are unclear but include variability of the pathophysiology of the disease process between individuals, influence of environmental factors on the development and progression of the disease, inability of a patient to effectively access health care, and inability of the patient to comply with the recommended treatment regimen. These various factors associated with asthma morbidity depend also on geographic locale and race. For example, asthma morbidity in Hispanics in the United States is highest in the Northeast (39).

PATHOPHYSIOLOGY

Asthma is characterized by variable and recurring symptoms, airflow obstruction, bronchial hyperresponsiveness, and inflammation. Inflammation lies at base of the pathophysiology (15). Airflow limitation is caused by a variety of factors including bronchoconstriction, airway hyperresponsiveness, and airway edema. Asthma involves the interplay between host factors (e.g., innate immunity, genetics) and environmental factors (e.g., airborne allergens and viral respiratory infections) (16, 62) (table 18.1).

Table 18.1 Factors Influencing the Development and Expression of Asthma

Host factors	Environmental factors
Genetic	Allergens
Obesity	Indoor (e.g., mites, domestic animals, cockroach allergens, fungi, molds)
Gender	Outdoor (e.g., pollens, molds, yeasts, fungi)
	Infections (e.g., viral)
	Tobacco smoke (passive and active)
	Air pollution (outdoor and indoor)
	Diet

The mechanisms influencing the development and expression of asthma are complex and interactive. While the clinical spectrum is highly variable, the presence of airway inflammation is persistent even though the symptoms are episodic. This inflammation is most pronounced in the medium-sized bronchi. Over 100 cellular mediators (e.g., cytokines, cysteinyl leukotrienes, histamine, nitric oxide, and prostaglandin D2) are responsible for the development of the inflammation associated with asthma (9, 37). The **CD4 lymphocyte** (Th2 subgroup) is currently believed to promote inflammation by the **eosinophils** and **mast cells** (38), with subsequent infiltration of these cells into the airway wall resulting in edema formation (46).

Structural changes in the airways can occur on top of this inflammatory response (38, 45, 55, 78). The airways may become remodeled with subepithelial fibrosis that results from the deposition of collagen fibers under the basement membrane. This is seen in most asthmatic patients even before the onset of symptoms. Thickening of the airway walls occurs secondary to hypertrophy and hyperplasia of the airway smooth muscle. Angiogenesis (proliferation of new blood vessels) occurs as well due to greater expression of vascular endothelial growth factor. Other structural alterations of the airway occur secondary to an increase in the number of goblet cells in the airway epithelium and an increased size of submucosal glands. These latter lead to increased mucus hypersecretion (9, 25, 69). While the disease is often reversible, in some instances this remodeling of the airways results in incomplete reversibility.

Airway hyperresponsiveness, the characteristic functional abnormality of asthma, results in airway narrowing (46). Linked to both inflammation and airway remodeling (figure 18.1), the overall mechanisms behind the hyperresponsiveness are not completely understood. The airway narrowing then results in the development of an individual patient's clinical symptoms.

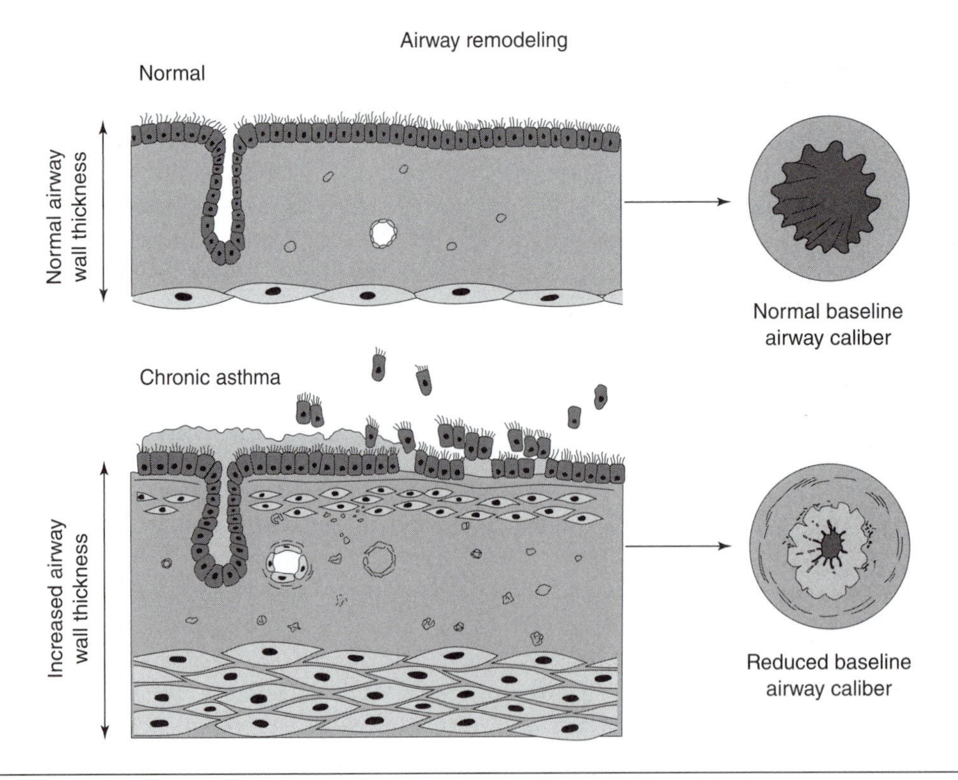

Figure 18.1 Illustrations of normal tissue, swelling, and remodeling.

CLINICAL CONSIDERATIONS

Establishing a correct diagnosis for a person who is suspected to have asthma is key. Using the signs and symptoms associated with the disease as a starting point, the clinician should perform a complete history and physical examination. This is followed by various types of laboratory testing. This sequence of evaluation should help to ensure that an accurate diagnosis is made and the correct treatment afforded to the given patient.

Signs and Symptoms

The diagnosis of asthma should be established from a clinical and laboratory perspective. Symptoms of recurrent episodes of airflow obstruction or airway hyperresponsiveness (that are at least partially reversible) should be present and alternative diagnoses excluded. The determination of airflow obstruction that is partially reversible should be made by spirometry. Reversibility is determined by an increase in FEV_1 greater than 200 ml and a 12% or greater increase from baseline measured after inhalation of a short-acting β-agonist (50, 58).

A clinical diagnosis of asthma is prompted by symptoms such as episodic wheezing, breathlessness, cough, and chest tightness. Asthma symptoms may be intermittent, and their significance may be overlooked by both patients and physicians (34, 35). This is particularly true in children. The symptoms may occur after an accidental exposure to allergens or in relation to seasonal rhinitis. They may also be precipitated by exposure to dust mites, smoke, strong fumes, cold air, or exercise (6). Variable clinical presentations occur from one patient to the next. Up to 15% of patients with asthma fail to notice any type of discomfort following a 20% decrease in the forced expiratory volume in 1 s (FEV_1) (70). In some patients, the clinical symptoms may develop only after exercise (exercise-induced bronchospasm) and may be manifest only as a decrease in the person's exercise tolerance without other symptoms. Death from asthma can occur even in patients who have mild asthma (68).

There are many diseases whose clinical symptoms mimic those seen in asthma. In infants and children, upper airway diseases (allergic rhinitis, sinusitis) and lower airway diseases (e.g., foreign body aspiration, vocal cord dysfunction, laryngotracheomalacia, vascular rings, cystic fibrosis, viral bronchiolitis) must be considered in the differential diagnosis. In adults, chronic obstructive pulmonary disease (COPD), congestive heart failure, pulmonary embolism, mechanical airway obstruction from tumor, vocal cord dysfunction, and cough secondary to medications are alternative causes for symptoms that need to be considered.

History and Physical Examination

A detailed medical history for a patient who is known or thought to have asthma should address several key factors. The presence of cough, wheezing, shortness of

Practical Application 18.1

CLIENT–CLINICIAN INTERACTION

To appropriately determine the level of exercise capability that a patient with asthma might be able to attain, the examiner should review the following features of the clinical assessment:

1. Presence or absence of symptoms (e.g., cough, wheezing, shortness of breath) at rest or with exercise. A history of any exercise limitation (attributable to these symptoms) in an otherwise asymptomatic patient should alert the examiner to the possibility of exercise-induced bronchospasm.
2. Use of medication before exercise (e.g., inhaled β-agonist, inhaled corticosteroid, oral leukotriene modifier).
3. Correct use of the medication (particularly with a metered dose inhaler).
4. Correct use of warm-up and cool-down periods during exercise.

A correct diagnosis of asthma (or exercise-induced bronchospasm) must be made and followed with appropriate use of medications before exercise to allow the patient to optimize his exercise capabilities.

breath, chest tightness, and sputum production should be assessed. Other important key history items are the pattern of symptoms (e.g., seasonal, weekly, nightly, exercise associated), precipitating and aggravating factors, development and progression of the disease, family history, social history, history of exacerbations, and assessment of the patient's and family's perception of the disease).

As the symptoms are often variable, the physical examination of the chest may be variable as well (7). The exam may be normal depending on the degree of airflow obstruction that is present at any one point in time. Wheezing confirms the presence of airflow limitation and is the most usual abnormal physical finding. In some, it may be possible to reproduce wheezing only if the patient forcefully exhales. In a patient with severe asthma, wheezing may be absent due to the extreme decrease in airflow, while other signs (such as cyanosis, drowsiness, difficulty speaking, tachycardia, and the use of the accessory muscles of respiration) may be present.

Diagnostic Testing

While the diagnosis of asthma is often made based on the presence of the characteristic symptoms, pulmonary function tests (in particular, spirometry) assist in the diagnostic evaluation. Lung function measurements provide an assessment of the severity of the airflow limitation, its variability, and its reversibility, as well as confirmation of the diagnosis.

Spirometry is the recommended method of measuring airflow limitation and reversibility to help establish a diagnosis of asthma. Measurements of FEV_1 and forced vital capacity (FVC) are performed during a forced expiration by the patient. Airflow limitation is defined by a decrease in the FEV_1 to less than 80% of predicted and a decrease in the FEV_1/FVC to less than 65%. The degree of reversibility was addressed earlier in the chapter. The flow–volume loops measured during spirometry (figure 18.2) can be helpful to differentiate airway obstruction secondary to asthma (which will show an improvement in flow rates following bronchodilator administration) or emphysema (which will not show an improvement in flow rates following bronchodilator administration). The flow–volume loops can also be helpful to determine if vocal cord dysfunction may be present (65).

Peak expiratory flow measurements, made using a peak flow meter, can be helpful in both the diagnosis and monitoring of asthma. These meters are very inexpensive, portable, and ideal for patients to use in their home and everyday surroundings. Measurements of peak expiratory flow are not interchangeable with other measurements of lung function (such as FEV_1). Peak expiratory flow can underestimate the degree of airflow limitation, particularly as airflow limitation worsens. Values for peak expiratory flow obtained with different peak flow meters vary, and the range of predicted values is wide. Peak flow measures should be compared

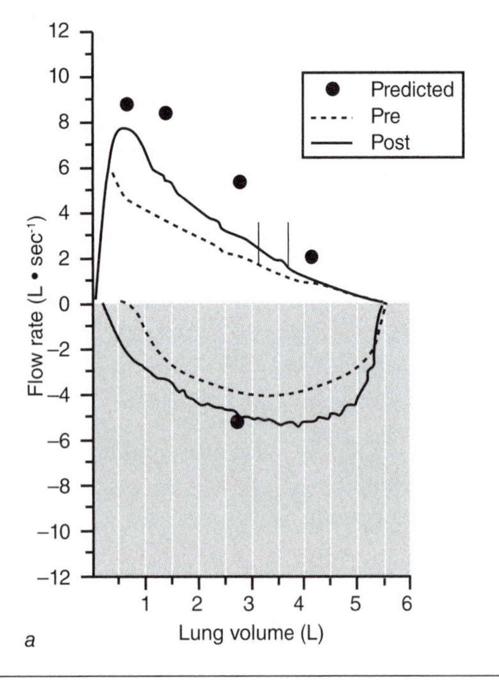

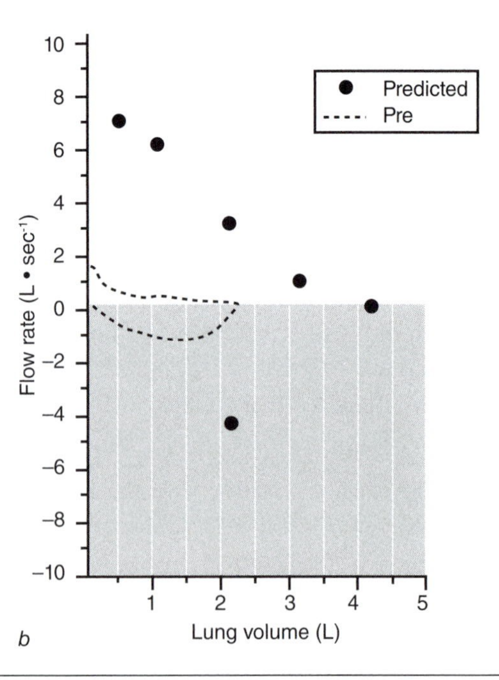

Figure 18.2 Flow–volume tracings of (*a*) a patient with asthma, and (*b*) a patient with emphysema.

to the patient's own previous best measurement and can serve as a reference value for monitoring the effects of treatment. Since many patients have a poor perception of symptoms, the use of peak flow measurements can help to improve the control of asthma for an individual patient.

In some instances, normal spirometry may be present and airflow limitation may be seen only after bronchial provocation testing or after exercise. Nonspecific airway irritants, such as methacholine or histamine, can be administered in aerosol form to determine if there is any decline in the FEV_1 or FVC. A greater than 20% decline in FEV_1 following administration of the irritant is abnormal and indicates the likely presence of asthma. Exercise challenge may also be used to uncover airflow limitation occurring after exercise. A greater than 20% decline in FEV_1 following exercise indicates the presence of exercise-induced bronchospasm (21).

Other studies may be helpful to substantiate the diagnosis of asthma and rule out other causes for the patient's symptoms. The chest roentgenogram may show hyperinflation of the lung (increase in the retrosternal airspace, diaphragm flattening). It may also be helpful for determining that other causes for the patient's symptoms (e.g., pneumonia, pneumothorax, congestive heart failure) are not present. Airway inflammation associated with asthma may be evaluated via examination of sputum produced by the patient for eosinophilic or neutrophilic inflammation. In addition, levels of exhaled nitric oxide (FeNO) and carbon monoxide (FeCO) may be used as noninvasive markers of airway inflammation (40, 43, 44, 66). Neither sputum eosinophilia nor FeNO has been prospectively evaluated as an aid in the diagnosis of asthma, but the latter is being studied as a means of monitoring the response to treatment (75).

Exercise Testing

In patients who have had asthma diagnosed clinically and confirmed by spirometry, exercise testing is generally not performed unless there is a decline in exercise tolerance that is out of proportion to the patient's symptoms or degree of airflow limitation. In some instances, for patients with such significant exercise limitation, standard exercise testing should be performed via a symptom-limited incremental test. Measurement of oxyhemoglobin saturation by pulse oximetry and monitoring of cardiac rhythm should be performed. Measurement of oxygen consumption, carbon dioxide production, and anaerobic threshold using a metabolic cart during progressive incremental exercise will provide a detailed assessment of an individual's response to maximal symptom-limited exercise and is helpful for further decisions regarding an exercise prescription.

For most patients with asthma, there are few contraindications to exercise. Should a patient have acute bronchospasm, chest pain, or an increased level of shortness of breath above that usually experienced, the exercise testing should be withheld. If a patient has severe exercise deconditioning or other comorbid conditions (e.g., unstable angina, orthopedic limitations), it may not be possible to do exercise testing.

Exercise-induced bronchospasm (EIB) occurs in a significant number of patients who have asthma (44, 77). The incidence of EIB was studied in 134 asthmatic and 102 nonasthmatic atopic children and compared to that in 56 nonatopic children. The incidence among those with asthma and nonatopic asthma was 63% and 41%, respectively (44). The pathophysiology behind EIB is thought to relate to the consequences of heating and humidifying large volumes of air during exercise. Cooling and drying of the airways lead to inflammatory mediator release. The airways are narrowed by the bronchial smooth muscle and cellular abnormalities similar to those discussed previously (4, 5). The symptoms of EIB are similar to those of asthma but are associated with short periods of intense physical activity. The typical response of a patient who has EIB involves a 10 min period of bronchodilation at the beginning of exercise followed by progressive bronchospasm peaking at 10 min following completion of exercise. Spontaneous resolution of EIB symptoms occurs over the ensuing 60 min (54).

It is important to confirm the presence of EIB in patients who have symptoms primarily during or following exercise and have minimal, if any symptoms, while at rest. Exercise testing in this situation should be performed for patients who have the relevant symptoms only during or after exercise and are suspected to have EIB. The exercise test is performed to a symptom-limited maximum (on either a bicycle or a treadmill). The work intervals should be short (e.g., 2 min stages) with an increase in workload of approximately 1 metabolic equivalent between stages. Ideally a protocol should be chosen that will elicit a patient's maximal effort between 8 and 12 min. Immediately following exercise, spirometry should be performed and then repeated at 15 and 30 min to determine whether airflow limitation has developed. Should the patient develop bronchospasm, an inhaled β-agonist (e.g., albuterol sulfate) should be administered.

Treatment

The general goals of asthma therapy are to control the overall disease process in order to reduce impairment (defined as the frequency and intensity of symptoms and functional limitations the patient is experiencing currently or has recently experienced) and reduce ongoing risks (defined as the likelihood of either asthma exacerbations, progressive decline in lung function, or risk of adverse effects from medication) associated with the disease. Reducing impairment includes prevention of chronic and bothersome symptoms, maintenance of normal (or near normal) pulmonary function, infrequent use of short-acting inhaled medications, maintenance of normal activity levels, and satisfaction on the part of the patient and family members that the disease is under control. Reducing risk includes prevention of recurrent exacerbations with minimization of the need for emergency department visits or hospitalizations, prevention of the loss of lung function, and provision of optimal pharmacotherapy with minimal or no side effects from such therapy.

In the past it was felt that classifying asthma severity based on the level of symptoms, airflow limitation, and lung function variability into four categories (intermittent, mild persistent, moderate persistent, or severe persistent) was useful in determining the initial and ongoing treatment (table 18.2). Classification of severity is useful when decisions are being made to determine the initial treatment regimen but are no longer recommended as the basis for ongoing treatment (32, 34). The classification has poor value with regard to predicting what treatment is required and what the patient's response to the given treatment might be.

The newer focus regarding asthma therapy involves monitoring of asthma control as the goal for asthma therapy and distinguishing between classifying asthma severity and monitoring asthma control. Severity is defined as the intrinsic intensity of the disease process and should be used as a basis to initiate therapy (table 18.2). Control is defined as the degree to which the manifestations of the disease are minimized by therapeutic interventions and to which the goals of therapy are met, and should be assessed and monitored to adjust therapy.

Monitoring of control includes daytime symptoms, limitations of activities of daily living, nocturnal symptoms or awakenings, need for reliever or rescue treatment, and lung function abnormalities (14). Treatment should be aimed at complete control of all of these clinical features of the disease. Periodic assessment of this level of control is important in the ongoing management of the patient.

The various treatment options for patients with asthma include medications delivered via the inhaled, oral, or parenteral routes; the majority are administered via the inhaled route. In this instance the medication is delivered directly into the airways, producing higher local drug concentrations with much less risk of systemic side effects. Medications are classified as either long-term control medications or quick-relief mediations.

Long-term control medications are used daily to achieve and maintain control of persistent asthma. The most effective are those that reduce the underlying inflammation. Such medications include corticosteroids

Table 18.2 Components of Asthma Severity by Clinical Features Before Treatment

	Days with symptoms	Nocturnal awakenings	Lung function	Interference with normal activity	Short-acting β_2-agonist use for symptom control
Intermittent	<2 d/wk	Fewer than two times per month	Normal FEV_1 between exacerbations; FEV_1 or PEF >80% predicted; FEV_1/FVC normal	None	<2 d/wk
Mild persistent	>2 d/wk but not daily	Three or four times per month	FEV_1 or PEF >80% predicted; FEV_1/FVC normal	Minor limitation	>2 d/wk but not daily, and not more than one time on any day
Moderate persistent	Daily	More than one time per week but not nightly	FEV_1 or PEF 60% to 80% predicted; FEV_1/FVC reduced 5%	Some limitation	Daily
Severe persistent	Throughout the day	Seven times per week	FEV_1 or PEF <60% predicted; FEV_1/FVC reduced >5%	Extreme limitation	Several times per day

(fluticasone, budesonide, beclomethasone, flunisolide, mometasone, triamcinolone), cromolyn and nedocromil, immunomodulators (e.g., omalizumab), leukotriene modifiers (montelukast, zileuton), long-acting β-agonists (e.g., salmeterol, formoterol), and methylxanthines (32, 60). Quick-relief medications are those used to treat the acute symptoms and exacerbations associated with the disease. They include anticholinergics (ipratropium), short-acting β-agonists (albuterol, levalbuterol, pirbuterol), and systemic corticosteroids (either oral or parenteral).

One must take into consideration various safety issues pertaining to the agents used to treat patients with asthma. Inhaled corticosteroids at the currently recommended doses in adults have been associated with few side effects (e.g., oral candidiasis, sore throat, dysphonia). Their use has been associated with reduced bone density and potential for reduced linear growth in children (1, 42, 73, 74). There is also some evidence that overall height is not affected in children who use inhaled corticosteroids (24).

In a large trial comparing a long-acting β-agonist (salmeterol) or placebo added to usual asthma therapy, there was an increased risk of death in the group treated with salmeterol. In addition, an increased number of severe exacerbations associated with formoterol administration have been noted in trials. Use of a long-acting β-agonist (e.g., salmeterol, formoterol) for treatment of acute symptoms or exacerbations is not currently recommended. These agents should also not be used as monotherapy for long-term control (29, 71).

Patients should not stop their inhaled corticosteroid therapy while taking a long-acting β-agonist even though their symptoms have improved.

A stepwise approach to the initial and ongoing management of a patient with asthma is presented in the National Asthma Education and Prevention Program Expert Panel Report 3 (30). This approach includes four components of care: assessment of severity to initiate therapy or assessment of control to monitor and adjust therapy; patient education; environmental control measures and management of comorbid conditions at every step; and selection of medication. As asthma is a variable disease process, an increase or a decrease in the administered therapy may be necessary. The overarching goal of therapy is to identify the minimum medication necessary to maintain adequate control.

For patients with EIB, bronchospasm can be prevented through a variety of means. An appropriate warm-up period before the exercise itself is an important nonpharmacological method that can be used. A warm-up period of 15 min of continuous exercise at 60% of maximal oxygen consumption can significantly decrease postexercise bronchoconstriction in moderately trained athletes. Interval warm-up may be used, but it has been noted in some instances (based on eight 30 s runs at 100% maximal oxygen consumption with a 1.5 min rest in between trials) that it does not significantly reduce postexercise bronchoconstriction. Thus at least 15 min of moderate-intensity exercise should precede significant exercise for active persons with asthma (23). A mask or scarf over the mouth and nose may be helpful to reduce cold-induced EIB.

Pharmacological treatment for EIB should be directed at the prevention of the EIB episodes for those patients who fail to respond to the nonpharmacologic measure just outlined. Both short-acting and long-acting β-agonists can be used. Short-acting β-agonists used shortly before exercise can be helpful for up to 3 h, while long-acting β-agonists can be protective for up to 12 h. Leukotriene receptor antagonists (e.g., montelukast), which have an onset of acting hours after administration, have also been shown to be helpful to reduce EIB (26, 49, 60). Inhaled corticosteroids and cromolyn sodium have been shown to be beneficial as well (36, 48). Given the widely variable pathophysiological components of EIB, therapy must be individually tailored. Table 18.3 provides a summary of medications or therapies for the treatment of various pulmonary disorders.

EXERCISE PRESCRIPTION

Given the great variability of the pathophysiological processes among patients with asthma, the response to exercise varies widely as well (31, 63, 76). Some patients may be able to exercise at an Olympic level while others may be unable to walk across the room without significant shortness of breath. The individual patient's exercise ability may vary from time to time depending on the current level of control of the disease. This is particularly true during an exacerbation, when the individual's exercise ability may be extremely limited (51, 63). The mode, frequency, time, intensity, and progression for health and fitness benefits should be similar to those for a patient who does not have lung disease unless other comorbid illnesses, as described earlier, are present.

Comprehensive supervised rehabilitation is helpful for patients with asthma. Appropriate control of the disease process is of primary importance. Once it is under maximal medical control, improvements in aerobic capacity, muscle strength, and endurance can be maximized. The exercise prescription should be based on objective

Table 18.3 Medications or Therapies for the Treatment of Various Pulmonary Disorders

Medications or therapies	Class and primary effects	Exercise effects	Special considerations
Albuterol (short-acting β-agonist) Bitolterol (short-acting β-agonist) Pirbuterol (short-acting β-agonist) Terbutaline (short-acting β-agonist) Salmeterol (long-acting β-agonist) Ipratropium (anticholinergic) Theophylline (methylxanthine)	Bronchodilators—decrease bronchospasm by opening up the airways	Improved ventilation during exercise, especially in patients with bronchospasm. Short-acting and long-acting β-agonists may cause cardiovascular stimulation. Salmeterol can cause skeletal muscle tremor and hypokalemia. Ipratropium may increase or have no effect on heart rate. Theophylline may produce premature ventricular contractions on the ECG at rest and during exercise.	Timing of administration before exercise affects gains. Short-acting β-agonists and ipratropium are used for rescue relief. Salmeterol and theophylline are controller medications. Salmeterol and albuterol are used in prevention of exercise-induced bronchospasm. Salmeterol is not used to treat acute symptoms or exacerbations. Ipratropium is treatment of choice for bronchospasm because of β-blockers. Side effect of ipratropium is dry mouth. Theophylline has gastrointestinal side effects.
Prednisone (systemic steroid) Prednisolone (systemic steroid) Methylprednisolone (systemic steroid) Beclomethasone (inhaled steroid) Budesonide (inhaled steroid) Flunisolide (inhaled steroid) Fluticasone (inhaled steroid) Triamcinolone (inhaled steroid) Zafirlukast (leukotriene modifier) Zileuton (leukotriene modifier) Cromolyn sodium (cromone) Nedocromil (cromone) Omalizumab (anti-IgE) Ibuprofen (NSAID)	Anti-inflammatories—decrease inflammatory component of the disease	None	Steroids may increase risk of osteoporosis, hypertension, diabetes, adrenal suppression, obesity, cataracts, muscle weakness, bruising, skin thinning, and growth suppression. Inhaled steroids can additionally cause oral thrush, cough, and dysphonia. Inhaled steroids are used as controller medications. Systemic steroids can be used as controller and rescue therapy. Leukotriene modifiers are used as controller therapy. Drug interactions and liver toxicity are possible. Cromolyn sodium and nedocromil are controller medications and can be used as preventive treatment before exposure to exercise or known allergen. Side effects are cough and sore throat. Omalizumab is used in patients with elevated IgE level who are uncontrolled on inhaled steroids.
A variety of oral or intravenous antibiotics (e.g., amoxicillin)	Antibiotics—prevent, treat, or decrease bacterial load	None	
Tobramycin Gentamicin Colistin Polymyxin B	Inhaled antibiotics—prevent, treat, or decrease bacterial load	None	May induce bronchospasm, vocal alteration, or hoarseness.
Nebulized hypertonic saline DNase N-acetylcysteine	Mucolytics—reduce viscosity of pulmonary secretions or sputum	Improved mucociliary clearance; enhanced cough; some improvement in pulmonary function	May induce bronchospasm or cough. N-acetylcysteine has unpleasant taste and odor.
Loratadine Cetirizine Fexofenadine	Antihistamines	Improved ventilation with exercise	Timing of administration before exercise affects gains.
Airway clearance techniques Mucus clearance techniques			Lack of appropriate techniques could result in decreased exercise ability.

measurement of exercise capabilities and individualized for each patient. A variety of training modalities are appropriate, including treadmill, stationary bicycle, walking, and swimming. If weakness of a specific group of muscles is noted, exercises to address that particular muscle group should be offered.

For cardiorespiratory fitness, the exercise training should be at least 20 to 30 min in duration on 2 to 3 d/wk (2). The mode of exercise should take into consideration the patient's interests, past exercise experience, and availability of equipment while acknowledging the effects of the surrounding environment. Exposure to cold air, low humidity, or air pollutants should be minimized. Intermittent exercise or lower-intensity sports performed in warm, humid air are generally better tolerated. There is, however, no consensus on the optimal intensity level at which a patient with asthma should train (18, 20). The intensity prescription should be based on the clinical and exercise test data in conjunction with the patient's goals. If a maximal oxygen consumption during exercise has been established using a metabolic cart, training can begin at an intensity level of 50% to 85% of the heart rate reserve (maximal heart rate minus resting heart rate). Again, this intensity is below anaerobic threshold for most individuals (3). For patients with more limiting asthma, a target intensity based on perceived dyspnea (such as a Borg scale) may be more appropriate (13).

Regular moderate-intensity exercise can be performed safely in children with mild to moderate asthma. In 62 children with mild to moderate asthma (mean age 10.4 ± 2.1 yr) who were randomly assigned to exercise or usual care, the group who underwent exercise had a significant improvement in both exercise capacity and reduction of symptoms (11). Regular exercise should be encouraged in patients who have asthma under good medical control to help improve exercise capacity, improve quality of life, and reduce symptoms (2, 3).

EXERCISE TRAINING

A wide variety of physiological outcomes might be expected as a result of exercise training for patients with asthma. No effect, either adverse or beneficial, has been reported to occur with regard to static lung function measurements (spirometry) including bronchial hyperresponsiveness related to exercise training (22). Several physiological changes have been observed, however, following a training program. These include increases in maximal oxygen uptake, oxygen pulse, and anaerobic threshold. Significant reductions in blood lactate level, carbon dioxide production, and minute ventilation at maximal exercise have also been demonstrated. Subjective responses also have been noted, particularly a reduction in perceived breathlessness at equivalent workloads, following exercise training. This latter response could be attributable to a central nervous system "desensitizing" effect, a decrease in minute ventilation at submaximal workloads, or an increase in the endorphin levels without a concomitant reduction in ventilatory chemosensitivity.

A variety of cardiopulmonary and metabolic responses to exercise in patients with asthma have been reported. From a cardiopulmonary perspective, treadmill exercise for patients with EIB without prior treatment increases ventilation and perfusion inequality, physiological dead space, and arterial blood lactate levels (6). From a metabolic response perspective, a "blunted" sympathoadrenal response to exercise (8, 79), an alteration in potassium homeostasis, and an excessive secretion of growth hormone have all been demonstrated. The role that each of these metabolic responses may play in the exercise limitation in patients with asthma is unknown. Again, given the wide variety of pathophysiological processes in each patient who has asthma, one can expect a wide variety of cardiopulmonary responses to exercise.

One of the most confounding variables in patients with asthma who are attempting to exercise is the effect of dyspnea on exercise capability. The decision to exercise is often weighted against discontinuation of exercise because of the increasing levels of dyspnea experienced by the patient. In one study, "harmful anticipation" significantly increased the perception of visceral changes associated with exercise (56). A wide variability between the degree of airway obstruction, exercise tolerance, and the severity of breathlessness has been noted in several studies (52, 64), but this accounts for up to only 63% of the variance in breathlessness that asthmatics noted during progressive incremental exercise. This complexity concerning the development of symptoms and exercise tolerance might be one reason that the diagnosis of EIB is obscured. Given the variable responses to exercise from a cardiopulmonary, metabolic, and symptom perspective, no unified conclusions can be drawn regarding such effects across patients with asthma. Individual assessment of each patient is thus important when one is trying to determine the degree (and thus subsequent effects) of exercise intolerance.

A variety of training schedules have been used for patients with asthma (practical application 18.2). Various types of exercise, including gym, games, distance running, swimming, cycling, altitude training, and treadmill running, have all been shown to improve exercise capability (10). The frequency of exercise training varied from study to study, ranging from once weekly to daily for periods of time ranging from 20 min up to 2 h per training session. Training periods of 6 to 8 wk were generally used. The intensity of exercise also varied, from a gradual increase in exercise endurance to a short, heavy increase in exercise endurance (19). In one study, 26 adults with mild to moderate asthma (FEV_1 63%) underwent a 10 wk supervised rehabilitation program with emphasis on individualized physical training. Daily exercise (swimming) for 2 wk was followed by twice-weekly exercise. Exercise training intensity was measured by a target heart rate during the first 2 wk and then by perceived exertion as measured by a Borg scale (1 to 10 scale) during the latter 8 wk. Each subject was encouraged to exercise to a Borg level of 7 or 8. All subjects were able to perform high-intensity exercise (80%-90% of their maximum predicted heart rate, and improvements in cardiovascular conditioning and walk distance were observed after the program. Decrease in asthma symptoms and decrease in anxiety were noted following the training period (28).

Ongoing exercise following the initial training program has been shown to be effective for patients with asthma. Of 58 patients who had previously undergone a 10 wk rehabilitation program, 39 reported continuation of regular exercise. Cardiovascular conditioning (as measured by a 12 min walk distance) and lung function values remained unchanged in all patients. There was, however, a significant decrease in the number of emergency department visits over the 3 yr period compared with the year before entry into the rehabilitation program in these 39 patients. A decrease in asthma symptoms was noted only in a subgroup of patients ($n = 26$) who exercised one or two times per week. Continued exercise following a supervised rehabilitation program is helpful for patients with mild to moderate asthma (27).

Breathing exercises to strengthen the respiratory muscles have been used by some, but their overall effectiveness is controversial. Deep diaphragmatic breathing was used in 67 patients with asthma and significantly decreased the use of medical services and the intensity of asthma symptoms. However, there was no significant

Practical Application 18.2

EXERCISE PRESCRIPTION SUMMARY

Mode	Frequency	Intensity	Duration	Special considerations
Treadmill (aerobic)	Five times per week	Just below anaerobic threshold	20 to 30 min per session	Optimize medication therapy before exercise.
Walking/running (track, sidewalk)	Five times per week	Just below anaerobic threshold	20 to 30 min per session	Optimize medication therapy before exercise.
Swimming	Five times per week	Just below anaerobic threshold	20 to 30 min per session	Optimize medication therapy before exercise.

1. Assess patient's underlying respiratory status and goals for exercise.
2. Assess maximum level of exercise.
3. If maximum level of exercise has been determined by measurement of oxygen consumption and carbon dioxide production (cardiopulmonary exercise testing), begin exercise prescription at an initial intensity level just below the anaerobic threshold.
4. If such measurements are unavailable, begin exercise at a level of exercise at which the patient is comfortable performing for 5 min.
5. Instruct the patient to continue exercise for 20 to 60 min per session.
6. Have the patient perform sessions three to five times per week.
7. Increase exercise intensity by 5% with each session.
8. When maximal level of intensity is attained, increase exercise duration by 5%.

change in overall physical activity as measured by an inventory scale (33). Inspiratory muscle training that used a threshold inspiratory muscle training device in a double-blind sham trial showed a significant increase in inspiratory muscle strength (as expressed by the maximum inspiratory pressure measured at residual volume) and respiratory muscle endurance. The training group also had a significant reduction in the number of asthma symptoms, number of hospitalizations, and absence from work or school compared with the sham group (80).

The overall effects of physical training in patients with asthma were reviewed in 13 studies involving 455 participants. Physical training (at least 20-30 min of exercise two or three times per week for a minimum of 4 wk) had no effect on resting lung function or the number of days a wheeze was present. Physical training improved cardiopulmonary fitness as measured by an increase in maximum oxygen uptake of 5.4 ml · kg^{-1} · min^{-1} (95% confidence interval, 4.2 to 6.6) and maximum expiratory ventilation of 6.0 L · min^{-1} (95% confidence interval, 1.5 to 10.4). Thus in people with asthma, physical training can improve cardiopulmonary fitness without adverse effects on lung function or symptoms of wheeze (67).

Patients with EIB should be encouraged to undergo exercise training as well. Instruction on preventive strategies allowing adequate control of airway inflammation and bronchoconstriction is important in the management of these patients. For each patient, the clinician should consider triggers for the development of asthma under such situations as being outside and exercising on a day with a high ozone concentration or high allergen counts in the atmosphere, or the development of symptoms following ingestion of certain foods within an hour or two before exercise (e.g., milk products, vegetables). As discussed previously, an adequate warm-up period and use of inhaled or oral medications before exercise should be stressed.

Comprehensive rehabilitation programs for patients with asthma include much more than just exercise training. Components of the initial patient assessment should include patient interview, medical history, diagnostic testing, symptoms and physical assessment, nutritional evaluation, activities of daily living assessment, educational and psychosocial history, and goal development. Program content should include education regarding the disease process, triggers of asthma, self-management of the disease (medication use, warning signs and symptoms associated with exacerbations, peak flow monitoring, metered dose inhaler technique, importance of exercise warm-up and cool-down), activities of daily living, psychosocial intervention, and dietary intake and nutrition counseling. Follow-up and evaluation of outcomes are of vital importance as part of the rehabilitation process. Questionnaires used to assess the asthma patient's quality of life (measuring variables such as symptoms, emotions, exposure to environmental stimuli, and activity limitation) have been well validated and should be used as part of this assessment process (41). In addition, one can assess cost of medications and equipment, time lost from work or school, and use of health care resources (e.g., emergency department visits, calls to the patient's physician) as part of the follow-up. Not all patients with asthma are candidates for such comprehensive rehabilitation programs.

There are potential risks to exercise in patients who have asthma. High-intensity exercise may trigger EIB through increasing minute ventilations and respiratory heat and water losses. This may lead to a greater drop in FEV_1 than would otherwise be expected (61). Some sports (e.g., cross-country skiing) expose individuals to dry, cool air, setting up a potential for the development of airway inflammation and subsequent airflow limitation (82). Sport-related deaths due to asthma have been reported in younger individuals (12).

The widespread acceptance, over many years, that patients with asthma cannot and should not exercise has led to many recommendations that these patients avoid exercise. Many parents have unnecessarily restricted children with asthma from exercise because of the fear that exercise may make the asthma "worse." Clinicians must attempt to more completely educate patients and their families about the importance of exercise and about how a child with asthma can safely perform exercise. The use of β-agonist or leukotriene modifier therapy before exercise, as well as the avoidance of conditions known to precipitate that person's asthma, are important mainstays of the treatment and should be used aggressively when asthma plays a role in exercise intolerance.

Most patients with asthma can be managed quite effectively with a combination of medications and general exercise recommendations. In patients who have moderate or severe persistent asthma, those who have failed medical therapy and have had a significant decline in their performance of the activities of daily living, or those in whom the disease process has drastically adversely affected lifestyle, comprehensive rehabilitation offers an effective means to improve overall quality of life.

CONCLUSION

Asthma represents a complex process involving airway narrowing secondary to airway inflammation and bronchoconstriction. Environmental risk factors (such as indoor allergens, viral infections) or other triggers (such as exercise, cold air) can initiate an allergic response, resulting in airway inflammation and airway hyperresponsiveness. Airflow limitation occurs and the patient develops symptoms such as chest tightness, wheezing, and shortness of breath. Exercise limitations and decreased levels of fitness are frequently noted in patients with asthma but in many instances are not considered important for some time following the initial development of symptoms. Exercise limitations and fitness levels can be improved in patients treated with an appropriate medication and exercise regimen.

Key Terms

airway hyperresponsiveness (p. 337)
CD4 lymphocyte (p. 338)
eosinophils (p. 338)
mast cells (p.338)
peak expiratory flow rate (PEFR) (p. 348)

CASE STUDY

MEDICAL HISTORY

Ms. JR is a 22 yr old Caucasian college senior. Throughout high school she was active in competitive sports including soccer, swimming, and field hockey. On occasion throughout high school she would develop an increase in shortness of breath and a cough. Her primary care physician told her that she had bronchitis and that she should not worry about it. After entering college, she continued with competitive soccer and swimming. At the end of a long run during soccer games she would note an increase in cough and slight wheeze. She did not note any symptoms following her swimming practice. She continued to exercise but noticed an increase in coughing and wheezing over the ensuing year.

DIAGNOSIS

Her parents became concerned about her discomfort and tried to convince her not to exercise because "it makes you feel much worse and could be dangerous." With ongoing symptoms, she withdrew from soccer. She sought advice of the college physician, who told her that she might have asthma given the symptoms of wheezing. Spirometry revealed an FEV_1 of 3.09 L (96% predicted), an FVC of 3.54 L (95% predicted), a **peak expiratory flow rate (PEFR)** of 6.97 L (95% predicted), and an FEV_1/FVC ratio of 87%. Given these results, showing "normal" pulmonary function, the patient was told that she did not have asthma but rather bronchitis and was advised to continue her exercise (after a course of antibiotics). She continued to swim but would note that at the end of a training session she was slightly more short of breath than usual and had anterior chest heaviness.

EXERCISE TEST RESULTS

Ms. JR sought the advice of another physician, who ordered an exercise test with the measurement of expired gases during progressive incremental bike exercise. Spirometry was performed at 15, 30, and 60 min following the exercise test. Maximal oxygen consumption was 3.13 $L \cdot min^{-1}$ (52.2 $ml \cdot kg^{-1} \cdot min^{-1}$).

Flow rates were as follows:

- FEV_1 (L): preexercise = 3.09; 15 min postexercise = 2.87; 30 min postexercise = 2.20; 60 min postexercise = 2.24
- FVC (L): preexercise = 3.54; 15 min postexercise = 3.32; 30 min postexercise = 2.97; 60 min postexercise = 3.03
- PEFR (L/s): preexercise = 6.97; 15 min postexercise = 6.00; 30 min postexercise = 5.25; 60 min postexercise = 5.26

EXERCISE PRESCRIPTION

As a result of these studies, a diagnosis of asthma (exercise induced) was made. The patient was started on a short-acting β-agonist (albuterol sulfate) administered 30 min before exercise. She was instructed to warm up for 15 min with low- to moderate-intensity exercise or swimming before starting a high-intensity swim practice. Exercise tolerance subsequently improved while exercise-associated symptoms became rare (for the most part abated).

DISCUSSION QUESTIONS

1. Why was the initial diagnosis of asthma not entertained?
2. How was the actual diagnosis of asthma (exercise induced) made? What tests should be useful in this determination?
3. How did the recommendations improve the patient's exercise tolerance? Why was swimming initially better tolerated than soccer?
4. Discuss the intensity, frequency, and duration of exercise training for patients with asthma.
5. How would the development of asthma symptoms at the end of a 3 h practice session influence the choice of medication (e.g., short-acting vs. long-acting β-agonist)?

Chapter 19

Cystic Fibrosis

Michael J. Danduran, MS
Julie Biller, MD

A common genetic disorder, cystic fibrosis results in severe mucus blockage that predominantly affects multiple organs and drastically reduces life expectancy. Exercise as part of a therapeutic plan has been shown to provide several physiological and psychological benefits including enhanced quality of life.

DEFINITION

Cystic fibrosis (CF) is a genetic disorder that affects the respiratory, digestive, and reproductive systems. Excessively viscid mucus causes obstruction of passageways including pancreatic and bile ducts, intestines, and bronchi. In addition, the sodium and chloride contents of sweat are increased.

SCOPE

Cystic fibrosis is the most common life-shortening genetic disease in the Caucasian population. Currently, more than 30,000 patients in the United States (70,000 worldwide) have CF, and nearly 1,000 new patients are diagnosed each year (25). Cystic fibrosis is inherited as an autosomal recessive disorder that affects approximately 1 in 2,500 live births in the Caucasian population and 1 in 17,000 in the African American population with a carrier rate of 1 in 25. Individuals with other ethnic or racial backgrounds are also affected but with less frequency. Fifty-three percent of all people with CF are younger than 18 yr of age. The remaining 47% make up the growing adult population. This growth in the adult-based CF population is further emphasized by the increased diagnosis of **adult variant CF** (71). The median survival age continues to improve, measuring in the mid-30s (37.4 yr). The estimated total cost to treat CF in the United States continues to be noteworthy at more than $900 million per year, representing a cost per CF patient of almost $40,000 per year (25).

PATHOPHYSIOLOGY

The gene for CF is located on chromosome 7 and results in the altered production of a protein called the **cystic fibrosis transmembrane conductance regulator (CFTR)**; the protein functions as a chloride channel regulated by cyclic adenosine triphosphate. More than 1,800 unique mutations of the dF508 CF gene have been identified; more than 87% of people with CF have at least one dF508 mutation. The primary role of CFTR appears to be as a chloride channel, although other functions for CFTR have been documented; most importantly it regulates reabsorption of sodium and water along the respiratory epithelium. The abnormal CFTR leads to abnormal sodium chloride and water movement across the cell membrane. When this occurs in the lungs, abnormal thick and dry mucus ensues, resulting in bronchial airway obstruction, bacterial infection, and inflammation. As this vicious cycle continues, lung tissue is progressively destroyed, leading to eventual respiratory failure (figure 19.1). Lung disease accounts for more than 95% of the morbidity and mortality associated with CF. Through aggressive intervention and early diagnosis, however, survival has been extended, with adults living well into their 30s and 40s. The average age at time of diagnosis is approximately 6 mo (25), but because many states are using newborn screening, the diagnosis can be made much earlier. Typically, one or more symptoms lead to diagnosis, including symptoms in the respiratory, gastrointestinal, sinus, and sweat gland systems. The underlying theme in all these systems is the cellular abnormality of ion transport necessary for proper function of epithelial structures.

Respiratory System

At birth, the lungs are normal on a histological basis. As the vicious cycle of infection, inflammation, and impaired mucus clearance ensues, the lungs become colonized with bacteria, and nearly 80% of all patients with CF grow *Pseudomonas aeruginosa* later in life (25). Bacteria such as *Staphylococcus aureus* occur in more than 50% of patients, with other bacteria occurring less frequently. When infants are not diagnosed by newborn screening, a young child most commonly presents with failure to thrive or recurrent respiratory tract infections. Most children present with signs of chronic infection including cough, sputum production, wheeze, fever, and failure to thrive at the time of diagnosis. Many infants or young children with CF have been previously misdiagnosed as having asthma, bronchitis, allergies, pneumonia, or bronchiolitis. Chest radiographs may indicate the presence of acute or chronic changes such as infiltrates, **bronchiectasis** (irreversibly irregular and dilated airways), or hyperlucency. When pulmonary function is assessed in older children (>5 yr) at the time of diagnosis, evidence of airway obstruction—reduced forced expiratory volume in 1 s (FEV_1) or forced expiratory flow between 25% and 75% of forced vital capacity (FEF_{25-75})—or hyperinflation (elevated residual volume and residual volume/total lung capacity ratio) may exist. In CF, the lower the FEV_1, the more severe the lung disease (>90%, normal; 70-89%, mild; 40-69%, moderate; and <40%, severe). New technologies have allowed for infant assessment of pulmonary function, inviting therapeutic interventions to occur early on in patients more severely affected with CF. Exercise tolerance may become significantly compromised as judged against normative values. Ultimately, the progressive loss of lung tissue and airway obstruction lead to respiratory failure. The time course for this progression is variable. Some adults with CF experience little lung damage, and some children experience extensive lung disease. This is thought to be due to modifier genes.

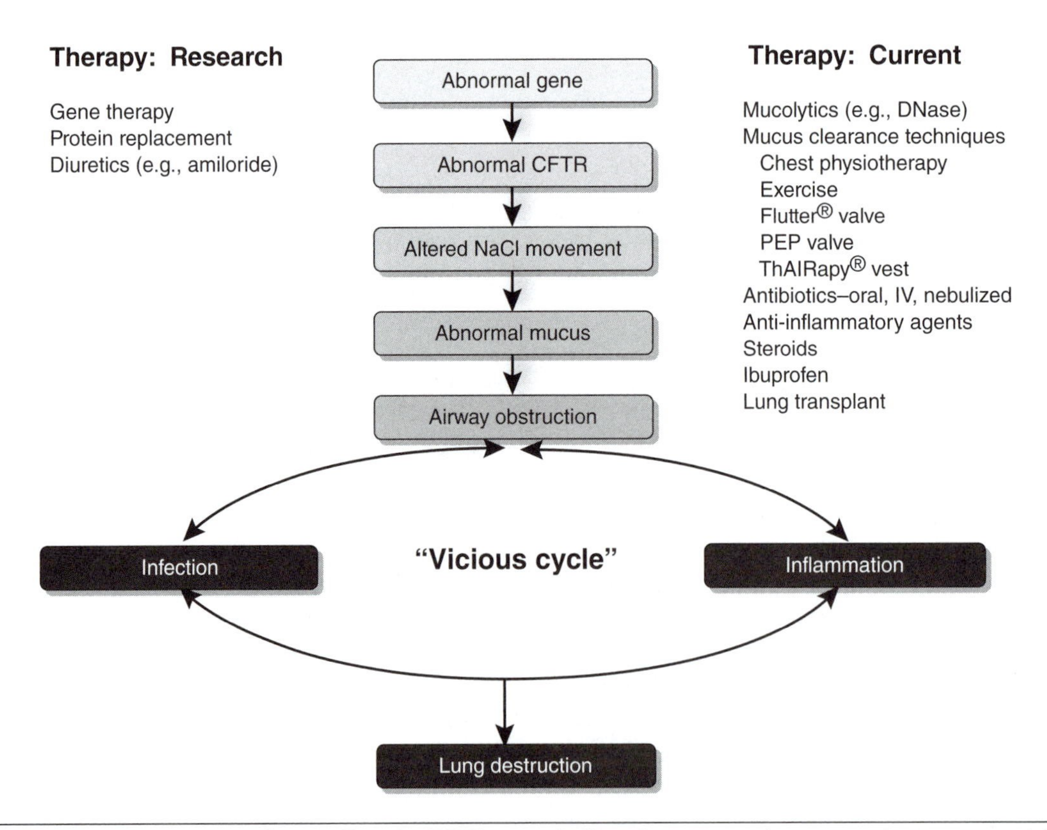

Figure 19.1 The vicious cycle of cystic fibrosis. CFTR = cystic fibrosis transmembrane conductance regulator.

Gastrointestinal and Nutritional Systems

In approximately 85% of individuals with CF, exocrine pancreatic insufficiency is present, resulting in malabsorption of important nutrients including fat and protein. Malabsorption can lead to frequent fatty stools (steatorrhea), malodorous stools, and abdominal pain. The combination of the need for increased caloric intake (attributable to increased resting energy expenditure, cough, and infection) and poor utilization of nutrients with malabsorption often leads to malnutrition or a constant struggle to maintain body weight. Maintaining a desirable body mass index (>22 women, >23 men) has been associated with improved pulmonary outcomes in the CF population (25). In patients with more advanced nutritional issues, a gastrostomy tube placed in the abdomen allows for alternate ways to offer food. Additionally, other organs can be affected, resulting in liver disease, endocrine pancreatic insufficiency (CF-related diabetes mellitus), and gallbladder disease.

Metabolic System

As individuals with CF age, they incur an increasing risk of developing **cystic fibrosis–related diabetes** (currently 30% of adults with CF) (25). Scarring of the pancreas, which produces insulin, often occurs in CF patients. This progressive scarring frequently prevents insulin from entering the bloodstream and can result in diabetes. In patients who develop CF-related diabetes, careful monitoring of blood sugar should occur, especially during times of increased activity. It is important to be able to recognize symptoms of hypoglycemia or hyperglycemia.

Bone Disease

Recent data suggested that approximately 20% of individuals with CF had signs of bone disease. The risk of developing bone disease increases significantly in the second decade of life, with further increases observed over time (25). Early identification of reduced bone mass may permit early intervention and help prevent the development of osteoporosis (70).

Psychosocial

Depression affects more than 21% of adults with CF. Although it is not uncommon in individuals with chronic disease, people with CF as well as their families and caregivers need to be cognizant of the increased risk of depression, and early treatment and intervention is recommended. Depression becomes even more prominent later in life, seen in more than 25% of patients greater than 35 yr of age. Depression is found in approximately 2% of young children with CF.

Sinuses

The development of pansinusitis and **nasal polyposis** is common in people with CF. This finding may be inconsequential, although some individuals may experience difficulty in breathing through the nose. Additionally, pansinusitis with associated bacterial colonization may contribute to the extent of lung disease. Some people require aggressive medical intervention (e.g., antibiotics, nasal irrigation, and endoscopic surgery).

Sweat Glands

All epithelial cells demonstrate the chloride transport defect. This defect in the sweat gland has been turned into a diagnostic test for CF. The basis of the **sweat test** (i.e., pilocarpine iontophoresis analysis) rests on the presence of extremely high salt content in the sweat of individuals with CF. A sweat chloride concentration greater than 60 $mEq \cdot dl^{-1}$ is highly suggestive of diagnosis of CF.

CLINICAL CONSIDERATIONS

The clinical manifestations of CF are variable, with differing involvement of the pulmonary and gastrointestinal organ systems. Comprehensive evaluation that includes assessment of the signs and symptoms, diagnostic studies, and pulmonary function testing helps determine the severity of disease.

Signs and Symptoms

Cystic fibrosis is usually diagnosed by the presence of classic signs and symptoms (table 19.1). Because of the expansive nature of the disease, many systems are affected. Patient care must often be coordinated by a CF care team that may include pulmonologists, gastroenterologists, nurses, respiratory therapists, physical therapists or exercise clinicians, a social worker, a nutritionist, a psychologist, a genetic counselor, and a pulmonary function technologist. Improved survival and a larger number of new patients diagnosed in adulthood have led many care teams to form adult care practices designed to address adult-specific needs. Nevertheless, respiratory and gastrointestinal support is the mainstay of therapy for patients with CF.

Table 19.1 Clinical Signs and Symptoms of Cystic Fibrosis

System	Signs and symptoms
Respiratory	Chronic productive cough, pneumonia, wheezing, hyperinflation, exercise intolerance, *Pseudomonas aeruginosa* bronchitis
Gastrointestinal and nutritional	Steatorrhea, failure to thrive, biliary cirrhosis, intestinal obstruction, abdominal pain, bone disease resulting in osteoporosis
Sinuses	Chronic sinusitis, nasal polyps
Metabolic	CF-related diabetes—pancreatic scarring inhibits insulin secretion
Sweat glands	Salty taste, recurrent dehydration, chronic metabolic acidosis
Other	Depression, infertility, pubertal delay, digital clubbing, family history

History and Physical Exam

A thorough medical history is necessary to identify potential risk factors that may limit exercise performance in individuals with CF. The history should focus on factors that may be present and can alter the pulmonary–cardiovascular–peripheral systems necessary for effective oxygen delivery and utilization during exercise. The most important consideration before testing a patient with CF is to determine the patient's level of pulmonary disease. Prior pulmonary function data can help predict which patients are likely to experience oxyhemoglobin desaturation with exercise testing. An FEV_1 less than 50% of predicted or a low resting **oxyhemoglobin saturation** places the person with CF at much greater risk of oxygen desaturation during exercise (37, 48, 60). A history of wheezing, chest tightness, or chest pain during exercise may indicate the presence of exercise-induced bronchoconstriction, which is seen in 22% to 55% of CF patients (54, 94, 100). Additional considerations, such as a history of pneumothorax or **hemoptysis** (coughing up blood) at rest or during exercise, should be reviewed. A history of nocturnal headaches or cyanosis may suggest advanced lung disease with associated **hypoxemia**. Because exercising at altitude may exaggerate hypoxemia, whether the exercise testing or training program will occur at altitude should be determined (8, 87). Few cardiovascular limitations exist for people with CF. History of pulmonary hypertension or cor pulmonale requires consultation with a cardiologist before testing. Signs and symptoms of right-side heart failure should be sought (e.g., edema, venous congestion, hypoxemia). Peripheral factors such as **scoliosis**, **kyphosis**, and tight hamstrings are commonly present in individuals with CF and may reduce mechanical efficiency during exercise.

Other organ systems can be affected by CF and become an issue during acute exercise. Liver disease with associated ascites (abdominal distension) may interfere with respiratory muscle effectiveness, whereas liver-related bleeding disorders may be exacerbated with increased blood pressure during exercise. Cystic fibrosis–related diabetes may result in abnormal blood sugars with exercise and can manifest as clinical symptoms associated with hyper- or hypoglycemia, including excessive fatigue, confusion, or dizziness (102). Increased incidence of bone disease in patients with CF has been noted, with progression occurring from the second decade of life on. Exercise may assist in retarding the progressive loss of bone minimal density over time (70). Cystic fibrosis patients are also at increased risk of dehydration state, because excessive salt loss with physical exertion is commonly seen in this disease (76, 77). Clinical signs of early dehydration should be discussed with those who plan to exercise. These signs include light-headedness, heat intolerance, flushed skin, decreased urine output, concentrated (dark yellow) urine production, nausea, headaches, and muscle cramps. Adequate hydration needs to be stressed for those who will exercise in warm, humid climates. Consumption of 4 oz (120 ml) of fluid every 20 min is a good general rule. For children who cannot readily quantify fluid amounts, eight gulps of fluid equals approximately 4 oz (120 ml) (57).

The extent of malnutrition and body composition (e.g., lean muscle mass) should be noted, because these considerations may alter the mechanical load applied during exercise testing. Finally, before developing a precise prescription, the exercise professional can use validated physical activity questionnaires or diaries to help determine how physically active the individual is. Options for assessing activity vary in format (e.g., recall questionnaires, activity diaries). Some tools may not be appropriate for younger individuals, but some are designed specifically for use in the pediatric population (e.g., previous-day physical activity recall) (80). The sensitivity of these tools in this population or in children in general has been questioned, especially when recall is required. In adults, monitoring activity may be less complicated. The section "Special Exercise Considerations" details other common conditions associated with CF and more specifically how they affect the exercise clinician's decisions.

Diagnostic Testing

Many states have instituted newborn screening for CF, allowing for earlier detection. In utero diagnosis has also become available. A positive sweat test or genetic mutation analysis can confirm the diagnosis of CF. Additional laboratory testing should be performed when the diagnosis of CF is considered:

- Sputum culture (positive for *P. aeruginosa* or other CF bacteria)
- Chest radiograph
- Sinus computed tomography
- Static and dynamic lung assessment if age appropriate
- Blood sampling for complete cell count
- Liver function
- Nutritional parameters (e.g., total protein, albumin)
- Renal function (e.g., blood urea nitrogen, creatinine)
- Fat-soluble vitamins A, D, E, and K (prothrombin time/international normalized ratio [PT/INR] for vitamin K)
- Glucose

Assessment of static pulmonary function, as defined by the properties of the lung at rest or baseline, is essential in the acute and chronic management of individuals with CF. Simple spirometry, as well as assessment of lung volumes, diffusion capacity, and bronchodilator responsiveness, assists in the detection of an acute **pulmonary exacerbation**. Although some individuals with CF have mild lung disease, most demonstrate varying degrees of airway obstruction with signs of hyperinflation. Additionally, one-third of people with CF will demonstrate signs of airway hyperreactivity when exposed to a bronchodilator. Declines in FEV_1 or indices of smaller-airway function (e.g., FEF_{25-75}, FEF_{50}, FEF_{75}) over time may serve as warning signs of acute or chronic lung deterioration.

In conjunction with static pulmonary function assessment, **dynamic pulmonary function**, as defined by lung function in response to changing physiological state (e.g., work, exercise, physiologic stress), also plays an important role as a diagnostic tool and in monitoring the patient's clinical condition. In fact, a single measure of aerobic exercise tolerance has been strongly correlated with long-term survival in patients with CF (2, 69, 74). Moreover, researchers studied patients who had undergone repeated exercise evaluations over a 5 yr period to look at mortality in years to follow. Individuals who maintained exercise tolerance had greater long-term survival when compared with those who showed a decline in exercise capacity (81). When health care workers assess the lungs under measurable stress (e.g., exercise), ventilatory limitations, as well as impairment in other parameters such as oxygen saturations that depend on dynamic lung function, may become apparent that are not noted when the patient is at rest. Regular assessment of exercise tolerance in patients with CF is an integral component of their medical care. The exercise clinician who treats and assesses individuals with CF should comprehensively understand both dynamic and static lung function measurements and the role that exercise testing plays in the management of this population.

Exercise Testing

The importance of performing a complete exercise evaluation before developing an exercise prescription for individuals with CF cannot be overestimated. Guidelines to exercise testing are summarized in table 19.2. This evaluation can help identify potential risk factors for this specific population. Although many people with CF have limitations in exercise performance, physical activity remains a vital part of the therapeutic management plan. An effective exercise program should help optimize all aspects of fitness including overall well-being, both physical and psychological. The exercise clinician plays an important role in conjunction with the patient, parents, and CF medical team in establishing realistic goals and developing an achievable exercise program to enhance the individual's quality of life.

A baseline maximal aerobic exercise challenge should be administered with monitoring of pulse oximetry and electrocardiogram. If the patient completes the challenge without desaturation (defined as a value <90% on room air), then supplemental oxygen is not required. Should an individual desaturate to less than 90% during the challenge, the point at which this occurs becomes critical. Because most aerobic exercise prescriptions use submaximal intensity levels (60-75% of maximal), the exercise clinician should determine whether the patient desaturated before reaching this submaximal level. If so, supplemental oxygen may be desired for subsequent exercise training. If supplemental oxygen is required, a

Table 19.2 Exercise Testing

Test type	Mode	Protocol specifics	Clinical measures	Clinical implications	Special considerations
Cardiovascular endurance	Treadmill Bicycle 6 min walk	Treadmill: Bruce, Balke, Naughton Bicycle: James, Godfrey	HR, BP, ECG, $\dot{V}O_2$ SaO_2	Assessment of endurance, PWC, risk of desaturation, extent of ventilatory limitations	FEV_1<50% = increased risk of desaturation In severe patients: reduced peak HR, increased $\dot{V}O_2$, RR
Muscular strength and endurance	Bicycle 1RM	Bicycle: Wingate anaerobic test 1RM, grip strength Respiratory muscle function	Peak and mean anaerobic power Force production PImax, PEmax	Assessment of muscular power and endurance Inspiratory and respiratory muscle strength	Measures significantly affected by nutritional status Muscular and respiratory strength typically reduced
Flexibility	Stretching Sit-and-reach	Range of motion testing using goniometer Sit-and-reach	Joint range of motion measured in degrees or inches (cm) in the case of classic sit-and-reach	Thoracic kyphosis develops as disease severity progresses	Early detection of inflexibility can lead to stabilization of the abnormality May enhance chest wall mechanics with exercise
Body composition	Height, weight BMI Body composition	Calculation of BMI Triceps skinfold assessment Three-site skinfold assessment	Stature, mass, BMI, % body fat, lean body mass	Desired BMI: Males >23 Females >22	Significant correlation between nutrition and exercise performance as well as long-term prognosis

Note. BMI = body mass index; 1RM = one-repetition maximum (to assess strength); HR = heart rate; BP = blood pressure; ECG = electrocardiogram (3-lead or 12-lead based on equipment); $\dot{V}O_2$ = oxygen consumption; SaO_2 = oxygen saturation (measured in percent); PImax = maximal inspiratory pressure; PEmax = maximal expiratory pressure; PWC = peak work capacity; $\dot{V}O_2max$ = maximum oxygen consumption during 1 min of exercise; FEV_1 = forced expiratory volume in 1 s; $\dot{V}O_2$= minute ventilation; RR = respiratory rate.

repeat exercise challenge should be administered after an appropriate recovery period to document that the patient remains normoxic during exercise. The level of oxygen supplementation required should be recorded.

Special considerations for the person requiring oxygen for exercise are warranted. The choice of activity may need to be modified to allow for the presence of oxygen tanks. Activities using stationary modalities (e.g., treadmills, bicycles) may be more appropriate for this group. Many health clubs can accommodate people with these special needs.

Individuals with advanced lung disease can obtain the beneficial effects of exercise training. Improved gas exchange, ventilation, aerobic tolerance, peripheral muscle adaptations, and sense of well-being have all been documented following exercise training (47, 65). Appropriate exercise prescriptions can be developed for even the most debilitated person with CF. This population has not traditionally had the benefit of interacting with an exercise clinician. One could argue that individuals with severe CF have the most to gain from exercise that would help reestablish functional ability. By understanding the particular needs of this population, the exercise clinician can play a major role in achieving this goal.

Management of CF may require varying types and amounts of medications. Table 19.3 lists the common pharmacologic agents used in patients with CF and includes special considerations with regard to exercise.

Table 19.3 Pharmacology

Medications	Primary effects	Exercise effects	Special considerations
BRONCHODILATORS			
Examples: β-Agonist (albuterol) Anticholinergics (ipratropium, tiotropium)	Relax muscles in the airways and increase airflow into the lungs Prevent bronchospasm, or narrowing in the lungs in patients with chronic lung disease	Improve airway patency, allowing for increased ventilation during exercise	β-agonist may cause tachycardia at rest, pounding or premature heartbeats, feeling jittery or nervous. Phase 3 trials have been completed for tiotropium to evaluate efficacy in CF, FDA approved for use in COPD.
ANTIBIOTICS			
Examples: Tobramycin inhalation solution (TOBI) Aztreonam inhalation solution (Cayston) Colistimethate	Treat all severities of infections, including those of the airways and lower respiratory tract Inhaled into the lungs using a nebulizer; used to treat lung infections that are commonly seen in patients with CF	No effects on physiologic responses to exercise May cause bronchospasm after use	Exercise participation during times of severe infection should be limited. Bronchospasm after use may affect exercise ability. Several other antibiotics are currently in phase 3 trials for patients with CF, including levofloxacin, ciprofloxacin, and amikacin.
AIRWAY HYDRATORS			
Examples: Hypertonic saline (Hypersal) Mannitol (Bronchitol)	Counteract the dehydration of airway surfaces associated with defective mucus clearance	Can cause cough or bronchospasm that may be exacerbated with exercise	Encouraging coughing and sputum production during exercise is important as an airway clearance technique.
ANTI-INFLAMMATORY			
Examples: Azithromycin Prednisone Inhaled corticosteroids Combination inhaled corticosteroids	In the patient with CF, used to treat chronic respiratory infections by preventing the release of substances that cause inflammation (many brands are in use)	May improve chronic inflammation of airways, resulting in improved ventilation during exercise	Predisone is typically not for routine use. Examples of inhaled corticosteroids are LABA, Symbicort, and Advair. Example of a combination corticosteroid is Symbicort, which contains budesonide to reduce inflammation and formoterol to relax the airways.
MUCOLYTICS			
Dornase alfa (rhDNase) (Pulmozyme)	Break down the excess DNA in the pulmonary secretions of patients with CF Improve lung function by thinning the pulmonary secretions and reducing the risk of respiratory infections	May improve airway obstruction with improvement in sputum mobilization, enhancing tolerance during a bout of exercise	Mucolytics may help maximize the effects of exercise as an airway clearance technique. The use of exercise in combination with other airway clearance techniques has been shown to be effective in mobilizing secretions.
Pancreatic enzymes	Maximize utilization of important nutrients that may go unabsorbed due to exocrine pancreatic insufficiency	No effects on physiologic response to exercise	Lack of patient compliance in taking pancreatic enzymes negatively affects nutritional status and may affect exercise tolerance and lung function.

Contraindications

No absolute contraindications to exercise testing exist within the CF population. However, special considerations need to be observed for individuals with a history of pulmonary hypertension, acute hemoptysis, pneumothorax, oxygen dependence, exercising at altitude, a bleeding disorder secondary to liver disease, and severe malnutrition. Monitoring during testing should include continuous pulse oximetry, electrocardiogram, and vital sign assessment. Supplemental oxygen and a short-acting bronchodilator (e.g., albuterol by inhaler) should be available to all patients with CF during or following testing as needed.

Recommendations

Following a complete history, one can assess exercise capacity. Typical cardiopulmonary responses in CF patients are listed in table 19.4. The precise protocol and testing location depend on several factors, including the desired goals for training, the age of the person, the medical considerations pertaining to the disease, and the resources available. Finally, goals for testing should be clearly defined so that individualized exercise programs can be designed. Such goals may include determining the heart rate at which oxyhemoglobin desaturation occurs, monitoring for improvement in response to medical therapy, or comparison to prior performance.

Cardiovascular Exercise

Numerous reproducible exercise protocols exist for testing maximal exercise performance in children and adults: the Godfrey, McMaster, and James protocols for bicycle testing and the Bruce, modified Bruce, and Balke protocols for treadmill testing (9, 67, 84). Monitoring of pulse oximetry, electrocardiogram, and blood pressure response with exercise should be considered in all patients with CF, especially those with more advanced disease. Despite minimal risk associated with testing, standard practices for emergency management should be followed, including access to a crash cart and supplemental oxygen as well as personnel trained in advanced life support and cardiopulmonary resuscitation.

Table 19.4 Cardiopulmonary Parameter Changes Among Individuals With Cystic Fibrosis

Parameter	Change
STATIC PULMONARY FUNCTION (REST OR BASELINE)[a]	
Spirometry: FEV_1, FEF_{25-75}, tidal volume	Decreased
Lung volumes: RV, FRC, RV/TLC	Increased
Diffusion capacity: DL_{CO}	Decreased
Oxyhemoglobin saturation: SpO_2	Decreased
DYNAMIC CARDIAC AND PULMONARY FUNCTION (IN RESPONSE TO EXERCISE OR STRESS)[b]	
Aerobic capacity: PWC, $\dot{V}O_2max$	Decreased
Breathing response: $\dot{V}O_2$, $\dot{V}O_2$/MVV, RR	Increased
Gas exchange: $\dot{V}O_2$, $EtCO_2$	Increased
Blood pressure response	Normal
Heart rate at rest	Increased
Heart rate during peak exercise	Decreased

Note. FEV_1 = forced expiratory volume in 1 s; FEF_{25-75} = forced expiratory flow between 25% and 75% of the forced vital capacity; RV = residual volume; FRC = functional residual capacity; DL_{CO} = diffusion capacity of the lung by the carbon monoxide technique; SpO_2 = pulse oximetry; PWC = peak work capacity; $\dot{V}O_2max$ = oxygen consumption; $\dot{V}O_2$ = minute ventilation; $\dot{V}O_2$/MVV = ratio of minute ventilation to maximal voluntary ventilation; RR = respiratory rate; $\dot{V}O_2$ = carbon dioxide production; $EtCO_2$ = end-tidal carbon dioxide.

[a]Static pulmonary function is decreased and declines with advancing lung disease.

[b]Abnormal parameters tend to follow extent of lung disease; aerobic performance is weakly correlated with static lung function parameters.

The comprehensive information provided by a maximal aerobic test allows the exercise clinician to develop an appropriate exercise prescription. Measures of maximal oxygen consumption, blood parameters, and 12-lead electrocardiograms have been performed as a part of clinical and research evaluations and have been extremely useful in defining the now-known limitations that exist in patients with CF. However, this form of testing requires sophisticated exercise equipment and highly trained technical staff and can result in a significant financial cost to the patient. A submaximal aerobic test may be useful in determining whether exercise desaturation or breathlessness occurs and can help verify the effectiveness of exercise prescriptions established from maximal tests. Submaximal assessment is used infrequently but may be easier to perform for the young child or adult with significant ventilatory limitations. A treadmill or bicycle similar to that used during a maximal test is the ideal equipment for the submaximal test. Traditional submaximal protocols require workloads consistent with 75% of the age-predicted maximal heart rate. Individuals with CF, however, may not reach the theoretical age-predicted maximal heart rate secondary to ventilatory limitations and should be allowed to terminate the exam short of reaching these physiologic criteria if symptoms or marked desaturation occurs (73). Heart rate, blood pressure, and pulse oximetry should be monitored during submaximal testing. The lowered technical demands and financial costs and the ease of repeat testing make the submaximal test an attractive alternative to the maximal test.

Lab-based exercise tests, whether maximal or submaximal, are not always convenient for individuals with CF. Several walking and running tests have been developed in an attempt to mimic real life more accurately and offer simple-to-administer testing protocols. Walk tests for 2, 6, and 12 min have been used for people with CF (41, 43, 86). These protocols allow patients to walk over a set period at their own pace while heart rate and oxygen saturations are monitored. Total distance traveled, the development of exercise breathlessness, and oxygen desaturation are recorded and can be compared over time with prior tests. Outcome variables have been relatively well correlated to standardized maximal tests for individuals with mild to severe CF lung disease. Serial walk tests may provide a simple yet valuable assessment of the usefulness of supplemental oxygen or pulmonary rehab in the patient severely affected with CF. Shuttle tests have also been used for people with CF (15, 16). In one version, the patient walks (or runs) at increasing speeds (set by an audio signal) over a set course until voluntary exhaustion occurs. Heart rate, pulse oximetry, and distance traveled are monitored during the test. Although both the walk tests and shuttle tests are relatively simple to perform, a limitation is that they depend on patient effort, which makes motivation by the test administrator essential. A 3 min step test has been developed as a modification of the Master two-step exercise test used in adult cardiac testing (9). The total number of steps can be tabulated along with change in heart rate, oxygen desaturation, and sensation of breathlessness (82). All these noninvasive tests are easy to perform and do not require sophisticated equipment. Their utility for assessment of aerobic fitness in patients with milder CF remains unknown, and they are generally reserved for patients with more severe disease.

Muscular Endurance

Although many tests have been proposed to assess muscular endurance in healthy individuals (101), relatively few protocols have been used for individuals with CF. The Wingate anaerobic test (WAnT) has been used to assess both short-term mechanical power or strength and leg muscle endurance over a brief, intense period of time. The WAnT was designed to measure nonoxidative muscle function, with peak power indicative of muscle power and mean power providing information about anaerobic muscle endurance. The test consists of a 30 s all-out sprint on a cycle ergometer against fixed resistance. Determination of resistance depends on lean muscle mass, but a standard starting point is 75 g of resistance per kilogram of body weight (50). The test is demanding, but patients with CF have been able to complete it (14, 17). Although sophisticated equipment is available to perform the WAnT, a mechanical cycle ergometer (e.g., Monark or Fleisch) can be adapted for this test. Alternative protocols for testing muscle endurance in individuals with CF include use of an isokinetic cycle ergometer and cycling at supramaximal levels. Both of these protocols have been used in the research setting for testing children with CF (58, 93) but are not readily accessible outside the academic exercise laboratory. Measurements of respiratory and peripheral muscle fatigue have been used as research tools to assess both respiratory and peripheral muscle function in individuals with CF (55, 59). Many school-based fitness testing (push-ups, pull-ups, long jump, and high jump) measures can be performed by CF patients with mild to moderate disease because they generally do not have restrictions or limitations to any activities. Finally, alternative measures, including balancing, performance accuracy, standing vertical jump, and timed exercises, have been used to further quantify fitness ability either as a one-time measure or as an outcome variable (39). These field tests may provide a fun and effective alternative for assessing both muscular strength and muscular endurance in young patients with CF.

Muscular Strength

Strength of both respiratory muscle and peripheral skeletal muscle groups has been assessed in people with CF. Peak inspiratory pressure determination specifically measures the muscles used for inspiration and consists of the patient's inspiring a breath of air at residual volume against an occluded airway. The greatest inspiratory subatmospheric pressure that can be developed is recorded. Similarly, peak expiratory pressure measures the strength of the abdominal and accessory muscles of breathing and consists of the patient's exhaling forcefully against an occluded airway, usually at total lung capacity. Inspiratory muscle training when corrected for workload can be effective in improving inspiratory muscle function and work capacity (34). These maneuvers are relatively easy to perform (85). The equipment required is usually part of a standard body plethysmography system. Alternatively, handheld direct-reading manometers or electronic pressure transducers and recorders can be used.

Peripheral skeletal muscle has been shown to respond to training in individuals with CF. Home-based strength training programs, as well as inpatient rehabilitation efforts, have resulted in increased strength and physical work capacity (39, 45, 78). Standard techniques, including use of dynamometers, cable tensiometers, isokinetic muscle testing, and free weights, can be applied to test specific muscle groups. The age of the patient often determines the choice of test. For very young children, the child's own body weight can be used as a resistance tool, according to the testing criteria of the President's Council on Physical Fitness and Sport (e.g., push-ups, pull-ups) (31). Expected muscular endurance and strength responses in children with CF are listed in table 19.5.

Body Composition

Individuals with CF tend to be lower in both body weight and height than those without CF. From a clinical perspective, monitoring body mass index as well as body composition is an important part of the nutritional assessment. Typical body composition responses in CF relative to age- and gender-matched peers are listed in table 19.5. Clinically, the desirable body mass index for males with CF is greater than 23 and for females with CF is greater than 22. Patients below these values tend to have decreased clinical standing and poor long-term prognosis when compared with those above the desired value (26). For the exercise clinician, documenting body composition is essential in determining testing workloads as well as for scaling absolute exercise data per muscle mass. Many techniques exist for determining fat distribution in healthy adults. Some of the underlying assumptions of these techniques are in question for children, for individuals with chronic lung disease, and in conditions associated with electrolyte disturbances. Currently, single-site (triceps) or multiple-site skinfold assessment is the most commonly used technique for monitoring body fat in children with CF. Skinfold calipers (e.g., Harpenden, Lange) are an inexpensive means for determining body composition. Use of pediatric reference equations for a child is mandatory (62). An alternative technique for body composition assessment uses bioelectrical impedance analysis through commercially available systems. Both skinfold and bioelectrical impedance assessments are easy to perform, inexpensive, and reproducible in the hands of a trained technician. A common practice in some CF specialty centers is to use both techniques as a means of establishing internal reliability of measurements.

Flexibility

Table 19.5 illustrates the typical changes in muscle endurance and strength, body composition, and flexibility observed in patients with CF. For most people with CF, flexibility is not a major limiting factor in exercise performance. As lung disease advances, thoracic kyphosis ensues, and associated mechanical inefficiencies are seen with exercise (51). Some of these postural changes are associated with tight hamstrings, leading to potential exercise limitations and injury (51). Early identification of these abnormalities through routine assessment of large muscle group range of motion, as part of an exercise assessment, can lead to establishment of stretching programs and stabilization of abnormal posture.

Anticipated Responses

Individuals with CF have impaired exercise tolerance as demonstrated by reduced maximal oxygen consumption and peak work capacity compared with healthy persons (35, 36, 63, 96). The ratio of minute ventilation to the **maximal voluntary ventilation**, a marker of ventilatory limitation, may exceed 100% (normal 70-80%) and worsen as CF lung disease progresses (18, 24). People with CF demonstrate expiratory airflow limitation as evidenced by tidal loop analysis during exercise (5). End-tidal carbon dioxide, another marker of ventilatory limitation, increases with exercise and is related to the severity of lung disease (21, 24, 64). Alveolar ventilation appears normal in patients with mild lung disease with a compensatory increase in the tidal volume (21). But as disease severity increases, alveolar hypoventilation becomes evident as the tidal volume approaches and is limited by the vital capacity (61). As this occurs, breathing frequency increases as a compensatory factor but does not provide the minute ventilation necessary for increased exercise intensity.

Table 19.5 Typical Muscle Endurance and Strength, Body Composition, and Flexibility Levels for Individuals With Cystic Fibrosis

Parameter	Change
MUSCULAR ENDURANCE[a]	
WAnT mean power; isokinetic cycle ergometry power	Decreased
Muscle efficiency	Decreased
MUSCULAR STRENGTH[b]	
Respiratory muscle strength (PImax, PEmax)	Decreased
Peripheral muscle strength	Decreased
BODY COMPOSITION[c]	
Body weight	Decreased
Lean muscle mass	Decreased
Body mass index	Decreased
Percent body fat (BIA, skinfold assessment)	Decreased
FLEXIBILITY[d]	
Peripheral muscle flexibility (hamstrings, quadriceps)	Decreased
Posture—extent of kyphosis	Increased

Note. WAnT = Wingate anaerobic test; PImax = peak inspiratory pressure; PEmax = peak expiratory pressure; BIA = bioelectric impedance analysis.

[a]Muscle endurance decreases as lung disease progresses and may reflect impaired nutritional status and intrinsic cellular deficiencies.

[b]Decreases as disease progresses and may reflect nutritional status and loss of mechanical efficiency.

[c]Decreases in body composition reflect increased caloric expenditure with advanced lung disease, poor oral intake, and release of cachectic mediators.

[d]Reflects deconditioning associated with decreased activity or advanced disease.

Gas exchange can also be compromised as evidenced by the lack of increase in the diffusion capacity of the lungs following exercise (103). In adult patients with moderate to severe disease, dead space ventilation increases. This appears to be secondary to reduced tidal volume in conjunction with increased respiratory rates resulting in less gas exchange with each breath (98). Finally, the phase II component of oxygen kinetics (increased ventilation secondary to a return of deoxygenated blood from muscles) appears to be slowed in patients with CF, resulting in peripheral adaptations (44). Risk of oxyhemoglobin desaturation was seen to increase significantly in patients with FEV_1 below 50% of predicted or in those demonstrating a reduced baseline saturation, below 90% (37, 48, 60).

The cardiovascular system is generally able to keep up with the oxygen demands of the exercising muscle and becomes compromised only with advanced disease. Heart rate and blood pressure responses to exercise appear normal (19, 40, 90), although a lower peak heart rate is seen as the disease progresses (35). Muscular efficiency in children with CF can be reduced by up to 25%. The reduced efficiency may reflect altered aerobic pathways at the mitochondrial level (30). Studies of muscular strength demonstrate that CF patients have reduced muscle strength when compared with healthy controls (27, 38, 49, 78, 91). Results regarding the trainability of the muscle systems in patients with CF are mixed. Some studies have reported strength gains associated with training programs (45, 49, 56, 78, 92), while others suggest that these responses are seen only in compromised patients with the most to gain. The authors of a recent investigation stated that patients with CF who are in good general status do not show improvement in muscle testing regardless of type of training (88).

Studies using the WAnT have demonstrated decreased anaerobic performance in individuals with CF that is related to muscle mass quantity (14). Overall oxygen cost of work appears elevated during exercise for people with CF (29). Additionally, it appears that energy metabolism during exercise is abnormal in children who have CF (12, 101).

Treatment

Current treatment for CF is complex. Specialized CF care centers that offer the multiple-specialty care necessary for these individuals have emerged. There are currently 122 Cystic Fibrosis Foundation–accredited centers across the United States that approve the multidisciplinary care required. Both preventive and acute management are required to optimize health for people in this group. Respiratory and gastrointestinal support are the mainstay of therapy for children with CF. The complexity of care increases in adult patients and in those with severe lung disease. Treatment of the pulmonary component can be best viewed in terms of addressing the vicious cycle of infection and inflammation (figure 19.1).

Current strategies are designed to intervene in this process at multiple levels, thus minimizing the progressive loss of lung tissue. A combination of mucolytic agents and daily mucus clearance techniques can help maintain good pulmonary hygiene. Because bacterial colonization leads to a brisk inflammatory response (e.g., cough, sputum production, increased work of breathing), use of antibiotics becomes necessary. The choice of oral, nebulized, or intravenous antibiotics is determined by the severity of the acute exacerbation. New therapy based on genotype is on the cusp of becoming available.

Exercise as part of the therapeutic medical regimen is standard care in most specialized CF centers. Approximately 6% of patients with CF use exercise as their only form of airway clearance (ACT), but a much larger percentage use exercise as an adjunct to other techniques such as high-frequency chest wall compressions (HFCWC: The Vest), which is the most common form of ACT in patients with CF (63% usage) (25). The Vest is an easy and portable airway clearance device for use in children and adults with CF. The Vest is similar to wearing a life jacket. An air pulse generator rapidly fills and deflates gently compressing and releasing the chest wall assisting in dislodging mucus from the airway. Although not the only technique, the Vest allows for greater independence and eliminates the need for special positioning associated with standard chest percussion. As progressive lung destruction and deterioration occur, the final choice of therapy is lung transplantation. According to the most recent registry information, 207 patients with CF underwent lung transplantation in 2009. The approximate 3 yr survival rate continues to be poor and is estimated at 60% (25).

Besides considering the pulmonary aspects of CF, the exercise clinician must give attention to the nutritional aspects. For the 85% of individuals with CF who are pancreatic insufficient, the use of pancreatic enzymes can help in the utilization of nutrients. Fat-soluble vitamins are also given as supplements. The use of a high-fat, high-calorie diet to ensure the consumption of sufficient calories is standard care for most patients with CF. Cystic fibrosis–diabetes is also on the rise. Careful monitoring of blood sugars during times of increased physical exertion should be a common practice in patients with known diabetes. Furthermore, recognition of the signs and symptoms of hyper- or hypoglycemia is essential. Fluid management during exercise poses unique challenges. People with CF tend to lose more salt in their sweat per surface area than do their non-CF counterparts. Thus, the sodium and chloride levels in the bloodstream often decrease after exercise, whereas these levels are maintained in those without CF (76). In addition, individuals with CF tend to underestimate their fluid needs during exercise. In one study, patients with CF lost twice as much body weight as healthy subjects did when they drank fluid only when thirsty (10). Although children with CF have a tendency to lose salt while exercising, especially in extremely hot, humid weather, most children consume sufficient salt. Ready access to a salt shaker or salty snacks (e.g., pretzels, potato chips), along with liberal fluid intake, usually suffices.

Anthropometric data including height, weight, body mass index (BMI), and percent body fat are vital markers of nutritional status of the patient with CF. Nutritional growth is highly correlated to prognosis. Desirable levels for BMI in both males and females have been established to assist in the nutritional maintenance of these patients. Exercise is routinely recommended for all people with CF, regardless of pulmonary status. In conjunction with regular chest physiotherapy, exercise can enhance clearance of mucus from the bronchial tree (7). Exercise alone, however, does not appear to be as effective as standard chest physiotherapy. Nevertheless, exercise therapy with either unsupervised or supervised pulmonary rehabilitation should be recommended because its positive physiologic outcomes have been proven even in patients severely affected with CF. Furthermore, with 21% of adult patients and 2% of children suffering from depression, the psychological effects of exercise also appear to be an important benefit of regular exercise (53).

EXERCISE PRESCRIPTION

As lung disease progresses, lung function may become significantly compromised. After the high-risk individual is identified, specific precautions may be needed before initiation of an exercise prescription. Use of supplemental oxygen during exercise for those who are prone to oxyhemoglobin desaturation has been beneficial in allowing successful exercise training and recovery (23, 65, 93). Family and medical personnel can encourage those who use supplemental oxygen to participate in exercise training. Although the clinical guidelines for exercise testing offered by the American College of Sports Medicine discourage maximal exercise testing in individuals with FEV_1 less than 60% of predicted (104), with appropriate medical direction patients with significant CF-associated lung disease have safely undergone clinical evaluation using a 6 min walk test or alternate submaximal evaluation. Participation in a formal exercise program or pulmonary rehabilitation can be established after the level of dyspnea and need for supplemental oxygen have been determined.

Additional concerns exist for the child who will be exercising at high altitude. Assessment may be warranted at sea level to determine the risk of desaturation at high altitude (8). Because individuals with CF are susceptible to dehydration, especially when exercising in warm weather, fluid intake should be carefully monitored and encouraged. Fluid intake every 20 to 30 min should suffice.

Special Exercise Considerations

Conditions associated with CF such as CF-related diabetes, exercise-induced asthma, and liver disease may require special consideration. Case-specific guidelines from the CF physician should be developed for these conditions before the beginning of an exercise program. Hypoglycemia and acute bronchospasm can be relatively easy to prevent. Patients who have a gastrostomy tube for additional nutritional support may require special exercise precautions for a brief period of time after placement so as to prevent damage or infection; subsequently they can resume normal activities provided that general care guidelines are followed. Although severe CF-related liver disease is uncommon, its presence can result in a bleeding tendency. People who have either enlarged visceral organs (e.g., liver, spleen) or a bleeding tendency should avoid contact sports. Prescribing exercise in CF patients with severe lung disease can be challenging. Specific application issues in individuals with advanced lung disease are discussed in practical application 19.1.

Patients with CF often develop secondary conditions that can alter the exercise prescription or affect the training program that is being managed by the exercise clinician. Aside from the obvious pulmonary issues caused by CF, *depression*, *CF-related diabetes*, and *malnutrition* secondary to malabsorption of nutrients are the most common secondary conditions that pose possible exercise concerns.

Approximately one in five adults with CF suffers from depression. It is believed that exercise and physical conditioning positively influence individuals with depression. In patients with CF, especially those with advanced disease, exercise may reinforce limitations secondary to their disease, leading to compliance issues and decreased feelings of well-being and a greater sense of their own mortality. The exercise clinician should recognize this possibility and create exercise sessions aimed at achieving desired improvements without magnifying limitations. Alternatives to classic exercises such as low-intensity yoga or Pilates may be desirable.

Patients with CF-related diabetes should be handled like any other patient with known diabetes. Recognizing blood glucose levels preexercise, in addition to the symptoms associated with hypo- or hyperglycemia, is important in working with diabetics. It may be necessary to have snacks readily available for patients with low blood glucose. Finally, if blood sugars can't be normalized, cancellation of the exercise session may be warranted.

Maintaining good nutritional status is a major concern for patients with CF. Malabsorption of nutrients is common in the presence of exocrine pancreatic insufficiency and results in decreased body weight. It is well recognized that patients who maintain an appropriate BMI have improved pulmonary outcomes. A common concern among patients with CF, as well as their families, is that exercise will result in greater loss of calories and thus undesired weight loss. When prescribing exercise, it is important for the exercise clinician to address this concern with the patient. Consultation with the CF team's nutritionist about supplementing calories through food or a gastrostomy tube can help ease these concerns. It is important to explain that the benefits of exercise in the patient with CF far outweigh the risk of expending calories.

Exercise Recommendations

The goals of an exercise program for individuals with CF should include enhancing physical fitness, reducing the severity or recurrence of disease, and ensuring safe and enjoyable participation. To maximize compliance with any exercise program, activity should be carefully selected to enhance cardiopulmonary fitness and other exercise goals as previously determined. The number of adults with CF is now approaching nearly 50% of the total CF population; but despite the size of this group, the majority of people with CF are still in the pediatric and adolescent age group and require special considerations based solely on age.

Children, especially those under the age of 8, generally do not respond well to formal structured exercise

Practical Application 19.1

EXERCISE RECOMMENDATIONS FOR INDIVIDUALS WITH ADVANCED LUNG DISEASE

Traditionally, patients with severe obstructive pulmonary disease have not received the attention of exercise clinicians because of an extremely conservative approach to their exercise participation. Fears of exercise-induced hypoxia leading to pulmonary hypertensive episodes, cardiac ischemia, and dyspnea, as well as the perception of limited beneficial effects of training, have all been deterrents to regular physical activity. Although concerns about hypoxia exist for individuals with advanced CF, appropriate exercise prescriptions can be safely administered. A consideration for a person with advanced lung disease who wishes to participate in regular physical activity is to determine whether exercise will induce oxyhemoglobin desaturation. A measure of blood oxygenation, using the partial pressure of arterial oxygenation (PaO_2) or percent saturation ($\%SaO_2$), should be obtained during initial testing. Oximetry is also recommended during initial exercise sessions to evaluate exercise-induced desaturation.

programs. Children do well when the program matches their muscular development, strength, and coordination with age-appropriate activities. Additionally, a gradual progression in the level of physical activity should allow for attainment of exercise goals while minimizing the risk of injury and noncompliance. This concept of gradual progression depends on the fitness parameter being addressed and is particular to different ages as well as disease severity. The use of "play" consisting of games and diversionary tactics may be most beneficial in meeting these criteria while maintaining compliance and teaching an active lifestyle. Reducing nonschool sedentary time by incorporating outside activities or tasks can be helpful. Older children (>8 yr) may be able to undergo a more structured program based on the mode, intensity, duration, and frequency of exercise. When prescribing exercise in the younger age groups, including the parent in the development of an exercise program is vital to success because parents often falsely perceive negative consequences of exercise for their child (e.g., weight loss) (26).

In adults with CF, many of the same compliance issues seen in the general population exist. Time management issues related to career and family are compounded by daily treatment regimes and therapies designed to maintain disease stability. Exercise progression in this group should not only follow the guidelines of the American College of Sports Medicine but also emphasize a feeling of well-being and quality of life. Finally, recent research suggests that individuals with CF who are new to an exercise training program are similar to any other first-time participants in that they stand to see the greatest gains from involvement in physical activity (40).

Recommendations for exercise prescription are presented in table 19.6. The guidelines have been adapted for adults with CF as well as for the pediatric population. This chapter does not address specifics of traditional exercise prescription; see chapter 5 for a discussion of classic training principles. Exercise compliance in the child or adolescent with CF depends largely on motivation and encouragement as discussed in practical application 19.2.

Cardiorespiratory Exercise

The main objective of cardiorespiratory training is to improve aerobic capacity. Higher levels of aerobic fitness have been associated with better quality of life and improved survival rates. This section reviews the components of the exercise prescription to optimize the client's cardiorespiratory exercise training program.

Mode

No specific activity has been identified as optimal for patients with CF and those with more severe lung dysfunction. Choice of modality depends on the patient's personal preference and need not be costly. Cardiopulmonary benefits have been seen with a multitude of activities, some of which require little or no equipment (e.g., walking, jogging). Treadmills, bicycles, or alternative aerobic modalities (e.g., elliptical trainers) can all be incorporated into a successful exercise program. For patients who experience desaturation, the need to exercise in an environment where supplemental oxygen is available may limit some choices but should not prohibit participation in an exercise program. In adults, specialized classes offered through community centers, health clubs, or YMCA programs may lend a social aspect to the conditioning program that adults may enjoy while

Practical Application 19.2

CLIENT–CLINICIAN INTERACTION

Motivation may be the most powerful factor in determining whether an exercise program will succeed.

Motivation is unique for each person. Because more than 50% of patients with CF are children or teenagers, motivational issues are especially important given adolescent issues of self-image, fitting in with peer groups, and establishing physical abilities. Adult patients with CF require motivation to maintain appropriate levels of fitness despite increasing time constraints associated with both career and daily life. Appropriate client–clinician interaction is important in ensuring successful exercise testing and satisfaction with a program that addresses both the physical and the emotional needs of the client.

Situational Motivational Tips for Clinical Testing

- Make the testing experience fun for younger children by creating a gamelike scenario, with cheering and enthusiasm throughout.
- Explain all procedures in detail. Forewarn children about what they will experience during testing (e.g., breathlessness, muscle fatigue, cough).
- Listen to the child's questions and concerns.
- Select apparatus (treadmill, bike) that the person believes will allow him the greatest success.
- Introduce equipment to children in a way that is fun and easily understood. The pulse oximeter might be described as the "ET light" or as a secret spy decoder that introduces the patient by his fingerprint.
- Use positive motivational phases such as "You can do it," "You're almost there," "We are so proud of you!" or "Only one more minute!" Try to avoid using phrases that influence decisions, such as "Do you need to stop?" or "Do you have to quit?"
- It is important to motivate adult patients as well, by ensuring their understanding that the exercise evaluation is a key component in the evaluation and treatment process. For many adult patients, performing exercise tests or assessments of lung function can be very stressful and emotional, as they understand that poor results are a sign of increasing disease and may see them as a reminder of their mortality. Proper explanation of testing results by the qualified CF team is warranted.

Tips for Exercise Adherence

- Allow children to play an active role in planning. Establish a partnership, and develop the exercise program with the child.
- Address the child's or parents' concerns (e.g., increased weight loss, not being able to keep up with friends, poor body image, presence of a gastrointestinal feeding tube) associated with exercise programs and facilities.
- Have the individual (adult or child) assist in setting exercise goals.
- In adults, acknowledge the concerns and barriers to adherence to the program and work to find alternative strategies.
- Understand that what is successful will be unique for each individual with CF (e.g., completing a marathon, being able to enjoy activities with family, being physically prepared for lung transplantation).
- In both children and adults, communicate frequently and address concerns before they lead to noncompliance, understanding that some days will be better than others.

Cystic fibrosis patients appear to benefit significantly from exercise programming. Pulmonary function tends to either improve or deteriorate more slowly following training. Muscle strength and endurance improvements are well documented. Additionally, body weight can be successfully maintained or increased during exercise intervention, and patient psychological well-being typically improves.

Table 19.6 Exercise Prescription

	DISEASE SEVERITY: MILD TO MODERATE[a]			DISEASE SEVERITY: SEVERE[b]		
	Aerobic	**Anaerobic**	**Flexibility**	**Aerobic**	**Anaerobic**	**Flexibility**
Mode	Any enjoyable aerobic activity—swimming, biking, walking, jogging, sports Elimination of nonschool sedentary time	Sprinting, push-ups, sit-ups, plyometrics, age-appropriate weight training	Stretching, yoga; Pilates can be used for gains as well as relaxation	Supplementation of O_2 if needed, which may limit choices; stationary ergometers work well	Increasing disease state increases risk; light weights can be used to maintain tone	Stretching, yoga as tolerated for control of breathing, relaxation
Frequency	3 to 5 d per week	3 to 5 d per week	2 to 7 d per week	3 to 5 d per week	1 or 2 d per week	2 to 7 d per week
Intensity	70% to 80% of measured maximum	10 to 12 repetitions, low resistance Inspiratory muscle training 80% of PImax	10 to 30 s for each stretch, pain free	Measuring a true max is vital because pulmonary factors limit peak HR to 70% to 80%	Very light resistance; limit activities that would induce Valsalva maneuver	10 to 30 s for each stretch, pain free
Duration	30 to 60 min	20 to 30 min	10 min	30 to 60 min	10 to 20 min	10 min
Progression	No more than 10% in any given 2 wk period	Work upper and lower body, progress when 10 to 12 reps are no longer challenging	Natural progression as flexibility improves	No more than 10% in any given 2 wk period	Minimal progression; maintain high reps with low resistance	Natural progression as flexibility improves
Goals	Improve aerobic function, increase lung function	Increase respiratory strength, assist in increasing body mass	Reduce risk of injury, maintain or enhance chest wall mobility	May be formalized pulmonary rehabilitation program to move toward or prepare for transplant	Increase respiratory strength, improve performance of tasks of daily living	Maintain or enhance chest wall mobility in advancing disease
Special considerations	Retest yearly Data suggest that exercise may retard disease progression.	Standard strength assessments have been used in CF with little difficulty.	Flexibility issues are similar to those for general population.	Formalized pulmonary rehabilitation may be best suited for clinical gains.	Alternative measures of strength assessment are warranted.	Exercise prescription should be an adjunct to regular CF treatments.

[a]Mild to moderate disease severity: minimal risk of desaturation, no sport or activity restrictions; forced expiratory volume in 1 s (FEV_1) >50%.

[b]Severe disease: increased risk of desaturation, which may require supplemental oxygen; FEV_1 <50%.

still providing structure. In children, participation in sports and activities, as permitted by the CF team, is recommended.

Frequency

In general, physical activity should be encouraged daily in people with CF. But in formal exercise programs to ensure improvements in cardiorespiratory conditioning, 3 to 5 d of exercise per week appears optimal. Intense exercise beyond five times per week may lead to increased risk of injury. If people want to engage in activity more than five times per week, cross-training (e.g., strength training, stretching) is advised to allow adequate muscle recovery. Signs of increased fatigue and staleness may be a result of overtraining and should prompt a reduction in exercise frequency. Maintenance of body weight should be a priority. Any reduction as a result of training should prompt increased nutritional support or modification of exercise frequency. For adults with CF, 5 d of exercise per week is optimal for enhanced fitness.

Intensity

Exercise intensity should range from 70% to 85% of the measured maximum heart rate, but lower intensities should be used for beginners. If a maximum cardiopulmonary test has not been performed, the general rule of using a maximum heart rate of 200 beats $\cdot$ min^{-1} for intense running and 195 for cycling can be applied for children and adolescents. After a child has completed puberty, the formula of 220 minus age can be applied as it would be for adults. Thus, a heart rate of 140 to 170 beats $\cdot$ min^{-1} is a reasonable estimate of the heart rate that should be attained for optimal cardiopulmonary benefit. The use of a steady-state protocol (e.g., treadmill or cycle ergometer) can ensure establishment of the appropriate workload. It is easy to meet this objective by choosing a submaximal workload and having the patient exercise for 10 min while measuring heart rate throughout. This intensity of exercise is necessary for obtaining gains in cardiorespiratory fitness. In patients with more progressive CF disease, predictions of maximal heart rate may not apply. A maximal exercise test to assess true maximum may be helpful. For the patient with severe CF, standard perceived exertion scales are an alternative way to establish intensities.

Duration

Exercise sessions should last 20 to 60 min. Two abbreviated sessions are an alternative and can provide similar benefits. Attention span may play a role in the child's ability to perform an activity for longer than 10 min. For the younger patient, varying the exercise sessions by interspersing different activities may minimize boredom and enhance compliance (e.g., 5 min of bike riding, followed by 5 min of jumping rope, followed by 5 min on the treadmill at various speeds and grades).

Progression

Because too rapid a progression in the exercise dose may cause patients to lose enthusiasm for the activity, special attention should be given to advances in the exercise prescription. As in adults, no more than a 10% increase in activity duration should occur after any 2 wk period during the program. Frequency can be gradually increased from three times a week for the beginner to the preferred five times per week over a 3 mo period. With each progression, recognize that the patient may have increased cough and sputum production. This should be encouraged, as it will enhance airway clearance. Finally, the progression of an individual's program should be based on the desired goals and the individual's particular needs.

Muscular Strength and Endurance Training

Although the magnitude of potential gains in strength has been disputed in the literature, there is no question that people with CF may benefit from resistance training through both a generalized increase in muscle strength and a decrease in residual air trapped in pulmonary dead space. Furthermore, strength training in older adolescents and adult patients may enhance body image and promote self-esteem. Researchers have attempted to quantify the added benefits of resistance training in individuals with CF and special considerations are discussed next.

Mode

Strength training programs using both supervised and home-based activities have been explored. Free weights, weight machines, and resistance against body weight can all be used to enhance peripheral muscle strength and endurance, provided that proper instruction is given. Anaerobic activities that mimic the way children play, such as plyometrics, sprinting, cycling, and other modalities that require high-intensity, short-burst duration, can also develop muscle strength and endurance. More recent investigations have looked at specific skill sets such as balance, sports accuracy, flexibility, and walking time in response to an anaerobic conditioning program as a way to quantify gains that may not be represented by true increases in muscular strength. Other modalities that address multiple muscle groups are ideal for enhancing muscular strength and endurance. Respiratory muscle training should take place in consultation with a clinician trained in respiratory disorders.

Frequency

A frequency of three to five times a week is usually appropriate. Care should be taken to allow for adequate muscle recovery. Alternating major muscle groups during training minimizes muscle injury while maximizing training effects. Subtle signs of overuse injuries (e.g., muscle soreness, joint pain) should prompt a reduction in frequency.

Intensity

Muscle strength and endurance can be optimized through high-repetition, low-intensity resistance training or through other modalities described earlier. The American Academy of Pediatrics Committee on Sports Medicine does not recommend high-intensity resistance training for children because of the potential of musculoskeletal injury, epiphyseal fractures, ruptured intervertebral discs, and growth plate injury before a child reaches full maturation (3). But strength training programs that use lower-intensity weights and modalities can be permitted if the planned program is appropriate for the child's stage of maturation (3). Whether adults with CF can safely participate in weightlifting is controversial. Lifting heavy weights can be associated with a Valsalva maneuver that results in increased thoracic pressures. In the susceptible patient with CF (e.g., history of **pneumothorax**, advanced lung disease), this increased thoracic pressure may result in a spontaneous pneumothorax. Consultation with a CF clinician before the start of a weightlifting program is strongly encouraged. After an individual is cleared for participation, a resistance of 50% to 60% of one-repetition maximum is generally used. Three sets of 12 or more repetitions should produce strength gains.

In addition, inspiratory muscle training can significantly improve lung function in many CF patients. Suggested training intensities that produce significant improvements may be at approximately 80% of maximal inspiratory effort.

Duration

In children and adolescents with CF, the duration of each session depends on the number of muscle groups exercised. Generally, 10 to 30 min of properly performed activities can increase muscular strength and endurance. In adults, 30 min of strength training should be sufficient when combined with inspiratory muscle training in addition to routine CF treatments.

Progression

In individuals with CF, the progression of a strength program should be slow. Repetitions or resistance should be increased only when the muscle has adapted to the current workload. The progression for increased resistance should occur when the individual is able to perform 8 to 12 repetitions without fatigue to the muscle. For non-weightlifting activities, activity should progress by no more than 10% during each 2 wk period.

Flexibility Training

Generalized flexibility exercises should be an adjunct to any exercise program and can benefit patients with CF in ways similar to those for the general population. Adequate range of motion is essential for minimizing risk of skeletal injury and ensuring healthy aging. The increasing popularity of alternative exercise options such as yoga and Pilates has resulted in enhanced flexibility while increasing aerobic and anaerobic fitness; these may be attractive alternatives or adjuncts to a regular fitness routine. At low intensity, these activities may also serve as "centering" or calming exercise options that allow the patient with anxiety or depression to relax or enhance mood.

Mode

Because stretching can be performed with little or no equipment, from a logistics standpoint it is one of the easiest aspects of fitness to address. Stretching may provide a protective mechanism against injury and can be a source of tension release and relaxation. Stretching exercises should focus on large muscle groups and should be included before and after activity as part of an effective warm-up and cool-down.

Frequency

Stretching exercises should be considered a routine component of any exercise program. Stretching can occur before, during, and after an exercise session depending on the activity chosen. A stretching program of 2 or more days per week can yield positive results such as decreased tightening of the hamstrings and quadriceps and more efficient use of respiratory muscles during exercise. Stretching for relaxation or tension relief can be performed daily. Individuals with specific flexibility issues (e.g., posture abnormalities, tight hamstrings) may need a more comprehensive stretching program. Finally, stretching can and should be used before and after aerobic and anaerobic activities as part of a normal warm-up and cool-down.

Intensity

Proper technique will help ensure that an appropriate intensity is used for stretching. A proper stretch often feels like a gentle pull in the muscle. Stretching should not be forced. Proper breathing, including exhaling before the stretch and inhaling afterward, helps to minimize injury. Finally, slowly releasing a stretch back into a neutral position allows the muscle to recover.

Duration

A stretch should last between 10 and 30 s. The use of progressive stretching that includes a 10 to 30 s stretch followed by an additional 10 to 30 s has been proposed as well.

Progression

The flexibility of a person gradually improves as the muscle adapts to an increased stretch. Slow progression of a stretching program should occur over a 5 wk period. The length at which a stretch is held can be gradually progressed from an initial 10 to 30 s to 40 to 50 s by the end of 5 wk.

EXERCISE TRAINING

Despite the pathophysiological manifestations, a number of people with CF accomplish many of the athletic endeavors pursued by their non-CF counterparts. The short-term benefits of exercise include the therapeutic aspects of enhanced mucus clearance, with 6% of CF patients using exercise as their main airway clearance technique. Other effects of exercise include improved cardiopulmonary fitness, maintenance of bone mineral density, and psychological well-being. The incidence of depression has risen in both adolescents and adults with CF, and the documented psychosocial benefits associated with regular physical activity are important. Many aspects of fitness appear to show benefits in conjunction with an exercise training program. Because many of the training programs for people with CF have varied in duration, intensity, and modality, establishing causal relationships between training and disease progression is difficult. The long-term benefits are more difficult to define, because the natural course of CF is complex and multifactorial. However, exercise tolerance and long-term survival have been correlated with one another even though no causal relationship has been established (74).

Cardiorespiratory Exercise

Some benefits in **static pulmonary function** have been seen after exercise training. Increases in FEV_1 were seen in response to exercise in studies that used both aerobic and resistance training techniques in both inpatient and outpatient settings (47, 88, 92). Additionally, improvements were observed in forced vital capacity (FVC), FEV_1, FEF_{25-75}, and peak expiratory flow following intensive exercise during a 17 d elective stay in a pediatric rehabilitation hospital in the mountains of Austria (103).

Despite these findings, most exercise training programs have not shown increases in spirometric indices. Multiple studies taking place in settings ranging from outpatient programs to formalized inpatient rehabilitation clinics have failed to demonstrate significant improvements in lung function but rather have resulted in a slower deterioration of lung function compared with that in a nonexercising cohort (56, 69, 75, 90). Improvements in dynamic lung function as well as parameters dependent on dynamic lung function following exercise training are well documented, as evidenced by a lower resting heart rate, increased maximal heart rate, improved maximal oxygen consumption, increased physical work capacity, enhanced ventilatory threshold, and improved maximal minute ventilation. Exercise programs during a hospitalization for an infectious exacerbation can serve as an adjunct to traditional modalities including chest physiotherapy and bronchial drainage, and have been associated with improvements in peak oxygen consumption and peak work capacity (1, 20, 45, 83). Formal supervised training programs including running, cycling, and swimming programs, as well as structured camps, have helped to maximize compliance with exercise. Length of participation and intensity of training vary in these studies, and greater training effects were seen in the more intense programs or in programs that included increases in intensity at points throughout (32, 33, 40, 42, 75, 99). More recent studies have shown that the greatest improvements in dynamic function are seen in patients who by report were the most sedentary at the beginning of the exercise intervention, suggesting that much like untrained healthy individuals, they have the most to gain from a formal program. The patients who entered formalized training programs with normal to above-normal baseline endurance showed minimal or no gains in dynamic pulmonary function (40, 99).

Inefficiencies in the mechanism by which individuals with CF accomplish exercise are apparent. Recent investigations have suggested that much of the ventilatory compromise, especially in patients more severely affected with CF, may be attributed to low tidal volume and resultant hyperventilation (66, 98). This marked increase in the respiratory rate decreases the air actively participating in gas exchange. Furthermore, subtle slowing in the phase II oxygen kinetics has been observed in individuals with CF as deoxygenated blood returns from the periphery (44). Improvement in tidal volume through exercise training, in both an aerobic and an anaerobic fashion, may contribute to increased cardiorespiratory function and improve delayed oxygen kinetics.

Muscular Strength and Endurance

There has been an increase in investigations exploring the benefits of anaerobic exercise or resistance training in recent years. The benefits of exercise training on anaerobic function have been mixed. A program that included upper body strength training demonstrated increased strength and physical work capacity as well as good compliance throughout, as more than 85% of the subjects completed the program (78). Weight training, home cycling for 6 mo, and interventions including sport participation have increased muscle strength in individuals with CF (42, 45, 97). Increased $\dot{V}O_2$ has been reported along with increased leg strength (56), in addition to peak and mean anaerobic power (45, 56, 92). Studies that have attempted to combine aerobic and anaerobic training principles, ranging from 6 wk to 6 mo, have resulted in gains in FEV_1 and in some measures of sport-related skills, such as balance, flexibility, and vertical jump, without significant increases in pure strength (39, 40, 88).

Exercise programs focused on respiratory muscle training have also shown training adaptations as demonstrated by increased peak inspiratory pressures (4, 89). The intensity of inspiratory muscle training has become better defined. High-intensity training (80% of maximum) was shown to improve muscle function significantly more than lower-intensity training did (34).

Body Composition and Nutrition

Children with CF who undergo regular exercise training (e.g., swimming, biking, running, weightlifting) are capable of increasing body mass despite increased caloric needs (6, 20, 46). Nutritional supplementation has been associated with improved aerobic exercise tolerance and respiratory muscle strength in some small case reports (22, 95). More recent randomized control studies are mixed in their findings; some showed no difference in anthropometric data (% body fat, BMI) in either a resistance program or a combined program using both aerobic and anaerobic methods (56, 68), while others demonstrated increases in body weight, fat free mass, and BMI (40, 92).

Psychological Well-Being

The long-term psychosocial benefits of exercise have been fairly well established in children with CF. The Quality of Well-Being Scale, designed to measure daily functioning, has been shown to correlate with exercise capacity in individuals with CF (52, 79). Furthermore, improvements in self-concept and well-being have been shown to be associated with involvement in CF summer camps (53, 103). The current rate of depression among adults with CF is now greater than 21%. As this number continues to grow, research in the area of quality of life and positive patient perceptions continues to flourish, and multiple studies cite improvements in both (28, 56, 90, 92).

Exercise Influence on Cystic Fibrosis

The role of exercise in preventing the deterioration of lung function or occurrence of complications associated with CF continues to be unclear. Although studies have linked exercise tolerance with long-term prognosis, the preventive benefits of exercise have not been established (69, 74). One study has attempted to examine the rate of exercise decline as a predictor of mortality. Measures of oxygen consumption were made annually over a 5 yr period. The rate of decline was calculated along with the subsequent 8 yr survival. On average, oxygen consumption fell by 2.1 $ml \cdot kg^{-1} \cdot min^{-1}$ per year. Furthermore, those with an oxygen consumption less than 32 $ml \cdot kg^{-1} \cdot min^{-1}$ had an 8 yr mortality of almost 60%, and those above 45 $ml \cdot kg^{-1} \cdot min^{-1}$ showed no mortality at all. This improved prognosis associated with exercise tolerance in CF has not been completely explained. Another study examined how an exercise group that performed a 3 yr home exercise program, consisting of three 20 minutes sessions a week, compared with a control group. Despite the increased activity, there was no significant difference between groups in the rate of annual decline in $\dot{V}O_2max$, peak heart rate, maximal minute ventilation, peak work capacity, body weight, or incidence of acute exacerbations. There were, however, significant differences in pulmonary function; a decreased rate of decline was seen in the exercising group, as well as increased quality of life and compliance with airway clearance techniques (90). This study is discussed in greater detail in Practical Application 19.3. The potential of an enhanced immune

Practical Application 19.3

RESEARCH FOCUS: A RANDOMIZED CONTROLLED TRIAL OF A 3-YEAR HOME EXERCISE PROGRAM IN CYSTIC FIBROSIS

Methods. This was a computer-randomized controlled study with a 3 yr parallel design in patients with CF. The two groups were an exercise group ($n = 30$, 13.4 ± 3.9 yr; performed a minimum of 20 min of aerobic exercise at a heart rate of approximately 150 beats $\cdot$ min^{-1}, three times a week, including a 5 min warm-up and cool-down) and a control group ($n = 35$, 13.3 ± 3.6 yr; maintained their usual physical activity practices). Both groups maintained their usual schedule of medications and airway clearance techniques. Clinic visits with the CF team were according to the usual schedule (approximately four or five a year) and included routine clinical evaluation (i.e., pulmonary function testing, anthropometrics). Yearly assessment of exercise tolerance was performed for both groups, with activity logs maintained throughout the investigation. Written questionnaires were completed regarding patient attitudes toward physical activity, as well as the perceived feasibility of such a program, at yearly intervals. Compliance was measured throughout, with annual incentives provided.

Results. Annual rates of decline were assessed for multiple variables that are included in the table (adapted from Schneiderman-Walker et al. [90]). Though not always significantly, the rate of disease progression was decreased in the exercise intervention group, most notably in the airway function indices FVC and FEV_1 ($p < 0.02$, and 0.07, respectively). Furthermore the study demonstrated a strong correlation between pulmonary function and exercise tolerance at baseline that was strengthened with 3 progressive years of exercise (see table). Positive attitudes toward exercise participation were noted throughout. Compliance to the program was also not an issue.

Variable	Exercise group (n = 30)	Control group (n = 35)	*P* value
Percent of ideal weight for height	0.48 ± 2.52	−0.04 ± 2.75	0.43
FVC (% predicted)	−0.25 ± 2.81	−2.42 ± 4.15	0.02
FEV_1 (% predicted)	−1.46 ± 3.55	−3.47 ± 4.93	0.07
FEF_{25-75} (% predicted)	−3.07 ± 5.34	−3.87 ± 7.00	0.61
Maximal heart rate (beats · min^{-1})	0.51 ± 3.68	−0.59 ± 4.33	0.28
Maximal minute ventilation (L · min^{-1})	3.93 ± 8.31	1.84 ± 6.57	0.26
$\dot{V}O_2max$ (ml · kg^{-1} · min^{-1})	−1.80 ± 2.21	−1.85 ± 2.51	0.93
Peak work capacity (% predicted)	−1.68 ± 5.16	−2.50 ± 6.08	0.56

	PEAK WORK			$\dot{V}O_2MAX$		
	R	*n*	*P* value	*R*	*n*	*P* value
FEV_1 (% PREDICTED)						
Baseline	.34	65	0.005	.35	65	0.005
Year 1	.55	65	0.0001	.42	64	0.0006
Year 2	.52	60	0.0001	.44	58	0.0006
Year 3	.59	63	0.0001	.46	62	0.0002
FVC (% PREDICTED)						
Baseline	.29	65	0.02	.28	65	0.03
Year 1	.54	65	0.0001	.37	64	0.0003
Year 2	.48	60	0.0001	.37	58	0.004
Year 3	.59	63	0.0001	.41	62	0.0009

Conclusions. Schneiderman-Walker and colleagues present a well-designed study that highlights the benefits of regular aerobic exercise over a substantial period of time (3 yr) in patients with CF. The findings include a significantly slower rate of decline in indices of pulmonary function, the foremost marker of disease severity. Furthermore, increasing correlations were observed between exercise and pulmonary status at each year of follow-up. The study also demonstrates that consistent compliance with a home program is attainable. This is important, as inpatient programs or formal rehabilitation programs in youth may be difficult to coordinate. In addition, self-reported positive attitudes toward exercise participation were observed, making a program such as this an attractive adjunct to traditional CF therapies.

(continued)

Practical Application 19.3 *(continued)*

Finally, this investigation helps to define the role of the exercise clinician in the chronic care of patients with CF. The exercise clinicians were responsible for education of the families and patients regarding the requirements of a physical fitness program. They provided safe and accurate testing and data collection for each patient. They ensured compliance through regular telephone contact as well as face-to-face counseling at outpatient visits, addressed potential barriers to participation, and provided incentives to maintain motivation throughout. The exercise clinician is a vital member of the CF care team, as demonstrated nicely by this investigation.

Reference: Schneiderman-Walker J, Pollock AL, Corey M, Wilkes DD, Canny GJ, Pedder L, Reisman JJ. A randomized controlled trial of a 3-year home exercise program in cystic fibrosis. *J Pediatr* 2000;136:304-310.

modulation is an attractive but speculative theory. The ability of exercise to alter the immune system is now well recognized (72). Although some studies have shown a beneficial effect of chronic exercise on the immune system, this relationship has not been established in people with CF (11, 13).

Table 19.7 provides a review of exercise training guidelines for patients with CF.

Clinical Exercise Pearls

- A significant benefit of exercise in this population is that it enhances airway clearance, induction of coughing and sputum production, and is often used as an adjunct to more traditional therapies.
- Maintaining exercise performance has been linked to improved long-term prognosis in the CF patient.
- Maintaining an appropriate nutritional status (BMI >23 in males, >22 in females) has resulted in a significant improvement in pulmonary outcomes. Body weight can be maintained or significantly increased through exercise.
- Oxyhemoglobin desaturation is more common in patients with FEV_1 <50% predicted and in patients with more severe disease as represented by lower resting O_2 saturation.
- Improved survival has resulted in a greater percentage of adult patients with CF (just below 50%). This has led to an increased incidence of depression, bone diseases such as osteoporosis, and CF-related diabetes. Clinicians should consider all of these when creating or maintaining an exercise prescription.
- The median age of survival in patients with CF is approximately 36 yr and has increased significantly in the past two decades. Over this time period, exercise has increasingly been shown to have significant effects in retarding disease progression.
- Although no absolute contraindications exist, clinicians should take care in the use of exercise in a CF patient with a history of pulmonary hypertension, acute hemoptysis, pneumothorax, oxygen dependence, exercising at altitude, bleeding disorders secondary to liver disease, and severe malnutrition.
- In patients with severe disease, maximal exercise testing may result in decreased cardiac responses (lower peak HR and BP) as the patient exhibits increasing ventilator limitations. It is important to take care when creating exercise prescriptions using intensities based on percentage of predicted peak heart, as these may exceed patient abilities.
- Assessment of body composition in the CF population is typically via skinfold measures or bioelectrical impedance. Use of techniques that correct for lung parameters (underwater weighing) tend to be inaccurate due to hyperinflation of the lungs.
- In children and adolescents with mild to moderate CF disease, traditional principles of exercise prescription can be followed (see information in chapter 5).

Table 19.7 Exercise Training Review in Patients With CF

Parameter	Training specifics
Cardiorespiratory endurance	Static (rest): Spiromtery values (FEV_1, FEF_{25-75}, tidal volume, DL_{CO}) are reduced with increased air trapping as evidenced by increased residual volume and RV/TLC. This lung impairment affects endurance activities through impaired ventilation. Lower resting oxygen saturation is also common. Dynamic (exercise): Aerobic capacity is decreased with an increased breathing response ($\dot{V}O_2$, $\dot{V}O_2$/MVV, RR, $PetCO_2$). Peak work capacity and maximal oxygen consumption are reduced. Although the HR may be elevated at rest, the HR with exercise is typically blunted, making prescriptions based on predicted maximal HRs difficult to use. Moderate to severe patients may desaturate with exercise, resulting in the need for supplemental O_2. This may limit exercise choices. Endurance training is critical. The benefit of exercise as an adjunct to traditional airway clearance techniques has been well documented.
Skeletal muscle strength	Peripheral and respiratory muscular strength is reduced as demonstrated by lower values in both 1RM at the periphery and decreased inspiratory and expiratory muscle strength (PImax, PEmax) at the lung. Strength training has been shown in enhance body image and self-confidence. The physical benefits of strength training have been argued. Standard strength training programs have been used. Caution is warranted in children and adolescents who are still undergoing physical development. This development can be delayed in patients with CF. The use of child-appropriate activities such as push-ups, sit-ups, and pull-ups in which one uses one's own body mass may be desirable in younger patients. In advanced disease, avoid the Valsalva maneuver, as this increases thoracic pressures.
Skeletal muscle endurance	Muscular endurance is reduced, as is muscular efficiency. Reduced lean body mass significantly affects both muscular strength and endurance. Strength training programs should follow the same guidelines as for individuals without CF with gradual progression. Respiratory muscle training should be incorporated into the program. Severe patients should avoid the Valsalva maneuver, which can increase thoracic pressures. Patients with a history of pneumothorax should not perform resistance training.
Flexibility	Peripheral muscle strength has been shown to be reduced (hamstrings, quadriceps). Poor posture, barrel chest, and kyphosis are common findings. Increases in flexibility are generally seen in individuals with CF using standard stretching techniques. Focusing on upper body range of motion and chest wall mobility may enhance breathing mechanics with exercise.
Body composition	Body weight, lean body mass, BMI, and % body fat are all decreased in individuals with CF. Maintaining proper nutrition is essential, as it is strongly correlated with long-term survival. Although it is well documented that individuals with CF can show positive weight gain with exercise training, caution is warranted to ensure proper nutrition. Working with the CF team's clinical nutritionist during exercise training is vital.

Note. FEV_1 = forced expiratory volume in 1 s; FEF_{25-75} = forced expiratory flow between 25% and 75%; DL_{CO} = diffusion capacity as measured by carbon monoxide technique; RV/TLC = ratio of residual volume to total lung capacity; $\dot{V}O_2$ = minute ventilation; $\dot{V}O_2$/MVV = ratio of minute ventilation to maximal voluntary ventilation; RR = respiratory rate; $PetCO_2$ = end-tidal carbon dioxide; PImax = maximal inspiratory pressure; PEmax = maximal expiratory pressure; BMI = body mass index.

CONCLUSION

People with CF appear to benefit from exercise training programs. Both static and dynamic pulmonary function either improve or have a significantly slower rate of deterioration following training programs. Reports on muscle strength and endurance following exercise programs have presented varying results regarding their effectiveness. Despite these varying results, data are sufficient to support the use of programs that target improvements in muscular strength and endurance in patients with CF. As nutrition is a major concern in patients with CF, it is critical to keep in mind that body weight can be successfully maintained or increased during an exercise intervention. The psychosocial implications of exercise are considerable, especially in the adult population, in whom the rate of depression is greater than 21%. Improvements in quality of life and sense of well-being have been well documented and may outweigh the physiologic benefits of exercise in some cases. Finally, the potential effect of exercise on improving patient prognosis makes exercise an attractive therapeutic modality.

Key Terms

adult variant CF (p. 351)
bronchiectasis (p. 352)
cystic fibrosis—related diabetes (p. 353)
cystic fibrosis transmembrane conductance regulator (CFTR) (p. 351)
dynamic pulmonary function (p. 355)
hemoptysis (p. 354)
hypoxemia (p. 354)
kyphosis (p. 354)
maximal voluntary ventilation (p. 360)
nasal polyposis (p. 353)
oxyhemoglobin saturation (p. 354)
pneumothorax (p. 368)
pulmonary exacerbation (p. 355)
scoliosis (p. 354)
static pulmonary function (p. 369)
sweat test (p. 353)

CASE STUDY

MEDICAL HISTORY

Ms. MH is a 16 yr old Caucasian female who was diagnosed with CF at 6 mo of age secondary to recurrent respiratory infections and failure to thrive. She has done relatively well with intermittent respiratory infections that require antibiotic and hospital therapy. Because of her inability to consume adequate calories, a gastrostomy tube was placed to allow supplemental nocturnal nutrition. She is uncomfortable with the tube and is currently self-conscious about her appearance; she has started to avoid social situations due to a declining self-esteem. Both Ms. MH and her parents are protective of her gastrostomy tube and consciously avoid activities that would either highlight its presence or potentially cause difficulties to the area.

DIAGNOSIS

An exercise evaluation was performed as part of Ms. MH's medical care. An activity questionnaire revealed that Ms. MH enjoys recreational activities but does not wish to participate in sports because she has difficulty keeping up with her peers, especially in prolonged aerobic activities. She owns a bicycle and in-line skates and says she uses them sporadically throughout the year, as she lives in a place where weather is typically not an issue. She enjoys dance and has been involved with tap, jazz, and ballet for most of her childhood through a local noncompetitive dance studio. Even though both of her parents are active, they neither encourage nor discourage activity on her part. She has an older sister who does not have CF. Before exercise testing, the following pulmonary function tests were obtained:

- Pulmonary function tests: FEV_1 = 60% of predicted; FEF_{25-75} = 40% of predicted
- Resting pulse oximetry = 95% on room air
- Residual volume = 195% of predicted
- Diffusing capacity of the lungs = 86% of predicted

EXERCISE TEST RESULTS

Maximal graded ergometry test (Godfrey protocol) results:

- Physical work capacity = 100 W (73% of predicted)
- $\dot{V}O_2peak = 33.2\ ml \cdot kg^{-1} \cdot min^{-1}$ (79% of predicted)

- Peak end-tidal CO_2 = 36 mmHg
- Lowest exercise oxygen saturation = 90%
- Ratio of minute ventilation to maximal voluntary ventilation = 96%
- Resting heart rate = 92 beats · min^{-1}
- Peak heart rate = 186 beats · min^{-1}
- Body composition: weight = 40.9 kg (<5th percentile for age); height = 158 cm (10th percentile for age)
- Body mass index = 16.49 kg · m^{-1} (<5th percentile for age)
- Percent body fat (bioelectrical impedance analysis) = 18%
- Flexibility: good posture; no muscle issues

EXERCISE PRESCRIPTION

Overall, Ms. MH has moderate obstructive pulmonary disease with mild ventilatory limitations to exercise. Although oxygen saturations decrease with exercise, she can safely participate in aerobic activities without adverse effect. Overall, nutrition is compromised; lower than expected lean body mass and body mass index are already being addressed with a gastrostomy tube supplying nocturnal feeds. From a psychological perspective, Ms. MH appears to be struggling with some body image issues and lower self-esteem and is starting to recognize some of the limitations of having CF. Her exercise tolerance remains adequate and is near normal. The biggest challenge will be to maintain or enhance her fitness as she ages. To that end, the following program was developed.

- Ms. MH was encouraged to select individual activities based on enjoyment, taking advantage of the warmer climate she lives in. Additionally, she was advised to seek out group classes such as yoga, Pilates, or Zumba at the local YMCA that will provide for aerobic conditioning while using her dancing background. Daily activities were encouraged, and intensity was set at self-limiting exertion (rating of perceived exertion 6 to 7 on a 0 to 10 scale) because heart rate prescription can be difficult and a burden in young children and adolescents.
- Incorporating a brief warm-up and cool-down into each exercise session was discussed.
- Her parents were asked to help in motivating Ms. MH and were given strategies to enhance compliance. In addition, Ms. MH and her parents were counseled about selecting activities that would not call attention to or affect her having a gastrostomy tube.
- Ms. MH will begin to see a counselor appointed through her CF care team regarding her negative self-image to improve her self-esteem and help to prevent depression later in life.
- The exercise clinician suggested formal reevaluation yearly to monitor progress, establish new goals, and offer motivation.

DISCUSSION QUESTIONS

1. How does Ms. MH's gastrostomy tube pose a challenge in designing an exercise prescription?
2. What is the role of the exercise clinician in optimizing compliance with the exercise program?
3. Discuss the potential challenges in prescribing exercise to adolescent children in general.
4. What role do the parents play in the success or failure of an exercise program in a child? How is this role complicated by a child's having a chronic disease?
5. What anticipatory counseling should be offered before initiation of an exercise program?
6. What steps should the exercise clinician take in the event that Ms. MH comes back for her yearly evaluation and demonstrates a significant decline in exercise capacity?

PART V

Oncology and the Immune System

A chief compliant among patients with a chronic disease is loss of exercise tolerance, or fatigue. In patients with cancer and those who are HIV positive, this common symptom may be related to the disease itself or it may be brought on (or worsened) by the medications or other therapies used to treat the disorder. Part V highlights the important role that regular exercise can have in helping to attenuate or reverse a patient's loss of exercise tolerance, regardless of the cause. And although an ever-increasing body of research describes the benefits of exercise in patients being treated for cancer, those who are cancer survivors, or those who are HIV positive, there continues to be vast opportunities for one who is interested in becoming a clinical research scientist who can help address the many exercise-related treatment questions that surround disease-related fatigue and clinical outcomes.

Chapter 20 focuses on cancer. In the late 1960s and early 1970s the beneficial role of exercise testing and training in the diagnosis and treatment of patients with heart disease was formalized and took hold. Today we stand at a similar threshold, this time regarding the beneficial role of exercise in the care of patients with cancer. An ever growing body of research evidence suggests that exercise helps reverse exercise intolerance and improves mood and quality of life. Still, many questions remain regarding safety, dose, rate of progression, clinical outcomes, and the optimal timing to interject exercise into a patient's treatment plan. This chapter summarizes what we know today about exercise in the prevention and treatment of cancer and causes us to think ahead, just like the exercise pioneers interested in preventive cardiology did more than four decades ago. Clearly, great potential awaits the use of exercise in yet another important health disorder.

Chapter 21 discusses human immunodeficiency virus. Over the last decade the medicines used to treat patients who are HIV positive have done much to attenuate viral load, maintain CD4+ count, and lessen disease-related morbidity and mortality. Unfortunately, however, like so many other health disorders that require complex medical therapies, the prescribed treatments often lead to other ailments that are themselves debilitating or increase the risk of other diseases. For those who are HIV positive these include the loss of skeletal muscle size, strength, and endurance; increased risk for cardiovascular disease; and body fat redistribution that require independent treatments. Exercise training is an important aspect of the treatment process for those who are HIV positive. Importantly, these patients can also present in cardiac rehabilitation programs due to their increased risk of coronary artery disease. Therefore, the clinical exercise physiologist who either works directly with patients who are HIV positive, or who may encounter these patients in a cardiac rehabilitation setting, must understand not only principles of exercise testing and training but also the pathophysiology, evaluation, and therapies unique to the disorder.

Chapter 20

Cancer

Dennis J. Kerrigan, PhD
John R. Schairer, DO
Lee W. Jones, PhD

Each of you has been touched by cancer. Either someone close to you or you yourself have heard the frightening words "You have cancer." For many confronted with this new diagnosis, the mind races ahead to thoughts of chemotherapy, illness, fatigue, and feelings of one's own mortality. With nearly 1.5 million new cases of cancer diagnosed in 2011 in the United States, scientists constantly strive to find new ways to prevent, diagnose, and treat it, as well as to lessen the often burdensome side effects that commonly accompany conventional care. To this end, much clinical research over the past 15 yr has focused on the role of regular exercise in the prevention and treatment of cancer.

DEFINITION

The words *cancer* and *malignancy* are commonly used for the medical term **neoplasm**, meaning an abnormal growth of tissue. In neoplasm, tissue grows through unregulated cellular proliferation, shows partial or complete lack of structural organization, lacks functional coordination with the normal tissue, and usually forms a distinct mass that may be either benign or malignant (3). Malignant neoplasms are generally fast growing, have the ability to invade host tissue, are associated with large areas of necrosis because the tumor outgrows its blood supply, and can **metastasize** to other parts of the body. They are eventually fatal if untreated. Benign neoplasms, on the other hand, are generally slower growing, have well organized and well differentiated cells, and usually are not fatal. Cancer is a unique disease because it can originate in any organ system, can spread to other organ systems, and has multiple etiologies. Cancer affects all nationalities, races, and ages, as well as both men and women. The treatment varies with the cancer type and location and includes chemotherapy, radiation therapy, biotherapy, and surgery, individually or in combination (2).

The therapeutic role of exercise for patients with cancer is multifaceted. Physical inactivity is a known risk factor for developing certain cancers, hence exercise may have a role in primary prevention. With respect to the adverse affects of anti-cancer treatments as well as the tumor burden itself, exercise training is now increasingly used as an adjunctive treatment to counteract commonly reported symptoms such as fatigue and muscle weakness. Additionally, preliminary data shows an inverse association between regular exercise following diagnosis and cancer recurrence, cancer-specific mortality, and all-cause mortality, suggesting a secondary prevention role for exercise. In this discussion, we focus primarily on the four most common cancers: lung, breast, prostate, and colorectal.

SCOPE

Cancer is one of the leading causes of morbidity and mortality throughout the world (2, 7). In the United States, approximately 570,000 Americans die annually, ranking it as the second leading cause of death, behind cardiovascular disease. Forty percent of Americans will develop cancer during their lifetime, approximately 1.5 million new cases occur each year, and approximately 13 million Americans have a history of cancer (2).

Relative to economics, the annual burden to care for the 100 million Americans who will develop cancer in their lifetime is estimated to be more than $263 billion: approximately $74 billion for hospital costs and physician services, $17.5 billion in lost productivity, and $118 billion in lost productivity due to premature death (11). But the story is not completely bleak. With continued improvement in diagnosis and treatments, overall death rates due to cancer have decreased in both men and women over the past 20 years (2). The 5 yr survival rate for all cancers now approaches 68%, leading to more than 13 million cancer survivors in the United States today (2).

Cancer is found throughout the world, with about a threefold difference between countries with the highest frequency of cancer and those with the least. Geographic variation can be as much as 100-fold for specific cancers. For example, the death rate from upper gastrointestinal cancer is extremely high in South Africa, China, Japan, and Iran. This type of cancer is much less frequent in the United States, except in areas that have a high incidence of alcoholism. The most common cancers in Western countries are lung, large bowel, and breast, whereas in southeast China, nasopharyngeal cancer is the most common malignancy.

Haenszel (35) and Muir and colleagues (55) demonstrated that migrating populations tend to acquire the cancer incidence profile of their new country of residence, suggesting that genetics is less important in the genesis of cancer than environmental influences. Results from epidemiological studies show that more than two-thirds of cancer deaths might be prevented through lifestyle modification (5, 28). One-third of cancer deaths are due to cigarette smoking, and another one-third are attributed to alcohol use, specific sexual practices, pollution, and dietary factors.

There are differences between men and women in the type and frequency of cancer and the likelihood of dying of cancer (1). For nearly all cancers, the incidence rates are higher in men than in women, with the exceptions of thyroid, gallbladder, and, of course, breast and uterine cancer. About 50% of men and 33% of women will develop cancer during their lifetime. Among men, the most common cancers in order of prevalence are prostate, lung, and colon or rectum. With women, the most common cancers are breast, colon or rectum, and lung. If death from cancer, is considered, the order of the cancers changes. More men die of lung cancer, followed by colon or rectum cancer, and finally prostate cancer. Among women, lung cancer is now responsible for more deaths than breast cancer and colon or rectum cancer.

Race affects cancer incidence and cancer death rates. African American males have the highest cancer incidence and death rates, followed by white men. The incidence and death rates are similar between African American women and white women. Asian and Pacific Islanders, American Indians, and Hispanics have lower incidences and death rates than does the white population.

Age also affects the distribution and frequency of cancer. Among patients under the age of 15, the most common cancers are leukemia, brain, and endocrine. For patients over age 75, the most common cancers are lung, colon, and breast. Cancer incidence increases with age. Between birth and age 39 yr, 1 in 58 men and 1 in 52 women will develop cancer. The incidence of cancer in men and women between ages 60 and 79 yr is 1 in 3 and 1 in 4, respectively.

PATHOPHYSIOLOGY

The final common pathway in virtually every instance is a cellular genetic mutation that converts a well-behaved cellular citizen of the body into a destructive renegade that is unresponsive to the ordinary checks and balances of the normal community of cells. (14, p. 1335)

The normal growth and proliferation of cells within the body is under genetic control. The stem cell theory is the model developed to explain the orderly proliferation of cells, specialization to perform discrete functions, and cell death within an organ (figure 20.1). The stem cell is pluripotent, which means that it is an uncommitted cell with various developmental options still open. The process by which the **pluripotent stem cell** is able to develop special functions and structures within an organ system is called differentiation. Thus, some stem cells are triggered to differentiate and become hair cells, and some cells become cardiac myocytes. The pluripotent stem cell also has the capacity for self-renewal. But after a stem cell becomes committed to a cell line such as a hair cell or cardiomyocyte, it no longer has pluripotent and self-renewal properties and is destined to develop along its specialized pathway of differentiation. The best example is a pluripotent hematopoietic stem cell, with its capacity to form both red blood cells and white blood cells (e.g., neutrophils, basophils). After it commits to a specific cell line, it can no longer differentiate into other cell types or divide to form new cells.

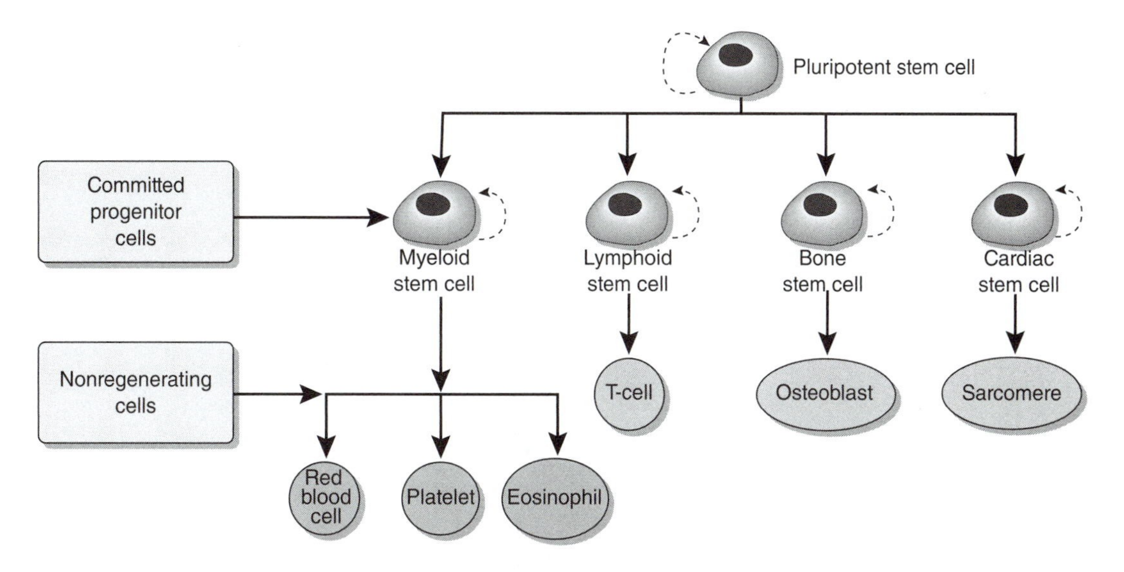

Figure 20.1 Stem cell sequence theory. This schematic presentation of the stem cell theory depicts how the uncommitted pluripotent stem cell develops or differentiates into several committed, nonregenerating cells with specialized functions.

In cancer, the cellular DNA of normal stem cells is damaged, leading to disordered cell growth and specialization. The stem cell model for cancer proposes that tumors arise from carcinogenic-causing events occurring within the normal stem cells of a particular tissue. A cancer-causing insult is believed to produce a defect in the control of normal stem cell function, resulting in abnormalities in self-renewal, differentiation, and proliferation. In other words, the normal quality and quantity control for cell function and growth is lost.

The carcinogenic event for many cancers is unknown, but five broad categories have been identified.

1. Environment
2. Heredity
3. Oncogenes
4. Hormones
5. Impaired immune system function

Cancers most likely arise when a factor from one or more of these categories is present (e.g., radiation exposure from the environment), causing cellular mutations that go unrecognized because of injury to the immune system as well.

Environmental factors are implicated in more than 60% of cancers, partly because of the known association between certain agents and the development of cancer. Common environmental factors that contribute to cancer death include tobacco (25-30%), obesity, and a sedentary lifestyle (25%), infections (15-20%), radiation (up to 10%), stress, and pollutants (8, 12) (figure 20.2). For example, cigarette smoking, a lifestyle choice, increases the risk for lung cancer because tobacco smoke contains carcinogens. Other lifestyle choices such as a diet high in fat, obesity, and a sedentary lifestyle are associated with an increased likelihood of developing breast and colon cancers. Excessive exposure to sunlight is responsible for a higher incidence of skin cancer in farmers. Lung cancer is more prevalent among miners and those who work with asbestos, chromate, or uranium. Exposure to certain solvents is associated with leukemia.

Some hormones are thought to possess carcinogenic potential. For example, the ovary, breast, and uterus are hormonally sensitive; therefore, estrogen may play a role in the cancers of these organs in women. The role that hormones may play in breast cancer is discussed in the literature review in the practical application 20.1.

While genetics alone account for a small number of cancers, neoplastic pathology often involves mutations from environmental carcinogens on specific genes responsible for cell growth, apoptosis, and regulation (36). Two of the most well-known types of genes associated with cancer are **proto-oncogenes** and **tumor suppressor genes**. Proto-oncogenes are genes that act at the cellular level to promote normal cell growth and development. When attacked by specific carcinogens (e.g., radiation, chemicals, viruses), proto-oncogenes are mutated into **oncogenes**, which can lead to uncontrolled proliferation resulting in tumorigenesis (36). However, tumor suppressor genes oppose such unregulated growth through the activation of anticancer pathways, which eventually leads to cellular apoptosis or inhibition of

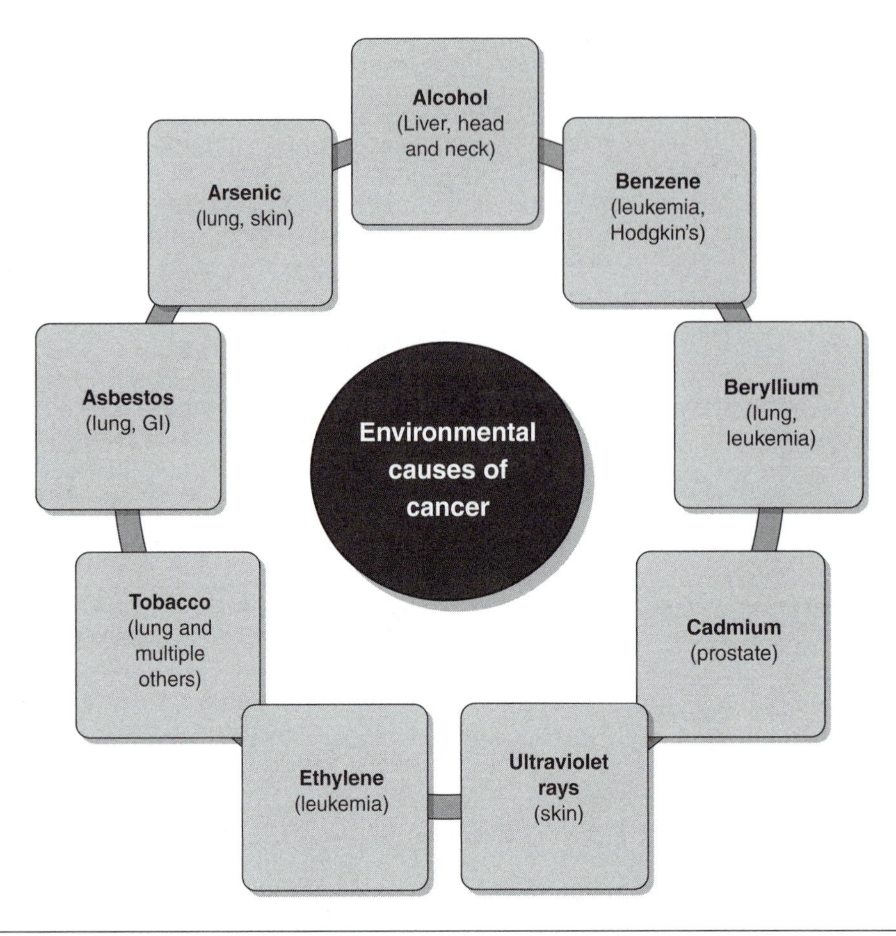

Figure 20.2 Environmental causes of cancer.

tumor growth (36). Interestingly, mutations to the tumor suppressor gene that regulates the p53 protein, a well-known tumor suppressor, have been found in many cancers (36).

The immune system is responsible for mediating the interaction between an individual's internal and external environment. The immune system is divided into two major categories: innate and acquired responses (table 20.1). The innate immune system response is nonspecific and immediate, beginning within minutes of an insult. The response occurs without "memory" for the eliciting stimulus. This process is called inflammation. The innate immune system represents our first line of defense against cancer.

The adaptive or acquired immune system is characterized by an antigen-specific response to a foreign antigen or pathogen and generally takes several days or longer to activate. A key feature of acquired immunity is memory for the antigen, such that subsequent exposure leads to a more rapid and often more vigorous response.

The overall function of the immune system is to rid the body of foreign agents such as bacteria, viruses, and malignant cells. The immune system recognizes infectious agents and malignant cells because they contain abnormal antigens in their cell membranes. The immune system can also inhibit subsequent formation of a tumor by countering factors responsible for its growth (table 20.2).

Table 20.1 Components of the Immune System

Immune system component	Description
Innate immune system Monocytes Macrophages Neutrophils Natural killer cells	Nonspecific response Nonspecific killing response of tumor cells by phagocytosis and cytolysis
Acquired or adaptive immune system Cytotoxic T lymphocytes	Antigen-specific response Requires tumor antigens in association with class I major histocompatibility antigens

Table 20.2 Effects of Exercise on the Immune System

INNATE IMMUNE SYSTEM		
Component	**Effect of acute exercise**	**Effect of chronic exercise**
NK cells	Immediate increase in cell count and cytolytic activity	NK cell count and activity increase, both in blood and (depressed for 2-24 h postexercise) in spleen.
Macrophages	Immediate increase in monocyte and macrophage count Adherence unchanged Increased phagocytosis with moderate activity	Response is unclear. Resting monocyte is count unchanged. May cause adaptations that alter exercise response.
Neutrophils	Large and sustained increase with moderate exercise Most PMN functions decrease significantly after strenuous exercise	Function is suppressed during periods of strenuous activity.
ACQUIRED IMMUNE SYSTEM		
Component	**Effect of acute exercise**	**Effect of chronic exercise**
T lymphocytes	Moderate activity enhances cell proliferation, with depressed levels 30 min postexercise Vigorous activity causes a transient decrease in proliferation	Regular, moderate exercise enhances cell proliferation.

Note. NK = natural killer; PMN = polymorphonuclear leukocytes.

Practical Application 20.1

EXERCISE AND PRIMARY CANCER PREVENTION

Data show that higher levels of physical activity are associated with lower overall cancer mortality (61, 63, 83). In a 2011 report from The American Cancer Society, both physical activity and dietary interventions were identified as strategies to successfully help reduce the overall incidence, morbidity, and mortality from certain kinds of cancer (2). In fact, for the nonsmoker, dietary and physical activity interventions are the most important modifiable determinants of cancer risk.

The role of physical activity, either leisure-time activity or occupational physical activity, in reducing overall cancer risk and site-specific cancer risk is gradually being defined (30, 47, 85). The relationship between colorectal cancer and physical activity has been studied the most. Forty-eight studies with 40,000 patients (85) demonstrated a 10% to 70% reduction in risk for colon cancer in physically active individuals. Decreased bowel transit time caused by physical activity may explain the observation that colon cancer frequency is reduced in physically active individuals whereas no change occurs in the frequency of rectal cancer.

Forty-one studies including 108,321 women (85) evaluated breast cancer and possible risk reduction with physical activity. Twenty-six studies demonstrated that both occupational and leisure-time activity reduce breast cancer risk by about 30%. In general, the results for breast cancer are less conclusive than those for colon cancer. Of the 12 studies in the literature regarding physical activity and the risk for developing endometrial cancer, eight demonstrated a 20% to 80% reduction in risk. In the studies that evaluated the possible protective effect of exercise in preventing ovarian, prostate, and testicular cancer, the data are generally favorable but inconsistent. The association of physical activity with lung cancer has been reported in 11 studies. Six of these support a protective effect of physical activity on lung cancer.

How much physical activity is needed to reduce the risk of cancer? A definitive answer is not currently available. Data from Blair and colleagues (10) indicate that the reduction in cancer risk occurred primarily between

(continued)

the very low fitness group and the moderately fit group, with no further decrease in risk among the more fit subjects. Paffenbarger and colleagues (62) reported a decrease in all-cause mortality for alumni who expended 1,500 or more kcal/wk, but the authors did not break out cancer deaths in this group. Data show that 1,000 kcal/wk (4 h of moderate activity per week or 3 h of vigorous activity) has a protective effect for colon cancer and breast cancer (78, 87). Lee and colleagues (51) also reported that exercising at moderate intensity, greater than 4 to 5 METs for 4 h per week, decreased the incidence of lung cancer.

Determining a relationship between physical activity and cancer risk is complex because the mechanism(s) through which exercise acts to lower cancer risk is not well understood. It is not known whether risk reduction occurs through lifelong exercise, exercise at certain stages in the life of the patient, moderate activity, or vigorous activity. Exercise may help block initiators of cancer, in which case exercise done consistently at a relatively young age may be most beneficial. This is the rationale behind studies that investigate whether participation in high school and college athletics reduces the risk of cancer during adulthood. Alternatively, exercise may counter the promoters of cancer cell replication, so that exercise during a later phase of the neoplastic process may be preferred to decrease the development of clinically significant disease (76, 81). Therefore, the time point at which exercise occurs during a person's life span may be an important factor relative to its effect on cancer development.

One good example is breast cancer. Estrogen plays a role during four phases of a woman's life: menarche, first pregnancy, menopause, and postmenopause. Several studies indicate that breast cancer risk is directly related to the cumulative number of ovulatory menstrual cycles (38, 50, 64, 86, 88). Intense exercise delays menarche, which can be thought of as favorable, because the risk of breast cancer increases twofold in women who experience menarche before age 12 versus at age 13 or older. Epidemiological data indicate that for every year menarche is delayed, breast cancer risk decreases by 5% to 15% (29, 42). Moderate levels of activity have been shown to increase the likelihood for anovulatory cycles threefold. Delayed menarche and anovulatory cycles decrease the woman's exposure to estrogen and progesterone (9, 31, 88). The first full pregnancy induces differentiation of the breast and may change the sensitivity of the breast to both endogenous and exogenous risk factors (17, 67). And women who experience natural or artificially induced menopause before the age of 45 have a markedly reduced risk of breast cancer compared with women whose menopause occurs after the age of 55 (82). Postmenopausal women generally have an increased body mass index, which is also a significant independent risk factor for breast cancer (39, 63). In postmenopausal women, fat tissue is the primary source of estrogen. Therefore, excess adiposity increases the woman's exposure to estrogen. Currently, it is not certain at which point or points in a woman's life cycle exercise exerts its greatest anticancer effect.

Another possible explanation for the role exercise plays in primary prevention of cancer is the potential effect on the immune system. Several excellent review articles discuss the effect of exercise on immune system function (40, 58, 76, 77, 90). The working hypothesis in the field of exercise immunology is the inverted-J hypothesis (figure 20.2), which suggests that enhanced immune system function occurs with a chronic moderate exercise training program, but that immune system function is depressed (even below that of sedentary individuals) after chronic exhaustive exercise. Clinically, moderate to high levels of physical activity seem to be associated with decreased incidence or mortality rates for some cancers (48, 61, 75, 81), and overtraining or intense competition may lead to immunosuppression as evidenced by an increased incidence of upper respiratory infections in people who train intensely (59, 65). To this point, no cancers have been shown to be associated with exhaustive exercise programs. Table 20.2 describes the effects of acute and chronic bouts of exercise on the immune system. Moderate exercise seems to enhance the body's immune system, but the significance of this finding in preventing and treating cancer is unknown.

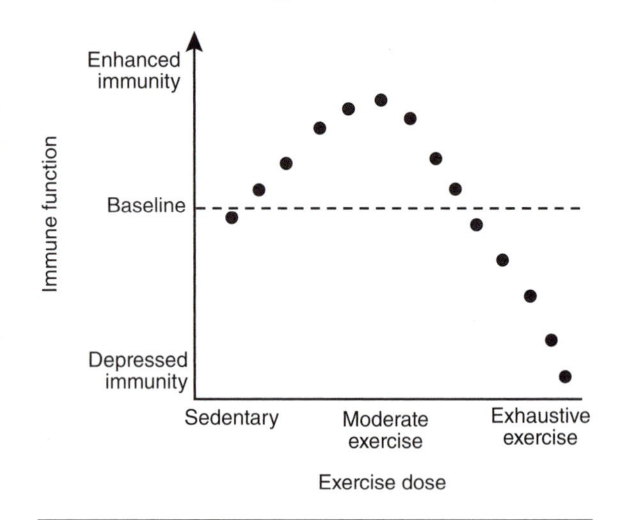

Figure 20.3 The inverted-J hypothesis.

Cancer therapies such as chemotherapy, radiation therapy, and cancer surgery are known to be immune suppressive. The role that this suppression of the immune system plays in the cure and recurrence rates during and after treatment of cancer is unknown. The important role that exercise plays in modulating the immune system has only recently been studied. Practical application 20.1 discusses the beneficial effects of acute and chronic exercise on immune function.

MEDICAL AND CLINICAL CONSIDERATIONS

There are four types of cancer: carcinoma, sarcoma, leukemia, and lymphoma. Carcinomas are cancers of epithelial tissues and include cancers of the skin, digestive tract, genitourinary tract, pulmonary system, and so on. Cancer of the breast, colon and rectum, lung, and prostate are examples of carcinoma. Carcinomas represent 90% of all cancers. Sarcomas are tumors of the connective tissue and include cancer of the bone, muscles, and cartilage. Leukemias are cancers of the white blood cells. Lymphomas are cancers of the lymphatic system. Because of the diversity of organ system involvement, no single part or segment of the history and physical examination focuses only on cancer. As a result, the physician or physician extender must perform a complete and accurate history and physical examination. Reviewing this information, the clinical exercise physiologist needs to understand the historical and physical aspects of the specific cancer that the patient has in order to help assess progress and identify any exercise-related concerns.

Signs and Symptoms

Early on, the symptoms of cancer are usually nonspecific, such as weight loss, fatigue, nausea, and malaise. Only an astute clinician can make the diagnosis of cancer at this time, yet early detection is key to maximizing the patient's chance for survival.

Later, the patient develops symptoms specific to the involved organ, such as shortness of breath in lung cancer or jaundice attributable to biliary obstruction in pancreatic carcinoma. By this time, however, prognosis is much poorer. Social history is also important, revealing occupational exposure to carcinogens or habits such as smoking or ethanol ingestion. The family history may reveal familial predisposition to a cancer, one that requires closer surveillance in the future. The review of systems may reveal symptoms indicating that the cancer has already metastasized.

As just mentioned, detecting cancer during a physical examination is extremely difficult. Any such examination is usually oriented toward examining for enlargement of an organ, such as an enlarged lymph node in the case of lymphoma or a testicle in testicular cancer. Because most organs (e.g., lung, pancreas, kidney) are deep within the body, the yield is low. Although examining for masses is important, a mass is often a later sign of cancer that occurs after the cancer has metastasized.

Diagnostic Testing

Because prevention of cancer is not always possible, the earliest detection of the disease is the next best strategy to reduce cancer mortality rates. To help accomplish this, the American Cancer Society recommends a series of screening procedures and evaluations depending on age, gender, and risk for specific cancers. When cancer is suspected, the first diagnostic principle is that adequate tissue must be obtained to establish the diagnosis. Because the therapy used for each type and subtype of cancer is often particular to that type or subtype, every effort must be made to obtain appropriate tissue samples even if treatment is delayed for a short time. The process of obtaining a sample of tissue is called a **biopsy**.

The second diagnostic principle is to determine the extent of spread of the cancer, also known as **staging**. In leukemia, staging can be accomplished through routine history and physical examination, laboratory tests, chest X-ray, and bone marrow biopsy. With solid tumors, computed tomography (CT) and magnetic resonance imaging (MRI) in conjunction with a biopsy are often needed to determine the size of the tumor and the extent of its spread. The degree to which the cancer has spread is reflected in its stage, which guides the type of treatment most appropriate for the patient. An example of a simplified staging system is shown in table 20.3. Each cancer has a staging system unique to itself, one that takes into consideration pathogenic features, the modes of spread, and the curability of the disease.

Treatment

There are four treatment options for cancer. The selection of which to use depends on the type, location, and stage of the cancer. Treatment options include the following:

- Surgery
- Radiation therapy
- Chemotherapy
- Biotherapy

Table 20.3 Tissue-Nodal-Metastasis (TNM) Classification System for Breast Cancer

Tumor size (T)	Nodal involvement (N)	Metastasis (M)
Ts = in situ	NO = no nodal metastasis	MO = no distant metastasis
T1 = <2 cm	N1 = movable axillary nodes	M1 = distant metastasis
T2 = 2 to 5 cm	N2 = fixed axillary nodes	
T3 = >5 cm	N3 = internal mammary nodes	

These TNM categories are combined to give the stage (e.g., stage 1 = T1 NO MO).

Surgery is the oldest and most definitive treatment for cancer. The two types of surgery are curative and palliative. **Curative surgery** is the primary treatment for about one-third of cancers that are small and have not yet metastasized. If the tumor is removed along with a small amount of surrounding normal tissue, the chance for a cure is good.

In **palliative surgery**, a large tumor mass is removed to make the patient more comfortable, to relieve obstruction of vital organs, and to reduce the tumor burden. For example, a colon cancer may be removed to prevent bowel obstruction, or an ovarian cancer may be removed to prevent obstruction of a ureter. Decreasing tumor burden may also make the tumor more susceptible to radiation or chemotherapy. Palliative surgery does not usually change overall chances for survival.

Radiation therapy is thought to stop the growth of malignant cells by damaging the DNA within the cell. Radiation can be applied either from implanted internal sources (brachytherapy) or from external machines. Most radiation therapy is applied in small fractions, usually between 180 and 250 rad/d. Doses above 4,000 rad over 4 wk increase the likelihood of developing radiation pericarditis. Radiation therapy can be used alone or in conjunction with surgery or chemotherapy. Radiation is good for a localized tumor that has not metastasized or for tumors that are difficult to reach with surgery (e.g., within the brain).

Cells that grow the fastest are best treated by chemical agents that interfere with cell replication. Known as **chemotherapy**, these agents frequently result in a cure. Some cancer cells, however, may become resistant to chemotherapeutic agents. Using several drugs at the same time (i.e., combination chemotherapy) is one way to minimize the resistance.

Cancer cells possess distinct surface protein antigens that are targets for antibody-directed or cell-mediated immunity. **Biotherapy** stimulates the immune response of the body to these protein antigens. Biotherapy also involves the production of antibodies outside the body, which are then administered to the cancer patient in an attempt to destroy the tumor. Biotherapy includes bone marrow transplantation and the use of cytokines such as interferon alpha and interleukin-5.

Exercise Testing

Assessment of cardiorespiratory fitness, and thus use of exercise testing, is not routinely performed in the clinical management of cancer patients other than to determine the preoperative physiologic status or risk (i.e., operability) of patients with pulmonary malignancies. Formalized exercise testing guidelines have been issued by several national (e.g., American Thoracic Society/American College of Chest Physicians [ATS/ACCP]) and international (e.g., European Respiratory Society [ERS]) organizations, and the reader is referred to the ATS/ACCP recommendations for a comprehensive overview of exercise testing methodology for clinical populations.

Several methods are available to investigators that enable the objective determination of cardiorespiratory fitness in adult cancer patients (table 20.4). Results of a recent systematic review suggest that maximal and submaximal exercise testing are relatively safe procedures in cancer survivors. The ATS and ACCP report that the risk of death and life-threatening complications during exercise testing is 2 to 5 per 100,000 tests (4).

Although supporting evidence is currently lacking, theoretically a cancer diagnosis and the use of conventional and novel therapies may increase the risk of exercise testing–related complications. The hypothesized increased risk associated with a cancer diagnosis is probably highly dependent on the type of cancer, the therapeutic management, and the stage of disease. However, a prior cancer diagnosis does not necessarily require referral to a physician or other health professional. The qualified exercise physiologist can use available tools such as the modified Physical Activity Readiness Questionnaire (PAR-Q) to determine preexercise risk in a client with cancer. Initial use of the PAR-Q should be undertaken when a client with cancer opts to do any of the following: (1) undergo a fitness assessment, (2) join a health club or sport team, (3) work with a personal trainer, or (4) become much more physically active than is the case currently (i.e., significantly increase habitual [daily] physical activity level or adopt a structured physical activity or exercise program).

It is not possible to screen or provide cancer-specific recommendations for all known cancer types; however, for certain forms of cancer, knowing which organ system(s) are involved may be immediately informative. For example, patients with lung cancer may be at particularly high risk for an adverse event, given that both the pathophysiology of the disease and concomitant comorbid conditions are often associated with a history of smoking (45, 46). These clients are considered high risk, and referral to a physician or other allied health professional is required for electrocardiogram, exercise testing, and other tests, as appropriate. If testing is unremarkable, clients are cleared for physical activity. If testing is remarkable, depending on the result, clients may be cleared for supervised exercise training in a rehabilitation program with experienced staff.

Preexercise Evaluation

Besides the usual contraindications for exercise and exercise testing in patients free of cancer, additional considerations are unique to patients with cancer. The history should include both noncancer and cancer considerations. Noncancer considerations include comorbid conditions such as age, diabetes, hypertension, fitness level, and orthopedic problems. Cancer-related issues of importance include type and stage of cancer, type of treatment, side effects of therapy, psychological status, and timing of tests and therapy. Other considerations include nutritional status, metabolic issues such as electrolyte abnormalities, and fluid status. The physical examination should attempt to identify acute signs or symptoms that are cancer type specific and would prevent participation in exercise. For example, women who have recently undergone axillary lymph node biopsy or breast surgery may not be able to participate in upper body exercises for several weeks after surgery. Another physical finding might include bone tenderness indicative of metastatic lesions to bones of the pelvis, back, or legs.

Table 20.4 Summary of Exercise Testing for Patients With Cancer

Test type	Mode	Protocol specifics	Clinical measures	Clinical implications	Special considerations
Maximal cardiovascular	Treadmill Cycle ergometer Arm ergometer	Individualized protocols, based on PA history and comorbidities	Peak oxygen consumption METs Peak workload Heart rate Respiratory exchange ratio Treadmill time	Provides basis for determining starting point for exercise training. Used to stratify presurgical risk. Used to evaluate response to training program.	Cancer treatment may result in cardiomyopathy, pulmonary fibrosis, or neuropathy.
Submaximal cardiovascular	Cycle ergometer Walking path	6 min walk Constant workload	Distance Heart rate Perceived exertion	Used to evaluate response to training program.	Sensitivity to detect changes may be reduced in more fit individuals.
Muscle strength and endurance	Machine weights	One-repetition maximum Multiple-repetition maximum	Kilograms	Provides basis for determining starting point for exercise training. Used to evaluate response to training program.	Modify or avoid 1RM tests with lymphedema or recent surgery.
Flexibility	Goniometry	Active stretching	Degrees	Used to evaluate response to training program.	Avoid pain. Assess upper extremity range of motion postmastectomy.

Contraindications to Exercise for Patients With Cancer

Hemoglobin <10.0 g · dl^{-1}

White blood cells <3,000/ml

Neutrophil count <0.5 · 10^9 · ml^{-1}

Platelet count <50 · 10^9 · ml^{-1}

Fever >38 °C (100.4 °F)

Unsteady gait (ataxia)

Cachexia or loss of >35% of premorbid weight

Limiting dyspnea with exertion

Bone pain

Severe nausea

Extensive skeletal metastases

Other findings can include gait instability attributable to chemotherapy or central nervous system involvement, delayed wound healing, immune suppression, and bleeding that may occur as a result of bone marrow suppression or surgery. Finally, some complications may require modification of the exercise plan, for example nausea, vomiting, fatigue, and weakness. All these factors must be considered before an exercise test and when a patient enters an exercise program, and they must be continually reevaluated throughout the program. See "Contraindications to Exercise for Patients With Cancer."

EXERCISE PRESCRIPTION

Currently there is no evidence supporting a different training response to exercise in the patient with cancer from that in the general adult population. This is reflected in the current American College of Sports Medicine guidelines for cancer survivors (70). Thus an exercise program consisting of aerobic, resistance, and flexibility training is recommended, with the initial modality and intensity depending on several factors:

Practical Application 20.2

CLIENT–CLINICIAN INTERACTION

Introducing exercise to patients with cancer requires that you, as the clinical exercise professional, accomplish two things. First, become familiar with the type of therapy or therapies that your patient is receiving. You may have to do some extra reading when you encounter a therapy that you are unfamiliar with. Also, talk with oncologists and surgeons about the agents or interventions used to treat cancer. Such discussions should include the clinical presentation of expected side effects and the natural history of the disease. Ultimately, you will improve your ability to interact with patients in a learned fashion.

Second, develop your ability to ask questions about how well your patient is tolerating therapy and his disease. Such ability usually comes from working with many patients, thus improving your skills to evaluate the interaction between exercise, cancer, and the cancer treatment. Because part of any exercise program includes regular follow-up, either in an exercise program or by telephone to the patient's home, you are in a unique position to establish patient confidence and assess clinical status. Obviously, you can take any concerns to the patient's attending oncologist or primary care physician. You play an important role in long-term patient surveillance.

Clinical features that you should pay special attention to include sudden loss of exercise tolerance over several days to a week, increased shortness of breath with exertion, an inordinate increase in anxiety or depression as manifested by difficulty falling asleep or lack of interest in social contact, and sudden changes in nutritional status. These are issues that you can work in to the routine follow-up phone calls or clinic visits that you might be using to evaluate exercise compliance and progress. At times the information you gather may dictate that exercise be withheld for a period of time while a specific treatment protocol runs its course.

Because more evidence is needed to pinpoint how exercise lowers risk for cancer recurrence or future hospitalizations, the best approach may be to keep a patient's attention focused on how exercise can help her lessen fatigue, regain control, and improve quality of life. This means using exercise to help patients lead a more comfortable life, one that allows them to perform the activities that they enjoy with fewer symptoms or limitations.

1. Active treatment versus survivorship phase
2. Response to treatment (i.e., adverse reactions)
3. Preexisting comorbidities (e.g., heart disease, osteoarthritis)
4. Current level of activity
5. Functional capacity
6. Fatigue level

Beginning exercise intensity may need modification in patients with cancer especially during active treatment. Patients who are sedentary to begin with, or who complain of adverse symptoms with treatment (e.g., fatigue), should probably start at lower intensities such as 40% to 50% of heart rate reserve. Conversely, already active patients, those who have completed treatments, or those without any reported adverse effects may be able to start at higher intensities such as 60% to 80% of heart rate reserve. A sound approach when introducing a patient with cancer to exercise is to start slowly and progress intensity, duration, and frequency as tolerated. Patients who present with severely impaired fitness may initially benefit from interrupted programs that incorporate several bouts of shorter-duration exercise. In an excellent review, Maryl Winningham (89) discussed how to get a cancer patient started in an exercise program. When in doubt about where to start, the exercise professional should use the 50% rule: Ask the patient how far he can walk before becoming too tired, and start at half that distance or time.

The exercise program for cancer patients does not typically require electrocardiographic monitoring, although some supervision and instruction about heart rate monitoring, proper exercise techniques, and cancer-specific exercises should be included. Initially, the exercise prescription should be reviewed with the patient, and the patient should be instructed about proper intensity and recognizing common adverse symptoms to exercise.

Similar to current exercise guidelines for healthy adults, patients with cancer are also encouraged to participate in general (i.e., total body) resistance as well as flexibility exercises. See table 20.5 for a summary of the exercise prescription recommendations for patients with cancer.

Because cancer is a constellation of diseases, when working with a new patient it is important to learn about

Table 20.5 Summary of Exercise Prescription for Patients with Cancer

Training method	Mode	Intensity	Frequency	Duration	Progression	Goals	Special considerations during active treatment
Aerobic	Walking, stationary bike, or other exercises that use large muscle groups	50% to 85% of heart rate reserve or oxygen uptake reserve, or RPE 11 to 14	Three to five times per week	≥150 min/wk of moderate intensity or ≥75 min/wk of vigorous intensity	30 s to 2 min per day	Increase peak $\dot{V}O_2$, total work, endurance	Intensity may need to be adjusted during treatment to <50% HRR or RPE 9 to 11. If needed, divide exercise into two or three sessions per day and begin at 5 to 10 min.
Resistance	Free weights Machines Resistance bands	50% to 70% of 1RM for lifts involving the lower body 40% to 70% of 1RM for lifts involving the upper body	Two or three times per week	One set of 8 to 12 reps	Gradual increase in resistance (1.1-2.3 kg) following two consecutive symptom-free sessions	Increase muscle strength and endurance	Patients with metastatic bone disease should avoid excessive weight-bearing exercises and seek medical approval due to increased fracture risk.
Flexibility	Stretching Yoga	Stretch maximally but avoid pain, especially in joints	Before and after exercise	15 to 30 s per stretch Repeat one to three times for a total of 60 s per stretch	As tolerated	Increase flexibility and range of motion	Following approval from surgeon, special attention should be given to shoulder mobility stretches in breast cancer survivors.

the specific cancer type, as well as the common treatment regimens and their associated side effects. An example is a patient receiving radiation therapy (RT); because of the localized nature of RT, areas proximal to the tumor (e.g., skin, lungs, bones) may be adversely affected. However, depending on where the cancer is, the side effects from RT (e.g., scarring, pain, reduced flexibility, increased risk of fractures) are variable. Therefore, clinical exercise professionals prescribing exercise during active treatment should give attention to specific treatment side effects; and when possible it is advisable to contact the oncologist regarding limitations.

Practical Application 20.3

LITERATURE REVIEW

The diagnosis of cancer carries a heavy burden. However, even though it is the second leading cause of death in the United States, better detection and treatments have led to improved survival rates over the past decade for most cancers (2). With life expectancies improving, increased emphasis has been placed on mitigating the side effects of treatment and improving the quality of life in cancer survivors. Regular exercise has been purported to counteract many of the transient and chronic effects associated with cancer treatments, including fatigue, nausea, muscle weakness, stress, and anxiety (21, 25, 32, 53, 70). The American College of Sports Medicine has released guidelines on exercise in patients with cancer (70), which include the recommendation to avoid the propensity to cease activities while undergoing treatment. Unfortunately, the majority of patients diagnosed with cancer become less active and remain so years after treatment (16).

Fatigue

Cancer-related fatigue (CRF) is the most common side effect with this disease (6). It is defined as sustained physical, emotional, or cognitive exhaustion experienced during and after adjuvant cancer treatments that is not proportional to recent activities and interferes with usual functioning (6). Because CRF is frequently debilitating, it is also associated with reduced compliance to medical treatments, which may increase the risk of recurrence (66). Cancer-related fatigue is experienced in >70% of patients undergoing chemotherapy, with 30% still reporting persistent fatigue 5 yr or more from the time of completion (26).

Despite the prevalence of CRF and its adverse impact on patients with cancer, little is known about its underlying mechanism or mechanisms and the means to ameliorate them. Part of the reason may be the likely multifaceted etiology of CRF, which may include emotional distress, sleep disturbance, anemia, nutrition, a decreased level of physical activity, and comorbidities (54) (figure 20.3).

Regular exercise, which counteracts many of the potential causes of fatigue, has shown promise in mitigating this common condition both during and following treatment (13, 19, 25, 37, 53, 72, 84). In fact, a recent study in patients with lymphoma found that changes in fatigue were largely explained by changes in fitness (i.e., as peak $\dot{V}O_2$ increases, levels of fatigue go down) (22).

Dimeo and colleagues reported less fatigue in patients who performed 30 min of daily supine cycle ergometry during a several-week in-hospital stay for high-dose chemotherapy (26, 27). A subsequent study in female patients with breast cancer reported improvements in self-reported fatigue in those who performed daily walking at home while receiving adjuvant chemotherapy (53). Similarly, using a mild home-based chair aerobics program, Headley and colleagues reported that patients undergoing adjuvant chemotherapy for breast cancer had less fatigue by the third cycle compared to usual-care controls (37).

Despite the evidence for exercise as an adjunctive treatment for CRF, not all exercise studies have found a beneficial effect on fatigue in this population (21, 24, 84). While this discrepancy may be due to methodological differences (i.e., frequency, intensity, time, and type of exercise), the heterogeneity of cancer populations (e.g., age, treatment, cancer type, stage) may also explain why some studies have reported benefits while others did not. The timing of exercise in relation to when cancer treatments occur can also affect how a patient perceives fatigue. Under the backdrop of dramatic changes that can occur with some treatments (e.g., hair loss, anemia, anxiety), the impact of exercise on fatigue may at times be masked. In fact, a recent meta-analysis by Speck and colleagues reported a stronger effect size in exercise studies conducted after patients have completed cancer treatment (80).

Another aspect of the elusiveness in fatigue research is the fact that it is mainly a self-reported condition with no clear biomarkers. Recent research has examined markers of chronic inflammation as either biochemical signals of fatigue or the underlying cause of the fatigue itself (18, 71, 74, 79). Patients with cancer exhibit higher than average inflammatory markers due to local immune response to the neoplasm, tissue injury from adjuvant treatments (i.e., surgery, chemotherapy, radiotherapy), and the neoplasm itself (23). While many biomarkers are involved in the inflammation process, much attention has focused on the cytokines released during the inflammation response as chemical mediators of fatigue. Interleukin-6 and tumor necrosis factor-α, in particular, are ubiquitous proinflammatory cytokines that are elevated in patients with cancer.

Interestingly, both prospective and cross-sectional studies have found inverse relationships between regular exercise and proinflammatory cytokines in other chronic disease populations (e.g., people with heart failure, those who are elderly or obese) (33, 34, 49, 57). While this has yet to be shown in CRF, it does provide a theoretical explanation as to how exercise may alleviate this condition.

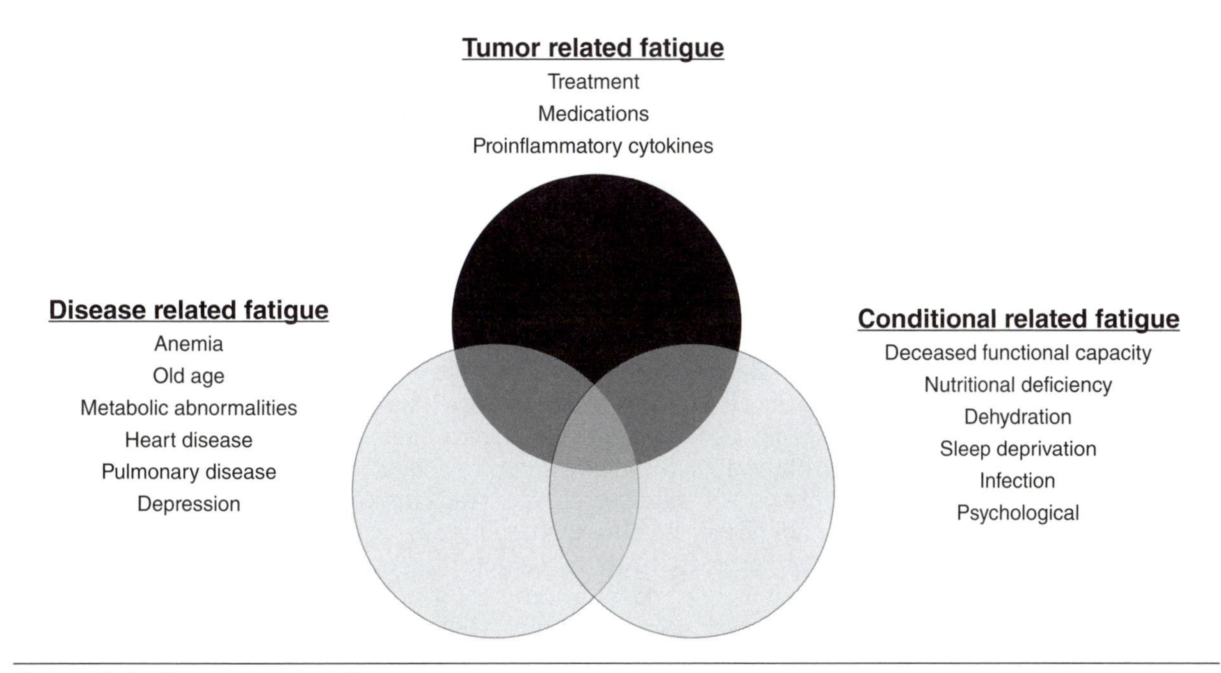

Figure 20.4 Determinants of fatigue.

Cancer Therapy Side Effects

Aside from fatigue, studies in patients with cancer have also shown exercise to improve side effects such as nausea and pain (25, 53, 56). However, these are often smaller or observational studies reporting positive outcomes cast from a "wide net" of secondary variables, not necessarily the main purpose of the study. Notwithstanding, while it is probable that exercise has an impact on several side effects associated with cancer treatments, the evidence for the most part is too preliminary to state this with any degree of certainty. Two areas in which the evidence for exercise seems to be strong are health-related physical fitness and quality of life (19, 21, 24, 80). Several trials have reported improvements in cardiorespiratory fitness, muscular strength, and various measures of quality of life (i.e., physical functioning, psychosocial, self-esteem) across a wide variety of cancer populations (19, 21, 24). While this evidence is again stronger for exercise training after cancer treatment, it does appear that exercise during treatment can mitigate the reduction of physical functioning often associated with chemotherapy and radiation. A large prospective randomized trial of 242 patients with breast cancer revealed that regular aerobic exercise during chemotherapy (i.e., 3 d/wk at 60% to 80% peak $\dot{V}O_2$) can offset the loss of cardiorespiratory fitness, which was 5% lower in the control group (21).

Another common side effect that is mitigated by regular exercise is **lymphedema** (68, 69). Lymphedema is a condition in which the lymphatic system is blocked or damaged, resulting in fluid buildup and edema in the affected extremity. Historically cancer patients who had multiple lymph nodes removed were discouraged from exercising the affected limb for fear that exercise (specifically weight training) might induce or exacerbate lymphedema. However, across many exercise studies involving different modalities (i.e., dragon boat racing, yoga, and weight training), exercise appears safe in individuals who are at risk for lymphedema (16).

(continued)

Practical Application 20.3 *(continued)*

A large randomized trial involving a cohort of breast cancer survivors with stable lymphedema found that those who performed resistance training twice per week had a 50% reduction in the incidence of lymphedema exacerbations (69). Additionally, a follow-up study showed that resistance training in breast cancer patients without lymphedema who had had five or more lymph nodes removed resulted in a 15% lower risk of developing lymphedema (68).

Chemotherapy-induced cardiomyopathy is a serious side effect for which exercise may potentially be protective. Doxirobicin (Adrianmyocin) is a well-documented chemotherapy agent known to have dose-dependent cardiotoxic effects (73). While this has not been well studied in humans, animal models have provided preliminary evidence showing that exercise can attenuate doxorubicin-induced myocyte damage (practical application 20.4). See Scott and colleagues for a comprehensive review on exercise and cardiotoxicity (73).

Exercise and Cancer Recurrence

Of all the studies involving exercise and cancer, early observational studies linking increased regular exercise to reduction of cancer recurrence and mortality have perhaps yielded some of the most intriguing results (41, 44, 52). To date, preliminary data have demonstrated a potential mortality benefit of exercise in patients with existing breast, colon, prostate, ovarian, brain, and lung cancers; the majority of these studies are on breast cancer.

Holmes and colleagues and Meyerhart and colleagues reported reductions in cancer recurrence rates, as well as mortality, for individuals with breast and colon cancer who were the most active following diagnosis. Holmes and colleagues reported, in patients following breast cancer treatment (stage I-III) who averaged over 9 MET-h per week (the equivalent of walking briskly for 30 min, 5 d/wk), an average 50% reduction in recurrence compared to those who were sedentary. The incidence of recurrence was similar in the most active patients in Meyerhardt's colon cancer study; however, the threshold for these results was double, at 18 MET-h per week (52). Both studies controlled for initial activity level, allowing the investigators to compare the increases in physical activity after time of prognosis.

These results suggest that exercise might prevent recurrence and early death in patients with breast and colon cancer; but because the studies were observational, more research is needed before a definitive answer can be obtained. Furthermore, for health professionals and researchers to reach this conclusion, it is also important to establish what the protective mechanism of exercise is for this population.

Concerning the potential antineoplastic properties of exercise, there are many plausible theories, including favorable changes in body weight and sex hormones (e.g., estrogen) and effects on various growth factors (e.g., insulin-like growth factor-I), all of which are associated with decreased survival (43). Additionally, exercise may influence the tumor itself through biological pathways not yet established. Another potential mechanism by which exercise may mediate cancer-free survivorship is through improved tolerance to treatment. The more a patient is able to tolerate a particular treatment, the greater the log kill (i.e., cancer cell destruction). Courneya and colleagues reported that patients undergoing treatment for breast cancer who performed resistance training were able to tolerate higher doses of chemotherapy (20). However, this was only a single study, so more work is needed to confirm the findings.

EXERCISE TRAINING

While the goal for all cancer patients is to progress gradually to 150 min/wk (or more) of moderate-intensity aerobic training, one must appreciate that the path to this goal is not always linear. As mentioned previously, there are many factors to consider when one is exercise training a cancer survivor. The exercise professional needs to anticipate setbacks such as side effects of chemotherapy, radiation therapy, and additional surgical procedures. If setbacks occur, exercise goals need to be modified accordingly. At some point it may even be necessary to suspend exercise if the risk–benefit relationship no longer justifies exercise as a component of the treatment.

Circumstances in which an exercise prescription may need modification include **neutropenia, thrombocytopenia**, anemia, neuropathy, lymphedema, and dehydration. Ideally, in working with patients during chemotherapy, labs including platelet, red blood, and white blood counts should be checked on a regular basis to ensure patient safety (see "Contraindications to Exercise"). However, when lab reports are not accessible, it is important to monitor for clinical signs of the toxicities. For example, patients with chemotherapy-induced anemia can experience early fatigue, exercise dyspnea, and tachycardia resulting in exercise intolerance and higher heart rate responses while exercising at a previously well-tolerated workload.

Just as important as monitoring for toxicities related to cancer therapies is anticipating when they are likely to occur. For example, if a patient is on a chemotherapy regimen that causes neutropenia, instructing her to exercise at home—staying away from fitness centers and "gym germs" for at least a week following chemotherapy treatment—may help reduce the chance of her contracting opportunistic infections.

Some patients have cancers that involve the bones, in which case exercise of the affected extremity may cause pain or result in injury. Patients with metastatic disease to the pelvis or legs should avoid high-impact exercises and may benefit from exercise programs that allow them to sit down, such as stationary cycling or chair exercises. If a patient reports new-onset bone pain, this could be a sign of metastatic disease, and exercise should be stopped immediately until physician approval is given.

Since most cancer patients receive surgery, special considerations are also needed for the postoperative period (typically 4-6 wk). Gentle stretching and stationary bicycle activities are frequently used for patients who are recovering from breast or thoracic surgery so that they can maintain joint-specific range of motion and lower body conditioning while avoiding upper body exercise.

Another common treatment, especially in breast and prostate cancer patients, is hormone-directed therapy (e.g., aromatase inhibitors inhibit estrogen production in postmenopausal breast cancer patients who overexpress the estrogen receptor [ER]). While these treatments have been shown to improve survival, they are also associated with bone and muscle loss as well as central adiposity. Because of these side effects, an exercise prescription that addresses weight loss, bone health, and muscular strength is warranted.

Resistance training to maintain or enhance muscle strength is important not only for patients undergoing hormonal therapy but for most cancer patients. Individuals who are undergoing treatment or are unfamiliar with resistance training should start with resistance bands, light hand weights, or machines (or more than one of these) to avoid any potential for bruising or bone fractures. The recommendations are similar to those for healthy adults—one or two sets of 8 to 12 repetitions, 2 or 3 d/wk. In addition, special precautions are needed for patients with (or at risk for) lymphedema. The current recommendations are to begin a supervised program with very light resistance and progress resistance gradually after two consecutive "pain-free" sessions (68, 69, 70). As long as the patient tolerates the strength training program, there is no upper limit for resistance.

As your patient's clinical exercise physiologist, you should be sure to keep in mind the following:

- Many patients with cancer are inactive and experience mood disturbances such as anxiety and depression that are themselves associated with little interest in exercise. Be sure to increase exercise dose progressively in an individualized manner for every patient.
- For the first week or so of exercise training, patients may be a bit tired later in the day. To compensate for this, ask them to limit the amount of other, home-based activities for the first few days.
- Pool activities are not advised for patients with intravenous catheters or those who are receiving radiation therapy.
- A variety of treatment-related obligations and complications often interrupt exercise therapy for

Practical Application 20.4

RESEARCH FOCUS

The use of exercise to attenuate the effects of cardiotoxic drugs, particularly anthracycline-containing drugs (e.g., doxorubicin), is an area of increased interest. While few studies have examined this in humans, some very good basic animal research has been conducted and yielded promising results. Chicco and colleagues examined the cardioprotective effects of a low-intensity exercise training program (20 min/d, 5 d/wk) in Sprague-Dawley rats exposed to doxorubicin (DOX) (15). Compared to a nonexercise DOX group, the exercise DOX group displayed preserved left ventricular function. Other similar animal studies have also found exercise training to attenuate the cardiotoxic effects of doxorubicin. The exact mechanism(s) resulting from exercise remains elusive. Five potential cardioprotective mechanisms were proposed in a review by Scott and colleagues (73). These were (1) reduced oxidative stress, (2) protection from apoptosis, (3) improved protein synthesis, (4) favorable myocyte calcium handling, and (5) increased adenosine mono-phosphate (AMP)-activated protein kinase pathway activity. Identifying the potential mechanism(s) and performing translational studies in humans are the next steps in determining the applicability of exercise training as a therapy against cardiotoxic chemotherapy. If this is established, it may result in patients' receiving a greater therapeutic dose of these anticancer drugs without greater risk of heart failure.

Clinical Exercise Pearls

Chemotherapy-induced peripheral neuropathy (CIPN) is a common side effect that can increase patient anxiety levels as well as negatively affect exercise adherence. Often associated with taxane, vinca alkaloids, or platinum drugs, CIPN causes common symptoms that include pain or numbness in hands and feet, progressing to radiating pain in the limbs (60). While currently little evidence is available regarding the use of exercise to attenuate CIPN, as with diabetic neuropathy, this condition does not preclude participation. The use of non-weight-bearing exercises (e.g., cycle or recumbent stepper) can help with patient tolerance as well as reducing the risk of falling (due to sensory loss). We have also found that patients enjoyed performing supervised balance exercises using various stability balls and coordination drills (e.g., catching a ball while standing on one leg). While it is not always easy to quantify subtle improvements in balance, the use of these exercises as a distraction may also provide a therapeutic benefit. Finally, in our experience, the symptoms of CIPN improve in most patients the further removed they are from chemotherapy. Assuring patients that this condition often resolves over time can also help reduce anxiety.

patients with cancer. Before patients even start an exercise regimen, inform them that setbacks and interruptions are not uncommon. Instead of not exercising at all or stopping the exercise program, patients should plan around interruptions and continue to adhere to their programs whenever possible.

- Learn about the different types of treatments that your patients are receiving.
- Consider developing an exercise buddy system, which matches up cancer survivors who are already exercising with patients who are just starting out. Such group support from patients with similar medical problems may help improve short-term adherence.
- Ask patients daily how they are feeling and what barriers to exercise they may be experiencing.
- When in doubt regarding the safety of a specific exercise, consult with the patient's supervising physician.

CONCLUSION

Cancer is a constellation of diseases. It can begin in any organ and spread to other organ systems. Treatment includes surgery, radiation therapy, chemotherapy, and, more recently, receptor-targeted drug therapy. Both the disease and the treatment bring emotional and physical challenges to patients with cancer. Exercise benefits these patients primarily through improving (or maintaining) function, reducing fatigue, and countering some of the side effects of cancer therapy. The end result of many of these benefits is an improved quality of life. Additionally, evidence is beginning to mount regarding survival and postdiagnostic activity levels, but the data are still preliminary.

Clinical exercise physiologists are underused for patients with this condition; this is analogous to early efforts to integrate cardiac rehabilitation programs 30 yr ago. Just as important as educating the prospective patient is reaching out to oncologists and primary physicians to communicate the evidence-based benefits of exercise.

Key Terms

biopsy (p. 385)
biotherapy (p. 386)
chemotherapy (p. 386)
curative surgery (p. 386)
lymphedema (p. 391)
metastasize (p. 379)
neoplasm (p. 379)
neutropenia (p. 392)
oncogenes (p. 381)
palliative surgery (p. 386)
pluripotent stem cell (p. 380)
proto-oncogenes (p. 381)
radiation therapy (p. 386)
staging (p. 385)
thrombocytopenia (p. 392)
tumor suppressor genes (p. 381)

CASE STUDY

MEDICAL HISTORY

Mr. CB is a 68 yr old Caucasian male with metastatic renal cell carcinoma. He was a millwright for the city water department but is now retired. He has a medical history of dyslipidemia (total cholesterol = 178 $mg \cdot dl^{-1}$, high-density lipoprotein cholesterol = 24 $mg \cdot dl^{-1}$, triglyceride = 774 $mg \cdot dl^{-1}$), hypertension, and myocardial infarction in August 2003. He also has a history of atrial fibrillation. In January 2005 his body mass was 83 kg, his blood pressure was 110/70 mmHg, and his heart rate was 72 $beats \cdot min^{-1}$. Current medications are metoprolol, isosorbide dinitrate, Solu-cortef, fenofibrate, and aspirin.

DIAGNOSIS

In 2000 during a follow-up visit for impotency in the Department of Urology, urine tests detected microscopic hematuria. The following year, still having difficulties of impotency, the patient complained of gross hematuria and right flank discomfort. An intravenous pyelogram suggested a renal mass. Results of a computed tomography scan showed a right renal mass that invaded the right kidney and the inferior vena cava. He underwent a right radical nephrectomy in November 2001 for stage IIIA renal cell carcinoma. Pathology showed invasion of the renal capsule, as well as the renal vein and inferior vena cava.

A follow-up computed tomography scan in April 2003 showed metastases in the lower lobe of the left lung. These nodules were considered too small for biopsy. In May 2003 he complained of radiating pain from the right buttock to the knee. A bone scan identified widespread bone metastases. Magnetic resonance imaging showed a mass in the left posterior lateral aspect of the lumbar vertebrae. He underwent 14 d of radiotherapy in June 2003, during which his back discomfort improved. As part of a clinical trial, he also received two courses of interferon and chemotherapy.

In May 2004, Mr. CB underwent surgical decompression and excision of an L3 vertebrae tumor. He began immunotherapy with interferon alpha in July 2004. Repeat magnetic resonance imaging and bone scans through January 2005 showed the disease to be stable.

EXERCISE TEST RESULTS

Mr. CB's most recent electrocardiogram showed normal sinus bradycardia with evidence of a previous inferior wall myocardial infarction. Resting heart rate was 53 $beats \cdot min^{-1}$. An exercise stress test completed in August 2003 was mildly suggestive of ischemia, with a $\dot{V}O_2$peak of 17.6 $ml \cdot kg^{-1} \cdot min^{-1}$. Peak heart rate was 92 $beats \cdot min^{-1}$, and exercise was discontinued because of mild- to moderate-grade angina. His cardiac status is now stable with Canadian Cardiovascular Society grade 2 angina.

EXERCISE PRESCRIPTION

During chemotherapy in 2003, Mr. CB began an exercise program. He exercised 3 or 4 d/wk at a perceived exertion of 11 to 12 (Borg scale 6-20) and at a prescribed heart rate range of 72 to 82 $beats \cdot min^{-1}$. This heart rate range was free of any electrocardiographic evidence of ischemia. Exercise modalities included treadmill, dual-action bike, and rower. Exercise therapy was tolerated well, without complications or symptoms.

DISCUSSION QUESTIONS

1. What effect does metoprolol have on heart rate response to exercise? Would you alter how you go about guiding exercise intensity for patients taking this drug?
2. Fatigue and mood disturbances are common complaints of patients with cancer, often attributable both to the disease itself and to the treatments used to manage the disorder. Explain whether exercise should be used or withheld in and around those times when a patient is undergoing therapy.

(continued)

What clinical features and symptoms would lead you to consider (or not consider) exercise during this time? How might you quantify whether, in fact, a patient is or is not responding to an exercise regimen?

3. This patient asks you to help him design and start a resistance training program. Do his test results indicate that this type of training should be incorporated into his exercise regimen? Explain your answer and any concerns you have with respect to resistance training in patients with cancer.

Chapter 21

Human Immunodeficiency Virus

Edward Archer, PhD, MS
Helmut Albrecht, MD
Gregory A. Hand, PhD, MPH

Human immunodeficiency virus (HIV) and acquired immunodeficiency syndrome (AIDS) were first recognized by the U.S. Centers for Disease Control and Prevention (CDC) in the early 1980s (17). Since that time, HIV/AIDS has become a leading cause of morbidity and mortality worldwide (109). The global impact of HIV-related infections on social and economic conditions has been substantial (109).

Without treatment, in most persons, HIV infection causes a progressive depletion of cellular immune functions. As the disease progresses, infected persons become increasingly debilitated and prone to opportunistic infections (e.g., pneumocystis pneumonia, cryptosporidiosis), wasting, and other HIV-associated complications. Over time, physical activity becomes increasingly difficult, resulting in a significant loss of physical functioning. Exercise training has been shown to attenuate the progression of the disease as well as complications arising from the current pharmacological treatments. This chapter provides a review of the disease and summarizes the evidence for the use of exercise as a complementary modality in the treatment of HIV infection and AIDS.

DEFINITION

In 1982, the CDC reported on a cluster of Kaposi's sarcoma and *Pneumocystis carinii* pneumonia among young, previously healthy men who had sex with men in Los Angeles and Orange County, California. These were the earliest cases of what was to become the AIDS pandemic in the United States (17). The cause of this virulent disease was found to be a novel retrovirus, the human immunodeficiency virus (HIV). That same year, the first case definition was published by the CDC. Since then the case definition of AIDS has evolved, reflecting an enhanced understanding of the disease progression as well as the clinical manifestations of HIV infection.

HIV is a lentivirus of the Retroviridae family (28, 107). The earliest positive genetic identification of HIV is from Africa in the late 1950s (111). There is strong evidence that HIV began as a zoonotic disease in nonhuman primates. Simian immunodeficiency virus (SIV) strains gave rise to HIV with subsequent adaptation to the human host (54). In 1983, HIV was identified as the causative virus of AIDS (4). HIV infects human T lymphocytes that express the CD4+ glycoprotein on their surface. There are two predominant species of HIV: HIV-1 and HIV-2. HIV-1 is the better human-adapted strain; it is more virulent and more easily transmitted and therefore more prevalent (40). HIV infection results from the transfer of bodily fluids (e.g., blood, semen, vaginal secretions, breast milk) from an infected individual to an uninfected person. Once chronic infection is established, a complex array of pathogenic mechanisms leads to increase in viral load and a concomitant decrement in CD4+ lymphocytes below the normal range of 500 to 1,400 cells · μl^{-1}. As the number of CD4+ lymphocytes declines, persons infected with HIV become predisposed to opportunistic infections, malignancies, wasting, and other complications of AIDS. Until the mid-1990s, survival was usually less than 2 yr once the CD4+ lymphocyte count declined to <200 cells · μl^{-1}. With the advent of highly active antiretroviral therapy (**HAART**), survival has improved significantly. By 2000, mortality rates were 85% lower than just 5 yr earlier, and CD4+ T lymphocyte counts often increased enough to forestall most life-threatening opportunistic infections (18).

Once HIV was established as the cause of AIDS, HIV antibody tests were developed, becoming widely available by 1985. This development led to the expansion of the case definition to include clinical conditions less closely associated with immune suppression (23). In 1987, the clinical definition of AIDS was further expanded to accommodate changing diagnostic and treatment practices (24, 110). In 1993, the CDC issued a revised classification system for defining AIDS using three CD4+ cell ranges and three clinical categories (16). The most important changes were the inclusion in the AIDS case definition of all patients with fewer than 200 CD4+ cells · μl^{-1} or a relative CD4+ percentage of less than 14% of total lymphocytes. In January 2000, the CDC devised a combined surveillance case definition that combined reporting criteria for HIV infection and AIDS.

In 2008, the CDC issued the most recent revision to the HIV case definition (91). It retains the 24 clinical conditions in the AIDS surveillance case definition published in 1987 and highlights the central role of objective measures of **immunosuppression** (i.e., **CD4+ T lymphocyte** counts, percentages, or both). HIV infection is classified in one of four stages. Laboratory-confirmed evidence of HIV infection is now required to meet the surveillance case definition for HIV infection, including stage 3 HIV infection (AIDS) (table 21.1).

SCOPE

In 2010 the World Health Organization (WHO) estimated that over 60 million people had been infected with HIV and that 30 million had died as a result. The WHO also estimated that over 34 million people globally and approximately 1.1 million in the United States are currently infected and living with HIV (19, 104, 109). The global social and economic impact of HIV-related infections has been substantial. In 2002, the cost of new HIV infections in the United States alone was estimated at $36.4 billion, including $6.7 billion in direct medical costs and $29.7 billion in productivity losses (49). The CDC estimates that more than 50,000 people are infected with HIV every year in the United States and that approximately 20% of all HIV-infected individuals are unaware of their infection (22).

The demographic characteristics of those newly infected with HIV have changed since the early period of the epidemic. Initially mostly an epidemic among white gay males, HIV infection now disproportionately affects individuals of low socioeconomic status as well as the African American and Hispanic communities. Minority women currently account for a substantial majority of all newly reported HIV and AIDS cases among women. A number of factors play a role in the epidemiologic patterns of infection (e.g., delayed presentation, transmission risk factors, socioeconomic status, and other comorbid conditions) (32, 78). A detailed description of the demographic characteristics of the HIV-AIDS epidemic is available online at the CDC website, www.cdc.gov. For information on the global epidemic, access the WHO website at www.who.org.

PATHOPHYSIOLOGY

HIV infection results from the transfer of infected bodily fluids (e.g., blood, semen, vaginal secretions, breast milk)

Table 21.1 HIV Stage Progression

HIV infection stage	Characteristics
HIV infection, stage 1	No AIDS-defining condition and either CD4+ T lymphocyte count ≥500 cells · μl^{-1} or CD4+ T lymphocyte percentage of total lymphocytes ≥29.
HIV infection, stage 2	No AIDS-defining condition and either CD4+ T lymphocyte count of 200 to 499 cells · μl^{-1} or CD4+ T lymphocyte percentage of total lymphocytes of 14 to 28.
HIV infection, stage 3 (AIDS)	CD4+ T lymphocyte count <200 cells · μl^{-1} or CD4+ T lymphocyte percentage of total lymphocytes <14 or documentation of an AIDS-defining condition. Documentation of an AIDS-defining condition supersedes a CD4+ T lymphocyte count ≥200 cells · μl^{-1} and a CD4+ T lymphocyte percentage of total lymphocytes ≥14.
HIV infection, stage unknown	No information available on CD4+ T lymphocyte count or percentage and no information available on AIDS-defining conditions.

from an infected individual to an uninfected person. The three primary routes of infection are unprotected penetrative sexual contact, perinatal exposure, and intravenous (IV) drug use. Alternative modes of transmission (e.g., insect bites, saliva transfer, sweat) have been suggested but were found not to be involved in the transmission of the virus. The risk of infection from HIV is much lower than for many other viral diseases. For example, the rate of occupational transmission from an HIV-positive source (e.g., needle stick) is thought to be 0.3% (i.e., 3 in 1,000) for a percutaneous exposure and 0.09% for a mucous membrane exposure (i.e., <1 in 1,000) (14, 20). Comparatively, the rate of transmission from a hepatitis B virus–positive source to a nonimmune host is 6% to 24% and is 1% to 10% for exposure to hepatitis C virus (21).

The time course of HIV infection often begins with a primary infection syndrome that usually presents as a moderate to severe case of influenza with various associated signs and symptoms such as fever, sore throat, fatigue, lymphadenopathy, rash, myalgia, malaise, and oral or esophageal sores or both. This seroconversion illness (i.e., the time of unchecked viremia before the development of HIV antibodies) may last from a few days to several weeks. During this phase, rapid viral replication leads to a massive increase in HIV viral load concentration. This increase in viral load is often accompanied by a decline in CD4+ cell counts resulting in a decline of cell-mediated immunity. This early CD4 cell decrease is usually reversible. The seroconversion illness is followed by a long clinically asymptomatic phase. However, over many years, a decline in CD4 cell count typically occurs. CD4+ cell destruction may result from direct infection and apoptosis (i.e., programmed cell death) of the CD4+ cells by the virus or from the cytotoxic immune response to the viral infection (e.g., CD8+-mediated cell destruction). Many patients are unaware of their infection and can remain asymptomatic for many years; 10 yr is the average in untreated individuals (10, 22, 58). Stage 3 (AIDS) develops when CD4+ T lymphocyte counts drop below 200 cells · μl^{-1}, or CD4+ T lymphocyte percentage of total lymphocytes drops below 14%, or patients develop an AIDS-defining condition. At this point, the infected individual is susceptible to additional opportunistic infections or cancers, and without treatment, life expectancy is significantly shortened. Before the development of HAART treatment, opportunistic infections were the primary cause of mortality in AIDS patients, with *Pneumocystis jiroveci* pneumonia and Kaposi's sarcoma the leading diagnoses, affecting approximately 50% and 25% of AIDS patients, respectively.

Wasting

Currently, less than 10% of patients in appropriate treatment experience wasting (94), whereas early in the epidemic (i.e., the 1980s), wasting was a common life-threatening complication of HIV infection and a common AIDS-defining condition (56). In 1988, wasting was classified as an unintentional decrease of >10% of body weight. In the early 1990s, it was shown that even a 5% loss of body weight had a significant impact on mortality in HIV-infected patients. As a result, wasting was reclassified to include individuals who experienced an unintentional loss of 5% or more (16, 44). Wasting induces a loss of function and weakness via a decrement in protein stores (e.g., lean body mass). High-intensity progressive resistance training (PRT) can ameliorate the catabolic effects of AIDS-induced wasting and lead to improvements in strength, protein stores (e.g., muscle mass), and physical functioning (41, 66, 115). As a result of pharmacological treatments, wasting is less of a concern today than during the early years of the epidemic.

Metabolic Complications

Metabolic complications are among the most common problems associated with antiretroviral therapy. These pathological metabolic changes confer an increased risk of cardiovascular disease similar to that with metabolic syndrome and include alterations in glucose and lipid metabolism, insulin resistance, diabetes mellitus, dyslipidemia, lipoatrophy, and lipodystrophy syndrome (13).

Lipodystrophy

Lipodystrophy is an **iatrogenic** metabolic effect of HIV and certain antiretroviral medications resulting in a loss of subcutaneous fat depots in the arms, legs, and face with a concomitant increase in visceral fat (2, 12, 15, 105). There is strong evidence that both endurance training (ET) and progressive resistance training (PRT) can ameliorate the metabolic complications in HIV-infected patients with lipodystrophy (64, 102, 103, 113).

Cardiac Dysfunction

HIV infection is associated with a number of subclinical functional and structural cardiac abnormalities, such as a higher prevalence of left ventricular diastolic dysfunction, cardiovascular malignancy, pulmonary arterial hypertension, vasculitis, pericardial effusion, premature atherosclerosis, and arrhythmias (71, 72, 77). HAART has led to a 30% reduction in the prevalence of HIV-associated cardiomyopathy and myocarditis, as well as a significant decrement in infection-related and all-cause mortality in HIV-infected patients (3). Nevertheless, attendant with this decline has been a substantial increase in cardiovascular disease (CVD)–specific morbidity and mortality. Furthermore, patients treated with HAART may experience a number of iatrogenic cardiovascular complications including congestive heart failure, accelerated atherosclerosis, increased myocardial infarction risk, and torsade de pointes ventricular tachycardia.

The increase in CVD is hypothesized to be a result of an increased prevalence of traditional CVD risk factors (e.g., smoking, physical inactivity, obesity) in patients with HIV as well as chronic inflammation, direct viral effects, and factors associated with HAART (70). Increased physical activity, ET, and PRT ameliorate these effects (27, 37, 50, 72). Given that CVD is an important contributor to morbidity and mortality in HIV-infected patients, early detection is essential for effective treatment (72).

CLINICAL CONSIDERATIONS

An increasing number of health care professionals are recommending exercise as a complementary modality to prevent disease and mitigate the signs and symptoms associated with chronic disease states. Nonpharmacologic interventions such as exercise training and nutritional behavioral counseling have the potential to improve patient outcomes via the enhancement of health and the augmentation of more traditional medical treatments. Complementary modalities such as exercise are low-cost and efficacious additions to the management of chronic diseases.

The long-term goals of exercise interventions are to decrease morbidity and mortality and improve the quality of life of those infected with HIV (45, 46, 75, 112). These aims are accomplished by exercise interventions tailored to address the infected individual's signs and symptoms. The more proximal goals of exercise training are increases in cardiorespiratory fitness (CRF), decreases in CVD risk factors, improvements in body composition and metabolic functioning, and the enhancement of musculoskeletal function. These objectives are most effectively achieved via exercise prescriptions that include both progressive resistance and aerobic training. These modalities improve people's capacity for activities of daily living as well as enhancing their ability to remain as physically and mentally active to the extent their illness allows.

Starting exercise training has been shown to benefit the HIV-infected patient's physiological and psychological functioning. Exercise improves mood, quality of life, aerobic fitness, and immune indices and also decreases the incidence of comorbid chronic disease states in HIV-infected populations (29, 30, 46, 62, 67, 75, 81, 102). Additionally, exercise significantly increases strength, functionality, and endurance; improves body composition and lipid profiles; decreases resting heart rate; and reverses the nervous system disorders comorbid with HIV infection (34, 36, 75, 103, 113). Thus HIV infection should be an impetus for the initiation of an exercise program and not a deterrent.

Signs and Symptoms

The clinical exercise professional must be cognizant of the fact that HIV-infected patients may present from completely asymptomatic (stage 1) to critically ill (stage 3). Patterns of physiological, behavioral, and psychological changes are idiosyncratic. Therefore, all patients must be evaluated on an individual basis to assess the effects and development of the disease.

While some individuals with primary HIV infection are asymptomatic, many patients will develop symptoms of viral illness (e.g., fever, pharyngitis, malaise, myalgias, rash) within 1 mo postexposure. The duration of these symptoms varies, with an average of 2 to 4 wk (53). This initial immune response reduces the number of viral particles and marks the advent of the latency period during which infected individuals may experience few symptoms. Latency may last as long as a few weeks to more than two decades.

Most HIV-infected individuals in a health care setting are on some form of HAART. Thus the most common signs and symptoms are lipodystrophy, impaired glucose and lipid metabolism, and increased CVD risk factors (e.g., obesity). Fortunately, these manifestations of pharmacological treatment are amenable to exercise training and other changes in lifestyle (e.g., nutrition).

History and Physical Examination

A detailed history combined with general and musculoskeletal examinations will identify individuals with significant changes in body weight or other symptoms of metabolic disorders (e.g., wasting, lipodystrophy), motor abnormalities (e.g., hyperreflexia, loss of equilibrioception), and indicators of CVD (e.g., arrhythmias, edema). Chapter 4 provides general medical history and examination information. The presence of motor disturbances requires vigilance with all subsequent exercise testing. The primary signs to be aware of include immune status (CD4+ cell counts) and functional capacity and fitness. Patients may present with alterations of body composition (e.g., lipodystrophy), changes in body weight (e.g., wasting), loss of skeletal muscle mass, visceral adiposity, and low muscular strength as well as low CRF (80). Strength in the legs and arms should be quantitatively examined, as well as functional strength (e.g., the ability to rise from a chair). Peripheral neuropathy is a common complication of both HIV and HAART regimens. Therefore, both sensory and functional motor testing should be performed. Numerous behavioral and psychological symptoms may also be present, such as fatigue, malaise, depression, anxiety, social isolation, and a decreased quality of life (QOL) (52, 68). Of particular concern are the reduced levels of physical activity and exercise in HIV-infected patients. High levels of sedentary living combined with low levels of CRF increase the risk of CVD and exacerbate the metabolic complications associated with pharmacological treatments.

Diagnostic Testing

The purpose of HIV testing is to ascertain the presence of HIV infection via the presence of antibodies, viral RNA, or both. Modern HIV testing is highly accurate (25). HIV testing has evolved through four generations (classified by the antigen used); the first commercially available immunoassay became available in 1985. Since that time, each generation has improved the specificity (i.e., the percentage of individuals correctly identified as HIV+) or the sensitivity (the percentage of individuals correctly classified as not HIV+) of HIV testing or both. These are the commonly performed HIV tests:

- **ELISA:** Enzyme-linked immunosorbent assay (also known as an enzyme immunoassay or EIA). This tests for the presence of HIV antibodies and has a high sensitivity. Results are reported as a single value. Second-generation tests detected immunoglobulin (Ig)G antibodies; third generation detected both IgM and IgG antibodies, while fourth generation detects both HIV antibody and the p24 antigen, which emanates directly from the HIV.
- Western blot: An HIV antibody test that examines levels of specific HIV proteins and is often used to confirm the results from an ELISA. Initial western blot results that are indeterminant most often yield HIV+ status results with a later test.
- HIV nucleic acid amplification test (HIV NAAT), also known as viral load tests. These tests detect the presence of RNA (viral genetic material).
- Indirect fluorescent antibody (IFA): Antibody test that is often used to confirm the results from an ELISA.
- Rapid antibody tests: Qualitative immunoassays intended for point-of-care testing to aid in the diagnosis of HIV infection. These tests must be used in combination with patient history, clinical status, and risk factors.

Antibody tests may use blood, mucosal or oral fluid (not saliva), or urine for detection. The period from initial exposure and infection to detectable levels of HIV antigens is less than 3 wk and for detection of antibodies is approximately 25 d.

While some individuals with acute primary HIV infection are asymptomatic, many patients develop symptoms of viral illness (e.g., fever, sore throat, malaise, myalgias, rash). Because they are nonspecific, these symptoms may indicate a wide range of conditions. Therefore, the diagnosis of acute HIV infection requires a thorough assessment of HIV exposure risk and appropriate HIV-related laboratory tests. Individuals with acute infection may have elevated viral loads and be unaware of their infection, and are therefore more likely to transmit the virus to their sexual partners (83, 106). The ability to adequately and effectively discuss HIV related issues with patients is important. Practical application 21.1 provides information regarding client and clinician interactions.

Exercise Testing

Exercise training is safe in most HIV-infected individuals (9). There is consistent and strong evidence that exercise can improve CVD risk factors and attenuate the loss of functional capacity as well as the psychological symptoms associated with HIV infection. Nevertheless, HIV-infected individuals should undergo a medical evaluation before beginning an exercise program. Continual monitoring of the general health of the HIV-infected participant is critical, especially with individuals unaccustomed to physical exertion. The detection of conditions that contraindicate exercise is of paramount importance in immune-compromised individuals. A number of disabling orthopedic complications and pathologies associated with HIV and HAART may preclude exercise training (e.g., osteonecrosis, bone tumors, and rheumatic manifestations) (42, 69, 74, 100, 114). While screening of asymptomatic patients is not recommended, patients experiencing chronic and debilitating joint pain should be evaluated, and the evaluation should include radiologic exams.

Conditions such as extreme fatigue, fever, or chills may indicate an active infection, and exercise may exacerbate these conditions. Nausea from HAART may make specific exercises difficult, decrease compliance with exercise prescription, or both. As with many

Practical Application 21.1

CLIENT–CLINICIAN INTERACTION

HIV is not easily transmitted between individuals. The risk of infection from HIV is much lower than for many other infectious agents such as the hepatitis B or hepatitis C viruses (21). Thus application of the general guidelines for disease prevention outlined by the CDC will provide an adequate level of safety for the clinical exercise physiologist (CEP) in all exercise testing and prescriptive contexts. The Universal Precautions and Blood Safety guidelines are available at www.cdc.gov/mmwr/preview/mmwrhtml/00000039.htm (last accessed on April 4, 2013). Specific recommendations for the prevention of HIV transmission in health care settings are listed at www.cdc.gov/MMWR/preview/mmwrhtml/00023587.htm (last accessed on April 4, 2013).

The initial contact between a CEP and an HIV-infected individual is most often for the treatment of established CVD due to the confluence of two factors: HIV-infected patients as a group are quite sedentary, and the pharmacological regimens used to treat HIV increase CVD risk factors. In clinical settings that test and treat CVD, the vast majority of patients are not HIV+. Thus, HIV+ patients may be concerned regarding the dissemination of their health status. In non-health care settings, a stigma is attached to HIV infection. Therefore, in the clinical setting it is helpful to reassure patients that their medical information is confidential and that only those staff members who need to be cognizant of their HIV+ status will have that information.

Precautions

1. Hands should be washed before and after examining or testing each patient.
2. CEPs should routinely use appropriate barrier precautions to prevent skin and mucous membrane exposure when there is the potential for contact with blood or other body fluids. Gloves, goggles, or face shields and gowns provide an appropriate level of protection for the majority of exercise testing contexts (e.g., removing mouthpieces, skin prep for electrocardiogram [ECG] electrode placement).
3. Face shields or goggles and surgical masks should be used for blood sampling.
4. Saliva has not been implicated in the transmission of HIV. Given the potential for emergency mouth-to-mouth resuscitation, resuscitation bags, mouthpieces, or other ventilation devices should be available when the need for resuscitation is anticipated.

populations, HIV-infected individuals suffer from poor exercise prescription adherence and compliance (57, 67, 82).

Wasting per se is not a contraindication to exercise. Nevertheless, care must be taken to include PRT to offset the catabolic effects of HIV infection and the metabolic complications associated with HAART. Additionally, nutrition counseling may be indicated for individuals beginning exercise training who may be experiencing wasting or lipodystrophy (31, 41).

While HIV infection can be associated with cardiomyopathy as well as subclinical functional and structural cardiac abnormalities, most patients present with normal ECGs and systolic morphologic parameters despite significant diastolic dysfunction (77). Pulmonary hypertension and left atrial enlargement are the most predominant pathologies associated with HIV infection and HAART regimens (71). Careful cardiac screening is warranted for all patients who are being evaluated for or who are currently receiving pharmacological treatment for HIV infection. This is particularly true for those with significant CVD risk factors (3).

Assessment of Functional Status

Baseline assessments of functional status and fitness are necessary to complement the medical history and physical examination. Baseline measures provide a reference point for evaluating response to the exercise program as well as comparisons in subsequent evaluations. Disease status can be categorized into one of four categories for functional status and fitness testing:

1. Asymptomatic, medically stable, and physically active
2. Asymptomatic, medically stable, and physically inactive
3. Recovering from a medical or disease-related event
4. Symptomatic and suffering from acute illness

Depending on the patient's status, various testing protocols have been proposed to ascertain CRF, body composition, flexibility, strength, and physical functioning. Patients who are symptomatic or are suffering an acute illness should refrain from testing until their condition stabilizes.

Asymptomatic patients may not differ with respect to age-adjusted norms on graded exercise tests. They may, however, exhibit a reduction in exercise capacity due to a sedentary lifestyle. Symptomatic patients exhibit significantly reduced time on treadmill or bike, peak oxygen consumption ($\dot{V}O_2$peak), and ventilatory threshold and increased heart rate at submaximal work rates. AIDS patients demonstrate dramatically reduced time on treadmill or bike and peak oxygen consumption ($\dot{V}O_2$peak) and possible failure to reach ventilatory threshold. There is also an increased probability of abnormal neuroendocrine responses at moderate- and high-intensity test stages.

Table 21.2 provides a brief summary of recommendations for exercise testing and the assessment of physical fitness and function in an HIV-infected population. This summary is not exhaustive, and specific protocols are described elsewhere (1, 65).

As with most populations, the general exercise testing recommendations apply (see chapter 5). Exercise professionals who administer exercise tests or supervise training are at minimal risk of infection. Nevertheless, universal precautions for the transmission of blood pathogens should be followed. The Universal Precautions and Blood Safety guidelines may be found at www.cdc.gov/mmwr/preview/mmwrhtml/00000039.htm (last accessed on April 4, 2013).

Cardiovascular

Most HIV-infected patients exhibit normal cardiovascular values (e.g., blood pressure, ventilation) in response to a graded exercise test. Some HIV-positive individuals, however, suffer from peripheral neuropathy and are at increased risk for autonomic neuropathy and abnormal responses to testing (e.g., attenuated heart rate). Cycle ergometry may be a better choice than treadmill protocols with patients who exhibit poor balance or coordination secondary to fatigue or neuromuscular pathologies.

Many HIV-infected individuals are sedentary and have low CRF as evidenced by low average $\dot{V}O_2$peak values (11, 46, 55, 79). These values may be as much as 24% to 44% below age-predicted normal values (66, 67) and may be exhibited in the absence of opportunistic cardiopulmonary infections (51). Individuals on HAART regimens may have a diminished ability to extract and use oxygen in the working skeletal musculature and exhibit a decrement in peak a-$\bar{v}O_2$ difference values. This

Table 21.2 Exercise Testing and Physical Fitness and Function Review

	DISEASE STATUS		
	STABLE AND ACTIVE	STABLE AND INACTIVE	IN RECOVERY
Cardiovascular: cardiorespiratory fitness (CRF)	Maximal or submaximal graded exercise test[1] Rockport 1 mi fitness walk test	Submaximal graded exercise test[1] 6 min walk Rockport 1 mi fitness walk test	6 min walk
Resistance: muscular strength	Grip strength 1RM[1] or 10RM	Grip strength 1RM[1] or 10RM	Grip strength 10RM
Resistance: muscular endurance	Modified 1 min sit-up test[1]	Modified 1 min sit-up test[1]	Modified 1 min sit-up test[1]
Flexibility	Sit-and-reach test[1]	Sit-and-reach test[1]	Sit-and-reach test[1]
Body composition	Skinfolds Waist circumference Waist-to-hip ratio Dual X-ray absorptiometry	Skinfolds Waist circumference Waist-to-hip ratio Dual X-ray absorptiometry	Skinfolds Waist circumference Waist-to-hip ratio Dual X-ray absorptiometry
Physical functioning	Repeated chair stands[2]	Repeated chair stands[2]	Repeated chair stands[2]

[1]American College of Sports Medicine (ACSM) guidelines for exercise testing and prescription. 9th ed. Baltimore: Lippincott, Williams & Wilkins, 2014.

[2]www.cdc.gov/nchs/data/nhanes/nhanes3/cdrom/nchs/.../physical.pdf

appears to be a result of mitochondrial effects of antiretroviral therapy rather than true HIV pathogenesis (11). Nevertheless, the literature reviewed later suggests that the decrement in CRF is a result of increased sedentary living and a lack of physical activity and exercise rather than disease progression or its treatment.

Musculoskeletal

Assessing muscular strength and flexibility in HIV-infected individuals should not differ appreciably from the testing of noninfected individuals. This population is often untrained and therefore unfamiliar with PRT. Therefore the use of a 3-repetition maximum (RM) or greater (e.g., 10RM) protocol may be more appropriate than 1RM for assessment of muscular strength and the prevention of injury or delayed-onset muscle soreness (DOMS). Given that most HIV-infected individuals are deconditioned and lead sedentary lives, assessments should be performed on multiple occasions over the course of training. Some individuals may demonstrate substantial improvements during the early stages of exercise training, and reassessment is necessary to ensure the sufficiency of the training stimulus, the progression, or both.

Treatment

The Food and Drug Administration has approved 29 drugs for the treatment of HIV infection. The five classes of pharmaceuticals are categorized by function (table 21.3).

HIV treatment regimens typically combine drugs from multiple categories. These regimens, when employed with medications for comorbid conditions (e.g., opportunistic infections), lead to significant issues with drug interactions and iatrogenic effects. These adverse conditions are labile and must be continuously monitored when an exercise regimen is being prescribed. Fortunately, a number of these unfavorable conditions, including diabetes, hyperglycemia, lipodystrophy, and hyperlipidemia, are responsive to exercise training. As with other clinical populations, nutritional interventions are often employed in conjunction with exercise prescription. There are a number of empirically supported nutritional strategies for HIV infection and its comorbidities (41, 63). While behavioral or lifestyle modifications such as diet and exercise have been shown to ameliorate many HIV comorbid conditions and iatrogenic effects, there is a dearth of evidence on the interaction of exercise and the pharmaceuticals employed in the treatment of HIV infection and AIDS.

EXERCISE PRESCRIPTION

All HIV-infected individuals should obtain medical clearance before participating in an exercise regimen. The long-term goals of exercise interventions are to decrease morbidity and mortality and improve the QOL of those infected with HIV (45, 46, 75, 112). To achieve these aims, all exercise prescriptions should be tailored to each individual's unique situation. A detailed history inclusive of past and current patterns of physical activity, when combined with exercise testing, provides the information

Table 21.3 Pharmacology

Medication name and class	Primary effects	Exercise effects	Special considerations
Protease inhibitors (PIs)	Prevent the cellular processing of HIV genetic material	None known	Note that these considerations are general to all HIV medications: *Potential liver or kidney damage *Elevated blood lipid levels
Nucleoside reverse transcriptase inhibitors (NRTIs)	Inhibit retroviral replication by inhibiting HIV RNA conversion to DNA		
Non-nucleoside reverse transcriptase inhibitors (non-NRTIs)	Inhibit HIV RNA conversion to DNA		
Fusion inhibitors (FIs)	Inhibit HIV virus from attaching to targeted cells and hinder the initial infection		
Integrase inhibitors	Prevent the integration of HIV genetic material into the infected cell's DNA		

necessary for the development of an effective exercise prescription. Ultimately, the goals of exercise training are to improve people's capacity for the activities of daily living as well as enhance their ability to remain healthy and active to the extent that their illness allows. As such, exercise prescriptions should be targeted toward improvements of any existing comorbidities. Improvements in physiological and psychological functioning are seen as progress is exhibited on a number of objective outcomes. These include increases in CRF, decreases in CVD risk factors (e.g., blood lipids, obesity), improvements in body composition (e.g., decreased visceral fat depots), augmented metabolic functioning (e.g., insulin resistance), and the enhancement of musculoskeletal function (e.g., strength and mobility). These objectives will be achieved only via exercise prescriptions that include some forms of both progressive resistance and aerobic training. Modifications will need to be made for individuals suffering significant morbidity (e.g., peripheral neuropathy).

Adherence is an important consideration. Research has shown that many factors influence adherence and compliance with exercise recommendations and prescriptions. For example, substance abuse, inadequate transportation, poverty, infection, and the side effects of treatment all demonstrate a reduced compliance with exercise prescriptions (46). Nausea is a common side effect of medication that is known to limit exercise tolerance and compliance. Clinical exercise professionals should work with individuals on the scheduling of meals, medication, and exercise training to minimize the nauseating effects of medication. While short-term high-intensity combined training significantly improves a number of health parameters (86), high-intensity training of any type (i.e., AT or PRT) has been shown to cause a decrement in exercise prescription adherence (57, 67, 82). It is postulated that this is the result of delayed-onset muscle soreness, the risk of injury in an untrained population, or both (46). It is well known in both clinical and research settings that anxiety, depression, and depressive symptoms decrease adherence with many forms of treatment (e.g., pharmacologic or behavioral) and across disease states (39, 84, 85). Given that the prevalence of depression and depressive symptoms in the HIV population is high, issues with both pharmacologic and exercise compliance should be continually addressed (38). Supervision substantially improves outcomes across a wide range of variables and should be incorporated when possible (67).

The confluence of direct viral effects, idiosyncratic lifestyle factors (e.g., physical inactivity), and pharmacologic treatment renders the management of HIV infection and related comorbidities complex and multidimensional. Continuous evaluation of the progression of HIV infection and comorbid diseases is essential for determining appropriate treatment and intervention strategies. Research has demonstrated that immune status and viral load are predictive markers of the relative risk for the development of opportunistic diseases and disease progression. CD4+ cell counts, as well as plasma HIV RNA concentration (an indicator of viral load), are convenient surrogate markers for HIV progression (48, 76, 89, 98). Alterations in values of immune function and viral load are much more strongly associated with stage progression or the development of AIDS than baseline values (76); as such, they may be used to inform decisions regarding exercise prescription. Patients treated with antiretroviral medications often exhibit a rapid decrement in plasma HIV RNA levels or increase of CD4+ lymphocyte counts, or both, over time. The failure to display a substantial reduction in plasma HIV RNA concentration by 8 wk and undetectable viral loads at 6 mo is suggestive of nonadherence to the drug regimen, inadequate medication potency, viral resistance, or some combination of these.

Patients with HIV infection, in addition to undergoing HAART, may also be subject to psychotropic drugs, chemotherapeutic agents for cancer, or the prophylactic use of antimicrobial agents and vaccines (or more than one of these) to prevent or treat comorbid conditions such as opportunistic infections (e.g., cytomegalovirus, pneumocystic pneumonia). Clinical health professionals must be cognizant of the physiological effects of HAART regimens as well as other pharmacologic treatments and allow this perception and understanding to inform exercise prescription. The following list provides a brief summary of iatrogenic complications of HIV treatment.

Complications of HAART Regimens

- Lipodystrophy
- Impaired glucose metabolism
- Insulin resistance
- Hyperlipidemia or hyperlipoproteinemia
- Osteonecrosis or avascular necrosis
- Diarrhea, nausea, and other gastrointestinal tract disorders
- Increased risk of CVD and type 2 diabetes

Cardiovascular Exercise

Given the deconditioned status and sedentary lifestyle of many HIV-infected individuals, initial recommendations for cardiovascular training should be at a level below those presented by the American College of Sports Medicine (ACSM). Initial prescription should begin at lower intensities and volumes and should progress at a slower rate than with uninfected individuals. Nevertheless, while the goal is to meet or exceed current guidelines (47), the final prescription may need to be modified downward due to viral and iatrogenic effects.

Resistance Exercise

The development of a progressive resistance program should adhere to ACSM guidelines while taking into account disease stage, complications, and any limitations induced via medication. Particular attention must be paid to individuals suffering osteonecrosis or peripheral neuropathies. With osteonecrosis, the joint capsule may be compromised, and resistance training is contraindicated. Peripheral neuropathy may significantly affect coordination, and caution is advised with all weight-bearing exercise. It is advisable to allow participants to experience a varied selection of exercises that allow a pain-free range of motion and appropriate progression.

Range of Motion Exercise

As with any sedentary population, flexibility and range of motion in HIV-infected individuals may be compromised. Therefore, the initial prescription should be mild, and the progression should be slow but consistent. Refer to chapter 5 for general information on exercise prescription.

Table 21.4 is a summary of exercise prescription for HIV-infected individuals.

EXERCISE TRAINING

For optimal benefits, all exercise training must be systematic and progressive. One must consider a number of factors when designing an exercise prescription. Type and dose (i.e., intensity, frequency, duration, and volume) are the two most important facets. The type of training will depend on the desired outcome (i.e., CRF, functional strength, CVD risk reduction). The three main types of training are aerobic (e.g., running, biking, walking), resistance (e.g., weightlifting), and flexibility training (e.g., stretching, yoga). The initial dose is determined by the patient's baseline level of fitness and motivation.

Cardiovascular Exercise

Aerobic training results in significant benefits for HIV-infected individuals: improved CRF, body composition, and mood; decreased visceral fat depots; and reductions in the risk of CVD. A minimum level of intensity is nec-

Table 21.4 Exercise Prescription Review

Type	Mode	Intensity	Duration	Frequency	Progression	Goals	Special considerations
Aerobic and endurance training	Walking, cycling, jogging, swimming, for example	40% to < 60% $\dot{V}O_2$ reserve or HRR	150 min/wk	3 to 5 d/wk	Slow incremental progression to 65% to 85% of peak $\dot{V}O_2$	Improved CRF	Initial prescription should be below ACSM guidelines.
PRT: progressive resistance training	Machine and free weights, resistance bands	80% of 10RM or 60% of 1RM	Two to four sets per exercise, 8 to 10 repetitions per set, six to eight exercises	2 to 3 d/wk	Slow incremental progression to 75% to 90% of 1RM	Focus on major muscle group strength and endurance	Suspect osteonecrosis or bone tumors with chronic, unexplained musculoskeletal pain.
Flexibility and range of motion training	Yoga, static stretching of the major muscle groups	Relaxed, comfortable stretch; no stretching to the point of muscle spindle activation	Each stretch held for 30 s	3 to 5 d/wk	Stretch should always be relaxed and without discomfort	Increased range of motion	Mild progression is advised due to general deconditioning.

RM = repetition maximum; CRF = cardiorespiratory fitness; ACSM = American College of Sports Medicine
HRR = heart rate reserve

essary for productive cardiorespiratory adaptations to occur and depends on the patient's current level of fitness. Severely detrained individuals can improve while exercising at 40% to 50% of their $\dot{V}O_2$peak, while more active individuals may need to work at between 55% and 85% of their $\dot{V}O_2$peak. The most productive approach is to begin slowly and progress to the desired duration and intensity as the patient's CRF or tolerance of exercise improves. While the minimum duration of an exercise bout at any given intensity has yet to be determined, current recommendations necessitate at least 150 min/wk of moderate to vigorous physical activity (>50% of $\dot{V}O_2$peak) (47). This dose can be achieved with varying schedules of daily or three times weekly exercise. Improvements in functional aerobic capacity (FAC) and reductions in heart rate (HR) at absolute submaximal work rates (i.e., training-induced bradycardia) can be expected in 12 wk or less of training.

In populations that suffer from chronic morbidity, the incorporation of aerobic exercise training into the treatment or recovery plan provides substantial health benefits. Research has demonstrated improvements in both physiological and psychological functioning (5, 93, 96). For people struggling with conditions such as HIV infection, AIDS, cancer, and type 2 diabetes, endurance or aerobic training may attenuate many symptoms and ameliorate the side effects associated with the disease state and its treatment (26, 67, 95).

A number of disease-related and iatrogenically induced signs and symptoms associated with HIV infection and AIDS are amenable to improvement with exercise. One of the most commonly reported symptoms in HIV-infected populations is fatigue (52). Individuals with HIV are more sedentary than the general public (46), and fatigue may play a role. The etiology of HIV-induced sedentary living is not clear, but low levels of physical activity lead to decrements in physical functioning and a lower QOL (80). Nevertheless, exercise training can reverse the morbidity induced by sedentary living and a lack of physical activity. Numerous studies have shown significant benefits for HIV-infected individuals, including improvements in physiological parameters such as CRF, immune status, and metabolic activity, as well as in psychological variables like depressive symptoms and anxiety (46, 60, 62, 73, 90, 102, 103).

While the long-term benefits of chronic AT have not been fully investigated in this population, the aforementioned benefits have been observed in as little as 6 wk of training (46). Given that CVD is a significant contributor to morbidity and mortality in HIV-infected individuals, reductions in CVD risk factors may be the strongest argument for the inclusion of AT in the treatment plan.

Table 21.5 provides a summary of the research examining the physiological and psychological outcomes of aerobic exercise in HIV-infected populations.

Resistance Exercise

Progressive resistance training results in significant benefits for HIV-infected individuals. Improved body composition (e.g., increased lean body mass), strength, and muscular function are a few of the demonstrated benefits. Many patients find that their activities of daily living are more easily performed due to higher levels of functional strength. Resistance exercises should incorporate both the lower and upper extremities and focus on the larger muscle groups (e.g., hip and knee flexors). The exercises that may be performed are myriad, and the equipment available to the patient will be a limiting dimension (e.g., free weights, resistance bands, plate-loaded machines). The frequency of training should be 2 to 4 d per week and should allow for at least 1 d of recovery between sessions involving the same musculature.

Most research on the effects of exercise has concentrated on aerobic training or on aerobic training in combination with resistance training (i.e., concurrent training). The few studies that have examined resistance training as a stand-alone intervention have focused mainly on changes in strength and muscle mass rather than overall fitness or immune function. Nevertheless, these studies strongly suggest that PRT has substantial physiological and many potential psychological benefits for HIV-infected individuals.

The hypokinesis that is characteristic of individuals with HIV leads to a loss of muscle mass, strength, and function (41). PRT has been demonstrated to offset these decrements in HIV-infected individuals with and without wasting (87, 113, 115). The increases in lean body mass from PRT have been associated with a decrease in mortality (108) and may lead to improved psychological functioning through enhanced body image and a reduction in depressive symptoms and anxiety (29, 68).

Table 21.6 provides a summary of the research on the physiological outcomes of resistance training in HIV-infected populations.

Combined Aerobic and Resistance Training

A substantial evidence base supports the use of concurrent aerobic and resistance training (27, 37, 45, 50). The

Table 21.5 Exercise Training Review: Cardiovascular

Authors	Participants	Type and dose of intervention	Outcomes
Schlenzig et al., 1989	Total: 28 males Exercise (*n* = 15) Control (*n* = 13)	Aerobic exercise: 8 wk of two sessions per week for 60 min at moderate intensity	Significant reductions in anxiety and depression
LaPerriere et al., 1990	Total: 23 males	Aerobic exercise: 45 min on a stationary bicycle at 80% age-predicted HRmax	Significant reductions in anxiety and depression scores on the POMS*
MacArthur et al., 1993	Total: 25 At 24 wk: 6 High intensity (*n* = 3) Low intensity (*n* = 3)	Aerobic exercise: 24 wk of three sessions per week for 40 min at 50% to 60% $\dot{V}O_2max$ (low) or 24 min at 75% to 85% $\dot{V}O_2max$ (high)	Significant increases in $\dot{V}O_2max$, minute ventilation, and oxygen pulse for those compliant in both groups Significant reductions in perceived stress for both groups
Stringer et al., 1998	Total: 34 (F = 4, M = 30) High intensity (*n* = 9) Moderate intensity (*n* = 9) Control (*n* = 8)	Aerobic exercise: 6 wk of three sessions per week for 60 min at 60% $\dot{V}O_2max$ (moderate) or 40 min at 75% $\dot{V}O_2max$ (high)	Significant improvements in quality of life and aerobic fitness for both exercise groups
Perna et al., 1999	Total: 28 Exercise (*n* = 18) Control (*n* = 10)	Aerobic exercise: 12 wk of three sessions per week for 45 min at 70% to 80% HRmax	Significant increases in $\dot{V}O_2max$, oxygen pulse, and max tidal volume
Terry et al., 1999	Total: 21 (F = 17, M = 4) High intensity (*n* = 11) Moderate intensity (*n* = 10)	Aerobic exercise: 12 wk of three sessions per week for 30 min at 60% HRmax (moderate) or 84% HRmax (high)	Significant improvements in functional capacity in both groups
Smith et al., 2001	Total: 49 (F = 8, M = 41) Exercise (*n* = 19) Control (*n* = 30)	Supervised aerobic exercise: 12 wk of three sessions per week for 30 min at 60% to 80% $\dot{V}O_2max$	Exercise subjects increased time to fatigue; lost weight; and decreased BMI, subcutaneous fat, and abdominal girth
Thoni et al., 2002	Total: 17 (F = 5, M = 12) No control group	Aerobic exercise: 4 mo of two sessions per week for 45 min at a light intensity	Significant increases in $\dot{V}O_2max$ and HDL Significant decreases in total abdominal adipose tissue, total cholesterol, and triglycerides
Neidig et al., 2003	Total: 30 (F = 8, M = 22) Exercise (*n* = 30) Control (*n* = 30)	Aerobic exercise: 12 wk of three sessions per week for 60 min at 60% to 80% $\dot{V}O_2max$	Significant reductions in depressive symptoms and depressed mood
Terry et al., 2006	Total: 30 (F = 10, M = 20) Exercise (*n* = 15) Control (*n* = 15)	Aerobic exercise: 12 wk of three sessions per week for 30 min at 70% to 85% HRmax	Significant increases in $\dot{V}O_2max$ and decreased BMI, waist-to-hip ratio, body density, and body fat

HR = heart rate; M = male; F = female.

*POMS = Profile of Mood States: A 64-item questionnaire that measures six mood states: tension/anxiety, depression/dejection, anger, vigor, fatigue, and confusion.

Reprinted, by permission, from G.A. Hand et al., 2009, "Impact of aerobic and resistance exercise on the health of HIV-infected persons," *American Journal of Lifestyle Medicine* 3: 489-499.

Table 21.6 Exercise Training Review: Resistance

Authors	Participants	Type and dose of intervention	Outcomes
Bhasin et al., 2000	Total: 61 males Placebo, no exercise (n = 14) Testosterone, no exercise (n = 17) Placebo, exercise (n = 15) Testosterone, exercise (n = 15)	PRT: wk 1 to 4: three sets, 12 to 15 reps, 60% 1RM, three times a week wk 5 to 10: four sets; four to six reps; 90%, 80%, 70% 1RM; three times a week wk 11 to 16: five sets, four to six reps, ↑7% upper body and 12% lower body	↑ body weight: 2.2 kg exercise (EX) and 2.6 kg testosterone (TEST) ↑ max leg press, leg curl, bench, lats for EX, TEST, and EX+TEST ↑ thigh muscle volume: TEST and EX ↑ LBM: TEST and EX+TEST
Yarasheski et al., 2001	Total: 18 males	Resistance exercise: Three upper body and four lower body exercises done 1 to 1.5 h/d, four times a week, for 64 sessions Exercises not specifically identified	↑ whole-body lean mass (1.4 kg) ↑ thigh cross-sectional area (5-7 cm^2) ↑ strength (23-38%), all exercises ↓ fasting serum TG (281-204 $mg \cdot dl^{-1}$)
Roubenoff and Wilson, 2001	Total: 25 males and females Wasting (n = 6) Nonwasting (n = 19)	PRT: Three times a week for 8 wk: 50% 1RM at first session, 60% at second session, and 75% to 80% for remainder of sessions Three sets, eight reps: double-leg press, leg extension, seated chest press, seated row	↑ 1RM: 44% nonwasted and 60% wasting ↑ lean body mass: 2.3% nonwasting and 5.3% wasting ↑ physical function

PRT = progressive resistance training; RM = repetition maximum; LBM = lean body mass

Reprinted, by permission, from G.A. Hand et al., 2009, "Impact of aerobic and resistance exercise on the health of HIV-infected persons," *American Journal of Lifestyle Medicine* 3: 489-499.

advantage to combining AT and PRT is obvious in that HIV-infected individuals suffer from deconditioning and exhibit low CRF, low muscle strength and function, and altered body composition. Improvements in CRF from AT and the increased strength and muscle mass from PRT have beneficial effects on both physiological and psychological capacity and functioning (27, 37, 50). Additionally, the improvements in body composition (e.g., reduced visceral fat depots, increased lean tissue mass), metabolic function, blood lipids, and insulin resistance from AT and PRT are associated with reductions in CVD risk and all-cause mortality (6, 35). The literature strongly demonstrates that AT and PRT provide myriad benefits that lead to improvements in both overall health and QOL. Given the substantial evidence base from empirical research, the inclusion of both AT and PRT should be standard practice in HIV treatment unless contraindicated.

Table 21.7 is a summary of research examining the physiological and psychological outcomes of combined cardiovascular and resistance training in HIV-infected populations.

Range of Motion Exercise

Flexibility and range of motion in sedentary populations are often compromised. Enhanced joint mobility via flexibility training improves functional outcomes and patient QOL. The inclusion of flexibility and range of motion exercises is necessary to offset the detrimental effects of physical inactivity, and improvements in chronic flexibility can be achieved in as little as 12 wk (88).

Exercise Training and Immune Function

Early in the HIV epidemic, many health care professionals were concerned that exercise might further compromise the already impaired immune function of HIV-infected individuals. The potential immunosuppressive effects of high-intensity bouts of exercise prejudiced many physicians against exercise and led to its underuse as a preventive or therapeutic modality (9, 92). While HIV-seropositive individuals have decreased CD4+ counts and an impaired ability to mobilize neutrophils, natural killer

Table 21.7 Exercise Training Review: Combined Cardiovascular and Resistance

Authors	Participants	Type and dose of intervention	Outcomes
Grinspoon et al., 2000	Total: 43 males Testosterone, no exercise (n = 10) Testosterone, exercise (n = 11) Placebo, no exercise (n = 12) Placebo, exercise (n = 10)	PRT: three times a week for 12 wk wk 1 and 2: two sets at 60% 1RM wk 3 to 6: two sets at 70% 1RM wk 7 to 9: two sets at 70% and one set at 80% 1RM wk 10 to 12: three sets at 80% 1RM Aerobic: 20 min, three times a week, 60% to 70% HRmax	↑ muscle cross-sectional area in trained versus nontrained and in testosterone (TEST) versus placebo: arm and leg ↓ HDL: TEST versus placebo ↑ HDL: trained versus nontrained
Fairfield et al., 2001	Total: 43 males Testosterone, no exercise (n = 10) Testosterone, exercise (n = 11) Placebo, no exercise (n = 12) Placebo, exercise (n = 10)	PRT: three times a week for 12 wk wk 1 and 2: two sets at 60% 1RM wk 3 to 6: two sets at 70% 1RM wk 7 to 9: two sets at 70% and one set at 80% 1RM wk 10 to 12: three sets at 80% 1RM Aerobic: 30 min, three times a week, 60% to 70% HRmax	↑ muscle attenuation with testosterone and training (p = 0.03)
Fillipas et al., 2006	Total: 40 males Experimental (n = 20) Control (n = 20)	Aerobic: two times a week, 20 min, 60% to 75% HRmax PRT: multiple exercises, three sets, 10 reps, 60% to 80% 1RM	↑ self-efficacy ↑ cardiovascular fitness through reduced heart rate ↑ overall health ↑ cognitive function
Engelson et al., 2006	Total: 18 females	Diet: 5,024 kJ (1,200 kcal), hypoenergetic Exercise 90 min, three times a week, supervised: Aerobic: 30 min, 70% to 80% HRmax Resistance: 10 exercises, three sets, 8 to 10 reps	↓ daily food intake and body weight (95.5% fat) ↓ resting energy expenditure ↑ strength, fitness, and QOL ↔ CD4, viral load, fasting glucose, insulin, insulin sensitivity, fasting lipids, TPA, PAI-1
Dolan et al., 2007	Total: 40 females	16 wk home-based combined regimen: Aerobic: 20 min, 60%, in wk 1 and 2; 30 min, 75%, in wk 3 to 16 PRT, multiple exercises: wk 1 and 2: 60% 1RM, three sets, 10 reps wk 3 and 4: 70% 1RM, four sets, eight reps wk 5 to 16: 80% 1RM, four sets, eight reps	↑ $\dot{V}O_2$max, endurance ↑ strength: knee extension, bench press, knee flexion, shoulder abduction, ankle plantar flexion, and elbow flexion ↑ total muscle area and attenuation
Robinson et al., 2007	Total: 5 males and females	16 wk preceded by 2 wk phase-in period: Aerobic: 20 min, three times a week, 70% to 80% $\dot{V}O_2$max Resistance: two times a week, one set, 8 to 10 reps at 80% 1RM—lat, seated row, shoulder press, bench press, leg press, calf press, seated leg curl	↓ total and trunk fat mass
Hand et al., 2008	Total: 40 males Experimental (n = 20) Control (n = 20)	Resistance training: two times a week for 6 wk at 60% 3RM Aerobic: 30 min two times a week, approximately 60% $\dot{V}O_2$peak	↑ $\dot{V}O_2$max, endurance ↓ HR at submaximal absolute workload ↑ strength: knee extension, bench press, knee flexion, and elbow flexion ↑ total muscle mass and reduced fat mass

PRT = Progressive resistance training; ↑ = increase, ↓ = decrease, ↔ = no change.

Reprinted, by permission, from G.A. Hand et al., 2009, "Impact of aerobic and resistance exercise on the health of HIV-infected persons," *American Journal of Lifestyle Medicine* 3: 489-499.

(NK), and lymphokine-activated killer (LAK) cells into the circulation (92), exercise training does not appear to negatively affect immune function or disease progression (29, 30, 62, 67, 102). Conversely, exercise training potentially has a number of positive effects on immune status in that it attenuates psychological stress and reduces depressive symptoms, perceived stress, and symptoms of anxiety, as well as improving QOL (38, 62, 67, 73, 90, 99). There is some evidence that increasing levels of physical activity have direct beneficial effects on immune function or viral load in HIV-infected individuals (8, 62).

Table 21.8 provides a summary of research on the immunological outcomes of exercise training in HIV-infected populations.

Table 21.8 Exercise Training Review: Immunologic Outcomes

Authors	Participants	Type and dose of intervention	Outcomes
LaPerriere et al., 1990	Total: 23 males Exercise (*n* = 10) Control (*n* = 13)	Interval training: three sessions per week for 5 wk of stationary bike at 70% to 80% HRmax, 45 min per session	Control: 61 cells · mm^{-3} decrease in CD4+ count Exercise: 38 cells · mm^{-3} increase in CD4+ count
LaPerriere et al., 1991	Total: 39 males Exercise (*n* = 23) Control (*n* = 16)	Interval training: three sessions per week for 5 wk of stationary bike at 70% to 80% HRmax, 45 min per session	Significant increases in exercise group CD4+ cells and CD45RA+CD4+ cells
MacArthur et al., 1993	Total: 25 males 6 finished at 24 wk High intensity (*n* = 3) Low intensity (*n* = 3)	Aerobic training: 24 wk of three sessions per week for 40 min at 50% to 60% $\dot{V}O_2max$ (low) or 24 min at 75% to 85% $\dot{V}O_2max$ (high)	No significant change in CD4+ cells
LaPerriere et al., 1994	Total: 14 males Exercise (*n* = 7) Control (*n* = 7)	Aerobic training: 10 wk of three sessions per week on stationary bike for 45 min at 70% to 80% HRmax	Significant increases in CD2+, CD4+, CD8+, CD45RA+CD4+, and CD20+ cells
Stringer et al., 1998	Total: 34 (4 females, 30 males) High intensity (*n* = 9) Moderate (*n* = 9) Control (*n* = 8)	Aerobic training: 6 wk of three sessions per week for 60 min at 60% $\dot{V}O_2max$ (moderate) or 40 min at 75% $\dot{V}O_2max$ (high)	No significant changes in viral load or CD4+ cells in all groups
Terry et al., 1999	Total: 21 (17 females, 4 males) High intensity (*n* = 11) Moderate intensity (*n* = 10)	Aerobic training: 12 wk of three sessions per week for 30 min at 60% HRmax (moderate) or 84% HRmax (high)	No significant changes in CD4+ cells, CD8+ cells, leukocytes, or lymphocytes in either group
Terry et al., 2006	Total: 30 (10 females, 20 males) Exercise (*n* = 15) Control (*n* = 15)	Aerobic training: 12 wk of three sessions per week for 30 min at 70% to 85% HRmax	No significant changes in CD4+ cells or viral load in either group

HR = heart rate.

Reprinted, by permission, from G.A. Hand et al., 2009, "Impact of aerobic and resistance exercise on the health of HIV-infected persons," *American Journal of Lifestyle Medicine* 3: 489-499.

Practical Application 21.2

RESEARCH FOCUS

Currently a number of ongoing research projects are examining the effects of exercise on people living with AIDS (PLWHA). The major concerns when one is working with HIV-infected patients depend on the demographics and characteristics of the population. Prevalence rates and the vectors for new infections vary by geographic location. Specific to current research being conducted in the United States is a focus on participants who are primarily heterosexual African American females and males who have sex with men from low socioeconomic

(continued)

backgrounds. Challenges in work with this segment of the population include a lack of transportation, homelessness, and other concerns concomitant with poverty. Furthermore, there is a tremendous psychological burden due to the stigma of being HIV+. Depression and anxiety are common and influence compliance and adherence to pharmacological and lifestyle regimens. As with many facets of HIV infection, the immense psychological burden varies as a function of social support (e.g., familial or community based).

In addition to the environmental and personal challenges, the iatrogenic effects of antiretroviral medications (e.g., lipodystrophy, impaired glucose metabolism) often cause this population to respond differently to an exercise regimen than the general population and other clinical populations. Health status and the presence of opportunistic infections vary with the stage of illness and pharmacological regimens. Therefore, continuous monitoring of general health is critical. Many HIV+ individuals are severely deconditioned and are at increased risk for opportunistic infection. Therefore clinical exercise physiologists should emphasize to participants the importance of reporting feelings of excessive effort or exhaustion, shortness of breath, gastrointestinal complications (e.g., nausea, diarrhea), or difficulty completing everyday tasks. Furthermore, with an untrained individual who is HIV+, one should expect functional aerobic impairment (i.e., low CRF), increased circulating blood lipids, decreased muscle mass, abnormal fat displacement (i.e., lypodystrophy), and possibly increased blood pressure. Many participants present with $\dot{V}O_2$peak values well below 20 ml · kg^{-1} · min^{-1}. It is recommend that new exercise regimens be started below current recommendations. Even modest amounts of increased activity have significant health benefits. It is not uncommon to see dramatic increases in as little as 6 wk in severely deconditioned HIV+ persons.

A current ongoing study at the University of South Carolina examines the effectiveness of home-based exercise programs on risk factors for CVD. The sample population exhibits a number of the metabolic and cardiovascular complications induced via the confluence of sedentary lifestyles, HIV, and HAART (i.e., decreased functional aerobic impairment, obesity, abnormal fat distribution, increased blood lipids, insulin resistance). For example, 69% of the sample had a waist circumference >80 cm, with some >150 cm. Fasting glucose was >5.6 mmol · L^{-1} in 55% of the sample, and 41% exhibited prehypertension with blood pressures >130/85 mmHg. A number of participants were medically excluded due to diastolic pressures in excess of 130 mmHg.

It is well documented that AIDS-related infections are no longer the primary cause of death in this population. Mortality in PLWHA is now due to the same chronic diseases experienced in the general aged population. Fortunately, research at the University of South Carolina has demonstrated substantial benefits and dramatic reductions in CVD risk factors among PLWHA in as little as 6 wk. The prescribed regimen combines moderate-intensity aerobic and resistance training for 30 min a day, 2 d per week. This is substantially less than current public health recommendations. Nevertheless, because the population is sedentary and deconditioned, promising results have been achieved. However, the long-term health benefits of exercise with this population may only be speculated upon, since there is a dearth of research investigations lasting longer than 3 mo.

Clinical Exercise Pearls

1. Exercise is a complementary modality to prevent disease and mitigate the signs and symptoms associated with HIV and other chronic disease states.
2. Exercise interventions decrease morbidity and mortality and improve the QOL of those infected with HIV.
3. Exercise reduces insulin resistance, lipodystrophy, hyperlipidemia, hyperlipoproteinemia, and the cardiovascular risk factors associated with HIV and HAART.
4. Combined AT and PRT provide the greatest benefit with respect to health status, the reduction of CVD risk factors, and productive changes in body composition.
5. Exercise training has a number of positive effects on immune status via reductions in psychological stress, depressive symptoms, and symptoms of anxiety.

CONCLUSION

Exercise training is a safe and effective complementary modality for the treatment of HIV infection, AIDS, and the complications arising from pharmacological treatment. The scientific literature provides compelling evidence that exercise training confers numerous benefits for HIV-infected persons. With the advent of HAART regimens, HIV-infected individuals are living longer and more productive lives (45). Unfortunately, they also experience a large number of complications from the progression of the disease as well as the effects associated with antiretroviral therapy (e.g., increased risk of CVD). Research has demonstrated that both AT and PRT can attenuate, and in some cases reverse, these detrimental effects so that patients experience improvements in functional capacity (e.g., increased strength and CRF), psychological functioning, and QOL. As a group, HIV-infected persons are sedentary and as a result experience an increased risk of CVD, obesity, insulin resistance, and mood disorders. This makes exercise training a vitally important component of every patient's treatment plan in all stages of HIV infection and AIDS.

While some health care professionals still work under the erroneous assumption that exercise training may compromise immune function in HIV-infected persons, the scientific evidence strongly suggests otherwise. Therefore HIV-positive status should not be a deterrent to starting an exercise program but rather should provide the impetus to begin.

Key Terms

CASE STUDY

MEDICAL HISTORY AND PHYSICAL EXAMINATION

Mr. BP, a 42 yr old African American with a seropositive HIV status since 2005, was recently placed on HAART after being hospitalized with bouts of thrush. The HAART regimen consists of Truvada and Kaletra. Shortly after beginning his new pharmacological regimen, the patient began experiencing more fatigue, muscle weakness, and nausea. He came to you seeking help in gaining strength and energy to "get my life back."

DIAGNOSIS

At the time of hospital admission, the patient's weight was 187 lb (85 kg), and his blood pressure was 132/90 mmHg. His laboratory work revealed the following:

- 254 CD4+ cells · mm^{-3}
- 89,650 HIV RNA copies · ml^{-1}
- Total cholesterol = 201 mg · dl^{-1}
- High-density lipoprotein cholesterol = 33 mg · dl^{-1}
- Triglycerides = 140 mg · dl^{-1}

Six months after beginning HAART, his laboratory work was as follows:

- 420 CD4+ cells · mm^{-3}
- 1,009 HIV RNA copies · ml^{-1}
- Total cholesterol = 280 mg · dl^{-1}
- High-density lipoprotein cholesterol = 34 mg · dl^{-1}
- Triglycerides = 160 mg · dl^{-1}

(continued)

EXERCISE TESTING RESULTS

At the time of the exercise evaluation, Mr. BP's resting vitals were as follows:

- Height = 5 ft 10 in. (1.78 m)
- Weight = 170 lb (77.27 kg)
- BMI = 24.38 $kg \cdot m^{-2}$
- Heart rate = 72 $beats \cdot min^{-1}$
- Blood pressure = 146/98 mmHg
- Respiratory rate = 22/min

The graded exercise evaluation resulted in the following values:

- Peak HR = 184 $beats \cdot min^{-1}$
- Peak BP = 270/96 mmHg
- Peak respiratory rate = 44/min
- Peak RPE = 19 out of 20
- Peak oxygen consumption = 26 $ml \cdot min^{-1} \cdot kg^{-1}$

His resting ECG revealed normal sinus rhythm with nonspecific ST-segment changes. No abnormal ECG changes or arrhythmias were noted during exercise. Before beginning his exercise program, the patient was advised to revisit his primary care provider and have his blood pressure reevaluated. The physician added a diuretic to the patient's medical regime. When he returned to the exercise facility the next week, his resting blood pressure was 128/90 mmHg.

EXERCISE PRESCRIPTION

- Mode: stationary cycling or walking/jogging.
- Frequency: 3 to 5 d/wk.
- Duration: 30 min, not including warm-up and cool-down. Patient may have to begin at 10 min each day and build up to 30.
- Intensity: 40% heart rate reserve, eventually increasing to 60% heart rate reserve.
- Exercise progression: As cardiovascular and muscular adaptations occur in response to training, the workload should be adjusted to maintain the heart rate within the initially prescribed range. Additionally the patient should try increasing the duration by 5 min every other week.

DISCUSSION QUESTIONS

1. What type of exercise increases lean body mass in HIV-infected patients experiencing skeletal muscle wasting?
2. What are the essential elements in an exercise prescription for someone with HIV?
3. What are the common complications of HAART? How can exercise attenuate these effects?
4. What factors are responsible for low CRF in persons with HIV infection?
5. Give examples of AIDS-defining conditions. How might these affect exercise testing and prescription?
6. What factors may limit exercise capacity in HIV-infected individuals?
7. What are stages of HIV infection? How are they defined? What are the stage-specific implications for exercise testing and training?
8. What are the common misconceptions about working with persons infected with HIV? What is the risk of contracting HIV during an exercise evaluation?

9. Case study: Compare the CD4+ cell counts from the time of admission to 6 mo. What information does this provide regarding the efficacy of the drug regimen?
10. Case study: What are the patient's triglyceride levels? HDL (high-density lipoprotein) levels? How might the exercise prescription alter these markers over time?
11. Case study: Compare the exercise prescription with the current ACSM guidelines. Do they differ? If so, how?

PART VI

Disorders of the Bone and the Joints

Part VI contains three chapters specific to orthopedic and musculoskeletal conditions. These include a chapter on arthritis, a chapter on osteoporosis, and a chapter on low-back discomfort. These chapters were selected based on the large numbers of people living with these conditions. With the aging of the population both in the United States, as well as world-wide, one can only anticipate that these conditions will increase in prevalence. Certainly, the practicing clinical exercise physiologist is likely to deal with these conditions on an almost daily basis.

Chapter 22 focuses on arthritis. Arthritis currently afflicts up to 50 million people in the United States. It is the leading cause of disability in the elderly and costs to treat arthritis are substantial. Although arthritis predominantly afflicts older individuals it can also affect those who are middle-aged. This chapter provides a comprehensive review of the various types of arthritis (over 100) and the strategies of using exercise for the treatment of this condition.

Chapter 23 examines osteoporosis. Although one typically thinks of women when considering osteoporosis, these bone conditions afflict an increasing number of older men. This chapter reviews these conditions and emphasizes the importance of exercise training in the long-term prevention of osteoporosis. The chapter also provides information regarding the development and implementation of exercise-training programs for individuals with established osteoporosis.

Chapter 24 focuses on low-back pain. Issues of low-back pain tend to affect people at younger ages than does either arthritis or osteoporosis. Low-back pain is also the most common reason for work disability in those under age 45. The clinical exercise physiologist should understand that exercise can play an important role in both the prevention and the long-term treatment of low-back pain. Issues such as improved cardiovascular conditioning and abdominal strength and their relationship to restoration of function and reduction in disability are presented.

Chapter 22

Arthritis

Andrew B. Lemmey, PhD

Despite traditional beliefs that individuals with arthritis should avoid vigorous physical activity for fear of exacerbating joint damage, research findings and clinical experience overwhelmingly demonstrate that regular and appropriate exercise reduces disability and improves joint function, general health, and quality of life (QOL), without aggravating symptoms in these individuals (28, 40, 65, 66, 98).

While broadly considering the various forms of arthritis, this chapter focuses mostly on osteoarthritis (OA), rheumatoid arthritis (RA), ankylosing spondylitis (AS), and the effects and benefits of exercise.

DEFINITION

Arthritis is a generic term for conditions that involve inflammation of one or more joints. There are more than 100 different forms of arthritis, each characterized by varying degrees of joint damage, restriction of movement, functional limitation, and pain.

SCOPE

In the United States, chronic arthritis affects 22% (50 million) of adults (36), is the leading cause of disability, and annually costs the economy $128 billion (222). Globally, the economic burden of arthritis is estimated to account for the equivalent of 1% to 2.5% of the gross national product of Western nations. Prevalence is higher for women (26%) than men (18%) and increases with age. For those over 65, arthritis is the most prevalent medical condition in both women and men (24, 218), with 50% of this age group affected (36). Disturbingly, due to the aging of the population and the increasing incidence of obesity, arthritis is only going to increase as a problem; the projection is that by 2030, 67 million adult Americans will be afflicted with these conditions (96). Exercise is identified as one of the principal ways of reducing the incidence of arthritis and attenuating the associated disability and economic burden (27).

With regard to the specific forms of arthritis, OA is the most common form, affecting 12.5%, or 27 million adult Americans (88). Rheumatoid arthritis occurs in approximately 0.7% (1.5 million) of Americans (148). Other common forms of arthritis include gout (1.4%, 3.0 million) (88) and **spondylarthropathies** (0.6-2.4 million) such as AS, psoriatic arthritis (PsA), reactive arthritis, and enteropathic arthritis. Arthritis is also associated with connective tissue diseases such as systemic lupus erythematosus (SLE), dermatomyositis, and systemic sclerosis (SSc; scleroderma). Although arthritis is most prevalent in the aged, an estimated 294,000 American children (under 18 yr) (i.e., 1 in 250) have some form of the condition (184).

Arthritis adversely affects both physical and psychosocial functioning and is the leading cause of disability in later life (125). The effects of arthritis on physical function are described in detail later in this chapter. Often ignored, however, are the consequences of arthritis on social functioning, which can be dramatic; that is, 25% of people with arthritis never leave their home or do so only with help, 18% never participate in social activities (16), and these individuals report a significantly worse QOL as compared to those who do not have arthritis (141). Additionally, arthritis is strongly associated with major depression (attributable risk of 18.1%) (54), and 6.6% of individuals with arthritis report severe psychological stress (95). These factors, combined with pain, fatigue, and the increased energy cost of performing activities of daily living (ADLs) with increasing impairment (72, 159), contribute to prolonged physical inactivity. In turn, the inactivity negatively affects health by increasing the risks of cardiovascular disease, dyslipidemia, hypertension, diabetes, obesity, and osteoporosis (145).

PATHOPHYSIOLOGY

While the etiologies, presentation, and clinical manifestations of the various forms of arthritis are generally distinct, they share some common features that negatively affect the following:

- Exercise tolerance (68, 169)
- Muscle strength
- Aerobic capacity
- Range of movement (ROM)
- Biomechanical efficiency
- Proprioception (99)

Consequently, all forms of arthritis are characterized by functional limitation and disability.

Osteoarthritis involves degradation of joints, which usually develops gradually and particularly affects the articular cartilage and the subchondral bone. Initially, the cartilage becomes pitted, rough, and brittle. In response to this, and to reduce the load on the cartilage, the underlying bone thickens. As a consequence, the synovial membrane swells and increases production of synovial fluid, and the joint capsule and surrounding ligaments thicken and contract (29, 168). These adaptations lead to a narrowing of the joint space, which, in advanced OA, can result in loss of cartilage, bone rubbing on bone, and ligaments becoming strained and weakened. Symptomatically, the consequences of this are joint pain, stiffness, and deformity; and patients can present with **crepitus**, locking, **effusion**, and bony growths at the edge of joints that form bone spurs (**osteophytes**).

The most commonly affected joints are those of the hands, feet, and spine and the large weight-bearing joints (hips and knees). As also with the other forms of arthritis, the exact etiology of OA is not known, although, as with the other forms, there is a clear genetic contribution (9, 59, 134, 190). The condition is more prevalent in women, and its incidence for both sexes increases with aging. Additionally, obesity and joint injury or trauma are known to predispose an individual to OA (9, 29, 59, 180). As the joint depends on dynamic loading for maintenance of joint function and joint nutrition (87, 102), chronic insufficient loading is deleterious to joints. However, so too is chronic excessive loading, as evidenced by the strong association between obesity and OA of the weight-bearing joints (9, 29, 59). As a result of reduced movement secondary to joint pain and stiffness, local muscles atrophy and ligaments may become lax. Thus, muscle weakness and joint instability are well-recognized consequences of OA. This loss of strength, in turn, is a major contributor to the disability associated with OA (e.g., muscle weakness) and is the best-established correlate of lower limb functional limitation in people with knee OA (160). Osteoarthritis is also characterized by loss of joint ROM (91, 103, 160), which exacerbates the reduced function and the disability produced by pain and muscle weakness (204).

Rheumatoid arthritis is more prevalent in females than males, with an age of onset between 40 and 50 yr. It is a chronic **autoimmune** disorder, characterized by systemic inflammation and symmetrical polyarthritis, that affects various tissues and organs but principally targets **synovial joints**. The resultant inflammatory response in joints (**synovitis**) is a consequence of synovial cell hyperplasia, excessive production of synovial fluid, and the development of **pannus**. In time, synovitis results in erosion of articular cartilage and marginal bone, with subsequent joint destruction and **ankylosis**. The joints most typically affected are the small joints of the hands and feet, followed, in order, by the larger joints of the wrists, elbows, shoulders, and knees, although any joint with a synovial lining is susceptible. In addition to joint pain and stiffness, RA also has extra-articular effects. Some are specific to RA, such as rheumatoid nodules, while nonspecific effects include muscle loss, adiposity, fatigue, and exacerbated cardiovascular disease (CVD) and osteoporosis risk (14, 46, 107, 140, 149, 157, 181).

Like RA, ankylosing spondylitis is a chronic inflammatory arthritis and an autoimmune disease. In contrast to RA and OA, though, the condition occurs more commonly in males (3:1 occurrence ratio), and onset is usually in younger individuals (20-40 yr). The condition primarily affects the spine and sacroiliac joint. Initially, the ligaments of the lower spine become inflamed at the **entheses**. This process in turn stimulates bone to grow within the ligaments. Gradually these bony growths form bony bridges between adjacent vertebrae, which eventually can cause the vertebrae to fuse, with resultant low back pain and immobility. Additionally, most AS patients suffer synovitis in the larger peripheral joints (especially the hips and knees), and common extra-articular features include fatigue, eye inflammation (uveitis or iritis), and inflammatory bowel disease (IBD). Unlike individuals with other forms of arthritis, AS patients have traditionally been encouraged to be physically active, as back pain, which is often severe at rest, usually improves with physical activity. Also, in the past it was observed that spinal fusion was accelerated by bed rest and immobilization.

Traditionally, arthritis disease stages have been classified at three levels:

1. Acute: reversible signs and symptoms in the joint related to synovitis
2. Chronic: stable but irreversible structural damage brought on by the disease process
3. Chronic with acute exacerbation of joint symptoms: increased pain and decreased ROM and physical function

Each of these stages has disease-specific presentations, treatment considerations, and goals. Table 22.1 provides specifics about the stages of OA, RA, and AS.

CLINICAL CONSIDERATIONS

The various arthritides can be differentiated based on whether symptoms arise from the joint or from a periarticular location, the number of joints involved, their location, whether the distribution is symmetric or asymmetric, the **chronicity** of disease, and extra-articular features (20, 122). Pharmacological treatment of OA, RA, AS, and the other forms of arthritis varies from condition to condition, between individuals with the same condition, and even for the same individual over time. Despite this disparity, all these conditions should feature exercise in their routine management.

Signs and Symptoms

In the evaluation of individuals with musculoskeletal complaints, the history and physical examination are the most informative elements. Restricted movement of a joint and tenderness to palpation along the axis of joint movement are indicative of arthritis. This contrasts with tenderness around the joint, which is more indicative of periarticular soft tissue problems. The signs and symptoms of arthritis are as follows:

- Pain
- Stiffness
- Joint effusion
- Synovitis
- Deformity
- Crepitus

Joint pain can arise from pathological changes in the joint capsule and periarticular ligaments, **intraosseous** hypertension, muscle weakness, subchondral microfractures, **enthesopathy**, and **bursitis**, and may be exacerbated by psychosocial factors including depression (26). Pain does not emanate, however, from articular cartilage directly, as cartilage is **aneural**. Stiffness in people with inflammatory arthropathies, such as RA and AS, varies with disease activity (i.e., the severity of inflammation). The prognosis of recent-onset arthritis is aided by a determination of whether the duration of symptoms has exceeded 4 to 6 wk (97).

Table 22.1 Types of Arthritis, Stages, and Related Impairments

Type of arthritis	Disease stage	Related impairments
Osteoarthritis	Acute joint pain	Often insidious
Chronic radiographic joint disease	Chronic with exacerbation	Increased joint pain and swelling, muscle weakness, and exacerbated functional impairment
Rheumatoid arthritis	Acute disease in multiple joints with pain, limited range of motion, and worsened functional impairment	Joint stiffness, adverse body composition changes ("rheumatoid cachexia"; muscle loss and fat gain), muscle weakness, fatigue, and increased cardiovascular disease risk
Chronic irreversible joint deformity	Chronic with exacerbation	Increased joint pain and swelling, which may occur in one or more joints
Ankylosing spondylitis	Acute spinal pain and stiffness without significant decrease in mobility	Muscle loss, muscle weakness, and fatigue
Chronic spinal ankylosis predominant with decreased spinal and thoracic mobility	Chronic with exacerbation	Increased pain and stiffness of the back or peripheral joints

History and Physical Examination

The medical history is essential for determining the duration, location, extent, and severity of musculoskeletal symptoms. And because of the genetic component to OA, RA, and especially AS, identifying the presence of this condition in the family greatly assists diagnosis of specific conditions. Obtaining information on the current level of functioning and any previous or ongoing efforts at an exercise intervention, including any barriers or facilitators to exercise, is also very useful for designing appropriate exercise interventions. The physical examination provides the majority of the information required for establishing an appropriate diagnosis and for recording specific information about any abnormalities of joint ROM, alignment, or function. Additionally, the physical examination indicates the presence of extra-articular features (e.g., rheumatoid nodules in RA, eye disease in AS, and skin disease in psoriatic arthritis), which greatly facilitates correct diagnosis. The physical examination is also used to assess joints for the four cardinal signs of inflammation: redness, swelling, pain, and heat.

Diagnostic Testing

The American College of Rheumatology has developed diagnostic criteria for the classification of hip, knee, and hand OA, RA, and AS (2-5, 182). Table 22.2 summarizes these criteria.

Table 22.2 Distinguishing Characteristics and American College of Rheumatology (ACR) Diagnostic Criteria for Arthritis

Arthritis type	Distinguishing characteristics	Presentation	ACR criteria
OA	Joint pain Crepitus Gel phenomenon	Affects hands, hips, knees, and lumbar and cervical spine Pain worsens throughout the day Affects any traumatized joint	Knee clinical: Knee pain and three of the following: a. Age >50 yr b. Morning stiffness <30 min c. Crepitus d. Bony tenderness e. Bony enlargement f. No warmth Knee clinical and radiographic: Knee pain and one of the following: Clinical criterion a, b, or c (see above) Osteophytes on knee X-ray Knee clinical and laboratory: Knee pain and five of the following: Clinical criteria a through f (see above) ESR <44 mm · h^{-1} RF <1:40 Synovial fluid compatible with OA Hip combined clinical, laboratory, and radiographic: Hip pain and one of the following: ESR <20 mm · h^{-1} Osteophytes on hip X-ray Joint space narrowing on hip X-ray Hand clinical: Hand pain or stiffness and three of the following: Bony enlargement of two or more DIPs Bony enlargement of two or more of 2nd and 3rd DIPs, 2nd and 3rd PIPs, 1st CMC Fewer than three swollen MCPs Deformity of at least one of 2nd and 3rd DIPs, 2nd and 3rd PIPs, 1st CMC

Arthritis type	Distinguishing characteristics	Presentation	ACR criteria
RA	Hand pain Swelling Fatigue Prolonged morning stiffness	Affects wrists, MCPs, and PIPs Symmetric	A score of >6 out of 10 based on the following: a. Joint involvement[‡] Two to 10 large joints*: 1 One to three small joints** (with or without involvement of large joints): 2 Four to 10 small joints (with or without involvement of large joints): 3 More than 10 joints (at least one small joint): 5 b. Serology Low-positive RF or low-positive ACPA: 2 High-positive RF or high-positive ACPA: 3 c. Acute-phase reactants Abnormal[†] CRP or ESR: 1 d. Duration of symptoms ≥6 wk: 1
AS	Low back pain Low back stiffness	Early: discovertebral bone erosion Late: vertebral fusion and sacro-iliac joint fusion (via bone formation, ossification)	Clinical: Inflammatory back pain (improves with activity, associated with stiffness and worsening with inactivity) Limited lumbar spine mobility Limited chest expansion At least one of the preceding and at least one of the following: Radiographic Grade 2 bilateral sacroiliitis Grade 3 or 4 unilateral sacroiliitis

*Present for at least 6 wk.

Note. ACR criteria are based on references 1-4 and 191. OA = osteoarthritis; RA = rheumatoid arthritis; AS = ankylosing spondylitis; ESR = erythrocyte sedimentation rate; RF = rheumatoid factor; DIP = distal interphalangeal joint; PIP = proximal interphalangeal joint; CMC = carpometacarpal joint; MCP = metacarpophalangeal joint; MTP = metatarsophalangeal joint.

[‡]Joint involvement refers to any swollen or tender joint on examination; *"large joints" refers to shoulders, elbows, hips, knees, and ankles; **"small joints" refers to the MCPs, PIPs, 2nd through 5th metatarsophalangeal joints, thumb interphalangeal joints, and wrists; ACPA = anti-citrullinated protein antibody; [†]"abnormal" is determined by local laboratory standards.

Currently, although no definitive diagnostic tests or markers of arthritis exist, several specific serum and synovial fluid tests are available that greatly assist in differentiating the various arthritic conditions (187). In combination with joint imaging, these tests contribute to specifying the arthritis diagnosis. For RA, a test for the presence of anti-citrullinated protein antibodies (ACPA; also called anticyclic citrullinated protein antibodies [anti-CCP]) has recently become part of routine practice. This test has high specificity (95%) and sensitivity (68%) for RA and for the future development of RA (15). Another blood test for RA determines the presence of rheumatoid factor (RF), which is positive in about 80% of RA patients. As with ACPA/anti-CCP, however, RF positivity is not expressed by all people with RA; thus a negative test should not exclude RA as a possible diagnosis. For AS, since the HLA-B27 genotype is present in approximately 90% of patients, a positive test for this genotype assists the diagnosis. However, as only 5% of individuals with the HLA-B27 gene develop AS, a positive test for this marker alone is not sufficient to justify a diagnosis of AS. Estimation of disease activity for arthritic conditions is aided (and often defined) by nonspecific measures of systemic inflammation, such as the erythrocyte sedimentation rate (ESR) and C-reactive protein (CRP). These inflammatory markers are high in active RA and are typically normal or only marginally elevated when the disease is controlled. In contrast, while mild elevations in ESR and CRP are often apparent in active AS and severe OA (161, 164, 196), these markers can also be normal in these conditions despite a significant amount of inflammation.

When synovial fluid is aspirated (i.e., removed) to relieve joint swelling, synovial fluid analysis is also often helpful for determining the type of arthritis. For example, the leukocyte count typically increases in synovial fluid from a normal value of 500 cells · mm^{-3} to 2,000 cells · mm^{-3} in OA and to 5,000 to 15,000 cells · mm^{-3} in RA and AS. Additionally, macromolecules originating from joint structures and measured in blood, synovial fluid, or urine reflect arthritic processes taking place locally in the joint (187).

Joint imaging, including radiographs (X-rays), ultrasound (US), and magnetic resonance imaging (MRI), is routinely used to confirm a particular arthritis diagnosis; for example, involvement of the small joints of the hands and feet usually indicates RA, whereas skeletal changes in the lower back are suggestive of AS. The plain radiograph, which detects bony changes, is the traditional imaging modality. However, since radiological changes are usually evident only in established or advanced disease, MRI and US may be the preferred imaging format for suspected early-onset RA and AS. Typical radiologic features of OA include osteophyte formation, joint space narrowing, cysts, and **subchondral sclerosis**. For RA, typical features are joints that have developed erosions at the joint margins due to the invasion of synovial tissue at the intersection of cartilage and bone. For early AS, there is a characteristic squaring of the corners of the vertebrae. Later in the disease process there is evidence of bone formation, ossification, or thin vertically oriented outgrowths that bridge the disc space and limit spinal motion (189).

In summary, from the information gained from the patient's history, physical examination, serum and synovial fluid tests, and joint imaging, the clinician discerns patterns that are consistent with the diagnosis of a particular condition (97).

Exercise Testing

This section reviews special exercise testing considerations for the arthritic population. Modes and protocols may need adjusting depending on the level of disability or the negative physical effect of an individual's arthritis.

Cardiovascular

Individuals with RA have an increased risk of CVD, which largely accounts for their excess mortality (14). This heightened risk, which is largely attributed to the inflammatory burden of RA (107), is approximately double that of the general population (165), making RA comparable to diabetes as a CVD risk factor (157, 197, 215). In view of this increased risk, the European League Against Rheumatism (EULAR) task force has advocated that risk scores determined by CVD calculators such as the Framingham and the Systemic Coronary Risk Evaluation (SCORE) methods should be multiplied by 1.5 when RA patients fulfill two out of three of the following criteria: RA disease duration ≥10 yr, presence

Practical Application 22.1

CLIENT–CLINICIAN INTERACTION

As part of the exercise evaluation, individuals with arthritis should be screened for factors that will affect the exercise prescription. Suggestions for questions directed to the patient include the following:

1. Individual's age, level of function, medications, personal goals, and lifestyle
2. Names and contact information of health care providers including primary care physician, rheumatologist, orthopedist, clinical exercise physiologist, and physical therapist
3. Musculoskeletal disease diagnosis
4. Pattern of joint involvement: (a) **symmetric** or **asymmetric**, (b) upper or lower extremity involvement, (c) joints affected
5. Severity of disease activity (acute, chronic, or chronic with acute exacerbation)
6. Comorbidities: other medical conditions, including cardiovascular risk factors (hypertension, dyslipidemia, insulin resistance or diabetes, obesity), pulmonary disease, fibromyalgia, **Raynaud's phenomenon**, **Sjogren's syndrome**, osteoporosis
7. Surgical history including joint replacements
8. Previous treatment (whether successful or not)
9. Presence of fatigue
10. Adequacy of footwear

of extra-articular features, or RF or ACPA/anti-CCP positivity (165). As evidence is emerging that AS is also associated with increased CVD risk, EULAR recommends that this condition also be regarded as a CVD risk factor (165). Moreover, since people with arthritis tend to be more deconditioned than sedentary individuals without arthritis, the risk of CVD is further exacerbated (140). In those with a high CVD risk profile who wish to commence exercise training, a symptom-limited exercise test should be considered, both to screen for the presence of CVD and to assist with developing the exercise prescription (7). An exercise test is also commonly used to assess cardiovascular status for surgical risk before joint replacement. Joint symptoms and fatigue, however, may adversely affect performance on an exercise test and may prohibit maximal testing (21, 43).

Arthritis and musculoskeletal conditions are often listed as relative contraindications to graded exercise testing, but one study found that 95% of those with severe end-stage hip or knee arthritis attributable to either OA or RA were capable of performing a symptom-limited exercise test using cycle ergometry methods (167). The majority also achieved a respiratory exchange ratio (RER) greater than 1.0, indicating a metabolically maximal test. Approximately two-thirds of these subjects were capable of completing tests by pedaling with their legs, and the remainder performed the same task using their arms. However, should directly measuring maximal aerobic capacity prove difficult, assessment of peak $\dot{V}O_2$ or estimating $\dot{V}O_2max$ by submaximal tests provides a viable alternative for developing appropriate exercise prescriptions (7, 25, 147; chapter 5). For all testing procedures, standard contraindications for exercise testing should be adhered to (see chapter 5). Exercise testing procedures for people with arthritis are similar to protocols recommended for elderly and deconditioned people (7). These tests should have small incremental changes in workload (e.g., increments of 10 to 15 W/min on the cycle ergometer or the modified Naughton protocol with use of a treadmill) (25). Cycle ergometry is generally the preferred mode for testing because of the high frequency of lower extremity impairment in patients with arthritis. This is the case because it is nonweight bearing and requires less reliance on balance, and also because most individuals with arthritis are able to achieve maximal cardiovascular effort using cycle ergometry (167). However, equations for estimating $\dot{V}O_2$ by cycle ergometry have not been validated for arthritis patients. In the instance of severe lower extremity joint pain and limitation, and in those with significant deformities that contraindicate cycling (e.g., **varus** or **valgus** deformity of the knee, which might cause the leg to strike the flywheel), testing by arm ergometry may be necessary. In contrast, the treadmill may be used in those with minimal or no functional disability.

The following equation was developed to predict $\dot{V}O_2peak$ in seniors with knee OA or CVD: $\dot{V}O_2$ $(ml \cdot min^{-1} \cdot kg^{-1}) = 0.0698 \times speed\ (m \cdot min^{-1}) + 0.8147 \times grade\ (\%) \times speed\ (m \cdot min^{-1}) + 7.533\ ml \cdot min^{-1} \cdot kg^{-1}$. This equation, validated in both men and women, requires that participants use front handrails for support during the exercise test (25). This is an advantageous method of testing an arthritic individual with lower extremity disability in whom standard nonhand support methods of treadmill testing may be hazardous. Recommendations for cardiovascular exercise testing are outlined in table 22.3.

Musculoskeletal

Recommendations for musculoskeletal testing are outlined in table 22.4.

Flexibility

Recommendations for flexibility testing are outlined in table 22.4.

Treatment

A comprehensive treatment strategy for arthritis should strive to counteract inactivity, restore a healthier body composition (i.e., increase muscle mass and reduce fat mass), and reduce disability. This approach should include appropriate medication to control disease activity and minimize symptoms, as well as exercise.

In the past, the traditional standard of treatment for an arthritic joint was rest (30). Current practice has shifted toward joint mobilization because of conclusive evidence of the beneficial effects and safety of exercise for individuals with arthritis (1, 7, 8, 17, 23, 39, 40, 45, 65, 66, 78, 98, 117). As depicted in figure 22.1, exercise interventions can interact at each stage of arthritis pathology and can help to mitigate the effects of the disease process on physical function and disability, as well as negative changes in body composition (118, 127).

The implementation of exercise as part of a comprehensive therapeutic management strategy is a challenge. Physicians and allied health professionals rarely recommended exercise to their arthritis patients, and when they do they usually provide little instruction (50). And they mostly advocate low-intensity (e.g., ROM) rather than moderate- to high-intensity exercise despite the proven efficacy and safety of moderate- to high-intensity exercise and the repeatedly demonstrated ineffectiveness of low-intensity exercise (146). This general failing by clinicians

Table 22.3 Cardiovascular Exercise Testing in Arthritis

Mode	Protocol specifics	Clinical measures	Clinical implications	Special considerations
Use a treadmill for those with minimal to mild joint impairment.	Use protocols with small increment increases (i.e., modified Naughton or a ramp protocol) unless disease activity and severity are minimal.	• Assess type of arthritis, degree of activity, and impairment. • Assess comorbidities, past surgical and medical history.	Standard peak $\dot{V}O_2$ prediction equations may overestimate functional capacity as they were developed on healthy (nonarthritic) populations.	With patient using handrails for support, use equation[a] to predict $\dot{V}O_2$max.
Use cycle ergometry for those with mild to moderate lower extremity impairment.	Use protocols with small increment increases (i.e., 10-15 W/min) or ramping protocols.	• Assess type of arthritis, degree of activity and impairment. • Assess comorbidities, past surgical and medical history.	Standard peak $\dot{V}O_2$ prediction equations may overestimate functional capacity as they were developed on healthy (nonarthritic) populations	Additional investigations are needed to improve prediction of peak $\dot{V}O_2$.
Use arm ergometry for those with severe lower extremity impairment.	Use arm ergometry–specific protocols with small increment increases or ramping protocols.	• Assess type of arthritis, degree of activity and impairment. • Assess comorbidities, past surgical and medical history.	Standard peak $\dot{V}O_2$ prediction equations may overestimate functional capacity as they were developed on healthy (nonarthritic) populations	Additional investigations are needed to improve prediction of peak $\dot{V}O_2$. Consider submaximal testing in those with severe impairment.

[a]$\dot{V}O_2$ ($ml^{-1} \cdot min^{-1} \cdot kg$) = 0.0698 × speed ($m \cdot min^{-1}$) + 0.8147 × grade (%) × speed ($m \cdot min^{-1}$) + 7.533 $ml^{-1} \cdot min^{-1} \cdot kg$ (14).

to encourage and provide informed instruction on how best to exercise is a contributor to the physical inactivity that characterizes the arthritic patient (114). Additionally, people with arthritis are typically poorly motivated

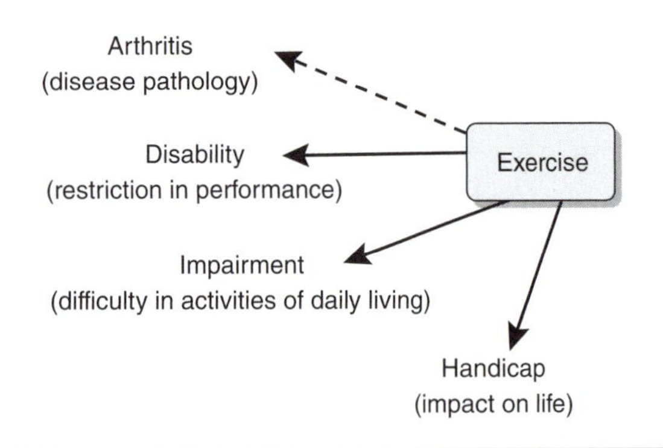

Figure 22.1 The World Health Organization classification of impairments, disabilities, and handicaps demonstrates the potential benefit of exercise as an interactive and mitigating factor in arthritis disease process. Strong evidence exists for benefits of exercise at the level of disability, impairment, and handicap with arthritis. While much remains to be learned about the effects of exercise on disease pathology, the overwhelming majority of studies show no worsening of arthritis with exercise, and quite a few suggest a beneficial effect on aspects of the disease.

Adapted from R. Shephard and P. Shek, 1997, "Autoimmune disorders, physical activity, and training, with particular reference to rheumatoid arthritis," *Exercise Immunology Review* 3:53-67.

to perform physical activity on their own. During leisure time, the period most amenable to efforts to increase physical activity, adults and children with arthritis are less active than healthy individuals (12, 108, 193, 194). For example, 69% of RA patients in the United States report doing no regular physical activity, and only 17% report achieving the recommended three or more bouts of physical activity per week (194). Additionally, despite the known benefits of community-based arthritis exercise programs such as the Arthritis Foundation YMCA Aquatics Program (AFYAP) and People with Arthritis Can Exercise (PACE), less than 1% of eligible persons with arthritis access these programs (96). Thus, individuals with arthritis require education and encouragement to increase and maintain appropriate physical activities.

In addition to exercise, multiple nonpharmacological interventions are used in the routine management and rehabilitation of people with chronic arthritis (42):

- Education
- Physical and occupational therapy
- Braces and bandages
- Canes and other walking aids
- Shoe modification and orthotics
- Ice and heat modalities
- Weight reduction

Table 22.4 Strength, Range of Motion, and Balance Testing in Arthritis

Test type	Mode	Protocol specifics	Clinical measures	Clinical implications*
Lower extremity	Dynamometer	All testing in supine position except knee flexion and extension (while seated)	Reference ranges for 50 to 79 yr olds (11)	Often (up to 50%) decreased in persons with arthritis
	30 s chair sit-to-stand test (30 s SST)	Number of stands completed in 30 s, without using arms, from a chair with a seat height of 17 in.	Reference ranges (175)	
	8RM (8-repetition maximum)	The maximum resistance that can be moved through the full range in a controlled manner for 8 repetitions (8RM; e.g., leg press, knee extension)	Reference ranges (7)	
Upper extremity and grip	Hydraulic dynamometer (Jamar)	In seated position with unsupported arm flexed 90° at elbow	Reference ranges for position 2 (129)	Often (up to 50%) decreased in persons with arthritis
	Electronic dynamometer (Grippit)	Peak grip force Average sustained force	Reference ranges (129, 153)	Usually (up to 90%) decreased in persons with hand arthritis
	30 s arm curl test	Total number of arm curls in 30 s with 5 lb dumbbell for women and 8 lb dumbbell for men	Reference ranges (175)	
	8RM	The maximum resistance that can be moved through the full range in a controlled manner for 8 repetitions (8RM; e.g., bench press)	Reference ranges (7)	Often (up to 50%) decreased in persons with arthritis
Range of motion	Goniometer	Align device fulcrum with joint fulcrum	Reference ranges	Usually (up to 90%) decreased in persons with arthritis
Balance	Figure-eight walking*	Useful in those with limited or mild impairments* Track width = 150 mm Inner diameter = 1.5 m Outer diameter = 1.8 m		
	Berg balance scale*	Useful in those with moderate to severe impairments* Includes 14 single tasks beginning with sitting unsupported and progressing to standing on one leg	More than two oversteps during two circuits suggestive of decreased balance Median ranges for RA functional classes (155)	Often (up to 50%) decreased in persons with arthritis

*Based on investigations in persons with RA (57, 155).

- Avoidance of repetitive-motion occupations
- Joint irrigation and joint surgery (in select circumstances)

Although joint surgery (arthroplasty) is usually performed only when conservative measures have failed, by 2006 more than 700,000 joint (knee and hip) replacements were being done every year in the United States (35, 110), 70% of which were for OA (61). Although joint replacement is very effective in resolving pain, physical function typically remains suboptimal 12 mo after surgery. This incomplete rehabilitation is generally attributed to postsurgical rehabilitative exercise that was inadequate (in terms of both volume and intensity) to restore muscle mass, strength, and aerobic capacity (170, 209).

Pharmacological therapies for arthritis vary according to the form of arthritis, and from individual to individual according to response. Unfortunately there are no effective cures for OA, and the mainstays of treatment remain lifestyle modification (exercise and weight loss) and analgesics and anti-inflammatories to reduce pain. Frequently used oral agents include nonopioid analgesics such as paracetamol and aspirin, as well as nonsteroidal anti-inflammatory drugs (NSAIDs) such as ibuprofen and naproxen. Topical agents include capsaicin cream and NSAID gels and creams, but the efficacy of these is unproven. Dosing of an oral NSAID 1 h before exercise may aid exercise compliance by reducing pain and stiffness during activity. Intra-articular injection of corticosteroid is commonly used and is effective in treating OA flares. Following injection, exercise should be avoided for 24 h, as should weight bearing when the injection is into weight-bearing joints; and it is generally advised that a patient receive no more than three intra-articular injections into the same joint during any 12 mo period. Glucosamine and chondroitin sulfate are naturally occurring constituents of **articular** cartilage. Data regarding the efficacy of taking oral supplements of glucosamine and chondroitin are conflicting (37, 90), but there are reports that this practice may reduce OA pain and slow progression of the disease, with minimal side effects (163, 173, 219). In persons with OA, being overweight is associated with increased pain and disability (10, 44, 60) and may worsen disease activity (56). Thus the inclusion of dietary intervention in combination with exercise is optimal when weight loss is a goal (139, 192).

In addition to the analgesics and anti-inflammatories mentioned, a variety of **disease-modifying antirheumatic drugs (DMARDs)** are available for treating RA and AS. For the autoimmune inflammatory arthropathies, such as RA and AS, the clinician aims to control disease activity by suppressing immune function. Currently, the favored DMARDs for RA include oral corticosteroids, methotrexate, leflunomide, sulfasalazine, hydroxychloroquine, and **biologics** such as TNF-α inhibitors (etanercept, infliximab, adalimumab, certolizumab pegol, and golimumab), anti-T cell (anti-CD28) therapy (abatacept), anti-B cell therapy (rituximab), anti-interleukin-1 (anti-IL-1) therapy (anakinra), and anti-interleukin-6 receptor (anti-IL-6R) agents (tocilizumab). Additionally, disease flares are often treated by intravenous or intra-articular corticosteroid injections. For AS, the DMARDs are oral corticosteroids, methotrexate, sulfasalazine, and cyclosporin, and the biologics are the anti-TNFs—etanercept, infliximab, adalimumab, certolizumab pegol, and golimumab.

Exercise is recommended in U.S. (6, 7, 92, 93) and European (39, 104, 223, 224) treatment guidelines for OA, RA, and AS due to its numerous benefits. These include increased aerobic capacity, muscle strength, and flexibility; improved physical function, weight control, self-efficacy, and mood; enhanced QOL; reduced pain; and decreased CVD risk (17, 23, 49, 89, 98, 101, 149, 179). It is important to point out that contrary to popular belief, exercise does not increase risk for OA, nor does it exacerbate joint damage in RA patients. These points are discussed in detail later in the chapter.

Table 22.5 provides a review of common arthritis medications.

EXERCISE PRESCRIPTION

The following are goals of exercise specific for the treatment of arthritis:

- Maintain or improve physical function by maintaining or improving muscle strength, cardiovascular fitness, and ROM
- Improve body composition (i.e., restore muscle mass and reduce fat mass) and, when appropriate, reduce body weight
- Reduce the risk of comorbidities such as CVD and osteoporosis
- Reduce inflammation and pain
- Prevent contractures and deformities

Immobilization and inactivity amplify the negative systemic and psychological manifestations that accompany arthritis (67). The effects of inactivity include rapid reductions in strength (about 3-8% per week) and aerobic capacity, muscle loss, reduced bone mass, and loss of cartilage matrix components (22, 105, 113). Because cartilage is **avascular**, it depends on normal repetitive loading of the joint for its nutrition and normal physiological function (94). Moreover, joints with effusions may develop synovial ischemia attributable to elevated intra-articular pressure. Walking and cycling increase synovial blood flow in inflamed knees (102). And in both healthy joints and those affected by OA, the intra-articular oxygen partial pressure increases during joint movement, although the increases are less in arthritic joints (142).

Supervised exercise training for individuals with arthritis most often occurs in a group setting (vs. personal training), and this is well supported as a treatment modality (38, 56, 85, 128, 146, 149, 211). A person in an exercise program, whether supervised or unsupervised, requires education, skill acquisition and reinforcement, and regular monitoring by health professionals (128). A supervised group setting may be beneficial because it provides peer social support in a positive environment and facilitates training compliance. Access to a clini-

Table 22.5 Pharmacology

Medication name and class	Primary effects	Exercise effects	Special considerations
OSTEOARTHRITIS			
Analgesics (e.g., nonopioid analgesics such as paracetamol, aspirin)	Reduce pain	Not reported	Allergy (hypersensitivity) to aspirin is relatively common (e.g., gastrointestinal [GI] irritation, bronchospasm).
Nonsteroidal anti-inflammatory drugs (NSAIDs; e.g., diclofenac, ibuprofen, naproxen, and selective COX-2 inhibitors such as celecoxib and etoricoxib)	Reduce inflammation by inhibiting the enzyme cyclooxygenase, which in turn reduces prostaglandin synthesis	Not reported	NSAIDs are associated with GI irritation, nausea, diarrhea, and occasionally ulceration. Consequently, coincident use of proton pump inhibitors is advised, and should be used with caution in the elderly. COX-2 inhibitors have reduced GI intolerance, but are associated with increased risk of thrombotic events (e.g., MI, stroke).
Intra-articular corticosteroid injection	Reduce inflammation and consequently improve pain and mobility	No exercise should be performed for 24 h following injection	Each joint should receive no more than three injections per year.
Glucosamine	Thought to stimulate cartilage growth, but the actual mechanism of action is unknown Has been shown to provide mild symptomatic relief	Not reported	GI side effects are occasionally reported.
Topical capsaicin	Reduces symptoms in mild hand and knee OA by counterirritation (i.e., pain in joint is relieved by irritation of the skin)	Not reported	Is associated with a burning sensation.
RHEUMATOID ARTHRITIS AND ANKYLOSING SPONDYLITIS			
NSAIDs	See above	See above	See above
Disease-modifying antirheumatic drugs (DMARDs; e.g., methotrexate, azathioprine, leflunomide, sulfasalazine, hydroxychloroquine) and the biologics (e.g., etanercept, infliximab, adalimumab, golimumab, certolizumab pegol, abatacept, rituximab, tocilizumab, and anakinra)	DMARDs are immunosuppressants and thus reduce inflammation. The biologics are specific cytokine or T cell or B cell modulators	Not reported	Immunosuppressants reduce the body's response to infection. Additionally each of the NSAIDs has recognized side effects and toxicities. Thus, careful monitoring is essential.
Systemic corticosteroids	Immunosuppression	Not reported	In addition to reduced response to infection, chronic corticosteroid use is associated with osteoporosis and, in high doses (>7.5 mg · d^{-1}), muscle loss.
Intra-articular corticosteroid injection	See above	See above	See above

cal exercise physiologist who has previously evaluated the individual may help decrease anxiety and improve compliance (200). Additionally, the clinical exercise physiologist may play a key role by encouraging regular attendance at sessions early in the treatment regimen, because session attendance in the initiation phase is the strongest predictor of session attendance in the later stages of the intervention (213). These are key considerations, as compliance is associated with training response, including improved physical function (128,

214). Regular monitoring by clinical exercise professionals also helps to ensure safety and appropriate progression of the exercise.

Supplementing supervised classes with home-based exercise may boost improvements in pain and function (40, 51, 133, 177). In addition, participants who do exercise partly at home during the maintenance stage of the intervention are more likely to adhere to the exercise program than peers who exercise only at a facility with supervision (213). Thus, a component of unsupervised and independent exercise appears to be important in modifying an individual's behavior so that regular exercise becomes a part of daily routine.

Many exercise options are available for unsupervised programs. These include land-based cardiovascular and strength training programs as well as independent water exercise programs, including water walking and joint ROM exercises. Videotapes of exercise programs suited to people with arthritis are available through the Arthritis Foundation (www.arthritis.org). Other activities and exercises that can be performed throughout the day include chin tucks, corner pectoral stretches in a doorway, abdominal tightening, checking posture throughout the day in the mirror, and extending walking time by taking the stairs or parking farther from a destination. For individuals with RA and severe morning stiffness, active ROM exercises performed at night within 15 min of going to bed can decrease morning stiffness (32). Walking is an excellent option for cardiovascular exercise. Performing a variety of cardiovascular exercises may help maintain interest and compliance and reduces the likelihood of overuse injuries. Range of motion exercises in a pool can be performed as a component of a supervised or unsupervised exercise program. Following the Arthritis Foundation's recreational aquatic program (AFYAP), available nationally in YMCAs and in private facilities, increases hip ROM, isometric strength, and flexibility when performed two or three times a week over 6 to 8 wk (206, 207). The AFYAP exercises, however, may not be of sufficient intensity to increase strength and ROM in joints not affected by arthritis (207) and should be supplemented by more intense strength training and active joint ROM activity performed on land.

Specific Exercise Prescription Considerations

In establishing an exercise prescription, the exercise professional should consider the success of various treatment modalities for particular joint impairments, as well as the individual's affected areas, level of fitness, surgical history, comorbidities, medications, age, personal goals, and lifestyle. Inflammation and joint degeneration associated with the disease process cause a cycle of decreasing function and increasing impairment. It is not yet clear whether therapeutic exercise alters the pathological process of arthritides, although it is clear that exercise—even long-term, high-intensity training—does not exacerbate disease activity; and reports of reductions in pain, inflammation, and even reduced joint damage are abundant in the literature.

Exercise can occasionally induce muscle and joint pain, which is the primary complaint and disabling factor related to arthritis. If this occurs, it can become difficult to motivate a person to maintain an exercise program. In an arthritic population that is generally older, is typically sedentary, and may be using systemic corticosteroids, avoidance of injury and pain is important to maintaining exercise compliance (18). The clinical exercise physiologist's role is to make recommendations that minimize these symptoms. For example, deconditioned patients should commence training programs at a low intensity and progress gradually to reduce the likelihood of muscle soreness (117). Additionally, attention to affected joints is important because of their decreased ROM, instability, reduced muscle strength and flexibility, poor joint proprioception, and increased pain. Consequently, the clinical exercise physiologist should ask the patient which joint sites are affected. This record may be refined with information from the individual's rheumatologist, primary care physician, or physical therapist. Some site-specific exercise recommendations are listed in table 22.6.

Impaired balance and increased fatigue are additional factors that must be considered in developing an exercise program. To increase or maintain patient motivation to commence or continue exercising, it is also essential that patients be aware that reduction in arthritis pain is an anticipated consequence of regular exercise (17, 89, 98, 101), and that improved strength and aerobic capacity will reduce the difficulty, and consequent fatigue, of performing ADLs.

Disease Staging

A consideration in developing an appropriate exercise prescription for a person with arthritis is the disease stage. The focus of exercise therapy for chronic stages of arthritis should be to maintain or improve function while minimizing or avoiding exacerbations. Most people referred for exercise therapy are experiencing arthritis in a chronic stage. Table 22.7 lists the arthritis stages and their associated exercise-related particulars. Table 22.8 provides specific recommendations for the clinical exercise professional to follow when developing an exercise prescription for arthritic patients with disease-specific skeletal conditions.

Table 22.6 Arthritis Site-Specific Recommendations

Site	Condition	Presentation	Recommendation
LOWER EXTREMITY			
Foot	Hallux valgus	Lateral deviation of the first digit of the foot leading to bursitis at the metatarsophalangeal joint (bunion)	Use padding between first and second toe. Use prescribed footwear that decreases need for extension of the joint during walking. If moderate to severe, avoid walking for exercise.
Foot	Plantar fasciitis or heel spur	Sharp foot pain on weight bearing, especially in heel; worse with first steps upon wakening and with walking	Avoid weight-bearing activities that increase pain. Wear shoes with supportive arch; may need soft or semirigid foot orthoses or heel cups. Stretch calf muscles throughout the day. Wear night splints to maintain stretch in dorsiflexion.
Leg	Musculotendinous inflammation of the anterior or posterior tibialis muscle (shinsplints) or compartment syndrome	Aching pain in anterior or lateral or posterior medial leg	Avoid painful activities (usually walking, and especially stair climbing) for up to 2 wk. Improved footwear or orthotics may be needed. Evaluate training routine for sudden change that preceded onset. Acute onset with excruciating pain and area hard to the touch requires immediate medical attention—send to emergency room.
Hip or knee	Hip OA, knee OA	Gait deviation Pain with weight bearing or stair climbing	Patellofemoral OA, hip or groin pain without gait deviation or balance problems: rearward walking. Patellofemoral OA: leg press; no leg extension. Hip or groin pain: hip bridging, free-speed walking, stationary cycling. If gait deviation is caused by pain or decreased joint range of motion (ROM), use of cane or rolling walker in the hand opposite the affected limb may be necessary.
Hip	Bursitis	Lateral hip pain, may extend to lateral knee Lying or single-leg stance on affected side not tolerated	Control inflammation—ice lateral hip. Patient should see doctor to rule out other problems and for medication to control inflammation. Patient should see physical therapist.
Knee	Valgus or varus deformity	Valgus deformity often called knock-knees; varus deformity often called bow-legged	Avoid weight-bearing exercise. Perform open-chain strengthening. Prescription for wedged insoles may be needed.
UPPER EXTREMITY			
Shoulder	Shoulder pathology, possibly adhesive capsulitis, tendinitis, or bursitis; may be secondary to OA	Pain with overhead or end-range motion	Perform shoulder ROM in pain-free range in pool with upper extremity submerged. If pain disturbs function or sleep, patient should see doctor. Physical therapy would be beneficial.
Hand	OA of carpometacarpal joint of the thumb	Pain in hand proximal to thumb	Avoid gripping activity during exercise. Enlarge grips.
Hand	Ulnar deviation in RA	Deviation of body of hand and fingers to the small-digit side of the hand	Avoid gripping activity during exercise. Use large muscles and joints for functional activities.

(continued)

Table 22.6 *(continued)*

Site	Condition	Presentation	Recommendation
		AXIAL SKELETON	
Lumbar spine	Spinal stenosis	Flexed low back during walking, standing, and sitting Symptoms increased with extension (standing, looking overhead) Often presents with claudication-type pain with walking	Flexion exercises and seated cardiovascular exercise (recumbent bicycle or stair stepper) can be tolerated. Perform aquatic exercise. Use rolling walker for household or community ambulation.
Cervical spine	Atlantoaxial subluxation in patients with RA	Facial sensory loss Vertigo with cervical extension Numbness or tingling of hands or feet Difficulty walking Loss of control of bowel or bladder Transient loss of consciousness with extension of cervical spine May be asymptomatic	Avoid any passive or heavy resistive neck ROM. Immobilization is needed if unstable (many symptoms). Surgery is indicated if neurological signs progress. Perform gentle active ROM, low repetitions, no extension.
Cervical or lumbar spine	Nerve compression secondary to OA	Gradual, recurrent pain or pain after activity Numbness, tingling, or pain in the extremities, sometimes only with certain movements	Avoid activity that results in numbness, tingling, or pain in the extremities.

Other Factors

Compliance with an exercise program can be a challenge for people with arthritis. Efforts on the part of the clinical exercise professional to prevent musculoskeletal injury from exercise and to appreciate the specialized concerns of the individual with arthritis will facilitate overall enjoyment and compliance with the exercise program. Some special considerations for people with arthritis are discussed next.

Preventing Musculoskeletal Injury Secondary to Exercise Because cardiovascular exercise involves high repetition of joint motion, risk of overuse injuries is present. Fortunately, injuries attributable to supervised exercise are infrequent. Estimates are that 2.2 minor injuries occur per 1,000 h of exercise, and that major injuries (i.e., those necessitating a reduction or discontinuation of exercise for at least 1 wk) occur at a rate of only 0.48 per 1,000 h of exercise (38). People with arthritis may minimize overuse of the soft tissue and bone by performing interval or cross-training during endurance exercise. Examples include the following:

1. Alternating cycle ergometry between 25% and 75% of the maximum work rate performed during a graded exercise test
2. Alternating between water walking and joint ROM exercises in a pool setting
3. Walking and weight training
4. Walking and higher-intensity cardiovascular exercise such as recumbent stair stepping or cycling

The development of strong knee extensors with quadriceps-strengthening exercise decreases impulse

Table 22.7 Arthritis Stages, General Signs and Symptoms, and Exercise Prescription-Related Considerations

Stage	Signs and symptoms	Exercise considerations
Acute	Fatigue Joint pain Reduced joint tissue tensile strength attributable to inflammation Reduced joint nutrition	Avoid activities that exacerbate joint pain.
Chronic	Permanent joint damage Pain at end of normal ROM Stiffness after rest Poor posture and ROM Joint deformities Pain with weight bearing Abnormal gait Weakness Contractures or adhesions Reduced aerobic endurance	Perform aerobic, strengthening, and ROM exercises. Perform exercises and intensities during resistance training that don't cause joint pain, but are still sufficient to ensure gains in strength and muscle mass. Initiate walking and perform in water if necessary to reduce pain. Perform low back flexion and abdominal strengthening exercise. Avoid trunk extension (especially with spinal stenosis). Maintain neutral spine position. To reduce risk of osteoporosis and ligament laxity, avoid long-term oral corticosteroids.
Chronic with acute exacerbation	Inflammation and joint size greater than normal Joint tenderness, warmth, swelling Joint pain at rest and with motion Stiffness Functional limitations Hips and spine affected	Normalize gait. Same recommendations as for acute phase.

Note. ROM = range of motion.

loading of the lower limb during walking by slowing the deceleration phase that occurs before heel strike. Therefore, adequate quadriceps strength may help prevent knee injury and slow the progression of knee OA.

People may have laxity in the structures that support a joint because of the rheumatic process or because of the use of corticosteroids. Under these circumstances, the joint should be protected when the person performs exercise or normal activities. Cautiously stretching to avoid extending beyond the functional ROM affords protection. Vigorous stretching or manipulative techniques are contraindicated (106). In many cases, providing external support to a joint may be necessary. To protect smaller joints during activities and exercises, larger muscles and joints should be used.

Fatigue Fatigue is common in those with rheumatic disease and can profoundly affect QOL. Fatigue beginning in the afternoon and lasting until evening, and morning stiffness lasting for 3 or 4 h, are common symptoms in people with inflammatory arthropathies, leaving only a few hours during midday when stiffness or fatigue is not a problem. Fatigue is a complex phenomenon related to exertion, deconditioning, depression, or a combination of these factors (22). A person who is fatigued is less able to exercise for a long duration, tends to be less motivated, and may become frustrated. Although relatively few studies have investigated the effects of exercise training on fatigue levels in arthritis patients, there is evidence to support a benefit (85, 149). And this is consistent with expectations that increased aerobic capacity and strength would allow performance of ADLs at a lower percentage of functional capacity, with less resultant fatigue.

Previous Joint Replacement Total joint replacement surgery is common in people with arthritis (110). Since individuals who undergo total knee or hip arthroplasty are typically deconditioned and overweight, exercise training is essential if they are to retain physical function and reduce the risk of comorbid conditions (170, 209). Appropriate exercise rehabilitation programs should be commenced soon after surgery, as this has been shown to improve strength and function, elicit muscle hypertrophy, reduce pain and stiffness, and enhance

Table 22.8 Exercise Prescription Review

Training method	Mode	Intensity	Frequency	Duration	Progression	Goals	Special considerations
Cardiovascular	Activities using large muscle groups with repetitive motion: walking, cycling, dancing, aquatics	30% to 75% HRR; 50% to 75% $\dot{V}O_2max$; RPE 12 to 16 (Borg 6 to 20 scale) or 3 to 6 (Borg 1 to 10 scale)	Begin at 2 or 3 d/wk and increase to 5 to 7 d/wk, more frequently for lower-intensity activities	20 to 60 min	Gradually increase duration and frequency of exercise, then increase intensity; begin at 60% MHR and progress up to 85% MHR	Improve cardiovascular fitness, function, and quality of life and reduce pain Reduce fat mass and, when appropriate, body weight	See text regarding the following: Injury Fatigue Joint replacement Time of day Aquatic therapy Footwear Cardiovascular and pulmonary manifestations Corticosteroids Body composition
Resistance	Isotonic	Begin at 30% to 60% 1RM and progress to 80% 1RM	2 or 3 d/wk, with at least 24 h between each session	8 to 10 exercises, 8 to 12 repetitions, one to three sets	Gradually increase the volume of training, then increase the intensity	Improve strength and function and reduce pain Improve body composition (increase muscle mass and reduce fat mass)	Perform in pain-free range. Whenever possible, use functional movement patterns. Include all major muscle groups, even with unilaterally distributed joint dysfunction.
Range of motion	Static and active stretching[a] (see chapter 5) Functional activities, for example sit to stand, stairs	To mild discomfort	Daily	10 to 30 s, three to five repetitions	Gradually increase ROM of stretching exercise Gradually increase volume and intensity of functional activities (e.g., sit to stand: reduce, then eliminate use of arms)	Improve range of motion and function, prevent deformity, and reduce pain	Target shortened muscle groups. Perform active ROM within the normal range for all joints, even with unilaterally distributed joint dysfunction.

Note. ROM = range of motion; HRR = heart rate reserve; RPE = rating of perceived exertion; OA = osteoarthritis; RA = rheumatoid arthritis; RM = repetition maximum; MHR = maximum heart rate.

[a]Passive ROM involves no muscle work by the individual while an outside force (another person or a passive motion machine) moves the body part through a range of motion. Active range of motion is movement of a body part by the individual performing the exercise without outside forces. Active assisted ROM involves partial assistance with motion by an outside force, whereby a portion of the motion of the limb may be provided by a mechanical device, another limb, or another person.

QOL (100, 116, 205). Individuals with lower extremity joint replacement should avoid high-impact activity. And those who have had hip replacements should not flex their hip past 90° or adduct or internally rotate the hip past neutral in the initial postoperative months.

Time of Day Morning stiffness is a problem for individuals with arthritis. The clinical exercise professional must be sensitive to the daily variability of symptoms, the difficulty of arising and performing ADLs, and the consequent difficulty of early morning activity. Moreover, a change in ability to perform exercise during periods of inclement weather is frequently reported by individuals with rheumatologic conditions. A drop in barometric pressure along with an increase in humidity can increase pain, as can cold conditions. For those with inflammatory arthritis characterized by prolonged morning stiffness, exercise should be prescribed for the late morning or early afternoon (67).

Aquatic Therapy Rheumatoid arthritis is commonly associated with Raynaud's phenomenon and Sjogren's syndrome. Raynaud's phenomenon is a **vasospastic** problem presenting as blanching or cyanosis of the hands and feet when exposed to cold or emotional stress. It can result in pitting scars and, in extreme cases, gangrene. Individuals with Raynaud's phenomenon should avoid cool air and water and should wear protective clothing including noncotton gloves, shirt, pants, and shoes. The choice of exercise modality should be dictated by the symptoms attributable to arthritis and Raynaud's phenomenon. Sjogren's syndrome is an autoimmune condition characterized by dry mouth and eyes caused by lymphocyte infiltration of salivary and lacrimal glands. People with this condition may find chlorinated water and the air surrounding pools especially irritating to the eyes, and should always wear goggles when in a pool.

Footwear Use of appropriate footwear can reduce the risk of injury related to poor lower extremity mechanics and repeated shock. Lightweight commercial athletic shoes that include hindfoot control, a supportive midsole of shock-absorbing materials, a continuous sole, and forefoot flexibility can improve shock attenuation and biomechanics. Individuals with OA, RA, and AS with biomechanical faults in the lower extremity may need custom-made rigid or semirigid orthotics from a podiatrist, orthotist, or physical therapist. People with lower extremity arthritis should be advised to wear pool shoes to assist mobility and protect feet from injury during aquatic exercise.

Cardiovascular and Pulmonary Manifestations of Rheumatic Disease Cardiovascular disease risk is exacerbated in RA and AS (107, 165). This increased risk is estimated to be 1.5 to 2-fold relative to that in the general population, making inflammatory arthritis comparable to diabetes as a CVD risk factor (157). Pulmonary disease can also be associated with arthritic conditions. For example, RA is associated with interstitial lung disease (10), and those with AS often have restrictive lung disease, caused by impairment of chest expansion, and upper lobe bilateral pulmonary fibrosis. In most cases, cardiovascular and pulmonary manifestations of rheumatologic disease are not contraindications to exercise, although a vital capacity of 1 L or less should be a relative contraindication to participating in pool therapy because of the restrictive effects of hydrostatic pressure on the chest wall. Chapters 12 and 13 in part III of this book review exercise for patients with CVD, and chapters 17 and 18 in part IV review exercise for those with pulmonary limitations.

Ankylosing Spondylitis With AS, the bony fusion that occurs in the spine and sacroiliac joint cannot be prevented, but rehabilitative strategies can improve spinal mobility and physical function (40). Since individuals with AS tend to be younger and more active at diagnosis, exercise for AS can generally be started at a higher relative intensity than for RA and OA (63, 171). When peripheral joints are involved, disease pathology is similar to that for RA, and thus exercise recommendations specific for RA should apply. A phenomenon that can occur is called the last-joint syndrome. With this, bridging ossification between vertebral bodies occurs at every level except one (86). This sole mobile segment is exposed to considerable stresses during exercise and can present with localized pain and **discitis**. In this circumstance, bracing or surgical fusion may be necessary.

Corticosteroids Systemic corticosteroids are a common treatment for RA, and chronic use is associated with bone loss and muscle atrophy. Muscle wasting from steroid-induced myopathy has been shown to contribute to reduced muscle strength in patients with RA (43).

Body Composition Obesity is a modifiable factor that negatively affects various arthritides. Obesity is also a strong risk factor for OA **incidence**, progression, and disability (138, 139) because of the increase in mechanical stress on weight-bearing joints (62, 137). Hence, it is estimated that a 5 kg weight loss decreases the risk (by 50%) of developing knee OA within 10 yr (62). In a large-scale intervention study, the Arthritis, Diet and Activity Promotion Trial (ADAPT) (138), significant improvements in physical function and pain were observed for overweight or obese knee OA patients when exercise and diet restriction were combined. The impressive effects of the combined treatment on function and pain were in contrast to the modest effects of exercise alone and the lack of effect of diet alone. These findings led the researchers to conclude that without exercise, dietary weight loss was ineffective in improving function, mobility, and pain in these patients.

Rheumatoid arthritis is characterized by "rheumatoid cachexia," a condition that features reduced muscle mass and elevated adiposity. Significant muscle loss is evident in approximately two-thirds of RA patients, including those with controlled disease (117, 181). Obesity is even more evident, with a prevalence of around 80% (117, 199). Due to the coincident loss of muscle and accretion of fat mass, body weight often remains stable, rendering body mass index (BMI) a misleading indicator of obesity in this population (198). In studies featuring high-intensity progressive resistance training (PRT), reversal of rheumatoid cachexia (i.e., increased muscle mass and decreased fat mass) has been achieved without exacerbating disease activity (76, 118, 127). Additionally, a study of obese subjects with RA found that a program of moderate physical training, reduced dietary energy intake, and a high-protein, low-energy supplement was successful in achieving a significant weight loss without loss of lean tissue (55). Muscle loss (cachexia) has also been shown to be a feature of AS; and as expected, this muscle loss is associated with reduced strength and physical function (126). Thus, interventions aimed at inducing muscle hypertrophy, such as PRT, would also appear to be needed for individuals with this condition.

General Exercise Prescription Recommendations

The sequencing of exercises for individuals with arthritis is similar to that for the general population, beginning with a warm-up and ending with a cool-down. A warm-up should be performed to increase tissue temperature throughout the body. As decreased stiffness and greater ROM evolve, people should be taught to judge whether increasing the range through which they are exercising is safe (191). Strengthening and cardiovascular conditioning exercises should be performed after a warm-up and should be followed by a cool-down period. Flexibility exercises should be performed during the cool-down. Laliberte and colleagues (111) provide extensive and useful examples and illustrations of various sequences of exercise appropriate for people with arthritis.

Exercise intensity or the time or repetition variable of a specific exercise should be progressed when the exercise is not as challenging as previously and when symptoms do not increase for two or three consecutive sessions. Conservative increments are recommended. In general, after 1 wk of consistent exercise without an increase in symptoms, the training demands should be increased toward the maximum recommended. Ways to progress exercises each session include increasing the volume (i.e., total duration of exercise, the number of exercises performed, or the time spent exercising with the use of shorter rest periods) or the intensity (e.g., increasing the % heart rate reserve for aerobic exercise, or the % one-repetition maximum for PRT). For both aerobic and resistance training, increase in training volume should precede increase in training intensity. An increase in symptoms may require lowering the intensity or volume of exercise, especially for the affected joint. The "2 h pain rule" is a helpful maxim for regulating exercise intensity. A localized increase in pain that lasts more than 2 h after an activity suggests the need to decrease the exercise intensity or volume for the next training session.

Collecting outcomes data serially to objectively monitor response to therapy is useful. Table 22.4 lists a variety of physical function measures appropriate to arthritis patients that can be used to evaluate strength, aerobic capacity, physical function, ROM, balance, and body composition. In addition, a number of standardized and validated instruments (questionnaires) are available for assessing arthritis pain, stiffness, and subjective function, as well as response to therapy, for OA, RA, and AS (21, 120, 166, 183, 185).

Cardiovascular Exercise

A key to cardiovascular exercise therapy for arthritic conditions is to manipulate the intensity and duration. Cardiovascular exercise intensity should be guided by heart rate (i.e., percent heart rate reserve, %HRR) or rating of perceived exertion (RPE) (64, 132). Since people with arthritis can achieve normal training heart rate ranges (135), standard training heart rate range development can be used (see chapter 5). A cardiovascular exercise intensity of 12 to 16 on the 6- to 20-point Borg scale or 3 to 6 on the 10-point Borg scale is recommended.

Aerobic training is ideally achieved with use of a mode that minimizes the magnitude and rate of joint loading (208). The best types include walking, cycling, and pool exercise (56). Free-speed walking produces less hip joint contractile pressure than isometric or standing dynamic hip exercises (208). If walking is uncomfortable, if pain lasts more than 2 h after walking, or if the individual has complicated biomechanics of the lower extremity, an alternative cardiovascular exercise should be used such as cycle ergometry, recumbent stair stepping, upper body ergometry, or water walking (38, 143). Cycle ergometry should be conducted with the seat height and crank length adjusted to limit knee flexion and minimize pedal load, which in turn decreases knee joint stress (123).

Water is a good medium for cardiovascular work, ROM exercise, and low-level strengthening. In both OA and RA, water-based, or aquatic, exercise moderately improves physical function and reduces pain (19, 98). The buoyant quality of water can help patients perform passive and active joint ROM exercises. Strengthening exercise can also be performed in the water, as water offers resistance to motion; and the faster a limb or body moves in water, the greater the resistance and consequently the workload. Many people with arthritis are able to tolerate longer and more vigorous workouts in the water than on land. With regard to the pool conditions; compliance with aquatic exercise decreases with water temperatures colder than 84 °F (28.9 °C), and cardiovascular stress increases with temperatures above 98 °F (36.7 °C) (31). Contraindications to hydrotherapy include a history of uncontrolled seizures, incontinence of bowel or bladder, pressure sores or contagious skin rashes, and cognitive impairments that would jeopardize the patient's safety. If a great deal of assistance is needed with dressing, or if changing clothes causes fatigue or joint pain, then pool therapy should not be used.

Resistance Exercise

Isotonic exercise is preferred over isometric exercise for dynamic strength training (159) because it more closely corresponds to everyday activities and therefore promotes improved daily function. Low-intensity isometric exercise, however, may be appropriate for muscle strengthening during the acute arthritic stage because it produces low articular pressures. Instructions should be to perform a submaximal isometric contraction for 6 s while exhaling. Isometric exercise should be targeted at one muscle group at a time.

For resistance training, intensity is determined by the percentage of the one-repetition maximum (1RM) that a load (weight) corresponds to. While increases in strength and muscle mass can be achieved in previously untrained individuals with loads of 50% 1RM, greater effects are seen with heavier loads, with 80% 1RM appearing to be optimal (109). At the commencement of resistance training, the intensity should be approximately 60% 1RM (~15 reps), and progression to 80% 1RM (~8 reps) should occur gradually over about 6 wk (118).

Range of Motion Exercise

Regular ROM exercise greatly facilitates maintenance of the degree of joint motion necessary for easy performance of ADLs (7). Thus, 5 to 10 min of active (i.e., executed without assistance) ROM exercises, performed on most or all days of the week, is recommended (7). However, while ROM exercises alone are effective in improving function and reducing pain and depression in AS patients (121), these exercises in isolation typically lack the intensity required to elicit functional improvements in most patients with arthritis (47, 77, 78, 79, 118). Therefore they should supplement aerobic and strength training.

It should be appreciated that considerable improvements in ROM are unlikely in those with severe joint destruction (e.g., limited or no joint space) and very restricted joint mobility. Under no circumstances should forceful stretching be employed.

EXERCISE TRAINING

Exercise training for cardiovascular, muscular, and ROM improvement is very important for the person afflicted with arthritis. Research demonstrates positive improvements with each of these types of exercise training in this population. The following section reviews the exercise training literature that applies to persons with OA, RA, and RS.

Osteoarthritis

The finding that the degree of severity of knee OA is associated with the level of cardiovascular deconditioning supports the concept that regular aerobic exercise should be performed by people with OA (176). Exercise is also recommended to assist with weight reduction. In this regard, resistance training may be more efficacious than aerobic training. And while aerobic training, by augmenting daily energy expenditure, has been shown to be an effective adjunct to restricted energy intake for weight loss in young adults, its effectiveness in middle-aged and elderly individuals is questioned. This is the case because sedentary elderly individuals, particularly those with chronic disease, are usually so deconditioned that they are unable to perform exercise of sufficient duration and intensity to significantly augment energy expenditure (58). In contrast, following 12 wk of PRT, resting metabolic rate (RMR) was increased 15% in elderly men and women as a consequence of increased lean body mass (34). Given that RMR typically accounts for 60% to 75% of daily energy expenditure, an increase of 15% is highly relevant to weight loss.

Reviews of exercise (aerobic or strength training or both) for OA conclude that training has positive effects on pain and disability and is safe (19, 49, 65, 66, 89, 101, 179). Recent Cochrane reviews of randomized controlled trials (RCTs) on land-based exercise report small improvements in pain and physical function for knee OA (65) and in pain for hip OA (66). Similarly, a Cochrane review on aquatic exercise for knee and hip

OA (19) concludes that this form of training has small to moderate benefit for function and QOL and a minor beneficial effect on pain. A large RCT of exercise for knee OA compared education with exercise interventions (home-based aerobic and resistance exercise with limited supervision) for 15 mo following an initial 3 mo of supervised exercise (56). In this trial, both aerobic and resistance training improved disability by arresting functional decline. A review of RCTs showed that aerobic walking and quadriceps-strengthening programs provide comparable positive effects on pain and disability in OA patients (178). Besides reducing pain and disability, therapeutic programs to strengthen knee extension and hip and ankle flexion in people with OA have resulted in decreased stiffness, as well as improved strength, mobility, balance, gait, independence, and physical function (113, 188). With respect to aerobic exercise, relatively low-level exercise intensity (at 40% of heart rate reserve) may be as effective as high-intensity cycling (at 70% of heart rate reserve) in improving function, gait, and aerobic capacity and decreasing pain in OA patients (123).

In addition to exercise interventions, weight loss has been shown to reduce knee joint forces in OA (137), and a recent RCT suggested that a combination of exercise and diet produced superior improvements in pain, function, and mobility compared to either exercise or diet interventions alone (138). Alternative modes of exercise such as yoga and tai chi also appear to improve pain, flexibility, and function for patients with OA (172, 195). In some instances, sport participation may be associated with an increased risk of injury, which in turn increases OA risk (112).

For OA, the MOVE consensus (179) concluded that group exercise and home-based exercise were equally beneficial and therefore that patient preference needs to be considered in the design of exercise programs. Not surprisingly, this group also concluded that exercise adherence was an important predictor of exercise benefit, and that adherence is aided by maintenance of an exercise diary, telephone contact, support from family and friends, provision of personal trainers, or some combination of these.

The benefits of exercise training for people with arthritis are listed in table 22.9.

Rheumatoid Arthritis

With RA, combating disability, deleterious body composition changes, and exacerbations of CVD and osteoporosis risk should be primary considerations in the design of appropriate exercise therapies. Thus, both resistance training and aerobic conditioning should be featured in an exercise prescription. These forms of training in RA patients have been shown to improve muscle strength, aerobic capacity, functional capacity, pain, and QOL without having detrimental effects on disease activity or joint damage; and in some cases they may even decrease disease activity (17, 74, 98, 117). The safety of exercise in RA patients is apparent in even high-intensity, long-term (≥2 yr) interventions (45, 47, 48, 78, 79, 151, 152). Initially, reports from the Rheumatoid Arthritis Patients in Training (RAPIT) trial, which featured 150 patients performing high-intensity aerobic and strength training twice weekly for 2 yr, raised concerns that this exercise program was accelerating joint damage progression in large joints with extensive preexisting damage (47, 147). However, when additional data were collected, these fears proved to be unfounded; and the authors concluded that long-term, intense weight-bearing exercise was safe for all joints, including large joints already extensively damaged (45). This verdict is in agreement with the findings of other studies (73, 78, 79, 152) and with the conclusion of a recent meta-analysis on RCTs of aerobic exercise interventions in RA (17).

Whole-body dynamic exercise is preferable to static or isometric exercise, as the former elicits greater improvements in body composition, strength, aerobic capacity, and function and is more relevant to the performance of ADLs (73, 81, 201). Additionally, higher-intensity exercise consistently produces significantly better gains than low-intensity ROM exercises (47, 77, 78, 79, 118). An example comes from the RCT conducted by Lemmey and colleagues (118), in which established RA patients were randomized to perform 24 wk of high-intensity PRT (two sessions per week, eight exercises, three sets of eight reps at 80% 1RM). These patients showed significant mean gains in muscle mass (1.5 kg [3.3 lb]), substantial reductions in total fat mass (2.3 kg [5.1b]) and trunk fat mass (2.5 kg [5.5 lb]), and improvements in strength (119%) and objective tests of physical function (17-30%). In contrast, the patients randomized to ROM exercise demonstrated no changes in body composition or physical function despite good compliance to training. Similarly, in the 2 yr RAPIT trial, combined high-intensity aerobic and resistance training produced significant improvements in functional ability, aerobic fitness, muscle strength, and emotional health and a decline in bone mineral density loss as compared with usual care, that is, ROM exercises (46, 47). This program demonstrated a 78% adherence and satisfaction rate after 2 yr (146). Another finding from the intervention reported by Lemmey and colleagues (118) was that after 24 wk of high-intensity PRT, RA patients who had

Table 22.9 Exercise Training Review

Cardiorespiratory endurance	Skeletal muscle strength	Skeletal muscle endurance	Flexibility	Body composition
		OSTEOARTHRITIS		
Aerobic exercise improves cardiorespiratory endurance, pain, depression, fatigue, function, health status, and gait and helps to reduce fat mass.	High-intensity resistance training improves strength, muscular endurance, function, health status, pain, and stiffness and helps to reduce fat and increase muscle mass.	Low-intensity resistance training improves strength, muscular endurance, function, health status, pain, and stiffness and helps to reduce fat and maintain muscle mass.	Dynamic exercise improves joint mobility, pain, and function. Aquatic exercise improves knee and hip range of motion, pain, and function. AFYAP and PACE programs improve flexibility and isometric strength (206).	Combined diet and resistance and aerobic training produces weight loss; improves function, mobility, and pain to a greater extent than diet or exercise alone.
		RHEUMATOID ARTHRITIS		
Aerobic exercise improves cardiorespiratory endurance, pain, and function, and, with resistance training, mood. Aerobic training improves fitness without worsening disease activity.	High-intensity resistance training improves strength, muscular endurance, function, and mobility; increases muscle mass; and reduces fat mass. Hand strengthening may improve dexterity and grip strength.	Low-intensity resistance training improves strength, muscular endurance, function, and mobility; maintains muscle mass; and reduces fat mass. Hand strengthening may improve dexterity and grip strength.	Joint mobility improves with dynamic exercise training. AFYAP and PACE programs improve flexibility and isometric strength (206).	Combination of aerobic and resistance training (RT) or RT alone increases muscle mass and decreases fat mass without significant weight loss; also slows bone mineral density loss.
		ANKYLOSING SPONDYLITIS		
Supervised physical therapy improves fitness, mobility, and function; improvements in fitness and stiffness may mediate improvements in global health.	Few investigations are relevant, but resistance training aimed at restoring muscle mass and improving function is warranted.	Few investigations are relevant, but resistance training aimed at restoring muscle mass and improving function is warranted.	Home or supervised exercise can improve spinal mobility, physical function, and (supervised only) patient global assessment Some evidence suggests improvements in pain and depression. Relative to conventional physical therapy, flexibility and strengthening exercises that target specific muscle groups involved in AS improve axial mobility and function.	Relevant studies have not been performed but are warranted, as AS is characterized by muscle loss and increased osteoporotic fracture risk.

Note. AFYAP = Arthritis Foundation YMCA Aquatics Program; PACE = People With Arthritis Can Exercise (land-based community program).

previously been moderately disabled achieved levels of physical function that were equivalent to or better than those of age- and sex-matched healthy controls. A similar restoration of normal physical function was observed in an earlier PRT intervention (127). Most investigations evaluating the efficacy of exercise interventions for RA patients use aerobic exercise at an intensity of 50% to 85% of maximum heart rate and strengthening beginning at 50% to 60% and progressing to 80% of maximum. These studies provide the basis for recommendations regarding exercise prescription in RA (7, 203).

Nontraditional modalities such as dance and tai chi chuan may have beneficial effects on depression, anxiety, fatigue, tension, and lower extremity ROM (83, 105, 154, 212). Additionally, water exercises, including seated immersion, improve aerobic capacity and other physical and psychological measures in patients with RA (82). When performed at low speeds, exercise in water results in lower heart rates and $\dot{V}O_2$ than exercise of a comparable speed performed on land because of the buoyancy effect of water; but at higher speeds, the resistance of water results in higher heart rates and $\dot{V}O_2$ than does land-based exercise (81). These consequences of adapting movement speed can be used to vary the training demands (i.e., intensity).

Although earlier investigations of home-based exercise programs failed to demonstrate improvements in physical function (41, 200), Hakkinen and colleagues (77, 78) showed significant improvements in strength, subjectively assessed physical function, and disease activity, as well as trends toward decreased pain and increased bone mineral density (BMD) in RA patients performing high-intensity home-based strength training compared to patients performing ROM exercises at home. These findings demonstrate that if the exercise intervention is appropriately designed, exercising at home is effective and beneficial.

As with any patient population, training effects are maintained only as long as exercise continues, and these benefits wane rapidly on termination of exercise (75, 119). Independent training following a supervised program sustains these beneficial effects (78).

In terms of training responses for RA patients, the changes in body composition, strength, and aerobic capacity are similar in magnitude to those generally reported for healthy middle-aged or elderly individuals (76). This similarity in response to training is consistent with recent reports that muscle quality is unaffected in RA, even in those with pronounced muscle wasting (130, 131). The finding that rheumatoid muscle is normal both qualitatively and in its response to chronic exercise is an important consideration for those prescribing exercise for people with RA.

Ankylosing Spondylitis

Investigations of exercise training effects in patients with AS are relatively few, although exercise was among the 10 key recommendations for the management of AS developed from a combination of research-based evidence and expert consensus, as part of a collaboration between the ASsessment in AS (ASAS) International Working Group and the European League Against Rheumatism (EULAR) (224). The literature review from which these recommendations arose suggested that different types of exercise-based intervention could affect disease outcomes in AS (225). Given the goals of maintaining posture and functional ability in this disease (in which disability is mainly related to effects in the spine), most interventions use flexibility and muscle-strengthening programs. Achieving functional ROM of the hip joints should be emphasized, because a lack of such capability can be extremely disabling (33, 158). The goal of muscle strengthening in this population is to maintain or approximate a neutral spine over the long term. A program to strengthen the back and hip extensors, as well as general strengthening, can be performed on land or in water. During strength training, the person should be initially supervised with a goal of independently maintaining proper posture during spinal extension (121).

Daily exercise is considered vital to maintenance of spinal mobility, but long-term effects have not been studied (40). Since rheumatoid cachexia (i.e., muscle wasting) occurs in AS, and since this muscle loss is associated with impaired strength and physical function, the high-intensity PRT advocated in RA patients for improving muscle mass and function would also appear appropriate for patients with AS (126).

Significant short-term improvement in spinal and hip ROM of AS patients enrolled in intensive physical therapy has been demonstrated, and the performance of regular moderate (2-4 h per week) exercise is associated with functional improvement for patients with AS (121, 186). Additionally, relative to weekly group physical therapy, improvements in pain, function, and patient global assessment were noted with combination spa-exercise therapy consisting of group exercises, walking, posture exercises, hydrotherapy, sports, and thermal treatment visits (i.e., sauna) for 3 wk followed by 37 additional weeks of combination group exercises, hydrotherapy, and sports (40, 217). This combined spa-exercise therapy also is cost-effective (216).

A recent Cochrane review, which evaluated 11 randomized and quasi-randomized trials in subjects with AS (40), concluded that home-based programs improved spinal mobility and physical function, although the gains in patient global assessment and spinal mobility were less than those observed following supervised group exercise programs (40). This is a common finding with home-based training, and as discussed earlier, relates to the lower training intensity that is characteristic of unsupervised training sessions. The mode and sequencing of exercise are important considerations for people with AS. An exercise program for an individual with AS should start as soon as possible after diagnosis of the condition, beginning with exercises to improve spinal and peripheral joint motion before commencement of a strengthening program (121). High-impact activities should be avoided because they are stressful to the spinal and sacroiliac joints. Swimmers with limited spinal and neck motion should use a mask, snorkel, and fins to avoid trunk and neck rotation. Sports that encourage extension are preferred over activities that require flexion (40). Contact sports are contraindicated for those with cervical spine or peripheral joint involvement.

Practical Application 22.3

FIBROMYALGIA

Fibromyalgia is an increasingly recognized chronic pain syndrome affecting around 5 million adult Americans (115). Although its etiology is unknown, the condition is possibly related to central neuromodulatory dysregulation (135). Fibromyalgia is not a form of arthritis but may easily be confused with arthritis because of its associated widespread musculoskeletal pain and so-called trigger or tender points, often with a periarticular location. Diagnosing fibromyalgia is additionally complicated by its high coexistence (25-65%) with other rheumatic conditions such as RA, AS, and SLE. It affects at least 2% of adult Americans, women more than men (ratio 7:1) (70), and is currently diagnosed using 1990 American College of Rheumatology (ACR) criteria (221). These criteria include widespread pain in combination with tenderness at 11 or more of the 18 specific tender point sites. The tender points occur in characteristic locations and are associated with fatigue and exercise intolerance (70), but laboratory tests and joint radiography are normal. Alternative ACR diagnostic criteria have recently been proposed that do not require a physical or tender point examination (220), but this method of evaluation has yet to be widely adopted.

Multiple investigations demonstrate that muscle and joint inflammation or damage is not associated with fibromyalgia, but that a lower pain threshold and altered pain-processing pathways exist, which provides clues to the underlying pathogenesis of this complicated syndrome (69). Fibromyalgia is also characterized by fatigue, inadequate and nonrestorative sleep, and cognitive symptoms (e.g., problems with thinking or remembering) and is often associated with depression (162).

Treatment of fibromyalgia requires a multidisciplinary approach employing exercise, education, and both pharmacologic and behavioral therapies for depression and sleep (70). Evidence shows that aerobic exercise is beneficial for persons with fibromyalgia and can improve aerobic capacity, physical function, and global well-being; possibly pain and tenderness (31, 71); and possibly sleep, fatigue, depression, and cognition (31, 71). For example, in an RCT conducted by Richards and Scott (174), fibromyalgia patients were allocated to either twice-weekly graded aerobic exercise training (walking and cycling at moderate intensity for progressively longer periods) or relaxation and flexibility classes for 12 wk. Significant improvements were noted for both groups in tender joint counts, pain scores, fatigue scores, and self-reported disability scores at 3 mo. More in the aerobic exercise group (24/69; 35%) rated themselves as "much" or "very much" better relative to the relaxation group (12/67; 18%) ($p = 0.03$). The benefits of aerobic training were even more apparent at 1 yr follow-up, when fewer of the exercise group fulfilled the fibromyalgia criteria compared to the relaxation group subjects (31/69 vs. 44/67, $p = 0.01$). Strength training appears to provide the best means of increasing muscular strength in persons with fibromyalgia, and also reduces pain, tender points, and depression. Aquatic exercise may provide superior benefits over land-based training in improving mood in fibromyalgia patients (13).

With respect to exercise testing, procedures outlined in chapter 5 can be followed in persons with fibromyalgia. One consideration is that, compared to age- and sex-matched controls, persons with fibromyalgia report higher levels of perceived exertion for the same relative workloads during exercise testing (150). Additionally, people with fibromyalgia are often deconditioned and may not achieve maximal effort (156, 210). In these cases, ventilatory threshold rather than peak $\dot{V}O_2$ is recommended as an indication of fitness levels (210). As they do for persons with arthritis, American College of Sports Medicine (ACSM) prediction equations appear to overestimate peak $\dot{V}O_2$ in persons with fibromyalgia (52). Although slight underestimations and overestimations, respectively, were noted with the FAST equation depicted in table 22.3 and the FOSTER equation (with handrail support: $\dot{V}O_2[\text{ml} \cdot \text{min}^{-1} \cdot \text{kg}^{-1}] = 0.694 \times \text{ACSM predicted} + 3.33$), these equations appear to provide clinically acceptable estimations of peak $\dot{V}O_2$ using a Duke–Wake Forest testing protocol (52).

For persons with fibromyalgia, exercise training goals include improving fitness and function. Potential modes of exercise include walking, cycling, dancing, and water aerobics. Pool options, such as the Arthritis Foundation aquatics program for fibromyalgia or FIT (Fibromyalgia Interval Training), may be ideal for individuals with significant complaints of pain with exercise or concerns that exercise will worsen pain symptoms (71). For people with fibromyalgia, as with other deconditioned individuals, a prudent approach should be taken to training initiation and progression. This includes beginning at a low intensity and progressing slowly and

(continued)

gradually to a moderate intensity (20-55% of heart rate reserve) for 20 to 30 min (31, 71). Some people with fibromyalgia may eventually progress to high-intensity exercise. Although exercise does not worsen self-reported pain in fibromyalgia, the clinical exercise physiologist should educate people about transient increases in pain with exercise bouts and about methods of adjustment of exercise protocols. Such education will enhance adherence and long-term benefits of exercise training and should be an essential component of the exercise prescription for people with fibromyalgia (124). Recommendations for minimizing pain with strength training include limiting eccentric exercises, performing upper and lower extremity training on alternate days, and resting between repetitions (71). As yet, it is not known whether combining different types of exercise training, or combining exercise with education, biofeedback, or medication (or more than one of these), is optimal for treating fibromyalgia.

Practical Application 22.4

RESEARCH FOCUS

This section emphasizes a recent clinical exercise physiology study of patients with RA. The purpose of the investigation was to determine the effects of participation in 12 wk of low-impact aerobic exercise, which was either supervised and class based or unsupervised and home based, on function, fatigue, pain, depression, and disease activity in RA patients. A convenience sample of 310 adult RA patients was randomized to either class exercise (CE: $n = 102$), home exercise using a videotape (HE: $n = 103$), or control (C: $n = 105$). Data were analyzed on the 220 subjects who completed the interventions and assessments at baseline and week 12 (CE: $n = 68$; HE: $n = 79$; C =73). The completers were aged 40 to 70 yr (median: 55.5 yr) and were predominantly white (85%) and female (83%). There were no significant group differences in sex, age, disease duration, comorbidities, or medication at baseline.

Training for the class exercise and home exercise groups involved performing 1 h of low-impact (i.e., no jumping or running) aerobic and strengthening exercises three times per week for 12 wk. The target HR during exercise was 60% to 80% maximum heart rate (MHR). The control group subjects were asked to maintain their baseline physical activity level. At 12 wk, the CE group, but not the HE group, had significantly reduced overall disease symptoms (fatigue, pain, and depression) relative to the controls ($p < 0.05$). The treatment groups (i.e., CE and HE combined) improved their physical function (50 ft walk time and grip strength) more than the nonexercising controls ($p < 0.005$), with the improvement being more apparent in the CE participants, albeit not significantly. Consistent with this trend, aerobic capacity (estimated $\dot{V}O_2max$) increased 12%, 10%, and 7% for the CE, HE, and C groups, respectively, over the course of the interventions. Although the time spent exercising was similar for the CE and HE groups, the intensity of training was higher for the CE subjects. For the last 6 wk of the intervention, 72% of the CE patients were exercising at 60% MHR and 16% were exercising at 80% MHR, whereas only 45% of the HE subjects achieved a training intensity of 60% MHR and just 4% achieved the 80% MHR level. Thus, the benefits of exercise training were in part determined by training intensity.

This study also assessed the factors that predicted exercise behavior and showed that lower fatigue and more positive perceptions of the relative benefits and barriers to exercise were associated with more minutes of exercise performed. The authors concluded from the investigation that their findings "reinforce the need for health providers to educate patients with RA about the many benefits of exercise, how to overcome barriers to exercise, and ways to manage fatigue" (p. 949).

Reference: Neuberger GB, Aaronson LS, Gajewski B, Embretson SE, Cagle PE, Loudon JK, Miller PA. Predictors of exercise and effects of exercise on symptoms, function, aerobic fitness, and disease outcomes of rheumatoid arthritis. *Arthritis Rheum* 2007;57(6):943-952.

Clinical Exercise Pearls

- Loss of strength and aerobic capacity, leading to impaired physical function and disability, is characteristic of arthritic conditions. Appropriate exercise substantially improves function and diminishes disability.
- RA and AS both worsen CVD risk. This increased risk (approximately twofold) is comparable to that conferred by diabetes.
- RA and AS are characterized by muscle loss, and RA and OA by obesity. Thus, improving body composition should be a goal of exercise programs for these patients.
- Exercise training, including long-term high-intensity training, does not exacerbate joint damage, pain, or inflammation in arthritic conditions.

CONCLUSION

The available data indicate that properly performed exercise is safe and effective for individuals with OA, RA, and AS. In the short term, exercise increases strength, aerobic capacity, and ROM; improves body composition; enhances physical function, thus decreasing or even removing disability; and attenuates stiffness and often pain. These benefits are in addition to the other well-accepted benefits of exercise to the general population. Precautions must be taken, however, to ensure that exercise is appropriate (i.e., safe and tolerable) for people with arthritis.

Key Terms

aneural (p. 421)
ankylosis (p. 420)
articular (p. 428)
asymmetric (p. 424)
autoimmune (p. 420)
avascular (p. 428)
biologics (p. 428)
bursitis (p. 421)
chronicity (p. 421)
crepitus (p. 420)
discitis (p. 435)
disease-modifying antirheumatic drugs (DMARDs) (p. 428)
effusion (p. 420)
entheses (p. 420)
enthesopathy (p. 421)
fibromyalgia (p. 441)
incidence (p. 435)
intraosseous (p. 421)
osteophytes (p. 420)
pannus (p. 420)
Raynaud's phenomenon (p. 424)
Sjogren's syndrome (p. 424)
spondylarthropathies (p. 419)
subchondral sclerosis (p. 424)
symmetric (p. 424)
synovial joints (p. 420)
synovitis (p. 420)
valgus (p. 425)
varus (p. 425)
vasospastic (p. 435)

CASE STUDY

MEDICAL HISTORY

At age 42 yr, Mrs. CJ, a schoolteacher, mother of two young children, and social tennis player who was generally physically active, started experiencing pain and stiffness in her hands, feet, and left knee. Accompanying this joint pain was profound fatigue. Over the ensuing 3 mo as her condition deteriorated, she found it progressively more difficult to perform ADLs, and she was forced to abandon her normal recreational pursuits. This was particularly the case in the mornings when her joint stiffness was at its worst, coinciding with when she needed

(continued)

to prepare her children for school. An inability to fulfill her role as a working mother and to enjoy her usual recreation activities led over time to feelings of inadequacy and of not being able to cope. Eventually, the combination of fatigue, diminished physical function, and depression forced Mrs. CJ to take extended leave from work.

At the insistence of her family, Mrs. CJ booked an appointment with a rheumatologist.

DIAGNOSIS

A rheumatologic evaluation revealed swollen proximal interphalangeal, metacarpophalangeal, and metatarsophalangeal (PIP, MCP, and MTP, respectively) joints and effusion of the left knee. Blood lab values included ESR = 65 mm · h^{-1}; C-reactive protein = 52; positivity for rheumatoid factor; ACPA (anti-CCP) = 92 units · ml^{-1} (normal: <5 units · ml^{-1}); and hemoglobin (Hb) = 11.0 g · dl^{-1}. A diagnosis of RA was made based on this information, and Mrs. CJ was prescribed combination therapy of etanercept (anti-TNF-α) and methotrexate (ETN + MTX). The rheumatologist also advised that she remain off work and continue taking things easy.

At 8 wk follow-up, the joint pain and swelling had mostly resolved (two slightly swollen MCPs); blood tests showed ESR = 16 mm · h^{-1}, C-reactive protein = 9, and Hb = 12.2 g · dl^{-1}; and X-rays gave no indication of significant joint damage. The rheumatologist was satisfied that the prescribed DMARDs were effectively controlling disease activity, and consequently this pharmaceutical treatment regime was continued.

While Mrs. CJ was pleased that her joint pain and stiffness had diminished and that her physical function had been partially restored, she was still experiencing difficulties coping with the demands of daily living, including the worry of lost income if she was unable to return to work. Additionally, she was upset at her inability to resume her usual recreational activities and attributed this reduced physical activity to the increase in body weight (5 lb) and adiposity she had experienced since being diagnosed with RA. Overall, the inability to perform activities she was accustomed to doing had caused a loss of self-esteem, persisting depression, and a reduced QOL.

Specifically to address these consequences of impaired physical functioning, the rheumatologist referred Mrs. CJ to a clinical exercise physiologist with training in exercise prescription.

EXERCISE TESTING RESULTS

A submaximal treadmill walk test was performed and predicted a peak aerobic capacity (estimated $\dot{V}O_2max$) of 27 ml O_2 · min^{-1} · kg^{-1}. Objective physical function tests (30 s seated step test (SST), 50 ft walk, 30 s arm curl test) indicated performance levels significantly poorer than for healthy sedentary reference women of the same age (174). Assessment of body composition showed Mrs. CJ to have a % body fat of 38% and to be significantly muscle wasted relative to population norms.

EXERCISE PRESCRIPTION

To improve body composition, aerobic capacity, and strength, Mrs. CJ was prescribed 8 wk of supervised gym-based combined aerobic and progressive resistance training (PRT) three times a week. Aerobic training initially involved cycling for 10 min at 30% HRR, but this gradually progressed to 20 min continuous cycling at 65% HRR. Resistance training involved six different exercises (upper and lower body, using large muscle groups) and commenced at 60% 1RM (one set of 15 repetitions). The intensity progressed gradually, initially to two sets of 15 reps at 60% 1RM, before progressing to two sets of 12 reps at 70% 1RM, and then, at the start of week 6, to two sets of eight reps at 80% 1RM. When supervision was withdrawn after 8 wk, Mrs. CJ was encouraged to continue training at the gym and to supplement this training with regular walking (i.e., replacing short driving trips with walking) and taking a long walk (30 min or more) 1 d/wk.

When reassessed at 6 mo after the start of exercise training, Mrs. CJ's estimated $\dot{V}O_2max$ was 36 ml O_2 · min^{-1} · kg^{-1}; she had gained 3 lb (1.4 kg) in lean body mass and lost 6 lb (2.7 kg) in fat mass (including 5 lb [2.3 kg] in trunk fat mass). Performance in the objective physical function tests had improved from 21% to 40%, and strength, as assessed by 1RMs, had increased on average 119%. The improvement in objectively assessed physical function indicated that she was now performing these tests as well as or better than age-matched, healthy reference women.

DISCUSSION

As a consequence of her improved aerobic capacity and strength, Mrs. CJ was now able to perform ADLs without difficulties and had returned to playing midweek social tennis. With the resumption of most of her previous activities and the knowledge that she was able to diminish, and even remove, some of the consequences of her chronic disease, Mrs. CJ's self-esteem, self-efficacy, and enjoyment of life had been largely restored. On the other hand, even though her fatigue had attenuated, periodically it remained debilitating, and maintenance on immune suppressors was notable for a slight increased risk of opportunistic infections (e.g., upper respiratory tract infections). The challenge for Mrs. CJ will be sustaining her motivation to exercise and thus maintaining the benefits of training that she has acquired.

DISCUSSION QUESTIONS

1. What exercise advice should the rheumatologist have given Mrs. CJ when she was initially diagnosed with RA?
2. Which comorbid conditions are likely in an individual such as Mrs. CJ, and why?
3. What plans could be put in place to help this patient achieve good long-term compliance to exercise?
4. What are the anticipated benefits of exercise training for Mrs. CJ?

Chapter 23

Osteoporosis

David L. Nichols, PhD
Andjelka Pavlovic, MS

Osteoporosis is an increasing health problem and should be a concern of everyone, although the disease primarily affects women. As the population ages, the cost and problems associated with osteoporosis will continue to increase. But osteoporosis should not be considered an inevitable consequence of aging. It is a preventable disease, but prevention must begin at an early age, perhaps even before puberty. Osteoporosis or osteopenia can be diagnosed with the use of dual-energy X-ray absorptiometry (DXA) technology, and bone density measurements should be seriously considered in anyone with existing risk factors for osteoporosis. Drug and exercise therapies are available to treat osteoporosis and its related problems, but when bone density is low enough to be considered osteoporotic, most of that lost density cannot be regained. Therefore, prevention should be the primary focus.

DEFINITION

During the prepubertal and adolescent growth years, **bone mineral density (BMD)** (also known as bone mass) increases in healthy individuals. Once people move past the age of 40, though, small amounts of bone mass are lost each year. In women, this bone loss accelerates during a 3 to 5 yr period after menopause. Because women typically have lower bone mass than men do and because of the accelerated decline following menopause, they tend to be more prone to **osteoporosis**. Osteoporosis has classically been defined, for older women and men, as a pathological condition associated with increased loss of bone mass caused by increased bone resorption. **Osteopenia** is a less severe form of the disease in which bone mass has declined below normal levels, but not to the extent seen in osteoporosis. As BMD declines, individuals are at increased risk for skeletal fractures. Osteoporosis is a major public health problem that results in significant morbidity, mortality, and economic burden. Fractures result in impairments that increase the risk of other underlying diseases, such as coronary artery disease, since with fracture, activity level will decline.

SCOPE

In 1994, the World Health Organization (WHO) defined osteoporosis as a BMD measurement more than 2.5 standard deviations below the young adult mean (32). Smaller amounts of bone loss, BMD 1 to 2.5 standard deviations below the young adult mean, were defined as osteopenia. However, these criteria were based solely on epidemiological data in postmenopausal women (35). Recent position statements from the International Society of Clinical Densitometry (8, 35), the American College of Sports Medicine (40), and others (30) have presented more appropriate criteria for identifying "low bone density for age" in populations other than postmenopausal women. The consequences of poor bone health have continued to escalate in recent years, despite advances in technology for screening and diagnosis and expanded treatment options. It is estimated that fracture risk at the hip is 2.5 times higher for each standard deviation decrease in hip

BMD (32). Osteoporosis now affects almost one of every two women at some point in life. Although osteoporosis is thought of primarily as a disease of women, prevalence rates in men can be as high as 15% (14, 30).

Medical costs for osteoporosis are estimated at $19 billion per year (12); the estimated overall cost of a single hip fracture for a patient is approximately $45,000 (49). The mortality associated with fractures is substantial, although the mortality rate depends greatly on the fracture site (34) as well as sex, as men have double the mortality rate of women following a hip fracture (23). A recent meta-analysis estimated a fivefold increase in all-cause mortality in the 3 mo following a hip fracture in older adults (23).

PATHOPHYSIOLOGY

Changes in bone result from the continual process of **bone resorption** and **bone formation**—known as bone remodeling. Bone remodeling does the following:

- Maintains the architecture and strength of the bone
- Regulates calcium levels
- Prevents fatigue damage

Remodeling is also important during periods of growth, puberty, and adolescence, when the majority of adult bone mass is laid down (44, 50). Bone modeling and remodeling often occur simultaneously, and distinctions between them are not always apparent. During growth, an increase in both length and size of the bone is accomplished by bone modeling in which bone formation occurs without prior bone resorption. The processes of both bone remodeling and modeling are complex; but in the simplistic view, bone resorption is carried out by bone cells known as osteoclasts, whereas bone cells called osteoblasts form new bone (50). The complete remodeling cycle takes several months. Modeling and remodeling are influenced by both hormonal and nutritional status; remodeling can also be affected by mechanical strain such as exercise.

During both childhood and adult life, if increases in mechanical strain are sufficient, the bone remodeling cycle can be altered and increases in bone mass and strength will occur (44). However, increases in mechanical strain do not affect the bone modeling cycle. On the other hand, if during the growth period severe deficits in mechanical strain occur, as with immobilization, a loss in bone mass and reduced growth in bone length will take place. Nutritional deficits have differing effects depending on the stage of life. If caloric intake is insufficient during childhood, when bone modeling is still occurring, the body will sacrifice increases in length of bone to maintain bone strength (18, 24). Nutritional deficits cause decreases in bone mass and strength after late adolescence when bone growth has ceased.

Before puberty, the effects of sex hormones on the skeleton are small, with changes in both bone mass and length influenced primarily by growth hormone and insulin-like growth factors. After puberty, estrogen deficiency in women for any reason at any stage of life results in rapid bone loss. In men, the role of sex hormones in bone loss is less clear, but an increase in sex hormone–binding globulin is likely a major factor (13).

When there is a disruption in the remodeling cycle of resorption and formation, bone loss occurs. In young adulthood, the cycles of resorption and formation are balanced and bone loss is minimal; peak bone mass is attained by approximately the end of the second decade (figure 23.1). It appears that some age-related bone loss (approximately 0.5-1.0% per year) is experienced by both men and women after approximately age 40 yr. Over her lifetime, a woman loses as much as 50% of her peak **trabecular bone** mass (5), and most of that loss is attributable to an estrogen deficiency. Although this loss occurs mostly after menopause, bone loss has been seen in the perimenopausal years (48), but the decline in bone mass during this time of life may be the result of increases in follicle-stimulating hormone rather than decreases in estrogen.

The two most important factors in the development of osteoporosis are the amount of peak bone mass attained and the rate of bone loss (15, 28, 48), although age of attainment of peak bone mass may alter both factors (7, 28). **Peak bone density** (or peak bone mass), for most purposes, is defined as the highest amount of bone mass attained during life. Although this definition is accurate, the term *peak bone mass* is often confused with the term *maximal bone mass*. Maximal bone mass should be defined as the highest bone mass that a person could possibly achieve. Maximal bone mass would, theoretically, be controlled solely by genetic factors. Peak bone mass is also influenced to a certain extent by genetics, but the difference between maximal bone mass and peak bone mass is controlled by other factors such as physical activity, diet, and hormonal balance. It is unlikely that anyone ever reaches maximal bone mass.

CLINICAL CONSIDERATIONS

Unfortunately, because osteopenia and even osteoporosis are asymptomatic, the disease may go undetected in many people until a fracture occurs. A risk factor assessment can be conducted to determine the likelihood of the disease. Various sophisticated measurement techniques

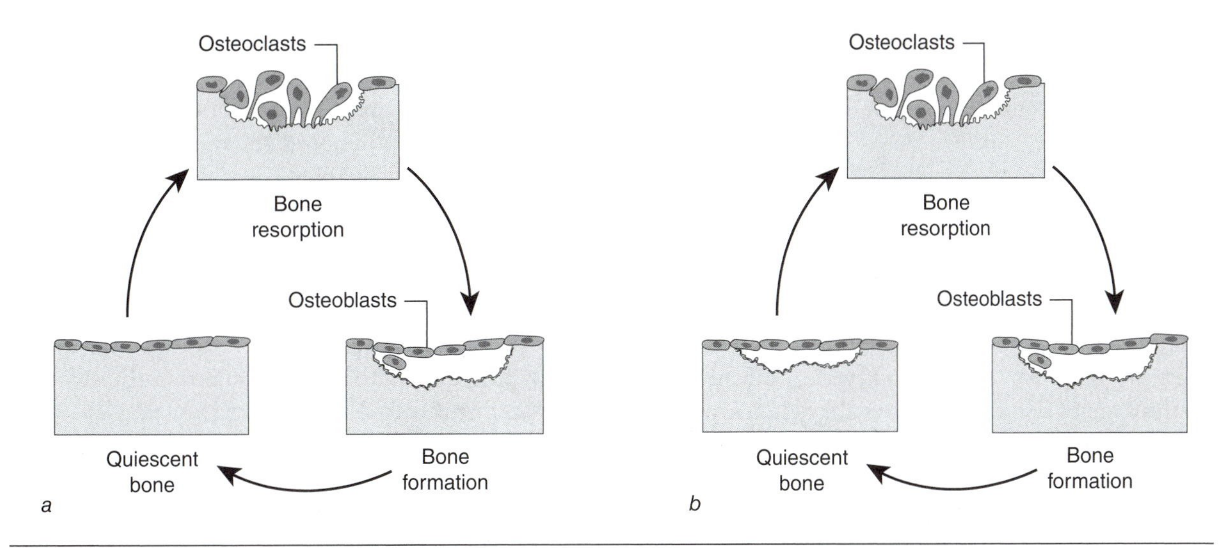

Figure 23.1 Bone modeling and remodeling.

have evolved for assessing bone density that may be useful for disease diagnosis.

Signs and Symptoms

Numerous risk factors, both inherited and environmental, exist for osteoporosis (table 23.1). Obtaining the patient's medical history can identify these risk factors. To date, the screening tools available to predict the presence of osteoporosis, other than measurement of BMD, are not always accurate. Risk factors that may be easily measured (e.g., age, gender, height, weight, inactivity, smoking) have a much lower predictability than actual measures of BMD (42). However, for a postmenopausal (or amenorrheic) Caucasian woman with low body weight who is not on hormone replacement therapy, the presence of any other risk factor would strongly suggest that this woman has low bone mass (40, 48). Furthermore, the risk of osteoporotic fractures is strongly related to an individual's BMD, and to a slightly smaller degree to fall frequency and bone geometry. Unfortunately, a fracture is often the first sign that a person has osteoporosis. Osteopenia occurs without outward symptoms, and by the time a fracture occurs, a patient may have lost as much as 30% or more of peak bone density. Some available treatments can increase BMD as much as 12% (42), but unfortunately, the loss of bone mass in an osteoporotic patient is so great that treatments are unlikely to reverse the disease. This holds true even in young amenorrheic women, who, after having been amenorrheic for an extended period, never regain sufficient bone mass to return to normal, even with the resumption of regular menstrual periods (40).

History and Physical Examination

A thorough understanding of the bone status of each osteoporosis patient is critical. Obtaining a history of fractures, age, and amount of BMD can help determine the level of caution that should be taken to minimize inappropriate mechanical stress on high-risk joints (hip,

Table 23.1 Risk Factors for Osteoporosis or Osteopenia

Inherited factors	Environmental factors
Caucasian or Asian	Below-normal weight
Female	Loss of menstrual function
Osteoporotic fracture in first-degree relative	Low calcium intake
Height <67 in. (170 cm)	Inactivity
Weight <127 lb (58 kg)	Prolonged corticosteroid use
Early menopause, before the age of 45	Smoking
Estrogen deficiency	Excessive alcohol intake
Hyperthyroidism	Caffeine
	Amenorrhea

spine, or wrist). The risk factors mentioned in table 23.1 may raise suspicions of the disease, but diagnostic testing is necessary for confirmation.

Diagnostic Testing

The primary means of assessing bone health is to measure BMD. **Dual-energy X-ray absorptiometry (DXA)** is the most commonly used technology for measuring BMD. DXA uses low-dose X-ray to emit photons at two different energy levels. Bone mineral density is calculated based on the amount of energy attenuated by the body. DXA measures the amount of **bone mineral content** per unit area; thus bone density measured by DXA is reported in $g \cdot cm^{-2}$. DXA is capable of differentiating between bone and soft tissue and can also be used to measure regional and total body composition. The advantages of DXA are that it is capable of measuring small changes in BMD over time, has a precision of 0.5% to 2.0%, requires short exam times (5-10 min), and provides low radiation exposure. Bone mineral density is most often measured in the spine, hip (femoral neck), and wrist because these are the most common sites for fracture in osteoporosis. The accuracy of predicting BMD and fracture risk at one site based on measurement of BMD at another site is low (≤50%) (3, 26, 29). Thus, a measurement at all three sites is preferable, especially the spine and hip, because the consequences of fracture are the greatest at these two sites (16).

Quantitative computed tomography is the only method currently available that provides an actual measurement of volumetric bone density. Because quantitative computed tomography has the advantage of providing a three-dimensional image, thus allowing a separation of trabecular and **cortical bone**, it would seem to be the measurement of choice. But this is a much more costly procedure and has a higher radiation exposure than DXA and thus is used less in clinical practice. The information provided with a bone density measurement is not only the absolute value in terms of grams of bone mineral but also a comparison of that value to established normal values. A DXA image of a femoral neck scan, along with its accompanying printout, is presented in figure 23.2. The subject's BMD is compared with reference standards.

Although DXA is widely available, measurement of BMD and diagnoses of osteoporosis or "low bone density for age" are not a simple matter. A multitude of factors can affect not only the individual's bone density but also the accuracy of the bone density measurement (20). In addition, although measurements of BMD account for much of the risk associated with fracture, there are other

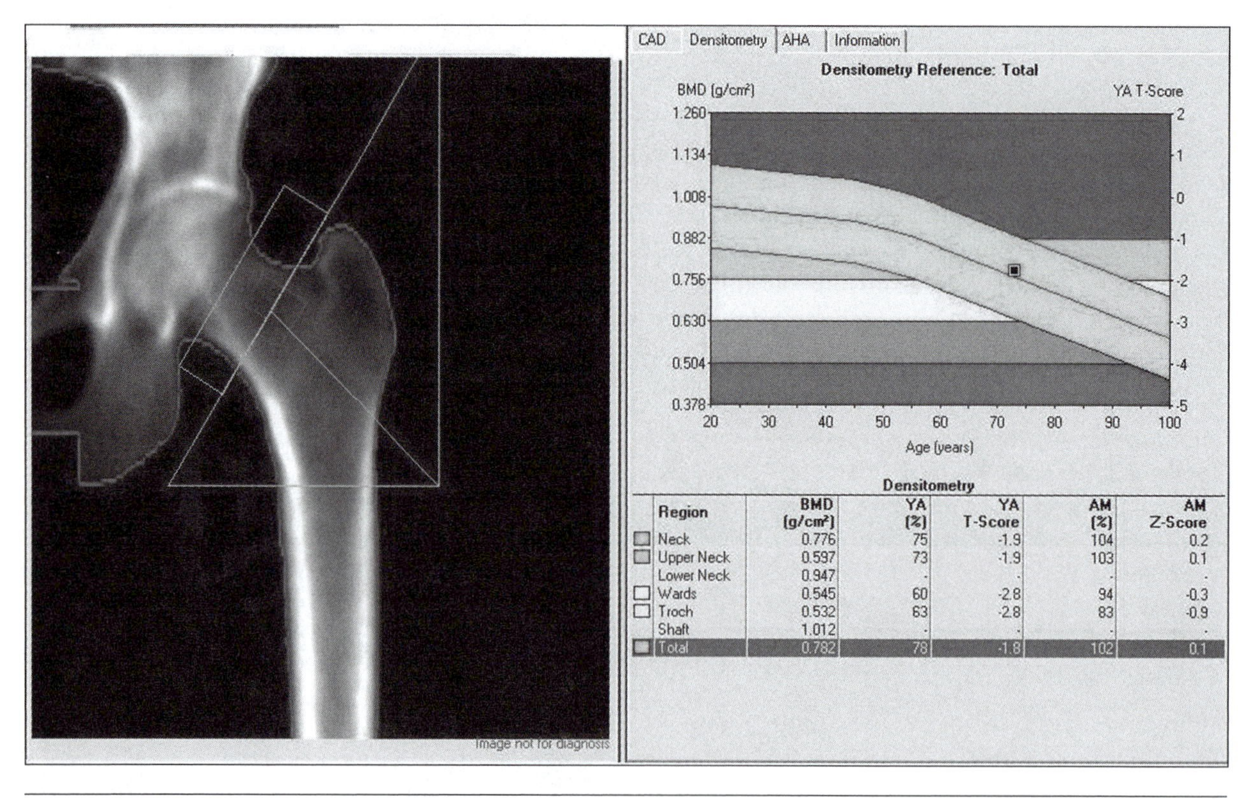

Region	BMD (g/cm²)	YA (%)	YA T-Score	AM (%)	AM Z-Score
Neck	0.776	75	-1.9	104	0.2
Upper Neck	0.597	73	-1.9	103	0.1
Lower Neck	0.947	-	-	-	-
Wards	0.545	60	-2.8	94	-0.3
Troch	0.532	63	-2.8	83	-0.9
Shaft	1.012	-	-	-	-
Total	0.782	78	-1.8	102	0.1

Figure 23.2 Femoral neck scan. Image of a proximal femur scan from a dual-energy X-ray absorptiometer.

Courtesy of Andrew B. Lemmey.

properties of bone (bone structure, quality of trabecular and cortical bone) that cannot be measured with DXA technology. Measurement of such properties along with BMD should provide a better assessment of fracture risk (20). Several very recent advances in tools to evaluate bone health have been emerging, and a thorough review of the diagnostic and evaluation procedures can be found elsewhere (20). Of importance is the new Fracture Risk Assessment tool (FRAX) developed by the WHO (31), which combines BMD with other readily obtainable risk factors (age, sex, and so on) to provide a 10 yr estimate of fracture risk. The clinician should be familiar with the critical anatomical locations where fractures typically occur and should have a fundamental understanding of the modalities used to evaluate BMD. After osteoporosis has been confirmed by low BMD and frequent fractures, the appropriate treatment should follow.

Exercise Testing

Typically, the primary purposes of exercise testing are to aid in the diagnosis of coronary artery disease and to determine an appropriate exercise prescription. In a patient with osteoporosis, the usefulness of exercise testing must be carefully evaluated to make certain that any potential benefits outweigh the risks. However, exercise testing can be of great benefit in generating an exercise prescription for this population. Additionally, most individuals at risk for osteoporosis (e.g., postmenopausal women, sedentary individuals, smokers) are at a greater risk for heart disease, which could potentially be diagnosed with a stress test.

Contraindications

The impact associated with exercise testing could lead to fractures, even in low-intensity protocols such as walking or bicycling. Therefore, individuals with osteoporosis or osteopenia should avoid exercise testing that involves high-impact skeletal loading such as jumping and stepping. In addition, strength testing should avoid twisting movements of the spine and neck. However, the American College of Sports Medicine does not specifically state that osteoporosis is an absolute or relative contraindication to exercise testing (1).

Recommendations and Anticipated Responses

During exercise testing for cardiorespiratory fitness, a protocol using a bicycle is preferred as it involves the least impact and trauma to the bones. During the bicycle protocol, the patient should maintain an upright posture at all times because any sort of spinal flexion or rotation is contraindicated in people with osteoporosis. Treadmill protocols should be avoided due to a loss in height or spinal deformation (or both) associated with vertebral compression fractures in osteoporosis patients. This may compromise ventilatory capacity, which can result in a shift of the center of gravity. These factors can affect balance during treadmill walking (1). Though treadmill protocols are not recommended, they may be used if necessary. However, care should be taken to ensure that the patient does not trip or fall. Within the osteoporotic population, exercise testing should not be avoided as long as the patient can tolerate the exercise with no pain. It is important to note, however, that osteoporosis can sometimes mask the presence of coronary artery disease by preventing the participant from achieving a heart rate and blood pressure necessary for accurate diagnosis. Specifically, the patient may not be able to complete a symptom-limited graded exercise test due to uncomfortable skeletal pain. In instances such as these, pharmacological tests may be a better alternative as they will aid in diagnosis of coronary artery disease without the use of exercise. A summary of exercise testing guidelines is presented in table 23.2.

Treatment

Exercise can be useful to help increase, or at least maintain, bone mass in patients with low BMD. A number of studies in postmenopausal women have shown that exercise can increase bone density or prevent further bone loss compared with values in nonexercising controls (36, 41). But these studies have also pointed out that exercise without other concomitant therapy will generally result in further bone loss. The same situation can be seen in young amenorrheic athletes who continue to lose bone mass, despite their exercise training, as long as they continue without their menstrual cycles (40). Exercise should still be one of the first choices in the treatment of osteoporosis for both men and women because it has the potential not only to increase bone mass but also to increase muscle strength and balance (which may help decrease falls). Several nonpharmacological and pharmacological agents are available to increase, or slow the loss of, bone mass:

- Calcium supplementation
- Vitamin D supplementation
- Estrogen (or hormone) replacement therapy
- Selective estrogen receptor modulators (SERMs)
- Bisphosphonates
- Parathyroid hormone (PTH)
- Denosumab
- Calcitonin

Table 23.2 Exercise Testing

Test type	Mode	Protocol specifics	Clinical measures	Clinical implications	Special considerations
Cardiovascular	Bicycle (treadmill walking if necessary)	Standard test (YMCA bike test, Standard Balke)	$\dot{V}O_2$, METs, RPE	Anticipated physiological responses likely to be similar to those in a healthy population, although no specific research has been done on osteoporotic populations regarding cardiovascular response to exercise.	Minimize inappropriate mechanical stress on high-risk joints (hip, spine) to avoid potential injury or fracture. Severe kyphosis can affect gait, balance, center of gravity, and decrease vital capacity.
Strength	10RM (repetition maximum); 1RM not appropriate due to risk of fracture	Dynamometer (e.g., Biodex isokinetic dynamometer)			May be used to determine intensity of strength training program; but evidence for its usefulness is lacking, and it may be contraindicated in some patients with osteoporosis.
Range of motion		Sit-and-reach, goniometer	Distance		May not be suitable for patients with known or suspected osteopenia.

Calcium is crucial for bone health in the growing years, but data on the usefulness of calcium as therapeutic intervention for either prevention or treatment of osteoporosis are equivocal. Many experts, though, still recommend calcium supplementation for patients with osteopenia or osteoporosis (47). Calcium supplementation may be problematic according to a recent meta-analysis indicating that calcium supplementation may be associated with an increased risk of cardiovascular events (10). Initial results from the Women's Health Initiative, though, indicated that calcium supplementation along with vitamin D did not increase risk for cardiovascular disease (27). However, a recent reanalysis of the data from the Women's Health Initiative suggested that an increased risk for coronary disease may still exist with calcium and vitamin D supplements (11). Thus, given that data are lacking to support efficacy of calcium supplementation as an intervention for the patient with low bone mass, and that calcium supplementation may increase risk of cardiovascular disease, a more careful assessment of all risk factors of a patient should be completed before a patient is advised to take calcium supplements. One important point is that increasing intake of calcium through dietary measures (such as increased low-fat dairy products) has not been associated with increased risk for cardiovascular disease. Thus, for patients with low calcium intake, achieving higher calcium intake through the diet is the best option. Another important point is that vitamin D supplementation may aid in reducing the risk of falls independent of any changes in bone mass (39).

Most of the current drugs with Food and Drug Administration (FDA) approval for osteoporosis are considered antiresorptive therapy (table 23.3). They halt the loss of bone or even increase bone mass by inhibiting bone resorption, while having no effect on bone formation. The only currently FDA-approved drug that increases bone formation is teriparatide (9, 17). Available drugs and their mechanism of action are discussed in chapter 3. An important point to note is that none of the current drug therapies or those under current investigation have been studied in the premenopausal woman with low bone mass, and they do not have FDA approval for use in this population. Thus, for the premenopausal woman with low bone mass, treatment should always focus on the underlying cause, which is most likely an estrogen deficiency. This person should be counseled on ways to regain her menstrual cycles by increasing energy intake. If this fails, or if she refuses to comply, then some form of drug therapy may need to be considered to offset the inevitable bone loss seen with amenorrhea. Insufficient data are available, however, to allow any specific pharmacology treatment recommendations for the amenorrheic population (40).

EXERCISE PRESCRIPTION

Although studies have shown that several forms of exercise training have the potential to increase BMD as well as bone strength (cortical thickness, trabecular architecture, and cross-sectional moment of inertia), the optimal training program for improving or maintaining skeletal

Table 23.3 Pharmacology

Medication name and class	Primary effects	Exercise effects	Special considerations
Hormone replacement Estrogen Estrogen + progesterone	Decrease osteoclast activity, thus decreasing bone resorption	Acute vasodilator but not known to alter exercise responses	Chronic use associated with increased risk of cardiovascular disease
SERMs Raloxifene	Decrease osteoclast activity	May increase vasodilation but no known exercise effects	
Bisphosphonates Alendronate Ibandronate Risedronate Zoledronic acid	Decrease osteoclast activity, thus decreasing bone resorption	Initial use may increase atrial fibrillation but no known exercise effects; data are limited	Can cause GI distress
Calcitonin	Decreases osteoclast activity	None known, but data are lacking	
Denosumab	Inhibits RANK-L, decreasing formation of new osteoclasts	None known, but data are lacking	
Teriparatide	Increases bone formation	None known, but data are lacking	Available only by injection

Note. SERMs = selective estrogen receptor modulators. RANK-L = receptor activator of nuclear factor-kappaB ligand

integrity has yet to be determined. Some studies have found significant correlations between BMD changes and exercise intensity while others have shown no differences in bone response to different intensities of training (6, 21). However, a consistent exercise program may have the greatest benefit on bone. A recent study examining exercise compliance and bone showed that women who were most compliant (accumulating >66 exercise sessions over a period of 12 mo) had 0.5% to 2.5% greater gain in cortical thickness, bone strength, and bone size at the proximal tibia as compared to the least compliant quartile (accumulating <19 exercise sessions) (51). Current experimental knowledge indicates the following with respect to an osteogenic exercise regimen (43):

- It should include load-bearing activities at high magnitude with few repetitions.
- It should create variable strain distributions throughout the bone structure (load the bone in directions to which it is unaccustomed).
- Bone responds to loading in a site-specific manner (exercise should load joints that are at greatest risk for fracture, such as hip, wrist, lower back).
- It should be long-term and progressive.
- Added benefit may result from dispersing loading activities throughout the day rather than completing the exercise all at one time.

Resistance training appears to be the best opportunity to meet these criteria on an individual basis (33, 37), partly because it requires the least skill and has the added advantage of being highly adaptable to changes in both magnitude and strain distribution.

Practical application 23.1 reviews the literature regarding whole-body vibration and bone health.

Resistance training combined with some sort of cardiorespiratory training (bicycling or walking) is probably the best exercise program for an individual with osteoporosis (21). Such a program will not only increase overall fitness, BMD, and bone strength, but will also greatly reduce the risk of falling, which is one of the primary causes of fracture in osteoporosis. However, alternative methods of exercise are available and can be used to increase or maintain BMD (e.g. whole-body vibration).

The exercise prescription, as well as goals of an exercise program for someone with osteoporosis, should not be substantially different from those for age-matched individuals without osteoporosis. For a person just beginning an exercise program, those goals should include an increase in cardiovascular fitness, increased muscular strength, and an increase in both BMD and bone strength (2). Heart disease remains the number-one killer of men and women by a wide margin. Therefore, the goal of both women and men should be to increase physical activity and to decrease their risk of heart disease. Recent guidelines from the American College of Sports Medicine recommend approximately 30 to 60 min of moderate physical activity each day such that the cumulative total for the week is 150 to 300 min (2). Examples of these activities are 30 min of brisk walking, 15 min of running,

Practical Application 23.1

RESEARCH FOCUS

Research has shown that an exercise program consisting of vigorous aerobic and strength training is an effective method for increasing bone strength (21). However, within the older population this type of exercise regimen is lacking with respect to long-term compliance and may in some instances increase risk of fracture. Recently, whole-body vibration (WBV) has been introduced as an alternative method of exercise, and it is less time-consuming than a typical exercise program. During WBV training, the participant performs various exercises on a vibrating platform. Ultimately, the sinusoidal vibrations stimulate mechanical loading and reflexive muscular contractions (46).

A recent study examined musculoskeletal effects of WBV training in postmenopausal women (53). One hundred fifty-one participants volunteered to participate in the study and were randomly assigned to one of three groups: a conventional training group (CT; $n = 50$), a conventional training + whole-body vibration group (WBV; $n = 50$), and the control group (CON; $n = 51$). Both CT and WBV groups trained four times weekly for the duration of the 18 mo study. Both CT and WBV groups performed 20 min of aerobic dance, 5 min of balance training, isometric strength training alternated with stretching exercises, and three sets of upper body resistance exercise using Thera-Bands. In addition, both CT and WBV groups performed dynamic exercises on the vibration platform; for the CT group, the platform did not vibrate, whereas for the WBV group the platform vibrated at a frequency of 25 Hz with 1.7 mm of displacement for a total of 6 of the 15 min. The CON group was instructed to refrain from any training. The researchers measured both lumbar and total hip bone density using DXA and also assessed the number of falls during the study period. A significant increase in lumbar BMD of 2.1% and 1.5% was found in the CT and WBV group, respectively; the control group did not change. Total hip BMD did not change in any group. The number of falls in the WBV group (0.70 falls per person) was significantly lower than in the CON group (1.50 falls per person). So, while the addition of WBV to conventional training did not provide an additional benefit to BMD, vibration training did significantly decrease the number of falls compared to no training, a benefit not found for training without vibration. Therefore, the findings suggest that WBV training may be an effective method for maintaining BMD and decreasing fall risk in postmenopausal women.

and 45 min of volleyball (noncompetitive). This goal would be worthwhile for all people, including those with osteoporosis. However, if the patient with osteoporosis is just beginning an exercise program, the duration of exercise might need to be shortened initially to allow time for adjustment to the activity to reduce the likelihood of musculoskeletal injury. As the person's fitness level increases, the amount of time exercising can be increased as well. For an elderly or osteoporotic individual, a simple walking program should provide the needed benefit along with the important component of safety. Practical application 23.2 presents exercise prescription guidelines for the patient with osteoporosis. Practical application 23.3 provides some guidelines on client–clinician interaction.

Practical Application 23.2

EXERCISE PRESCRIPTION GUIDELINES

In designing an exercise program for the osteoporotic patient, the clinical exercise physiologist should bear in mind two primary goals:

1. Increase overall fitness (cardiovascular, muscular strength, balance, and flexibility)
2. Increase or at least maintain BMD and bone strength

Most individuals with osteoporosis are postmenopausal women or elderly men in whom the risk of cardiovascular disease is significant. To improve cardiorespiratory fitness, this population should use a walking program because it has the potential to improve or maintain BMD. Bicycling, although beneficial for improving overall fitness, is less likely to improve BMD because it is not weight bearing. The walking program should begin at a slow pace and gradually progress over time. Ultimately, the goal would be to walk five times a week for 30 to 45 min at a speed of at least 3 mph (4.8 kph) or faster if possible (but never reaching a jogging speed). The clinical exercise physiologist should ensure that the patient can comfortably handle the speed and distance used for the first 2 wk. These components can then be increased gradually. The best choice is to increase only speed on a couple of days a week and increase the distance walked on the other days of the week.

A resistance training program can help increase BMD, bone strength, and muscular strength in elderly people. Although increases in bone density and strength are generally only modest, strength gains can be much greater and will help reduce the risk of falling, which as mentioned earlier is a major cause of hip fractures. Initially, the resistance should be set low enough that the person can easily complete 15 repetitions of an exercise without undue strain. Exercises should target all the major muscle groups, especially for the legs and back. After the first 2 wk, during which the client does one set of 15 repetitions for each exercise, a second set for each exercise should be added. After a couple of weeks of two sets per exercise, resistance can be added to increase the intensity of the program. The resistance should be increased enough that the client can complete no more than 10 to 12 repetitions on the second set. After that, the weight should be increased gradually as needed to maintain the 10- to 12-repetition limit on the second set. Depending on the patient, a clinical exercise physiologist may use any of the following modes of resistance training when developing and exercise program: weight machines, elastic bands, dumbbells, kettle bells, or any combination of these. Exercises should always be performed with slow, controlled movements. Table 23.4 provides a summary of exercise prescription guidelines.

Table 23.4 Exercise Prescription

Training method	Mode	Intensity	Frequency	Duration	Progression	Goals	Special considerations
Aerobic	Walking or bicycling	Moderate intensity (40-70% of HRR)	5 d/wk	Accumulation of 30 min per day	Very gradually, increase speed and resistance after 2 wk	30 to 45 min at ≥3.0 mph (4.8 kph)	No jogging; avoid activities that increase risk of falling (e.g., step aerobics). Patients with severe kyphosis may be limited to stationary cycling.
Strength	Resistance training (e.g., weight machines, elastic bands, dumbbells, kettle bells)	15 reps of 8 to 10 exercises (may require a less strenuous program initially)	2 d/wk	One or two sets up to 30 to 60 min	After an initial 2 wk, add a set	3 to 4 d/wk at 10 to 12 repetitions per set	Avoid spinal flexion and rotation; use slow and controlled movements (avoid jerky movements). Target legs and back as these sites are most commonly affected by osteoporosis.
Range of motion	Stretching	Stretch to maximal range of motion without pain	5 to 7 d/wk	15 to 20 min		Increase or maintain range of motion	Stretching exercises involving spinal flexion should be avoided.

Practical Application 23.3

CLIENT–CLINICIAN INTERACTION

Osteoporotic patients can fracture with little or no trauma, so special issues should be considered in the development of an exercise program for a patient with osteoporosis or osteopenia. High-impact activities such as running, jumping, or high-impact aerobics should be avoided in both populations. Another activity that absolutely must not be done by people with osteoporosis or osteopenia is spinal flexion, which drastically increases the forces on the spine and increases the likelihood of a fracture; exercises to be avoided include sit-ups and toe touches. Other activities that should be avoided are those that may increase the chance of falling, such as trampoline exercise, step aerobics, or exercising on slippery floors (table 23.5).

Table 23.5 The Osteoporotic Patient and Exercise

Beneficial exercises	Exercises to avoid
Modified sit ups in which only the head and shoulders come off the ground	Any type of abdominal crunch that involves the spine coming off the ground
Lying leg lifts—the person lies flat on a firm surface with legs straight out in front and then lifts legs 6 in (15 cm) off the floor	Any movement involving spinal flexion such as toe touches or rowing on machines
Back extension exercises	Jogging or running
Walking to help increase cardiovascular fitness and increase bone mineral density	High-impact aerobics or other activities that may jar the spine or hip
Strength training exercises using either dumbbells or resistance exercise machines; pay special attention to hips, thighs, back, and arms	Leg adduction or abduction or squat exercises with any significant resistance
Exercises to improve balance or agility such as standing on one foot (with assistance if needed) for ≥15 s	Exercises or activities that increase the chance of falling, such as using trampolines or exercising on slippery floors

EXERCISE TRAINING

Exercise and physical activity are essential for bone mass in all phases of the life cycle. They may reduce the risk of osteoporotic fractures by maximizing bone during growth and development as well as slow the rate of bone loss with aging. However, it should be noted that excess exercise training can sometimes be detrimental to bone health, especially if it leads to amenorrhea, as is often seen in female athletes. Table 23.6 provides a brief review of considerations for exercise training.

Cardiovascular Training

Studies evaluating cardiovascular adaptations in patients with osteopenia or osteoporosis are lacking. Furthermore, studies that have examined the effects of endurance activities on bone density in older populations have typically found that aerobic activities have smaller effects on either femoral neck or lumbar spine BMD than resistance training (37, 38). However, a recent study showed that adding aerobic exercise to a weight loss program increased BMD at the femoral neck by 2% in overweight postmenopausal women (45). In older individuals, prolonged cardiorespiratory training (> 16 wk) can increase fitness levels by 10% to 30% (2). Because endurance training can decrease cardiovascular disease risk factors such as hypertension and cholesterol, it should be incorporated into an exercise regimen for osteoporotic women and men. Weight-bearing endurance activities (e.g., walking) may be most beneficial for retaining bone density.

Resistance Training

Resistance training offers the most benefits for muscular strength and bone density. Current recommendations suggest a single set of 12 to 15 repetitions of 8 to 10 exercises involving the major muscle groups 2 d/wk (2). This

Table 23.6 Exercise Training

Cardiorespiratory endurance	Skeletal muscle strength	Skeletal muscle endurance	Flexibility	Body composition
Weight-bearing activities may be most beneficial for retaining bone density. Endurance training can decrease cardiovascular disease risk factors.	Increases are seen in muscle strength and bone density with resistance training.	Benefits seen in healthy populations are similar to those expected in the osteoporotic patient.	Benefits seen in healthy populations are similar to those expected in the osteoporotic patient.	Benefits seen in healthy populations are similar to those expected in the osteoporotic patient.

is a worthwhile goal for the person with osteoporosis, but a less strenuous program may be needed initially. Care should be taken to avoid the exercises previously mentioned that are dangerous for people with osteoporosis.

Balance and Agility Training

Neuromuscular training, which incorporates balance and agility, has been shown to be effective at preventing falls if performed two or three times per week (2, 4, 19, 54). General guidelines from the American College of Sports Medicine include (1) a gradual reduction of the base of support through postural changes (e.g., progressing from a two-legged stand to a one-legged stand), (2) dynamic movements that challenge the center of gravity (e.g., tandem walk), (3) stressing stabilizing muscle groups (e.g., toe stands), and (4) reducing sensory input (e.g., standing with eyes closed). Furthermore, the exercise prescription should focus on strengthening the quadriceps muscles because poor strength in this muscle group has been linked to risk of falling (25). Squats with free weights should be avoided, however, because of the excess load applied to the spine as well as the potential for spinal flexion during the squat lift. The osteoporotic patient should also be encouraged to do spine extension (but not spinal flexion) exercises. Spine extension exercises can be performed in a chair. These exercises can strengthen the back muscles, which should help reduce the development of a dowager's hump and possibly reduce the risk of vertebral fracture. Patients with osteoporosis should perform these and all exercises with slow and controlled movements; they should avoid jerky rapid movements.

Whole-Body Vibration Training

Recent research has explored the possibility of added bone health benefits through whole-body vibration training (46). In postmenopausal women, 6 mo of vibration training resulted in a 0.93% increase in BMD (52). Furthermore, a study comparing vibration training to a walking program in postmenopausal women found that whole-body vibration elicited a 4.3% increase in BMD at the femoral neck as compared to walking (22). Though vibration training appears to have a positive effect on bone, a detailed exercise prescription has not been presented. Perhaps a progressive overload model can be applied to this mode of exercise with gradual increases in vibration intensity and duration.

Clinical Exercise Pearls

- Low bone mass can be assumed to exist in postmenopausal Caucasian women with low body weight, and therefore certain precautions should be taken.
- Currently, osteoporosis is neither an absolute nor a relative contraindication to exercise.
- Osteoporotic fractures can occur with little or no trauma; therefore, high-impact activities such as running, jumping, or high-impact aerobics should be avoided within this population.
- Spinal flexion should be avoided at all costs in the osteoporotic population (i.e., sit-ups and toe touches).
- The exercise physiologist should also avoid any activities that may increase the patient's chance of falling (i.e., trampolines, step aerobics, BOSU ball, or exercising on a slippery floor).

CONCLUSION

As discussed throughout this chapter, exercise can have a multitude of health benefits, but a clinician caring for an osteoporosis patient must observe caution regarding the type of exercise program to be used and the specific exercises to be performed. Patients with severe osteoporosis who are just beginning an exercise program should probably be supervised until it is determined that they can properly perform the exercises without danger to themselves. Nevertheless, with proper guidance and education, the osteoporosis patient can self-sufficiently sustain a viable level of physical activity.

Key Terms

bone formation (p. 448)
bone mineral content (p. 450)
bone mineral density (BMD) (p. 447)
bone resorption (p. 448)
cortical bone (p. 450)
dual-energy X-ray absorptiometry (DXA) (p. 450)
osteopenia (p. 447)
osteoporosis (p. 447)
peak bone density (p. 448)
trabecular bone (p. 448)

CASE STUDY

MEDICAL HISTORY

Ms. PW, a 58 yr old Caucasian woman, is referred by her physician for the development of an exercise prescription. She is 6 yr postmenopausal and has never been on hormone replacement therapy. In 1987, she was diagnosed with hyperthyroidism (overactive thyroid gland), which was treated with radioactive iodine. Since the treatment, Ms. PW has been on synthroid, a drug used when the thyroid gland does not produce enough thyroid hormones. She has mild hypertension, which is regulated by blood pressure–altering medications. Ms. PW is 5 ft 5 in. (165 cm) tall with a weight of 128 lb (58 kg), and her body mass index is in the normal range. She does not follow a regular exercise regimen but reports that she enjoys biking.

DIAGNOSIS

A few months ago, Ms. PW was diagnosed with osteopenia. Considering her medical history, she is likely to become osteoporotic. Her risk factors are age, sedentary lifestyle, being Caucasian, being petite in stature, loss of menstrual function, and lack of thyroid hormone production.

EXERCISE TEST RESULTS

She recently underwent an exercise stress test that was normal, but no physiological testing was done other than regular blood chemistries.

EXERCISE PRESCRIPTION

A walking program is surely the best choice because it has the potential of improving not only her cardiorespiratory fitness, but also her bone density. However, Ms. PW is particularly fond of bicycling, and therefore the exercise prescription will incorporate some biking activities, as that is likely to increase her adherence to exercise. These cardiorespiratory activities will start slow and gradually progress over time. Ultimately, her goal would be to walk (3.5 mph [5.6 kph] or faster) or bike 30 to 45 min five times a week. In order to reach this goal, Ms. PW may need to start walking or biking at a slower pace and for a shorter distance, but she should strive to exercise five times a week. After it is determined that she can comfortably handle the speed and distance used for the first 2 wk, these two components should be gradually increased. The best choice is to increase the speed a couple of days a week and increase the distance covered on the other days of the week.

A resistance training program should also be implemented, and may be more beneficial with respect to bone than the cardiorespiratory program. Resistance training should also start slowly, with the frequency beginning at 2 d/wk and moving up to 3 or 4 d/wk. Initially, resistance should be low enough that she can easily complete 12 repetitions of an exercise without undue strain. Exercises that target all the major muscle groups should be incorporated, especially for the legs and back. However, exercises that involve spinal flexion should be avoided. In order to strengthen abdominal muscles, Ms. PW can do leg lifts or modified sit-ups, in which only the head is being lifted off the floor. Upon completion of the first 2 wk, during which the patient does one set of 12 repetitions of each exercise, she can add a second set. This same progression can be used after Ms. PW completes another 2 wk. At this point, the patient should start increasing the intensity by slowly increasing the resistance. An adequate resistance at this point would allow the patient to complete no more than 10 repetitions on the second set. All exercises should be performed with slow, controlled movements. Based on Ms. PW's medical history, this exercise prescription should be well tolerated and should provide optimal health benefits. Individual cases vary, however, and for a woman who has suffered fragility fractures as a result of osteoporosis, even the minimal exercise described here may not be tolerated initially. Such a patient will need a more gradual, less intense program.

DISCUSSION QUESTIONS

1. What other information should be obtained from Ms. PW?
2. What other physiological tests might be recommended?
3. Explain the difference in benefits achieved through cardiorespiratory training versus resistance training for this patient. Why is it always important to incorporate both?

Chapter 24

Nonspecific Low Back Pain

Jan Perkins, PT, PhD
J. Tim Zipple, PT, DSc

Most people in modern society experience nonspecific low back pain (NSLBP). This complicated and poorly understood phenomenon involves the interaction of a wide range of physical, social, and psychological factors. For an individual, the effect may be devastating; for society, the costs are enormous. Although exercise is widely used in the management of NSLBP, experts do not agree on the optimal type or dose of exercise. Furthermore, exercise prescription is complicated by differences between acute, subacute, and chronic NSLBP and by substantial variability in recommended management strategies.

DEFINITION

NSLBP is an umbrella term that includes pain in the lumbosacral area caused by a variety of somatic (musculoskeletal) dysfunctions. A specific origin of pathology, such as herniated disc, vertebral fracture, congenital malformation, or general medical condition, has not been identified. Theories about causation appear to be as plentiful as the number of structures in the back. Indeed, "the actual origin of the pain is more of a philosophic statement of the training of the practitioner than hard, scientific fact" (4, p. 63). Also, NSLBP is not a single entity but rather a syndrome with pain and disability as the most important symptoms (53). The best definition of NSLBP may be simply pain experienced in the lumbosacral region in the absence of major identifiable pathology. The pain is typically diffuse and located in a region that includes the areas of the back below the ribs and above the distal fold of the buttocks (53).

This chapter focuses primarily on management of subacute, recurrent, or chronic NSLBP. Management of the acute phase of an episode of NSLBP mainly involves medical screening to exclude serious conditions, reassurance as to the benign nature of the problem, and simple symptom-based treatment (53, 101). The minority of back pain caused by serious pathologies or requiring urgent medical or surgical intervention is not covered, and postsurgical rehabilitation routines are also excluded. The primary care medical screening for serious pathology is also not covered, although a recent overview for primary care physicians may be of interest (54). For information on typical postsurgical rehabilitation, recovery paths, and reviews of surgical outcomes, refer to other resources (e.g., references 24, 30, 83), which should be used in consultation with the health care workers caring for individual clients. Therapeutic exercise has not been shown to be any better than other treatments in promoting the resolution of an acute episode of NSLBP (52, 53, 103). The mainstay of prescribed treatment is to avoid prolonged bed rest and stay as physically active (without aggravating the condition) as possible (36, 37, 53, 97). As pain improves and the condition becomes subacute, exercise can be more helpful (53, 103). Exercise treatment may possibly help prevent recurrences and is suggested most strongly for chronic low back pain (52); but more research is needed, and there is insufficient evidence on which forms of exercise should be used (15, 52).

SCOPE

Most people, in all societal groups, will experience at least one episode of NSLBP during their lifetime (22, 27). Back problems are "the most expensive musculoskeletal affliction and industrial injury and the most common cause of disability for Americans under the age of 45" (7, p. 192). Reported rates of back pain vary widely (18). A recent systematic review found annual rates between 14% and 93% in individuals with a history of back pain (40), whereas a population-based cohort study in North America found an annual **incidence** of 18.6%, with most episodes being mild and with a recurrence rate of 28.7% within 6 mo (10). Another study found an annual incidence of back pain of 49.1% in working-age adults (16). The variability likely reflects different locations, definitions, populations, and survey techniques (18). But all experts agree that NSLBP is a frequent aspect of life. Many people experience distress and some degree of temporary or mild disability, but only a small percentage become seriously disabled or go on to experience chronic back pain (53, 101). Another problem is that the prevalence of disabling low back pain appears to have increased over the past couple of decades (26). The incidences of obesity and depression are also both increasing (11, 14, 17, 43), and it has been suggested that these factors are linked (14, 17, 26)

Up to a point, NSLBP increases with age. A cohort study that examined individuals at age 14 and used survey follow-up at age 38 found that **prevalence** of NSLBP increased as the participants aged (with lifetime prevalence at 70% by age 38) and that rates for men and women were similar (34). Many others have noted an increase in NSLBP with age, with a peak between age 45 and 60, followed by a decline in reported pain (36, 47, 58). This finding may be attributable to the inherent stability of the spine associated with the loss of elasticity and increased stiffness seen with aging. Older people may also be more cautious with lifting. Younger people are more likely than older people to experience complete resolution of an episode of pain, whereas older people who continue to have pain have more frequent and prolonged pain (10). But back pain in childhood and adolescence tracks into adulthood, suggesting that we need to pay more attention to first-time episodes of NSLBP in young people to determine whether **prevention** at that stage can decrease later, more severe problems (39). Onset in childhood is associated with participation in competitive sport, with incidence increasing as weekly time in sport increases (84). The relatively high rates of NSLBP in young adults and adolescents in cohort studies, and the association with sport in both genders (15, 39, 84), match our clinical experience: Many adults describe a first episode of NSLBP in high school, frequently precipitated by athletic competition, heavy lifting, or trauma. The cost of managing the minority of NSLBP patients whose acute episode becomes chronic is enormous. Chronic back pain direct cost estimates in the United States range from $12.2 to $90.6 billion each year (32).

The average adult can expect between one and three episodes of NSLBP in a year. In most cases, a single episode resolves spontaneously without producing major disability (10, 53). In a small percentage of cases, acute episodes of NSLBP lead to chronic NSLBP (10, 53).

PATHOPHYSIOLOGY

NSLBP is a syndrome with many causes and consequently many poorly understood physiologic abnormalities. Up to 85% of low back pain has no identifiable cause (93). In our clinical experience, the common presenting diagnosis of "low back pain" in not really a diagnosis, but a symptom. In this chapter the focus is on this type of NSLBP, for which precise pathologic descriptions are impossible. Opinions as to the structure causing pain have varied over time and may indeed be more reflective of fashions in health care and contemporary technologies than actual pathologies.

An impressive list of structures with pain receptors could cause NSLBP, including the anterior and posterior ligaments, interspinous ligament, yellow ligament, posterior annular fibers of the disc, intervertebral joint capsules, vertebral fascia, blood vessel walls, and paravertebral muscles (18, 104, 107). Mechanical stimulation during microsurgery has been used to test tissues for pain sensitivity. Fascia, supraspinous and interspinous ligaments, spinous process, muscle, lamina, ligamentum flavum, facet capsule, facet synovium, annulus fibrosus, and nucleus pulposus were tested (42). Pain was "always" or "usually" caused with nerve root compression and with stimulation of the outer rim of the annulus fibrosus, vertebral endplate, anterior dura, and posterior longitudinal ligament (42). All other structures tested rarely or never produced pain consistently (42).

Many experts consider the intervertebral disc the most frequently implicated structure in nontraumatic NSLBP, and popular treatments are often based on theories relating to disc pathologies (4, 98). Others disagree and consider the importance of the disc as the source of NSLBP to have been overrated (104). It makes intuitive sense to suppose that NSLBP may be caused by multiple anatomic structures that may all respond similarly to the appropriate regimen. An approach that focuses on risk factors and prevention rather than pathologies is best.

CLINICAL CONSIDERATIONS

If an individual's medical history includes NSLBP, a prudent approach is to determine whether it may be associated with a more serious pathology that would require further physician evaluation before exercise testing and training. Medical screening is important for those with new-onset NSLBP for the identification of the minority of people with potentially serious pathology (18, 52-54, 96).

A number of risk factors are associated with NSLBP, but the strongest predictors of recurring NSLBP are the length of time between episodes of NSLBP (10, 47, 88) and a history of back pain (39, 53). Numerous medical diagnostic tests can be used to assess NSLBP. These tests generally are not beneficial, however, unless it is thought that the NSLBP is associated with a serious spinal pathology, such as a space-occupying lesion (neoplastic growth), **cauda equina syndrome**, or another underlying medical condition. Once serious pathology has been ruled out, noninvasive treatment rather than surgery is strongly suggested in most cases. An exception is a situation in which precise diagnosis is possible, such as a disc **herniation** with progressive neuromuscular deficits. In this case, surgical treatment has a high rate of success.

As a patient goes through treatment for acute to subacute NSLBP, consideration should be given to **secondary** prevention to decrease recurrence rates. In addition, because of the high incidence of NSLBP in our society, **primary** prevention is strongly encouraged, especially in occupational, household, and recreational pursuits with a high incidence of NSLBP. Many recent studies suggest that physicians and other health care providers should help patients change their perception of NSLBP as a serious disorder that results in permanent disability. They suggest shifting the emphasis back onto the patient with a convincing argument that self-care produces effects comparable to those with traditional medical care (15, 77, 100). Only 5% of people who experience an acute episode of NSLBP go on to chronic pain and disability (53). Increasing use of interventions for NSLBP has not had the intended effect on back pain disability, leading experts to question their suitability for widespread use (21, 27).

Signs and Symptoms

The person presenting with NSLBP may complain of localized or generalized lumbosacral region pain of variable intensity, duration, and frequency. Radiating pain with a specific distribution of sensory changes, numbness, or lower extremity weakness can be associated with more serious pathology and can indicate specific tissue involvement. The client may have NSLBP with weight-bearing activities and complain of increasing pain with certain lumbar motions and postures. NSLBP may cause nocturnal discomfort that awakens the client when changing positions in bed. Symptoms are usually decreased with rest and anti-inflammatory medication. Any client presenting with **red flags** or other symptoms such as those listed in table 24.1, or with new undiagnosed symptoms, chest pain, heart palpitations, shortness of breath, hernia, or unremitting spinal pain that is not relieved by rest, should consult with a physician before beginning an exercise program. Clinicians should also be alert to signs and symptoms that may indicate an inflammatory arthritic disorder such as **ankylosing spondylitis**, because a disorder of that type will require further medical evaluation and treatment.

A number of established risk factors pertain to NSLBP and disability. Prevalence increases with age into midlife. NSLBP is lower in those with greater endurance in back extensor muscles, and people who have had previous episodes of NSLBP are more likely to experience additional back pain. This risk increases as the interval since the last episode shortens. People with recurrent or persistent back pain have decreased flexibility of hamstring and back extensor muscles and lower trunk muscle strength. The best predictor of an episode of back pain (and the only one sufficiently discriminative to be valuable in job selection) is a history of previous episodes, with risk of recurrence being higher the more recent the previous episode (10, 47, 88). Other risk factors for back pain include obesity (55), smoking (31), and whole-body vibration such as with prolonged motor vehicle driving (75, 99). Sedentary occupations are also a possible risk factor (75, 99, 104). Heavy lifting, or lifting with twisting regardless of the weight, is also a risk factor for back pain (75, 99). Research is now beginning to clarify the influence of genetics on degenerative disc disease (67).

Note that the associations do not necessarily imply causation. For example, although a modest correlation exists between back pain and obesity, it is not found in all studies, and some experts believe that the association is unlikely to be causal (53, 56). Instead, obesity may play a part in simple NSLBP's becoming chronic or recurrent.

Psychosocial and work environment factors are more predictive of back pain and disability than are physical examination findings or mechanical stress at work (94, 99, 100, 102). Psychological distress, dissatisfaction with employment, low levels of physical activity, poor self-rated health, and smoking status are, along with poor spinal movement, associated with persistent back pain after an acute episode (94, 100, 101). For many years, calls have been made for a shift from treatment of pain to management of disability (100, 102). This proposal may

Table 24.1 Indications of Possible Severe Spinal or General Pathology

Red flags	• Age under 20 yr or over 55 yr **Significant medical history, such as the following:** • Carcinoma • Systemic steroids • Human immunodeficiency virus (HIV) • Other major medical pathology • High-impact trauma such as a fall from a height or a motor vehicle accident (or low-impact trauma in a person with known or suspected osteoporosis) • Constant or progressive nonmechanical pain (not related to particular times or actions) • Spontaneous or persistent pain at night or with lying supine • Undiagnosed thoracic pain in addition to the low back pain **Systemic unwellness or with constitutional signs such as the following:** • Unexplained weight loss • Fever • Nausea • Vomiting • Current or recent infection • Widespread neurologic signs or symptoms such as major or progressive motor weakness or reflex changes • Anesthesia around anus, perineum, or genitals, or difficulties with bladder or bowel function (fecal incontinence, urinary retention, or incontinence) • Severe and lasting restriction of flexion
Inflammatory disorders such as ankylosing spondylitis	• Insidious onset in young adulthood • Prolonged morning stiffness • Limitation of spinal movement in all directions • Peripheral joint inflammation • Family history of ankylosing spondylitis or related disorders • Recurrent sacroiliac inflammation may be the first clinical sign • Psoriasis • Iritis

Based on Bigos et al. 1994; Bigos and Davis 1996; Bogduck 2005; Koes et al. 2001; Koes, van Tulder, and Thomas 2006; van Tulder 2006; Waddell 2004.

be the key to cost control, because disability costs and ability to return to work seem to be more strongly influenced by modifiable psychosocial factors than by physical attributes (38, 100). Failure to address these issues will lead to inadequate management of individual cases and continued high costs to society (29, 38, 100, 102).

In attempting to predict long-term disability in the small percentage (1%) of people who don't recover from NSLBP, Waddell and Burton (102) describe "yellow flags" for risk of chronicity. These yellow flags are primarily associated with several categories of the biopsychosocial model and represent obstacles to recovery as elaborated in table 24.2. Yellow flags are found in most modern guidelines for back pain management, but there is some variation in what is considered a yellow flag (52).

Psychosocial factors are important factors in management of subacute and chronic NSLBP, and most authorities suggest that they be addressed as a part of conservative management (52, 74). Studies found that strong encouragement by the practitioner and selection of an enjoyable general exercise program are equally effective and potentially more cost-effective than a management strategy that employs specific back exercises. Most current guidelines recommend general aerobic exercise (swimming, walking, cycling), stretching, and strengthening exercises as part of a multidisciplinary management plan that also addresses psychosocial factors (22, 36, 37, 41, 50, 52, 53, 77, 85, 90, 96, 97).

Conventional wisdom has long argued that the natural history of a single episode of acute NSLBP is fairly

Table 24.2 Yellow Flags Indicating a Risk of Developing Chronic Pain

Personal and psychosocial obstacles	Work preparedness obstacles	Environmental and social obstacles
Dysfunctional attitudes, beliefs, and expectations about pain and disability	Physical or mental demands of work	Inappropriate medical information and advice about work
Inappropriate attitudes, beliefs, and expectation about health care	Occupational "stress"	Medical leave practices that sometimes reinforce illness behaviors
Uncertainty, anxiety, fear, avoidance	Job dissatisfaction	Lack of occupational health provisions
Depression, distress, low mood, negative emotions	Lack of social support at work, relationships with coworkers and employer	Employers' lack of understanding of common health problems and their modern management; assuming that they automatically mean sickness absence
Passive or negative coping strategies (e.g., catastrophizing)	Attribution of health condition to work-related activities	Coworkers' unhelpful attitudes and behaviors
Lack of motivation and readiness to change, failure to take personal responsibility for rehabilitation, awaiting a "fix," lack of effort	Beliefs that work is harmful and that return to work will cause further damage or be unsafe	Belief by many employers that symptoms must be "cured" before they can risk permitting return to work, for fear of reinjury and liability
Illness behavior	Self-perceptions of current and future "work-ability"	Loss of contact and lack of communication between worker, employer, and health professionals
	Beliefs about being too sick or disabled to contemplate return to work	Lack of suitable policies and practices for sickness absence, return to work, and so forth
	Beliefs that one cannot or should not become fully active or return to work until the health condition is completely "cured"	Rigidity of rules of employment, duties, and sick pay
	Expectation of increased pain or fatigue if work is resumed	Lack of modified work
	Low self-efficacy	Organizational size and structure; poor organizational culture
	Low expectations about return to work	Impending downsizing, termination of employment
	Beliefs and expectations about (early) retirement	Detachment and distance from the labor market

Adapted from *Best Practice & Research Clinical Rheumatology* Vol. 19, G. Waddell and A.K. Burton, "Concepts of rehabilitation for the management of low back pain," pgs. 655-670, Copyright 2005, with permission from Elsevier.

well defined, and 75% to 90% of people recover within a few weeks (100). Unfortunately, recent epidemiological studies have found that although improvement is usual, complete resolution is less typical and recurrences are common (40, 51, 100). The risk of additional episodes following a brief acute episode of back pain is 60% to 75% within a year (40, 95). In line with the traditional perception that most acute NSLBP will improve on its own, aggressive intervention is often reserved for the minority of people who do not improve in 2 mo. The increasing awareness of the recurrent nature of the problem has led some experts to reevaluate their approach. They now suggest that exercise therapy and self-management strategies be initiated in the subacute stage in the hopes of decreasing the risks of recurrence and persistence of pain. Recent work has shown that individuals with subacute back pain have more modifiable psychosocial risk factors than those with chronic pain, making this population a key one to target with psychosocial interventions (38).

History

Medical evaluation should have cleared the individual referred with NSLBP for major pathologies, but a clinical exercise physiologist should be alert to the possibility that serious pathologies have been missed. Table 24.1 lists key red flag findings that indicate when medical evaluation is required.

The history of an individual with NSLBP should first focus on this screening for possible serious pathology. Questions should cover all the areas indicated in table 24.1. Following this, the interview should focus on the mechanisms of injury, both initial and for recurrences, because this information may guide the practitioner in selecting management strategies. For example, recurrences may be associated with specific situations such as spinal flexion with lifting, prolonged postures as with driving, or particular sporting activities. Careful questioning and analysis of such patterns can guide physical examinations, suggest postures and activities to avoid or use in early rehabilitation, and indicate what education is needed during the rehabilitation process. For example, a patient with a history of pain recurrences following long periods of driving with the spine in flexion may benefit from exercises that avoid this posture, from modifying motor vehicle seating, and from education on incorporating breaks for stretching and brief walking into any long drives in the future. A history of usual vocational and recreational activities can also offer valuable hints for education on self-management strategies after resolution of the current episode. Asking about the benefit of treatments tried for previous episodes is also important in directing treatments for the current episode of NSLBP.

Physical Examination

Again, the physical examination in NSLBP is primarily one of exclusion. An examiner should do a quick scan of posture and general range of motion. A neurologic screening should check for any abnormalities of sensation, motor function, or reflexes that may indicate serious pathologies. Assuming that these scans are clear, the physical examination should include range of motion and flexibility testing that specifically looks at common deficits seen in NSLBP. Hamstring and hip flexor tightness are common in people with NSLBP, and spinal flexion and particularly extension are frequently limited. As a minimum, these should be tested. Palpation of paraspinal muscles may reveal increased muscle tone and tenderness.

Ideally, the examination would incorporate aerobic testing and spinal and abdominal muscle strength testing. Back extensor and abdominal muscle weakness is common in individuals with NSLBP, and those who have a long history of recurrences may be considerably deconditioned even when compared with their sedentary peers without back pain. Unfortunately, if the client is being seen early after a flare-up of pain, strength and aerobic testing may be difficult and the results may be invalid because of pain limitations. In this case, formal testing of these functional capacities may have to be deferred. The examination should instead clear the patient for general safety for exercise using American College of Sports Medicine (ACSM) guidelines.

Diagnostic Testing

Diagnosis of specific low back pathology is fraught with difficulty. Despite the use of highly sophisticated tests, experts are often unable to give definitive diagnoses, and it is argued that the diagnosis given relates more to the specialist who is consulted than to the patient's back (12, 101). Radiographic imaging identifies loss of disc space, malalignment (e.g., **scoliosis**, **spondylolisthesis**), and osteoarthritic changes such as **osteophytes** or stenosis. Often these do not correlate with the severity of signs and symptoms. Abnormal findings on imaging studies are common and do not predict back pain recurrence (45, 46). Some have suggested that imaging is currently overused and should be reserved for cases in which serious pathology is suspected, or possibly when conservative management has failed (13, 19, 21, 45, 46, 54, 68, 100). **Magnetic resonance imaging (MRI)** is better at soft tissue examination than plain X-ray is, but neither technology can identify the pain receptors responsible for the reported pain. Use of diagnostic injections or neurolytic injections has shown some promise at identifying pain-sensitive structures and a specific cause of LBP.

Most people who see physicians for back pain do not have serious pathology. In primary care, approximately 1% of patients have back pain from serious spinal pathology (100). Careful screening for red flag symptoms that indicate serious pathology is an essential part of primary health care management (6, 7, 52-54, 96, 100) (table 24.1). For most cases in which an underlying systemic disease or clearly definable injury is not present, diagnosis may be less important than management.

For determining appropriate examination tests and measures used by clinical exercise physiologists, consult chapter 5 of this text. Besides being able to measure and record vital signs, the exercise physiologist is able to perform baseline postural observations and basic gait deviations related to velocity, cadence, and base of support. Exercise physiologists can perform palpation on common superficial anatomical landmarks to determine soft tissue irritability and help locate the source of pathology. For a more thorough neuromusculoskeletal examination, a physician, physical therapist, or other suitable practitioner may be the appropriate referral.

Exercise Testing

NSLBP is a diagnostic category that does not in itself indicate a need for a graded exercise test (GXT). Maximal or submaximal testing, although useful for prescription, is not required in general clinical practice unless history indicates possible coronary artery disease or other medical conditions that would normally require a GXT before the formulation of an exercise prescription (81). The ACSM guidelines suggest that older adults (men ≥45 or women ≥55) should have medical clearance, including exercise testing and follow-up, if vigorous exercise is planned (2, 49). ACSM suggests similar precautions for those at increased risk for cardiovascular events and for those with known pulmonary, cardiac, or metabolic disease (2, 49). People with NSLBP should be screened routinely to identify individuals in any of these categories. For formal GXT, any submaximal protocol that does not exacerbate pain may be used. The selection may be made based on the client's history. For example, those who find the spinal loading stress of walking painful may do better with a cycle ergometer test. Submaximal testing is suggested because pain may prevent maximal testing. For muscle strength testing, submaximal tests are suggested. Similarly, testing for flexibility can be performed according to usual protocols. In all cases, information on whether a particular movement is known to trigger pain should be solicited and assessment modified accordingly. For example in a cycle test, people may find that a forward lean position aggravates their pain, and a modification to minimize this would increase the accuracy of the test. Education for the client is important in light of the emphasis on learning lifelong self-management and prevention strategies. Practical application 24.1 describes client–clinician interaction for patients with NSLBP.

Contraindications

A brief period of rest is commonly used for the first few days of an acute flare-up of back pain. During this period, GXT is inadvisable. Moreover, practical concerns relating to the patient's pain pattern and presentation

Practical Application 24.1

CLIENT–CLINICIAN INTERACTION

Many people with NSLBP have not had adequate education about their condition. Consequently, they may have fears about exercise that prevent them from participating fully in exercise programs. They also have often been given testing results that mislead them into believing that serious pathology is present. After evaluation has ensured that the condition is NSLBP, a clinician can use the following points as the core of client education, both at the initial evaluation and as the client goes through the slow and often frustrating process of learning to manage NSLBP.

- Radiographic evidence of pathology does not correlate with level of spinal pain or disability.
- Studies provide evidence that adherence to a specific, progressive exercise program reduces the incidence and frequency of episodic NSLBP and improves function.
- Ergonomic adaptations and postural awareness reduce the incidence of episodic NSLBP.
- Minor lifestyle changes will permit compliance with self-management strategies for episodic NSLBP.
- Improvement in pain will not be instantaneous. Muscle strength can take 4 to 6 wk to improve, and loss of pain may not be noted until the spine is stable and has had time to adjust to muscle and ligament changes.
- After the initial rehabilitation stage, continued general exercise is suggested.

Clinicians should also consider qualitative research findings as they interact with their clients. Individuals with back pain seek clinicians who listen to them, who learn of their individual exercise and activity preferences and experiences, and take these into consideration as they prescribe exercise therapies (86, 87). Taking these steps will increase exercise adherence and client willingness to participate (86, 87).

can affect GXT administration and evaluation. Because both upper extremity and lower extremity ergometers and treadmills require coordinated trunk mobility and stability, pain may interfere with the individual's ability to perform the test or reach maximal exercise levels. Selection of the means of testing should be tailored to the individual. For some, the seated position could be most painful and ambulation may be preferred, whereas for others a stationary bicycle could provoke less pain than treadmill testing does. In any case, pain may prevent the person from reaching maximal exercise levels, and aggressive testing protocols could produce considerable posttesting soreness. A submaximal testing protocol using the exercise modality least likely to cause a flare-up in the given person's pain may be most appropriate for individuals who report pain that is easily aggravated by activity.

Recommendations and Anticipated Responses

No specific GXT protocols have been developed for individuals with NSLBP. Any of the standard testing procedures recommended by the ACSM (2, 49) and in chapter 5 may be suitable. As discussed previously, however, if someone is currently having an acute episode of NSLBP, GXT is not recommended because the person will be limited by pain and unable to expend the necessary effort. When the medical history reveals previous NSLBP, the mode of exercise testing should be carefully determined to avoid further exacerbation. An individual who has experienced multiple episodes of NSLBP and who reports interference with recreation or vocational activities may be deconditioned beyond the level usually seen in sedentary individuals, and this circumstance should be taken into consideration in planning testing. High-impact test protocols, protocols that place the individual in sustained spine flexion, or maximal testing may provoke pain and limit validity of the results.

Treatment

Many medical management strategies are used for NSLBP. Common medical management includes a wide range of medications; prescription of exercise or passive modalities such as heat, massage, or **spinal traction**; facet joint injections; and surgeries such as **spinal discectomy**, **spinal decompression**, and **spinal fusion**. Current emphasis in primary care management of NSLBP is early return to activity, avoidance of needless surgery or use of unnecessary diagnostic tests, and ultimately cost-effective medical and self-management of back pain.

Surgical rates and types of surgery vary widely across geographic areas. Outcomes show no corresponding variation, and agreement is lacking on indications for surgery and on successful outcomes. Surgical rates have increased dramatically, but no decrease in disability has occurred (21, 27, 100). Acceptable clinical trials provide little evidence that the expensive option of surgery offers any benefit over nonsurgical management in most cases of back pain (100). Unequivocal disc herniation is an important exception to this general statement. When strict criteria are adhered to in selecting surgical patients, the success rate can be high (103). Unfortunately, the wide variation in surgical rates nationally and globally attests to the lack of strict application of criteria by all practitioners. The actual surgical management of disc pathology varies from partial excision of the offending protruding disc material, to **laminectomy**, to significant disc excision with a stabilizing fusion using bone fragments from other parts of the body or metal spacers between the vertebrae.

The management of an isolated episode of acute back pain is less controversial than that for subacute, recurrent, or chronic pain. Current medical practice advocates conservative care for acute back pain in the absence of any red flag findings. Bed rest is discouraged and, if used at all, is limited to 24 to 48 h; passive treatment modalities (such as hot packs, **transcutaneous electrical nerve stimulation (TENS)**, **ultrasound**, and traction) are used sparingly; and early resumption of normal activities with or without additional exercise training is advocated (6, 7, 15, 18, 20, 52, 96, 100, 103). Table 24.3 summarizes generally accepted principles in management strategies for acute and chronic NSLBP.

Many patients with NSLBP need pharmacologic intervention only briefly; over-the-counter medications are the most often used. Generally, acetaminophen is the safest medication for low back pain. Nonsteroidal anti-inflammatory drugs (NSAIDs), including aspirin and ibuprofen, are also effective at decreasing pain but tend to cause gastrointestinal irritation and ulcers. They also may cause renal pathology, blood thinning, or an allergic reaction. Either acetaminophen or NSAIDs, or a combination, is the typical first-line choice for NSLBP, and guidelines consistently recommend these as first-line pharmacologic treatment (18, 52, 54).

Muscle relaxants may also be prescribed to people with NSLBP. Although they may reduce pain, a number of side effects limit their use. They carry the adverse effects of drowsiness, impaired motor function, and increased reaction times, creating potential risks with operation of motor vehicles and power tools or machinery. Although some physicians still prescribe narcotic analgesics or opioids for relief of musculoskeletal pain, these appear

Table 24.3 Treatments for Acute and Chronic Low Back Pain

Effectiveness	Acute LBP	Chronic LBP
Beneficial	Advice to stay active NSAIDs	Exercise therapy Intensive multidisciplinary treatment programs
Trade-off*	Muscle relaxants	Muscle relaxants
Likely to be beneficial	Spinal manipulation Behavior therapy Multidisciplinary treatment programs	Analgesics Acupuncture Antidepressants Back schools Behavior therapy NSAIDs Spinal manipulation
Unknown	Analgesics Acupuncture Back schools Epidural steroid injections Lumbar supports Massage Multidisciplinary treatment TENS Traction Thermal modalities EMG biofeedback	Epidural steroid injections EMG biofeedback Lumbar supports Massage TENS Traction Local injections
Unlikely to be beneficial	Specific back exercises	
Ineffective or harmful	Prolonged bed rest	Facet joint injections

*Trade-off refers to the balance between beneficial effects and undesirable side effects.

to have no greater benefit than safer analgesics and are generally considered a poor choice for NSLBP. Narcotics can be addicting, and patients tend to become dependent on these medications during long-term use. As with muscle relaxants, a greater risk of injury occurs with the use of machinery or with driving under the influence of narcotics. Narcotics should never be taken with alcoholic beverages, which can compound their effects. Clients should be warned of the potential side effects listed in table 24.4.

The limited studies that look at primary and secondary prevention (discussed subsequently) make it possible to suggest that some intervention, preferably in a community or worksite setting, may be beneficial in avoiding recurrence of back pain after an initial nonspecific injury. Low-stress aerobic activity is usually considered safe within 2 wk, and "conditioning," especially of back extensors, is begun thereafter (6). Aerobic exercise—fast walking is particularly suitable for people recovering from a NSLBP episode (69)—should continue as long as there is evidence that the patient is making functional progress (23). Few patients have insurance coverage for this type of program in a rehabilitation center, because the current focus of the health care system is on short-term secondary prevention and tertiary prevention. Instead, people may need to continue exercise programs under the guidance of fitness professionals at health clubs and fitness or wellness centers. Many community wellness centers have an important role in both secondary and primary prevention of nonspecific back pain. The health care specialist providing rehabilitation for the client should work collaboratively with the community wellness trainer to create a reasonable long-term maintenance program for the client. This combination returns people to the community setting quickly and encourages them to learn active self-management skills.

Primary prevention is avoidance of a condition for which a person is at risk. Intervening before an injury or disease develops and preventing it from appearing

Table 24.4 Pharmacological Agents Commonly Used With NSLBP

Drug categories and common generic names	Trade names	Side effects	Contraindications
Nonnarcotic analgesics		Gastrointestinal (GI) distress, GI ulcerations, allergic reactions, renal dysfunction	Patients with history of allergic reactions or GI distress with these medications
Aspirin	• Bufferin, Empirin		
Acetaminophen	• Tylenol, others		
Ibuprofen	• Motrin, Rufen, Nuprin		
Naproxen	• Naprosyn, Anaprox		
Piroxicam	• Feldene		
Diflunisal	• Dolobid		
Narcotic (opiate) analgesics		Poor tolerance, gastrointestinal disturbances, sleep disturbances, drowsiness, increased reaction time, clouded judgment, misuse or abuse and dependence issues	Patients with history of poor tolerance or allergic reactions, dependent personality types
Meperidine	• Demerol		
Hydromorphone	• Dilaudid		
Methadone	• Dolophine		
Codeine	• Codeine		
Morphine	• Avinza, Roxicodone		
Oxycodone	• Oxycontin, OxyIR, Tylox, Percodan, Percocet		
Muscle relaxants		Central nervous system depressant, drowsiness, tachycardia, hives, mental depression, shortness of breath, skin rash, itching	Allergies, blood disease caused by an allergy or reaction to any other medicine, drug abuse or dependence, kidney or liver disease, porphyria, epilepsy
Cyclobenzaprine	• Flexeril		
Carisoprodol	• Soma		
Metaxalone	• Skelaxin		

would be the optimal form of preventive health care. For traumatic back pain, frequently caused by vehicle accidents or falls, primary prevention includes standard safety precautions in household, recreational, and occupational situations (4). Safety equipment such as seat belts should be used in all appropriate situations, and watchful caution should be practiced during activities that place people at risk of falling or encountering moving vehicles. For NSLBP in general, leisure-time physical activity for at least 3 h per week is associated with decreased incidence of back pain (33). In addition, a number of "best practices" in work and leisure activities can reduce the risk of sustaining a back injury (69).

A considerable body of literature exists regarding risk factors for NSLBP, but to summarize, primary prevention strategies include the following:

- Use safety equipment in work and leisure activities.
- Address risk factors of smoking, poor general fitness, obesity, stress, and poor seating.
- Perform balanced exercise programs that include both spinal flexion and extension.
- Avoid long periods in one position—take breaks to move the spine out of fixed positions, and balance postures of flexion and extension.
- Avoid lifting with twisting and preserve the neutral lordotic spine curve during lifting.
- Avoid lifting activities immediately after rising or prolonged spinal flexion positions without breaks to move into extension.
- Be physically active and keep fit.
- Vary postures during prolonged sitting.
- Balance activity in flexion with activity in extension.

The more common nontraumatic low back injuries are thought to result from a series of cumulative events in a person with risk factors such that a relatively minor incident precipitates symptoms (4). The best options for primary prevention of nontraumatic back pain come from addressing risk factors such as smoking, obesity, psychosocial stress, and poor seating for those exposed to occupational risk factors like truck driving (4). Any awareness of early signs of dysfunction such as joint stiffness, minor aches and pains, or difficulties straightening the spine can minimize progression. Physical activity that promotes both cardiovascular and musculoskeletal fitness is beneficial, as are a balance of spinal flexion and extension activities and avoidance of prolonged loading of the spine in either flexion or extension. The spine is not meant for static positioning but for movement, and sustained postures or repetitive movements in any direction should be balanced with movement in the opposite direction.

In secondary prevention, treatment is used early in the course of a condition to cure that condition or prevent or slow its progression. In back pain, the aim is to prevent recurrences and to prevent acute and subacute back pain from becoming chronic.

Secondary prevention strategies for back pain include the following:

- Catch problems early before an injury becomes disabling.
- Avoid or minimize bed rest.
- Encourage early return to activity, even in the presence of some pain.
- Avoid aggressive spinal loading exercises during early rehabilitation.
- Avoid loading the spine throughout the full range of motion, particularly in the general (nonathletic) population.
- Keep use of passive modalities to a minimum or use them to assist with a more active program.
- Be willing to adapt a program to individual needs.
- Use behavioral strategies to encourage participation.

After the resolution of the NSLBP episode, continuing to follow the commonsense principles of primary prevention is important. Preferred management strategies include emphasizing early return to activity even with some continuing pain; weaning from pain medications; avoiding dependence on passive modalities such as heat, rest, transcutaneous electrical nerve stimulation, or massage; and strongly emphasizing self-management (4, 6, 77, 88, 90). After recovery, a program that incorporates the measures suggested for primary prevention, with extra consideration of the individual's particular at-risk activities, will enhance outcomes.

EXERCISE PRESCRIPTION AND TRAINING

Most programs of exercise for people with back pain include a combination of several forms of exercise and educational advice regarding lifestyle factors and general back care. With a single episode of back pain, it is not clear whether exercise will have any effect on the anticipated natural history of that episode. Exercise intervention after back pain has occurred is aimed at reducing risk factors and minimizing recurrences. With chronic and recurrent back pain, the evidence for more aggressive intervention is stronger. Increasing function and decreasing the severity and frequency of back pain episodes should be the goal.

The clinical exercise physiologist may need to emphasize endurance over strength in the early training, adjusting repetitions and resistance to reflect this emphasis. Exercise usually starts with low levels of exercise done frequently and progresses through a system of exercising to quota. The focus should be on promoting function improvements rather than reducing pain. A **periodization** schedule based on estimated maximums from low-intensity testing and use of ACSM guidelines for testing and progression (1-3, 49) may be appropriate. In the absence of specific restrictions imposed by pathology, the exercise program should be designed to correct specific impairments found in initial comprehensive evaluation (82).

Adherence to exercise is a problem for many people with recurrent or chronic NSLBP. Although function can improve quickly, individuals with chronic back pain may need more than 2 mo of training to experience significant pain relief (65). Discontinuing exercise after the completion of a program is as common among individuals with back pain (23, 24) as it is in the general population. Support, encouragement, and easy availability of follow-up programs in the community may be especially important until an exercise habit is well established and whenever a program has to be restarted after a lapse for any reason.

As a general guideline for exercise in people with NSLBP, watch for the following pain provocation complaints:

- Pain that is severe enough to halt exercise
- Pain that persists for more than 3 h after cessation of exercise
- Pain that results in several days of disability or sleep disturbances
- Pain that initiates, exacerbates or extends the distribution of radiating pain

Exercise should be generally tolerable during the activity and should result only in mild musculoskeletal discomfort associated with delayed-onset muscle soreness (DOMS) postexercise. Pain similar to the examples in the preceding list or pain that results in referred or radiating pain into the posterior thigh or down into the lower extremity, indicates that the exercise is too aggressive or irritating to sensitive neuromuscular structures in the lumbar spine. Radiating pain into the lower extremity, in itself, is cause for concern and usually indicates irritation of lumbar nerve roots. Because a common etiology of nerve root irritation is disc herniation that compresses the nervous tissue, or loss of intervertebral disc space that results in narrowing of the space where the nerve root must pass, recurrent **radicular** pain should be referred to the appropriate health care practitioner.

Clients coming for treatment of NSLBP may have a long history of flare-ups of back pain and experience with a wide number of traditional and alternative approaches to manual therapy and exercise. Background information on several of these is given in practical application 24.2. Gradually, theory-based research is developing improved guidelines for management and practical clinical protocols.

Supervision is valuable, particularly in the early stages, to ensure use of correct form and encourage exercise adherence. In one study, chronic back pain patients were randomly assigned to either supervised or unsupervised training for flexibility, muscle strength, and aerobic conditioning according to a periodization schedule (82). After 6 mo, marked differences were seen in the experimental groups. The group given independent exercise averaged only 31.95 of a planned 96 sessions. The other group, assigned a certified strength and conditioning specialist who supervised them, completed an average of 90.75 sessions. Because long-term follow-up was not done, it is not known whether benefits and exercise participation continued, but this work supports the common-sense idea that supervision enhances adherence. Other studies have noted the benefits of supervised exercise as part of a comprehensive management model, particularly when pain has become chronic (47, 53, 63). The client is held accountable and experiences the positive benefits of social interaction with a health care provider, who acts as a cheerleader for the patient. Although not all studies have found a significant effect of supervision (35), enough have to suggest that it be used when feasible. Of course the quality of the interaction between the clinician doing the supervising and the client is an important factor, and can either help or hinder participation (86, 87).

Practical Application 24.2

LITERATURE REVIEW

Exercises have been used to manage back pain for more than 100 yr, yet well-controlled trials on the benefits of exercise are lacking. "Some form of exercise is probably the most commonly prescribed therapy for patients recovering from low-back pain" (103, p. 363). Types of exercise favored have varied widely (67). Lumbar flexion exercises, usually based on the Williams flexion routine (105, 106), have had their vogue, as have hyperextension exercises focusing on increasing paravertebral muscle strength and endurance and reducing the posterior displacement of the intervertebral discs. At present, a balanced approach that involves strengthening most spinal musculature and lower extremity muscles is arguably the most popular strategy, usually in combination with stretching exercises and recommendations for some form of aerobic conditioning.

Exercise is one of the most frequently used conservative modalities for management of back pain (25, 26), but high-quality research is still not available to determine which form of exercise is optimal (52, 78, 89). Given the heterogeneity of NSLBP, it is reasonable to expect that there are subgroups who may respond differently to different forms of exercise. There have been calls for research in this area (52), and clinical trials are starting to explore the effects of individually prescribed exercise based on movement testing (59-61, 89).

When prescribed based on an individual evaluation, exercises are hypothesized to have multiple beneficial effects, including reduction of disc herniation; improved joint mobility; strengthening of weakened muscles; and stretching of adaptively shortened ligaments, capsules, and muscles. Increasingly, behavioral approaches to management are also included to address aspects of the back pain syndrome other than the simple physical limitation.

Popular approaches in rehabilitation have included variations on the Swedish Back School approach (41), which mainly involves education on back care; functional restoration and work-hardening using individually prescribed exercise combined with aggressive disability management (79, 85); and a wide range of manual therapy approaches.

Although these manual therapies are based on widely different theories of causation, many use surprisingly similar positions and techniques. The McKenzie approach, which uses treatments based on symptom response to movement and a theory of disc pathology (28, 72), has become one of the more common manual therapy approaches in rehabilitation. Health care and wellness practitioners also encounter people who have used a wide range of alternative approaches to therapeutic exercise and body work. The Manual Therapy and Therapeutic Exercise Approaches table describes selected exercise approaches that are frequently encountered.

One approach to using exercise in a clinical setting focuses on initially establishing normal movement patterns and then moving on to stretching tight structures, strengthening with early progression through increasing

Manual Therapy and Therapeutic Exercise Approaches

	Founder and history	Treatment philosophies
	MANUAL THERAPY PHILOSOPHIES	
Chiropractic approach	Founder: Daniel David Palmer; treated misalignments (subluxations) of the spine with manipulations (adjustments). Chiropractors began treatment of the spine in the late 1800s.	Many chiropractors also use a variety of spinal manipulative techniques and joint oscillations to realign spinal segments that are disrupting normal neurological function.
Osteopathic approach	Founder: Andrew Taylor Still, a medical doctor in the United States in 1874	Treatment of somatic dysfunctions with a variety of gentle muscle contractions and joint mobilizations, as well as high-velocity, low-amplitude thrust techniques (HVLA).
Australian approach	Credited to Geoffrey Maitland, a physiotherapist from Australia	Treatment emphasis on subjective complaints and quality of pain, as well as behavior of pain and response to positions and mobilization techniques. Advocates use of comparable signs (pre- and posttreatment assessment) but avoids "diagnostic titles."
New Zealand approach	Credited to Robin McKenzie and Stanley Paris, two physiotherapists from New Zealand	Approach emphasizes self-management strategies. Three predisposing factors to low back pain: sitting postures, limited lumbar extension, and predominance of flexion activities in society. Divides back problems into postural, dysfunction, and derangement syndromes.
Norwegian approach	Credited to Oddvar Holten, a physiotherapist in Oslo, Norway. Introduced in the United States by Freddy Kaltenborn and Olaf Eventh.	Uses a biomechanical approach to assess mobility of spinal and extremity joints. Uses Maitland and Kaltenborn oscillatory mobilization grading system to normalize passive joint mobility and improve function.
Craniosacral therapy	Early work of William Garner Sutherland in early 1900s, popularized in the 1970s by John Upledger, an osteopathic physician in the United States	Treatment involves manipulation of the cranial sutures and sacrum to adjust the flow of cerebral spinal fluid, which oscillates caudally and cranially (craniosacral rhythm). Disruptions in the rhythm lead to a variety of autonomic and musculoskeletal dysfunctions.
Variety of soft tissue approaches	Myofascial release: Fred Mitchell Sr., DO Rolfing: Ida P. Rolf, PhD Hellerwork and SOMA: Joseph Heller and Bill Williams, PhD Bindegewemassage: Elisabeth Dicke, Swedish remedial techniques	Soft tissue approaches run the gamut of gentle myofascial stretching to vigorous mobilization of deep tissues and internal body cavities. Realignment and restructuring of tissues lead to normalization of posture and muscle function.
	THERAPEUTIC EXERCISE APPROACHES	
Pilates	Joseph Pilates, developed while working as a nurse in World War I	Use of specialized equipment to promote balance, strength, proper posture, and agility. Currently in vogue as an exercise approach.
Tai chi	Passed on from many Chinese generations including Wu style, Yang style, Ch'en style, and Chuan style	Use of slow, balanced movements of extremities around stable trunk through a series of 108 postures. Uses concept of power centering during standing postures, balancing ying and yang. Slow form used for strengthening and meditation; fast form used as defense.
Feldenkrais	Developed by Moshe Feldenkrais, a physicist, judo expert, and athlete	Observed unnatural movement patterns and created a system of movement awareness exercises. Recognition of mental and emotional activity that perturb all aspects of human performance. Uses visual, auditory, and kinesthetic cues to alter movement.

(continued)

Practical Application 24.2 *(continued)*

Manual Therapy and Therapeutic Exercise Approaches *(continued)*

	Founder and history	Treatment philosophies
THERAPEUTIC EXERCISE APPROACHES		
Alexander technique	Developed by Frederick M. Alexander during late 1800s	Use of specific movements of the body to promote bodily awareness and health. Self-awareness of harmful movement patterns allows subject to correct and move in harmony with breathing pattern.
Trager approach	Milton Trager, MD, former boxer and acrobat. He developed the techniques for 50 yr and began teaching in the 1970s and 1980s.	Concentration, repetition, and refinement of movement, targeted through the unconscious mind. Use of active effortless movements called Mentastics.
Yoga	Origins in Egypt and India 5,000 yr ago. Popular in the 1960s and 1970s, it has made a resurgence in the last 10 yr.	Uses a series of stretching positions that liberate the natural flow of energy in the body, with slow meditative assumption of the poses and strong emphasis on deep relaxation. Improves flexibility, posture, and body awareness. Improves circulation.

repetitions, and adding weighted training exercises (23). Great emphasis is placed on form and maintenance of good muscle control throughout the entire range of movement. This approach is compatible with the functional restoration philosophy of treatment and, in our opinion, is a logical dynamic approach to back care.

Some excellent attempts have been made to validate exercise selection and to link experimental research and clinical practice. One interesting technique combines **electromyography** and computer modeling to estimate tissue loads during activities (48, 69-71). In a text specifically written for clinicians, McGill (69) synthesizes research evidence and makes specific suggestions for exercise and prevention. Although emphasizing that any program must be tailored to the specifics of the individual patient, he suggests that the ideal early exercise program for a patient with back pain would avoid loading the spine through range of motion yet would provide sufficient challenge to allow a muscle-conditioning effect. Whereas a healthy athlete is able to load the spine through full range of motion, McGill's approach is safer for the less trained individual. Other treatments that are currently in vogue unload the spine during early rehabilitation and then add progressive loading and resistance training as recovery occurs. Examples include some programs that incorporate aquatic therapy and use unloading harnesses for treadmill ambulation (88). Many vendors in the United States offer unloading equipment, but the basic premise with the use of the equipment is that reducing compressive forces on the lumbosacral region allows pain-free ambulation on a treadmill or during the performance of functional activities for the low back such as partial squats, heel raises, weight shifting, and step-ups onto a platform or step. This approach, however, is not considered traction for the spine, a common misconception among the uninformed. Figure 24.1 shows a typical unloading system found in an increasing number of U.S. clinics.

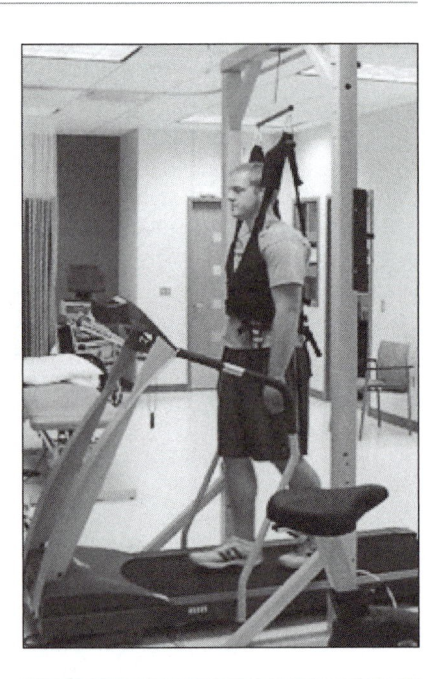

Figure 24.1 Typical unloading system.
Photo courtesy of J. Tim Zipple.

Special Exercise Considerations

After a person has been cleared by a medical professional, few contraindications to an exercise program exist. But many patients with recurrent or chronic pain may be so deconditioned and fearful of exercise or movement that progression should be slow and initial exercise levels low.

Special exercise considerations for patients with NSLBP include the following:

- The patient should obtain medical clearance for exercise.

- Patients experience a fear-avoidance behavior pattern when they believe that movement and activity will further damage or injure the spine.
- The exercise specialist should monitor for red flag findings.
- Deconditioning may be greater in these patients than in sedentary healthy individuals. This deconditioning may be more severe in smaller phasic muscle groups than in larger tonic muscle groups.
- The exercise specialist should use caution if loading through the spine and should consider unloading in some cases for pain management.
- Progression should be slow and initial exercise levels low.
- Smokers may need a slower progression.

One caution that should apply is to avoid overstressing or overtraining. With overtraining, large muscle groups are allowed to compensate or substitute for smaller, deconditioned muscle groups. An emphasis on form and evaluation of postexercise response are important in determining exercise level. The goal is to provide adequate stress for a training effect but to avoid stressing any tissue beyond tolerable levels, and to do this without using abnormal movement patterns (44). The clinical exercise specialist must remember that a person with NSLBP may have been extremely inactive for a considerable time and so will need to start with a lower-intensity program than would a sedentary but healthy individual. Persons with chronic NSLBP have been shown to have selective atrophy of type II muscle fibers in the back muscles. This condition has consequences for fatigue resistance and may help explain the poor back muscle endurance seen in people with back pain (76).

In general, the best approach is to avoid exercises that provide compression loading of the spine in injured individuals until late in rehabilitation. Early exercise focus should be on safe performance of rhythmic exercises with low resistance and an emphasis on correct form that avoids substitution patterns. Initially, exercises such as squats or calf raises may be done with less than full body weight. Progression can then occur to full weight bearing and finally to additional resistance. For most people, the use of handheld weights may be more appropriate for adding resistance than a bar behind the neck and across the shoulders for exercises like lunges, squats, and calf raises. Similarly, leg presses that rely on adding weight through the shoulders and spine should usually be avoided in early treatment. Early resistance training is more easily tolerated with the use of pulley-assisted equipment (performing squats with a pulley assist) or reduced body weight exercises (e.g., Total Gym squats or heel raises). A system such as a Total Gym machine allows the resistance progression to be objectified and provides a stabilized position for the spine during resistance exercises for the upper and lower extremities.

Unless they can be performed with supervision, deadlifts should be avoided or used with extreme caution with patients with a history of back pain. Some clinicians, however, use these in their rehabilitation programs for patients with back pain (23). The argument in favor of deadlifts states that they are an important means of strengthening the trunk extensor muscles. Correct training in technique can help avoid future exacerbations and permit retraining for stressful activities. This argument has validity, but the exercise remains controversial because of the high potential for reinjury. Unless an exercise advisor is confident of his ability to instruct people in this lift and is certain that the client will continue to use correct form, the deadlift is best left out of an exercise program.

Given the association between smoking and back pain, caution should be used in exercise prescription with smokers. Although most published work takes a judgmental approach to the issue of smoking or merely documents the association, a better approach may be to acknowledge the fact that some people will not be able to quit and to use a slower exercise progression for smokers than for nonsmokers (23). Becoming judgmental with patients who smoke often leads to feelings of resentment that impair the client–provider relationship. Clinical experience shows that only those who are committed to quitting smoking benefit from active cessation programs.

Exercise Recommendations

Most people with NSLBP should finish rehabilitation with a general exercise prescription that matches ACSM guidelines for healthy adults (1-3). Adoption of these guidelines is appropriate in the early stages of exercise training to allow for common problems such as significantly decreased exercise tolerance, lack of flexibility, poor neuromuscular coordination, and pain with movement or loading through the spine. Practical application 24.3 includes general guidelines for prescribing exercise and anticipated benefits of exercise for the population with NSLBP. Although there is some suggestion that patient-specific programs based on expert biomechanical evaluation may be better than general exercise, this form of evaluation and treatment requires more research and access to specialized practitioners (28). For individuals with persistent pain and disability, a referral to such practitioners for initial treatment may be useful.

ACSM recommends moderate-intensity aerobic exercise done at least 5 d/wk, or vigorous exercise at least 3 d/wk, or a combination of moderate and vigorous exercise (1, 2). For this population, at least initially, moderate exercise is more likely to be acceptable with less risk of flaring back pain than vigorous.

Cardiovascular Exercise

Aerobic exercise is often used in rehabilitation of NSLBP, although evidence is not conclusive (9). Despite this, most experts still recommend aerobic exercise with subacute and chronic back pain because of its known benefits in

Practical Application 24.3

EXERCISE PRESCRIPTION SUMMARY

Exercise Prescription Review

Training method	Mode	Intensity	Frequency	Duration	Progression	Goals	Special considerations and comments
Cardiovascular	• Brisk walking with arm movement ideal • Cycling (preferably recumbent) • Swimming • Elliptical training	• Moderate, 40 to <60% peak $\dot{V}O_2$, to high, ≥60% peak $\dot{V}O_2$ • Start at lower range, particularly if deconditioned • Borg 12 to 16	• 3 to 5 d/wk minimum, 5 or more for moderate exercise • Preferably most or all days in early stages when at lower intensity	• Build up to 20 to 60 min • May do multiple 10 min bouts throughout day • Begin with several 2 to 5 min bouts if very deconditioned • Anticipate 4 to 6 wk program for substantial improvement	• Increase bouts (if used) to at least 10 min until goal is reached • ACSM recommended progression for sedentary low-risk individuals unless comorbidities increase restrictions	• At least 30 min per day or 150 min per week for moderate exercise • 20 to 25 min per day or 75 min a week for vigorous exercise	• Low impact best initially
Resistance	• Free weights • Machines • Resistance tubing or bands	• Submaximal suggested • 12 to 13 on Borg at start • Progress to 15 to 16	• 2 or 3 nonconsecutive days a week • Separate training of same muscle group by at least 48 h	• Sessions of less than 1 h	• Follow ACSM guidelines for healthy sedentary individuals	• Avoid low repetitions and high intensity; use more repetitions and lower resistance to maximize endurance • Include back extensor and latissimus dorsi endurance training	• Avoid unstable surfaces (e.g., exercise balls) early in training • Later, unstable surfaces may enhance training • Emphasize good form • Limit spine loading through full range

Training method	Mode	Intensity	Frequency	Duration	Progression	Goals	Special considerations and comments
Range of motion	• Static stretching exercises or propriocep-tive neuro-muscular facilitation (PNF) technique stretches	• Comfort-able tightness without pain	• 2 to 7 d/wk; higher number preferable with this population • Four or more stretches per key muscle group	• Each static stretch held 15 to 60 s • Session total of at least 10 min	• Dictated by intensity	• Return ROM to normal	• Balance flexion and extension stretches of spine • Ensure ham-strings are stretched • Select other stretches based on pa-tient deficits • Avoid flexion stretches soon after rising • Avoid stand-ing toe touch

Exercise Testing Review

Test type	Mode	Protocol specifics	Clinical measures	Clinical implications	Special considerations
Cardiovascular	• Treadmill • Cycle ergometer • Stepping	• Ramp • Incremental	• BP • HR • Others as indi-cated by comor-bidities or other ACSM guidelines • Subjective ratings of intensity and pain	• May be decon-ditioned below healthy sedentary level	• Difficulties with prolonged flexion may make recum-bent bike more suitable than upright • Testing soon after rising not recom-mended
Resistance	• Free weights or machines	• Submaximal testing	• BP • HR • Others as indi-cated by comor-bidities or other ACSM guidelines • Subjective ratings of intensity and pain	• Many experience some discomfort, especially early in testing and train-ing	• Ensure that spinal support and good posture are used at all times
Range of motion	• Standard flexibil-ity testing	• Ensure testing of trunk flexors and extensors, hamstrings, quadriceps, and iliotibial band flexibility	• Spinal and ex-tremity joint mo-tion falls within normative ranges for specific popu-lations	• Many experience some discomfort, especially early in testing and train-ing • Avoid forcing into uncomfortable range	• Emphasize neutral spine while test-ing extremity flex-ibility • Avoid spinal flex-ion in first 1 to 2 h after rising

(continued)

Practical Application 24.3 *(continued)*

Exercise Training Review

Cardiorespiratory endurance	Skeletal muscle strength	Skeletal muscle endurance	Flexibility	Body composition
• Improves performance of functional activities • Enhances the effect of back exercise program • Benefits general health	• Increases tolerance of activities of daily living (ADLs), recreational activities, and occupational performance	• Back extensor and latissimus dorsi improvements may protect against recurrences	• Return to normal values (with associated strength gains) may decrease risk of recurrence	• Improves muscle mass and lowers BMI, reducing compressive loads on joints and risk of cardiovascular disease

other areas of health, hypothesized psychological benefits, or the theoretical rationale regarding back pain. General fitness is considered desirable for many reasons and has several hypothesized benefits in general pain management.

Given known benefits of aerobic exercise for general health and mental well-being, including a graduated exercise program with an individualized exercise prescription is reasonable. Maximal or submaximal testing, although useful for prescription, is not required in general clinical practice unless history indicates possible coronary artery disease (60) or other risk factors, according to ACSM guidelines, that make exercise testing advisable (2, 49). Any aerobic activity that interests the client may be used, but in general, high-impact activities such as jogging or exercise requiring sustained spinal flexion (as in some bicycling) are considered poor choices. But the mode of activity that will aggravate NSLBP differs for each person. People with NSLBP who start an aerobic exercise program should start slowly and should be sensitive to activities that precipitate NSLBP. Otherwise, aerobic exercise training should follow the ACSM guidelines for promoting health and fitness. Goals should be to return clients to exercise routines suggested for the general adult population. People with NSLBP may need to begin training with short bouts at lower than usual intensity until they achieve low-level exercise tolerance. Evidence of progressive low back pain during and after exercise, residual postexercise muscle soreness lasting greater than 48 h, impairment in tolerance for activities of daily living, or evidence of neuropathy requires reevaluation by a qualified health care practitioner before exercise is resumed. Modifications in the parameters of resistance or fitness training will be prescribed by the appropriate health care provider.

Resistance Exercise

Available evidence suggests that the more important component of resistance training, at least for the typical NSLBP and for prevention, is endurance (5, 62, 71). **Dynamic endurance** may be more critical than **static endurance** (73). A well-designed program considers both strength and endurance, although literature and clinical expertise support a somewhat heavier emphasis on endurance. Many people with back pain have difficulty tolerating positions for more than a brief period. Poor endurance can predispose them to injuries at relatively low loads. If sustained positions are required for vocational or avocational activities, the clinical exercise physiologist should consider training for these specific positions (23).

Most programs encourage strengthening of back and abdominal muscles. To strengthen all muscles involved, a variety of exercises are required. The following specific suggestions, from the series of articles by McGill and colleagues and the text by McGill (48, 69-71), focus on safety through minimizing spinal loading while providing adequate stimulus for muscle training. Based on electromyography studies for iliopsoas muscle activity and intervertebral disc pressure, no evidence exists to support the use of bent-knee sit-ups over straight-leg sit-ups. Instead of sit-ups, various types of curl-ups or crunches are suggested to strengthen rectus abdominis muscles, including curl-ups with one leg straight and hands used to maintain a neutral lumbar curve (figure 24.2*a*). Others believe that the people who perform crunches (figure 24.2*b*) should independently stabilize the pelvis to promote lumbopelvic neuromuscular control. This neutral curve may be most important in early recovery.

For strengthening the lateral oblique muscles, the horizontal side support exercise shown in figure 24.2,

Figure 24.2 Strengthening exercises commonly used in low back rehabilitation.

c and *d,* is suggested (48, 69-71). This exercise activates the lateral oblique muscles without causing high lumbar compressive loading. The exercise has the additional advantage of training quadratus lumborum, which is considered important in lumbar stabilization.

Back extensions are suggested for strengthening the erector spinae muscles (23). Hyperextensions (of back with fixed legs or legs with fixed spine) are an example of a once-popular exercise that is not indicated and that has not been shown to benefit patients. Evaluating individual responses to end-range exercise may help minimize problems and increase responsiveness to treatment. It is important to avoid loading the spine through range of motion in clients who are beginning a program (69-71).

Extension of both back and legs simultaneously in prone lying (figure 24.3) is often suggested (7), but high lumbar spine compression is produced (69-71, 104). Back extensors can be strengthened in other ways. Alternatives include resisted limited-range back extension with fixed pelvis or less demanding exercises involving extension of one arm, one leg, or both in kneeling (figure 24.2, *e* and *f*). Equipment that isolates the lumbar spine from the pelvis and legs may produce greater strength improvement than more traditional Roman chair exercises (81). But most programs do without such sophisticated equipment, and lack of access to equipment should not be a reason for avoiding lumbar extension training. A simple exercise to start a program that does not need special equipment and that can be used to teach form is the "good morning exercise" which is a bowing motion done with the pelvis blocked by a bench or table (figure 24.2*g*).

Upper extremity and lower extremity muscle strengthening should be part of any general prevention program. In the upper extremities, latissimus dorsi exercises are important because this muscle is involved in back protection and some movement initiation (23). Anatomical connections link lumbodorsal fascia with gluteals, hamstrings, and latissimus dorsi (8), connecting them

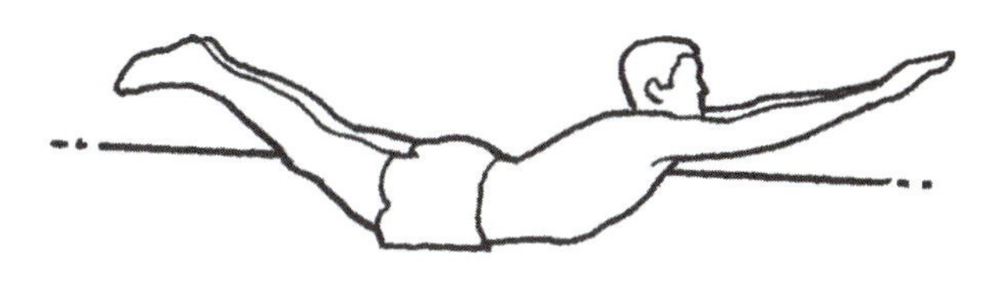

Figure 24.3 Lumbar hyperextension exercise.

to overall back function. Exercises such as the lunge and squat (with handheld weights providing resistance) provide training for trunk stabilization as well as desirable lower extremity strengthening (23).

We concur with the suggestions from the literature given previously but would also consider specific rotator muscle strengthening. As suggested earlier, a potential source of back pain may be the selective loss of strength and endurance of the smaller muscles of the low back such as the rotatores and multifidi. Specific exercises for these muscles may be helpful. Figure 24.2*f* shows an exercise that encourages activation of the thoracolumbar rotators, but this exercise may allow substitution of the large muscle groups. A simple seated exercise (later progressed to standing to promote lumbopelvic coordination) uses a pulley, rubber band, or tubing as light resistance (figure 24.4).

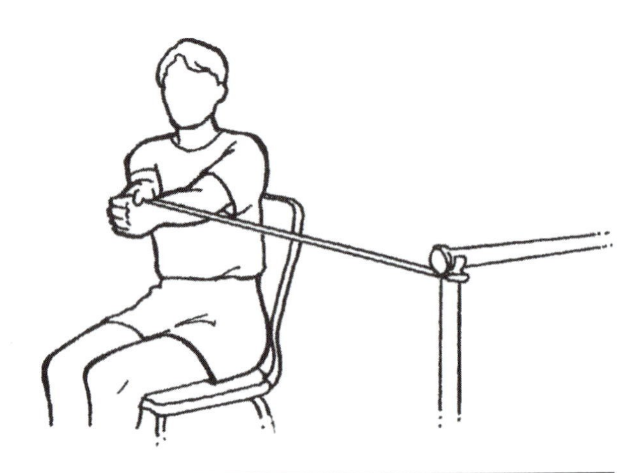

Figure 24.4 Seated resisted trunk rotation exercise.

Having made the previous suggestions for specific back-strengthening exercises, we are obliged to indicate that many of the current studies recommending therapeutic exercise for patients with NSLBP suggest a more generalized approach to exercise. General aerobic exercises such as walking, swimming, or light sport-related activities that the patient self-selects tend to be comparable to or even more effective with respect to positive outcomes than a specific back training exercise program. A program that is enjoyable may be more likely to be followed long-term, which is an important consideration for people with NSLBP. This recommendation seems to be related to the shift in philosophy in management strategies for patients with NSLBP to a more active approach, with simple back school education and reliance on a self-care program.

Range of Motion Exercise

Providing general recommendations for flexibility exercises to someone with low back pain is difficult (69-71). A stretching program must be developed based on the patient's history and physical examination findings. The limited data available do not support the view that greater flexibility of the spine prevents injury. Maintaining adequate flexibility at hips and knees for lifting is important; and if stretching of the spine is done, it is probably best done in an unloaded position (70). Deficits in lower extremity flexibility should be addressed individually. Spinal flexibility should not be emphasized until the person has the muscular strength and endurance to control a mobile spine (71). We recommend starting with low-resistance, high-repetition strengthening to reeducate neuromuscular control of increased spinal movement that occurs with stretching. An important caveat is that flexion stretching after sleep or prolonged sitting is not recommended (69, 91, 92).

Individuals with a history of back pain have reduced back and hamstring flexibility (5). This finding is associated with recurrence or persistence of NSLBP. On the other hand, those with very mobile spines are at increased risk of experiencing NSLBP during the following year (5). Apart from this, there is no demonstrated link between spinal flexibility and back pain.

Most programs incorporate flexibility exercises into their exercise routines. In the absence of clear evidence, it seems reasonable to continue to do so unless hypermobility or instability is a particular problem. When in doubt, the best approach is to have the person evaluated for abnormal flexibility before advocating spinal stretching exercises. Nonspinal stretches, usually calf muscles and hamstrings stretches (23), are part of most back programs. Other areas and muscles that are frequently tight include the iliotibial band, hip adductors, hip flexors, and quadriceps. Specific muscle length tests should be administered by qualified individuals to establish an individualized stretching program.

A reasonable approach would be to evaluate for tightness and to design an exercise program that includes stretches to correct any impairments noted. A few commonly used stretches for the low back and lower extremities are shown in figure 24.5. Many variations are equally suitable. Any exercise that involves repeated flexion should be balanced with stretches into extension. Caution should be used with people who have known disc pathology, because repeated flexion exercises can shift disc nuclear material posteriorly, potentially aggravating symptoms. Passive hyperextension stretches (such as pushing up from prone by using the arms with the hips staying in the floor) may be a better option than active extension stretches. A balance in stretching movements and inclusion of extension as well as flexion stretches are important. Patients can be instructed to move to a comfortable range of movement and hold a prolonged (15-30 s) stretch in the end portion of the range where tissue tension is developed. Sometimes, in clients with NSLBP, the addition of strategically timed breathing can allow relaxation of the spinal muscles, enhancing the stretching effect.

Frequency

Only continued adherence to an exercise program is likely to offer lasting benefit (64, 65). In a study using intensive training of people with chronic back pain, all subgroups improved with the 3 mo training program, but a year later only those who continued training at least once a week had remained significantly improved (64, 65). Training at least twice a week is usually suggested (69, 81). Addressing adherence at home or in the community is important, as it has been shown that self-efficacy and adherence to home programs are more predictive of success than therapy attendance (66).

Given the population targeted in this chapter, two or three sessions a week is appropriate for a maintenance program. In early rehabilitation, daily exercise is often suggested. Frequency can decrease as intensity increases and as exercise tolerance approaches population norms. Clients who are empowered during their rehabilitation phase of recovery are more likely to adhere to a maintenance program after their subacute or chronic pain has dissipated. Having patients select exercises that they

Figure 24.5 Stretches commonly used in low back rehabilitation.

enjoy will provide motivation to continue compliance in self-care of the back after they have completed rehabilitation with a health care provider.

Intensity

Few guidelines are specific to the management of NSLBP regarding intensity selection. Those available usually suggest that exercise needs to be more intense than normally provided for patients. Sometimes a very thin line separates exercise intensity sufficient to be effective from exercise that is too intense and leads to worsening of symptoms. Self-selected intensity or exercise to pain tolerance often leads to inadequate exercise levels. Although pain may not improve for several months in many people, an intensive program will result in greater functional and psychological benefit than a less aggressive approach. Some advocate a quota approach to prescribing exercise intensity to prevent underexercising and suggest the use of operant conditioning behavioral tactics (57, 82).

In setting exercise intensity, we suggest a trial-and-error approach, such as the DeLorme method of determining maximal resistance, to select resistance weights that allow 20 to 30 repetitions with good neuromuscular control in a pain-free or minimal range of motion. This initial program will promote endurance and control of movement. As the person progresses, resistance should increase while the number of repetitions decreases to 8 to 12, compatible with ACSM training guidelines (1-3).

Duration

To complete a prescribed program, weight training may require 30 to 60 min per session. More than this may lead to decreased adherence. The total resistance training program includes intermittent rest periods between intensive sets to recuperate.

CONCLUSION

NSLBP is one of society's most common problems. Details of suitable exercise programs are limited, but research shows that most people with nonspecific back pain can use exercise to restore function and decrease disability. Programs should start out with modifications to accommodate individual impairments and should progress to maintenance programs that are as close as possible to those suggested for general fitness with healthy populations.

Although people with back pain may need extra encouragement to begin and continue a general exercise program, those who are able to do so can hope for considerable improvement in function and quality of life. Reassurance, education, and encouragement with self-management are important components of care. Individuals too rarely receive this information, and open communication with people who have back pain can be a key aspect of management. For most people, back pain is nonspecific and can be best managed with conservative yet active treatment strategies.

Key Terms

ankylosing spondylitis (p. 463)
cauda equina syndrome (p. 463)
dynamic endurance (p. 478)
electromyography (p. 474)
herniation (p. 463)
incidence (p. 462)
laminectomy (p. 468)
magnetic resonance imaging (MRI) (p. 466)
osteophytes (p. 466)
periodization (p. 471)
prevalence (p. 462)
prevention (p. 462)
primary (p. 463)
radicular (p. 472)
red flags (p. 463)
scoliosis (p. 466)
secondary (p. 463)
spinal decompression (p. 468)
spinal discectomy (p. 468)
spinal fusion (p. 468)
spinal traction (p. 468)
spondylolisthesis (p. 466)
static endurance (p. 478)
transcutaneous electrical nerve stimulation (TENS) (p. 468)
ultrasound (p. 468)

CASE STUDY

MEDICAL HISTORY

Mrs. AB is a 28 yr old Caucasian bank worker. She has had recurrent back pain since the age of 16. At that time she had an awkward fall while playing softball. She was taken to her local emergency department, where she was given the diagnosis of muscle strain and treated with painkillers and muscle relaxants. Although her back improved quickly, she believes that the pain never completely resolved. As she continued through high school she noticed that although she had continuing periods of pain, they seemed to be less severe when she was physically active.

At age 21 she experienced another acute episode of back pain that began suddenly when she sat down on a couch. This time she did not seek medical help but treated herself with over-the-counter pain medication. The pain slowly resolved over several months.

DIAGNOSIS

Two years ago over the July 4th weekend, Mrs. AB was camping with her husband and daughter. On the morning when they were packing up to go home, she bent over to help her husband pick up the tent and had a sudden onset of severe back pain. The pain was in her back and right buttock. She saw both her physician and chiropractor for treatment. Her X-ray results were normal, but she was told that her MRI showed a bulging disc between the fourth and fifth lumbar vertebrae. All other medical screening was clear, and she was told that she did not have any serious medical problems. She was given a short course of muscle relaxants and stayed off work for 3 d. Her condition slowly improved but continued to bother her for the next several months.

EXERCISE TEST RESULTS

In October the patient's chiropractor suggested that she start fitness exercise. The previous January she had joined her town's fitness center, but her attendance had been sporadic. She had completely stopped working out before she injured her back while camping, and the manager of the facility had put her membership on hold while she was receiving treatment for her back pain.

Following her chiropractor's advice, Mrs. AB returned to her fitness facility for a reevaluation and treatment program. Her physician had cleared her to begin an exercise program, and the fitness facility cleared her based on PAR-Q. The facility manager attempted a graded cycle ergometer test of aerobic capacity, but Mrs. AB was unable to complete it because of back pain. Similarly, muscle strength testing of back extensors and abdominal muscles was not possible. She was able to extend her hips only to neutral, and an attempt to test hamstring flexibility by reaching for her toes in sitting yielded a 13 in. (33 cm) distance from her fingertips to her toes.

She was able to do five repetitions of a trunk extension exercise from midflexion range to neutral spine position using a resistance of 30 lb (14 kg). She attempted latissimus pull-downs with 20 lb (9 kg) of resistance and a leg press that loaded through her shoulders, but she found the exercises too painful to continue to a formal strength evaluation. She was able to walk for 15 min on a treadmill with 0° incline at 3 mph (4.8 kph). She found this moderately painful.

EXERCISE PRESCRIPTION

Under the guidance of the fitness center manager, she began a program of treadmill walking to tolerance and resumed a modified version of the program of resistance exercise she had started when she joined the facility. Her walking started with 15 min at 3 mph (4.8 kph), and her goal was to increase to 4 mph (6.4 kph) and a distance of 2 mi (3.2 km). She did curl-ups using an exercise ball for the abdominal muscles, beginning with one set of 10 repetitions. She also started with one set of 10 repetitions of the following exercises: chest press (two varieties), seated rowing, leg curl, triceps pull-down, and knee extension. All were done on weight equipment rather than with free weights, and initial resistance was determined via selection of a weight that

(continued)

did not increase her back pain and allowed her to do the required number of repetitions without loss of form. On the back extension machine she started with one set of five repetitions with 30 lb (14 kg) of resistance in a very limited range.

The fitness center is adjacent to the bank where Mrs. AB works, and she now goes there after work several days a week. Her employer is a major corporate sponsor of the fitness center, so she is able to take advantage of a reduced membership rate. Currently, she experiences some back pain at least weekly. She does not take time off work for pain but may modify her activities slightly. On days when pain bothers her, she takes over-the-counter pain medication. With the severe episodes she had experienced in the past, she had returned to work despite considerable pain within 3 to 5 d of the episode. Her supervisor insisted on buying an ergonomic chair for her use at work, and Mrs. AB finds the chair helpful. Her job allows her to change position frequently, and she is never required to either sit or stand for long periods. She has noticed that any prolonged posture aggravates her back pain for several days.

Similarly, she has found that beginning any new sporting activity increases pain. Last summer she coached her daughter's softball team and found that the frequent squatting activities caused her back to flare up considerably. Depending on the activity, days or weeks may pass before the pain settles down to the usual level. She continues to see her chiropractor once a month.

Mrs. AB normally exercises three or four times a week at the wellness center. She has found that her episodes of back pain worsen and the frequent low levels of pain increase in severity whenever she fails to exercise regularly. She believes that her back symptoms are still improving but extremely slowly. Her goal continues to be complete elimination of pain.

In the first 6 mo after her injury, Mrs. AB increased her exercise routine to include the following exercises: chest press (two varieties), seated rowing, leg curl, triceps pull-down, and knee extension. These exercises are all done on weight equipment rather than with free weights and she normally does three sets of 12 to 15 repetitions. Additionally she does two sets of 30 abdominal crunches. For the chest presses and rowing she now uses 40 lb (18 kg). The triceps pull-down is 20 lb (9 kg); the leg curls are 30 lb (14 kg); and the knee extension is 35 lb (16 kg). Her back extension exercises have increased to three sets of 20 reps with a resistance of 80 lb (36 kg), with extension only to neutral. She now walks 2 mi (3.2 km) on the treadmill at a speed of 4 mph (6.4 kph) and does light stretching of arms and legs before her workout. The flexibility of her spine has improved somewhat, but she is cautious of allowing extension of her back much beyond a straightened position with any exercise. Before the episode she had been using a leg press machine that loaded through the shoulders and a lat pull-down. Both were extremely painful after the injury, and she dropped them from her program. She has not attempted them since. Although she has continued to work out regularly, the weights have been the same on the machines for about a year. She uses the fitness center treadmill for a walking program. She still has tight hamstrings, fingertips 6 in. (15 cm) from toes, and reduced active and passive back extension.

DISCUSSION QUESTIONS

1. How well does Mrs. AB fit the profile of the typical person with nonspecific back pain?
2. In her presentation are there any red or yellow flags?
3. What characteristics in her work and recreational situation have helped her deal with her pain?
4. Do you think her current back program will produce the desired outcome (pain-free spinal mobility and physical tolerance of all work-related tasks)?
5. Do you believe, based on the recurrent nature of her symptoms, that present management strategies are adequate?
6. What specific recommendations for changes in her exercise program could you make for her based on the information provided?
7. List general principles of exercise that Mrs. AB should follow to decrease her risk of recurrence.

8. What advice would you give her for a suitable aerobic conditioning program?
9. Write an outline of advice that you would give her for those days when her job requires a lot of sitting activity.
10. Would you reintroduce the partial squats and lat pull-downs in her program? If so, what modifications could help her resume those exercises?
11. She wants to continue coaching her daughter's softball team. What strategies should she use to prevent flare-ups of her back pain from this activity?
12. How can she safely stretch into extension to regain some of the lost mobility in her spine?

PART VII

Selected Neuromuscular Disorders

Part VII contains four chapters that cover the major neuromuscular conditions likely to be encountered by the practicing clinical exercise physiologist. Although most of these conditions require some type of specialized care or rehabilitation, many persons who are minimally afflicted or significantly recovered will seek exercise-training advice and programs. The clinical exercise physiologist should be aware of these conditions and know how to design and implement exercise programs that address patients' needs.

Chapter 25 centers on spinal cord injury. This chapter addresses the degrees of spinal cord injury, the special care needed immediately after injury, and the barriers that these people face for the rest of their lives. Those who recover well and maintain a degree of function that allows them to perform regular exercise training may seek services at facilities staffed by clinical exercise physiologists. This chapter discusses the needs of these patients and their specific exercise issues.

Chapter 26 investigates multiple sclerosis (MS), a disease that typically affects grown individuals. This chapter reviews the many presentations of MS and the degree of functioning along the course of the disease. Lesser affected, higher functioning people may not have much need to alter a regular exercise-training routine, whereas individuals with greater disability may need significant adaptations to their exercise program.

Chapter 27 explores cerebral palsy. Cerebral palsy (CP) afflicts people at birth or in early childhood. Degrees of deficit are certainly present in these individuals, and this chapter provides a summary of specific issues that they face. Because aging can compound these deficits, the chapter provides important information with respect to designing, implementing, and adjusting exercise-training programs for these individuals.

Chapter 28 focuses on stroke. Although one might argue that stroke should be placed in the cardiovascular section because it is a vascular issue, the long-term effects are often related to neuromuscular function. For instance, a stroke that affects the functioning of limbs becomes an important issue in exercise training. This chapter discusses the important concerns for these patients that the clinical exercise physiologist may encounter.

Chapter 25

Spinal Cord Injury

David R. Gater, Jr., MD, PhD
Stephen F. Figoni, PhD

Spinal cord injury (SCI) profoundly affects the body's usual responses to exercise due to somatic and autonomic nervous system disruption, and requires the clinical exercise physiologist to carefully plan and monitor exercise testing and training. Many body systems are affected; the degree of muscle paralysis, sensory loss, and autonomic impairment can vary greatly among individuals with SCI. In addition to usual contraindications and exercise limitations based on age and gender, the client with SCI will likely have neurogenic hypotension, circulatory hypokinesis, adaptive myocardial **atrophy**, and subsequently diminished cardiac output and cardiac reserve, mixed restrictive and obstructive pulmonary dysfunction, severe osteoporosis, and marked **sarcopenia** with obesity and metabolic syndrome (obesity, glucose intolerance, insulin resistance, hypertension, and dyslipidemia) despite "normal" body weight. This chapter reviews the most common issues encountered by the clinical exercise physiologist during exercise planning, prescription, and training for persons with SCI. It also provides a brief update on current exercise research and makes recommendations for prescribing exercise. Discussion with the client's SCI specialty physician is strongly encouraged should any questions remain.

DEFINITION

The spinal cord serves as the major conduit for motor, sensory, and autonomic neural information transmission between the brain and the body. A SCI affects conduction of neural signals across the site of the injury or lesion. Spinal cord injury is classified by the lowest segment of the spinal cord with normal sensory and motor function on both sides of the body and may be defined as complete (without sensory function in the lowest sacral segment) or incomplete (partial preservation of sensory or motor function below the neurological level, including the lowest sacral segment). **Tetraplegia** (the term preferred over quadriplegia) is "the impairment or loss of motor and/or sensory function in the cervical segments of the spinal cord due to damage of neural elements within the spinal canal" (3). Tetraplegia is distinguished from paraplegia in that it includes dysfunction of the arms, whereas both tetraplegia and paraplegia involve impairment of function of the trunk, legs, and pelvic organs. Table 25.1 lists the American Spinal Injury Association's (ASIA's) definitions of five degrees (completeness) of SCI (3); table 25.2 lists ASIA's specific clinical syndromes as they relate to injuries to specific spinal cord locations.

SCOPE

Approximately 237,000 to 301,000 U.S. citizens have a SCI (123). The incidence of SCI approaches 40 new injuries per million U.S. persons per year. Almost 40% die before reaching a hospital. Hence, approximately 7,400 Americans will survive a new SCI each year, and more than 90% will return to a private residence following rehabilitation. Men are affected four times as frequently as women, and most SCIs occur in 16 to 30 yr olds. Most of these injuries occur in young men involved in motor vehicle collisions. Acts of violence, falls (mostly elderly),

Table 25.1 American Spinal Injury Association Impairment Scale

Degree of impairment	Conditions of impairment
A	Complete: No motor or sensory function is preserved in the sacral segments S4-S5.
B	Incomplete: Sensory but not motor function is preserved below the neurological level and includes the sacral segments S4-S5.
C	Incomplete: Motor function is preserved below the neurological level, and more than half of key muscles below the neurological level have a muscle grade of <3.
D	Incomplete: Motor function is preserved below the neurological level, and at least half of key muscles below the neurological level have a muscle grade of ≥3.
E	Normal: Motor and sensory function is normal.

Adapted, by permission, from American Spinal Injury Association, 2006, *International Standards for Neurological Classification of Spinal Cord Injury*, revised 2006 (Chicago, IL: American Spinal Injury Association).

Table 25.2 Causes and Impairments of Spinal Cord Injury Clinical Syndromes

Syndrome	Cause	Functional deficits
Central cord syndrome	Incomplete cervical injury with cord damage	Weakness, sensory deficits in upper extremities (less in lower)
Brown-Sequard syndrome	Unilateral cord lesion	**Ipsilateral** proprioceptive and motor deficit; contralateral pain impairment and temperature deficit below level of injury
Anterior cord syndrome	Anterior cord ischemia (T10-L2)	Below level of injury, pain impairment and temperature deficit
Conus medullaris syndrome	Upper and lower motor neuron damage	Bowel, bladder, lower extremity **areflexia**, and **flaccidity**; preserved or facilitated reflexes
Cauda equina syndrome	Lumbosacral nerve root injury	Bowel, bladder, and lower extremity areflexia and severe **dysesthetic** pain

and sport injuries (mostly diving) are other primary causes of SCI. Pediatric SCI is often congenital, presenting at birth with incomplete formation and closure of neural and skeletal bony elements, referred to as spina bifida or **myelomeningocele**. Recent advances in acute care have significantly reduced the number of complete SCIs, particularly in cervical injuries. Since 2000, 34% of SCIs have resulted in **incomplete tetraplegia**, whereas only 18% are considered complete tetraplegia at the time of discharge from a rehabilitation facility. In the same period, 19% of SCIs resulted in **incomplete paraplegia**, whereas 23% still involved **complete paraplegia** at the time of discharge. A small percentage of SCIs (<1%) completely resolve by the time of hospital discharge, without neurological deficits. The medical costs associated with SCI are staggering. Average lifetime medical costs for those with complete tetraplegia (levels C5-C8) acquired at age 25 exceed $1.7 million per person, and average lifetime foregone earnings and fringe benefits per case may exceed $2.1 million (119). Even with the advent of managed care, the initial hospitalization and rehabilitation length of stay will typically exceed 60 d and $80,000 for tetraplegia and 40 d and $50,000 for thoracic paraplegia. As SCI care and technology have improved, life expectancy has likewise improved. A 20 yr old man with newly acquired C5-C8 tetraplegia now has an additional life expectancy of 40 yr or more, whereas the 20 yr old with paraplegia may expect to live an additional 45 yr or more (123).

PATHOPHYSIOLOGY

The spinal cord, a portion of the central nervous system, links the conscious and subconscious functions of the brain with the peripheral and **autonomic nervous systems**. It extends through and is protected by the spinal column, a flexible segment of interdigitating bones and discs arranged to maximize mobility and reduce risk of injury. There are 33 total vertebrae (7 cervical, 12 thoracic, 5 lumbar, 5 sacral, and 3 to 5 coccygeal). A pair of nerves arises from each vertebral segment. Figure 25.1 depicts these segments and nerves. The spinal cord is about 25% shorter than the spinal column; thus subsequent spinal nerves exit from the cord as the **cauda equina**. The vascular supply to the cord comes from one anterior and two posterior spinal arteries at each vertebral level, supplying the anterior two-thirds and posterior one-third of the cord, respectively. Neural tracts of the central nervous system and cell bodies of the peripheral nervous system are susceptible to primary and secondary injury. Primary injury can damage neural tracts, cell bodies, and vascular structures that supply the cord. Secondary injury occurs because of hemorrhage and local edema within the cord, which compromise vascular supply, resulting in local ischemia. Infarction of the gray matter occurs within 4 to 8 h after injury if blood flow cessation remains. Inevitably, necrosis, or cell death, occurs and can enlarge over the one or two vertebral levels above and below the area of trauma. **Astrocytic gliosis** and syringomyelia may form during the next several months. Formation of fibrous and glial scarring is the final phase of the injury process. Injury to the spinal cord obstructs the transmission of neural messages through the cord and results in the loss of somatic and autonomic control over trunk, limbs, and viscera below the site of the lesion. Systemic responses to exercise seen in nondisabled individuals are blunted in persons with SCI. This diminishes their ability to

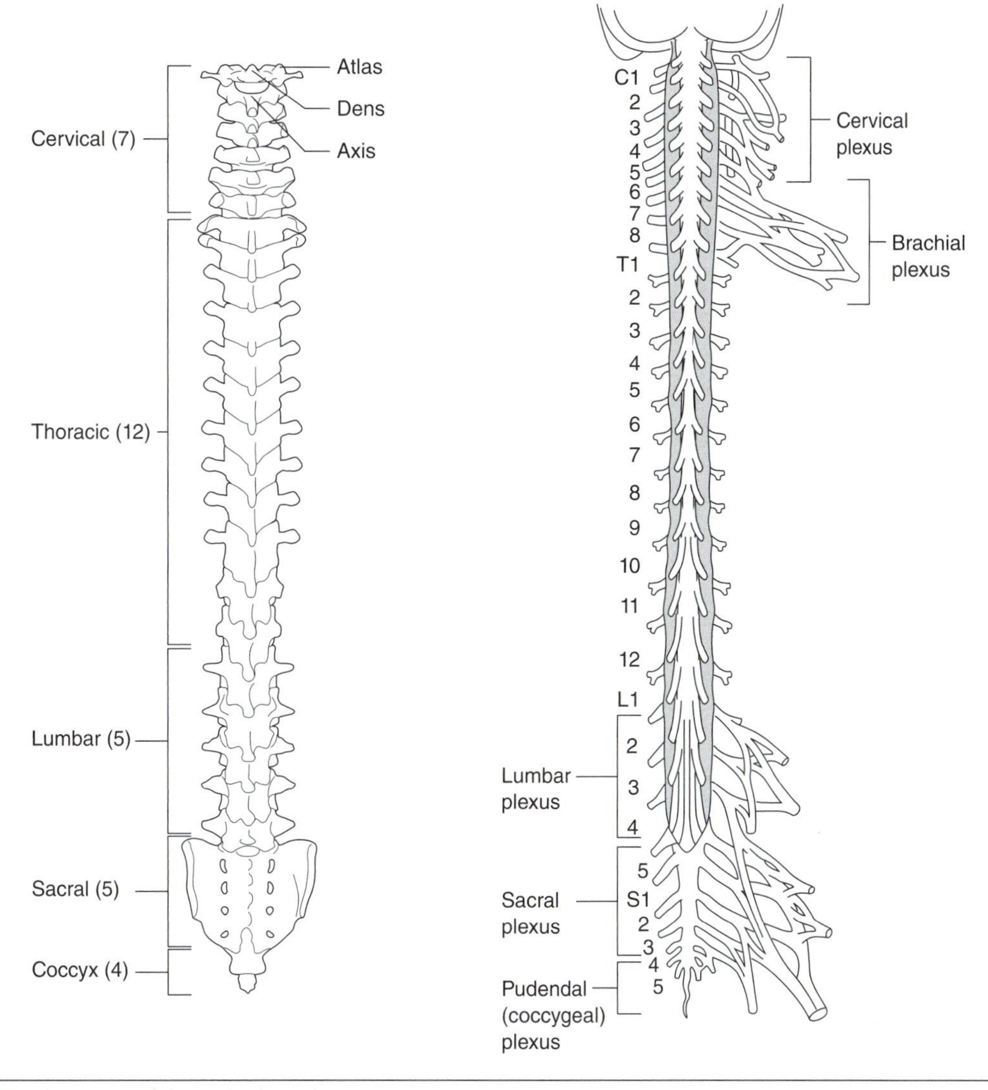

Figure 25.1 Anatomy of the spinal cord.

Reprinted, by permission, from R. Behnke, 2001, *Kinetic anatomy* (Champaign, IL: Human Kinetics), 130, 171.

perform physical activity and exercise, at both the conscious (somatic) and the subconscious (autonomic) level.

Somatic Nervous System Disruption

The somatic nervous system consists of motor (efferent) and sensory (afferent) neural pathways connecting the brain to the body by the spinal cord. Complete SCI interrupts transmission of these signals, and voluntary movement and sensory perception are absent below the lesion. Neurological classification of SCI is standardized as seen in figure 25.2.

Autonomic Nervous System Disruption

The autonomic nervous system coordinates automatic life-sustaining processes and organizes visceral responses to somatic reactions. It is composed of sympathetic and parasympathetic divisions, which regulate the action of smooth muscle and glands. Essential functions of the autonomic nervous system during exercise include modulating heart rate, stroke volume, blood pressure, blood flow, ventilation, thermoregulation, and metabolism. The **sympathetic nervous system** and **parasympathetic nervous system** make up the autonomic nervous

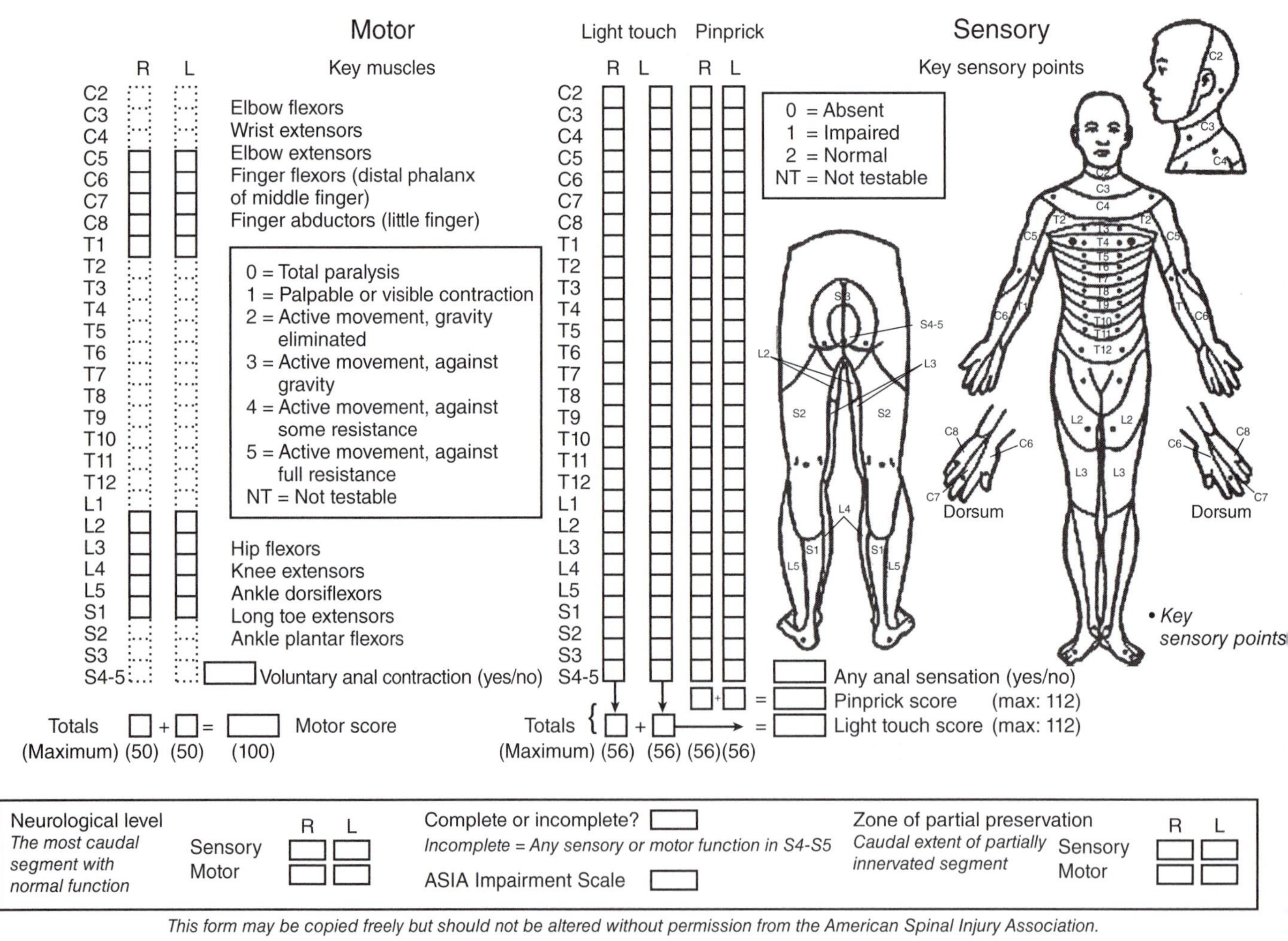

ASIA

Standard neurological classification of spinal cord injury

Motor

Key muscles

R L

C2 C3 C4 C5 C6 C7 C8 T1 T2 T3 T4 T5 T6 T7 T8 T9 T10 T11 T12 L1 L2 L3 L4 L5 S1 S2 S3 S4-5

Elbow flexors
Wrist extensors
Elbow extensors
Finger flexors (distal phalanx of middle finger)
Finger abductors (little finger)

0 = Total paralysis
1 = Palpable or visible contraction
2 = Active movement, gravity eliminated
3 = Active movement, against gravity
4 = Active movement, against some resistance
5 = Active movement, against full resistance
NT = Not testable

Hip flexors
Knee extensors
Ankle dorsiflexors
Long toe extensors
Ankle plantar flexors

Voluntary anal contraction (yes/no)

Totals ☐ + ☐ = ☐ Motor score
(Maximum) (50) (50) (100)

Light touch Pinprick
R L R L

Sensory

Key sensory points

0 = Absent
1 = Impaired
2 = Normal
NT = Not testable

Any anal sensation (yes/no)
Pinprick score (max: 112)
Light touch score (max: 112)

Totals ☐ + ☐
(Maximum) (56) (56) (56)(56)

Neurological level
The most caudal segment with normal function
Sensory R ☐ L ☐
Motor R ☐ L ☐

Complete or incomplete? ☐
Incomplete = Any sensory or motor function in S4-S5
ASIA Impairment Scale ☐

Zone of partial preservation
Caudal extent of partially innervated segment
Sensory R ☐ L ☐
Motor R ☐ L ☐

This form may be copied freely but should not be altered without permission from the American Spinal Injury Association.

Figure 25.2 Tool used to classify the neurological extent of a spinal cord injury.

Reprinted from American Spinal Injury Association, 2006, *International standards for neurological classification of spinal cord injury*, revised (Chicago, IL: American Spinal Injury Association).

system. Table 25.3 presents the effects of level of SCI on function.

Systemic Adaptation

In this section we review the systemic effects of and adaptations to SCI.

Cardiovascular: In response to acute cervical and upper thoracic SCI, bradycardia is common and is attributable to loss of sympathetic nervous system influences, with no effect on the parasympathetic nervous system (84, 85). This situation usually resolves in 2 to 6 wk (39, 101). Reduced sympathetic influence on peripheral and splanchnic vascular beds reduces peripheral vascular resistance, which enhances venous pooling and **orthostatic hypotension**.

Autonomic dysreflexia: For persons with SCI above T6, an uncontrolled outflow of sympathetic activity in response to stimuli below the SCI (e.g., distended bowel or bladder, lacerations, fractures, pressure sores, sunburn) may occur, resulting in life-threatening paroxysmal hypertension (19, 30, 100). Reflex bradycardia and vasodilation, manifested as flushing, headache, hyperhydrosis, piloerection, pupillary dilation, and blurred vision, may occur (19). Such symptoms suggest an autonomic crisis that can lead to intracerebral hemorrhage, seizure, arrhythmia, and death if not immediately and appropriately treated.

Pulmonary: Ventilation is impaired in most SCI patients because of paralysis of rib cage and abdominal musculature, reduced pulmonary compliance, and reduced diaphragmatic excursion. Thus, some patients may require ventilator assistance. Tetraplegia below C5 typically spares voluntary control of the diaphragm, although expiration remains impaired. Ventilation worsens as the level of disability increases (16, 27), although aggressive spirometry and exercise training may improve it (29).

Bowel and bladder function: Cervical, thoracic, and lumbar spinal cord lesions increase the risk of gastric

Table 25.3 Effects of Level of Injury on Somatic and Autonomic Function

Level of injury	Somatic: movements preserved include all above each level affected	Autonomic: functions preserved include all above each level affected
C1-3	Chin	Cranial-level parasympathetic nerves are not affected by SCI.
C4	Shoulder shrug and head turn	
C5	Shoulder movement and elbow flexion	
C6	Wrist extension	
C7	Elbow, wrist, finger extension	
C8	Lowest level of tetraplegia; finger flexion	
T1-10	Begins level of paraplegia; variable sensory deficits; intercostal control	Sympathetic nerves are from T1 to L5; SCI above this level affects smooth muscles, organs, and glandular sympathetic-mediated effects (e.g., heart rate, stroke volume, ventilation, sweating, splanchnic vasoconstriction, skeletal muscle vasodilation).
T6-12	Trunk stabilizers	
T1-S1	Paraspinal muscles	
L2	Hip flexion	
L3	Knee extension	
L4	Ankle dorsiflexion	
L5	Toe extension, hip abduction	
S1	Ankle plantar flexors, hip extensors	Affects lower digestive structures, gallbladder, and bladder function.
S2-S5	Bowel, bladder sphincter control	

Reprinted from American Spinal Injury Association, 2006, *International standards for neurological classification of spinal cord injury*, revised (Chicago, IL: American Spinal Injury Association).

and duodenal ulcers, increase bowel motility and hyperreflexia, and eliminate voluntary control of defecation (129). Hyperreflexia of the bladder wall and sphincter muscles results in greater risk of vesicoureteral reflux, hydronephrosis, and acute renal failure if not appropriately managed. Bladder spasms and loss of voluntary sphincter control may lead to urinary incontinence. Urinary tract infections, as well as renal or bladder stones, are more frequent in SCI patients than in the nondisabled population.

Hyperreflexia: Spasticity, seen in up to 60% of individuals with SCI (95), is attributable to central disinhibition of spinal reflex arcs (141). Some people with SCI are able to use their spasticity to assist them in performing mobility and **activity of daily living (ADL)** tasks, but many find that the spasticity is painful, disrupts sleep, interferes with function, causes muscle contractures, and can lead to shear- or pressure-induced skin breakdown.

Thermoregulation: The interruption of autonomic pathways in tetraplegia results in partially **poikilothermic** responses to thermal stress and reduced ability to dissipate environmental and internally generated heat by sweating (106, 117) and skin vasodilation. This occurs because most of the sweat response is limited to regions above the level of SCI.

Endocrine: Autonomic dysfunction and somatic paralysis in SCI may also affect metabolic and hormonal function, including the sympathomedullary response and the adrenocortical–pituitary axis, resulting in flattened circadian rhythms and poorly regulated corticosteroid responses (135). Glucose intolerance often occurs and is frequently accompanied by **hyperinsulinemia** (1). Although thyroid function may be acutely altered in SCI, thyroid function tests are generally normal in healthy SCI adults (12). Conversely, testosterone and free testosterone levels in men with SCI are often reduced (12, 128), whereas growth hormone release is blunted and chronically depressed (11). Following the acute phase of SCI, ovulatory menstrual cycles are fairly well preserved (112).

Osteopenia: Neurogenic osteopenia in complete tetraplegia results from the withdrawal of stress and strain on bone (140). Bone mineral loss is rapid during the first 4 mo of SCI (52). Homeostasis at 67% of original bone mass is achieved at about 16 mo after injury. Thus, the risk of fractures increases.

CLINICAL CONSIDERATIONS

The clinical management of SCI is complex and beyond the scope of the clinical exercise professional who is working with these patients on long-term fitness management. See table 25.4 on management issues during acute hospitalization and rehabilitation in SCI patients. Comorbidities that may affect the long-term rehabilitation of these patients include brain injury, thoracic contusions and fractures, intra-abdominal trauma, upper and lower extremity fractures, **plexopathies**, and peripheral neuropathies. Typically, medical management issues take precedence over fitness concerns during the first 3 mo postinjury, and the altered physiology must constantly be considered as the person with SCI is reintegrated into community and recreational pursuits.

Signs and Symptoms

The common signs and symptoms of chronic SCI are listed in table 25.5. Each of these poses potential issues during the daily life of the patient. The clinical exercise professional working with these patients should be aware of these signs and symptoms and able to assist when needed.

History and Physical Examination

Comorbid medical issues common among patients with SCI that should be considered in relation to exercise testing and training include respiratory complications, coronary artery disease, peripheral arterial disease, **circulatory hypokinesis** leading to hypotensive responses, obesity, type 2 diabetes mellitus, pressure sores, joint contractures, and osteopenia. Incidence of smoking (35%) and upper extremity overuse problems is high. Finally, medications should be listed and assessed for their effect on exercise tolerance. Commonly prescribed medications used in SCI, their indications, and common side effects are listed in table 25.6. Heart disease is the second leading cause of death in patients with SCI, accounting for approximately 22% of all deaths (37, 62). Unfortunately, silent myocardial ischemia caused by disrupted visceral afferent fibers in higher levels of SCI may prevent an individual from recognizing symptoms such as angina (12). Amputations for peripheral arterial disease are sometimes required because of diminished healing and the development of ulcers. Obesity is a problem attributable to reduced (12-54%) basal energy expenditure. As a result, the risk for glucose intolerance and type 2 diabetes increases (56). Pressure ulcers occur as the result of shear, friction, and unrelieved pressure, usually over a bony prominence, which damages overlying skin tissue. This problem is exacerbated in persons with tetraplegia because they are unable to feel the sensations of tingling, discomfort, and pain. Appropriately prescribed wheelchair-seating systems with scheduled pressure relief

Table 25.4 Management Issues During Acute SCI Hospitalization and Rehabilitation

ACUTE HOSPITALIZATION
Spine management: imaging of cervical, thoracic, lumbar, sacral spine
Surgical or orthotic stabilization of unstable spinal column injuries and spinal cord decompression
Range of motion limitations to allow complete bony and soft tissue healing of spinal elements
RESPIRATORY
Assisted ventilation often required for high cervical injuries
Secretion management essential because of impaired cough and increased parasympathetic nervous system influence on pulmonary secretions
Assisted cough required for SCI above T6 attributable to intercostal and abdominal muscle paralysis
CARDIOVASCULAR
Relative bradycardia attributable to impaired sympathetic nervous system in SCI above T6; occasionally requires pacemaker placement
Hypotension attributable to systemic vasodilation resulting from impaired sympathetic drive
Venous stasis can result in deep venous thrombosis or pulmonary embolism
FUNCTIONAL MOBILITY
Upper extremity range of motion, strengthening, and endurance within limitations of orthotics and medical management
Bed mobility (including side to side, supine to prone to supine, supine to sit)
Wheelchair mobility (including forward and backward propulsion, turning, uneven terrain, curbs, ramps, hills)
Transfers (including bed to wheelchair to bed, wheelchair to toilet to wheelchair, wheelchair to bath to wheelchair, wheelchair to floor to wheelchair, wheelchair to car to wheelchair)
Activities of daily living including feeding, grooming, dressing, bathing, toileting
Bladder management training, typically with intermittent catheterization or alternative
Bowel management training, typically with suppositories and digital stimulation or alternative
Skin management training with monitoring and pressure relief techniques
Equipment evaluation for personal care, mobility, and public accessibility
Home and vehicle evaluation for accessibility
Psychological and social adjustment to SCI
Introduction to vocational and recreational opportunities for persons with SCI

Table 25.5 Signs and Symptoms of Spinal Cord Injury

Signs	Symptoms
Motor paralysis (BLOI)	Impaired or absent voluntary motor function
Sensory loss (BLOI)	Impaired or absent sensation
Hyperreflexia (UMN lesion)	Brisk DTRs, spasticity, spasms, clonus
Flaccidity (LMN lesion)	Flaccid paralysis with absent DTRs
Hypotension	Dizziness or loss of consciousness
Pulmonary dysfunction	Accessory muscles of respiration required
Neurogenic bladder	Urinary incontinence, urinary tract infection
Neurogenic bowel	Fecal incontinence, constipation

Note. BLOI = below neurological level of SCI; UMN = upper motor neuron; LMN = lower motor neuron; DTR = deep tendon reflexes.

Adapted from A.H. Ropper, 1994, Trauma of the head and spine. In *Harrison's principles of internal medicine*, edited by K.J. Isselbacher et al. (St. Louis: McGraw Hill).

every 15 to 30 min should reduce the risk of pressure ulcers in the exercising person with SCI (42). Phantom or neuropathic pain (e.g., burning, tingling, electrical sensations) may occur from a region below the SCI and can adversely affect exercise ability (20).

Upper extremity overuse injuries in wheelchair-reliant individuals occur most frequently at the shoulder (31, 44, 126). Hip, knee, and plantar flexion contractures may result from unbalanced muscle forces in wheelchair-reliant individuals over time, but these do not usually affect the person's function unless he desires to stand or walk (incomplete SCI). Heterotopic ossification (bony overgrowth within the joint space) can occasionally limit range of motion to such an extent that the individual loses the ability to transfer and perform certain ADLs. Management is with gentle range of motion exercises, nonsteroidal anti-inflammatory agents, bisphosphonates, and occasionally surgical resection (125).

Spasticity can affect exercise ability. Treatment includes the removal of any stimuli inducing increased tone, daily prolonged stretch of affected muscle groups, and pharmacological or surgical management (83). Understanding the level of SCI and associated medical problems can help the clinical exercise professional appreciate the physical limitations of each patient. Table 25.7 lists ADLs and functional ability by level of SCI.

Men 40 or more years of age and women 50 or more years of age who have SCI should be considered at moderate risk for untoward events during exercise, independent of traditional coronary artery disease risk. Persons with SCI should obtain medical clearance from a physician knowledgeable in SCI care before performing regular exercise.

Diagnostic Testing

The spine is considered unstable when two or more of the spinal columns (36) are damaged (involving soft tissue or bony elements) and surgical stabilization is warranted (figure 25.3). Note, however, that spinal cord damage may occur in the presence of apparent spine stability and that neurological dysfunction does not always occur in the presence of an unstable spine. Several examples of spinous fractures and dislocations are provided in table 25.8.

The diagnosis of SCI is largely based on the physical examination according to the ASIA criteria listed in table 25.1. Motor function sensory levels (pinprick and light touch) are assessed, and the injury is listed as complete if no motor or sensory function is spared at the S4 to S5 level; any preservation of function spared at these sacral levels denotes an incomplete SCI. The use of electrodiagnostic studies facilitates the diagnosis of SCI, most notably **somatosensory evoked potentials**

Table 25.6 Pharmacology: Medications Commonly Used in Spinal Cord Injury

Generic/brand names (class)	Primary effects	Exercise effects	Special considerations
Amitriptyline/Elavil (tricyclic antidepressant)	Decrease neuropathic pain	Dysrhythmia, hypotension	Dry mouth, constipation
Carbamazepine/Tegretol (anticonvulsant)	Control seizures	Decrease neuropathic pain, dizziness	Drowsiness, aplastic anemia, agranulocytosis
Diazepam/Valium (skeletal muscle relaxant)	Anxiety relief, decrease spasticity	Spasticity, dizziness, hypotension	Drowsiness, dry mouth
Enoxaparin/Lovenox (anticoagulant)	Anticoagulation, prevention of DVT and PE		Bleeding, bruising
Gabapentin/Neurontin (anticonvulsant)	Decrease neuropathic pain	Dizziness, hypotension	Drowsiness, dry mouth
Imipramine/Tofranil (tricyclic antidepressant)	Decrease bladder incontinence, depression and anxiety	Dysrrhythmia, hypotension	Dry mouth, constipation
Lioresal/Baclofen (skeletal muscle relaxant)	Decrease spasticity	Dizziness, hypotension	Drowsiness, hallucination, seizure
Nortriptyline/Pamelor (tricyclic antidepressant)	Decrease neuropathic pain	Dysrhythmia, hypotension	Dry mouth, constipation
Oxybutynin/Ditropan (anticholinergic)	Decrease bladder spasm	Dizziness, hypotension	Drowsiness, dry mouth

Generic/brand names (class)	Primary effects	Exercise effects	Special considerations
Phenoxybenzamine/Dibenzyline (α-antagonist)	Facilitate bladder emptying, decrease autonomic dysreflexia	Dizziness, hypotension	
Prazosin/Minipress (α-blocker)	Facilitate bladder emptying, decrease autonomic dysreflexia	Dizziness, hypotension	
Terazosin/Hytrin (α-blocker)	Facilitate bladder emptying, decrease autonomic dysreflexia	Dizziness, hypotension	
Tizanidine/Zanaflex (skeletal muscle relaxant)	Decrease spasticity	Dizziness, hypotension	Drowsiness, dry mouth
Tolterodine/Detrol (antimuscarinic)	Decrease bladder spasm	Dizziness, hypotension	Drowsiness, dry mouth
Warfarin/Coumadin (anticoagulant)	Anticoagulation, prevent DVT and PE	None	Bleeding, bruising

Note. DVT = deep venous thrombosis; PE = pulmonary embolism.

Table 25.7 Functional Ability by Level of Spinal Cord Injury

Activity	C3-C4	C5	C6	C7	C8-T1	T2-T6	T7-T10	T11-L2	L3-S3
Feeding	D	Modified I	I	I	I	I	I	I	I
Grooming	D	Modified I	I	I	I	I	I	I	I
Upper extremity dressing	D	Min A	Modified I	Modified I	I	I	I	I	I
Lower extremity dressing	D	Modified I	Min A	Modified I	I	I	I	I	I
Bathing	D	D	Modified I	Modified I	I	I	I	I	I
Pulmonary hygiene	D	Assisted cough	Assisted cough	Assisted cough	Assisted cough	Assisted cough	I	I	I
Bowel management	D	D	Modified I	Modified I	I	I	I	I	I
Bladder management	D	D	Mod A	Modified I	I	I	I	I	I
Bed mobility	D	Mod A	Modified I	I	I	I	I	I	I
Wheelchair propulsion	D	Modified I	Modified I	I	I	I	I	I	I
Wheelchair transfers	D	Max A	Modified I	I	I	I	I	I	I
Pressure relief	D	D	I	I	I	I	I	I	I
Driving	D	Modified I	Modified I	Modified I	Modified I	Modified I	Modified I	Modified I	Modified I
Ambulation	NA	NA	NA	NA	Exercise only	KAFO + Loft	KAFO + Loft	KAFO + Loft	AFO

Note. D = totally dependent; Min A = minimal (25%) assistance required from another person; Mod A = moderate (50%) assistance required from another person; Max A = maximal (75%) assistance required from another person; Modified I = independent with modified equipment; I = independent; NA = not applicable; KAFO = knee–ankle–foot orthosis (bracing) required; AFO = ankle–foot orthosis (bracing) required; Loft = Lofstrand (forearm) crutches required (130).

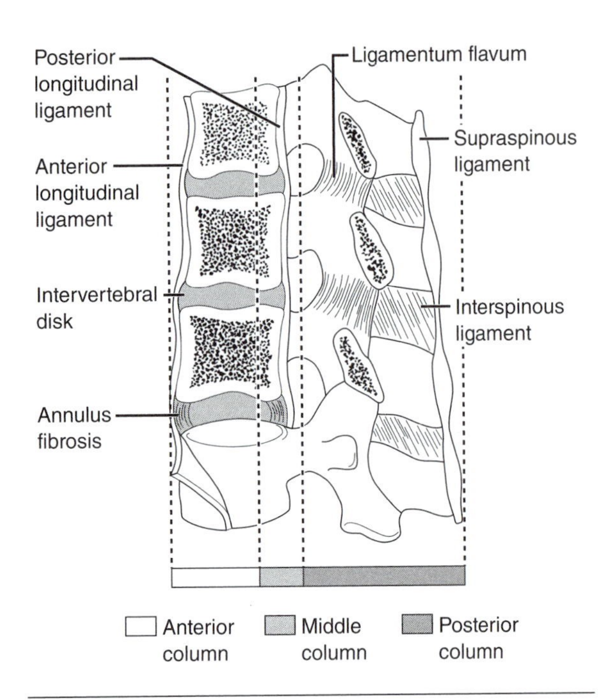

Figure 25.3 Denis three-column model.

Adapted, by permission, from F. Denis, 1983, The three column spine and its significance in the classification of acute thoracolumbar spinal injuries, *Spine* 8(8): 817-831.

(23). When consulting the exercise literature, the exercise professional should note the motor and impairment levels for SCI subjects, because the level and completeness of the injury significantly affect the degree to which the somatic and autonomic nervous systems contribute to exercise responses. The **Wheelchair and Ambulatory Sports, USA (WASUSA)** classification system for track events may help the clinical exercise professional determine the degree of impairment as it relates to exercise (table 25.9).

Body composition analysis is important in this population because body mass index (BMI) is not representative and obesity is a common problem. Although recently touted as the gold standard for determining body composition in SCI (124), dual-energy X-ray absorptiometry (DXA) introduces significant error because it does not account for hydration status. **Four-compartment body composition modeling** (fat, protein, water, mineral) should be used to determine percent body fat in persons with SCI until regression-based equations specific to DXA in SCI are established (10, 67). Additional testing appropriate for the person with SCI before exercise training may include pulmonary function testing, quantified strength and flexibility measures, DXA to determine bone mineral density, radiographs of paralyzed extremities to exclude asymptomatic fractures, lipid profiles, high-sensitivity C-reactive protein, and HbA1c to rule out glucose intolerance and diabetes.

Exercise Testing

Exercise testing to evaluate cardiorespiratory, musculoskeletal, and flexibility fitness can help to define areas of need with respect to exercise training. Table 25.10 provides a review of exercise testing specifics related to cardiovascular, skeletal muscle, and flexibility fitness.

Cardiovascular

Graded exercise testing can be used to assess aerobic fitness or training effects in asymptomatic or athletic populations, screen individuals at risk for heart disease, determine progress in rehabilitation, demonstrate maximal strength and power capacities, and assist with the exercise prescription (121). Additional benefits to persons with SCI include the opportunity to establish a

Table 25.8 Types and Causes or Results of Vertebral Fractures and Dislocations

Type of injury	Common cause and result of fracture
Atlantoaxial dislocation	Rheumatoid arthritis; can result in respiratory failure
Atlanto-occipital dislocation	Most common in children; typically causes death
Jefferson's fracture	Burst ring of atlas from descending force on vertex of skull (e.g., diving accident)
Hangman's fracture	Hyperextension and longitudinal distraction of upper cervical spine (e.g., chin striking steering wheel in vehicular accident)
Teardrop fracture	Vertebral body compression with anterior bony fragment
Compression or burst fracture	Retropulsed bony fragment into spinal cord
Hyperextension of cervical spine	With cervical stenosis, central cord syndrome
Thoracolumbar fracture	High-impact spinal cord injury associated with rib fracture

Adapted from Ropper 1994.

Table 25.9 The Wheelchair and Ambulatory Sports, USA System for Competition in Track Events for Participants With Spinal Cord Injury

Class	Neurological Level	Anatomical capability
T51	C6	Have functional elbow flexors and wrist extensors Have no functional elbow extensors or wrist palmar flexors May have shoulder weakness
T52	C7-C8	Have functional pectoral muscles, elbow flexors and extensors, wrist extensors, and some to all wrist flexors May have finger flexors or extensors Do not have the ability to perform finger abduction and adduction
T53	T1-T7	Have normal or nearly normal upper limb function Have no abdominal muscle function May have weak upper spinal extension
T54	T8-S1	Have abdominal muscles as well as back extension, which usually includes both upper and lower extensors

Adapted, by permission, from "International Paralympic Committee (IPC) Classification information in the WTFUSA 2013-2014 Rule Book for Track Event Participants with Spinal Cord Injury". http://www.wasusa.org/WTFUSA/2013%20WTFUSA%20Rulebook_FINAL_2_10_2013.pdf

Table 25.10 Exercise Testing Review

Test type	Mode	Protocol specifics	Clinical measures	Clinical implications	Special considerations
Cardiovascular	Arm or leg ergometry or both, including wheelchair ergometry	• Warm-up • Start at 5 to 25 W • Increase by 5 to 10 W every 2 to 3 min • Pause for BP	• Heart rate • Blood pressure • Rate of perceived exertion • Symptoms • 12-lead ECG if heart disease is suspected or documented • Monitor through 5 to 10 min recovery	• Signs and symptoms reveal exercise tolerance and source of limitation	• Empty bowel and bladder • Handgrip and postural stability • Skin protection • Adjust loads according to degree of training and paralysis and other impairments
Strength	Combined or isolated joint motions with free weights or machines	• Warm-up • Three trials of manual muscle testing for weaker muscles with grades ≥3 • Three trials of isometric or isotonic testing for stronger muscles • One trial of upper body anaerobic power testing for athletes	• Manual muscle testing for weakest muscles, if expecting neurological recovery • Determine maximal isometric force production with handheld dynamometer or 1- to 10RM • Peak or average anaerobic power during 30 s maximal effort	• Sufficient strength in proximal muscles for independent mobility • Balance of strength around joints to prevent contractures, rotator cuff injury, and joint degeneration	• Same as above • Higher joint angular velocities may be impeded by spasticity
Range of motion	Active and passive assisted stretching	• Warm-up • Evaluation of major joints • Focus on joints with spasticity or contracture	• Range of motion measured using a goniometer or other device	• Emphasize joints with spasticity or contracture • Support midshaft of long bones with osteopenia	• Spasticity • Contracture • Osteopenia

RM = repitition maximum
W = watts

relationship between fitness and posttraumatic return to gainful employment and to determine how the fitness level of a person with SCI changes over time (33).

Field testing is the easiest, least expensive, and most mobility-specific method of evaluation in selected wheelchair users (51). Recent reports, however, have failed to show significant correlations of field testing with actual $\dot{V}O_2$peak, possibly because of variability in terrain, wind speed, temperature, and humidity (134). Arm crank ergometry is the most often used test mode with SCI (28). Wheelchair ergometry is mobility specific for most SCI patients. Several systems have been developed and tested, including wheelchairs mounted on a motorized treadmill (133), low-friction rollers (94), and specialized devices to simulate overground propulsion (26, 60, 88). When compared with arm crank ergometry, wheelchair ergometry results in similar or greater $\dot{V}O_2$peak responses with lower peak power output (59), indicating reduced mechanical efficiency.

Several devices are available to assess all-extremity oxygen consumption (93), which may be appropriate for persons with incomplete SCI and for monitoring aerobic fitness of those using combined upper extremity and lower extremity **functional electrical stimulation (FES)**. Inclusion of the lower body muscle pump increases venous return, stroke volume, and cardiac output (99, 108).

Protocols should employ graded increments in resistance or power output requirement with periodic discontinuance for blood pressure, heart rate, and electrocardiogram determination (59, 86). A typical arm ergometry protocol employs an initial resistance of 5 to 25 W, with 5 to 10 W increases every 2 to 3 min to symptom-limited fatigue. People with SCI commonly achieve only 40 to 100 W at peak exercise. Population-specific prediction equations based on heart rate should be used for estimating $\dot{V}O_2$peak (73). Standard test termination rationale should be applied (see chapter 5).

For people undergoing testing to rule out ischemic heart disease, postexercise echocardiography (87) or nuclear imaging studies (9) may improve the sensitivity of the exercise stress test. In a study using standard exercise testing, only 5 of 13 subjects with known myocardial ischemia had ST-segment changes indicative of ischemia (9). The arm crank or wheelchair ergometer should be adjusted appropriately to allow optimal efficiency and reduce musculoskeletal injuries at the shoulder, elbow, and wrist. Straps applied to the torso improve trunk stability. Wheelchair gloves or flexion mitts with Velcro straps can prevent blisters, lacerations, and abrasions, especially for those with tetraplegia whose hands and fingers are **insensate** or unable to grasp sufficiently. Velcro straps and cuffed weights are commonly used for resistance training equipment modification. Abdominal binders and leg wraps may improve pulmonary dynamics and venous return, which reduce the risk of hypotension.

Because of sympathetic impairment, peak heart rate rarely exceeds 120 beats · min^{-1} in those with complete tetraplegia and T1-T3 **paraplegia**. Although variable responses occur in T4-T6 paraplegia, most persons with SCI below T7 are able to reach their age-adjusted peak heart rate. Similar trends are reported for blood pressure responses. In general, $\dot{V}O_2$peak and peak power output are significantly diminished in patients with SCI (7, 16, 76, 80, 116, 118). But the lower the injury, the less the impairment. $\dot{V}O_2$peak values range from 12 ml · kg^{-1} · min^{-1} for tetraplegic individuals to more than 30 ml · kg^{-1} · min^{-1} in low-level injury paraplegic persons. In these same groups, peak power output ranges from less than 30 W to more than 100 W, respectively.

Persons with SCI and a resting systolic blood pressure below 100 mmHg should be closely monitored during exercise. As people in this group approach peak exercise, the risk of a hypotensive response increases despite the use of leg wraps and abdominal binders. Should symptomatic hypotension occur, exercise testing should be halted and the person should be tilted back in the wheelchair to elevate the lower extremities above the level of the heart, promoting venous return.

Practical application 25.1 provides some guidelines for interacting with persons with SCI.

Musculoskeletal

In general, testing for muscular strength and endurance for normally innervated muscles in persons with SCI is very similar to that in persons without SCI. These normal muscles, unaffected by the SCI, can be tested with 1- to 10RM (repetition maximum) on resistance machines or free weights, provided that the trunk is stabilized, balance is ensured, and excessive axial loading of the spine or weight-bearing skin is avoided. Wingate-type anaerobic power testing with arm ergometry can be performed with younger fit individuals and athletes.

With muscle groups affected by paralysis, manual muscle testing may be necessary to evaluate strength relative to the ability to move the joint through the range of motion with or without gravity-neutral positioning. Muscles with spastic paralysis can be strengthened through the use of FES exercise or elicitation of spastic movement patterns, but caution is advised to avoid aggravating problem spasticity or joint contracture or increasing muscle imbalance. Muscles with flaccid paralysis and extreme atrophy cannot be stimulated to contract for testing or training.

Practical Application 25.1

CLIENT–CLINICIAN INTERACTION

Persons with a SCI must overcome tremendous physical, emotional, spiritual, and intellectual obstacles to succeed in life. They must wake up early to bathe, dress, groom, and perform bowel and bladder care. Seated pressure relief must be performed every 15 to 20 min. They must manage bodily functions according to a timed regimen. Obstacles are everyday occurrences. As a clinical exercise professional, you will find often that the person with SCI who consults you is one of the most disciplined, motivated, and enthusiastic clients you will have. Prepare to be surprised! You should understand the goals and obstacles of the person with SCI. To do this, you must have direct interaction with the patient. Whenever possible, speak at eye level with the client. Gain an understanding of the client's daily routine, environmental and transportation barriers that must be overcome, and concerns that the person may have about an exercise routine. Provide empathetic listening. Be prepared to discuss the application of exercise benefits to the client's functional abilities. Become familiar with accessible facilities in the community. You might consider spending a 24 h day without standing. During this time, perform community mobility and seated ADLs (including car, tub, and toilet transfers) from a wheelchair to gain greater appreciation for your SCI client's perspective and needs. This experience can be eye-opening, allowing you a glimpse into the world of the SCI patient. This activity may be helpful when you are considering the daily life of SCI patients and how they might incorporate an exercise or physical activity regimen.

Flexibility

In general, testing for joint flexibility in persons with SCI is very similar to that in persons without SCI. Joint range of motion can be tested with goniometry by a trained assistant. Joints should be moved passively, with assistance, or actively, but always slowly, gently, and painlessly, especially if spasticity, tightness, or contracture is present. Support long bones to avoid overstress to bone midshafts that are at risk of osteoporosis and osteopenia.

Treatment

The care and management of the person with SCI have significantly advanced, extending longevity with improved quality of life but also unmasking the problems associated with physical inactivity, blunted autonomic and hormonal responses, and upper extremity overuse syndromes common to chronic survivors of SCI. Clinical practice guidelines developed by the Consortium for Spinal Cord Medicine and published by the Paralyzed Veterans of America (www.pva.org) describe expected outcomes for SCI and discuss the prevention and management of autonomic dysreflexia, thromboembolism, bowel problems, depression, and pressure ulcers in SCI. Future clinical practice guidelines must address the application and prescription of exercise including a 12-lead electrocardiogram and risk profile assessment before performance of an exercise test. Standard contraindications for exercise testing noted in chapter 5 should be applied to the SCI population. The following are common contraindications in patients with SCI. Without adequate assessment and treatment, these should be considered absolute reasons not to perform an exercise test:

- Autonomic dysreflexia resulting from recent fracture (may precipitate spasms or increase the risk of fatty emboli, hypertensive crisis, or cerebrovascular events)
- Orthostatic hypotension, with the risk of syncope
- Recent deep vein thrombosis or pulmonary embolism
- Pressure ulcers, which increase the risk of autonomic dysreflexia during exercise

Common relative contraindications include the following:

- Active tendinitis (e.g., rotator cuff, elbow flexors, wrist flexors and extensors)
- Chronic heterotopic ossification
- Peripheral neuropathy
- Pressure ulcers of grade 2 or less
- Spasticity

Bladder and bowel evacuation should be implemented immediately before the graded exercise test to minimize the risk of exertional incontinence or autonomic dysreflexia. Manual rectal stimulation or **disimpaction** is required to maintain fecal continence; and in some cases, colostomy is warranted for bowel management. Bladder management is most often performed via intermittent catheterization, although indwelling catheters and bladder diversion techniques are sometimes warranted. Environmental considerations are also necessary because of the SCI patient's difficulty with regulation of body temperature.

EXERCISE PRESCRIPTION AND TRAINING

The exercise prescription for a person with SCI should focus on the typical parameters (see chapter 5). Because of potential complexities, however, the prescription should ideally be developed by a team composed of an exercise physiologist (physiological responses), a physical therapist (orthopedic limitations), and a physician (medical concerns and oversight). The best approach is for the person with complete SCI to begin an exercise training program under the supervision of either a physical therapist or a clinical exercise physiologist. A review of exercise prescription variables is provided in table 25.11.

Special Exercise Considerations

Because of their reduced body temperature regulatory ability, persons with tetraplegia should exercise only in a mildly temperate climate (outside during appropriate weather or inside in a climate-controlled area) to avoid the risk of hypo- and hyperthermia. Appropriate seating and positioning are necessary to reduce the risk of pressure sores, autonomic dysreflexia, spasticity, and musculoskeletal trauma.

Few commercial fitness centers are able to accommodate the needs of the person with SCI, including wheelchair access. A survey of physical fitness facilities in a major city demonstrated that none of the 34 facilities reviewed met all of the 1990 Americans with Disabilities Act (ADA) requirements for accessibility (47). Other potential barriers include individual constraints attributable to transportation availability and required assistance. Finances may also be a concern for many patients. These issues have a direct effect on compliance.

Adapted or adaptable equipment, some of which is outlined in the section on exercise testing, is required for proper and safe exercise. Other recommendations include the use of upper armbands or Coban tape to prevent abrasions at the medial upper arm with wheelchair propulsion. Abdominal binders and leg wraps may be used to facilitate improved pulmonary dynamics and greater venous return. FES-LCE (functional electrical stimulation–leg cycling exercise) and hybrid systems are now commercially available but remain expensive to purchase and maintain as compared to arm crank ergometry. Velcro straps and cuffed weights are commonly used for resistance training to improve or create a grip.

Individuals with complete lesions at or above C4 are limited to using FES-LCD or hybrid exercise equipment because of arm, trunk, and leg paralysis. Manual or other resistive exercise can be performed for inspiration (diaphragm). Range of motion exercise can be provided by an exercise professional or trained family member. The person with C5 tetraplegia requires exercises that do not entail active wrist extension, elbow extension, or grasp. Conversely, those with C6 lesions have active wrist extension but little or no elbow extension or grasp. Table 25.3 can be used to determine the effects of the level of SCI on the ability to exercise. Additionally, persons with incomplete SCI who have relative sparing of sensation in the lower extremities may not be able to tolerate FES-LCE or hybrid exercise modes because of pain.

Hybrid exercise (FCE-LCE combined with voluntary arm exercise) is a promising exercise mode for cardiovascular training because it elicits higher levels of metabolic and cardiopulmonary responses than either arm or leg exercise alone (66).

The initial stages of an aerobic exercise program should focus more on developing the habit of exercise than on the intensity and duration, because exercise adherence is poor in this population. Succinct, precise, and quantifiable goals optimize chances of a successful outcome. Having the patient make the informed choice of exercise mode positively affects patient compliance. Also, individuals should be told to expect delayed-onset muscle soreness (DOMS). The onset and duration of DOMS are similar to those for nondisabled people. Other reviews have ably summarized the many exercise recommendations for persons with SCI (43, 46, 49, 77, 92).

Cardiovascular Exercise Training

It may be misleading to assume that aerobic exercise will train the cardiovascular system in tetraplegia. Central cardiovascular adaptations, such as increased peak stroke volume, cardiac contractility, or cardiac output, have not yet been documented. Such observed adaptations may well result from at least two mechanisms: (a) increased active muscle mass that exhausts a larger fraction of the cardiac reserve, hence producing higher peak cardiovas-

Table 25.11 Exercise Prescription Review

Training method	Mode	Intensity	Frequency	Duration	Progression	Goals	Special considerations and comments
Aerobic	ACE Wheelchair ergometry Arm crank cycling Community wheeling Seated aerobics Aquatics Wheelchair recreation FES and leg cycling ergometry ACE and FES leg ergometry	RPE 11 to 14 50% to 85% $\dot{V}O_2$peak or peak power output 30% to 80% HR reserve 60% to 90% HRpeak Talk test	3 to 7 d/wk	20 to 60 min, continuous or interval	Slow (<5% per week)	Improved functional capacity Reduction in activity-affected cardiovascular disease risk factors	Avoid exertional hypotension. May initially require multiple sets of 5 to 10 min duration. Monitor for autonomic dysreflexia. Avoid thermal stress. Include warm-up and cool-down.
Resistance	Elastic bands Wrist weights Body weight Dumbbells Free weights Wheelchair-accessible machines FES isokinetic	8 to 12 reps at 60% to 75% 1RM	One to three sets on 2 or 3 d/wk	30 to 60 min	Increased resistance when 12 reps are achieved	Improved strength Improved ability to ambulate using arms	Avoid Valsalva maneuver. Provide spotter. Use seat belt and chest strap for balance. Use adaptive grip and mitts. Do not exceed stress limits of wheelchair.
Range of motion	Active assisted (anterior shoulder, pectoral, rotator cuff) Passive assisted (hip flexors, knee flexors, plantar flexors) Standing in standing frame if medically cleared	Two sets, 30 to 60 s per set; gentle, slow, painless	7 d/wk	5 to 15 min	As tolerated	Prevention of contracture and reduction of spasticity Improved joint ROM for affected and unaffected joints Sufficient flexibility to accomplish ADLs Degrees of normal range of motion that are missing	Stretch to strain, no pain. Avoid Valsalva maneuver. Don't overstress insensate joints. Provide midshaft support for long osteopenic limbs.

Note. ACE = arm crank ergometry; FES = functional electrical stimulation; RPE = rating of perceived exertion (6 to 20 scale); RM = repetition maximum; HR = heart rate; ROM = range of motion; reps = repetitions; ADLs = activities of daily living.

cular responses; or (b) external lower body compression or supine posture that reduces the lower body venous pool and improves venous return and (via baroreflex) reduces heart rate and increases stroke volume and cardiac output. Overall, training-induced increases in $\dot{V}O_2$peak may not be caused by central cardiovascular limitations but by increased active muscle, venous return, and peripheral O_2 extraction in trained muscles (43).

The general pattern of cardiovascular acute responses in persons with SCI during upright exercise is "hypokinetic circulation" or "circulatory hypokinesis" (33, 43, 70, 76). This is characterized by a lower increase in cardiac output per unit increase in $\dot{V}O_2$ during exercise. This is achieved by a relatively high heart rate and relatively limited increases in cardiac contractility, stroke volume, and arterial blood pressure and decreases in peripheral vascular resistance.

Persons with SCI are at an increased risk for cardiovascular disease. Therefore, during cardiovascular exercise, care must be taken to reduce the risk of an adverse cardiovascular event. In 2006, Spinal Cord Injury Rehabilitation Evidence (SCIRE) evaluated the strength of the evidence that aerobic exercise training can modify cardiovascular risk factors (109, 110, 136, 139). The rationale for the evidence strength is provided in table 25.12.

The management of SCI, as reported by SCIRE, using voluntary upper body aerobic exercise performed at a moderate-to-vigorous intensity for 20 to 30 minutes per day, 3 days per week for at least 8 weeks positively affects many risk factors related to cardiovascular disease. These include the following with a SCIRE strength of evidence of 1 (table 25.12): improved exercise tolerance, $\dot{V}O_2$peak, and power output; reduced LDL and total cholesterol, and serum triglycerides; and enhanced insulin sensitivity and glucose tolerance. The level of research evidence was also strong (SCIRE score of 2) for improved cardiac output and stroke volume, reduced peripheral vascular resistance, and increased HDL cholesterol. Submaximal and maximal heart rate, as well as cellular adaptations for oxidative metabolism, also likely improve; however the level of strength of the evidence is lower (SCIRE score of 4). The management of cardiovascular risk factors has also been assessed in those with SCI by the use of functional electrical stimulation (FES) training. Although each of the same parameters noted for the upper body ergometry were positively influenced by FES, the strength of the research evidence currently is all SCIRE level 4.

SCIRE concluded that "Marked physical inactivity appears to play a central role in this increased risk . . . the relationship between increasing physical activity and health status of SCI has not been evaluated adequately to date. Based on preliminary evidence, it would appear that various exercise modalities (including arm ergometry, resistance training, body weight supported treadmill training, and FES) may attenuate and/or reverse abnormalities in glucose homeostasis, lipid lipoprotein profiles, and cardiovascular fitness. As such, exercise training appears to be an important therapeutic intervention for reducing the risk for cardiovascular disease and multiple comorbidities (such as type II diabetes, hypertension, obesity) in individuals with SCI. Well-designed randomized controlled trials are required in the future to firmly establish the primary mechanisms by which exercise interventions elicit these beneficial changes. Similarly, further research is required to evaluate the effects of lesion level and injury severity on exercise prescription, such that exercise programs can be developed that address the varied needs of persons with SCI. Moreover, long-term follow-up investigations are required to determine whether training-induced changes in risk factors for cardiovascular disease translate directly into a reduced incidence of cardiovascular disease and premature mortality in persons with SCI" (p. 136).

The most appropriate method for monitoring and prescribing intensity of aerobic exercise for the individual with SCI is controversial. Heart rate responses in people with tetraplegia and high-level paraplegia are lower than in nondisabled people because of reduced active muscle mass and impaired sympathetic response (113). Although variable, 30% to 80% of heart rate reserve appears to correspond to 50% to 85% of $\dot{V}O_2$peak in those with

Table 25.12 Exercise Training Review

Cardiorespiratory endurance	Skeletal muscle strength	Skeletal muscle endurance	Flexibility	Body composition
• Increased exercise tolerance • Increased $\dot{V}O_2$peak • Reversal of myocardial disuse atrophy • Increased cardiac output • Increased stroke volume • Decreased submaximal exercise HR • Decreased total peripheral vascular resistance • Potential for increased daily energy expenditure	• Increased strength • Increased peak and mean anaerobic power output • Increased functional mobility	• Increased fatigue resistance • Increased functional mobility	• Maintenance of joint flexibility, in absence of heterotopic ossification and contracture • Temporary decrease in spasticity	• Decreased % body fat • Increased % lean body mass • Increased energy expenditure • Potential for treatment of obesity, especially with diet modification

high-level paraplegia and tetraplegia (79, 113). Using a percentage of peak power output is also recommended (113). This method is cumbersome, however, as continual reevaluation is required to maintain optimal training intensity because peak power output increases with training. Rating of perceived exertion (RPE) may also be used to guide exercise training intensity. This method is used successfully in cardiac transplant patients who, like patients with SCI, have reduced peak heart rates secondary to cardiac denervation (84). Although not investigated in patients with SCI, RPE values of 11 to 14 may be used during exercise training because this level corresponds to approximately 60% to 90% of $\dot{V}O_2$peak in nondisabled individuals. Another possible method is the talk test. Nondisabled people able to speak during exercise without feeling short of breath are typically at appropriate exercise intensity. Again, however, this method has not been validated in the SCI population or other groups with sympathetic nervous system dysfunction (2).

The duration of a single aerobic exercise bout will vary depending on fitness level, but a goal is to follow the recommendations for the general population (2, 65). Exercise progression should occur as tolerated to a maximum of 60 min duration at the prescribed intensity range.

Aerobic exercise should be performed no fewer than 3 d/wk to maintain fitness but may occur up to 5 d/wk for optimal gains without negative consequences, particularly when performed at relatively low intensity levels (2, 64). Persons with SCI may require increased exercise frequency to optimize caloric expenditure for weight management but should be judiciously monitored to reduce the incidence of upper extremity overuse syndromes. People with SCI who have $\dot{V}O_2$peak less than 10.5 ml $\cdot$ kg^{-1} $\cdot$ min^{-1} (<3 METs) may require multiple bouts of exercise daily, each lasting 5 to 15 min, until they are able to tolerate 20 to 30 min sessions. SCI Action Canada recommends at least 20 min of moderate-to-vigorous aerobic activity two times a week, for example wheeling, arm cycling, sports, partial body weight–supported treadmill training (PBWSTT), leg cycling, recumbent stepping, and water exercise (57, 119).

Resistance Exercise

Minimally, scapular stabilization and rotator cuff resistance exercises should be used in all SCI patients capable of voluntary control of these muscles. Initial intervention should include two sets of 10 repetitions, with 6 s isometric contractions for shoulder protractors, retractors, elevators, and depressors, as well as for internal and external shoulder rotators. As the patient tolerates, progression to dynamic exercise should occur. Resistance band exercises are useful initially, but a plateau in gains can be expected because of the limitation in resistance of this device. Although dumbbells and free weights may be used under close supervision, paralyzed lower extremities and truncal musculature significantly reduce a person's ability to balance even small objects when lying supine or when seated without significant truncal support. When the person is using free weights or isotonic or isokinetic machines, wheelchair brakes should be set before lifting, and care should be taken not to exceed the weight and stress limitations of the wheelchair as provided by the manufacturer. Standard recommendations for intensity and progression should be followed as discussed in chapter 5. SCI Action Canada recommends strength training exercises two times a week, consisting of three sets of 8 to 10 repetitions of each exercise for each major muscle group, for example, free weights, elastic resistance bands, cable pulleys, weight machines, and FES exercise (57, 119).

Range of Motion Exercise

Range of motion exercise should be performed daily and should focus on all major joints, especially those with contracture and spasticity. Both active and passive assisted methods of static stretching can be used. During passive stretching, carefully working through the range of motion in joints lacking sensation is important because the individual cannot determine when the maximal range of the joint has been reached. Supporting the midshaft of long bones in the patient who has osteopenia may be important to reduce the chance of fracture. If the person is medically cleared for full weight bearing, standing in a standing frame can be used to maintain or increase range of motion of the spinal extensors and hip, knee, and ankle plantar flexors.

EXERCISE TRAINING

Tasks such as feeding, grooming, hygiene, dressing, bathing, transfers, and toileting are referred to as activities of daily living (ADLs), whereas tasks relating to community mobility include traversing sidewalks, stairs or ramps, paths, and environmental barriers (e.g., curbs, speed bumps). Patients with longstanding (>20 yr) SCI require greater physical assistance as they age. The percentage of maximal heart rate during performance of ADLs and community mobility tasks is higher in people with tetraplegia than in those with paraplegia, and an inverse relationship exists between physical capacity and physical strain (32, 78). In one study, only 29% of SCI subjects with $\dot{V}O_2$peak less than 15 ml $\cdot$ kg^{-1} $\cdot$ min^{-1} were able to perform independent ADLs (71). Table 25.12 presents the general adaptations that can be expected from exercise training in those with SCI.

Upper Extremity Aerobic Training

Critical review of upper extremity aerobic conditioning studies demonstrates variability in exercise prescription and results, partially attributable to the level of SCI (90). For instance, Gass and colleagues (53) trained seven subjects (four with C5-C6 lesions, three with T1-T4 lesions) five times weekly to exhaustion on a graded exercise test protocol using a wheelchair on a treadmill ergometer. After 7 wk of training with this ergometry system, the mean $\dot{V}O_2$peak had increased from 9.5 to 12.7 ml · kg^{-1} · min^{-1} and endurance time had increased by 4.4 min, suggesting considerable change in functional capacity (53). Knutsson and colleagues (86) evaluated 20 SCI inpatients with complete and incomplete SCI between C5 and L1. For the 10 persons assigned to the training group, a 40% increase in peak work rate (40-57 W) was reported, although no significant change was noted in the three subjects with SCI above T6. Taylor and colleagues (127) assessed the effects of arm crank ergometry (ACE) performed 30 min/day, 5 d/wk for 8 wk in 10 individuals with paraplegia. The trained group significantly improved $\dot{V}O_2$peak from 22.8 to 26.3 ml · kg^{-1} · min^{-1} without significant changes in maximal heart rate, postexercise lactate, or body fat. Davis and Shephard (34) assessed four patterns of ACE training (50% or 70% $\dot{V}O_2$peak, 20 or 40 min per session, three sessions per week, 24 wk) in inactive subjects with paraplegia. Cardiac stroke volume during submaximal exercise and $\dot{V}O_2$peak increased significantly during ACE tests, except in the nonexercise control group and the group with the lowest training intensity and duration. McLean and Skinner (96) matched 14 tetraplegic subjects, by peak power output, to either a supine or seated exercise training regimen to assess for the effect of changes in postural position on stroke volume, cardiac output, and exercise capacity. Their subjects performed arm crank ergometry exercise in either a sitting or a supine position at 60% of their $\dot{V}O_2$peak three times a week for 10 wk, with progressive increments in either duration or resistance; no significant differences were found in stroke volume or cardiac output, although absolute $\dot{V}O_2$peak increased from 720 to 780 ml · min^{-1}, suggesting peripheral adaptations.

Locomotor Training

Recent research indicates that locomotor training can greatly improve human walking ability after SCI (4, 5, 14, 15, 18, 38). When therapists provide body weight support and manual assistance to SCI subjects during treadmill stepping, task-specific motor learning occurs and improves gait control and mechanics. Studies on patients with chronic incomplete SCI indicated that approximately 80% of wheelchair-reliant individuals became independent walkers after locomotor training (137). Dobkin and colleagues (40) conducted a multicenter clinical trial for subjects with acute incomplete SCI to validate the effectiveness of locomotor training in clinical practice. Although improvements were noted, **partial body weight–supported treadmill training (PBWSTT)** was not significantly better than "usual" rehabilitation in the acute setting, likely because of neural recovery often seen in the acute phase. Despite its possible benefits, only a handful of clinics in the United States currently offer manually assisted locomotor training with partial body weight support for gait rehabilitation after SCI, presumably because of the high cost of therapist labor and equipment for locomotor training. Reports on overall health parameters including aerobic fitness, lipid profiles, body composition changes, insulin sensitivity and glucose tolerance, and lower extremity bone mineral density are promising but sparse to date (68, 103, 107, 136).

Resistance Exercise

Because of the large number of shoulder and upper extremity musculoskeletal problems encountered by persons with SCI, a prophylactic, structured, and progressively resistive strengthening program that focuses on scapular, rotator cuff, and pectoral muscles is likely to increase strength and reduce the risk for overuse injury (44). Such a program will probably improve the ability of these people to perform functional tasks in the community. In a RCT, Mulroy and colleagues (98) compared the efficacy of an exercise movement optimization program to an attention control intervention for decreasing shoulder pain in people with paraplegia from SCI. The 12 wk home-based intervention consisted of shoulder strengthening and stretching exercises, along with recommendations on how to optimize the movement technique of transfers, raises, and wheelchair propulsion. Shoulder pain, as measured with the Wheelchair User's Shoulder Pain Index, significantly decreased to one-third of baseline levels after the intervention in the exercise movement optimization group but remained unchanged in the attention control group. Peak torques in scapular elevation and in shoulder adduction and internal and external rotation improved in the exercise movement optimization group (18-32%) but not in the attention control group. Improvements were maintained at the 4 wk follow-up assessment. Self-reported physical activity did not change in either group.

Increasing information is available on the effects of resistance training on strength, power, muscle mass, or functional abilities in persons with SCI. Clinical studies have recommended resistance training for many years. Nilsson and colleagues (105) reported increases in dynamic strength (16%) and endurance (80%) when comparing bench press before and after a 7 wk combined arm crank ergometry and resistance training program in adults with paraplegia. Chawla and colleagues (24) reported that a resistance training program including bench press, incline press, lateral raises, incline curls, lat pulls, and triceps stretch improved ADL function in 10 patients with SCI. Unfortunately, specific ADLs and quantitative measures of strength were not reported. From these limited studies, it appears promising that specific strength training will benefit patients with SCI. Increasing information from RCTs is available on the effects of resistance training on strength, power, muscle mass, and muscle morphology.

Five research groups have used RCTs to test the efficacy of resistance training to improve muscular strength and endurance (61, 64, 69, 74, 104). Needham-Shropshire and colleagues (104) found more significant increases in upper extremity (e.g., triceps) muscle strength (by manual muscle testing) following 8 wk of FES-assisted arm ergometry than after either voluntary arm ergometry or FES-assisted arm ergometry plus voluntary arm ergometry. Hicks and colleagues (69) trained 11 participants with SCI twice weekly for 90 to 120 min per session over 9 mo using progressive resistance exercise and arm ergometry. Participants significantly improved 1RM upper body muscle strength by 19% to 34% and submaximal arm ergometry power output by 81%. Jacobs (74) trained two groups of nine clients with paraplegia over 12 wk, one with arm ergometry and the other with resistance circuit training (three sessions per week, six exercises, three sets of 10 reps per session, 60-70% of 1RM intensity). Muscular strength significantly increased for all exercises in the resistance training group with no changes in the arm ergometry group. Mean arm crank anaerobic power increased in both groups by 5% to 8%. Peak arm crank anaerobic power increased significantly by 16% in the resistance training group and 3% in the arm ergometry group. Glinsky and colleagues (61) used progressive resistance exercise to train 32 participants with C5-C6 tetraplegia during three sessions a week for 8 wk. They showed no clinically significant increases in strength and endurance (8% and 11%, respectively) in paretic wrist extensor and flexor muscle groups. Hartkopp and colleagues (64) used high- and low-resistance FES exercise to train 12 participants with C5-C6 tetraplegia in paretic wrist extensors (30 min per session, five sessions per week, 12 wk). Compared to the untrained arm, the trained arm increased strength significantly with the high-resistance protocol, but not the low-resistance protocol. Muscular endurance increased with both

Practical Application 25.2

RESEARCH FOCUS

Resting metabolism in persons with SCI is diminished in association with reductions in fat-free body mass (13, 21, 22). Subsequently, energy expenditure during many daily activities is also reduced, contributing to a net positive energy balance in most cases (22, 35). A compendium of physical activity energy expenditures for persons with SCI has recently been published, and may offer additional insights into the weight management issues faced by most individuals with SCI (25). Most clinicians are only beginning to recognize obesity as a significant issue in SCI. Recent investigations have demonstrated a significant discrepancy between BMI and percent body fat in persons with SCI (21, 81), suggesting that the incidence of overfat and obesity in persons with SCI is greatly underestimated (50, 55, 91) and may be at epidemic proportions in our society (54). An investigation is currently under way to develop a quick clinical tool to assess body composition in persons with SCI based on the criterion four-compartment model. Since adiposity is now considered one of the major causal factors of the metabolic syndrome (obesity, glucose intolerance or insulin resistance, hypertension, dyslipidemia), a relationship between adiposity and the metabolic syndrome is being sought in the SCI population (54, 55). Several cross-sectional SCI studies have reported that the incidence of metabolic syndrome is at least as high as that reported in the non-SCI population (54). Most previous exercise research in SCI has focused on physiological improvements in fitness levels, endurance capacity, and strength. The role of energy expenditure in the exercise prescription needs immediate consideration, as it may affect overall body composition and health parameters more than changes in aerobic fitness or strength.

protocols by 41% and 42%, respectively. Although paradigms varied greatly across trials, most studies report increases in muscular strength and endurance compared to values in control groups.

Functional Electrical Stimulation

In 1987, medical guidelines were developed for patient participation in FES rehabilitation, including medical criteria for inclusion and exclusion. Computerized FES is a neuromuscular aid used to restore purposeful movement of limbs paralyzed by upper motor neuron lesions. Numerous reviews are devoted to FES and its potential to stimulate beneficial exercise adaptations in patients with SCI (48, 107, 108). Briefly, FES of the lower extremities can be used to do the following:

- Stimulate skeletal muscle strength (41, 58, 96, 111, 114) and endurance (53, 59, 72, 97, 114)
- Increase energy expenditure and increase stroke volume (42, 59, 72, 75)
- Increase total body peak power, $\dot{V}O_2$peak, and ventilatory rate (6, 59, 72, 97)
- Reverse myocardial disuse atrophy (102)
- Increase high-density lipoprotein levels and improve body composition (8)
- Improve self-perception (122)
- Increase lower extremity bone mineral density (17, 63, 97)

Despite these encouraging findings, functional gains in upper extremity strength, aerobic capacity, **community mobility**, and ADLs were not evaluated in response to FES lower extremity training. After 8 wk of inactivity that followed a 12 wk training program, the 45% improvement in $\dot{V}O_2$peak attained by FES plus leg ergometry training was reduced by approximately 50%, whereas power output returned to pretraining levels. Submaximal heart rate, peak heart rate, and peak ventilatory volume did not change at any time point.

A logical and intuitive progression in the development of FES lower extremity exercise training has been the combined use of concurrent arm crank ergometry and FES leg cycle ergometry (LCE), termed **hybrid** exercise. As expected from the combination of upper and lower extremity exercise, peak power, $\dot{V}O_2$peak, stroke volume, and cardiac output significantly increase during hybrid exercise bouts with SCI subjects (45, 99) and during combined upper extremity rowing plus lower extremity FES (89).

Range of Motion Exercise

Katalinic and colleagues (82) completed a meta-analysis of 35 studies involving 1,391 participants on the efficacy of stretching for the purpose of treating or preventing contractures. In people with neurological conditions, moderate- to high-quality evidence indicated that stretch does not have clinically important immediate (mean difference 3°; 95% confidence interval [CI] 0 to 7), short-term (mean difference 1°; 95% CI 0 to 3), or long-term (mean difference 0°; 95% CI –2 to 2) effects on joint mobility. No study performed stretch for more than 7 mo. The results were similar for people with nonneurological conditions. The authors concluded that for all conditions, there is little or no effect of stretch on pain, spasticity, activity limitation, participation restriction, or quality of life. Therefore, the value of stretching may be limited to maintenance of joint range of motion and prevention of contractures.

Other interventions to improve range of motion such as splinting, continuous passive motion, standing in a standing frame, or neuromuscular facilitation techniques have not been as rigorously evaluated. Many fitness and clinically based stretching protocols are available online:

Lower body

http://calder.med.miami.edu/pointis/lower.html (131)

Upper body

http://calder.med.miami.edu/pointis/upper.html (132)

Clinical Exercise Pearls

- Maximize active muscle mass during cardiovascular or aerobic exercise testing to try to exhaust cardiovascular reserve before peripheral muscle fatigue limits test.
- Preserve balanced strength and range of motion about the shoulder girdle to prevent impingement syndrome and rotator cuff pathology.
- For maximal long-term exercise training success, emphasize improving general health and prevention of secondary conditions that will impede ability to train and remain physically active.
- Promote sustainable physical activities and energy expenditure for prevention of metabolic syndrome, for long-term health, and for functional independence.

Physicians and rehabilitative therapists should be consulted when muscle imbalance, spasticity, or contracture is present.

CONCLUSION

The patient with SCI presents with unique obstacles and considerations that the clinical exercise professional must be familiar with to provide safe and optimal exercise testing and training oversight. People with SCI tend to be sedentary and are excellent candidates for regular exercise training. Because of the many potential medical and possible exercise-related problems, a team approach to evaluation, exercise prescription development, and exercise training guidance is recommended.

Key Terms

activity of daily living (ADL) (p. 494)
areflexia (p. 490)
astrocytic gliosis (p. 491)
atrophy (p. 489)
autonomic dysreflexia (p. 493)
autonomic nervous systems (p. 490)
cauda equina (p. 491)
circulatory hypokinesis (p. 494)
community mobility (p. 508)
complete paraplegia (p. 490)
disimpaction (p. 502)
dysesthetic (p. 490)
endocrine (p. 494)
flaccidity (p. 490)
four-compartment body composition modeling (p. 498)
functional electrical stimulation (FES) (p. 500)
hybrid (p. 508)
hyperinsulinemia (p. 494)
hyperreflexia (p. 494)
incomplete paraplegia (p. 490)
incomplete tetraplegia (p. 490)
insensate (p. 500)
ipsilateral (p. 490)
myelomeningocele (p. 490)
orthostatic hypotension (p. 493)
osteopenia (p. 494)
paraplegia (p. 500)
parasympathetic nervous system (p. 492)
partial body weight–supported treadmill training (PBWSTT) (p. 506)
plexopathies (p. 494)
poikilothermic (p. 494)
sarcopenia (p. 489)
somatosensory evoked potentials (p. 496)
spasticity (p. 494)
spinal cord injury (SCI) (p. 489)
sympathetic nervous system (p. 492)
tetraplegia (p. 489)
Wheelchair and Ambulatory Sports, USA (WASUSA) (p. 498)

CASE STUDY

MEDICAL HISTORY

Ms. BF is a 28 yr old white female with thoracic paraplegia since age 19 caused by a motor vehicle accident. She played volleyball in high school and was athletic before the accident. Since her rehabilitation, she has been wheelchair reliant for community mobility but is otherwise independent with respect to ADLs, including bowel and bladder management and driving with hand controls. She received an MBA this year and has a full-time job as an accountant for a law firm, but is otherwise quite sedentary. Her body weight has remained stable at 150 lb (68 kg) for the past 4 yr. Her BMI is 31. Other exam findings include blood pressure (BP) of 100/60mmHg; heart rate (HR) of 85 beats · min^{-1}; electrocardiogram (ECG) with normal sinus rhythm; normal heart, lung, and bowel sounds; intact skin; mildly reduced range of motion at the shoulder (extension, internal and external rotation); 5° hip, knee, and plantar flexion contractures bilaterally; 5/5 upper extremity with 0/5 lower extremity motor strength; absent pinprick and light touch sensation everywhere below the xyphoid process; and 3+ (brisk) reflexes at both the knees and ankles. She reports poor sleep quality and excessive fatigue and intermittently notes pounding headaches and sweating associated with an overfull bladder. Her medications include Detrol for bladder spasms and Lioresal for lower extremity spasms and spasticity.

(continued)

DIAGNOSIS

Ms. BF has T6 American Spinal Injury Association grade A paraplegia with neurogenic bowel and bladder, spasticity, and occasional autonomic dysreflexia. She likely has impaired pulmonary function attributable to lower intercostal, paraspinal, and abdominal muscle paralysis. Although she is fully functional with her upper extremities, she has poor to fair sitting balance because of paraspinal, abdominal, and lower extremity paralysis. Her elevated BMI is probably the result of her sedentary lifestyle. The fact that she has completed a graduate degree and holds a full-time job suggests that she has adjusted well to her SCI.

EXERCISE TEST RESULTS

Ms. BF will most likely want to know how exercise can affect her current lifestyle and health, and whether the benefits will be worth investing time already committed to a busy and productive life. Encouraging her to share short- and long-term life goals will provide an opportunity to educate her about the acute and chronic benefits of exercise, particularly as they can improve her community mobility and modify her elevated risks for chronic diseases, including coronary artery disease, obesity, diabetes, osteoporosis, and upper extremity overuse syndromes. As an initial step to increasing Ms. BF's physical activity level, an exercise evaluation was suggested. Ms. BF's exercise evaluation was performed on an upper body ergometer using a discontinuous protocol with an initial workload of 24 W at a constant 50 rpm and 6 W increments every 3 min. Her peak exercise performance yielded the following results: peak HR = 172 beats $\cdot$ min^{-1}; BP = 130/50mmHg; $\dot{V}O_2$peak = 16.4 ml $\cdot$ kg^{-1} $\cdot$ min^{-1}; peak power output = 42 W. ECG revealed isolated premature ventricular contractions (PVCs) with no evidence of ST-segment depression. Ms. BF stopped the test herself because of shoulder fatigue and dizziness, associated with a 10 mmHg decrease in systolic blood pressure. She experienced some exertional dyspnea but no other symptoms. In the past, Ms. BF's physician has asked her to try FES. She was initially evaluated and thought to be a good candidate but never seriously considered FES.

EXERCISE PRESCRIPTION

Ms. BF is interested in improved community mobility, losing weight, and maintaining good health. She is motivated to begin a regular physical activity program. She will begin under the guidance of an outpatient physical therapist. Her goal is to perform her routine at a medical fitness center under the guidance of a clinical exercise physiologist. Her maximal heart rate response on the exercise test will be used to set a target heart rate zone based on her heart rate reserve, and she will be introduced to a wheelchair ergometer as well as the upper body ergometer. Her goal is to perform between 20 and 60 min of aerobic exercise 3 to 5 d each week, although her initial workouts may be divided into two or three 10 min sessions, with 10 to 15 min recovery between sessions to reduce her risk of hypotension. Ms. BF will also begin a resistance training program that emphasizes shoulder stabilization with resistance bands 2 to 3 d per week, progressing to machines or free weights as accessible. She will perform range of motion exercises daily to improve sitting posture and shoulder biomechanics as well as to prepare for an eventual trial with FES exercise.

DISCUSSION QUESTIONS

1. What is your interpretation of the medical history?
2. What might you discuss with this patient regarding her examination findings and lifestyle, and how would you do this?
3. Discuss the exercise test with respect to mode, protocol, and her physiological responses. Considering these results, what are your recommendations for exercise?
4. Do you think that FES might benefit Ms. BF? Why or why not?
5. At what heart rate should she perform aerobic exercise?
6. What specific precautions should be considered for people with SCI who perform exercise?
7. What types of strength and range of motion exercises should she perform?
8. How might you keep Ms. BF motivated?

Chapter 26

Multiple Sclerosis

Linda H. Chung, PhD
Jane Kent-Braun, PhD

Individuals with multiple sclerosis (MS) have varying levels of disability. The level may remain stable for years or even for the rest of one's life. But in other cases, disability may gradually or suddenly increase. Exercise is an important aspect of treatment for those with MS. This chapter provides information about using exercise as part of the treatment in people with this disease.

DEFINITION

Multiple sclerosis is an inflammatory autoimmune disease of the central nervous system. The etiology of MS is unknown; however, MS is characterized by nerve **demyelination**. This process may result in multiple plaques (scleroses) in the white matter of the brain and spinal cord (110). These plaques can develop into permanent scars (104) that impair nerve transmission (40, 117, 119) and lead to an array of symptoms, such as muscle weakness, fatigue, and motor function difficulties (58).

SCOPE

Approximately 400,000 people are living with MS in the United States (65). Diagnosis of MS generally occurs between the ages of 20 and 50, although young children and older adults also may receive this diagnosis (65). Women appear to be affected at nearly twice the rate as men and make up more than 60% of those diagnosed with the disease (21, 42).

Susceptibility to MS appears to be complex (44), possibly resulting from an interaction of genetic, infectious, and environmental factors. Several genes (HLA [human leukocyte antigen] class II, ApoE [alipoprotein E], IL [interleukin]-1ra, IL-1β, and TGF [transforming growth factor]-β1) are reported to be associated with the course and severity of MS (32, 44, 96). However, no single identifiable gene is associated with all aspects of MS. A genetic component of MS is supported by studies of twins, which report concordance rates of about 26% in identical twins in contrast with only about 2% in other siblings (93). Additionally, approximately 20% of people with MS have a close relative with the disease (94). Some scientists believe that MS may be triggered by infectious agents such as herpes simplex virus (24) or the Epstein-Barr virus (3). However, no virus has been proven to be involved in the initiation of MS. Susceptibility to MS may be due to environmental factors, as the prevalence of MS is strongly associated with latitude (106). Higher prevalence of MS has been observed in northern Europe, the northern United States, Canada, and southern Australia and New Zealand (93). Notably, MS is most common in Caucasians of northern European descent compared with other ethnicities (e.g., Inuit, Norwegian Lapps, Australian Aborigines, New Zealand Maoris) living in the same latitude (65). Geographic clusters (or outbreaks) of MS are observed, but the cause or significance of these clusters is not yet known (38).

Multiple sclerosis can be personally devastating, as demonstrated by the fact that within 10 yr of diagnosis

Acknowledgment: Much of the writing in this chapter was adapted from the first and second editions of *Clinical Exercise Physiology*. As a result, we wish to gratefully acknowledge the previous efforts of and thank Chad C. Carroll, PhD, and Charles P. Lambert, PhD.

more than 50% of individuals with MS will become unemployed (41), primarily due to lower functional capacity (39). Lifetime cost of MS health care can exceed $1,000,000 per afflicted person (8), and the cost increases with the severity of neurological dysfunction (7).

PATHOPHYSIOLOGY

The pathophysiologic hallmark of MS is the demyelination of neurons in the central nervous system, due to an inflammatory autoimmune response. Autoreactive T cells are thought to initiate an immune response against myelin (35, 72, 88). These activated T cells cross the blood–brain barrier, proliferate, and secrete lymphokines or cytokines, which recruit microglia, macrophages, and other immune cells to participate in oligodendrocyte death and myelin destruction (14, 22, 88). At this time, it is not clear whether macrophages or T cells are the primary cells that mediate demyelination (72, 90). As the myelin sheath deteriorates, plaques form, and the end result may be axonal destruction (103). The impairment of normal nerve conduction can lead to an array of symptoms that affect people living with MS, including their ability to carry out typical activities of daily living. Furthermore, some symptoms may limit exercise performance or temporarily be worsened by exercise.

Multiple **sclerosis** can follow at least four courses of clinical progression (table 26.1) (55). A fifth course, *relapsing-progressive*, has been defined; but it overlaps other defined clinical courses, and some recommend that this term not be used (55). The most prevalent course of MS is *relapsing-remitting* (RRMS), affecting ~80% of the population with MS (72). The remaining ~20% of the population with MS is diagnosed initially with the *primary progressive* course. More than 50% of individuals initially diagnosed with RRMS develop a steady, progressive form of MS within 10 yr (i.e., convert to secondary progressive) (65). Many people with MS have occasional attacks or exacerbations, but they frequently recover and may never incur any permanent disability (64). In other cases, individuals may have frequent attacks; and although they do not completely recover, they may retain sufficient function for typical activities of daily living.

CLINICAL CONSIDERATIONS

This section reviews the medical and clinical issues that persons with MS face and the general methods used to evaluate these individuals.

Signs and Symptoms

Signs and symptoms vary from person to person with MS and are associated with the areas of the central nervous system that are demyelinated to the greatest degree (table 26.2). Symptoms of MS can negatively affect daily activities, physical and mental function, and quality of life (125). Common early indicators of MS include visual impairments (e.g., **optic neuritis, nystagmus**), motor function difficulties, and paresthesia (64). For many people with MS, disease progression is slow, and deterioration occurs over 10 to 25 yr. Ambulation can become increasingly difficult (11), although proper treatment can help delay or diminish disability.

Muscle weakness (2, 9, 13, 53, 67, 69, 81, 98, 101, 115) is a common problem in persons with MS, and generally affects the lower extremity muscles more than upper extremity muscles (98). Muscle **spasticity** may reduce range of motion about a joint and may limit voluntary movement during high-speed flexion or extension (or both) by coactivating antagonist muscles (15). Symptomatic fatigue, defined as an overwhelming sense of tiredness (51), is common and often rated as the worst symptom by people with MS (4, 23, 26). Individuals with MS commonly report that fatigue prevents sustained physical functioning, worsens with heat, interferes with responsibilities, and can come on

Table 26.1 Clinical Courses of Multiple Sclerosis

Type	Characteristic
Relapsing-remitting	Characterized by disease relapses with either a full recovery or a deficit after recovery; no progression of disease symptoms in recovery stage
Primary progressive	Disease progression from onset with infrequent plateaus and only temporary, small improvements; clinical status continuously worsens with no distinctive remissions
Secondary progressive	Begins as relapsing-remitting but progresses either with or without infrequent relapses, plateaus, and remissions
Progressive-relapsing	Progressive from onset with short, definite relapses with or without full recovery

Data from Lublin and Reingold (55).

Table 26.2 Common Signs and Symptoms of Multiple Sclerosis

Symptoms	Signs
Muscle weakness	Optic neuritis
Symptomatic fatigue	Nystagmus
Numbness	Paresthesia
Visual disturbances	Spasticity
Walking, balance, and coordination problems	
Bladder dysfunction	
Bowel dysfunction	
Cognitive dysfunction	
Dizziness and vertigo	
Depression	
Emotional changes	
Sexual dysfunction	
Pain	

easily (51). Cooling is reported to improve symptomatic fatigue (51).

Walking, balance, and coordination difficulties are prevalent in MS, and the incidence of falls is ~52%. Approximately 45% of those with MS use assistive devices for mobility (25). Although compromised sensory or motor systems (or both) are likely explanations for slowed walking and poor balance control in MS, muscle weakness, spasticity, and lower physical activity level may also contribute to declines in physical function.

Psychological effects such as an impaired cognitive ability and memory loss can occur early after disease onset (83). A strong correlation exists between the level of physical disability and a defect of cognition (59). Depression is also common in MS (64) and can lead to reduced participation in work, social activities, and other endeavors (83).

History and Physical Examination

Many of the signs and symptoms associated with MS (table 26.2) are difficult for untrained individuals to recognize. Therefore, the clinical exercise physiologist should consult with a patient's physician before exercise testing and training. The clinical course, current symptoms, and medications should be considered before any exercise bout.

Some people with MS exhibit problems with vision, and issues related to sight during exercise should be addressed in these cases. Those with cognitive deficits may have impaired ability to provide informed consent, maintain focus during an exercise test, or follow an exercise prescription. In such cases, a caregiver or support person should accompany the patient. Loss of muscle strength, numbness, symptomatic fatigue, poor flexibility, poor balance and coordination, and gait abnormalities can affect the patient's ability to perform certain types of exercise. These conditions must be considered on an individual basis before any exercise bout.

A review of the medical history and medications that an individual is using should be conducted. For example, medications to treat muscle spasticity often result in fatigue, which may limit maximal exercise capacity. Skeletal muscle fatigue may occur before cardiorespiratory fatigue, which may limit the ability to diagnose coronary artery disease (i.e., reduce the predictive value of the test) using a standard exercise test. Likewise, persons with MS who have autonomic dysfunction may not display the typical heart rate or blood pressure responses to graded exercise, rendering those measures less useful for monitoring test progression.

The level of MS-related disability based on the medical evaluation by a neurologist (or trained clinician) can be classified with use of the Kurtzke Functional Systems and the Kurtzke Expanded Disability Status Scale (table 26.8) (52). In addition, simple functional assessments before and during an exercise training intervention can provide useful insight as to the patient's functional level over time (table 26.4). With this information, the clinical exercise physiologist can objectively rate an individual's ability to perform certain types of exercise.

Table 26.3 Kurtzke Functional Systems

Category	Rating scale range
Pyramidal	0-6
Bowel and bladder	0-6
Cerebellar	0-5
Visual (optic)	0-6
Brainstem	0-5
Cerebral (mental)	0-5
Sensory	0-6
Other[a]	0-3

Note. Ratings range from normal function (0) to signs only without disability (1) to increasing levels of disability (2-6).

[a]Any other neurological findings attributed to multiple sclerosis (e.g., fatigue).

Adapted from J.A. Ponichtera-Mulcare et al., 1994, "Maximal aerobic exercise in ambulatory and semi-ambulatory patients with multiple sclerosis," *Medicine and Science in Sports and Exercise* 26: S29.

Table 26.4 Functional Assessments

	Assessments	Evaluations
MS Functional Composite (MSFC)	Timed 25 ft walk 9-hole peg test (9-HPT) Paced Auditory Serial Addition Test (PASAT)	Leg function and ambulation Arm and hand function Cognitive function
Modified Fatigue Impact Scale (MFIS)	Self-report questionnaire	Physical function Cognitive function Psychosocial function
Health Status Questionnaire (SF-36)	Self-report questionnaire	Physical component Mental component

Diagnostic Testing

The International Panel on the Diagnosis of Multiple Sclerosis has recently updated its criteria for establishing a diagnosis of MS (80). The aim of the revised criteria is to provide physicians with clearer definitions of *definite, possible,* and *not* MS for patients who demonstrate a clinically isolated syndrome or one that is progressive. Diagnostic tools to detect MS include a review of medical history, neurological exam, and various tests (magnetic resonance imaging, evoked potential response, cerebrospinal fluid analysis, blood tests).

Diagnosis initially focuses on the medical history and neurological examination. A medical history provides information regarding past and present symptoms that may become relevant upon diagnosis. Neurological examination provides insight into the health of the nervous system. Evoked potential tests measure the electrical response to sensory stimulation (e.g., visual, auditory, somatosensory); a lower magnitude and longer response time may indicate the presence of lesions along the nerve. **Magnetic resonance imaging (MRI)** has become an important diagnostic tool for MS, particularly in demonstrating the important criterion of spatial and temporal dissemination of lesions (figure 26.1). Cerebrospinal fluid, obtained by spinal tap, is analyzed to detect levels of immunoglobulins and determine the presence of oligoclonal bands, which can support an elevated immune

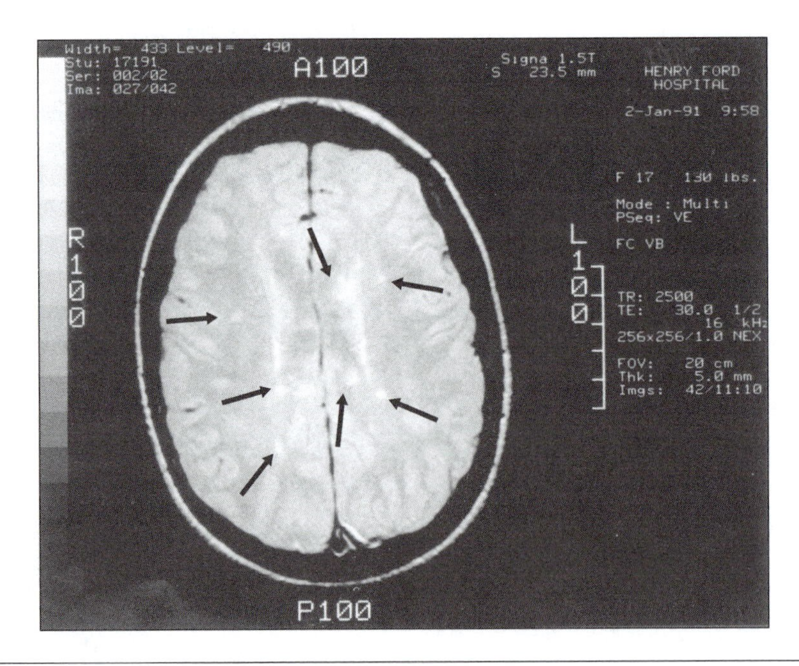

Figure 26.1 Magnetic resonance image depicting multiple (arrows) areas of demyelination within the white matter of the brain in a person with MS.

Reprinted, by permission, from C.A. Brawner and J.R. Schairer, 2000, "Multiple sclerosis; Case report from Henry Ford Hospital," *Clinical Exercise Physiology* 2(1): 17.

response. Although there are no specific blood tests for MS, blood tests are useful to rule out other diseases with similar signs and symptoms.

Exercise Testing

Exercise testing is useful in persons with MS to determine current state of fitness and individual responses to a bout of exercise. It is also helpful for evaluating the effectiveness of exercise training. Each person should have an evaluation before testing, including a review of the medical history and list of medications, and a functional assessment. Information resulting from an exercise test can be useful in developing an exercise prescription.

Cardiovascular

To assess cardiovascular fitness, the clinical exercise physiologist should carefully consider the mode of exercise used, because each mode has specific advantages. A leg cycle ergometer is most commonly used as it requires little balance and can be used by persons with ambulatory impairment (e.g., foot drop, need for assistive devices for mobility, **ataxia**). Additionally, the resistance of a cycle ergometer can be adjusted to a level that is within most patients' capabilities. An excellent option for exercise testing is an ergometer with arm poles (e.g., Schwinn Airdyne or a Nu-Step), which allows the use of all four limbs, thereby providing the opportunity for individuals with weakness in one limb to continue exercising by compensating with the other limbs. People with little ambulatory impairment can use a treadmill, whereas nonambulatory persons with adequate upper body function can use an arm cycle ergometer. The risk of falling should be considered; the choice can be quickly evaluated during the walk from the waiting room to the testing room or with balance tests. These types of functional assessments are useful for understanding any limitations a person may have before beginning an exercise intervention. In addition to baseline measures, these assessments are useful for monitoring changes in function over time. Clinical exercise physiologists may find the standardized functional tests for MS listed in table 26.4 useful.

The manuals and forms for the assessments provided in table 26.4, as well as for other functional tests, are available at the National Multiple Sclerosis Society website (65). The Berg Balance Scale, Timed Up and Go (TUG), rating of perceived exertion (RPE), and a 10 cm visual analogue fatigue scale may provide additional information with regard to the functional limitations of a person with MS.

Persons with MS should be closely monitored for any signs of paresis, fatigue, or overheating as exercise intensity increases. The use of toe straps with the leg ergometer is recommended for persons with sensory deficits, and an ankle–foot orthotic may be needed for patients in whom foot drop may develop during exercise.

A low-level protocol beginning with a 1 to 2 min warm-up period is recommended. Workload increases should occur at 10 to 25 W for leg ergometry (89, 97) and 8 to 12 W for arm ergometry and should be no more than 2 metabolic equivalents (METs) on the treadmill. Each stage should last approximately 2 to 3 min or until a steady state is achieved. Some patients with little disability may tolerate a higher-level protocol. Table 26.5 reviews the physiological responses to exercise in persons with MS. Table 26.6 provides a summary of exercise testing specifics. Standard graded exercise testing contraindications reviewed in chapter 5 should be followed (46); no specific contraindications are common among persons with MS. Because of the highly variable nature of the disease, each person may respond somewhat differently to testing, so close monitoring is required.

Individuals with MS may have an attenuated systolic blood pressure response during exercise (78, 100) that may be related to autonomic dysfunction (111), a reduced skeletal muscle metabolic response that affects the skeletal muscle chemoreflex (67), or both. For this reason, the RPE score is often used during graded exercise testing to monitor exercise intensity (6). This approach, developed and used at the Can Do MS wellness programs (formerly the Jimmie Heuga Center), gives the person with MS a means of adjusting exercise intensity to match current clinical status. Individuals with MS may also have abnormal temperature regulation and sweat response (16),

Table 26.5 Physiological Responses During Exercise

Physiological response	Response of MS relative to non-MS individuals
Submaximal oxygen consumption during treadmill walking	Increased (74)
Submaximal and maximal arterial blood pressure	Same or decreased (77)
Temperature	Same or increased (10, 71)
Skeletal muscle fatigue	Earlier (49, 101)

Table 26.6 Exercise Testing Review

	Cardiorespiratory endurance	Skeletal muscle strength	Flexibility
Mode	Cycle ergometry	Machine weights or manual muscle testing may be used in more disabled patients	Goniometer
Protocol specifics	1 to 2 min of light warm-up; 2 to 3 min stages, 10 to 25 W workload increases per stage	1RM testing of upper and lower body muscle groups	Ankle, knee, hip, shoulder, elbow
Clinical measures	BP, HR, $\dot{V}O_2$, RPE	Weight lifted, number of repetitions	Degrees
Clinical implications	Persons with MS may have increased CV risk, respiratory dysfunction, very low fitness levels, low physical activity level, or some combination of these	Muscle weakness may contribute to reduced mobility and functional impairment	Poor flexibility may contribute to poor ambulation and performance of ADLs
Special considerations	Attenuated BP and HR responses possible; impaired thermoregulation; muscle weakness and fatigue	Poor flexibility; spasticity; muscle fatigue	Contractures, spasticity

Note. BP = blood pressure; HR = heart rate; $\dot{V}O_2$ = oxygen consumption; CV = cardiovascular; RM = repetition maximum; ADL = activities of daily living; RPE = rating of perceived exertion.

increasing the risk of hyperthermia during exercise. Use of electric fans to improve evaporative and convective cooling, use of cold neck packs, attention to fluid replacement, and control of climate (i.e., room temperature 72-76 °F [22.2-24.4 °C], low humidity) are recommended to reduce the risk of heat-related symptoms, as well as symptomatic fatigue. Skeletal muscle fatigue may occur before patients reach peak cardiovascular levels. This circumstance may be caused by an inability to recruit additional skeletal muscle or by impaired skeletal muscle metabolism (48, 49). A higher energy cost of walking in persons with MS (28, 74, 75) may also affect their ability to reach peak cardiovascular levels.

Musculoskeletal

Standard testing procedures can be used to determine muscle strength and endurance in persons with MS. Due to the heterogeneity of lesion location and impact, some muscle groups may be quite weak while others exhibit normal strength. A one-repetition maximum (1RM) is typically used to measure muscle strength. However, if a person with MS is weak, a modified 1RM may be more appropriate to minimize the risk of injury. One can assess muscle endurance by determining the number of repetitions a person can perform before volitional fatigue while using a light resistance weight.

Flexibility

Range of motion can be assessed with a goniometer. Because flexibility often is reduced in persons with MS, especially in muscles affected by spasticity, stretches should be slow and gentle (without bouncing). Patients should perform stretches while seated or lying down to minimize the risk of falling.

Information regarding important issues to discuss with those with MS is presented in practical application 26.1.

Treatment

There is no cure for MS, but several disease-modifying drugs (table 26.7) have been approved by the Food and Drug Administration to treat MS. These medications are designed to minimize the frequency of exacerbations, reduce the number of lesions in the central nervous system (56, 60, 76, 105, 120), and slow the progression of disability (43, 54, 91). Currently, only one disease-modifying drug has been approved for secondary progressive MS, and none have been approved for primary progressive course of MS. An exacerbation is the result of an increased inflammatory response that promotes demyelination and causes slowing or blocking of neural signaling. Exacerbations are characterized by a mild to severe worsening of symptoms that interferes with a person's functioning and lasts for more than 24 h. Corticosteroid treatment is typically used to ameliorate the inflammation and reduce the recovery time from an exacerbation (14). Side effects of corticosteroid treatments are bone softening, high blood pressure, and transient weight gain.

Practical Application 26.1

CLIENT–CLINICIAN INTERACTION

The symptomatic fatigue noted by individuals with MS varies from day to day. The clinical exercise physiologist must anticipate the need to reduce exercise training volume (i.e., frequency, intensity, duration) if a person is fatigued. Doing this requires daily assessment for indicators of fatigue. Verbal communication is the best means of assessment. Appropriate questions include "How do you feel today?" and "Are you tired today?" Use of a visual analogue fatigue scale (51, 99) can also be helpful in ascertaining symptomatic fatigue before exercise. Intensity of exercise can be adjusted, and duration monitored, by use of the RPE score. As the RPE approaches the "hard" and "very hard" zones and the individual begins to fatigue despite reductions in exercise pace, their exercise session should be ended. Exercise above a moderate intensity, particularly on days when significant symptomatic fatigue is reported, should be avoided.

The clinical exercise physiologist must also be aware that depression is somewhat common in persons living with MS. Indicators of stress may be related to depression. These signs include poor sleep habits, noncompliance with lifestyle change, and elevated scores on standardized questionnaires such as the SF-36 and Beck Depression Inventory. The exercise clinician should share information about the potential positive effects of exercise on various psychological variables (e.g., mood, depression, anxiety). People who exhibit signs of depression or are concerned about it should be referred to the mental health professional on their MS management team.

The clinical exercise physiologist should counsel individuals with MS about the risk factors associated with a sedentary lifestyle, including coronary artery and cardiovascular disease, obesity, and type 2 diabetes. Any steps the person can take to include or increase daily activity levels will help with many health outcomes.

The clinical exercise physiologist should help motivate individuals with MS to exercise regularly. Because some people with MS have cognitive problems, the exercise clinician may need to repeat instructions, clarify explanations, or present the training plan in an easy-to-follow format. A common reason for nonadherence to exercise training is fatigue or a prolonged recovery from exercise. The clinical exercise physiologist must stress that the person must regulate exercise training intensity each day to avoid excessive fatigue; the best way to do this is to use the RPE. Additionally, one can make the point that regular resistance exercise may strengthen the skeletal muscles and allow people to perform activities of daily living with less overall symptomatic fatigue. This type of information can be helpful for maintaining motivation.

Persons with MS suffer an array of symptoms, as previously discussed (table 26.2). Symptoms can be managed through medication (table 26.7), rehabilitation, and use of assistive devices. Rehabilitation is beneficial for maintaining daily function and quality of life. Physical therapy can be important for improving mobility. Occupational therapy is useful in improving motor function and reasoning abilities and for learning to compensate for disabilities (e.g., mitigating fatigue). Speech pathologists and cognitive remediation specialists can help persons with MS improve problems with verbal communication, memory, and cognition. Complementary treatments (e.g., yoga, Chinese medicine, naturopathy, relaxation techniques) may provide additional benefits through stress management strategies, nutritional recommendations, exercise, and lifestyle changes, although evidence for efficacy of these treatments is often lacking. Exercise in persons with MS improves impaired bladder and bowel function (78), positively affects psychological health and quality of life (29, 77, 78, 97), improves **muscle weakness** (17, 114, 123), and potentially reduces symptomatic fatigue (1). Anecdotally, flexibility improves with exercise, but outcome measures of range of motion are not generally included in exercise training studies in persons with MS. The use of assistive devices (e.g., cane, Canadian crutch, ankle–foot orthosis) is an effective strategy to maintain balance control and allow mobility-challenged persons with MS to continue their regular activities of daily living.

EXERCISE PRESCRIPTION

The ultimate goal of exercise in persons with MS is to maintain or reachieve the ability to do habitual tasks of daily living, from household chores and walking (without risk of falling) to participation in recreational activities. Lower physical activity levels are observed in persons

Table 26.7 Common Medications Used for Persons With MS

Generic name	Brand name	Primary effects	Potential side effects relevant to exercise
Interferon β-1a	Avonex Rebif	Disease modifying	Flu-like symptoms, dyspnea, headache, fatigue Novantrone, specifically, may cause nausea and arrhythmias
Interferon β-1b	Betaseron Extavia		
Glatiramer acetate	Copaxone		
Fingolimod	Gilenya		
Mitoxantrone	Novantrone		
Natalizumab	Tysabri		
Onabotulinumtoxin A	Botox	Spasticity	Muscle weakness, dizziness, nausea, hypotension, headache, hypotonia, fatigue, ataxia, blurred vision, diplopia, dry mouth
Dantrolene sodium	Dantrium		
Baclofen (intrathecal)	Lioresal		
Diazepam	Valium		
Clonazepam	Klonopin		
Tizanidine hydrochloride	Zanaflex		
Amantadine hydrochloride	Symmetrel	Symptomatic fatigue	Nausea, dizziness, headache
Modafinil	Provigil		
Fluoxetine hydrochloride	Prozac		
Dalfampridine	Ampyra	Walking problems	
Clonazepam	Klonopin	Tremor	Peripheral neuropathy
Isoniazid	Laniazid Nydrazid		
Desmopressin	DDAVP nasal spray or tablets	Bladder dysfunction	Dry mouth, dizziness, headache
Tolterodine	Detrol		
Oxybutynin chloride	Ditropan (XL or LA) Oxytrol		
Darifenacin	Enablex		
Tamsulosin hydrochloride	Flomax		
Terazosin hydrochloride	Hytrin		
Prazosin hydrochloride	Minipress		
Propantheline bromide	Pro-Banthine		
Trospium chloride	Sanctura		
Imipramine hydrochloride	Tofranil		
Solifenacin succinate	VESIcare		
Duloxetine hydrochloride	Cymbalta	Depression	Nausea, dizziness, asthenia, sweating, dry mouth, headache
Venlafaxine hydrochloride	Effexor		
Paroxetine hydrochloride	Paxil		
Fluoxetine hydrochloride	Prozac		
Bupropion hydrochloride	Wellbutrin		
Sertraline hydrochloride	Zoloft		

Generic name	Brand name	Primary effects	Potential side effects relevant to exercise
Duloxetine hydrochloride	Cymbalta	Pain	Nausea, nystagmus, ataxia, dizziness
Phenytoin sodium	Dilantin		
Amitriptyline hydrochloride	Elavil		
Clonazepam	Klonopin		
Gabapentin	Neurontin		
Nortriptyline hydrochloride	Pamelor		
Carbamazepine	Tegretol		
Imipramine hydrochloride	Tofranil		

Note. All disease-modifying drugs are used to treat relapsing forms of MS with the exception of mitoxantrone, which treats progressive or worsening forms of MS.

with MS compared to non-MS individuals (68). Inactivity leads to muscle weakness, lower bone density, lower cardiovascular fitness, and an increase in fatigue. Exercise or increased physical activity in general has been shown to benefit persons with MS by reducing symptoms (1) and improving overall quality of life (29, 78, 97). Increased activity also may lower risk factors for cardiovascular and metabolic disease in this generally sedentary population.

Currently, there are no specific exercise prescription standards for persons with MS. As long as the exercise is safe and does not exacerbate symptoms, persons with MS can perform standardized (or slightly modified) aerobic and muscle-strengthening activities. When deciding about appropriateness of exercise and level of supervision, the clinical exercise physiologist can use the functional disability systems presented in tables 26.3 and 26.8. In addition, due to the varying nature of some symptoms of MS, the clinical exercise physiologist should have up-to-date knowledge of symptoms before any exercise bout and adjust the exercise prescription accordingly. For example, when someone with MS reports increased symptomatic fatigue, strenuous activities should be avoided and fatigue levels monitored regularly using a 10 cm visual analogue fatigue scale during the training session. Those with known impairment of heart rate or blood pressure response during exercise may require monitoring. Persons with MS, most commonly in those with relapsing–remitting MS, should avoid strenuous exercise during an exacerbation to prevent further worsening of symptoms.

Because people with MS can have problems with thermoregulation and because symptomatic fatigue can be brought on by increased body core temperature, cooling with an electric fan and controlling room temperature are important. Good hydration and fluid replacement during and after exercise at the rate at which fluid is lost from the body are recommended. Although time-consuming, precooling by body immersion in cool water has been successful in improving exercise tolerance and may be useful in some cases (121). Cold immersion can also be helpful following exercise in warm conditions. Cold neck packs may be useful during and after exercise to minimize the effects of symptomatic fatigue. Because of the potential for heat-related illness or temporary worsening of symptoms, persons with MS should be educated about the need to take steps to avoid temperature problems during exercise training. Persons with MS experiencing an exacerbation should avoid exercise training until symptoms improve or stabilize.

A training program that involves intermittent exercise may be necessary for a person with MS in order to avoid excessive buildup of fatigue and heat stress (122). Alternating exercise with short rest periods is recommended, especially for people who have low cardiovascular fitness. Clinical exercise physiologists should monitor RPE and fatigue levels during each exercise bout, along with the percentage of peak $\dot{V}O_2$ or peak heart rate (HR), to determine a person's target exercise intensity. Moderate intensity is the ideal target intensity when symptoms are not present (122).

Cardiovascular Exercise

Persons with MS can attain the same types of improvement (e.g., increased peak $\dot{V}O_2$, improved psychological variables, decreased muscle fatigue, and increased skeletal muscle metabolism) with cardiovascular training as non-MS individuals do, and at a similar rate (17, 29, 34, 73, 78, 87, 89, 97, 112, 116, 123). The risk of disease exacerbation is small (77), and symptomatic improvements are observed following training (1, 17, 29, 78, 97, 114, 123). Table 26.9 provides a review of the exercise prescription.

Table 26.8 Kurtzke Expanded Disability Status Scale

Rating	Disability	Functional limitations
0	Normal neurological exam	None
1.0	None	None
1.5	None	None
2.0	Minimal	Affects one FS
2.5	Minimal	Affects two FS
3.0	Moderate	Affects three or four FS
3.5	Moderate, fully ambulatory	Affects three or four FS
4.0	Mildly severe disability: fully ambulatory without aid; self-sufficient up to 12 h per day	Affects one or more FS
4.5	Moderately severe disability: same as 4.0 with some limitation of ADLs or need for minimal assistance	Affects one or more FS
5.0	Severe: walks only 200 m without rest or aid; impaired ADLs	Affects one or more FS
5.5	Severe: walks only 100 m without rest or aid; impaired ADLs	Affects one or more FS
6.0	Severe: needs intermittent or unilateral aid to walk 100 m	Affects two or more FS
6.5	Very severe: needs constant aid (cane, crutches, braces) to ambulate 20 m	Affects two or more FS
7.0	Extremely severe: unable to walk 5 m; needs aid 12 or more hours per day; can transfer self from chair	Affects more than one FS
7.5	Extremely severe: takes only a few steps; cannot wheel self; full day in wheelchair	Affects more than one FS
8.0	No ambulation; restricted to bed or chair; retains self-care and can use arm	Affects many FS
8.5	Bedridden; minimal arm use; some self-care	Affects many FS
9.0	Bedridden; no self-care; can talk and eat	Affects all FS
9.5	Bedridden; cannot communicate, eat, or swallow	Affects all FS
10.0	Death attributable to multiple sclerosis	

Note. FS = functional systems; ADL = activities of daily living.

To improve cardiovascular function, aerobic activity should be performed at a moderate intensity (122) for ≥30 min on 3 to 5 d each week (36). If needed, the duration of aerobic exercise can be accumulated with three 10 min bouts interspersed with rest. Intensities should range between 40% and 70% of $\dot{V}O_2$reserve. Persons with MS who have heightened symptoms, autonomic nervous system dysfunction, or poor fitness may benefit from a lower exercise intensity at the outset. Use of Borg's RPE score and the visual analogue fatigue scale is highly recommended. These scales can easily be applied throughout an exercise session and should be considered in conjunction with the percentage of peak $\dot{V}O_2$ or peak HR for determining a person's exercise intensity.

Stationary cycling with legs or arms (or both), walking, swimming, and aquatic exercise are common methods used to improve aerobic fitness in persons with MS, and may be the modality of choice for those with ambulatory limitations. High-functioning persons with MS may opt to use a treadmill or elliptical trainer instead. Progression over time should be individualized and based on the percentage of peak exercise test heart rate or peak $\dot{V}O_2$, as well as the ability to tolerate the exercise (i.e., perceived exertion and symptomatic fatigue levels).

Resistance Exercise

Muscle strength is critically important for mobility and independence. Resistance training should focus on the major muscle groups and use resistive movements (e.g., weightlifting, stair climbing, calisthenics, elastic bands) 2 or more days each week (36). Resistance should be set

Table 26.9 Exercise Prescription Review

Training method	Mode	Intensity	Frequency	Duration	Progression	Goals	Special considerations and comments
Aerobic	Cycle ergometry, walking, swimming, treadmill, elliptical trainer	Moderate or 40% to 70% of $\dot{V}O_2$reserve*, RPE 11 to 14, fatigue <6	Three to five times per week	30 min or three 10 min bouts progressing to 60 min as tolerated	Careful progression based on the individual's responses to the exercise; avoid excessive fatigue	Improving cardiorespiratory function and exercise capacity, increasing quality of life, reducing fatigue	Balance on some equipment; consider seated modes or handrail use on treadmill; temperature regulation; monitor perceived exertion and fatigue levels; attenuated HR or BP response.
Resistance	Machine weights, elastic bands, calisthenics, stair climbing, aquatic training	60% to 80% of 1RM, 8 to 15 repetitions for 1 to 2 sets	Two or more times per week	About 30 min If fatigued, increase between set rest time to 2 to 5 min	Careful progression based on the individual's abilities; otherwise follow standard PRT principles	Increasing muscle strength and decreasing muscle fatigue, increasing ability to perform ADLs	Unilateral weakness; balance; muscle spasticity; excessive fatigue; work to correct imbalances; rest at least 1 d between sessions; avoid free weights unless spotter is used; use handrail for balance control during stair climbing.
Balance	Standing on one foot	10 to 15 repetitions, twice on each foot	Two or more times per week	10 s stance	Exercises performed on a hard, flat surface; challenging exercises can be performed on rubber foam or uneven surface or with interrupted sensory system(s)	Increasing lower body strength and coordination, increasing ability to perform ADLS without fear of falling	Use handrail or back of chair with one hand for safety and extra sensory support; as balance improves over time, use a finger, then no hands; for walking tests, have arms raised sideways at shoulder level, choose a spot ahead, and focus on it.
	Walking heel to toe	20 steps		1 s pause at each step			
	Balance walk	20 steps		Lift back leg without flexing and pause for 1 s, then step forward			
	Back leg raises	10 to 15 repetitions, twice with each leg		Lift leg straight back			
	Side leg raises	10 to 15 repetitions, twice with each leg		Lift leg sideways			
Flexibility	Slow, gentle stretching; may be assisted by a partner or using a device such as a rope	To the point of tight or mild discomfort	Most days of the week; before and after each training session	Each stretch held for 30 to 60 s 2 to 4 repetitions with short rest between	Gradual stretch without bouncing to the end of comfort range	Increasing ROM, reducing or avoiding contractures	Balance; avoid excessive joint range and muscle spasticity aggravation; perform while seated or on the floor; perform after either warm-up or cool-down.

Note. HR = heart rate; 1RM = one-repetition maximum; NA = not applicable; BP = blood pressure; PRT = progressive resistance training; ADL = activities of daily living.

*Largely dependent on patient's ability; assess on an individual basis.

at a weight that allows for 8 to 15 repetitions, resulting in volitional fatigue, for each exercise (36). If a person with MS is weak, lower resistance should be used initially. Seated weightlifting is recommended to avoid the potential risk of falls in people who have problems with balance and coordination. In general with MS, the best approach may be to alternate strength training and cardiovascular training days to avoid excessive fatigue. However, persons with MS who are clinically stable and have only minor side effects from the disease may be able to perform resistance training following an aerobic training session. Modifications, such as training muscle groups unilaterally because of differences in strength between the limbs or because of range of motion limitation, may be required. These limitations need to be considered on an individual basis.

Aquatic exercise is another resistance training option for persons with MS who are weak or who use a wheelchair. Aquatic exercise minimizes gravitational effects, provides a greater sense of balance, and helps dissipate body heat (in water temperatures at 27 to 29 °F) (122). Limb devices can be used to increase water resistance to movement and strengthen muscles (122).

Simple upright balance exercises may also be used to improve strength and coordination. The National Institutes of Health recommends five exercises for individuals who are at risk for falls. Balance exercises include standing on one foot, walking heel to toe, balance walking, leg raises to the back, and leg raises to the side. These exercises are designed to increase lower body strength and enhance coordination and are simple enough to perform at home. Individuals with good balance control can perform the exercises at an increased level of difficulty, with supervision, by doing them while standing on rubber foam, on uneven surfaces, or with one or more of their sensory systems perturbed (e.g., having eyes closed, leaning). Table 26.9 reviews the exercise prescription for resistance training.

Range of Motion Exercise

Because persons with MS may have reduced range of motion in some joints, a general flexibility program is recommended. Flexibility exercises can counteract spasticity, increase muscle length, and improve posture and balance (122). Stretches should be gentle, slow, prolonged (30-60 s), and not painful and should focus on all major upper and lower body muscle groups (122). They should be performed before and after a strength or cardiovascular training session, as well as during brief periods throughout every day (122).

Some persons with MS may need assistance with passive stretching. The use of a rope or the participation of a trained clinician may be necessary for persons with spasticity. Light massages, yoga, and other relaxation techniques may also complement stretching (122). Chapter 5 provides general information about range of motion training.

EXERCISE TRAINING

Exercise training and daily physical activity are important in persons with MS in order to maintain cardiovascular and bone health, prevent type 2 diabetes, retain the capacity to perform activities of daily living, and minimize some common symptoms such as symptomatic fatigue (86, 92, 109). The following sections provide a brief review of the physiological and functional changes observed in people with MS and the effects of training on those variables. Table 26.10 provides a brief exercise training review.

Cardiovascular Exercise

Peak oxygen consumption (62, 89) and respiratory muscle function (12, 27, 95) are lower in persons with MS. These declines may be in part explained by lower levels of physical activity in this population (68). Lower cardiovascular fitness and physical activity levels may place persons with MS (62) at a greater risk for diseases related to sedentary lifestyle. As found in the landmark study by Petajan and colleagues (78), endurance exercise can have a significant positive impact on a range of health-related variables in people with MS. Increased leisure-time physical activity can result in a smaller

Table 26.10 Exercise Training Review

Cardiorespiratory endurance	Skeletal muscle strength	Skeletal muscle endurance	Flexibility	Balance	Body composition
Increased (78, 82, 87) which improves cardiovascular health	Increased (31, 114, 123) via neural adaptation and fiber hypertrophy	Increased (31, 113, 114) possibly by a reduction of neural activation failures	• Likely increased to similar degree as in healthy individuals • May reduce spasticity	Improved (37) with possible reduced fall risk	Smaller waist circumference (108); benefits likely similar to those for healthy individuals

waist circumference, lower blood triglycerides, and lower blood glucose values in this group (108). Increased $\dot{V}O_2$peak (78, 82, 87), decreased serum triglyceride levels (78), increased fatigue tolerance (73, 112), improved health perception and quality of life (50, 78, 97), and increased walking speed and endurance (82, 97, 116) have all been reported in response to aerobic exercise training in this population. Overall, aerobic exercise is important in maintaining cardiovascular health and minimizing secondary illnesses related to inactivity in persons with MS.

Resistance Exercise

Isometric strength (67, 69, 70, 98, 101) and dynamic power (2, 9, 13, 53, 81, 115) generally are lower in persons with MS compared with non-MS individuals. A number of potential physiological changes in MS may explain reduced strength. Compared to those without MS, persons with MS may have decreased central activation (69, 85, 101) and lower motor unit discharge rates (20, 85). Although some studies have shown alterations in contractile function (69, 101) and smaller muscle size ((47, 69) in persons with MS, others reported no change in these variables (18, 19). Given the heterogeneity of this disease, variability in strength loss is to be expected. In addition to variation due to clinical status, it is possible that some of these physiological changes may be secondary to decreased physical activity (68). Spasticity, characterized by a velocity-dependent increase in tonic stretch reflexes, may also contribute to lower dynamic power. Coactivation of antagonist muscles, possibly due to spasticity, may work against an agonist muscle contraction. Physical inactivity may also contribute to muscle weakness in persons with MS (63, 68).

Resistance training improves isometric strength and dynamic power via neural adaptation (short-term) and muscle hypertrophy (long-term) (61). Several studies have evaluated the effects of resistance training in persons with MS (17, 34, 37, 89, 113, 114, 123, 124). Increases in isometric strength (123) and power (17, 114) are observed following a resistance training program. Functional gains, such as decreased Timed Up and Go test (17), faster walking speed (89), increased number of steps over 3 min (123), greater muscular endurance (113, 114), improved balance (37), and improved gait kinematics (34) also are observed following resistance training. Symptomatic fatigue has been shown to be reduced following resistance training (34, 123). Interestingly, White and colleagues (123) observed increases in isometric strength but not muscle size following an 8 wk resistance training (two times a week for 30 min). The lack of muscle hypertrophy may be in part due to the low intensity and volume of the training sessions. Thus, strength gains may be explained by neural adaptation. A 10 wk aquatic exercise program resulted in improvements in power, muscular endurance, and total work in both the upper and lower extremities (31). Studies using supported treadmill training in persons with MS with high levels of disability have shown increases in walking speed and a reduction in the percentage of body weight support needed (31, 79). Overall, there are physiological and functional benefits to resistance training in persons with MS.

Balance Exercise

In persons with MS, walking speed often is slower (9, 13, 57, 74, 98, 115) and demands a high amount of energy expenditure (28, 74, 75). Slower walking coincides with shorter stride lengths, decreased cadence, and prolonged double-support phase during the gait cycle (5, 57). Gait initiation is altered in people with MS; smaller posterior shifts in the center of pressure and minimal displacement of the center of mass have been observed (84). A higher energy cost during walking is associated with spasticity (74).

Balance control is also compromised in MS. Greater postural sway is observed when one or more sensory systems (visual, somatosensory, vestibular, proprioception) are perturbed (66). Postural sway is faster compared to that in others when persons with MS stand quietly on stable and foam rubber support surfaces (33). Center-of-pressure displacements are lower during leaning (45, 118) and reaching tasks (118). Higher center-of-pressure variability and asymmetric limb loading during quiet stance have been observed in persons with MS compared with non-MS individuals (13). When vision was limited, subjects with MS showed increased postural sway while leaning backward and greater limb-loading asymmetry during quiet stance (118). Increasing levels of fatigue also had an effect on postural control during balance tasks (118). Differences in power across limbs are observed and are associated with center-of-pressure variability during quiet stance (13). Thus, a functional adaptation occurs to maintain balance control. The strategies used by persons with MS appear to keep the body within the individual's stability boundaries in an effort to minimize the risk of falling.

Resistance training in people with MS improves gait kinematics (34). Improvements in the Berg Balance Test have also been observed in this group following resistance training (37). Additional research with objective measures is needed to advance our understanding of the benefits of resistance training on gait kinematics and postural control in MS.

Practical Application 26.2

RESEARCH FOCUS: MUSCULAR FATIGUE IN MS

Skeletal muscle fatigue may be more pronounced in MS patients than in others (18, 98, 101, 107), although this is not always the case (48, 67). Kent-Braun and colleagues (48) reported that the half-time of skeletal muscle phosphocreatine (PCr) resynthesis after exercise-induced depletion was slower in people with MS compared with non-MS controls, suggesting that a poor capacity to supply energy oxidatively may contribute to fatigue in this population. The impaired PCr resynthesis rate was consistent with a finding of lower oxidative enzyme activities in those with MS (47). The skeletal muscle of patients with MS may also contain a greater proportion of type II muscle fibers compared with controls (30, 47). Greater reliance on these more "fatigable" type II muscle fibers could explain a portion of the muscle fatigue reported in MS patients. The lower skeletal muscle oxidative capacity and fiber type changes are consistent with lower physical activity levels in this population (63, 68). It is possible that skeletal muscle fatigue may result more from activation failure than from metabolic factors known to induce fatigue (49). Sheean and colleagues (102) observed greater central activation failure following a 45 s sustained maximal voluntary contraction in persons with MS compared with non-MS controls. Recently, Skurvydas (107) reported greater central fatigue than peripheral fatigue in subjects with MS during a 2 min sustained maximal voluntary contraction, indicating that lower central activation of the muscle may result in a smaller demand within the periphery. Resistance training can elicit neural adaptations (61), which may reduce skeletal muscle fatigue. Therefore, exercise training may be an effective therapy for decreasing fatigue in these patients. Further research is necessary to provide definitive findings.

Clinical Exercise Pearls

1. If people with MS are encouraged to be as active as possible, on as many days as possible, they will take charge of their overall health and reap the physical, psychological, and social benefits of regular physical activity.
2. Teaching people with MS to use the RPE scale to set the intensity of each exercise session will allow for daily adjustments as their signs and symptoms wax and wane. This will lead to greater consistency in incorporating activity or exercise into their lifestyle.
3. Avoiding overheating is critically important to a normal recovery from an exercise bout. The cost of overheating can include both short-term (2-4 h) exhaustion and increased symptoms and longer-term (24-48 h) fatigue.
4. Regular practice of activities such as yoga or tai chi can provide multiple benefits, including muscle strength, endurance, balance control, flexibility, and mental relaxation. These are great activities for individuals with or without MS. Many communities offer classes for individuals with mobility limitations.

CONCLUSION

It is best for people with MS to begin an exercise routine and increase or maintain daily physical activity levels as early in the disease process as possible. There is ample evidence that exercise, when performed properly, can limit the degree of physical disability and ameliorate symptoms. While specific exercise training recommendations do not yet exist for the MS population, persons with MS may gain cardiovascular and strength benefits using a modified exercise prescription similar to that for adults without MS. It is prudent, however, to adjust the exercise prescription when symptoms are heightened by lowering exercise intensity, frequency, duration, or some combination of these. The ultimate goal of exercise training in persons with MS is to maintain physical and psychological health, functional capacity, mobility, and quality of life.

Key Terms

ataxia (p. 515)
demyelination (p. 511)
diplopia (p. 525)
magnetic resonance imaging (MRI) (p. 514)
multiple sclerosis (p. 511)
nystagmus (p. 512)
optic neuritis (p. 512)
sclerosis (p. 512)
spasticity (p. 512)
white matter (p. 525)

CASE STUDY

MEDICAL HISTORY AND PHYSICAL EXAM

Ms. NB, a 29 yr old Caucasian woman, was diagnosed with MS 6 yr ago. At that time she had problems with ataxia and **diplopia**. She has had one or two exacerbations of MS per year since that time. An increase in disease stability was noted after she started recombinant interferon β-1b (Betaseron) 5 yr ago. She stopped the interferon therapy briefly during a pregnancy 4 yr ago, and after the pregnancy she had three exacerbations within 12 mo despite being back on Betaseron. With each exacerbation, her symptoms of ataxia, vertigo, and diplopia worsened. These symptoms involve primarily the left side of her body, and she has had muscle cramps in her left extremities on occasion. Overall, she functions reasonably well. She takes care of her 4 yr old daughter. She rests daily while her daughter goes to preschool. She had to quit working within a year of her diagnosis because of problems with ataxia and fatigue.

An MRI of her brain indicated multiple **white matter** lesions consistent with MS. A spinal fluid examination showed elevated immunoglobulin G synthesis rate and oligoclonal bands, again consistent with the diagnosis of MS.

Ms. NB is taking no other medication on a regular basis and has not been involved in a regular exercise program. She occasionally gets a urinary tract or upper respiratory infection that necessitates antibiotics. Cramping in her extremities has not been bad enough to warrant a muscle relaxant on a regular basis. She notes that when she walks more than a couple of blocks, she feels weakness in her left leg and often needs to rest for a few minutes before walking farther. She also thinks that at times her left leg may give out. Examination shows mild incoordination and diffuse hyperreflexia, which is more pronounced in the left-side extremities. She also has a few beats of nystagmus.

DIAGNOSIS

Her diagnosis is relapsing–remitting MS, grade 4.0 on the Kurtzke Expanded Disability Status Scale (table 26.8). Grade 4.0 means that the individual is fully ambulatory without aid, is self-sufficient, is up and about some 12 h a day despite relatively severe disability, and is able to walk at least 500 m without aid or rest, as well as indicating that the disease affects one or more functional systems.

EXERCISE TEST RESULTS

An exercise test was performed to assess functional ability and to rule out a cardiac origin of her arm weakness during walking. Ms. NB performed a cycling protocol of 3 min stages at 25, 50, and 75 W. No signs of electrocardiographic abnormalities were observed. Exercise was discontinued because of volitional fatigue.

Resting Values

Heart rate = 92 beats · min^{-1}

Blood pressure = 100/60 mmHg

Electrocardiogram: normal sinus rhythm and within normal limits

(continued)

Peak Exercise Values

Heart rate = 188 beats · min^{-1}

Blood pressure = 150/50 mmHg

Rating of perceived exertion = 19 (6 to 20 scale)

Peak $\dot{V}O_2$ = 17.5 ml · kg^{-1} · min^{-1}

Peak METs = 5.0

EXERCISE PRESCRIPTION

Because Ms. NB is just beginning an exercise program and because of the ataxia and fatigue that she experiences, she should start out at low intensity and duration for both aerobic and strength training. Limb strength is reduced on the left side because of increased MS-related symptoms.

Cardiovascular training: She should begin aerobic training on a cycle ergometer at a light intensity (RPE ~11 or ~40% of peak $\dot{V}O_2$) for up to 10 min, three times per week, progressing very gradually to 30 min at a moderate intensity (RPE 13 or 14, ~60% peak $\dot{V}O_2$reserve). Each cardiovascular training session should start with a 5 min warm-up at a light intensity and end with a 5 min cool-down. The intensity of the exercise portion of each session should be moderate, within an RPE of 13 or 14 on the Borg scale. If 60% of peak $\dot{V}O_2$ reserve generates an RPE >14, then a lower percentage should be prescribed. The intensity should be increased gradually, as tolerated, over 3 to 4 wk based on the RPE to an upper limit of 70% of $\dot{V}O_2$reserve.

Strength training: Ms. NB should perform resistance training two times per week, gradually building up to two sets of 8 to 15 repetitions at 40% of 1RM for each of the major muscle groups of the upper and lower body. This training will require using hand weights, dumbbells, or resistance training machines. A warm-up of five repetitions should be performed at a low intensity before any resistance training of a muscle group. The volume should be increased gradually (2-5% increase) when she is able to complete 15 repetitions in good form in consecutive sessions. Aquatic resistance training may also be used to increase muscle strength. Intensity of the aquatic exercise may be monitored using the RPE scale and should be limited to somewhat hard (RPE ~13 or 14).

Muscular endurance training: An endurance training program for skeletal muscles can be administered on the days resistance exercise is performed. A suggestion is two times per week, at 40% to 50% of 1RM initially, and progressing to 80% of 1RM. She should do one set to muscular failure for the gastrocnemius, quadriceps, hamstrings, biceps and triceps brachii, chest, and abdominal muscles.

Flexibility training: She should perform stretching on a daily basis, in addition to before or after each exercise session. Stretches should focus on the ankles, knees, hips, lower back, shoulders, wrists, and neck. Static stretches should be held for 30 to 60 s and performed at least twice for each muscle group. Yoga may also be a good form of flexibility training. However, care should be taken to prevent or minimize the risk of falling, and yoga sessions should be carefully monitored by the clinical physiologist.

Balance training: Ms. NB should perform balance exercises at least twice a week. She should stand on one foot for 10 s and repeat 10 to 15 times. She should try to walk heel to toe, then walk in a straight line, lifting the back leg for 1 s before bringing it forward. She should perform leg raises to the back and to the side, gradually increasing the number of repeats, up to 10 to 15 times. For all of these exercises, support should be provided to maintain balance; for example, she can lightly hold on to the back of a chair or touch the wall.

DISCUSSION QUESTIONS

1. What can you say about the ability of the graded exercise test to measure cardiovascular fitness in a person with MS?
2. How should the prescribed exercise be modified for a person with more disability?

3. How should the prescribed exercise be modified for a person with less disability?
4. What modes of aerobic and resistance exercise might a person with less disability use?
5. What is an advantage of using non-weight-bearing exercise in a person with MS who has a high degree of disability?
6. What physiological variables need to be monitored closely in a person with high degree of disability? Why? Are there other variables that need to be monitored as well?
7. Do you think people with MS should exercise during an exacerbation of disease activity? Why?
8. What are the potential benefits of exercise for people with MS in terms of functional capacity?

Chapter 27

Cerebral Palsy

Amy E. Rauworth, MS
James H. Rimmer, PhD

Cerebral palsy is a condition that afflicts people throughout their lives. It certainly can affect functional ability. Regular exercise training can be beneficial in people with this condition. This chapter reviews cerebral palsy and methods of using exercise to help individuals live healthier and more functional lives.

DEFINITION

Cerebral palsy is "a generic term for various types of nonprogressive motor dysfunction present at birth or beginning in early childhood. Risk factors are both hereditary and acquired; depending upon cause, classified as intrauterine, natal, and early postnatal; motor disturbances include **diplegia, hemiplegia, quadriplegia, choreoathetosis,** and ataxia" (105).

Dr. William Little, an English surgeon, was the first to document the symptoms of what is now known as spastic diplegia, in the early 1860s. He described infants who had difficulty grasping, walking, or crawling. Other symptoms included greater involvement of the legs compared to the arms and no progression from the initial degree of involvement. For many years, this form of cerebral palsy was referred to as Little's disease and was linked to difficult births associated with asphyxiation. Later, Dr. Sigmund Freud suggested that the etiology of this disease might stem from problems with the development process of the brain during gestation (42). Despite this conclusion, complication during birth was the primary identifiable cause of cerebral palsy for many years.

In 1958, a definition was presented at the First International Study Group on Child Neurology and Cerebral Palsy (64). The study group defined cerebral palsy as "a persisting qualitative motor disorder due to non-progressive interference with development of the brain occurring before the growth of the central nervous system is complete." This description led to a consensus definition: "Cerebral palsy is a persistent but not changing disorder of movement and posture, appearing in the early stages of life and due to a non-progressive disorder of the brain, the result of interference during its development" (64). Bax annotated this definition in 1964 to read "a disorder of movement and posture due to a defect or lesion of the immature brain" (14).

More recently, advances in the understanding of the developmental process of infants with early brain damage led to the definition of cerebral palsy as "an umbrella term covering a group of non-progressive, but often changing motor impairment syndromes secondary to lesions or anomalies of the brain arising in the early stages of its development" (75, p. 549). The heterogeneous nature of cerebral palsy evokes controversy even today with respect to defining the condition. A proposed definition and classification were presented in 2005 (94, p. 480): "Cerebral palsy describes a group of developmental disorders of movement and posture, causing activity restriction or disability that is attributed to disturbances occurring in the fetal or infant brain. The motor impairment may be accompanied by a seizure disorder and by impairment of sensation, cognition, communication and/or behaviour."

This definition was established to broaden the classic definition and to include the common secondary conditions associated with cerebral palsy.

SCOPE

In the United States, cerebral palsy (CP) occurs in 1.5 to 2.5 per 1,000 live births (2). There are 500,000 to 764,000 persons in the United States with CP. The incidence of CP has remained stable or may have increased slightly because of the survival of premature infants with extremely low birth weight (83). The prevalence of CP worldwide varies from 1.6 per 1,000 children under the age of 7 in China (60) to 2.0 to 2.5 per 1,000 live births in Australia (87). Cerebral palsy and low birth weight are also more prevalent in populations of lower socioeconomic status (78).

According to the Centers for Disease Control and Prevention, the lifetime cost of persons born in the year 2000 with CP is estimated to reach $11.5 billion (23). This figure emphasizes the importance of primary and secondary prevention in managing CP.

PATHOPHYSIOLOGY

The most common risk factors associated with the development of CP are premature birth and low birth weight (32). Twin and higher-order multiple-birth infants are at increased risk for developing CP (47, 76). Although twinning, prematurity, and low birth weight are interrelated, these factors have been found to be independent risk factors of CP (127). Stanley and colleagues (104) have suggested a "causal pathways" concept of interdependent events such as twinning, prematurity, and low birth weight that could result in CP. Maternal or fetal abnormalities in coagulation, such as thrombophilia, and the presence of factor V Leiden mutation increase the risk for complications during pregnancy and birth that can lead to CP (9, 56, 109). Maternal infection is also highly correlated with the development of CP and has been directly related to brain lesions in low-birth-weight infants (129). Maternal fever during term labor has been shown to increase the risk of CP ninefold.

Other prenatal risk factors associated with increased risk of CP include maternal abuse of drugs and cigarettes, congenital brain malformations, and genetic conditions (34). Complications during birth that cause anoxia or hemorrhage with direct brain damage and kernicterus result in increased risk for CP. Estimates are that approximately 75% of all cases of CP are a result of prenatal events (87). Postnatal events that occur before the age of 2 yr (e.g., viral or bacterial meningitis; traumatic head injury from vehicular accidents or abuse; anoxia as a result of near drowning, cerebral vascular accidents, tumors, or surgery; and toxins that cause heavy metal encephalopathy) are also risk factors for CP (34). In developing countries, common causes of CP include septicemia, meningitis, and conditions such as malaria (87).

CLINICAL CONSIDERATIONS

Patients with CP face a host of medically related issues. The exercise physiologist should have a general familiarity with these issues to improve basic assessment skills when working with these patients. This section reviews some of the basic medical and clinical issues for patients with CP.

Signs and Symptoms

The hallmark sign of the group of disorders identified as CP is an impairment of voluntary motor control. Cerebral palsy frequently involves one or more limbs and the trunk musculature. It is classified by the type of motor abnormality (spasticity, athetosis, ataxia), the area of the body that is affected (**monoplegia**, diplegia, hemiplegia, **triplegia**, **tetraplegia**), and the location of the lesion in the brain (**pyramidal**, **extrapyramidal**, mixed) (34, 98). Figure 27.1 provides details on the classification of CP.

Classification and Clinical Manifestation

Classification by movement disorder is often difficult because of the appearance of characteristics that represent multiple movement disorders. Abnormalities in movement include spasticity, athetosis, **chorea**, dystonia, and ataxia (see table 27.1 for clinical symptoms of muscle impairment associated with spastic paresis). The clinical manifestations of CP can be related to the location of the lesions within the brain. Injury to the cortical system (pyramidal type) results in spasticity. Spasticity is present in approximately 75% of people with CP (74). Injuries of the extrapyramidal type are lesions within the **basal ganglia** and **cerebellum**. This form of injury produces **dyskinetic** cerebral palsy, which includes movement disorders such as athetosis, chorea, ataxia, and **rigidity**. Having clinical signs and symptoms of mixed forms of CP is also common.

Cerebral palsy can also be classified by systems developed specifically for athletes with CP. Disabled Sports and the Cerebral Palsy International Sports and Recreation Association (CPISRA) created eight classes to determine movement quality of people with CP (24). Class 1

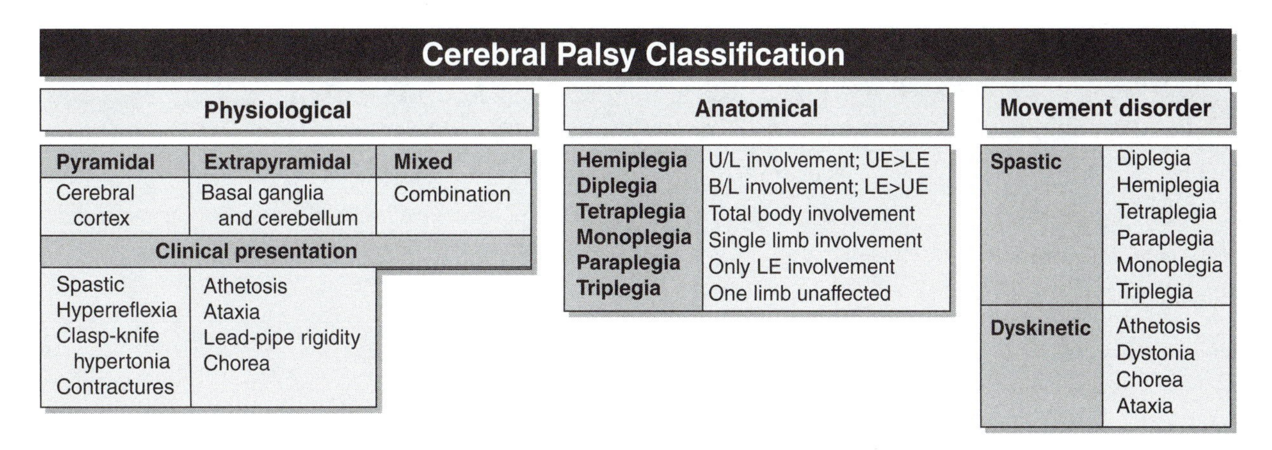

Figure 27.1 Cerebral palsy classification. U = upper; L = lower; E = extremity.

Adapted from Andersson and Mattsson 2001.

delineates severe involvement, whereas class 8 describes minimal involvement. Classes 1 through 4 are for people who use wheelchairs for mobility, and classes 5 through 8 are for individuals who are ambulatory. See table 27.2 for a complete description of all classes. Additional classification systems have been developed, such as the Quantitative Sports and Functional Classification (QSFC) (52). This classification system determines potential functional ability and allows evaluation of improvement, which may be useful to the clinical exercise physiologist. This assessment consists of 26 items and involves skills such as transfers, touching behind the head, combination movement, kicking or thrusting with the foot, control of the wheelchair footrest, and leg abduction.

Associated Conditions

Besides the neuromuscular impairments present in CP, associated conditions such as impaired cognition, sensory deficits, seizure disorders, feeding problems, behavioral dysfunction, and emotional problems have been identified (37). (See table 27.3 for a list of associated conditions and general exercise guidelines.) The prevalence of intellectual disabilities varies with the type and severity of CP and increases with the presence of **epilepsy** (78). Seizures occur in approximately half of all children with CP. When seizures occur without a direct trigger, the condition is labeled epilepsy. Visual disturbances frequently occur in CP. The most common type is **strabismus**, a condition in which a discrepancy is present between the right and left eye muscles. Individuals with hemiparesis often have an impairment in or loss of sight in the normal field of vision of one eye, referred to as **hemianopia**. This condition can also occur in both eyes and affect the same area of the visual field. Approximately 90% of people with hemiplegic CP display significant bilateral sensory deficits (78).

Speech impairments such as **dysarthria** and **aphasia** occur most commonly in individuals with dyskinetic and tetraplegic CP (39). Dental problems such as malocclusion are prevalent in children with CP, resulting in a significant increased overjet (an acute angle of the mandible) (41). **Scoliosis** is commonly seen in preteens, teens, and young adults. Scoliosis occurs most frequently in people with spastic CP (95). Surgical management is sometimes necessary to prevent further progression of the spinal curvature. For detailed information on the prevalence of **associated conditions**, see Odding and colleagues (78).

Secondary Conditions

The clinical symptoms and **secondary conditions** that result from CP lead to a functional decline and are often exacerbated by the aging process (8, 43). (See table 27.4 for a complete list of secondary conditions.) Secondary conditions have been identified as viable portions of health conditions for people who are disabled (66). They often begin in late childhood and progress throughout adulthood (80). Due to the increased risk for developing low bone mineral density (BMD) and low-impact fractures, children with severe CP should have BMD and vitamin D monitored, in addition to the optimization of calcium intake. Interventions that increase muscle mass are advisable in this population (69).

The decline in health over time can be related to the effects of clinical signs and symptoms, such as abnormal movement patterns that stress the structural stability of the orthopedic and musculoskeletal systems; **contractures** resulting in decreased range of motion and changes in posture such as spinal kyphosis; prolonged use of medication with undesired side effects such as osteoporosis and fractures; and poor nutrition as a result of gastrointestinal complications (8, 43, 53, 74).

Table 27.1 Summary of Clinical Symptoms of Impaired Muscle Function in Spastic Paresis

IMPAIRMENT OF MUSCLE ACTIVATION	
DEFICIT SYMPTOMS (SIGNS OF REDUCTION OR LOSS OF NORMAL VOLUNTARY MUSCLE ACTIVATION)	
Paresis	Inadequate force
Loss of selective motor control	Impaired ability to activate and control selective or isolated movements across specific joints
Loss of dexterity of movement	Impaired ability to coordinate temporal and spatial activation of many muscles
Enhanced fatigability	Inadequate sustained force
EXCESS SYMPTOMS (SIGNS OF ABNORMAL INVOLUNTARY MUSCLE ACTIVATION)	
PASSIVE MOVEMENT	
Hypertonia (tonic stretch reflex)	Non-velocity-dependent resistance to passive movement experienced by the examiner as increased muscle tone, caused by continuous muscle activation (tonic stretch reflex activity)
Spasticity	Velocity-dependent resistance to passive movement experienced by the examiner as increased muscle tone, caused by an increase in stretch reflex activity
Clonus	Involuntary rhythmic muscle contraction
ACTIVE MOVEMENT	
Mirror movements	Simultaneous contralateral, involuntary, identical movements that accompany voluntary movements
Associated abnormal postures	Involuntary abnormal muscle activity related to posture or the performance of any task
Cocontraction	Simultaneous involuntary contraction of the antagonist at voluntary contraction of the agonist
ABNORMAL REFLEXES	
Hyperreflexia or tendon jerks	Raised reaction to tendon tap, extension of the reflexogenic zone, spread of reflexes
Abnormal nociceptive flexion reflexes	Abnormal pathological reflexes to painful stimuli (e.g., Babinski reflex)
Abnormal musculocutaneous reflexes	Abnormal pathological reflexes to cutaneous stimuli (e.g., adductor reflex)
CHANGES IN BIOMECHANICAL PROPERTIES OF MUSCLE AND OTHER CONNECTIVE TISSUES	
Contractures	Fixed shortening of muscle and tendon, resulting in a reduced range of motion
Hypertonia	Non-velocity-dependent resistance to passive stretch experienced by examiner as increased muscle tone, caused by biomechanical changes within muscle itself

Reprinted, by permission, from B.A. Scholtes et al., 2006, "Clinical assessment of spasticity in children with cerebral palsy: a critical review of available instruments," *Developmental Medicine and Child Neurology* 48: 64-73.

Deconditioning is an important secondary condition observed in persons with CP. Ando and Ueda and others (8, 43, 53, 74) have reported that 35% of adults with CP have reduced ability to perform activities of daily living over time. The time sequence for this decline could be as short as 5 yr. Strauss and colleagues (106) identified a decline in ambulatory function after the age of 60 with a subsequent increase in mortality. Secondary conditions such as poor joint alignment and overuse syndromes may also contribute to the loss of independent walking (74). Fatigue as a limiting factor for ambulation has been reported in several studies (49, 118). Murphy and colleagues reported that approximately 75% of their cohorts discontinued ambulation by the age of 25 because of fatigue and inefficient movement (74). This cohort also reported pain in weight-bearing joints that decreased ambulation around age 45. Pain is commonly reported among people with CP. Seventy-nine percent of parents of children with CP reported that their child had severe to moderate pain. Of those who reported higher levels of pain, poorer health as identified by difficulty in feeding, seizures, and severe motor or intellectual impairments increased the risk of familiar stress (82). Turk and colleagues reported that 84% of the adult women with CP identified pain as a deterrent to participation in certain activities of daily living (114). As adults with CP age, inactivity levels appear to be directly related to pain. Health care utilization studies suggest that adults with CP seek preventive and rehabilitative services less often than others and do not consult physicians regarding pain (122). The pain level for an adult with CP is typically in the moderate to intense range (101).

Table 27.2 Cerebral Palsy International Sport and Recreation Association Classes (CPISRA)

Classification	Description
	WHEELCHAIR USERS
CP1	Athletes use power wheelchairs or assistance for mobility. They are unable to propel a wheelchair.
CP2	Athletes are able to propel a wheelchair but have very poor useful strength in their arms, legs, and trunk.
CP3	Athletes show fair trunk movement when pushing wheelchairs, but forward trunk movement is limited during forceful pushing.
CP4	Athletes have minimal limitations or control problems in their arms and trunk while pushing wheelchairs.
	SELF-AMBULATORY
CP5	Athletes may need assistive devices for walking but not in standing or throwing. Athletes may have sufficient function to run but demonstrate poor balance.
CP6	Athletes do not have the capacity to remain still, and they show involuntary movements with all four limbs affected. They usually walk without assistive devices.
CP7	Athletes have movement and coordination problems on one half of the body. They have good ability in the dominant side of the body.
CP8	Athletes have minimal involvement in one limb only.

CP = cerebral palsy

Reprinted, by permission, Cerebral Palsy International Sports and Recreation Association (CP-ISRA) *Classification and sports rule manual*, 9th ed. By permission of California Diabetes Program.

Table 27.3 Associated Conditions Observed in Persons With Cerebral Palsy

Associated conditions	Causative and related factors	Exercise guidelines
Bowel and bladder problems	• Incontinence • Neurogenic bladder	• Bladder should be emptied before exercise session.
Cognitive and behavioral problems	• Intellectual disabilities About one-third of children who have CP have mild intellectual disabilities; one-third have moderate or severe impairments, and the remaining third do not have an intellectual disability. Mental impairment is even more common among children with spastic quadriplegia. • Learning disabilities • ADD (attention deficit disorder) • ADHD (attention deficit hyperactivity disorder) • Sensory integration disorder	• Ask questions that require brief answers. • Allow time for accommodation to a new task; repetition is important. • Use precise language and simple words. • Limit the number of instructions given at one time.
Musculoskeletal problems	• Bone loss • Fractures • Hip dislocation • Hypertonia • Leg length discrepancies • Pelvic obliquity • Spasticity	• Incorporate weight-bearing exercise when appropriate. • Assess balance to choose appropriate exercise modality to prevent falls. • Be cautious when stretching or strengthening the hip musculature if history of dislocations is present. • Decrease cadence of exercise activity to avoid increase in spasticity.

(continued)

Table 27.3 *(continued)*

Associated conditions	Causative and related factors	Exercise guidelines
Oral motor dysfunction	• Dysphagia • Gastroesophageal reflux disease, or GERD • Oral hypersensitivity • Dysphagia, difficulty in swallowing • Drooling • Dysarthria • Poor motor control of the mouth • Feeding problems	• Provide a towel for possible drooling. • Make adaptations to the mouthpiece during exercise testing or use a face mask.
Respiratory problems	• Blocked or obstructed airways • Airway inflammation • Increased mucus production • Chemical or bacterial pneumonia • Lung damage • Diminished ability to clear secretions • Weakened pulmonary defenses	• Use breathing exercises (pursed-lip breathing, diaphragmatic breathing) to increase or maintain lung function.
Seizures or epilepsy	A seizure occurs when the normal, orderly pattern of electrical activity in the brain is disrupted by uncontrolled bursts of electricity. When seizures recur without a direct trigger, such as fever, the condition is called epilepsy.	• Consider the effects of medication on exercise and patient's endurance level.
Sensory and perception impairments	• Impaired perception of pain and pressure • Sensory integration dysfunction • Stereognosis • Visuo-auditory and visuo-spatial processing impairments	• Break down instructions and cues.
Vision deficits	• Strabismus • Hemianopia	• Speak to someone in the appropriate field of vision. • Provide an obstacle-free environment to prevent falls.

GERD = gastroesophagal reflux disease

Data from Gajdosik et al. and Krigger.

The occurrence of pain is most commonly identified in the lower extremities, back, shoulders, and neck (74, 91, 114). In relation to pain in these areas, limited range of motion is commonly reported in the ankle, hip, and shoulder joints (97). Sandstrom and colleagues (97) reported a high incidence of pain in the lower extremities of persons who could walk with or without assistance. The life span of people with CP who do not have significant comorbidities approaches that of the general population (106).

History and Physical Examination

The clinical exercise physiologist must consider all associated and secondary conditions related to CP when developing an exercise program. Because of the heterogeneous nature of CP, the exercise program should be tailored to the needs of the individual. When necessary, the clinical exercise physiologist should consult with a team of rehabilitative specialists, which may include physical therapists, occupational therapists, speech therapists, neurologists, or physiatrists, to develop an appropriate program. Regular reporting of progress to the team of specialists who work with the client is important. Accurate and detailed SOAP (subjective, objective, assessment, plan) notes are critical to facilitate quality communication among all health care providers who serve the client. The SOAP notes will also assist the client in tracking the occurrence of secondary conditions and may provide motivation and self-empowerment in

Table 27.4 Secondary Conditions in Cerebral Palsy

Secondary condition	Causative or related factors	Exercise prescription
Cardiovascular and cardiopulmonary conditions	• Inactivity • Weakness • Postural instability	Aerobic exercises: at least 3 d/wk; start with low duration, gradually increasing to 20 to 60 min as tolerated. Examples: ergometers, chair aerobics, dancing.
Communication disorders	• Learning disability • Perception issues • Speech difficulties (dysphasia) • Intellectual disability	Ask yes and no questions, or provide a pad and paper to assist in the communication process.
Depression	• Communication difficulties • Low self-esteem • Social isolation	• Use motivational coaching techniques. • Use short-term goals to illustrate progress. • Provide opportunity for group activity.
Fatigue and decreased endurance	• Overuse syndromes • Nerve entrapments	• Begin with intermittent exercise sessions with frequent rest intervals (three 10 min sessions) and slowly progress exercise duration. • Increase rest intervals between sets and between exercises to avoid muscle fatigue (2-3 min after each set).
Mobility impairments	• Balance and coordination problems • Contractures • Hypertonicity	Incorporate balance, stability, and fall prevention exercises.
	Muscle weakness and atrophy	• Use high volume (resistance times reps) and low intensity. • Work large muscles before smaller muscles. Use multi- and single-joint exercises. Concentrate on eccentric muscle actions. Muscles should be moved in a slow, controlled manner. • Start slowly with resistance training program. Examples: chair aerobics, elastic resistance bands.
	Contractures and musculoskeletal deformities (scoliosis, hip dislocation, pelvic obliquity, foot and ankle conditions such as clubfoot and flat foot)	• Improve flexibility and ROM. • Use modified interval training (changes in intensity as well as direction). • Use static technique for stretching, a minimum of 2 or 3 d/wk to a position of mild discomfort; hold for 15 to 30 s; repeat three or four times for each stretch. • Use proprioceptive neuromuscular facilitation (PNF) and active assistive stretching to help achieve a full range of motion and help relax muscles in contraction. Examples: yoga, tai chi, ergometers.
	Decreased bone density and strength	• Use weight-bearing exercise. Example: dancing.
Obesity	• Food used as a reward • Decreased caloric expenditure • May lead to joint pain	• Incorporate aerobic exercises. • Encourage client to increase general physical activity levels beyond structured exercise. • Pedometer may be useful to motivate those who ambulate. • With appropriate adaptations, pedometer can be used by people in wheelchairs.

(continued)

Table 27.4 *(continued)*

Secondary condition	Causative or related factors	Exercise prescription
Pain	• Arthritis • Contractures • Musculoskeletal deformities • Nerve entrapments • Overuse syndromes • Osteoarthritis • Osteoporosis • Postural instability • Pressure ulcers	• Incorporate non-weight-bearing activities for training (swimming, hand cycling). • Use elastic bands, balance balls, and hydrotherapy. • Record pain scale in SOAP notes before and after exercise session.
Poor dental health and hygiene	• Cavities • Gingival and periodontal disease • Occlusal disorders • Hyperactive gag reflex • Hypoplasia (incomplete calcification) of the primary teeth, reflecting an early disturbance in the development of enamel and dentin • Temporomandibular contractures may be present • Can cause hypoxemia	• Make adaptations to mouthpiece for gas exchange testing; make sure fit is secure and seal is tight. • Use face mask for $\dot{V}O_2$ testing.
Pressure sores or poor skin integrity	• Due to impaired sensory perception • Chronic infections • Skin disorders	• Take precautions with skin care, positioning, and duration of activity. • If person is in a wheelchair, use wheelchair push-ups for 20 to 30 min per day in addition to other forms of resistance training to avoid pressure sores. • Check placement of orthotics before and after exercise to ensure skin integrity.

Data from 43, 57, and 115.

relation to the person's overall health promotion program. Detailed SOAP notes allow the clinical exercise physiologist to determine the most effective rate of exercise progression. Pain levels, fatigue, and any medication changes should be noted before execution of an exercise session. See figure 27.2 for a sample SOAP note format.

Diagnostic Testing

The diagnosis of CP is often a result of the failure of a child to reach developmental milestones, the presence of abnormal **muscle tone**, and persistence of primitive reflexes or qualitative differences in movement patterns (93). A physical examination, complete history, and specialized tests are required to diagnose CP accurately (55). Tests such as the Movement Assessment of Infants (MAI), Bayley Motor Scale, and the Test of Infant Motor Performance (TIMP) are used to determine the diagnosis (22). Specialized technology such as cranial ultransonography, magnetic resonance imaging (MRI), and computed tomography can assist in determining the extent of the insult to the central nervous system (55). Experimental technology such as functional MRI and transcranial magnetic stimulation are currently used only for research purposes, but these modalities show promise in the diagnosis of CP. According to Samson and colleagues, an accurate diagnosis can be made at 6

PARTICIPANT S.O.A.P. NOTE

Participant name: ____________________ ID #: __________

Interviewer: ____________________ Date: __________

	Preexercise	Postexercise
Heart rate (BPM)		
Blood pressure (mmHg)		
Blood glucose if diabetic (mg • dl $^{-1}$)		
Activities pain index if indicated (0-10)		

QUESTIONS (Yes/No):

_______ Any illnesses or changes in health? __________

_______ Eat breakfast/lunch today?

_______ Take medications today?

_______ Any changes or adjustments to medications? __________

Subjective statements made by participant: __________

Objective modalities or activities: __________

Assessment of participant progress and compliance: __________

Plan for next visit: __________

Additional comments, including complications or significant events: __________

Figure 27.2 Sample SOAP note format.

mo of age except in the mildest forms of CP (96). It has been suggested that because of the appearance of the progression of signs and symptoms in the early stages of an infant's life, the extent of the condition may not be identified until 2 or 3 yr of age or later (32).

The exercise professional should understand the functional levels of the client and any other factors that may influence the person's health behavior, such as measures of quality of life. See "Functional Instruments and Health-Related Quality of Life Measures" for a list of tests to determine those factors.

When working with an adult with CP, it is important to have a comprehensive understanding of the factors that influence exercise testing and prescription, such as gait, **spasticity,** and balance.

Gait and Posture Assessment

The ability to ambulate and the extent to which spasticity, contractures, and deformities affect gait are variable among individuals with CP. During ambulation and activity, abnormalities in movement can be clearly identified. Clinical exercise physiologists who are knowledgeable about the types of deformities, contractures, and gait deviations associated with CP can develop a precise exercise prescription to accommodate these issues and assist in the prevention of secondary conditions. Commonly observed postures include (1) shoulder internal rotation, elbow flexion, forearm pronation, wrist flexion, finger flexion, and thumb in palm; and (2) hip flexion, hip abduction, knee flexion, ankle equinus, **hindfoot valgus**, and toe

Functional Instruments and Health-Related Quality of Life Measures

Global Function
- Pediatric Evaluation of Disability Inventory
- Functional Independence Measure for Children

Arm Functional Assessment Tests and Scales
- Assessment of quality of movement for unilateral upper limb function (Melbourne)
- Quality of Upper Extremity Skills Test
- House classification

Leg Functional Assessment Instruments
- Gross Motor Performance Measure
- Gross Motor Functional Measure
- Gross Motor Functional Classification System (GMFCS)
- Physician rating scale (lower extremity rating scale)

Spasticity Assessment Instruments
- Ashworth Scale
- Modified Ashworth Scale
- Tardieu Scale
- Burke-Rahn-Marsden Scale of Dystonia
- Barry-Albright Dystonia Scale

Health-Related Quality of Life in CP: Generic Instruments
- Child Health Questionnaire
- Pediatric musculoskeletal-functional health questionnaire
- PedsQL 4.0

Health-Related Quality of Life in CP: Disease-Specific Instruments

Caregiver questionnaire

Utility measures
- Health Utilities Index Mark 3

Adapted from Koman, Smith, and Shilt 2004.

flexion (55). Common gait abnormalities include toe walking, crouched gait, jump gait, and scissoring. Tables 27.5 and 27.6 describe the common contractures and the resulting deformities, gait deviations, and orthotics. Tables 27.7 and 27.8 present the types of contracture and muscles that should be stretched and strengthened. Motion analysis is a tool used to understand gait abnormalities. Markers are placed on select body points to define body segments. Multiple cameras capture the motion, which is then analyzed according to specific kinematic parameters. Motion analysis is used to determine the appropriate surgical procedure and to determine the effect of interventions that involve treatments such as orthotics, botulinum, physical therapy, and selected dorsal rhizotomy (72).

Practical application 27.1 presents the use of a classification system that can be used to individualize the exercise prescription as it is designed for the person with CP.

Spasticity Assessment

Spasticity is an involuntary velocity-dependent increase in muscle tone. Spasticity is often measured by tests such as the Ashworth Scale, Modified Ashworth Scale, Tardieu Scale, and Modified Tardieu Scale. These tests differ in how they measure spasticity. The Ashworth Scale measures muscle tone by scoring the resistance encountered

Table 27.5 Common Contractures of the Lower Extremities

Joints	Common types of contractures	Types of deformities	Gait deviations	Orthotics
Hip	Flexion	Coxa vara	Scissoring gait	Hip–knee–ankle–foot-orthosis (HKAFO)
	Adduction			
	Internal rotation			
	Flexion	Coxa valga	Abducted gait Wide-based gait	Hip–knee–ankle–foot–orthosis
	Abduction			
	External rotation			
Knee	Flexion	Genu valgum	Crouched gait	Knee–ankle–foot orthosis (KAFO): medial T-strap
		Genu varum		Knee–ankle–foot orthosis: lateral T-strap
Ankle	Plantar flexion	Pes equinus		Ankle–foot orthosis (AFO) Medial wedge
	Pronation			
	Eversion	Hindfoot valgus		
	Plantar flexion	Pes equinus	Equinus gait	Ankle–foot orthosis Lateral wedge
	Supination	Clubfoot		
	Inversion	Hindfoot varus		

Table 27.6 Common Contractures of the Upper Extremities

Joints	Common types of contractures	Orthotics
Shoulder	Flexion	Orthotics not commonly used
	Adduction	
	Internal rotation	
Elbow	Flexion	Orthotics not commonly used
	Pronation	
Wrist, MCP, IP, and CMC (thumb)	Flexion	• Resting splints: • Wrist in 30° extension • MCP: 60° flexion • IP: neutral • Thumb: opposition • Functional splints: cock-up splint • Wrist: 30° extension • Thumb: abduction

Note. MCP = metacarpophalangeal; IP = interphalangeal; CMC = carpometacarpal.

in a specific muscle group with passive movement of the limb through its range of motion on a 5-point scale (the Modified Ashworth Scale uses a 6-point scale) (100). The Tardieu Scale measures spasticity by passive movement of the joints at three specified velocities; the intensity and duration of the muscle reaction to the stretch is rated on a 5-point scale and the angle recorded. The Modified Tardieu Scale is less time-consuming and requires the clinician to define the moment of the "catch" (the first occurrence or "showing" of resistance or spasticity when the muscle is stretched) that is experienced in the range of motion at a fast, passive stretch (100). Although the clinical exercise physiologist may not use these tests, they are important clinical assessments that should be

Table 27.7 Stretching and Strengthening of Muscles Associated With Common Contractures in the Lower Extremities

Joint	Types of contracture	Direction of stretch*	Muscle strengthening
Hip	Flexion, adduction, internal rotation	Extension, abduction, external rotation	• Extensors: gluteus maximus, biceps femoris, semimembranosus, semitendinosus • Abductors: gluteus maximus, gluteus medius, gluteus minimus, tensor fasciae latae, sartorius • External rotators: gluteus maximus, sartorius
	Flexion, abduction, external rotation	Extension, adduction, internal rotation	• Extensors: gluteus maximus, biceps femoris, semimembranosus, semitendinosus • Adductors: adductor longus, adductor magnus, adductor brevis, gracilis, pectineus • Internal rotators: tensor fasciae latae, adductor longus, adductor magnus, adductor brevis, pectineus
Knee	Flexion	Extension	• Extensors: quadriceps femoris
Ankle and foot	Plantar flexion and eversion	Dorsiflexion and inversion	• Dorsiflexors: tibialis anterior, extensor digitorum longus, extensor hallucis longus • Invertors: tibialis posterior, flexor digitorum longus, flexor hallucis longus, popliteus
	Plantar flexion and inversion	Dorsiflexion and eversion	• Dorsiflexors: tibialis anterior, extensor digitorum longus, extensor hallucis longus • Evertors: peroneus longus, peroneus brevis

*Stretch the antagonist muscles.

Table 27.8 Stretching and Strengthening of Muscles Associated With Common Contractures in the Upper Extremities

Joint	Types of contracture	Direction of stretch*	Muscle strengthening
Shoulder	Flexion, adduction, internal rotation	Extension, abduction, external rotation	• Extensors: pectoralis major, latissimus dorsi, posterior deltoid • Abductors: middle deltoid, supraspinatus • External rotators: infraspinatus, teres minor
Elbow	Flexion	Extension	• Extensors: triceps, anconeus
Wrist	Flexion, ulnar deviation	Extension, radial deviation	• Extensors: extensor carpi radialis longus and brevis, extensor digitorum, extensor carpi ulnaris • Radial deviators: flexor carpi radialis, extensor carpi radialis longus and brevis
MCP, IP, CMC	Flexion at MCP, IP Adduction at CMC	Extension at MCP, IP Abduction at CMC	• MCP extensors: extensor indices • IP extensors: extensor indices, lumbricals • CMC abductors: abductor pollicis longus and brevis

Note. MCP = metacarpophalangeal; IP = interphalangeal; CMC = carpometacarpal.

*Stretch the antagonist muscles.

Practical Application 27.1

USE OF THE ICF TO PREPARE INDIVIDUALIZED EXERCISE PRESCRIPTIONS FOR PERSONS WITH CP

The International Classification of Functioning, Disability and Health (ICF) provides a comprehensive framework to assess the relationship among an individual's function, activities, and participation while also considering environmental and personal factors that influence the person's overall health. According to the World Health Organization, the overall aim of the ICF classification is to provide a unified and standard language and framework for the description of health and health-related states. The classification defines components of health and some health-related components of well-being (such as education and labor). The domains in the ICF are referred to as health domains and health-related domains. These domains are described from the perspective of the body, the individual, and society in two basic categories: (1) body functions and structures (system level) and (2) activities and participation (person level and person–environment interaction). The ICF can be used as a clinical tool in exercise physiology to conduct a needs assessment or as an outcome evaluation (48). It allows the clinical exercise physiologist to identify the barriers and facilitators that affect the health of the client with CP (figure 27.3) (80). Adults with CP have higher than expected rates of medical problems such as deterioration of functional ambulation, increased contractures, bowel and bladder dysfunction, osteopenia and fractures, fatigue, and pain. As a result of the lack of mobility, they are less likely to find meaningful employment. When developing interventions to improve the health of individuals with CP, it is prudent to focus on the optimization of health during the transitional years between adolescence and adulthood. The ICF model can assist in the development of interventions that will optimize the health and well-being of adults with CP (59).

For detailed information on application of the ICF model to cardiovascular and resistance training, see figure 27.4.

ICF Structure

Body functions are the physiological functions of body systems (including psychological functions).

Body structures are anatomical parts of the body, such as organs, limbs, and their components.

Impairments are problems in body function or structure such as a significant deviation or loss.

Activity is the execution of a task or action by an individual.

Participation is involvement in a life situation.

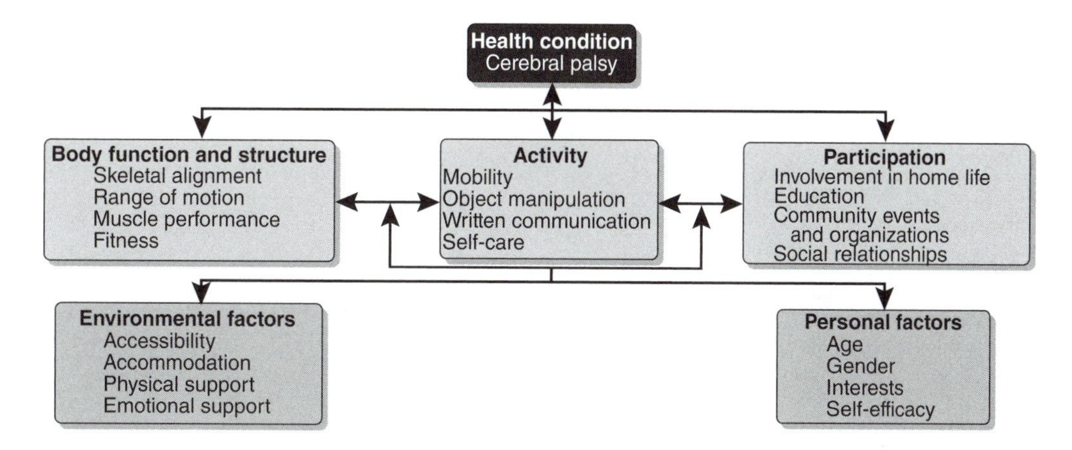

Figure 27.3 Relationships among components of health and contextual factors in children with cerebral palsy. The conceptual framework is the International Classification of Functioning, Disability and Health.

Reprinted from *Seminars in Pediatric Neurology*, Vol.11, R.J. Palisano, L.M. Snider and M.N. Orlin, "Recent advances in physical and occupational therapy for children with cerebral palsy from seminars in pediatric neurology," pg. 67, Copyright 2004, with permission from Elsevier.

(continued)

Activity limitations are difficulties an individual may have in executing activities.

Participation restrictions are problems an individual may experience in involvement in life situations.

Environmental factors make up the physical, social, and attitudinal environment in which people live and conduct their lives.

Components

Part I. Functioning and Disability

1. Body functions and structures
2. Activities and participation

Part II. Contextual Factors

1. Environmental factors
2. Personal factors (*note:* not classified in ICF, but included here to show contribution and impact)

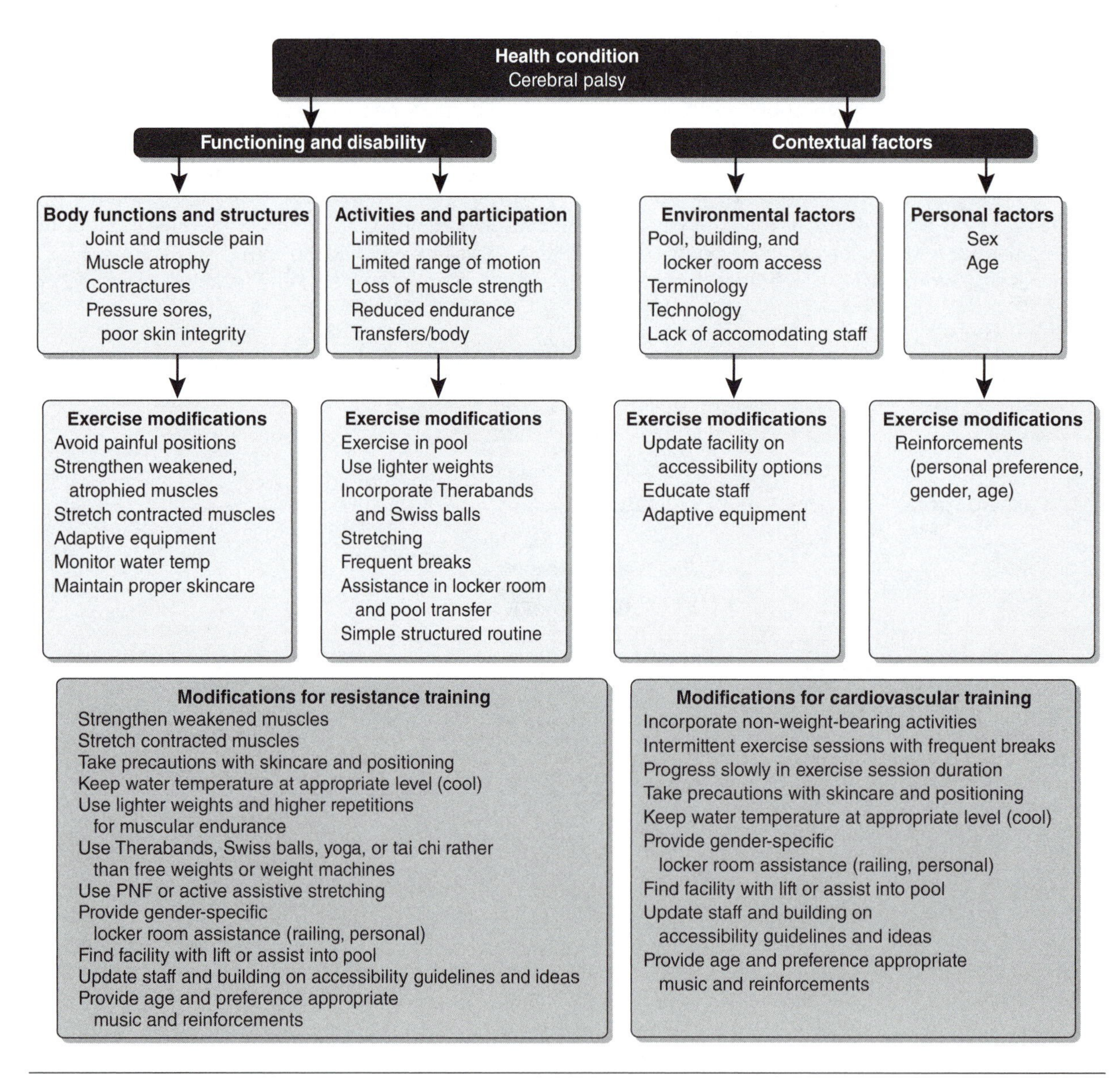

Figure 27.4 International Classification of Functioning, Disability and Health model and cardiovascular and resistance training for cerebral palsy.

understood. The clinical exercise physiologist must know the level of spasticity that a client experiences to allow detection of any increases in spasticity that may be exercise related. Spasticity levels should be monitored following each exercise bout. The exercise physiologist should observe the client's range of motion during functional activities, such as reaching tasks, before and after exercise. Inability to complete a task or decreased range of motion may indicate an increase in spasticity.

Balance Assessment

People with CP often have poor balance because of the initial insult to the cerebellum, which produces **ataxia,** abnormal involuntary movements, altered postural alignments, and gait abnormalities. It is important to assess the client's seated and standing balance along with static and dynamic balance control. Balance can be assessed by using tests such as the Functional Reach Test or by observing the client stand from a seated position (36). If a client's balance is poor, alterations to exercise testing and prescription should be made to ensure safety. Other tests used to determine balance include the Clinical Test of Sensory Interaction and Balance (CTSIB), the Tinetti Balance Test or the Performance-Oriented Mobility Assessment (POMA), Timed Get Up and Go, Berg Balance Scale, and the Physical Performance Test (16, 26, 67, 86, 88, 111).

Practical application 27.2 provides a patient's perspective on ways to work and with those with CP.

Practical Application 27.2

CLIENT–CLINICIAN INTERACTION

June Kailes, author of *Health, Wellness and Aging with Disability*, is a well-known national disability policy consultant. She is in her 50s and has been diagnosed with ataxic CP. June uses a scooter to ambulate and to manage musculoskeletal pain. Her job requires long hours on the computer, on the phone, on airplanes, and in meetings. She makes physical activity a priority because her job and lifestyle are sedentary. She works out regularly and can provide insight to the clinical exercise physiologist from both a personal and a professional standpoint.

June's primary exercise goals include pain reduction, prevention of secondary conditions, maintenance of function, and weight control. June works with a personal trainer once a week on average and exercises independently two to four times per week. June's exercise program includes walking on the treadmill for 40 min followed by a comprehensive set of flexibility exercises. June needs a treadmill that can operate at 1.3 mph (2.1 kph). She uses a balance ball to perform strength training exercises that also challenge her balance. Using the treadmill for support, she performs standing exercises to improve her balance. In the evenings, June does floor exercises that focus on strength and flexibility for her back and core muscles.

When asked what motivates her most to continue her exercise program, June suggested that concerns about losing function and preventing increased pain keep her exercise regimen a high priority in her lifestyle. June said, "In the past we didn't age, we just died! We are the first generation to live this long, so the question is not, 'Will we live?' but 'How well will we live?'" She also stated, "Many of us will probably live longer than we think, so we have to think about what we can influence or change in terms of the quality of our years as we age." June advises the clinical exercise physiologist not to minimize the symptoms that she is experiencing and simply relate them to the aging process. "What's coming into sharp focus for many of us is that the changes brought about by regular aging (yet to be clearly defined) can play havoc with a person's ability to function." Like many of her fellow baby boomers, she states that she expects more, not less, with age. When asked what she enjoys the most about exercise, June said, "When it's over . . . a sense of accomplishment!"

June also encourages the exercise physiologist to be specific about the type and amount of exercise recommended. She believes that the more she knows about why she is doing an exercise and what benefits it will bring, the more likely she will be to include it in her activities. She advises exercise physiologists to speak in nonclinical terms, to be creative in exercise design, to ask for the patient's suggestions or thoughts, and to challenge people with CP. She stresses the importance of understanding the secondary and associated conditions of CP and the importance of tailoring the program to the patient's wants and needs. June often travels and sometimes encounters barriers such as a lack of accessible fitness centers in hotels or equipment that is not accommodating. When fitness centers lack accessible options, June walks the hotel halls and does flexibility and strength exercises in her room. June expects that exercise will always be part of her life. She sees the benefits and believes that the investment of time is more important for her than for people without disabilities.

Exercise Testing

Exercise testing for those with CP may be required if they have cardiovascular disease risk factors or for exercise prescription purposes. Because of the disorder and its complications, some modifications to the typical stress test may be required. These are reviewed in the next several sections. For general information regarding exercise testing, see chapter 5.

Cardiovascular

Cardiovascular exercise testing can help the clinical exercise physiologist determine cardiovascular fitness and create an appropriate exercise prescription. Researchers reported significantly lower levels of aerobic capacity in children and adolescents who ambulated independently (GMFCS I or II), especially in girls as they increased in age, when compared to typically developing controls (121). The principle of specificity is important when one is choosing a modality for performing the graded exercise test.

Wheelchair ergometry involves the use of the individual's own wheelchair and a wheelchair roller, which locks the chair in place and provides a stationary means of propelling the wheels. Calculating or controlling workload using wheelchair ergometry with a mechanism to measure power is often difficult. An arm ergometer allows control of the workload and still provides an accurate measure of oxygen consumption. Testing individuals with CP can be difficult because the test may increase spasticity as a result of the increase in the speed of movement and temporarily affect coordination. The recommended protocol for arm ergometry begins with a starting power output of 0 to 15 W at 30 to 50 rpm, increasing 5 to 10 W every 1 to 2 min until volitional fatigue is reached (5). The exercise physiologist must make sure the client is in a stable position during testing and that the arm ergometer is secured into a fixed position. If the client has hemiplegia (a contracture or weakness in the hands), straps or an adaptive glove may be needed to secure the hand to the handlebar of the ergometer (5, 81). The exercise physiologist must identify sufficient range of motion in the hand and wrist before attaching it to the ergometer with any assistive devices. If the person lacks adequate range of motion to perform this movement, an injury can result.

Leg cycle ergometry has been used effectively in this population and provides a stable testing method if balance is an issue. The protocol for cycle ergometry varies from 25 to 50 W at 50 to 60 rpm with an increase of 15 to 25 W for each 2 min increment until volitional fatigue (3, 5). As a person with CP reaches muscular fatigue, she may have difficulty continuing to pedal on a cycle ergometer because foot control and knee alignment may prevent performance of full revolutions (81). Stationary cycling can also be effective in individuals with CP who have varying degrees of gross motor function limitations that render then non-ambulant (126). Other testing devices such as the Nu-Step recumbent stepper, Schwinn Airdyne, or treadmill can be used as testing modalities for individuals with CP, depending on their functional ability. An advantage of the Nu-Step is the greater recruitment of musculature as compared with cycling and the ability to increase workload without increasing cadence. The Schwinn Airdyne cannot increase workload without increasing cadence (rpm), which may increase spasticity and affect coordination. Studies in the nondisabled population show that peak $\dot{V}O_2$ values achieved during treadmill walking correlate strongly with all-extremity exercise testing values and do not differ significantly from those with nonwalking modalities (61). The treadmill is a useful testing device, but may be limited in some persons with CP because of limitations in balance and coordination. If a treadmill is used, the exercise professional must protect against increased spasticity and use a spotter. The client should be allowed to practice the testing methods to allow accommodation and therefore produce a more accurate test (50). During $\dot{V}O_2$ tests, it may be necessary to adjust the mouthpiece or use a face mask because of oral motor dysfunctions and occlusal disorders. Table 27.9 reviews specifics for cardiovascular exercise testing.

Musculoskeletal

According to the American College of Sports Medicine (ACSM), the best format for determining dynamic strength testing (involving movement of the body or an external load) is the one-repetition maximum (1RM) (4). In some cases, a 1RM test is contraindicated, so a 6RM or 10RM test is used. But extrapolating those data and estimating a person's true 1RM are often problematic because of the marked variation in the number of repetitions that can be performed at a fixed percentage of a 1RM for different muscle groups (e.g., leg press vs. bench press). Use of the Holton curve takes the 10RM of an individual and adjusts the score to estimate the 1RM. According to ACSM, an 8RM or 25RM can be used to test muscular strength and endurance of individuals with CP (3, 5). An alternative method of testing is to determine the number of repetitions that can be completed in 1 min. Performance level is graded by individual progression from one testing period to the next.

Simple field tests can be used to evaluate the endurance of certain muscle groups. For example, the number of abdominal crunches that can be performed without any rest provides information regarding the endurance of the abdominal muscles. Certain resistance training

Table 27.9 Cardiovascular Exercise Testing Review*

Mode	Protocol specifics	Clinical measures	Clinical implications	Special considerations
Wheelchair ergometry using rollers		ECG, HR, BP, consider $\dot{V}O_2$	Difficult to calculate or control workload	• Allow practice time before testing, but do not trigger fatigue by allowing excessive practice. • Make sure that client is in a stable position. • Be cautious if using straps to attach the client's hand to the handle. Be sure that client has adequate range of motion to perform the exercise.
Arm ergometry	Begins with a starting power output of 0 to 15 W at 30 to 50 rpm and increases 5 to 10 W every 1 to 2 min until volitional fatigue is reached	ECG, HR, BP, consider $\dot{V}O_2$	Easier to control workload	• Allow practice time before testing, but do not trigger fatigue by allowing excessive practice. • Make sure that client is in a stable position. • Be cautious if using straps to attach the client's hand to the handle. Be sure that client has adequate range of motion to perform the exercise.
Cycle ergometry; Airdyne-type cycle; Nu-Step recumbent stepper	Varies from 25 to 50 W at 50 to 60 rpm with an increase of 15 to 25 W for each 2 min increment until volitional fatigue is reached	ECG, HR, BP, consider $\dot{V}O_2$	As a person with CP reaches muscular fatigue, he may have difficulty continuing to pedal on a cycle ergometer because foot control and knee alignment may prevent performance of full revolutions Greater recruitment of musculature; can increase workload without increasing cadence Easier to control workload	• Allow practice time before testing, but do not trigger fatigue by allowing excessive practice. • Make sure that client is in a stable position. • Be cautious if using straps to attach the client's hand to the handle. Be sure that client has adequate range of motion to perform the exercise.
Treadmill	RPE (6 to 20) $\dot{V}O_2$max, METs 12-lead ECG	ECG, HR, BP, consider $\dot{V}O_2$	Cannot increase workload without increasing cadence (rpm), which may increase spasticity and impair coordination	• Allow practice time before testing, but do not trigger fatigue by allowing excessive practice. • May be limited in persons with CP because of limitations in balance and coordination. • Protect against increased spasticity and use a spotter at all times.
Metabolic data (e.g., peak $\dot{V}O_2$)			None	• For metabolic data, mouthpiece may be a problem; use mask but be sure to maintain a tight seal.

*Some tests may increase spasticity and affect coordination. *Note.* ECG = electrocardiogram; HR = heart rate; BP = blood pressure; METs = metabolic equivalents; RPE = rating of perceived exertion; rpm = revolution per minute.

equipment can be adapted to measure muscular endurance if one chooses a submaximal level of resistance and then measures the number of reps the person can perform before fatigue sets in. As an example, the YMCA bench press involves performing standardized reps at a rate of 30 lifts per minute. Women are tested using a 35 lb (16 kg) barbell, and men are tested with an 80 lb (36 kg) barbell. Individuals are scored based on the number of successful repetitions. The weight suggested in the YMCA bench press may not be appropriate for everyone and should be determined by individual strength levels.

Other modalities that can be used to test strength include handheld dynamometers and isokinetic dynamometers such as the Cybex II (10, 107). Primary assessments used to assess the gait function response to progressive resistance exercise training include functional tests such as the 6 min walk test. Additionally secondary assessment tests such as the 10 m walk test, stair climbing, the Timed Get Up and Go Test, and isokinetic muscle strength of the quadriceps are also used to assess the resistance training response (65).

When testing persons with CP, the exercise professional should remember several specific issues that can affect the safety and effectiveness of the testing (3, 5):

1. Cocontraction may offset strength in tested muscle groups (agonists).
2. Measure range of motion (ROM) in tested muscle groups to determine the safety of exercise testing and specific exercise prescription.
3. Test muscle groups unilaterally (there may be more spasticity on one side or a significant strength difference).
4. Focus on stability, coordination, ROM, and timing.
5. Adaptations include wide benches, low seats, and trunk and pelvic strapping.
6. Machines are safer than free weights and provide greater fluidity to the movement.
7. Use a metronome to ensure appropriate fluidity. Be sure to use a slow cadence to decrease spasticity.
8. Use nonslip handgrips and gloves if necessary.
9. Always provide adequate practice before testing.

Table 27.10 presents an overview of muscular strength and endurance testing.

Flexibility

Measuring the range of motion for each joint is crucial to designing an effective and safe exercise program for a person with CP. Maintaining flexibility of joints facilitates movement; tissue damage can easily occur if an activity moves the structures of a joint beyond its shortened range of motion. Because flexibility is joint specific, the clinical exercise physiologist must be aware of any contractures that are present to prevent injury. Tables 27.7 and 27.8 show which muscles should be stretched and which muscles should be strengthened. Common devices to measure flexibility include goniometers, electrogoniometers, and tape measures. The range of each joint can be measured by the use of a goniometer. This is most commonly done through passive motion but can be done through active motion as well. The measurement is taken throughout one plane of motion at a time. Placing the joint in the zero position is important. The goniometer must be aligned with the bony prominence to measure the range of motion accurately. Other forms of flexibility tests include the sit-and-reach test (which measures low back and hip joint flexibility) and Apley's Scratch Test. Some of these tests may not be appropriate for all individuals with CP. Table 27.11 presents a review of range of motion assessment.

Treatment

The medical management of CP includes nonpharmacological, pharmacological, and surgical interventions. See table 27.12 for a list of nonpharmacologic treatment options along with the characteristic functions or outcomes. Table 27.13 provides a review of pharmacologic treatments.

Nonpharmacological Treatments

The following sections provide an overview of several nondrug treatment regimens for patients with CP.

Neuromuscular electrical stimulation Neuromuscular electrical stimulation can be used functionally to cause a muscular contraction. If this method is effective, an increase in strength can be seen in 1 to 2 wk (55). As children with CP enter their teen years, a progressively crouched gait and loss of ambulation occurs, which has been linked to gastrocnemius insufficiency (44). The use of serial casting and pharmacological interventions such as botulinum toxin A has been shown to increase ankle dorsiflexion and improve functional gait (18).

Assistive technology devices (ATDs) Orthotics are used commonly in the management of clinical signs and symptoms of CP. The main goal of an orthotic is to prevent deformity of the joints and musculature and increase function (12). The solid ankle–foot orthosis (AFO) is frequently used to hold the foot in a neutral position and reduce equinus caused by a spastic gastrocsoleus (34). If the orthotic is removed during strengthening or stretching exercises, it must be replaced correctly. Improper positioning can cause skin degradation.

People with CP commonly use posterior walkers to facilitate ambulation. This device allows the individual to maintain better stability and balance while ambulating. The clinical exercise physiologist should be cognizant that ambulation with assistive devices requires greater oxygen demands independent of the presence or absence of a physical disability (68). The increased energy expenditure in ambulation should be considered in the management of symptoms such as fatigue.

Table 27.10 Strength Exercise Testing Review

Mode	Protocol specifics	Clinical measures	Clinical implications	Special considerations (may apply to all strength testing modes)
Free weights, weight machines	1RM = the heaviest weight that can be lifted or moved only once using proper exercise form	8RM or 25RM; alternatively, determine the number of reps in 1 min	None	• Machines are safer than free weights and provide greater "fluidity" to the movement. • Cocontraction may offset strength in tested muscle groups (agonists). • Measure ROM in tested muscle groups. • Test muscle groups unilaterally (more spasticity may be present on one side). • Focus on stability, coordination, ROM, and timing. • Adaptations: wide benches, low seats, and trunk and pelvic strapping. • Use a metronome to ensure appropriate fluidity; be sure to use a slow cadence to decrease spasticity. • Use nonslip handgrips and gloves (if necessary). • Always provide adequate practice before testing.
Field tests (e.g., abdominal crunches performed without rest)	Simple field tests can be used to evaluate muscular endurance of certain muscle groups	Determine the number of reps in 1 min	None	
Handheld dynamometers and isokinetic dynamometers (e.g., Cybex II)		The dynamometer measures the amount of grip strength used to depress the handle in pounds or kilograms	None	
Hydraulic resistance			None	

Note. RM = repetition maximum; ROM = range of motion.

Table 27.11 Flexibility Testing Review

Mode	Protocol specifics	Clinical measures	Clinical implications	Special considerations
Goniometry, electrogonimeters, and tape measures	• Align goniometer with the bony prominence to measure available range of motion. • Measure during passive or active range of motion throughout available ROM.	Joint angles at full flexion or extension	None	ROM may be limited because of spasticity, athetosis, or contractures.
Sit-and-reach test Apley's Scratch Test				Tests may not be appropriate for all individuals.

Table 27.12 Treatment Options for the Management of Cerebral Palsy

Treatment options		Characteristics and functions
NONPHARMACOLOGIC TREATMENTS		
Therapeutic intervention	Physical therapy	Strength training
	Occupational therapy	Endurance training
	Speech therapy	Neurodevelopmental techniques (NDTs)
		Patterning
		Conductive education
		Hippotherapy
		Aquatics
		Vojta techniques
		Pediatric massage
		Adeli, or Polish, Suit program
		Constraint-induced movement therapy (CIMT)
		Inhibit spastic agonist muscles
		Improve motor development
Therapeutic modalities	Electrotherapeutic modalities of neuromuscular electrical stimulation: Therapeutic NMES Functional NMES	Maintain or improve ROM
		Delay or prevent deformity
		Provide joint stability
	Splinting	Maintains resting musculotendinous length
	Serial casting	Strengthen weak muscles
	Orthotics Assistive technology and adaptive devices	Improve functionality by using ATDs Facilitate community ambulation and integration
	Spinal bracing	
Hyperbaric oxygen		Experimental
SURGICAL		
Surgical techniques to reduce spasticity	Selective dorsal rhizotomy	Decreases spasticity by balancing spinal cord–mediated facilitatory and inhibitory control
Orthopedic surgery	Neurectomy	
	Tenotomy	
	Arthrodesis	
	Osteotomy	
	Ostectomy	
	Tendon transfer	
	Tendon lengthening	
	Fractional myotendinous lengthening	
	Multisegmental spinal fusion	

Note. NMES = neuromuscular electrical stimulation; ROM = range of motion; ATD = assisted technology device.

Table 27.13 Pharmacology Review

Medication name and class	Primary effects	Exercise effects	Special considerations
Antispasmodics and muscle relaxants: Valium, clonidine, baclofen, dantrolene, tizanidine, alcohol, phenol	Work to inhibit the reflexes that cause increased muscle tone	Sedation, nausea, hypotonicity, dizziness, loss of seizure control	These medications act systemically; therefore, an already low-tone trunk may be compromised in its ability to act as a base of support, thus further limiting extremity function.
Anticonvulsants: phenobarbital, phenytoin, valproic acid, gabapentin, topiramate	Work through many pathways to decrease transmission of the brain activity	Some are toxic to the liver; fatigue, ataxia, dizziness, decreased cognitive function.	Monitor BP carefully (phenytoin). Check pulse regularly, and if possible, watch for abnormalities on ECG screen (phenobarbital, phenytoin, valproic acid). Watch for signs of dyspnea (phenobarbital, valproic acid, gabapentin) and apnea (valproic acid). Maintain proper hydration to avoid kidney stones (toprimate).
Antiseizure: phenobarbital, phenytoin, carbamazepine	Block the overload of signals to the brain to control seizures	Some antiseizure medications have a depressant effect on the central nervous system, thus possibly blunting the physiologic responses to exercise.	Carbamazepine has the fewest side effects, which may include mental confusion or irritability, dizziness, nausea, weight loss, and sensitivity to sunburn. In some instances, the paradoxical effect of hyperactivity may be a result of these medications.
Anticholinergics/dopaminergics: benztropine mesylate, carbidopa-levodopa, glycopyrrolate, trihexyphenidyl	Work to block cholinergic nerve impulses that affect the muscles in the arms, legs, and other parts of the body and reduce uncontrollable body movements for those with dystonic CP or those who drool frequently.	Dry mouth, constipation, agitation, and dysuria	Exercise may increase these side effects; monitor clients carefully and have them sit or lie down during initial signs of syncope. Maintain proper hydration.

BP = blood pressure; ECG = electrocardiogram.

Data from 2, 11, 33, 53, 57, 58, and 119.

Wheelchair use may be necessary to facilitate community participation. Using a wheelchair results in a decrease in energy expenditure that can allow a person with CP to participate fully and maximize functional outcomes. A manual wheelchair should be encouraged for individuals with adequate upper body strength and optimal endurance levels. Power wheelchairs or scooters can be used for people with secondary conditions related to musculoskeletal and cardiovascular deconditioning.

Pharmacological Treatments

The management of spasticity and seizures is a primary goal for the majority of the pharmacological treatments used with clients who have CP. Table 27.13 lists commonly used drugs along with the potential side effects and exercise implications. Many of the medications cause dizziness, weakness, and fatigue (2, 11, 33, 53, 57, 58, 119). It is important to know if the client has taken the appropriate medications before exercise and to monitor any changes in medication that may lead to undesired side effects during exercise. Tone-reducing medications such as baclofen, dantolene sodium, and diazepam can be administered orally. Baclofen serum levels peak 2 to 3 h after oral administration, which necessitates multiple daily doses (2). Continuous intrathecal baclofen (CIB) is administered through a subcutaneous pump and has been found to be an effective treatment for severe spasticity; it also alleviates some forms of generalized dystonia. An overdose of CIB can cause unresponsiveness and profound hypotonia requiring ventilation (2).

Neuromuscular blocking agents such as botulinum toxins, phenol, and alcohol are injected into the spastic muscle to reduce tone. The side effects of phenol and alcohol include pain on injection; possible permanent muscle fibrosis; and painful sensations such as burning, numbness, or tingling lasting for several weeks (53). Botulinum has fewer side effects and greater reversibility than alcohol or phenol. The exercise physiologist should know whether the client has taken the appropriate medications before exercise and should monitor any changes in medication that may cause undesired side effects during exercise.

Surgical Treatments

Surgical procedures are used to treat spasticity, to lengthen muscle, or to decrease the effects of contractures. Selective dorsal rhizotomy interrupts the afferent limb of the reflex arc, which reduces activity in the circuits that facilitate spasticity (83). Because of this procedure, transient muscle weakness is observed and requires orthotics and extensive physical therapy (55). Orthopedic surgical interventions are symptom dependent and consider age, disease severity, comorbidities, and overall well-being (55).

EXERCISE PRESCRIPTION

The frequency, intensity, and duration of physical activity used in an exercise prescription for the nondisabled population are well known; refer to chapter 5 for details. Those guidelines serve as the basis for exercise training. But the specifics for frequency, intensity, and duration of physical activity necessary for improving fitness and functional health outcomes in persons with CP are less clear. Therefore, it is necessary to make certain modifications to the training program while considering both associated and secondary conditions (89).

Cardiovascular Exercise

To minimize joint and muscle pain, the use of non-weight-bearing activities is recommended. These activities reduce pressure on joints while still enabling the individual to receive the same benefits from the exercise routine. Good examples include swimming or other water activity and hand cycling rather than running or walking on a hard surface. Other modalities include recumbent stepping and dual-action cycling. Individuals with hemiparesis or those who use a wheelchair can use recumbent steppers. Adaptive straps may be needed to secure the feet in place. Interval training entails working for intense bouts of short duration and alternating them with relief periods. For a person with CP, interval training can be used as well, but modifications may have to be made. For example, the level of intensity for the short-duration bouts should not be so great that the entire workout session has to be lessened because of secondary conditions (reduced endurance and joint and muscle pain). Training on an arm ergometer should include directional changes that work both agonistic and antagonistic upper body muscles (e.g., 5 min forward on the arm ergometer followed by 5 min backward on the arm ergometer). The speed of training may be changed throughout (slow 5 min, fast 3 min, slow 5 min, moderate 7 min, and so on). The person should maintain a cadence that does not increase spasticity.

Because of the reduced endurance levels of individuals with CP, exercise may have to be broken down into intermittent sessions rather than performed in one continuous session. Instead of one 30 min session of cardiovascular exercise, shorter bouts of exercise with frequent rest intervals (3 to 10 min bouts) may be necessary. These exercise sessions can be performed throughout the day and should progress slowly in duration.

To avoid pressure sores, precautions must be taken with prolonged sitting periods during exercise. Proper positioning is important as well. Trunk stability may be inadequate to provide proper support and ensure safety. Therefore, additional support may be required. If aquatic training is used, maintaining proper skin care is important to ensure that the integrity of the skin is not deteriorating, which could lead to increased risk of pressure sore development.

Resistance Exercise

People with CP can benefit from incorporating resistance training using devices such as elastic bands and balance balls, as well as hydrotherapy. Elastic bands and balance balls provide a variety of resistance exercise options throughout all multijoint movements (e.g., bicep curls, knee extensions, trunk rotations) without the added risk of injury from using free weights. Elastic bands offer many levels of resistance, and the individual should progress as with any other training program (least to most resistant: yellow, red, green, blue, black, and silver). If **athetosis** is present, elastic bands may be inappropriate because of the involuntary movement and decrease in stability of the exercise band. The most appropriate exercises for people with slow, writhing movements are cuffed weights or exercise machines that can provide fluidity of movement. Active assistive exercise may be needed for smooth performance of the motion. The use of aquatic therapy enables the individual to exercise in a non-weight-bearing environment while still obtaining exercise benefits. This modality is extremely useful in decreasing the effect of musculoskeletal pain.

According to ACSM, many program variables are needed to produce appropriate progression of resistance exercise for the given individual's fitness level. With the secondary conditions in mind, these variables need to be adjusted to meet each individual's needs and capabilities. In general, a high volume (resistance times reps) is warranted, using lower resistance and increasing the number of repetitions to account for muscle atrophy. The exercise professional must be aware of fatigue that may occur rapidly because of reduced muscular endurance. Large muscle group work should be completed before small muscle group work (e.g., quadriceps before calf muscles). The use of both multi- and single-joint exercises is not contraindicated (e.g., squat vs. knee extension). Each muscle should move in a slow, controlled manner for each exercise. To ensure safety, the client should start slowly with the resistance training program and progress at a comfortable pace. Resistance training programs should also focus on enhancing or maintaining good range of motion in the affected limbs.

A common type of CP that results in weakness or **paralysis** to the right or left side of the body is spastic hemiplegia. This condition often requires greater attention to developing strength on the weaker side of the body. The amount of improvement that can be seen on the hemiplegic side depends on the amount of damage that was sustained to that part of the central nervous system. If the person has complete paralysis on one side of the body, flexibility training should be substituted for resistance training.

The strong pull of the hip adductors experienced by many persons with CP requires a resistance training program that places greater emphasis on strengthening the hip abductors. This does not necessarily mean that the hip adductors do not need to be strengthened. Although the adductor muscles are often extremely tight because of spasticity, they may also be very weak. Therefore, both sets of muscle must be strengthened, although the abductors might have to receive more emphasis. Make sure that clients have not had a hip dislocation before you have them work these muscle groups. If the client has a history of hip dislocation, seek advice from his primary care physician or health provider to determine whether hip exercises can be conducted safely.

Because balance is often affected in ambulatory persons with CP, the exercise physiologist must protect clients from injury by developing safe resistance training programs that do not expose them to high risk of injury. Some clients are able to work on strength exercises in a standing position with physical assistance from the instructor, whereas others have to perform the exercise routines from a chair. Before developing the resistance training program, the clinical exercise physiologist should measure the client's static and dynamic balance to determine whether standing exercises are safe.

Range of Motion Exercise

Active and passive range of motion should be incorporated into a well-balanced flexibility program. When possible, functional activities that use all joints throughout their full range of motion can be performed in conjunction with passive stretching. Each exercise session should begin and end with flexibility exercises. Active assistive stretching and proprioceptive neuromuscular facilitation (PNF) are effective stretching techniques that assist the client in achieving full range of motion. These methods also assist in the relaxation of muscles in contraction. Chapter 5 provides details about how to perform PNF stretching. Also, refer to tables 27.7 and 27.8 to determine the appropriate muscles to stretch when contractures are present. Before prescribing range of motion exercise, the exercise physiologist should know whether the client has leg length discrepancy, scoliosis, hip dislocations, or pelvis abnormalities. In some instances, flexibility exercises may exacerbate the problem. For example, a spastic muscle should not be overstretched. Tai chi and yoga are also good exercise choices to increase or maintain flexibility in individuals with CP.

Tables 27.14 and 27.15 provide reviews of exercise prescription and benefits for individuals with CP.

EXERCISE TRAINING

The heterogeneous nature of CP creates a complex array of signs and symptoms that vary between individuals. Knowledge of the primary, secondary, and associated conditions related to CP is necessary for understanding the beneficial effects of exercise for people with CP.

Turk defines the primary condition as the fundamental source of disability (113). In the case of CP, the etiology is not always identifiable, but in most cases it is a result of an initial insult to the brain or disruption in normal development. The primary disability does not progress. The signs and symptoms, however, can be transient, and progression of secondary and associated conditions may occur. Associated conditions are aspects of the pathology of the primary condition (113). Associated conditions are not necessarily preventable, but they can often be managed through medical intervention. An example of an associated condition of CP is spasticity. See table 27.3 for a list of associated conditions. Secondary conditions are defined as an injury, impairment, functional limitation, or disability that occurs because of the primary condition

Table 27.14 Exercise Prescription Review

Training method	Mode	Frequency	Intensity	Duration	Progression	Goals	Special considerations and comments
Cardiovascular	• Non-weight-bearing activities, including swimming, hand cycling, stationary cycles, and recumbent steppers • Schwinn Airdyne • Any upper and lower limb ergometer • Arm ergometer	3 to 5 d/wk Interval training	Moderate intensity 40% to 50% $\dot{V}O_2$ reserve or HRR for severe CP 50% to 80% $\dot{V}O_2$ reserve of HRR as tolerated	• Two or three 10 min sessions • Intermittent exercise sessions alternating with relief periods (3 to 10 min bouts)	Slow progression in duration and intensity	Increase aerobic capacity and cardiovascular endurance	Use proper positioning and avoid prolonged sitting position to avoid pressure sores. Provide proper trunk support. If aquatic training is used, maintain proper skin care.
Resistance	Free weights, elastic resistance bands, balance balls, exercise machines, or hydrotherapy	2 d/wk	To fatigue Ideally, lower resistance is used	• Three sets of 8 to 12 reps • Increased number of reps	Based on individual's abilities and needs	Improve overall muscle strength	Strengthen hip and shoulder abductors. Check for history of hip dislocation, osteopenia, and osteoporosis. If using hydrotherapy, keep water temperature at appropriate level.
Range of motion	Stretching, PNF, or active assistive stretching	3 to 5 d/wk Used as warm-up and cool-down for aerobic and strengthening exercise segments	Maintain stretch to point just before discomfort occurs	30 to 60 s per stretch	Progress as tolerated up to 20 min for contracted muscles	Improve ROM directly related to ADLs	Check for leg length discrepancy, scoliosis, hip dislocations, or pelvis abnormalities when client performs range of motion exercise.

Note. PNF = proprioceptive neuromuscular facilitation; ADLs = Activities of daily living.

Table 27.15 Benefits of Exercise for Individuals With Cerebral Palsy

Cardiorespiratory endurance	Skeletal muscle strength	Skeletal muscle endurance	Flexibility	Body composition
Improved aerobic capacity and endurance Improvement in HR, BP Increased maximal aerobic power	Improved functional strength, improved capacity to perform ADLs	Improved muscle endurance, decreased fatigue or later onset of muscle fatigue, improved capacity to perform functional activities and ADLs Improved work capacity	Increased ROM, improved coordination and skill of movements; improved capacity to perform ADLs	Lower levels of body fat, improved BMIs

Note. ROM = range of motion; ADL = activities of daily living; BMI = body mass index; HR = heart rate; BP = blood pressure.

or pathology. Secondary conditions include physical problems, social concerns, and mental health difficulties. Secondary conditions also can develop when the primary disability interferes with the delivery of standard health care for the treatment or prevention of a health condition. See table 27.4 for a list of secondary conditions.

Persons with CP report a decline in functional abilities with age (8, 54, 97). The decline with age may include symptoms such as fatigue (49, 79), pain (8, 43, 73, 79), progressive musculoskeletal deformity and dysfunction (43, 73, 79), and a decrease in functional ambulation (8, 73, 118).

Practical Application 27.3

RESEARCH FOCUS

The purpose of this uncontrolled trial was to examine the effectiveness of a home- and Internet-based upper limb training program for adults with hemiplegic CP (18). While significant declines in locomotor function in young adults with CP have been reported, no study had examined upper limb coordination and the potential impact of reach and grasp coordination deficits on activities of daily living. The researchers developed and implemented a sensorimotor training program specifically designed for the upper limb. The Upper Limb Training and Assessment (ULTrA) program incorporated a variety of arm reaching movements, as well as hand manipulation and tactile discrimination tasks. It had several unique features, including real-time remote data collection and counseling of participants regarding progress, as well as the ability to incorporate more challenging tasks to maximize gains. Furthermore, it did not require intensive, one-on-one professional contact that generally involves travel and disruption of school or work schedules. The researchers hypothesized that this unique sensorimotor training program would lead to improved upper limb reaching ability, hand manipulation skills, and tactile sensibility in adults with hemiplegic CP.

Twelve adults (four males, eight females) with predominantly unilateral involvement of the upper limb and age ranging from 21 to 57 yr completed the study. All participants reported difficulty in the ability to use the affected arm and hand in activities of daily living. Upper limb function was assessed using the Motor Activity Log, which measures self-reported amount of daily limb use and the quality of limb function, 1 to 2 wk before and 1 wk following the home training program. In addition, the Nine-Hole Peg Test was used to quantify fine hand dexterity, and grip strength was measured using handheld dynamometry.

Participants completed a series of arm reaching and hand manipulation tasks in their homes 5 d/wk for 8 wk, each training session lasting 40 to 45 min. Tasks were chosen that reflected components of everyday activities and were designed by a physical therapist, and occupational therapist with feedback from adults with CP. Custom-designed computer software was used to guide participants through a series of warm-ups, followed by task-specific sensorimotor training activities and finally cool-down exercises. Webcams were used by trained researchers to observe performance two or three times per week and to provide verbal feedback regarding training performance; this information was used to make progression decisions.

Results showed that no significant change in spasticity, range of motion, or performance on the Nine-Hole Peg Test occurred following the intervention for either arm. Hand grip force increased for the affected hand but not for the less affected hand. Performance on the Motor Activity Log improved significantly for both amount of use and quality of movement. All participants showed improvements in reaching with their affected arm following the intervention, as well as in tasks requiring grasping or manipulating objects (or both) and tactile discrimination. This was the first known study to demonstrate the effectiveness of a home-based training program aimed at improving upper limb function in adults with hemiplegic CP, as well as illustrating the value of using the Internet to monitor progress and providing quantitative information about performance throughout the intervention period.

As demonstrated in this research study, the clinical exercise physiologist can work with other health care providers to create programs that support activities of daily living. Using technologies such as the Internet or programs that include virtual coaching can reduce barriers such as lack of transportation that participants with CP experience and provide in-home programming that can greatly improve the participant's quality of life.

Reference: Brown SH, Lewis CA, McCarthy JM, Doyle ST, Hurvitz EA. The effects of internet-based home training on upper limb function in adults with cerebral palsy. *Neurorehabil Neural Repair* 2010;24(6):575-583.

A systematic review of the literature suggests that children with CP may benefit from improved exercise programs that focus on lower extremity muscle strength, cardiovascular fitness, or a combination of the two. A need still exists to determine the efficacy of exercise programs that provide the most benefit for children with CP with regard to the daily activity and participation levels of this cohort (120). Research has not yet yielded a specific set of exercise guidelines and safety considerations for children with CP. However the findings suggest that children with CP are able to train to elicit positive physiological outcomes (90).

Exercise can play an important role in the prevention of many secondary conditions and in the delay or prevention of the functional decline associated with aging. Most exercise research has focused on children and adolescents with CP. The specific intensity, frequency, and duration that are most appropriate and effective for adults with CP are still unknown (89). Nevertheless, the clinical exercise physiologist can use the successful studies as a basis for developing an appropriate exercise prescription. The literature supports preventive and promotional health strategies to be used in this population to lead to better health, increased independence, and fewer secondary conditions (40). Higher levels of physical activity are correlated to better overall health and greater levels of community participation in adults with mobility impairments (27).

Cardiovascular Exercise

The physical work capacity of adults with CP is significantly lower than that of the general population (112). Studies suggest that persons with CP require more muscle activity to perform tasks than people without disabilities (62, 63). Higher oxygen uptake, oxygen pulse, and respiratory exchange ratio occur at given workloads for individuals with CP (91, 116). Studies show that the $\dot{V}O_2$max for people with CP is similar to, or slightly lower than, that of people without disabilities (63, 91). In conjunction with having a greater level of energy expenditure, children with CP often are smaller, have a higher body fat percentage, and are less physically active than children without disabilities (117). Other research reports poor fitness levels in children and adults with CP (38, 92).

Cardiovascular exercise can be beneficial for individuals with CP. Aerobic exercise increases lower extremity strength, decreases energy required to ambulate, and improves gross motor function for adolescents with spastic CP (99). A systematic review of the effectiveness of treadmill training for children with CP suggests that this mode is safe and feasible and may provide benefits in walking speed over short distances and in general gross motor skills (21). The largest effect was found for children with full body involvement and greater difficulty in ambulation at baseline (128). Data suggests that evidence exists to support the use of cardiorespiratory training to improve aerobic function (21). However, there is little carryover to the amount of activity that children engage in on a daily basis. Short-term cardiorespiratory training (2 to 4 mo) increased aerobic fitness by 18% to 22%; long-term training (8 to 9 mo) led to an improvement of 26% to 41% (21).

With use of treadmills the techniques of spotting, gait belts, or weight-supported treadmill training systems may be required. In a systematic review, Damiano and colleagues hypothesized that the use of lower extremity reciprocal exercise devices, such as cycles, water-based treadmills, or elliptical trainers, may offer benefits for individuals with multiple disabilities with less repetitive joint stress. However, further research is needed to strengthen the evidence and to identify the optimal training protocol for body weight–supported treadmill training in CP and other central motor disorders (29).

An 8 wk training program using a bicycle ergometer at 40% to 70% of peak $\dot{V}O_2$ twice weekly produced a 12% increase in cardiorespiratory fitness (64). This program elicited an increase in physical work capacity as evaluated by a Schwinn Airdyne ergometer (85). Other studies using cycle ergometry report significant increases in maximal heart rate and aerobic capacity in children with CP (15, 103). The stationary cycle provides a safe and effective means of exercise for young people with CP who are nonambulatory, in addition to producing valuable improvements in functional ability that promote sustained independence in activity (122).

Thorpe and Reilly used water walking in conjunction with lower extremity resistance exercise and found improved gross motor function for an adult male with CP (110). This study was conducted over a 10 wk period with an exercise frequency of three times per week. A review of aquatic exercise for children with CP suggests that this form of exercise may be beneficial for children with CP, but more research is needed to determine the effect on fitness (51).

Resistance Exercise

Muscle weakness, common in people with CP, directly affects functional ability (125). Several studies have elicited enhanced ambulation through resistance training. Andersson and Mattson found that individuals with CP who were physically active reported improved ability to walk (7). Further research suggests that a 6 to 10 wk progressive strength training program improves muscle strength and walking ability without increasing spasticity (6, 28). Besides improving walking mobility, a progressive resistance training program that used a 6RM for upper body exercises three times a week increased muscular strength and wheelchair performance in children with CP (77). Progressive resistance training in children with CP has demonstrated positive outcomes in strength and function with no adverse effects (71). A systematic review of progressive strength training in youth and adolescents with CP reported no adverse events and showed that isotonic rather than isokinetic training improved function and gait in younger subjects (70).

Research suggests that a therapeutic goal of health professionals should be to encourage active standing and other forms of weight-bearing activities to promote bone mineral accrual and decrease skeletal fragility in children with spastic CP (25). This, however, should involve a well-balanced approach that does not advance the secondary conditions experienced by individuals with CP. As shown in a clinically feasible home-based training program that focused on lower limb strength and physical activity in people with spastic diplegia, a short (6 wk) program can lead to lasting changes in the strength of key lower limb muscles that affect daily function of young people with CP (35). Exercises that used virtual reality for practice of isometric exercise focusing on ankle dorsiflexion were successful in children (20). The range of motion and hold times increased with the virtual reality exercise, increasing compliance over that with the conventional home exercise program. Programs that can successfully use technology to provide in-home exercise opportunities can eliminate the barriers of transportation and cost of the typical therapist-based supervised exercise session conducted in a clinical environment (19).

According to a recent study by Weightman and colleagues, an important feature in children with CP is impairment of upper limb movements through weakness, spasticity, and loss of selective muscle activation (123). These impairments, in addition to sensory deficits, hinder the development of the smooth, coordinated movements required for activities of daily living. Children with such difficulties are naturally reluctant to use the paretic limb, leading to further restrictions in the child's exploratory learning, acquisition of basic manual skills, personal independence, education, and play. Current methods of undertaking upper limb exercise are principally limited by the amount of appropriate exercise that can be performed; and from a child's perspective, formal exercises are often unexciting. It is therefore difficult to maintain enthusiasm in the absence of a therapist. For these reasons, interest is growing in technologies that offer the possibility of providing programs of simpler exercises for children to pursue independently. To address this need, Weightman and colleagues developed and evaluated a powered joystick system linked to a computer exercise workspace that provided controlled guidance to the child's arm to enable performance of the exercises. For children with poor voluntary arm movement, the joystick offers physical guidance and in this sense is different from a computer gaming system (123).

Community-based exercise has been shown to be effective in increasing strength and motor activity in adults over the age of 40 with CP (108). Strength increases ranged from 17% to 22% over the 10 wk program. An added benefit was social interaction within the group. Resistance training in children with spastic CP demonstrated positive functional outcomes in individuals with a wide array of involvement (30). This study elicited strength increases in the hemiplegic side and an increase in walking pace. Dodd and colleagues completed a systematic review of the strength training literature for individuals with CP (35). The authors concluded that most training programs were conducted three times per week for 6 to 10 wk with a progressive adjustment in resistance to provide sufficient intensity as suggested by ACSM (4). A short program (4 wk of exercise class conducted for 1 h twice a week) of task-specific strengthening exercise in the form of circuit training resulted in improved strength and functional performance that was maintained over time (8 wk postintervention) (17). Functional physical training programs have been demonstrated to improve the aerobic endurance and functional walking ability of children with CP (45). In one study, children trained for half an hour twice weekly for 9 wk, performing circuit training activities. The circuit training activities, on four workstations, were (1) jumping on a trampoline; (2) combinations of bending, jumping, running, and climbing; (3) a basketball circuit; and (4) a run circuit. Studies are inconclusive on the benefits of strengthening intervention for youth with CP; however, the research demonstrates no increase in spasticity levels (102). Additional reviews report a general increase in muscle force production without an increase in spasticity (31). Ahlborg and colleagues found that an 8 wk intervention of whole-body vibration or resistance training can increase muscle strength without negative effect on spasticity in adults with CP (1).

Range of Motion Exercise

Individuals with CP experience musculoskeletal changes such as decreased muscle strength, soft tissue shortening, and abnormalities in bony structures. All these secondary conditions can cause a loss of joint range of motion, gait alterations, and a decrease in overall functional movement. In a clinical setting, therapeutic stretching techniques are used to (1) maintain current range of motion, (2) increase range of motion for functional tasks, (3) defer or avoid surgery, and (4) attain full range of motion (124). Strategies to elicit these goals include passive stretching, active stretching, and prolonged positioning. These strategies are often used in conjunction with other interventions such as spasticity-reducing medications, splints and orthoses, serial casting, and surgery. However, the mechanisms and etiology of muscle contractures in children and adults with CP are not well understood. Clinical research evaluating the effectiveness of these therapeutic strategies is inconclusive and cannot guide practice through sound evidence (84, 124). The increased muscle tone and diminished selective motor control may also contribute to a decrease in voluntary movement. It is assumed that this decrease in voluntary movement frequency and variety elicits an adaptive response of the muscle that reduces the number of in-series sarcomeres; this in turn results in a decrease in the muscle belly length. There is also speculation that the shortening of the muscle belly may be due to muscle fiber atrophy. The reduction in the number of in-series sarcomeres supports the use of stretching; however, the theory of muscle fiber atrophy supports the use of resistance training to prevent or reduce contractures.

Currently, fitness and therapeutic interventions focus on activity and participation level and can be developed or evaluated with use of the ICF. The ICF cautions against assuming a direct relationship between factors related to the body functions and structures component and changes at the activities and participation level (124). For example, increasing flexibility in the gastrocnemius medialis and the soleus may not make it easier for a child to participate in a traditional physical education class or climb stairs at home. Despite the emphasis on functional outcomes, most research on flexibility has focused on impairment level outcomes such as spasticity and joint range of motion (124). For example, Pin reported weak evidence to support the effectiveness of passive stretching in children with spastic CP (84). The author further stated that of the limitations of inadequate rigorousness of the research design and the small number of participants, there is some evidence that sustained stretching is more effective than manual stretching of short duration in improving range of motion and reducing spasticity. The inclusion of community participation as well as fitness and functional goals in the contemporary approaches to rehabilitation for children with CP provides an ideal opportunity for the clinical exercise physiologist—to work with physical therapists to increase opportunities for community participation in fun activities that use physical activity to enhance the level of physical fitness achievable by children with CP. Yoga, tai chi, ballet, horseback riding, and swimming can allow the participant a chance to stretch the muscle fibers in a more functional context. Wiart and colleagues state that community fitness programs are a viable alternative to medically oriented therapy programs (124). In a study published by Zhao and associates in 2011, the use of technology in the intelligent ankle stretching device to promote passive stretching and active movement in a highly engaging environment motivates children through interactive games and provides a convenient and low-cost treatment for muscle stiffness or tightness and the prevention of contractures (130).

Clinical Exercise Pearls

- Research suggests that fitness levels of persons with CP are generally poor.
- Pain and fatigue are prevalent, particularly in aging.
- Use non-weight-bearing machines if pain is an issue.
- Walking may increase pain.
- Do not confuse speech impairment with cognitive disability.
- Be aware of possible exercise-related side effects of medications:
  - Drowsiness
  - Dizziness
  - Weakness
  - Fatigue
  - Ataxia
  - Dry mouth
  - Blurred vision
  - Depression

CONCLUSION

Because people with disabilities and chronic health conditions are spending significantly less time in inpatient and outpatient rehabilitation facilities than they did 15 yr ago, clinical exercise physiologists play a greater role in responding to the health and exercise needs of these populations. As a result, exercise physiologists assist in the prevention of many secondary conditions, which improves the quality of life for people with CP and other disabilities.

Key Terms

aphasia (p. 531)
associated conditions (p. 531)
ataxia (p. 543)
athetosis (p. 550)
basal ganglia (p. 530)
cerebellum (p. 530)
chorea (p. 530)
choreoathetosis (p. 529)
contractures (p. 531)
diplegia (p. 529)
dysarthria (p. 531)
dyskinetic (p. 530)
epilepsy (p. 531)
extrapyramidal (p. 530)
hemianopia (p. 531)
hemiplegia (p. 529)
hindfoot valgus (p. 537)
monoplegia (p. 530)
muscle tone (p. 536)
paralysis (p. 551)
pyramidal (p. 530)
quadriplegia (p. 529)
rigidity (p. 530)
scoliosis (p. 531)
secondary conditions (p. 531)
spasticity (p. 537)
strabismus (p. 531)
tetraplegia (p. 530)
triplegia (p. 530)

CASE STUDY

MEDICAL HISTORY AND PHYSICAL EXAMINATION

At age 7, Mr. AT suffered a series of seizures, and after a physical exam, lab tests, and electroencephalography, he was diagnosed with epilepsy and put on epileptic medication. He has gained 25 lb (11 kg) over the last 2 yr. He complains about excessive postural sway and loss of balance while walking and therefore uses his father's cane. He has no history of smoking or alcohol consumption, but he has a familial history of diabetes. At a recent physical exam his low-density lipoprotein cholesterol (LDL-C) was 178 mg · dl^{-1}, high-density lipoprotein (HDL) was 61 mg · dl^{-1}, fasting glucose was 88 mg · dl^{-1}, and waist circumference was 42 in. (107 cm). He is currently on Carbatrol (carbamazepine) and baclofen.

DIAGNOSIS

Mr. AT, now 28 yr old, was born prematurely along with his twin sister at 34 wk. His mother experienced a severe infection early in her pregnancy and was hospitalized for several days; upon birth he had an Apgar score of 4. At age 1 he was diagnosed with spastic hemiplegia.

EXERCISE TEST RESULTS

Because Mr. AT is ambulatory but has balance difficulties, leg cycle ergometry was chosen as the method of aerobic exercise testing. His resting blood pressure and heart rate in a seated position were 134/80 mmHg and 82 beats · min^{-1}, respectively. A small face mask was used to provide an adequate seal. A toe clip on his right side was used to maintain proper foot positioning on the pedals, and an adaptive glove was used to help his right hand grip the handlebar throughout the test. Mr. AT performed a 2 min warm-up without resistance. The leg cycle ergometry test began at 25 W at a cadence of 50 to 60 rpm. The protocol consisted of 2 min stages with an increase of 15 W per stage. The following data were collected during the exercise test.

The exercise test was terminated at 5:15 because of increased hip adductor spasticity and inability to maintain speed of 50 rpm, and by request of the participant because of shortness of breath. No signs of electrocardiogram abnormalities were present during the test. His $\dot{V}O_2$ at 5:00 was 16 ml · kg^{-1} · min^{-1}. The RER was <1.00 and the heart rate and $\dot{V}O_2$ failed to reach peak. This was not a valid peak $\dot{V}O_2$ test.

Because of the termination of the clinical leg cycle ergometry test, a more functional approach was chosen. The Timed Get Up and Go test was used as a baseline measure. This test used the following protocol.

(continued)

Test Instructions

1. Starting position: Sit upright against the chair.
2. Stand up from the chair (time start).
3. Move quickly along the line toward the opposite end.
4. Touch the end and pivot back toward the chair.
5. Move quickly all the way back to the chair.
6. Pivot.
7. Sit back down (time end).

The total distance was 10 ft (3.05 m). The test was performed three times. Mr. AT's scores were 42 s, 38 s, and 40 s. (A score over 30 s suggests a higher risk of falls.) Mr. AT used his assistive device during all testing.

EXERCISE PRESCRIPTION

Using the information presented in the medical history and exercise testing sections and in tables 27.9 through 27.11, develop an exercise prescription that encompasses aerobic, strength, flexibility, and neuromuscular training. Consider the different modes for each of the exercise training domains and the frequency, duration, intensity, and progression of each.

DISCUSSION QUESTIONS

1. How would the clinical exercise physiologist best determine the appropriate resistance levels to prescribe for this person?
2. What factors represented in the ICF model should the clinical exercise physiologist take into consideration to increase the adherence level and success of this client?
3. What fall prevention strategies should be incorporated into the exercise prescription for an individual who is at increased risk for falls?
4. Is the assistive device described in this case study appropriate for use by Mr. AT?
5. How should the clinical exercise physiologist incorporate weight loss strategies into the goals for this client?

Chapter 28

Stroke

Christopher J. Womack, PhD

Stroke is a leading cause of both death and disability, making stroke patients a key population for the clinical exercise professional. Paradoxically, stroke patients are often referred only to physical therapy without consultation regarding long-term exercise prescription from either their physical therapist or a clinical exercise physiologist. Unfortunately, standard physical therapy does not typically improve functional capacity in this population, and incorporation of a regular exercise program can be a determinant of whether the patient can live independently or perform standard daily tasks without assistance. This chapter provides background on the pathophysiology and diagnosis of stroke, illuminates the importance of exercise for this population, and presents considerations for exercise testing and prescription.

DEFINITION

A stroke is the loss of blood flow to a region of the brain. This loss can occur because of a manifestation of cardiovascular disease, characterized by the buildup of atherosclerotic plaque in cerebrovascular arteries, ultimately resulting in an ischemic stroke. In most ischemic strokes, a blood clot ultimately seals off the narrowing artery. Strokes can also occur because of excessive bleeding in a cerebral artery, also known as a hemorrhagic stroke. The excessive bleeding and swelling in the brain prevent blood from flowing to brain cells downstream of the hemorrhage. Most strokes are ischemic, accounting for approximately 87% of all strokes (20). When a stroke occurs, neurons in the brain die, and the accompanying brain damage is the main cause of subsequent disability in stroke survivors. The brain damage can impair voluntary muscle movement, speech, vision, and judgment. Strokes can also occur in the form of silent cerebral infarctions. These strokes are not coupled with clinically apparent acute symptoms.

SCOPE

Cardiovascular disease is the number-one cause of mortality in U.S. adults (20). Of the causes of death from cardiovascular disease, stroke is the second leading cause, encompassing 17% of all deaths attributable to cardiovascular disease (20). This figure amounts to 43.6 deaths per 100,000 people, making stroke the third leading cause of death in the United States, behind coronary artery disease and cancer (20). The incidence of stroke is such that one occurs every 40 s in the United States; thus as of 2010, over 6 million American adults had had a stroke (20). Each year, almost 800,000 Americans will experience a stroke, and the vast majority of these will be new stroke events (20). Clearly, there are many survivors of stroke. As evidence of this, the number of noninstitutionalized stroke survivors increased from 1.5 to 2.4 million from the early 1970s to the early 1990s (26). Although this is a positive trend in that stroke is taking fewer lives, a larger number of people now have a stroke-related disability. The effect that stroke has on quality of life is large because 20% of stroke survivors require institutional care 3 mo after onset, and 15% to 30% of stroke survivors are permanently disabled (20). In a study evaluating patients after onset of stroke, 26% of individuals were institutionalized, and 30% were unable to walk without some assistance 6 mo after stroke onset (15). In total, over 1 million American adults reported functional limitations because of stroke (2). Because of the high rate of mortality and disability associated with this condition, indirect and direct costs for stroke are high; they have been estimated at $73.7 billion for 2010 (20).

Gender

Overall, stroke incidence in men between 55 to 64 yr of age is 1.25 times that for women, but these differences become smaller with increasing age, to the point where incidence is higher in women over the age of 85 (20). In fact, approximately 1 million more women than men have had strokes (20). The larger number for women is mainly due to the longer life span in women, but it also points out that stroke is an important health concern for women as well as men. The increased risk of stroke is likely due to the increased cardiovascular disease risk associated with menopause. Furthermore, estrogen replacement therapy can increase risk of cardiovascular disease and stroke. Estrogen alone increased risk of stroke by 39% in women posthysterectomy (4). Estrogen plus progestin has been shown to increase risk of ischemic stroke in postmenopausal females (38).

Although menopause represents the primary state that places females at increased risk for stroke, pregnancy also temporarily increases risk. Pregnant women are at nearly a ninefold increased risk for stroke during the 6 wk postpartum period (16). This risk is greater in African American women (12). A study on postpartum stroke showed that 4.1% of women with postpartum stroke died in the hospital, whereas 22% of survivors died at home after discharge (12).

Ethnicity

Table 28.1 summarizes the distinct racial influences on the incidence and mortality rate from stroke. As evidenced by recent data, 3.9% of adult American Indians

Table 28.1 Stroke Incidence and Mortality for Selected Ethnic Groups (20)

	Prevalence (% of population)	Mortality rate
CAUCASIAN		
Males	2.3	41.7
Females	3.1	41.1
Total	2.7	
AFRICAN AMERICAN		
Males	3.8	67.1
Females	4.3	57.0
Total	4.1	
HISPANIC		
Males	*	35.9
Females	*	32.3
Total	2.6	
AMERICAN INDIAN AND ALASKAN NATIVE		
Males	*	25.8
Females	*	30.9
Total	3.9	
ASIAN		
Males	*	39.8
Females	*	34.9
Total	1.8	

*Data not available. Mortality rate given is per 100,000 deaths within a specified group.

Based on D. Lloyd-Jones et al., 2010, "Heart disease and stroke statistics—2010 update: A report from the American Heart Association," *Circulation* 121: e46–e215.

Based on T. Thom et al., 2006, "Heart disease and stroke statistics—2006 update: a report from the American Heart Association Statistics Committee and Stroke Statistics Subcommittee," Circulation 113: e85–151.

and Alaska Natives have had a stroke, contrasted with 4.1% of African Americans, 2.7% of Caucasians, and 1.8% of Asians (20). The death rate from stroke is correspondingly higher in African American males and females compared with Caucasians (67.1 deaths per 100,000 for black males, 57.0 for black females, 41.7 for white males, and 41.1 for white females) (20). Furthermore, African Americans experience a higher degree of functional limitation following stroke than Caucasians do (1), although the exact cause of these differences is unknown. Residents of southeastern states have a higher prevalence of stroke (1), which likely reflects cultural differences regarding diet and physical activity. The additive effects of these geographical and racial influences combine to make African American males in southeastern states the population at greatest risk for stroke in the United States (1).

PATHOPHYSIOLOGY

The atherosclerotic process that causes cerebrovascular disease, and ultimately an ischemic stroke, proceeds in the same fashion as plaque progression in coronary artery disease (CAD). For a more detailed explanation of this process, see chapter 12. For this reason, the same traditional and nontraditional risk factors that are related to the development and progression of CAD and peripheral arterial disease (PAD) are associated with the development of ischemic cerebrovascular disease. **Ischemic strokes** can be further categorized as thrombotic, embolic, and hemodynamic. In the case of **thrombotic strokes**, in which an occlusive thrombus develops in or outside an ulcerated plaque, hypercoagulable states due to increased coagulation potential or decreased fibrinolytic potential are particularly important risk factors. Emboli that cause **embolic strokes** are typically from the carotid or other arteries. In these cases, the thrombus is not large enough to occlude the large vessel, but the embolus that breaks off ultimately lodges in a smaller cerebral artery or arteriole. Often, major strokes are preceded by transient ischemic attacks (TIAs), which are considered a major predictor of impending stroke (13). In a study of patients who reported to an emergency room with a TIA, approximately 10% experienced a stroke within 90 d. Perhaps even more compelling is the fact that 5% experienced a stroke within 2 d (13). All older populations should be screened for possible prior TIA because about 50% of patients who experience a TIA do not report it to a clinician (13).

Hypertension is the major risk factor for hemorrhagic stroke, which makes up approximately 10% of strokes (35). **Hemorrhagic strokes** can also be caused by aneurysm, drug use, brain tumor, congenital arteriovenous malformations, and anticoagulant medication. Hemorrhagic strokes are classified as either intracerebral, which refers to bleeding inside the brain, or subarachnoid, which refers to bleeding in and around the spaces that surround the brain (35). Unfortunately, there is usually little warning for a hemorrhagic stroke. Acute signs or symptoms include altered consciousness, headache, vomiting, and large elevations in blood pressure. Additionally, patients with subarachnoid hemorrhage may develop neck stiffness (35).

CLINICAL CONSIDERATIONS

Medical and clinical considerations for stroke include signs and symptoms specific to whether damage occurs on the right or left side of the brain. The clinician needs to be aware of the acute signs of the stroke and the clinical manifestation that occur after the event. Finally, the method of determining the definitive diagnosis of stroke is another important clinical decision.

Signs and Symptoms

Memory loss and paralysis are two of the more consistent symptoms of stroke. In the case of paralysis, the brain damage causes paralysis on the opposite side of the body (i.e., right-brain damage causes left-side paralysis). Furthermore, right-brain damage can result in vision problems and awkward or inappropriate behavior, whereas left-brain damage causes speech and language problems and slow or cautious behavior. A patient suffering from acute stroke can have one of the following symptoms: (1) numbness or weakness of the face, arm, or leg; (2) confusion, speech problems, and cognitive defects; (3) impaired bilateral or unilateral vision; (4) impaired coordination and walking; and (5) headache.

History and Physical Exam

One of the main changes for the clinical exercise physiologist to be aware of is the hemiplegic gait of stroke patients. Concomitant risk factors for cardiovascular disease such as hypertension and diabetes are frequently present. Furthermore, underlying CAD is often present (24), so resting electrocardiogram (ECG) changes and symptoms for ischemia should be evaluated. Most stroke patients develop mental depression during the poststroke period (39), so a psychological referral may be necessary.

Diagnostic Testing

Ultrasound, magnetic resonance imaging (MRI), and angiography are the main diagnostic tests used to assess impending occlusions that could cause an ischemic stroke. The major diagnostic tool for determining hemorrhagic stroke is noncontrast computerized tomography (CT) (34). CT can also be used to diagnose ischemic stroke, although the sensitivity of this technique varies across research studies (34). Recently, diffusion-weighted MRI has been shown to be more effective than CT for diagnosing ischemic stroke (34), making this a promising diagnostic tool.

Exercise Testing

Contraindications for exercise testing are the same as those for all patient populations. Because hypertension is the major risk factor for hemorrhagic stroke, particular attention should be directed toward ensuring that preexercise resting blood pressure is below the contraindicative values, systolic pressure of 200 mmHg and diastolic pressure of 110 mmHg. Furthermore, because ischemic stroke is highly associated with CAD, screening should ensure that symptoms of CAD such as unstable angina are not present.

Considerations for exercise testing in stroke patients are summarized in table 28.2. Not all stroke patients are capable of traditional treadmill graded exercise tests to determine functional capacity. Some researchers have even suggested that submaximal measurements such as lactate or ventilatory threshold or oxygen pulse ($\dot{V}O_2$/HR) should be the criterion measures of cardiorespiratory fitness in stroke patients because these markers are easier to obtain and put the patients at a lower risk (19). Recent research has shown, however, that if patients achieve a self-selected walking speed of at least 0.5 mph (0.8 kph) during 30 ft (9 m) of floor walking, they are capable of performing a graded treadmill exercise test with handrail support as needed (24). Out of 30 patients with hemiparetic stroke, 29 were able to meet this criterion and achieve an average of 84% of age-predicted maximal heart rate. This intensity was sufficient to allow detection of asymptomatic myocardial ischemia in 29% of patients without previously determined CAD. The treadmill test was at a self-selected walking velocity with grade increases of 2% every 2 min (24).

If a patient is unable to perform treadmill or cycle ergometry work, protocols are available that do not require a large amount of leg muscle mass. Tsuji and colleagues (37) developed an incremental protocol that uses an increasing rate of "bridging," which involves elevating the pelvis to a point of maximal hip extension. The protocol uses 4 min stages and increases the rate of bridges per minute from a starting point of three to six, with six bridge per minute increases thereafter up to 24. About 89% of patients were able to complete this protocol, and the test–retest intraclass coefficients were over 0.90 for HR, over 0.70 for $\dot{V}O_2$, and 0.98 for the oxygen pulse. However, the peak oxygen consumption does not appear to be higher than 2 metabolic equivalents (METs) for this protocol (37), suggesting that it should be used only for extremely deconditioned patients. Although 6 min walk distance is commonly used to assess cardiorespiratory fitness in several clinical populations, it has been observed (27) that $\dot{V}O_2$peak obtained during cycle

Table 28.2 Exercise Testing Considerations for Stroke Patients

Test type	Mode	Protocol specifics	Clinical measures	Clinical implications	Special considerations
Cardiovascular	Treadmill, cycle ergometry	Self-selected speed, 2% increase in grade every 2 min; begin at 20 W and increase 10 W/min	ECG, $\dot{V}O_2 \cdot HR^{-1}$, ventilation	Myocardial ischemia can be detected	Patients unable to achieve at least 0.5 mph (0.8 kph) during floor walking can use cycle ergometry, arm ergometry, basic activities protocol, or bridging protocol.
Strength	Free weights, exercise machines, dynamometry	10RM testing, handgrip	NA	NA	Assessment should be made before testing on functional ability to perform motion on paretic side.
Range of motion	Goniometer, sit-and-reach	Range of motion in affected joints	NA	NA	NA

Based on T. Thom et al., 2006, "Heart disease and stroke statistics—2006 update: a report from the American Heart Association Statistics Committee and Stroke Statistics Subcommittee," Circulation 113: e85–151.

ergometry and 6 min walk distance are not associated in stroke patients. Rather, 6 min walk performance is primarily dictated by balance, knee extensor strength, and degree of muscle spasticity (27). Therefore, the 6 min walk may be a good addition to the battery of tests in that it is a functional outcome of these impairments, but it should not be viewed as a measure of cardiorespiratory fitness per se.

Treatment

For advanced atherosclerosis (70-99% occlusion) in the carotid arteries, carotid endarterectomy is a common surgery (35). In this procedure, most of the plaque is physically removed from the artery wall by the surgeon. However, it has recently been observed that stenting with an embolic protection device is as effective in both the short and long term (11).

Pharmacological treatment depends on the type of stroke. As shown in table 28.3, this may include anticoagulants, antiplatelet medications, and antihypertensives; the latter are especially important for patients with hemorrhagic stroke. Antihypertensives such as labetolol, enalapril, and clonidine are the most common drugs for subarachnoid hemorrhage (35). Patients with ischemic stroke are often on medication, which could potentially include statins and diabetes medications, to control associated risk factors. Nitroglycerin or other vasodilators may be prescribed following an ischemic, hemodynamic stroke to lessen the chance of vasospasm. Patients with subarachnoid hemorrhage are typically given nimodipine to reduce vasospasm every 4 h until symptoms subside (35). It has recently been observed that amphetamines enhance recovery of motor control during therapy in stroke patients, making this a potential future area of pharmacological intervention (28).

Standard rehabilitation for stroke patients usually includes physical therapy. The primary aim of this treatment is to restore balance, movement, and coordination. Basic strengthening exercises, passive movements to increase range of motion, assisted and unassisted walking, and functional movements such as chair stands and transferring from bed to chair are commonly prescribed. Because of the hemiparesis, patients must often relearn daily activities such as dressing and bathing, particularly if they are often able to perform these tasks with one hand. Therapy will also include instruction and practice using assistive walking devices such as walkers and canes. Speech therapy is commonly necessary because of the effects of the stroke on speech control. Nutritional consultations with a registered dietitian may be necessary if the patient is overweight or obese. Additionally, because depression is common in patients who have had a stroke, psychological referrals may be employed.

Practical application 28.1 provides guidelines for interacting with stroke victims.

Table 28.3 Common Medications for Stroke Patients and Their Respective Effects

Medications	Class and primary effects	Exercise effects	Other effects	Special considerations
Warfarin	Anticoagulant	None		Patients should avoid activities with high risk for trauma or injury.
Ticlopidine, clopidogrel, aspirin	Antiplatelet	None		
Ramipril, enalapril	Angiogensin-converting enzyme inhibitor	None	Decrease in resting and exercise BP	
Nimodipine	Calcium channel blocker	Increased exercise capacity in patients with angina	Possible increase in resting and exercise HR; decrease in resting and exercise BP	
Hydrochlorothiazide	Diuretic	None	Decrease in resting and exercise BP	

Based on T. Thom et al., 2006, "Heart disease and stroke statistics—2006 update: a report from the American Heart Association Statistics Committee and Stroke Statistics Subcommittee," Circulation 113: e85–151.

Practical Application 28.1

CLIENT–CLINICIAN INTERACTION

Dealing with stroke patients presents some unique challenges that one can best address by emphasizing good communication with the patient and associated family members or caregivers. Previously mentioned consequences include depression, inappropriate behavior, or slow or cautious behavior. The clinician should be sensitive to these considerations, establish appropriate boundaries for interaction with the patient, and, if necessary, make appropriate referrals to mental health professionals. Communication with family members or caregivers can help determine whether depression or other psychological conditions may be a concern. Furthermore, because many of the desired outcomes of rehabilitation are improved activities of daily living, some of the outcome data may be qualitative information received from family members or caregivers about the level of independence that the patient exhibits when completing daily tasks.

EXERCISE PRESCRIPTION

The average functional capacity of a stroke patient is approximately 14.4 $ml \cdot kg^{-1} \cdot min^{-1}$ (23), which is of great concern because 20 $ml \cdot kg^{-1} \cdot min^{-1}$ has been suggested as the minimum necessary for independent living (8). Therefore, a major goal of stroke rehabilitation should be to increase functional capacity. Standard therapy, however, does not provide enough of an aerobic stimulus to achieve an increase in cardiorespiratory fitness (22). In support of this, only 12% of patients receiving physical therapy and 2% of patients receiving occupational therapy received therapy directed toward improving this parameter (8). Therefore, supplemental aerobic exercise should be prescribed for stroke patients. The clinical exercise physiologist should be aggressive about implementing this therapy in the patient's rehabilitation regimen. Patient education is important, because many patients may believe that their standard physical therapy provides a sufficient amount of exercise.

Besides the decrease in functional capacity, profound decreases occur in muscular strength and endurance, and these should be addressed. A particular challenge in terms of exercise prescription for a stroke patient who may have led a sedentary lifestyle is the perceived volume of work per week necessary to engender appropriate changes. One suggested approach is that of Rimmer and colleagues (31), who observed significant improvements in body fat percentage, 10RM (repetition maximum) strength, $\dot{V}O_2$peak, and hamstring and lower back flexibility in predominantly African American patients poststroke. The program consisted of 3 d per week for 60 min each session (30 min of cardiovascular training using a variety of modalities, 20 min of strength training on commercial strength training machines, and 10 min of flexibility training). Therefore, impressive gains can be realized with a total of 3 h per week of training, a time commitment that should not place an excessive burden on patients. Information on all aspects of exercise prescription is presented in practical application 28.2.

Cardiovascular Exercise

Most studies that have shown improvements in patients with stroke employed a frequency, intensity, and duration similar to those included in the American College of Sports Medicine (ACSM) guidelines for healthy populations. Modes of training are listed in practical application 28.2 and can include walking (on ground or on a treadmill), water exercise, and cycle ergometry. A mixture of these modalities is warranted because some have specific supplementary benefits. Employing some weight-bearing exercise is highly recommended, because these activities have been shown to maintain bone mineral density in patients poststroke (28). Treadmill exercise not only increases cardiorespiratory endurance but also may aid in improving balance, coordination, and gait abnormalities (17, 29). Water-based exercise training has been shown to increase $\dot{V}O_2$peak by approximately 22% and to cause concomitant increases in isokinetic strength (5). Water exercises can also be a relatively safe way for these patients to train because risk of falls is avoided. Cycle ergometry training results improved in stair-climbing ability (14), suggesting that this form of training can lead to nonspecific improvements in performance related to ambulation. If either the patient or the clinician is concerned about falls on the treadmill, supported treadmill exercise, as discussed in chapter 25, can be employed (9). By using supported

Practical Application 28.2

EXERCISE PRESCRIPTION SUMMARY

Training method	Mode	Intensity	Frequency	Duration	Progression	Goals	Special considerations and comments
Aerobic	Floor and treadmill walking, cycle ergometry, Nu-Step, water exercise	40% to 80% HRR	3 to 5 d/wk	15 to 30 min	Progress from low to high intensity and to longer durations. Because of biomechanical limitations, heart rate recommendations should be superseded by perceived effort.	To increase functional capacity above 20 $ml \cdot kg^{-1} \cdot min^{-1}$	
Resistance	Elastic resistance bands, body weight exercises, sandbags, active motion, water exercise, commercial strength training equipment	As tolerated, up to 80% of 1RM	3 to 5 d/wk	30 to 45 min	As tolerated	To improve gait-related parameters and activities of daily living	
Range of motion	Passive movement, proprioceptive neuromuscular facilitation (PNF)	Below point of discomfort	3 to 5 d/wk	10 to 20 min	As tolerated		Emphasis on stretching muscles on the paretic side, particularly in muscle groups experiencing spasticity

Based on T. Thom et al., 2006, "Heart disease and stroke statistics—2006 update: a report from the American Heart Association Statistics Committee and Stroke Statistics Subcommittee," Circulation 113: e85–151.

treadmill exercise, patients can train at the highest velocity possible without stumbling and with minimal risk. Stroke patients who perform 10 s intervals at these higher speeds experience changes in 10 m walking speed, stride length, and functional ambulation that are superior to those produced with standard treadmill training (29).

Resistance Training and Flexibility

Because of the neuromuscular compromise caused by stroke, resistance training is an important part of the rehabilitative exercise prescription. ACSM recommendations for strength training could be easily followed with possible modification in the mode of training. Because of lowered muscular strength and endurance in these patients, strength training exercise can consist of low-resistance modalities such as elastic bands, body weight exercises, and sandbags. For patients who are extremely weak, exercises against gravity (e.g., shoulder and leg abductions), either with or without assistance, may be necessary until they are able to use any of the other recommended modalities. Rimmer and colleagues (31) successfully employed commercial resistance training equipment in poststroke patients, suggesting that these devices can be used provided that the resistance can be decreased enough to accommodate the specific demands in this population. Whatever mode is selected, both concentric and eccentric movements should be included, because concentric-only training does not increase isokinetic strength in stroke patients (10). A portion of the exercise prescription should focus on exercises designed to improve activities of daily living. Examples include chair stands, stair climbing, ball kicking, balance beam step-ups, and walking through obstacles. These activities require combinations of strength, endurance, and balance. These patients can perform traditional flexibility exercises as tolerated. A raised platform with a stretching mat can be useful because it greatly assists with lying down and returning to a standing position following the exercises. The focus of the flexibility program should be on the paretic limbs, particularly muscle groups in which the patient is experiencing a large degree of muscle spasticity.

EXERCISE TRAINING

Functional capacity can be dramatically improved in stroke patients. The improved $\dot{V}O_2$peak, combined with the lowered oxygen cost of ambulation because of improved economy, results in a profound decrease in the relative intensity of normal ambulation. As an example, Macko and coworkers (25) observed that 6 mo of treadmill aerobic exercise resulted in a 20% reduction in the metabolic cost of ambulating at a constant, submaximal intensity because of both increased $\dot{V}O_2$peak and decreased $\dot{V}O_2$ during submaximal walking. This study also suggested that the improvements in exercise economy are maximized within a 3 mo period, while $\dot{V}O_2$peak continues to improve throughout the course of training (25). More recent data further suggest that these benefits are accompanied by increased brain activity in the cerebellum and midbrain that correlates with improvements in walking velocity. Exercise rehabilitation also improves functional ambulation, as evidenced by consistent increases in 6 min walk distances found throughout the literature (3, 7, 9, 33). As stated earlier, 6 min walk performance is largely influenced by balance, strength, and spasticity of the lower paretic limb in stroke patients (27, 28). Therefore, the degree of improvement in the 6 min walk is likely due to improvement in these factors either in place of or in conjunction with improved cardiorespiratory endurance. Table 28.4 presents an overview of the anticipated physiologic adaptations to exercise training in these patients.

Table 28.4 Adaptations to Exercise Rehabilitation in Stroke Patients

Cardiorespiratory endurance	Skeletal muscle strength	Skeletal muscle endurance	Flexibility	Body composition
• Increased functional capacity • Improved walking economy • Increased oxygen pulse • Decreased submaximal blood pressure • Improved 6 min walk time • Increased voluntary physical activity	• Increased isokinetic strength • Improved upper extremity function	• Improved Timed Get Up and Go • Increased number of chair stands • Improved upper extremity function • Increased number of steps in 15 s	• Improved range of motion • Decreased spasticity in muscles on the paretic side	• Minimal decrease in body fat percentage • Large decreases in overweight or obese patients who receive nutritional counseling in addition to exercise

Based on T. Thom et al., 2006, "Heart disease and stroke statistics—2006 update: a report from the American Heart Association Statistics Committee and Stroke Statistics Subcommittee," Circulation 113: e85–151.

Because of the aforementioned relationship between functional capacity and the ability to live independently (6), these adaptations are extremely meaningful to the quality of life of stroke patients. A direct relationship also exists between improvements in sensorimotor function because of exercise training and the magnitude of improvement in $\dot{V}O_2$peak, which again points to the importance of improving functional capacity. Because blood pressure is an important risk factor for both ischemic and hemorrhagic strokes, another important clinical adaptation is a decrease in blood pressure in response to acute exercise (30).

Because most studies provide a holistic intervention that includes aerobic and strength training, discerning the exact contribution that resistance training makes to the outcome variables is difficult. Nonetheless, rehabilitation programs that incorporate resistance training typically yield impressive changes in both laboratory and functional evaluations. Isokinetic quadriceps and ankle plantar flexion strength can increase in the paretic limb without increases in muscle spasticity (36). Similar to the aerobic adaptations, strength improvements appear to increase performance in functional activities. Short-distance (5, 10, and 30 m) walking speed (7, 33, 36), Timed Get Up and Go (7), number of steps performed in 15 s (7), and force production in the paretic limb when going from a seated to standing position (7) all increase with programs that incorporate resistance exercises, especially those designed to increase functional performance. Combining aerobic and resistance training can drastically improve quality of life and activities of daily living (36). The latter is an especially important finding given that physical activity is associated with risk for stroke (18).

Clinical Exercise Pearls

- Physical therapy alone is not sufficient to increase functional capacity in stroke patients.
- The majority of hemiparetic stroke patients can participate in treadmill exercise.
- The inclusion of strength training, aquatic exercises, or both can be important in improving functional capacity.

CONCLUSION

Stroke remains a significant cause of death and disability. Exercise testing and training are important in both the prevention and treatment of patients after stroke. The clinical exercise physiologist will likely become an increasingly demanded allied health professional in the care of patients at risk for stroke or after stroke. This is especially true for poststroke patients with only minimal related loss of motor function.

Practical Application 28.3

RESEARCH FOCUS

Although it is well known that stroke compromises brain activation patterns, it was not established until recently whether functional improvements following exercise training are accompanied by improvements in brain activation patterns in stroke patients. Luft and colleagues (21) performed a randomized, controlled trial on 71 stroke patients with hemiparesis. Thirty-seven of the patients completed a 6 mo treadmill training program; the remaining 34 patients served as a control group and performed 6 mo of a stretching program. The treadmill exercise group increased their maximal treadmill walking speed during a graded exercise test by 51%, their $\dot{V}O_2$peak by 19%, and their walking velocity during a 6 min walk test by 19%. These findings confirmed that exercise training improves both physiological and functional parameters in these patients. However, the investigators further observed that brain activation was altered as an adaptation to exercise. Functional MRI was performed on all subjects to assess brain activation during movement of both the paretic and nonparetic legs. Treadmill training increased brain activation in the posterior cerebellum and midbrain during movement of the paretic leg, while no such changes were observed in the control group. The increase in activation of the posterior cerebellum was correlated ($R = 0.54$) with the increase in maximal walking speed during the treadmill test, suggesting that ambulatory improvements in stroke patients may be at least partially mediated by improvements in brain activation.

Key Terms

embolic strokes (p. 561)
hemorrhagic strokes (p. 561)
ischemic strokes (p. 561)
thrombotic strokes (p. 561)

CASE STUDY

MEDICAL HISTORY

The patient is a 69 yr old male with a history of obesity, type 2 diabetes, and hypertension. He is 3 mo poststroke and has gone through basic physical therapy. He has right-side hemiplegia, speech impediment, and depression. His resting BP is 150/90.

Medications: Ticlopidine, hydrochlorothiazide, and glucophage.

DIAGNOSIS

The patient reported to his physician with numbness, dizziness, motor function impairment on the right side of his body, and vision problems. The patient was referred, and stroke was confirmed and diagnosed by MRI.

EXERCISE TEST RESULTS

The patient achieved 6 METs on a progressive cycle ergometer test to exhaustion with no significant ECG changes. Peak HR was 140, and peak BP was 210/100.

EXERCISE PRESCRIPTION

The patient was prescribed aerobic exercise 3 d per week, progressing up to 5 d per week, and resistance exercises 3 d per week. The initial intensity for aerobic exercise was set at 40% to 50% heart rate reserve and a 12 to 15 rating of perceived exertion on the Borg scale, progressing every 3 wk as tolerated. The resistance exercises were to be performed as tolerated for 10 to 15 repetitions.

Initial duration for the aerobic exercise was 15 min, progressing every 3 wk as tolerated up to 1 h per session. Resistance exercises were to start with one set of exercises for 3 wk, progressing up to three sets. The initial mode of training was a mix of water-based aerobic exercises and supported treadmill exercises. Support in the latter is preferably in the form of handrails, but a support device can be used if it enhances patient comfort. These modes serve to increase cardiorespiratory fitness and other factors related to walking performance such as balance and muscular strength. As frequency and duration progress, cycle ergometry exercise can be added. The patient's ability to perform resistance exercises with elastic bands and available strength training equipment should be evaluated. Initial exercises may need to include several movements against gravity and assisted movements.

Because of comorbid conditions (diabetes, obesity, depression), the lower end of the ACSM-recommended range for intensity and duration should be used initially. But because caloric expenditure and improved insulin sensitivity are secondary goals, frequency should progress to the upper end of the ACSM-recommended range. A variety of modes should be employed to enhance patient interest, an important consideration because of depression. Tests for $\dot{V}O_2$peak should be repeated, and the testing battery should include at least two functional tests (e.g., 40 m walk speed, 6 min walk, Timed Get Up and Go, chair stands) so that the patient and the clinician can progress ambulation. To assess progress for body fat percentage and weight, an estimate of body fat percentage (skinfolds, girths, or impedance) can be used. Because this patient exhibits characteristics of metabolic syndrome, waist-to-hip ratio should be determined regularly.

DISCUSSION QUESTIONS

1. What are the main factors that determine the mode of training in this patient population?
2. What common associated conditions (e.g., obesity, diabetes, hypertension) may alter the exercise prescription for stroke patients?
3. What are the factors that determine the type of assessments performed?

PART VIII

Special Populations

Part VIII discusses various special populations that have also been found to benefit greatly from appropriate exercise therapy for the purpose of preventing health problems and improving quality of life. The clinical exercise physiologist needs to be knowledgeable in many areas related to these special populations in order to maximize exercise-related benefits and minimize any potential risks.

Chapter 29 focuses on children. Our children have a great opportunity to avoid chronic diseases if they start living a healthy lifestyle at an early age. But with the rise in obesity observed in our children and the relationship of obesity with other CVD risk factors, concern has increased about the ramifications that may occur over time. To have a favorable influence on our children's health as it pertains to chronic diseases, intervention programs need to begin at an early age. This is supported by the fact that early blood vessel changes are observed in children who have multiple CVD risk factors. The clinical exercise physiologist should be knowledgeable about developing safe and effective lifestyle modification programs that address major factors that are negatively influencing the health of our children.

Chapter 30 deals with the aging process with a particular focus on those over the age of 65. Since the publication of the 2nd edition of this text the 'baby boomer' generation, or those born between 1946 and 1964, has begun to reach the age of 65. Additionally, there are more and more people living over the age of 90. To meet the medical, social, and other demands of an aging society, we must address chronic health concerns. We must encourage and educate our elderly population to stay physically active so that they can maintain activities of daily living and live an independent lifestyle for as long as possible. When an elderly person plans to begin an exercise program, physician clearance is important because many older people will have some disease-related risk factor that calls for specific modifications of the exercise prescription. Medical clearance is a greater concern when an older person is interested in exercising above moderate intensity. The clinical exercise physiologist should be aware of health issues that affect the elderly and the process for developing and maintaining an exercise training process for this population.

Chapter 31 reviews depression. Depression is the most prevalent psychiatric disease or disorder in the United States. Patients with depression may require more attention by the clinical exercise physiologist than do patients not suffering from depression. Having the knowledge to recognize depression-related symptoms is important for the exercise physiologist who often has the opportunity to interact with these patients more than other health care providers. Depression is not uncommon among patients having recently survived or experienced a major medical event and for these patients, and others suffering from chronic depression, exercise has been shown to be an effective adjunctive or alternate treatment to medication.

Chapter 32 is a new chapter that addresses individuals with an intellectual disability (e.g. Down's syndrome) that affects intellectual functioning and adaptive behavior. These individuals are generally inactive, poorly fit, and often experience associated risk factors such as obesity, which itself increases the risk for diabetes and vascular diseases. Other co-existing health problems are common as well, including genetic heart abnormalities. This chapter will address this special population relative to both the scope and pathophysiology of the disorder, as well as the important and emerging role for exercise testing and exercise training.

Chapter 29

Children

Timothy J. Michael, PhD
William A. Saltarelli, PhD

Registered clinical exercise physiologists and other clinical practitioners who evaluate, diagnose, and treat individuals with disease or disorders will more than likely see children in their practices. As seen in adult populations, exercise can be used as a diagnostic tool, as an evaluative tool, and as a treatment for clinical conditions. However, we do need to remind ourselves that children are different than adults in a variety of ways, physically, psychologically, and emotionally. Children's responses to exercise may be different than those of adults, and their adaptations to exercise are predicated, to a certain degree, on growth and maturation. Their maturity also influences their ability to take instruction and to give full efforts when needed. Physical size and fitness may influence test modality and protocol choices (i.e., one size does not fit all).

DEFINITION

Pediatrics is the branch of medicine concerned with children and their diseases. **Children** are the population between infancy and adolescence.

SCOPE

Although children do not, in and of themselves, constitute a clinical population, it is important for clinicians to know about the exercise responses of children and how they may influence exercise testing, exercise training, and exercise prescription. Children can present with many different clinically relevant signs and symptoms, diseases, and disorders. In recent years the number of children who are overweight or obese has increased dramatically. These children are now being diagnosed with chronic diseases that we often see in adult clinical populations, such as type 2 diabetes, hypertension, dyslipidemia, and metabolic syndrome. It is not unusual to see the many different clinical conditions in children that we would in adults, ranging from cardiovascular to pulmonary, musculoskeletal, and neurological conditions. It is therefore appropriate for clinicians to use exercise with children similarly to the way they do adult clinical populations, with an understanding of the underlying differences between these two groups.

This chapter does not discuss all clinical conditions that afflict children; readers can refer to the other chapters in this text for information on many of these. The aim of the chapter is to bring to light the differences that children exhibit in relation to adults when exercise is used for evaluation, diagnosis, and treatment. The chapter also addresses the adjustments that may be needed to accommodate children in the clinical exercise laboratory.

CLINICAL CONSIDERATIONS

Children may develop a variety of conditions that warrant an exercise test or an exercise program intervention. This section deals with some of the considerations that clinicians need to take into account when conducting exercise testing and prescribing an exercise program for a child.

Signs and Symptoms

Children may present with a variety of signs and symptoms that would warrant the use of exercise for evaluation, as an aid to diagnosis, or as a part of a treatment plan. The signs and symptoms may not be very different from what we see in adults: shortness of breath; chest pain; faintness, dizziness, syncope; unusual fatigue; and exercise intolerance. Paridon and colleagues for the American Heart Association (21) have listed the common reasons why a child should have a stress test.

Common Reasons for Pediatric Stress Testing

1. To evaluate specific signs that are induced or aggravated by exercise
2. To assess or identify abnormal responses to exercise in children with cardiac, pulmonary, or other organ disorders, including the presence of myocardial ischemia and arrhythmias
3. To assess efficacy of specific medical or surgical treatments
4. To assess functional capacity for recreational, athletic, and vocational activities
5. To evaluate prognosis, including both baseline and serial testing measurements
6. To establish baseline data for institution of cardiac, pulmonary, or musculoskeletal rehabilitation

From Paridon et al. 2006.

History and Physical Examination

The first step, as with adults, is to obtain a complete medical history and physical exam, along with any pertinent laboratory findings (2). The information gleaned from this will help the clinician evaluate the child for risk of exercise through stratification for level of risk and determine contraindications to exercise testing and exercise participation. The goal is to make sure that, in the case of exercise testing, the benefit of the testing outweighs information to be gained by having the child attempt to complete the exercise test (table 29.1) (21).

Children have many conditions that may elicit a referral for exercise testing, including, but not limited to, asthma, cystic fibrosis, diabetes, obesity, and various forms of heart disease (2, 12, 20, 27, 32).

Exercise Testing

Helge Hebestreit (11) lists three reasons that clinical exercise testing of children may be difficult:

- Children have small body size in relation to testing equipment.
- They show poor peak performance.
- They have a short attention span and usually poor motivation during exercise testing; this is especially true with longer exercise protocols.

Furthermore, Hebestreit says, very young children and those with chronic conditions are much more of a challenge than others when one is attempting to have them complete a clinical exercise test.

The American College of Sports Medicine (ACSM) (2) offers the following guidelines with regard to exercise testing for children:

- Exercise testing for children is not indicated unless there is a health concern.
- The testing protocol should be based on the reason the test was requested and the capabilities of the child.
- Children should always be familiarized with the test modality and protocol before the actual test; this will help to reduce test anxiety as well as increase the chances for success of the exercise test.
- A treadmill and a cycle ergometer should each be available for exercise testing.
- Children are psychologically and emotionally immature and may require more motivation and support than older people to complete the exercise test.

General Considerations for Conducting Exercise Tests With Children

This section considers issues relevant to exercise testing specifically in children. Adjustments often need to be made to fit the equipment to the given child as well as to accommodate the child's condition, maturity (physical and emotional), and fitness.

Modalities

The mode of exercise testing for children in a clinical setting is most often either a treadmill or a cycle ergometer. These modes are used to assess functional capacity and to determine exercise prescriptions for aerobic-based exercise plans. Treadmill testing tends to elicit greater oxygen consumption than does cycle ergometer testing. It is particularly important to have an ergometer that can be adjusted appropriately to the child's size (2, 12, 20, 27, 32).

Although the treadmill and the cycle ergometer are the most commonly used modalities for clinical exercise testing of children, other laboratory-based tests can be used

Table 29.1 Relative Risks for Stress Testing

Lower risk	Higher risk
Symptoms during exercise in an otherwise healthy child who has a normal CVS exam and ECG	Patients with pulmonary hypertension
Exercise-induced bronchospasm studies in the absence of severe resting airways obstruction	Patients with documented long-QTc syndrome
Asymptomatic patients undergoing evaluation for possible long-QTc syndrome	Patients with dilated/restrictive cardiomyopathy with CHF or arrhythmia
Asymptomatic ventricular ectopy in patients with structurally normal hearts	Patients with a history of hemodynamically unstable arrhythmia
Patients with unrepaired or residual congenital cardiac lesions who are asymptomatic at rest, including these: Left to right shunts (ASD, VSD, PDA, PAPVR) Obstructive right heart lesions without severe resting obstruction (TS, PS, ToF) Obstructive left heart lesions without severe resting obstruction (cor triatriatum, MS, AS, CoA) Regurgitant lesions regardless of severity	Patients with hypertrophic cardiomyopathy with the following: Symptoms Greater than mild LVOTO Documented arrhythmia
Routine follow-up of asymptomatic patients at risk for myocardial ischemia, those with the following: Kawasaki disease without giant aneurysms or known coronary stenosis After repair of anomalous LCA After arterial switch procedure	Patients with greater than moderate airway obstruction on baseline pulmonary function tests
Routine monitoring in cardiac transplant patients not currently experiencing rejection	Patients with Marfan syndrome and activity-related chest pain in whom a noncardiac cause of chest pain is suspected
Patients with palliated cardiac lesions without uncompensated CHF, arrhythmia, or extreme cyanosis	Patients suspected to have myocardial ischemia with exertion
Patients with a history of hemodynamically stable SVT	Routine testing of patients with Marfan syndrome
Patients with stable dilated cardiomyopathy without uncompensated CHF or documented arrhythmia	Unexpected syncope with exercise

CVS = cardiovascular system; ASD = atrial septal defect; VSD = ventricular septal defect; PDA = patent ductus arteriosus; PAPVR = partial anomalous pulmonary venous return; TS = tricuspid stenosis; PS = pulmonary stenosis; ToF = tetralogy of Fallot; MS = mitral stenosis; AS = aortic stenosis; CoA = coarctation of aorta; LCA = left coronary artery; SVT = supraventricular tachycardia; CHF = congestive heart failure; LVOTO = left ventricular outflow tract obstruction.

Reprinted, by permission, from S.M. Paridon et. al., 2006, "Clinical stress testing in the pediatric age group: A statement from the American Heart Association Council on Cardiovascular Disease in the Young, Committee on atherosclerosis, hypertension, and obesity in youth," *Circulation* 113: 1905-1920.

to assess factors other than aerobic capacity. Laboratory tests can assess muscular strength and anaerobic power and capacity, as well as flexibility. Again, it is essential that whatever modality is used, it must be appropriately adapted to the child.

In 2006, Chang and coworkers reviewed the use of the treadmill versus the cycle ergometer in clinical pediatric exercise testing. As shown in figure 29.1, approximately 24% of pediatric cardiology and pulmonology centers use cycle ergometers and 76% prefer to use the treadmills (6).

Protocols

Protocols for adults can usually be used for children. Occasionally adjustments will be needed to accommodate the earlier onset of fatigue in children (i.e., stage time may be shortened), or the large increases in metabolic equivalent (MET) values from stage to stage may be lessened to accommodate children in cases in which large jumps in METs could be discouraging and affect motivation to complete the test. Stephens and Paridon reported that one should determine the protocol based on

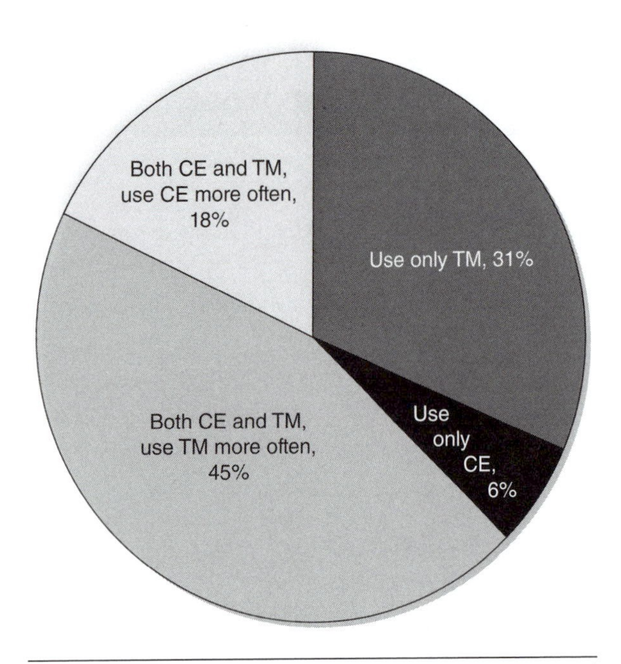

Figure 29.1 Percentage of respondents who reported using cycle ergometer (CE), treadmill (TM), or both.

Springer and *Pediatric Cardiology*, 2006, page 111, Current practice of exercise stress testing among pediatric cardiology and pulmonology centers in the United States," R.K.R. Chang, figure 1, with kind permission from Springer Science+Business Media B.V.

the kind of information desired by the person requesting the test (2, 12, 20, 27, 32). Table 29.2 reviews principles of exercise testing in pediatric populations.

Chang and colleagues, as part of the study mentioned previously, also reviewed the use of stress echocardiography (6). Figure 29.2 shows the results. Kimball (13) encourages the use of stress echocardiography as a means of extending physical examinations for children.

If general health or fitness assessment is all that is needed, use of the American Alliance for Health, Physical Education, Recreation and Dance FITNESSGRAM is suggested. Most of the assessments in this battery are field-based tests but should be used from only a functional, not a diagnostic, point of view. In addition, participation in these field-type tests should be limited to children who are free of contraindications to exercise participation or testing as listed in *ACSM's Guidelines for Exercise Testing and Prescription, Eighth Edition* (2).

Contraindications to Exercise Testing

The safety of the child is always the first priority. As with adults, strict compliance with the contraindications to exercise testing in *ACSM's Guidelines for Exercise Testing and Prescription, Ninth Edition,* is warranted (2, 12, 20, 27, 32).

In addition, the criteria for testing termination specified in the ACSM guidelines should be strictly adhered to. Paridon and colleagues for the American Heart Association (21) list three general reasons for terminating a clinical exercise test in children:

- In cases in which diagnostic findings have been established—continuing the test would not produce any further information
- Failure of monitoring equipment
- Appearance of signs or symptoms indicating that continuation of the test would put the patient at potential risk for an adverse event

Practical application 29.1 provides special consideration for the exercise physiologist when dealing with children.

Table 29.2 Exercise Testing

Test	Mode	Protocol	Variables	Special considerations
Aerobic (cardiopulmonary)	Treadmill (TM) Cycle ergometer (CE)	Bruce (TM) Balke (TM) James (CE) Godfrey(CE)	HR, RPE, BP, signs, symptoms, ECG, SaO_2, $\dot{V}O_2$, anaerobic threshold	Arrhythmia evaluation Ischemic evaluation Echocardiography
Anaerobic	Cycle ergometer	Wingate (CE)	Peak power output, mean power output, fatigue index	Exercise–induced bronchospasm
Strength	Hand grip dynamometer Isokinetic dynamometer		Force (kg) Torque (kg)	
Flexibility (ROM)	Inclinometer Flexometer Goniometer		Degree (ROM)	

Reprinted from *Pediatric Clinics of North America*, Vol. 51, P. Stephens Jr. and S.M. Paridon, "Exercise testing in pediatrics," pgs. 1569-1587, copyright 2004, with permission from Elsevier.

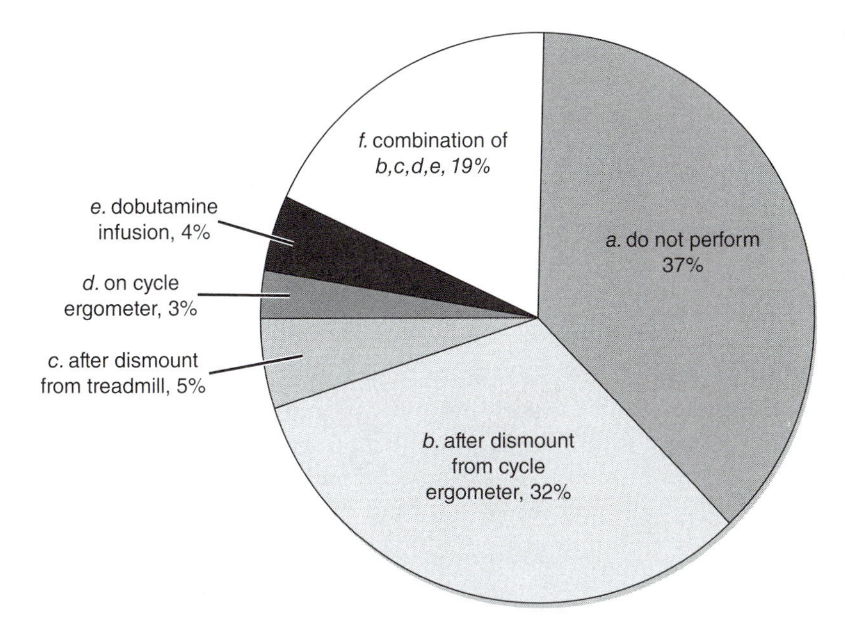

Figure 29.2 Responses of cardiology programs regarding performance of stress echocardiography.

Springer and *Pediatric Cardiology*, 2006, page 111, Current practice of exercise stress testing among pediatric cardiology and pulmonology centers in the United States," R.K.R. Chang, figure 1, with kind permission from Springer Science+Business Media B.V.

Practical Application 29.1

CLIENT–CLINICIAN INTERACTION

When referring to responses and adaptations to exercise, the pediatric exercise physiologist Oded Bar-Or (4) often stated in his presentations that "children are not small adults." This is important to keep in mind when one is evaluating the exercise performance of children. Following is a brief review of special considerations that one must take into account when testing children.

Laboratory Environment

- The lab must be safe and well illuminated.
- Staff should be trained and have warm, friendly personalities that will help them establish a positive relationship with children as soon as they enter the facility.

Particular Safety Issues

- Two testers are essential to ensure constant visual and verbal contact with the child.
- During treadmill testing, two spotters should be used—one at the subject's side and one behind.

Pretest Protocols

- Establish a relationship with the child as he enters the facility.
- Completely explain the test to both parents and children. Be sure that both know exactly what will take place and what the child will be asked to do.
- Following the explanation of the procedure, obtain parental consent and complete the child's assent documents. A child's assent document (for children age 6 and older) should be written in age-appropriate language and be short and to the point.
- Although offering an incentive to the child is controversial, rewards can be extremely helpful in eliciting a maximal effort.

Laboratory Equipment

- When appropriate, testing equipment should be modified for the size and maturity of the subject.

EXERCISE PRESCRIPTION

This section reviews the exercise prescription guidelines relevant to children, specifically those for aerobic exercise, resistance exercise, and flexibility training, and concludes with a review of the response to exercise training. Special attention is given to the differences between children's and adults' responses to exercise.

Aerobic

A review of the literature revealed very little in the way of specific details for aerobic exercise prescriptions. General guidelines for aerobic training are found in a variety of publications and are summarized in the following list and in table 29.3. Note that the guidelines in the table are divided into three objectives: basic health-related fitness, intermediate health-related fitness, and athletic performance. This approach allows for differences in abilities, interests, and fitness objectives of children (2, 19, 29).

- The type, intensity, and duration of exercise activities need to be based on the maturity of the child, medical status, and previous experiences with exercise.
- Regardless of age, the exercise intensity should start low and progress gradually.
- Because of the difficulty in monitoring heart rates with children, modified Borg or OMNI scales are practical methods of monitoring exercise intensity in children.
- Children are involved in a variety of activities throughout the day. A specific time should be dedicated to sustained aerobic activities.
- Adolescents should be physically active daily, or nearly every day, as part of play, games, sports, work, transportation, recreation, physical education, or planned exercise, in the context of family, school, and community activities.
- The intensity or duration of the activities is probably less important than the fact that energy is expended and a habit of daily activity is established.
- Adolescents should engage in three or more sessions per week of activities that last 20 min or more and that require moderate to vigorous levels of exertion. Moderate to vigorous activities are those that require at least as much effort as brisk or fast walking.

Table 29.3 Aerobic Training Prescription for Children

Goals	Mode	Intensity	Frequency	Duration	Overload and progression	Special considerations
Basic health-related fitness	Walking/jogging Dancing Recreational biking and swimming Active games	1. HR 50% to 60% HRmax 2. Moderate (beginning to sweat) 3. RPE Borg 4 to 5 or OMNI 2 to 5	Three times per week	30 min	Get children moving. It is not necessary to overload. Use minimal progression.	Make activity fun and part of an active lifestyle.
Intermediate-health-related fitness	Jogging, running, biking Fitness-based games Intramural and local league sports	1. HR 60% to 75% HRmax 2. Vigorous (breathing hard, sweating) 3. RPE Borg 6 to 9 or OMNI 6 to 9	Three to five times per week	40 to 60 min	Introduce training principles and small-progression overload by increasing tempo or decreasing rest periods.	Be aware of heat stress and overtraining.
Athletic performance	Training programs similar to those for adults (running, cycling, aerobics, school and community sports)	1. HR 65% to 90% HRmax 2. Vigorous (breathing hard, sweating) 3. RPE Borg 6 to 9 or OMNI 6 to 9	Five or six times per week	60 to 120 min	Use structured training programs stressing variable intensities and durations.	Be aware of heat stress and overtraining.

- Children and adolescents should do 60 min or more of physical activity each day.

An important omission in the ACSM guidelines is setting proper intensity. Rowland (25, 26) and others have suggested the use of adult guidelines to increase aerobic fitness, and studies have shown higher intensity to be effective.

Vigorous physical activities are rhythmic, repetitive activities that require the use of large muscle groups to elicit a heart response of 60% or more of a subject's maximum heart rate adjusted for age. An exercise heart rate of 60% of maximum heart rate for age is sufficient for cardiorespiratory conditioning. Alternative methods of calculating optimal heart rate in children are available. Most methods of calculating optimal aerobic training zones for children require an estimate of maximum heart rate. Adult formulas such as 220 minus age are usually used for children. As stated by Rowland, however, maximal heart rate determined by treadmill and cycle studies remains constant across the pediatric years (26). Therefore, adult formulas strongly dependent on age may not be appropriate. Direct maximal heart rate is best for determining exercise prescriptions. An alternative formula suggested by Tanaka and colleagues (33) is max HR = 208 – (.7 × age) in years. This formula is slightly less dependent on age. It was developed from adult data and requires further study in children (30).

Resistance Training

As recently as the late 1970s, pediatric exercise experts and medical doctors believed that prepubescent children could not benefit from strength or resistance training because this developmental group lacked the prerequisite circulatory hormones (34). Furthermore, many believed that the stress imposed by this type of training was not safe and could injure bones, especially at the growth plates. Since that time, numerous controlled studies have provided compelling evidence that strength or resistance training produces strength gains in both prepubescent girls and boys (22, 31, 35).

As with adults, the effectiveness of training appears to depend on intensity, volume, and duration. But these specific factors have not been established with certainty in children. An excellent reference is a document from a conference sponsored by the American Orthopedic Society for Sports Medicine (5), which concludes that strength training for the prepubescent

- improves muscular strength and endurance,
- improves motor skills,
- protects against injury (sports),
- has positive psychological benefits, and
- provides a forum for the introduction of safe and proper training.

Training prescriptions for children show great variability but seem generally to follow adult prescriptions with the exception of lower resistance and higher repetitions. Supervision to ensure proper technique is most important when one is prescribing resistance training to children (28). Table 29.4 and practical application 29.2 review training principles for children with respect to improving muscular strength and endurance for basic health and athletic performance.

Flexibility and Range of Motion

Flexibility is defined as the ability to move joints through a full range of motion. A widely accepted concept is that children are extremely flexible and that therefore flexibility should not be a priority in activities or training. Exceptions seem to be children in sports such as gymnastics and dancing, in which flexibility is required and its importance is appreciated. Nevertheless, flexibility training is recommended at all ages to ensure safe activity. Most research studies show a decline in flexibility as children become older. Clark (8) concluded that boys tend to lose flexibility after age 10 and girls after age 12. In fact, Milne and colleagues (18) found that flexibility in both boys and girls declined between kindergarten and 2nd grade. Girls as a group are usually more flexible than boys, which may reflect the activities in which girls participate. In addition, Youth Risk Behavior Surveillance System data showed that girls were more apt to engage in flexibility training than boys (7). Contraindications and specific medical concerns are absent unless the child has preexisting musculoskeletal problems. Table 29.5 summarizes range of motion training guidelines.

EXERCISE TRAINING

Pediatric exercise physiologists are continually challenged to explain basic exercise physiology of children, that is, their responses and adaptations to bouts of exercise. We have learned from Oded Bar-Or that with respect to these responses, children are not small adults. However, despite the desirability of fully understanding children's responses to exercise, exercise research has been hindered by many obstacles that lie in the

Table 29.4 Resistance Training Prescription for Children

Goals	Mode	Intensity	Frequency	Duration	Overload and progression	Special considerations
Basic health-related fitness	Use of body weight for major muscle groups (sit-ups, push-ups) Stretch bands	Very light, <40% of projected maximal effort	Two or three times per week	One or two sets of 6 to 12 reps	Overload and progression are not essential unless child has interest.	Main objective is introducing resistance training and correct movement patterns.
Intermediate-health-related fitness	Resistance exercise with machines such as leg press, curls, shoulder press Pull-ups	Light to moderate, 50% of projected maximal effort	Three or four times per week; alternating upper and lower body segments allows for consecutive days	One to three sets of 6 to 15 reps	Introduce one or two components of overload one or two times per week.	Be sure child wants to be involved and avoid maximal lifts
Athletic performance	Advanced sport-specific multijoint lifts (clean pull, power press, Olympic-style lifts)	Match specific load requirements with sport Max should be <70% projected maximal effort	Four or five times per week for training activities	Three to five sets of 5 to 20 reps specific to sport	Program design should stress variable intensity and duration to cause overload two or three times per week.	Be sure child wants to be involved and avoid maximal lifts

Practical Application 29.2

A SIMPLE METHOD OF ESTIMATING THE ONE-REPETITION MAXIMUM (1RM) FOR CHILDREN

1. The child warms up with 5 to 10 repetitions at 50% of the estimated 6RM.
2. After 1 min of rest and some stretching, the child performs six repetitions at 70% of the estimated 6RM.
3. The child repeats step 2 at 90% of the estimated 6RM.
4. After about 2 min of rest, depending on the effort needed to perform the 90% set, the child performs six repetitions with 100% or 105% of the estimated 6RM.
5. If the child successfully completes six repetitions in step 4, add 2.5% to 5% of the resistance used in step 4 and have the child attempt six repetitions after 2 min of rest. If the child does not complete six repetitions in step 4, subtract 2.5% to 5% of the resistance used in step 4 and have the child attempt six repetitions after 2 min of rest.
6. If the first part of step 5 is successful (the child lifts 2.5% to 5% more resistance than used in step 4), retest the child starting with a higher resistance after at least 24 h of rest, because fatigue will greatly affect performance of more sets. If the second part of step 5 is successful (the child lifts 2.5% to 5% less than the resistance used in step 4), this is the child's 6RM. If the second part of step 5 is not successful (the child does not lift 2.5% to 5% less than the resistance used in step 4), retest the child after at least 24 h of rest, starting with less resistance.

Adapted, by permission, from W.J. Kraemer and S.J. Fleck, 1993, *Strength training for young athletes* (Champaign, IL: Human Kinetics).

Table 29.5 Range of Motion Prescription for Children

Goals	Mode	Intensity	Frequency	Duration	Overload and progression	Special considerations
Basic health-related fitness	Static stretches for major muscle groups	To mild tension or slight muscular discomfort	Before and after each activity or exercise session or three times per week	Hold stretch 10 to 15 s, two times per stretch	Overload is not necessary. Start easy into stretch with minimal applied resistance.	Objective is to get children into the habit of stretching. Be sure child is active before beginning stretches.
Intermediate health-related fitness	Use static stretches and introduce dynamic stretches	To mild tension or slight muscular discomfort	Before and after each activity or exercise session or three times per week	Hold stretch 10 to 15 s, three times per stretch	Overload is not necessary. Start with easy multijoint dynamic movements, progress to more resistive.	Stretch major core muscles first and move to extremities. Introduce dynamic flexibility.
Athletic performance	Usually dynamic or ballistic; major muscle groups and sport-specific stretches	To mild tension or slight muscle discomfort at a level appropriate for sport participation	Before and after each activity or training session	Depending on static, dynamic, or ballistic (usually conducted by qualified trainer)	Overload is not necessary. Start with easy multijoint dynamic movements, progress to more resistive.	Progression could include moderate static or proprioceptive neuromuscular facilitation stretching.

way of adequately explaining these responses through experiments. For example, ethical issues prevent us from taking multiple muscle biopsies to determine adenosine triphosphate (ATP) use, and the fact that training studies are complicated by developmental changes has limited our knowledge. Often we simply have to rely on what we know about adults and hope it is similar in children. What science has told us about the differences between children and adults with respect to exercise responses and their training implications is summarized in table 29.6. This table also includes exercise prescription implications where appropriate.

Aerobic Training Response in Children

The plasticity of aerobic fitness in children is currently being debated and studied. Plasticity refers to the extent to which normal growth-related changes (improvements) in maximal aerobic capacity can be altered by changes in physical activity. To state this more succinctly, plasticity refers to whether increases in the level of physical activity or physical training improve aerobic fitness ($\dot{V}O_2$max). Plasticity, therefore, can be thought of as trainability. Numerous studies have established that improvement in $\dot{V}O_2$max after a period of aerobic training is less substantial in children than it is in adults. Rowland (25) reviewed 13 studies that attempted to correlate habitual activity in children with their level of aerobic capacity ($\dot{V}O_2$max). Only five of the studies concluded that a significant correlation existed between levels of physical activity and aerobic capacity. All these studies measured habitual physical activity of children and were not training studies.

A possible explanation for the lack of support for a significant training effect for habitual physical activity of children can be found in the literature. For example, studies assessing the intensity of physical activity in children have consistently shown that only a small percentage of children meet the guidelines that call for at least 20 min of sustained activity eliciting between 60% and 90% of maximal heart rate. Specifically, Armstrong and colleagues (3) found that only 13% of boys and 6.5% of girls attained heart rates over 160 beats · min^{-1} for a 20 min period during a 3 d assessment period.

If physical activity is not a major contributor to $\dot{V}O_2$max, then what about formal endurance training? Adult training using ACSM guidelines for intensity, duration, mode, and frequency (large muscle groups, rhythmic activities, 20 to 60 min, 3 to 5 d per week, equivalent to 65% to 90% maximum heart rate) usually produces an improvement between 5% and 35% in 12 wk (2). Mahon (16) reviewed three controlled training studies on children less than 8 yr of age. In these studies, the experimental group showed a 12.5% increase in

$\dot{V}O_2$max, whereas the control group increased only 7.5%. Studies with children 8 to 13 yr old showed an average 13.8% increase in $\dot{V}O_2$max, and the controls increased only 0.7%

For adolescents 13 yr of age or older, an increase of 6.8% in $\dot{V}O_2$max was found, and the control group exhibited no change in $\dot{V}O_2$max (17). These data show that children and adolescents can adapt to training by increasing $\dot{V}O_2$max, but at a much lower rate than adults. When considering $\dot{V}O_2$max changes in children, we must recognize that initial fitness and genetic endowment can also influence responses. In the previously mentioned training studies, the average frequency was 3 d per week, the average duration was 30 min, and the intensities were greater than 160 beats · min^{-1} and as high as 85% maximal heart rate (HRmax). The modes were continuous running, weight training, aerobics, and jumping rope. Clearly, children as young as 8 yr old can increase $\dot{V}O_2$max with training, but their increase will not be as great as that of adults.

Anaerobic Training Response in Children

Little information is available concerning exercise prescription and trainability of anaerobic systems in children using short-burst activities. Rowland (25) indicated that the major reason for this lack of information is that an accurate, noninvasive method of assessing anaerobic metabolism similar to $\dot{V}O_2$max in aerobic systems does not exist.

Consequently, short-burst activities, which are common in the habitual activity of children, are poorly understood. The limited available research indicates that children can increase their anaerobic power by training. For example, studies by Grodjinovsky and colleagues (10), Rotstein and colleagues (24), and Sargeant and colleagues (30) have shown that children improve anaerobic performance from 4% to 14% following interval-type training. In the 6 wk training study by Grodjinovsky and coworkers, one group rode a cycle ergometer for five 10 s all-out bouts followed by three 30 s all-out bouts. The other group performed three all-out 40 m runs followed by three all-out 150 m runs during each training session. Both groups trained 3 d per week. Improvement in anaerobic performance was about 4% in each group (10). Although this study shows that anaerobic performance can be improved by training, it also gives some information on duration, intensity, and frequency of training. Table 29.6 summarizes the physiological and training responses of children, highlighting the similarities and differences with adults.

CONCLUSION

The purpose of this chapter was to bring to light some of the considerations that clinicians need to take into account when working with children. Particularly important considerations have to do with exercise testing protocols, testing modalities, and exercise responses, all of which affect the exercise prescription for children in various clinical conditions. It is imperative for clinicians to keep in mind that children are not small adults and to treat children appropriately in order to maximize results from exercise testing and exercise programming.

Table 29.6 Physiological Responses of Exercising Children Compared to Adults, Their Training Responses, and Prescription Implications

Function	Exercise responses of children compared to adults	Training response and exercise prescription implications
CARDIORESPIRATORY		
$\dot{V}O_2$max (ml · kg^{-1} · min^{-1})	Similar	
$\dot{V}O_2$max (L · min^{-1})	Lower values due to smaller body size and mass	Training response is age dependent, increasing less in younger children. Children improve (1-16%, mean 9.7%) Training intensity may need to be greater in children to achieve improvements similar to those seen in adults.
Exercise HR	Similar to adult values, lower at all submax levels	As children train, HR intensity should be increased to account for lower response.

Function	Exercise responses of children compared to adults	Training response and exercise prescription implications
Max HR	Similar to adult values, that is, no change or slight decrease	
Resting HR	Higher	Training lowers resting HR as in adults. Exercise prescription using Karvonen formula and resting HR should be considered.
Submaximal oxygen demand (economy)	Cycling responses similar; walking and running less efficient with higher metabolic cost	Running and walking are less economic in children. Consider lower exercise intensity until efficiency improves.
Maximal cardiac output (Qmax) at a given $\dot{V}O_2$	Lower because of smaller cardiac dimensions	Training increases Q due to increased SV and possible myocardial hypertrophy.
Maximal stroke volume (SVmax)	Lower because of size and heart volume difference	Training increases SV, reflecting greater blood volume, venous return, or myocardial contractility similar to adults.
Maximal heart rate (HRmax)	Higher due to smaller size	Training may slightly decrease HR max. HR intensities are therefore higher in children at any % max.
Heart rate at submax work	Higher at a given power output and relative metabolic load	Higher heart rate compensates for lower stroke volume. Training prescription intensity may need to be adjusted lower.
Oxygen-carrying capacity	Blood volume and total hemoglobin lower	Training increases blood volume and total hemoglobin.
$a\text{-}\bar{v}O_2$ difference	Somewhat higher	Training does not seem to change $a\text{-}\bar{v}O_2$ difference.
Blood flow to active muscle	Higher	Training does not seem to change blood flow.
Systolic and diastolic pressures	Lower maximal and submaximal pressures	Training does not change systolic BP but may increase diastolic BP.
PULMONARY		
Maximal minute ventilation ($\dot{V}O_2$max, $L \cdot min^{-1}$)	Lower	Training increases $\dot{V}O_2$max.
$\dot{V}O_2$max ($ml \cdot kg^{-1} \cdot min^{-1}$)	Same as in adolescents and young adults	Training increases $\dot{V}O_2$max.
$\dot{V}O_2$, submax	Higher $\dot{V}O_2$ at any given $\dot{V}O_2$	Training decreases submax $\dot{V}O_2$ as seen in adults.
Respiratory rate, submax	Marked by higher rate (tachypnea) and shallow breathing response	Training decreases respiratory rate, allowing increased intensity for progression.
THERMOREGULATORY AND PERCEIVED EXERTION		
Sweating rate	About 40% lower in children, greater increase in core temperature required to begin sweating	Training does not affect sweating rate. Great risk of heat-related illness on hot, humid days because of reduced capacity to evaporate sweat, lower tolerance time in extreme heat. Exercise prescription should be modified in hot environments.

(continued)

Table 29.6 *(continued)*

Function	Exercise responses of children compared to adults	Training response and exercise prescription implications
THERMOREGULATORY AND PERCEIVED EXERTION		
Acclimatization to heat	Longer and more gradual program of acclimatization required; special attention required during early stages of acclimatization	Training improves heat tolerance somewhat in children. With exercise in hot weather, caution is warranted in prescribing exercise intensity. This is especially true for obese, diabetic, and unfit children.
Body cooling in water	Faster cooling because of higher surface-to-volume ratio and lower thickness of subcutaneous fat	Training does not appear to improve cold tolerance. Caution is needed in prescribing exercise in cold water with children, as potential for hypothermia increases.
Body core heating during dehydration	Greater	For prolonged activity, children must hydrate well before and take in fluid during activity.
Perception (RPE)	Exercising at a given physiological strain perceived to be easier	When exercise intensity is governed by RPE, care must be taken to observe HR and general fatigue to avoid overexertion in children. The OMNI Scale (23) is suggested as an alternative to the Borg scale (4).
ANAEROBIC		
Glycogen stores	Lower concentration and rate of utilization of muscle glycogen	Training increases glycogen stores. With interval prescription, lower glycogen stores may initially require smaller number of reps and longer rest intervals.
Phosphofructokinase (PFK)	Glycolysis limited because of lower level of PFK	Training increases PFK. Ability of children to perform intense anaerobic tasks that last 10 to 90 s is distinctly lower than that of adults. Consider lower number of reps and longer rest intervals.
Phosphagen stores	CP stores lower, ATP stores the same; breakdown of ATP and CRP the same	Training increases high-energy phosphate stores. Consider lower number of reps and longer rest intervals until adaptation.
Oxygen transient	Reach steady state faster and develop smaller oxygen deficit; shorter half-time of oxygen increase	During GXT, stage duration can be shorter (i.e., 2 min) in children. Children also recover faster and therefore are well suited to intermittent aerobic activities.
LAmax	Lower maximal blood lactate levels and lower lactate at given percent of $\dot{V}O_2max$	Training does not appear to increase LAmax. This may be why children perceive a given workload as easier.
Heart rate at lactate threshold (LT)	Same or higher	Training may lower LT. Because HR at LT is higher in children, exercise HR should be adjusted accordingly.

Data from American College of Sports Medicine, ACSM's Certified News, 8: 1, p.2 and Bar-Or O, Rowland TW. *Pediatric Exercise Medicine.* Champaign, IL: Human Kinetics, 2004.

Practical Application 29.3

RESEARCH FOCUS

The use of a cardiopulmonary exercise test to stratify dilated cardiomyopathy patients for heart transplantation has been studied and used for adults. However, very few studies have addressed whether cardiopulmonary exercise testing is useful in children with dilated cardiomyopathy to establish exercise intolerance related to poor short-term prognosis as a means of stratifying pediatric patients for transplantation. A group from Great Ormond Street Hospital for Children in London, United Kingdom (9), reviewed cardiopulmonary exercise test and echocardiography data from 2001 to 2009 for children (n = 82) with dilated cardiomyopathy. The following variables were all related to adverse outcomes: left ventricular shortening fraction, peak heart rate, peak oxygen consumption, peak systolic blood pressure, and ventilator efficiency. Based on the results of this study, the researchers indicated that cardiopulmonary exercise testing (using a cycle ergometer) in children with dilated cardiomyopathy is possible and that peak oxygen consumption is related to poor outcomes in this clinical population; that is, children with peak $\dot{V}O_2 \leq 62\%$ of predicted had more adverse events in a 24 mo time frame than children with a peak $\dot{V}O_2 > 62\%$ of predicted. In addition, peak $\dot{V}O_2$ was the only variable that was associated with the study endpoint of death without transplantation or listing for urgent transplantation.

Clinical Exercise Pearls

- Children perceive and experience exercise differently than do adults.
- Adult rating of perceived exertion (RPE) scales, such as the original Borg 15-point scale, can pose problems for children.
- Because of the differences between adults and children with regard to RPE, a new scale, the OMNI RPE Scale, was developed (23). Specific OMNI RPE scales include scales for running and cycling (figures 29.3 and 29.4).
- Clinicians are advised to use the OMNI RPE scales for children during exercise testing and as a means to prescribe exercise intensities.

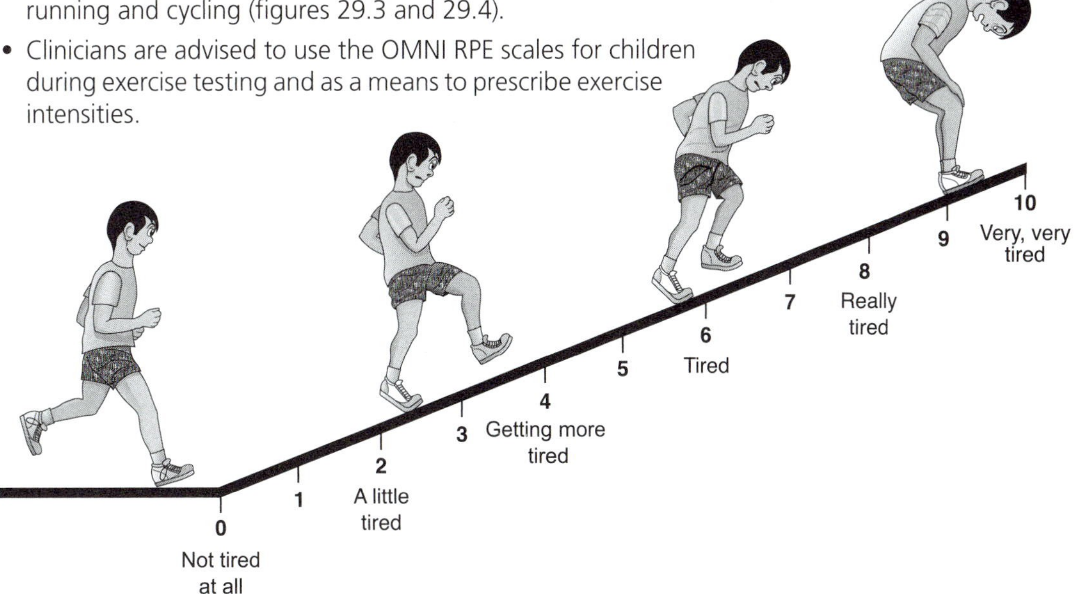

Figure 29.3 OMNI Scale of Perceived Exertion: Child, Walking to Running Format.

Reprinted, by permission, from R.J. Robertson, 2004, *Perceived exertion for practitioners: Rating effort with the Omni Picture System* (Champaign, IL: Human Kinetics), 146.

(continued)

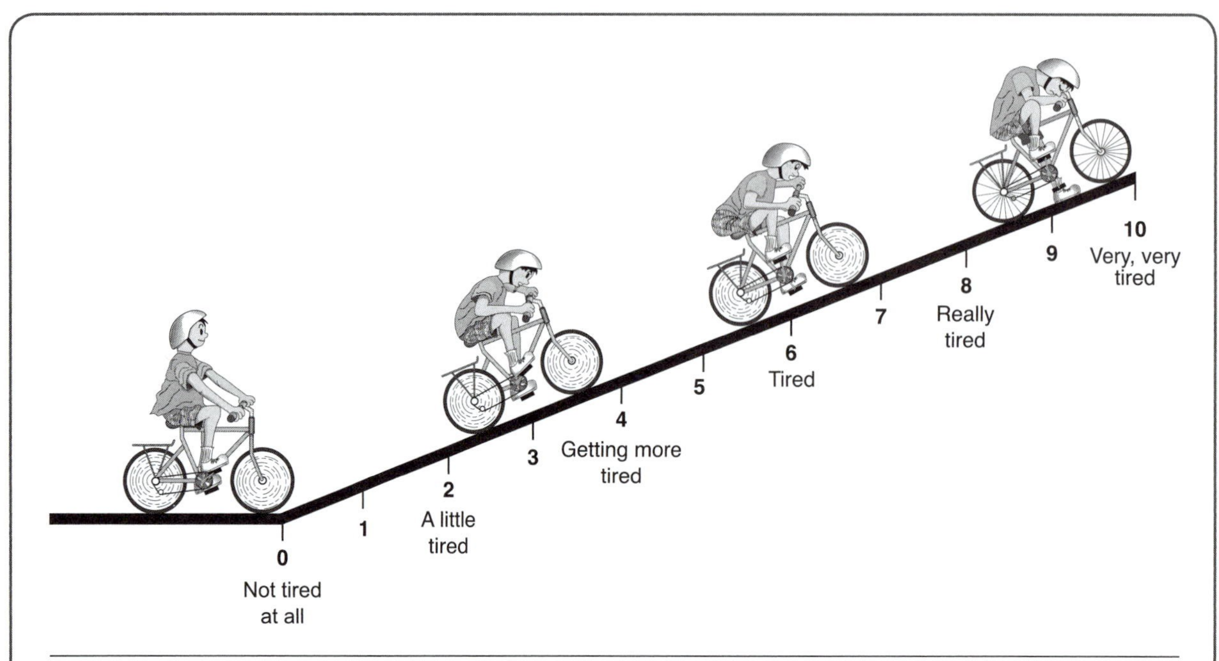

Figure 29.4 OMNI Scale of Perceived Exertion: Child, Cycle Format.

Reprinted, by permission, from R.J. Robertson, 2004, *Perceived exertion for practitioners: Rating effort with the Omni Picture System* (Champaign, IL: Human Kinetics), 145.

Key Terms

children (p. 573)

pediatrics (p. 573)

CASE STUDY

MEDICAL HISTORY

JT is a 10 yr old white male. He is sedentary most of the time and would rather play computer games than play outdoors or participate in sports. His mother has noticed that he has shortness of breath when exercising, and she is worried about his health. She took JT to see his pediatrician due to these concerns.

DIAGNOSIS

The pediatrician completed a physical exam and referred JT for an exercise test.
The results were as follows:

BMI = 31

Maximal functional capacity = 8 METs

Blood lipids and glucose:

- Total cholesterol = 200 $mg \cdot dl^{-1}$
- High-density lipoprotein = 32 $mg \cdot dl^{-1}$

- Triglycerides = 190 $mg \cdot dl^{-1}$
- Low-density lipoprotein = 220 $mg \cdot dl^{-1}$
- Glucose = 119 $mg \cdot dl^{-1}$

Resting blood pressure = 130/86 mmHg

Resting HR = 90

DISCUSSION QUESTIONS

1. Is JT at risk with respect to body mass index?
2. Is JT at risk with respect to his aerobic capacity?
3. Is JT at risk with respect to blood lipids, glucose, or blood pressure?
4. What would be an appropriate exercise test for JT (mode, protocol) and why?
5. As you sit down with JT and his mother following testing, how would you approach this patient regarding his results?
6. What suggestions would you give JT to improve his results? If you suggest increasing physical activity, how would you justify this to him? Explain why physical activity would or would not be beneficial for JT.
7. Write an exercise prescription for JT to improve the following: aerobic endurance, muscular endurance and strength, and range of motion.

Chapter 30

Aging

Daniel E. Forman, MD
Jerome L. Fleg, MD

With the continued increases in longevity, there are more older Americans and individuals worldwide than ever before. And this trend is likely to continue for the foreseeable future, particularly in the United States with the baby boom generation beginning to reach retirement age. An important aspect of **aging** is **frailty** and reduced physical functioning. This can lead to a loss of independent living and an increased risk of a variety of chronic diseases. Exercise and physical activity have the potential to improve functional capacity and reduce the risk of disease.

DEFINITION

Gerontology is the study of the aging process and typically focuses on the period from maturation through old age and death. Most physiological processes and organ systems change gradually with age over the entire adult age span. The rates of change vary between organ systems in a given individual and can vary substantially among individuals depending on intrinsic biological differences, lifestyle differences, and superimposed disease.

Geriatrics is the branch of clinical medicine that involves the diagnosis and management of individuals ages 65 yr and older. This chapter emphasizes aging changes in the cardiovascular (CV) system and other organ systems that are most prominent after age 75, that is, a time when the most significant physiological transitions of advanced aging are manifest. Whereas adults 65 to 75 yr respond to exercise testing and training similarly to younger adults, those aged 75 yr and older require a modified approach to exercise testing and training.

SCOPE

By current estimates, there are 35 million Americans age 65 yr and above (58.5% women), of whom 9.2 million are ≥80 yr old (65.7% women). By 2030, there will be ~70 million adults ≥65 yr old, representing ~20% of the U.S. population, with anticipated proportions even greater in the ensuing decades (60). The subset of adults aged 85 yr and older is the fastest growing in the U.S. population. While the aging of the baby boom generation accounts for some of this growth, the key determinant of this demographic shift is increased longevity that has resulted from improved public health measures as well as improved health care. Many other developed countries throughout the world are experiencing a similar "graying" of their populations.

The societal, public health, and economic impacts of the explosive growth of the **elderly** population are profound. Given that the average 65 yr old American has a further life expectancy of 16 to 19 yr, the ideal is for extended life to be a period of fulfillment and independence, not one of progressive functional decline, frailty, and illness. However, the fact remains that aging fundamentally predisposes to multiple morbidities that compound one another, often to the detriment of function and well-being. As a result, older adults disproportionately use health care resources. In 2004, total health care expenditures for persons 65 yr and older averaged ~4.5-fold those for younger individuals (27).

The number of CV risk factors and their duration both increase as a function of aging, with cumulative injurious effects that predispose to cardiovascular disease (CVD). For example, hypertension affects ~70% of older adults (14) and is a major risk factor for coronary heart disease, heart failure, atrial fibrillation, stroke, and peripheral artery disease, all of which increase disproportionally with age. Diabetes, dyslipidemia, and obesity are also highly prevalent in the elderly, adding to CV risk. High ambient inflammation on a cellular level is a factor linked to aging itself and to cumulative CV risk factors (10). Unfortunately, most CVDs increase the likelihood of a sedentary lifestyle, which itself is a CV risk factor and further reduces mobility due to the effects of deconditioning.

Along with CVD, aging is associated with multiple **comorbidities** that also may reduce functional capacity and add to the complexity of management. Chronic obstructive lung disease, anemia, and other systemic processes exacerbate functional limitations attributable to CV abnormalities. Arthritis and other musculoskeletal disorders also increase in prevalence among older adults, further reducing mobility and ability to perform daily activities. Loss of skeletal muscle mass, termed sarcopenia, and loss of bone mass accelerate with advanced age, predisposing to loss of strength, falls, and fractures (25, 58). Age-associated sarcopenia involves both the loss of muscle fibers and reduction in their size (11). By age 75, muscle mass typically represents ~15% of body weight compared to 30% in a young adult (39). The associated loss of strength parallels the loss of mass, resulting in a much greater effort to perform activities that require lifting, pushing, pulling, and so on (34). Deconditioning exacerbates the impact of age-associated physiologic changes and diseases. This accumulation leads to severe functional limitations. Only 20% to 25% of adults ≥65 yr old report participation in exercise at least 30 min five times per week (6).

PATHOPHYSIOLOGY

This section focuses on CV pathophysiology because CVD is the leading cause of morbidity and mortality in this population. Although moderate- and high-intensity physical activities require the integration of multiple organ systems, the CV and musculoskeletal systems are especially critical to such activity. Major age-associated physiological changes in these systems are summarized in table 30.1.

In industrialized societies, aging is typically accompanied by stiffening of the arterial tree, due largely to degeneration of elastic fiber and deposition of nondistensible collagen and calcium in the walls of larger arteries (37). This arterial stiffening is manifest clinically by increased systolic blood pressure (BP) without increases in diastolic BP. Thus, isolated systolic hypertension is the dominant form of hypertension in the elderly and constitutes a major risk factor for CVD in this age group (3).

Partially in response to the higher systolic arterial pressure that occurs with age, as well as due to intrinsic cellular aging changes independent of BP, the left ventricle (LV) undergoes a modest concentric wall thickening (26, 30). Although resting systolic LV function appears to be well preserved throughout the life span (19), a prominent reduction of early diastolic LV filling rate occurs with aging (54). This reduced early LV filling rate, which approximates 50% between the third and eighth decades, is compensated for by increased late LV filling from a more vigorous left atrial contraction (12). In this way, LV end-diastolic volume is preserved. However, the greater dependence on left atrium-mediated LV filling in older adults means that they are more likely than younger adults to develop signs and symptoms of heart failure when they develop atrial fibrillation. Figure 30.1 is a schematic representation of common age-associated changes in CV structure and resting function (16).

It has been recognized for over 50 yr that peak aerobic capacity (peak $\dot{V}O_2$) declines approximately 8% to 10% per decade in cross-sectional studies (17, 59). This reduction in peak $\dot{V}O_2$ is due to reduction in maximal heart rate of ~1 beat · min^{-1} per year and to reduced arteriovenous oxygen difference (a-$\bar{v}O_2$ difference). Although earlier cross-sectional studies assumed a linear age-associated decline of peak $\dot{V}O_2$, recent data suggest that the decline of aerobic capacity accelerates with age, with reductions of 20% or more per decade in persons 70 yr and older (18, 33) (figure 30.2). These age-associated declines in peak $\dot{V}O_2$ in healthy older adults, magnified by deconditioning and by CV and musculoskeletal diseases, have profound implications for exercise testing and training in this age group.

In addition to the significant reduction in aerobic capacity with aging, important hemodynamic changes also occur during exercise. Because of the age-associated increases in large artery stiffness, systolic BP generally rises more briskly in older adults than in young adults, especially at fixed external work rates. In contrast to the striking reduction in maximal heart rate with age, exercise stroke volume is well maintained in healthy older adults (19). However, the maintenance of exercise stroke volume in the elderly depends more on use of the Frank-

Table 30.1 Physiologic Aging Effects on Organ Systems

Organ system	Effects of aging	Clinical significance
Skeletal muscle	↓ mass of ~1.2 kg/decade from fifth to ninth decade ↓ muscle strength, contractile speed, and power Greater loss of fast-twitch fibers Relative ↑ in slow myosin isoform	↓ ability to perform strenuous activities
Bone Cartilage and connective tissue	Loss of calcium, leading to ↓ bone mass and density, especially in women ↓ thickness, elasticity, and tensile strength; degenerative changes	↑ fracture risk ↑ joint and tendon injury Arthritis
Body composition	↓ lean mass and total body water ↑ % body fat	↓ volume of distribution of water-soluble drugs
Cardiovascular	Arterial stiffening and thickening ↓ vasodilator capacity ↑ left ventricular wall thickness ↓ early left ventricular diastolic filling rate ↓ maximal heart rate and arteriovenous oxygen difference ↓ peak aerobic capacity	Hypertension Diastolic heart failure ↓ ability to perform strenuous activities
Respiratory	↑ chest wall stiffening ↓ vital capacity ↑ residual volume and dead space ↓ maximal voluntary ventilation	Chronic lung disease
Metabolic	↓ resting metabolic rate ↓ insulin sensitivity ↓ glucose tolerance ↓ liver size and blood flow	Obesity Diabetes ↓ metabolism of many drugs
Thermoregulation	↓ thirst sense ↓ skin blood flow ↓ sweat production, sweat gland	↑ risk of dehydration and heatstroke
Renal	Glomerulosclerosis ↓ kidney size ↓ renal blood flow ↓ Glomerular filtration rate	↓ renal excretion of drugs Electrolyte disturbances
Central nervous system	↓ β-adrenergic sensitivity ↓ cholinergic sensitivity ↓ brain volume Degenerative brain changes ↓ balance, coordination, hearing, and vision	↓ maximal heart rate ↓ heart rate variability ↓ cognitive function and memory ↑ falls

↑ = increase; ↓ = decrease.

Starling mechanism, that is, increased LV end-diastolic volume, to compensate for a blunted ability to reduce LV end-systolic volume and increase ejection fraction (19). These age-associated changes in exercise hemodynamics occur despite exaggerated increases in plasma catecholamines (21), suggesting that they are mediated by reduced β-adrenergic responsiveness, which has been confirmed pharmacologically (20). The following list summarizes salient age-associated changes in maximal exercise hemodynamics (16).

Normal Changes in Maximal Aerobic Capacity and Its Determinants Between Ages 20 and 80 Years

Oxygen consumption ↓ 50%
AV oxygen difference ↓ 20%
Cardiac output ↓ 30%
Heart rate ↓ 30%
LV stroke volume: no change
LV end-diastolic volume ↑ 30%
LV end-systolic volume ↑ 100%
LV ejection fraction ↓ 15%
LV contractility ↓ 60%
Systemic vascular resistance ↑ 30%
Plasma catecholamines ↑
CV β-adrenergic responses ↓

Abbreviations: AV = arteriovenous; CV = cardiovascular; LV = left ventricular. ↑ = increase; ↓ = decrease.

Based on Fleg et al. 1995.

CLINICAL CONSIDERATIONS

As already noted, essentially all of the common major CV disorders affect the elderly disproportionately. These include hypertension, coronary heart disease, peripheral artery disease, cerebrovascular disease, heart failure, and atrial fibrillation; most of these are discussed in detail in separate chapters. Chronic pulmonary disease, obesity, arthritis, and other musculoskeletal disorders are also common causes of exercise limitation in the elderly. Prior stroke and other neurological and cognitive disorders may pose additional barriers. A screening questionnaire administered before exercise testing should identify significant disorders that warrant careful evaluation during the history and physical examination.

Signs and Symptoms

Symptoms of aging are varied and are often dependent on specific physical conditions and chronic diseases. Elderly people often have pain associated with their joints, which is typically related to arthritis. It is also typical for these individuals to have exertional dyspnea, which may be

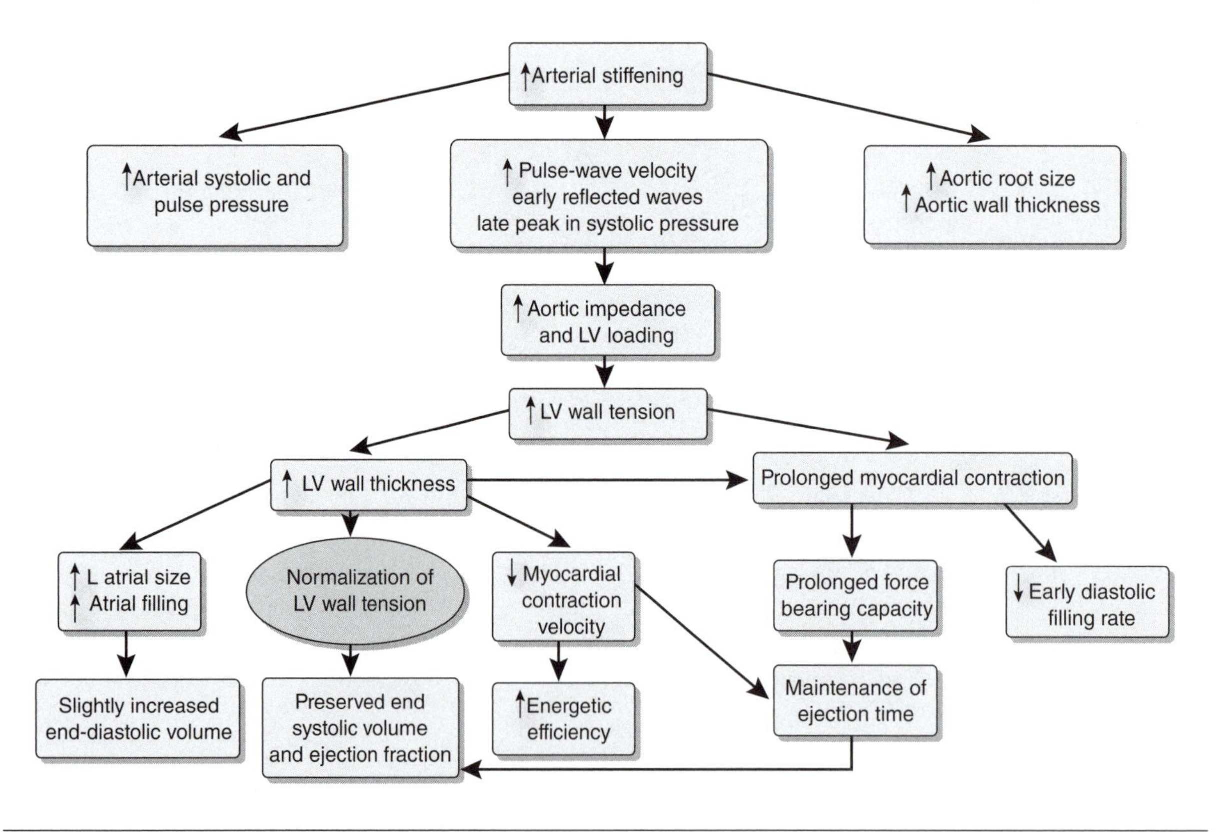

Figure 30.1 Conceptual framework of the major cardiovascular structural and resting functional changes that occur with aging in apparently healthy individuals.

Reprinted, by permission, from J.L. Fleg and E.G. Lakatta, 2008, Normal aging of the cardiovascular system. In *Cardiovascular disease in the elderly*, 4th ed., edited by W.S. Aronow, J.L. Fleg, and M.W. Rich (New York: Informa Healthcare USA, Inc.), 1-43.

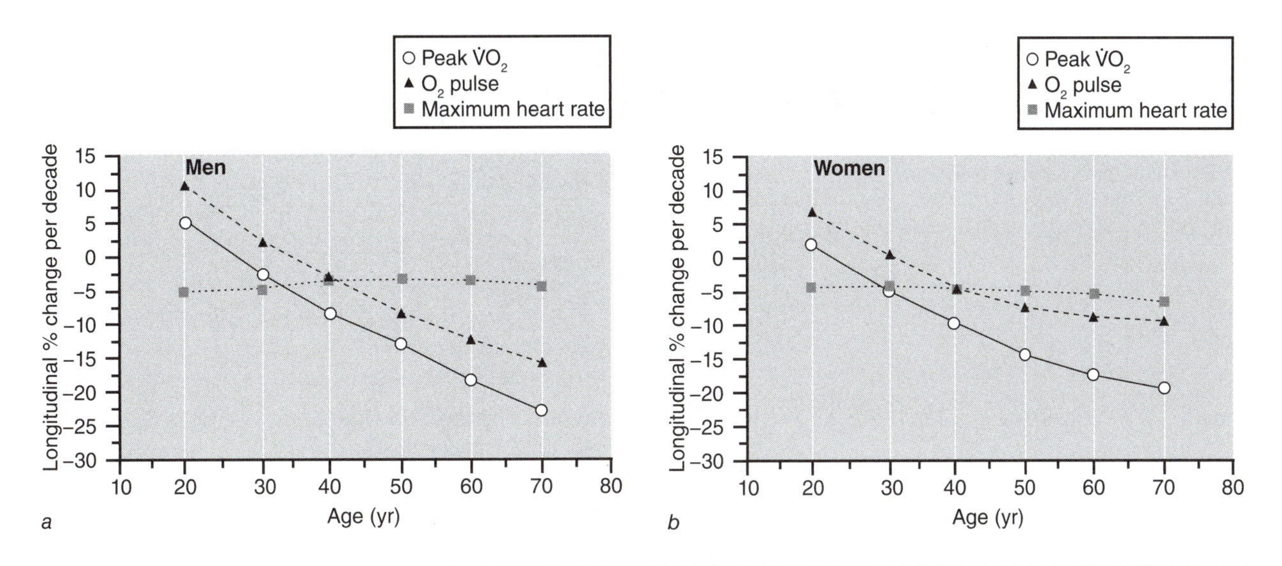

Figure 30.2 Longitudinal changes in peak $\dot{V}O_2$, maximal heart rate, and oxygen (O_2) pulse in healthy volunteers during treadmill exercise. Although the decline in maximal heart rate across age remains relatively constant at ~5% per decade, an acceleration in the decline in oxygen pulse with age parallels the decline in peak $\dot{V}O_2$.

Data from Fleg and Lakatta 1988.

related to deconditioning or to cardiac or lung disease. Since cardiac disease is prevalent in elderly people, they may also experience angina during exertion. It is important in this population to follow up specific symptoms with a medical history and exam when indicated.

History and Physical Examination

Although important in assessing all individuals before exercise testing and training, a careful medical history and physical examination are especially critical in older adults because of their much higher likelihood of significant CVD and other comorbidities that may affect their evaluation and exercise prescription. The history and physical exam should be performed by an experienced provider, with special attention to the CV and musculoskeletal systems. The patient's typical physical activity pattern and exercise-induced symptoms such as chest pain, marked dyspnea, palpitations, lightheadedness, or claudication should be sought. A history of such symptoms or prior myocardial infarction, coronary or peripheral revascularization, and other cardiac disorders or interventions places the patient in a higher risk category. A complete list of medications should be obtained, including use of heart rate–slowing β-blockers and calcium channel blockers.

The physical examination should focus on BP, cardiac and lung auscultation for evidence of heart failure or significant valvular disease, and evaluation of peripheral pulses. Significant gait disorders may be elicited by a brief corridor walk. Severe cognitive deficits may preclude formal exercise testing or require special precautions to ensure understanding and safety. See chapter 4 for more about physical examinations.

Diagnostic Testing

A 12-lead electrocardiogram should be done on all elderly individuals. This is often performed immediately before an exercise test. In such cases, it should be reviewed before the exercise test for acute ST-segment changes or significant arrhythmias that contraindicate exercise testing. Additional CV diagnostic testing to determine cardiac structure and function can be performed either before or after exercise testing, depending on the clinical situation. A detailed pretraining evaluation such as the one outlined in table 30.2 is useful in assessing gait, balance, range of motion, and coordination in frail older adults.

Exercise Testing

Older adults with well-controlled BP and no other clinical CVD can generally begin a low- to moderate-intensity exercise program such as walking (3-6 metabolic equivalents [METs] or 40-60% of estimated aerobic capacity) without a formal exercise test. For older individuals at higher risk or those planning to begin a high-intensity training program, a formal exercise test is generally

Table 30.2 Preexercise Evaluation for Frail Older Adults

Test	Measurement	Outcome	Indicators of risks with testing
Chair stand	Stand from a chair of standard height, unaided and without using arms	Ability Time required	Inability >2.0 s
Step-ups	Step-ups onto a single 23 cm step in 10 s	Ability Number of times	Inability Less than three in 10 s
Walking speed	5 m walk	Time Gait abnormalities such as asymmetry	<0.65 to 0.8 m · s^{-1}
Tandem walk	Walking along a 2 m line, 5 cm wide	Number of errors (off line, touching examiner or another object)	More than eight errors
One-leg stand	Stand on one leg	Ability Time	<2 s
Functional reach	Maximal distance a person can reach forward beyond arm's length while maintaining a fixed base of support in the standing position	Inches	
Timed Get Up and Go	Stand up from standard chair, walk distance of 3 m, turn, walk back to chair, and sit down again	Time	>10 s
Range of motion	Goniometer used to assess shoulder abduction (SA), flexion (SF), extension (SE); elbow flexion (EF), extension (EE); hip flexion (HF), extension (HE); knee flexion (KF), extension (KE); ankle dorsiflexion (DF), plantar flexion (PF)	Degrees	<90° (SA), <150° (SF), <20° (SE), <140° (EF), <20° (EE), <90° (HF), within 10° (HE), <90° (KF), not within <10° full KE, inability to perform DF and PF

Adapted from E.F. Binder et al., 1999, "Peak aerobic power is an important component of physical performance in older women," *Journal of Gerontology* 54A: M353-356; W.J. Chodzko-Zajko and K.A. Moore, 1994, "Physical fitness and cognitive function in aging," *Exercise in Sport and Science Review* 22: 195-220; R.J. Kuczmarski et al., 1994, "Increasing prevalence of overweight among U.S. adults," *Journal of the American Medical Association* 272: 205-211; M.C. Nevitt et al., 1989, "Risk factors for recurrent non-syncopal falls," *JAMA* 261: 2663-2668; M.L. Pollock et al., 2000, "Resistance exercise in individuals with and without cardiovascular disease: Benefits, rationale, safety, and prescription. An advisory from the committee on exercise, rehabilitation, and prevention, council on clinical cardiology, and the American Heart Association," *Circulation* 101: 828-833.

recommended. For persons planning to incorporate resistance training, strength and range of motion testing should also be performed (table 30.3). Assessment of balance may be particularly useful in older adults with a history of falling.

Cardiovascular

It should be emphasized that most older adults can perform an exercise test rather than being referred routinely for pharmacological stress testing, which provides no information on aerobic capacity. However, particular attention to the mode of exercise testing and the protocol is imperative in older adults. Those with gait disturbances or instability may be better suited for cycle ergometry than for treadmill exercise. A modest downside of cycle exercise is the lower aerobic capacity and peak heart rate generally achieved relative to treadmill exercise, which increases the likelihood of submaximal CV responses (42). Regardless of the mode of exercise or protocol used, the patient should be encouraged to perform a maximal effort to optimize detection of ischemia, arrhythmias, and other abnormalities. Moreover, even a test deemed submaximal by heart rate criteria may provide useful information regarding the individual's conditioning status and ability to exercise safely to the level achieved during the test.

With treadmill exercise, a protocol using modest increments of ~1 MET between successive stages, such as the modified Naughton, Cornell, or Balke protocol, should be employed. These modest increments help instill confidence in the participant and reduce the likelihood of falling or premature termination. A ramp protocol employing frequent, very small increments in work rate is a viable alternative on either a treadmill or a cycle ergometer. Regardless of the protocol used, the patient's general well-being, BP, and ST-segment response should be closely monitored. Because exercise-induced arrhythmias are more common and complex in the elderly (7, 41), cardiac rhythm should be continuously monitored. The indications for cardiopulmonary

Table 30.3 Exercise Testing Review

Test type	Mode	Protocol specifics	Clinical measures	Clinical implications	Special considerations
Cardiovascular	Treadmill or ergometer	Low initial intensity with small increments in work rate	Estimated METs or peak $\dot{V}O_2$, heart rate, blood pressure, ECG	Greater vigilance for age-related limitations	High incidence of undiagnosed CV disease. Gait deficits may affect safety and ability to test.
Strength	Weight machines	Modified 1RM with focus on muscles used for ADLs Low initial load with small increases	Weight lifted, repetitions	Avoid injury by close supervision	Agility, balance, and coordination may affect safety and ability to test.
Range of motion	Goniometer	Hip, ankle, knee, shoulder, and elbow; lower back and hamstrings	Degrees of motion	Limited ROM in these joints related to limitations in performing ADLs	Mental and intellectual impairment may affect ability to perform testing.
Balance	One-leg stand Tandem walk		Ability, duration, number of errors	Increased risk for falls	Useful for assessing risk of falls.

ROM = range of motion; ADLs = activities of daily living; RM = repetition maximum.

exercise (CPX) testing are generally similar to those in younger adults. By quantifying aerobic capacity, respiratory efficiency, and level of effort, CPX testing is especially useful in determining the severity of exercise intolerance and whether it is due to cardiac, pulmonary, or motivational factors.

For older adults in whom an assessment of exercise capacity is desired without the need for heart rate, BP, or electrocardiogram (ECG) monitoring, a 6 min walk test may be useful (49). The test involves traversing a 20 to 30 m course as many times as possible in 6 min. A variation of this test is a timed 400 m walk (56). These tests generally correlate well with measured peak $\dot{V}O_2$ and can be useful in designing a walking program. Their nonintimidating nature, minimal cost, and ease of administration make them particularly attractive in assessing older adults.

Musculoskeletal

Formal strength testing is not always necessary in healthy older adults planning to undertake a low-resistance strength training program. However, strength assessment can often be useful in identifying and quantifying the typical loss that occurs in older age. Assessing for strength may provide additional motivation for the individual with respect to improving function. In frail individuals or those planning a more intensive strength training regimen, a modified or general one-repetition maximum (1RM) test can be used. Due to the greater loss of fast-twitch compared to slow-twitch muscle fibers with age, reduction in muscular power—the ability to perform rapid muscular contractions—is typically greater than the loss of strength (57). These losses in power can be quantified by machines that measure both strength and speed of contraction. Muscle endurance, the ability to perform repetitive muscle contractions, also declines with age. Because everyday activities involve not only muscle strength, but also power and endurance, an assessment of all three components may provide more useful information than a 1RM test alone. Special resistance machines, such as those using pneumatic pressure as the basis of resistance, are particularly well suited for older adults and help minimize the risk of injury. For frail elderly persons, strength testing should focus on the muscles used in typical activities of daily living, including the hip and shoulder flexors, extensors, and abductors and the knee and ankle flexors and extensors.

Flexibility

In older adults, the flexibility of the large joints, including the lower back, hips, shoulders, elbows, and knees, is often reduced due to arthritis, muscle weakness, and disuse. Because many routine activities require reasonable flexibility of these joints, range of motion (ROM) testing is particularly useful in elderly persons. Those joints with especially limited ROM can be targeted in a personalized training program.

Treatment

As already noted, today's expanding older population is particularly susceptible not only to cardiovascular disease (CVD) but to multiple chronic diseases in combination with CVD (22). Typically, these diseases compound one another such that senior patients are inherently more unstable and management decisions more complex. For example, it is not uncommon for an older patient to develop heart failure in the context of kidney disease, diabetes, atrial fibrillation, and anemia—a combination of factors that not only predispose to heart failure but also make it more difficult to treat (5). Even diseases that may seem distinct from one another may have considerable bearing on a patient's management choices; for example, the common occurrence of lumbar stenosis and mild dementia in an older adult might totally confound considerations for surgical interventions for coronary artery disease (28). Polypharmacy is a related concern, as many elderly people are prescribed multiple medications for multiple disease states simultaneously, resulting in greater likelihood of unintended iatrogenic consequences, especially in the context of the altered drug metabolism and elimination that occur in old age (35).

Faced with such age- and disease-related limitations, many older adults constrict their living patterns, fearing situations that seem likely to threaten their fragile equilibrium. Ironically, these preferences tend to make the health issues worse (64); that is, sedentary lifestyles tend to worsen medical instability and accelerate progression into frailty and associated susceptibility to weakness, falls, diminished independence, and reduced quality of life (24). The challenge to care providers is to increase physical activity among seniors, particularly those with CVD and multiple morbidities. Vigilance and encouragement are required, as exercise may initially contribute to instability (e.g., arthritic pain, glucose fluctuations, and increased fatigue) but in the long term become beneficial (48). Proper nutrition, adequate sleep, and proper footwear are also fundamental.

Exercise training serves as a means to potentially forestall age-related morbidity and prevent disease in seniors (43). It also plays a key role in secondary prevention, modifying symptoms, increasing function, and moderating pathophysiology (e.g., plaque stabilization, enhanced

Practical Application 30.1

CLIENT–CLINICIAN INTERACTION

Exercise training evokes complex feelings for many older adults. For those who have never trained, the idea of developing exercise behaviors in old age is often seen as daunting and unappealing, especially in the midst of fears about illness, cost, and mortality. For many other adults who were vigorous athletes in their youth, the notion of exercise training in old age evokes unrealistic expectations, making it a priority to teach them about moderation, self-control, and the efficacy of lower-intensity regimens. Therefore, as part of his job, the exercise physiologist must perceive the needs and orientation of the patient. Teaching is implicitly personal, that is, a tailoring of the training details to the circumstances of each individual.

For most elderly persons, particularly those who are frail, initiating exercise routines, learning safe exercise techniques, and safely advancing exercise duration and intensity require close supervision and teaching. For the majority of frail elders, most exercise is experienced as "high intensity" since each modality is challenging. Therefore, the exercise physiologist must employ as much vigilance and urgency with a 2 lb weight and slow walking in a frail elder as with a 50 lb weight and jogging in an elder who is robust. Motivation, adherence, efficacy, and safety depend critically on these important dynamics.

A related point is for the exercise physiologist to be vigilant regarding the likelihood that patients are struggling with multiple morbidities that confound even the best efforts to exercise. Joint or back pain, CV limitations, breathing limitations, concerns about urinary or fecal continence, and depression or anxiety are among myriad impediments that are more typical than exceptional. It is incumbent on the physiologist to be resourceful in helping patients overcome or work around these limitations to still achieve a successful exercise training routine. This does not mean belittling medical concerns, but working with the rest of a patient's caregiving team (physicians, nurses, family) to ensure safety and efficacy as exercise proceeds. A key point of emphasis is that an effective exercise physiologist can play a vital role in facilitating exercise therapy amid medical clinical complexities that might otherwise exclude healthful exercise behaviors.

vasodilation, improved autonomic responses) in those with established disease (9). In addition to secondary prevention benefits for CVD, exercise therapy improves cognition, respiration, bone strength, gastrointestinal motility, sleep, mood, and multiple other factors (9, 32, 62). In a very real sense it is a wonder drug, improving morbidity, mortality, and, perhaps most importantly, quality of life. Information about interacting appropriately and effectively with older adults is provided in practical application 30.1.

EXERCISE PRESCRIPTION

Exercise recommendations for older adults were once primarily extrapolated from recommendations oriented toward younger adults (47). However, in 2007, the American College of Sports Medicine (ACSM) and the American Heart Association (AHA) published physical activity and exercise recommendations specifically geared toward seniors (45) (table 30.4). Moreover, in 2009, the ACSM published a position stand on exercise that extensively elaborated on these principles oriented to exercise benefits in elderly (9). In general, these efforts emphasize the importance of regular physical activity, including aerobic, strength, stretching, and balance activities, as part of healthy aging. Parallel literature from the AHA emphasizes the utility of exercise training as part of secondary prevention of CVD (3).

In general, an active lifestyle preserves and enhances skeletal muscle mass, strength and endurance, flexibility, and cardiorespiratory fitness, which otherwise would diminish and increase chronic disease risk with advancing age. Goals are to develop exercise interventions for older adults that stimulate CV and autonomic responses and that also increase strength, ROM, and balance. Increased physical activity also helps increase energy use (moderating weight gain and other CV risk factors) and preserve functional independence.

Exercise Training Caveats Specific to the Elderly

While seen in any population, the heterogeneity of older individuals is particularly striking; that is, while some are quite robust, others can be very frail (9, 45, 53). Such heterogeneity plays an important role in exercise prescription for older adults. The notion of exercise intensity is also relative to each person's capacity, such that lifting a 2 lb weight might be high intensity for someone who is

Table 30.4 Exercise Prescription Review

- Older adults may exceed the recommended minimum amounts of physical activity if they desire to improve their fitness (supervision and close monitoring are often vital in these situations).
- If chronic conditions preclude activity at the recommended minimum amount, older adults should perform physical activities as tolerated to avoid being sedentary.

Mode	Frequency	Intensity	Duration	Special considerations
AEROBIC TRAINING				
• More often: walking, cycling, seated recumbent (Nu-Step), pool activity, and seated aerobics. • Incorporating aerobic activities into daily routines can help achieve these goals. • Less often: jogging, swimming laps, rowing, aerobic dance.	• Moderate-intensity exercise five times per week or vigorous-intensity exercise three times per week. • General physical activity should be performed daily.	• ADLs at comfortable pace. • Moderate intensity at 50% to 70% HRR or 5 to 6 on the 10-point exertion scale. • Vigorous intensity at >70% HRR or 7 to 8 on the 10-point exertion scale.	• Moderate-intensity exercise is 30 to 60 min intervals, but may be as short as 10 min, targeting 150 to 300 min per week. • Vigorous-intensity exercise is 20 to 30 min for a total of 75 to 100 min a week.	• For adults who are highly deconditioned, are functionally limited, or have chronic conditions that affect their ability to perform physical tasks, intensity and duration of physical activity should be low at the outset, with progression tailored to each person's tolerance and preferences. • For those who are very limited at the outset, small amounts of exercise may be tantamount to high intensity and should be accorded corresponding duration, frequency, and monitoring. • Muscle-strengthening activities or balance training (or both) may need to precede aerobic training among very frail individuals. • Comorbidities such as arthritis, osteoporosis, and heart disease need to be considered. • See relevant chapters for further special considerations.

(continued)

Table 30.4 *(continued)*

Mode	Frequency	Intensity	Duration	Special considerations
STRENGTH TRAINING				
• Multistation machine-type (e.g., Keiser pneumatic) equipment provides greatest safety. • Elastic bands and free weights are less costly and often more available.	Two or more times per week	5 to 6 (moderate) or 7 to 8 (vigorous) on the 10-point exertion scale.	• 8 to 12 repetitions for each muscle group, usually targeting 8 to 10 different muscle groups. • Increase number of repetitions (up to 20) for greater endurance. • 20 to 30 min per session.	• Strength training regimens depend on maximum resistance as well as endurance and velocity; supervision and monitoring are often helpful. • Proper breathing is important; avoiding the Valsalva maneuver is a safety priority. • Frail or deconditioned adults critically benefit from strength training. • Free weights may be difficult for a frail person to balance, so instruction and assistance should always be available. • Focus on major muscle groups used to perform ADLs (shoulders, legs). • See chapters that review heart diseases, osteoporosis, and arthritis for further special considerations.
RANGE OF MOTION TRAINING				
Static stretching: See chapter 5 for examples.	Minimally two times per week	• Moderate intensity (5 to 6 on a 0 to 10 scale). • Subject should feel a mild stretch without inducing pain. • Progress range of stretch based on lack of discomfort experienced.	• 5 to 30 min total, with two 30 s bouts for each muscle group. • Involve multiple muscle groups (neck, shoulders, arms, lower back, quadriceps, hamstrings, calves, ankles).	• Avoid ballistic movements and the Valsalva maneuver during the stretching routine. • A brief routine performed before or after aerobic or resistance exercise need focus only on the muscle groups used and need last as little as 5 min. • Consider using alternative methods such as seated movements and use of a towel or elastic band to assist those with difficulties performing standard flexibility training. • See chapters that review osteoporosis and arthritis for further special considerations.
BALANCE TRAINING				
There are currently no specific recommendations regarding frequency, intensity, or type of balance exercises for older adults. ACSM exercise prescription guidelines recommend using activities that include the following: (1) progressively difficult postures that gradually reduce the base of support (e.g., two-legged stand, semitandem stand, tandem stand, one-legged stand); (2) dynamic movements that perturb the center of gravity (e.g., tandem walk, circle turns); (3) stressing postural muscle groups (e.g., heel stand, toe stand); or (4) reducing sensory input (e.g., standing with eyes closed). Yoga, tai chi, and Bosu ball exercises may be beneficial.				

Note. ADL = activities of daily living; HRR = heart rate reserve; RM = repetition maximum.

frail. Therefore, exercise goals need to be tailored to a patient's unique circumstances. For frail elders, priorities often center on building strength, stability, and balance before progressing to aerobic modalities. When frailty or a chronic condition prohibits specific activities, exercise should be tailored to those capacities that are preserved. For some frail adults, exercises may be restricted to arm movements, seated machines, or water activities because disabilities preclude other options.

It is also critical to appreciate that most successful exercise regimens correspond to an individual's specific needs and preferences. Exercise adherence often ultimately relates to an individual's affinity for the activities themselves. Therefore, it is sometimes helpful to identify a specific activity that a patient prefers and then begin a regimen involving strength or aerobic performance–enhancing exercises that complement this preference. For example, one study showed that a regimen of dancing activity was more effective than traditional aerobic training for older heart failure patients (36).

Warm-up and cool-down activities are particularly important for elderly individuals. A warm-up activity allows gradual increases in heart and breathing rates and limb blood flow before higher-intensity exercises are undertaken. Warm-up for strength training commonly involves performing the same movements with less weight. A cool-down should follow exercise as in younger adults (36, 53).

Specific Training Recommendations

The ACSM/AHA special report on physical activity for older adults recommends aerobic exercise activity 5 d per week for 30 min at moderate intensity (<70% heart rate reserve or 5 to 6 on a 10-point intensity scale) or at vigorous intensity on 3 d per week for 20 min (45). It also recommends 2 d per week strength training (in 8-10 maneuvers), as well as biweekly flexibility training and exercises to improve balance. While to many sedentary elderly people, these goals may seem intimidating, even small amounts of exercise have benefit. Recent work by Sattelmair and colleagues corroborates that even low-intensity exercise provides substantial CV benefit (52). Therefore, for many seniors, only 10 min of light activity may be realistic, with goals to slowly increase frequency, duration, and intensity as tolerated. Furthermore, for adults who are frail, exercises to build strength and balance often need to precede aerobic exercise training. This perspective is critical with regard to the guidance of older adults, as low-intensity activities should never be disparaged.

Light- to moderate-intensity protocols are generally recommended for patients who are exercising on their own, thereby providing a greater margin of safety than high-intensity regimens, as well as for seniors who are primarily trying to maintain their current capacity. Exercise progression, however, is particularly important for frail older adults. In these cases, higher-intensity training goals with greater supervision may be most beneficial. In this connection it is important to acknowledge that activities that may seem like low-intensity routines (e.g., lifting one's own body weight as part of so-called warm-up activity) may actually be high-intensity training for the frail senior patient (thus requiring careful supervision and monitoring) and demand careful awareness with regard to quantity, recovery time, and so on.

Cardiovascular Exercise

Aerobic activity entails repeated movements of large muscle groups (e.g., large-muscle rhythmic activity such as walking, swimming, and bicycling). In a gymlike setting, this usually entails walking (treadmill or indoor track), bicycle ergometry, or recumbent stepping. However, many physical activities provide similar exercise benefits, including walking (outdoors), cycling, gardening, swimming (laps or water aerobics), and golfing. Those who are unable to perform ambulatory activities may be candidates to perform seated aerobic activities including chair and water exercises.

Low- to moderate-intensity exercise programs can be performed daily. Higher-intensity or vigorous exercise sessions (>70% heart rate reserve or 7 to 8 on a 10-point exertion scale) are recommended only 3 to 5 d per week, allowing for increased recovery time between exercise sessions and thereby minimizing musculoskeletal risks.

Older individuals with low exercise capacity often best tolerate (and enjoy) multiple daily sessions of short duration, whereas seniors who are relatively more robust may prefer exercise bouts performed once per day (9, 53, 64). Often, older adults are advised to exercise at a level that feels "somewhat hard" for aerobic and strength training (i.e., a level that usually corresponds to moderate exercise intensity and therefore leads to many key physiologic and clinical benefits). These individuals can be counseled to exercise below a threshold at which they become short of breath as a means to maximize efficacy as well as safety and comfort.

High-intensity interval training (short bouts of vigorous exercise separated by periods of rest or active recovery) is one form of high-intensity exercise that incorporates physiologic principles well suited to many elderly people. The conceptual advantage is that the higher intensity maximizes physiologic benefit, and, when coupled with a rest interval, maximizes recovery and stabilization (63).

Resistance Exercise

In general, muscle-strengthening exercises are oriented to the legs, hips, chest, back, abdomen, shoulders, and arms, as these muscle groups are all highly engaged during activities of daily living. Strength training sets usually entail 10 to 15 repetitions of moderate to heavy intensity. While one set of exercises (a series of flexion–extension repetition movements oriented to each of a targeted group of muscles) twice a week is beneficial, achieving two or three sets per session has greater physiological impact. Just as with aerobic training, physiological benefits must be counterbalanced by musculoskeletal risks and behavioral inertia. Strength training sessions are usually separated by 48 h to facilitate recovery.

The recommended intensity level, using the 10-point exertion scale, should range from 5 to 6 for moderate efforts and 7 to 8 for vigorous efforts. Strength training sessions generally last 20 to 30 min. Progress is carefully monitored, and progression of intensity is usually considered every 2 to 3 wk, depending on the patient and the training goals.

Strength training technique also varies in relation to specific training goals. In general, increasing capacity to tolerate greater resistance is the primary training endpoint (i.e., adaptations in muscle size and force). However, dimensions of muscle power and endurance are related concepts that may be associated with even greater clinical relevance. Power implies a dimension of velocity integrated with force, which can provide a critical advantage in reducing risk of falls. Higher-velocity training protocols have been demonstrated to better diminish falling risks than strength alone (15).

Muscle endurance is the ability of muscle to generate force and power over an extended period of time, and often plays a critical role in determining a senior adult's independence and quality of life. Studies suggest that higher-intensity strength training regimens are more likely to achieve significant improvements in endurance (9).

While strength machines generally provide the advantage of structured movements that minimize musculoskeletal risk, free weights are less expensive and more readily available, particularly for a home environment. Milk jugs or even cans filled with sand are even more available and less expensive. Nonetheless, exercise with free weights entails inherent instability and potential for injury in some elderly persons. Clinical exercise physiologists must be particularly proactive in guiding use of free weights (or the equivalents), since proper technique substantially reduces the risk of injury. Correct breathing patterns should also be emphasized such that participants avoid the **Valsalva maneuver** (i.e., breath holding), which can reduce venous return and increase BP.

In persons who are too weak to sustain aerobic activities, strength training may be a critical first step toward increasing minimum strength and balance needed for aerobic capacity. In such circumstances, training may begin with using a patient's own body weight for resistance, then slowly progress to strength training that uses resistance tubing and bands, ankle weights, and other weights over many weeks or months. As strength increases, the best way to advance resistance training is to increase the number of repetitions before increasing the resistance. Such activity is progressively integrated with daily walking and other aerobic activities as they become tolerated. Supervision, monitoring, teaching, and encouragement are essential, with the recognition that long adaptation periods may be needed as deconditioned elderly persons begin a more active lifestyle.

Range of Motion Exercise

Both flexibility (ROM) and stretching exercises are recommended in the ACSM/AHA report and play a critical role in mitigating some of the natural bone, joint, and muscle changes that occur with aging. The recommendation for these exercises is at least 2 d per week for at least 10 min each time, but frequency can also be greater. Generally, the exercises consist of static stretches (ballistic or bouncing stretches are not recommended) lasting 10 to 30 s per stretch, with three or four repetitions each. Range of motion exercises usually focus on all the major muscle and tendon groups (i.e., all the major joints of the body including hip, back, shoulder, knee, trunk, and neck region). Performing them in conjunction with aerobic or resistance training is usually well tolerated. In fact, before performing a ROM routine, the participant should usually begin with a warm-up period of light-intensity aerobic-type exercise. The degree of stretch slowly increases as tolerated. Chapter 5 reviews general recommendations for ROM exercises.

Balance Training

Balance training complements ROM training with a critical capacity to mitigate or reverse falling risks and forestall frailty. Usually balance training is recommended three times per week, but formal dosing recommendations remain ambiguous. Balance training options include various walking activities (backward, sideways, heel to toe), as well as standing activities (one-legged stand, heel stand, toe stand) and dynamic movements (circle turns) (34). Tai chi is often touted for its utility in inducing balance training benefits in association with strength and aerobic training effects (38). Integrating a Bosu ball into training activities is also a useful balance training technique.

EXERCISE TRAINING

Exercise training-induced benefits in older adults tend to be similar to those achievable in younger adults.

Cardiovascular Exercise

Increased aerobic capacity, cardiac output, peripheral oxygen utilization, vascular responsiveness and perfusion, neuromuscular responses, muscle mass and strength, proprioception and balance, glucose metabolism, and reduced inflammation have all been described in connection with aerobic exercise (2, 8, 9, 13, 32, 53). While the absolute magnitude of exercise training benefit may be greater among elderly people who are robust and healthy, even frail and infirm seniors can derive gains that may translate into enormous clinical benefits, including enhanced independence, confidence, and well-being (9, 23, 50, 55).

Resistance Exercise

Strength training is often particularly important in the elderly since the natural susceptibility to sarcopenia and osteoporosis frequently limits the capacities of frail elders to begin aerobic-type training. Resistance training programs increase muscle strength and mass through muscle hypertrophy and neuromuscular adaptation, facilitating improvements in gait, balance, and overall functional capacity and resistance to falls (1, 9, 53). Strength training also increases bone mineral density and content (40, 46, 61), increases metabolic rate (31, 51), assists with maintenance of body weight by decreasing fat mass and increasing lean mass, and improves insulin sensitivity (31).

Home- and Community-Based Exercise Training

Home-based exercise takes place in an individual's residence, is not supervised, and is often conducted

Practical Application 30.2

RESEARCH FOCUS: CARDIAC REHABILITATION AND SURVIVAL IN OLDER CORONARY PATIENTS

This study was a hallmark analysis of older cardiac patients aimed at assessing the benefits of cardiac rehabilitation (CR) as a life-prolonging therapy. The authors found that CR users had 5 yr mortality rates 21% to 34% lower than nonusers, depending on the statistical analysis employed. Whereas many smaller studies have demonstrated the efficacy of CR in older CV patients, a repeated criticism has been implicit selection bias and the related likelihood that confounders such as socioeconomics, complexity of disease, and extent of revascularization predictably detract from the strength of the favorable conclusions. Furthermore, given the progressive improvements in management of coronary artery disease, acute myocardial infarction, and CV surgery, many have expressed doubt that CR adds further life-prolonging benefits, particularly for older patients.

A major strength of this study is that it relied on a large Medicare database (601,099 adults). Furthermore, the statistical methods were robust. The researchers used propensity analysis to balance for many confounders that have been associated with CR use in earlier analyses. Socioeconomic factors, disease complexity, comorbidities, extent of cardiac injury and revascularization, and other clinical features were carefully matched between the 70,040 CR users and an equal number of nonusers.

In addition, regression modeling was used to assess the life-prolonging benefit of CR, with the same risk factors as used in the propensity analysis, but applied to the entire cohort of 601,099 Medicare beneficiaries. Instrumental variables were incorporated in this model to protect against residual confounding effects (e.g., distance to the CR facility from the center of the patient's residence zip code and density of facilities per 10,000 population age 65 yr or older by state).

Overall, 5 yr mortality rates were 21% to 34% lower in CR users than nonusers in this socioeconomically and clinically diverse, older population after extensive analyses to control for potential confounding. These results are similar in magnitude to those observed in published randomized controlled trials and meta-analyses of CR in younger, more selected populations. Furthermore, CR users who participated in 25 or more sessions had 19% lower mortality than matched CR users with fewer sessions. Of note, another recent analysis in 30,161 Medicare patients showed a strong protective dose–response effect relating the number of CR sessions attended to the 4 yr risk of death and myocardial infarction (29). However, only 12% of the eligible Medicare population participated in CR at all, indicating a major obstacle to reaping these life-prolonging benefits.

Whereas the benefits of CR include both exercise training and behavior modification and education, this study highlights the pivotal role of exercise therapy as part of composite care in the management of older CV patients. The implicit value of the exercise physiologist to facilitate exercise training is an elemental part of excellent CV care.

Reference: Suaya JA, Stason WB, Ades PA, Normand ST, Shepard DS. Cardiac rehabilitation and survival in older coronary patients. *J Am Coll Cardiol* 2009;54:25-33.

alone. As noted earlier, training intensity is typically moderated to mitigate risks. While this is conducive to safety and adherence, potential physiological gains are also usually smaller. Furthermore, home-based exercises present particular challenges to those who are frail. Exercise is inherently likely to be "high intensity" since their baseline capacities are so limited. Frail older adults typically require personal instruction and supervision to best ensure the safety and efficacy of activity, but such guidance is difficult to facilitate at home. Furthermore, lack of emergency care for home-based exercise compounds risk and apprehension among patients as well as their families.

Community-based exercise generally takes place in a health or fitness center, is usually supervised, and allows several individuals to exercise together or at the same time. In addition to incorporating components of assessment, monitoring, and safety, community-based programs provide opportunities for social reinforcement and higher adherence. Community-based supervised resistance training has more consistently achieved greater increases in lean mass and muscle strength than home-based programs (44). Furthermore, community-based programs are better suited for older adults with baseline frailty or concomitant medical conditions, such as neurologic, CV, orthopedic, or pulmonary diseases.

CONCLUSION

Whatever the structure of the exercise program, the success of exercise training usually entails an important relationship between the patient and clinician. Initial assessments are essential, both for determining an appropriate training program and for assessing CV risk and all the other conditions, medications, and circumstances that likely affect exercise goals and safety.

The optimal regimen and dose of exercise must be carefully selected relative to each person's circumstances. Ideally, each person's regimen should be personalized and coordinated by an exercise provider. For most elderly people, particularly those who are frail, the capacity to initiate exercise routines, to learn safe exercise techniques, and to advance exercise duration and intensity safely all require close supervision and teaching. Motivation and adherence depend critically on these important dynamics.

Clinical Exercise Pearls

- Aging is typically accompanied by multiple changes that affect central (heart and major vessels) and peripheral (skeletal muscle, bones, joints, and balance) performance, predisposing to constricted functional capacities and frailty. However, exercise therapy can attenuate the progression of these changes and can reverse some of those that have already occurred, thereby improving aerobic, strength, flexibility, and balance capabilities.
- Most elders referred for exercise training are struggling with a range of medical issues and related therapies. While many assume that chronic medical issues trump exercise training priorities, the reverse is often the case. Exercise training is a critical component of therapy, helping to maintain function, independence, and high quality of life despite the effects of diseases.
- Although many older adults conceptualize exercise only in terms of aerobic activities, it is often critical to implement strength training as the first step in facilitating aerobic performance. Typical aging entails progression of sarcopenia and muscle deconditioning, which erode performance capacity but which can be counterbalanced by strength training.
- Balance and flexibility training, though important at any age, may be especially useful in the elderly to reduce falls and enhance their ability to perform **activities of daily living (ADLs)**, respectively.

Key Terms

activities of daily living (ADL) (p. 602)
aging (p. 589)
comorbidities (p. 590)
elderly (p. 589)
frailty (p. 589)
valsalva maneuver (p. 600)

CASE STUDY

MEDICAL HISTORY

Mrs. HK is an 84 yr old active woman with a history of hypertension and hyperlipidemia for which she takes hydrochlorothiazide (12.5 mg), simvastatin (10 mg), and 81 mg aspirin daily. She has been experiencing progressive weakness, particularly in her legs, and associated breathlessness when she is active. She states, "I can't seem to do as much as I could two years ago." While still independent in all ADLs, she is no longer comfortable walking with her friends and tends to stay home watching television. Last month, while she was carrying groceries to the car, she became so weary that she almost fell as she stepped from the curb. Her weight has been stable. She has no other constitutional complaints. She is compliant with her medications, eats a well-balanced diet, and has been sleeping reasonably well.

DIAGNOSIS

Recently, on the advice of her children, Mrs. HK went to see her internist. Her exam was normal, including BP 142/70 mmHg on her medications, heart rate 72 beats $\cdot$ min^{-1}, and BMI 22 kg $\cdot$ m^{-2}. The lipid profile showed total cholesterol of 170 mg $\cdot$ dl^{-1}, with LDLc 102 mg $\cdot$ dl^{-1}, TG 80 mg $\cdot$ dl^{-1}, and HDL 32 mg $\cdot$ dl^{-1}. The internist noted a normal physical exam, though the ECG showed left ventricular hypertrophy with a strain pattern. He ordered an exercise stress test to explore the possibility that cardiac ischemia was the cause of her symptoms. Since the baseline ECG was abnormal, he ordered that sestamibi radionuclide imaging be included to ensure better diagnostic specificity.

EXERCISE TEST RESULTS

Mrs. HK walked 6 min on a modified Bruce protocol, achieving 5 METs. Peak HR was 152 beats $\cdot$ min^{-1}, and peak BP was 170/70 mmHg. There were no additional ECG changes and no significant ventricular or supraventricular rhythm abnormalities. However, radionuclide imaging showed a small area of reversible inferior wall ischemia.

Mrs. HK's doctor added metoprolol (12.5 mg twice daily) to her regimen and raised her simvastatin from 10 to 20 mg. He told her that if things did not improve, he would likely recommend a heart catheterization. Over the next few weeks she experienced more fatigue plus myalgias and lapsed into even greater inactivity. Despite some initial apprehension, she is now inclined to undergo a heart catheterization since she believes it will help improve her exercise capacity, quality of life, and presumably her longevity.

DEVELOPMENT OF EXERCISE PRESCRIPTION

This is a common scenario that highlights many typical aspects of geriatric medicine. Mrs. HK has exercise intolerance, most likely related to aspects of normal aging that progressively erode functional capacity. While morbidities such as coronary artery disease are likely to occur in the elderly, especially in the context of chronic hypertension and hypercholesterolemia, her progressive exercise decline does not necessarily indicate that coronary artery disease is the cause or that she will benefit from revascularization. Moreover, the recently published COURAGE trial (4) indicates that even if patients like Mrs. HK are revascularized percutaneously by coronary stents, they will not live longer.

Instead, Mrs. HK will likely benefit from an exercise training program, one that is inclusive of strength training in association with aerobic, flexibility, and balance training. While issues of coronary artery disease as well as iatrogenic medication effects can complicate training goals, they do not preclude training. In Mrs. HK's case, side effects of β-blockers and statins are often temporary, but they need to be carefully monitored over time, especially in association with an exercise training regimen.

(continued)

An exercise physiologist will need to be cognizant of Mrs. HK's complex dynamics, that is, the components of normal aging, disease, medications, and the inevitable dynamics of her own hopes and frustrations and those of her children. Exercise must begin at low levels and progress in small increments to ensure that the regimen is well tolerated and that it addresses the spectrum of needs underlying her functional limitations and health requirements.

DISCUSSION QUESTIONS

1. What is an appropriate exercise activity and exercise intensity for Mrs. HK?
2. What are the major concerns for Mrs. HK during exercise activity?
3. Based on Mrs. HK's medical history and living situation, what are important barriers to her continued participation in the physical activity program?
4. What special precautions were likely taken during her exercise test?
5. What motivational strategies can be used to ensure her exercise adherence?
6. What are the major risks of Mrs. HK's participation in an exercise program?
7. What other tests might you perform on Mrs. HK (diagnostic or functional)?
8. What might be appropriate recommendations for advancing Mrs. HK to walking three times a week for 30 min or on a daily basis?

Chapter 31

Depression

Krista A. Barbour, PhD
Benson M. Hoffman, PhD
James A. Blumenthal, PhD

Depression is the most prevalent psychiatric disease or disorder in the United States. The term *depression* is often used to refer to varying levels of psychological distress, ranging from a dysphoric mood state to diagnosis of a clinical disorder, such as **major depressive disorder** (MDD). This chapter focuses on the clinical syndrome of MDD and includes discussion of subclinical depressive symptoms where appropriate.

DEFINITION

Because depression often results from or co-occurs with many of the chronic diseases described in this book, we believe it important that the student interested in clinical exercise physiology appreciate the challenges clinicians face when treating this disorder. For example, the clinical exercise physiologist must feel comfortable working with not only patients who have diabetes, but also patients with diabetes who are experiencing symptoms of depression as well.

In the United States, psychiatric disorders are currently diagnosed using criteria outlined in the *Diagnostic and Statistical Manual of Mental Disorders, Fourth Edition* (DSM-IV) (1). The DSM-IV diagnostic criteria for MDD are presented in the following sidebar. A revision of the DSM-IV is currently under way but was not available at the time of this writing. To our knowledge, there will not be any substantive changes in the diagnostic criteria for MDD. A diagnosis of MDD requires endorsement of at least five symptoms, one of which must be either depressed mood or diminished interest or pleasure, that have been present during the same 2 wk period.

SCOPE

A recent estimate of the lifetime prevalence of MDD in community-based adults was reported as 16.6%, and a 12 mo prevalence estimate was 6.7% (44). Kessler and colleagues (45) found that of those individuals meeting criteria for MDD, 80% were classified as having moderate or severe episodes. Thus, MDD is a prevalent disorder that people often experience at a significant level of severity. If not treated, most episodes of depression last for several months before remitting (49). Furthermore, MDD tends to be episodic; many individuals who recover from an episode of MDD experience a recurrence of the disorder (28).

MDD is consistently found more commonly in women than in men, and this gender difference holds across cultures (85). Studies examining differences in prevalence of MDD between ethnic minorities and Caucasians have yielded mixed results, but both African Americans and Latinos tend to report less depression compared with Caucasians (45). Subgroups of ethnic minorities may be at increased risk for MDD because of difficulties in meeting basic needs, such as food and shelter (65); and it has been suggested that depression tends to be underrecognized and undertreated in minorities with

Summary of DSM-IV Diagnostic Criteria for Major Depressive Episode

Five (or more) of the following symptoms are present during the same 2 wk period and represent a change from previous functioning (note that either number 1 or number 2 must be one of the symptoms).

1. Depressed mood most of the day, nearly every day
2. Diminished interest or pleasure in all or most activities
3. Significant change in weight or appetite
4. Insomnia or hypersomnia
5. Psychomotor agitation or retardation
6. Fatigue or loss of energy
7. Feelings of worthlessness or guilt
8. Diminished ability to think or concentrate
9. Recurrent thoughts of death or suicide

Note. The symptoms cause clinically significant distress or impairment, are not caused by the effects of a substance or general medical condition, and are not better accounted for by bereavement or another psychiatric disorder.

coronary heart disease (CHD), particularly African Americans (82). Besides the great personal cost of MDD, the public health burden is enormous. For example, MDD has been linked to greater health care utilization, decreased quality of life, missed time at work, and increased rates of attempted suicide (42). When compared with patients who have chronic medical disease (e.g., diabetes, hypertension), people with depression report similar or worse functioning (87). For the year 2000, the economic burden of depression in the United States (including direct treatment costs, mortality costs of depression-related suicide, and indirect costs associated with depression in the workplace) were estimated at approximately $44 billion (34). Depression ranked fourth on the list of conditions with the greatest global disease burden for the year 2000 (80). Clearly, MDD is a prevalent, recurrent disorder associated with significant morbidity and economic costs.

PATHOPHYSIOLOGY

Depression is widely viewed as an interaction between genetics and the environment. As a result, both neurochemical and psychosocial variables have been implicated as contributing to the pathophysiology of depression (10). Specifically, three neurotransmitter systems—norepinephrine, serotonin, and dopamine—have been theorized to play a significant role in the onset and maintenance of depression. These theories of neurotransmitter dysregulation are complex but generally suggest neurotransmitter deficiencies in individuals with depression. In addition, dysregulation in the hypothalamus–pituitary–adrenal (HPA) axis (a marker of reaction to stressors) has been implicated in the pathophysiology of depression (10). HPA axis dysregulation, in turn, has been associated with CHD risk factors such as truncal obesity, hypercholesterolemia, hypertriglyceridemia, and hypertension (30, 52, 69). Depression has been associated with other pathophysiologic processes as well, such as autonomic nervous system dysregulation as evidenced by reduced heart rate variability (19), increased inflammation (36, 51), increased platelet reactivity (17), and endothelial dysfunction (72). Additional psychosocial contributions to the etiology of depression involve the role that negative cognitions play in causing depressed mood. Exercise training may serve to improve depressive symptomatology through alterations in neurotransmitter functioning or improvement in negative evaluations of self (e.g., enhanced self-efficacy and self-worth). Low social support is also implicated in the occurrence of depression, and may represent an independent risk factor for CHD. Because exercise frequently occurs in the context of group exercise sessions, the presence of other participants in facility-based exercise programs may also improve depressive symptoms.

CLINICAL CONSIDERATIONS

The gold standard for the assessment of depression is a structured diagnostic interview administered by a trained mental health professional, but an extensive assessment of that kind is impractical in most exercise settings. Fortunately, a number of tools or measures exist that allow exercise physiologists to screen for depressive symptoms and to refer patients to the appropriate mental health professional for further assessment when necessary.

Signs and Symptoms

Besides helping to administer these pen-and-paper screening tests, exercise physiologists should be cognizant of the cardinal signs and symptoms that might alert them to the fact that intervention by a trained mental health professional is urgent. Examples include self-reported symptoms of hopelessness and suicidal ideations. Patients who report these symptoms should be carefully monitored, and the appropriate health care professional should be consulted (e.g., the patient's primary care provider).

History and Physical Exam (or Diagnostic Testing)

A number of self-report measures are available to assess the presence and severity of depressive symptoms. These questionnaires generally consist of a series of written questions with multiple-choice responses that patients complete on their own. Widely used examples include the Beck Depression Inventory-II (BDI-II) (8), the Center for Epidemiological Studies Depression Scale (CES-D) (66), and the Hospital Anxiety and Depression Scale (HADS) (91). These measures are used to obtain information about the type and severity of depressive symptoms. For additional depression management strategies, see practical application 31.1.

Practical Application 31.1

CLIENT–CLINICIAN INTERACTION

Exercise professionals who work with patients with depression may find the experience both challenging and rewarding. As noted elsewhere in the chapter, symptoms of depression such as fatigue and loss of interest in people and activities may adversely affect a patient's ability to adhere to an exercise program.

Patients with depression often require more attention by the exercise professional than do nondepressed individuals. The exercise professional's consideration may be especially important during the initiation phase of an exercise program. For example, many patients with depression feel a lack of self-worth and may not believe in their ability to participate successfully in an exercise program. The loss of energy that occurs with depression may further contribute to a patient's reluctance to begin exercising. Thus, depressed patients may need more reassurance and positive reinforcement (even for just showing up to an exercise training session) when they begin the program. In addition, the cognitive symptoms of depression, such as indecisiveness and difficulty in concentrating, may lead to problems with recalling the exercise prescription or remembering to set exercise goals. Occasional reminders of such exercise information can be helpful (presenting information in both oral and written forms is a good idea).

Depression often is accompanied by social problems such as family conflict and unemployment. The exercise professional should be aware that such stressful life situations may present barriers to participation in an exercise program. For example, a depressed patient who is going through a divorce and is between jobs may be less likely to consider exercise a priority.

When clinicians encounter patients with depressive symptoms that interfere significantly with exercise participation, the appropriate approach may be to refer them to a mental health provider for treatment to improve quality of life and, potentially, to achieve motivation to exercise. Several treatments for depression have shown success, including antidepressant medication and cognitive–behavioral therapy. Patients with depression should also be informed that exercise training has been shown to reduce depressive symptoms in both healthy and medically ill populations.

In summary, depressed patients may require greater attention to ensure continued adherence to the exercise prescription compared with nondepressed patients. Because depression is associated with increased risk of dropout from exercise programs, staff should closely monitor depressed clients and refer them for specialized treatment when appropriate.

Self-report questionnaires are not designed to diagnose MDD and should not be substituted for a diagnostic clinical interview conducted by a trained mental health professional. Instead, these screening questionnaires enable clinical exercise physiologists and other clinicians to triage each case and develop a plan for action. For example, one patient's depressive symptoms may be mild enough to manage within the exercise setting, whereas another may benefit from referral to a mental health provider for further evaluation of mood. Often it is not practical for a mental health specialist to assist in the interpretation of these questionnaires, so specific screening questions have been developed for use by primary care physicians. For example, the depression module of the Patient Health Questionnaire (PHQ-9) is a nine-item symptom checklist designed to quickly screen for clinical depression and to measure depressive symptom severity (77). In a multisite validation study, PHQ-9 scores ≥10 had a sensitivity and specificity of 88% for MDD (47). Patients typically complete the PHQ-9 in less than 1 min, and Spanish and English versions of the PHQ-9 can be downloaded at no cost (www.phqscreeners.com).

Finally, in the assessment of depression, it is important to recognize that several of the symptoms (especially those within the somatic realm, such as weight loss and fatigue) are also symptoms of a variety of medical conditions. Thus, the diagnosis of depression can be complicated in patients who also are medically ill (20).

Treatment of Depression

Many people with depression do not seek treatment. In a recent household survey of 9,282 adults aged 18 yr and older (84), 6.7% of respondents met criteria for MDD. Of these individuals, 56.8% had sought treatment for MDD within the previous 12 mo period. Unfortunately, only 37.5% of those who sought treatment received treatment that was considered even minimally adequate (i.e., based on current evidence-based treatment guidelines). These results suggest that MDD is untreated or inadequately treated for most individuals. Indeed, the undertreatment of MDD has been recognized as a significant public health problem in the United States (38), particularly among ethnic minority groups (62).

Except in cases of severe depression (for which care is usually rendered by a psychiatrist, psychologist, or other mental health professional), most depressed patients who seek treatment are treated by primary care practitioners (88). Patients who present with depression in the primary care setting are more likely to be offered medication than psychotherapy or a combination of medication and psychotherapy (68).

Antidepressant Medication

Antidepressant medication is the most widely used treatment for depression in the United States (60, 62); see table 31.1 for a list of commonly used antidepressant medications. Results from numerous randomized clinical trials (RCTs) have provided evidence for efficacy of antidepressant medications, both for the treatment of acute depression and for the prevention of relapse (35, 41, 73), although there is uncertainty about which patients stand to benefit most from which treatments. RCTs generally conclude that less than one in two clinically depressed adults responds to treatment with one antidepressant (40), and less than two in three depressed adults respond to treatment with more than one antidepressant (71, 78). Furthermore, two recent meta-analyses determined that the effects of antidepressant medications were superior to placebo for severe depression, but were comparable to placebo for mild to moderate depression (29, 47). In addition, adherence to antidepressant medication regimens often is inadequate relative to treatment guidelines (25), and early discontinuation of treatment is associated with recurrence of MDD (59). These findings underscore the importance of identifying alternatives to pharmacotherapy in the treatment of depression.

Depression is a particularly important issue for patients with CHD. As many as 50% of CHD patients exhibit elevated depressive symptoms (83). Furthermore, depression appears to be an important and independent risk factor for fatal and nonfatal events in CHD patients. In one meta-analysis, patients with CHD and depression were almost twice as likely to die within 2 yr compared to nondepressed CHD patients (6). CHD also complicates the treatment of depression. The use of tricyclic antidepressants (TCAs) and monoamine oxidase inhibitors (MAOIs) by cardiac patients has been discouraged due to potential cardiovascular safety concerns (53).

To date, only a handful of RCTs have evaluated the effect of depression treatment among cardiac patients. The Sertraline Antidepressant Heart Attack Randomized Trial (SADHART) determined that the selective serotonin reuptake inhibitor (SSRI) sertraline was safe for patients with MDD and either a recent myocardial infarction (MI) or unstable angina. Sertraline was effective at reducing depressive symptoms compared to placebo, but only for a subset of patients with a history of recurrent depression and more severe depression (32). In two additional RCTs, sertraline was observed to be safe for patients with CHD (81) and chronic heart failure (61), but no more effective than placebo. The Canadian Cardiac Randomized Evaluation of Antidepressant and

Psychotherapy Efficacy (CREATE) trial demonstrated that the SSRI citalopram combined with clinical management was more effective than placebo and clinical management in the reduction of depressive symptoms among CHD patients (50). Because patients received 20 min of clinical management as well as the medication, the benefit of the medication as it might be prescribed and monitored in actual clinical practice is uncertain. These trials suggest that SSRIs appear to be safe for depressed cardiac patients, although their effectiveness for such patients has not been definitively established.

Because many patients either do not respond to or cannot tolerate antidepressants, it is important to identify alternative approaches to depression treatment. Two such treatments that have received empirical support are psychotherapy and aerobic exercise.

Psychotherapy

There is considerable empirical support for the efficacy of psychotherapies for the treatment of MDD (21). Two of the most well-studied and widely used psychotherapies for depression are cognitive–behavioral therapy (CBT) and interpersonal therapy (IPT). CBT emphasizes the important influence that negative thoughts have on emotion and behavior (7). For people with MDD, treatment focuses on modifying maladaptive thoughts as well as addressing deficits in behavior (e.g., unassertiveness, isolating oneself from others) that lead to and maintain depression. For example, a depressed patient undergoing CBT might be taught to identify cognitive distortions associated with her depression, to challenge those distortions, and to replace them with more realistic thoughts. Because patients with depression often lack motivation, another therapeutic exercise might involve planning a daily schedule of activities in which to engage. Following the schedule may result in decreased boredom and loneliness as well as increased motivation.

IPT is a structured and time-limited psychotherapy that emphasizes the interpersonal context in which depressive symptoms occur (e.g., social isolation, interpersonal conflicts, role transition, or loss of a loved one) (86). The IPT therapist forms a supportive relationship with the depressed patient, helps the patient identify interpersonal problems, and uses strategies to help the patient develop and adapt new interpersonal behaviors (e.g., assertive communication, reaching out for social support). CBT, IPT, and other empirically validated psychotherapies have been associated with large effect sizes both in RCTs and in clinical practice (5), and there is little reason to believe that one empirically validated psychotherapy is necessarily superior to another (22). Table 31.2 provides cognitive–behavioral therapy (CBT) resources.

To date, only three RCTs have evaluated the effects of psychotherapy for depression in CHD patients. In the largest of these three trials, Enhancing Recovery in Coronary Heart Disease (ENRICHD), 2,481 post-MI patients with MDD, minor depression, or low social support were randomly assigned either to CBT or to an education control group (11). In addition, patients with severe depression or those who were nonresponsive to CBT

Table 31.1 Pharmacology: Commonly Prescribed Antidepressant Medications

Medication	Brand name	Class	Exercise effect	Other effects
Bupropion	Wellbutrin	Misc.	Rapid heartbeat, dizziness	Restlessness, insomnia, headache, tremor, dry mouth, confusion, nausea, constipation, menstrual complaints
Citalopram	Celexa	SSRI	None identified	Nausea, nervousness, insomnia, diarrhea, rash, agitation, sexual side effects
Duloxetine	Cymbalta	SNRI		
Escitalopram	Lexapro	SSRI		
Fluoxetine	Prozac	SSRI		
Paroxetine	Paxil	SSRI		
Sertraline	Zoloft	SSRI		
Venlafaxine	Effexor	SNRI		
Mirtazapine	Remeron	NaSSA		

Note. Bupropion does not share a chemical structure with other types of antidepressant medications and as a result is considered a miscellaneous or "other" antidepressant medication. SSRI = selective serotonin reuptake inhibitor; SNRI = serotonin and norepinephrine reuptake inhibitor; NaSSA = noradrenergic and specific serotonergic antidepressant.

Table 31.2 Cognitive–Behavioral Therapy (CBT) Resources

Resource	Description
Feeling Good: The New Mood Therapy	Self-help book by David Burns that uses the basic principles of CBT (18)
Association for Behavioral and Cognitive Therapies website (www.aabt.org)	Website that can be used to locate psychotherapists by geographical location
Cognitive Therapy: Basics and Beyond	Book by Judith Beck that describes the cognitive model and therapy in detail (9)
International Society for Interpersonal Psychotherapy (www.interpersonalpsychotherapy.org)	Website that provides a concise overview of interpersonal psychotherapy

were considered candidates for augmentation with the antidepressant sertraline. At 6 mo, CBT was associated with greater improvement in depression severity scores compared with education control. The aforementioned CREATE trial randomly assigned 284 CHD patients to receive clinical management and either (1) IPT plus citalopram, (2) IPT plus placebo pill, (3) citalopram, or (4) placebo pill. Clinical management was associated with improvement, and IPT did not appear to confer any additional benefit beyond clinical management. However, clinical management was conducted by IPT therapists and therefore may have masked any beneficial effects of IPT. Finally, the Coronary Psychosocial Evaluation Study (COPES) assigned 157 patients with acute coronary syndrome and persistently elevated depressive symptoms to receive either usual care or their choice of antidepressant treatment (23). At 9 mo, depression severity scores were considerably lower for the treatment group, 81% of whom chose to receive psychotherapy. Taken together, these results suggest that psychotherapy may be effective for reducing depressive symptoms in CHD patients, although additional studies are needed to establish the benefits of psychotherapy in reducing depressive symptoms and improving clinical outcomes.

Exercise Testing

Intuitively, several of the key symptoms of depression might be expected to affect performance during exercise testing. For example, a patient with psychomotor retardation and fatigue may demonstrate poorer performance relative to nondepressed patients. Using the PRIME-MD and the BDI-II, Lavoie and colleagues (48) assessed depression in 1,367 patients referred for exercise stress testing. Thirteen percent of the sample was found to meet criteria for MDD. Compared with nondepressed patients, individuals with MDD exhibited a lower peak heart rate when expressed as a percentage of age-predicted maximum, a lower peak metabolic equivalent (MET) level, and shorter total exercise duration. Similar results were found in a sample of outpatient veterans with stable coronary artery disease (70). Thus, some researchers (48) recommend routine depression screening of patients who are undergoing exercise testing. For patients with significant depression, pharmacological stress testing may be preferred as a way to minimize the effect of motivational factors on test performance.

Exercise as a Treatment for Depression

A number of studies have reported that exercise is an effective treatment for depression (56, 57, 75). A recent meta-analysis summarized the results of RCTs of exercise on depression. Of the 25 studies included, 23 trials randomly assigned a total of 907 participants to either an exercise condition or to a control or no treatment condition. On average, participants assigned to an exercise condition had depression severity scores that were 0.82 standard deviations lower than in participants assigned to control or no treatment conditions. This is considered a large effect (58). Although the authors cautioned that many of the studies included in the meta-analysis suffered from methodological limitations (e.g., absence of follow-up data, insufficient randomization procedures), they also concluded that exercise is a reasonable recommendation for patients with MDD.

To date, only two published RCTs have examined the use of exercise compared with antidepressant medication as a treatment for MDD. In the Standard Medical Intervention and Long-term Exercise (SMILE) study (13), 156 middle-aged and older adults diagnosed with MDD were randomized to supervised exercise, sertraline, or a combination of exercise and sertraline. The 16 wk exercise treatment consisted of three weekly sessions of aerobic activity. By the end of the treatment period, each of the three treatment groups experienced a significant reduction in their levels of depression. The treatments did not differ significantly from one another in efficacy, suggesting that exercise may be a viable alternative to medication in the treatment of MDD. A follow-up assessment completed 6 mo after the beginning of treatment (4) showed that those individuals assigned to exercise alone experienced lower rates of depression than did those who received medication or a combination of exercise and medication. In addition, only 9% of remitted participants in the exercise group relapsed, compared with more than 30% of participants in the medication and combination groups. Another finding was that 64% of participants who received the exercise treatment continued to exercise following completion of the program. Self-reported exercise among all participants was associated with a 50% reduction in risk of depression 6 mo after the study ended. However, this study did not include a placebo control condition, and participants exercised in a group setting. Consequently, the possibility that the observed treatment effects were due to the nonspecific effects of staff attention, group support, or regression to the mean could not be ruled out.

These methodologic limitations were addressed in a second RCT, known as SMILE II, in which 202 clinically depressed adults were assigned to supervised exercise, home-based exercise, sertraline, or placebo pill (12). After 4 mo of treatment, depression severity scores improved for all groups. In addition, MDD remission rates tended to be higher for the active treatment groups (supervised exercise = 45%; home-based exercise = 40%; sertraline = 47%) compared with placebo (31%; $p = .057$). One year after the completion of treatment, the rate of MDD remission increased to 66%, although remission rates no longer differed by treatment group assignment (39). However,

Exercise, Depression, and Sleep

Disturbed sleep is a defining feature of depression, and the presence of disturbed sleep among depressed patients predicts poor response to treatment and future depressive episodes (55). Furthermore, many medications used to treat depression may also lead to sleep disruption (3). There is now evidence to suggest that exercise, known to be an effective treatment for depression, may also improve disturbed sleep.

Epidemiological investigations support a significant link between engaging in exercise and better sleep (89). Several interventional studies have demonstrated that acute bouts of exercise are generally associated with small, but significant, improvement in total sleep time among "normal" sleepers (89, 90). Montgomery and Dennis (60) set out to conduct a meta-analytic review of *chronic* exercise interventions for insomnia among nondepressed older adults. However, they were able to identify only one study that met their inclusion criteria (46). In that study of 48 male and female community-dwelling adults, participants assigned to a 16 wk exercise intervention (four 30-40 min aerobic exercise sessions per week) scored better on a standardized sleep scale than wait-list controls, and exercise was associated with an average improved sleep duration of 42 min per night. In a recent study of 17 nondepressed adults with insomnia, participants who completed a 16 wk aerobic exercise program plus sleep hygiene reported that their sleep duration improved by an average of 75 min per night, compared to 13 min for the attention control plus sleep hygiene condition (67).

Singh and colleagues reported that an 8 wk strength training intervention was associated with greater improvement in both depressive symptoms and sleep, compared to an attention control (74), and that a high-intensity 8 wk strength training intervention was associated with greater improvement in both depression and sleep compared to a low-intensity strength training intervention and a usual-care condition (76). However, data regarding aerobic or resistance training exercise as a treatment for disturbed sleep among depressed adults are limited; ultimately, there is no definitive evidence that regular aerobic or resistance training exercise improves sleep among depressed persons.

Evening exercise has been associated with sleep disturbance in some studies (79), but this issue has not been investigated among patients with clinical depression.

In summary, there is encouraging evidence that exercise may be an effective treatment for disturbed sleep among depressed adults. When formulating an exercise prescription for depressed patients with sleep disturbance, the exercise physiologist is encouraged to prescribe daytime exercise, as evening exercise may disrupt sleep for some patients.

patients who exercised during the 1 yr follow-up period had lower depression severity scores and were more likely to be classified as MDD remission compared to patients who did not exercise.

In summary, exercise may be as effective as antidepressant medication in the treatment of MDD. In addition, maintaining a long-term exercise program may extend the short-term benefits of exercise and may augment the benefits of antidepressant use. To date, no RCTs have examined the effects of exercise on depression in cardiac patients with MDD. Understanding Prognostic Benefits of Exercise and Antidepressant Therapy (UPBEAT) is an ongoing RCT that will randomly assign stable CHD patients with elevated depressive symptoms to 4 mo of treatment with aerobic exercise, sertraline, or placebo (14). Primary endpoints will include depressive symptom severity, as well as intermediate biomarkers of disease risk.

A number of potential mechanisms may be responsible for the reduction in depressive symptoms associated with exercise (16). For example, the central monoamine theory suggests that exercise corrects dysregulation of the central monoamines believed to lead to depression. Psychological factors also may be responsible for exercise-related improvements in mood. One hypothesis is that exercise reduces depression through increases in self-esteem and self-efficacy. Other potential psychological mechanisms include the distraction from negative emotion provided by exercising as well as the behavioral activation occurring with exercise, which is also an important component of CBT for MDD. The issue of how much exercise is needed to achieve an antidepressant effect is also an important topic, one that is discussed in practical application 31.2.

EXERCISE PRESCRIPTION AND TRAINING

An exercise prescription for people with depression will likely differ little from the prescription used for healthy individuals. Clinicians should be aware, however, that several symptoms of depression (e.g., loss of interest, fatigue) may interfere with participation in exercise, and that **anxiety** disorders can further complicate matters. Concerning anxiety, although its relationship with exercise is not as well studied as that between depression

Practical Application 31.2

HOW MUCH EXERCISE?

In general, studies have demonstrated the effectiveness of exercise in treating depression. One question that remains unanswered is the dose of exercise required to obtain an antidepressant effect. Specifically, what frequency, intensity, and duration are most beneficial in treating patients with depression? Dunn and colleagues (27) completed a trial designed to address this issue. In this trial, 80 sedentary adults diagnosed with MDD were randomized to undergo 12 wk of one of five exercise training treatment conditions: low energy expenditure and 3 d of exercise training per week, high energy expenditure and 3 d of exercise training per week, low energy expenditure and 5 d of exercise training per week, high energy expenditure and 5 d of exercise training per week, or stretching and flexibility control.

Results of the trial showed that exercise conducted at the high energy expenditure dose (consistent with public health recommendations) was effective in reducing depressive symptoms (47% from baseline measurement) over the 12 wk treatment period. In contrast, although participants randomized to the low energy expenditure dose did experience some reduction in depression over the treatment period (30% from baseline measurement), they did not respond significantly better than participants in the control condition did (29% from baseline measurement). Regarding frequency of training, no significant difference in treatment response was found between participants who exercised 3 d per week relative to those who exercised 5 d per week. This result identifies a higher energy expenditure as the key aspect of exercise dose related to remission of MDD, regardless of days per week of exercise.

In summary, this trial demonstrated that an exercise training dose consistent with public health recommendations for energy expenditure is effective for the treatment of MDD in previously sedentary adults. Importantly, an exercise prescription that includes lower energy expenditure does not appear to be beneficial in reducing depressive symptoms. Thus, exercise professionals should encourage depressed patients to achieve at least the minimum recommended levels of energy expenditure.

and exercise, evidence suggests that anxiety does affect exercise participation and compliance rates (see "Anxiety and Exercise" sidebar).

For individuals whose depression is considered stable enough to allow starting an exercise program (e.g., they are not suicidal; their depression is not so severe that it prevents active participation), the question arises how to manage depressive symptoms that may affect the course of participation in exercise. First, recognize that significant comorbidity exists between depression and other chronic diseases; people with depression are unlikely to present with depression alone. Furthermore, depression can affect the course of chronic disease. For example, the presence of depression is associated with more complications and increased mortality in patients with diabetes (24, 43). Evidence also identifies depression as a powerful and independent risk factor for cardiac outcomes in patients with coronary heart disease (such as recurrent MI and mortality) (52). For cancer patients, untreated depression has been linked to poorer treatment adherence, increased hospital stays, and mortality (64).

Besides often being present as a comorbidity in patients with another chronic disease, depression is associated with unhealthy lifestyle behaviors such as tobacco and alcohol use (2), poor diet, and physical inactivity (15).

An exercise prescription for patients with depression appears in practical application 31.3.

Depression has been found to be associated with decreased adherence to treatment regimens across a wide variety of medical illnesses. In a quantitative review of 12 studies that included patients with renal failure,

Anxiety and Exercise

In patients with depression, anxiety may be the single best predictor of both treatment dropout and failure to benefit from treatment. For example, participants enrolled in an exercise treatment program who also rated themselves as high in anxiety during the screening period were at greater risk of dropping out of the treatment program prematurely. In addition, if patients with comorbid anxiety and depression completed the treatment program, they were less likely to experience remission of depression (37).

Exercise professionals may encounter several types of anxiety:

Social phobia: This disorder is characterized by an intense fear of making mistakes or looking foolish in public. This fear often leads to avoidance of certain people, places, or social events.

Posttraumatic stress disorder (PTSD): PTSD arises in response to experiencing or witnessing a traumatic event. Although many people exposed to trauma temporarily experience stress-related symptoms, those with PTSD continue to struggle with intrusive thoughts and nightmares, as well as increased arousal (e.g., anger) and avoidance of reminders of the trauma.

Panic disorder: People with panic disorder experience sudden, unexpected periods of extreme fear known as panic attacks. Symptoms of panic attacks include sweating, heart palpitations, a feeling of choking, dizziness, and a fear of dying.

Obsessive–compulsive disorder (OCD): OCD is characterized by disturbing, uncontrollable thoughts (obsessions). To alleviate these thoughts, individuals engage in repetitive behaviors (compulsions) in an effort to prevent some feared situation (e.g., excessive hand washing to prevent contamination).

Generalized anxiety disorder (GAD): People with GAD experience excessive worry across a number of life domains (e.g., family, work) that is difficult to control. The worry is associated with insomnia, muscle tension, and restlessness.

Exercise professionals should familiarize themselves with the symptoms of anxiety disorders. Remember that occasional, minimal symptoms of anxiety are a normal part of life. To meet criteria for a disorder, symptoms must result in impaired functioning in an aspect of a person's life. If a patient appears to be experiencing significant anxiety, the exercise professional may wish to discuss this with him and make a referral for treatment as needed. Both CBT and anxiolytic medication have been shown to be effective in treating anxiety disorders.

Practical Application 31.3

SUMMARY OF EXERCISE PRESCRIPTION

Mode	Frequency	Intensity	Duration	Special considerations
Gross motor activities such as walking or biking	Five times per week	60% to 75% of heart rate reserve	40 to 60 min	Untreated or undertreated depression can negatively affect adherence to exercise. Facilitate care to prevent depression symptoms from interfering with exercise participation.

cardiovascular disease, arthritis, and cancer, patients with coexisting depression were three times more likely to be nonadherent to treatment recommendations as nondepressed patients (26). This finding has been replicated in more recent studies of patients with diabetes (54) and coronary heart disease (31), as well as with individuals referred to exercise programs. For example, results of studies of depressed patients enrolled in cardiac rehabilitation have demonstrated that depression status at program entry is predictive of number of sessions attended as well as dropping out (33). Thus, patients with depression may find it more difficult to stay engaged in an exercise program compared with patients who are not depressed, and specific symptoms of depression such as fatigue and a loss of interest in people and activities may interfere with adherence to an exercise regimen (see "Strategies for Improving Exercise Adherence in Patients With Depression").

Strategies for Improving Exercise Adherence in Patients With Depression

- Most importantly, work to establish good rapport with patients. Positive feedback and empathy from exercise staff can go a long way toward promoting adherence.
- At the beginning of an exercise program, review with patients their unique barriers to participation (e.g., work responsibilities, family issues). These barriers should then be discussed with patients to find ways to overcome or minimize these obstacles.
- Educate patients about the benefits of exercise for physical health and depression. Elicit from patients other benefits of exercise specific to them and periodically remind them of those benefits.
- Patients are more likely to adhere to exercise training if the experience is enjoyable. Work with patients to increase their satisfaction with the program (e.g., switching equipment used, varying the time of day).
- Help patients develop realistic exercise goals (e.g., gradual increase in number of exercise sessions per week).
- Encourage patients to reward themselves for participation in exercise. Emphasize the importance of positive reinforcement for accomplishments. Even simple rewards (e.g., a new book, a pleasant dinner out) can be powerful motivators.
- Recommend to patients that they talk to family members and friends about their exercise program and goals. Such people are often a valuable resource in offering encouragement for patients' participation in exercise.
- Remember that untreated or undertreated depression is likely to have a negative effect on adherence to exercise. Encourage patients to seek treatment for depression if symptoms appear to interfere with exercise participation.

O'Neal and colleagues (63) have offered recommendations for working with depressed people in the supervised exercise setting. First, they emphasize that with depressed individuals, nonadherence should be expected. Exercise professionals should avoid judging or blaming the patient for her depression, because doing so will likely lead to guilt and a sense of failure that may cause the person to drop out from an exercise program. Instead, when nonadherence occurs, it should be viewed as a learning opportunity. That is, lapses in exercise participation can be used to identify an individual's unique barriers to adherence. The exercise professional can then assist her in finding ways around these obstacles. Finally, when working in exercise settings, the exercise professional should be familiar with the symptoms of depression and have some knowledge of treatment options. When depression is identified, the exercise professional should express warmth and empathy toward the patient while taking care to preserve an appropriate clinician–client boundary. That is, the exercise professional should not attempt to be the patient's psychotherapist but should instead have referral sources available.

CONCLUSION

Depression is a prevalent condition that can affect exercise testing as well as level of participation in exercise programs. Screening for depressive symptoms that may interfere with exercise and subsequent referral for treatment are essential. The clinical exercise physiologist should recognize that the depressed person who exercises is at risk for nonadherence, that depression is common in patients with CHD and other chronic diseases, and that depressive symptoms (e.g., loss of interest, lack of self-confidence, fatigue) may interfere with the enjoyment of exercise and motivation to fully engage in the rehabilitation process. People with depression will likely require increased attention from exercise professionals to ensure adequate adherence to the exercise prescription.

Fortunately, several treatments have shown success in the treatment of depression, including antidepressant medication, cognitive–behavioral therapy, and exercise training. Patients who exhibit significant depressive symptoms should be approached in an empathetic manner and encouraged to seek treatment to improve quality of life and gain maximal benefit from exercise.

Key Terms

anxiety (p. 612)

major depressive disorder (p. 605)

CASE STUDY

MEDICAL HISTORY

Ms. TH is a 62 yr old woman who joined your medical fitness center with the intention of reinitiating an exercise program. She is ambivalent about exercise even though her primary care provider recommended that she increase her activity level because of recent weight gain. Ms. TH revealed that she has struggled with her mood for the past year and that her primary care physician recently diagnosed her with major depression.

DIAGNOSIS

Her physician started her on sertraline to manage her symptoms. She noted that before becoming depressed she enjoyed walking around her neighborhood several times per week but has been sedentary. Her medical history is significant for hypothyroidism, for which she takes daily levothyroxine.

EXERCISE TEST RESULTS

Ms. TH recently underwent exercise testing; the results showed that she exercised for 6 min (Bruce protocol, approximately 7 METs) and achieved a maximum heart rate of 136 beats · min^{-1} (resting = 73 beats · min^{-1}) and blood pressure of 165/76 mmHg (resting = 118/76). No ischemia, chest pain, or arrhythmia was present.

(continued)

It was noted that although she was not experiencing symptoms, she required much prompting throughout to continue the test.

EXERCISE PRESCRIPTION

Ms. TH's exercise prescription would not differ significantly from that for other healthy but sedentary adults. Specifically, over several weeks and as tolerated, the prescription should progressively increase her activity levels from inactivity to exercise 3 to 5 d per week. During walking or cycling, she should try to achieve an intensity that is progressively increased from 70% of heart rate reserve to up to 85% of heart rate reserve, which is 117 to 126 beats $\cdot$ min^{-1}.

DISCUSSION QUESTIONS

1. Based on her exercise test results, how might you interact with Ms. TH as she begins her exercise program?
2. What do you note in Ms. TH's history that is promising in terms of starting an exercise program?
3. What can you tell Ms. TH about the effects of exercise on depression?
4. What symptoms should you monitor to decide whether Ms. TH's depression is worsening while she is participating at your center? How might you handle this situation?
5. Given her age, we would predict a peak heart rate around 158 beats $\cdot$ min^{-1}. Assuming that she is not on a β-adrenergic blocking agent, what might account for a peak heart rate that appears to be submaximal?

Chapter 32

Intellectual Disability

Bo Fernhall, PhD
Tracy Baynard, PhD

Individuals with an intellectual or developmental disability may have one of several disorders, all of which are dynamic over time yet have limitations in mental function and personal skills (e.g., ambulation, communication, self-care, social adjustment) in common. An intellectual disability is not a disease and cannot be communicated. It's also not a type of mental illness, like anxiety or depression. There is no cure for intellectual disabilities; and although most children with an intellectual disability can learn to engage in a variety of skills and activities, mastering them takes more time and effort than it does for others.

DEFINITION

The American Association on Intellectual and Developmental Disabilities (AAIDD) (16) defines intellectual disability (ID) as "a disability characterized by significant limitations both in intellectual functioning and in adaptive behavior, which covers many everyday social and practical skills. This disability originates before the age of 18." In general, a person with ID has reduced intellectual functioning (e.g., lower intelligence), which includes reductions in learning, reasoning, and problem solving. Intellectual functioning is usually determined from standardized intelligence tests such as the Wechsler Adult Intelligence Scale or the Stanford-Binet scale. These tests are also used to determine limitations in adaptive behavior by focusing on conceptual, social, and practical skills. The definition from AAIDD has been adopted by the Social Security Administration Revised Medical Criteria for Mental Disorders (16). However, this definition differs slightly from that in the proposed *Diagnostic and Statistical Manual of Mental Disorders V* (DSM-V) guidelines, which will require two or three deficits in adaptive behavior; the AAIDD definition requires a deficit in only one of the adaptive behaviors.

Individuals with ID usually need support services in one or more of the areas of adaptive behavior; the intensity of support needed is used to plan services but is not part of the definition or diagnosis of ID. Intellectual disability is not a static, nonchangeable condition. Instead, it is a fluid condition, and early intervention may help some individuals progress to the point where they are no longer classified as having ID (5).

SCOPE

The estimated prevalence of ID in industrialized society is 3% of the total population. If this estimate is correct, there are approximately 9 to 10 million individuals with ID in the United States; over 90% of all persons with ID are classified with mild ID (5, 19). Most of these individuals live in the community, either at home or in community-based group homes. As a consequence of the deinstitutionalization movement, very few individuals with ID now live in state-supported institutions. People with more severe forms of ID live in community-based facilities that offer more specialized and intensive support (20). Mortality has traditionally been higher in people with ID than in the general population, with estimates varying from 1.5 to 4 times higher than expected in populations without ID (43, 74). Mortality is typically linked to the low IQ and poorer self-care skills of persons with ID. However, low levels of physical activity, and especially movement ability (which may be related to self-care and low IQ as well), may also contribute to the higher mortality and morbidity of individuals with ID (26, 43). Except in persons with **Down syndrome** (DS), cardiovascular and pulmonary disorders are the most common medical problems in persons with ID (43, 64). For individuals with DS, leukemia, infections, and the early development of Alzheimer's disease are the most frequent causes of both mortality and morbidity (18, 62).

Life expectancy has been increasing rapidly in individuals with ID who do not have DS, to the point where it is now approaching that of the general population (62). In individuals with DS, life expectancy is still lower than in others with ID, but it has been increasing rapidly and is now getting close to 60 yr of age, with some people living into their 80s (62). The drastic change in life expectancy is likely due to better medical care, better living conditions, and, in the case of individuals with DS, better survival following corrective cardiac surgery. Considering that almost all individuals with ID live in supported living situations, the cost of ID is likely substantial, but little or no information is available on these costs.

PATHOPHYSIOLOGY

The pathophysiology, or cause of ID, is often difficult to identify, as it includes any condition that negatively affects brain development before, during, or after birth until the age of 18. Several hundred causes of ID have been identified or suspected, but in half to one-third of individuals with ID, the cause is unknown. The three major causes are **fetal alcohol syndrome**, **fragile X syndrome**, and DS (77). In general, causes of ID usually fit in one of the following general categories: (1) genetic, (2) problems during pregnancy, (3) problems at birth, (4) problems after birth, and (5) poverty and cultural deprivation. These categories are summarized in table 32.1.

Genetic conditions that cause ID typically result from either inherited gene disorders such as fragile X syndrome or gene disorders caused during pregnancy. Down syndrome, or trisomy 21, is the most common genetic cause for ID and in more than 90% of individuals with DS an extra or third (instead of just two) chromosome is found on chromosome 21. Other causes of DS include disjunction (24 chromosomes in one haploid cell and 22 in the other) and translocation (two chromosomes grown together as one but containing the genetic material of both) (49). Maternal age is the primary risk factor for DS, and up to 50% of all children born with DS are born with congenital heart disease. Although **congenital heart disease** used to be a major cause of early mortality in patients with DS, advances in surgical techniques and early diagnosis have substantially improved the prognosis of these children (5, 20).

The most common cause of ID when a problem occurs during pregnancy is maternal alcohol or drug use. In fact, excessive alcohol consumption is the leading preventable cause of ID. In contrast to the genetic causes of ID, many of the factors during pregnancy that result in ID are preventable (4, 77).

Premature delivery and low birth weight are the leading causes of ID associated with problems at birth. It is also possible for fetal oxygen deprivation during birth to cause ID, in addition to certain birth injuries (4, 77). Later on, several childhood diseases, such as measles, whoop-

Table 32.1 Summary of Potential Causes of Intellectual Disabilities and Related Factors

Potential causes	Factors associated with the potential causes					
Genetic	Maternal age	Infection(s)	Phenylketonuria		Trisomy 21 (Down syndrome)	
Problems during pregnancy	Maternal substance abuse	Rubella	Toxoplasmosis	Syphilis	Maternal malnutrition	Maternal exposure to environmental toxins
Problems at birth	Premature delivery, low birth weight	Fetal oxygen deprivation	Certain injuries during birth			
Problems after birth	Meningitis	Encephalitis	Other causes of brain damage (e.g., near drowning)		Environmental toxins (e.g., lead)	
Poverty and cultural deprivation	Malnutrition	Inadequate health care	Low educational and cultural stimulation			

ing cough, and chicken pox, can lead to meningitis and encephalitis, which may produce brain damage and cause ID. Poverty increases the risk of malnutrition, inadequate health care, and deprivation of educational and cultural experiences, which can also lead to ID over time. Thus, many of the known causes of ID are preventable, but it is important to remember that in up to half of all cases, the cause is unknown (4, 77).

Several features common in individuals with ID affect the physiologic responses during exercise. For example, obesity is more prevalent in persons with ID than in the general population (44, 73), and persons with ID are more likely to present with more advanced stages of obesity. Obesity is even more prevalent in certain subcategories of ID, such as DS and **Prader-Willi syndrome** (6, 39), with many of these individuals categorized as having severe obesity. Most individuals with ID exhibit low or very low levels of cardiorespiratory fitness, often coupled with obesity (see sidebar). It is not unusual for people with ID in their mid-20s to have a peak oxygen uptake ($\dot{V}O_2peak$) of only 25 to 30 ml · kg^{-1} · min^{-1}, which is approximately 25% to 35% below what would be expected (28). However, more recent information suggests that it may be mostly individuals with DS who present with exceptionally low levels of cardiovascular fitness; $\dot{V}O_2peak$ in patients with ID that is not due to DS have close to normal levels of cardiovascular fitness (6). It is not unusual for individuals with DS between 18 and 30 yr of age to have a $\dot{V}O_2peak$ between 15 and 20 ml · kg^{-1} · min^{-1}. These low levels of cardiovascular fitness are not the result of the high incidence of congenital heart disease in this population, as all published studies have been conducted in individuals free of congenital heart disease. Thus, the influence of congenital heart disease on cardiorespiratory fitness among individuals with DS is currently unknown.

A primary cause for the lower levels of cardiorespiratory fitness in individuals with DS, and potentially in individuals with ID who do not have DS, is a lower maximal heart rate. Although most individuals with ID exhibit maximal heart rates below expected levels, for persons without DS the reduction is not severe enough to classify their exercise heart rate response as chronotropic incompetence (6, 23, 28). However, it is well documented that almost all individuals with DS exhibit chronotropic incompetence (6, 22, 23, 28, 40, 41, 53). In fact, the average maximal heart rate can be expected to be ~30 beats · min^{-1} less than in age-matched individuals without disabilities (23). This reduction in peak heart rate leads to a reduced peak cardiac output and subsequently to the reduced $\dot{V}O_2peak$. It is likely that autonomic dysfunction is a cause of the reduced maximal heart rate in people with DS, as they exhibit reduced vagal withdrawal to handgrip exercise (24, 36). However, the primary cause for the reduced maximal heart rate during maximal treadmill exercise appears to be an inability to increase plasma catecholamines during exercise (22). Since catecholamines are a major contributor to increases in heart rate at high exercise intensities (above the ventilatory threshold), this partly explains the lower maximal heart rates in DS.

Regardless of how muscle strength has been measured, across all age groups, strength in persons with ID is 30% or more below that of age-matched nondisabled individuals (14, 25, 42, 45, 46, 63, 65, 66, 78). People with DS exhibit even lower levels of muscle strength, often 30% to 40% below their peers with ID and less than 50% of the expected strength levels of their nondisabled peers (11, 12, 42, 65). These low levels of muscle strength are present in childhood (25, 45, 66) and persist into adulthood (14, 46). Interestingly, in individuals with ID who were extremely active and exhibited very high aerobic capacities, muscle strength was still ~25% below the expected values for age- and activity-matched subjects without ID (37); this suggests a possible underlying abnormality in skeletal muscle function. The lower levels of muscle strength are important in persons with ID, because muscle strength has been related to both cardiorespiratory fitness (11, 63) and the ability to perform tasks of daily living (11). Furthermore, many individuals with ID

Physiologic Responses to Acute Exercise in Individuals With Intellectual Disability

- Cardiovascular responses to aerobic exercise
  - Lower maximal heart rate; in persons with DS, ~30 beats · min^{-1} lower than expected
  - Lower cardiac output
  - Lower peak aerobic capacity ($\dot{V}O_2$)
  - Low plasma catecholamine levels in persons with DS
- Responses to resistance exercise
  - 30% to 50% reductions in expected strength
  - Reduced strength even in very active individuals
  - Cardiac autonomic dysfunction

exhibit poor coordination, as well as reduced walking economy and walking stability, which have also been associated with muscle strength (2, 3, 55).

CLINICAL CONSIDERATIONS

Individuals are diagnosed with ID based on intelligence and mental ability and adaptive skills. Compared to individuals free of an ID, people with an ID have a reduced ability to think and solve problems and to adapt and function independently. Once a diagnosis of ID has been made, an individual's strengths and weakness can be evaluated in order to tailor the amount of support or help needed to function at home, in school, and in the community.

Signs and Symptoms

The major signs and symptoms are usually related to delayed developmental stages in infants and children. This can include delays in sitting up, crawling, and walking. Children may also experience delays in or difficulty with talking. Most developmental milestones are delayed, but the amount of delay depends on the severity and cause of the disability. Later in childhood, a child with ID may have difficulties understanding directions and experience problems with logical thinking and problem solving. It is not unusual for children with ID to be unable to understand social rules or the consequences of their actions (77).

History and Physical Examination

The history and physical examination should follow standard formats. It is likely that a history needs to be acquired from a significant caregiver, as individuals with ID may not accurately remember their history. For people with DS, it important to obtain information on congenital heart disease and any joint problems, such as instability of the atlantoaxial joint (**atlantoaxial instability**) in the upper neck (i.e., where the base of the skull meets the neck).

Diagnostic Testing

Chromosomal microarrays are being used increasingly as a first-tier genetic test among individuals with unexplained IDs (57). G-banded **karyotyping** should be reserved for individuals with obvious chromosomal problems or a family history of genetic abnormalities (57). This is a rapidly developing area in relation to ID, and our knowledge and understanding of genetic links to various types of ID will expand substantially in the coming years.

The level of ID is diagnosed through the use of two types of tests: standardized tests of intelligence (e.g., intelligence quotient [IQ] tests) and adaptive behavior (16, 20). The cutoff score used for both is generally 2 standard deviations below the mean for the particular assessment instrument. Well-known IQ tests include the Wechsler Adult Intelligence Scale, Wechsler Intelligence Scale for Children, Stanford-Binet, Woodcock-Johnson Tests of Cognitive Abilities, Raven's Progressive Matrices, and the Kaufman Assessment Battery for Children. Intellectual disability is diagnosed with an overall score at or below 70 points on most IQ tests. It is important to note that IQ scores are fluid, as intelligence and adaptive behavior can change over the course of a life span, in either a positive or a negative direction. Therefore it is recommended that intelligence and adaptive behavior tests be repeated through the life span.

Adaptive behavioral tests address conceptual, social, and practical skills (16, 20). Conceptual skills have to do with issues of language and literacy, money, time, numerical concepts, and self-direction (16). Skills related to the social being include "interpersonal skills, social responsibility, self-esteem, gullibility, naiveté, social problem solving, and the ability to follow rules, obey laws and avoid being victimized" (16). Lastly, practical skills involve areas related to personal care, travel and transportation, job skills, health care, maintaining a schedule or routine, safety, and using money and the telephone. The AAIDD is slated to release its new Diagnostic Adaptive Behavior Scale in 2012, which will provide a comprehensive standardized assessment of adaptive behavior (16). It will be used with individuals 4 to 21 yr old and provide specific diagnostic information around the cutoff at which someone is deemed to have "significant limitations" regarding adaptive behavior (16). Current tests available include the Woodcock-Johnson Scales of Independent Behavior (used for children), the Vineland Adaptive Behavior Scale, and the 2010 version of AAIDD's Diagnostic Adaptive Behavior Scale. The Vineland Scale is used for testing social skills in persons from birth through 19 yr of age and is administered to the individual's caregiver.

A diagnosis of ID is important for several reasons. It establishes eligibility for special education services, home- or community-based services, and Social Security benefits, as well as allowing for special treatment within the criminal justice system. Additional diagnostic tests should be considered for individuals with ID, in particular those with DS. These include tests for **hypothyroidism** and congenital heart disease, neck X-ray to determine possible atlantoaxial instability, and hearing and vision tests. Individuals with DS have higher rates of infection and respiratory problems; therefore regular blood tests should be performed to detect any abnormalities with the immune system (62, 74).

Exercise Testing

Exercise testing in individuals with ID appears to be largely safe, and safety considerations with respect to cardiovascular complications are likely not different from those for the general population (21, 26). However, it is important to keep in mind that while few reports, if any, exist on exercise-induced complications in persons with ID, there is no body of current scientific evidence to suggest whether exercise testing in this population is indeed safe or not. Certainly this is an area that needs scientific study. Common to a diagnosis of ID is a concern about the person's ability to follow and understand instructions and cooperate with the testing procedures, which is a concern regarding any type of test in this population. In spite of these concerns, standard laboratory exercise tests appear to be valid in persons with ID (21, 26). Individuals with ID can undergo a full complement of exercise-based tests, which should include tests of aerobic capacity, muscle strength and endurance, and body composition. A summary of exercise testing recommendations is provided in table 32.2.

Several key points are essential with regard to valid outcomes for exercise testing (20, 21, 32, 34, 35):

1. Pretest screening procedures should follow American College of Sports Medicine (ACSM) guidelines (1). The exception is for individuals with DS. Because nearly half of all individuals with DS have congenital heart disease, a careful health history and physical examination is necessary in this population, regardless of age. Further, there is a relatively high incidence of atlantoaxial instability, which also warrants a physical examination.
2. Appropriate familiarization with both personnel and testing procedures is necessary before performance of any "real" tests. The level of ID will determine the amount of familiarization needed to ensure test validity. Often two or three sessions are necessary, depending on the person and the test being performed. These sessions should include adequate time for the personnel to demonstrate the task(s) and for the individual with ID to practice them. Regular and positive feedback will enhance results.
3. Provide explicit and simple one-step instructions. Allow the individual to move along with each step

Table 32.2 Summary of Exercise Testing for Individuals With Intellectual Disabilities

Test type	Mode	Protocol specifics	Clinical measures	Clinical implications	Special considerations
Cardiorespiratory endurance	Treadmill Airdyne cycle 20 m shuttle run	Individualized walking speeds on treadmill Airdyne—use both arms and legs, 25 W stages	METs HR $\dot{V}O_2$	To assess progress in a training program	Familiarization is critical. See text for full list of considerations.
Skeletal muscle strength and endurance	1RM on weight machines Isokinetic testing Isometric maximal voluntary contraction (MVC)	RM testing: progress slowly Isokinetic testing: standard protocols, three or four maximal contraction trials; select best one or average of the three Perform 2 min of isometric contraction at 30% MVC	Kilograms	Same as above	Watch range of motion. Be careful to fit subject to machine, due to short statures, adjustments will need to be made. With MVC, verbal encouragement is needed.
Flexibility	Sit-and-reach Joint-specific goniometry	Active stretching	Distance Degrees	Same as above	Do not allow hyperextension; be watchful of range of motion.
Body composition	Air displacement plethysmography Skinfolds Waist circumference		% body fat Centimeters	Same as above	Estimate lung volume with air displacement plethysmography.

before proceeding to the next step (e.g., "Step up on the treadmill"; "Hold on"; "Begin walking").

4. Have extra personnel assist with testing on the treadmill, especially in persons with DS, due to the balance issues that are common.
5. Individualize testing protocols as needed, but do so in as systematic way as possible. For example, on a graded exercise treadmill test, start at an individualized comfortable walking speed. Then ramp up speed in similar increments between subjects. Then increase the incline similarly between subjects. Finally, increase speed to a jog or run.
6. Be careful with use of an age-predicted maximal heart rate (HR) formula in individuals with ID, especially those with DS (see practical application 32.2). A population-specific formula has been developed for DS (23) and is useful as a guide during exercise testing, but it should not be used for exercise prescription: estimated maximal HR = 210 − 0.56(age) − 15.5(x) [x = 1 for non-DS and 2 for DS].
7. There are several tests that are not recommended for use in individuals with ID due to poor reliability, validity, or both. For cardiovascular testing, these include treadmill running protocols, arm ergometry, and the 1 or 1.5 mi runs. Cycle ergometry protocols are also not recommended, with the exception of a dual-action cycle ergometer. In testing for muscle strength or muscle endurance, the one-repetition maximum (1RM) with free weights, push-ups, and the flexed arm hang are not recommended for this population.
8. When field tests are used to estimate cardiovascular fitness, it is necessary to apply population-specific formulas, which are outlined in table 32.3.
9. While cardiovascular field tests are reliable, they are not valid for predicting peak aerobic capacity in persons with DS.

Exercise Testing Review

Similar to the exercise testing of any other individual with a physical disability (e.g., osteoarthritis) or metabolic disorder (e.g., diabetes), the exercise testing of individuals with an ID requires that the personnel performing the testing be highly familiar with the various medical aspects of the specific disability, common secondary treatments, and the unique physiologic adjustments to exercise. In patients with ID, exercise assessments are often modified to accommodate the skills and abilities of the individual.

Cardiovascular

Cardiovascular testing largely comprises individualized treadmill walking protocols. Fernhall and colleagues (26, 28, 32-34) have demonstrated that individualized walking protocols are valid in this population. In general, after proper familiarization, the individual begins the treadmill protocol walking at a speed between 0.89 and 1.56 m/s (2-3.5 mph) for 4 min at 0% grade. Holding speed constant, grade is increased by 2.5% every 4 min until 7.5% grade is achieved. After this, grade is increased further every 2 min by 2.5% until 12.5% grade is reached; at this point, speed is increased ~0.22 m/s (0.5 mph) until the person reaches exhaustion. At these higher speeds, people will need to jog if they are able. Participants are allowed to hold on to the handrail as needed to maintain balance, especially at the higher speeds. Considering that many individuals with ID are poorly coordinated, holding on to the handrails is usually recommended. Importantly, this means that work capacity cannot be accurately predicted from speed and grade on the treadmill, and actual measurement of $\dot{V}O_2$ is recommended.

Dual-action cycle ergometers that require individuals to use both their arms and legs have also been demonstrated to be valid in this population (67, 68). Stages increase by 25 W, and it is important to focus on keeping the revolutions per minute in a reasonable range in order to obtain an accurate work rate. Field tests are another

Table 32.3 Formulas for Predicting $\dot{V}O_2$max From Field Test Performance in Individuals With ID

Field test	Formula
20 m shuttle run (11)	$\dot{V}O_2$max (ml · kg^{-1} · min^{-1}) = 0.35(no. of 20 m laps) − 0.59(BMI) − 4.5(gender; 1 = boys, 2 = girls) + 50.8
600 yd run/walk (11)	$\dot{V}O_2$max (ml · kg^{-1} · min^{-1}) = −5.24 (600 yd run time in minutes) − 0.37(BMI) − 4.61(gender; 1 = boys, 2 = girls) + 73.64
1-mile Rockport walk fitness test (18)	$\dot{V}O_2$max (L · min^{-1}) = −0.18 (walk time in minutes) + 0.03 (body weight in kilograms) + 2.90

BMI = body mass index.

alternative, and some are valid and reliable in estimating aerobic capacity in persons with ID. Because these tests are submaximal and do not require expensive equipment, and also because several people can be tested at the same time, they are an attractive alternative. The 20 m shuttle run, the 600 yd run/walk, and the 1 mi Rockport walking fitness test have been validated for use and deemed reliable in both children and adolescents with ID (27, 29, 30, 33, 81). Prediction formulas for estimating $\dot{V}O_2$peak with these tests are displayed in table 32.3. Only these population-specific formulas should be used, as they differ from the formulas used in nondisabled individuals. Furthermore, due to the large variability associated with field tests in persons with ID, it is important to keep in mind that individual estimates may be less accurate even though group data are very accurate (27, 29).

Leg strength contributes to endurance run performance in this population (25), which is quite different from what is observed in nondisabled individuals. Leg strength is also an independent contributor to field test performance in children with ID, which is an important issue for exercise training in this population.

Resistance Exercise

Use of 1RM testing with free weights is not advised in individuals with ID. However, weight machines are a perfect alternative, and 1RM testing can be achieved in this population across six to eight major muscle groups. Following standard protocols for 1RM testing is recommended. Verbally encourage proper breathing and provide physical spotting to ensure proper form. Ensure that overextension or overflexion is avoided.

Isokinetic testing can be performed for measures of strength, endurance, or both in this population. While no alterations need to be made to the computer-based protocols on these devices, proper fitting to the device is essential. This may be especially difficult among individuals with DS due to their generally short stature. Isometric testing can be used in this population, especially for determining cardiovascular responsiveness to the adrenergic stress. Visual feedback is important for this task, allowing the individual to make adjustments to the strength necessary to maintain the target resistance. Generally, three or four trials of isometric maximal voluntary contraction are needed; two or three of the trials should be within 1 or 2 kg of each other. Use either a mean of the three trials (assuming they are close enough to each other) or use the highest value obtained as the maximal voluntary contraction. For handgrip testing, 30% of maximal voluntary contraction is commonly set as the target resistance, and this level is generally held for 2 min in individuals with ID.

Flexibility

Both the sit-and-reach test and joint-specific goniometry are used in testing flexibility in this population. A possible concern is joint laxity in individuals with DS.

Treatment

There is no treatment for ID per se. The AAIDD states that the primary reason for evaluation and classification of individuals with ID is to individualize support specific to each individual, whereby detailed strategies and services are outlined over a sustained time period (16). The purpose of tailoring the support system is to maximize people's functioning within "their own culture and environment in order to lead a more successful and satisfying life" (16). This helps to increase the individual's sense of self-worth, well-being, pride, and social engagement (16). Interestingly, increased physical activity or formal exercise programming (or both) has now been recognized as important aspect of maintaining health in persons with ID (44). Finally, conditions, symptoms, or behaviors that are either directly or indirectly associated with ID are treated as needed. For instance, many individuals with ID are treated for symptoms of depression with antidepressive medications, and many with DS are treated for thyroid disorders as well.

Pharmacology

Individuals with ID may take various antidepressant medications. Many also take anticonvulsive agents. These medications help control inappropriate behavior and physiological symptoms caused by the ID, but do not affect ID per se. In severe cases, hypnotics may be used to control psychotic behavior.

It is not unusual for individuals with DS to have hypothyroid conditions, which are usually diagnosed during adolescence or early adulthood. These patients are on thyroxine replacement therapy. Large weight gains should be evaluated in this population, even in persons who have already been diagnosed with hypothyroidism; further titration of the thyroxine dose may be necessary (74).

Antidepressant medications can have a minor effect on heart rate or blood pressure, but generally have little effect on how these variables respond to acute or chronic exercise. These agents may also affect the ability of an individual with ID to understand and follow directions, as well as motivation to perform high-intensity exercise (1, 20). Thyroid medications may also increase both heart rate and blood pressure and are potentially arrhythmogenic. Use of β-blockers is not uncommon, particularly in persons without DS (1). β-blockers reduce heart rate and blood pressure both at rest and in response to exercise (31). Table 32.4 provides a summary of the pharmacology issues in persons with ID.

The involvement of individuals with ID in exercise testing or training is increasing; thus clinical exercise physiologists are likely to encounter and work with this population more frequently (see practical application 32.1). Consequently, it is important to understand the unique aspects of conducting exercise testing and training in persons with ID. The major differences apply primarily to exercise testing, as exercise training and exercise prescriptions are for the most part similar to those for populations without disabilities.

EXERCISE PRESCRIPTION

The exercise prescription for people with ID is quite similar to that used for nondisabled adults (38). Due to the generally high level of overweight and obesity and low physical activity levels, it is important to develop an exercise prescription that progressively increases both physical activity and muscle strength as part of a healthy lifestyle. The exercise prescription is outlined in table 32.5. The following are important special considerations for persons with ID.

1. More positive encouragement will likely be necessary in individuals with ID than in others. Because motivation can be problematic, and given the likelihood of a short attention span, it is important not to ask leading questions (e.g., "Are you tired?"—people are likely to answer yes to such a question even if they have not begun their workout). Also, in planning exercise sessions, consider other tasks or duties that will impose a great attentional demand during the day.
2. Exercise training sessions need to be carefully supervised at all times, particularly in the beginning of an exercise regimen to help individuals gain familiarization.
3. Individualize the training as much as feasible. Group exercise is not as effective in individuals with ID.
4. Simple games and music can be incorporated easily into an exercise training program, as can participation in activities akin to the Special Olympics, for increased motivation.

Table 32.4 Pharmacology: Major Classes of Medications Used In Individuals With ID and Possible Exercise-Related Effects

Medication class or name	Primary effects	Exercise effects	Special considerations
Antidepressants	Used to alleviate mood disorders.	Minor to negligible effects on HR and BP during acute or chronic exercise	May produce false-positive or false-negative stress test results; may affect individual's understanding of test instructions and motivation
Antiepileptics and anticonvulsants	Primary molecular targets are voltage-gated sodium or calcium channels and parts of the GABA system, which help reduce hyperirritability in parts of the brain's cortex.	Appears safe to use in conjunction with exercise	May cause hyperinsulinemia and increased appetite, possibly hormonal changes as well (e.g., testosterone)
Levothyroxine; thyroid replacement hormone	Used to treat hypothyroidism and to treat or suppress goiter (enlarged thyroid) or both.	Normalizes (increases) HR and BP response to exercise compared to that in hypothyroid condition Normalizes (increases) LVEF during exercise compared to that in hypothyroid condition	Small risk of provoking an arrhythmia, and because of increased cardiac contractility, some individuals may experience brief episodes of angina
β-blockers	β-adrenergic receptor antagonist; reduces heart rate and blood pressure at rest.	Reduced heart rate and blood pressure during exercise	

HR = heart rate, BP = blood pressure, GABA = gamma-aminobutyric acid, LVEF = left ventricular ejection fraction.

Practical Application 32.1

CLIENT–CLINICIAN INTERACTION

Due to the many types and various degrees of ID, one of the most important qualities in the clinical exercise physiologist is patience. Individuals with ID may have difficulties understanding the context surrounding what a clinical exercise physiologist is trying to accomplish and therefore may not understand directions given. It is also likely that task persistence, attention span, and willingness to try something new (as in exercise testing or training) are reduced. It is not unusual for people with ID to also exhibit various forms of behavioral problems, which can make exercise testing and training more challenging.

Allocate sufficient time in the schedule for exercise testing or training sessions, and be flexible about the degree of availability that individuals with ID have. For instance, because people in this group live primarily in group- or community-based homes, they often rely on caregivers for transportation and therefore their schedules are not uniquely their own. Creativity is often key in successful planning. Furthermore, it is helpful for the clinical exercise physiologist just entering the field to gain experience with this population by working closely with a more experienced and senior CEP before independently supervising either exercise training or testing sessions. While this chapter includes "tricks of the trade," nothing can replace actual hands-on experience and the opportunity to learn from an experienced individual or group. Finally, as with all patients or clients seen by the clinical exercise physiologist for exercise testing or training, it is important to preserve modesty and treat the individual with ID respectfully, as well as to provide a positive experience for not only the individual but also the caregiver. This often requires more attention and time than are usually allocated to patients with other health problems and frequently involves some "creative" scheduling. It should go without saying that being short with a patient or "talking down" is not effective. Individuals with ID enjoy and expect being treated as an equal.

Table 32.5 Summary of Exercise Prescription for Individuals With Intellectual Disabilities

Training method	Mode	Intensity	Frequency	Duration	Progression	Goals	Special considerations
Cardiovascular	Walking primarily, with possible progression to running when used intermittently Swimming Combined arm and leg ergometry Exercise to music (e.g., modified aerobics)	40% to 80% HRR RPE may not work	3 to 7 d/wk, with 3 or 4 d/wk comprising moderate to vigorous exercise; emphasis on increased activity on the remaining days	30 to 60 min/d; accumulated exercise may be used, particularly in the beginning of an exercise program	Increase speed, duration, or both gradually after the initial 2 to 3 wk.	45 to 60 min at a moderate to vigorous intensity	Do not base exercise prescription on predicted maximal heart rate; always use measured maximal heart rate. RPE will likely not work. See text for additional comments.
Resistance	Machines: six to eight major muscle groups	12 reps at 15- to 20RM for 1 or 2 wk	2 or 3 d/wk	Two or three sets, 1 to 2 min rest in between	Progress to 8- to 12RM after second week.	2 or 3 d/wk at 12 reps using 8- to 12RM	Closely supervise for the first 3 mo. See text for additional comments.
Range of motion	Stretching		5 to 7 d/wk	10 to 20 min	Increase or maintain range of motion.		Be very careful of hyperextension and hyperflexion.

The following are additional special considerations for individuals with DS:

1. Exercises that may involve hyperflexion or hyperextension of the neck are contraindicated due to a high rate of atlantoaxial instability.
2. Individuals with DS often present with skeletal muscle hypotonia and joint laxity. To avoid hyperextension or hyperflexion, ensure that the individual is performing the exercise in a smooth and controlled manner and is not "jerking" through a given exercise with excessive momentum.
3. **Pulmonary hypoplasia** may also be present in individuals with DS. Possible modifications may be needed to the exercise prescription depending on the severity of the hypoplasia, especially if exercise is performed at altitude. A lower than normal intensity will likely be needed.
4. Specific physical features may limit exercise performance. These include short stature and short limbs and digits. These characteristics may be coupled with additional characteristics such as malformed feet and toes, a small mouth or small nasal cavities or both, and a large or protruding tongue. Poor vision and balance may also contribute to reduced exercise performance.

Cardiovascular Exercise

It is important to find an exercise modality that is truly enjoyable for individuals with ID. This will ensure their attention and increase the likelihood of their continued participation. Acceptable modes of exercise include walking, with the possibility of some jogging; swimming; combined arm and leg ergometry (e.g., elliptical, Airdyne cycle ergometer); and exercising to music or using games. Since balance is an issue, ski ergometry is not recommended.

Because obesity is prevalent in this population, particularly in people with DS, some modes of exercise may be more difficult (e.g., orthopedic limitations) or less enjoyable, such as jogging. Proper progression is critical, and suggested exercise intensities for cardiovascular-based exercise are between 40% and 80% of heart rate reserve. Due to the lower maximal heart rates in persons with DS, this may make programming challenging and also requires that maximal heart rates be measured (vs. predicted using a standard age-dependent formula). Using a rating of perceived exertion scale is not recommended. However, if the clinical exercise physiologist does use a rating of perceived exertion scale in higher-functioning individuals, it is important to concomitantly use an additional marker as well (e.g., heart rate). Exercise should take place most days of the week and should be between 30 and 60 min/d. Accumulated exercise is highly recommended, especially in the beginning of a cardiovascular exercise program. The advice is to slowly increase intensity of the workout by adjusting speed, grade, duration, or more than one of these after the initial 2 to 3 wk. The goal should be for individuals with ID to be able to exercise 45 to 60 min/d at a moderate to vigorous intensity. If feasible, it may be helpful for them to experience a variety of cardiovascular exercises to increase enjoyment and participation.

Resistance Exercise

Weight machines are recommended for resistance training in individuals with ID. Two advantages of resistance machines are that they require less attention to proper form, though this is not absent entirely, and that balance is not as necessary. A majority of individuals with ID benefit from exercising six to eight major muscle groups and may not be as interested in many of the specialized machines available at fitness facilities. Using a load of 15- to 20RM for the first few weeks, it is suggested that people lift 2 or 3 d/wk, for 12 repetitions and two or three sets with 1 to 2 min rest in between sets. After the first 1 or 2 wk, they can progress to a load of 8- to 12RM once they are comfortable with the machines and their technique and when any delayed-onset muscle soreness has largely disappeared. Close supervision is necessary to ensure proper breathing and provide spotting. This level of supervision may be necessary for 3 mo or longer.

Range of Motion Exercise

People with ID are encouraged to engage in stretching exercises most days of the week, for 10 to 20 min per session. Yoga or similar types of exercise may be especially helpful for individuals suffering from some level of depression. Group exercises may not be ideal in this population and would need to be highly individualized for the person depending on attention span and physical characteristics. Also it is important to avoid hyperextension and hyperflexion in individuals with DS with associated joint laxity.

EXERCISE TRAINING

Although a multitude of exercise training studies aimed at improving cardiorespiratory fitness in individuals with ID have been conducted, very few have employed appropriate methodologies. Most early studies did not employ either validated tests or control groups, and very few have used a randomized controlled design. Nevertheless, the information in the literature is remarkably uniform. For instance, it was commonly found that both children and adults with ID improved field test or submaximal exercise performance following a standard endurance training program (7, 8, 82, 83). However, since none of these studies used validated test protocols, it is difficult to evaluate whether the findings are indicative of the actual training response among individuals with ID.

Cardiovascular Training

Several early studies also showed substantial improvements in $\dot{V}O_2$peak following endurance training in individuals with ID. However, most of these studies had a small number of subjects and did not include a control group, although the uniform response suggested that improvements in $\dot{V}O_2$peak were real (48, 70, 75). Pitetti and Tan (69) provided more conclusive evidence when they evaluated individuals with ID before, after 16 wk of endurance training using Airdyne ergometry, and then again after a 7 mo of no organized training. Improvements in $\dot{V}O_2$peak between 16% and 43% were shown in these studies with use of standard recommended exercise prescriptions. A recent study also showed that a 16 wk community-based swim training program improved body composition in eight adolescents with ID, but no control was included (10).

Two recent studies using more appropriate study designs have shown disparate results on the effect of exercise training in children or adolescents with ID. Ozmen and colleagues (60) evaluated the effect of a 10 wk school-based cardiovascular fitness training program in 30 children and adolescents randomized to training and control (8-15 yr of age). The training group exercised for 1 h at 60% to 80% of peak heart rate three times per week. Following the 10 wk program, the training group significantly improved 20 m shuttle run performance; however, $\dot{V}O_2$peak was not measured. Conversely, Elmahgoub and colleagues (17) found that 10 wk of combined strength and endurance exercise (three times per week, 50 min per session) did not change $\dot{V}O_2$peak, even though 6 min walk performance increased and muscle strength improved in the exercise training group. However, only 20 min per session was devoted to endurance training.

These studies included only persons with ID who did not have DS. Two early studies in persons with ID with DS suggested that $\dot{V}O_2$peak does not improve with endurance training in DS. Millar and colleagues (56) reported no change in $\dot{V}O_2$peak following 10 wk of endurance training; but the training group improved their work capacity as shown by increased treadmill time, whereas the control group did not change either $\dot{V}O_2$peak or work capacity. Varela and colleagues (86) presented similar results following a rowing training program, in which the exercise group improved the amount of work done on the rowing ergometer but not $\dot{V}O_2$peak. These investigations suggested that $\dot{V}O_2$peak is immutable to change in response to a standard endurance program. However, more recent studies have obtained different results. Significant improvements in $\dot{V}O_2$peak were reported by Tsimaras and coworkers (85) following a 12 wk aerobic training regimen, and by Rimmer and colleagues (71) following combined aerobic and resistance exercise training in older individuals with DS (mean age 39 yr). Unfortunately, Rimmer and colleagues used maximal cycle ergometry to evaluate $\dot{V}O_2$peak, which has not been validated and is not reliable in persons with ID (51).

More recently, two studies by Mendonca and coworkers (52, 53) provided support for increases in aerobic capacity with endurance training in individuals with ID with DS. These findings were supported by a meta-analysis showing large effect sizes for improvements in both aerobic capacity and work capacity in persons with ID with DS . Changes in $\dot{V}O_2$peak following endurance training in individuals with DS range from 0% to 27%, with most recent studies demonstrating increases similar to those expected in populations without disabilities (53). Combined aerobic and resistance exercise training also improved both strength and walking economy in both individuals with DS and persons without disabilities. This was especially important for individuals with DS who initially exhibited poor walking economy and low levels of strength. However, exercise training did not normalize muscle strength and walking economy, as both of these variables were lower in the group with DS compared to individuals without disabilities both pre- and posttraining. Based on the available evidence, individuals with ID, both with and without DS, appear to improve aerobic capacity and work capacity with appropriate endurance training. However, many studies had very small sample sizes; several had no control groups; and most did not used randomized controlled designs.

Resistance Exercise Training

Early studies of the effect of resistance training in individuals with ID used field tests including sit-ups, push-ups, and pull-ups to measure muscle performance. These are primarily tests of muscle endurance and do not provide information on muscle strength. Also, neither push-up nor flexed arm hang (a version of pull-up) is reliable with persons with ID. Nevertheless, substantial improvements (up to 58%) in sit-ups and chin-ups following 3 to 28 wk of training, in both children and adults with ID, have been seen (35, 47, 58). Exercise training using surgical tubing has also been shown to improve muscle strength in individuals with ID (13). Although handgrip strength is often tested in individuals with ID, it does not appear to improve with exercise training (58). This is not surprising, since handgrip is not normally used as an exercise in these types of programs. In summary, nontraditional muscle strength and endurance programs can improve muscle strength and endurance (primarily endurance) in persons with ID.

Standard circuit training types of resistance programs varying in length from 5 wk to 3 mo, using two or three training sessions per week, also improved muscle strength in individuals with ID to an extent similar to that in persons without disabilities (50, 72, 78-80). The improvements in muscle strength varied between 8% and 82% depending on the muscle group tested and the type of program. There does not appear to be a difference in findings between studies with and without control groups. In addition, Suomi and colleagues (78, 79) reported excellent compliance rates (~90%) even though participants had to provide their own transportation to the exercise facility. Furthermore, these investigators showed that individuals with ID could continue to improve muscle strength through participation in a self-directed 9 mo program following the initial 12 wk supervised program. These findings have important implications, as they show that self-motivated individuals with mild ID who have learned how to safely conduct strength exercises can maintain their strength gains on their own without supervision.

Because individuals with DS have very low levels of muscle strength, and because muscle strength is associated with activities of daily living and aerobic capacity in persons with DS, inventions designed to improve muscle function may be especially important (11). Interestingly, muscle strength has been improved using exercise regimens other than resistance training in persons with DS. These have included high-intensity sport training and training using jumping and balance activities in young individuals (61, 84), as well as a treadmill walking program in older participants (age ~63 yr) (9). These programs improved muscle strength 7% to 25%, probably reflecting very low starting levels of strength.

Several trials have investigated the effect of progressive resistance training on muscle strength in persons

Practical Application 32.2

RESEARCH FOCUS

In a recent study (22), epinephrine and norepinephrine responses were evaluated at rest and immediately following maximal treadmill exercise in individuals with DS compared to a control group of individuals without disabilities. The rationale for the study was that maximal heart rate is much lower in persons with DS than in the general population, but the reason for the low maximal heart rate is unclear. Prior research had suggested reduced vagal withdrawal during isometric exercise in persons with DS; however, this did not appear to be the case with treadmill exercise. Since plasma catecholamine concentrations are partly responsible for further increases in heart rate during higher exercise intensities, it is possible that reduced levels of plasma epinephrine and norepinephrine contribute to the low maximal heart rate.

Twenty participants with DS (age ~24 yr) and 21 participants without disabilities (age ~26 yr) completed a maximal treadmill exercise protocol with heart rate measured and oxygen uptake measured using indirect open-circuit spirometry. Blood samples were collected before and immediately after exercise and analyzed for plasma epinephrine and norepinephrine. As expected, both maximal heart rate and $\dot{V}O_2$peak were significantly lower in the group with DS (HRmax 170 vs. 189 beats · min^{-1}; $\dot{V}O_2$peak 27 vs. 41 ml · kg^{-1} · min^{-1}). There was no difference in resting catecholamine concentrations between groups; but plasma epinephrine did not change from rest to maximal exercise in individuals with DS whereas it increased almost 900% in the control group. Similarly, plasma norepinephrine doubled from rest to exercise in persons with DS but increased almost 10-fold in the control group. The changes in catecholamines were also associated with maximal heart rate and $\dot{V}O_2$peak.

This study demonstrated that individuals with DS exhibit chronotropic incompetence and low aerobic capacity and that the attenuation is due, in part, to the lack of an increase in plasma epinephrine and norepinephrine during maximal exercise. These findings highlight the importance of measuring maximal heart rate and oxygen uptake in this population, and provides a physiologic explanation for the low maximal heart rate in persons with DS.

with DS. Two of these trials (53, 71) used a combined aerobic and resistance regimen but still found substantial increases in muscle strength (reviewed earlier in the section on endurance training). Two other trials also showed substantial improvements in muscle strength, but did not include a control a group (15, 87). Shields and colleagues (76), found a significant improvement in muscle endurance in the exercise group (no change in controls), but not in muscle strength. There was also no change in tests of daily activities. Conversely, Cowley and colleagues (12) found significant improvement in leg strength in the exercise group but not in the control group. Furthermore, the exercise group improved stair climbing ascending and descending performance, showing that the resistance training program positively influenced a functional activity of daily living. Interestingly, although leg strength improved significantly, there was no change in $\dot{V}O_2$peak. This suggests that the association between leg strength and $\dot{V}O_2$peak in persons with DS does not imply a causal relationship.

Overall, it appears that appropriately conducted resistance exercise training programs will improve muscle strength in persons with ID, including those with DS. It has been suggested that programs using training frequencies of 3 d/wk are required (54), but the recent study by Cowley and colleagues (12) showed significant improvement with 2 d/wk. Thus, it would appear that it is appropriate to follow standard ACSM guidelines to elicit desired resistance training responses in persons with ID.

CONCLUSION

As mentioned at the beginning of this chapter, an ID is not a disease and there is no cure. However, just like people with other chronic diseases or disabilities, individuals with an ID can both participate in and enjoy the benefits derived from a regular exercise program. This is a unique and special population of individuals that are well deserving of the skills and abilities of the practicing clinical exercise physiologist.

Clinical Exercise Pearls

- Familiarizing the individual with the testing or training is of the utmost importance; repetition, practice, and clear instructions can help with this. It also important for individuals with ID to become familiar and comfortable with testing and training staff, as this will facilitate their ability to pay attention to instructions.
- Do not treat individuals with ID like children, and be careful not to condescend. Individuals with ID understand when this happens and often "act out" in response.
- Be patient; individuals with ID may have difficulties understanding directions and often have a short attention span. Behavioral problems are more frequent than in the general population.
- For most individuals with ID, expect low work and aerobic capacities, maximal HR, and muscle strength, together with a higher prevalence of overweight and obesity. Very low aerobic capacities and muscle strength levels are often observed in individuals with DS. Measured values are approximately 50% lower than expected for age and sex. Further, maximal heart rates are ~30 beats · min^{-1} lower than expected for age.
- Individuals with DS may present with congenital heart disease, atlantoaxial instability, or both.
- Testing may not always go as planned; tests often have to be repeated on a different day.

Key Terms

atlantoaxial instability (p. 620)
chromosomal microarrays (p. 620)
congenital heart disease (p. 618)
Down syndrome (p. 618)
fetal alcohol syndrome (p. 618)
fragile X syndrome (p. 618)
hypothyroidism (p. 621)
karyotyping (p. 620)
Prader-Willi syndrome (p. 619)
pulmonary hypoplasia (p. 626)

CASE STUDY

MEDICAL HISTORY AND PHYSICAL EXAM

Mr. RK is a 45 yr old male who has mild ID and also Down syndrome. He lives in a community group home with 24 h support and assistance. He works 6 h/d at a local fast food restaurant. His favorite activity is to watch TV, and he enjoys eating popcorn while watching movies.

He does not have any history of heart disease or other serious medical conditions. He is 5 ft 6 in. (168 cm) and 240 lb (109 kg), with a BMI of 38.8 $kg \cdot m^{-2}$. His medical record states that his total cholesterol is 240 $mg \cdot dl^{-1}$, with high-density lipoprotein (HDL) cholesterol of 35 $mg \cdot dl^{-1}$. There is no information on triglycerides or low-density lipoprotein (LDL) cholesterol. His blood pressure was 110/70 mmHg. His case worker has noted that over the past year he has experienced increased shortness of breath when walking up the stairs to his bedroom. However, at his last physical examination there was no note on any suggested pulmonary problems. He has started to display the early stage of Alzheimer's disease and demonstrates abnormal laxity of the left knee. Mr. RK cannot walk or jog for any extended period of time without pain. Other findings on the physical examination were unremarkable. Both his physician and case worker have encouraged him to become more physically active, but at present he performs no physical activity outside of work. He is not currently taking any medications.

DIAGNOSIS

Mr. RK is diagnosed with mild ID with DS and with an early stage of Alzheimer's disease. He is obese and presents with several cardiovascular disease risk factors.

EXERCISE TEST RESULTS

He completed 4 min on a standard Bruce treadmill protocol. His maximal heart rate was 148 $beats \cdot min^{-1}$(85% of predicted), and his maximal work capacity was predicted from treadmill time to be 4 METs. Oxygen uptake was not measured. His maximal blood pressure was 150/80 mmHg. He exhibited no ECG abnormalities, and the test was interpreted as negative for exercise-induced myocardial ischemia; but it was noted that maximal effort may not have been reached as evidenced by the low maximal heart rate achieved.

EXERCISE PRESCRIPTION

Since Mr. RK has a problem with knee instability and knee pain, he was prescribed a cycling program. He was advised to exercise three to six times per week, starting at 30 min of comfortable cycling. After 2 wk he was advised to increase his exercise intensity to ensure training for 30 min in a heart rate range of 100 to 126 $beats \cdot min^{-1}$. Following another 2 wk he was to increase the exercise duration by 5 min each week until he was training between 45 and 60 min per session. His exercise was preceded by 5 min warm-up and followed by a 5 min cool-down period consisting of easy pedaling. Resistance exercise training was not included because of lack of staff resources.

DISCUSSION

Although Mr. RK has an abnormal blood lipid profile, this is not a major concern for individuals with DS. People with DS often have elevated blood lipids, but they normally do not develop atherosclerotic heart disease. (A few early autopsy studies showed no significant atherosclerotic plaque in the arteries of individuals with DS, in contrast to similarly aged individuals without disabilities. These findings led to the suggestion that DS is an atheroma-free model, but the reason for this is unknown [59, 88].) The main concerns are obesity, joint instability with associated pain, and Alzheimer's disease. Low work capacity is also a concern, but this could be due to the protocol (i.e., Bruce) used for testing. Certainly, an individualized protocol would have been preferred, but such protocols are not often employed in clinical testing settings. Mr. RK's low maximal heart rate is actually normal for a person with DS and would suggest that he provided a maximal effort. Since people with DS often have lower blood pressures than the general population, his resting blood pressure and his response to exercise were expected.

The exercise prescription may be too general. Many individuals with DS cannot ride a bicycle, so stationary cycle ergometry should have been specified. However, some persons with DS have difficulty pedaling any bike; thus it is important to ensure that this patient will be able to do so before prescribing cycling. It is also unlikely that a person with DS would be able to exercise at a specified heart rate range; considerable time and effort would be needed to teach him how to use a heart rate monitor. It is important to educate the case worker about the exercise prescription, because eventually the case worker will likely be the person supervising the exercise. Thus, a general exercise prescription is indeed appropriate for Mr. RK, but much more detail is needed to operationalize it and provide a reasonable chance of success.

Considering Mr. RK's obesity, it is important to devise nutritional and dietary strategies aimed at decreasing body weight and, at minimum, preventing further weight gain. It is unlikely that his exercise program alone will have an appreciable effect on his body weight. Adhering to dietary strategies targeting weight loss will be difficult since his meals are usually cooked by staff and he has free access to the kitchen for snacks.

DISCUSSION QUESTIONS

1. What other information from the history and physical would be helpful to have?
2. What factors should you take into consideration before developing an exercise program for Mr. RK?
3. How can you develop a more specific exercise program using a specific exercise prescription? How long should the program last, and what outcomes can you expect?
4. Would you involve any other health care professionals when working with Mr. RK? Who and why?
5. How would you evaluate progress? What type of follow-up would you recommend?

Glossary

α_1-antitrypsin (AAT)—Protein produced in the liver and found in the lungs that inhibits neutrophil elastase.

α_1-antitrypsin deficiency—A genetic disorder characterized by abnormally low levels of α_1-antitrypsin, thereby predisposing an individual to emphysema.

abdominal obesity—Condition characterized by excessive fat on the trunk, also known as android obesity. Increased fat in the abdominal region increases the risk for development of hypertension, type 2 diabetes, dyslipidemia, coronary artery disease, and premature death compared to gynoid obesity (increased fat in the hip and thigh).

absolute contraindications for test termination—Conditions that occur during a GXT necessitating that the test be terminated.

absolute oxygen uptake—Oxygen uptake expressed in liters per minute ($L \cdot min^{-1}$).

acquired immunodeficiency syndrome (AIDS)—A disease caused by HIV. See text for the CDC case definition.

activities of daily living (ADLs)—Bathing, dressing, grooming, toileting, feeding, and transferring.

acute MI—The initial stages of an evolving myocardial infarction (MI).

adult variant CF—Adult diagnosis of cystic fibrosis secondary to presentation of respiratory symptoms. The population with this diagnosis has resulted in the increasing numbers of adults with CF.

afterload—The resistance against which the pumping chamber or ventricle of the heart works.

aging—The process of growing old.

agonist—A drug or agent that stimulates or enacts the biologic response for a given receptor.

airflow limitation—The blockage of the flow of air out of the lung that can occur secondary to narrowing of the airway lumen.

airway hyperresponsiveness—The ability of the airway wall to be sensitive to various inhalants.

akinetic—Denoting loss of movement of a left ventricular wall during the normal cardiac cycle.

amenorrhea—Absence of normal menses; for most studies, a woman is considered amenorrheic if she has fewer than three menses per year.

anabolic steroids—Testosterone derivatives or steroid hormones resembling testosterone that stimulate the building up of body tissues.

androgenic—Having masculinizing effects, that is, stimulation of male sex characteristics and male hair characteristics.

anemia—A decrease in the red blood cells that carry hemoglobin, resulting in reduced oxygen-carrying capacity.

aneural—Characterized by absence of nerve fibers.

aneurysm—Dilation of an artery that is connected with the lumen of the artery or cardiac chamber. Usually occurs because of a congenital or acquired (e.g., myocardial infarction) weakness in the wall of the artery or chamber. Forms of aneurysms include true, dissecting, and false.

angina pectoris—Constricting chest pain or pressure, often radiating to the left shoulder or arm, back, or neck and jaw regions, caused by ischemia of the heart muscle.

angina threshold—Point at which the supply of oxygen is less than the demand, leading to ischemia and producing symptoms of angina pectoris. Generally observed during physical or mental exertion in patients with significant coronary artery disease.

angiography—Medical imaging technique used to assess the arterial anatomy, often to confirm the findings of noninvasive studies as to the site of a blockage. Broadly includes the following modalities: invasive angiography, computed tomography angiography, and magnetic resonance angiography.

angioplasty—Reconstitution or recanalization of a blood vessel; may involve balloon dilation, mechanical stripping of intima, forceful injection of fibrinolytics, or placement of a stent.

angiotensin-converting enzyme inhibitor—Medication that prevents the conversion of angiotensin 1 to angiotensin 2, which ultimately decreases blood vessel vasoconstriction.

ankle–brachial index (ABI)—Noninvasive peripheral artery disease screening test that uses a blood pressure cuff and Doppler device to assess the systolic blood pressures of the right and left brachial, posterior tibial, and dorsalis pedis arteries. An index value is computed by dividing the highest ankle pressure for each lower extremity by the higher of the two brachial pressures; a value less than 0.90 is considered abnormal.

ankylosing spondylitis—A chronic rheumatic disease that causes inflammation, stiffness, and pain in the spine; sacroiliac joints (often an early indicator); and in some cases, the neck, hips, jaw, and rib cage. This disease may be accompanied by fever, loss of appetite, and heart and lung problems. It may cause spinal deformities and eventually causes the spinal segments to fuse (ankylose), with the result that the back assumes a rigid fixed posture.

ankylosis—Immobility and consolidation of a joint due to disease, injury, or surgery.

anorexia nervosa—Loss of appetite associated with intense fear of becoming obese. Can lead to life-threatening weight loss, disturbed body image, hyperactivity, and amenorrhea.

anovulatory—Not accompanied by the discharge of an ovum.

antagonist—A drug or agent that reduces, blocks, or inhibits the biologic response of a given receptor.

anterior cord syndrome—Symptoms of anterior SCI (T10-L2), including pain, impairment, and temperature deficit below level of injury. Symptoms are the result of an anterior cord lesion.

anticoagulation therapy—Pharmacological delaying or preventing of blood coagulation (clotting).

antioxidant—An agent that inhibits oxidation and thus prevents the deterioration of other materials through oxidative processes.

antiresorptive therapy—The group of drug therapies currently available for treatment of osteoporosis. The term originates from the fact that the therapies halt the loss of bone by inhibiting bone resorption.

antiretroviral medications—Medications demonstrated to be effective against the HIV virus, which is a retrovirus.

anuria—Suppression or arrest of urinary output, resulting from impairment of renal function or from obstruction in the urinary tract.

anxiety—A complex psychophysiological response to an environmental stressor, disaster, or trauma; more often manifested in people who are genetically vulnerable to the disorder.

APGAR—A noninvasive clinical test, designed by Dr. Virginia Apgar (1953), carried out immediately on a newborn. The name is also an acronym for activity (muscle tone), pulse, grimace (reflex irritability), appearance (skin color), and respiration. A score is given for each sign at 1 min and 5 min after birth.

aphasia—Partial or total loss of the ability to articulate ideas or comprehend spoken or written language, resulting from damage to the brain caused by injury or disease.

apolipoproteins—Proteins associated with lipoproteins; these proteins act to stabilize the lipid portion of the lipoprotein in the circulation. Apolipoproteins serve as ligands for cell receptor binding and as cofactors in enzyme reactions.

areflexia—Absence of neurologic reflexes, usually a sign of peripheral nerve damage affecting muscular, bowel, bladder, and sexual function.

arterial intimal layer—The innermost layer of an artery, composed of endothelial cells.

arthrodesis—The surgical immobilization of a joint, intended to result in bone fusion.

articular—Relating to a joint.

ASCVD—Atherosclerotic cardiovascular disease; cardiovascular disease resulting from the accumulation of atherosclerotic plaque.

associated conditions—Conditions that accompany a primary disability but are not necessarily preventable. They can, however, be controlled with medication, surgery, or medical devices.

asthma—A continuum of disease processes characterized by inflammation of the airway wall.

astrocyte—Star-shaped neural cell that provides nutrients, support, and insulation for neurons of the central nervous system.

astrocytic gliosis—Proliferation of astrocytes in damaged areas of the central nervous system, forming scar tissue.

asymmetric—Denoting lack of symmetry between two or more like parts.

ataxia—Inability to coordinate voluntary muscle movements; unsteady movements and staggering gait or loss of ability to coordinate muscular movement most often caused by disorders of the cerebellum or the posterior columns of the spinal cord; may involve the limbs, head, or trunk.

atherectomy—Procedure used for revascularization of an obstructed coronary artery; uses a catheter tipped with either a metal burr that grinds a calcified atheroma (rotational atherectomy), or a rotating cup-shaped blade housed in a windowed cylinder that cuts or shaves the atheroma (directional atherectomy).

atherogenic—Having the capacity to initiate, increase, or accelerate the process of atherosclerosis.

atheroma—An accumulation of lipid in vascular walls. Atheromas are also called fatty streaks and atherosclerotic lesions.

atherosclerosis—An extremely common form of arteriosclerosis in which deposits of yellowish plaques (atheroma) containing cholesterol, lipid material, and lipophages are formed within the intima and inner media of large and medium-sized arteries.

atherosclerotic—Relating to or characterized by atherosclerosis.

athetosis—A constant succession of slow, writhing, involuntary movements of flexion, extension, pronation, and supination of fingers and hands and sometimes of toes and feet.

atlantoaxial instability—Increased flexibility in the atlantoaxial joint, which is actually a composition of three joints: two lateral and one median atlantoaxial joint, the joint where the base of the skull meets the neck. Because of its proximity to the brain stem and importance in stabilization, fracture or injury at this level can be catastrophic. Due to ligament laxity, instability is not uncommon in Down syndrome

atrophy—Partial or complete wasting away of a part of the body, as from disuse.

auscultation—Listening to the sounds made by various body structures as a diagnostic method.

autoimmune—Referring to cells or antibodies that arise from and are directed against the person's own tissues, as in autoimmune disease.

automatic internal cardioverter defibrillator (AICD or ICD)—An implantable battery-powered generator used in patients who are at risk for sudden death due to ventricular fibrillation or tachycardia. The device can detect life-threatening cardiac arrhythmias and can deliver a jolt of electricity designed to stop the arrhythmia.

autonomic dysreflexia—Sudden, exaggerated reflex increase in blood pressure in persons with a spinal cord injury above T6, sometimes accompanied by bradycardia, in response to a noxious stimulus originating below the level of a spinal cord injury.

autonomic nervous system—Components of the nervous system responsible for coordinating life-sustaining processes and organization of visceral responses to somatic reactions.

avascular—Denoting absence of blood vessels.

A-V synchrony—The sequence and timing of the atria and ventricles during systole.

balance—The ability to make adjustments to maintain body equilibrium.

ballistic flexibility (stretching)—Stretching using active muscle movement with a bouncing action.

bariatric surgery—Surgical intervention using one of several methods designed to assist morbidly obese people in losing weight.

basal ganglia—The caudate and lentiform nuclei of the brain and the cell groups; all of the large masses of gray matter at the base of the cerebral hemisphere; the large masses of gray matter at the base of the brain that, if damaged, would impair motor abilities.

behavior therapy—Strategies, based on learning principles, that provide tools for overcoming barriers to compliance.

β-agonist—A drug or hormone capable of combining with β-receptors to initiate drug actions.

β-blocker—Medication used to block β-receptors in the myocardium, which decreases myocardial work by decreasing heart rate and myocardial contractility.

β-receptor—A cell receptor that is activated by a β-agonist such as epinephrine, norepinephrine, or dopamine.

bifurcation—Point at which an artery branches to form two arteries.

biologics—A recently developed class of drugs that interfere with biologic substances causing or exacerbating inflammation; cytokine or T cell or B cell modulators used to suppress the rheumatic disease process.

biopsy—A surgical procedure whereby a sample of tissue is obtained.

biotherapy—Stimulation of the body's immune response system to cancer-specific protein antigens.

blood pressure (BP)—The force of circulating blood on the walls of the blood vessels as it circulates throughout the body. Chronic elevated blood pressures (systolic blood pressure ≥140 mmHg or diastolic blood pressure ≥90 mmHg) are associated with increased risk for cardiovascular disease.

body mass index (BMI)—Relative weight for height; weight in kilograms divided by height in meters squared ($kg \cdot m^{-2}$).

bone formation—Also called bone remodeling; the process by which new bone is formed and deposited within the existing bone matrix. Bone formation is accomplished primarily by bone cells called osteoblasts.

bone geometry—The overall cross-sectional area of the bone; the cross-sectional area of the outer cortex, the number of cross-links in the trabecular bone, and other related factors. With reference to the femoral neck, also includes the angle that the neck of the femur makes with the shaft of the femur.

bone mineral content—A measurement of the total amount of hydroxyapatite (calcium phosphate crystal) of bone, expressed as $g \cdot cm^{-2}$; synonymous with bone mass.

bone mineral density (BMD)—Relative amount of bone mineral per measured bone width. Values are expressed as $g \cdot cm^{-2}$.

bone modeling—Alterations in the shape of the bone such as changes in length.

bone remodeling—A constant state of formation and resorption.

bone resorption—The process of eroding old bone from the existing bone matrix so that new bone can be formed in its place. Resorption is accomplished primarily by bone cells called osteoclasts. Bone resorption is greatly increased in estrogen-depleted women.

botulinum toxin—A neurotoxin that blocks the release of acetylcholine from the motor endplates of the lower motor neuron at the myoneural junction, thereby preventing muscle contraction.

bradycardia—A heart rate less than 60 beats $\cdot$ min^{-1}.

bradypnea—A respiratory rate less than 8 breaths $\cdot$ min^{-1}.

bromocriptine—Ergot derivative that suppresses secretion of prolactin; used to stimulate ovulation in the galactorrhea–amenorrhea syndrome.

bronchi—Large airways of the lungs.

bronchiectasis—Chronic dilation of a bronchus usually associated with secondary bacterial infection.

bronchioles—Small airways of the lungs.

bronchodilator—A drug that relaxes the smooth muscles surrounding the bronchi and bronchioles.

bronchoprovocation—A type of pulmonary function testing in which a particular medication (e.g., methacholine) is aerosolized to induce bronchospasm.

bronchospasm—Spasmodic contraction of the smooth muscle of the bronchi, as occurs in asthma.

Brown-Sequard syndrome—Symptoms of a unilateral spinal cord injury, including ipsilateral proprioceptive and motor deficit, as well as contralateral pain impairment and temperature deficit below the level of injury.

bruits—Acquired sounds of venous or arterial origin caused by turbulent blood flow, heard by auscultation.

bulimia—Disorder that includes recurrent episodes of binge eating, self-induced vomiting and diarrhea, excessive exercise, strict diet, and exaggerated concern about body shape.

bursitis—Inflammation of one of the fluid-filled sacs located at sites of friction surrounding the joint.

calcitonin—Hormone that is responsible for calcium regulation and inhibits bone resorption.

cardiac activation—The biological processes resulting in the formation of electrical impulses that regulate the heartbeat, either normal or abnormal.

cardiac resynchronization therapy (CRT)—A feature of an implanted pacemaker that can pace both the septal and lateral walls of the left ventricle so that the heart can contract in a more synchronized manner to improve cardiac function. The failure of the opposing walls of the left ventricle to contract in synchrony occurs in about 25% to 50% of patients with heart failure.

cardiac tamponade—Accumulation of fluid in the pericardial sac that may compress the ventricles and markedly reduce cardiac output.

cardiogenic shock—Lack of cardiac and systemic oxygen supply resulting from a decline in cardiac output secondary to serious heart disease; typically follows a myocardial infarction.

cardiometabolic risk factors—Collection of health-related variables that increase one's risk for both cardiovascular and metabolic diseases (hypertension, obesity, high triglycerides, glucose intolerance).

cardiorespiratory fitness—Also known as aerobic capacity; the ability of the body to perform high-intensity activity for a prolonged period without undue physical stress or fatigue.

cardiovascular autonomic neuropathy—Neural damage to the autonomic nerves of the cardiovascular system, which can result in a high resting and low peak exercise heart rate and severe orthostatic hypotension.

cardiovascular disease (CVD)—A group of disease conditions that affect the heart, arteries, or veins of the circulatory system or more than one of these; the term most often refers to the atherosclerotic process occurring in the coronary arteries of the heart.

catecholamines—Chemicals released in the body that are major elements in the response to stress and exercise. Two catecholamines of interest are epinephrine and norepinephrine. Both exert, among other effects, a positive inotropic and chronotropic effect on cardiac function.

cauda equina—Lumbosacral spinal nerve roots forming a cluster at the terminal region of the spinal cord that resembles a horse's tail.

cauda equina syndrome—Severe compression of the cauda equina, resulting in loss of bowel or bladder function, loss of sensation in the buttocks and groin, and weakness in the legs. Cauda equina syndrome requires an emergency surgical intervention to prevent permanent damage.

CD4+ cells—A membrane receptor found on T-helper lymphocytes (or T4 cells); the preferred target of HIV.

CD4 lymphocyte—A type of white blood cell that is part of the body's immunological system.

CD4+ T lymphocytes—Immune cells that express the CD4+ glycoprotein on their surface. Without treatment, HIV infection leads to a progressive decline in T cells expressing the CD4+ glycoprotein.

central cord syndrome (CNS)—Symptoms of incomplete SCI including weakness and sensory deficits in the upper extremities (less than in the lower extremities).

central nervous system—The brain and the spinal cord.

cerebellum—A trilobed structure of the brain, lying posterior to the pons and medulla oblongata and inferior to the occipital lobes of the cerebral hemispheres; responsible for the regulation and coordination of complex voluntary muscular movement as well as maintenance of posture and balance.

cerebral cortex—The thin, convoluted surface layer of gray matter of the cerebral hemispheres that consists of the frontal, parietal, temporal, and occipital lobes.

chemotherapy—Use of chemical agents to kill rapidly growing cancer cells.

children—Persons between infancy and adolescence.

cholesterol—A fatlike, waxy substance found throughout the body. High levels of cholesterol in the blood ($>200 \text{ mg} \cdot \text{dl}^{-1}$) increase the risk of developing cardiovascular disease.

cholesterol ester transfer protein (CETP)—Protein in the circulation that transfers nonpolar lipids, cholesterol esters and triglycerides, between lipoproteins.

chorea—State of excessive, spontaneous movements, irregularly timed; causes nonrepetitive and abrupt motions and inability to maintain voluntary muscle contraction.

choreoathetosis—A type of athetosis marked by extreme range of motion, jerky involuntary movements that are more proximal than distal, and muscle tone fluctuating from hypotonia to hypertonia.

chromosomal microarray—A molecular–cytogenetic method for the analysis of copy number changes (gains and losses) in the content of a given subject's DNA.

chronic bronchitis—Disease characterized by the presence of a productive cough on most days during 3 consecutive months in each of 2 successive years.

chronicity—The state of being chronic, or long in duration.

chronic kidney disease (CKD)—A chronic disease condition that is associated with kidney damage.

chronic MI—The latest phase of a myocardial infarction, during which the heart is stable.

chronic obstructive pulmonary disease (COPD)—Presence of airflow obstruction attributable to either chronic bronchitis or emphysema.

chronotropic assessment exercise protocol—Treadmill protocol used to determine whether heart rate response is appropriate throughout the length of the exercise test.

chronotropic incompetence—Lack of an appropriate increase in heart rate with physical exertion. Considered an abnormal response if peak heart rate does not reach 2 standard deviations of the person's age-predicted maximum heart rate, assuming that the patient was highly motivated and not on medications to blunt heart rate response (i.e., β-blockers, calcium channel blockers).

chylomicron—Relatively large, triglyceride-rich lipoprotein secreted by the intestine after digestion and absorption of food.

circulatory hypokinesis—Insufficient vascular tone resulting in hypotension; occurs when increased metabolic demands of upper extremity exertion are not matched by appropriate hemodynamic responses.

clasp-knife hypertonia—Increased muscle resistance to passive movement of a joint followed by a sudden release of the muscle; commonly seen in individuals with spasticity.

claudication—Limping, lameness, and pain that occur in individuals who have an ischemia response in the muscles of the legs, which is brought on with physical activity (e.g., walking). A scale can be used to determine the severity of claudication.

claudication onset time or distance—The time point or distance at which the patient first experiences claudication pain in the calves, thighs, or buttocks during walking. Often used as a secondary outcome measure to evaluate walking performance in peripheral artery disease studies.

clonus—An abnormality in neuromuscular activity characterized by rapidly alternating muscular contraction and relaxation; a form of movement marked by contractions and relaxations of a muscle, occurring in rapid succession, after forcible extension or flexion of a part. Also called clonospasm.

clubbing—Rounding and enlargement of the distal-most parts of the fingers, usually most prominently under the fingernails.

clubfoot—Also known as talipes equinovarus, a congenital deformity affecting one or both feet characterized by the heel pointing downward and the forefoot turning inward. The heel cord (Achilles tendon) is tight, causing the heel to be drawn up toward the leg.

cognitive behavioral therapy—For patients with major depressive disorder, treatment that focuses on modifying maladaptive thoughts and addresses deficits in behavior (e.g., unassertiveness, isolating oneself from others).

community mobility—Locomotion and transportation of individuals through their community.

comorbidity—A concomitant but unrelated pathologic or disease process, usually with reference to the coexistence of two or more disease processes.

complete paraplegia—Motor and sensory dysfunction of the trunk, legs, and pelvic organs resulting from SCI, without motor or sensory sparing below the level of the injury.

complete tetraplegia—Motor and sensory dysfunction of the arms, trunk, legs, and pelvic organs resulting from SCI, without motor or sensory function sparing below the level of the injury.

computed axial tomography (CAT or CT scan)—Tomography (moving X-ray tube and film) whereby a pinpoint radiographic beam sweeps transverse planes of tissue and a computerized analysis of the variance in absorption produces a precise image of that area.

conduction block—A disease of the electrical system of the heart that, depending on the severity, impedes or completely blocks the conduction of the electrical impulses that initiate the contraction of heart muscle.

congenital heart disease—A defect in the structure of the heart and great vessels present at birth.

congestive or chronic heart failure (CHF)—The symptom complex associated with shortness of breath, edema, and exercise intolerance attributable to abnormal left ventricular function, resulting in congestion of blood in other bodily organs.

contracture—An abnormal, often permanent shortening, as of muscle or scar tissue, that results in distortion or deformity, especially of a joint of the body. Shortening of a muscle group and tendon is usually observed in persons with spasticity.

conus medullaris syndrome—Symptoms of upper and lower motor neuron damage to the conus medullaris, including bowel, bladder, and lower extremity areflexia and flaccidity and preserved or facilitated reflexes.

coronary dissection—Separation of tissue within the lining of a coronary artery.

coronary revascularization—Establishment of restored blood flow through a stenosed coronary artery via catheter intervention (angioplasty, stenting, thrombectomy) or coronary artery bypass graft surgery.

cortical bone—One of the two main types of bone tissue; hard, compact bone found mainly in the shafts of long bones. The other type is trabecular.

coxa valga—A hip deformity produced when the angle of the head of the femur with the shaft exceeds 120°. The greater the degree of coxa valga, the longer the resulting limb length.

coxa vara—A hip deformity produced when the angle made by the head of the femur with the shaft is below 120°. In coxa vara, it may be 80° to 90°. Coxa vara occurs in rickets or may result from bone injury. The affected leg appears shortened, resulting in a limp.

C-reactive protein (CRP)—A β-globulin found in the serum of persons with certain inflammatory, degenerative, and neoplastic diseases. CRP levels are often detectable in the blood of individuals with metabolic syndrome, suggesting chronic inflammation.

creatinine—End product of creatine metabolism excreted in the urine at a constant rate—a blood marker of renal function.

creatinine clearance—An index of the glomerular filtration rate, calculated by multiplying the concentration of creatinine in a timed volume of excreted urine by the milliliters of urine produced per minute and dividing the product by the plasma creatinine value.

crepitus—Crackling from the joint palpated on examination.

critical limb ischemia—The most severe form of peripheral artery disease; generally affects the tibial and pedal arteries of the lower extremity. Can cause lower extremity rest pain, nonhealing ulcer wounds in the foot, and gangrene. Critical limb ischemia may ultimately lead to amputation of the lower extremities.

cross-training—The concept of training in one mode that allows for the development of physiology that will have a carryover effect to another mode; for example, resistance training is often performed to develop sport-specific strength.

CRT (cardiac resynchronization therapy)—The use of a biventricular pacemaker to restore the coordinated (or synchronized) pumping action of the ventricles when electrical conduction is delayed by bundle branch block, a common feature of chronic heart failure.

culprit lesion—The primary obstruction responsible for decreased blood flow through a coronary artery.

curative surgery—Surgery aimed at complete removal of tumor along with a small amount of surrounding normal tissue.

cystic fibrosis–related diabetes—Diabetes associated with scarring of the pancreas that prevents insulin from entering the bloodstream and results in abnormal blood glucose levels.

cystic fibrosis transmembrane conductance regulator (CFTR)—A protein that is altered secondary to CF; the gene for CF that is located on chromosome 7 and results in altered production of CFTR, a protein that functions as a chloride channel regulated by cyclic adenosine triphosphate. This protein causes abnormal sodium chloride and water movement across the cell membrane, resulting in thick, dry mucus.

cysts—Abnormal sacs containing gas, fluid, or a semisolid material, with a membranous lining.

cytokines—Nonantibody proteins, released by one cell population on contact with a specific antigen, that act as intercellular mediators.

cytomegalovirus—One of a group of highly host-specific herpes viruses.

dementia—A progressive decline in mental function, memory, and acquired intellectual skills.

demyelination—The loss of the myelin covering that insulates the nerve tissue.

diabetic ketoacidosis—A type of metabolic acidosis caused by accumulation of ketone bodies in diabetes mellitus.

dialysate—The fluid that passes through the membrane in dialysis and contains the substances with greater diffusibility in solution.

dialysis—A method used to remove excess fluid and uremic toxins from the extracellular fluid when the kidneys are no longer capable of performing this function adequately.

diastolic—Referring to the pressure remaining in the arteries after cardiac contraction.

diastolic dysfunction—Most often with reference to left ventricular function, a stiff or less compliant chamber that is partially unable to relax and expand as blood flows in during diastole.

differential diagnosis—The determination of which of two or more diseases with similar symptoms is the one from which the patient is suffering, through systematic comparison and contrasting of the clinical findings.

digoxin—A cardioactive steroid glycoside used to increase myocardial contractility.

diplegia—Paralysis of corresponding parts on the two sides of the body. Fine motor function in the upper extremities may be affected and the trunk may be slightly affected, but primarily the legs are affected.

diplopia—Double vision caused by a disorder of the nerves that innervate the extraocular muscles or by impaired function of the muscles themselves.

disability—Loss of physical function.

discitis—Inflammation of an intra-articular disc.

disease-modifying antirheumatic drugs (DMARDs)—A category of otherwise unrelated drugs that suppress the rheumatic disease process by diminishing the immune response. Can improve not only the symptoms of inflammatory joint disease but also some of the extra-articular manifestations, such as vasculitis.

disimpaction—Manual removal of fecal material from the rectal vault.

disordered eating—Inappropriate eating behaviors leading to insufficient energy intake.

distal—Away from the origin or center line, as opposed to proximal.

Doppler ultrasonography—Application of the Doppler effect in ultrasound to detect movement of scatterers (usually red blood cells) through analysis of the change in frequency of the returning echoes.

dorsal rhizotomy—A surgical procedure used to treat spasticity, particularly in young children, usually between 2 and 8 yr old with cerebral palsy; often referred to as selective dorsal rhizotomy. This surgical procedure permanently reduces spasticity by selectively cutting the abnormal sensory nerve rootlets.

double product—The value obtained by multiplying the heart rate and systolic blood pressure; an estimate of myocardial oxygen demand that is reproducible, for example at the ischemic threshold; used to determine confidence of results of a diagnostic exercise evaluation (i.e., value should be >24,000 for highest predictive confidence).

Down syndrome—A chromosomal condition caused by the presence of all or part of an extra 21st chromosome. It is named after a British physician, John Langdon Down, who first described the syndrome in 1866.

drug-eluting stent—Stent that slowly releases a drug (sirolimus), resulting in a reduction of restenosis rates.

dual-chamber pacemaker—Pulse generator that can pace or sense in the atrium or ventricle.

dual-energy X-ray absorptiometry (DXA)—A method for measuring bone mineral density and bone mineral content. It is based on the amount of radiation absorption, or attenuation, in body tissues. When bone mass is measured, the higher the attenuation of radiation by the bone, the greater the mass. Radiation exposure is minimal (<5 mR) compared with that from a chest X-ray (100 mR) or lumbar X-ray (600 mR).

dual-photon absorptiometry—A method similar to DXA for measuring bone density but one that relies on a radionuclide source as opposed to X-ray. The photon intensity is not as great as with DXA, and precision is therefore reduced.

Duke nomogram—Five-step tool to estimate a person's prognosis (5 yr survival or average annual mortality rate) following completion of a maximal GXT.

dynamic endurance—Classification of exercise in which concentric–eccentric shifting occurs until muscular fatigue is induced. An example of dynamic endurance is performing biceps curls until fatigue occurs and the subject is unable to continue full motion against resistance.

dynamic flexibility (stretching)—Slow and constant stretch held for a period of time.

dynamic pulmonary function—Assessment of pulmonary status during exercise. Dynamic pulmonary function includes measures of oxygen consumption, minute ventilation, and ventilator equivalents. These measures help to identify ventilator limitations to stress.

dysarthria—Difficulty speaking because of impairment of the tongue and other muscles essential for speech.

dysesthetic—Referring to an abnormal unpleasant spontaneous or evoked sensation or pain, caused by lesions of the peripheral or central nervous system; involves sensations such as burning, wetness, itching, electric shock, and "pins and needles."

dyskinetic (pertaining to cerebral palsy)—Characterized by an abnormal amount and type of involuntary motion with varying amounts of tension, normal reflexes, and asymmetric involvement.

dyskinetic (pertaining to the heart)—Denoting an outward or bulging movement of the myocardium during systole; often associated with aneurysm.

dyslipidemia—Plasma lipid disorders resulting in abnormal lipid profiles.

dyslipoproteinemia—Abnormally elevated or reduced lipoprotein concentrations.

dysmenorrhea—Pain in association with menstruation.

dyspareunia—Pain in the labial, vaginal, or pelvic areas during or after sexual intercourse.

dyspepsia—Stomach discomfort, including symptoms such as heartburn, gas, and acid reflux.

dysphagia—Difficulty in swallowing.

dyspnea—Shortness of breath or labored or difficult breathing that is perceived by an individual at rest or with exertion (also referred to as dyspnea on exercise, DOE). A scale can be used to determine the severity of dyspnea.

dystonia—Sustained muscle contractions that result in twisting and repetitive movements or abnormal posture.

eccentric lesion—A blockage that is equal distance away from the center of the artery—around the lining of the artery.

echocardiogram—An investigation of the heart and great vessels with ultrasound technology as a means to diagnose cardiovascular abnormalities.

echocardiography—Use of ultrasound images to evaluate the heart and great vessels.

economy—The rate of oxygen uptake necessary to perform a given activity.

ectopic adiposity—Accumulation of fat in various non-adipose tissue depots, such as the liver and skeletal muscle.

ectopic pregnancy—Implantation of the fertilized ovum outside of the uterine cavity.

edema—A condition in which body tissue contains an excessive amount of fluid.

effective insulin—Insulin available for use by body tissues.

effusion—Excess synovial fluid within a joint.

ejection fraction (EF)—Percentage of blood that is ejected from the left ventricle per beat (normal 55-60%); EF = [(EDV − ESV)/EDV] × 100, where EDV = end-diastolic volume and ESV = end-systolic volume. Decreases are noted with systolic heart failure to values below 35% to 40%.

elastin—Structural protein found in the walls of the alveoli.

elderly—Past middle age; relating to later life.

electromechanical dissociation (pulseless electrical activity)—Any heart rhythm observed on the electrocardiogram that does not result in a pulse.

electromyography—A diagnostic neurological test to study the potential (electrically measured activity) of a muscle at rest, the reaction of muscle to contraction, and the response to muscle insertion of a needle. The test is an aid in ascertaining whether a patient's illness is directly affecting the spinal cord, muscles, or peripheral nerves.

ELISA—Enzyme-linked immunosorbent assay. The most commonly used test for the presence of HIV antibodies.

embolic stroke—Emboli that cause embolic strokes are typically from the carotid or other arteries. In these cases, the thrombus is not large enough to occlude the large vessel, but the embolus that breaks off ultimately lodges in a smaller cerebral artery or arteriole.

embolism—Obstruction of a blood vessel by foreign substances or a blood clot.

emphysema—Disease characterized by abnormal permanent enlargement of the respiratory bronchioles and the alveoli.

endocrine—Referring to glands that secrete hormones into the bloodstream.

endothelial cell—One of the squamous cells forming the lining of serous cavities, blood, and lymph vessels and the inner layer of the endocardium.

endothelial-derived relaxing factors—Diffusible substances produced by endothelial cells that cause vascular smooth muscle relaxation; nitric oxide is one such substance.

endothelium—A thin layer of cells that line the inner surface of blood vessels.

endovascular revascularization—Percutaneous, catheter-based interventions used to treat obstructive blood vessel disease. Endovascular treatment options include but are not limited to stenting, angioplasty, and atherectomy procedures.

end-stage renal disease (ESRD)—The final stage of chronic kidney disease in which kidney function has deteriorated so drastically that patients must either be dialyzed or receive a kidney transplant or expire from uremia.

enteral—Referring to a route for administration of a drug that is through the gastrointestinal tract.

entheses—Sites where ligaments, tendons, or joint capsules are attached to bone.

enthesopathy—Inflammation at entheses.

environmental factors—Physical and social factors that can influence participation in physical activity (e.g., vehicular traffic, inclement weather, and unsafe neighborhoods).

eosinophil—A granular leukocyte that contains vasoactive amines.

epilepsy—The paroxysmal transient disturbances of brain function that may be manifested as episodic impairment or loss of consciousness, abnormal motor phenomena, psychic or sensory disturbances, or perturbation of the autonomic nervous system.

epistaxis—Bleeding from the nose.

erythrocyte sedimentation—The sinking of red blood cells in a volume of drawn blood.

erythropoiesis—Stimulation of red blood cell production.

estrogen replacement therapy—Therapy useful for protecting against bone loss in postmenopausal women.

etiology—Cause.

evidence based—Referring to use of the best available clinical research to guide treatment.

evoked response testing—Test in which brain electrical signals are recorded as they are elicited by specific stimuli of the somatosensory, auditory, and visual pathways.

evolving MI—Period of time after the acute onset of an MI when the myocardial tissue is transforming from ischemic to necrotic tissue.

exacerbation—A period of worsening symptoms.

excess body weight—Condition that results when too

few calories are expended and too many consumed for individual metabolic requirements; overweight (>25) and obesity (≥30) as defined by body mass index ($kg \cdot m^{-2}$).

exercise intolerance—Condition in which the individual is unable to perform physical exercise at the level that would be expected for the person's age, gender, and comorbid conditions. It is not a disease in itself but a symptom, either of worse than expected underlying disease or of deconditioning.

exercise prescription (ExRx)—A specific plan of exercise activities that form the basis of an exercise program. The term prescription implies that the plan is based on objective criteria from an exercise test and that the activities are designed to restore health, for example after a cardiac event, or prevent disease, for example by reducing cardiac risk factors.

exercise testing—Measuring the body's reaction to increases to an exercise challenge. Most commonly, testing the cardiovascular response to exercise in terms of ability to perform work; blood pressure, heart rate, and electrocardiographic responses; and development of symptoms.

exertional ischemia—Myocardial ischemic response produced by exerting oneself physically.

extrapyramidal—Denoting the area of the brain that includes the basal ganglia and the cerebellum.

fatigue—That state, following a period of mental or bodily activity, characterized by a lessened capacity for work and reduced efficiency of accomplishment, usually accompanied by a feeling of weariness, sleepiness, or irritability; may also supervene when, from any cause, energy expenditure outstrips restorative processes, and may be confined to a single organ.

fatty liver disease—Also known as steatorrhoeic hepatosis. This condition is characterized by the accumulation of triglycerides in the cells of the liver. Fatty liver disease is most commonly associated with excessive alcohol intake or obesity (nonalcoholic fatty liver disease).

fetal alcohol syndrome—A pattern of mental and physical defects that can develop in a fetus in association with high levels of alcohol consumption during pregnancy. Often leads to intellectual disability and behavioral problems.

fibrinolysis—The process of dissolving a coronary artery thrombosis with either an intrinsic thrombolytic peptide or a thrombolytic medication.

fibrinolytic—Causing fibrinolysis, which is the breakdown of fibrin, a blood-coagulating protein.

fibromyalgia—Condition featuring chronic widespread pain and diffuse tenderness at discrete anatomical sites. Generally associated with mood and sleep disturbances and debilitating fatigue.

flaccidity—Lacking muscle tone (opposite of spasticity).

foam cells—Smooth muscle cells that take up intimal lipid when it accumulates in the cytoplasm and have a bubbly appearance when observed microscopically.

follicle-stimulating hormone—Hormone produced by the anterior pituitary to stimulate the growth of the follicle in the ovary and spermatogenesis in the testes.

forced expiratory volume in 1 s (FEV_1)—Marker of airway obstruction; the maximum amount of air that can be exhaled in 1 s; may be expressed as an absolute value, a percentage of the forced vital capacity, or a percentage of a predicted value.

forced vital capacity (FVC)—The maximum amount of air that can be exhaled forcefully after a maximal inspiration.

four-compartment body composition modeling—Method of studying body composition by determining the contents of four body compartments: fat, protein, water, and mineral.

fragile X syndrome—Also called Martin–Bell syndrome or Escalante's syndrome. This is a genetic condition that is the most commonly known single-gene cause of autism and the most common inherited cause of intellectual disability. Results in physical, intellectual, emotional, and behavioral limitations.

frail—Having delicate health.

functional aerobic impairment (FAI)—Percentage of an individual's observed functional capacity that is below that expected for the person's sex, age, and conditioning level. $\%FAI = (\text{predicted } \dot{V}O_2 - \text{observed } \dot{V}O_2)/(\text{predicted } \dot{V}O_2) \times 100$.

functional capacity—A person's maximum level of oxygen consumption; can be measured at maximal effort with the use of a metabolic cart or predicted based on the maximum workload achieved.

functional electrical stimulation (FES)—Externally applied electrical stimulation of neuromuscular elements to activate paralyzed muscles in a precise sequence and at a precise intensity to restore muscular function.

galactorrhea—Continuation of flow of milk after cessation of nursing.

gangrene—Necrosis of body tissues caused by obstruction, loss, or diminution of blood supply.

gastroepiploic artery—An artery with its origin in the stomach region used for coronary revascularization surgery.

gel phenomenon—The sensation of difficulty moving a joint after a period of joint rest or immobility.

genotype—The resultant expression of specific genes.

genu valgum—More commonly referred to as knock-knee deformity; a condition in which the knees angle in and touch when the legs are straightened.

genu varum—More commonly referred to as bowleg deformity; a condition in which the medial angulation of the leg in relation to the thigh results in an outward bowing of the legs.

geriatrics—A branch of medicine that deals with the problems and diseases associated with elderly people (>65 yr) and the aging process.

gestational diabetes—Carbohydrate intolerance of variable severity with onset or first recognition during pregnancy.

ghrelin—A potent appetite-increasing gut hormone.

gliosis—Excess of astroglia in damaged areas of the central nervous system.

glomerular filtration rate—The amount of fluid that the kidneys filter each minute.

glomerulonephritis—An acute, subacute, or chronic, usually bilateral, diffuse inflammatory kidney disease that primarily affects the glomeruli.

glucagon—A hormone produced by the pancreas that stimulates the liver to release glucose, causing an increase in blood glucose levels and thus opposing the action of insulin.

glucose intolerance—A transitional state between normoglycemia and diabetes. Diagnosed when fasting blood glucose levels are $\geq$100 mg $\cdot$ dl^{-1} but <126 mg $\cdot$ dl^{-1}, or when glucose levels are between 140 and 199 mg $\cdot$ dl^{-1} 2 h after a 75 g oral glucose tolerance test.

GLUT 4—Insulin-regulated glucose transporter responsible for the removal of glucose from blood and delivery to the inner cell membrane.

glycemic goals—A goal range for blood glucose concentration.

Golgi tendon organ—A proprioceptive sensory nerve ending embedded among the fibers of a tendon, often near the musculotendinous junction; it is compressed and activated by any increase of the tendon's tension, caused either by active contraction or by passive stretch of the corresponding muscle.

gonadotropin-releasing hormone—Hormone produced in the hypothalamus that acts on the pituitary and causes the release of gonadotropic substances, luteinizing hormone, and follicle-stimulating hormone.

graded exercise testing (GXT)—Testing that uses a gradual increase in exercise workload to a predetermined point or until volitional fatigue, unless symptoms occur before this point. Generally completed on a treadmill or bicycle ergometer.

growth factor—A category of hormones responsible for stimulating the process of tissue growth.

HAART—Highly active antiretroviral therapy. A broad category of pharmacologic agents used in HIV treatment regimens.

half-life (t-1/2)—Pertaining to a drug, the time it takes for one-half of the drug concentration to be eliminated.

hazard ratio—Multiplicative measure of association. Exposure to a certain risk factor or certain characteristic is associated with a fixed instantaneous risk compared with the hazard in the unexposed.

HbA1c—Glycosylated hemoglobin. This form of hemoglobin is primarily used to identify the plasma glucose concentration. A very high HbA1c (i.e., >7%) represents poor glucose control.

HDL cholesterol—High-density lipoprotein cholesterol. Binds to cholesterol in the arteries and carries the cholesterol back to the liver. Low levels of HDL cholesterol (<40 mg $\cdot$ dl^{-1}) increase the risk of cardiovascular disease.

health belief model—Theory proposing that only psychological variables influence health behaviors.

heart and lung bypass—Device for maintaining the functions of the heart and lungs while either or both are unable to continue to function adequately.

heart failure (HF of CHF)—The pathophysiological state in which an abnormality of cardiac function is responsible for failure of the heart to pump blood at a rate commensurate with the requirements of metabolizing tissues.

heart rate reserve (HRR)—The difference between a person's resting heart rate and maximal heart rate.

hematocrit (Hct)—The percentage by volume of packed red blood cells in a sample of blood.

hematuria—Red blood cells in the urine.

hemianopia—Loss of vision for one-half of the visual field in one or both eyes.

hemiparesis—Paralysis affecting only one side of the body.

hemiparetic—Referring to paralysis affecting only one side of the body.

hemiplegia—Paralysis that affects only one side of the body.

hemodialysis—A method of dialysis in which the patient's blood is pumped through an artificial kidney, external to the body. As the blood passes through the dialyzer, its composition is favorably altered.

hemolysis—Alteration or destruction of red blood cells.

hemophilia—A hereditary hemorrhagic diathesis caused by deficiency of coagulation factor VIII. Characterized by spontaneous or traumatic subcutaneous and intramuscular hemorrhages.

hemoptysis—Expectoration of blood arising from the respiratory system; for people with CF, this occurrence reflects further infection or advancing disease.

hemorrhagic—Of or relating to excessive bleeding.

hemorrhagic stroke—Hypertension is the major risk factor for hemorrhagic stroke. They can also be caused by aneurysm, drug use, brain tumor, congenital arteriovenous malformations, and anticoagulant medication. Hemorrhagic strokes are classified as either intracerebral, which refers to bleeding inside the brain, or subarachnoid, which refers to bleeding in and around the spaces that surround the brain. There is usually little warning for a hemorrhagic stroke.

hepatic lipase (HL)—Lipase produced in the liver. Hepatic lipase activity results in the liver uptake of fatty acids by hydrolyzing triglycerides and phospholipids of VLDL and HDL. Hepatic lipase activity also contributes to cholesterol delivery to the liver.

herniation—Development of an abnormal protrusion or projection of an intervertebral disc.

high tone—Excess tone in a muscle group, often referred to as spasticity or hypertonicity.

hindfoot valgus—Excessive lateral deviation of the talocalcaneal complex relative to the trochanteric knee ankle (TKA) line.

hindfoot varus—Excessive medial deviation of the talocalcaneal complex relative to the trochanteric knee ankle (TKA) line.

histoplasmosis—Infection resulting from inhalation or, infrequently, ingestion of fungal spores. May cause pneumonia.

HIV negative—Denoting an individual without antibodies to HIV viral proteins. An individual who has been recently infected with HIV may not have had the opportunity to have developed antibodies to the virus.

HIV positive—Denoting an individual with antibodies to HIV viral proteins; often refers to individuals who are infected but who have not yet developed an AIDS-defining condition or whose CD4+ cell count is greater than 200 cells $\cdot$ mm^{-3}.

homocysteine—A homolog of cysteine.

human immunodeficiency virus (HIV)—The pathogen that leads to AIDS.

hybrid—Referring to concurrent use of both upper extremity exercise and lower extremity functional electrical stimulation ergometry.

hydronephrosis—Kidney condition characterized by dilation of the renal pelvis and collecting system attributable to ureteral obstruction or backflow (reflux) from the bladder.

hyperandrogenemia—Elevated androgen hormone levels in the plasma.

hypercapnia—An increased arterial carbon dioxide content.

hyperemia—Increased amount of blood in a part or organ.

hyperglycemia—An abnormally high concentration of glucose in the circulating blood, seen especially in people with diabetes mellitus.

hyperinflation—Overinflation of the lung, resulting in a greater functional residual capacity and total lung capacity.

hyperinsulinemia—Condition characterized by excess levels of circulating insulin in the blood. Also known as prediabetes, insulin resistance, and syndrome X.

hyperkalemia—Excess concentrations of potassium in the bloodstream.

hyperlipidemia—Elevated blood lipid levels that include elevated cholesterol and triglyceride concentrations.

hyperlipoproteinemia—Elevated lipoprotein concentrations.

hyperparathyroidism—A state produced by increased function of the parathyroid glands; results in dysregulation of calcium.

hyperpnea—More rapid and deeper breathing than normal.

hyperreflexia—A condition in which the deep tendon reflexes are exaggerated and are defined by overactive or overresponsive reflexes, which may include twitching and spastic tendencies.

hypertension—A condition in which blood pressure is chronically elevated. Diagnosed by resting systolic blood pressure ≥140 mmHg, resting diastolic blood pressure ≥90 mmHg, or both on two separate occasions.

hypertonia—Increased rigidity, tension, and spasticity of the muscles.

hypertriglyceridemia—Elevated blood triglyceride concentrations.

hypertrophy—An increase in cell size.

hyperuricemia—Presence of high levels of uric acid in the blood. Diagnosed when uric acid levels are ≥7 mg $\cdot$ dl^{-1} in men and are ≥6 mg $\cdot$ dl^{-1} in women.

hypoestrogenic—Referring to decreased plasma estrogen levels.

hypoglycemia—Abnormally small concentration of glucose in the circulating blood; symptoms resulting from low blood glucose (normal glucose range 60-100 mg $\cdot$ dl^{-1}, or 3.3-5.6 mmol $\cdot$ L^{-1}) that are either autonomic or neuroglycopenic.

hypokalemia—Extreme potassium depletion in the circulating blood.

hypotension—Abnormally low blood pressure; typically associated with symptoms.

hypothyroidism—A condition in which the thyroid gland does not make enough thyroid hormone and the person becomes thyroid deficient.

hypotonic—Having a lesser degree of tension.

hypovolemia—Diminished blood volume.

hypoxemia—Insufficient oxygenation of the blood; assessed by arterial blood gas or pulse oximetry.

hypoxia—A state of oxygen deficiency.

iatrogenic—Denoting decrements in health status due to medical treatment. HAART induces numerous adverse (i.e., iatrogenic) effects, such as lipodystrophy, insulin resistance, osteonecrosis, and an increased risk of cardiovascular disease (CVD).

idiopathic—Denoting a disease of unknown cause.

IMAT—Intermuscular adipose tissue; fat that infiltrates the skeletal muscle.

immune system—System that mediates the body's interaction between internal and external environments. Helps rid the body of infectious agents and malignant cells.

immunosuppression—Suppression of immune responses produced primarily by a variety of immunosuppressive agents.

incidence—The frequency of occurrence of any event or condition over a period of time and in relation to the population in which it occurs.

incomplete paraplegia—Incomplete motor and sensory dysfunction of the trunk, legs, and pelvic organs resulting from SCI.

incomplete tetraplegia—Incomplete motor and sensory dysfunction of the arms, trunk, legs, and pelvic organs resulting from SCI.

incontinence—Lack of control of urination or defecation.

infarct-related artery—The completely, or nearly completely, stenosed coronary artery that is responsible for the interruption of blood flow and the subsequent myocardial infarction.

insensate—Lacking sensation.

inspiratory resistive loading—The act of inspiring air against a resistance greater than normal.

inspiratory threshold loading—The act of inspiring after attaining and proceeding at a predetermined inspiratory pressure (threshold point).

insulin resistance—A condition in which normal amounts of insulin secreted by the pancreas are inadequate to produce a normal insulin response in the muscle or liver. As a result, the pancreas secretes additional insulin, thereby elevating insulin levels in the plasma. High levels of insulin in the plasma often lead to the development of type 2 diabetes or metabolic syndrome.

intermittent claudication—Recurrent cramping symptoms, usually of the calf muscles, thighs, or buttocks at regular intervals during walking.

internal mammary artery—An artery with its origin in the chest region used for coronary revascularization surgery.

intima—The inner layer of blood vessels containing endothelial cells.

intradialytic exercise—Exercise performed during a hemodialysis treatment session.

intraosseous—Within bone.

ipsilateral—On the same side of the body.

iritis—Inflammation of the iris of the eye.

ischemia—Deficiency of blood flow, attributable to functional constriction or actual obstruction of a blood vessel.

ischemic—Referring to a sustained deficiency in oxygen delivery.

ischemic heart disease—A pathological condition in which blood flow to the myocardium is reduced below the demand, resulting in a lack of oxygen delivery to cardiac tissue (i.e., coronary atherosclerosis).

ischemic stroke—Associated with the development of ischemic cerebrovascular disease, they can be further categorized as thrombotic, embolic, and hemodynamic.

isocapnia—Normal arterial carbon dioxide levels.

isokinetic—Denoting the condition in which muscle fibers shorten at a constant speed in such a manner that the tension developed may be maximal over the full range of joint motion.

isometric—Denoting the condition in which the ends of a contracting muscle are held fixed so that contraction produces increased tension at a constant overall length.

isotonic—Denoting the condition in which muscle fibers shorten with varying tension as the result of a constant load.

joint contractures—Reduced passive range of motion at a joint caused by shortened tendons, typically associated with unbalanced spasticity.

joint effusion—Increased fluid in synovial cavity of a joint.

joule (J)—A unit of energy; the heat generated, or energy expended, by an ampere flowing through an ohm for 1 s; equal to 107 erg and to a newton meter. The joule is an approved multiple of the SI fundamental unit of energy, the erg, and is intended to replace the calorie (4.184 J).

Kaposi's sarcoma—Firm, subcutaneous, brown-black or purple lesions usually observed on the face, chest, genitals, oral mucosa, or viscera.

karyotyping—A technique used to quantify the DNA copy number on a genomic scale.

ketone—A substance with the carbonyl group linking two carbon atoms.

kilocalorie (kcal)—A unit of heat content or energy; the amount of heat necessary to raise 1 g of water from 14.5 °C to 15.5 °C times 1,000.

kyphosis—Excessive angulation of the spine resulting in increased anteroposterior diameter of the chest cavity; humpback; may reflect chronic pulmonary disease.

laminectomy—A surgical procedure in which the posterior arch of a vertebra is removed to relieve pressure on the spinal cord or on the nerve roots that emerge from the spinal canal. The procedure may be used to treat a herniated disc or spinal stenosis.

lead-pipe rigidity—Diffuse muscular rigidity resembling

the resistance to bending of a thin-walled metal pipe. Lead-pipe rigidity persists throughout the range of movement of a particular joint, as distinct from clasp-knife rigidity, which varies.

lecithin-cholesterol acyltransferase (LCAT)—An enzyme that hydrolyzes a fatty acid from lecithin and the subsequent esterification of the fatty acid with cholesterol. The products of this reaction are cholesterol esters and lysolecithin. LCAT is essential for the function of HDL and the maintenance of reverse cholesterol transport.

left ventricular dysfunction (LVD)—Abnormal function of the left ventricle (i.e., poor wall motion).

left ventricular ejection fraction (LVEF)—The percentage of the end-diastolic volume ejected per beat; an index of systolic function.

leptin—A protein messenger from adipose tissue to the satiety center in the hypothalamus involved in regulating appetite.

lifestyle-based physical activity—Home- or community-based participation in forms of activity that include much of a person's daily routine (e.g., transport, home repair and maintenance, yard maintenance).

lipoprotein—Macromolecule consisting of proteins, phospholipids, cholesterol, and triglycerides that serve to transport lipids in aqueous mediums found in the blood, interstitium, and lymph.

lipoprotein lipase (LPL)—An enzyme, found in skeletal muscle and adipose tissues, that hydrolyzes triglycerides into fatty acids. The fatty acids are taken up by these tissues for energy use or storage.

low-calorie diet—A hypocaloric diet, 1,200 kcal per day or less.

low energy availability—Also known as energy deficit, energy drain, or negative energy balance; results from low dietary energy intake and high energy expenditure.

lower rate limit—The rate at which the pulse generator begins pacing in the absence of intrinsic activity.

low tone—Often referred to as flaccidity or hypotonia; decreased amount of tone in a muscle group.

luteinizing hormone—Hormone secreted by the anterior lobe of the pituitary to stimulate the development of the corpus luteum.

LV diastolic dysfunction—Clinically, diagnosis is less exact than for systolic dysfunction. Diagnosis is often made when the clinical syndrome of congestive heart failure (fatigue, dyspnea, and orthopnea) requires hospitalization in the presence of a relatively normal ejection fraction.

LV systolic dysfunction—Ejection fraction reduced below 45% (severe considered <30%) as measured by echocardiogram or another quantitative measure.

lymphedema—Swelling as a result of obstruction of lymphatic vessels, resulting in fluid buildup and edema in the affected extremity.

lymphocytes—Any of the mononuclear, nonphagocytic leukocytes found in the blood, lymph, or lymphoid tissues that are the body's immunologically competent cells.

lymphocytopenia—A reduction in the number of lymphocytes in the circulating blood.

macrovascular disease—Atherosclerosis that affects large vessels such as the aorta, femoral artery, and carotid artery.

magnetic resonance imaging (MRI)—The diagnostic test that uses principles of magnetism to generate an electromagnetic field around the body, causing certain atoms in the nucleus of the body cells to "line up." Then, by sending and receiving radio signals, which are fed into a computer, the device records the position of those atoms, providing a distinct picture of the tissues being investigated. The patient lies inside a large, tunnel-like tube for 30 to 60 min while the images are formulated by the computer. This diagnostic imaging technique has been found to have certain advantages over radiographs and CT scans in the diagnosis of spinal disorders.

major depressive disorder—A diagnosis requiring endorsement of at least five symptoms, one of which must be either depressed mood or diminished interest or pleasure, that have been present during the same 2 wk period.

maladaptation—Adaptation to a progressive stimulus (e.g., exercise) that results in an overload to the system to the degree that performance is reduced and the risk of injury is increased.

mast cells—Connective tissue cells that are important in the defense mechanisms of the body needed during injury or infection.

maximal bone mass—The highest bone mass that a person could possibly achieve.

maximal oxygen consumption—The maximum amount of oxygen consumed (or used) by the body, usually measured under conditions of maximal exercise.

maximal voluntary ventilation (MVV)—Amount of air maximally breathed in, expressed as liters per minute.

maximum sensor rate—The maximum rate for a rate-responsive pacemaker that can be achieved under sensor control.

maximum tracking rate—The maximum rate at which the pulse generator will respond to atrial events.

medical nutrition therapy (MNT)—The use of nutrition as a treatment for a clinical condition or disease.

medication reconciliation—The formal process that compares a patient's current medications to those in the patient's record or medication orders.

menarche—The beginning of menstrual function.

mentation—The process of reasoning and thinking.

metabolic equivalent (MET)—An expression of the rate of energy expenditure at rest. 1 MET = 1 kcal $\cdot$ kg^{-1} $\cdot$ h^{-1} = 3.5 ml $\cdot$ kg^{-1} $\cdot$ min^{-1}.

metabolic syndrome—A constellation of insulin resistance characterized by central obesity, elevated triglycerides, suppressed HDL cholesterol, hypertension, or prediabetes.

metastasize—To spread from one part of the body to another, as when neoplasms appear in parts of the body remote from the site of the primary tumor.

microalbuminuria—A condition in which the kidneys leak a small amount of albumin into the urine. Diagnosed with 24 h urine collections ($\geq$20 μg $\cdot$ min^{-1}) or when levels are $\geq$30 μg $\cdot$ min^{-1} on two separate occasions.

microvascular disease—Atherosclerosis that affects small blood vessels such as those of the kidney, eye, heart, and brain.

moderate-intensity physical activity—Activities that cause small increases in breathing and heart rate; ~50% to 70% $\dot{V}O_2$peak.

mode switching—A programmed feature of dual-chamber pacemakers that prevents tracking or matching every atrial impulse with a ventricular pacing pulse; purpose is to prevent tracking of rapid atrial rates to the ventricle.

monoplegia—Paralysis of a single limb, muscle, or muscle group.

monounsaturated fat—Dietary fatty acid that contains one double bond along the main carbon chain.

morbidity—Manifestations of disease other than death.

morphology—Configuration or shape (e.g., shape of the ST segment: downsloping, upsloping, or horizontal).

mortality—Death.

motion artifact—Incidental activity that is picked up on an ECG during body movement.

multiple sclerosis (MS)—A debilitating disease characterized by multiple areas of scar tissue replacing myelin around axons in the central nervous system.

muscle spindle receptors—A fusiform end organ in skeletal muscle in which afferent and a few efferent nerve fibers terminate; this sensory end organ is particularly sensitive to passive stretch of the muscle in which it is enclosed.

muscle tone—Amount of tension in a muscle or muscle group at rest.

muscle weakness—Condition in which skeletal muscle lacks strength and power-generating capability.

muscular strength and endurance—The ability of skeletal muscles to perform hard or prolonged work.

musculoskeletal flexibility—The range of motion in a joint or sequence of joints.

myalgia—Pain in a muscle or muscles.

***Mycobacterium avium* complex**—Complex that consists of two predominant species, *M. avium* and *M. intracellulare.* More than 95% of infections in patients with AIDS are caused by *M. avium,* whereas 40% of infections in immunocompetent patients are caused by *M. intracellulare.*

myelography—Radiographic inspection of the spinal cord and nerve roots by use of a radiopaque contrast medium (a substance that causes the absorbing tissues to appear darker or lighter on a radiograph) injected into the intrathecal space. Air or oil dye may be used as contrasting agents.

myelomeningocele—Congenital open neural tube defect with disruption of skin, bone, and neural elements; usually involves spinal cord dysfunction despite surgical closure.

myocardial ischemia—Temporary lack of oxygen to the heart or myocardium, due to an imbalance between oxygen supply and demand.

myocardium—The heart muscle.

myoglobinuria—Excretion of myoglobin in the urine resulting from muscle degeneration.

nasal polyposis—Growths of tissue in the nose that may block air passage through the nostril; not life threatening.

neoplasm—Abnormal tissue that grows by cellular proliferation more rapidly than normal and continues to grow after the stimuli that initiated the growth cease. Structural organization and function of neoplastic tissue are partially or completely different from what is seen in normal tissue.

neurectomy—Partial or total excision or resection of a nerve.

neurogenic—Controlled by the nervous system, as in neurogenic bladder, bowel, or sexual function.

neuroglycopenic hypoglycemia—Symptoms of hypoglycemia that include feelings of dizziness, confusion, tiredness, difficulty speaking, headache, and inability to concentrate.

neuromuscular—Of, relating to, or affecting both nerves and muscles.

neuropathy—A disease involving the cranial nerves or the peripheral or autonomic nervous system.

neuropeptide Y (NPY)—Central nervous system appetite stimulant.

neutral spine—The position in which the trunk and neck, and therefore the joints of the spine, are in neither flexion nor extension.

neutropenia—The presence of abnormally small numbers of neutrophils in the blood.

neutrophilia—An increase in neutrophilic leukocytes in blood or tissue.

nitroglycerine—Medication used to promote vasodilation in patients with angina pectoris.

nonischemic cardiomyopathy—Disease process involving cardiac muscle that is not related to ischemic heart disease; may be attributable to viral infection or alcohol abuse.

normal-weight obesity—A condition in which an individual is classified in the normal range for body mass or BMI, but has elevated adiposity or risk factors for metabolic syndrome, atherosclerotic cardiovascular disease, diabetes, or more than one of these.

nuclear perfusion—Radioactive isotope that has the ability to perfuse through tissue so that select organs can be imaged.

nucleoside reverse transcriptase inhibitor (NRTI)—A specific type of antiretroviral medication.

nystagmus—Rhythmic, involuntary movements of the eyes.

obesity—A condition in which the proportion of body fat is abnormally high; defined as a body mass index greater than 30 $kg \cdot m^{-2}$.

obstructive sleep apnea—Collapse of the airway during sleep resulting in snoring, poor sleep quality, and intermittent complete lack of breathing (apnea).

occipital—Referring to the posterior part of the skull.

occlusion—The complete blockage of an artery, resulting in decreased blood flow through the artery to the peripheral muscles of the legs.

old age—Between 65 and 74 yr of age.

oldest old—Older than 85 yr of age.

oligomenorrhea—Scanty or infrequent menstrual flow.

oliguria—A diminution in the quantity of urine excreted; specifically, less than 400 ml in a 24 h period.

omega-3 fatty acids—Long-chain polyunsaturated fatty acids that contain a double bond in the n-3 position.

oncogene—A mutated proto-oncogene that may foster unregulated or malignant cell growth.

opportunistic infections—Infections most commonly seen in people who are immunocompromised, such as individuals with late or advanced HIV-1 disease, cancer, or other immunocompromising conditions.

optic neuritis—Inflammation of the optic nerve.

orthopnea—Labored or difficult breathing while lying flat or supine.

orthostatic—Relating to upright or erect posture.

orthostatic hypotension—Decrease of at least 20 mmHg in systolic blood pressure when an individual moves from a supine position to a standing position.

ostectomy—Surgical excision of a bone or a portion of one.

osteoarthritis (OA)—Erosion of articular or joint cartilage that leads to pain and loss of function.

osteogenic—Increasing bone mass.

osteopenia—Reduced bone mineral density, defined as between 1 and 2.5 standard deviations below the young adult mean.

osteophyte—A bony excrescence or outgrowth, usually branched in shape.

osteoporosis—A pathological condition associated with increased susceptibility to fracture and decreased bone mineral density, more than 2.5 standard deviations below the young adult mean.

osteoporotic fracture—Broken bone caused by a reduction in the mass of the bone per unit of volume.

osteotomy—Operation for cutting through a bone to improve alignment or correct deformities.

overweight—A body mass index of 25.0 to 29.9 $kg \cdot m^{-2}$.

oxygen uptake (consumption) ($\dot{V}O_2$)—A measure of a person's ability to take in and use oxygen.

oxyhemoglobin saturation (SaO_2)—Percentage of hemoglobin bound to oxygen; assessed noninvasively by pulse oximeter or invasively by arterial blood gas sampling.

pacemaker—Implantable medical device that uses electrical impulses delivered by electrodes to the heart muscle to regulate the beating of the heart. Modern pacemakers can serve multiple functions, such as maintaining an adequate heart rate or improving the synchronization of the chambers of the heart. Some implantable devices also combine a pacemaker with an automatic internal cardioverter defibrillator. A temporary pacemaker is not implanted and serves primarily to regulate the heart rate.

pacemaker sensor—Sensor incorporated into the pulse generator that detects a physiological stimulus to control the heart rate to match physiological demands.

pacing system—A medical device that uses electrical impulses to regulate the beating of the heart when the heart's natural pacemaker is diseased. The pacing system consists of the pacemaker, or generator, which is the component of the system containing the circuitry that senses when the pacemaker needs to initiate a response, and the battery that powers the device. The pacemaker lead wires are the components of the pacing system that are attached to the pacemaker and deliver the electrical impulses to the heart muscle.

pack-years—Number of packs of cigarettes smoked per day multiplied by the number of years the person smoked; for example, two packs a day for 20 yr equals 40 pack-years.

palliative surgery—Surgery aimed at removal of a tumor to make the patient more comfortable, relieve organ obstruction, or reduce tumor burden.

pancreas—A gland lying behind the stomach that secretes pancreatic enzymes into the duodenum and secretes insulin, glucagon, and somatostatin into the bloodstream.

pancreatic insufficiency—Inadequate exocrine function of the pancreas resulting in little or no production of pancreatic enzymes needed for digestion (i.e., lipase, amylase, protease); results in nutrient malabsorption.

pannus—Inflammatory exudates overlying the lining layer of synovial cells on the inside of a joint.

pansinusitis—Chronic inflammation and infection involving all sinuses; commonly seen in individuals with CF.

papilledema—Swelling of the optic disc in the eye caused by severe hypertension.

paralysis—Loss or impairment of the ability to move a body part, usually as a result of damage to its nerve supply, loss of sensation over a region of the body, or complete loss of motor control.

paraplegia—Motor and sensory dysfunction of the trunk, legs, and pelvic organs resulting from a spinal cord injury.

parasympathetic nervous system—Craniosacral portion of the central nervous system that promotes anabolic activity and energy conservation.

parathyroid hormone—A peptide hormone formed by the parathyroid glands; raises the serum calcium when administered parenterally by causing bone resorption.

parenchyma—The essential or primary tissue of the lungs.

parenchymal infiltrates—Deposition or diffusion in lung tissue of substances not normal to it.

parenteral—Relating to the route of administration of a drug that is through the skin or a mucosal membrane.

paresis—Slight or partial paralysis, or partial weakness in one or more limbs.

paresthesia—A subjective feeling such as numbness, "pins and needles," or tingling.

paroxysmal nocturnal dyspnea—Sudden awakening caused by labored or difficult breathing.

partial body weight-supported treadmill training (PBWSTT)—Mode of exercise training with the person's body weight partially supported over the treadmill by a mechanical suspension system, a therapist, or both.

patency—Of the arterial lumen, the condition of being open, unobstructed.

peak bone density—The highest amount of bone mass achieved by a person during his lifetime. It is assumed that peak bone mass is achieved in the second or third decade of life. The age at which peak bone mass is achieved also varies based on which bone site is being measured.

peak expiratory flow rate (PEFR)—The highest flow rate (exhalation of gas from the lung) that a person can generate during a forceful expiration.

peak walking time or distance—Most typically, the time point or distance at which a patient cannot walk any farther due to severe claudication pain; the most commonly used outcome measure for assessing walking performance in peripheral artery disease studies.

pediatrics—A branch of medicine that is concerned with children and their diseases.

percussion—A diagnostic method that uses short, sharp tapping of the body surface to produce different reflected sounds from the underlying organs.

pericardial effusion—Increased amounts of fluid within the pericardial sac, usually attributable to inflammation.

perineum—The area between the scrotum and anus in males and between the vulva and anus in females.

periodization—A system of fractioning larger periods of muscle training into smaller phases or cycles. Intensity, frequency, sets, repetitions, and rest periods are altered to reduce the risk of overtraining and minimize uncomfortable responses.

peripheral neuropathy—Damage to the nerves of the legs or arms resulting in a loss of sensation (e.g., touch, temperature).

peripheral nervous system—Sensory and motor components of the nervous system that have extensions outside the brain and spinal cord.

peripheral vascular disease (PVD, or peripheral artery deisease PAD)—Disease of the vascular system that can be found in the periphery (i.e., commonly observed in the legs, which leads to claudication with physical exertion).

peritoneal dialysis—Dialysis performed through introduction of fluid into the peritoneal cavity. Dialysis fluid can be cycled through the peritoneal cavity by a machine over a 10 to 12 h period daily (intermittent peritoneal dialysis) or exchanged every 4 h, with the fluid staying in the peritoneal cavity between exchanges (continuous ambulatory peritoneal dialysis). The fluid is introduced through a catheter (tube) placed in the abdomen.

pes equinus—A condition marked by walking without touching the heel to the ground.

pharmacodynamic phase—The science that pertains to the effect of a drug on the body.

pharmacokinetic phase—The science that pertains to the movement of a drug's molecules through the body; encompasses four discrete subphases—absorption, distribution, metabolism, and excretion.

physical activity—Any bodily movement produced by skeletal muscles that results in caloric expenditure.

physiological pacing—Setting a pacemaker to regulate the heartbeat to mimic natural physiological processes, such as timed stimuli between the chambers of the heart, to best meet physiological demands. Modern pacemakers can be programmed to provide multiple functions, depending the indication for the need for the device.

plaque (pertaining to an artery)—A yellow area or swelling on the intimal surface of an artery, produced by the atherosclerotic process of lipid deposition.

plaques (pertaining to the central nervous system)—Scarring of axons in the central nervous system attributable to demyelination.

plasticity—The extent to which normal maturation of maximal aerobic power can be altered by changes in physical activity.

platelet aggregation—The congregation of platelets, which are disc-shaped fragments found in the peripheral blood and involved in the clotting process.

plegia—Greater involvement of one or more limbs than paresis; often associated with paralysis.

plexopathies—Disorders affecting a network of nerves, blood vessels, or lymph vessels. The regions of nerves affected are at the brachial or lumbosacral plexus. Symptoms include pain, loss of motor control, and sensory deficits.

pluripotent stem cell—Uncommitted cell with various developmental options pending.

***Pneumocystis carinii* pneumonia (PCP)**—An AIDS-defining condition caused by the parasite *P. carinii*.

pneumothorax—An acute collection of air in the pleural space; results in collapse of the affected lung; common in advancing CF lung disease.

poikilothermic—Having body temperature that varies with the environment.

point of maximal cardiac impulse (PMI)—Point identified by palpation and inspection of the chest wall during physical examination as the most prominent location for cardiac apical impulse.

polycystic kidney disease—Hereditary bilateral cysts distributed throughout the renal parenchyma, resulting in markedly enlarged kidneys and progressive renal failure.

polycystic ovarian syndrome—Endocrine disturbance associated with primary anovulation and polycystic ovaries.

polycythemia—An abnormally elevated level of red cells in the blood.

polydipsia—Excessive thirst that is relatively prolonged.

polymorphonuclear leukocytosis—An elevation in neutrophilic leukocyte (white blood cell) count.

polyunsaturated fats—Dietary fatty acids that contain two or more double bonds along the main carbon chain.

polyuria—Excessive excretion of urine.

postprandial—Referring to the time period 1 to 2 h following a meal.

postprandial lipemia—Exaggerated levels of triglycerides in the blood and failure to return to baseline levels within 8 to 10 h after consumption of dietary fat.

Prader-Willi syndrome—A rare genetic disorder in which seven genes, or a subset of seven genes, on chromosome 15 are deleted or unexpressed. Characterized by excessive weight gain and obesity.

prevalence—The number of cases of a disease present in a specified population at a given time. This number may be given at one identified time (point prevalence) or during a specified period, such as 2 wk or a year (period prevalence).

prevention—Three general categories of intervention strategies to limit the effect of potential or established disease in the population.

primary—Referring to an intervention geared toward removing or reducing the risk factors of disease.

primary amenorrhea—Delay of menarche beyond age 18.

protease inhibitor—A specific type of antiretroviral medication.

proteinuria—The presence of abnormal amounts of protein in the urine.

prothrombotic—Condition or agent that increases the risk of formation or presence of a thrombus.

proto-oncogene—A gene involved in regulation of normal cell growth or proliferation.

pulmonary exacerbation—An episode of worsening lung disease caused by increased infection and inflammation, resulting in increased patient infections, symptoms, and limitations to activities of daily living.

pulmonary hypoplasia—An incomplete development of the lungs, resulting in an abnormally low number or size of bronchioles or alveoli; primarily a congenital condition.

pulmonary rehabilitation—An evidence-based, multidisciplinary, and comprehensive intervention for patients with chronic respiratory diseases who are symptomatic and often have decreased daily life activities.

pulsatile—Characterized by a rhythmical pulsation.

pyelonephritis—The disease process from the immediate and late effects of bacterial and other infections of the parenchyma and the pelvis of the kidney.

pyramidal—Denoting an area of the brain that consists of the cortical system, which controls all voluntary movements.

Q angle—Acute angle formed by a line from the anterior superior iliac spine of the pelvis through the center of the patella and a line from the tibial tubercle through the patella.

quadriplegia—Paralysis of all four limbs; also called tetraplegia.

quality of life—Perception of life satisfaction.

quantitative computed tomography—The only method currently available that provides an actual measurement of volumetric bone density.

quantitative ultrasound—A device that measures structural properties of bone with sound waves. Unlike densitometry devices, it uses no ionizing radiation.

radiation therapy—Therapy meant to stop growth of malignant cells by damaging RNA within the cells.

radicular—Relating to pain caused by compression or injury of the root of a nerve.

radionuclide agent—Isotope (natural or artificial) that exhibits radioactivity. Used in nuclear cardiology medicine to image the myocardium for potential ischemia.

radionuclide imaging—A type of cardiac imaging that can detect ischemia and wall motion; uses an injected radioisotope (i.e., thallium 201 or technetium-99m sestamibi) that is scanned using X-ray.

ramping protocol—Continuous gradual increase in workload (treadmill, speed and grade; cycle ergometer, watts) over a select period of time.

range of motion (ROM)—The total degrees of movement that a joint can move through.

rate-pressure product (RPP)—Indirect indication of how hard the heart is working. RPP = systolic BP × HR.

rate-responsive pacing—Function of a pacemaker that changes the rate by sensing a physiological stimulus. This type of pacemaker is also described as modulated, adaptive rate, or sensor driven.

rating of perceived exertion (RPE)—A person's perception of how hard she is working physically. Currently two scales are commonly used to assess RPE (6 to 20 and 0 to10).

Raynaud's phenomenon—Intermittent, bilateral attacks of ischemia of the toes and fingers (and sometimes ears or nose). The classical features are episodic pallor of digits, followed by cyanosis (i.e. blue color due to deoxygenation) and then redness, pain, and tingling.

red flags—Generally, signs and symptoms indicative of serious pathology that is usually beyond the capabilities of the treating health care practitioner to treat; these indicators of disease usually mean that an immediate referral to a physician is warranted. Examples include constitutional signs like unexplained nausea, shortness of breath or diaphoresis, unexplained weight loss or gain, spine pain with eating, nocturnal spine pain, and constant spine pain that is not modified by change in position or activity.

refractory—Resilient or resistant to treatment.

rejection—Immune response to foreign tissue (transplanted organ).

relapse—Reversion to an active disease process in multiple sclerosis after a remission.

relapse prevention model—A model used to help new exercisers anticipate problems with adherence. Factors that contribute to relapse include negative emotional or physiological states, limited coping skills, social pressure, interpersonal conflict, limited social support, low motivation, high-risk situations, and stress.

relative contraindications for test termination—Conditions occurring during a GXT that require strong clinical judgment concerning the safety of continuing the exercise test.

relative oxygen uptake—Oxygen uptake expressed in milliliters of oxygen per kilogram of body weight per minute ($ml \cdot kg^{-1} \cdot min^{-1}$).

remission—Recovery period from the active disease process.

renal replacement therapy—A type of therapy used to replace the functioning of failing kidneys. Medical technologies that serve as substitutes for renal function include hemodialysis, peritoneal dialysis, and transplantation. Without this therapy, the patient with no renal function would die.

reocclusion—Closing again; reclosure.

reperfusion—The process of reinstituting blood flow to an area of tissue previously deprived of normal blood flow.

repetition maximum (RM)—The number of times a weight can be lifted; 1RM is the maximal amount of weight that can be lifted one time only.

resolving MI—The phase of an MI in which necrotic tissue forms a scar.

respiratory failure—Failure of the respiratory system to keep gas exchange at an acceptable level.

restenosis—The recurrence of a narrowing or restriction.

retroviruses—Viruses containing both RNA-dependent and DNA-synthesizing material.

revascularization—Restoration of blood flow to a body part.

rhabdomyolysis—Acute, potentially fatal disease of the skeletal muscle; entails destruction of the muscle as evidenced by myoglobinuria.

rheology—The study of the deformation and flow of liquids and semisolids.

rheumatoid arthritis (RA)—Inflammation of the joints attributable to autoimmune attack; leads to pain and loss of function.

rheumatoid cachexia—Loss of body cell mass, predominantly skeletal muscle, that is characteristic of inflammatory arthritis. Thought to be cytokine driven, the condition is also characterized by increased fat mass; consequently reduced body weight (or BMI) is uncommon (<10%).

rigidity—Stiffness or inflexibility that may result in the inability to bend or be bent.

roentgenogram—A photograph made with X-rays.

sarcopenia—Degenerative loss of skeletal muscle mass and strength associated with aging.

sclerosis—Tissue hardening that occurs because scar tissue replaces lost myelin around axons in the central nervous system.

scoliosis—Abnormal lateral curvature (side-bending and rotational components) of the spine that may be congenital or acquired by extremely poor posture, disease, or muscular weakness. Usually the curvature consists of two curves, the original abnormal curve and a second compensatory curve in the opposite direction (also referred to as an S-curve).

secondary—Referring to an intervention that promotes early detection and treatment of disease with the goals of preventing recurrences or progression, promoting recovery, and avoiding complications.

secondary amenorrhea—Cessation of menses in a woman who has previously menstruated.

secondary condition—An injury, impairment, functional limitation, or disability that occurs as a result of the primary condition or pathology. Secondary conditions include physical problems, social concerns, and mental health difficulties. Secondary conditions also can develop when the primary disability interferes with the delivery of standard health care for the treatment or prevention of a health condition.

selective estrogen receptor modulators (SERMs)—Antiresorptive agents that have fewer side effects than estrogen replacement therapy and may be a good alternative to ERT for the woman with a history of breast cancer.

selective serotonin reuptake inhibitors—A class of antidepressant medications commonly used to treat depression.

self-efficacy—A person's belief in his capability to perform a behavior and the perceived incentive to do so.

sensitivity—The proportion of affected people who give a positive test result for the disease that the test is intended to reveal.

set-point theory—A metabolic theory that postulates stability in weight in both overfeeding and calorie-restriction situations.

sick sinus syndrome—Syndrome in which the sinus node is not functioning at an appropriate rate, leading to sinus bradycardia, pauses, arrest, or exit block. Syncopal episodes can be caused by this abnormality.

significant Q wave—Wave that depicts a prior MI on an ECG. For Q waves to be considered significant, they must be either greater than or equal to 0.04 s wide and one-third the height of the associated R wave.

single-chamber pacemaker—Pulse generator that can pace or sense in the atrium or ventricle.

single-photon absorptiometry—A method for determining bone mineral content through measurement of the absorption by bone of a monoenergetic photon beam.

Sjogren's syndrome—A chronic, autoimmune disease of unknown etiology, usually occurring in middle-aged or older women. Features lymphocyte infiltration of exocrine glands (i.e., tear and salivary glands), with resultant dryness of the eyes and mouth, and the presence of a connective tissue disease—usually rheumatoid arthritis but sometimes systemic lupus erythematosus, scleroderma, or polyarthritis.

social cognitive theory—Theory that behavior change is affected by environmental influences, personal factors, and attributes of the behavior itself.

social support—Support and encouragement that a person receives from others to maintain behavior change.

soft tissue—The total amount of tissue in the body minus bone mass as determined by DXA.

somatic nervous system—Neural elements over which a person has conscious awareness and control.

somatosensory evoked potentials—A noninvasive diagnostic test to assess the speed of electrical conduction across the spinal cord. The technique involves applying electrical stimulus at specific nerves in the arms and legs and measuring the impulses generated by the stimulus at various points in the body.

spasticity—Of, relating to, or characterized by spasms, an involuntary increase in muscle tone.

spastic paresis—A posture- and movement-dependent tone regulation disorder. The clinical symptoms are the loss or absence of tone in lying and an increase in tone in sitting, standing, walking, or running, depending on the degree of involvement.

specificity—The proportion of people with negative test results for the disease that the test is intended to reveal.

spina bifida—Congenital neural tube defect with varying degrees of skin, bone, and neural element involvement.

spinal cord injury (SCI)—Damage involving the spinal cord.

spinal decompression—Surgical intervention to excise bony or soft tissue structures that exert pressure on neural tissues in the spine.

spinal discectomy—Surgical intervention to excise the portion of a herniated disc that is causing compression on neural tissue. The extent of the tissues removed is based on the extent of the intervertebral disc herniation.

spinal fusion—Surgical intervention to fixate unstable hypermobile vertebral segments by the use of metal plates, screws, wires, and autologous bony transplants.

spinal traction—Use of specialized harness systems and electronic winch or manually applied distractive forces on the spine in a variety of spinal and bodily positions. The purpose is to separate vertebrae and stretch the associated soft tissues, thus decompressing nerve roots and relieving symptoms.

spirometry—Measurement of the breathing capacity of the lungs.

spondyloarthropathy—A type of inflammatory arthritis involving ligament or tendon insertion sites (enthuses), leading to spinal and peripheral joint arthritis, usually in human lymphocyte antigen B27–positive individuals.

spondylolisthesis—Forward subluxation (malalignment) of superior lumbar vertebra on an inferior vertebra, leading to traction or compression of nerve roots and intervertebral supportive soft tissues and causing associated irritation and nociceptive input. This condition may be benign, depending on the amount of the slippage.

staging—A system used to classify the extent and spread of cancer.

static endurance—Classification of exercise in which isometric contractions of muscle groups lead to anaerobic exhaustion. An example is a trunk extension position prolonged until fatigue occurs and the subject is unable to hold the position.

static pulmonary function—Assessment of pulmonary status at rest. Static pulmonary function includes measures of pulmonary function including FEV_1, FVC, RV/TLC, and diffusion capacity.

steatosis—Pathologic adipose depots.

stenosis—Constriction or narrowing of a passage or orifice. In spinal stenosis, congenital or degenerative narrowing of the intervertebral or vertebral foramen (opening) is present, leading to compressive forces on the nerve roots that travel through these openings.

stent—A stainless steel bridge, expanded by a balloon-tipped catheter, designed to hold open an area of stenosis within an artery.

sternotomy—The operation of cutting through the sternum.

strabismus—A deviation of the eye that the individual cannot overcome. The visual axes assume a position relative to each other different from that required by the physiological conditions. The various forms of strabismus are termed tropias. Their direction is indicated by the appropriate prefix (e.g., cyclotropia, esotropia, exotropia, hypertropia, and hypotropia). Also called cast, heterotropia, manifest deviation, and squint.

strategies to promote adherence—Techniques commonly used among exercise professionals to improve initiation of and compliance to a structured exercise regimen.

stratify—To separate individuals or samples into subcategories based on variables of interest (e.g., sex, age, number of risk factors, symptoms).

strength—Maximal voluntary contractile force of a given muscle group or groups.

stress echocardiogram—Combination of an exercise test and an echocardiogram. Resting and postexercise echocardiogram images are compared for wall motion abnormalities that can suggest an ischemia response.

subarachnoid—Pertaining to space in the brain under the arachnoid membrane containing cerebrospinal fluid.

subchondral sclerosis—Thickening of the bone beneath the cartilage layer of an arthritic joint.

subendocardial—Referring to the endocardial surface of the heart.

subendocardial ischemia—Myocardial ischemic response beneath the endocardium.

subepithelial fibrosis—The structural changes noted beneath the epithelial layer of the bronchus resulting in scar tissue formation in this area.

sweat test—Diagnostic test used for CF; usually involves stimulation of the skin's sweat glands by chemical (i.e., pilocarpine) and electrical means (i.e., iontophoresis); an elevation of greater than 60 $mEq \cdot dl^{-1}$ is highly suggestive of CF.

symmetric—Referring to quality of or correspondence in the form of parts on the opposite sides of any body.

sympathetic nervous system—Lumbosacral portion of the central nervous system that promotes the classic fight-or-flight response to a given stimulus.

syncope—Loss of consciousness caused by diminished cerebral blood flow.

synovial joint—A joint in which the opposing bony surfaces are covered with a layer of hyaline cartilage or fibrocartilage and which is nourished and lubricated by synovial tissue.

synovitis—Swelling within a joint attributable to inflammation of the synovial lining.

syringomyelia—Chronic syndrome characterized pathologically by cavitation and gliosis of the spinal cord (usually cervical or thoracic), medulla, or both.

systemic—Referring to the arterial system supplying the body.

systolic—Referring to the pressure generated in the arteries by contraction of the heart muscle.

systolic dysfunction—Most often with reference to left ventricular function, an inability or lesser ability of the cardiac myofibrils to shorten or contract against a load.

tachycardia—Heart rate greater than 100 $beats \cdot min^{-1}$.

tachypnea—Respiratory rate greater than 20 $breaths \cdot min^{-1}$.

technetium-99m sestamibi—A radioisotope, introduced into the bloodstream by a catheter, that tags red blood cells and when imaged using a gamma camera can provide a measure of ventricular volume, ejection fraction, and regional ventricular wall motion at rest and during exercise. Used to depict myocardial ischemia.

tenotomy—Irreversible surgical section of severely contracted tendons attached to muscles that do not respond to any other type of spasticity control and are causing intractable pain and skin complications related to lack of physical movement.

tertiary—Characterizing an intervention designed to reduce the functional effect of an illness or disability.

tetraplegia—Motor and sensory dysfunction of the arms, trunk, legs, and pelvic organs resulting from SCI. The term is preferred over "quadriplegia."

thalamotomy—Destruction of a selected portion of the thalamus for the relief of pain and involuntary movements.

thallium 201—A white metallic substance with radioactivity, introduced into the bloodstream by a catheter, that is perfused into the myocardium; used in conjunction with stress testing (exercise and pharmacological) to image the myocardium to detect transient ischemia and tissue necrosis.

T-helper lymphocyte—Lymphocytes whose secretions and other activities coordinate the cellular and humoral immune responses.

theory of planned behavior—Theory that adds to the theory of reasoned action with the concept of perceived control over the opportunities, resources, and skills necessary to perform a behavior.

theory of reasoned action—Theory that performance of a given behavior (e.g., exercise) is primarily determined by the person's attitude toward the behavior and the influence of the person's social environment or subjective norm (i.e., beliefs about what other people think the person should do as well as the person's motivation to comply with the opinions of others).

therapeutic index—Indicator of a drug's safety; the ratio between lethal dose$_{50}$ (LD_{50}) and effective dose$_{50}$ (ED_{50}), where LD_{50} equals the dose of drug that is lethal in 50% of the animals tested and ED_{50} is the dose of the drug needed to be therapeutically effective in 50% of a like population of animals.

thrombocytopenia—A condition in which the amount of platelets in the blood is abnormally small.

thrombolytic—Referring to agents that degrade fibrin clots by activating plasminogen, a naturally occurring modulator of hemostatic and thrombotic processes.

thrombosis—The formation, development, or existence of a clot or thrombus within the vascular system.

thrombotic stroke—A stroke where an occlusive thrombus develops in or outside an ulcerated plaque. Hypercoagulable states due to increased coagulation potential or decreased fibrinolytic potential are particularly important risk factors.

thromboxane—Vasoconstrictor and platelet activation substance.

thrombus—An aggregation of blood factors, primarily platelets and fibrin with entrapment of cellular elements, that frequently causes vascular obstruction at the point of formation.

trabecular bone—One of the two main types of bone tissue, also known as cancellous or spongy bone. Trabecular bone is made up of interlacing plates of bone tissue and is found mainly at the ends of long bones and within the vertebrae.

tracking—The concept that risk factors or other conditions expressed in childhood will persist and be expressed in adulthood.

transcutaneous electrical nerve stimulation (TENS)—Use of small battery-operated or plugged-in devices for delivery of electrical current across the skin to provide patients with pain relief, artificial contraction of muscles, fatigue of spastic muscles, and pulsations to decrease swelling in a joint. The stimulation is given through electrode pads placed directly on the skin over the muscles selected for stimulation or inhibition of nociceptive input, or in areas determined by nerve supply or acupuncture points. The underlying theories are based on the gate theory of pain control, in which sensory stimulation inhibits pain transmission at the spinal cord level, or stimulation of Aδ and C fibers, to cause the release of endogenous opiates.

transmural—Referring to effects on all tissue layers of the heart.

transmural ischemia—Myocardial ischemic response that occurs throughout the myocardial wall.

transtheoretical model of behavior change—Model wherein behavior change is conceptualized as a five-stage process or continuum related to a person's readiness to change: precontemplation, contemplation, preparation, action, and maintenance. People are thought to progress through the stages.

tremor—Repetitive, often regular, oscillatory movements caused by alternate, or synchronous, but irregular contraction of opposing muscle groups.

triglycerides—Chemical storage form of fat in the body. Hypertriglyceridemia ($\geq 150\ mg \cdot dl^{-1}$) increases the risk of cardiovascular disease.

triplegia—A form of cerebral palsy that affects three limbs. The most common pattern is for both legs and one arm to be affected. Triplegia is sometimes thought of as hemiplegia overlapping with diplegia because the primary motor difficulty is with the legs.

true maximal heart rate—The highest heart rate as measured at, or near, the end of an exercise test, as opposed to peak heart rate estimated by the equation 220 – age.

tumor suppressor genes—Genes encoding proteins that normally serve to restrain cell proliferation.

type 2 diabetes mellitus—The most common form of diabetes, affecting approximately 90% to 95% of all those with diabetes. With type 2 diabetes, abnormal blood glucose regulation is often a result of insulin resistance of the peripheral tissues and defective insulin secretion.

ultrasound—Use of sonic wave energy, created by a vibrating quartz crystal, to deliver heat or medication to healing musculoskeletal structures. A variety of machines deliver the ultrasonic waves through a transducer rubbed directly over the skin using gel or water as the transmitting medium.

upper rate limit—The highest rate at which ventricular pacing will track 1:1 each sensed atrial event.

uremia—The medical state that occurs when kidney function is extremely low. The manifold symptoms include confusion, lethargy leading to coma, loss of appetite, and prurititis (itchy skin). Complications include gastrointestinal bleeding, pericarditis, pulmonary edema, and ultimately death.

valgus—Bowlegged deformity.

Valsalva maneuver—Forced exhalation with the glottis, nose, and mouth closed, resulting in increased intrathoracic pressure, slowing of the heart rate, and decreased return of blood to the heart.

variant angina—Angina pectoris occurring during rest; not necessarily preceded by exercise or an increase in heart rate.

varus—Knock-kneed deformity.

vascular pathologies—Manifestations of disease in blood vessels.

vasospastic—Referring to contraction or spasm of the muscular coats of the blood vessels.

ventilatory-derived lactate threshold (V-LT)—The point where a nonlinear increase in blood lactate occurs during exercise; when determined with ventilatory parameters, it is sometimes referred to as ventilatory threshold.

ventilatory muscle training—Specific exercises that are used to increase respiratory muscular strength.

ventricular hypertrophy—Muscle thickening in a pumping chamber of the heart.

vertigo—A sensation of spinning or whirling motion.

very old age—Between 75 and 84 yr of age.

vesicoureteral reflux—Backflow of urine from the bladder into the upper urinary tracts.

vigorous-intensity physical activity—Activities that result in large increases in breathing and heart rate; >70% $\dot{V}O_2$peak.

viremia—Viral particles in the blood.

virion—A single, encapsulated piece of viral genetic material.

visceral fat—One of three compartments of abdominal fat. Others are retroperitoneal and subcutaneous.

wall motion—Movement of the left ventricular segments of the heart; used to describe normal or abnormal movement during contraction and to calculate ejection fraction during two-dimensional echocardiography or some types of nuclear imaging.

wasting—Involuntary loss of more than 10% of body weight.

Wheelchair and Ambulatory Sports, USA (WASUSA)—A national 501(c)3 organization, consisting of grassroots programs arranged into 14 Regional Sports Organizations (RSOs) to support sport competition for athletes with disabilities in the United States.

white matter—Regions of the brain and spinal cord that are largely or entirely composed of nerve fibers and contain few or no neuronal cell bodies or dendrites.

xanthomas—Usually yellow, soft raised bumps that are caused by cholesterol deposits under the skin.

References

CHAPTER 1

1. American Association of Cardiovascular and Pulmonary Rehabilitation. www.aacvpr.org.
2. American Council on Exercise. www.acefitness.org.
3. American Council on Exercise. Advanced Health & Fitness Specialist Certification. www.acefitness.org/getcertified/certification_ahfs.aspx [accessed April 25, 2013].
4. American College of Sports Medicine. www.acsm.org.
5. American College of Sports Medicine. ACSM Certified Exercise Specialist. http://certification.acsm.org/acsm-certified-clinical-exercise-specialist [accessed April 25, 2013].
6. American College of Sports Medicine. ACSM Registered Clinical Exercise Physiologist. http://certification.acsm.org/acsm-registered-clinical-exercise-physiologist [accessed April 25, 2013].
7. American Physical Therapy Association of Massachusetts, Inc. Bills of interest to APTA of MA in the 2011-2012 legislative cycle. www.aptaofma.org/Content/Legislative.asp [accessed April 25, 2013].
8. American Society for Exercise Physiologists. www.asep.org.
9. Berry R, Verrill D. Point: Licensure for clinical exercise physiologists—an argument in favor of this proposal. *J Clin Exerc Physiol* 2012;1(1):36-39.
10. Canadian Society of Exercise Physiology. www.csep.ca.
11. Canadian Society of Exercise Physiology. CSEP Certified Exercise Physiologist. www.csep.ca/english/view.asp?x=739 [accessed April 25, 2013].
12. Clinical Exercise Physiology Association. Clinical exercise physiology. www.acsm-cepa.org/i4a/pages/index.cfm?pageid=3393 [accessed April 25, 2013].
13. Commission on Accreditation of Allied Health Education Programs (CAAHEP). www.caahep.org/Content.aspx?ID=40 [accessed April 25, 2013].
14. Cooper KH. *Aerobics*. New York: Evans, 1968.
15. Foster C, Porcari JP. Counterpoint: Licensure for clinical exercise physiologists—an elusive outcome. *J Clin Exerc Physiol* 2012;1(1):40-42.
16. 187th General Court of the Commonwealth of Massachusetts, The. Bill S.1072. www.malegislature.gov/Bills/187/Senate/S01072 [accessed April 25, 2013].
17. Haskell WL, Lee I-M, Pate RR, et al. Physical activity and public health: Updated recommendation for adults from the American College of Sports Medicine and the American Heart Association. *Circulation* 2007;116:1081-1093.
18. Nelson ME, Rejeski WJ, Blair SN, et al. Physical activity and public health in older adults: Recommendations from the American College of Sports Medicine and the American Heart Association. *Circulation* 2007;116:1094-1105.
19. Saltin B, et al. Response to submaximal and maximal exercise after bed rest and training. *Circulation* 1968;38(Suppl. 7).
20. U.S. Department of Health and Human Services. *Physical Activity and Health: A Report of the Surgeon General*. Atlanta: U.S. Department of Health and Human Services, Centers for Disease Control and Prevention, National Center for Chronic Disease Prevention and Health Promotion, 1996.
21. U.S. Department of Labor, Bureau of Labor Statistics. Standard Occupational Classification. www.bls.gov/soc/2010/soc291128.htm [accessed April 25, 2013].
22. Warburton DE, Bredin SS, Charlesworth SA, Foulds HJ, McKenzie DC, Shepherd RJ. Evidence-based risk recommendations for best practices in the training of qualified exercise professionals working with clinical populations. *App Physiol Nutr Metab* 2011;36:S232-S265.

CHAPTER 2

1. Ajzen I. *Attitudes, Personality, and Behavior*. Chicago: Dorsey Press, 1988.
2. Ajzen I, Fishbein M. *Understanding Attitudes and Predicting Social Behavior*. Englewood Cliffs, NJ: Prentice Hall, 1980.
3. American College of Sports Medicine. Position stand on the recommended quantity and quality of exercise for developing and maintaining cardiorespiratory, musculoskeletal, and neuromotor fitness in apparently healthy adults: Guidance for prescribing exercise. *Med Sci Sports Exerc* 2011;43:1334-1359.
4. Bandura A. Self-efficacy: Toward a unifying theory of behavioral change. *Psychol Rev* 1977;84:191-215.
5. Bandura A. *Social Foundations of Thought and Action: A Social–Cognitive Theory*. Englewood Cliffs, NJ: Prentice Hall, 1986.
6. Bandura A. *Social Learning Theory*. Englewood Cliffs, NJ: Prentice Hall, 1977.
7. Baranowski T, Lin LS, Wetter DW, Resnicow K, Hearn MD. Theory as mediating variables: Why aren't community interventions working as desired? *Ann Epidemiol* 1997;S7:S89-S95.
8. Blamey A, Mutrie N, Aitchison T. Health promotion by encouraged use of stairs. *BMJ* 1995;311:289-290.
9. Brownell K, Stunkard AJ, Albaum J. Evaluation and modification of exercise patterns in the natural environment. *Am J Psychiatry* 1980;137:1540-1545.
10. Cady LD, Bischoff DP, O'Connell ER, Thomas PC, Allan JH. Strength and fitness and subsequent back injuries in firefighters. *J Occup Med* 1979;21:269-272.
11. Calfas KJ, Long BJ, Sallis JF, Wooten WJ, Pratt M, Patrick K. A controlled trial of physician counseling to promote the adoption of physical activity. *Prev Med* 1996;25:225-233.
12. Calfas KJ, Sallis JF, Zabinski BA, Wilfley BE, Rupp R, Prochaska JJ, Thompson T, Pratt M, Patrick K. Preliminary evaluation of a multicomponent program for nutrition and physical activity change in primary care: PACE+ for Adults. *Prev Med* 2002;34:153-161.

13. Caspersen CJ, Powell KE, Christenson GM. Physical activity, exercise, and physical fitness: Definitions and distinctions for health-related research. *Public Health Rep* 1985;100:126-130.

14. Dishman RK. Increasing and maintaining exercise and physical activity. *Behav Ther* 1991;22:345-378.

15. Dishman RK, Buckworth J. Increasing physical activity: A quantitative synthesis. *Med Sci Sports Exerc* 1996;28:706-719.

16. Dunn AL, Marcus BH, Kampert JB, et al. Comparison of lifestyle and structured interventions to increase physical activity and cardiorespiratory fitness: A randomized trial. *JAMA* 1999;281:327-334.

17. Epstein LH, Wing RR. Aerobic exercise and weight. *Addict Behav* 1990;5:371-388.

18. Haskell WL, Lee IM, Pate RR, Powell KE, Blair SN, Franklin BA, Macera CA, Heath GW, Thompson PD, Bauman A. Physical activity and public health: Updated recommendation for adults from the American College of Sports Medicine and the American Heart Association. *Circulation* 2007;116:1-13.

19. Hassman PR, Ceci R, Backman L. Exercise for older women: A training method and its influences on physical and cognitive performance. *Eur J Appl Physiol* 1992;64:460-466.

20. Heath GW, Brownson RC, Kruger J, Miles R, Powell KE, Ramsey LT. The effectiveness of urban design and land use and transport policies and practices to increase physical activity: A systematic review. *J Phys Act Health* 2006;1:S55-S71.

21. Heath GW, Parra DC, Sarmiento OL, Andersen LB, Owen N, Goenka S, Montes F, Brownson RC. Evidence-based intervention in physical activity: Lessons from around the world. Lancet 2012; 380:272-81.

22. Kahn EB, Ramsey LT, Brownson RC, Heath GW, Howze EH, Powell KE, Stone EJ, Rajab MW, Corso P. The effectiveness of interventions to increase physical activity: A systematic review. *Am J Prev Med* 2002;22(4S):73-107.

23. King AC. Intervention strategies and determinants of physical activity and exercise behavior in adult and older adult men and women. *World Rev Nutr Diet* 1997;82:148-158.

24. King AC, Haskell WL, Taylor CB, Kraemer HC, DeBusk RF. Group- vs. home-based exercise training in healthy older men and women. *JAMA* 1991;266:1535-1542.

25. King AC, Haskell WL, Young DR, Oka RK, Stefanick ML. Long-term effects of varying intensities and formats of physical activity on participation rates, fitness, and lipoproteins in men and women aged 50 to 65 years. *Circulation* 1995;91:2596-2604.

26. Kriska AM, Bayles C, Cauley JA, et al. Randomized exercise trial in older women: Increased activity over two years and the factors associated with compliance. *Med Sci Sports Exerc* 1986;18:557-562.

27. Lee IM, Paffenbarger RS, Hsieh CC. Physical activity and risk of developing colorectal cancer among college alumni. *J Natl Cancer Inst* 1991;83:1324-1329.

28. Lee IM, Shiroma EJ, Lobelo F, Puska P, Blair SN, Katzmarzyk PT. Effect of physical inactivity on major non-communicable diseases worldwide: An analysis of burden of disease and life expectancy. *Lancet,* in press.

29. Linenger JM, Chesson CV 2nd, Nice DS. Physical fitness gains following simple environmental change. *Am J Prev Med* 1991;7:298-310.

30. Long BJ, Calfas KJ, Patrick K, Sallis JF, Wooten WJ, Goldstein M, Marcus B, Schwenck T, Carter R, Torez T, Palinkas L, Heath G. Acceptability, usability, and practicality of physician counseling for physical activity promotion: Project PACE. *Am J Prev Med* 1996;12:73-81.

31. Marcus BH, Banspach SW, Leffebvre RC, Rossi JS, Carleton RA, Abrams DB. Using the stages of change model to increase adoption of physical activity among community participants. *Am J Health Promot* 1992;6:424-429.

32. Marcus BH, Stanton AL. Evaluation of relapse prevention and reinforcement interventions to promote exercise adherence in sedentary females. *Res Q Exerc Sport* 1993;64:447-452.

33. Marlatt GA, George WH. Relapse prevention and the maintenance of optimal health. In: Shumaker SA, Schron EB, Ockene J, eds. *The Handbook of Health Behavior Change* (pp. 44-63). New York: Springer, 1990.

34. Martin JE, Dubbert PM, Katell AD, et al. The behavioral control of exercise in sedentary adults: Studies 1 through 6. *J Consult Clin Psychol* 1984;52:795-811.

35. McAuley E, Courneya DS, Rudolph DL, Lox CL. Enhancing exercise adherence in middle-aged males and females. *Prev Med* 1994;23:498-506.

36. McLeroy KR, Bibeau D, Steckler A, Glanz K. An ecological perspective on health promotion programs. *Health Educ Q* 1988;15:351-377.

37. New South Wales Physical Activity Task Force. *Simply Active Every Day: A Discussion Document From the NSW Physical Activity Task Force on Proposals to Promote Physical Activity in NSW, 1997-2002.* Summary report. Sydney, Australia: New South Wales Health Department, 1997.

38. Ntoumanis N, Biddle SJ. A review of motivational climate in physical activity. *J Sports Sci* 1999;17:643-665.

39. Paffenbarger RS, Hyde RT, Wing AL. Physical activity as an index of heart attack risk in college alumni. *Am J Epidemiol* 1978;108:161-175.

40. Paffenbarger RS Jr, Hyde RT, Wing AL, Hsieh CC. Physical activity, all-cause mortality, and longevity of college alumni. *N Engl J Med* 1986;314:605-613.

41. Passmore R. The regulation of body weight in man. *Proc Nutr Soc* 1971;30:122-127.

42. Physical Activity Guidelines Advisory Committee. *Physical Activity Guidelines Advisory Committee Report, 2008.* Washington, DC: U.S. Department of Health and Human Services, 2008.

43. Pollock ML. Prescribing exercise for fitness and adherence. In: Dishman RK, ed. *Exercise Adherence* (pp. 259-277). Champaign, IL: Human Kinetics, 1995.

44. Prochaska JO, DiClemente CC. *The Transtheoretical Approach: Crossing Traditional Boundaries of Change.* Homewood, IL: Dorsey Press, 1984.

45. Rosenstock IM. The health belief model: Explaining health behavior through expectancies. In: *Health Behavior and Health Education. Theory, Research, and Practice* (pp. 39-62). San Francisco: Jossey-Bass, 1990.

46. Sallis JF, Bauman A, Pratt M. Environmental and policy interventions to promote physical activity. *Am J Prev Med* 1998;15:379-396.

47. Sallis JF, Hovell MF. Determinants of exercise behavior. *Exerc Sport Sci Rev* 1990;18:307-330.

48. Sallis JF, Hovell MF, Hofstetter CR. Predictors of adoption and maintenance of vigorous physical activity in men and women. *Prev Med* 1992;21:237-251.

49. Sallis JF, Hovell MF, Hofstetter CR, et al. A multivariate study of determinants of vigorous exercise in a community sample. *Prev Med* 1989;18:20-34.

50. Shephard RJ. Exercise and aging: Extending independence in older adults. *Geriatrics* 1993;48:61-64.

51. Sidney KH, Shephard RJ, Harrison JE. Endurance training and body composition of the elderly. *Am J Clin Nutr* 1977;30:326-333.

52. Smith DM, Khairi MRA, Norton J, et al. Age and activity effects on rate of bone mineral loss. *J Clin Invest* 1976;58:716-721.

53. Stokols D. Establishing and maintaining healthy environments: Toward a social ecology of health promotion. *Am Psychol* 1992;47:6-22.

54. Taylor CB, Miller NH, Smith PM, DeBusk RF. The effect of home-based, case-managed, multifactorial risk reduction program on reducing psychological distress in patients with cardiovascular disease. *J Cardiopulm Rehabil* 1997;17:157-162.

55. Tipton CH. Exercise, training, and hypertension: An update. *Exerc Sport Sci Rev* 1991;19:447-505.

56. Trends in strength training—United States 1998-2004. *MMWR* 2006;55(28, July 21):769-772.

57. U.S. Department of Agriculture and U.S. Department of Health and Human Services. *Dietary Guidelines for Americans, 2010.* 7th ed. Washington, DC: U.S. Government Printing Office, December 2010.

58. U.S. Department of Health and Human Services. Healthy people 2020. www.healthypeople.gov/2020/topicsobjectives2020/default.aspx [accessed November 2, 2011].

59. U.S. Department of Health and Human Services. Physiologic responses and long-term adaptations to exercise. In: *Physical Activity and Health: A Report of the Surgeon General* (pp. 61-73). Atlanta: U.S. Department of Health and Human Services, Centers for Disease Control and Prevention, National Center for Chronic Disease Prevention and Health Promotion, 1996.

60. Wankel LM, Yardley JK, Graham J. The effects of motivational interventions upon the exercise adherence of high and low self-motivated adults. *Can J Appl Sport Sci* 1985;10:147-156.

61. West Virginia University, Department of Safety and Health Studies and Department of Sports and Exercise Studies. *Physical Fitness and the Aging Driver: Phase I.* Washington, DC: AAA Foundation of Traffic Safety, 1988.

62. Wood PD, Stefanick ML, Dreon DM, Frey-Hewitt D, Garay BC, Williams PT, Superko HR, Fortmann SP, Albers JJ, Vranizan KM, Ellworth NM, Terry RB, Haskell WL. Changes in plasma lipids and lipoproteins in overweight men during weight loss through dieting as compared with exercise. *N Engl J Med* 1988;319:1173-1179.

63. Young DR, Haskell WL, Taylor CB, Fortmann SP. Effect of community health education on physical activity knowledge, attitudes, and behavior: The Stanford Five-City Project. *Am J Epidemiol* 1996;144:264-274.

CHAPTER 3

1. American College of Sports Medicine. *ACSM's Guidelines for Exercise Testing and Prescription.* 9th ed. Philadelphia: Lippincott Williams & Wilkins, 2013.

2. Balkrishnan R, Rajagopalan R, Camacho FT, Huston SA, Murray FT, Anderson RT. Predictors of medication adherence and associated health care costs in an older population with type 2 diabetes mellitus: A longitudinal cohort study. *Clin Ther* 2003;25:2958-2971.

3. Barat I, Andreasen F, Damsgaard A. Drug therapy in the elderly: What doctors believe and patients actually do. *Br J Clin Pharmacol* 2001;51:615-622.

4. Barnsteiner JH. Medication reconciliation. In: *Patient Safety and Quality: An Evidence-Based Handbook for Nurses.* Rockville, MD: Agency for Healthcare Research and Quality, April 2008. AHRQ publication 08-0043. www.ahrq.gov/legacy/qual/nurseshdbk [accessed April 25, 2013].

5. Etchells E, Matlow A, Coffey M, Cornish P, Koonthanam T. Implementation of admission medication reconciliation at two academic health science centres: Challenges and success factors. *Healthcare Q* 2009;12:102-109.

6. The Joint Commission. Medication reconciliation. Sentinel event alert. 2006;35. www.jointcommission.org.SentinelEvents/SentinelEventsAlert/sea_35.htm [accessed December 7, 2011].

7. Kaplan RC, Bhalodkar NC, Brown EJ, White J, Brown DL. Race, ethnicity, and sociocultural characteristics predict noncompliance with lipid lowering medications. *Prev Med* 2004;39:1249-1255.

8. Kee JL, Hayes ER, McCuistion LE. *Pharmacology: A Nursing Process.* 7th ed. St. Louis: Elsevier Saunders, 2012.

9. Kohn LT, Corrigan JM, Donaldson MS, eds. *To Err Is Human: Building a Safer Health System.* Washington, DC: National Academies Press, 2000.

10. McKenry LM, Salerno E. *Mosby's Pharmacology in Nursing.* 21st ed. St. Louis: Mosby, 2003.

11. Miller LG, Matson CC, Rogers JC. Improving prescription documentation in the ambulatory setting. *Fam Pract Res J* 1992;12:421-429.

12. Nielsen-Bohlman L, Panzer AM, Kindig DA, Institute of Medicine Committee on Health Literacy. *Health Literacy: A Prescription to End Confusion.* Washington, DC: National Academies Press, 2004.

13. Provost P, Weast B, Schwartz M, Wyskiel RM, Prow D, Milanovich SN, Berenholtz S, Dorman T, Lipsett P. Medication reconciliation: A practical tool to reduce the risk of medication errors. *J Crit Care* 2003;18:201-205.

14. Roach SS, Ford SM. *Introductory Clinical Pharmacology.* 8th ed. Philadelphia; Lippincott Williams & Wilkins, 2008.

15. Rozich JD, Howard RJ, Justeson JM, Macken PD, Lindsay ME, Resar RK. Standardization as a mechanism to improve safety in health care. *Jt Comm J Qual Saf* 2004;30:5014.

16. Vik SA, Maxwell CJ, Hogan DB. Measurement, correlates, and health outcomes of medication adherence among seniors. *Ann Pharmacother* 2004;38:303-312.

CHAPTER 4

1. Ades PA, Savage PD, Brawner CA, Lyon CE, Ehrman JK, Bunn JY, Keteyian SJ. Aerobic capacity in patients entering cardiac rehabilitation. *Circulation* 2006;113:2706-2712.

2. Campeau L. Grading of angina pectoris. *Circulation* 1976;54:522-523.

3. Carhart RL, Ades PA. Gender differences in cardiac rehabilitation. *Cardiol Clin* 1998;16:37-43.

4. Centers for Disease Control and Prevention. Diabetes prevalence among American Indians, Alaskan Natives, and the overall population—United States. *MMWR* 2003;52:702.

5. Coutinho T, Goel K, Correa de Sa D, et al. Central obesity and survival in subjects with coronary artery disease: A systematic

review of the literature and collaborative analysis with individual subject data. *J Am Coll Cardiol* 2011;57:1877-1886.

6. Diamond GA, Forrester JS. Analysis of probability as an aid in the clinical diagnosis of coronary-artery disease. *N Engl J Med* 1979; 300:1350-1358.
7. Hirsch HT, Haskal ZJ, Hertzer NR, Bakal CW, Creager MA, Halperin JL, Hiratzka LF, et al. ACC/AHA 2005 practice guidelines for the management of patients with peripheral arterial disease (lower extremity, renal, mesenteric and abdominal aortic): Executive summary: A collaborative report from the American Association for Vascular Surgery/Society for Vascular Surgery, Society for Cardiovascular Angiography and Interventions, Society for Vascular Medicine and Biology, Society of Interventional Radiology, and the ACC/AHA Task Force on Practice Guidelines (Writing Committee to Develop Guidelines for the Management of Patients with Peripheral Arterial Disease). *Circulation* 2006;113:1474-1547.
8. Kloner, RA. *The Guide to Cardiology*. Greenwich, CT: Le Jacq Communications, 1995.
9. Linzer M, Yang EH, Estes NA 3rd, Wang P, Vorperian VR, Kapoor WN. Diagnosing syncope. Part 1: Value of history, physical examination, and electrocardiography. Clinical Efficacy Assessment Project of the American College of Physicians. *Ann Intern Med* 1997;126:989-996.
10. Parish JM, Somers VK. Obstructive sleep apnea and cardiovascular disease. *Mayo Clin Proc* 2004;79:1036-1046.
11. Lloyd-Jones DM, Larson MG, Beiser A, Levy D. Lifetime risk of developing coronary heart disease. *Lancet* 1999;353:89-92.
12. Rankin SL, Briffa TG, Morton AR, Hung J. A specific activity questionnaire to measure the functional capacity of cardiac patients. *Am J Cardiol* 1996;77:1220-1223.
13. Smith DW. Changing causes of death of elderly people in the United States, 1950–1990. *Gerontology* 1998;44:331-335.
14. Swartz MH. *Textbook of Physical Diagnosis: History and Examination*. Philadelphia: Saunders, 1998.
15. Thom T, Haase N, Rosamond W, Howard VJ, Rumsfeld J, Manolio T, Zheng ZJ, et al. Heart disease and stroke statistics—2006 update: A report from the American Heart Association Statistics Committee and Stroke Statistics Subcommittee. *Circulation* 2006;113:e85-151.

CHAPTER 5

1. Adams GM. *Exercise Physiology: Laboratory Manual*. 3rd ed. Boston: McGraw-Hill, 1998.
2. American College of Sports Medicine. *ACSM's Guidelines for Exercise Testing and Prescription*. 9th ed. Baltimore: Lippincott Williams & Wilkins, 2013.
3. American College of Sports Medicine. Exercise and physical activity for older adults. *Med Sci Sports Exerc* 2010;42:1510-1530.
4. American College of Sports Medicine. Position stand. Exercise and hypertension. *Med Sci Sports Exerc* 2004;36:533-553.
5. American College of Sports Medicine. Position stand. The quantity and quality of exercise for developing and maintaining cardiorespiratory, musculoskeletal, and neuromotor fitness in apparently healthy adults: Guidance for prescribing exercise. *Med Sci Sports Exerc* 2011;43:1334-1359.
6. American College of Sports Medicine and the American Diabetes Association: Joint position statement. Exercise and type 2 diabetes. *Med Sci Sports Exerc* 2010;42:2282-2303.
7. American Thoracic Society/American College of Chest Physicians. ATS/ACCP statement on cardiopulmonary exercise testing. *Am J Respir Crit Care Med* 2003;167:211-277.
8. Amon KW, Richards KL, Crawford MH. Usefulness of the postexercise response of systolic blood pressure in the diagnosis of coronary artery disease. *Circulation* 1984;70:951-956.
9. Balady GJ, Arena R, Siestsema K, et al. Clinician's guide to cardiopulmonary exercising testing in adults: A scientific statement from the American Heart Association. *Circulation* 2010;122:191-225.
10. Bartels R, Billings CE, Fox EL, Mathews DK, O'Brien R, Tanz D, Webb W. AAHPER Con 1968;123.
11. Blair SN, Kohl HW III, Paffenbarger RS Jr, Clark DG, Cooper KH, Gibbons LW. Physical fitness and all-cause mortality: A prospective study of healthy men and women. *JAMA* 1989;262:2395-2401.
12. Brawner CA, Ehrman JK, Schairer JR, Cao JJ. Predicting maximum heart rate among patients with coronary heart disease receiving beta-adrenergic blockade therapy. *Am Heart J* 2004;148:910-914.
13. Brawner CA, Keteyian SJ, Ehrman JK. The relationship of heart rate reserve to $\dot{V}O_2$ reserve in patients with heart disease. *Med Sci Sports Exerc* 2002;34(3):418-422.
14. Brubaker PH, Rejeski WJ, Law HD, et al. Cardiac patients' perception of work intensity during graded exercise testing: Do they generalize to field testing? *J Cardiopulm Rehabil* 1994;14:127-133.
15. Butler RM, Belerwalters WH, Rodger FJ. The cardiovascular response to circuit weight training in patients with cardiac disease. *J Cardiac Rehabil* 1987;7:402-409.
16. Chapman, CB. Edward Smith (1818-1874), physiologist, human ecologist, reformer. *J Hist Med Allied Sci* 1967;22:1-26.
17. Cornelius WL, Craft-Hamm K. Proprioceptive neuromuscular facilitation flexibility techniques: Acute effects on arterial blood pressure. *Phys Sportsmed* 1988;16(4):152-161.
18. Coyle EF, Martin WH, Sinacore DR, Joyner MJ, Hagberg JM, Holloszy JO. Time course of loss of adaptations after stopping prolonged intense endurance training. *J Appl Physiol* 1984;57(6):1857-1864.
19. DeMichele PD, Pollock ML, Graves JE, et al. Isometric torso rotation strength: Effect of training frequency on its development. *Arch Physiol Med Rehabil* 1997;78:64-69.
20. Etnyre BR, Abraham LD. Antagonist muscle activity during stretching: A paradox re-assessed. *Med Sci Sports Exerc* 1988;20:285-289.
21. Etnyre BR, Lee JA. Chronic and acute flexibility of men and women using three different stretching techniques. *Res Q Exerc Sport* 1988;59:222-228.
22. Feigenbaum MS, Pollock ML. Strength training: Rationale for current guidelines for adult fitness programs. *Phys Sportsmed* 1997;25:44-64.
23. Fleck SJ, Kraemer WJ. *Designing Resistance Training Programs*. 2nd ed. Champaign, IL: Human Kinetics, 1997.
24. Foss ML, Keteyian SJ, eds. *Fox's Physiological Basis for Exercise and Sport*. 6th ed. Boston: McGraw-Hill, 1998.
25. Frontera WR, Meredith CN, O'Reilly KP, Evans WJ. Strength conditioning in older men: Skeletal muscle hypertrophy and improved function. *J Appl Physiol* 1988;68:1038-1044.

26. Garber CE, Blissmer B, Deschenes MR, et al. Quantity and quality of exercise for developing and maintaining cardiorespiratory, musculoskeletal, and neuromotor fitness in apparently healthy adults: Guidance for prescribing exercise. *Med Sci Sports Exerc* 2011;43:1334-1359.

27. Gettman LR, Pollock ML. Circuit weight training: A critical review of its physiological benefits. *Phys Sportsmed* 1981;9:44-60.

28. Gettman LR, Pollock ML, Durstine JL, Ward A, Ayres J, Linnerud AC. Physiological responses of men to 1, 3, and 5 days per week training program. *Res Q* 1976;47:638-646.

29. Ghilarducci LE, Holly RG, Amsterdam EA. Effects of high resistance training in coronary artery disease. *Am J Cardiol* 1989;64:866-870.

30. Gibbons RJ, Balady GJ, Bricker JT, et al. ACC/AHA 2002 guideline update for exercise testing: Summary article: A report of the American College of Cardiology/American Heart Association Task Force on Practice Guidelines (Committee to Update the 1997 Exercise Testing Guidelines). *Circulation* 2002;106:1883-1892.

31. Gordon NF, Kohl HW. Exercise testing and sudden cardiac death. *J Cardiopulm Rehabil* 1993;13:381-386.

32. Graves JE, Pollock ML, Jones AE, et al. Specificity of limited range of motion variable resistance training. *Med Sci Sports Exerc* 1989;21:84-89.

33. Graves JE, Pollock ML, Leggett SH, Braith RW, Carpenter DM, Bishop LE. Effect of reduced training frequency on muscular strength. *Int J Sports Med* 1988;9:316-319.

34. Hermansen L, Saltin B. Oxygen uptake during maximal treadmill and bicycle exercise. *J Appl Physiol* 1969;26:31-37.

35. Hurley BF, Seals DR, Ehsani AA, Cartier LJ, Dalsky GP, Hagberg JM, Holloszy JO. Effects of high-intensity strength training on cardiovascular function. *Med Sci Sports Exerc* 1984;16(5):483-488.

36. Jackson J, Sharkey B, Johnston L. Cardiorespiratory adaptations to training at specified frequencies. *Res Q* 1968;39:295-300.

37. Karvonen M, Kentala K, Mustala O. The effects of training heart rate: A longitudinal study. *Ann Med Exp Biol Fenn* 1957;35:307-315.

38. Keteyian SJ, Isaac D, Thadani U, et al. Safety of symptom-limited cardiopulmonary exercise testing in patients with chronic heart failure due to left ventricular systolic dysfunction. *Am Heart J* 2009;158:S72-S77.

39. Lauer MS. Exercise electrocardiogram testing and prognosis: Novel markers and predictive instruments. In: Balady GJ, ed. *Cardiol Clin* 2001;19:401-414.

40. Lee IM, Hsieh C, Paffenbarger RS. Exercise intensity and longevity in men. *JAMA* 1995;273:1179-1184.

41. Mancini DM, Eisen H, Kussmaul W, et al. Value of peak exercise oxygen consumption for optimal timing of cardiac transplantation in ambulatory patients with heart failure. *Circulation* 1991;83:778-786.

42. Mark DB, Shaw L, Harrell FE, et al. Prognostic value of a treadmill exercise score in outpatients with suspected coronary artery disease. *N Engl J Med* 1991;325:849-853.

43. Master MA. Two-step test of myocardial function. *Am Heart J* 1934;10:495-501.

44. McHam SA, Marwick TH, Pashkow FJ, Lauer MS. Delayed systolic blood pressure recovery after graded exercise: An independent correlate of angiographic coronary disease. *J Am Coll Cardiol* 1999;34:754-759.

45. Miyamura M, Kitamura K, Yamada A, Matsui H. Cardiorespiratory responses to maximal treadmill and bicycle exercise in trained and untrained subjects. *J Sports Med Phys Fitness* 1978;18:25-32.

46. Morgan RE, Adamson GT. *Circuit Weight Training*. London: G Bell and Sons, 1961.

47. Morgan W, Borg G. Perception of effort in the prescription of physical activity. In: Nelson T, ed. *Mental Health and Emotional Aspects of Sports* (pp. 126-129). Chicago: American Medical Association, 1976.

48. Murphy MH, Blair SN, Murtagh EM. Accumulated versus continuous exercise for health benefit: A review of empirical studies. *Sports Med* 2009;39(1):29-43.

49. Myers J. Effect of exercise training on abnormal ventilatory responses to exercise in patients with chronic heart failure. *Congest Heart Fail* 2000; 6:243-249.

50. Myers J, Arena R, Franklin B, et al. Recommendations for clinical exercise laboratories: A scientific statement from the American Heart Association. *Circulation* 2009;119:3144-3161.

51. Myers J, Madhavan R. Exercise testing with gas exchange analysis. In: Balady GJ, ed. *Cardiol Clin* 2001;19:433-446.

52. Nostratian F, Froelicher VF. ST elevation during exercise testing: A review. *Am J Cardiol* 1989;63:986-988.

53. Paffenbarger RS, Hyde RT, Wing AL, Hsieh CC. Physical activity, all-cause mortality, and longevity of college alumni. *N Engl J Med* 1986;314:605-613.

54. Pate RR, Pratt M, Blair SN, et al. Physical activity and public health: A recommendation from the Centers for Disease Control and Prevention and the American College of Sports Medicine. *JAMA* 1995;273:402-407.

55. Pollock ML, Cureton TK, Greninger L. Effects of frequency of training on working capacity, cardiovascular function, and body composition of adult men. *Med Sci Sports* 1969;1:70-74.

56. Pollock ML, Leggett SH, Graves JE, Jones A, Fulton M, Cirulli J. Effect of resistance training on lumbar extension strength. *Am J Sports Med* 1989;17:624-629.

57. Ramos-Barbon D, Fitchett D, Gibbons WJ, et al. Maximal exercise testing for the selection of heart transplantation candidates: Limitation of peak oxygen consumption. *Chest* 1999;115:410-417.

58. Rodgers GP, Ayanian JZ, Balady G, et al. American College of Cardiology/American Heart Association clinical competence statement on stress testing. A report of the American College of Cardiology/American Heart Association/American College of Physicians/American Society of Internal Medicine Task Force on Clinical Competence. *Circulation* 2000;102:1726-1738.

59. Sady SP, Wortman M, Blanke D. Flexibility training: Ballistic, static or proprioceptive neuromuscular facilitation? *Arch Phys Med Rehabil* 1982;63:261-263.

60. Saltin B, Blomqvist G, Mitchell JH, Johnson RL, Wildenthal K, Chapman CB. Response to exercise after bed rest and after training. *Circulation* 1968;38(Suppl. VII):VII-1–VII-55.

61. Scherer D, Kaltenbach M. Frequency of life-threatening complications associated with exercise testing. *Dtsch Med Wochenschr* 1979;33:1161-1165.

62. Shetler K, Marcus R, Froelicher VF, et al. Heart rate recovery: Validation and methodologic issues. *J Am Coll Cardiol* 2001;38:1980-1987.

63. Sidney KH, et al. In: Taylor AW, ed. *Training: Scientific Basis and Application.* Springfield, IL: Charles C Thomas, 1972.

64. Squires RW, Allison TG, Johnson BD, Gau GT. Non-physician supervision of cardiopulmonary exercise testing in chronic heart failure: Safety and results of a preliminary investigation. *J Cardiopulm Rehabil* 1999;19:249-253.

65. Squires RW, Muri AJ, Anderson LJ, et al. Weight training during phase II (early outpatient) cardiac rehabilitation: Heart rate and blood pressure responses. *J Cardiac Rehabil* 1991;11:360-364.

66. Stuart RJ, Ellestad MH. Upsloping S-T segments in exercise stress testing: Six year follow-up study of 438 patients and correlation with 248 angiograms. *Am J Cardiol* 1976;37:19-22.

67. Swain D, Leutholtz B. Heart rate reserve is equivalent to %$\dot{V}O_2$ reserve, not to %$\dot{V}O_2$max. *Med Sci Sports Exerc* 1997;29:410-414.

68. Swain D, Leutholtz B, King M, Haas L, Branch J. Relationship between % heart rate reserve and % $\dot{V}O_2$ reserve in treadmill exercise. *Med Sci Sports Exerc* 1998;30:318-321.

69. Tanaka H, Monahan KD, Seals DR. Age-predicted maximal heart rate revisited. *J Am Coll Cardiol* 2001;37:153-156.

70. U.S. Department of Health and Human Services. *Physical Activity and Health: A Report of the Surgeon General.* Atlanta: U.S. Department of Health and Human Services, Centers for Disease Control and Prevention, National Center for Chronic Disease Prevention and Health Promotion, 1996.

71. U.S. Department of Health and Human Services. *2008 Physical Activity Guidelines for Americans.* [Online]. Washington, DC: U.S. Department of Health and Human Services. ODPHP publication U0036. [accessed 2011 October 19]. www.health.gov/paguidelines/pdf/paguide.pdf.

72. Verstappen FTJ, Huppertz RM, Snoeckx LHEH. Effect of training specificity on maximal treadmill and bicycle ergometer exercise. *Int J Sports Med* 1982;3:43-46.

73. Will PM, Walter JD. Exercise testing: Improving performance with a ramped Bruce protocol. *Am Heart J* 1999;138:1033-1037.

74. Wilmore J, Davis JA, O'Brien RS, Vodak PA, Wolder G, Amsterdam EA. Physiological alterations consequent to 20-week conditioning programs of bicycling, tennis, and jogging. *Med Sci Sports Exerc* 1980;12:1-8.

75. Wisloff U, Stoylen A, Loennechen JP, et al. Superior cardiovascular effect of aerobic interval training versus moderate continuous training in heart failure patients: a randomized study. *Circulation* 2007;115:3086-3094.

CHAPTER 6

1. Ackermann RT, Finch EA, Brizendine E, Zhou H, Marrero DG. Translating the Diabetes Prevention Program into the community: The DEPLOY Pilot Study. *Am J Prev Med* 2008;35:357-363.

2. Action to Control Cardiovascular Risk in Diabetes (ACCORD) Study Group. Effects of intensive glucose lowering in type 2 diabetes. *New Engl J Med* 2008;358:2545-2559.

3. American College of Sports Medicine. *Resource Manual for Exercise Testing and Prescription.* 6th ed. Baltimore: Williams & Wilkins, 2009.

4. American College of Sports Medicine. Thompson WR, Gordon NF, Pescatello LS, eds. *ACSM's Guidelines for Exercise Testing and Prescription.* 9th ed. Baltimore: Williams & Wilkins, 2013.

5. American College of Sports Medicine, Durstine JL, Moore GE, Painter PL, Roberts SO. *ACSM's Exercise Management for Persons with Chronic Diseases and Disabilities.* 3rd ed. Champaign, IL: Human Kinetics, 2009.

6. American Diabetes Association. Ruderman N, Devlin J, Schneider S, Kriska A, eds. *Handbook of Exercise in Diabetes.* Alexandria, VA: American Diabetes Association, 2002.

7. American Diabetes Association. Diagnosis and classification of diabetes mellitus. *Diabetes Care* 2011;34:S62-S69.

8. American Diabetes Association. Economic costs of diabetes mellitus in the U.S. in 2007. *Diabetes Care* 2008;31:596-615.

9. American Diabetes Association. National standards for diabetes self-management education. *Diabetes Care* 2011;34:S89-S96.

10. American Diabetes Association. Nutrition recommendations and interventions for diabetes. *Diabetes Care* 2008;31:S61-S78

11. American Diabetes Association. Physical activity/exercise and diabetes. *Diabetes Care* 2004;27:S58-S62.

12. American Diabetes Association. Standards of medical care in diabetes—2011. *Diabetes Care* 2011;34:S11-S61.

13. Andersson DKG, Svaardsudd K. Long-term glycemic control related to mortality in type II diabetes. *Diabetes Care* 1995;18:1534-1543.

14. Araiza P, Hewes H, Gashetewa C, Vella CA, Burge MR. Efficacy of a pedometer-based physical activity program on parameters of diabetes control in type 2 diabetes mellitus. *Metabolism* 2006;55:11382-1387.

15. Bantle JP, Swanson JE, Thomas W, Laine DC. Metabolic effects of dietary sucrose in type II diabetic subjects. *Diabetes Care* 1996;19:1249-1256.

16. Björntorp P. Portal adipose tissue as a generator of risk factors for cardiovascular disease and diabetes. *Arteriosclerosis* 1990;10:493-496.

17. Boule NG, Haddad E, Kenny GP, Wells GA, Sigal RJ. Effects of exercise on glycemic control and body mass in type 2 diabetes mellitus: A meta-analysis of controlled clinical trials. *JAMA* 2001;286:1218-1227.

18. Boule NG, Kenny GP, Haddad E, Wells GA, Sigal RJ. Meta-analysis of the effect of structured exercise training on cardiorespiratory fitness in type 2 diabetes mellitus. *Diabetologia* 2003;46:1071-1081.

19. Castaneda C. Diabetes control with physical activity and exercise. *Nutr Clin Care* 2003;6:89-96.

20. Castaneda C, Layne JE, Munoz-Orians L, et al. A randomized controlled trial of resistance exercise training to improve glycemic control in older adults with type 2 diabetes. *Diabetes Care* 2002;25:2335-2341.

21. Cauza E, Hanusch-Enserer U, Strasser B, et al. The relative benefits of endurance and strength training on the metabolic factors and muscle function of people with type 2 diabetes mellitus. *Arch Phys Med Rehabil* 2005;86:1527-1533.

22. Centers for Disease Control and Prevention. www.cdc.gov/diabetes. [accessed April 25, 2013].

23. Centers for Disease Control and Prevention. National diabetes fact sheet: National estimates and general information on diabetes and prediabetes in the United States, 2011. Atlanta: U.S. Department of Health and Human Services, Centers for Disease Control and Prevention, 2011.

24. Christ-Roberts CY, Pratipanawatr T, Pratipanawatr W, et al. Exercise training increases glycogen synthase activity and GLUT 4 expression but not insulin signaling in overweight non-diabetic and type 2 diabetic subjects. *Metabolism* 2004;53:1233-1242.

25. Cohen ND, Dunstan DW, Robinson C, Vulikh E, Zimmet PZ, Shaw JE. Improved endothelial function following a 14-month

resistance exercise training program in adults with type 2 diabetes. *Diabetes Res Clin Pract* 2008;79:405-411.

26. Colberg SR. *Diabetic Athlete's Handbook: Your Guide to Peak Performance.* Champaign, IL: Human Kinetics, 2009.

27. Colberg SR, Albright A, Blissmer B, Brau B, Chasen-Taber L, Fernhall B, Regensteiner JG, Rubin RR, Sigal RJ for the American College of Sports Medicine and American Diabetes Association. Exercise and type 2 diabetes *Med Sci Sports Exerc* 2010;42:2282-2303.

28. Colberg SR. Exercise prescription. In: Mensing C, McLaughlin S, Halstenson C, eds. *The Art and Science of Diabetes Self-Management Education Desk Reference.* 2nd ed. Chicago: American Association of Diabetes Educators, 2011.

29. Cornelissen VA, Fagard RH. Effect of resistance training on resting blood pressure: A meta-analysis of randomized controlled trials. *J Hypertens* 2005;23:251-259.

30. Craft LL, Perna FM. The benefits of exercise for the clinically depressed. *Prim Care Companion J Clin Psychiatry* 2004;6:104-111.

31. Diabetes Control and Complications Trial Research Group. The effect of intensive treatment of diabetes on the development and progression of long-term complications in insulin-dependent diabetes mellitus. *New Engl J Med* 1993;329:977-986.

32. Diabetes Prevention Program Research Study Group. Achieving weight loss and physical activity goals among diabetes prevention program lifestyle participants. *Obes Res* 2004;12:1426-1434.

33. Diabetes Prevention Program Research Study Group. Reduction in the incidence of type 2 diabetes with lifestyle intervention or metformin. *N Engl J Med* 2002;346:393-403.

34. Diabetes Prevention Program Research Study Group. 10-year follow-up of the diabetes incidence and weight loss in the Diabetes Prevention Program Outcomes Study. *Lancet* 2009;374:1677-1686.

35. Donnelly JE, Blair SN, Jakicic JM, Manore MM, Rankin JW, Smith BK, American College of Sports Medicine. American College of Medicine position stand: Appropriate physical intervention strategies for weight loss and prevention of weight regain for adults. *Med Sci Sports Exerc* 2009;41:459-471.

36. Dowse GK, Zimmet PZ, Gareeboo H, Alberti KGMM, Tuomilehto J, Finch CF, Chitson P, Tulsidas H. Abdominal obesity and physical inactivity are risk factors for NIDDM and impaired glucose tolerance in Indian, Creole, and Chinese Mauritians. *Diabetes Care* 1991;14:271-282.

37. Duncan GE, Perri MG, Theriaque DW, Hutson AD, Eckel RH, Stacpoole PW. Exercise training, without weight loss, increases insulin sensitivity and postheparin plasma lipase activity in previously sedentary adults. *Diabetes Care* 2003;26:557-562.

38. Duncan JJ, Gordon NF, Scott CB. Women walking for health and fitness: How much is enough? *JAMA* 1991;266:3295-3299.

39. Dunstan DW, Daly RM, Owen N, et al. High-intensity resistance training improves glycemic control in older patients with type 2 diabetes. *Diabetes Care* 2002;25:1729-1736.

40. Ewing DJ, Boland O, Neilson JM, Cho CG, Clarke B. Autonomic neuropathy, QT interval lengthening, and unexpected deaths in diabetic autonomic neuropathy. *J Clin Endocrinol Metab* 1991;54:751-754.

41. Featherstone JF, Holly RG, Amsterdam EA. Physiologic responses to weight lifting in coronary artery disease. *Am J Cardiol* 1993;71:287-292.

42. Feskens EJ, Loeber JG, Kromhout D. Diet and physical activity as determinants of hyperinsulinemia: The Zutphen elderly study. *Am J Epidemiol* 1994;140:350-360.

43. Frisch RE, Wyshak G, Albright TE, Albright NL, Schiff I. Lower prevalence of diabetes in female former college athletes compared with nonathletes. *Diabetes* 1986;35:1101-1105.

44. Fujimoto WY, Leonetti DL, Kinyoun JL, Shuman WP, Stolov WC, Wahl PW. Prevalence of complications among second-generation Japanese-American men with diabetes, impaired glucose tolerance or normal glucose tolerance. *Diabetes* 1987;36:730-739.

45. Fujioka S, Matsuzawa Y, Tokunaga K, Kawamoto T, Kobatake T, Keno Y, Kotani K, Yoshida S, Tarui S. Improvement of glucose and lipid metabolism associated with selective reduction of intra-abdominal visceral fat in premenopausal women with visceral fat obesity. *Int J Obes* 1991;15:853-859.

46. Ghilarducci LE, Holly RG, Amsterdam EA. Effects of high resistance training in coronary artery disease. *Am J Cardiol* 1989;64:866-870.

47. Gillison FB, Skevington SM, Sato A, Standage M, Evangelidou S. The effects of exercise interventions on quality of life in clinical and healthy populations: A meta-analysis. *Soc Sci Med* 2009;68:1700-1710.

48. Goodpaster BH, Katsiaras A, Kelley DE. Enhanced fat oxidation through physical activity is associated with improvements in insulin sensitivity in obesity. *Diabetes* 2003;52:2191-2197.

49. Gordon BA, Benson AC, Bird SR, Fraser SF. Resistance training improves metabolic health in type 2 diabetes: A systematic review. *Diabetes Res Clin Pract* 2009;83:157-175.

50. Gordon LA, Morrison EY, McGrowder DA, et al. Effects of exercise therapy on lipid profile and oxidative stress indicators in patients with type 2 diabetes. *BMC Complement Altern Med* 2008;8:21.

51. Helmrish SP, Ragland DR, Leung RW, Paffenbarger RW. Physical activity and reduced occurrence of non-insulin-dependent diabetes mellitus. *New Engl J Med* 1991;325:147-152.

52. Holten MK, Zacho M, Gaster M, Juel C, Wojtaszewski JF, Dela F. Strength training increases insulin-mediated glucose uptake, GLUT 4 content, and insulin signaling in skeletal muscle in patients with type 2 diabetes. *Diabetes* 2004;53:294-305.

53. Jakicic JM, Marcus BH, Gallagher KI, Napolitano M, Lang W. Effect of exercise duration and intensity on weight loss in overweight, sedentary women: A randomized trial. *JAMA* 2003;290:1323-1330.

54. Johansen KL. Exercise and chronic kidney disease: Current recommendations. *Sports Med* 2005;35:485-499.

55. Kadoglou NP, Iliadis F, Angelopoulou N, et al. The anti-inflammatory effects of exercise training in patients with type 2 diabetes mellitus. *Eur J Cardiovasc Prev Rehabil* 2007;14:837-843.

56. Kahn JK, Sisson JC, Vinik AI. Prediction of sudden cardiac death in diabetic autonomic neuropathy. *J Nucl Med* 1988;29:1605-1606.

57. Kahn JK, Zola B, Juni J, Vinik A. Decreased exercise heart rate and blood pressure response in diabetic subjects with cardiac autonomic neuropathy. *Diabetes Care* 1986;9:389-394.

58. Kahn SE, Hull RL, Utzschneider KM. Mechanisms linking obesity to insulin resistance and type 2 diabetes. *Nature* 2006;444:840-846.

59. Kelley GA, Kelley KS. Effects of aerobic exercise on lipids and lipoproteins in adults with type 2 diabetes: A meta-analysis of randomized-controlled trials. *Public Health* 2007;12:643-655.

60. Kelley GA, Kelley KS. Progressive resistance exercise and resting blood pressure: A meta-analysis of randomized controlled trials. *Hypertension* 2000;35:838-843.

61. Kelley GA, Kelley KS, Tran ZV. Aerobic exercise and resting blood pressure: A meta-analytic review of randomized, controlled trials. *Prev Cardiol* 2001;4:73-80.

62. Kim SH, Lee SJ, Kang ES, et al. Effects of lifestyle modification on metabolic parameters and carotid intima-media thickness in patients with type 2 diabetes. *Metabolism* 2006;55:1053-1059.

63. Kjaer M, Hollenbeck CB, Frey-Hewitt B, Galbo H, Haskell W, Reaven GM. Glucoregulation and hormonal responses to maximal exercise in non-insulin-dependent diabetes. *J Appl Physiol* 1990;68:2067-2074.

64. Klem ML, Wing RR, Mcguire MT, Seagle HM, Hill JO. A descriptive study of individuals successful at long-term maintenance of substantial weight loss. *Am J Clin Nutr* 1997;66:239-246.

65. Koh KP, Fassett RG, Sharman JE, Coombes JS, Williams AD. Effect of intradialytic versus home-based aerobic exercise training on physical function and vascular parameters in hemodialysis patients: A randomized pilot study. *Am J Kidney Dis* 2010;55:88-99.

66. Kohner EM, Aldington SJ, Stratton IM, Manley SE, Holman RR, Matthews DR, Turner RC, for the United Kingdom Prospective Diabetes Study. Diabetic retinopathy at diagnosis of non-insulin-dependent diabetes mellitus and associated risk factors. *Arch Ophthalmol* 1998;116:297-303.

67. Lawrence RH. The effects of exercise on insulin action in diabetes. *Better Med J* 1926;1:648-652.

68. Loghmani E, Rickard K, Washburne L, Vandagriff J, Fineberg N, Golden M. Glycemic response to sucrose-containing mixed meals in diets of children with insulin-dependent diabetes mellitus. *J Pediatr* 1991;119:531-537.

69. Loimaala A, Groundstroem K, Rinne M, et al. Effect of long-term endurance and strength training on metabolic control and arterial elasticity in patients with type 2 diabetes mellitus. *Am J Cardiol* 2009;103:972-977.

70. Look AHEAD Research Group. Long-term effects of a lifestyle intervention on weight and cardiovascular risk factors in individuals with type 2 diabetes mellitus. *Arch Intern Med* 2010;170:1566-1575.

71. Manson JE, Nathan DM, Krolewski AS, Stampfer MJ, Willett WC, Hennekens CH. A prospective study of exercise and incidence of diabetes among US male physicians. *JAMA* 1992;268:63-67.

72. Manson JE, Rimm EB, Stampfer MJ, Colditz GA, Willett WC, Krolewski AS, Rosner B, Hennekens CH, Speizer FE. Physical activity and incidence of non-insulin-dependent diabetes mellitus in women. *Lancet* 1991;338:774-778.

73. Marliss EB, Vranic M. Intense exercise has unique effects on both insulin release and its roles in glucoregulation: Implications for diabetes. *Diabetes* 2002;51:S271-S283.

74. Marrero DG, Sizemore JM. Motivating patients with diabetes to exercise. In: Anderson BJ, Ruben RR, eds. *Practical Psychology for Diabetes Physicians: How to Deal With Key Behavioral Issues Faced by Health Care Teams.* Alexandria, VA: American Diabetes Association, 1996.

75. Mayer-Davis EJ, D'Agostino R, Karta AJ, Haffner SM, Rewers MJ, Saad M, Bergman RN. Intensity and amount of physical activity in relation to insulin sensitivity. *JAMA* 1998;279:669-674.

76. McEwen LN, Kim C, Haan M, Ghosh D, Lantz PM, Mangione CM, Safford MM, Marrero D, Thompson TJ, Herman WH for the TRIAD Study Group. Diabetes reporting as a cause of death. *Diabetes Care* 2006;29:247-253.

77. Meigs JB, Cupples LA, Wilson PWF. Parental transmission of type 2 diabetes. The Framingham Offspring Study. *Diabetes* 2000;49:2201-2207.

78. Minuk HL, Vranic M, Marliss EB, Hanna AK, Albisser AM, Zinman B. Glucoregulatory and metabolic response to exercise in obese non-insulin-dependent diabetes. *Am J Physiol* 1981;240:E458-E464.

79. Morrato EH, Hill JO, Waytt HR, Ghushchyan V, Sullivan PW. Physical activity in U.S. adults with diabetes and at risk for developing diabetes, 2003. *Diabetes Care* 2007;30:203-209.

80. Mourier A, Gautier J-F, Dekerviler E, Biagard AX, Villette J-M, Garnier JP, Duvallet A, Guezennec CY, Cathelineau G. Mobilization of visceral adipose tissue related to the improvement in insulin sensitivity in response to physical training in NIDDM. *Diabetes Care* 1997;20:385-392.

81. O'Gorman DJ, Karlsson HK, McQuaid S, et al. Exercise training increases insulin-stimulated glucose disposal and GLUT4 (SLC2A4) protein content in patients with type 2 diabetes. *Diabetologia* 2006;49:2983-2992.

82. Olefsky JM. Insulin resistance. In: Porte D, Sherwin RS, eds. *Ellenberg and Rifkin's Diabetes Mellitus: Theory and Practice.* 5th ed. New Hyde Park, NY: Medical Examination Publishing, 151-178.

83. Pan X-P, Li G-W, Hu Y-H, Wang J, Yang W, Hu Z-X, Lin J, Xiao J-Z, Cao H-B, Liu P, Jiang X-G, Jiang Y-Y, Wang J-P, Zheng H, Zhang H, Bennet PH, Howard BV. Effects of diet and exercise in preventing NIDDM in people with impaired glucose tolerance. *Diabetes Care* 1997;20:537-544.

84. Panagiotakos DB, et al. The implications of obesity and central fat on markers of chronic inflammation: The ATTICA Study. *Atherosclerosis* 2005;183:308-315.

85. Pereira M, Kriska A, Joswiak M, Dowse G, Collins V, Zimmet P, Gareeboo H, Chitson P, Hemraj F, Purran A, Fareed D. Physical inactivity and glucose intolerance in the multi-ethnic island of Mauritius. *Med Sci Sports Exerc* 1995;27:1626-1634.

86. Perry I, Wannamethee S, Walker M, et al. Prospective study of risk factors for development of non-insulin-dependent diabetes in middle aged British men. *BMJ* 1995;310:560-564.

87. Pi-Suner X, Blackburn G, Brancati F, et al. Reduction in weight and cardiovascular disease risk factors in individuals with type 2 diabetes: One-year results of the Look AHEAD trial. *Diabetes Care* 2007;30:1374-1383.

88. President's Council on Physical Fitness. Definitions: Health, fitness, and physical activity. *Research Digest* 2000.

89. Reaven GM. Pathophysiology of insulin resistance in human disease. *Physiol Rev* 1995;75:473-486.

90. Regensteiner JG, Shetterly SM, Mayer EJ, Eckel RH, Haskell WL, Baxter J, Hamman RF. Relationship between habitual physical activity and insulin area among individuals with impaired glucose tolerance. *Diabetes Care* 1995;18:490-497.

91. Rice B, Janssen I, Hudson R, Ross R. Effects of aerobic exercise and/or diet on glucose tolerance and plasma levels in obese men. *Diabetes Care* 1999;22:684-691.

92. Ronnemaa T, Marniemi J, Puukka P, Kuusi T. Effects of long-term physical exercise on serum lipids, lipoproteins and lipid metabolizing enzymes in type 2 (non-insulin-dependent) diabetic patients. *Diabetes Res* 1988;7:79-84.

93. SEARCH for Diabetes in Youth Study Group. The burden of diabetes mellitus among US youth: Prevalence estimates from the SEARCH for diabetes in youth study. *Pediatrics* 2006;118:1510-1518.

94. Sigal RJ, Kenny GP, Boule G, et al. Effects of aerobic training, resistance training, or both on glycemic control in type 2 diabetes: A randomized trial. *Ann Intern Med* 2007;147:357-369.

95. Sigal RJ, Kenny GP, Wassserman DH, Castaneda-Sceppa C. Physical activity/exercise and type 2 diabetes. *Diabetes Care* 2004;27:2518-2539.

96. Sigal RJ, Kenny GP, Wassserman DH, Castaneda-Sceppa C, White RD. Physical activity/exercise and type 2 diabetes: A consensus statement from the American Diabetes Association. *Diabetes Care* 2006;29:1433-1438.

97. Snowling NJ, Hopkins WG. Effects of different modes of exercise training on glucose control and risk factors for complications in type 2 diabetic patients: A meta-analysis. *Diabetes Care* 2006;29:2518-2527.

98. Sothmann MS, Horn TS, Hart BA, Gustafson AB. Comparison of discrete cardiovascular fitness groups on plasma catecholamine and selected behavioral responses to psychological stress. *Psychophysiology* 1987;24:47-54.

99. Spijkerman AMW, Dekker JM, Nijpels G, Adriaanse MC, Kostense PJ, Ruwaard D, Stehouwer CDA, Bouter LM, Heine RJ. Microvascular complications at time of diagnosis of type 2 diabetes are similar among diabetic patients detected by targeted screening and patients newly diagnosed in general practice. The Hoorn Screening Study. *Diabetes Care* 2003;26:2604-2608.

100. Stellingwerff T, Boon H, Gijsen AP, Stegen JH, Kuipers H, van Loon LJ. Carbohydrate supplementation during prolonged cycling exercise spares muscle glycogen but does not affect intramuscular lipid use. *Pflugers Arch* 2007;45:635-647.

101. Tudor-Locke C, Bell RC, Myers AM, et al. Controlled outcome evaluation of the First Step Program: A daily physical activity intervention for individuals with type II diabetes. *Metab Disord* 2004;28:113-119.

102. Tuomilehto JLJ, Eriksson JG, Valle TT, Hamalainin H, Ilanne-Parikka P, Keinanen-Kiukaanniemi S, Laakso M, Louheranta A, Rastas M, Salminen V, Uusitupa M. Prevention of type 2 diabetes mellitus by changes in lifestyle among subjects with impaired glucose tolerance. *N Engl Med* 2001;344:1343-1392.

103. UKPDS Group. UK Prospective Diabetes Study 33: Intensive blood-glucose control with sulphonylureas or insulin compared with conventional treatment and risk of complications in patients with type 2 diabetes. *Lancet* 1998;352:837-853.

104. Uusitupaa MIJ, Niskanen LK, Siitonen O, Voutilainen E, Pyorala K. Ten year cardiovascular mortality in relation to risk factors and abnormalities in lipoprotein composition in type 2 (non-insulin-dependent) diabetic and non-diabetic subjects. *Diabetologia* 1993;18:1534-1543.

105. Vinik AI, Ziegler D. Diabetic cardiovascular autonomic neuropathy. *Circulation* 2007;115:387-397.

106. Wagner H, Degerblad M, Thorell A, et al. Combined treatment with exercise training and acarbose improves metabolic control and cardiovascular risk factor profile in subjects with mild type 2 diabetes. *Diabetes Care* 2006;29:1471-1477.

107. Wang Y, Simar D, Fiatarone Singh MA. Adaptations to exercise training within skeletal muscle in adults with type 2 diabetes or impaired glucose tolerance: A systematic review. *Diabetes Metab Res Rev* 2009;25:13-40.

108. Watts NB, Spanheimer RG, Digirolamo A, Gebhart SSP, Musey VC, Siddiq K, Phillips LS. Prediction of glucose response to weight loss in patients with non-insulin-dependent diabetes mellitus. *Arch Intern Med* 1990;150:803-806.

109. West KM. *Epidemiology of Diabetes and Its Vascular Lesions*. New York: Elsevier, 1978.

110. Williamson DA, Rejeski J, Lang W, et al. Impact of a weight management program on health-related quality of life in overweight adults with type 2 diabetes. *Arch Intern Med* 2009;169:163-171.

111. Wing RR. Behavioral strategies for weight reduction in obese type 2 diabetic patients. *Diabetes Care* 1989;12:139-144.

112. Wing RR, Koeske R, Epstein LH, Nowalk MP, Gooding W, Becker D. Long-term effects of modest weight loss in type II diabetic patients. *Arch Intern Med* 1987;147:1749-1753.

113. Winnick JJ, Sherman WM, Habash DL, et al. Short-term aerobic exercise training in obese humans with type 2 diabetes mellitus improves whole-body insulin sensitivity through gains in peripheral, not hepatic insulin sensitivity. *J Clin Endorinol Metab* 2008;93:771-778.

CHAPTER 7

1. Adams T, et al. Long-term mortality after gastric bypass surgery. *N Engl J Med* 2007;357:753-761.

2. Ades P, Savage PD. The obesity paradox: Perception vs knowledge. *Mayo Clin Proc* 2010;85(2):112-114.

3. Ades PA, Savage PD, Toth MJ, et al. High-caloric expenditure exercise: A new approach to cardiac rehabilitation for overweight coronary patients. *Circulation* 2009;119(20):2671-2678.

4. American College of Sports Medicine. *ACSM's Guidelines for Exercise Testing and Prescription*. 9th ed. American College of Sports Medicine. Baltimore: Lippincott Williams & Wilkins, 2013.

5. American Heart Association. Obesity Information. www.heart.org/HEARTORG/GettingHealthy/WeightManagement/Obesity/Obesity-Information_UCM_307908_Article.jsp [accessed April 25, 2013].

6. Anderson JW, Greenway FL, Fujioka K, et al. Bupropion SR enhances weight loss: A 48-week double-blind, placebo-controlled trial. *Obes Res* 2002;10:633-641.

7. Anderson JW, Vichitbandra S, Oian W, Kryscio RJ. Long-term weight maintenance after an intensive weight-loss program. *J Am Coll Nutr* 1999;18:620-627.

8. Andreyeva T, Sturm R, Ringel JS. Moderate and severe obesity have large differences in health care costs. *Obes Res* 2004;12:1936-1943.

9. Angulo P. Nonalcoholic fatty liver disease. *N Engl J Med* 2002;346:1221-1231.

10. Ballor DL, Katch VL, Becque MD, Marks CR. Resistance weight training during caloric restriction enhances lean body weight maintenance. *Am J Clin Nutr* 1988;47:19-25.

11. Behme MT. Leptin: Product of the obese gene. *Nutr Today* 1996;31:138-141.

12. Blackburn GL, Dwyer J, Flanders WD, Hill JO, Keller CH, Pi-Sunyer FX, St Jeart ST, Willett WC. Report of the American Institute of Nutrition (AIN) Steering Committee on Healthy Weight. *J Nutr* 1994;124:2240-2243.

13. Blair SN, Kohl HW, Barlow CE, et al. Changes in physical fitness and all-cause mortality: A prospective study of healthy and unhealthy men. *JAMA* 1995;273:1093-1098.

14. Bouchard C, Despres JP, Mauriege P. Genetic and nongenetic determinants of regional fat distribution. *Endocr Rev* 1993;14(1):72-93.

15. Bouchard C, Tremblay A, Despres JP, Nadeau A, Lupien PJ, Theriault G, Dussault J, Moorjani S, Pinault S, Fournier G. The response to long-term overfeeding in identical twins. *N Engl J Med* 1990;322(21):1477-1482.

16. Bray GA. Obesity: A time bomb to be defused. *Lancet* 1998;352(9123):160-161.

17. Bray GA, York DA. Clinical review 90: Leptin and clinical medicine: A new piece in the puzzle of obesity. *J Clin Endocrinol Metab* 1997;82:2771-2776.

18. Brownell, KD. *The LEARN Program for Weight Control.* Dallas: American Health, 2002.

19. Calle EE, Thun MJ, Petrelli JM, Rodriguez C, Heath CW Jr. Body-mass index and mortality in a prospective cohort of U.S. adults. *N Engl J Med* 1999;341(15):1097-1105.

20. Cassidy SB, Schwartz S. Prader-Willi and Angelman syndromes. Disorders of genomic imprinting. *Medicine (Baltimore)* 1998;77:140.

21. Centers for Disease Control and Prevention. Obesity: Halting the Epidemic by Making Health Easier. www.cdc.gov/chronicdisease/resources/publications/aag/pdf/2010/AAG_Obesity_2010_Web_508.pdf [accessed April 25, 2013].

22. Christou NV, Sampalis JS, Liberman M, et al. Surgery decreases long-term mortality, morbidity, and health care use in morbidly obese patients. *Ann Surg* 2004;240:416-423.

23. Church TS, Earnest CP, Skinner JS, Blair SN. Effects of different doses of physical activity on cardiorespiratory fitness among sedentary, overweight or obese postmenopausal women with elevated blood pressure: A randomized controlled trial. *JAMA* 2007;297(19):2081-2091.

24. Colditz GA, Willett WC, Rotnitzky A, Manson JE. Weight gain as a risk factor for clinical diabetes mellitus in women. *Ann Intern Med* 1995;122(7):481-486.

25. Cummings DE, Weigle DS, Frayo RS, Breen PA, Ma MK, Dellinger EP, Purnell JQ. Plasma ghrelin levels after diet-induced weight loss or gastric bypass surgery. *N Engl J Med* 2002;346(21):1623-1630.

26. Curtis JP, Selter JG, Wang Y, et al. The obesity paradox: Body mass index and outcomes in patients with heart failure. *Arch Intern Med* 2005;165:55-56.

27. Davidson MH, Hauptman J, DiGirolamo M, et al. Weight control and risk factor reduction in obese subjects treated for 2 years with orlistat. A randomized controlled trial. *JAMA* 1999;281:235-242.

28. De Fillipis E, Cusi K, Ocampo G, et al. Exercise-induced improvement in vasodilatory function accompanies increased insulin sensitivity in obesity and type 2 diabetes mellitus. *J Clin Endocrinol Metab* 2006;91:4903-4910.

29. De Gonzalez AB, Hartge P, Cerhan JR, et al. Body-mass index and mortality among 1.46 million white adults. *N Engl J Med* 2010;363:2211-2219.

30. Dixon JB, O'Brien PE. Changes in comorbidities and improvements in quality of life after LAP-BAND placement. *Am J Surg* 2002;184:51S.

31. Donnelly JE, Blair SN, Jakicic JM, et al. Appropriate physical activity intervention strategies for weight loss and prevention of weight regain for adults. *Med Sci Sports Exerc* 2009;41(2):459-471.

32. Duncan GE, Perri MG, Theriaque DW, et al. Exercise training, without weight loss, increases insulin sensitivity and postheparin plasma lipase activity in previously sedentary adults. *Diabetes Care* 2003;26(3):944-945.

33. Dunn CL, Hannan PJ, Jeffery RW, Sherwood NE, Pronk NP, Boyle R. The comparative and cumulative effects of a dietary restriction and exercise on weight loss. *Int J Obes* 2006;30:112-121.

34. Fagard RH. Physical activity in the prevention and treatment of hypertension in the obese. *Med Sci Sports Exerc* 1999;31(Suppl.):s624-s630.

35. Felson DT, Anderson JJ, Naimark A, et al. Obesity and knee osteoarthritis. The Framingham study. *Ann Int Med* 1998;109:18-24.

36. Finkelstein E. Annual medical spending attributable to obesity: Payer and service specific estimates. *Health Aff* 2009 28(5):w822-w831.

37. Finkelstein EA, Fiebelkorn IC, Wang G. National medical spending attributable to overweight and obesity: How much, and who's paying? *Health Aff* 2003;W3:219-226.

38. Flegal KM, Graubard BI, Williamson DF, Gail MH. Cause-specific excess deaths associated with underweight, overweight, and obesity. *JAMA* 2007;298:2028-2037.

39. Flegal KM, Graubard BI, Williamson DF, Gail MH. Excess deaths associated with underweight, overweight, and obesity. *JAMA* 2005;293(15):1861-1867.

40. Fletcher GF, Balady GJ, Amsterdam EA, et al. Exercise standards for testing and training: A statement for healthcare professionals from the American Heart Association. *Circulation* 2001;104(14):1694-1740.

41. Fogelholm M, Kukkonin-Harjula K, Nenonen A, Pasanen M. Effects of walking training on weight maintenance after a very low energy diet in premenopausal obese women: A randomized controlled trial. *Arch Intern Med* 2000;160(14):2177-2184.

42. Folsom AR, Rasmussen ML, Chambless LE, et al. Prospective associations of fasting insulin, body fat distribution, and diabetes with risk of ischemic stroke. *Diabetes Care* 1999;22:1077-1083.

43. Fontaine KR, Redden DT, Wang C, et al. Years of life lost due to obesity. *JAMA* 2003;289(2):187-193.

44. Foreyt JP, et al. Diet, genetics and obesity. *Food Tech* 1997;51:70-73.

45. Fornitano LD, Godoy MF. Exercise testing in individuals with morbid obesity. *Obes Surg* 2010;20:583-588.

46. Foster GD, Wadden TA, Vogt RA, Brewer G. What is a reasonable weight loss? Patients' expectations and evaluations of obesity treatment outcomes. *J Consult Clin Psychol* 1997;65:79-85.

47. Franz MJ, VanWormer JJ, Crain AL, Boucher JL, et al. Weight-loss outcomes: A systematic review and meta-analysis of weight-loss clinical trials with a minimum 1-year follow-up. *J Am Diet Assoc* 2007;107:1755-1767.

48. Gallagher D, Heymsfield SB, Heo M, et al. Healthy percentage body fat ranges: An approach for developing guidelines based on body mass index. *Am J Clin Nutr* 2000;72:694-701.

49. Gearhardt AN. Neural correlates of food addiction. *Arch Gen Psych* 2011;68(8):808-816.

50. Gortmaker SL, Must A, Perrin JM, et al. Social and economic consequences of overweight in adolescence and young adulthood. *N Engl J Med* 1993;329:1008-1012.

51. Gray L, Lee I-M, Sesso HD, Batty GD. Body weight in early and mid-adulthood in relation to subsequent coronary heart disease mortality: 80-year follow-up in the Harvard Alumni Study. *Arch Intern Med* 2011;171(19):1768-1770.

52. Gregg EW, Cheng YJ, Cadwell BL, Imperatore G, Williams DE, Flegal KM, Narayan KM, Williamson DF. Secular trends in cardiovascular disease risk factors according to body mass index in US adults. *JAMA* 2005;293(15):1868-1874.

53. Grundy SM, Cleeman JI, Daniels SR, et al. Diagnosis and management of the metabolic syndrome: An American Heart Association/National Heart, Lung, and Blood Institute scientific statement. *Circulation* 2005;112:2735-2752.

54. Gulati M, Pandey DK, Arnsdorf MF, et al. Exercise capacity and the risk of death in women: The St James Women Take Heart Project. *Circulation* 2003;108:1554-1559.

55. Guo SS, Wu W, Chumlea WC, et al. Predicting overweight and obesity in adulthood from body mass index values in childhood and adolescence. *Am J Clin Nutr* 2002;76(3):653-658.

56. Gurm HS, Whitlow PL, Kip KE. The impact of body mass index on short- and long-term outcomes in patients undergoing coronary revascularization: Insights from the bypass angioplasty revascularization investigation (BARI). *J Am Coll Cardiol* 2002;39:834-840.

57. Hall KD. Quantification of the effect of energy imbalance on bodyweight. *Lancet* 2011;378:826-837.

58. Haslam D, Sattar N, Lean M. ABC of obesity. Obesity—time to wake up. *BMJ* 2006;333:640-642.

59. Hastie CE, Padmanabhan S, Slack R, et al. Obesity paradox in a cohort of 4880 consecutive patients undergoing percutaneous coronary intervention. *Eur Heart J* 2010;31:222-226.

60. Hensrud DD, Klein S. Extreme obesity: A new medical crisis in the United States. *Mayo Clin Proc* 2006;81(10, Suppl.):S5-S10.

61. Himpens J, et al. Long-term outcomes of laproscopic adjustable gastric banding. *Arch Surg* 2011;146:802-807.

62. Hortobagyi T, Herring C, Pories W, et al. Massive weight loss-induced mechanical plasticity in obese gait. *J Appl Physiol* 2011; ePub ahead of print.

63. Jakicic JM, Marcus BH, Gallagher KI, Napolitano M, Lang W. Effect of exercise duration and intensity on weight loss in overweight, sedentary women. *JAMA* 2003;290:1323-1330.

64. Keith SW, Redden DT, Katzmarzyk PT, et al. Putative contributors to the secular increase in obesity: Exploring the roads less traveled. *Int J Obes* 2006;30:1585-1594.

65. Klem ML, Wing RR, McGuire MT, Seagle HM, Hill JO. A descriptive study of individuals successful at long term maintenance of substantial weight loss. *Am J Clin Nutr* 1997;66:239-246.

66. Knowler WC, Barrett-Connor E, Fowler SE, Hamman RF, Lachin JM, Walker EA, Nathan DM. Reduction in the incidence of type 2 diabetes with lifestyle intervention or metformin. *N Engl J Med* 2002;346(6):393-403.

67. Kuczmarski RJ, Flegal KM. Criteria for definition of overweight in transition: Background and recommendations for the United States. *Am J Clin Nutr* 2000;72:1074-1081.

68. Leibel RL, Rosenbaum M, Hirsch J. Changes in energy expenditure resulting from altered body weight. *N Engl J Med* 1995;332(10):621-628.

69. Leibowitz SF, Hoebel BG. Behavioral neuroscience of obesity. In: Bray GA, Bouchard C, James WPT, eds. *Handbook of Obesity* (p. 313). New York: Marcel Dekker, 1997.

70. Lewis CE, Jacobs DR Jr, McCreath H, et al. Weight gain continues in the 1990s: 10-year trends in weight and overweight from the CARDIA study. Coronary Artery Risk Development in Young Adults. *Am J Epidemiol* 2000;151:1172-1181.

71. Li Z, Hong K, Wong E, Maxwell M, Heber D. Weight cycling in a very low-calorie diet programme has no effect on weight loss velocity, blood pressure and serum lipid profile. *Diabetes Obes Metab* 2007;9:379-385.

72. Look AHEAD Research Group. The Look AHEAD study: A description of the lifestyle intervention and the evidence supporting it. *Obesity* 2006;14(5):737-752.

73. Lucotti P, Monti LD, Setola E, et al. Aerobic and resistance training effects compared to aerobic training alone in obese type 2 diabetic patients on diet treatment. *Diabetes Res Clin Pract* 2011;2; ePub.

74. Lundgren JD. Remission of metabolic syndrome following a 15-week low-calorie lifestyle change program for weight loss. *Int J Obes* 2009;33:144-150.

75. MacLean LD, Rhode BM, Sampalis J, Forse RA. Results of the surgical treatment of obesity. *Am J Surg* 1993;165:155-160.

76. Matsushita Y, Nakagawa T, Yamaoto T, et al. Associations of visceral and subcutaneous fat areas with the prevalence of metabolic risk factor clustering in 6,292 Japanese individuals: The Hitachi health study. *Diabetes Care* 2010;33(9):2117-2119.

77. McAuley PA, Kokkinos PF, Oliveira RB, et al. Obesity paradox and cardiorespiratory fitness in 12,417 male veterans aged 40 to 70 years. *Mayo Clin Proc* 2010;85(2):115-121.

78. McCullough PA, Gallagher MJ, Dejong AT, Sandberg KR, Trivax JE, Alexander D, Kasturi G, Jafri SM, Krause KR, Chengelis DL, Moy J, Franklin BA. Cardiorespiratory fitness and short-term complications after bariatric surgery. *Chest* 2006;130(2):517-525.

79. Meeuwsen S, Horgan GW, Elia M. The relationship between BMI and percent body fat, measured by bioelectrical impedance, in a large adult sample is curvilinear and influenced by age and sex. *Clin Nutr* 2010;29:560-566.

80. Metropolitan Life Insurance Company. 1983 Metropolitan height and weight tables. *Stat Bull Metropol Life Insur Co* 1983;64:1-19.

81. Miller WCD, Koceja DM, Hamilton EJ. A meta-analysis of the past 25 years of weight loss research using diet, exercise, or diet plus exercise intervention. *Int J Obes* 1997;21:941-947.

82. Miller WR, Rollnick S. *Motivational Interviewing.* New York: Guilford Press, 2002.

83. Mokdad AH, Marks JS, Stroup DF, Gerberding JL. Correction: Actual causes of death in the United States, 2000. *JAMA* 2005;293:293-294.

84. Mun EC, Blackburn GL, Matthews JB. Current status of medical and surgical therapy for obesity. *Gastroenterology* 2001;120:669-681.

85. Mustajoki P, Pekkarinen T. Very low energy diets in the treatment of obesity *Obes Rev* 2001;2:61-72.

86. Najjar MF, Rowland M. *Anthropometric Reference Data and Prevalence of Overweight, United States, 1976–80.* Washington, DC: U.S. Government Printing Office, 1987. Series 11, 238; DHHS publication (PHS) 87-1688.

87. National Center for Health Statistics. *Health, United States, 2010: With Special Feature on Death and Dying.* Hyattsville, MD: National Center for Health Statistics, 2010.

88. National Heart, Lung and Blood Institute (NHLBI) Obesity Education Initiative Expert Panel on the Identification, Evaluation, and Treatment of Overweight and Obesity in Adults. Clinical guidelines on the identification, evaluation, and treatment of overweight and obesity in adults—the evidence report. *Obes Res* 1998;6:51S-209S.

89. National Institutes of Health. Clinical guidelines on the identification, evaluation, and treatment of overweight and obesity in adults. *Obes Res* 1998;6(Suppl. 2):51S-209S.

90. Niessner A, Richter B, Penka M, et al. Endurance training reduces circulating inflammatory markers in persons at risk of coronary events: Impact on plaque stabilization? *Atherosclerosis* 2006;186:160-165.

91. Ogden CL, Carroll MD, Curtin LR, et al. Prevalence of overweight and obesity in the United States, 1999–2004. *JAMA* 2006;295:1549-1555.

92. Olshansky SJ, Passaro DJ, Hershow RC, Layden J, Carnes BA, Brody J, Hayflick L, Butler RN, Allison DB, Ludwig DS. A potential decline in life expectancy in the United States in the 21st century. *N Engl J Med* 2005;352(11):1138-1145.

93. Orpana HM, Berthelot JM, Kaplan MS, et al. BMI and mortality: Results from a national longitudinal study of Canadian adults. *Obesity* 2010;18:214-218.

94. Padilla J, Wallace JP, Park S. Accumulation of physical activity reduces blood pressure in pre- and hypertension. *Med Sci Sports Exerc* 2005;37:1264-1275.

95. Park W, Ramachandran J, Weisman P, Jung ES. Obesity effect on male active joint range of motion. *Ergonomics* 2010;53(1):102-108.

96. Parr RB, Capozzi L. Counseling patients about physical activity. *J Am Acad Phys Assist* 1997;10:45-49.

97. Peeters A, Barendregt JJ, Willekens F, Mackenbach JP, Al Mamun A, Bonneux L. Obesity in adulthood and its consequences for life expectancy: A life-table analysis. *Ann Intern Med* 2003;138(1):24-32.

98. Poirier P, Giles TD, Bray GA, Hong Y, et al. Obesity and cardiovascular disease: Pathophysiology, evaluation, and effect of weight loss. An update of the 1997 American Heart Association scientific statement on obesity and heart disease from the Obesity Committee of the Council on Nutrition, Physical Activity, and Metabolism. *Circulation* 2006;113:898-918.

99. Potteiger JA, Jacobsen DJ, Donnelly JE, Hill JO. Glucose and insulin responses following 16 months of exercise training in overweight adults: The Midwest Exercise Trial. *Metabolism* 2003;52(90):1175-1181.

100. Rao G, Burke LE, Spring BJ, et al. New and emerging weight management strategies for busy ambulatory settings: A scientific statement from the American Heart Association. *Circulation* 2011;124:1182-1203.

101. Ray CS, Sue DY, Bray JE, et al. Effects of obesity on respiratory function. *Am Rev Respir Dis* 1983;128:501-506.

102. Rexrode KM, Carey VJ, Hennekens CH, et al. Abdominal adiposity and coronary heart disease in women. *JAMA* 1998;280:1843-1848.

103. Rimm EB, Stamfer MJ, Giovannucci E, et al. Body size and fat distribution as predictors of coronary heart disease among middle-aged and older US men. *Am J Epidemiol* 1995;141:1117-1127.

104. Ross R, Dagnone D, Jones PJ, et al. Reduction in obesity and related comorbid conditions after diet-induced weight loss or exercise-induced weight loss in men: A randomized, controlled trial. *Ann Intern Med* 2000;133:92-103.

105. Shiwaku K, Anuurad E, Enkhmaa B, et al. Overweight Japanese with body mass indexes of 23.0–24.9 have higher risks for obesity-associated disorders: A comparison of Japanese and Mongolians. *Int J Obes Relat Metab Disord* 2004;28(1):152-158.

106. Sjostrom L. Impact of body weight, body composition, and adipose tissue distribution of morbidity and mortality. In: Stunkard AJ, Wadden TA, eds. *Obesity: Theory and Therapy* (pp. 13-41). New York: Raven Press, 1993.

107. Sjostrom L, Lindroos AK, Peltonen M, et al. Lifestyle, diabetes, and cardiovascular risk factors 10 years after bariatric surgery. *N Engl J Med* 2004;351:2683-2693.

108. Skelton NK, Skelton WP. Medical implications of obesity. *Postgrad Med* 1992;92:151-152.

109. Snow V, Barry P, Fitterman N, et al.; Clinical Efficacy Assessment Subcommittee of the American College of Physicians. Pharmacologic and surgical management of obesity in primary care: A clinical practice guideline from the American College of Physicians. *Ann Intern Med* 2005;142:525-531.

110. Stafford RS, Farhat JH, Misra B, Schoenfeld DA. National patterns of physician activities related to obesity management. *Arch Fam Med* 2000;9:631-638.

111. Stampfer MJ, Maclure KM, Colditz GA, et al. Risk of symptomatic gallstones in women with severe obesity. *Am J Clin Nutr* 1992;55:652-658.

112. Stunkard AJ, Harris JR, Pedersen NL, McClearn GE. The body-mass index of twins who have been reared apart. *N Engl J Med* 1990;322(21):1483-1487.

113. Tataranni PA, Young JB, Bogardus C, Ravussin E. A low sympathoadrenal activity is associated with body weight gain and development of central adiposity in Pima Indian men. *Obes Res* 1997;5(4):341-347.

114. Tsai AG, Wadden TA. An evaluation of major commercial weight loss programs in the United States. *Ann Intern Med* 2005;142:55-66.

115. Tsai AG, Wadden TA. Treatment of obesity in primary care practice in the United States: A systematic review. *J Gen Intern Med* 2009;24(9):1073-1079.

116. U.S. Department of Agriculture and U.S. Department of Health and Human Services. *Dietary Guidelines for Americans, 2010.* 7th ed. Washington, DC: U.S. Government Printing Office, December 2010. [Online]. http://health.gov/dietaryguidelines/dga2010/DietaryGuidelines2010.pdf [accessed April 25, 2013].

117. U.S. Department of Health and Human Services. Healthy people 2020. www.healthypeople.gov/2020/default.aspx [accessed April 25, 2013].

118. U.S. Department of Health and Human Services. *Physical Activity and Health: A Report of the Surgeon General.* Atlanta: U.S. Department of Health and Human Services, Centers for Disease Control and Prevention, National Center for Chronic Disease Prevention and Health Promotion, 1996.

119. U.S. Department of Health and Human Services. Decision Memo for Intensive Behavioral Therapy for Obesity (CAG-00423N). www.cms.gov/medicare-coverage-database/details/nca-proposed-decision-memo.aspx?NCAId=253&ver=2&NcaName=Intensive+Behavioral+Therapy+for+Obesity&TimeFrame=7&DocType=All&bc=AgAAYAAAIAAA& [accessed April 25, 2013].

120. Venditti EM, Bray GA, Carrion-Petersen ML, et al. First versus repeat treatment with a lifestyle program: Attendance and weight loss outcomes. *Int J Obes* 2008;32(10):1537-1544.

121. Vismara L, Menegoni F, Zaina F, et al. Effect of obesity and low back pain on spinal mobility: A cross sectional study in women. *J Neuroeng Rehabil* 2010;7:3.

122. Wadden TA. Treatment of obesity by moderate and severe caloric restriction. Results of clinical research trials. *Ann Intern Med* 1993;119(7 Pt 2):688-693.

123. Wadden TA, Berkowitz RI, Womble LG, et al. Randomized trial of lifestyle modification and pharmacotherapy for obesity. *N Engl J Med* 2005;353:2111-2120.

124. Wadden TA, Neiberg RH, Wing RR, et al. Four-year weight losses in the Look AHEAD Study: Factors associated with long-term success. *Obesity* 2011;19(10):1987-1998.

125. Wadden TA, Vogt RA, Anderson RE, et al. Exercise in the treatment of obesity: Effects of four interventions on body composition, resting energy expenditure, appetite, and mood. *J Consult Clin Psychol* 1997;65(2):269-277.

126. Wadden TA, West DS, Neiberg RH, et al. One-year weight losses in the Look AHEAD Study: Factors associated with success. *Obesity* 2009;17(4):713-722.

127. Wang JS. Exercise prescription and thrombogenesis. *J Biomed Sci* 2006;13:753-761.

128. Wang TJ, Parise H, Levy D, et al. Obesity and the risk of new-onset atrial fibrillation. *JAMA* 2004;292:2471-2477.

129. Wei M, Kampert JB, Barlow CE, et al. Relationship between low cardiorespiratory fitness and mortality in normal-weight, overweight, and obese men. *JAMA* 1999;282(16):1547-1553.

130. Weiss EP, Racette SB, Villareal DT, et al. Lower extremity muscle size and strength and aerobic capacity decrease with caloric restriction but not with exercise-induced weight loss. *J Appl Physiol* 2007;102:634-640.

131. Whitaker RC, Wright JA, Pepe MS, Seidel KD, Dietz WH. Predicting obesity in young adulthood from childhood and parental obesity. *N Engl J Med* 1997;337:869-873.

132. Willett WC, Manson JE, Stampfer MJ, et al. Weight, weight change, and coronary heart disease in women. Risk within the "normal" weight range. *JAMA* 1995;273:461-465.

133. Wing RR. Physical activity in the treatment of the adulthood overweight and obesity: Current evidence and research issues. *Med Sci Sports Exerc* 1999;31(Suppl. 11):s547-s552.

134. Wing RR, Phelan S. Long-term weight loss maintenance. *Am J Clin Nutr* 2005;82(Suppl.):222S-225S.

135. Wing R, Tate D, Gorin A, et al. A self-regulation program for maintenance of weight loss. *N Engl J Med* 2006;355:1563-1571.

136. World Health Organization. www.who.int/mediacentre/factsheets/fs311/en/index.html [accessed April 25, 2013].

137. World Health Organization. Obesity: Preventing and managing the global epidemic. Report of a WHO consultation (p. i). WHO Technical Report Series 894, 2000.

138. Yusuf S, Hawken S, Ounpuu S, et al. Obesity and the risk of myocardial infarction in 27,000 participants from 52 countries: A case-control study. *Lancet* 2005;366:1640-1649.

CHAPTER 8

1. Allison TG, Cordeiro MAS, Miller TD, Daida HD, Squires RW, Gau GT. Prognostic significance of exercise induced systemic hypertension in healthy subjects. *Am J Cardiol* 1999;83:371-375.

2. American College of Cardiology and American Heart Association. ACC/AHA 2002 guideline update for exercise testing: A report of the American College of Cardiology/American Heart Association Task Force on Practice Guidelines. *Circulation* 2002;106:1883-1892.

3. American College of Sports Medicine. Position stand. Exercise and hypertension. *Med Sci Sports Exerc* 2004;36:533-553.

4. American College of Sports Medicine. Position stand. The recommended quantity and quality of exercise for developing and maintaining cardiorespiratory and muscular fitness, and flexibility in healthy adults. *Med Sci Sports Exerc* 1998;30:975-991.

5. American College of Sports Medicine. Thompson W, Gordon NF, Pescatello LS, eds. *ACSM's Guidelines for Exercise Testing and Prescription.* 9th ed. Philadelphia: Lippincott Williams & Wilkins, 2013.

6. American Heart Association. Dietary approaches to prevent and treat hypertension: A scientific statement from the American Heart Association. *Circulation* 2006;47:296-308.

7. American Heart Association. Heart disease and stroke statistics—2011 update: A report from the American Heart Association. *Circulation* 2011;123:e18-e209.

8. American Heart Association. The importance of population-wide sodium reduction as a means to prevent cardiovascular disease and stroke: A call to action from the American Heart Association. *Circulation* 2011;123:1138-1143.

9. American Heart Association. Resistance exercise in individuals with and without cardiovascular disease: 2007 update. *Circulation* 2000;101:828-833.

10. American Heart Association. Resistant hypertension: Diagnosis, evaluation, and treatment. A scientific statement from the American Heart Association Professional Education Committee of the Council for High Blood Pressure Research. *Circulation* 2008;51:1403-1419.

11. American Heart Association. Treatment of hypertension in the prevention and management of ischemic heart disease: A scientific statement from the American Heart Association Council for High Blood Pressure Research and the Councils on Clinical Cardiology and Epidemiology and Prevention. *Circulation* 2007;115:2761-2788.

12. Aronow WS, Fleg JL, Pepine CJ, Artinian NT, Bakris G, Brown AS, Ferdinand KC, Forciea MA, Frishman WH, Jaigobin C, Kostis JB, Mancia G, Oparil S, Ortiz E, Reisin E, Rich MW, Schocken DD, Weber MA, Wesley DJ. ACCF/AHA expert consensus document on hypertension in the elderly: A report of the American College of Cardiology Foundation Task Force on Clinical Expert Consensus Documents. *Circulation* 2011;123:2434-2506.

13. Augeri AL, Tsongalis GJ, Van Heest JL, Maresh CM, Thompson PD, Pescatello LS. The endothelial nitric oxide synthase -786 T>C polymorphism and the exercise-induced blood pressure and nitric oxide responses among men with elevated blood pressure. *Atherosclerosis* 2009;204:e28-e34.

14. Barnard RJ, Wen SJ. Exercise and diet in the prevention and control of the metabolic syndrome. *Sports Med* 1994;18:218-228.

15. Benbassat J, Froom P. Blood pressure response to exercise as a predictor of hypertension. *Arch Intern Med* 1986;146:2053-2055.

16. Blair SN, Kohl HW III, Barlow CE, Gibbons LW. Physical fitness and all-cause mortality in hypertensive men. *Ann Med* 1991;23:307-312.

17. Braith R, Stewart KJ. Resistance exercise training. Its role in the prevention of cardiovascular disease. *Circulation* 2006;113:2642-2650.

18. Brown MD, Moore GM, Korytkowski MT, McCole SD, Hagberg JM. Improvement of insulin sensitivity by short-term exercise

training in hypertensive African-American women. *Hypertension* 1997;30:1549-1553.

19. Chobanian AV, Bakris GL, Black HR, Cushman WC, Green LA, Izzo JL Jr, Jones DW, Materson BJ, Oparil S, Wright JT Jr, Roccella EJ. The seventh report of the Joint National Committee on Prevention, Detection, Evaluation, and Treatment of High Blood Pressure: The JNC 7 report. *JAMA* 2003;289:2560-2572.
20. Church TS, Earnest CP, Skinner JS, Blair SN. Effects of different doses of physical activity on cardiorespiratory fitness among sedentary, overweight or obese postmenopausal women with elevated blood pressure: A randomized controlled trial. *JAMA* 2007;297:2081-2091.
21. Cornelissen VA, Fagard RH. Effect of resistance training on resting blood pressure: A meta-analysis of randomized controlled trials. *J Hypertens* 2005;23:251-259.
22. Cortez-Cooper MY, DeVan AE, Anton MM, Farrar RP, Beckwith KA, Todd JS, Tanaka H. Effects of high intensity resistance training on arterial stiffness and wave reflection in women. *Am J Hypertens* 2005;18:930-934.
23. Dengel DR, Hagberg JM, Pratley RE, Rogus EM, Goldberg AP. Improvements in blood pressure, glucose metabolism, and lipoprotein lipids after aerobic exercise plus weight loss in obese, hypertensive middle-aged men. *Metabolism* 1998;47:1075-1082.
24. Dickinson HO, Mason JM, Nicolson DJ, Campbell F, Beyer FR, Cook JV, Williams B, Ford GA. Lifestyle interventions to reduce raised blood pressure: A systematic review of randomized controlled trials. *J Hypertens* 2006;24:215-233.
25. Ebrahim S, Smith GD. Lowering blood pressure: A systematic review of sustained effects of non-pharmacological interventions. *J Public Health Med* 1998;20:441-448.
26. Fields LE, Burt VL, Cutler JA, Hughes J, Roccella EJ, Sorlie P. The burden of adult hypertension in the United States, 1999–2000: A rising tide. *Hypertension* 2004;44:398-404.
27. Filipovsky J, Ducimetiere P, Safar ME. Prognostic significance of exercise blood pressure and heart rate in middle-aged men. *Hypertension* 1992;20:333-339.
28. Franklin BF, Gordon NF. *Contemporary Diagnosis and Management in Cardiovascular Exercise.* Newtown, PA: Handbooks in Healthcare, 2009.
29. Franz IW. Blood pressure measurement during ergometric stress testing. *Z Kardiol* 1996;85(Suppl. 3):71-75.
30. Gordon NF. Hypertension. In: Durstine JL, Moore GE, Painter PL, Roberts SO, eds. *ACSM's Exercise Management for Persons with Chronic Diseases and Disabilities* (pp. 107-113). 3rd ed. Champaign, IL: Human Kinetics, 2009.
31. Gordon NF, Scott CB, Wilkinson WJ, Duncan JJ, Blair SN. Exercise and mild essential hypertension: Recommendations for adults. *Sports Med* 1990;10:390-404.
32. Gordon NF, Salmon RD, Franklin BA, Sperling LS, Hall L, Leighton RF, Haskell WL. Effectiveness of therapeutic lifestyle changes in patients with hypertension, hyperlipidemia, and/or hyperglycemia. *Am J Cardiol* 2004;94:1558-1561.
33. Guidry MA, Blanchard BE, Thompson PD, Maresh CM, Seip RL, Taylor AL, Pescatello LS. The influence of short and long duration on the blood pressure response to an acute bout of dynamic exercise. *Am Heart J* 2006;151:1322.e5-12.
34. Hagberg JM, Ferrell RE, Dengel DR, Wilund KR. Exercise training-induced blood pressure and plasma lipid improvements in hypertensives may be genotype dependent. *Hypertension* 1999;34:18-23.
35. Harris KA, Holly RG. Physiological response to circuit weight training in borderline hypertensive subjects. *Med Sci Sports Exerc* 1987;19:246-252.
36. Izzo JL Jr, Sica DA, Black HR. *Hypertension Primer.* 4th ed. Dallas: American Heart Association, 2008.
37. Kaplan NM. The deadly quartet. *Arch Intern Med* 1989;149:1514-1520.
38. Kaplan NM. Systemic hypertension: Mechanisms and diagnosis. In: Braunwald E, ed. *Heart Disease: A Textbook of Cardiovascular Medicine.* Philadelphia: Saunders, 1997.
39. Kelley GA, Kelley KS. Progressive resistance exercise and resting blood pressure: A meta-analysis of randomized controlled trials. *Hypertension* 2000;35:838-843.
40. LaFontaine T. Resistance training for patients with hypertension. *Strength Cond J* 1997;19:5-9.
41. Lewis GD, Gona P, Larson MG, Plehn JF, Benjamin EJ, O'Donnell CJ, Levy D, Vasan RS, Wang TJ. Exercise blood pressure and the risk of incident cardiovascular disease (from the Framingham Heart Study). *Am J Cardiol* 2008;101:1614-1620.
42. Matthews CE, Pate RR, Jackson KL, Ward DS, Macera CA, Kohl HW, Blair SN. Exaggerated blood pressure response to dynamic exercise and risk of future hypertension. *J Clin Epidemiol* 1998;51:29-35.
43. Miyachi M, Kawano H, Sugawara J, Takahashi K, Hayashi K, Yamazaki K, Tabata I, Tanaka H. Unfavorable effects of resistance training on central arterial compliance: A randomized intervention study. *Circulation* 2004;110:2858-2863.
44. Mundal R, Kjeldsen SE, Sandvik L, Erikssen G, Thaulow E, Erikssen J. Exercise blood pressure predicts cardiovascular mortality in middle-aged men. *Hypertension* 1994;24:56-62.
45. Nemoto K, Gen-no H, Masuki S, Okazaki K, Nose H. Effects of high-intensity interval walking training on physical fitness and blood pressure in middle-aged and older people. *Mayo Clin Proc* 2007;82:803-811.
46. Rakobowchuk M, McGowan CL, de Groot PC, Bruinsma D, Hartman JW, Phillips SM, MacDonald MJ. Effect of whole body resistance training on arterial compliance in young men. *Exp Physiol* 2005;90:645-651.
47. Reaven GM. Role of insulin resistance in human disease. *Diabetes* 1988;37:1595-1607.
48. Reisin E. Nonpharmacologic approaches to hypertension. Weight, sodium, alcohol, exercise, and tobacco considerations. *Med Clin North Am* 1997;81:1289-1303.
49. Sacks FM, Svetkey LP, Vollmer WM, Appel LJ, Bray GA, Harsha D, et al. Effects on blood pressure of reduced dietary sodium and the Dietary Approaches to Stop Hypertension (DASH) diet. DASH-Sodium Collaborative Research Group. *N Engl J Med* 2001;344:3-10.
50. Singh JP, Larson MG, Manolio TA, O'Donnell CJ, Lauer M, Evans JC, Levy D. Blood pressure response during treadmill testing as a risk factor for new-onset hypertension. The Framingham Heart Study. *Circulation* 1999;99:1831-1836.
51. Smith SC, Benjamin EJ, Bonow RO, Braun LT, Creager MA, Franklin BA, Gibbons RJ, Grundy SM, Hiratzka LF, Jones DW, Lloyd-Jones DM, Minissian M, Mosca L, Peterson ED, Sacco RL, Spertus J, Stein JH, Taubert KA. AHA/ACCF secondary prevention and risk reduction therapy for patients with coronary and other atherosclerotic vascular disease: 2011 update: A guideline from the American Heart Association and American College of Cardiology Foundation. *Circulation* 2011;124.

52. *Stedman's Medical Dictionary*. 27th ed. Baltimore: Lippincott Williams & Wilkins, 2000.

53. Tanaka H, Reiling MJ, Seals DR. Regular walking increases peak limb vasodilatory capacity of older hypertensive humans: Implications for arterial structure. *J Hypertens* 1998;16:423-428.

54. Vasan RS, Beiser A, Seshadri S, Larson MG, Kannel WB, D'Agostino RB, Levy D. Residual lifetime risk for developing hypertension in middle-aged women and men: The Framingham Heart Study. *JAMA* 2002;287:1003-1010.

55. Williams GH. Approach to the patient with hypertension. In: Braunwald E, Fauci A, Hauser S, Jameson J, Kasper D, Longo D, eds. *Harrison's Principles of Internal Medicine*. New York: McGraw-Hill, 1998.

56. Writing Group of the PREMIER Collaborative Research Group. Effects of comprehensive lifestyle modification on blood pressure control. Main results of the PREMIER clinical trial. *JAMA* 2003;289:2083-2093.

57. Xin X, He J, Frontini MG, Ogden LG, Motsamai OI, Whelton PK. Effects of alcohol reduction on blood pressure: A meta-analysis of randomized controlled trials. *Hypertension* 2001;38:1112-1117.

CHAPTER 9

1. Alhassan S, Reese KA, Mahurin AJ, et al. Blood lipid responses to plant stanol ester supplementation and aerobic exercise training. *Metabolism* 2006;55:541-546.

2. Altena TS, Michaelson JL, Ball SD, et al. Lipoprotein subfraction changes after continuous or intermittent exercise training. *Med Sci Sports Exerc* 2006;38:367-372.

3. American College of Sports Medicine. *ACSM's Guidelines for Exercise Testing and Prescription*. Philadelphia: Lippincott Williams & Wilkins, 2013.

4. Anum EA, Adera T. Hypercholesterolemia and coronary heart disease in the elderly: A meta-analysis. *Ann Epidemiol* 2004;14:705-721.

5. Artinian NT, Fletcher GF, Mozaffarian D, et al. Interventions to promote physical activity and dietary lifestyle changes for cardiovascular risk factor reduction in adults: A scientific statement from the American Heart Association. *Circulation* 2010;122:406-441.

6. Athyros VG, Elisaf M, Papageorgiou AA, et al. Effects of statins versus untreated dyslipidemia on serum uric acid levels in patients with coronary heart disease: A subgroup analysis of the Greek atorvastatin and coronary heart disease evaluation (GREACE) study. *Am J Kidney Dis* 2004;43:589-599.

7. Athyros VG, Mikhailidis DP, Papageorgiou AA, et al. Effects of statins versus untreated dyslipidemia on renal function in patients with coronary heart disease: A subgroup analysis of the Greek atorvastatin and coronary heart disease evaluation (GREACE) study. *J Clin Pathol* 2004;57:728-734.

8. Athyros VG, Papageorgiou AA, Athyrou VV, et al. Atorvastatin and micronized fenofibrate alone and in combination in type 2 diabetes with combined hyperlipidemia. *Diabetes Care* 2002;25:1198-1202.

9. Austin MA, Hokanson JE, Edwards KL. Hypertriglyceridemia as a cardiovascular risk factor. *Am J Cardiol* 1998;81:7B-12B.

10. Avramoglu RK, Basciano H, Adeli K. Lipid and lipoprotein dysregulation in insulin resistant states. *Clin Chem Acta* 2006;368:1-19.

11. Avramoglu RK, Qiu W, Adeli K. Mechanisms of metabolic dyslipidemia in insulin resistant states: Deregulation of hepatic and intestinal lipoprotein secretion. *Front Biosci* 2003;8:d464-d476.

12. Babirak SP, Iverius PH, Fujimoto WY, et al. Detection and characterization of the heterozygote state for lipoprotein lipase deficiency. *Arteriosclerosis* 1989;9:326-334.

13. Ballantyne CM, Olsson AG, Cook TJ, et al. Influence of low high-density lipoprotein cholesterol and elevated triglyceride on coronary heart disease events and response to simvastatin therapy in 4S. *Circulation* 2001;104:3046-3051.

14. Bautista MC, Engler MM. The Mediterranean diet. Is it cardioprotective? *Prog Cardiovasc Nurs* 2005;20:70-76.

15. Berensen GS, Srinivasan SR, Bao W, et al. Association between multiple cardiovascular risk factors and atherosclerosis in children and young adults: The Bogalousa Heart Study. *N Engl J Med* 1998;338:1650-1656.

16. Blaha MJ, Bansal S, Rouf R, et al. A practical "ABCDE" approach to the metabolic syndrome. *Mayo Clin Proc* 2008;83:932-943.

17. Bond DS, Phelan S, Leahey TM, et al. Weight-loss maintenance in successful weight losers: Surgical vs non-surgical methods. *Int J Obes* 2009;33:173-180.

18. Borgia M, Medici F. Perspectives in the treatment of dyslipidemias in the prevention of coronary heart disease. *Angiology* 1998;49:339-348.

19. Bounds RG, Martin SE, Grandjean PW, et al. Diet and short-term plasma lipoprotein-lipid changes after exercise in trained men. *Int J Sport Nutr Exerc Metab* 2000;10:114-127.

20. Breckenridge W. Deficiencies of plasma lipolytic activities. *Am Heart J* 1987;113:567-573.

21. Brewer HB, Gregg RE, Hoeg JM, et al. Apolipoproteins and lipoproteins in human plasma: An overview. *Clin Chem* 2004;34:B4-B8.

22. Briel M, Ferreira-Gonzalez I, You JJ, et al. Association between change in high density lipoprotein cholesterol and cardiovascular disease morbidity and mortality: A systematic review and meta-regression analysis. *BMJ* 2009;338:1-8.

23. Brown BG, Zhao XQ, Chait A. Simvastatin and niacin, antioxidant vitamins, or the combination for the prevention of coronary disease. *N Engl J Med* 2001;345:1583-1592.

24. Brunzell JD, Albers JJ, Chait A, et al. Plasma lipoproteins in familial combined hyperlipidemia and monogenic familial hypertriglyceridemia. *J Lipid Res* 1983;24:147-155.

25. Burke GL, Bertoni AG, Shea S, et al. The impact of obesity on cardiovascular disease risk factors and subclinical vascular disease: The multi-ethnic study of atherosclerosis. *Arch Intern Med* 2008;168:928-935.

26. Capell J, Jonas S, Kaplan-Liss E, et al. *ACSM's Exercise is Medicine. A Clinician's Guide to Exercise Prescription*. Philadelphia: Wolters Kluwer/Lippincott Williams & Wilkins, 2009.

27. Capewell S, Ford ES. Why have total cholesterol levels declined in most developed countries? *BMC Med* 2011;11:1-5.

28. Carlson LA, Mossfeldt F. Acute effects of prolonged, heavy exercise on the concentration of plasma lipids and lipoprotein in man. *Acta Physiol Scand* 1964;62:51-59.

29. Carroll MD, Lacher DA, Sorlie PD, et al. Trends in serum lipids and lipoproteins of adults, 1960-2002. *JAMA* 2005;294:1773-1781.

30. Carroll S, Dudfield M. What is the relationship between exercise and metabolic abnormalities? A review of the metabolic syndrome. *Sports Med* 2004;34:371-418.

31. Chapman MJ, Ginsberg HN, Amarenco P, et al. Triglyceride-rich lipoproteins and high-density lipoprotein cholesterol in patients at high risk of cardiovascular disease: Evidence and guidance for management. *Eur Heart J* 2011;32:1345-1361.

32. Chobanian AV, Bakris GL, Black HR, et al. National Heart, Lung, and Blood Institute Joint National Committee on Prevention, Detection, Evaluation, and Treatment of High Blood Pressure; National High Blood Pressure Education Program Coordinating Committee. The seventh report of the Joint National Committee on Prevention, Detection, Evaluation, and Treatment of High Blood Pressure: The JNC 7 report. *JAMA* 2003;289:2560-2572.

33. Chodzko-Zajko W, Proctor D, Singh M, et al. Exercise and physical activity for older adults. *Med Sci Sports Exerc* 2009;41:1510-1530.

34. Colhoun HM, Betteridge DJ, Durrington PN, et al. Primary prevention of cardiovascular disease with atorvastatin in type 2 diabetes in the Collaborative Atorvastatin Diabetes Study (CARDS): Multicentre randomized placebo-controlled trial. *Lancet* 2004;364:685-696.

35. Crouse SF, Cockrill SL, Grandjean PW, et al. LDL and HDL densities after exercise in postmenopausal women with normal and high cholesterol. *Med Sci Sports Exerc* 1999;31:S370. [Abstract]

36. Crouse S, O'Brien B, Grandjean P, et al. Effects of exercise training and a single session of exercise on lipids and apolipoproteins in hypercholesterolemic men. *J Appl Physiol* 1997;83:2019-2028.

37. Crouse S, O'Brien B, Rohack J, et al. Changes in serum lipids and apolipoproteins after exercise in men with high cholesterol: Influence of intensity. *J Appl Physiol* 1995;79:279-286.

38. Cullinane E, Lazarus B, Thompson P, et al. Acute effects of a single exercise session on serum lipids in untrained men. *Clin Chim Acta* 1981;109:341-344.

39. Daniels TF, Killinger KM, Michal JJ, et al. Lipoproteins, cholesterol homeostasis and cardiac health. *Int J Biol Sci* 2009;5:474-488.

40. Davidson MH. Combination therapy for dyslipidemia: Safety and regulatory considerations. *Am J Cardiol* 2002;90:50K-60K.

41. Davignon J, Cohn JS, Mabile L, et al. Apolioprotein E and atherosclerosis: Insight from animal and human studies. *Clin Chem Acta* 1999;286:115-143.

42. de Koning EJP, Rabelink TJ. Endothelial function in the postprandial state. *Atheroscler Suppl* 2002;3:16.

43. De Lorgeril M, Salen P, Martin J, et al. Mediterranean diet, traditional risk factors, and the rate of cardiovascular complications after myocardial infarction: Final report of the Lyon Diet Heart Study. *Circulation* 1999;99:779-785.

44. Di Angelantonio E, Sarwar N, Perry P, et al. Major lipids, apolipoproteins, and risk of vascular disease. *JAMA* 2009;302:1993-2000.

45. Djousse L, Arnett DK, Eckfeldt JH, et al. Alcohol consumption and the metabolic syndrome: Does the type of beverage matter? *Obes Res* 2004;12:1375-1385.

46. Dominiczak MH, Caslake MJ. Apolipoproteins: Metabolic role and clinical biochemistry applications. *Ann Clin Biochem* 2011;48:498-515.

47. Donnelly J, Blair S, Jakicic J, et al. Appropriate physical activity intervention strategies for weight loss and prevention of weight regain for adults. *Med Sci Sports Exerc* 2009;41:459-471.

48. Durstine JL, Grandjean PW, Cox CA, et al. Lipids, lipoproteins, and exercise. *J Cardiopulm Rehabil* 2002;22:385-398.

49. Durstine JL, Grandjean PW, Davis PG, et al. Blood lipid and lipoprotein adaptations to exercise: A quantitative analysis. *Sports Med* 2001;31:1033-1062.

50. Eckel RH, Grundy SM, Zimmet PZ. The metabolic syndrome. *Lancet* 2005;365:1415-1428.

51. Enger S, Stromme S, Refsum H. High density lipoprotein cholesterol, total cholesterol, triglycerides in serum after a single exposure to prolonged heavy exercise. *Acta Med Scand* 1981;645:57-64.

52. Evans M, Rees A. Effects of HMG-CoA reductase inhibitors on skeletal muscle: Are all statins the same? *Drug Safety* 2002;25:629-663.

53. Evans M, Roberts A, Davies S, et al. Medical lipid-regulating therapy: Current evidence, ongoing trials and future developments. *Drugs* 2004;64:1181-1196.

54. Fappa E, Yannakoulia M, Pitsavos C, et al. Lifestyle intervention in the management of metabolic syndrome: Could we improve adherence issues? *Nutrition* 2008;24:286-291.

55. Ferguson M, Alderson N, Trost S, et al. Effects of four different single exercise sessions on lipids, lipoproteins, and lipoprotein lipase. *J Appl Physiol* 1998;85:1169-1174.

56. Flegal KM, Carroll MD, Ogden CL, et al. Prevalence and trends in obesity among US adults, 1999-2008. *JAMA* 2010;303:235-241.

57. Fletcher B, Berra K, Ades P, et al. Managing abnormal blood lipids: A collaborative approach. *Circulation* 2005;112:3184-3209.

58. Fletcher GF, Blair S, Blumenthal C, et al. Statement on exercise: Benefits and recommendations for physical activity programs for all Americans. A statement for health professionals by the Committee on Exercise and Cardiac Rehabilitation of the Council on Clinical Cardiology, American Heart Association. *Circulation* 1996;94:857-862.

59. Foger B, Wohlfarter T, Ritsch A, et al. Kinetics of lipids, apolipoproteins and cholesteryl ester transfer protein in plasma after a bicycle marathon. *Metabolism* 1994;43:633-639.

60. Fojo SS, Brewer HB. Hypertriglyceridemia due to genetic defects in lipoprotein lipase and apolipoprotein C-II. *J Intern Med* 1992;231:669-677.

61. Frederickson D, Lees R. A system for phenotyping hyperlipoproteinemia. *Circulation* 1965;31:321-326.

62. Friedewald W, Levy R, Frederickson D. Estimation of the concentration of low-density lipoprotein cholesterol in plasma without the use of preparative centrifugation. *Clin Chem* 1972;18:499-502.

63. Gale EAM. The myth of the metabolic syndrome. *Diabetologia* 2005;48:1679-1683.

64. Genser B, Marz W. Low density lipoprotein cholesterol, statins and cardiovascular events: A meta-analysis. *Clin Res Cardiol* 2006;95:393-404.

65. Ginsberg HN. Lipoprotein physiology. *Endocrinol Metab Clin North Am* 1998;27:503-519.

66. Goldstein J, Brown M. Familial hypercholesterolemia. In: Scriver C, Beaudet A, Sly W, et al., eds. *The Metabolic Basis of Inherited Disease* (pp. 1215-1250). New York: McGraw-Hill, 1989.

67. Goodyear L, Van Houten D, Fronsoe M, et al. Immediate and delayed effects of marathon running on lipids and lipoproteins in women. *Med Sci Sports Exerc* 1990;22:588-592.

68. Gould AL, Davies GM, Alemao E, et al. Cholesterol reduction yields clinical benefits: Meta-analysis including recent trials. *Clin Ther* 2007;29:778-794.

69. Grandjean PW, Alhassan S. Essential laboratory methods for blood lipid and lipoprotein analysis. In: Moffatt R, Stamford B, eds. *Lipid Metabolism and Health* (pp. 117-146). Boca Raton, FL: CRC Press, 2005.

70. Grandjean PW, Crouse SF, Rohack JJ. Influence of cholesterol status on blood lipid and lipoprotein enzyme responses to aerobic exercise. *J Appl Physiol* 2000;89:472-480.

71. Grandjean PW, Crouse SF, Rohack JJ. Lipid responses to a single bout of exercise in type IIa and IIb hypercholesterolemic men. *Med Sci Sports Exerc* 2000;32:S1877. [Abstract]

72. Grundy SM. A constellation of complications: The metabolic syndrome. *Clin Cornerstone* 2005;7:36-45.

73. Grundy SM. Metabolic syndrome: Therapeutic considerations. *Handb Exp Pharmacol* 2005;170:107-133.

74. Grundy SM, Cleeman JI, Daniels SR, et al. Diagnosis and management of the metabolic syndrome. An American Heart Association/National Heart, Lung, and Blood Institute Scientific statement. *Circulation* 2005;112:2735-2752.

75. Grundy SM, Cleeman JI, Merz NB, et al. Implications of recent clinical trials for the National Cholesterol Education Program Adult Treatment Panel III guidelines. *Circulation* 2004;110:227-239.

76. Guyton JR, Blazing MA, Hagar J. Extended release niacin vs. gemfibrozil for the treatment of low levels of high-density lipoprotein cholesterol. *Arch Intern Med* 2000;160:1177-1184.

77. Halbert J, Silagy C, Finucane P, et al. Exercise training and blood lipids in hyperlipidemic and normolipidemic adults: A meta-analysis of randomized controlled trials. *Eur J Clin Nutr* 1999;53:514-522.

78. Halverstadt A, Phares DA, Wilund KR, et al. Endurance exercise training raises high-density lipoprotein cholesterol and lowers small low-density lipoprotein and very low-density lipoprotein independent of body fat phenotypes in older men and women. *Metabolism* 2007;56:444-450.

79. Hardman A, Lawrence J, Herd SL. Postprandial lipemia in endurance-trained people during short interruption to training. *J Appl Physiol* 1998;84:1895-1901.

80. Harper CR, Jacobson TA. Managing dyslipidemia in chronic kidney disease. *J Am Coll Cardiol* 2008;51:2375-2384.

81. Haskell WL, Lee I, Pate R, et al. Physical activity and public health: Updated recommendation for adults from the American College of Sports Medicine and the American Heart Association. *Med Sci Sports Exerc* 2007;39:1423-1434.

82. Hasler CM, Kundrat S, Wool D. Functional foods and cardiovascular disease. *Curr Atheroscler Rep* 2000;2:467-475.

83. Havel R, Kane J. Introduction: Structure and metabolism of plasma lipoproteins. In: Scriver C, Beaudet A, Sly W, et al., eds. *The Metabolic Bases of Inherited Disease,* vol. 2 (pp. 1841-1851). New York: McGraw-Hill, 1995.

84. Havel RJ, Rapaport E. Management of primary hyperlipidemia. *N Engl J Med* 1995;332:1491-1498.

85. Hokanson JE, Austin MA. Plasma triglyceride level as a risk factor for cardiovascular disease independent of high-density lipoprotein cholesterol level: A meta-analysis of population-based prospective studies. *J Cardiovasc Risk* 1996;3:213-219.

86. Holloszy J, Skinner J, Toro G, et al. Effects of a six month program of endurance exercise on lipids in middle aged men. *Am J Cardiol* 1964;14:753-760.

87. Hyre AD, Munter P, Menke A, et al. Trends in ATP-III-defined high blood cholesterol prevalence, awareness, treatment, and control among U.S. adults. *Ann Epidemiol* 2007;17:548-555.

88. Ip S, Lichtenstein AH, Chung M, et al. Systematic review: Association of low-density lipoprotein subfractions with cardiovascular outcomes. *Ann Intern Med* 2009;150:474-484.

89. Jackevicius CA, Chou MM, Ross JS, et al. Generic atorvastatin and health care costs. *N Engl J Med* 2012;366:201-204.

90. Janiszewski PM, Ross R. The utility of physical activity in the management of global cardiometabolic risk. *Obesity* 2009;17:S3-S14.

91. Johansen CT, Kathiresan S, Hegele RA. Genetic determinants of plasma triglycerides. *J Lipid Res* 2011;52:189-206.

92. Kantor M, Cullinane E, Sady S, et al. Exercise acutely increases high density lipoprotein-cholesterol and lipoprotein lipase activity in trained and untrained men. *Metabolism* 1987;36:188-192.

93. Keevil JG, Cullen MW, Gangnon R, et al. Implications of cardiac risk and low-density lipoprotein cholesterol distributions in the United States for the diagnosis and treatment of dyslipidemia. *Circulation* 2007;115:1363-1370.

94. Kelley GA, Kelley KS. Aerobic exercise and lipids and lipoproteins in men: A meta-analysis of randomized controlled trials. *J Mens Health Gend* 2005;3:61-70.

95. Kelley GA, Kelley KS. Impact of progressive resistance training on lipids and lipoproteins in adults: Another look at a meta-analysis using prediction intervals. *Prev Med* 2009;49:473-475.

96. Kelley GA, Kelley KS, Franklin BA. Aerobic exercise and lipids and lipoproteins in patients with cardiovascular disease. A meta-analysis of randomized controlled trials. *J Cardiopulm Rehabil* 2006;26:131-139.

97. Kelley GA, Kelley KS, Roberts S, et al. Efficacy of aerobic exercise and a prudent diet for improving selected lipids and lipoproteins in adults: A meta-analysis of randomized controlled trials. *BMC Med* 2011;9:1-15.

98. Kelley GA, Kelley KS, Tran ZV. Aerobic exercise, lipids and lipoproteins in overweight and obese adults: A meta-analysis of randomized controlled trials. *Int J Obes* 2005;29:893.

99. Kelley GA, Kelley KS, Tran ZV. Walking and non-HDL-C in adults: A meta-analysis of randomized controlled trials. *Prev Cardiol* 2005;8:102-107.

100. Kelley GA, Kelley KS, Tran ZV. Walking, lipids, and lipoproteins: A meta-analysis of randomized controlled trials. *Prev Med* 2004;38:651-661.

101. Kirkeby K, Stromme S, Bjerkedal I, et al. Effects of prolonged, strenuous exercise on lipids and thyroxine in serum. *Acta Med Scand* 1977;202:463-467.

102. Kolovou GD, Mikhailidis DP, Kovar J, et al. Assessment and clinical relevance of non-fasting and postprandial triglycerides: An expert panel statement. *Curr Vasc Pharmacol* 2011;9:258-270.

103. Kolovou GD, Mikhailidis DP, Nordestgaard BG, et al. Definition of postprandial lipemia. *Curr Vasc Pharmacol* 2011;9:292-301.

104. Kones R. Primary prevention of coronary heart disease: Integration of new data, evolving views, revised goals, and role of rosuvastatin in management. A comprehensive survey. *Drug Des Devel Ther* 2011;5:325-380.

105. Kraus WE, Houmard JA, Duscha BD, et al. Effects of the amount and intensity of exercise on plasma lipoproteins. *N Eng J Med* 2002;347:1483-1492.

106. Krauss RM, Eckel RH, Howard B, et al. AHA dietary guidelines: Revision 2000: A statement for healthcare professionals from the Nutrition Committee of the American Heart Association. *Circulation* 2000;102:2284-2299.

107. Kwiterovich PO. Clinical relevance of the biochemical, metabolic, and genetic factors that influence low-density lipoprotein heterogeneity. *Am J Cardiol* 2002;90:30i-47i.

108. Lairon D, Defoort C. Effects of nutrients on postprandial lipemia. *Curr Vasc Pharmacol* 2011;9:309-312.

109. Lamon-Fava S, McNamara J, Farber H, et al. Acute changes in lipid, lipoprotein, apolipoprotein, and low-density lipoprotein particle size after an endurance triathalon. *Metabolism* 1989;38:921-925.

110. Levin BE. Why some of us get fat and what we can do about it. *J Physiol* 2007;583:425-430.

111. Li C, Balluz LS, Okoro CA, et al. Surveillance of certain health behaviors and conditions among states and selected local areas — Behavioral Risk Factor Surveillance System, United States, 2009. *MMWR Surveill Summ* 2011;60:1-250.

112. Liberopoulos EN, Mikhailidis DP, Elisaf MS. Diagnosis and management of the metabolic syndrome in obesity. *Obes Rev* 2005;6:283-296.

113. Liu M, Bergholm R, Makimattila S, et al. A marathon run increases the susceptibility of LDL to oxidation in vitro and modifies plasma antioxidants. *Am J Physiol* 1999;276:E1083-E1091.

114. Lokey E, Tran Z. Effects of exercise training on serum lipid and lipoprotein concentrations in women: A meta-analysis. *Int J Sports Med* 1989;10:424-429.

115. Lyznicki JM, Young DC, Riggs JA, et al. Obesity: Assessment and management in primary care. *Am Fam Phys* 2001;63:2185-2196.

116. MacEneaney OJ, O'Gorman DJ, O'Connor PL. Effect of prior exercise on postprandial lipemia and markers of inflammation and endothelial activation in normal weight and overweight adolescent boys. *Eur J Appl Physiol* 2009;106:721-729.

117. Mansbach CM, Siddiqi SA. The biogenesis of chylomicrons. *Annu Rev Physiol* 2010;72:315-333.

118. Mestek ML, Garner JC, Plaisance EP, et al. Blood lipid responses after continuous and accumulated aerobic exercise. *Int J Sport Nutr Exerc Metab* 2006;16:245-254.

119. Mestek ML, Plaisance EP, Ratcliff LA, et al. Aerobic exercise and postprandial lipemia in men with metabolic syndrome. *Med Sci Sports Exerc* 2008;40:2105-2111.

120. Mihas C, Kolovou GD, Mikhailidis DP, et al. Diagnostic value of postprandial triglyceride testing in healthy subjects: A meta-analysis. *Curr Vasc Pharmacol* 2011;9:271-280.

121. Miranda PJ, DeFronzo RA, Califf RM, et al. Metabolic syndrome: Definition, pathophysiology, and mechanisms. *Am Heart J* 2005;149:33-45.

122. Mooradian AD. Dyslipidemia in type 2 diabetes mellitus. *Nat Clin Pract Endocrinol Metab* 2009;5:150-159.

123. Murphy MH, Blair SN, Murtagh EM. Accumulated versus continuous exercise for health benefit: A review of empirical studies. *Sports Med* 2009;39:29-43.

124. National Cholesterol Education Program. Third report of the National Cholesterol Education Program (NCEP) Expert Panel on Detection, Evaluation and Treatment of High Cholesterol in Adults (Adult Treatment Panel III). Final report. *Circulation* 2002;106:3143-3421.

125. National Institutes of Health. Clinical guidelines on the identification, evaluation, and treatment of overweight and obese adults: The evidence report. *Obes Res* 1998;6:51S-209S.

126. Negi S, Ballantyne CM. Insights from recent meta-analysis: Role of high-density lipoprotein cholesterol in reducing cardiovascular events and rates of atherosclerotic disease progression. *J Clin Lipidol* 2010;4:365-370.

127. Nelson M, Rejeski J, Blair S, et al. Physical activity and public health in older adults: Recommendations from the American College of Sports Medicine and the American Heart Association. *Med Sci Sports Exerc* 2007;39:1435-1445.

128. Newman WPI, Freedman DS, Voors AW, et al. Relation of serum lipoprotein levels and systolic blood pressure to early atherosclerosis: The Bogalousa Heart Study. *N Engl J Med* 1986;314:138-144.

129. Nordestgaard BG, Freiberg JJ. Clinical relevance of non-fasting and postprandial hypertriglyceridemia and remnant cholesterol. *Curr Vasc Pharmacol* 2011;9:281-286.

130. Ooi TC, Nordestgaard BG. Methods to study postprandial lipemia. *Curr Vasc Pharmacol* 2011;9:302-308.

131. Oscai L, Patterson J, Bogard D, et al. Normalization of serum triglycerides and lipoprotein electrophoretic patterns by exercise. *Am J Cardiol* 1972;30:775-780.

132. Oscai L, Tsika R, Essig D. Exercise training has a heparin-like effect on lipoprotein lipase activity in muscle. *Can J Physiol Pharmacol* 1992;70:905-909.

133. Pate R, Pratt M, Blair S, et al. Physical activity and public health: A recommendation from the Centers for Disease Control and Prevention and the American College of Sports Medicine. *JAMA* 1995;273:402-407.

134. Pauciullo P, Mancini M. Treatment challenges in hypercholesterolemia. *Cardiovasc Drugs Ther* 1998;12:325-337.

135. Perez-Martinez P, Garcia-Rios A, Delgado-Lista J, et al. Nutrigenetics of the postprandial lipoprotein metabolism: Evidences from human intervention studies. *Curr Vasc Pharmacol* 2011;9:287-291.

136. Plaisance EP, Grandjean PW, Mahurin AJ. Independent and combined effects of aerobic exercise and pharmacological strategies on serum triglyceride concentrations: A qualitative review. *Phys Sportsmed* 2009;37:1-9.

137. Plaisance EP, Mestek ML, Mahurin AJ, et al. Postprandial triglyceride response to aerobic exercise and extended-release niacin. *Am J Clin Nutr* 2008;88:30-37.

138. Pollock M, Franklin B, Balady G, et al. AHA scientific statement: Resistance exercise in individuals with and without cardiovascular disease: Benefits, rationale, safety, and prescription. *Circulation* 2000;101:828-833.

139. Poobalan A, Aucott L, Smith WC, et al. Effects of weight loss in overweight/obese individuals and long-term lipid outcomes. A systematic review. *Obes Rev* 2004;5:43-50.

140. Porier P, Giles TD, Bray GA, et al. The ezetimibe controversy: Implications for clinical practice. *Atheroscler Thromb Vasc Biol* 2011;26:968-976.

141. Pronk N, Crouse S, O'Brien B, et al. Acute effects of walking on serum lipids and lipoproteins in women. *J Sports Med Phys Fitness* 1995;35:50-58.

142. Pyorala K, Ballantyne CM, Gumbiner B, et al. Reduction of cardiovascular events by simvastatin in non-diabetic coronary heart disease patients with and without the metabolic syndrome: Subgroup analyses of the Scandinavian Simvastatin Survival Study (4S). *Diabetes Care* 2004;27:1735-1740.

143. Ratner R, Goldberg R, Haffner S, et al. Impact of intensive lifestyle and metformin therapy on cardiovascular disease risk factors in the Diabetes Prevention Program. *Diabetes Care* 2005;28:888-894.

144. Reaven GM. The metabolic syndrome: Time to get off the merry-go-round? *J Intern Med* 2010;269:127-136.

145. Reaven GM, Abbasi F, McLaughlin T. Obesity, insulin resistance, and cardiovascular disease. *Recent Prog Horm Res* 2004;59:207-223.

146. Renaud S, De Lorgeril M. Dietary lipids and their relation to ischemic heart disease: From epidemiology to prevention. *J Intern Med* 1989;225:39-46.

147. Rizos CV, Elisaf MS, Liberopoulos EN. Effects of thyroid dysfunction on lipid profiles. *Open Cardiovasc Med J* 2011;5:76-84.

148. Roger VL, Go AS, Lloyd-Jones DM, et al. Executive summary: Heart disease and stroke statistics—2012 update. A report from the American Heart Association. *Circulation* 2012;125:188-197.

149. Rotondo D, Davidson J. Genetics and molecular biology: Identification of dyslipidemia genes. *Curr Opin Lipidol* 2010;21:548-549.

150. Rudkowska I, Jones PJ. Functional foods for the prevention and treatment of cardiovascular diseases: Cholesterol and beyond. *Expert Rev Cardiovasc Ther* 2007;5:477-490.

151. Rumawas ME, Meigs JB, Dwyer JT, et al. Mediterranean-style dietary pattern, reduced risk of metabolic syndrome traits, and incidence in the Framingham Offspring Cohort. *Am J Clin Nutr* 2009;90:1608-1614.

152. Sady S, Thompson P, Cullinane E, et al. Prolonged exercise augments plasma triglyceride clearance. *JAMA* 1986;256:2552-2555.

153. Sarwar N, Danesh J, Eiriksdottir G, et al. Triglycerides and the risk of coronary heart disease: 10,158 incident cases among 262,525 participants in 29 Western prospective studies. *Circulation* 2007;115:450-458.

154. Schober SE, Carroll MD, Lacher DA, et al. High serum total cholesterol—an indicator for monitoring cholesterol lowering efforts; U.S. adults, 2005-2006 (pp. 1-8). *NCHS Data Brief* 2. Hyattsville, MD: National Center for Health Statistics, 2007.

155. Shepherd J. Lipoprotein metabolism: An Overview. *Ann Acad Med* 1992;21:106-113.

156. Sirtori CR, Galli C, Anderson JW, et al. Functional foods for dyslipidaemia and cardiovascular risk prevention. *Nutr Res Rev* 2009;22:244-261.

157. Smutok A, Reece C, Kokkinos P, et al. Aerobic versus strength training for risk factor intervention in middle-aged men at high risk for coronary heart disease. *Metabolism* 1993;42:177-184.

158. Stolar M. Metabolic syndrome: Controversial but useful. *Cleve Clin J Med* 2007;74:199-208.

159. Thompson P, Buchner D, Pina I, et al. AHA scientific statement: Exercise and physical activity in prevention and treatment of atherosclerotic heart disease: A statement from the Council on Clinical Cardiology (Subcommittee on Exercise, Rehabilitation, and Prevention) and the Council on Nutrition, Physical Activity, and Metabolism (Subcommittee on Physical Activity). *Circulation* 2003;107:3109-3116.

160. Thompson PD, Clarkson P, Karas RH. Statin-associated myopathy. *JAMA* 2003;289:1690.

161. Thompson P, Cullinane E, Henderson O, et al. Acute effects of prolonged exercise on serum lipids. *Metabolism* 1980;29:662-665.

162. Tomkin GH. Atherosclerosis, diabetes and lipoproteins. *Expert Rev Cardiovasc Ther* 2010;8:1015-1029.

163. Tomkin GH. Targets for intervention in dyslipidemia in diabetes. *Diabetes Care* 2008;31:S241-S248.

164. Tonstad S, Despres J-P. Treatment of lipid disorders in obesity. *Expert Rev Cardiovasc Ther* 2011;9:1069-1080.

165. Tota-Maharaj R, Defilipps AP, Blumenthal RS, et al. A practical approach to the metabolic syndrome: Review of current concepts and management. *Curr Opin Clin Cardiol* 2010;25:502-512.

166. Tran ZV, Weltman A. Differential effects of exercise on serum lipid and lipoprotein levels seen with changes in body weight. *JAMA* 1985;254:919-924.

167. U.S. Department of Health and Human Services. *2008 Physical Activity Guidelines for Americans,* 2008. ODPHP publication U0036.

168. van der Velde AE. Reverse cholesterol transport: From classical view to new insights. *World J Gastroenterol* 2010;16:5908-5915.

169. Van Gaal LF, Mertens IL, De Block CE. Mechanisms linking obesity with cardiovascular disease. *Nature* 2006;444:875-880.

170. Vega GL, Ma PTS, Cater NB. Effects of adding fenofibrate (200 mg/dy) to simvastatin (10 mg/dy) in patients with combined hyperlipidemia and metabolic syndrome. *Am J Cardiol* 2003;91:956-960.

171. Visich PS, Goss FL, Gordon PM, et al. Effects of exercise with varying energy expenditure on high-density lipoprotein cholesterol. *Eur J Appl Physiol* 1996;72:242-248.

172. Wagh A, Stone NJ. Treatment of metabolic syndrome. *Expert Rev Cardiovasc Ther* 2004;2:213-228.

173. Wallace JP, Johnson B, Padilla J, et al. Postprandial lipemia, oxidative stress and endothelial function: A review. *Int J Clin Pract* 2010;64:389-403.

174. Weiss EC, Galuska DA, Kettel KL, et al. Weight regain in U.S. adults who experienced substantial weight loss, 1999-2002. *Am J Prev Med* 2007;33:34-40.

175. Wessel TR, Arant CB, Kelsey SF, et al. Relationship of physical fitness vs body mass index with coronary artery disease and cardiovascular events in women. *JAMA* 2004;292:1179-1187.

176. Williams P, Krauss R, Vranizan K, et al. Changes in lipoprotein subfractions during diet-induced and exercise-induced weight loss in moderately overweight men. *Circulation* 1990;81:1293-1304.

177. Williams P, Krauss R, Vranizan K, et al. Effects of exercise induced weight loss on low-density lipoprotein subfractions in healthy men. *Atherosclerosis* 1989;9:623-632.

178. Williams P, Krauss R, Wood P, et al. Lipoprotein subfractions of runners and sedentary men. *Metabolism* 1986;35:45-52.

179. Williams P, Stefanick M, Vranizan K, et al. The effects of weight loss by exercise or by dieting on plasma high density lipoprotein (HDL) levels in men with low, intermediate, and normal-to-high HDL at baseline. *Metabolism* 1994;43:917-924.

180. Wood P, Haskell W, Blair S, et al. Increased exercise level and plasma lipoprotein concentrations: A one year randomized, controlled study in sedentary middle-aged men. *Metabolism* 1983;32:31-39.

181. Wood P, Stefanick M, Dreon D, et al. Changes in plasma lipids and lipoproteins in overweight men during weight loss through dieting as compared with exercise. *N Engl J Med* 1988;319:1173-1179.

182. Wooten JS, Biggerstaff KD, Anderson C. Response of lipid, lipoprotein-cholesterol, and electrophoretic characteristics of lipoproteins following a single bout of aerobic exercise in women. *Eur J Appl Physiol* 2008;104:19-27.

183. Wooten JS, Biggerstaff KD, Ben-Ezra V. A single 1-h session of moderate-intensity aerobic exercise does not modify lipids and lipoproteins in normolipidemic obese women. *Appl Physiol Nutr Metab* 2011;36:715-722.

184. Zhu S, St-Onge MP, Heshka S, et al. Lifestyle behaviors associated with lower risk of having the metabolic syndrome. *Metabolism* 2004;53:1503-1511.

185. Zivkovic AM, German B, Sanyal AJ. Comparative review of diets for the metabolic syndrome: Implications for nonalcoholic fatty liver disease. *Am J Clin Nutr* 2007;86:285-300.

CHAPTER 10

1. *ACSM's Guidelines for Exercise Testing and Prescription*. 9th ed. Philadelphia: Lippincott Williams & Wilkins, 2013.

2. American College of Sports Medicine and the American Diabetes Association: Joint position statement. Exercise and type 2 diabetes. *Med Sci Sports Exerc* 2010;42:2282-2303.

3. Clinical Guidelines on the Identification, Evaluation, and Treatment of Overweight and Obesity in Adults—The Evidence Report. National Institutes of Health. *Obes Res* 1998;6 Suppl. 2:51S-209S.

4. Diagnosis and classification of diabetes mellitus. *Diabetes Care* 2011;34 Suppl. 1:S62-69.

5. Executive Summary of the Third Report of the National Cholesterol Education Program (NCEP) Expert Panel on Detection, Evaluation, and Treatment of High Blood Cholesterol in Adults (Adult Treatment Panel III). *JAMA* 2001;285:2486-2497.

6. Third Report of the National Cholesterol Education Program (NCEP) Expert Panel on Detection, Evaluation, and Treatment of High Blood Cholesterol in Adults (Adult Treatment Panel III) final report. *Circulation* 2002;106:3143-3421.

7. Vital signs: State-specific obesity prevalence among adults—United States, 2009. *MMWR* 2010;59:951-955.

8. Abate N, Chandalia M. Ethnic differences in the metabolic syndrome. In: Grundy SM, ed. *Atlas of Atherosclerosis and Metabolic Syndrome, 5th ed.* New York: Springer, 2011.

9. Abbasi F, Brown BW Jr, Lamendola C, McLaughlin T, Reaven GM. Relationship between obesity, insulin resistance, and coronary heart disease risk. *J Am Coll Cardiol* 2002;40:937-943.

10. ADA. Economic costs of diabetes in the U.S. in 2007. *Diabetes Care* 2008;31:596-615.

11. Aguiari P, Leo S, Zavan B, Vindigni V, Rimessi A, Bianchi K, Franzin C, Cortivo R, Rossato M, Vettor R, Abatangelo G, Pozzan T, Pinton P, Rizzuto R. High glucose induces adipogenic differentiation of muscle-derived stem cells. *Proc Natl Acad Sci U S A* 2008;105:1226-1231.

12. Alberti KG, Eckel RH, Grundy SM, Zimmet PZ, Cleeman JI, Donato KA, Fruchart JC, James WP, Loria CM, Smith SC Jr. Harmonizing the metabolic syndrome: A joint interim statement of the International Diabetes Federation Task Force on Epidemiology and Prevention; National Heart, Lung, and Blood Institute; American Heart Association; World Heart Federation; International Atherosclerosis Society; and International Association for the Study of Obesity. *Circulation* 2009;120:1640-1645.

13. Alberti KG, Zimmet PZ. Definition, diagnosis and classification of diabetes mellitus and its complications. Part 1: Diagnosis and classification of diabetes mellitus provisional report of a WHO consultation. *Diabet Med* 1998;15:539-553.

14. Alberti KG, Zimmet P, Shaw J. The metabolic syndrome—a new worldwide definition. *Lancet* 2005;366:1059-1062.

15. Avogaro P, Crepaldi G, Enzi G, Tiengo A. Associazione di iperlipidemia, diabete mellito e obesità di medio grado. *Acta Diabetol Lat* 1967;4:36-41.

16. Avogaro P, Crepaldi G, Enzi G, Tiengo A. Metabolic aspects of essential obesity. *Epatologia* 1965;11:226-238.

17. Ayyad C, Andersen T. Long-term efficacy of dietary treatment of obesity: A systematic review of studies published between 1931 and 1999. *Obes Rev* 2000;1:113-119.

18. Bastard JP, Jardel C, Delattre J, Hainque B, Bruckert E, Oberlin F. Evidence for a link between adipose tissue interleukin-6 content and serum C-reactive protein concentrations in obese subjects. *Circulation* 1999;99:2221-2222.

19. Beasley LE, Koster A, Newman AB, Javaid MK, Ferrucci L, Kritchevsky SB, Kuller LH, Pahor M, Schaap LA, Visser M, Rubin SM, Goodpaster BH, Harris TB. Inflammation and race and gender differences in computerized tomography-measured adipose depots. *Obesity (Silver Spring)* 2009;17:1062-1069.

20. Benson AC, Torode ME, Fiatarone Singh MA. The effect of high-intensity progressive resistance training on adiposity in children: A randomized controlled trial. *Int J Obes (Lond)* 2008;32:1016-1027.

21. Benson AC, Torode ME, Singh MA. Muscular strength and cardiorespiratory fitness is associated with higher insulin sensitivity in children and adolescents. *Int J Pediatr Obes* 2006;1:222-231.

22. Berenson GS, Srinivasan SR, Bao W, Newman WP 3rd, Tracy RE, Wattigney WA. Association between multiple cardiovascular risk factors and atherosclerosis in children and young adults. The Bogalusa Heart Study. *N Engl J Med* 1998;338:1650-1656.

23. Biddinger SB, Emanuelli B. Insulin Resistance in the Metabolic Syndrome. In: Ahima RS, ed. *Metabolic Basis of Obesity* (pp. 391). New York: Springer, 2011.

24. Biddinger SB, Kahn CR. From mice to men: Insights into the insulin resistance syndromes. *Annu Rev Physiol* 2006;68:123-158.

25. Birnbaum HG, Mattson ME, Kashima S, Williamson TE. Prevalence rates and costs of metabolic syndrome and associated risk factors using employees' integrated laboratory data and health care claims. *J Occup Environ Med* 2011;53:27-33.

26. Borch-Johnsen K, Wareham N. The rise and fall of the metabolic syndrome. *Diabetologia* 2010;53:597-599.

27. Brochu M, Tchernof A, Dionne IJ, Sites CK, Eltabbakh GH, Sims EA, Poehlman ET. What are the physical characteristics associated with a normal metabolic profile despite a high level of obesity in postmenopausal women? *J Clin Endocrinol Metab* 2001;86:1020-1025.

28. Byrne SM. Psychological aspects of weight maintenance and relapse in obesity. *J Psychosom Res* 2002;53:1029-1036.

29. Cali AM, Caprio S. Ectopic fat deposition and the metabolic syndrome in obese children and adolescents. *Horm Res* 2009;71 Suppl.:1:2-7.

30. Cauza E, Hanusch-Enserer U, Strasser B, Kostner K, Dunky A, Haber P. Strength and endurance training lead to different post exercise glucose profiles in diabetic participants using a continuous subcutaneous glucose monitoring system. *Eur J Clin Invest* 2005;35:745-751.

31. Cawley J, Meyerhoefer C. The medical care costs of obesity: An instrumental variables approach. *J Health Econ* 2012; 31:219-30.

32. Chomentowski P, Coen PM, Radikova Z, Goodpaster BH, Toledo FG. Skeletal muscle mitochondria in insulin resistance: Differences in intermyofibrillar versus subsarcolemmal subpopulations

and relationship to metabolic flexibility. *J Clin Endocrinol Metab* 2011;96:494-503.

33. Chung HY, Cesari M, Anton S, Marzetti E, Giovannini S, Seo AY, Carter C, Yu BP, Leeuwenburgh C. Molecular inflammation: Underpinnings of aging and age-related diseases. *Ageing Res Rev* 2009;8:18-30.
34. Church TS, Blair SN, Cocreham S, Johannsen N, Johnson W, Kramer K, Mikus CR, Myers V, Nauta M, Rodarte RQ, Sparks L, Thompson A, Earnest CP. Effects of aerobic and resistance training on hemoglobin A1c levels in patients with type 2 diabetes: A randomized controlled trial. *JAMA* 2010;304:2253-2262.
35. Coker RH, Williams RH, Yeo SE, Kortebein PM, Bodenner DL, Kern PA, Evans WJ. The impact of exercise training compared to caloric restriction on hepatic and peripheral insulin resistance in obesity. *J Clin Endocrinol Metab* 2009;94:4258-4266.
36. Cook S, Auinger P, Li C, Ford ES. Metabolic syndrome rates in United States adolescents, from the National Health and Nutrition Examination Survey, 1999-2002. *J Pediatr* 2008;152:165-170.
37. Curtis LH, Hammill BG, Bethel MA, Anstrom KJ, Gottdiener JS, Schulman KA. Costs of the metabolic syndrome in elderly individuals: Findings from the Cardiovascular Health Study. *Diabetes Care* 2007;30:2553-2558.
38. de Ferranti SD, Gauvreau K, Ludwig DS, Neufeld EJ, Newburger JW, Rifai N. Prevalence of the metabolic syndrome in American adolescents: Findings from the Third National Health and Nutrition Examination Survey. *Circulation* 2004;110:2494-2497.
39. Dela F, Larsen JJ, Mikines KJ, Ploug T, Petersen LN, Galbo H. Insulin-stimulated muscle glucose clearance in patients with NIDDM. Effects of one-legged physical training. *Diabetes* 1995;44:1010-1020.
40. Delmonico MJ, Harris TB, Visser M, Park SW, Conroy MB, Velasquez-Mieyer P, Boudreau R, Manini TM, Nevitt M, Newman AB, Goodpaster BH. Longitudinal study of muscle strength, quality, and adipose tissue infiltration. *Am J Clin Nutr* 2009;90:1579-1585.
41. Despres J-P, Lemieux I. Abdominal obesity and metabolic syndrome. *Nature* 2006;444:881-887.
42. Ding EL, Smit LA, Hu FB. The metabolic syndrome as a cluster of risk factors: Is the whole greater than the sum of its parts?: Comment on "The metabolic syndrome, its component risk factors, and progression of coronary atherosclerosis." *Arch Intern Med* 2010;170:484-485.
43. Ditschuneit HH, Flechtner-Mors M, Johnson TD, Adler G. Metabolic and weight-loss effects of a long-term dietary intervention in obese patients. *Am J Clin Nutr* 1999;69:198-204.
44. Donnelly JE, Blair SN, Jakicic JM, Manore MM, Rankin JW, Smith BK. American College of Sports Medicine position stand. Appropriate physical activity intervention strategies for weight loss and prevention of weight regain for adults. *Med Sci Sports Exerc* 2009;41:459-471.
45. Dunstan DW, Daly RM, Owen N, Jolley D, De Courten M, Shaw J, Zimmet P. High-intensity resistance training improves glycemic control in older patients with type 2 diabetes. *Diabetes Care* 2002;25:1729-1736.
46. Einhorn D, Reaven GM, Cobin RH, Ford E, Ganda OP, Handelsman Y, Hellman R, Jellinger PS, Kendall D, Krauss RM, Neufeld ND, Petak SM, Rodbard HW, Seibel JA, Smith DA, Wilson PW. American College of Endocrinology position statement on the insulin resistance syndrome. *Endocr Pract* 2003;9:237-252.
47. Eisenmann JC. On the use of a continuous metabolic syndrome score in pediatric research. *Cardiovasc Diabetol* 2008;7:17.
48. Eisenmann JC, Laurson KR, DuBose KD, Smith BK, Donnelly JE. Construct validity of a continuous metabolic syndrome score in children. *Diabetol Metab Syndr* 2010;2:8.
49. Ekelund U, Brage S, Froberg K, Harro M, Anderssen SA, Sardinha LB, Riddoch C, Andersen LB. TV viewing and physical activity are independently associated with metabolic risk in children: The European Youth Heart Study. *PLoS Med* 2006;3:e488.
50. Ervin RB. Prevalence of metabolic syndrome among adults 20 years of age and over, by sex, age, race and ethnicity, and body mass index: United States, 2003-2006. *Natl Health Stat Report* 2009:1-7.
51. Fenicchia LM, Kanaley JA, Azevedo JL Jr, Miller CS, Weinstock RS, Carhart RL, Ploutz-Snyder LL. Influence of resistance exercise training on glucose control in women with type 2 diabetes. *Metabolism* 2004;53:284-289.
52. Festa A, D'Agostino R Jr, Howard G, Mykkanen L, Tracy RP, Haffner SM. Chronic subclinical inflammation as part of the insulin resistance syndrome: The Insulin Resistance Atherosclerosis Study (IRAS). *Circulation* 2000;102:42-47.
53. Finkelstein EA, Trogdon JG, Cohen JW, Dietz W. Annual medical spending attributable to obesity: Payer-and service-specific estimates. *Health Aff (Millwood)* 2009;28:w822-831.
54. Flegal KM, Carroll MD, Ogden CL, Curtin LR. Prevalence and trends in obesity among US adults, 1999-2008. *JAMA* 2010;303:235-241.
55. Ford ES. Prevalence of the metabolic syndrome defined by the International Diabetes Federation among adults in the U.S. *Diabetes Care* 2005;28:2745-2749.
56. Ford ES, Giles WH, Dietz WH. Prevalence of the metabolic syndrome among US adults: Findings from the third National Health and Nutrition Examination Survey. *JAMA* 2002;287:356-359.
57. Ford ES, Giles WH, Mokdad AH. Increasing prevalence of the metabolic syndrome among U.S. adults. *Diabetes Care* 2004;27:2444-2449.
58. Ford ES, Li C, Zhao G. Prevalence and correlates of metabolic syndrome based on a harmonious definition among adults in the US. *J Diabetes* 2010;2:180-193.
59. Foreyt JP. Need for lifestyle intervention: How to begin. *Am J Cardiol* 2005;96:11E-14E.
60. Franco OH, Massaro JM, Civil J, Cobain MR, O'Malley B, D'Agostino RB Sr. Trajectories of entering the metabolic syndrome: The Framingham Heart Study. *Circulation* 2009;120:1943-1950.
61. Franks PW, Hanson RL, Knowler WC, Sievers ML, Bennett PH, Looker HC. Childhood obesity, other cardiovascular risk factors, and premature death. *N Engl J Med* 2010;362:485-493.
62. Frohlich M, Imhof A, Berg G, Hutchinson WL, Pepys MB, Boeing H, Muche R, Brenner H, Koenig W. Association between C-reactive protein and features of the metabolic syndrome: A population-based study. *Diabetes Care* 2000;23:1835-1839.
63. Galassi A, Reynolds K, He J. Metabolic syndrome and risk of cardiovascular disease: A meta-analysis. *Am J Med* 2006;119:812-819.
64. Gallagher D, Kelley DE, Yim JE, Spence N, Albu J, Boxt L, Pi-Sunyer FX, Heshka S; Ancillary Study Group of the Look AHEAD Research Group. Adipose tissue distribution is different in type 2 diabetes. *Am J Clin Nutr* 2009;89:807-814.

65. Gami AS, Witt BJ, Howard DE, Erwin PJ, Gami LA, Somers VK, Montori VM. Metabolic syndrome and risk of incident cardiovascular events and death: A systematic review and meta-analysis of longitudinal studies. *J Am Coll Cardiol* 2007;49:403-414.

66. Gaziano JM. Fifth phase of the epidemiologic transition: The age of obesity and inactivity. *JAMA* 2010;303:275-276.

67. Gilsanz V, Kremer A, Mo AO, Wren TA, Kremer R. Vitamin D status and its relation to muscle mass and muscle fat in young women. *J Clin Endocrinol Metab* 95:1595-1601.

68. Giorgino F. Adipose tissue function and dysfunction: Organ cross-talk and metabolic risk. *Am J Physiol Endocrinol Metab* 2009;297(5):E975-976.

69. Goldberg L, Elliot DL, Schutz RW, Kloster FE. Changes in lipid and lipoprotein levels after weight training. *JAMA* 1984;252:504-506.

70. Goodman E, Daniels SR, Meigs JB, Dolan LM. Instability in the diagnosis of metabolic syndrome in adolescents. *Circulation* 2007;115:2316-2322.

71. Goodpaster BH, Kelley DE, Thaete FL, He J, Ross R. Skeletal muscle attenuation determined by computed tomography is associated with skeletal muscle lipid content. *J Appl Physiol* 2000;89:104-110.

72. Goodpaster BH, Wolf D. Skeletal muscle lipid accumulation in obesity, insulin resistance, and type 2 diabetes. *Pediatr Diabetes* 2004;5:219-226.

73. Gorgey AS, Dudley GA. Skeletal muscle atrophy and increased intramuscular fat after incomplete spinal cord injury. *Spinal Cord* 2007;45:304-309.

74. Greco AV, Mingrone G, Giancaterini A, Manco M, Morroni M, Cinti S, Granzotto M, Vettor R, Camastra S, Ferrannini E. Insulin resistance in morbid obesity: Reversal with intramyocellular fat depletion. *Diabetes* 2002;51:144-151.

75. Grundy S. The Metabolic Syndrome. In: Grundy SM, ed. *Atlas of Atherosclerosis and Metabolic Syndrome, 5th ed* (pp. 1-26). New York: Springer, 2011.

76. Grundy SM. Metabolic syndrome pandemic. *Arterioscler Thromb Vasc Biol* 2008;28:629-636.

77. Grundy SM, Cleeman JI, Daniels SR, Donato KA, Eckel RH, Franklin BA, Gordon DJ, Krauss RM, Savage PJ, Smith SC Jr, Spertus JA, Costa F. Diagnosis and management of the metabolic syndrome: An American Heart Association/National Heart, Lung, and Blood Institute scientific statement. *Circulation* 2005;112:2735-2752.

78. Grundy SM, Hansen B, Smith SC Jr, Cleeman JI, Kahn RA. Clinical management of metabolic syndrome: Report of the American Heart Association/National Heart, Lung, and Blood Institute/American Diabetes Association conference on scientific issues related to management. *Circulation* 2004;109:551-556.

79. Guo SS, Wu W, Chumlea WC, Roche AF. Predicting overweight and obesity in adulthood from body mass index values in childhood and adolescence. *Am J Clin Nutr* 2002;76:653-658.

80. Haas JT, Biddinger SB. Dissecting the role of insulin resistance in the metabolic syndrome. *Curr Opin Lipidol* 2009;20:206-210.

81. Haller H. Epidemiology and associated risk factors of hyperlipoproteinemia. *Z Gesamte Inn Med* 1977;32:124-128.

82. Hayes L, Pearce MS, Firbank MJ, Walker M, Taylor R, Unwin NC. Do obese but metabolically normal women differ in intra-abdominal fat and physical activity levels from those with the expected metabolic abnormalities? A cross-sectional study. *BMC Public Health* 2010;10:723.

83. Hillier TA, Rousseau A, Lange C, Lepinay P, Cailleau M, Novak M, Calliez E, Ducimetiere P, Balkau B. Practical way to assess metabolic syndrome using a continuous score obtained from principal components analysis: The D.E.S.I.R. Cohort. *Diabetologia* 2006;49:1528-1535.

84. Hood DA, Irrcher I, Ljubicic V, Joseph AM. Coordination of metabolic plasticity in skeletal muscle. *J Exp Biol* 2006;209:2265-2275.

85. Hurley BF, Hagberg JM, Goldberg AP, Seals DR, Ehsani AA, Brennan RE, Holloszy JO. Resistive training can reduce coronary risk factors without altering VO2max or percent body fat. *Med Sci Sports Exerc* 1988;20:150-154.

86. Itani SI, Ruderman NB, Schmieder F, Boden G. Lipid-induced insulin resistance in human muscle is associated with changes in diacylglycerol, protein kinase C, and IkappaB-alpha. *Diabetes* 2002;51:2005-2011.

87. Jornayvaz FR, Samuel VT, Shulman GI. The role of muscle insulin resistance in the pathogenesis of atherogenic dyslipidemia and nonalcoholic fatty liver disease associated with the metabolic syndrome. *Annu Rev Nutr* 2010;30:273-290.

88. Juhan-Vague I, Alessi MC, Morange PE. Hypofibrinolysis and increased PAI-1 are linked to atherothrombosis via insulin resistance and obesity. *Ann Med* 2000;32 Suppl. 1:78-84.

89. Kahn R, Buse J, Ferrannini E, Stern M. The metabolic syndrome: Time for a critical appraisal. *Diabetologia* 2005;48:1684-1699.

90. Kahn R, Buse J, Ferrannini E, Stern M. The metabolic syndrome: Time for a critical appraisal: Joint statement from the American Diabetes Association and the European Association for the Study of Diabetes. *Diabetes Care* 2005;28:2289-2304.

91. Kaplan NM. The deadly quartet. Upper-body obesity, glucose intolerance, hypertriglyceridemia, and hypertension. *Arch Intern Med* 1989;149:1514-1520.

92. Kelley DE, Slasky BS, Janosky J. Skeletal muscle density: Effects of obesity and non-insulin-dependent diabetes mellitus. *Am J Clin Nutr* 1991;54:509-515.

93. Kelley GA, Kelley KS. Progressive resistance exercise and resting blood pressure: A meta-analysis of randomized controlled trials. *Hypertension* 2000;35:838-843.

94. Kern PA, Ranganathan S, Li C, Wood L, Ranganathan G. Adipose tissue tumor necrosis factor and interleukin-6 expression in human obesity and insulin resistance. *Am J Physiol Endocrinol Metab* 2001;280:E745-751.

95. Kewalramani G, Bilan PJ, Klip A. Muscle insulin resistance: Assault by lipids, cytokines and local macrophages. *Curr Opin Clin Nutr Metab Care* 2010;13:382-390.

96. Klimcakova E, Polak J, Moro C, Hejnova J, Majercik M, Viguerie N, Berlan M, Langin D, Stich V. Dynamic strength training improves insulin sensitivity without altering plasma levels and gene expression of adipokines in subcutaneous adipose tissue in obese men. *J Clin Endocrinol Metab* 2006;91:5107-5112.

97. Krebs NF, Himes JH, Jacobson D, Nicklas TA, Guilday P, Styne D. Assessment of child and adolescent overweight and obesity. *Pediatrics* 2007;120 Suppl. 4:S193-228.

98. Kuk JL, Katzmarzyk PT, Nichaman MZ, Church TS, Blair SN, Ross R. Visceral fat is an independent predictor of all-cause mortality in men. *Obesity (Silver Spring)* 2006;14:336-341.

99. Kylin E. Studien uber das hypertonie-hyperglykamie-hyperurikamiesyndrom. *Zentralbl f innere Med Leipz* 1923;81:105-127.

100. Lalonde L, Gray-Donald K, Lowensteyn I, Marchand S, Dorais M, Michaels G, Llewellyn-Thomas HA, O'Connor A, Grover SA.

Comparing the benefits of diet and exercise in the treatment of dyslipidemia. *Prev Med* 2002;35:16-24.

101. Lara-Castro C, Garvey WT. Intracellular lipid accumulation in liver and muscle and the insulin resistance syndrome. *Endocrinol Metab Clin North Am* 2008;37:841-856.

102. Larson-Meyer DE, Redman L, Heilbronn LK, Martin CK, Ravussin E. Caloric restriction with or without exercise: The fitness versus fatness debate. *Med Sci Sports Exerc* 2010;42:152-159.

103. Leroy-Willig A, Willig TN, Henry-Feugeas MC, Frouin V, Marinier E, Boulier A, Barzic F, Schouman-Claeys E, Syrota A. Body composition determined with MR in patients with Duchenne muscular dystrophy, spinal muscular atrophy, and normal subjects. *Magn Reson Imaging* 1997;15:737-744.

104. Levy-Marchal C, Arslanian S, Cutfield W, Sinaiko A, Druet C, Marcovecchio ML, Chiarelli F. Insulin resistance in children: Consensus, perspective, and future directions. *J Clin Endocrinol Metab* 2010;95:5189-5198.

105. Malik S, Wong ND, Franklin SS, Kamath TV, L'Italien GJ, Pio JR, Williams GR. Impact of the metabolic syndrome on mortality from coronary heart disease, cardiovascular disease, and all causes in United States adults. *Circulation* 2004;110:1245-1250.

106. Manini TM, Clark BC, Nalls MA, Goodpaster BH, Ploutz-Snyder LL, Harris TB. Reduced physical activity increases intermuscular adipose tissue in healthy young adults. *Am J Clin Nutr* 2007;85:377-384.

107. Marangos P, Okamoto L, Caro J. Economic burden of the components of the metabolic syndrome. In: Preedy VR and Watson RR, eds. *Handbook of Disease Burdens and Quality of Life Measures.* New York: Springer, 2011.

108. Marques-Vidal P, Pecoud A, Hayoz D, Paccaud F, Mooser V, Waeber G, Vollenweider P. Normal weight obesity: Relationship with lipids, glycaemic status, liver enzymes and inflammation. *Nutr Metab Cardiovasc Dis* 2010;20:669-675.

109. Martinez-Gomez D, Tucker J, Heelan KA, Welk GJ, Eisenmann JC. Associations between sedentary behavior and blood pressure in young children. *Arch Pediatr Adolesc Med* 2009;163:724-730.

110. Mauras N, Delgiorno C, Kollman C, Bird K, Morgan M, Sweeten S, Balagopal P, Damaso L. Obesity without established comorbidities of the metabolic syndrome is associated with a proinflammatory and prothrombotic state, even before the onset of puberty in children. *J Clin Endocrinol Metab* 2010;95:1060-1068.

111. Misra R, Patel T, Kotha P, Raji A, Ganda O, Banerji M, Shah V, Vijay K, Mudaliar S, Iyer D, Balasubramanyam A. Prevalence of diabetes, metabolic syndrome, and cardiovascular risk factors in US Asian Indians: Results from a national study. *J Diabetes Complications* 2010;24:145-153.

112. Morino K, Petersen KF, Shulman GI. Molecular mechanisms of insulin resistance in humans and their potential links with mitochondrial dysfunction. *Diabetes* 2006;55 Suppl. 2:S9-S15.

113. Morrison JA, Friedman LA, Gray-McGuire C. Metabolic syndrome in childhood predicts adult cardiovascular disease 25 years later: The Princeton Lipid Research Clinics Follow-up Study. *Pediatrics* 2007;120:340-345.

114. Mottillo S, Filion KB, Genest J, Joseph L, Pilote L, Poirier P, Rinfret S, Schiffrin EL, Eisenberg MJ. The metabolic syndrome and cardiovascular risk a systematic review and meta-analysis. *J Am Coll Cardiol* 2010;56:1113-1132.

115. Nair KS, Bigelow ML, Asmann YW, Chow LS, Coenen-Schimke JM, Klaus KA, Guo ZK, Sreekumar R, Irving BA. Asian Indians have enhanced skeletal muscle mitochondrial capacity to produce ATP in association with severe insulin resistance. *Diabetes* 2008;57:1166-1175.

116. Nassis GP, Papantakou K, Skenderi K, Triandafillopoulou M, Kavouras SA, Yannakoulia M, Chrousos GP, Sidossis LS. Aerobic exercise training improves insulin sensitivity without changes in body weight, body fat, adiponectin, and inflammatory markers in overweight and obese girls. *Metabolism* 2005;54:1472-1479.

117. Nguyen QM, Srinivasan SR, Xu JH, Chen W, Berenson GS. Changes in risk variables of metabolic syndrome since childhood in pre-diabetic and type 2 diabetic subjects: The Bogalusa Heart Study. *Diabetes Care* 2008;31:2044-2049.

118. Ogden CL, Carroll MD, Curtin LR, Lamb MM, Flegal KM. Prevalence of high body mass index in US children and adolescents, 2007-2008. *JAMA* 2010;303:242-249.

119. Okorodudu DO, Jumean MF, Montori VM, Romero-Corral A, Somers VK, Erwin PJ, Lopez-Jimenez F. Diagnostic performance of body mass index to identify obesity as defined by body adiposity: A systematic review and meta-analysis. *Int J Obes (Lond)* 34:791-799.

120. Okosun IS, Lyn R, Davis-Smith M, Eriksen M, Seale P. Validity of a continuous metabolic risk score as an index for modeling metabolic syndrome in adolescents. *Ann Epidemiol* 2010;20:843-851.

121. Ortega FB, Ruiz JR, Sjostrom M. Physical activity, overweight and central adiposity in Swedish children and adolescents: The European Youth Heart Study. *Int J Behav Nutr Phys Act* 2007;4:61.

122. Pagel-Langenickel I, Bao J, Pang L, Sack MN. The role of mitochondria in the pathophysiology of skeletal muscle insulin resistance. *Endocr Rev* 2010;31:25-51.

123. Park YW, Zhu S, Palaniappan L, Heshka S, Carnethon MR, Heymsfield SB. The metabolic syndrome: Prevalence and associated risk factor findings in the US population from the Third National Health and Nutrition Examination Survey, 1988-1994. *Arch Intern Med* 2003;163:427-436.

124. Pedersen BK. The anti-inflammatory effect of exercise: Its role in diabetes and cardiovascular disease control. *Essays Biochem* 2006;42:105-117.

125. Petersen KF, Dufour S, Befroy D, Garcia R, Shulman GI. Impaired mitochondrial activity in the insulin-resistant offspring of patients with type 2 diabetes. *N Engl J Med* 2004;350:664-671.

126. Peterson M, Liu D, Gordish-Dressman H, Hubal M, Pistilli E, Angelopoulos T, Clarkson P, Moyna N, Pescatello L, Seip R, Visich P, Zoeller R, Thompson P, Devaney J, Hoffman E, Gordon P. Adiposity attenuates muscle quality and the adaptive response to resistance exercise in non-obese, healthy adults. *Int J Obes (Lond)* 2011;35: 1095-1103.

127. Peterson MD, Liu D, IglayReger HB, Saltarelli WA, Visich PS, Gordon PM. Principal Component Analysis Reveals Gender-Specific Predictors of Cardiometabolic Risk in 6th Graders. *Cardiovasc Diabetol,* 2012;11:146.

128. Phielix E, Meex R, Moonen-Kornips E, Hesselink MK, Schrauwen P. Exercise training increases mitochondrial content and ex vivo mitochondrial function similarly in patients with type 2 diabetes and in control individuals. *Diabetologia* 2010;53:1714-1721.

129. Phielix E, Schrauwen-Hinderling VB, Mensink M, Lenaers E, Meex R, Hoeks J, Kooi ME, Moonen-Kornips E, Sels JP, Hesselink MK, Schrauwen P. Lower intrinsic ADP-stimulated mitochondrial respiration underlies in vivo mitochondrial dysfunction in muscle of male type 2 diabetic patients. *Diabetes* 2008;57:2943-2949.

130. Pladevall M, Singal B, Williams LK, Brotons C, Guyer H, Sadurni J, Falces C, Serrano-Rios M, Gabriel R, Shaw JE, Zimmet PZ, Haffner S. A single factor underlies the metabolic syndrome: A confirmatory factor analysis. *Diabetes Care* 2006;29:113-122.

131. Rabol R, Svendsen PF, Skovbro M, Boushel R, Haugaard SB, Schjerling P, Schrauwen P, Hesselink MK, Nilas L, Madsbad S, Dela F. Reduced skeletal muscle mitochondrial respiration and improved glucose metabolism in nondiabetic obese women during a very low calorie dietary intervention leading to rapid weight loss. *Metabolism* 2009;58:1145-1152.

132. Raffaitin C, Feart C, Le Goff M, Amieva H, Helmer C, Akbaraly TN, Tzourio C, Gin H, Barberger-Gateau P. Metabolic syndrome and cognitive decline in French elders: The Three-City Study. *Neurology* 2011;76:518-525.

133. Raffaitin C, Gin H, Empana JP, Helmer C, Berr C, Tzourio C, Portet F, Dartigues JF, Alperovitch A, Barberger-Gateau P. Metabolic syndrome and risk for incident Alzheimer's disease or vascular dementia: The Three-City Study. *Diabetes Care* 2009;32:169-174.

134. Ramos RG, Olden K. The prevalence of metabolic syndrome among US women of childbearing age. *Am J Public Health* 2008;98:1122-1127.

135. Reaven GM. Banting lecture 1988. Role of insulin resistance in human disease. *Diabetes* 1988;37:1595-1607.

136. Rey-Lopez JP, Vicente-Rodriguez G, Biosca M, Moreno LA. Sedentary behaviour and obesity development in children and adolescents. *Nutr Metab Cardiovasc Dis* 2008;18:242-251.

137. Richter EA, Derave W, Wojtaszewski JF. Glucose, exercise and insulin: Emerging concepts. *J Physiol* 2001;535:313-322.

138. Ridker PM, Hennekens CH, Buring JE, Rifai N. C-reactive protein and other markers of inflammation in the prediction of cardiovascular disease in women. *N Engl J Med* 2000;342:836-843.

139. Roger VL, Go AS, Lloyd-Jones DM, Adams RJ, Berry JD, Brown TM, Carnethon MR, Dai S, de Simone G, Ford ES, Fox CS, Fullerton HJ, Gillespie C, Greenlund KJ, Hailpern SM, Heit JA, Ho PM, Howard VJ, Kissela BM, Kittner SJ, Lackland DT, Lichtman JH, Lisabeth LD, Makuc DM, Marcus GM, Marelli A, Matchar DB, McDermott MM, Meigs JB, Moy CS, Mozaffarian D, Mussolino ME, Nichol G, Paynter NP, Rosamond WD, Sorlie PD, Stafford RS, Turan TN, Turner MB, Wong ND, Wylie-Rosett J. Heart disease and stroke statistics—2011 update: A report from the American Heart Association. *Circulation* 2011;123:e18-e209.

140. Rolland Y, Lauwers-Cances V, Pahor M, Fillaux J, Grandjean H, Vellas B. Muscle strength in obese elderly women: Effect of recreational physical activity in a cross-sectional study. *Am J Clin Nutr* 2004;79:552-557.

141. Romero-Corral A, Somers VK, Sierra-Johnson J, Korenfeld Y, Boarin S, Korinek J, Jensen MD, Parati G, Lopez-Jimenez F. Normal weight obesity: A risk factor for cardiometabolic dysregulation and cardiovascular mortality. *Eur Heart J* 2010;31:737-746.

142. Ross R, Dagnone D, Jones PJ, Smith H, Paddags A, Hudson R, Janssen I. Reduction in obesity and related comorbid conditions after diet-induced weight loss or exercise-induced weight loss in men. A randomized, controlled trial. *Ann Intern Med* 2000;133:92-103.

143. Rossi A, Zoico E, Goodpaster BH, Sepe A, Di Francesco V, Fantin F, Pizzini F, Corzato F, Vitali A, Micciolo R, Harris TB, Cinti S, Zamboni M. Quantification of intermuscular adipose tissue in the erector spinae muscle by MRI: Agreement with histological evaluation. *Obesity (Silver Spring)* 2010;18(12):2379-2384.

144. Sardinha LB, Andersen LB, Anderssen SA, Quiterio AL, Ornelas R, Froberg K, Riddoch CJ, Ekelund U. Objectively measured time spent sedentary is associated with insulin resistance independent of overall and central body fat in 9- to 10-year-old Portuguese children. *Diabetes Care* 2008;31:569-575.

145. Sattar N, McConnachie A, Shaper AG, Blauw GJ, Buckley BM, de Craen AJ, Ford I, Forouhi NG, Freeman DJ, Jukema JW, Lennon L, Macfarlane PW, Murphy MB, Packard CJ, Stott DJ, Westendorp RG, Whincup PH, Shepherd J, Wannamethee SG. Can metabolic syndrome usefully predict cardiovascular disease and diabetes? Outcome data from two prospective studies. *Lancet* 2008;371:1927-1935.

146. Schrager MA, Metter EJ, Simonsick E, Ble A, Bandinelli S, Lauretani F, Ferrucci L. Sarcopenic obesity and inflammation in the InCHIANTI study. *J Appl Physiol* 2007;102:919-925.

147. Schrauwen P, Schrauwen-Hinderling V, Hoeks J, Hesselink MK. Mitochondrial dysfunction and lipotoxicity. *Biochim Biophys Acta* 2010;1801:266-271.

148. Schrauwen-Hinderling VB, Kooi ME, Hesselink MK, Jeneson JA, Backes WH, van Echteld CJ, van Engelshoven JM, Mensink M, Schrauwen P. Impaired in vivo mitochondrial function but similar intramyocellular lipid content in patients with type 2 diabetes mellitus and BMI-matched control subjects. *Diabetologia* 2007;50:113-120.

149. Schultz AB, Edington DW. Metabolic syndrome in a workplace: Prevalence, co-morbidities, and economic impact. *Metab Syndr Relat Disord* 2009;7:459-468.

150. Selby JV, Newman B, Quiroga J, Christian JC, Austin MA, Fabsitz RR. Concordance for dyslipidemic hypertension in male twins. *JAMA* 1991;265:2079-2084.

151. Shah PK, Gregory CM, Stevens JE, Pathare NC, Jayaraman A, Behrman AL, Walter GA, Vandenborne K. Non-invasive assessment of lower extremity muscle composition after incomplete spinal cord injury. *Spinal Cord* 2008;46:565-570.

152. Shi H, Kokoeva MV, Inouye K, Tzameli I, Yin H, Flier JS. TLR4 links innate immunity and fatty acid-induced insulin resistance. *J Clin Invest* 2006;116:3015-3025.

153. Simmons RK, Alberti KG, Gale EA, Colagiuri S, Tuomilehto J, Qiao Q, Ramachandran A, Tajima N, Brajkovich Mirchov I, Ben-Nakhi A, Reaven G, Hama Sambo B, Mendis S, Roglic G. The metabolic syndrome: Useful concept or clinical tool? Report of a WHO Expert Consultation. *Diabetologia* 2010;53:600-605.

154. Simoneau JA, Colberg SR, Thaete FL, Kelley DE. Skeletal muscle glycolytic and oxidative enzyme capacities are determinants of insulin sensitivity and muscle composition in obese women. *FASEB J* 1995;9:273-278.

155. Sims EA. Are there persons who are obese, but metabolically healthy? *Metabolism* 2001;50:1499-1504.

156. Sinaiko AR, Steinberger J, Moran A, Hong CP, Prineas RJ, Jacobs DR Jr. Influence of insulin resistance and body mass index at age 13 on systolic blood pressure, triglycerides, and high-density lipoprotein cholesterol at age 19. *Hypertension* 2006;48:730-736.

157. Srikanthan P, Hevener A, Karlamangla A. Sarcopenia exacerbates obesity-associated insulin resistance and dysglycemia: Findings from the National Health and Nutrition Examination Survey III. *PloS One* 2010;5:e10805, doi:10810.11371/journal.pone.0010805.

158. Steinberger J, Daniels SR, Eckel RH, Hayman L, Lustig RH, McCrindle B, Mietus-Snyder ML. Progress and challenges in metabolic syndrome in children and adolescents: A scientific statement from the American Heart Association Atheroscler-

sis, Hypertension, and Obesity in the Young Committee of the Council on Cardiovascular Disease in the Young; Council on Cardiovascular Nursing; and Council on Nutrition, Physical Activity, and Metabolism. *Circulation* 2009;119:628-647.

159. Sturm R, Wells KB. Does obesity contribute as much to morbidity as poverty or smoking? *Public Health* 2001;115:229-235.

160. Thivel D, Malina RM, Isacco L, Aucouturier J, Meyer M, Duche P. Metabolic syndrome in obese children and adolescents: Dichotomous or continuous? *Metab Syndr Relat Disord* 2009;7:549-555.

161. Toledo FG, Menshikova EV, Azuma K, Radikova Z, Kelley CA, Ritov VB, Kelley DE. Mitochondrial capacity in skeletal muscle is not stimulated by weight loss despite increases in insulin action and decreases in intramyocellular lipid content. *Diabetes* 2008;57:987-994.

162. Vague J. La differenciation sexuelle, factuer determinant des formes de l'obesite. *Presse Medicale* 1947;30:339-340.

163. Van Der Heijden GJ, Wang ZJ, Chu Z, Toffolo G, Manesso E, Sauer PJ, Sunehag AL. Strength exercise improves muscle mass and hepatic insulin sensitivity in obese youth. *Med Sci Sports Exerc* 2010;42:1973-1980.

164. Venojarvi M, Puhke R, Hamalainen H, Marniemi J, Rastas M, Rusko H, Nuutila P, Hanninen O, Aunola S. Role of skeletal muscle-fibre type in regulation of glucose metabolism in middle-aged subjects with impaired glucose tolerance during a long-term exercise and dietary intervention. *Diabetes Obes Metab* 2005;7:745-754.

165. Visser M, Pahor M, Taaffe DR, Goodpaster BH, Simonsick EM, Newman AB, Nevitt M, Harris TB. Relationship of interleukin-6 and tumor necrosis factor-alpha with muscle mass and muscle strength in elderly men and women: The Health ABC Study. *J Gerontol A Biol Sci Med Sci* 2002;57:M326-332.

166. Wang X, You T, Yang R, Lyles MF, Demons J, Gong DW, Nicklas BJ. Muscle strength is associated with adipose tissue gene expression of inflammatory adipokines in postmenopausal women. *Age Ageing* 2010;39(5):656-659.

167. Wannamethee SG, Shaper AG, Lennon L, Morris RW. Metabolic syndrome vs Framingham Risk Score for prediction of coronary heart disease, stroke, and type 2 diabetes mellitus. *Arch Intern Med* 2005;165:2644-2650.

168. Wijndaele K, Beunen G, Duvigneaud N, Matton L, Duquet W, Thomis M, Lefevre J, Philippaerts RM. A continuous metabolic syndrome risk score: Utility for epidemiological analyses. *Diabetes Care* 2006;29:2329.

169. Wilson PW, D'Agostino RB, Parise H, Sullivan L, Meigs JB. Metabolic syndrome as a precursor of cardiovascular disease and type 2 diabetes mellitus. *Circulation* 2005;112:3066-3072.

170. Wing RR, Hill JO. Successful weight loss maintenance. *Annu Rev Nutr* 2001;21:323-341.

171. Yaffe K, Haan M, Blackwell T, Cherkasova E, Whitmer RA, West N. Metabolic syndrome and cognitive decline in elderly Latinos: Findings from the Sacramento Area Latino Study of Aging study. *J Am Geriatr Soc* 2007;55:758-762.

172. Zhang H, Park Y, Wu J, Chen X, Lee S, Yang J, Dellsperger KC, Zhang C. Role of TNF-alpha in vascular dysfunction. *Clin Sci (Lond)* 2009;116:219-230.

173. Zoico E, Rossi A, Di Francesco V, Sepe A, Olioso D, Pizzini F, Fantin F, Bosello O, Cominacini L, Harris TB, Zamboni M. Adipose tissue infiltration in skeletal muscle of healthy elderly men: Relationships with body composition, insulin resistance, and inflammation at the systemic and tissue level. *J Gerontol A Biol Sci Med Sci* 2010;65:295-299.

CHAPTER 11

1. American College of Sports Medicine. *ACSM's Guidelines for Exercise Testing and Prescription.* 9th ed. Philadelphia: Lippincott Williams & Wilkins, 2013.

2. Barnea N, Drory Y, Iaina A, et al. Exercise tolerance in patients on chronic hemodialysis. *Isr J Med Sci* 1980;16(1):17-21.

3. Beasley CR, Smith DA, Neale TJ. Exercise capacity in chronic renal failure patients managed by continuous ambulatory peritoneal dialysis. *Aust N Z J Med* 1986;16(1):5-10.

4. Bohannon RW. Comfortable and maximum walking speed of adults aged 20-79 years: Reference values and determinants. *Age Ageing* 1997;26(1):15-19.

5. Bohm CJ, Ho J, Duhamel TA. Regular physical activity and exercise therapy in end-stage renal disease: How should we move forward? *J Nephrol* 2010;23(3):235-243.

6. Brenner BM, Stein JH. *Chronic Renal Failure.* New York: Churchill Livingstone, 1981.

7. Carey S, Painter P. An exercise program for CAPD patients. *Nephrol News Issues* 1997;11(6):15-18.

8. Carlson L, Carey S. Staff responsibility to exercise. *Adv Renal Replace Ther* 1999;6(2):172-180.

9. Chang Y, Cheng S, Lin M, Gau F, Chao YC. The effectiveness of intradialytic leg ergometry exercise for improving sedentary life style and fatigue among patients with chronic kidney disease: A randomized clinical trial. *Int J Nurs Stud* 2010;47(11):1383-1388.

10. Cheema BSB, Singh MAF. Exercise training in patients receiving maintenance hemodialysis: A systematic review of clinical trials. *Am J Nephrol* 2005;25(4):352-364.

11. Deligiannis A, Kouidi E, Tassoulas E, Gigis P, Tourkantonis A, Coats A. Cardiac effects of exercise rehabilitation in hemodialysis patients. *Int J Cardiol* 1999;70(3):253-266.

12. DeOreo PB. Hemodialysis patient-assessed functional health status predicts continued survival, hospitalization, and dialysis-attendance compliance. *Am J Kidney Dis* 1997;30(2):204-212.

13. Gjertson DW. Determinants of long-term survival of adult kidney transplants: A 1999 UNOS update. *Clin Transplant* 1999;341-352.

14. Goldberg AP, Geltman EM, Hagberg JM, Gavin JR III, Delmez JA, Carney RM, Naumowicz A, Oldfield MH, Harter HR. Therapeutic benefits of exercise training for hemodialysis patients. *Kidney Int* 1983;24(16):S303-S309.

15. Headley S, Germain M, Mailloux P, et al. Resistance training improves strength and functional measures in patients with end-stage renal disease. *Am J Kidney Dis* 2002;40(2):355-364.

16. Johansen KL. Exercise and chronic kidney disease: Current recommendations. *Sports Med (Auckland, NZ)* 2005;35(6):485-499.

17. Johansen KL. Exercise and dialysis. *Hemodial Int* 2008;12(3):290-300.

18. Johansen KL. Physical functioning and exercise capacity in patients on dialysis. *Adv Renal Replace Ther* 1999;6(2):141-148.

19. Kempeneers G, Noakes TD, van Zyl-Smit R, et al. Skeletal muscle limits the exercise tolerance of renal transplant recipients: Effects of a graded exercise training program. *Am J Kidney Dis* 1990;16(1):57-65.

20. Koh KP, Fassett RG, Sharman JE, Coombes JS, Williams AD. Intradialytic versus home-based exercise training in hemodialysis patients: A randomised controlled trial. *BMC Nephrol* 2009; 10:2.

21. Koufaki P, Kouidi E. Current best evidence recommendations on measurement and interpretation of physical function in patients with chronic kidney disease. *Sports Med* 2010;40(12):1055-1074.

22. Leung R, Germain M, Manos T, Headley S. Effect of intradialytic exercise on urea kinetics. *Clin Exerc Physiol* 2001;3(3):144-153.

23. Levey AS, Tangri N, Stevens LA. Classification of chronic kidney disease: A step forward. *Ann Intern Med* 2011;154(1):65-67.

24. Lo CY, Li L, Lo WK, et al. Benefits of exercise training in patients on continuous ambulatory peritoneal dialysis. *Am J Kidney Dis* 1998;32(6):1011-1018.

25. Mercer TH, Naish PF, Gleeson NP, Wilcock JE, Crawford C. Development of a walking test for the assessment of functional capacity in non-anaemic maintenance dialysis patients. *Nephrol Dial Transplant* 1998;13(8):2023-2026.

26. Miller TD, Squires RW, Gau GT, Ilstrup DM, Frohnert PP, Sterioff S. Graded exercise testing and training after renal transplantation: A preliminary study. *Mayo Clin Proc* 1987;62(Sept):773-777.

27. Moore GE, Painter PL, Brinker KR, Stray-Gundersen J, Mitchell JH. Cardiovascular response to submaximal stationary cycling during hemodialysis. *Am J Kidney Dis* 1998;31(4):631-637.

28. National Institutes of Health, National Institute of Diabetes and Digestive and Kidney Diseases. *U.S. Renal Data System, USRDS 2010 Annual Data Report: Atlas of Chronic Kidney Disease and End-Stage Renal Disease in the United States.* Bethesda, MD: National Institutes of Health, National Institute of Diabetes and Digestive and Kidney Diseases, 2010.

29. Nissenson AR. Dialysis therapy in the elderly patient. *Kidney Int Suppl* 1993;40:S51-S7.

30. Oberly E. *Renal Rehabilitation: Bridging the Barriers.* Madison, WI: Medical Education Institute, 1994.

31. Padilla J, Krasnoff J, Da Silva M, et al. Physical functioning in patients with chronic kidney disease. *J Nephrol* 2008;21(4):550-559.

32. Painter P. Determinants of exercise capacity in CKD patients treated with hemodialysis. *Adv Chronic Kidney Dis* 2009;16(6):437-448.

33. Painter P. Exercise in chronic disease: Physiological research needed. *Exerc Sport Sci Rev* 2008;36(2):83-90.

34. Painter P. Implementing exercise: What do we know? Where do we go? *Adv Chronic Kidney Dis* 2009;16(6):536-544.

35. Painter P, Carlson L, Carey S, Paul SM, Myll J. Physical functioning and health-related quality of life changes with exercise training in hemodialysis patients. *Am J Kidney Dis* 2000;35(3):482-492.

36. Painter P, Hanson P, Messer-Rehak D, Zimmerman SW, Glass NR. Exercise tolerance changes following renal transplantation. *Am J Kidney Dis* 1987;10(6):452-456.

37. Painter P, Krasnoff JB, Kuskowski M, Frassetto L, Johansen KL. Effects of modality change and transplant on peak oxygen uptake in patients with kidney failure. *Am J Kidney Dis* 2011;57(1):113-122.

38. Painter PL, Luetkemeier MU, Moore GE, et al. Health-related fitness and quality of life in organ transplant recipients. *Transplantation* 1997;64(12):1795-1800.

39. Painter P, Messer-Rehak D, Hanson P, Zimmerman SW, Glass NR. Exercise capacity in hemodialysis, CAPD, and renal transplant patients. *Nephron* 1986;42(1):47-51.

40. Painter PL, Stewart AL, Carey S. Physical functioning: Definitions, measurement, and expectations. *Adv Renal Replace Ther* 1999;6(2):110-123.

41. Zabetakis PM, Gleim GW, Pasternack FL, Saraniti A, Nicholas JA, Michelis MF. Long-duration submaximal exercise conditioning in hemodialysis patients. *Clin Nephrol* 1982;18(1):17-22.

42. Zakliczynski M, Spiechowicz U, Krynicka A, et al. Fluctuations of exercise capacity in patients after kidney transplantation. *Transplant Proc* 2009;41(1):184-187.

43. Zelle DM, Corpeleijn E, Stolk RP, et al. Low physical activity and risk of cardiovascular and all-cause mortality in renal transplant recipients. *Clin J Am Soc Nephrol* 2011;6:898-905.

CHAPTER 12

1. Ades PA, Savage PD, Toth MJ, et al. High-calorie-expenditure exercise: A new approach to cardiac rehabilitation for overweight coronary patients. *Circulation* 2009;119:2671-2678.

2. American Association of Cardiovascular and Pulmonary Rehabilitation. *Guidelines for Cardiac Rehabilitation and Secondary Prevention Programs.* 4th ed. Champaign, IL: Human Kinetics, 2004.

3. American College of Sports Medicine. *ACSM's Guidelines for Exercise Testing and Prescription.* 9th ed. Philadelphia: Lippincott Williams & Wilkins, 2013.

4. American Heart Association. *Heart Disease and Stroke Statistics—2004 Update.* Dallas: American Heart Association, 2003.

5. Anderson JL, Adams CD, Antman EM, et al. 2011 ACCF/AHA focused update incorporated into the ACC/AHA 2007 guidelines for the management of patients with unstable angina/non-ST-elevation myocardial infarction: A report of the American College of Cardiology Foundation/American Heart Association task force on practice guidelines. *Circulation* 2011;123:e426-e579.

6. Arnett EN, Isner JM, Redwood DR. Coronary artery narrowing in coronary heart disease: Comparison of cineangiography and necropsy findings. *Ann Intern Med* 1979;91:350-356.

7. Balady GJ, Arena R, Sietsema K, et al. Clinician's guide to cardiopulmonary exercise testing in adults: A scientific statement from the American Heart Association. *Circulation* 2010;122:191-225.

8. Benz Scott LA, Ben-Or K, Allen JK. Why are women missing from outpatient cardiac rehabilitation programs? A review of multilevel factors affecting referral, enrollment, and completion. *J Women's Health* 2002;11:773-791.

9. Braunwald E, Sobel BE. Coronary blood flow and myocardial ischemia. In: Braunwald E, ed. *Textbook of Cardiovascular Medicine.* 3rd ed. Philadelphia: Saunders, 1988.

10. Canto JG, Goldberg RJ, Hand MM, et al. Symptom presentation of women with acute coronary syndromes. *Arch Intern Med* 2007;167:2405-2413.

11. Caplice NM, Bunch TJ, Stalboerger PG, Wang S, Simper D, Miller DV, Russell SJ, Litzow MR, Edwards WD. Smooth muscle cells in human coronary atherosclerosis can originate from cells administered at marrow transplant. *Proc Natl Acad Sci* 2003;100:4754-4759.

12. Chesebro JH, Zoldelyi P, Fuster V. Plaque disruption and thrombosis in unstable angina pectoris. *Am J Cardiol* 1991;68:9C-15C.

13. Corti R, Fuster V, Badimon JJ. Pathogenetic concepts of acute coronary syndromes. *J Am Coll Cardiol* 2003;41:7S-14S.

14. Cowers AW, Reidy MA, Clowers MM. Kinetics of cellular proliferation after arterial injury. I. Smooth muscle growth in the absence of endothelium. *Lab Invest* 1983;49:327-333.

15. Ellestad MH, Selvester RHS, Mishkin FS, James FW. *Stress Testing: Principles and Practice*. 5th ed. New York: Oxford University Press, 2003.

16. Evenson KR, Rosamond WD, Luepker RV. Predictors of outpatient cardiac rehabilitation utilization: The Minnesota Heart Survey Registry. *J Cardiopulm Rehabil* 1998;18:192-198.

17. Faggiotto A, Ross R, Harker L. Studies of hypercholesterolemia in the nonhuman primate. I. Changes that lead to fatty streak formation. *Arteriosclerosis* 1984;4:323-340.

18. Ferrari R. Metabolic disturbances during myocardial ischemia and reperfusion. *Am J Cardiol* 1995;76:17B-24B.

19. Franklin BA. Myocardial infarction. In: Durstine JL, Moore GE, eds. *ACSM's Exercise Management for Persons with Chronic Diseases and Disabilities* (pp. 24-31). 2nd ed. Champaign, IL: Human Kinetics, 2003.

20. Franklin BA, Bonzheim K, Gordon S, et al. Safety of medically supervised outpatient cardiac rehabilitation exercise therapy: A 16-year follow-up. *Chest* 1998;114:902-906.

21. Frink RJ, Ostrach LH, Rooney PA. Coronary thrombosis, ulcerated plaques and platelet/fibrin microemboli in patients dying with acute coronary disease. A large autopsy study. *J Invasive Cardiol* 1990;2:199-210.

22. Fuster V. Elucidation of the role of plaque instability and rupture in acute coronary events. *Am J Cardiol* 1995;76:24C-33C.

23. Fuster V, Chesebro JH. Atherogenesis: Pathogenesis, initiation, progression, acute coronary syndromes, and regression. In: Giuliani ER, Gersh BJ, McGoon MD, Hayes DL, Schaff HV, eds. *Mayo Clinic Practice of Cardiology*. 3rd ed. St Louis: Mosby, 1996.

24. Fuster V, Gotto AM, Libby P, Loscalzo J, McGill HC. 27th Bethesda conference: Matching the intensity of risk factor management with the hazard for coronary disease events. *J Am Coll Cardiol* 1996;27:964-976.

25. Garratt KN, Morgan JP. Coronary circulation. In: Giuliani ER, Fuster V, Gersh BJ, McGoon MD, McGoon DC, eds. *Cardiology: Fundamentals and Practice*. 2nd ed. St. Louis: Mosby Year Book, 1991.

26. Garratt KN, Morgan JP. Pathophysiology of myocardial ischemia and reperfusion. In: Giuliani ER, Fuster V, Gersh BJ, McGoon MD, McGoon DC, eds. *Cardiology: Fundamentals and Practice*. 2nd ed. St. Louis: Mosby Year Book, 1991.

27. Gersh BJ, Clements IP. Acute myocardial infarction: Diagnosis and prognosis. In: Giuliani ER, Gersh BJ, McGoon MD, Hayes DL, Schaff HV, eds. *Mayo Clinic Practice of Cardiology*. 3rd ed. St. Louis: Mosby, 1996.

28. Gibbons RJ, Balady GJ, Beasley JW, et al. ACC/AHA guidelines for exercise testing: A report of the American College of Cardiology/American Heart Association task force on practice guidelines (Committee on Exercise Testing). *J Am Coll Cardiol* 1997;30:260-315.

29. Gielen S, Schuler G, Adams V. Cardiovascular effects of exercise training: Molecular mechanisms. *Circulation* 2010;122:1221-1238.

30. Goel K, Lennon RJ, Tilbury RT, Squires RW, Thomas RJ. Impact of cardiac rehabilitation on mortality and cardiovascular events after percutaneous coronary intervention in the community. *Circulation* 2011;123:2344-2352.

31. Goel K, Shen J, Wolter AD, et al. Prevalence of musculoskeletal and balance disorders in patients enrolled in phase II cardiac rehabilitation. *J Cardiopulm Rehabil Prev* 2010;30:235-239.

32. Gumina RJ, Wright RS, Murphy JG. Right ventricular infarction. In: Murphy JG, Lloyd MA, eds. *Mayo Clinic Cardiology: Concise Textbook* (pp. 807-812). 3rd ed. Rochester: Informa Healthcare, 2007.

33. Guyton AC. *Textbook of Medical Physiology*. 7th ed. Philadelphia: Saunders, 1986.

34. Hammill BG, Curtis LH, Schulman LA, et al. Relationship between cardiac rehabilitation and long-term risks of death and myocardial infarction among elderly Medicare beneficiaries. *Circulation* 2010;121:63-70.

35. Helgerud J, Karlsen T, Kim WY, et al. Interval and strength training in CAD patients. *Int J Sports Med* 2011;32:54-59.

36. Hilliard AA, Kopecky SL. Acute coronary syndromes. In: Murphy JG, Lloyd MA, eds. *Mayo Clinic Cardiology: Concise Textbook* (pp. 781-793). 3rd ed. Rochester: Informa Healthcare, 2007.

37. Jolliffe JA, Rees K, Taylor RS, et al. Exercise-based cardiac rehabilitation for coronary heart disease. *Cochrane Database Syst Rev* 2001;CD001800.

38. Kannel WB. Cardiovascular disease: A multifactorial problem (insights from the Framingham study). In: Pollock ML, Schmidt DH, eds. *Heart Disease and Rehabilitation*. Boston: Houghton Mifflin, 1979.

39. Krumholz HM, Currie PM, Riegel B, et al. A taxonomy for disease management: A scientific statement from the American Heart Association disease management taxonomy group. *Circulation* 2006;114:1432-1445.

40. Kullo IJ, Gau GT, Tajik AJ. Novel risk factors for atherosclerosis. *Mayo Clin Proc* 2000;75:369-380.

41. Lavie CJ, Thomas RJ, Squires RW, et al. Exercise training and cardiac rehabilitation in primary and secondary prevention of coronary heart disease. *Mayo Clin Proc* 2009;84:373-383.

42. Lee LV, Foody JM. Women and heart disease. *Cardiol Clin* 2011;29:35-45.

43. Libby P. Molecular bases of the acute coronary syndromes. *Circulation* 1995;91:2844-2850.

44. Lie JT. Pathology of coronary artery disease. In: Giuliani ER, Fuster V, Gersh BJ, McGoon MD, McGoon DC, eds. *Cardiology: Fundamentals and Practice*. 2nd ed. Chicago: Mosby Year Book, 1991.

45. Linke A, Erbs S, Hambrecht R. Exercise and the coronary circulation—alterations and adaptations in coronary artery disease. *Prog Cardiovasc Dis* 2006;48:270-284.

46. Marmot MG. Epidemiologic basis for the prevention of coronary heart disease. *Bull WHO* 1979;57:331-347.

47. Maseri A. Myocardial ischemia in man: Current concepts, changing views and future investigation. *Can J Cardiol* 1986;Suppl. A:255A-259A.

48. Mayer EL, Jacobsen DW, Robinson K. Homocysteine and coronary atherosclerosis. *J Am Coll Cardiol* 1996;27:517-527.

49. McGoon MD, Fuster V. Coronary artery spasm and vasotonicity. In: Giuliani ER, Fuster V, Gersh BJ, McGoon MD, McGoon DC, eds. *Cardiology: Fundamentals and Practice*. 2nd ed. St. Louis: Mosby Year Book, 1991.

50. Mittleman MA, Maclure M, Sherwood JB, et al. Triggering of acute myocardial infarction onset by episodes of anger. *Circulation* 1995;92:1720-1725.

51. Motivational Interviewing. www.motivationalinterview.org/ [accessed July 5, 2011].

52. Muhlestein JB, Hammond EH, Carlquist JF, Radicke E, Thomson MJ, Karagounis LA, Woods ML, Anderson JJ. Increased incidence of chlamydia species within the coronary arteries of patients with symptomatic atherosclerosis versus other forms of cardiovascular disease. *J Am Coll Cardiol* 1996;27:1555-1561.

53. Murphy JG, Bresnahan JE, Lloyd MA, Reeder GS. Complications of acute myocardial infarction. In: Murphy JG, Lloyd MA, eds. *Mayo Clinic Cardiology: Concise Textbook* (pp. 827-842). 3rd ed. Rochester: Informa Healthcare, 2007.

54. Nesto RW, Kowalchuk GJ. The ischemic cascade: Temporal sequence of hemodynamic, electrocardiographic and symptomatic expression of ischemia. *Am J Cardiol* 1987;57:23C-30C.

55. O'Connor CM, Whellan DJ, Lee KL, et al. Efficacy and safety of exercise training in patients with chronic heart failure: HF-ACTION randomized controlled trial. *JAMA* 2009;301:1439-1450.

56. Otvos JD, Mora S, Shalaurova I, et al. Clinical implications of discordance between low-density lipoprotein cholesterol and particle number. *J Clin Lipid* 2011;5:105-113.

57. Owen N, Sparling PB, Healy GN, et al. Sedentary behavior: Emerging evidence for a new health risk. *Mayo Clin Proc* 2010;85:1138-1141.

58. Pasternak RC, Braunwald E, Sobel BE. Acute myocardial infarction. In: Braunwald E, ed. *Textbook of Cardiovascular Medicine.* 3rd ed. Philadelphia: Saunders, 1988.

59. Rahimtoola SH. From coronary artery disease to heart failure: Role of the hibernating myocardium. *Am J Cardiol* 1995;75:16E-22E.

60. Ridker PM. Clinical application of C-reactive protein for cardiovascular disease detection and prevention. *Circulation* 2003;107:363-369.

61. Roger VL, Go AS, Lloyd-Jones DM, et al. Heart disease and stroke statistics—2011 update: A report from the American Heart Association. *Circulation* 2011;123:e18-e209.

62. Ross R. Atherosclerosis—an inflammatory disease. *N Engl J Med* 1999;340:115-126.

63. Ross R. The pathogenesis of atherosclerosis. In: Braunwald E, ed. *Heart Disease: A Textbook of Cardiovascular Medicine.* 5th ed. Philadelphia: Saunders, 1997.

64. Ross R. The pathogenesis of atherosclerosis—an update. *N Engl J Med* 1986;314:488-500.

65. Ross R, Glomset J. Atherosclerosis and the smooth muscle cell. *Science* 1973;180:1332-1339.

66. Shah PK, Forrester JJ. Pathophysiology of acute coronary syndromes. *Am J Cardiol* 1991;68:16C-23C.

67. Shapiro BP, Jaffe AS. Cardiac biomarkers. In: Murphy JG, Lloyd MA, eds. *Mayo Clinic Cardiology: Concise Textbook* (pp. 773-779). 3rd ed. Rochester: Informa Healthcare, 2007.

68. Shub C. Angina pectoris and coronary heart disease. In: Giuliani ER, Fuster V, Gersh BJ, McGoon MD, McGoon DC, eds. *Cardiology: Fundamentals and Practice.* 2nd ed. St. Louis: Mosby Year Book, 1991.

69. Smith SC, Blair SN, Bonow RO, et al. AHA/ACC guidelines for preventing heart attack and death in patients with atherosclerotic cardiovascular disease: 2001 Update: A statement for healthcare professionals from the American Heart Association and the American College of Cardiology. *Circulation* 2001;104:1577-1579.

70. Squires RW. Are cardiac rehabilitation programs underutilized by patients with coronary heart disease? *Nat Clin Pract Cardiovasc Med* 2008;5:192-193.

71. Squires RW. *Exercise Prescription for the High-Risk Cardiac Patient* (pp. 1-20). Champaign, IL: Human Kinetics, 1998.

72. Squires RW. Mechanisms by which exercise training may improve the clinical status of cardiac patients. In: Pollock ML, Schmidt DH, eds. *Heart Disease and Rehabilitation* (pp. 147-160). 3rd ed. Champaign, IL: Human Kinetics, 1995.

73. Squires RW, Gau GT. Cardiac rehabilitation and cardiovascular health enhancement. In: Brandenburg RO, Fuster V, Giuliani ER, McGoon DC, eds. *Cardiology: Fundamentals and Practice* (pp. 1944-1960). Chicago: Yearbook Medical, 1987.

74. Squires RW, Montero-Gomez A, Allison TG, Thomas RJ. Long-term disease management of patients with coronary disease by cardiac rehabilitation program staff. *J Cardiopulm Rehabil Prev* 2008;28:180-186.

75. Stamatakis E, Hamer M, Dunstan DW. Screen-based entertainment time, all-cause mortality, and cardiovascular events. *J Am Coll Cardiol* 2011;57:292-299.

76. Strong JP, McGill HC Jr. The natural history of coronary atherosclerosis. *Am J Pathol* 1962;40:37-49.

77. Suaya JA, Shepard DS, Normand SL, Ades PA, Prottas J, Stason WB. Use of cardiac rehabilitation by Medicare beneficiaries after myocardial infarction or coronary bypass surgery. *Circulation* 2007;116:1653-1662.

78. Thomas RJ, King M, Lui K, et al. AACVPR/ACC/AHA 2007 performance measures on cardiac rehabilitation for referral to and delivery of cardiac rehabilitation/secondary prevention services. *Circulation* 2007;116:1611-1642.

79. Thomas RJ, Miller NH, Lamendola C, et al. National survey on gender differences in cardiac rehabilitation programs. Patient characteristics and enrollment patterns. *J Cardiopulm Rehabil* 1996;16:402-412.

80. Thygesen K, Alpert JS, Jaffe AS, et al; the Writing Group on behalf of the Joint ESC/ACCF/AHA/WHF Task Force for the Universal Definition of Myocardial Infarction. Third universal definition of myocardial infarction. Circulation 2012; 126:2020-2035.

81. Ting HH. Reperfusion strategy for ST-elevation myocardial infarction: Fibrinolysis versus percutaneous coronary intervention. In: Murphy JG, Lloyd MA, eds. *Mayo Clinic Cardiology: Concise Textbook* (pp. 843-853). 3rd ed. Rochester: Informa Healthcare, 2007.

82. Van Furth R. Current view on the mononuclear phagocyte system. *Immunobiology* 1982;161:178-185.

83. Wang TY, Nallamothu BK, Krumholz HM, et al. Association of door-in to door-out time with reperfusion delays and outcomes among patients transferred for primary percutaneous coronary intervention. *JAMA* 2011;305:2540-2547.

84. Wisloff U, Stoylen A, Loennechen JP, et al. Superior cardiovascular effect of aerobic interval training versus moderate continuous training in heart failure patients: A randomized study. *Circulation* 2007;115:3086-3094.

85. Witt BJ, Jacobsen SJ, Weston SA, et al. Cardiac rehabilitation after myocardial infarction in the community. *J Am Coll Cardiol* 2004;44:988-996.

86. Wright RS, Santrach PJ, Kopecky SL. Diagnosis of acute myocardial infarction. In: Murphy JG, ed. *Mayo Clinic Cardiology Review.* 2nd ed. Philadelphia: Lippincott, 2000.

87. Wright RS, Syed IS, Murphy JG. Adjunctive therapy in acute myocardial infarction. In: Murphy JG, Lloyd MA, eds. *Mayo Clinic Cardiology: Concise Textbook* (pp. 813-825). 3rd ed. Rochester: Informa Healthcare, 2007.

CHAPTER 13

1. 2009 Focused Updates: ACC/AHA guidelines for the management of patients with ST-elevation myocardial infarction (updating the 2004 Guideline and 2007 Focused Update) and ACC/AHA/SCAI guidelines on percutaneous coronary intervention (updating the 2005 Guideline and 2007 Focused Update). *Catheter Cardiovasc Interv* 2009;74:E25-E68.
2. ACC/AHA guidelines for the management of patients with acute myocardial infarction. A report of the American College of Cardiology/American Heart Association task force on practice guidelines. Writing committee to revise the 1999 guidelines for the management of patients with acute myocardial infarction. *J Am Coll Cardiol* 2004;44:671-719.
3. ACC/AHA 2004 guideline update for coronary artery bypass graft surgery: A report of the American College of Cardiology/American Heart Association Task Force on Practice Guidelines (Committee to Update the 1999 Guidelines for Coronary Artery Bypass Graft Surgery) Developed in Collaboration With the American Association for Thoracic Surgery and the Society of Thoracic Surgeons. *Circulation* 2004;110:e340-e437.
4. ACC/AHA/SCAI 2005 guidelines update for percutaneous coronary intervention. A report of the American College of Cardiology/American Heart Association task force on practice guidelines (ACC/AHA/SCAI writing committee to update the 2001 guidelines for percutaneous coronary intervention). *Circulation* 2006;113:156-175.
5. ACCF/ASE/ACEP/AHA/ASNC/SCAI/SCCT/SCMR 2008 appropriateness criteria for stress echocardiography. *Circulation* 2008;117:1478-1497.
6. ACCF/ASNC/ACR/AHA/ASE/SCCT/SCMR/SNM 2009 appropriate use criteria for cardiac radionuclide imaging. *Circulation* 2009;119:e561-e587.
7. Adams J, Cline MJ, Hubbard M, McCullough T, Hartman J. A new paradigm for post-cardiac event resistance exercise guidelines. *Am J Cardiol* 2006;97:281-286.
8. AHA scientific statement. Cardiac rehabilitation and secondary prevention of coronary heart disease. *Circulation* 2005;111:369-376.
9. American Association of Cardiovascular and Pulmonary Rehabilitation. *Guidelines for Cardiac Rehabilitation and Secondary Prevention Programs.* 4th ed. Champaign, IL: Human Kinetics, 2004.
10. American College of Sports Medicine. *ACSM's Exercise Management for Persons with Chronic Diseases and Disabilities.* 3rd ed. Champaign, IL: Human Kinetics, 2009.
11. American College of Sports Medicine. *ACSM's Guidelines for Exercise Testing and Prescription.* 9th ed. Lippincott Williams & Wilkins, 2013.
12. American College of Sports Medicine. *ACSM's Resource Manual for Guidelines for Exercise Testing and Prescription.* 5th ed. Lippincott Williams & Wilkins, 2006.
13. Arena R, Myers J, Williams MA, et al. Assessment of functional capacity in clinical and research settings. A scientific statement from the American Heart Association Committee on Exercise, Rehabilitation, and Prevention of the Council on Clinical Cardiology and the Council on Cardiovascular Nursing. *Circulation* 2007;116:329-343.
14. Aronson D. Revascularization for coronary artery disease in diabetes mellitus: Angioplasty, stents and coronary artery bypass grafting. *Rev Endocr Metab Disord* 2010;11:75-86.
15. Carlier M, Meier B, Finci L, Karpuz H, Nukta E, Righett A. Early stress tests after successful coronary angioplasty. *Cardiology* 1993;83:339-344.
16. Clinician's guide to cardiopulmonary exercise testing in adults: A scientific statement from the American Heart Association. *Circulation* 2010;122:191-225.
17. Core components of cardiac rehabilitation/secondary prevention programs: 2007 update: A scientific statement from the American Heart Association Exercise, Cardiac Rehabilitation, and Prevention Committee, the Council on Clinical Cardiology; the Councils on Cardiovascular Nursing, Epidemiology and Prevention, and Nutrition, Physical Activity, and Metabolism; and the American Association of Cardiovascular and Pulmonary Rehabilitation. *Circulation* 2007;115:2675-2682.
18. Dash H. Delayed coronary occlusion after successful percutaneous transluminal coronary angioplasty: Association with exercise testing. *Am J Cardiol* 1982;52:1143-1144.
19. Dendale P, Berger J, Hansen D, Vaes J, Benit E, Weymans W. Cardiac rehabilitation reduces the rate of major cardiac events after percutaneous coronary intervention. *Eur J Cardiovasc Nurs* 2005;4:113-116.
20. Di Valentino M, et al. Prognostic value of cycle exercise testing prior to and after outpatient cardiac rehabilitation. *Int J Cardiol* 2010;140:34-41.
21. Dolansk MA, Moore SM. Effects of cardiac rehabilitation on the recovery outcomes of older adults after coronary artery bypass surgery. *J Cardiopulm Rehabil* 2004;24:236-244.
22. Exercise for patients with coronary artery disease. American College of Sports Medicine position stand. *Med Sci Sports Exerc* 1994;26(3):i-v.
23. Faris J, Stotts N. The effect of percutaneous transluminal coronary angioplasty on quality of life. *Prog Cardiovasc Nurs* 1990;5:132-140.
24. Foley JB, Chisholm RJ, Common AA, Langer A, Armstrong PW. Aggressive clinical pattern of angina at restenosis following coronary angioplasty in unstable angina. *Am Heart J* 1992;124:1174-1180.
25. Franklin, BA. New insights in preventive cardiology and cardiac rehabilitation. *Curr Opin Cardiol* 2008;23:477-486.
26. Franklin BA, Berra K. The case for cardiac rehabilitation after coronary revascularization: Achieving realistic outcome assessments. *J Cardiopulm Rehabil* 2002;22:418-420.
27. Frobert O, et al. Differences in restenosis rate with different drug-eluting stents in patients with and without diabetes mellitus. *J Am Coll Cardiol* 2009;53:1660-1667.
28. Froelicher VF, Myers JN. *Manual of Exercise Testing.* 3rd ed. Philadelphia: Mosby-Elsevier, 2007.
29. Furze G, et al. "Prehabilitation" prior to CABG surgery improves physical functioning and depression. *Int J Cardiol* 132;2009:51-58.
30. Gardner JK, McConnell TR, Klinger TA, Herman CP, Hauck CA, Laubach CA Jr. Quality of life and self-efficacy. Gender and diagnosis considerations for management during cardiac rehabilitation. *J Cardiopulm Rehabil* 2003;23:299-306.
31. Gaw B. Motivation to change lifestyle following percutaneous transluminal coronary angioplasty. *Dimens Crit Care Nurs* 1992;11:68-74.
32. Gaw-Ens B, Laing GP. Risk factor reduction behaviors in coronary angioplasty and myocardial infarction patients. *Can J Cardiovasc Nurs* 1994;5:4-12.

33. Gillis C. The family dimension of cardiovascular care. *Can J Cardiovasc Nurs* 1991;2:3-7.

34. Goodman JM, Pallandi DV, Reading JR, Plyley ML, Liu PP, Kavanagh T. Central and peripheral adaptations after 12 weeks of exercise training in postcoronary artery bypass surgery patients. *J Cardiopulm Rehabil* 1999;19:144-150.

35. Greason, KL. Coronary artery bypass graft surgery (CABG) for patients with diabetes and multivessel coronary artery disease: Identifying patients who would benefit with CABG and understanding the potential mechanisms involved. *Coron Artery Dis* 2010 Nov;21(7):402-406.

36. Hambrecht R, Walther C, Mobius-Winkler S, Geilen S, Linke A, Conradi K, Erbs S, Kluge R, Kendziorra K, Sabri O, Sick P, Schuler G. Percutaneous coronary angioplasty compared with exercise training in patients with stable coronary artery disease. A randomized trial. *Circulation* 2004;109:1371-1378.

37. Hannan EL, Racz MJ, Arani DT, McCallister BD, Walford G, Ryan TJ. A comparison of short- and long-term outcomes for balloon angioplasty and coronary stent placement. *J Am Coll Cardiol* 2000;36:395-403.

38. Jensen K, Banwart L, Venhaus R, Popkess-Vawter S, Perkins S. Advanced rehabilitation nursing care of coronary angioplasty patients using self-efficacy theory. *J Adv Nurs* 1993;18:926-931.

39. Kavanagh T, Mertens DJ, Hamm LF, et al. Peak oxygen intake and cardiac mortality in women referred for cardiac rehabilitation. *J Am Coll Cardiol* 2003;42:2139-2143.

40. Kavanagh T, Mertens DJ, Hamm LF, et al. Prediction of long-term prognosis in 12169 men referred for cardiac rehabilitation. *Circulation* 2002;106:666-671.

41. Kelemen MH. Resistive training safety and assessment guidelines for cardiac and coronary prone patients. *Med Sci Sports Exerc* 1989;21:675-677.

42. Kestin AS, Ellis PA, Barnard MR, Errichetti A, Rosner BA, Michelson AD. Effects of strenuous exercise on platelet activation state and reactivity. *Circulation* 1993;88:1502-1511.

43. King SB III. Interventions in cardiology: What does and does not work. *Am J Cardiol* 2000;86:3H-5H.

44. Kligfield P, McCormick A, Chai A, Jaccobson A, Feuerstadt P, Hao S. Effect of age and gender on heart rate recovery after submaximal exercise during cardiac rehabilitation in patients with angina pectoris, recent acute myocardial infarction, or coronary bypass surgery. *Am J Cardiol* 2003;92:601-603.

45. Kubo H, Hirai H, Machii K. Exercise training and the prevention of restenosis after percutaneous transluminal coronary angioplasty (PTCA). *Ann Acad Med Singapore* 1992;21:42-46.

46. Lan C, Chen S, Chiu S, Hsu C, Lai J, Kuan P. Poor functional recovery may indicate restenosis in patients after coronary angioplasty. *Arch Phys Med Rehabil* 2003;84:1023-1027.

47. Lan C, Chen SY, Hsu CJ, Chiu SF, Lai JS. Improvement of cardiorespiratory function after percutaneous transluminal coronary angioplasty or coronary artery bypass grafting. *Am J Phys Med Rehabil* 2002;81:336-341.

48. Levine GN, Kern MJ, Berger PB, Brown DL, Klein LW, Kereiakes DJ, Sanborn TA, Jacobs AK, for the American Heart Association Diagnostic and Interventional Catheterization Committee and Council on Clinical Cardiology. Management of patients undergoing percutaneous coronary revascularization. *Ann Int Med* 2003;139:123-136.

49. Lissper J, Sudin O, Ohman A, Hofman-Bang C, Fydan L, Nygren A. Long-term effects of lifestyle behavior change in coronary artery disease: Effects on recurrent coronary events after percutaneous coronary intervention. *Health Psychol* 2005;24:41-48.

50. Marzolini S, et al. Aerobic and resistance training in coronary disease: Single versus multiple sets. *Med Sci Sports Exerc* 2008;40(9):1557-1564.

51. McConnell TR, Klinger TA, Gardner JK, Laubach CA Jr, Herman CE, Hauck CA. Cardiac rehabilitation without exercise tests for post-myocardial infarction and post-bypass surgery patients. *J Cardiopulm Rehabil* 1998;18:458-463.

52. Milani RV, Lavie CJ, Cassidy MM. Effects of cardiac rehabilitation and exercise training programs on depression in patients after major coronary events. *Am Heart J* 1996;132:726-732.

53. Moser DK, Dracup KA, Marsden C. Needs of recovering cardiac patients and their spouses: Compared views. *Int J Nurs Stud* 1993;30:105-114.

54. Newton K, Sivarajan E, Clarke J. Patient perceptions of risk factor changes and cardiac rehabilitation outcomes after myocardial infarction. *J Cardiac Rehabil* 1985;5:159-168.

55. Ong ATL, van Domburg RT, Aoki J, Sonnenschein K, Lemos PA, Serruys PW. Sirolimus-eluting stents remain superior to bare-metal stents at two years: Medium-term results from the Rapamycin-Eluting Stent Evaluated at Rotterdam Cardiology Hospital (RESEARCH) Registry. *J Am Coll Cardiol* 2006;47:1356-1360.

56. Oxford JL, Selwyn AP, Ganz P, Popma JJ, Rogers C. The comparative pathobiology of atherosclerosis and restenosis. *Am J Cardiol* 2000;86(Suppl.):6H-11H.

57. Pashkow FJ. Issues in contemporary cardiac rehabilitation. *J Am Coll Cardiol* 1993;21:822-834.

58. Patterson JM. 1989. Illness beliefs as a factor in patient spouse adaptation to treatment for coronary artery disease. *Fam Syst Med* 1989;7:428-443.

59. Pierson LM, Norton HJ, Herbert WG, Pierson ME, Ramp WK, Kiebzak GM, Fedor JM, Cook JW. Recovery of self-reported functional capacity after coronary artery bypass surgery. *Chest* 2003;123:1367-1374.

60. Przybojewski JZ, Welch HFH. Acute coronary thrombus formation after stress testing following percutaneous transluminal coronary angioplasty: A case report. *S Afr Med J* 1985;67:378-382.

61. Resistance exercise in individuals with and without cardiovascular disease: 2007 update: A scientific statement from the American Heart Association Council on Clinical Cardiology and Council on Nutrition, Physical Activity, and Metabolism. *Circulation* 2007;116:572-584.

62. Robertson D, Keller C. Relationships among health beliefs, self-efficacy, and exercise adherence in patients with coronary artery disease. *Heart Lung* 1992;21:56-63.

63. Rollman BL, et al. Telephone-delivered collaborative care for treating post CABG depression: A randomized controlled trial. *JAMA* 2009 Nov 18;302(19):2095-2103.

64. Samuels B, Schumann J, Kiat H, Friedman J, Berman DS. Acute stent thrombosis associated with exercise testing after successful percutaneous transluminal angioplasty. *Am Heart J* 1995;130:1210-1222.

65. Shaw DK, Deutsch DT, Schall PM, Bowling RJ. Physical activity and lean body mass loss following coronary artery bypass graft surgery. *J Sports Med Phys Fitness* 1991;31:67-74.

66. Sigwart U. Drug-eluting stents are safe and effective: Right or wrong? *J Am Coll Cardiol* 2006;47:1361-1362.

67. Soga Y, et al. Safety of early exercise training after elective coronary stenting in patients with stable coronary artery disease. *Eur J Cardiovasc Prev Rehabil* 2010;17(2):230-234.

68. Stewart K, Badenhop D, Brubaker PH, Keteyian SJ, King M. Cardiac rehabilitation following percutaneous revascularization, heart transplant, heart valve surgery, and for chronic heart failure. *Chest* 2003;123:2104-2111.

69. Sumide T, et al. Relationship between exercise tolerance and muscle strength following cardiac rehabilitation: Comparison of patients after cardiac surgery and patients with myocardial infarction. *J Cardiol* 2009;54(2):273-281.

70. Thompson PD, et al. Exercise and physical activity in the prevention and treatment of atherosclerotic cardiovascular disease: A statement from the Council on Clinical Cardiology (Subcommittee on Exercise, Rehabilitation, and Prevention) and the Council on Nutrition, Physical Activity, and Metabolism (Subcommittee on Physical Activity). *Circulation* 2003;107:3109-3116.

71. Valeur N. The prognostic value of pre-discharge exercise testing after myocardial infarction treated with either primary PCI or fibrinolysis: A DANAMI-2 substudy. *Eur Heart J* 2005;26:119-127.

72. Weisz G, Leon MB, Holmes DR Jr, Kereiakes DJ, Clark MR, Cohen BM, Ellis SG, Coleman P, Hill C, Shi C, Cutlip DE, Kuntz RE, Moses JW. Two-year outcomes after sirolimus-eluting stent implantation: Results from the Sirolimus-Eluting Stent in de Novo Native Coronary Lesions (SIRIUS) trial. *J Am Coll Cardiol* 2006;47:1350-1355.

73. Yu C, Lau C, Chau J, McGhee S, Kong S, Cheung BM. A short course of cardiac rehabilitation program is highly cost effective in improving long-term quality of life in patients with recent myocardial infarction or percutaneous coronary intervention. *Arch Phys Med Rehabil* 2004;85:1915-1922.

CHAPTER 14

1. Adams V, Yu J, Mobius-Winkler S, Linke A, Weigl C, Hilbrich L, et al. Increased nitric oxide synthase in skeletal muscle biopsies from patients with chronic heart failure. *Biochem Mol Med* 1997;61:152-160.

2. American Association of Cardiovascular and Pulmonary Rehabilitation. *Guidelines for Cardiac Rehabilitation and Secondary Prevention Programs.* 4th ed. Champaign, IL: Human Kinetics, 2004.

3. Arena R, Myers J, Abella J, Peberdy MA, Bensimhon D, Chase P, Guazzi M. Development of ventilatory classification system in patients with heart failure. *Circulation* 2007;115:2410-2417.

4. Barnard KL, Adams KJ, Swank AM, Kaelin M, Kushnik MR, Denny DM. Combined high intensity and aerobic training in congestive heart failure patients. *J Strength Cond Res* 2000;14:383-388.

5. Beaver WL, Wasserman K, Whipp BJ. A new method for detecting anaerobic threshold by gas exchange. *J Appl Physiol* 1986;60:2020-2027.

6. Belardinelli R, Georgiou D, Cianci G, Purcaro A. Randomized, controlled trial of long-term moderate exercise training in chronic heart failure. *Circulation* 1999;99:1173-1182.

7. Belardinelli R, Georgiou D, Scoccoo V, et al. Low intensity exercise training in patients with heart failure. *J Am Coll Cardiol* 1995;26:975-982.

8. Braith RW, Mills RM, Welsch MA, Keller JW, Pollock ML. Resistance training restores bone mineral density in heart transplant recipients. *J Am Coll Cardiol* 1996;28:1471-1477.

9. Braith RW, Welsch MA, Feigenbaum MS, et al. Neuroendocrine activation in heart failure is modified by endurance exercise training. *J Am Coll Cardiol* 1999;34:1170-1175.

10. Coats AJS, Adamopoulos S, Radaelli A. Controlled trial of physical training in chronic heart failure. *Circulation* 1992;85:2119-2131.

11. Cohen-Solal A, Chabernaud JM, Gourgon R. Comparison of oxygen uptake during bicycle exercise in patients with chronic heart failure and in normal subjects. *J Am Coll Cardiol* 1990;16:80-85.

12. Colucci WS, Ribeiro JP, Rocco MB. Impaired chronotropic response to exercise in patients with congestive heart failure. *Circulation* 1989;80:314-323.

13. Davies EJ, Moxham T, Rees K, et al. Exercise based rehabilitation for heart failure. *Cochrane Database Syst Rev* 2010;CD003331: 13.

14. de Jonge N, Kirkels H, Lahpor JR, et al. Exercise performance in patients with end-stage heart failure after implantation of a left ventricular assist device and after heart transplantation. *J Am Coll Cardiol* 2001;37:1794-1799.

15. Drexler H, Hayoz D, Munzel T, et al. Endothelial function in chronic congestive heart failure. *Am J Cardiol* 1992;69:1596-1601.

16. Dubach P, Myers J, Dziekan G, et al. Effect of exercise training on myocardial remodeling in patients with reduced left ventricular function after myocardial infarction. *Circulation* 1997;95:2060-2067.

17. Dubach P, Myers J, Dziekan G, et al. Effect of high intensity exercise training on central hemodynamic responses to exercise in men with reduced left ventricular function. *J Am Coll Cardiol* 1997;29:1591-1598.

18. Duscha BD, Kraus WE, Keteyian SJ, et al. Capillary density of skeletal muscle. *J Am Coll Cardiol* 1999;33:1956-1963.

19. Duscha BD, Schulz PC, Robbins JL, Forman DE. Implications of chronic heart failure on peripheral vasculature and skeletal muscle before and after exercise training. *Heart Fail Rev* 2008;13:21-27.

20. Ehrman JK, Keteyian SJ, Shepard R, et al. Cardiovascular responses of heart transplant recipients to graded exercise testing. *J Appl Physiol* 1992;73:260-264.

21. Erbs S, Hollriegel R, Linke A, et al. Exercise training in patients with advanced chronic heart failure (NYHA IIIb) promotes restoration of peripheral vasomotor function, induction of endogenous regeneration, and improvement of left ventricular function. *Circ Heart Fail* 2010;3:486-494.

22. Feiereisen P, Delagardelle C, Vaillant M, et al. Is strength training the more efficient training modality in chronic heart failure? *Med Sci Sports Exerc* 2007;39:1910-1917.

23. Flynn KE, Pina IL, Whellan DJ, et al. Effects of exercise training on health status in patients with chronic heart failure. *JAMA* 2009;301:1451-1459.

24. Foss ML, Keteyian SJ. *Fox's Physiological Basis for Exercise and Sport.* 6th ed. Boston: McGraw-Hill, 1998.

25. Gary RA, Sueta CA, Dougherty M, et al. Home-based exercise improves functional performance and quality of life in women with diastolic heart failure. *Heart Lung* 2004;33:210-218.

26. Gielen S, Adams V, Linke A, Erbs S, Mobius-Winkler S, Schubert A, Schuler G, Hambrecht R. Exercise training in chronic heart failure: Correlation between reduced local inflammation and improved oxidative capacity in the skeletal muscle. *Eur J Cardiovasc Prev Rehabil* 2005;12:393-400.

27. Gielen S, Adams V, Mobius-Winkler S, Linke A, S Erbs, Yu J, Kempf W, Schubert A, Schuler G, Hambrecht R. Anti-inflammatory effects of exercise training in the skeletal muscle of patients with chronic heart failure. *J Am Coll Cardiol* 2003;42:861-868.

28. Go AS, Mozaffarian D, Rober VL, Benjamin EJ, Berry JD, Borden WB, Bravata DM, Dai S, Ford ES, Fox CS, Franco S, Fullerton HJ, Gillespie C, Hailpern SM, Heit JA, Howard VJ, Huffman MD, Kissela BM, Kittner SJ, Lackland DT, Lichtman JH, Lisabeth LD, Magid D, Marcus GM, Marelli A, Matchar DB, McGuir DK, Mohler ER, Moy CS, Mussolino ME, Nichol G, Paynter NP, Schreiner PH, Sorlie PD, Stein J, Turan TN, Virani SS, Wong ND, Woo D, Turner MB; on behalf of the American Heart Association Statistics Committee and Stroke Statistics Subcommittee. Heart disease and stroke statistics—2013 update: a report from the American Heart Association. *Circulation,* 2013;127:e6-e245.

29. Griffith BP, Kormos RL, Borovetz HS, Litwak K, Antaki JF, Poirier VL, Butler KC. HeartMate II left ventricular assist system: From concept to first clinical use. *Ann Thorac Surg* 2001;71:S116-120.

30. Hambrecht R, Fiehn E, Weigl C, et al. Regular physical exercise corrects endothelial dysfunction and improves exercise capacity in patients with chronic heart failure. *Circulation* 1998;98:2709-2715.

31. Hambrecht R, Fiehn E, Yu J, et al. Effects of endurance training on mitochondrial ultrastructure and fiber type distribution in skeletal muscle of patients with stable chronic heart failure. *J Am Coll Cardiol* 1997;29:1067-1073.

32. Hambrecht R, Gielen S, Linke A, et al. Effects of exercise training on left ventricular function and peripheral resistance in patients with chronic heart failure. *JAMA* 2000;283:3095-3101.

33. Hambrecht R, Niebauer J, Fiehn E, et al. Physical training in patients with stable chronic heart failure: Effects on cardiorespiratory fitness and ultrastructural abnormalities of leg muscles. *J Am Coll Cardiol* 1995;25:1239-1249.

34. Hammill BG, Curtis LH, Schulman KA, et al. Relationship between cardiac rehabilitation and long-term risks of death and myocardial infarction among elderly Medicare beneficiaries. *Circulation* 2010;121:63-70.

35. Haykowsky MJ, Liang Y, Pechter D, et al. A meta-analysis of the effect of exercise training on left ventricular remodeling in heart failure patients. *J Am Coll Cardiol* 2007;49:2329-2336.

36. Higginbotham MB, Morris KG, Conn EH, et al. Determinations of variable exercise performance among patients with severe left ventricular dysfunction. *Am J Cardiol* 1983;51:52-60.

37. Hunt SA, Abraham WT, Chin MH, Feldman AM, Francis GS, Gantias TG, Jessup M, Konstam MA, Mancini DM, Michl K, Oats JA, Rahko PS, Silver MA, Stevenson LW, Yancy CW. ACC/AHA 2009 focused update incorporated into the ACC/AHA 2005 guidelines for the evaluation and management of heart failure in adults: A report of the American College of Cardiology Foundation/American Heart Association Task Force on Practice Guidelines. *J Am Coll Cardiol* 2009;53:e1-90.

38. Jakovljevic DG, George RS, Nunan D, Donovan G, Bougard RS, Yacoub MH, Birks EJ, Brodie DA. The impact of acute reduction of continuous-flow left ventricular assist device support on cardiac and exercise performance. *Heart* 2010;96:1390-1395.

39. Jaski BE, Kim J, Maly RS, Branch KR, Adamson R, Favrot LK, Smith SC Jr, Dembitsky WP. Effects of exercise during long-term support with a left ventricular assist device: Results of the Experience With Left Ventricular Assist Device With Exercise (EVADE) Pilot Trial. *Circulation* 1997;95:2401-2406.

40. Jaski BE, Lingle RJ, Kim J, Branch KR, Goldsmith R, Johnson MR, Lahpor JR, Icenogle TB, Piña I, Adamson R, Favrot LK, Dembitsky WP. Comparison of functional capacity in patients with end-stage heart failure following implantation of a left ventricular assist device versus heart transplantation: Results of the experience with left ventricular assist device with exercise trial. *J Heart Lung Transplant* 1999;18:1031-1040.

41. Kataoka T, Keteyian SJ, Marks CRC, et al. Exercise training in a patient with congestive heart failure on continuous dobutamine. *Med Sci Sports Exerc* 1994;26:678-681.

42. Keteyian SJ. Exercise training in congestive heart failure: Risks and benefits. *Prog Cardiovasc Dis* 2011;53:419-428.

43. Keteyian SJ, Brawner CA, Schairer JR, et al. Effects of exercise training on chronotropic incompetence in patients with heart failure. *Am Heart J* 1999;138:233-240.

44. Keteyian SJ, Duscha BD, Brawner CA, Green HJ, Marks CRC, Schachat FH, Annex BH, Kraus WE. Differential effects of exercise training in men and women with chronic heart failure. *Am Heart J* 2003;145:912-918.

45. Keteyian SJ, Ehrman J, Fedel F, Rhoads K. Heart rate-perceived exertion relationship during exercise in orthotopic heart transplant patients. *J Cardiopulm Rehabil* 1990;10:287-293.

46. Keteyian SJ, Fleg JL, Brawner CA, Pina IL. Role and benefits of exercise in management of patients with heart failure. *Heart Fail Rev* 2010;15:523-530.

47. Keteyian SJ, Isaac D, Thadani U, et al. Safety of symptom-limited cardiopulmonary exercise testing in patients with chronic heart failure due to left ventricular systolic dysfunction. *Am Heart J* 2009;158:S72-S77.

48. Keteyian SJ, Levine AB, Brawner CA, et al. Exercise training in patients with heart failure. A randomized, controlled trial. *Ann Intern Med* 1996;124:1051-1057.

49. Keteyian SJ, Pina IL, Hibner BA, Fleg JL. Clinical role of exercise training in the management of patients with chronic heart failure. *J Cardiopulm Rehabil Prev* 2010;30:67-76.

50. Khan MN, Pothier CE, Lauer MS. Chronotropic incompetence as a predictor of death among patients with normal electrograms taking beta blockers (metoprolol or atenolol). *Am J Cardiol* 2005;96:1328-1333.

51. Kiilavuori K, Sovijarvi A, Naveri H, et al. Effect of physical training on exercise capacity and gas exchange in patients with chronic heart failure. *Chest* 1996;110:985-991.

52. Kiilavuori K, Toivonen L, Naveri H, Leinonen H. Reversal of autonomic derangements by physical training in chronic heart failure assessed by heart rate variability. *Eur Heart J* 1995;16:490-496.

53. Kinugawa T, Ogino K, Kitamura H, et al. Response of sympathetic nervous system activity to exercise in patients with congestive heart failure. *Eur J Clin Invest* 1991;221:542-546.

54. Kitzman DW, Brubaker PH, Morgan TM, et al. Exercise training in older patients with heart failure and preserved ejection fraction. *Circ Heart Fail* 2010;3:659-667.

55. Kitzman K, Little WC, Brubaker PH, Anderson RT, Hundley WG, Marburger CT, Brosnihan B, Morgan TM, Stewart KP. Pathophysiological characterization of isolated diastolic heart failure in comparison to systolic heart failure. *JAMA* 2002;288:2144-2150.

56. Kobashigawa JA, Leaf DA, Lee N, et al. A controlled trial of exercise rehabilitation after heart transplantation. *N Engl J Med* 1999;340:272-277.

57. Kubo SH, Rector TS, Bank AJ, et al. Endothelial-dependent vasodilatation is attenuated in patients with heart failure. *Circulation* 1991;84:1586-1596.

58. Linke A, Schoene N, Gielen S, Hofer J, Erbs S, Schuler G, Hambrecht R. Endothelial dysfunction in patients with chronic heart failure: Systemic effects of lower-limb exercise training. *J Am Coll Cardiol* 2001;37:392-397.

59. Mancini DM, Walter G, Reichek N, et al. Contribution of skeletal muscle atrophy to exercise intolerance and altered muscle metabolism in heart failure. *Circulation* 1992;85:1364-1373.

60. Minotti JR, Christoph I, Oka R, Weiner MW, Wells L, Massie BM. Impaired skeletal muscle function in patients with congestive heart failure. *J Clin Invest* 1991;88:2077-2082.

61. Myers J, Buchanan N, Walsh D, et al. Comparison of the ramp versus standard exercise protocols. *J Am Coll Cardiol* 1991;17:1334-1342.

62. Myers J, Gullestad L, Vagelos R, et al. Cardiopulmonary exercise testing and prognosis in severe heart failure: 14 ml/kg/min revisited. *Am Heart J* 2000;139:78-84.

63. O'Connor CM, Whellen DJ, Lee KL, et al. Efficacy and safety of exercise training in patients with chronic heart failure. HF-ACTION randomized controlled trial. *JAMA* 2009;301:1439-1450.

64. Page E, Cohen-Solal A, Jondeau G, et al. Comparison of treadmill and bicycle exercise in patients with chronic heart failure. *Chest* 1994;106:1002-1006.

65. Passino C, Severino S, Poletti R, Piepoli MF, Mammini C, Clerico A, Gabutti A, Nassi G, Emdin M. Aerobic training decreases B-type natriuretic peptide expression and adrenergic activation in patients with heart failure. *J Am Coll Cardiol* 2006;47:1835-1839.

66. Robbins M, Francis G, Pashkow FJ, et al. Ventilatory and heart rate responses to exercise. *Circulation* 1999;100:2411-2417.

67. Rogers JG, Aaronson KD, Boyle AJ, Russell SD, Milano CA, Pagani FD, Edwards BS, Park S, John R, Conte JV, Farrar DJ, Slaughter MS. Continuous flow left ventricular assist device improves functional capacity and quality of life of advanced heart failure patients. *J Am Coll Cardiol* 2010;55:1826-1834.

68. Rose EA, Gelijns AC, Moskowitz AJ, Heitjan DF, Stevenson LW, Dembitsky W, Long JW, Ascheim DD, Tierney AR, Levitan RG, Watson JT, Ronan NS, Shapiro PA, Lazar RM, Miller LW, Gupta L, Frazier OH, Desvigne-Nickens P, Oz MC, Poirier VL, Meier P. Long-term use of a left ventricular assist device for end-stage heart failure. *New Engl J Med* 2001;345:1435-1443.

69. Roveda F, Middlekauff HR, Rondon MUPB, Reis SF, Souza M, Nastari L, Barretto ACP, Krieger EM, Negrao CE. The effects of exercise training on sympathetic neural activation in advanced heart failure. *J Am Coll Cardiol* 2003;42:854-860.

70. Smart N, Haluska B, Jeffriess L, et al. Exercise training in systolic and diastolic dysfunction: Effects on cardiac function, functional capacity, and quality of life. *Am Heart J* 2007;153:530-536.

71. Stehlik J, Edwards LB, Kucheryavaya AY, Christie JD, Kirk R, Dobbels F, Rahmel AO, Hertz MI. The registry of the International Society for Heart and Lung Transplantation: Twenty-seventh official adult heart transplant report—2010. *J Heart Lung Transplant* 2010;29:1089-1103.

72. Sullivan MJ, Green HJ, Cobb FR. Skeletal muscle biochemistry and histology in ambulatory patients with long-term heart failure. *Circulation* 1991;81:518-527.

73. Sullivan MJ, Knight DJ, Higginbotham MB, Cobb FR. Relation between central and peripheral hemodynamics during exercise in patients with chronic heart failure. *Circulation* 1989;80:769-781.

74. Toepher M, Meyer K, Maier P, et al. Influence of exercise training and restriction of activity on autonomic balance in patients with severe congestive heart failure. *Clin Sci* 1996;91(Suppl.):116.

75. Volterrani M, Clark AL, Ludman PF, et al. Predictors of exercise capacity in chronic heart failure. *Eur Heart J* 1994;15:801-809.

76. Williams MA, Haskell WL, Ades PA, et al. Resistance exercise in individuals with and without cardiovascular disease: 2007 update. *Circulation* 2007;116:572-584.

77. Wilson JR, Martin JL, Schwartz D, et al. Exercise intolerance in patients with chronic heart failure: Role of impaired nutritive flow to skeletal muscle. *Circulation* 1984;69:1079-1087.

78. Wisloff U, Stoylen A, Loennechen JP, et al. Superior cardiovascular effect of aerobic interval training versus moderate continuous training in heart failure patients: A randomized study. *Circulation* 2007;115:3086-3094.

79. Zhang YY, Wasserman K, Sietsmea KE, et al. O2 uptake kinetics in response to exercise. *Chest* 1993;103:735-741.

CHAPTER 15

1. Abdel-Latif A, Ziada K. The role of vascular ultrasonography in peripheral arterial disease. In: Casserly I, Sachar R, Yadav JS, eds. *Practical Peripheral Vascular Intervention* (pp. 18-35). Philadelphia: Lippincott Williams & Wilkins, 2011.

2. Albers M, Battistella VM, Romiti M, Rodrigues AA, Pereira CA. Meta-analysis of polytetrafluoroethylene bypass grafts to infrapopliteal arteries. *J Vasc Surg* 2003;37(6):1263-1269.

3. Albers M, Romiti M, Brochado-Neto FC, De Luccia N, Pereira CA. Meta-analysis of popliteal-to-distal vein bypass grafts for critical ischemia. *J Vasc Surg* 2006;43(3):498-503.

4. Allison MA, Ho E, Denenberg JO, Langer RD, Newman AB, Fabsitz RR, Criqui MH. Ethnic-specific prevalence of peripheral arterial disease in the United States. *Am J Prev Med* 2007;32(4):328-333.

5. Alpert JS, Larsen OA, Lassen NA. Exercise and intermittent claudication. Blood flow in the calf muscle during walking studied by the xenon-133 clearance method. *Circulation* 1969;39(3):353-359.

6. Altavilla D, Bitto A, Polito F, Marini H, Minutoli L, Di Stefano V, Irrera N, Cattarini G, Squadrito F. Polydeoxyribonucleotide (PDRN): A safe approach to induce therapeutic angiogenesis in peripheral artery occlusive disease and in diabetic foot ulcers. *Cardiovasc Hematol Agents Med Chem* 2009;7(4):313-321.

7. Andrew GM, Oldridge NB, Parker JO, Cunningham DA, Rechnitzer PA, Jones NL, Buck C, Kavanagh T, Shephard RJ, Sutton JR. Reasons for dropout from exercise programs in post-coronary patients. *Med Sci Sports Exerc* 1981;13(3):164-168.

8. Aronow WS. Management of peripheral arterial disease. *Cardiol Rev* 2005;13(2):61-68.

9. Ashworth NL, Chad KE, Harrison EL, Reeder BA, Marshall SC. Home versus center based physical activity programs in older adults. *Cochrane Database Syst Rev* 2005(1):CD004017.

10. Attenberger UI, Haneder S, Morelli JN, Diehl SJ, Schoenberg SO, Michaely HJ. Peripheral arterial occlusive disease: Evaluation of a high spatial and temporal resolution 3-T MR protocol with a low total dose of gadolinium versus conventional angiography. *Radiology* 2010;257(3):879-887.

11. Belch JJ, Bell PR, Creissen D, Dormandy JA, Kester RC, McCollum RD, Mizushima Y, Ruckley CV, Scurr JH, Wolfe JH. Randomized, double-blind, placebo-controlled study evaluating the efficacy and safety of AS-013, a prostaglandin E1 prodrug, in patients with intermittent claudication. *Circulation* 1997;95(9):2298-2302.

12. Bendermacher BL, Willigendael EM, Teijink JA, Prins MH. Medical management of peripheral arterial disease. *J Thromb Haemost* 2005;3(8):1628-1637.

13. Bendermacher BL, Willigendael EM, Teijink JA, Prins MH. Supervised exercise therapy versus non-supervised exercise therapy for intermittent claudication. *Cochrane Database Syst Rev* 2006(2):CD005263.

14. Bonow RO, Mann DL, Zipes DP, Libby P, eds. *Braunwald's Heart Disease: A Textbook of Cardiovascular Medicine*. 9th ed. Philadelphia: Elsevier, 2011.

15. Borg G. *An Introduction to Borg's RPE-Scale*. Ithaca, NY: Mouvement, 1985.

16. Brendle DC, Joseph LJ, Corretti MC, Gardner AW, Katzel LI. Effects of exercise rehabilitation on endothelial reactivity in older patients with peripheral arterial disease. *Am J Cardiol* 2001;87(3):324-329.

17. Brevetti G, Chiariello M, Ferulano G, Policicchio A, Nevola E, Rossini A, Attisano T, Ambrosio G, Siliprandi N, Angelini C. Increases in walking distance in patients with peripheral vascular disease treated with L-carnitine: A double-blind, cross-over study. *Circulation* 1988;77(4):767-773.

18. Brevetti G, Diehm C, Lambert D. European multicenter study on propionyl-L-carnitine in intermittent claudication. *J Am Coll Cardiol* 1999;34(5):1618-1624.

19. Brevetti G, di Lisa F, Perna S, Menabo R, Barbato R, Martone VD, Siliprandi N. Carnitine-related alterations in patients with intermittent claudication: Indication for a focused carnitine therapy. *Circulation* 1996;93(9):1685-1689.

20. Bui BT, Miller S, Mildenberger P, Sam A 2nd, Sheng R. Comparison of contrast-enhanced MR angiography to intraarterial digital subtraction angiography for evaluation of peripheral arterial occlusive disease: Results of a phase III multicenter trial. *J Magn Reson Imaging* 2010;31(6):1402-1410.

21. Casserly IP. Interventional management of critical limb ischemia in renal patients. *Adv Chronic Kidney Dis* 2008;15(4):384-395.

22. Cheetham DR, Burgess L, Ellis M, Williams A, Greenhalgh RM, Davies AH. Does supervised exercise offer adjuvant benefit over exercise advice alone for the treatment of intermittent claudication? A randomised trial. *Eur J Vasc Endovasc Surg* 2004;27(1):17-23.

23. Clement DL, Duprez D. Effect of ketanserin in the treatment of patients with intermittent claudication: Results from 13 placebo-controlled parallel group studies. *J Cardiovasc Pharmacol* 1987;10 Suppl. 3:S89-95.

24. Coffman JD. Intermittent claudication—be conservative. *N Engl J Med* 1991;325(8):577-578.

25. Coffman JD. Pathophysiology of obstructive arterial disease. *Herz* 1988;13(6):343-350.

26. Collins TC, Lunos S, Carlson T, Henderson K, Lightbourne M, Nelson B, Hodges JS. Effects of a home-based walking intervention on mobility and quality of life in people with diabetes and peripheral arterial disease: A randomized controlled trial. *Diabetes Care* 2011;34(10):2174-2179.

27. Criqui MH, Fronek A, Barrett-Connor E, Klauber MR, Gabriel S, Goodman D. The prevalence of peripheral arterial disease in a defined population. *Circulation* 1985;71(3):510-515.

28. Criqui MH, Vargas V, Denenberg JO, Ho E, Allison M, Langer RD, Gamst A, Bundens WP, Fronek A. Ethnicity and peripheral arterial disease: The San Diego population study. *Circulation* 2005;112(17):2703-2707.

29. da Cunha-Filho IT, Pereira DA, de Carvalho AM, Campedeli L, Soares M, de Sousa Freitas J. The reliability of walking tests in people with claudication. *Am J Phys Med Rehabil* 2007;86(7):574-582.

30. Dahllof AG, Bjorntorp P, Holm J, Schersten T. Metabolic activity of skeletal muscle in patients with peripheral arterial insufficiency. *Eur J Clin Invest* 1974;4(1):9-15.

31. Dawson DL, Cutler BS, Hiatt WR, Hobson RW 2nd, Martin JD, Bortey EB, Forbes WP, Strandness DE Jr. A comparison of cilostazol and pentoxifylline for treating intermittent claudication. *Am J Med* 2000;109(7):523-530.

32. De Backer T, Vander Stichele R, De Buyzere M, De Backer G, Van Bortel L. Silence of the limbs pharmacological symptomatic treatment of intermittent claudication. *Curr Vasc Pharmacol* 2010;8(3):383-387.

33. De Backer TL, Vander Stichele RH, Warie HH, Bogaert MG. Oral vasoactive medication in intermittent claudication: Utile or futile? *Eur J Clin Pharmacol* 2000;56(3):199-206.

34. Degischer S, Labs KH, Hochstrasser J, Aschwanden M, Tschoepl M, Jaeger KA. Physical training for intermittent claudication: A comparison of structured rehabilitation versus home-based training. *Vasc Med* 2002;7(2):109-115.

35. de Graaff JC, Ubbink DT, Kools EI, Chamuleau SA, Jacobs MJ. The impact of peripheral and coronary artery disease on health-related quality of life. *Ann Vasc Surg* 2002;16(4):495-500.

36. Delis KT, Nicolaides AN, Wolfe JH, Stansby G. Improving walking ability and ankle brachial pressure indices in symptomatic peripheral vascular disease with intermittent pneumatic foot compression: A prospective controlled study with one-year follow-up. *J Vasc Surg* 2000;31(4):650-661.

37. DeRubertis BG, Faries PL, McKinsey JF, Chaer RA, Pierce M, Karwowski J, Weinberg A, Nowygrod R, Morrissey NJ, Bush HL, Kent KC. Shifting paradigms in the treatment of lower extremity vascular disease: A report of 1000 percutaneous interventions. *Ann Surg* 2007;246(3):415-424.

38. Dettori AG, Pini M, Moratti A, Paolicelli M, Basevi P, Quintavalla R, Manotti C, Di Lecce C. Acenocoumarol and pentoxifylline in intermittent claudication. A controlled clinical study. The APIC study group. *Angiology* 1989;40(4 Pt 1):237-248.

39. de Vries SO, Hunink MG. Results of aortic bifurcation grafts for aortoiliac occlusive disease: A meta-analysis. *J Vasc Surg* 1997;26(4):558-569.

40. D'Hooge D, Lehert P, Clement DL. Naftidrofuryl in quality of life (NIQOL). A Belgian study. *Int Angiol* 2001;20(4):288-294.

41. Diehm C, Balzer K, Bisler H, Bulling B, Camci M, Creutzig A, Gruss JD, Horsch S, Odemar F, Piehler U, Rogatti W, Scheffler P, Spengel F, Treese N, Turowski A, Waldhausen P, Weber B, Weiss T. Efficacy of a new prostaglandin E1 regimen in outpatients with severe intermittent claudication: Results of a multicenter placebo-controlled double-blind trial. *J Vasc Surg* 1997;25(3):537-544.

42. Dobesh PP, Stacy ZA, Persson EL. Pharmacologic therapy for intermittent claudication. *Pharmacotherapy* 2009;29(5):526-553.

43. Dolmatch BL. Imaging peripheral arterial disease: A randomized controlled trial comparing contrast-enhanced MR angiography and multi-detector row CT angiography. *Perspect Vasc Surg Endovasc Ther* 2006;18(2):191-193.

44. Ekroth R, Dahllof AG, Gundevall B, Holm J, Schersten T. Physical training of patients with intermittent claudication: Indications, methods, and results. *Surgery* 1978;84(5):640-643.

45. Ericsson B, Haeger K, Lindell SE. Effect of physical training of intermittent claudication. *Angiology* 1970;21(3):188-192.

46. Ernst EE, Matrai A. Intermittent claudication, exercise, and blood rheology. *Circulation* 1987;76(5):1110-1114.

47. Fabsitz RR, Sidawy AN, Go O, Lee ET, Welty TK, Devereux RB, Howard BV. Prevalence of peripheral arterial disease and associated risk factors in American Indians: The Strong Heart Study. *Am J Epidemiol* 1999;149(4):330-338.

48. Feiring AJ, Wesolowski AA, Lade S. Primary stent-supported angioplasty for treatment of below-knee critical limb ischemia and severe claudication: Early and one-year outcomes. *J Am Coll Cardiol* 2004;44(12):2307-2314.

49. Fowkes FG, Housley E, Cawood EH, Macintyre CC, Ruckley CV, Prescott RJ. Edinburgh artery study: Prevalence of asymptomatic and symptomatic peripheral arterial disease in the general population. *Int J Epidemiol* 1991;20(2):384-392.

50. Gallasch G, Diehm C, Dorfer C, Schmitt T, Stage A, Morl H. [Effect of physical training on blood flow properties in patients with intermittent claudication]. *Klin Wochenschr* 1985;63(12):554-559.

51. Garber CE, Monteiro R, Patterson RB, Braun CM, Lamont LS. A comparison of treadmill and arm-leg ergometry exercise testing for assessing exercise capacity in patients with peripheral arterial disease. *J Cardiopulm Rehabil* 2006;26(5):297-303.

52. Gardner AW, Afaq A. Management of lower extremity peripheral arterial disease. *J Cardiopulm Rehabil Prev* 2008;28(6):349-357.

53. Gardner AW, Katzel LI, Sorkin JD, Killewich LA, Ryan A, Flinn WR, Goldberg AP. Improved functional outcomes following exercise rehabilitation in patients with intermittent claudication. *J Gerontol A Biol Sci Med Sci* 2000;55(10):M570-577.

54. Gardner AW, Montgomery PS, Flinn WR, Katzel LI. The effect of exercise intensity on the response to exercise rehabilitation in patients with intermittent claudication. *J Vasc Surg* 2005;42(4):702-709.

55. Gardner AW, Parker DE, Montgomery PS, Scott KJ, Blevins SM. Efficacy of quantified home-based exercise and supervised exercise in patients with intermittent claudication: A randomized controlled trial. *Circulation* 2011;123(5):491-498.

56. Gardner AW, Poehlman ET. Exercise rehabilitation programs for the treatment of claudication pain. A meta-analysis. *JAMA* 1995;274(12):975-980.

57. Gardner AW, Skinner JS, Bryant CX, Smith LK. Stair climbing elicits a lower cardiovascular demand than walking in claudication patients. *J Cardiopulm Rehabil* 1995;15(2):134-142.

58. Gardner AW, Skinner JS, Cantwell BW, Smith LK. Progressive vs single-stage treadmill tests for evaluation of claudication. *Med Sci Sports Exerc* 1991;23(4):402-408.

59. Gardner AW, Skinner JS, Smith LK. Effects of handrail support on claudication and hemodynamic responses to single-stage and progressive treadmill protocols in peripheral vascular occlusive disease. *Am J Cardiol* 1991;68(1):99-105.

60. Gardner AW, Skinner JS, Vaughan NR, Bryant CX, Smith LK. Comparison of three progressive exercise protocols in peripheral vascular occlusive disease. *Angiology* 1992;43(8):661-671.

61. Garg PK, Liu K, Tian L, Guralnik JM, Ferrucci L, Criqui MH, Tan J, McDermott MM. Physical activity during daily life and functional decline in peripheral arterial disease. *Circulation* 2009;119(2):251-260.

62. Girolami B, Bernardi E, Prins MH, Ten Cate JW, Hettiarachchi R, Prandoni P, Girolami A, Buller HR. Treatment of intermittent claudication with physical training, smoking cessation, pentoxifylline, or nafronyl: A meta-analysis. *Arch Intern Med* 1999;159(4):337-345.

63. Gupta MK, Verma S. Painful legs, aching heart: Reducing cardiovascular risk in patients with peripheral arterial disease with angiotensin-converting enzyme inhibition. *Can J Cardiol* 2005;21(2):194-195.

64. Guralnik JM, Simonsick EM, Ferrucci L, Glynn RJ, Berkman LF, Blazer DG, Scherr PA, Wallace RB. A short physical performance battery assessing lower extremity function: Association with self-reported disability and prediction of mortality and nursing home admission. *J Gerontol* 1994;49(2):M85-94.

65. Guyton JR. Benefit versus risk in statin treatment. *Am J Cardiol* 2006;97(8A):95C-97C.

66. Hankey GJ, Norman PE, Eikelboom JW. Medical treatment of peripheral arterial disease. *JAMA* 2006;295(5):547-553.

67. Hennrikus D, Joseph AM, Lando HA, Duval S, Ukestad L, Kodl M, Hirsch AT. Effectiveness of a smoking cessation program for peripheral artery disease patients: A randomized controlled trial. *J Am Coll Cardiol* 2010;56(25):2105-2112.

68. Heredero AF, Acin F, March JR, Utrilla F. Impact of endovascular surgery on management of critical lower-limb ischemia in a vascular surgery department. *Vasc Endovascular Surg* 2005;39(5):429-435.

69. Hiatt WR. Medical treatment of peripheral arterial disease and claudication. *N Engl J Med* 2001;344(21):1608-1621.

70. Hiatt WR. The US experience with cilostazol in treating intermittent claudication. *Atheroscler Suppl* 2005;6(4):21-31.

71. Hiatt WR, Creager MA, Amato A, Brass EP. Effect of propionyl-L-carnitine on a background of monitored exercise in patients with claudication secondary to peripheral artery disease. *J Cardiopulm Rehabil Prev* 2011;31(2):125-132.

72. Hiatt WR, Hirsch AT, Cooke JP, Olin JW, Brater DC, Creager MA. Randomized trial of AT-1015 for treatment of intermittent claudication. A novel 5-hydroxytryptamine antagonist with no evidence of efficacy. *Vasc Med* 2004;9(1):18-25.

73. Hiatt WR, Hirsch AT, Regensteiner JG, eds. *Peripheral Arterial Disease Handbook*. Boca Raton, FL: CRC Press, 2001.

74. Hiatt WR, Regensteiner JG, Creager MA, Hirsch AT, Cooke JP, Olin JW, Gorbunov GN, Isner J, Lukjanov YV, Tsitsiashvili MS, Zabelskaya TF, Amato A. Propionyl-L-carnitine improves exercise performance and functional status in patients with claudication. *Am J Med* 2001;10(8):616-622.

75. Hiatt WR, Regensteiner JG, Hargarten ME, Wolfel EE, Brass EP. Benefit of exercise conditioning for patients with peripheral arterial disease. *Circulation* 1990;81(2):602-609.

76. Hiatt WR, Regensteiner JG, Wolfel EE, Carry MR, Brass EP. Effect of exercise training on skeletal muscle histology and metabolism in peripheral arterial disease. *J Appl Physiol* 1996;81(2):780-788.

77. Hiatt WR, Wolfel EE, Meier RH, Regensteiner JG. Superiority of treadmill walking exercise versus strength training for patients with peripheral arterial disease. Implications for the mechanism of the training response. *Circulation* 1994;90(4):1866-1874.

78. Higgins D, Santamore WP, Walinsky P, Nemir P Jr. Hemodynamics of human arterial stenoses. *Int J Cardiol* 1985;8(2):177-192.

79. Hirsch AT, Criqui MH, Treat-Jacobson D, Regensteiner JG, Creager MA, Olin JW, Krook SH, Hunninghake DB, Comerota AJ, Walsh ME, McDermott MM, Hiatt WR. Peripheral arterial disease detection, awareness, and treatment in primary care. *JAMA* 2001;286(11):1317-1324.

80. Hirsch AT, Hartman L, Town RJ, Virnig BA. National health care costs of peripheral arterial disease in the Medicare population. *Vasc Med* 2008;13(3):209-215.

81. Hirsch AT, Haskal ZJ, Hertzer NR, Bakal CW, Creager MA, Halperin JL, Hiratzka LF, Murphy WR, Olin JW, Puschett JB, Rosenfield KA, Sacks D, Stanley JC, Taylor LM Jr, White CJ, White J, White RA, Antman EM, Smith SC Jr, Adams CD, Anderson JL, Faxon DP, Fuster V, Gibbons RJ, Hunt SA, Jacobs AK, Nishimura R, Ornato JP, Page RL, Riegel B. ACC/AHA 2005 guidelines for the management of patients with peripheral arterial disease (lower extremity, renal, mesenteric, and abdominal aortic): Executive summary. A collaborative report from the American Association for Vascular Surgery/Society for Vascular Surgery, Society for Cardiovascular Angiography and Interventions, Society for Vascular Medicine and Biology, Society of Interventional Radiology, and the ACC/AHA Task Force on Practice Guidelines (writing committee to develop guidelines for the management of patients with peripheral arterial disease) endorsed by the American Association of Cardiovascular and Pulmonary Rehabilitation; National Heart, Lung, and Blood Institute; Society for Vascular Nursing; Transatlantic Inter-Society Consensus; and Vascular Disease Foundation. *J Am Coll Cardiol* 2006;47(6):1239-1312.

82. Holm J, Dahllof AG, Bjorntorp P, Schersten T. Enzyme studies in muscles of patients with intermittent claudication. Effect of training. *Scand J Clin Lab Invest Suppl* 1973;128:201-205.

83. Igawa T, Tani T, Chijiwa T, Shiragiku T, Shimidzu S, Kawamura K, Kato S, Unemi F, Kimura Y. Potentiation of anti-platelet aggregating activity of cilostazol with vascular endothelial cells. *Thromb Res* 1990;57(4):617-623.

84. Johnson EC, Voyles WF, Atterbom HA, Pathak D, Sutton MF, Greene ER. Effects of exercise training on common femoral artery blood flow in patients with intermittent claudication. *Circulation* 1989;80(5 Pt 2):III59-72.

85. Jonason T, Ringqvist I, Oman-Rydberg A. Home-training of patients with intermittent claudication. *Scand J Rehabil Med* 1981;13(4):137-141.

86. Kakkos SK, Geroulakos G, Nicolaides AN. Improvement of the walking ability in intermittent claudication due to superficial femoral artery occlusion with supervised exercise and pneumatic foot and calf compression: A randomised controlled trial. *Eur J Vasc Endovasc Surg* 2005;30(2):164-175.

87. Kihara H, Koganei H, Hirose K, Yamamoto H, Yoshimoto R. Antithrombotic activity of AT-1015, a potent 5-HT(2A) receptor antagonist, in rat arterial thrombosis model and its effect on bleeding time. *Eur J Pharmacol* 2001;433(2-3):157-162.

88. Klinkert P, Post PN, Breslau PJ, van Bockel JH. Saphenous vein versus PTFE for above-knee femoropopliteal bypass. A review of the literature. *Eur J Vasc Endovasc Surg* 2004;27(4):357-362.

89. Kohda N, Tani T, Nakayama S, Adachi T, Marukawa K, Ito R, Ishida K, Matsumoto Y, Kimura Y. Effect of cilostazol, a phosphodiesterase III inhibitor, on experimental thrombosis in the porcine carotid artery. *Thromb Res* 1999;96(4):261-268.

90. Kruidenier LM, Nicolai SP, Willigendael EM, de Bie RA, Prins MH, Teijink JA. Functional claudication distance: A reliable and valid measurement to assess functional limitation in patients with intermittent claudication. *BMC Cardiovasc Disord* 2009;9:9.

91. Kudo T, Chandra FA, Kwun WH, Haas BT, Ahn SS. Changing pattern of surgical revascularization for critical limb ischemia over 12 years: Endovascular vs. open bypass surgery. *J Vasc Surg* 2006;44(2):304-313.

92. Kuhlmann MT, Klocke R, Nikol S. Therapeutic angiogenesis for peripheral artery disease: Cytokine therapy. *VASA* 2007;36(4):253-260.

93. Lamar Welch VL, Casper M, Greenlund K, Zheng ZJ, Giles W, Rith-Najarian S. Prevalence of lower extremity arterial disease defined by the ankle-brachial index among American Indians: The Inter-Tribal Heart Project. *Ethn Dis* 2002;12(1):S1-63-67.

94. Lampman RM. Exercise prescription for chronically ill patients. *Am Fam Physician* 1997;55(6):2185-2192.

95. Larsen OA, Lassen NA. Effect of daily muscular exercise in patients with intermittent claudication. *Lancet* 1966;2(7473):1093-1096.

96. Le Faucheur A, Abraham P, Jaquinandi V, Bouye P, Saumet JL, Noury-Desvaux B. Measurement of walking distance and speed in patients with peripheral arterial disease: A novel method using a global positioning system. *Circulation* 2008;117(7):897-904.

97. Leng GC, Fowler B, Ernst E. Exercise for intermittent claudication. *Cochrane Database Syst Rev* 2000(2):CD000990.

98. Levin DC, Rao VM, Parker L, Frangos AJ, Sunshine JH. The effect of the introduction of MR and CT angiography on the utilization of catheter angiography for peripheral arterial disease. *J Am Coll Radiol* 2007;4(7):457-460.

99. Lindgarde F, Jelnes R, Bjorkman H, Adielsson G, Kjellstrom T, Palmquist I, Stavenow L. Conservative drug treatment in patients with moderately severe chronic occlusive peripheral arterial disease. Scandinavian study group. *Circulation* 1989;80(6):1549-1556.

100. Lundgren F, Dahllof AG, Lundholm K, Schersten T, Volkmann R. Intermittent claudication—surgical reconstruction or physical training? A prospective randomized trial of treatment efficiency. *Ann Surg* 1989;209(3):346-355.

101. Makin A, Silverman S, Lip GY. Ethnic differences in peripheral vascular disease. *Int J Clin Pract* 2002;56(8):605-608.

102. Mangiafico RA, Fiore CE. Current management of intermittent claudication: The role of pharmacological and nonpharmacological symptom-directed therapies. *Curr Vasc Pharmacol* 2009;7(3):394-413.

103. Mannarino E, Pasqualini L, Innocente S, Scricciolo V, Rignanese A, Ciuffetti G. Physical training and antiplatelet treatment in stage II peripheral arterial occlusive disease: Alone or combined? *Angiology* 1991;42(7):513-521.

104. Mannarino E, Pasqualini L, Menna M, Maragoni G, Orlandi U. Effects of physical training on peripheral vascular disease: A controlled study. *Angiology* 1989;40(1):5-10.

105. Marso SP, Hiatt WR. Peripheral arterial disease in patients with diabetes. *J Am Coll Cardiol* 2006;47(5):921-929.

106. Matsuo H, Shigematsu H. Patient-based outcomes using the Walking Impairment Questionnaire for patients with peripheral arterial occlusive disease treated with Lipo-PGE1. *Circ J* 2010;74(2):365-370.

107. McDermott MM, Ades PA, Dyer A, Guralnik JM, Kibbe M, Criqui MH. Corridor-based functional performance measures correlate better with physical activity during daily life than treadmill measures in persons with peripheral arterial disease. *J Vasc Surg* 2008;48(5):1231-1237.

108. McDermott MM, Ades P, Guralnik JM, Dyer A, Ferrucci L, Liu K, Nelson M, Lloyd-Jones D, Van Horn L, Garside D, Kibbe

M, Domanchuk K, Stein JH, Liao Y, Tao H, Green D, Pearce WH, Schneider JR, McPherson D, Laing ST, McCarthy WJ, Shroff A, Criqui MH. Treadmill exercise and resistance training in patients with peripheral arterial disease with and without intermittent claudication: A randomized controlled trial. *JAMA* 2009;301(2):165-174.

109. McDermott MM, Ferrucci L, Liu K, Guralnik JM, Tian L, Kibbe M, Liao Y, Tao H, Criqui MH. Women with peripheral arterial disease experience faster functional decline than men with peripheral arterial disease. *J Am Coll Cardiol* 2011;57(6):707-714.

110. McDermott MM, Ferrucci L, Liu K, Guralnik JM, Tian L, Liao Y, Criqui MH. Leg symptom categories and rates of mobility decline in peripheral arterial disease. *J Am Geriatr Soc* 2010;58(7):1256-1262.

111. McDermott MM, Greenland P, Liu K, Guralnik JM, Criqui MH, Dolan NC, Chan C, Celic L, Pearce WH, Schneider JR, Sharma L, Clark E, Gibson D, Martin GJ. Leg symptoms in peripheral arterial disease: Associated clinical characteristics and functional impairment. *JAMA* 2001;286(13):1599-1606.

112. McDermott MM, Liu K, Ferrucci L, Tian L, Guralnik JM, Liao Y, Criqui MH. Decline in functional performance predicts later increased mobility loss and mortality in peripheral arterial disease. *J Am Coll Cardiol* 2011;57(8):962-970.

113. McDermott MM, Liu K, Greenland P, Guralnik JM, Criqui MH, Chan C, Pearce WH, Schneider JR, Ferrucci L, Celic L, Taylor LM, Vonesh E, Martin GJ, Clark E. Functional decline in peripheral arterial disease: Associations with the ankle brachial index and leg symptoms. *JAMA* 2004;292(4):453-461.

114. McDermott MM, Tian L, Ferrucci L, Liu K, Guralnik JM, Liao Y, Pearce WH, Criqui MH. Associations between lower extremity ischemia, upper and lower extremity strength, and functional impairment with peripheral arterial disease. *J Am Geriatr Soc* 2008;56(4):724-729.

115. McDermott MM, Tiukinhoy S, Greenland P, Liu K, Pearce WH, Guralnik JM, Unterreiner S, Gluckman TJ, Criqui MH, Ferrucci L. A pilot exercise intervention to improve lower extremity functioning in peripheral arterial disease unaccompanied by intermittent claudication. *J Cardiopulm Rehabil* 2004;24(3):187-196.

116. Menke J. Improving the image quality of contrast-enhanced MR angiography by automated image registration: A prospective study in peripheral arterial disease of the lower extremities. *Eur J Radiol* 2010;75(3):e1-8.

117. Mohler ER 3rd, Hiatt WR, Creager MA. Cholesterol reduction with atorvastatin improves walking distance in patients with peripheral arterial disease. *Circulation* 2003;108(12):1481-1486.

118. Mohler ER 3rd, Treat-Jacobson D, Reilly MP, Cunningham KE, Miani M, Criqui MH, Hiatt WR, Hirsch AT. Utility and barriers to performance of the ankle-brachial index in primary care practice. *Vasc Med* 2004;9(4):253-260.

119. Mondillo S, Ballo P, Barbati R, Guerrini F, Ammaturo T, Agricola E, Pastore M, Borrello F, Belcastro M, Picchi A, Nami R. Effects of simvastatin on walking performance and symptoms of intermittent claudication in hypercholesterolemic patients with peripheral vascular disease. *Am J Med* 2003;114(5):359-364.

120. Mosti MP, Wang E, Wiggen ON, Helgerud J, Hoff J. Concurrent strength and endurance training improves physical capacity in patients with peripheral arterial disease. *Scand J Med Sci Sports* 2011;21(6):e308-e314.

121. Mouser MJ, Zlabek JA, Ford CL, Mathiason MA. Community trial of home-based exercise therapy for intermittent claudication. *Vasc Med* 2009;14(2):103-107.

122. Mukherjee D, Yadav JS. Update on peripheral vascular diseases: From smoking cessation to stenting. *Cleve Clin J Med* 2001;68(8):723-733.

123. Murphy TP, Cutlip DE, Regensteiner JG, Mohler ER, Cohen DJ, Reynolds MR, Massaro JM, Lewis BA, Cerezo J, Oldenburg NC, Thum CC, Goldberg S, Jaff MR, Steffes MW, Comerota AJ, Ehrman J, Treat-Jacobson D, Walsh ME, Collins T, Badenhop DT, Bronas U, Hirsch AT. Supervised exercise versus primary stenting for claudication resulting from aortoiliac peripheral artery disease: Six-month outcomes from the claudication: Exercise versus endoluminal revascularization (CLEVER) study. *Circulation* 2012;125(1):130-139.

124. Nicolai SP, Hendriks EJ, Prins MH, Teijink JA. Optimizing supervised exercise therapy for patients with intermittent claudication. *J Vasc Surg* 2010;52(5):1226-1233.

125. Norgren L, Hiatt WR, Dormandy JA, Nehler MR, Harris KA, Fowkes FG. Inter-society consensus for the management of peripheral arterial disease (TASC II). *J Vasc Surg* 2007;45 Suppl. S:S5-67.

126. Norgren L, Hiatt WR, Dormandy JA, Nehler MR, Harris KA, Fowkes FG, Bell K, Caporusso J, Durand-Zaleski I, Komori K, Lammer J, Liapis C, Novo S, Razavi M, Robbs J, Schaper N, Shigematsu H, Sapoval M, White C, White J, Clement D, Creager M, Jaff M, Mohler E 3rd, Rutherford RB, Sheehan P, Sillesen H, Rosenfield K. Inter-society consensus for the management of peripheral arterial disease (TASC II). *Eur J Vasc Endovasc Surg* 2007;33 Suppl. 1:S1-75.

127. Ohara Y, Peterson TE, Sayegh HS, Subramanian RR, Wilcox JN, Harrison DG. Dietary correction of hypercholesterolemia in the rabbit normalizes endothelial superoxide anion production. *Circulation* 1995;92(4):898-903.

128. Oka RK, Altman M, Giacomini JC, Szuba A, Cooke JP. Abnormal cardiovascular response to exercise in patients with peripheral arterial disease: Implications for management. *J Vasc Nurs* 2005;23(4):130-138.

129. Parr BM, Noakes TD, Derman EW. Peripheral arterial disease and intermittent claudication: Efficacy of short-term upper body strength training, dynamic exercise training, and advice to exercise at home. *S Afr Med J* 2009;99(11):800-804.

130. Patterson RB, Pinto B, Marcus B, Colucci A, Braun T, Roberts M. Value of a supervised exercise program for the therapy of arterial claudication. *J Vasc Surg* 1997;25(2):312-319.

131. Pereira CE, Albers M, Romiti M, Brochado-Neto FC, Pereira CA. Meta-analysis of femoropopliteal bypass grafts for lower extremity arterial insufficiency. *J Vasc Surg* 2006;44(3):510-517.

132. Pipinos II, Judge AR, Selsby JT, Zhu Z, Swanson SA, Nella AA, Dodd SL. The myopathy of peripheral arterial occlusive disease: Part 2. Oxidative stress, neuropathy, and shift in muscle fiber type. *Vasc Endovascular Surg* 2008;42(2):101-112.

133. Porter JM, Cutler BS, Lee BY, Reich T, Reichle FA, Scogin JT, Strandness DE. Pentoxifylline efficacy in the treatment of intermittent claudication: Multicenter controlled double-blind trial with objective assessment of chronic occlusive arterial disease patients. *Am Heart J* 1982;104(1):66-72.

134. Prchal D, Holmes DT, Levin A. Nephrogenic systemic fibrosis: The story unfolds. *Kidney Int* 2008;73(12):1335-1337.

135. Prevention of Atherosclerotic Complications with Ketanserin Trial Group. Prevention of atherosclerotic complications: Controlled trial of ketanserin. *BMJ* 1989;298(6671):424-430.

136. Radack K, Wyderski RJ. Conservative management of intermittent claudication. *Ann Intern Med* 1990;113(2):135-146.

137. Raynor DA, Coleman KJ, Epstein LH. Effects of proximity on the choice to be physically active or sedentary. *Res Q Exerc Sport* 1998;69(1):99-103.

138. Regensteiner JG. Exercise in the treatment of claudication: Assessment and treatment of functional impairment. *Vasc Med* 1997;2(3):238-242.

139. Regensteiner JG, Hiatt WR, Coll JR, Criqui MH, Treat-Jacobson D, McDermott MM, Hirsch AT. The impact of peripheral arterial disease on health-related quality of life in the peripheral arterial disease awareness, risk, and treatment: New resources for survival (PARTNERS) program. *Vasc Med* 2008;13(1):15-24.

140. Regensteiner JG, Meyer TJ, Krupski WC, Cranford LS, Hiatt WR. Hospital vs home-based exercise rehabilitation for patients with peripheral arterial occlusive disease. *Angiology* 1997;48(4):291-300.

141. Regensteiner JG, Stewart KJ. Established and evolving medical therapies for claudication in patients with peripheral arterial disease. *Nat Clin Pract Cardiovasc Med* 2006;3(11):604-610.

142. Regensteiner JG, Ware JE Jr, McCarthy WJ, Zhang P, Forbes WP, Heckman J, Hiatt WR. Effect of cilostazol on treadmill walking, community-based walking ability, and health-related quality of life in patients with intermittent claudication due to peripheral arterial disease: Meta-analysis of six randomized controlled trials. *J Am Geriatr Soc* 2002;50(12):1939-1946.

143. Ritti-Dias RM, Basyches M, Camara L, Puech-Leao P, Battistella L, Wolosker N. Test-retest reliability of isokinetic strength and endurance tests in patients with intermittent claudication. *Vasc Med* 2010;15(4):275-278.

144. Ritti-Dias RM, Wolosker N, de Moraes Forjaz CL, Carvalho CR, Cucato GG, Leao PP, de Fatima Nunes Marucci M. Strength training increases walking tolerance in intermittent claudication patients: Randomized trial. *J Vasc Surg* 2010;51(1):89-95.

145. Robeer GG, Brandsma JW, van den Heuvel SP, Smit B, Oostendorp RA, Wittens CH. Exercise therapy for intermittent claudication: A review of the quality of randomised clinical trials and evaluation of predictive factors. *Eur J Vasc Endovasc Surg* 1998;15(1):36-43.

146. Robless P, Mikhailidis DP, Stansby GP. Cilostazol for peripheral arterial disease. *Cochrane Database Syst Rev* 2008(1):CD003748.

147. Rogers JH, Laird JR. Overview of new technologies for lower extremity revascularization. *Circulation* 2007;116(18):2072-2085.

148. Romano M, Mainenti PP, Imbriaco M, Amato B, Markabaoui K, Tamburrini O, Salvatore M. Multidetector row CT angiography of the abdominal aorta and lower extremities in patients with peripheral arterial occlusive disease: Diagnostic accuracy and interobserver agreement. *Eur J Radiol* 2004;50(3):303-308.

149. Rosinberg R, Kibbe MR. Hemodynamic evaluation of peripheral arterial disease. In: Casserly I, Sachar R, Yadav JS, eds. *Practical Peripheral Vascular Intervention* (pp. 8-17). Philadelphia: Lippincott Williams & Wilkins, 2011.

150. Rueda CA, Nehler MR, Perry DJ, McLafferty RB, Casserly IP, Hiatt WR, Peyton BD. Patterns of artery disease in 450 patients undergoing revascularization for critical limb ischemia: Implications for clinical trial design. *J Vasc Surg* 2008;47(5):995-1000.

151. Rutherford RB, Baker JD, Ernst C, Johnston KW, Porter JM, Ahn S, Jones DN. Recommended standards for reports dealing with lower extremity ischemia: Revised version. *J Vasc Surg* 1997;26(3):517-538.

152. Sacks D, Bakal CW, Beatty PT, Becker GJ, Cardella JF, Raabe RD, Wiener HM, Lewis CA. Position statement on the use of the ankle brachial index in the evaluation of patients with peripheral vascular disease. A consensus statement developed by the standards division of the Society of Interventional Radiology. *J Vasc Interv Radiol* 2003;14(9 Pt 2):S389.

153. Sallis JF, Hovell MF, Hofstetter CR, Elder JP, Hackley M, Caspersen CJ, Powell KE. Distance between homes and exercise facilities related to frequency of exercise among San Diego residents. *Public Health Rep* 1990;105(2):179-185.

154. Salmasi AM, Nicolaides A, Al-Katoubi A, Sonecha TN, Taylor PR, Serenkuma S, Eastcott HH. Intermittent claudication as a manifestation of silent myocardial ischemia: A pilot study. *J Vasc Surg* 1991;14(1):76-86.

155. Saltin B. Physical training in patients with intermittent claudication. In: Cohen LS, Mock MB, Ringqvist I, eds. *Physical Conditioning and Cardiovascular Rehabilitation* (pp. 181-196). New York: Wiley, 1981.

156. Sanne H, Sivertsson R. The effect of exercise on the development of collateral circulation after experimental occlusion of the femoral artery in the cat. *Acta Physiol Scand* 1968;73(3):257-263.

157. Savage P, Ricci MA, Lynn M, Gardner A, Knight S, Brochu M, Ades P. Effects of home versus supervised exercise for patients with intermittent claudication. *J Cardiopulm Rehabil* 2001;21(3):152-157.

158. Schlager O, Giurgea A, Schuhfried O, Seidinger D, Hammer A, Groger M, Fialka-Moser V, Gschwandtner M, Koppensteiner R, Steiner S. Exercise training increases endothelial progenitor cells and decreases asymmetric dimethylarginine in peripheral arterial disease: A randomized controlled trial. *Atherosclerosis* 2011;217(1):240-248.

159. Schoop W. Mechanism of beneficial action of daily walking training of patients with intermittent claudication. *Scand J Clin Lab Invest Suppl* 1973;128:197-199.

160. Schroder F, Diehm N, Kareem S, Ames M, Pira A, Zwettler U, Lawall H, Diehm C. A modified calculation of ankle-brachial pressure index is far more sensitive in the detection of peripheral arterial disease. *J Vasc Surg* 2006;44(3):531-536.

161. Selvin E, Erlinger TP. Prevalence of and risk factors for peripheral arterial disease in the United States: Results from the National Health and Nutrition Examination Survey, 1999-2000. *Circulation* 2004;110(6):738-743.

162. Shah AP, Klein AJ, Sterrett A, Messenger JC, Albert S, Nehler MR, Hiatt WR, Casserly IP. Clinical outcomes using aggressive approach to anatomic screening and endovascular revascularization in a Veterans Affairs population with critical limb ischemia. *Catheter Cardiovasc Intervent* 2009;74(1):11-19.

163. Skinner JS, Strandness DE Jr. Exercise and intermittent claudication. II. Effect of physical training. *Circulation* 1967;36(1):23-29.

164. Sorlie D, Myhre K. Effects of physical training in intermittent claudication. *Scand J Clin Lab Invest* 1978;38(3):217-222.

165. Spengel F, Brown TM, Poth J, Lehert P. Naftidrofuryl can enhance the quality of life in patients with intermittent claudication. *VASA* 1999;28(3):207-212.

166. Stein R, Hriljac I, Halperin JL, Gustavson SM, Teodorescu V, Olin JW. Limitation of the resting ankle-brachial index in symptomatic patients with peripheral arterial disease. *Vasc Med* 2006;11(1):29-33.

167. Sum CF, Lim SC, Tavintharan S. Peripheral arterial disease and diabetes foot care. *Singapore Med J* 2008;49(2):93-94.

168. Treat-Jacobson D, Halverson SL, Ratchford A, Regensteiner JG, Lindquist R, Hirsch AT. A patient-derived perspective of

health-related quality of life with peripheral arterial disease. *J Nurs Scholarsh* 2002;34(1):55-60.

169. Tsai JC, Chan P, Wang CH, Jeng C, Hsieh MH, Kao PF, Chen YJ, Liu JC. The effects of exercise training on walking function and perception of health status in elderly patients with peripheral arterial occlusive disease. *J Intern Med* 2002;252(5):448-455.

170. Tsuchikane E, Fukuhara A, Kobayashi T, Kirino M, Yamasaki K, Izumi M, Otsuji S, Tateyama H, Sakurai M, Awata N. Impact of cilostazol on restenosis after percutaneous coronary balloon angioplasty. *Circulation* 1999;100(1):21-26.

171. Utter AC, Robertson RJ, Green JM, Suminski RR, McAnulty SR, Nieman DC. Validation of the adult OMNI scale of perceived exertion for walking/running exercise. *Med Sci Sports Exerc* 2004;36(10):1776-1780.

172. Varu VN, Hogg ME, Kibbe MR. Critical limb ischemia. *J Vasc Surg* 2010;51(1):230-241.

173. Vavra AK, Kibbe MR. Women and peripheral arterial disease. *Women's Health* 2009;5(6):669-683.

174. Wang E, Helgerud J, Loe H, Indseth K, Kaehler N, Hoff J. Maximal strength training improves walking performance in peripheral arterial disease patients. *Scand J Med Sci Sports* 2010;20(5):764-770.

175. White CJ, Gray WA. Endovascular therapies for peripheral arterial disease: An evidence-based review. *Circulation* 2007;116(19):2203-2215.

176. Wiginton CD, Kelly B, Oto A, Jesse M, Aristimuno P, Ernst R, Chaljub G. Gadolinium-based contrast exposure, nephrogenic systemic fibrosis, and gadolinium detection in tissue. *AJR Am J Roentgenol* 2008;190(4):1060-1068.

177. Willigendael EM, Teijink JA, Bartelink ML, Kuiken BW, Boiten J, Moll FL, Buller HR, Prins MH. Influence of smoking on incidence and prevalence of peripheral arterial disease. *J Vasc Surg* 2004;40(6):1158-1165.

178. Wolosker N, Ritti-Dias RM, Camara LC, Garcia YM, Jacob-Filho W, Puech-Leao P. Treadmill test is limited in elderly patients with peripheral arterial disease. *VASA* 2010;39(3):237-241.

179. Womack CJ, Sieminski DJ, Katzel LI, Yataco A, Gardner AW. Improved walking economy in patients with peripheral arterial occlusive disease. *Med Sci Sports Exerc* 1997;29(10):1286-1290.

180. Wullink M, Stoffers HE, Kuipers H. A primary care walking exercise program for patients with intermittent claudication. *Med Sci Sports Exerc* 2001;33(10):1629-1634.

181. Wyatt MF, Stickrath C, Shah A, Smart A, Hunt J, Casserly IP. Ankle-brachial index performance among internal medicine residents. *Vasc Med* 2010;15(2):99-105.

182. Yamamoto K, Miyata T, Onozuka A, Koyama H, Ohtsu H, Nagawa H. Plantar flexion as an alternative to treadmill exercise for evaluating patients with intermittent claudication. *Eur J Vasc Endovasc Surg* 2007;33(3):325-329.

183. Young DF, Cholvin NR, Kirkeeide RL, Roth AC. Hemodynamics of arterial stenoses at elevated flow rates. *Circ Res* 1977;41(1):99-107.

184. Zetterquist S. The effect of active training on the nutritive blood flow in exercising ischemic legs. *Scand J Clin Lab Invest* 1970;25(1):101-111.

185. Zwierska I, Walker RD, Choksy SA, Male JS, Pockley AG, Saxton JM. Upper- vs lower-limb aerobic exercise rehabilitation in patients with symptomatic peripheral arterial disease: A randomized controlled trial. *J Vasc Surg* 2005;42(6):1122-1130.

CHAPTER 16

1. Alexander T, Friedman DB, Levine BD, Pawelczyk JA, Mitchell JH. Cardiovascular responses during static exercise. Studies in patients with complete heart block and dual chamber pacemakers. *Circulation* 1994;89(4):1643-1647.

2. Alt E, Combs W, Willhaus R, Condie C, Bambl E, Fotuhi P, Pache J, Schomig A. A comparative study of activity and dual sensor: Activity and minute ventilation pacing responses to ascending and descending stairs. *Pacing Clin Electrophysiol* 1998;21(10):1862-1868.

3. Alt E, Matula M. Comparison of two activity-controlled rate-adaptive pacing principles: Acceleration versus vibration. *Cardiol Clin* 1992;10(4):635-658.

4. Bacharach DW, Hilden TS, Millerhagen JO, Westrum BL, Kelly JM. Activity-based pacing: Comparison of a device using an accelerometer versus a piezoelectric crystal. *Pacing Clin Electrophysiol* 1992;15(2):188-196.

5. Barold SS, Barold HS. Optimal cardiac pacing in patients with coronary artery disease. *Pacing Clin Electrophysiol* 1998;21(2):456-461.

6. Bodenhamer RM, Grantham RN. Mode selection: The therapeutic challenge. In: *Adaptive-Rate Pacing* (pp. 19-52). St. Paul, MN: Cardiac Pacemakers, Inc., 1993.

7. Carmouche DG, Bubien RS, Kay GN. The effect of maximum heart rate on oxygen kinetics and exercise performance at low and high workloads. *Pacing Clin Electrophysiol* 1998;21(4 Pt 1):679-686.

8. De Marco T, Wolfel E, Feldman AM, Lowes B, Higginbotham MB, Ghali JK, Wagoner L, Kirlin PC, Kennett JD, Goel S, Saxon LA, Boehmer JP, Mann D, Galle E, Ecklund F, Yong P, Bristow MR. Impact of cardiac resynchronization therapy on exercise performance, functional capacity, and quality of life in systolic heart failure with QRS prolongation: COMPANION trial sub-study. *J Card Fail* 2008;14(1):9-18.

9. Epstein AE, DiMarco JP, Ellenbogen KA, Estes NA 3rd, Freedman RA, Gettes LS, Gillinov AM, Gregoratos G, Hammill SC, Hayes DL, Hlatky MA, Newby LK, Page RL, Schoenfeld MH, Silka MJ, Stevenson LW, Sweeney MO, Smith SC Jr, Jacobs AK, Adams CD, Anderson JL, Buller CE, Creager MA, Ettinger SM, Faxon DP, Halperin JL, Hiratzka LF, Hunt SA, Krumholz HM, Kushner FG, Lytle BW, Nishimura RA, Ornato JP, Riegel B, Tarkington LG, Yancy CW. ACC/AHA/HRS 2008 Guidelines for Device-Based Therapy of Cardiac Rhythm Abnormalities: A report of the American College of Cardiology/American Heart Association Task Force on Practice Guidelines (Writing Committee to Revise the ACC/AHA/NASPE 2002 Guideline Update for Implantation of Cardiac Pacemakers and Antiarrhythmia Devices): Developed in collaboration with the American Association for Thoracic Surgery and Society of Thoracic Surgeons. *Circulation* 2008;117(21):e350-408.

10. Greco EM, Guardini S, Citelli L. Cardiac rehabilitation in patients with rate responsive pacemakers. *Pacing Clin Electrophysiol* 1998;21(3):568-575.

11. Harper GR, Pina IL, Kutalek SP. Intrinsic conduction maximizes cardiopulmonary performance in patients with dual chamber pacemakers. *Pacing Clin Electrophysiol* 1991;14(11 Pt 2):1787-1791.

12. Hayes DL, Von Feldt L, Higano ST. Standardized informal exercise testing for programming rate adaptive pacemakers. *Pacing Clin Electrophysiol* 1991;14(11 Pt 2):1772-1776.

13. Holmes DR. Hemodynamics of cardiac pacing. In: Furman S, Hayes DL, Holmes DR, eds. *A Practice of Cardiac Pacing* (pp. 167-191). 2nd ed. Mount Kisco, NY: Futura, 1989.

14. Hoth KF, Nash J, Poppas A, Ellison KE, Paul RH, Cohen RA. Effects of cardiac resynchronization therapy on health-related quality of life in older adults with heart failure. *Clin Interv Aging* 2008;3(3):553-560.
15. Joglar JA, Hamdan MH, Welch PJ, Page RL. Interaction of a commercial heart rate monitor with implanted pacemakers. *Am J Cardiol* 1999;83(5):790-792, A10.
16. Leung SK, Lau CP, Tang MO, Leung Z, Yakimow K. An integrated dual sensor system automatically optimized by target rate histogram. *Pacing Clin Electrophysiol* 1998;21(8):1559-1566.
17. Medtronic. Therapies for Medical Conditions. [Online]. Medtronic, Inc., 1999 [updated September 3, 1999; cited 1999 11/26/99]. www.medtronic.com/brady/clinician/therapies/clinther.html.
18. Patwala AY, Woods PR, Sharp L, Goldspink DF, Tan LB, Wright DJ. Maximizing patient benefit from cardiac resynchronization therapy with the addition of structured exercise training: A randomized controlled study. *J Am Coll Cardiol* 2009;53(25):2332-2339.
19. Pereira de Sousa LA, Britto RR, Ribeiro AL, Baracho SM, da Costa Val Barros V, Carvalho VT, Parreira VF. Six-minute walk test in patients with permanent cardiac pacemakers. *J Cardiopulm Rehabil Prev* 2008;28(4):253-257.
20. Philippon F. Cardiac resynchronization therapy: Device-based medicine for heart failure. *J Card Surg* 2004;19(3):270-274.
21. Schlosshan D, Barker D, Pepper C, Williams G, Morley C, Tan LB. CRT improves the exercise capacity and functional reserve of the failing heart through enhancing the cardiac flow- and pressure-generating capacity. *Eur J Heart Fail* 2006;8(5):515-521.
22. Seidl K, Rameken M, Vater M, Senges J. Cardiac resynchronization therapy in patients with chronic heart failure: Pathophysiology and current experience. *Am J Cardiovasc Drugs* 2002;2(4):219-226.
23. Seifert M, Schlegl M, Hoersch W, Fleck E, Doelger A, Stockburger M, Butter C. Functional capacity and changes in the neurohormonal and cytokine status after long-term CRT in heart failure patients. *Int J Cardiol* 2007;121(1):68-73.
24. Sharp CT, Busse EF, Burgess JJ, Haennel RG. Exercise prescription for patients with pacemakers. *J Cardiopulm Rehabil* 1998;18(6):421-431.
25. Shukla HH, Flaker GC, Hellkamp AS, James EA, Lee KL, Goldman L, Orav EJ, Lamas GA. Clinical and quality of life comparison of accelerometer, piezoelectric crystal, and blended sensors in DDDR-paced patients with sinus node dysfunction in the mode selection trial (MOST). *Pacing Clin Electrophysiol* 2005;28(8):762-770.
26. Sparks PB. Cardiac resynchronisation therapy. *Heart Lung Circ* 2004;13 Suppl. 3:S56-59.
27. Steendijk P, Tulner SA, Bax JJ, Oemrawsingh PV, Bleeker GB, van Erven L, Putter H, Verwey HF, van der Wall EE, Schalij MJ. Hemodynamic effects of long-term cardiac resynchronization therapy: Analysis by pressure-volume loops. *Circulation* 2006;113(10):1295-1304.
28. Sulke N, Dritsas A, Chambers J, Sowton E. Is accurate rate response programming necessary? *Pacing Clin Electrophysiol* 1990;13(8):1031-1044.
29. Wilkoff B, Corey J, Blackburn G. A mathematical model of the chronotropic response to exercise. *J Electrophysiol* 1989;3:176-180.
30. Wilkoff BL, Miller RE. Exercise testing for chronotropic assessment. *Cardiol Clin* 1992;10(4):705-717.
31. Wood MA, Stambler BS, Ellenbogen KA. Patient management: Optimal programming of adaptive-rate pacemakers. In: *Adaptive-Rate Pacing* (pp. 86-110). St. Paul, MN: Cardiac Pacemakers, Inc., 1993.

CHAPTER 17

1. Abraham AS, Cole RB, Bishop JM. Reversal of pulmonary hypertension by prolonged oxygen administration to patients with chronic bronchitis. *Circ Res* 1968;23:147-157.
2. Abraham AS, Cole RB, Green ID, Hedworth-Whitty RB, Clarke SW, Bishop JM. Factors contributing to the reversible pulmonary hypertension of patients with acute respiratory failure studies by serial observations during recovery. *Circ Res* 1969;24:51-60.
3. Abraham AS, Hedworth-Whitty RB, Bishop JM. Effects of acute hypoxia and hypervolaemia singly and together, upon the pulmonary circulation in patients with chronic bronchitis. *Clin Sci* 1967;33:371-380.
4. Adams L, Chronos N, Lane R, Guz A. The measurement of breathlessness induced in normal subjects: Validity of two scaling techniques. *Clin Sci* 1985;69:7-16.
5. Agency for Health Care Policy and Research. *Clinical Classifications for Health Policy Research: Hospital Inpatient Statistics, 1995* (pp. 16-17). HCUP-3 Research Note. Rockville, MD: Agency for Health Care Policy and Research, 1999.
6. Agusti AGN, Noguera A, Sauleda J, Sala E, Pons J, Busquets X. Systemic effects of chronic obstructive pulmonary disease. *Eur Respir J* 2003;21:347-360.
7. American Association of Cardiovascular and Pulmonary Rehabilitation. *Guidelines for Cardiac Rehabilitation and Secondary Prevention Programs*. 4th ed. Champaign, IL: Human Kinetics, 2004.
8. American Association of Cardiovascular and Pulmonary Rehabilitation. *Guidelines for Pulmonary Rehabilitation Programs*. 2nd ed. Champaign, IL: Human Kinetics, 1998.
9. American College of Chest Physicians/American Association of Cardiovascular and Pulmonary Rehabilitation Pulmonary Rehabilitation Guidelines Panel. Pulmonary rehabilitation: Joint ACCP/AACVPR evidence-based guidelines. *Chest* 2007;131:4S-42S.
10. American College of Sports Medicine. *ACSM's Exercise Management for Persons with Chronic Diseases and Disabilities*. 3rd ed. Champaign, IL: Human Kinetics, 2009.
11. American College of Sports Medicine. *ACSM's Guidelines for Exercise Testing and Prescription*. 9th ed. Baltimore: Williams & Wilkins, 2013.
12. American Thoracic Society. Evaluation of impairment/disability secondary to respiratory disorders. *Am Rev Respir Dis* 1986;133:1205-1209.
13. American Thoracic Society. Standards for the diagnosis and care of patients with chronic obstructive pulmonary disease. *Am J Respir Crit Care Med* 1995;152:S77-S152.
14. American Thoracic Society and European Respiratory Society. Skeletal muscle dysfunction in chronic obstructive pulmonary disease. *Am J Respir Crit Care Med* 1999;159:S1-40.
15. American Thoracic Society Documents. An official American Thoracic Society public policy statement: Novel risk factors and the global burden of chronic obstructive pulmonary disease. *Am J Respir Crit Care Med* 2010;182:693-718.
16. American Thoracic Society/European Respiratory Society statement on pulmonary rehabilitation. *Am J Respir Crit Care Med* 2006;173:1390-1413.

17. Atkins CJ, Kaplan RM, Timms RM, Reinsch S, Lofback K. Behavioral exercise programs in the management of chronic obstructive pulmonary disease. *J Consult Clin Psychol* 1984;52:591-603.

18. Badgett RG, Tanaka DJ, Hunt DK, Jelley MJ, Feinberg LE, Steiner JF, Petty TL. Can moderate chronic obstructive pulmonary disease be diagnosed by historical and physical findings alone? *Am J Med* 1993;94:188-196.

19. Barnard KL, Adams KJ, Swank AM, Mann E, Denny DM. Injuries and muscle soreness during the one repetition maximum assessment in a cardiac rehabilitation population. *J Cardiopulm Rehabil* 1999;19:52-58.

20. Battaglia E, Fulgenzi A, Ferrero ME. Rationale of combined use of inspiratory and expiratory devices in improving maximal inspiratory pressure and maximal expiratory pressure of patients with chronic obstructive pulmonary disease. *Arch Phys Med Rehabil* 2009;90:913-918.

21. Beckerman M, Magadle R, Weiner R, Weiner P. The effects of 1 year of specific inspiratory muscle training in patients with COPD. *Chest* 2005;128:3177-3182.

22. Bernard S, LeBlanc P, Whittom F, Carrier G, Jobin J, Belleau R, Maltais F. Peripheral muscle weakness in patients with chronic obstructive pulmonary disease. *Am J Respir Crit Care Med* 1998;158:629-634.

23. Berry MJ, Adair NE, Sevensky KS, Quinby A, Lever HM. Inspiratory muscle training and whole-body reconditioning in chronic obstructive pulmonary disease. *Am J Respir Crit Care Med* 1996;153:1812-1816.

24. Berry MJ, Rejeski WJ, Adair NE, Zaccaro D. Exercise rehabilitation and chronic obstructive pulmonary disease stage. *Am J Respir Crit Care Med* 1999;160:1248-1253.

25. Borg GA. Psychophysical bases of perceived exertion. *Med Sci Sports Exerc* 1982;14:377-381.

26. Bowen JB, Votto JJ, Thrall RS, Haggerty MC, Stockdale-Woolley R, Bandyopadhyay T, ZuWallack RL. Functional status and survival following pulmonary rehabilitation. *Chest* 2000;118:697-703.

27. Bradley BL, Garner AE, Billiu D, Mestas JM, Forman J. Oxygen-assisted exercise in chronic obstructive lung disease. The effect on exercise capacity and arterial blood gas tensions. *Am Rev Respir Dis* 1978;118:239-243.

28. Braun SR, Keim NL, Dixon RM, Clagnaz P, Anderegg A, Shrago ES. The prevalence and determinants of nutritional changes in chronic obstructive pulmonary disease. *Chest* 1984;86:558-563.

29. Breyer M, Breyer-Kohansal R, Funk G, Dornhofer N, Spruit M, Wouters E, Burghuber O, Hartl S. Nordic walking improves daily physical activities in COPD: A randomized controlled trial. *Respir Res* 2010;11:1-9.

30. Burrows B, Bloom JW, Traver GA, Cline MG. The course and prognosis of different forms of chronic airways obstruction in a sample from the general population. *N Engl J Med* 1987;317:1309-1314.

31. Calverley PM. Modern treatment of chronic obstructive pulmonary disease. *Eur Respir J Suppl* 2001;34:60S-66S.

32. Cambach W, Chadwick-Straver RV, Wagenaar RC, van Keimpema AR, Kemper HC. The effects of a community-based pulmonary rehabilitation programme on exercise tolerance and quality of life: A randomized controlled trial. *Eur Respir J* 1997;10:104-113.

33. Casaburi R. Skeletal muscle function in COPD. *Chest* 2000;117:267S-271S.

34. Casaburi R, Patessio A, Ioli F, Zanaboni S, Donner CF, Wasserman K. Reductions in exercise lactic acidosis and ventilation as a result of exercise training in patients with obstructive lung disease. [See comment]. *Am Rev Respir Dis* 1991;143:9-18.

35. Celli BR, Rassulo J, Make BJ. Dyssynchronous breathing during arm but not leg exercise in patients with chronic airflow obstruction. *N Engl J Med* 1986;314:1485-1490.

36. Clark CJ, Cochrane LM, Mackay E, Paton B. Skeletal muscle strength and endurance in patients with mild COPD and the effects of weight training. *Eur Respir J* 2000;15:92-97.

37. Collins EG, Langbein WE, Fehr L, Maloney C. Breathing pattern retraining and exercise in persons with chronic obstructive pulmonary disease. *AACN Clin Issues* 2001;12:202-209.

38. Cotes JE, Gilson JC. Effect of oxygen on exercise ability in chronic respiratory insufficiency. *Lancet* 1956;1:872-876.

39. Couser JI Jr, Martinez FJ, Celli BR. Pulmonary rehabilitation that includes arm exercise reduces metabolic and ventilatory requirements for simple arm elevation. *Chest* 1993;103:37-41.

40. Dean NC, Brown JK, Himelman RB, Doherty JJ, Gold WM, Stulbarg MS. Oxygen may improve dyspnea and endurance in patients with chronic obstructive pulmonary disease and only mild hypoxemia. *Am Rev Respir Dis* 1992;146:941-945.

41. Debigare R, Maltais F, Mallet M, Casaburi R, LeBlanc P. Influence of work rate incremental rate on the exercise responses in patients with COPD. *Med Sci Sports Exerc* 2000;32:1365-1368.

42. Decramer M, de Bock V, Dom R. Functional and histologic picture of steroid-induced myopathy in chronic obstructive pulmonary disease. *Am J Respir Crit Care Med* 1996;153:1958-1964.

43. Decramer M, Lacquet LM, Fagard R, Rogiers P. Corticosteroids contribute to muscle weakness in chronic airflow obstruction. *Am J Respir Crit Care Med* 1994;150:11-16.

44. Degre S, Sergysels R, Messin R, Vandermoten P, Salhadin P, Denolin H, De Coster A. Hemodynamic responses to physical training in patients with chronic lung disease. *Am Rev Respir Dis* 1974;110:395-402.

45. Dekhuijzen PN, Folgering HT, van Herwaarden CL. Target-flow inspiratory muscle training during pulmonary rehabilitation in patients with COPD. *Chest* 1991;99:128-133.

46. 46 .Dillard TA. Ventilatory limitation of exercise. Prediction in COPD. *Chest* 1987;92:195-196.

47. Engelen MP, Schols AM, Baken WC, Wesseling GJ, EF Wouters. Nutritional depletion in relation to respiratory and peripheral skeletal muscle function in out-patients with COPD. *Eur Respir J* 1994;7:1793-1797.

48. Engelen MP, Schols AM, Does JD, Wouters EF. Skeletal muscle weakness is associated with wasting of extremity fat-free mass but not with airflow obstruction in patients with chronic obstructive pulmonary disease. *Am J Clin Nutr* 2000;71:733-738.

49. Epstein SK, Celli BR. Cardiopulmonary exercise testing in patients with chronic obstructive pulmonary disease. *Cleve Clin J Med* 1993;60:119-128.

50. Evans WJ. Exercise training guidelines for the elderly. *Med Sci Sports Exerc* 1999;31:12-17.

51. Falk P, Eriksen AM, Kolliker K, Andersen JB. Relieving dyspnea with an inexpensive and simple method in patients with severe chronic airflow limitation. *Eur J Respir Dis* 1985;66:181-186.

52. Farkas GA, Roussos C. Adaptability of the hamster diaphragm to exercise and/or emphysema. *J Appl Physiol* 1982;53:1263-1272.

53. Fiaccadori E, Del Canale S, Coffrini E, Vitali P, Antonucci C, Cacciani G, Mazzola I, Guariglia A. Hypercapnic-hypoxemic chronic

obstructive pulmonary disease (COPD): Influence of severity of COPD on nutritional status. *Am J Clin Nutr* 1988;48:680-685.

54. Fletcher GF, Balady G, Froelicher VF, Hartley LH, Haskell WL, Pollock ML. Exercise standards. A statement for healthcare professionals from the American Heart Association. *Circulation* 1995;91:580-615.
55. Gibbons LW, Mitchell TL, Gonzalez V. The safety of exercise testing. *Prim Care* 1994;21:611-629.
56. Goldstein RS, Gort EH, Stubbing D, Avendano MA, Guyatt GH. Randomised controlled trial of respiratory rehabilitation. *Lancet* 1994;344:1394-1397.
57. Gordon AM, Huxley AF, Julian FJ. The variation in isometric tension with sarcomere length in vertebrate muscle fibres. *J Physiol (Lond)* 1966;184:170-192.
58. Gosker HR, van Mameren H, van Dijk PJ, Engelen MP, van der Vusse GJ, Wouters EF, Schols AM. Skeletal muscle fibre-type shifting and metabolic profile in patients with chronic obstructive pulmonary disease. *Eur Respir J* 2002;19:617-625.
59. Gosselink R, Troosters T, Decramer M. Peripheral muscle weakness contributes to exercise limitation in COPD. *Am J Respir Crit Care Med* 1996;153:976-980.
60. Greenberg SB, Allen M, Wilson J, Atmar RL. Respiratory viral infections in adults with and without chronic obstructive pulmonary disease. *Am J Respir Crit Care Med* 2000;162:167-173.
61. Griffin SE, Robergs RA, Heyward VH. Blood pressure measurement during exercise: A review. *Med Sci Sports Exerc* 1997;29:149-159.
62. Guyatt G, Keller J, Singer J, Halcrow S, Newhouse M. Controlled trial of respiratory muscle training in chronic airflow limitation. *Thorax* 1992;47:598-602.
63. Hackett TL, Knight DA, Sin DD. Potential role of stem cells in management of COPD. *Int J COPD* 2010;5:81-88.
64. Hamilton AL, Killian KJ, Summers E, Jones NL. Muscle strength, symptom intensity, and exercise capacity in patients with cardio-respiratory disorders. *Am J Respir Crit Care Med* 1995;152:2021-2031.
65. Heaton RK, Grant I, McSweeny AJ, Adams KM, Petty TL. Psychologic effects of continuous and nocturnal oxygen therapy in hypoxemic chronic obstructive pulmonary disease. *Arch Intern Med* 1983;143:1941-1947.
66. Henke KG, Sharratt M, Pegelow D, Dempsey JA. Regulation of end-expiratory lung volume during exercise. *J Appl Physiol* 1988;64:135-146.
67. Hildebrand IL, Sylven C, Esbjornsson M, Hellstrom K, Jansson E. Does chronic hypoxaemia induce transformations of fibre types? *Acta Physiol Scand* 1991;141:435-439.
68. Hill K, Jenkins SC, Philippe DL, Cecins N, Shepard KL, Green DJ, Hillman DR, Eastwood PR. High-intensity inspiratory muscle training in COPD. *Eur Resp J* 2006;27:1119-1128.
69. Holford N, Black P, Couch R, Kennedy J, Briant R. Theophylline target concentration in severe airways obstruction—10 or 20 mg/L? A randomised concentration-controlled trial. *Clin Pharmacokinet* 1993;25:495-505.
70. Hunter AM, Carey MA, Larsh HW. The nutritional status of patients with chronic obstructive pulmonary disease. *Am Rev Respir Dis* 1981;124:376-381.
71. Jakobsson P, Jorfeldt L, Brundin A. Skeletal muscle metabolites and fibre types in patients with advanced chronic obstructive pulmonary disease (COPD), with and without chronic respiratory failure. *Eur Respir J* 1990;3:192-196.
72. Jones NL, Jones G, Edwards RH. Exercise tolerance in chronic airway obstruction. *Am Rev Respir Dis* 1971;103:477-491.
73. Kaelin ME, Swank AM, Adams KJ, Barnard KL, Berning JM, Green A. Cardiopulmonary responses, muscle soreness, and injury during the one repetition maximum assessment in pulmonary rehabilitation patients. *J Cardiopulm Rehabil* 1999;19:366-372.
74. Kelsen SG, Ference M, Kapoor S. Effects of prolonged undernutrition on structure and function of the diaphragm. *J Appl Physiol* 1985;58:1354-1359.
75. Knudson RJ, Slatin RC, Lebowitz MD, Burrows B. The maximal expiratory flow-volume curve. Normal standards, variability, and effects of age. *Am Rev Respir Dis* 1976;113:587-600.
76. Kongsgaard M, Backer V, Jorgensen K, Kjaer M, Beyer N. Heavy resistance training increases muscle size, strength and physical function in elderly male COPD patients: A pilot study. *Respir Med* 2004;98:1000-1007.
77. Koppers RJH, Vos PJE, Boot CRL, Folgering HTM. Exercise performance improves in patients with COPD due to respiratory muscle endurance training. *Chest* 2006;129:886-892.
78. Lacasse Y, Guyatt GH, Goldstein RS. The components of a respiratory rehabilitation program: A systematic overview. *Chest* 1997;111:1077-1088.
79. Lacasse Y, Wong E, Guyatt GH, King D, Cook DJ, Goldstein RS. Meta-analysis of respiratory rehabilitation in chronic obstructive pulmonary disease. *Lancet* 1996;348:1115-1119.
80. Lake FR, Henderson K, Briffa T, Openshaw J, Musk AW. Upper-limb and lower-limb exercise training in patients with chronic airflow obstruction. *Chest* 1990;97:1077-1082.
81. Lan C, Yang M, Lee C, Huang Y, Huang C, Huang K, Wu Y. Pulmonary rehabilitation improves exercise capacity and quality of life in underweight patients with chronic obstructive pulmonary disease. *Respirology* 2011;16:276-283.
82. Lange P, Groth S, Nyboe GJ, Mortensen J, Appleyard M, Jensen G, Schnohr P. Effects of smoking and changes in smoking habits on the decline of FEV_1. *Eur Respir J* 1989;2:811-816.
83. Larson JL, Covey MK, Wirtz SE, Berry JK, Alex CG, Langbein WE, Edwards L. Cycle ergometer and inspiratory muscle training in chronic obstructive pulmonary disease. *Am J Respir Crit Care Med* 1999;160:500-507.
84. Leith DE, Bradley M. Ventilatory muscle strength and endurance training. *J Appl Physiol* 1976;41:508-516.
85. Levine BE, Bigelow DB, Hamstra RD, Beckwitt HJ, Mitchell RS, Nett LM, Stephen TA, Petty TL. The role of long-term continuous oxygen administration in patients with chronic airway obstruction with hypoxemia. *Ann Intern Med* 1967;66:639-650.
86. Lewis MI, Sieck GC, Fournier M, Belman MJ. Effect of nutritional deprivation on diaphragm contractility and muscle fiber size. *J Appl Physiol* 1986;60:596-603.
87. Lisboa C, Munoz V, Beroiza T, Leiva A, Cruz E. Inspiratory muscle training in chronic airflow limitation: Comparison of two different training loads with a threshold device. *Eur Respir J* 1994;7:1266-1274.
88. Lisboa C, Villafranca C, Leiva A, Cruz E, Pertuze J, Borzone G. Inspiratory muscle training in chronic airflow limitation: Effect on exercise performance. *Eur Respir J* 1997;10:537-542.
89. Mador MJ, Bozkanat E, Aggarwal A, Shaffer M, Kufel TJ. Endurance and strength training in patients with COPD. *Chest* 2004;125:2036-2045.

90. Magadle R, McConnell AK, Beckerman M, Weiner P. Inspiratory muscle training in pulmonary rehabilitation program in COPD patients. *Respir Med* 2007;101:1500-1505.

91. Magnussen H, Richter K, Taube C. Are chronic obstructive pulmonary disease (COPD) and asthma different diseases? *Clin Exp Allergy* 1998;28(Suppl. 5):187-194.

92. Mahler DA, Rosiello RA, Harver A, Lentine T, McGovern JF, Daubenspeck JA. Comparison of clinical dyspnea ratings and psychophysical measurements of respiratory sensation in obstructive airway disease. *Am Rev Respir Dis* 1987;135:1229-1233.

93. Mahler DA, Weinberg DH, Wells CK, Feinstein AR. The measurement of dyspnea. Contents, interobserver agreement, and physiologic correlates of two new clinical indexes. *Chest* 1984;85:751-758.

94. Maltais F, LeBlanc P, Whittom F, Simard C, Marquis K, Belanger M, Breton MJ, Jobin J. Oxidative enzyme activities of the vastus lateralis muscle and the functional status in patients with COPD. *Thorax* 2000;55:848-853.

95. Maltais F, Simard AA, Simard C, Jobin J, Desgagnes P, LeBlanc P. Oxidative capacity of the skeletal muscle and lactic acid kinetics during exercise in normal subjects and in patients with COPD. *Am J Respir Crit Care Med* 1996;153:288-293.

96. Marciniuk DD, Gallagher CG. Clinical exercise testing in chronic airflow limitation. *Med Clin North Am* 1996;80:565-587.

97. Martinez FJ, Vogel PD, Dupont DN, Stanopoulos I, Gray A, Beamis JF. Supported arm exercise vs unsupported arm exercise in the rehabilitation of patients with severe chronic airflow obstruction. *Chest* 1993;103:1397-1402.

98. McCully KK, Faulkner JA. Length-tension relationship of mammalian diaphragm muscles. *J Appl Physiol* 1983;54:1681-1686.

99. McKeon JL, Turner J, Kelly C, Dent A, Zimmerman PV. The effect of inspiratory resistive training on exercise capacity in optimally treated patients with severe chronic airflow limitation. *Aust N Z J Med* 1986;16:648-652.

100. Medical Research Council Working Party. Long term domiciliary oxygen therapy in chronic hypoxic cor pulmonale complicating chronic bronchitis and emphysema. *Lancet* 1981;1:681-686.

101. Miller WF, Taylor HF. Exercise training in the rehabilitation of patients with severe respiratory insufficiency due to pulmonary emphysema. *South Med J* 1962;55:1216-1221.

102. Murray RP, Anthonisen NR, Connett JE, Wise RA, Lindgren PG, Greene PG, Nides MA. Effects of multiple attempts to quit smoking and relapses to smoking on pulmonary function. *J Clin Epidemiol* 1998;51:1317-1326.

103. National Heart, Lung, and Blood Institute. *Morbidity and Mortality: 1998 Chart Book on Cardiovascular, Lung and Blood Diseases.* Bethesda, MD: National Institutes of Health, 1998.

104. Nickerson BG, Sarkisian C, Tremper K. Bias and precision of pulse oximeters and arterial oximeters. *Chest* 1988;93:515-517.

105. Nocturnal Oxygen Therapy Trial Group. Continuous or nocturnal oxygen therapy in hypoxemic chronic obstructive lung disease: A clinical trial. *Ann Intern Med* 1980;93:391-398.

106. Noseda A, Carpiaux JP, Vandeput W, Prigogine T, Schmerber J. Resistive inspiratory muscle training and exercise performance in COPD patients. A comparative study with conventional breathing retraining. *Bull Eur Physiopathol Respir* 1987;23:457-463.

107. Oelberg DA, Medoff BD, Markowitz DH, Pappagianopoulos PP, Ginns LC, Systrom DM. Systemic oxygen extraction during incremental exercise in patients with severe chronic obstructive pulmonary disease. *Eur J Appl Physiol* 1998;78:201-207.

108. Orozco-Levi M. Structure and function of the respiratory muscles in patients with COPD: Impairment or adaptation? *Eur Respir J Suppl* 2003;46:41S-51S.

109. Ortega F, Toral J, Cejudo P, Villagomez R, Sanchez H, Castillo J, Montemayor T. Comparison of effects of strength and endurance training in patients with chronic obstructive pulmonary disease. *Am J Respir Crit Care Med* 2002;166:669-674.

110. Panton LB, Golden J, Broeder CE, Browder KD, Cestaro-Seifer DJ, Seifer FD. The effects of resistance training on functional outcomes in patients with chronic obstructive pulmonary disease. *Eur J Appl Physiol* 2004;91:443-449.

111. Pardy RL, Rivington RN, Despas PJ, Macklem PT. Inspiratory muscle training compared with physiotherapy in patients with chronic airflow limitation. *Am Rev Respir Dis* 1981;123:421-425.

112. Pereira A, Santa-Clara H, Pereira E, Simoes S, Remedios I, Cardoso J, Brito J, Cabri J, Fernhall B. Impact of combined exercise on chronic obstructive pulmonary patients' state of health. *Rev Port Pneumol* 2010;15:737-757.

113. Pierce AK, Paez PN, Miller WF. Exercise training with the aid of a portable oxygen supply in patients with emphysema. *Am Rev Respir Dis* 1965;91:653-659.

114. Polkey MI, Kyroussis D, Hamnegard CH, Mills GH, Green M, Moxham J. Diaphragm strength in chronic obstructive pulmonary disease. *Am J Respir Crit Care Med* 1996;154:1310-1317.

115. Preusser BA, Winningham ML, Clanton TL. High- vs low-intensity inspiratory muscle interval training in patients with COPD. *Chest* 1994;106:110-117.

116. Reardon J, Awad E, Normandin E, Vale F, Clark B, ZuWallack RL. The effect of comprehensive outpatient pulmonary rehabilitation on dyspnea. *Chest* 1994;105:1046-1052.

117. Rejbi I, Trabelsi Y, Chouchene A, Turkia W, Saad H, Zbidi A, Kerken A, Tabka Z. Changes in six-minute walking distance during pulmonary rehabilitation in patients with COPD and in healthy subjects. *Int J Chron Obstruct Pulmon Dis* 2010;5:209-215.

118. Ries AL, Ellis B, Hawkins RW. Upper extremity exercise training in chronic obstructive pulmonary disease. *Chest* 1988;93:688-692.

119. Ries AL, Kaplan RM, Limberg TM, Prewitt LM. Effects of pulmonary rehabilitation on physiologic and psychosocial outcomes in patients with chronic obstructive pulmonary disease. *Ann Intern Med* 1995;122:823-832.

120. Rodrigo C, Rodrigo G. Treatment of acute asthma. Lack of therapeutic benefit and increase of the toxicity from aminophylline given in addition to high doses of salbutamol delivered by metered-dose inhaler with a spacer. *Chest* 1994;106:1071-1076.

121. Rollier H, Bisschop A, Gayan-Ramirez G, Gosselink R, Decramer M. Low load inspiratory muscle training increases diaphragmatic fiber dimensions in rats. *Am J Respir Crit Care Med* 1998;157:833-839.

122. Rooyackers JM, Dekhuijzen PN, Van Herwaarden CL, Folgering HT. Training with supplemental oxygen in patients with COPD and hypoxaemia at peak exercise. *Eur Respir J* 1997;10:1278-1284.

123. Schols AM, Soeters PB, Dingemans AM, Mostert R, Frantzen PJ, Wouters EF. Prevalence and characteristics of nutritional depletion in patients with stable COPD eligible for pulmonary rehabilitation. *Am Rev Respir Dis* 1993;147:1151-1156.

124. Schwaiblmair M, Beinert T, Seemann M, Behr J, Reiser M, Vogelmeier C. Relations between cardiopulmonary exercise testing and quantitative high-resolution computed tomography associated in patients with alpha-1-antitrypsin deficiency. *Eur J Med Res* 1998;3:527-532.

125. Serres I, Gautier V, Varray A, Prefaut C. Impaired skeletal muscle endurance related to physical inactivity and altered lung function in COPD patients. *Chest* 1998;113:900-905.

126. Shahin B, Germain M, Kazem A, Annat G. Benefits of short inspiratory muscle training on exercise capacity, dyspnea, and inspiratory fraction in COPD patients. *Int J Chron Obstruct Pulmon Dis* 2008;3:423-427.

127. Shannon M. Predictors of major toxicity after theophylline overdose. *Ann Intern Med* 1993;119:1161-1167.

128. Sharratt MT, Henke KG, Aaron EA, Pegelow DF, Dempsey JA. Exercise-induced changes in functional residual capacity. *Respir Physiol* 1987;70:313-326.

129. Sherrill DL, Enright P, Cline M, Burrows B, Lebowitz MD. Rates of decline in lung function among subjects who restart cigarette smoking. *Chest* 1996;109:1001-1005.

130. Shuey CB Jr, Pierce AK, Johnson RL Jr. An evaluation of exercise tests in chronic obstructive lung disease. *J Appl Physiol* 1969;27:256-261.

131. Similowski T, Yan S, Gauthier AP, Macklem PT, Bellemare F. Contractile properties of the human diaphragm during chronic hyperinflation. *N Engl J Med* 1991;325:917-923.

132. Simpson K, Killian K, McCartney N, Stubbing DG, Jones NL. Randomised controlled trial of weightlifting exercise in patients with chronic airflow limitation. *Thorax* 1992;47:70-75.

133. Sin DD, McAlister FA, Man SFP, Anthonisen NR. Contemporary management of chronic obstructive pulmonary disease. *JAMA* 2003;290(17):2301-2312.

134. Smith J, Bellemare F. Effect of lung volume on in vivo contraction characteristics of human diaphragm. *J Appl Physiol* 1987;62:1893-1900.

135. Snider GL. Chronic obstructive pulmonary disease: A definition and implications of structural determinants of airflow obstruction for epidemiology. *Am Rev Respir Dis* 1989;140:S3-S8.

136. Snider GL, Faling LJ, Rennard SI. Chronic bronchitis and emphysema. In: Murray JF, Nadel JA, eds. *Textbook of Respiratory Medicine* (p. 1342). Philadelphia: Saunders, 1994.

137. Snider GL, Kleinerman J, Thurlbeck WM, Bengali ZK. The definition of emphysema. Report of a National Heart, Lung, and Blood Institute, Division of Lung Diseases workshop. *Am Rev Respir Dis* 1985;132:182-185.

138. Spruit MA, Gosselink R, Troosters T, De Paepe K, Decramer M. Resistance versus endurance training in patients with COPD and peripheral muscle weakness. *Eur Respir J* 2002;19:1072-1078.

139. Storer TW. Exercise in chronic pulmonary disease: Resistance exercise prescription. *Med Sci Sports Exerc* 2001;33:S680-S692.

140. Strijbos JH, Postma DS, van Altena R, Gimeno F, Koeter GH. A comparison between an outpatient hospital-based pulmonary rehabilitation program and a home-care pulmonary rehabilitation program in patients with COPD. A follow-up of 18 months. *Chest* 1996;109:366-372.

141. Stubbing DG, Pengelly LD, Morse JL, Jones NL. Pulmonary mechanics during exercise in subjects with chronic airflow obstruction. *J Appl Physiol* 1980;49:511-515.

142. Sue DY. Exercise testing in the evaluation of impairment and disability. *Clin Chest Med* 1994;15:369-387.

143. Sue DY, Wasserman K, Moricca RB, Casaburi R. Metabolic acidosis during exercise in patients with chronic obstructive pulmonary disease. Use of the V-slope method for anaerobic threshold determination. *Chest* 1988;94:931-938.

144. Sullivan P, Bekir S, Jaffar Z, Page C, Jeffery P, Costello J. Anti-inflammatory effects of low-dose oral theophylline in atopic asthma. [Erratum appears in *Lancet* 1994;343:1512]. *Lancet* 1994;343:1006-1008.

145. Tager IB, Segal MR, Speizer FE, Weiss ST. The natural history of forced expiratory volumes. Effect of cigarette smoking and respiratory symptoms. *Am Rev Respir Dis* 1988;138:837-849.

146. Tangri S, Woolf CR. The breathing pattern in chronic obstructive lung disease during the performance of some common daily activities. *Chest* 1973;63:26-127.

147. Thomason MJ, Strachan DP. Which spirometric indices best predict subsequent death from chronic obstructive pulmonary disease? *Thorax* 2000;55:785-788.

148. Thurlbeck WM. Pathophysiology of chronic obstructive pulmonary disease. *Clin Chest Med* 1990;11:389-403.

149. Thurlbeck WM, Muller NL. Emphysema: Definition, imaging, and quantification. *Am J Roentgenol* 1994;163:1017-1025.

150. Toshima MT, Kaplan RM, Ries AL. Experimental evaluation of rehabilitation in chronic obstructive pulmonary disease: Short-term effects on exercise endurance and health status. *Health Psychol* 1990;9:237-252.

151. Urbano FL, Pascual RM. Contemporary issues in the care of patients with chronic obstructive pulmonary disease. *J Manag Care Pharm* 2005;11(5):S2-S17.

152. Vogiatzis I, Nanas S, Roussos C. Interval training as an alternative modality to continuous exercise in patients with COPD. *Eur Respir J* 2002;26:65-72.

153. Wadell K, Henriksson-Larsen K, Lundgren R. Physical training with and without oxygen in patients with chronic obstructive pulmonary disease and exercise-induced hypoxaemia. *J Rehabil Med* 2001;33:200-205.

154. Wanke T, Formanek D, Lahrmann H, Brath H, Wild M, Wagner C, Zwick H. Effects of combined inspiratory muscle and cycle ergometer training on exercise performance in patients with COPD. *Eur Respir J* 1994;7:2205-2211.

155. Wasserman K, Hansen JE, Sue DY, Whipp BJ. *Principles of Exercise Testing and Interpretation*. Philadelphia: Lea & Febiger, 1986.

156. Wedzicha JA, Bestall JC, Garrod R, Garnham R, Paul EA, Jones PW. Randomized controlled trial of pulmonary rehabilitation in severe chronic obstructive pulmonary disease patients, stratified with the MRC dyspnoea scale. *Eur Respir J* 1998;12:363-369.

157. Wehr KL, Johnson RL Jr. Maximal oxygen consumption in patients with lung disease. *J Clin Invest* 1976;58:880-890.

158. Weiner P, Azgad Y, Ganam R. Inspiratory muscle training combined with general exercise reconditioning in patients with COPD. *Chest* 1992;102:1351-1356.

159. Weiner P, Weiner M. Inspiratory muscle training may increase peak inspiratory muscle flow in chronic obstructive pulmonary disease. *Respiration* 2006;73:151-156.

160. Whittom F, Jobin J, Simard PM, Leblanc P, Simard C, Bernard S, Belleau R, Maltais F. Histochemical and morphological characteristics of the vastus lateralis muscle in patients with chronic obstructive pulmonary disease. *Med Sci Sports Exerc* 1998;30:1467-1474.

161. Wijkstra PJ, Ten Vergert EM, van Altena R, Otten V, Kraan J, Postma DS, Koeter GH. Long term benefits of rehabilitation at home on quality of life and exercise tolerance in patients with chronic obstructive pulmonary disease. *Thorax* 1995;50:824-828.

162. Wilson DO, Rogers RM, Openbrier D. Nutritional aspects of chronic obstructive pulmonary disease. *Clin Chest Med* 1986;7:643-656.

163. Wilson DO, Rogers RM, Sanders MH, Pennock BE, Reilly JJ. Nutritional intervention in malnourished patients with emphysema. *Am Rev Respir Dis* 1986;134:672-677.

164. Wilson DO, Rogers RM, Wright EC, Anthonisen NR. Body weight in chronic obstructive pulmonary disease. The National Institutes of Health Intermittent Positive-Pressure Breathing Trial. *Am Rev Respir Dis* 1989;139:1435-1438.

165. Wright PR, Heck H, Langenkamp H, Franz KH, Weber U. Influence of a resistance training on pulmonary function and performance measures of patients with COPD. *Pneumologie* 2002;56:413-417.

166. Wuyam B, Payen JF, Levy P, Bensaidane H, Reutenauer H, Le Bas JF, Benabid AL. Metabolism and aerobic capacity of skeletal muscle in chronic respiratory failure related to chronic obstructive pulmonary disease. *Eur Respir J* 1992;5:157-162.

167. Zack MB, Palange AV. Oxygen supplemented exercise of ventilatory and nonventilatory muscles in pulmonary rehabilitation. *Chest* 1985;88:669-675.

168. Ziment I. Pharmacologic therapy of obstructive airway disease. *Clin Chest Med* 1990;11:461-486.

CHAPTER 18

1. Agertoft L, Pedersen S. Effect of long-term treatment with inhaled budesonide on adult height in children with asthma. *N Engl J Med* 2000;343:1064-1069.

2. American College of Sports Medicine. *ACSM's Guidelines for Exercise Testing and Prescription.* 9th ed. Baltimore: Lippincott Williams & Wilkins, 2013.

3. American College of Sports Medicine position stand. The recommended quantity and quality of exercise for developing and maintaining cardiorespiratory and muscular fitness, and flexibility in healthy adults. *Med Sci Sports Exerc* 1998;30:975-991.

4. Anderson SD, Daviskas E. The mechanism of exercise-induced asthma is. . . . *J Allergy Clin Immunol* 2000;106:453-459.

5. Anderson SD, Holzer K. Exercise induced asthma: Is it the right diagnosis in elite athletes? *J Allergy Clin Immunol* 2000;106:419-428.

6. Anderson SD, Silverman M, Walker SR. Metabolic and ventilatory changes in asthmatic patients during and after exercise. *Thorax* 1972;27:718-725.

7. Banzett RB, Dempsey JA, O'Donnell DE, Wambolt MZ. NHLBI workshop summary. Symptom perception and respiratory sensation in asthma. *Am J Respir Crit Care Med* 2000;162:1178-1182.

8. Barnes PJ, Brown MJ, Silverman M, et al. Circulating catecholamines in exercise and hyperventilation induced asthma. *Thorax* 1981;36:435-440.

9. Barnes PJ, Chung KF, Page CP. Inflammatory mediators of asthma: An update. *Pharmacol Rev* 1998;50:515-596.

10. Bar-Or O, Inbar O. Swimming and asthma. Benefits and deleterious effects. *Sports Med* 1992;14:397-405.

11. Basaran S, Guler-Uysal F, Ergen N, Seydaoglu G, et al. Effect of physical exercise on quality of life, exercise capacity, and pulmonary function in children with asthma. *J Rehabil Med* 2006;38:130-135.

12. Becker JM, Rogers J, Rossini G, Mirchandani H, D'Alonzo GE Jr. Asthma deaths during sports: Report of a 7-year experience. *J Allergy Clin Immunol* 2004;113:264-267.

13. Borg G. Perceived exertion as an indicator of somatic stress. *Scand J Rehabil Med* 1970;2:92-98.

14. Boulet LP, Boulet V, Milot J. How should we quantify asthma control? A proposal. *Chest* 2002;122:2217-2223.

15. Busse WW. Inflammation in asthma: The cornerstone of the disease and target therapy. *J Allergy Clin Immunol* 1998;102:S17-22.

16. Busse WW, Lemanske RF Jr. Asthma. *N Engl J Med* 2001;344:350-362.

17. Centers for Disease Control and Prevention. Forecasted state-specific estimates of self-reported asthma prevalence—United States. *MMWR* 1998;47:1022-1025.

18. Clark CJ, Cochrane LM. Assessment of work performance in asthma for determination of cardiorespiratory fitness and training capacity. *Thorax* 1988;43:745-798.

19. Clark CJ, Cochrane LM. Physical activity and asthma. *Curr Opin Pulmon Med* 1999;5:68-75.

20. Cochrane LM, Clark LJ. Benefits and problems of a physical training programme for asthmatic patients. *Thorax* 1990;45:345-351.

21. Cockroft DW. Bronchoprovocation methods: Direct challenges. *Clin Rev Allergy Immunol* 2003;24:19-26.

22. Counil FP, Varray A, Matecki S, et al. Training of aerobic and anaerobic fitness in children with asthma. *J Pediatr* 2003;142:179-184.

23. Cypcar D, Lemanske DF. Asthma and exercise. *Clin Chest Med* 1994;15:351-368.

24. Doull IJ. The effect of asthma and its treatment on growth. *Arch Dis Child* 2004;89:60-63.

25. Ebina M, Yaegashi H, Chiba R, et al. Hyperreactive site in the airway tree of asthmatic patients recorded by thickening of bronchial muscles: A morphometric study. *Am Rev Respir Dis* 1990;141:1327-1332.

26. Edelman JM, Turpin JA, Bronsky EA, et al. Oral montelukast compared with inhaled salmeterol to prevent exercise-induced bronchoconstriction. *Ann Intern Med* 2000;132:97-104.

27. Emtner M, Finne M, Stalenheim G. A three year followup of asthmatic patients participating in a 10-week rehabilitation program with emphasis on physical training. *Arch Phys Med Rehabil* 1998;78:539-544.

28. Emtner M, Herela M, Stalenheim G. High intensity physical training in adults with asthma. *Chest* 1996;109:323-330.

29. Ernst P, McIvor A, Ducharme FM, et al. Safety and effectiveness of long-acting inhaled beta-agonist bronchodilators when taken with inhaled corticosteroids. *Ann Intern Med* 2006;145:692-694.

30. Expert Panel Report 3 (EPR3): Guidelines for the Diagnosis and Management of Asthma. www.nhlbi.nih.gov/guidelines/asthma/asthgdln.htm [accessed December 31, 2012].

31. Garfinkel SK, Kesten S, Chapman KR, et al. Physiologic and nonphysiologic determinants of aerobic fitness in mild to moderate asthma. *Am Rev Respir Dis* 1992;145:741-745.

32. GINA: The Global Initiative for Asthma. 2011. http://ginasthma.org [accessed December 31, 2012].

33. Girodo M, Ekstrand KA, Metivier GJ. Deep diaphragmatic breathing: Rehabilitation exercise for the asthmatic patient. *Arch Phys Med Rehabil* 1992;73:717-720.

34. Global Strategy for Asthma Management and Prevention. 2006. www.ginaasthma.com.

35. Helenius I, Haahtela T. Allergy and asthma in elite summer sport athletes. *J Allergy Clin Immunol* 2000;106:444-452.

36. Henriksen JM, Dahl R. Effect of inhaled budesonide alone and in combination with low-dose terbutaline in children with exercise induced asthma. *Am Rev Respir Dis* 1983;128:993-997.

37. Holgate ST. The cellular and mediator basis of asthma in relation to natural history. *Lancet* 1997;350(Suppl. II):5-9.

38. Holgate ST, Davies DE, Lackie PM, Wilson SJ, Puddicombe SM, Lordan JL. Epithelial-mesenchymal interactions in the pathogenesis of asthma. *J Allergy Clin Immunol* 2000;105:193-204.

39. Homa DM, Mannino DM, Lara M. Asthma mortality in U.S. Hispanics of Mexican, Puerto Rican, and Cuban heritage, 1990-1995. *Am J Respir Crit Care Med* 2000;161:504-509.

40. Horvath I, Barnes PJ. Exhaled monoxides in asymptomatic atopic subjects. *Clin Exp Allergy* 1999;29:1276-1280.

41. Juniper EF, Guyatt GH, Ferrie PJ, et al. Measuring quality of life in asthma. *Am Rev Respir Dis* 1993;147:832-838.

42. Kelly HW, Nelson HS. Potential adverse effects of the inhaled corticosteroids. *J Allergy Clin Immunol* 2003;112:469-478.

43. Kharitonov S, Alving K, Barnes PJ. Exhaled and nasal nitric oxide measurement: Recommendations. The European Respiratory Society Task Force. *Eur Respir J* 1997;10:1683-1693.

44. Kawabori I, Pierson WE, Conquest LL, Bierman CW. Incidence of exercise induced asthma in children. *J Allergy Clin Immunol* 1996;58:447-455.

45. Laitinen A, Laitinen LA. Airway morphology: Endothelium/basement membrane. *Am J Respir Crit Care Med* 1994;150:514-517.

46. Laitinen LA, Laitinen A, Haahtela T. Airway mucosal inflammation even in patients with newly diagnosed asthma. *Am Rev Respir Dis* 1993;147:697-704.

47. Lang DM, Polansky M. Patterns of asthma mortality in Philadelphia 1969 to 1991. *N Engl J Med* 1994;331:1542-1546.

48. Lee TM, Brown MJ, Nagy L, et al. Exercise induced release of histamine and neutrophil chemotactic factor in atopic asthmatics. *J Allergy Clin Immunol* 1982;70:73-81.

49. Leff JA, Busse WW, Pearlman D, et al. Montelukast, a leukotriene-receptor antagonist, for the treatment of mild asthma and exercise induced bronchospasm. *N Engl J Med* 1998;339:147-152.

50. Levy ML, Fletcher M, Price PB, Hausen T, Halbert RJ, Yawn BP. International Primary Care Respiratory Group guidelines: Diagnosis of respiratory diseases in primary care. *Prim Care Respir J* 2006;15:20-34.

51. Ludwick SK, Jones JW, Jones TK, et al. Normalization of cardiopulmonary endurance in severely asthmatic children after bicycle ergometry therapy. *J Pediatr* 1986;109:446-451.

52. Mahler DA, Faryniarz K, Lentine T, et al. Measurement of breathlessness during exercise in asthmatics: Predictor variables, reliability, and responsiveness. *Am Rev Respir Dis* 1991;144:39-44.

53. Mannino DM, Homa DM, Pertowski CA, et al. Surveillance for asthma—United States, 1960-1995. *MMWR* 1998;47:1-27.

54. McFadden ER Jr, Gilbert FA. Exercise induced asthma. *N Engl J Med* 1994;330:1362-1367.

55. McParland BE, Macklem PT, Pare PD. Airway wall remodeling: Friend or foe? *J Appl Physiol* 2003;95:426-434.

56. Meyer R, Froner-Herwig B, Sporkel H. The effect of exercise and induced expectations on visceral perception in asthmatic patients. *J Psychosom Res* 1990;34:454-460.

57. Mosli M, Fabian D, Holt S, Beasley R. The global burden of asthma: Executive summary of the GINA Dissemination Committee report. *Allergy* 2004;59:469-478.

58. Murphy S. National Asthma Education and Prevention Program. *Expert Panel Report 2: Guidelines for the Diagnosis and Management of Asthma* (NIH publication 1997:97-4051). Bethesda, MD: U.S. Department of Health and Human Services, 1997.

59. Department of Health and Human Services. Summary Health Statistics for U.S. Adult: National Health Interview Survey, 2010. Series 10, Number 252, January 2012. DHHS Publication No. (PHS) 2012-1580. cdc.gov/nchs/data/series/S1_10/SR10_252.pdf [accessed December 31, 2012].

60. Nelson JA, Strauss L, Skowronski M, et al. Effect of long term salmeterol treatment on exercise induced asthma. *N Engl J Med* 1998;339:141-146.

61. Noviski N, Bar-Yishay E, Gur I, Godfrey S. Respiratory heat/water loss alone does not determine the severity of exercise-induced asthma. *Eur Respir J* 1988;1:253-256.

62. Ober C. Perspectives on the past decade of asthma genetics. *J Allergy Clin Immunol* 2005;116:274-278.

63. Orenstein DM, Reed ME, Grogan FT, et al. Exercise conditioning in children with asthma. *J Pediatr* 1985;106:556-560.

64. Peel ET, Soutar CA, Seaton A. Assessment of variability of exercise tolerance limited by breathlessness. *Thorax* 1988;43:960-964.

65. Pelligrino R, Viegi G, Brusasco B, Crapo RO, Burgos F, Casaburi R, et al. Interpretative strategies for lung function tests. *Eur Respir J* 2005;26:948-968.

66. Pizzichini MM, Popov TA, Efthimiadis A, Hussack P, Evans S, Pizzichini E, et al. Spontaneous and induced sputum to measure indices of airway inflammation in asthma. *Am J Respir Crit Care Med* 1996;154(4 Pt 1):866-869.

67. Ram FS, Robinson SM, Black PN, Picol J. Physical training for asthma. *Cochrane Database Syst Rev* 2005;Oct 19(4):CD001116.

68. Robertson CF, Rubinfeld AR, Bowes G. Pediatric asthma deaths in Victoria: The mild are at risk. *Pediatr Pulmonol* 1992;13:95-100.

69. Roche WR, Beasley R, Williams JH, Holgate ST. Subepithelial fibrosis in the bronchi of asthmatics. *Lancet* 1989;1:520-524.

70. Rubenfeld A, Pain M. Perception of asthma. *Lancet* 1976;1(7965):882-884.

71. Salpeter SR, Buckley NS, Ormiston TM, Salpeter EE. Meta-analysis: Effect of long-acting beta-agonists on severe asthma exacerbations and asthma-related deaths. *Ann Intern Med* 2006;144:904-912.

72. Senthilselvan A. Prevalence of physician diagnosed asthma in Saskatchewan, 1981-1990. *Chest* 1999;114:388-392.

73. Sharek PJ, Bergman DA. Beclomethasone for asthma in children: Effects on linear growth. *Cochrane Database Syst Rev* 2000;CD001282.

74. Sharek PJ, Bergman DA. The effect of inhaled steroids on the linear growth of children with asthma: A meta-analysis. *Pediatrics* 2000;106:E8.

75. Smith AD, Cowan JO, Brassett KP, Herbison GP, Taylor DR. Use of exhaled nitric oxide measurements to guide treatment in chronic asthma. *N Engl J Med* 2005;352:2163-2173.

76. Strunk RC, Rubin D, Kelly L, et al. Determination of fitness in children with asthma: Use of standardized tests for functional endurance, body fat composition, flexibility, and abdominal strength. *Am J Dis Child* 1988;142:940-944.

77. Tsanakas JN, Milner RD, Bannister OM, et al. Free running asthma screening test. *Arch Dis Child* 1988;63:261-265.

78. Wang L, McParland BE, Pare PD. The functional consequences of structural changes in the airways: Implications for airway hyperresponsiveness in asthma. *Chest* 2003;123(3 Suppl.):356S-362S.

79. Warren JB, Keynes RJ, Brown MJ, et al. Blunted sympathoadrenal response to exercise in asthmatic subjects. *Br J Dis Chest* 1982;76:147-150.

80. Weiner P, Axgad R, Ganam R, Weiner R. Inspiratory muscle training in patients with bronchial asthma. *Chest* 1992;102:1357-1361.

81. Zeiger FS, Dawson C, Weiss S. Relationships between duration of asthma and asthma severity among children in the childhood asthma management program (CAMP). *J Allergy Clin Immunol* 1999;103:376-387.

82. Zeitoun M, Wilk B, Matsuzaka A, KnOpfli BH, Wilson BA, Bar-Or O. Facial cooling enhances exercise-induced bronchoconstriction in asthmatic children. *Med Sci Sports Exerc* 2004;36:767-771.

CHAPTER 19

1. Alison JA, Donnelly PM, Lennon M, Parker S, Torzillo P, Mellis C, Bye PTP. The effect of a comprehensive, intensive inpatient treatment program on lung function and exercise capacity in patients with cystic fibrosis. *Phys Ther* 1994;74:583-593.

2. Alison J, Duong B, Robinson M, Regnis J, Donnelly P, Bye PTP. Level of aerobic fitness and survival in adults with cystic fibrosis. *Am J Respir Crit Care Med* 1997;155(Pt 2):A642.

3. American Academy of Pediatrics: Committee on Sports Medicine Policy Statement. Strength training, weight and power lifting, and body building by children and adolescents. *Pediatrics* 1990;86:801-803.

4. Asher MI, Pardy RL, Coates AL, Thomas E, Macklem PT. The effects of inspiratory muscle training in patients with CF. *Am Rev Respir Dis* 1982;126:855-859.

5. Babb TG. Mechanical ventilatory constraints in aging, lung disease, and obesity: Perspectives and brief review. *Med Sci Sports Exerc* 1999;31(Suppl. 1):S12-S22.

6. Bakker W. Nutritional state and lung disease in cystic fibrosis. *Neth J Med* 1992;41:130-136.

7. Baldwin DR, Hill AL, Peckham DG, Knox AJ. Effect of addition of exercise to chest physiotherapy on sputum expectoration and lung function in adults with cystic fibrosis. *Respir Med* 1994;88:49-53.

8. Balfour-Lynn IM, Carr SB, Madge S, Prasad A, MacAlister L, Laverty A, Dinwiddie R. *Effect of altitude on exercise testing in children with cystic fibrosis.* Proceedings of the 20th European Working Group on Cystic Fibrosis. Brussels, 1995.

9. Balfour-Lynn IM, Prasad SA, Laverty A, Whitehead BF, Dinwiddie R. A step in the right direction: Assessing exercise tolerance in children with cystic fibrosis. *Pediatr Pulmonol* 1998;25:278-284.

10. Bar-Or O, Blimkie CJ, Hay JA, MacDougall JD, Ward DS, Wilson WM. Voluntary dehydration and heat intolerance in cystic fibrosis. *Lancet* 1992;339:696-699.

11. Boas SR, Danduran MJ, McBride AL, McColley SA, O'Gorman MRG. Post exercise immune correlates in children with and without cystic fibrosis. *Med Sci Sports Exerc* 2000;32:1997-2004.

12. Boas SR, Danduran MJ, McColley SA. Energy metabolism during anaerobic exercise in children with cystic fibrosis and asthma. *Med Sci Sports Exerc* 1999;31:1242-1249.

13. Boas SR, Danduran MJ, McColley SA, Beeman K, O'Gorman MRG. Immune modulation following aerobic exercise in children with cystic fibrosis. *Int J Sports Med* 2000;21:294-301.

14. Boas SR, Joswiak ML, Nixon PA, Fulton JA, Orenstein DM. Factors limiting anaerobic performance in adolescent males with cystic fibrosis. *Med Sci Sports Exerc* 1996;28:291-298.

15. Bradley J, Howard J, Wallace E, Elborn S. Reliability, repeatability, and sensitivity of the modified shuttle test in adult cystic fibrosis. *Chest* 2000;117:1666-1671.

16. Bradley J, Howard J, Wallace E, Elborn S. Validity of a modified shuttle test in adult cystic fibrosis. *Thorax* 1999;54:437-439.

17. Cabrera ME, Lough MD, Doershuk CF, DeRivera GA. Anaerobic performance—assessed by the Wingate test—in patients with cystic fibrosis. *Pediatr Exerc Sci* 1993;5:78-87.

18. Canny GJ. Ventilatory response to exercise in cystic fibrosis. *Acta Paediatr Scand* 1985;74:451-452.

19. Cerny FJ. Relative effects of bronchial drainage and exercise for in-hospital care of patients with cystic fibrosis. *Phys Ther* 1989;69(8):633-639.

20. Cerny FJ, Cropp GJA, Bye MR. Hospital therapy improves exercise tolerance and lung function in cystic fibrosis. *Am J Dis Child* 1984;138:261-265.

21. Cerny FJ, Pullano TP, Cropp GJA. Cardiorespiratory adaptations to exercise in cystic fibrosis. *Am Rev Respir Dis* 1982;126:217-220.

22. Charge TD, Drury D, Pianosi P, Kopelman H, Coates AL. Nutritional rehabilitation and changes in respiratory strength, function, and maximal exercise capacity in cystic fibrosis. *Am Rev Respir Dis* 1991;143(Suppl.):A300.

23. Coates AL. Oxygen therapy, exercise, and cystic fibrosis. *Chest* 1992;101:2-4.

24. Cropp GJ, Pullano TP, Cerny FJ, Nathanson IT. Exercise tolerance and cardiorespiratory adjustments at peak work capacity in cystic fibrosis. *Am Rev Respir Dis* 1982;126:211-216.

25. Cystic Fibrosis Foundation. Cystic Fibrosis Foundation Patient Registry, 2009. Annual data report. Bethesda, MD, October 2010.

26. Danduran MJ, Boas SR. Fibrosi cistica ed esercizio fisico. *Il Fisioterapista* 2000;6:36-41.

27. Darbee J, Watkins M. Isokinetic evaluation of muscle performance in individuals with cystic fibrosis. *Pediatr Pulmonol* 1987;3(Suppl.):140-141.

28. de Jong W, Grevink RG, Roorda RJ, Kaptein AA, van der Schans CP. Effects of a home exercise training program in patients with cystic fibrosis. *Chest* 1994;105:463-468.

29. deMeer K, Gulmans VAM, van der Laag J. Peripheral muscle weakness and exercise capacity in children with cystic fibrosis. *Am J Respir Crit Care Med* 1999;159:748-754.

30. deMeer K, Jeneson JAL, Gulmans VAM, van der Laag J, Berger R. Efficiency of oxidative work performance of skeletal muscle in patients with cystic fibrosis. *Thorax* 1995;50:980-983.

31. Docherty D. Field tests and test batteries. In: Docherty D, ed. *Measurement in Pediatric Exercise Science* (pp. 285-327). Champaign, IL: Human Kinetics, 1996.

32. Dunlevy CL, Douce FH, Hill E, Baez S, Clutter J. Physiological and psychological effects of low-impact aerobic exercise on young adults with cystic fibrosis. *J Cardiopulm Rehabil* 1994;14:47-51.

33. Edlund LD, French RW, Herbst JJ, Ruttenberg HD, Ruhling RO, Adams TD. Effects of a swimming program on children with cystic fibrosis. *Am J Dis Child* 1986;140:80-83.

34. Enright S, Chatham K, Ionesco AA, Unnithan VB, Shale DJ. Inspiratory muscle training improves lung function and exercise capacity in adults with cystic fibrosis. *Chest* 2004;126:405-411.

35. Freeman W, Stableforth DE, Cayton RM, Morgan MDL. Endurance exercise capacity in adults with cystic fibrosis. *Respir Med* 1993;87:541-549.

36. Godfey S, Mearns M. Pulmonary function and response to exercise in cystic fibrosis. *Arch Dis Child* 1971;46:144-151.

37. Goldring RM, Fishman AP, Turino GM, Cohen HI, Denning CR, Andersen DH. Pulmonary hypertension and cor pulmonale in cystic fibrosis of the pancreas. *J Pediatr* 1964;65:501-524.

38. Griffiths DM, Miller L, Flack E, Connett GJ. Reduced grip strength in children with cystic fibrosis. *Neth J Med* 1999;54:S62.

39. Gruber W, Orenstein DM, Braumann KM, Huls G. Health-related fitness and trainability in children with cystic fibrosis. *Pediatr Pulmonol* 2008;43:953-964.

40. Gruber W, Orenstein DM, Braumann KM, Paul K, Huls G. Effects of an exercise program in children with cystic fibrosis: Are there differences between females and males? *J Pediatr* 2011;158:71-76.

41. Guillen MAJ, Posadas AS, Asensi JRV, Moreno RMG, Rodriguez MAN, Gonzalez AS. Reproducibility of the walking test in patients with cystic fibrosis. *An Esp Pediatr* 1999;51:475-478.

42. Gulmans VAM, deMeer K, Brackel HJL, Faber JAJ, Berger R, Helders PJM. Outpatient exercise training in children with cystic fibrosis: Physiological effects, perceived competence, and acceptability. *Pediatr Pulmonol* 1999;28:39-46.

43. Gulmans VAM, van Veldhoven NHMJ, de Meer K, Helders PJM. The six-minute walking test in children with cystic fibrosis: Reliability and validity. *Pediatr Pulmonol* 1996;22:85-89.

44. Hebestreit H, Hebestreit A, Trusen A, Hughson RL. Oxygen uptake kinetics are slowed in cystic fibrosis. *Arch Dis Child* 2004;89:928-933.

45. Hebestreit H, Kieser S, Junge S, Ballmann M, Hebestreit A, Schindler T, Schenk T, Posseit H-G, Kriemier S. Long-term effects of a partially supervised conditioning programme in cystic fibrosis. *Eur Respir J* 2010;35:578-583.

46. Heijerman HGM. Chronic obstructive lung disease and respiratory muscle function: The role of nutrition and exercise training in cystic fibrosis. *Respir Med* 1993;Suppl. B:49-51.

47. Heijerman HGM, Bakker W, Sterk PJ, Dijkman JH. Oxygen-assisted exercise training in adult cystic fibrosis patients with pulmonary limitation to exercise. *Int J Rehabil Res* 1991;14:101-115.

48. Henke KG, Orenstein DM. Oxygen saturation during exercise in cystic fibrosis. *Am Rev Respir Dis* 1984;129:708-711.

49. Hussy J, Gormley J, Leen G, Greally P. Peripheral muscle strength in young males with cystic fibrosis. *J Cyst Fibros* 2002;1(3):116-121.

50. Inbar O, Bar-Or O, Skinner JS. *The Wingate Anaerobic Test.* Champaign, IL: Human Kinetics, 1996.

51. Jenesma M, Concannon D, Gallagher CG. An evaluation of spinal posture and hamstring muscle flexibility in cystic fibrosis adults. *Pediatr Pulmonol* 1998;17:350.

52. Kaplan RM, Anderson JP, Wu AW, Mathews WC, Kozin F, Orenstein DM. The quality of well-being scale. Application in AIDS, cystic fibrosis, and arthritis. *Med Care* 1989;27(Suppl.):27-43.

53. Kaplan TA, McKey RM, Toraya N, Moccia G. Impact of CF summer camp. *Clin Pediatr* 1992;31:161-167.

54. Kaplan TA, Moccia G, McKey RM. Unique pattern of pulmonary function after exercise in patients with cystic fibrosis. *Pediatr Exerc Sci* 1994;6:275-286.

55. Keens TG, Krastius IRB, Wannemaker EM, Levison H, Crozier DN, Bryan AC. Ventilatory muscle endurance training in normal subjects and patients with cystic fibrosis. *Am Rev Respir Dis* 1977;116:853-860.

56. Klijn PH, Oudshoorn A, van der Ent CK, van der Net J, Kimpen JL, Helders PJ. Effects of anaerobic training in children with cystic fibrosis. A randomized controlled study. *Chest* 2004;125:1299-1305.

57. Kriemler S, Wilk B, Schurer W, Wilson WM, Bar-Or O. Preventing dehydration in children with cystic fibrosis who exercise in the heat. *Med Sci Sports Exerc* 1999;31:774-779.

58. Lands LC, Heigenhauser GJ, Jones NL. Analysis of factors limiting maximal exercise performance in cystic fibrosis. *Clin Sci* 1992;83:391-397.

59. Lands LC, Heigenhauser GJF, Jones NL. Respiratory and peripheral muscle function in cystic fibrosis. *Am Rev Respir Dis* 1993;147:865-869.

60. Lebecque P, Lapierre JG, Lamarre A, Coates AL. Diffusion capacity and oxygen desaturation effects on exercise in patients with cystic fibrosis. *Chest* 1987;91:693-697.

61. Levison H, Cherniack RM. Ventilatory cost of exercise in chronic obstructive pulmonary disease. *J Appl Physiol* 1968;25:21-27.

62. Lohman T. *Advances in Body Composition Assessment.* Champaign, IL: Human Kinetics, 1992.

63. Loutzenhiser JK, Clark R. Physical activity and exercise in children with cystic fibrosis. *J Pediatr Nurs* 1993;8:112-119.

64. Marcotte JE, Grisdale RK, Levison H, Coates AL, Canny GJ. Multiple factors limit exercise capacity in cystic fibrosis. *Pediatr Pulmonol* 1986;2:274-281.

65. Marcus CL, Bader D, Stabile M, Wang CI, Osher AB, Keens TG. Supplemental oxygen and exercise performance in patients with severe cystic fibrosis. *Chest* 1992;101:52-57.

66. McKone EF, Barry SC, FitzGerald MX. The role of arterial hypoxemia and pulmonary mechanics in exercise limitation in adults with cystic fibrosis. *J Appl Physiol* 2005;99(3):1012-1018.

67. McKone EF, Barry SC, FitzGerald MX, Gallagher CG. Reproducibility of maximal exercise ergometer testing in patients with cystic fibrosis. *Chest* 1999;116:363-368.

68. Moorcroft AJ, Dodd ME, Morris J, Webb AK. Individualized home exercise training in adults with cystic fibrosis: A 1 year randomized controlled study. *Thorax* 2004;59(12):1074-1080.

69. Moorcroft AJ, Dodd ME, Webb AK. Exercise testing and prognosis in adult cystic fibrosis. *Thorax* 1997;52:291-293.

70. Neri AS, Lori I, Festini F, Masi L, Brandi ML, Galici V, Braggion C, Taccetti G. Bone mineral density in cystic fibrosis patients under the age of 18 years. *Minerva Pediatrica* 2008;60(2):147-154.

71. Nick J, Rodman D. Manifestations of cystic fibrosis diagnosed in adulthood. *Curr Opin Pulm Med* 2005;11(6):513-518.

72. Nieman DC. The immune response to prolonged cardiorespiratory exercise. *Am J Sports Med* 1996;24:S98-S103.

73. Nixon PA, Orenstein DM. Exercise testing in children. *Pediatr Pulmonol* 1988;5:107-122.

74. Nixon PA, Orenstein DM, Kelsey SF, Doershuk CF. The prognostic value of exercise testing in patients with cystic fibrosis. *N Engl J Med* 1992;327:1785-1788.

75. Orenstein DM, Franklin BA, Doershuk CF, Hellerstein HK, Germann KJ, Horowitz JG, Stern RC. Exercise conditioning and cardiopulmonary fitness in cystic fibrosis: The effects of a three-month supervised running program. *Chest* 1981;80:392-398.

76. Orenstein DM, Henke KG, Costill DL, Doechuk CF, Lemon PJ, Stern RC. Exercise and heat stress in cystic fibrosis patients. *Pediatr Res* 1983;17:267-269.

77. Orenstein DM, Henke KG, Green CG. Heat acclimation in cystic fibrosis. *J Appl Physiol* 1984;57:408-412.

78. Orenstein DM, Hovell MF, Mulvhill M, Keating KK, Hofstetter CR, Kelsey S, Morris K, Nixon PA. Strength vs aerobic training in children with cystic fibrosis: A randomized controlled trial. *Chest* 2004;126:1204-1214.

79. Orenstein DM, Nixon PA, Ross EA, Kaplan RM. The quality of well-being in cystic fibrosis. *Chest* 1989;95:344-347.

80. Pate RR, Barabowski T, Dowda M, Trost S. Tracking of physical activity in young children. *Med Sci Sports Exerc* 1996;28:92-96.

81. Pianosi P, LeBlanc J, Almudevar A. Peak oxygen uptake and mortality in children with cystic fibrosis. *Thorax* 2005;60:50-54.

82. Prasad SA, Randall SD, Balfour-Lynn IM. Fifteen-count breathlessness score: An objective measure for children. *Pediatr Pulmonol* 2000;30:56-62.

83. Rachinsky SV, Kapranow NI, Tatochenko VK, Simonowa OI, Turina JE. Submaximal physical loads in cystic fibrosis. *Acta Univ Carol Med* 1990;36:198-200.

84. Rowland TM. Aerobic exercise testing protocols. In: Rowland TW, ed. *Pediatric Laboratory Exercise Testing: Clinical Guidelines* (pp. 19-41). Champaign, IL: Human Kinetics, 1991.

85. Ruppel GL. Spirometry and related test. In: Ruppel GL, ed. *Manual of Pulmonary Function Testing* (pp. 52-53). St. Louis: Mosby, 1998.

86. Ruter K, Staab D, Magdorf K, Kleinau I, Paul K, Hetzer R, Wahn U. The 12-minute walk test as assessment for lung transplantation in CF patients. *Neth J Med* 1999;54(Suppl.):S58.

87. Ryujin DT, Samuelson WM, Marshall BC. Oxygen saturation in adult cystic fibrosis patients during exercise at 1500 meters above sea level. *Pediatr Pulmonol* 1998;17(Suppl.):331.

88. Sahlberg M, Svantesson U, Magnusson Thomas E, Andersson BA, Saltin B, Strandvik B. Muscular strength after different types of training in physically active patients with cystic fibrosis. *Scand J Med Sci Sports* 2008;18:756-764.

89. Sawyer EH, Clanton TL. Improved pulmonary function and exercise tolerance with inspiratory muscle conditioning in children with cystic fibrosis. *Chest* 1993;104:1490-1497.

90. Schneiderman-Walker J, Pollock AL, Corey M, Wilkes DD, Canny GJ, Pedder L, Reisman JJ. A randomized controlled trial of a 3-year home exercise program in cystic fibrosis. *J Pediatr* 2000;136:304-310.

91. Selvadurai HC, Blimkie CJ, Cooper PJ, Mellis CM, Van Asperen PP. Gender differences in habitual activity in children with cystic fibrosis. *Arch Dis Child* 2004;89:928-933.

92. Selvadurai HC, Blimkie CJ, Meyers N, Mellis CM, Cooper PJ, Van Asperen PP. Randomized controlled study of in-hospital exercise training program in children with cystic fibrosis. *Pediatr Pulmonol* 2002;33:194-200.

93. Shah AR, Gozal D, Keens TG. Determinants of aerobic and anaerobic exercise performance in cystic fibrosis. *Am J Respir Crit Care Med* 1998;157:1145-1150.

94. Silverman M, Hobbs FDR, Gordon IRS, Carswell F. Cystic fibrosis, atopy, and airways lability. *Arch Dis Child* 1978;53:873-877.

95. Skeie B, Askanazi J, Rothkopf MM, Rosenbaum SH, Kvetan V, Ross E. Improved exercise tolerance with long term parenteral nutrition in cystic fibrosis. *Crit Care Med* 1987;15:960-962.

96. Stanghelle JK, Hjeltnes N, Michalsen H, Bangstad HJ, Skyberg D. Pulmonary function and oxygen uptake during exercise in 11 year old patients with cystic fibrosis. *Acta Paediatr Scand* 1986;75:657-661.

97. Strauss GD, Osher A, Wang C, Goodrich E, Gold F, Colman W, Stabile M, Dobrenchuk A, Keens TG. Variable weight training in cystic fibrosis. *Chest* 1987;92:273-276.

98. Thin AG, Dodd JD, Gallagher CG, Fitzgerald MX, Mcloughlin P. Effect of respiratory rate on airway deadspace ventilation during exercise in cystic fibrosis. *Respir Med* 2004;98:1063-1070.

99. Turchetta A, Salerno T, Lucidi V, Libera F, Cutrera R, Bush A. Usefulness of a program of hospital-supervised physical training in patients with cystic fibrosis. *Pediatr Pulmonol* 2004;38:115-118.

100. van Haren EHJ, Lammers JWJ, Festen J, van Herwaarden CL. Bronchial vagal tone and responsiveness to histamine, exercise and bronchodilators in adult patients with cystic fibrosis. *Eur Respir J* 1992;5:1083-1088.

101. Van Praagh E, Franca NM. Measuring maximal short term power output during growth. In: Van Praagh E, ed. *Pediatric Anaerobic Performance* (pp. 151-190). Champaign, IL: Human Kinetics, 1998.

102. Ward SA, Tomezsko JL, Holsclaw DS, Paolone AM. Energy expenditure and substrate utilization in adults with cystic fibrosis and diabetes mellitus. *Am J Clin Nutr* 1999;69:913-919.

103. Zach M, Oberwaldner B, Hausler F. Cystic fibrosis: Physical exercise versus chest physiotherapy. *Arch Dis Child* 1982;57:587-589.

104. Zwiren LD, Manos TM. Exercise testing and prescription considerations throughout childhood. In: Rottman JL, ed. *ACSM's Resource Manual for Guidelines for Exercise Testing and Prescription* (pp. 507-515). Baltimore: Williams & Wilkins, 1998.

CHAPTER 20

1. American Cancer Society. *Cancer Facts and Figures - 2006*. Atlanta: American Cancer Society, 2006.

2. American Cancer Society. *Cancer Facts & Figures 2011*. Atlanta: American Cancer Society, 2011.

3. *The American Heritage Stedman's Medical Dictionary*. Boston: Houghton Mifflin, 2002.

4. ATS/ACCP statement on cardiopulmonary exercise testing. *Am J Respir Crit Care Med* 2003;167:211-277.

5. Harvard Report on Cancer Prevention. Volume 1: Causes of human cancer. *Cancer Causes Control* 1996;7 Suppl. 1:S3-59.

6. The NCCN Clinical Practice Guidelines in Oncology (NCCN Guidelines). Cancer-Related Fatigue (Version 1.2110). National Comprehensive Cancer Network, Inc. http://www.nccn.org/professionals/physician_gls/f_guidelines.asp#fatigue.

7. U.S. Mortality Public use data tape 2003. National Center for Health Statistics. Centers for Disease Control and Prevention; 2006.

8. Anand P, Kunnumakara A, Sundaram C, Harikumar K, Tharakan S, Lai O, Sung B, Aggarwal B. Cancer is a preventable disease that requires major lifestyle changes. *Pharm Res* 2008;25:2097-2116.
9. Bernstein L, Ross RK, Lobo RA, Hanisch R, Krailo MD, Henderson BE. The effects of moderate physical activity on menstrual cycle patterns in adolescence: Implications for breast cancer prevention. *Br J Cancer* 1987;55:681-685.
10. Blair SN, Kohl HW 3rd, Paffenbarger RS Jr, Clark DG, Cooper KH, Gibbons LW. Physical fitness and all-cause mortality. A prospective study of healthy men and women. *JAMA* 1989;262:2395-2401.
11. Brown ML. The national economic burden of cancer: An update. *J Natl Cancer Inst* 1990;82:1811-1814.
12. Campbell KL, McTiernan A. Exercise and biomarkers for cancer prevention studies. *J Nutr* 2007;137:161S-169S.
13. Carlson LE, Smith D, Russell J, Fibich C, Whittaker T. Individualized exercise program for the treatment of severe fatigue in patients after allogeneic hematopoietic stem-cell transplant: A pilot study. *Bone Marrow Transplant* 2006;37:945-954.
14. Cecil RL, Bennett JC, Plum F. *Cecil Textbook of Medicine* (pp. 1335-1533). Philadelphia: Saunders, 1996.
15. Chicco AJ, Hydock DS, Schneider CM, Hayward R. Low-intensity exercise training during doxorubicin treatment protects against cardiotoxicity. *J Appl Physiol* 2006;100:519-527.
16. Cho TJ, Shreiner CD, Hwang SH, Moorefield CN, Courneya B, Godinez LA, Manriquez J, Jeong KU, Cheng SZ, Newkome GR. 5,10,15,20-Tetrakis[4'-(terpyridinyl)phenyl]porphyrin and its RuII complexes: Synthesis, photovoltaic properties, and self-assembled morphology. *Chem Commun (Camb)* 2007;4456-4458.
17. Colditz GA, Frazier AL. Models of breast cancer show that risk is set by events of early life: Prevention efforts must shift focus. *Cancer Epidemiol Biomarkers Prev* 1995;4:567-571.
18. Collado-Hidalgo A, Bower JE, Ganz PA, Cole SW, Irwin MR. Inflammatory biomarkers for persistent fatigue in breast cancer survivors. *Clin Cancer Res* 2006;12:2759-2766.
19. Courneya KS, Mackey JR, Bell GJ, Jones LW, Field CJ, Fairey AS. Randomized controlled trial of exercise training in postmenopausal breast cancer survivors: Cardiopulmonary and quality of life outcomes. *J Clin Oncol* 2003;21:1660-1668.
20. Courneya KS, Segal RJ, Gelmon K, Reid RD, Mackey JR, Friedenreich CM, Prouix C, Lane K, Ladha AB, Vallance JK, Liu Q, Yasui Y, McKenzie DC. Six-month follow-up of patient-rated outcomes in a randomized controlled trial of exercise training during breast cancer chemotherapy. *Cancer Epidemiol Biomarkers Prev* 2007;16:2572-2578.
21. Courneya KS, Segal RJ, Mackey JR, Gelmon K, Reid RD, Friedenreich CM, Ladha AB, Proulx C, Vallance JK, Lane K, Yasui Y, McKenzie DC. Effects of aerobic and resistance exercise in breast cancer patients receiving adjuvant chemotherapy: A multicenter randomized controlled trial. *J Clin Oncol* 2007;25:4396-4404.
22. Courneya KS, Sellar CM, Stevinson C, McNeely ML, Peddle CJ, Friedenreich CM, Tankel K, Basi S, Chua N, Mazurek A, Reiman T. Randomized controlled trial of the effects of aerobic exercise on physical functioning and quality of life in lymphoma patients. *J Clin Oncol* 2009;27:4605-4612.
23. Coussens LM, Werb Z. Inflammation and cancer. *Nature* 2002;420:860-867.
24. Daley AJ, Crank H, Saxton JM, Mutrie N, Coleman R, Roalfe A. Randomized trial of exercise therapy in women treated for breast cancer. *J Clin Oncol* 2007;25:1713-1721.
25. Dimeo F, Fetscher S, Lange W, Mertelsmann R, Keul J. Effects of aerobic exercise on the physical performance and incidence of treatment-related complications after high-dose chemotherapy. *Blood* 1997;90(9):3390-3394.
26. Dimeo FC, Stieglitz RD, Novelli-Fischer U, Fetscher S, Keul J. Effects of physical activity on the fatigue and psychologic status of cancer patients during chemotherapy. *Cancer* 1999;85:2273-2277.
27. Dimeo FC, Tilmann MH, Bertz H, Kanz L, Mertelsmann R, Keul J. Aerobic exercise in the rehabilitation of cancer patients after high dose chemotherapy and autologous peripheral stem cell transplantation. *Cancer* 1997;79:1717-1722.
28. Doll R, Peto R. The causes of cancer: Quantitative estimates of avoidable risks of cancer in the United States today. *J Natl Cancer Inst* 1981;66:1191-1308.
29. Francis K. Physical activity: Breast and reproductive cancer. *Compr Ther* 1996;22:94-99.
30. Friedenreich CM. Physical activity and cancer prevention: From observational to intervention research. *Cancer Epidemiol Biomarkers Prev* 2001;10:287-301.
31. Frisch RE, Wyshak G, Albright NL, Albright TE, Schiff I, Jones KP, Witschi J, Shiang E, Koff E, Marguglio M. Lower prevalence of breast cancer and cancers of the reproductive system among former college athletes compared to non-athletes. *Br J Cancer* 1985;52:885-891.
32. Galvao DA, Taaffe DR, Spry N, Joseph D, Newton RU. Combined resistance and aerobic exercise program reverses muscle loss in men undergoing androgen suppression therapy for prostate cancer without bone metastases: A randomized controlled trial. *J Clin Oncol* 2010;28:340-347.
33. Gielen S, Adams V, Mobius-Winkler S, Linke A, Erbs S, Yu J, Kempf W, Schubert A, Schuler G, Hambrecht R. Anti-inflammatory effects of exercise training in the skeletal muscle of patients with chronic heart failure. *J Am Coll Cardiol* 2003;42:861-868.
34. Goldhammer E, Tanchilevitch A, Maor I, Beniamini Y, Rosenschein U, Sagiv M. Exercise training modulates cytokines activity in coronary heart disease patients. *Int J Cardiol* 2005;100:93-99.
35. Haenszel W. Cancer mortality among the foreign-born in the United States. *J Natl Cancer Inst* 1961;26:37-132.
36. Hanahan D, Weinberg RA. Hallmarks of cancer: The next generation. *Cell* 2011;144:646-674.
37. Headley JA, Ownby KK, John LD. The effect of seated exercise on fatigue and quality of life in women with advanced breast cancer. *Oncol Nurs Forum* 2004;31:977-983.
38. Henderson BE, Ross RK, Judd HL, Krailo MD, Pike MC. Do regular ovulatory cycles increase breast cancer risk? *Cancer* 1985;56:1206-1208.
39. Hershcopf RJ, Bradlow HL. Obesity, diet, endogenous estrogens, and the risk of hormone-sensitive cancer. *Am J Clin Nutr* 1987;45:283-289.
40. Hoffman-Goetz L. Exercise, natural immunity, and tumor metastasis. *Med Sci Sports Exerc* 1994;26:157-163.
41. Holmes MD, Chen WY, Feskanich D, Kroenke CH, Colditz GA. Physical activity and survival after breast cancer diagnosis. *JAMA* 2005;293:2479-2486.
42. Hsieh CC, Trichopoulos D, Katsouyanni K, Yuasa S. Age at menarche, age at menopause, height and obesity as risk factors for breast cancer: Associations and interactions in an international case-control study. *Int J Cancer* 1990;46:796-800.

43. Irwin ML. Physical activity interventions for cancer survivors. *Br J Sports Med* 2009;43:32-38.

44. Irwin ML, Smith AW, McTiernan A, Ballard-Barbash R, Cronin K, Gilliland FD, Baumgartner RN, Baumgartner KB, Bernstein L. Influence of pre- and postdiagnosis physical activity on mortality in breast cancer survivors: The health, eating, activity, and lifestyle study. *J Clin Oncol* 2008;26:3958-3964.

45. Jones LW, Eves ND, Mackey JR, Peddle CJ, Haykowsky M, Joy AA, Courneya KS, Tankel K, Spratlin J, Reiman T. Safety and feasibility of cardiopulmonary exercise testing in patients with advanced cancer. *Lung Cancer* 2007;55:225-232.

46. Jones LW, Peddle CJ, Eves ND, Haykowsky MJ, Courneya KS, Mackey JR, Joy AA, Kumar V, Winton TW, Reiman T. Effects of presurgical exercise training on cardiorespiratory fitness among patients undergoing thoracic surgery for malignant lung lesions. *Cancer* 2007;110:590-598.

47. Kiningham RB. Physical activity and the primary prevention of cancer. *Prim Care* 1998;25:515-536.

48. Kohl HW, LaPorte RE, Blair SN. Physical activity and cancer. An epidemiological perspective. *Sports Med* 1988;6:222-237.

49. Lambert CP, Wright NR, Finck BN, Villareal DT. Exercise but not diet-induced weight loss decreases skeletal muscle inflammatory gene expression in frail obese elderly persons. *J Appl Physiol* 2008;105:473-478.

50. La Vecchia C, Decarli A, di Pietro S, Franceschi S, Negri E, Parazzini F. Menstrual cycle patterns and the risk of breast disease. *Eur J Cancer Clin Oncol* 1985;21:417-422.

51. Lee IM, Sesso HD, Paffenbarger RS Jr. Physical activity and risk of lung cancer. *Int J Epidemiol* 1999;28:620-625.

52. Meyerhardt JA, Giovannucci EL, Holmes MD, Chan AT, Chan JA, Colditz GA, Fuchs CS. Physical activity and survival after colorectal cancer diagnosis. *J Clin Oncol* 2006;24:3527-3534.

53. Mock V, Burke MB, Sheehan P, Creaton EM, Winningham ML, McKenney-Tedder S, Schwager LP, Liebman M. A nursing rehabilitation program for women with breast cancer receiving adjuvant chemotherapy. *Oncol Nurs Forum* 1994;21:899-907, discussion 908.

54. Morrow GR, Andrews PLR, Hickok JT, Roscoe JA, Matteson S. Fatigue associated with cancer and its treatment. *Support Care Cancer* 2002;10:389-398.

55. Muir C. *Cancer Incidence in Five Continents*, vol. V (p. xxxii). Lyon: International Agency for Research on Cancer, 1987.

56. Mustian KM, Griggs JJ, Morrow GR, McTiernan A, Roscoe JA, Bole CW, Atkins JN, Issell BF. Exercise and side effects among 749 patients during and after treatment for cancer: A University of Rochester Cancer Center Community Clinical Oncology Program study. *Support Care Cancer* 2006;14:732-741.

57. Nicklas BJ, Hsu FC, Brinkley TJ, Church T, Goodpaster BH, Kritchevsky SB, Pahor M. Exercise training and plasma C-reactive protein and interleukin-6 in elderly people. *J Am Geriatr Soc* 2008;56:2045-2052.

58. Nieman DC. Exercise, upper respiratory tract infection, and the immune system. *Med Sci Sports Exerc* 1994;26:128-139.

59. Nieman DC, Nehlsen-Cannarella SL. The immune response to exercise. *Semin Hematol* 1994; 31:166-179.

60. Pachman DR, Barton DL, Watson JC, Loprinzi CL. Chemotherapy-induced peripheral neuropathy: Prevention and treatment. *Clin Pharmacol Ther* 2011;90:377-387.

61. Paffenbarger RS Jr, Hyde RT, Wing AL. Physical activity and incidence of cancer in diverse populations: A preliminary report. *Am J Clin Nutr* 1987;45:312-317.

62. Paffenbarger RS Jr, Hyde RT, Wing AL, Hsieh CC. Physical activity, all-cause mortality, and longevity of college alumni. *N Engl J Med* 1986;314:605-613.

63. Paffenbarger RS Jr, Kampert JB, Chang HG. Characteristics that predict risk of breast cancer before and after the menopause. *Am J Epidemiol* 1980;112:258-268.

64. Pike MC, Krailo MD, Henderson BE, Casagrande JT, Hoel DG. "Hormonal" risk factors, "breast tissue age" and the age-incidence of breast cancer. *Nature* 1983;303:767-770.

65. Pyne DB, Baker MS, Fricker PA, McDonald WA, Telford RD, Weidemann MJ. Effects of an intensive 12-wk training program by elite swimmers on neutrophil oxidative activity. *Med Sci Sports Exerc* 1995;27:536-542.

66. Rao A, Cohen HJ. Symptom management in the elderly cancer patient: Fatigue, pain, and depression. *J Natl Cancer Inst Monogr* 2004;32:150-157.

67. Russo J, Russo IH. Role of differentiation in the pathogenesis and prevention of breast cancer. *Endocr Relat Cancer* 1997;4:7-21.

68. Schmitz KH, Ahmed RL, Troxel AB, Cheville A, Lewis-Grant L, Smith R, Bryan CJ, Williams-Smith CT, Chittams J. Weight lifting for women at risk for breast cancer-related lymphedema: A randomized trial. *JAMA* 2010;304:2699-2705.

69. Schmitz KH, Ahmed RL, Troxel A, Cheville A, Smith R, Lewis-Grant L, Bryan CJ, Williams-Smith CT, Greene QP. Weight lifting in women with breast-cancer-related lymphedema. *N Engl J Med* 2009;361:664-673.

70. Schmitz KH, Courneya KS, Matthews C, Demark-Wahnefried W, Galvao DA, Pinto BM, Irwin ML, Wolin KY, Segal RJ, Lucia A, Schneider CM, von Gruenigen VE, Schwartz AL. American College of Sports Medicine roundtable on exercise guidelines for cancer survivors. *Med Sci Sports Exerc* 2010;42:1409-1426.

71. Schubert C, Hong S, Natarajan L, Mills PJ, Dimsdale JE. The association between fatigue and inflammatory marker levels in cancer patients: A quantitative review. *Brain Behav Immun* 2007;21:413-427.

72. Schwartz AL, Mori M, Gao RL, Nail LM, King ME. Exercise reduces daily fatigue in women with breast cancer receiving chemotherapy. *Med Sci Sports Exerc* 2001;33:718-723.

73. Scott JM, Khakoo A, Mackey JR, Haykowsky MJ, Douglas PS, Jones LW. Modulation of anthracycline-induced cardiotoxicity by aerobic exercise in breast cancer: Current evidence and underlying mechanisms. *Circulation* 2011;124:642-650.

74. Seruga B, Zhang H, Bernstein LJ, Tannock IF. Cytokines and their relationship to the symptoms and outcome of cancer. *Nat Rev Cancer* 2008;8:887-899.

75. Shephard RJ. Physical activity and the healthy mind. *Can Med Assoc J* 1983;128:525-530.

76. Shephard RJ, Shek PN. Associations between physical activity and susceptibility to cancer: Possible mechanisms. *Sports Med* 1998;26:293-315.

77. Shephard RJ, Shek PN. Cancer, immune function, and physical activity. *Can J Appl Physiol* 1995;20:1-25.

78. Slattery ML, Potter J, Caan B, Edwards S, Coates A, Ma KN, Berry TD. Energy balance and colon cancer—beyond physical activity. *Cancer Res* 1997;57:75-80.

79. Sloan RP, Shapiro PA, Demeersman RE, McKinley PS, Tracey KJ, Slavov I, Fang Y, Flood PD. Aerobic exercise attenuates inducible TNF production in humans. *J Appl Physiol* 2007;103:1007-1011.

80. Speck RM, Courneya KS, Masse LC, Duval S, Schmitz KH. An update of controlled physical activity trials in cancer survivors: A systematic review and meta-analysis. *J Cancer Surviv* 2010;4(2):87-100.

81. Sternfeld B. Cancer and the protective effect of physical activity: The epidemiological evidence. *Med Sci Sports Exerc* 1992;24:1195-1209.

82. Thompson HJ. Effect of exercise intensity and duration on the induction of mammary carcinogenesis. *Cancer Res* 1994;54:1960s-1963s.

83. Thompson WG. Exercise and health: Fact or hype? *South Med J* 1994;87:567-574.

84. Thorsen L, Skovlund E, Stromme SB, Hornslien K, Dahl AA, Fossa SD. Effectiveness of physical activity on cardiorespiratory fitness and health-related quality of life in young and middle-aged cancer patients shortly after chemotherapy. *J Clin Oncol* 2005;23:2378-2388.

85. Thune I, Furberg AS. Physical activity and cancer risk: Dose-response and cancer, all sites and site-specific. *Med Sci Sports Exerc* 2001;33:S530-550, discussion S609-610.

86. Trichopoulos D, MacMahon B, Cole P. Menopause and breast cancer risk. *J Natl Cancer Inst* 1972;48:605-613.

87. Verloop J, Rookus MA, van der Kooy K, van Leeuwen FE. Physical activity and breast cancer risk in women aged 20-54 years. *J Natl Cancer Inst* 2000;92:128-135.

88. Vihko R, Apter D. Endocrine characteristics of adolescent menstrual cycles: Impact of early menarche. *J Steroid Biochem* 1984;20:231-236.

89. Winningham ML. Strategies for managing cancer-related fatigue syndrome: a rehabilitation approach. Cancer 2001; 92(S4), 988-997.]

90. Woods JA, Davis JM, Smith JA, Nieman DC. Exercise and cellular innate immune function. *Med Sci Sports Exerc* 1999;31:57-66.

CHAPTER 21

1. ACSM. *ACSM's Guidelines for Exercise Testing and Prescription.* 9th ed. Baltimore: Lippincott Williams & Wilkins, 2013.

2. Anuurad E, Bremer A, Berglund L. HIV protease inhibitors and obesity. *Curr Opin Endocrinol Diabetes Obes* 2010;17:478-485.

3. Barbaro G. Heart and HAART: Two sides of the coin for HIV-associated cardiology issues. *World J Cardiol* 2010;2:53-57.

4. Barre-Sinoussi F, Chermann JC, Rey F, Nugeyre MT, Chamaret S, Gruest J, Dauguet C, Axler-Blin C, Vezinet-Brun F, Rouzioux C, Rozenbaum W, Montagnier L. Isolation of a T-lymphotropic retrovirus from a patient at risk for acquired immune deficiency syndrome (AIDS). *Science* 1983;220:868-871.

5. Bergasa NV, Mehlman J, Bir K. Aerobic exercise: A potential therapeutic intervention for patients with liver disease. *Med Hypotheses* 2004;62:935-941.

6. Bevilacqua M, Dominguez LJ, Barbagallo M. Insulin resistance and the cardiometabolic syndrome in HIV infection. *J Cardiometab Syndr* 2009;4:40-43.

7. Bhasin S, Storer TW, Javanbakht M, Berman N, Yarasheski KE, Phillips J, Dike M, Sinha-Hikim I, Shen R, Hays RD, Beall G. Testosterone replacement and resistance exercise in HIV-infected men with weight loss and low testosterone levels. *JAMA* 2000;283:763-770.

8. Bopp CM, Phillips KD, Fulk LJ, Dudgeon WD, Sowell R, Hand GA. Physical activity and immunity in HIV-infected individuals. *AIDS Care* 2004;16:387-393.

9. Bopp CM, Phillips KD, Fulk LJ, Hand GA. Clinical implications of therapeutic exercise in HIV/AIDS. *J Assoc Nurses AIDS Care* 2003;14:73-78.

10. Buchbinder SP, Katz MH, Hessol NA, O'Malley PM, Holmberg SD. Long-term HIV-1 infection without immunologic progression. *AIDS* 1994;8:1123-1128.

11. Cade WT, Fantry LE, Nabar SR, Keyser RE. Decreased peak arteriovenous oxygen difference during treadmill exercise testing in individuals infected with the human immunodeficiency virus. *Arch Phys Med Rehabil* 2003;84:1595-1603.

12. Calmy A, Carey D, Mallon PW, Wand H, Law M, Cooper DA, Carr A. Early changes in adipokine levels and baseline limb fat may predict HIV lipoatrophy over 2 years following initiation of antiretroviral therapy. *HIV Med* 2008;9:101-110.

13. Calza L, Masetti G, Piergentili B, Trapani F, Cascavilla A, Manfredi R, Colangeli V, Viale P. Prevalence of diabetes mellitus, hyperinsulinaemia and metabolic syndrome among 755 adult patients with HIV-1 infection. *Int J STD AIDS* 2011;22:43-45.

14. Cardo DM, Culver DH, Ciesielski CA, Srivastava PU, Marcus R, Abiteboul D, Heptonstall J, Ippolito G, Lot F, McKibben PS, Bell DM. A case-control study of HIV seroconversion in health care workers after percutaneous exposure. Centers for Disease Control and Prevention Needlestick Surveillance Group. *N Engl J Med* 1997;337:1485-1490.

15. Carr A, Samaras K, Chisholm DJ, Cooper DA. Pathogenesis of HIV-1-protease inhibitor-associated peripheral lipodystrophy, hyperlipidaemia, and insulin resistance. *Lancet* 1998;351:1881-1883.

16. CDC. Centers for Disease Control and Prevention. 1993 revised classification system for HIV infection and expanded surveillance case definition for AIDS among adolescents and adults. *JAMA* 1993;269:729-730.

17. CDC. A cluster of Kaposi's sarcoma and Pneumocystis carinii pneumonia among homosexual male residents of Los Angeles and Orange Counties, California. *MMWR* 1982;31:305-307.

18. CDC. HIV/AIDS. www.cdc.gov/hiv/. 2010 [accessed April 25, 2013].

19. CDC. HIV prevalence estimates—United States, 2006. *MMWR* 2008;57:1073-1076.

20. CDC. Public Health Service guidelines for the management of health-care worker exposures to HIV and recommendations for postexposure prophylaxis. Centers for Disease Control and Prevention. *MMWR Recomm Rep* 1998;47:1-33.

21. CDC. Recommendations for follow-up of health-care workers after occupational exposure to hepatitis C virus. *MMWR* 1997;46:603-606.

22. CDC. Results of the Expanded HIV Testing Initiative—25 jurisdictions, United States, 2007-2010. *MMWR* 2011;60:805-810.

23. CDC. Revision of the case definition of acquired immunodeficiency syndrome for national reporting—United States. *MMWR* 1985;34:373-375.

24. CDC. Revision of the CDC surveillance case definition for acquired immunodeficiency syndrome. *MMWR* 1987;36:1S - 15S.

25. Chou R, Huffman LH, Fu R, Smits AK, Korthuis PT. Screening for HIV: A review of the evidence for the U.S. Preventive Services Task Force. *Ann Intern Med* 2005;143:55-73.

26. Dimeo FC, Thomas F, Raabe-Menssen C, Propper F, Mathias M. Effect of aerobic exercise and relaxation training on fatigue and physical performance of cancer patients after surgery. A randomised controlled trial. *Support Care Cancer* 2004;12:774-779.

27. Dolan SE, Frontera W, Librizzi J, Ljungquist K, Juan S, Dorman R, Cole ME, Kanter JR, Grinspoon S. Effects of a supervised home-based aerobic and progressive resistance training regimen in women infected with human immunodeficiency virus: A randomized trial. *Arch Intern Med* 2006;166:1225-1231.

28. Douek DC, Roederer M, Koup RA. Emerging concepts in the immunopathogenesis of AIDS. *Annu Rev Med* 2009;60:471-484.

29. Dudgeon WD, Phillips KD, Bopp CM, Hand GA. Physiological and psychological effects of exercise interventions in HIV disease. *AIDS Patient Care STDS* 2004;18:81-98.

30. Dudgeon WD, Phillips KD, Durstine JL, Burgess SE, Lyerly GW, Davis JM, Hand GA. Individual exercise sessions alter circulating hormones and cytokines in HIV-infected men. *Appl Physiol Nutr Metab* 2010;35:560-568.

31. Engelson ES, Agin D, Kenya S, Werber-Zion G, Luty B, Albu JB, Kotler DP. Body composition and metabolic effects of a diet and exercise weight loss regimen on obese, HIV-infected women. *Metabolism* 2006;55:1327-1336.

32. Ezzati M, Friedman AB, Kulkarni SC, Murray CJ. The reversal of fortunes: Trends in county mortality and cross-county mortality disparities in the United States. *PLoS Med* 2008;5:e66.

33. Fairfield WP, Treat M, Rosenthal DI, Frontera W, Stanley T, Corcoran C, Costello M, Parlman K, Schoenfeld D, Klibanski A, Grinspoon S. Effects of testosterone and exercise on muscle leanness in eugonadal men with AIDS wasting. *J Appl Physiol* 2001;90:2166-2171.

34. Farinatti PT, Borges JP, Gomes RD, Lima D, Fleck SJ. Effects of a supervised exercise program on the physical fitness and immunological function of HIV-infected patients. *J Sports Med Phys Fitness* 2010;50:511-518.

35. Feigenbaum K, Longstaff L. Management of the metabolic syndrome in patients with human immunodeficiency virus. *Diabetes Educ* 2010;36:457-464.

36. Fillipas S, Cherry CL, Cicuttini F, Smirneos L, Holland AE. The effects of exercise training on metabolic and morphological outcomes for people living with HIV: A systematic review of randomised controlled trials. *HIV Clin Trials* 2010;11:270-282.

37. Fillipas S, Oldmeadow LB, Bailey MJ, Cherry CL. A six-month, supervised, aerobic and resistance exercise program improves self-efficacy in people with human immunodeficiency virus: A randomised controlled trial. *Aust J Physiother* 2006;52:185-190.

38. Fulk LJ, Kane BE, Phillips KD, Bopp CM, Hand GA. Depression in HIV-infected patients: Allopathic, complementary, and alternative treatments. *J Psychosom Res* 2004;57:339-351.

39. Gensichen J, Petersen JJ, Karroum T, Rauck S, Ludman E, Konig J, Gerlach FM. Positive impact of a family practice-based depression case management on patient's self-management. *Gen Hosp Psychiatry* 2011;33:23-28.

40. Gilbert PB, McKeague IW, Eisen G, Mullins C, Gueye NA, Mboup S, Kanki PJ. Comparison of HIV-1 and HIV-2 infectivity from a prospective cohort study in Senegal. *Stat Med* 2003;22:573-593.

41. Glover EI, Phillips SM. Resistance exercise and appropriate nutrition to counteract muscle wasting and promote muscle hypertrophy. *Curr Opin Clin Nutr Metab Care* 2010;13:630-634.

42. Govender S, Harrison WJ, Lukhele M. Impact of HIV on bone and joint surgery. *Best Pract Res Clin Rheumatol* 2008;22:605-619.

43. Grinspoon S, Corcoran C, Parlman K, Costello M, Rosenthal D, Anderson E, Stanley T, Schoenfeld D, Burrows B, Hayden D, Basgoz N, Klibanski A. Effects of testosterone and progressive resistance training in eugonadal men with AIDS wasting. A randomized, controlled trial. *Ann Intern Med* 2000;133:348-355.

44. Guenter P, Muurahainen N, Simons G, Kosok A, Cohan GR, Rudenstein R, Turner JL. Relationships among nutritional status, disease progression, and survival in HIV infection. *J Acquir Immune Defic Syndr* 1993;6:1130-1138.

45. Hand GA, Lyerly GW, Jaggers JR, Dudgeon WD. Impact of aerobic and resistance exercise on the health of HIV-infected persons. *Am J Lifestyle Med* 2009;3:489-499.

46. Hand GA, Phillips KD, Dudgeon WD, Lyerly GW, Durstine JL, Burgess SE. Moderate intensity exercise training reverses functional aerobic impairment in HIV-infected individuals. *AIDS Care* 2008;20:1066-1074.

47. Haskell WL, Lee IM, Pate RR, Powell KE, Blair SN, Franklin BA, Macera CA, Heath GW, Thompson PD, Bauman A. Physical activity and public health: Updated recommendation for adults from the American College of Sports Medicine and the American Heart Association. *Med Sci Sports Exerc* 2007;39:1423-1434.

48. Hogg RS, Heath KV, Yip B, Craib KJ, O'Shaughnessy MV, Schechter MT, Montaner JS. Improved survival among HIV-infected individuals following initiation of antiretroviral therapy. *JAMA* 1998;279:450-454.

49. Hutchinson AB, Farnham PG, Dean HD, Ekwueme DU, del Rio C, Kamimoto L, Kellerman SE. The economic burden of HIV in the United States in the era of highly active antiretroviral therapy: Evidence of continuing racial and ethnic differences. *J Acquir Immune Defic Syndr* 2006;43:451-457.

50. Jaggers J, Phillips K, Dudgeon W, Lyerly GW, Duncan B, Burgess S, Durstine L, Hand G. Effects of combined aerobic and resistance exercise training on cardiovascular fitness in HIV-infected men: 1905: Board #69 May 29 3:30 PM - 5:00 PM, 2008, p. S329.

51. Johnson JE, Slife DM, Anders GT, Bailey SR, Blanton HM, McAllister CK, Latham RD. Cardiac dysfunction in patients seropositive for the human immunodeficiency virus. *West J Med* 1991;155:373-379.

52. Justice AC, Rabeneck L, Hays RD, Wu AW, Bozzette SA. Sensitivity, specificity, reliability, and clinical validity of provider-reported symptoms: A comparison with self-reported symptoms. Outcomes Committee of the AIDS Clinical Trials Group. *J Acquir Immune Defic Syndr* 1999;21:126-133.

53. Kahn JO, Walker BD. Acute human immunodeficiency virus type 1 infection. *N Engl J Med* 1998;339:33-39.

54. Kalish ML, Wolfe ND, Ndongmo CB, McNicholl J, Robbins KE, Aidoo M, Fonjungo PN, Alemnji G, Zeh C, Djoko CF, Mpoudi-Ngole E, Burke DS, Folks TM. Central African hunters exposed to simian immunodeficiency virus. *Emerg Infect Dis* 2005;11:1928-1930.

55. Keyser RE, Peralta L, Cade WT, Miller S, Anixt J. Functional aerobic impairment in adolescents seropositive for HIV: A quasiexperimental analysis. *Arch Phys Med Rehabil* 2000;81:1479-1484.

56. Kotler DP, Tierney AR, Wang J, Pierson RN Jr. Magnitude of body-cell-mass depletion and the timing of death from wasting in AIDS. *Am J Clin Nutr* 1989;50:444-447.

57. Kyser M, Buchacz K, Bush TJ, Conley LJ, Hammer J, Henry K, Kojic EM, Milam J, Overton ET, Wood KC, Brooks JT. Factors associated with non-adherence to antiretroviral therapy in the SUN study. *AIDS Care* 2011;23(5):601-611.

58. *Lancet.* Time from HIV-1 seroconversion to AIDS and death before widespread use of highly-active antiretroviral therapy: A collaborative re-analysis. Collaborative Group on AIDS Incubation and HIV Survival including the CASCADE EU Concerted Action. Concerted Action on SeroConversion to AIDS and Death in Europe. *Lancet* 2000;355:1131-1137.

59. LaPerriere A, Antoni MH, Ironson G, Perry A, McCabe P, Klimas N, Helder L, Schneiderman N, Fletcher MA. Effects of aerobic exercise training on lymphocyte subpopulations. *Int J Sports Med* 1994;15 Suppl. 3:S127-130.

60. LaPerriere AR, Antoni MH, Schneiderman N, Ironson G, Klimas N, Caralis P, Fletcher MA. Exercise intervention attenuates emotional distress and natural killer cell decrements following notification of positive serologic status for HIV-1. *Biofeedback Self Regul* 1990;15:229-242.

61. LaPerriere A, Fletcher MA, Antoni MH, Klimas NG, Ironson G, Schneiderman N. Aerobic exercise training in an AIDS risk group. *Int J Sports Med* 1991;12 Suppl. 1:S53-57.

62. LaPerriere A, Klimas N, Fletcher MA, Perry A, Ironson G, Perna F, Schneiderman N. Change in CD4+ cell enumeration following aerobic exercise training in HIV-1 disease: Possible mechanisms and practical applications. *Int J Sports Med* 1997;18 Suppl. 1:S56-61.

63. Leyes P, Martinez E, Forga Mde T. Use of diet, nutritional supplements and exercise in HIV-infected patients receiving combination antiretroviral therapies: A systematic review. *Antivir Ther* 2008;13:149-159.

64. Lindegaard B, Hansen T, Hvid T, van Hall G, Plomgaard P, Ditlevsen S, Gerstoft J, Pedersen BK. The effect of strength and endurance training on insulin sensitivity and fat distribution in human immunodeficiency virus-infected patients with lipodystrophy. *J Clin Endocrinol Metab* 2008;93:3860-3869.

65. Lipkin DP, Scriven AJ, Crake T, Poole-Wilson PA. Six minute walking test for assessing exercise capacity in chronic heart failure. *Br Med J (Clin Res Ed)* 1986;292:653-655.

66. Lox CL, McAuley E, Tucker RS. Aerobic and resistance exercise training effects on body composition, muscular strength, and cardiovascular fitness in an HIV-1 population. *Int J Behav Med* 1996;3:55-69.

67. MacArthur RD, Levine SD, Birk TJ. Supervised exercise training improves cardiopulmonary fitness in HIV-infected persons. *Med Sci Sports Exerc* 1993;25:684-688.

68. Malaspina L, Woods SP, Moore DJ, Depp C, Letendre SL, Jeste D, Grant I. Successful cognitive aging in persons living with HIV infection. *J Neurovirol* 2011;17:110-119.

69. Mazzotta E, Agostinone A, Rosso R, Di Biagio A, De Socio GV, Cappelletti A, Zicolella R, Polilli E, Bonfanti P, Di Matteo L, Manzoli L, Parruti G. Osteonecrosis in human immunodeficiency virus (HIV)-infected patients: A multicentric case-control study. *J Bone Miner Metab* 2011;29(3):383-388.

70. Miller LH, Coppola JT. Noninvasive assessment of HIV-related coronary artery disease. *Curr HIV/AIDS Rep* 2011;8(2):114-121.

71. Mondy KE, Gottdiener J, Overton ET, Henry K, Bush T, Conley L, Hammer J, Carpenter CC, Kojic E, Patel P, Brooks JT. High prevalence of echocardiographic abnormalities among HIV-infected persons in the era of highly active antiretroviral therapy. *Clin Infect Dis* 2011;52:378-386.

72. Nanavati KA, Fisher SD, Miller TL, Lipshultz SE. HIV-related cardiovascular disease and drug interactions. *Am J Cardiovasc Drugs* 2004;4:315-324.

73. Neidig JL, Smith BA, Brashers DE. Aerobic exercise training for depressive symptom management in adults living with HIV infection. *J Assoc Nurses AIDS Care* 2003;14:30-40.

74. Nguyen BY, Reveille JD. Rheumatic manifestations associated with HIV in the highly active antiretroviral therapy era. *Curr Opin Rheumatol* 2009;21:404-410.

75. O'Brien K, Nixon S, Tynan AM, Glazier R. Aerobic exercise interventions for adults living with HIV/AIDS. *Cochrane Database Syst Rev* 2010:CD001796.

76. O'Brien WA, Hartigan PM, Daar ES, Simberkoff MS, Hamilton JD. Changes in plasma HIV RNA levels and CD4+ lymphocyte counts predict both response to antiretroviral therapy and therapeutic failure. VA Cooperative Study Group on AIDS. *Ann Intern Med* 1997;126:939-945.

77. Oliviero U, Bonadies G, Bosso G, Foggia M, Apuzzi V, Cotugno M, Valvano A, Leonardi E, Borgia G, Castello G, Napoli R, Sacca L. Impaired diastolic function in naive untreated human immunodeficiency virus infected patients. *World J Cardiol* 2010;2:98-103.

78. Oramasionwu CU, Brown CM, Ryan L, Lawson KA, Hunter JM, Frei CR. HIV/AIDS disparities: The mounting epidemic plaguing US blacks. *J Natl Med Assoc* 2009;101:1196-1204.

79. Oursler KK, Katzel LI, Smith BA, Scott WB, Russ DW, Sorkin JD. Prediction of cardiorespiratory fitness in older men infected with the human immunodeficiency virus: Clinical factors and value of the six-minute walk distance. *J Am Geriatr Soc* 2009;57:2055-2061.

80. Paton NI, Elia M, Jebb SA, Jennings G, Macallan DC, Griffin GE. Total energy expenditure and physical activity measured with the bicarbonate-urea method in patients with human immunodeficiency virus infection. *Clin Sci (Lond)* 1996;91:241-245.

81. Perna FM, LaPerriere A, Klimas N, Ironson G, Perry A, Pavone J, Goldstein A, Majors P, Makemson D, Talutto C, Schneiderman N, Fletcher MA, Meijer OG, Koppes L. Cardiopulmonary and CD4 cell changes in response to exercise training in early symptomatic HIV infection. *Med Sci Sports Exerc* 1999;31:973-979.

82. Petroczi A, Hawkins K, Jones G, Naughton DP. HIV patient characteristics that affect adherence to exercise programmes: An observational study. *Open AIDS J* 2010;4:148-155.

83. Pilcher CD, Eron JJ Jr, Vemazza PL, Battegay M, Harr T, Yerly S, Vom S, Perrin L. Sexual transmission during the incubation period of primary HIV infection. *JAMA* 2001;286:1713-1714.

84. Pyne JM, Fortney JC, Curran GM, Tripathi S, Atkinson JH, Kilbourne AM, Hagedorn HJ, Rimland D, Rodriguez-Barradas MC, Monson T, Bottonari KA, Asch SM, Gifford AL. Effectiveness of collaborative care for depression in human immunodeficiency virus clinics. *Arch Intern Med* 2011;171:23-31.

85. Rao D, Feldman BJ, Fredericksen RJ, Crane PK, Simoni JM, Kitahata MM, Crane HM. A structural equation model of HIV-related stigma, depressive symptoms, and medication adherence. *AIDS Behav* 2012;16(3):711-716.

86. Robinson FP, Quinn LT, Rimmer JH. Effects of high-intensity endurance and resistance exercise on HIV metabolic abnormalities: A pilot study. *Biol Res Nurs* 2007;8:177-185.

87. Roubenoff R, Wilson IB. Effect of resistance training on self-reported physical functioning in HIV infection. *Med Sci Sports Exerc* 2001;33:1811-1817.

88. Sainz de Baranda P, Ayala F. Chronic flexibility improvement after 12 week of stretching program utilizing the ACSM recommendations: Hamstring flexibility. *Int J Sports Med* 2010;31:389-396.

89. Sanchez-Ramon S, Bellon JM, Resino S, Canto-Nogues C, Gurbindo D, Ramos JT, Munoz-Fernandez MA. Low blood CD8+ T-lymphocytes and high circulating monocytes are predictors of HIV-1-associated progressive encephalopathy in children. *Pediatrics* 2003;111:E168-175.

90. Schlenzig C, Jager H, Reider H. Supervised physical exercise leads to psychological and immunological improvement in pre-AIDS patients. In: *Proceedings of the 5th International AIDS Conference.* Montreal, Canada, 1989.

91. Schneider E, Whitmore S, Glynn KM, Dominguez K, Mitsch A, McKenna MT. Revised surveillance case definitions for HIV infection among adults, adolescents, and children aged <18 months and for HIV infection and AIDS among children aged 18 months to <13 years—United States, 2008. *MMWR Recomm Rep* 2008;57:1-12.

92. Shephard RJ. Exercise, immune function and HIV infection. *J Sports Med Phys Fitness* 1998;38:101-110.

93. Short KR, Vittone JL, Bigelow MK, Proctor DN, Rizza RA, Coenen-Schimke JM, Nair KS. Impact of aerobic exercise training on age-related changes in insulin sensitivity and muscle oxidative capacity. *Diabetes* 2003;52:1888-1896.

94. Siddiqui J, Phillips AL, Freedland ES, Sklar AR, Darkow T, Harley CR. Prevalence and cost of HIV-associated weight loss in a managed care population. *Curr Med Res Opin* 2009;25:1307-1317.

95. Sigal RJ, Kenny GP, Wasserman DH, Castaneda-Sceppa C, White RD. Physical activity/exercise and type 2 diabetes: A consensus statement from the American Diabetes Association. *Diabetes Care* 2006;29:1433-1438.

96. Smart N, Marwick TH. Exercise training for patients with heart failure: A systematic review of factors that improve mortality and morbidity. *Am J Med* 2004;116:693-706.

97. Smith BA, Neidig JL, Nickel JT, Mitchell GL, Para MF, Fass RJ. Aerobic exercise: Effects on parameters related to fatigue, dyspnea, weight and body composition in HIV-infected adults. *AIDS* 2001;15:693-701.

98. Stein DS, Korvick JA, Vermund SH. CD4+ lymphocyte cell enumeration for prediction of clinical course of human immunodeficiency virus disease: A review. *J Infect Dis* 1992;165:352-363.

99. Stringer WW, Berezovskaya M, O'Brien WA, Beck CK, Casaburi R. The effect of exercise training on aerobic fitness, immune indices, and quality of life in HIV+ patients. *Med Sci Sports Exerc* 1998;30:11-16.

100. Takhar SS, Hendey GW. Orthopedic illnesses in patients with HIV. *Emerg Med Clin North Am* 2010;28:335-342.

101. Terry L, Sprinz E, Ribeiro JP. Moderate and high intensity exercise training in HIV-1 seropositive individuals: A randomized trial. *Int J Sports Med* 1999;20:142-146.

102. Terry L, Sprinz E, Stein R, Medeiros NB, Oliveira J, Ribeiro JP. Exercise training in HIV-1-infected individuals with dyslipidemia and lipodystrophy. *Med Sci Sports Exerc* 2006;38:411-417.

103. Thoni GJ, Fedou C, Brun JF, Fabre J, Renard E, Reynes J, Varray A, Mercier J. Reduction of fat accumulation and lipid disorders by individualized light aerobic training in human immunodeficiency virus infected patients with lipodystrophy and/or dyslipidemia. *Diabetes Metab* 2002;28:397-404.

104. UNAIDS, WHO. AIDS epidemic update: December 2009: Geneva: Joint UN Programme on HIV/AIDS, World Health Organization, 2009.

105. Vassimon HS, Deminice R, Machado AA, Monteiro JP, Jordao AA. The association of lipodystrophy and oxidative stress biomarkers in HIV-infected men. *Curr HIV Res* 2010;8:364-369.

106. Wawer MJ, Gray RH, Sewankambo NK, Serwadda D, Li X, Laeyendecker O, Kiwanuka N, Kigozi G, Kiddugavu M, Lutalo T, Nalugoda F, Wabwire-Mangen F, Meehan MP, Quinn TC. Rates of HIV-1 transmission per coital act, by stage of HIV-1 infection, in Rakai, Uganda. *J Infect Dis* 2005;191:1403-1409.

107. Weiss RA. How does HIV cause AIDS? *Science* 1993;260:1273-1279.

108. Wheeler DA, Gibert CL, Launer CA, Muurahainen N, Elion RA, Abrams DI, Bartsch GE. Weight loss as a predictor of survival and disease progression in HIV infection. Terry Beirn Community Programs for Clinical Research on AIDS. *J Acquir Immune Defic Syndr Hum Retrovirol* 1998;18:80-85.

109. WHO. *Global Report: UNAIDS Report on the Global AIDS Epidemic*, 2010.

110. Whyte BM, Cooper DA. The surveillance definition of the acquired immunodeficiency syndrome and the clinical classification of infection with the human immunodeficiency virus type 1. *Med J Aust* 1988;149:368-373.

111. Worobey M, Gemmel M, Teuwen DE, Haselkorn T, Kunstman K, Bunce M, Muyembe JJ, Kabongo JM, Kalengayi RM, Van Marck E, Gilbert MT, Wolinsky SM. Direct evidence of extensive diversity of HIV-1 in Kinshasa by 1960. *Nature* 2008;455:661-664.

112. Yahiaoui A, McGough EL, Voss JG. Development of evidence-based exercise recommendations for older HIV-infected patients. *J Assoc Nurses AIDS Care* 2012;23(3):204-219.

113. Yarasheski KE, Tebas P, Stanerson B, Claxton S, Marin D, Bae K, Kennedy M, Tantisiriwat W, Powderly WG. Resistance exercise training reduces hypertriglyceridemia in HIV-infected men treated with antiviral therapy. *J Appl Physiol* 2001;90:133-138.

114. Yoo JJ, Chun SH, Kwon YS, Koo KH, Yoon KS, Kim HJ. Operations about hip in human immunodeficiency virus-positive patients. *Clin Orthop Surg* 2010;2:22-27.

115. Zinna EM, Yarasheski KE. Exercise treatment to counteract protein wasting of chronic diseases. *Curr Opin Clin Nutr Metab Care* 2003;6:87-93.

CHAPTER 22

1. Work group recommendations: 2002 Exercise and Physical Activity Conference, St. Louis, Missouri. Session V: Evidence of benefit of exercise and physical activity in arthritis. *Arthritis Rheum* 2003;49:453-454.

2. Aletaha D, Neogi T, Silman AJ, Funovits J, Felson DT, Bingham CO III, Birnbaum NS, Burmester GR, Bykerk VP, Cohen MD, et al. 2010 Rheumatoid arthritis classification criteria. An American College of Rheumatology/European League Against Rheumatism collaborative initiative. *Arthritis Rheum* 2010;62:2569-2581.

3. Altman R, Alarcon G, Appelrouth D, Bloch D, Borenstein D, Brandt K, Brown C, Cooke TD, Daniel W, Feldman D, et al. The American College of Rheumatology criteria for the classification and reporting of osteoarthritis of the hip. *Arthritis Rheum* 1991;34:505-514.

4. Altman R, Alarcon G, Appelrouth D, Bloch D, Borenstein D, Brandt K, Brown C, Cooke TD, Daniel W, Gray R, et al. The American College of Rheumatology criteria for the classification and reporting of osteoarthritis of the hand. *Arthritis Rheum* 1990;33:1601-1610.

5. Altman R, Asch E, Bloch D, Bole G, Borenstein D, Brandt K, Christy W, Cooke TD, Greenwald R, Hochberg M, et al. Development of criteria for the classification and reporting of osteoarthritis.

Classification of osteoarthritis of the knee. Diagnostic and Therapeutic Criteria Committee of the American Rheumatism Association. *Arthritis Rheum* 1986;29:1039-1049.

6. American College of Rheumatology Subcommittee on Rheumatoid Arthritis Guidelines. Guidelines for the management of rheumatoid arthritis: 2002 update. *Arthritis Rheum* 2002;46:328-346.
7. American College of Sports Medicine. Exercise prescription for other clinical populations: Arthritis. *ACSM's Guidelines for Exercise Testing and Prescription* (pp. 225-228). 8th ed. Philadelphia: Lippincott Williams & Wilkins, 2010.
8. American Geriatrics Society Panel on Exercise and Osteoarthritis. Exercise prescription for older adults with osteoarthritis pain: Consensus practice recommendations. A supplement to the AGS clinical practice guidelines on the management of chronic pain in older adults. *J Am Geriatr Soc* 2001;49:808-823.
9. Anderson JJ, Felson DT. Factors associated with osteoarthritis of the knee in the first national Health and Nutrition Examination Survey (HANES I). Evidence for an association with overweight, race, and physical demands of work. *Am J Epidemiol* 1988;128:179-189.
10. Anderson R. Rheumatoid arthritis. In: Schumacher HR Jr, ed. *Primer on Rheumatic Diseases*. Atlanta: Arthritis Foundation, 1993.
11. Andrews AW, Thomas MW, Bohannon RW. Normative values for isometric muscle force measurements obtained with hand-held dynamometers. *Phys Ther* 1996;76:248-259.
12. Anonymous. Prevalence of leisure-time physical activity among persons with arthritis and other rheumatic conditions—United States, 1990-1991. *MMWR* 1997;46:389-393.
13. Assis MR, Silva LE, Alves AM, Pessanha AP, Valim V, Feldman D, Neto TL, Natour J. A randomized controlled trial of deep water running: Clinical effectiveness of aquatic exercise to treat fibromyalgia. *Arthritis Rheum* 2006;55:57-65.
14. Avina-Zubieta JA, Choi HK, Sadatsafavi M, Etminan M, Esdaile JM, Lacaille D. Risk of cardiovascular mortality in patients with rheumatoid arthritis: A meta-analysis of observational studies. *Arthritis Rheum* 2008;59:1690-1697.
15. Avouac J, Gossec L, Dougados M. Diagnostic and predictive value of anti-CCP (cyclic citrullinated protein) antibodies in rheumatoid arthritis: A systematic literature review. *Ann Rheum Dis* 2006;65:845-851.
16. Badley EM. The impact of disabling arthritis. *Arthritis Care Res* 1995;8:221-228.
17. Baillet A, Zeboulon N, Gossec L, Combescure C, Bedin L-A, Juvin R, Dougados M, Gaudin P. Efficacy of cardiorespiratory aerobic exercise in rheumatoid arthritis: Meta-analysis of randomized controlled trials. *Arthritis Care Res* 2010;62:984-992.
18. Barry H. Activity for older persons and mature athletes. In: Safran M, McKeag D, Van Camp S, eds. *Manual of Sports Medicine* (pp. 184-189). New York: Lippincott-Raven, 1998.
19. Bartels EM, Lund H, Hagen KB, Dagfinrud H, Christensen R, Danneskiold-Samsoe B. Aquatic exercise for the treatment of knee and hip osteoarthritis (Review). *Cochrane Database Syst Rev* 2007;CD005523.
20. Barth WF. Office evaluation of the patient with musculoskeletal complaints. *Am J Med* 1997;102:3S-10S.
21. Bellamy N, Buchanan WW, Goldsmith CH, Campbell J, Stitt LW. Validation study of WOMAC: A health status instrument for measuring clinically important patient relevant outcomes to antirheumatic drug therapy in patients with osteoarthritis of the hip or knee. *J Rheumatol* 1988;15:1833-1840.
22. Belza B. The impact of fatigue on exercise performance. *Arthritis Care Res* 1994;7:176-180.
23. Bennell K, Hinman R. Exercise as a treatment for osteoarthritis. *Curr Opin Rheumatol* 2006;17:634-640.
24. Berman A, Studenski S. Musculoskeletal rehabilitation. *Clin Geriatr Med* 1998;14:641-659.
25. Berry MJ, Brubaker PH, O'Toole ML, Rejeski WJ, Soberman J, Ribisl PM, Miller HS, Afable RF, Applegate W, Ettinger WH. Estimation of $\dot{V}O_2$ in older individuals with osteoarthritis of the knee and cardiovascular disease. *Med Sci Sports Exerc* 1996;28:808-814.
26. Brandt K, Heilman D, Slemenda C, Mazzuca S, Braunstein E. Knee pain in elderly community subjects. Differences in lower extremity muscles strength, body weight, and depression scores among those with and without radiographic evidence of osteoarthritis. *Trans Orthop Res* 1999;24:222.
27. Brooks PM. The burden of musculoskeletal disease—a global perspective. *Clin Rheumatol* 2006;25:778-781.
28. Brosseau L, MacLeay L, Robinson V, Wells G, Tugwell P. Intensity of exercise for the treatment of osteoarthritis. *Cochrane Database Syst Rev* 2003;CD004259.
29. Bullough P, Cawston T. Pathology and biochemistry of osteoarthritis. In: Doherty M, ed. *Osteoarthritis* (pp. 29-58). London: Wolfe, 1994.
30. Bunning RD, Materson RS. A rational program of exercise for patients with osteoarthritis. *Semin Arthritis Rheum* 1991;21:33-43.
31. Busch AJ, Barber KA, Overend TJ, Peloso PMJ, Schachter CL. Exercise for treating fibromyalgia syndrome (Review). *Cochrane Database Syst Rev* 2007;CD003786.
32. Byers PH. Effect of exercise on morning stiffness and mobility in patients with rheumatoid arthritis. *Res Nurs Health* 1985;8:275-281.
33. Calin A, Elswood J, Rigg S, Skevington SM. Ankylosing spondylitis—an analytical review of 1500 patients: The changing pattern of disease. *J Rheumatol* 1988;15:1234-1238.
34. Campbell WW, Crim MC, Young VR, Evans WJ. Increased energy requirements and body composition changes with resistance training in older adults. *Am J Clin Nutr* 1994;60:167-175.
35. Centers for Disease Control and Prevention. Racial disparities in total knee replacement among Medicare enrollees—United States, 2000-2006. *MMWR* 2009;58:133-138.
36. Cheng YJ, Hootman JM, Murphy LB, Langmaid GA, Helmick CG. Prevalence of doctor-diagnosed arthritis and arthritis-attributable activity limitation—United States, 2007-2009. *MMWR* 2010;59:1261-1265.
37. Clegg DO, Reda DJ, Harris CL, Klein MA, O'Dell JR, Hooper MM, Bradley JD, Bingham CO 3rd, Weisman MH, Jackson CG, Lane NE, Cush JJ, Moreland LW, Schumacher HR Jr, Oddis CV, Wolfe F, Molitor JA, Yocum DE, Schnitzer TJ, Furst DE, Sawitzke AD, Shi H, Brandt KD, Moskowitz RW, Williams HJ. Glucosamine, chondroitin sulfate, and the two in combination for painful knee osteoarthritis. *N Engl J Med* 2006;354:795-808.
38. Coleman EA, Buchner DM, Cress ME, Chan BK, de Lateur BJ. The relationship of joint symptoms with exercise performance in older adults. *J Am Geriatr Soc* 1996;44:14-21.
39. Combe B, Landewe R, Lukes C, Bolosiu HD, Breedveld F, Dougados M, Emery P, Ferraccioli G, Hazes JMW, Klareskog L, et al. EULAR recommendations for the management of early arthritis: Report of a task force of the European Standing Com-

mittee for International Clinical Studies Including Therapeutics (ESCISIT). *Ann Rheum Dis* 2007;66:34-45.

40. Dagfinrud H, Hagen KB, Kvien TK. Physiotherapy interventions for ankylosing spondylitis (Review). *Cochrane Database Syst Rev* 2008;CD002822.
41. Daltroy LH, Robb-Nicholson C, Iversen MD, Wright EA, Liang MH. Effectiveness of minimally supervised home aerobic training in patients with systemic rheumatic disease. *Br J Rheumatol* 1995;34:1064-1069.
42. Daly MP, Berman BM. Rehabilitation of the elderly patient with arthritis. *Clin Geriatr Med* 1993;9:783-801.
43. Danneskiold-Samsoe B, Grimby G. The relationship between the leg muscle strength and physical capacity in patients with rheumatoid arthritis, with reference to the influence of corticosteroids. *Clin Rheumatol* 1986;5:468-474.
44. Davis MA, Ettinger WH, Neuhaus JM. Obesity and osteoarthritis of the knee: Evidence from the National Health and Nutrition Examination Survey (NHANES I). *Semin Arthritis Rheum* 1990;20:34-41.
45. de Jong Z, Munneke M, Kroon HM, van Schaardenburg D, Dijkmans BAC, Hazes JMW, Vliet Vlieland TP. Long-term follow-up of a high intensity exercise program in patients with rheumatoid arthritis. *Clin Rheumatol* 2009;28:663-671.
46. de Jong Z, Munneke M, Lems WF, Zwinderman AH, Kroon HM, Pauwels EK, Jansen A, Ronday KH, Dijkmans BA, Breedveld FC, Vliet Vlieland TP, Hazes JM. Slowing of bone loss in patients with rheumatoid arthritis by long-term high-intensity exercise: Results of a randomized, controlled trial. *Arthritis Rheum* 2004;50:1066-1076.
47. de Jong Z, Munneke M, Zwinderman AH, Kroon HM, Jansen A, Ronday KH, van Schaardenburg D, Dijkmans BA, Van den Ende CH, Breedveld FC, Vliet Vlieland TP, Hazes JM. Is a long-term high-intensity exercise program effective and safe in patients with rheumatoid arthritis? Results of a randomized controlled trial. *Arthritis Rheum* 2003;48:2415-2424.
48. de Jong Z, Munneke M, Zwinderman AH, Kroon HM, Ronday KH, Lems WF, Dijkmans BA, Breedveld FC, Vliet Vlieland TP, Hazes JM, Huizinga TW. Long term high intensity exercise and damage of small joints in rheumatoid arthritis. *Ann Rheum Dis* 2004;63:1399-1405.
49. Devos-Comby L, Cronan T, Roesch SC. Do exercise and self-management interventions benefit patients with osteoarthritis of the knee? A meta-analytic review. *J Rheumatol* 2006;33:744-756.
50. Dexter PA. Joint exercises in elderly persons with symptomatic osteoarthritis of the hip or knee. Performance patterns, medical support patterns, and the relationship between exercising and medical care. *Arthritis Care Res* 1992;5:36-41.
51. Deyle GD, Allison SC, Matekel RL, Ryder MG, Stang JM, Gohdes DD, Hutton JP, Henderson NE, Garber MB. Physical therapy treatment effectiveness for osteoarthritis of the knee: A randomized comparison of supervised clinical exercise and manual therapy procedures versus a home exercise program. *Phys Ther* 2005;85:1301-1317.
52. Dominick KL, Gullette EC, Babyak MA, Mallow KL, Sherwood A, Waugh R, Chilikuri M, Keefe FJ, Blumenthal JA. Predicting peak oxygen uptake among older patients with chronic illness. *J Cardiopulm Rehabil* 1999;19:81-89.
53. Dougados M, Gueguen A, Nguyen M, Thiesce A, Listrat V, Jacob L, Nakache JP, Gabriel KR, Lequesne M, Amor B. Longitudinal radiologic evaluation of osteoarthritis of the knee. *J Rheumatol* 1992;19:378-384.
54. Dunlop DD, Lyons JS, Manheim LM, Song J, Chang RW. Arthritis and heart disease as risk factors for major depression: The role of functional limitation. *Med Care* 2004;42:502-511.
55. Engelhart M, Kondrup J, Hoie LH, Andersen V, Kristensen JH, Heitmann BL. Weight reduction in obese patients with rheumatoid arthritis, with preservation of body cell mass and improvement of physical fitness. *Clin Exp Rheumatol* 1996;14:289-293.
56. Ettinger WH Jr, Burns R, Messier SP, Applegate W, Rejeski WJ, Morgan T, Shumaker S, Berry MJ, O'Toole M, Monu J, Craven T. A randomized trial comparing aerobic exercise and resistance exercise with a health education program in older adults with knee osteoarthritis. The Fitness Arthritis and Seniors Trial (FAST). *JAMA* 1997;277:25-31.
57. Eurenius E, Stenstrom CH. Physical activity, physical fitness, and general health perception among individuals with rheumatoid arthritis. *Arthritis Rheum* 2005;53:48-55.
58. Evans WJ. Exercise training guidelines for the elderly. *Med Sci Sports Exerc* 1999;31:12-17.
59. Felson DT. Developments in the clinical understanding of osteoarthritis. *Arthritis Res Ther* 2009;11:203-303.
60. Felson DT. The epidemiology of knee osteoarthritis: Results from the Framingham Osteoarthritis Study. *Semin Arthritis Rheum* 1990;20:42-50.
61. Felson DT. Weight and osteoarthritis. *Am J Clin Nutr* 1996;63:430S-432S.
62. Felson DT, Anderson JJ, Naimark A, Walker AM, Meenan RF. Obesity and knee osteoarthritis. The Framingham Study. *Ann Intern Med* 1988;109:18-24.
63. Fisher LR, Cawley MI, Holgate ST. Relation between chest expansion, pulmonary function, and exercise tolerance in patients with ankylosing spondylitis. *Ann Rheum Dis* 1990;49:921-925.
64. Frangolia DD, Rhodes EC. Metabolic responses and mechanisms during water immersion running and exercise. *Sports Med* 1996;22:38-53.
65. Fransen M, McConnell S. Exercise for osteoarthritis of the knee (Review). *Cochrane Database Syst Rev* 2008;CD004376.
66. Fransen M, McConnell S, Hernandez-Molina G, Reichenbach S. Exercise for osteoarthritis of the hip (Review). *Cochrane Database Syst Rev* 2009;CD007912.
67. Galloway MT, Jokl P. The role of exercise in the treatment of inflammatory arthritis. *Bull Rheum Dis* 1993;42:1-4.
68. Gecht MR, Connell KJ, Sinacore JM, Prochaska TR. A survey of exercise beliefs and exercise habits among people with arthritis. *Arthritis Care Res* 1996;9:82-88.
69. Giesecke T, Gracely RH, Grant MA, Nachemson A, Petzke F, Williams DA, Clauw DJ. Evidence of augmented central pain processing in idiopathic chronic low back pain. *Arthritis Rheum* 2004;50:613-623.
70. Goldenberg DL, Burckhardt C, Crofford L. Management of fibromyalgia syndrome. *JAMA* 2004;292:2388-2395.
71. Gowans SE, deHueck A. Effectiveness of exercise in management of fibromyalgia. *Curr Opin Rheumatol* 2004;16:138-142.
72. Gussoni M, Margonato V, Ventura R, Veicsteinas A. Energy cost of walking with hip joint impairment. *Phys Ther* 1990;70:295-301.
73. Hakkinen A, Hakkinen K, Hannonen P. Effects of strength training on neuromuscular function and disease activity in patients with recent-onset inflammatory arthritis. *Scand J Rheumatol* 1994;23:237-242.

74. Hakkinen A, Hannonen P, Nyman K, Lyyski T, Hakkinen K. Effects of concurrent strength and endurance training in women with early or longstanding rheumatoid arthritis: Comparison with healthy subjects. *Arthritis Rheum* 2003;49:789-797.

75. Hakkinen A, Malkia E, Hakkinen K, Jappinen I, Laitinen L, Hannonen P. Effects of detraining subsequent to strength training on neuromuscular function in patients with inflammatory arthritis. *Br J Rheumatol* 1997;36:1075-1081.

76. Hakkinen A, Pakarinen A, Hannonen P, Kautiainen H, Nyman K, Kraemer WJ, Hakkinen K. Effects of prolonged combined strength and endurance training on physical fitness, body composition and serum hormones in women with rheumatoid arthritis and in healthy controls. *Clin Exp Rheumatol* 2005;23:505-512.

77. Hakkinen A, Sokka T, Hannonen P. A home-based two year strength training period in early rheumatoid arthritis led to good long-term compliance: A five year followup. *Arthritis Rheum* 2004;51:56-62.

78. Hakkinen A, Sokka T, Kautiainen H, Kotaniemi A, Hannonen P. Sustained maintenance of exercise induced muscle strength gains and normal bone mineral density in patients with early rheumatoid arthritis: A 5 year follow up. *Ann Rheum Dis* 2004;63:910-916.

79. Hakkinen A, Sokka T, Kotaniemi A, Hannonen P. A randomized two-year study of the effects of dynamic strength training on muscle strength, disease activity, functional capacity, and bone mineral density in early rheumatoid arthritis. *Arthritis Rheum* 2001;44:515-522.

80. Hakkinen A, Sokka T, Lietsalmi AM, Kautiainen H, Hannonen P. Effects of dynamic strength training on physical function, Valpar 9 work sample test, and working capacity in patients with recent-onset rheumatoid arthritis. *Arthritis Rheum* 2003;49:71-77.

81. Hall J, Grant J, Blake D, Taylor G, Garbutt G. Cardiorespiratory responses to aquatic treadmill walking in patients with rheumatoid arthritis. *Physiother Res Int* 2004;9:59-73.

82. Hall J, Skevington SM, Maddison PJ, Chapman K. A randomized and controlled trial of hydrotherapy in rheumatoid arthritis. *Arthritis Care Res* 1996;9:206-215.

83. Han A, Robinson V, Judd M, Taixiang W, Wells G, Tugwell P. Tai chi for treating rheumatoid arthritis. *Cochrane Database Syst Rev* 2004;CD004849.

84. Hansen TM, Hansen G, Langgaard AM, Rasmussen JO. Long-term physical training in rheumatoid arthritis. A randomized trial with different training programs and blinded observers. *Scand J Rheumatol* 1993;22:107-112.

85. Harkcom TM, Lampman RM, Banwell BF, Castor CW. Therapeutic value of graded aerobic exercise training in rheumatoid arthritis. *Arthritis Rheum* 1985;28:32-39.

86. Haslock I. Ankylosing spondylitis: Management. In: Klippel J, Dieppe P, eds. *Rheumatology* (pp. 3.29.21–23.29.10). St Louis: Mosby, 1994.

87. Hasselbacher P. Joint physiology. In: Klippel JH, Dieppe PA, eds. *Rheumatology* (pp. 1.3.1-1.3.6). London: Mosby, 1994.

88. Helmick CG, Felson DT, Lawrence RC, Gabriel S, Hirsch R, Kwosh CK, Liang MH, Kremers HM, Mayes MD, Merkel PA, et al. National Arthritis Data Workshop. Estimates of the prevalence of arthritis and other conditions in the United States: Part I. *Arthritis Rheum* 2008;58:15-25.

89. Hernandez-Molina G, Reichenbach S, Zhang B, Lavalley M, Felson DT. Effect of therapeutic exercise for hip osteoarthritis pain: Results of a meta-analysis. *Arthritis Care Res* 2008;59:1221-1228.

90. Herrero-Beaumont G, Roman J, Trabado MC, Blanco FJ, Benito P, Martin-Mola E, Paulino J, Marenco J, Porto A, Laffon A, Araujo D, Figueroa M, Branco J. Effects of glucosamine sulfate on 6-month control of knee osteoarthritis symptoms versus placebo and acetaminophen: Results from the Glucosamine Unum in Die Efficacy (GUIDE) trial. *Arthritis Rheum* 2005;52:240.

91. Hicks JE. Exercise in patients with inflammatory arthritis and connective tissue disease. *Rheum Dis Clin North Am* 1990;16:845-870.

92. Hochberg MC, Altman RD, Brandt KD, Clark BM, Dieppe PA, Griffin MR, Moskowitz RW, Schnitzer TJ. Guidelines for the medical management of osteoarthritis. Part I. Osteoarthritis of the hip. American College of Rheumatology. *Arthritis Rheum* 1995;38:1535-1540.

93. Hochberg MC, Altman RD, Brandt KD, Clark BM, Dieppe PA, Griffin MR, Moskowitz RW, Schnitzer TJ. Guidelines for the medical management of osteoarthritis. Part II. Osteoarthritis of the knee. American College of Rheumatology. [See comments]. *Arthritis Rheum* 1995;38:1541-1546.

94. Hoffman DF. Arthritis and exercise. *Prim Care* 1993;20:895-910.

95. Hootman JM, Cheng WY. Psychological distress and fair/poor health among adults with arthritis: State-specific prevalence and correlates of general health status, United States, 2007. *Int J Public Health* 2009;54:75-83.

96. Hootman JM, Helmick CG. Projections of US prevalence of arthritis and associated activity limitations. *Arthritis Rheum* 2006;54:226-229.

97. Hubscher O. Pattern recognition in arthritis. In: Klippel J, Dieppe P, eds. *Rheumatology* (pp. 2.3.1-2.3.6). Philadelphia: Mosby, 1998.

98. Hurkmans E, van der Griesen FJ, Vliet Vlieland TPM, Schoones J, van den Ende ECHM. Dynamic exercise programs (aerobic and/or muscle strength training) in patients with rheumatoid arthritis (Review). *Cochrane Database Syst Rev* 2009;CD006853.

99. Hurley MV. The role of muscle weakness in the pathogenesis of osteoarthritis. *Rheum Dis Clin North Am* 1999;25:283-298, vi.

100. Husby VS, Helgerud J, Bjorgen S, Husby OS, Benum P, Hoff J. Early maximal strength training is an efficient treatment for patients operated with total hip arthroplasty. *Arch Phys Med Rehabil* 2009;90:1658-1667.

101. Iverson MD. Managing hip and knee osteoarthritis with exercise: What is the best prescription? *Ther Adv Musculoskel Dis* 2010;2:279-290.

102. James MJ, Cleland LG, Gaffney RD, Proudman SM, Chatterton BE. Effect of exercise on 99mTc-DTPA clearance from knees with effusions. *J Rheumatol* 1994;21:501-504.

103. Jokl P. Prevention of disuse muscle atrophy in chronic arthritides. *Rheum Dis Clin North Am* 1990;16:837-844.

104. Jordan KM, Arden NK, Doherty M, Bannwarth B, Bijlsma JW, Dieppe P, Gunther K, Hauselmann H, Herrero-Beaumont G, Kaklamanis P, Lohmander S, Leeb B, Lequesne M, Mazieres B, Martin-Mola E, Pavelka K, Pendleton A, Punzi L, Serni U, Swoboda B, Verbruggen G, Zimmerman-Gorska I, Dougados M. EULAR recommendations 2003: An evidence based approach to the management of knee osteoarthritis: Report of a task force of the Standing Committee for International Clinical Studies Including Therapeutic Trials (ESCISIT). *Ann Rheum Dis* 2003;62:1145-1155.

105. Kirsteins AE, Dietz F, Hwang SM. Evaluating the safety and potential use of a weight-bearing exercise, Tai-Chi Chuan,

for rheumatoid arthritis patients. *Am J Phys Med Rehabil* 1991;70:136-141.

106. Kisner C, Colby L. *Therapeutic Exercise: Foundations and Techniques* (p. 713). 2nd ed. Philadelphia: Davis, 1990.

107. Kitas GD, Gabriel SE. Cardiovascular disease in rheumatoid arthritis: State of the art and future perspectives. *Ann Rheum Dis* 2011;70:8-14.

108. Klepper SE. Effects of an eight-week physical conditioning program on disease signs and symptoms in children with chronic arthritis. *Arthritis Care Res* 1999;12:52-60.

109. Kraemer WJ, Ratamess NA. Fundamentals of resistance training: Progression and exercise prescription. *Med Sci Sports Exerc* 2004;36:674-688.

110. Kurtz S, Mowat F, Ong K, Chan N, Lau E, Halpern M. Prevalence of primary and revision total hip and knee arthroplasty in the United States from 1990 through 2002. *J Bone Joint Surg Am* 2005;87:1487-1497.

111. Laliberte R, Kraus V, Rooks D. *The Everyday Arthritis Solution* (p. 239). Pleasantville, NY: Reader's Digest, 2003.

112. Lane NE. Exercise: A cause of osteoarthritis. *J Rheumatol Suppl* 1995;43:3-6.

113. Lane NE, Buckwalter JA. Exercise: A cause of osteoarthritis? *Rheum Dis Clin North Am* 1993;19:617-633.

114. Law R-J, Breslin A, Oliver EJ, Mawn L, Markland DA, Maddison PJ, Thom JM. Perceptions of the effects of exercise on joint health in rheumatoid arthritis patients. *Rheumatology (Oxford)* 2010;49:2444-2451.

115. Lawrence RC, Felson DT, Helmick CG, Arnold LM, Choi H, Deyo RA, Gabriel S, Hirsch R, Hochberg MC, Hunder GC, et al. National Arthritis Data Workshop. Estimates of the prevalence of arthritis and other conditions in the United States: Part II. *Arthritis Rheum* 2008;58:26-35.

116. Leibs TR, Herzberg W, Ruther W, Haasters J, Russlies M, Hassenpfug J. Ergometer cycling after hip or knee replacement surgery: A randomized, controlled trial. *J Bone Joint Surg Am* 2010;92:814-822.

117. Lemmey AB. Efficacy of progressive resistance training for patients with rheumatoid arthritis and recommendations regarding its prescription. *Int J Clin Rheumatol* 2011;6:189-205.

118. Lemmey AB, Marcora SM, Chester K, Wilson S, Casanova F, Maddison PJ. Effects of high-intensity resistance training in patients with rheumatoid arthritis: A randomised controlled trial. *Arthritis Rheum* 2009;61:1726-1734.

119. Lemmey AB, Williams SL, Marcora SM, Jones J, Maddison PJ. Are the benefits of a high intensity resistance training program sustained in rheumatoid arthritis patients?: A 3 year follow-up study. *Arthritis Care Res* 2012, in press.

120. Lequesne MG, Mery C, Samson M, Gerard P. Indexes of severity for osteoarthritis of the hip and knee. Validation—value in comparison with other assessment tests. *Scand J Rheumatol Suppl* 1987;65:85-89.

121. Lim HJ, Moon YI, Lee MS. Effects of home-based daily exercise therapy on joint mobility, daily activity, pain, and depression in patients with ankylosing spondylitis. *Rheumatol Int* 2005;25:225-229.

122. Loeser RF Jr. Evaluation of musculoskeletal complaints in the older adult. *Clin Geriatr Med* 1998;14:401-415.

123. Mangione KK, McCully K, Gloviak A, Lefebvre I, Hofmann M, Craik R. The effects of high-intensity and low-intensity cycle ergometry in older adults with knee osteoarthritis. *J Gerontol A Biol Sci Med Sci* 1999;54:M184-190.

124. Mannerkorpi K. Exercise in fibromyalgia. *Curr Opin Rheumatol* 2005;17:190-194.

125. March LM, Bachmeier CJ. Economics of osteoarthritis: A global perspective. *Baillieres Clin Rheumatol* 1997;11:817-834.

126. Marcora SM, Casanova F, Williams E, Jones J, Elamanchi R, Lemmey AB. Preliminary evidence for cachexia in patients with well-established ankylosing spondylitis. *Rheumatology (Oxford)* 2006;45:1385-1388.

127. Marcora SM, Lemmey AB, Maddison PJ. Can progressive resistance training reverse cachexia in patients with rheumatoid arthritis? Results of a pilot study. *J Rheumatol* 2005;32:1031-1039.

128. Marks R, Allegrante JP. Chronic osteoarthritis and adherence to exercise: A review of the literature. *J Aging Phys Act* 2005;13:434-460.

129. Massy-Westropp N, Rankin W, Ahern M, Krishnan J, Hearn TC. Measuring grip strength in normal adults: Reference ranges and a comparison of electronic and hydraulic instruments. *J Hand Surg Am* 2004;29:514-519.

130. Matschke V, Murphy P, Lemmey AB, Maddison PJ, Thom JM. Muscle quality, architecture, and activation in cachectic patients with rheumatoid arthritis. *J Rheumatol* 2010;37:282-284.

131. Matschke V, Murphy P, Lemmey AB, Maddison PJ, Thom JM. Skeletal muscle properties in rheumatoid arthritis patients. *Med Sci Sports Exerc* 2010;42:2149-2155.

132. McCardle W, Katch F, Katch V. *Exercise Physiology: Energy, Nutrition, and Human Performance* (p. 696). 2nd ed. Philadelphia: Lea & Febiger, 1986.

133. McCarthy CJ, Mills PM, Pullen R, Roberts C, Silman A, Oldham JA. Supplementing a home exercise programme with a class-based exercise programme is more effective than home exercise alone in the treatment of knee osteoarthritis. *Rheumatology (Oxford)* 2004;43:880-886.

134. McCurdy D. Genetic susceptibility to the connective tissue diseases. *Curr Opin Rheumatol* 1999;11:399-407.

135. Mease PJ, Clauw DJ, Arnold LM, Goldenberg DL, Witter J, Williams DA, Simon LS, Strand CV, Bramson C, Martin S, Wright TM, Littman B, Wernicke JF, Gendreau RM, Crofford LJ. Fibromyalgia syndrome. *J Rheumatol* 2005;32:2270-2277.

136. Melton-Rogers S, Hunter G, Walter J, Harrison P. Cardiorespiratory responses of patients with rheumatoid arthritis during bicycle riding and running in water. *Phys Ther* 1996;76:1058-1065.

137. Messier SP, Gutekunst DJ, Davis C, DeVita P. Weight loss reduces knee-joint loads in overweight and obese older adults with knee osteoarthritis. *Arthritis Rheum* 2005;52:2026-2032.

138. Messier SP, Loeser RF, Miller GD, Morgan TM, Rejeski WJ, Sevick MA, Ettinger WH Jr, Pahor M, Williamson JD. Exercise and dietary weight loss in overweight and obese older adults with knee osteoarthritis: The Arthritis, Diet, and Activity Promotion Trial. *Arthritis Rheum* 2004;50:1501-1510.

139. Messier SP, Loeser RF, Mitchell MN, Valle G, Morgan TP, Rejeski WJ, Ettinger WH. Exercise and weight loss in obese older adults with knee osteoarthritis: A preliminary study. *J Am Geriatr Soc* 2000;48:1062-1072.

140. Metsios GS, Stavropoulos-Kalinoglou A, Veldhuijzen van Zanten JJCS, Treharne GJ, Panoulas VF, Douglas KMJ, Koutedakis Y, Kitas GD. Rheumatoid arthritis, cardiovascular disease and physical exercise: A systematic review. *Rheumatology (Oxford)* 2008;47:239-248.

141. Mili F, Helmick CG, Moriarty DG. Health related quality of life among adults reporting arthritis: Analysis of data from the Behavioral Risk Factor Surveillance System, US, 1996-1999. *J Rheumatol* 2003;30:160-166.

142. Miltner O, Schneider U, Graf J, Niethard FU. Influence of isokinetic and ergometric exercises on oxygen partial pressure measurement in the human knee joint. *Adv Exp Med Biol* 1997;411:183-189.

143. Minor MA. Exercise in the treatment of osteoarthritis. *Rheum Dis Clin North Am* 1999;25:397-415, viii.

144. Minor MA, Johnson JC. Reliability and validity of a submaximal treadmill test to estimate aerobic capacity in women with rheumatic disease. *J Rheumatol* 1996;23:1517-1523.

145. Minor MA, Lane NE. Recreational exercise in arthritis. *Rheum Dis Clin North Am* 1996;22:563-577.

146. Munneke M, de Jong Z, Zwinderman AH, Jansen A, Ronday HK, Peter WF, Boonman DC, van den Ende CH, Vliet Vlieland TP, Hazes JM. Adherence and satisfaction of rheumatoid arthritis patients with a long-term intensive dynamic exercise program (RAPIT program). *Arthritis Rheum* 2003;49:665-672.

147. Munneke M, de Jong Z, Zwinderman AH, Jansen A, Ronday HK, van Schaardenburg D, Dijkmans BAC, Kroon HM, Vliet Vlieland TP, Hazes JM. Effect of a high-intensity weight-bearing exercise program on radiologic damage progression of the large joints in subgroups of patients with rheumatoid arthritis. *Arthritis Rheum* 2005;53:410-417.

148. Myasoedova E, Crowson CS, Kremers HM, Therneau TM, Gabriel SE. Is the incidence of rheumatoid arthritis rising?: Results from Olmsted County, Minnesota, 1995-2007. *Arthritis Rheum* 2010;62:1576-1582.

149. Neuberger GB, Press AN, Lindsley HB, Hinton R, Cagle PE, Carlson K, Scott S, Dahl J, Kramer B. Effects of exercise on fatigue, aerobic fitness, and disease activity measures in persons with rheumatoid arthritis. *Res Nurs Health* 1997;20:195-204.

150. Nielens H, Boisset V, Masquelier E. Fitness and perceived exertion in patients with fibromyalgia syndrome. *Clin J Pain* 2000;16:209-213.

151. Nordemar R. Physical training in rheumatoid arthritis: A controlled long-term study. II. Functional capacity and general attitudes. *Scand J Rheumatol* 1981;10:25-30.

152. Nordemar R, Ekblom B, Zachrisson L, Lundqvist K. Physical training in rheumatoid arthritis: A controlled long-term study. I. *Scand J Rheumatol* 1981;10:17-23.

153. Nordenskiold UM, Grimby G. Grip force in patients with rheumatoid arthritis and fibromyalgia and in healthy subjects. A study with the Grippit instrument. *Scand J Rheumatol* 1993;22:14-19.

154. Noreau L, Moffet H, Drolet M, Parent E. Dance-based exercise program in rheumatoid arthritis. Feasibility in individuals with American College of Rheumatology functional class III disease. *Am J Phys Med Rehabil* 1997;76:109-113.

155. Noren AM, Bogren U, Bolin J, Stenstrom C. Balance assessment in patients with peripheral arthritis: Applicability and reliability of some clinical assessments. *Physiother Res Int* 2001;6:193-204.

156. Norregaard J, Bulow PM, Lykkegaard JJ, Mehlsen J, Danneskiold-Samsoe B. Muscle strength, working capacity and effort in patients with fibromyalgia. *Scand J Rehabil Med* 1997;29:97-102.

157. Nurmohamed MT, Kitas GD. Cardiovascular risk in rheumatoid arthritis and diabetes: How does it compare and when does it start? *Ann Rheum Dis* 2011;70:881-883.

158. O'Driscoll SL, Jayson MI, Baddeley H. Neck movements in ankylosing spondylitis and their responses to physiotherapy. *Ann Rheum Dis* 1978;37:64-66.

159. O'Grady M, Fletcher J, Ortiz S. Therapeutic and physical fitness exercise prescription for older adults with joint disease: An evidence-based approach. *Rheum Dis Clin North Am* 2000;26:617-646.

160. O'Reilly S, Jones A, Doherty M. Muscle weakness in osteoarthritis. *Curr Opin Rheumatol* 1997;9:259-262.

161. Otterness IG, Zimmerer RO, Swindell AC, Poole AR, Saxne T, Heinegard D, Ionescu M, Weiner E. An examination of some molecular markers in blood and urine for discriminating patients with osteoarthritis from healthy individuals. *Acta Orthop Scand* 1995;66(Suppl. 266):148-150.

162. Patten SB, Beck CA, Kassam A, Williams JV, Barbui C, Metz LM. Long-term medical conditions and major depression: Strength of association for specific conditions in the general population. *Can J Psychiatry* 2005;50:195-202.

163. Pavelka K, Gatterova J, Olejarova M, Machacek S, Giacovelli G, Rovati LC. Glucosamine sulfate use and delay of progression of knee osteoarthritis: A 3-year, randomized, placebo-controlled, double-blind study. *Arch Intern Med* 2002;162:2113-2123.

164. Penninx BW, Abbas H, Ambrosius W, Nicklas BJ, Davis C, Messier SP, Pahor M. Inflammatory markers and physical function among older adults with knee osteoarthritis. *J Rheumatol* 2004;31:2027-2031.

165. Peters MJL, Symmons DPM, McCarey D, Dijkmans BAC, Nicola P, Kvien TK, McInnes IB, Haentzschel H, Gonzalez-Gay MA, Provan S, et al. EULAR evidence-based recommendations for cardiovascular risk management in patients with rheumatoid arthritis and other forms of inflammatory arthritis. *Ann Rheum Dis* 2010;69:325-331.

166. Pham T, van der Heijde D, Altman RD, Anderson JJ, Bellamy N, Hochberg M, Simon L, Strand V, Woodworth T, Dougados M. OMERACT-OARSI initiative: Osteoarthritis Research Society International set of responder criteria for osteoarthritis clinical trials revisited. *Osteoarthritis Cartilage* 2004;12:389-399.

167. Philbin EF, Ries MD, French TS. Feasibility of maximal cardiopulmonary exercise testing in patients with end-stage arthritis of the hip and knee prior to total joint arthroplasty. *Chest* 1995;108:174-181.

168. Poole AR, Rizkalla G, Ionescu M, Reiner A, Brooks E, Rorabeck C, Bourne R, Bogoch E. Osteoarthritis in the human knee: A dynamic process of cartilage matrix degradation, synthesis and reorganization. *Agents Actions Suppl* 1993;39:3-13.

169. Rall LC, Roubenoff R. Body composition, metabolism, and resistance exercise in patients with rheumatoid arthritis. *Arthritis Care Res* 1996;9:151-156.

170. Rasch A, Bystrom AH, Dalen N, Martinez-Carranza N, Berg HE. Persisting muscle atrophy two years after replacement of the hip. *J Bone Joint Surg Br* 2009;91:583-588.

171. Rasmussen JO, Hansen TM. Physical training for patients with ankylosing spondylitis. *Arthritis Care Res* 1989;2:25-27.

172. Raub JA. Psychophysiologic effects of Hatha Yoga on musculoskeletal and cardiopulmonary function: A literature review. *J Altern Complement Med* 2002;8:797-812.

173. Reginster JY, Deroisy R, Rovati LC, Lee RL, Lejeune E, Bruyere O, Giacovelli G, Henrotin Y, Dacre JE, Gossett C. Long-term effects of glucosamine sulphate on osteoarthritis progres-

sion: A randomised, placebo-controlled clinical trial. *Lancet* 2001;357:251-256.

174. Richards SCM, Scott DL. Prescribed exercise in people with fibromyalgia: Parallel group randomized controlled trial. *BMJ* 2002;325:185, doi10.1136/bmj.325.7357.185 (published July 27, 2002).

175. Rickli RE, Jones CJ. *Senior Fitness Test Manual*. Champaign, IL: Human Kinetics, 2001.

176. Ries MD, Philbin EF, Groff GD. Relationship between severity of gonarthrosis and cardiovascular fitness. *Clin Orthop Relat Res* 1995;313:169-176.

177. Roddy E, Doherty M. Changing life-styles and osteoarthritis: What is the evidence? *Best Pract Res Clin Rheumatol* 2006;20:81-97.

178. Roddy E, Zhang W, Doherty M. Aerobic walking or strengthening exercise for osteoarthritis of the knee? A systematic review. *Ann Rheum Dis* 2005;64:544-548.

179. Roddy E, Zhang W, Doherty M, Arden NK, Barlow J, Birrell F, Carr A, Chakravarty K, Dickson J, Hay E, Hosie G, Hurley M, Jordan KM, McCarthy C, McMurdo M, Mockett S, O'Reilly S, Peat G, Pendleton A, Richards S. Evidence-based recommendations for the role of exercise in the management of osteoarthritis of the hip or knee—the MOVE consensus. *Rheumatology (Oxford)* 2005;44:67-73.

180. Roos EM. Joint injury causes knee osteoarthritis in young adults. *Curr Opin Rheumatol* 2005;17:195-200.

181. Roubenoff R, Roubenoff RA, Cannon JG, Kehayias JJ, Zhuang H, Dawson-Hughes B, Dinarello CA, Rosenberg IH. Rheumatoid cachexia: Cytokine-driven hypermetabolism accompanying reduced body cell mass in chronic inflammation. *J Clin Invest* 1994;93:2379-2386.

182. Rudwaleit M, Khan MA, Sieper J. The challenge of diagnosis and classification in early ankylosing spondylitis: Do we need new criteria? *Arthritis Rheum* 2005;52:1000-1008.

183. Ruof J, Sangha O, Stucki G. Comparative responsiveness of 3 functional indices in ankylosing spondylitis. *J Rheumatol* 1999;26:1959-1963.

184. Sacks JJ, Helmick CG, Luo Y-H, Ilowite NT, Bowyer S. Prevalence of and annual ambulatory health care visits for pediatric arthritis and other rheumatological conditions in the United States in 2001-2004. *Arthritis Care Res* 2007;57:1439-1445.

185. Salaffi F, Stancati A, Neri R, Grassi W, Bombardieri S. Measuring functional disability in early rheumatoid arthritis: The validity, reliability and responsiveness of the Recent-Onset Arthritis Disability (ROAD) index. *Clin Exp Rheumatol* 2005;23:S31-42.

186. Santos H, Brophy S, Calin A. Exercise in ankylosing spondylitis: How much is optimum? *J Rheumatol* 1998;25:2156-2160.

187. Saxne T. Differential release of molecular markers in joint disease. *Acta Orthop Scand* 1995;66(Suppl. 266):80-83.

188. Schilke JM, Johnson GO, Housh TJ, O'Dell JR. Effects of muscle-strength training on the functional status of patients with osteoarthritis of the knee joint. *Nurs Res* 1996;45:68-72.

189. Schweitzer M, Resnick D. Enthesopathy. In: Klippel J, Dieppe P, eds. *Rheumatology* (pp. 3.27.21–23.27.26). St Louis: Mosby, 1994.

190. Scott DL, Wolfe F, Huizinga TW. Rheumatoid arthritis. *Lancet* 2010;376:1094-1108.

191. Scully R, Barnes M, eds. *Physical Therapy*. Philadelphia: Lippincott, 1989.

192. Sevick MA, Miller GD, Loeser RF, Williamson JD, Messier SP. Cost-effectiveness of exercise and diet in overweight and obese adults with knee osteoarthritis. *Med Sci Sports Exerc* 2009;41:1167-1174.

193. Shih M, Hootman JM, Kruger J, Helmick CG. Physical activity in men and women with arthritis. National Health Interview Survey, 2002. *Am J Prev Med* 2006;30:385-393.

194. Sokka T, Hakkinen A, Kautiainen H, Maillerfert JF, Toloza S, Hansen TM, Calvo-Alen J, Oding R, Liveborn M, Huisman M, et al. The QUEST-RA Group. Physical inactivity in patients with rheumatoid arthritis: Data from twenty-one countries in a cross-sectional, international study. *Arthritis Rheum* 2008;59: 42-50.

195. Song R, Lee EO, Lam P, Bae SC. Effects of tai chi exercise on pain, balance, muscle strength, and perceived difficulties in physical functioning in older women with osteoarthritis: A randomized clinical trial. *J Rheumatol* 2003;30:2039-2044.

196. Spector TD, Hart DJ, Nandra D, Doyle DV, Mackillop N, Gallimore JR, Pepys MB. Low-level increases in serum C-reactive protein are present in early osteoarthritis of the knee and predict progressive disease. *Arthritis Rheum* 1997;40:723-727.

197. Stamatelopoulos KS, Kitas GD, Papamichael CM, Chryssohou E, Kyrkou K, Georgiopoulos G, Protogerou A, Panoulas VF, Sandoo A, Tendoloouris N, et al. Artherosclerosis in rheumatoid arthritis versus diabetes: A comparative study. *Arterioscler Thromb Vasc Biol* 2009;29:1702-1708.

198. Stavropoulos-Kalinoglou A, Metsios GS, Koutedakis Y, Nevill AM, Douglas KM, Jamurtas A, van Zanten JJCSV, Labib M, Kitas GD. Redefining overweight and obese in rheumatoid arthritis patients. *Ann Rheum Dis* 2007;66:1316-1321.

199. Stavropoulos-Kalinoglou A, Metsios GS, Panoulas VF, Nevill AM, Jamurtas AZ, Koutedakis Y, Kitas GD. Underweight and obese states both associate with worse disease activity and physical function in patients with established rheumatoid arthritis. *Clin Rheumatol* 2009;28:439-444.

200. Stenstrom CH. Home exercise in rheumatoid arthritis functional class II: Goal setting versus pain attention. *J Rheumatol* 1994;21:627-634.

201. Stenstrom CH. Therapeutic exercise in rheumatoid arthritis. *Arthritis Care Res* 1994;7:190-197.

202. Stenstrom CH, Lindell B, Swanberg E, Swanberg P, Harms-Ringdahl K, Nordemar R. Intensive dynamic training in water for rheumatoid arthritis functional class II—a long-term study of effects. *Scand J Rheumatol* 1991;20:358-365.

203. Stenstrom CH, Minor MA. Evidence for the benefit of aerobic and strengthening exercise in rheumatoid arthritis. *Arthritis Rheum* 2003;49:428-434.

204. Steultjens MP, Dekker J, van Baar ME, Oostendorp RA, Bijlsma JW. Range of joint motion and disability in patients with osteoarthritis of the knee or hip. *Rheumatology (Oxford)* 2000;39:955-961.

205. Suetta C, Magnusson SP, Rosted A, Aagaard P, Jakobsen AK, Larsen LH, Duus B, Kjaer M. Resistance training in the early postoperative phase reduces hospitalization and leads to muscle hypertrophy in elderly hip surgery patients—a controlled, randomized study. *J Am Geriatr Soc* 2004;52:2016-2022.

206. Suomi R, Collier D. Effects of arthritis exercise programs on functional fitness and perceived activities of daily living measures in older adults with arthritis. *Arch Phys Med Rehabil* 2003;84:1589-1594.

207. Suomi R, Lindauer S. Effectiveness of Arthritis Foundation Aquatic Program on strength and range of motion in women with arthritis. *J Aging Phys Act* 1997;5:341-351.

208. Tackson SJ, Krebs DE, Harris BA. Acetabular pressures during hip arthritis exercises. *Arthritis Care Res* 1997;10:308-319.

209. Trudelle-Jackson E, Emerson R, Smith S. Outcomes of total hip arthroplasty: A study of patients one year postsurgery. *J Orthop Sports Phys Ther* 2002;32:260-267.

210. Valim V, Oliveira LM, Suda AL, Silva LE, Faro M, Neto TL, Feldman D, Natour J. Peak oxygen uptake and ventilatory anaerobic threshold in fibromyalgia. *J Rheumatol* 2002;29:353-357.

211. van den Ende CH, Hazes JM, le Cessie S, Mulder WJ, Belfor DG, Breedveld FC, Dijkmans BA. Comparison of high and low intensity training in well controlled rheumatoid arthritis. Results of a randomised clinical trial. *Ann Rheum Dis* 1996;55:798-805.

212. Van Deusen J, Harlowe D. The efficacy of the ROM Dance Program for adults with rheumatoid arthritis. *Am J Occup Ther* 1987;41:90-95.

213. van Gool CH, Penninx BW, Kempen GI, Miller GD, van Eijk JT, Pahor M, Messier SP. Determinants of high and low attendance to diet and exercise interventions among overweight and obese older adults. Results from the arthritis, diet, and activity promotion trial. *Contemp Clin Trials* 2005.

214. van Gool CH, Penninx BW, Kempen GI, Rejeski WJ, Miller GD, van Eijk JT, Pahor M, Messier SP. Effects of exercise adherence on physical function among overweight older adults with knee osteoarthritis. *Arthritis Rheum* 2005;53:24-32.

215. van Halm VP, Peters MJ, Voskuyl AE, Boers M, Lems WF, Visser M, Stehouwer CDA, Spijkerman AMW, Dekker JM, Nijpels G, et al. Rheumatoid arthritis versus diabetes as a risk factor for cardiovascular disease: A cross-sectional study, the CARRE investigation. *Ann Rheum Dis* 2009;68:1395-1400.

216. Van Tubergen A, Boonen A, Landewe R, Rutten-Van Molken M, Van Der Heijde D, Hidding A, Van Der Linden S. Cost effectiveness of combined spa-exercise therapy in ankylosing spondylitis: A randomized controlled trial. *Arthritis Rheum* 2002;47:459-467.

217. van Tubergen A, Landewe R, van der Heijde D, Hidding A, Wolter N, Asscher M, Falkenbach A, Genth E, The HG, van der Linden S. Combined spa-exercise therapy is effective in patients with ankylosing spondylitis: A randomized controlled trial. *Arthritis Rheum* 2001;45:430-438.

218. Verbrugge LM, Patrick DL. Seven chronic conditions: Their impact on US adults' activity levels and use of medical services. *Am J Public Health* 1995;85:173-182.

219. Verbruggen G. Chondroprotective drugs in degenerative joint diseases. *Rheumatology (Oxford)* 2006;45:129-138.

220. Wolfe F, Clauw DJ, Fitzcharles M-A, Goldenberg DL, Katz RS, Mease P, Russell AS, Russell IJ, Winfield JB, Yunus MB. The American College of Rheumatology preliminary diagnostic criteria for fibromyalgia and measurement of symptom severity. *Arthritis Care Res* 2010;62:600-610.

221. Wolfe F, Smythe HA, Yunus MB, Bennett RM, Bombardier C, Goldenberg DL, Tugwell P, Campbell SM, Abeles M, Clark P, et al. The American College of Rheumatology 1990 criteria for the classification of fibromyalgia. Report of the Multicenter Criteria Committee. *Arthritis Rheum* 1990;33:160-172.

222. Yelin E, Cisternas M, Foreman A, Pasta D, Murphy L, Helmick C. National and state medical expenditures and lost earnings attributable to arthritis and other rheumatic conditions—United States, 2003. *MMWR* 2007;56:4-7.

223. Zhang W, Doherty M, Arden N, Bannwarth B, Bijlsma J, Gunther KP, Hauselmann HJ, Herrero-Beaumont G, Jordan K, Kaklamanis P, Leeb B, Lequesne M, Lohmander S, Mazieres B, Martin-Mola E, Pavelka K, Pendleton A, Punzi L, Swoboda B, Varatojo R, Verbruggen G, Zimmermann-Gorska I, Dougados M. EULAR evidence based recommendations for the management of hip osteoarthritis: Report of a task force of the EULAR Standing Committee for International Clinical Studies Including Therapeutics (ESCISIT). *Ann Rheum Dis* 2005;64:669-681.

224. Zochling J, van der Heijde D, Burgos-Vargas R, Collantes E, Davis JC Jr, Dijkmans B, Dougados M, Geher P, Inman RD, Khan MA, Kvien TK, Leirisalo-Repo M, Olivieri I, Pavelka K, Sieper J, Stucki G, Sturrock RD, van der Linden S, Wendling D, Bohm H, van Royen BJ, Braun J. ASAS/EULAR recommendations for the management of ankylosing spondylitis. *Ann Rheum Dis* 2006;65:442-452.

225. Zochling J, van der Heijde D, Dougados M, Braun J. Current evidence for the management of ankylosing spondylitis: A systematic literature review for the ASAS/EULAR management recommendations in ankylosing spondylitis. *Ann Rheum Dis* 2006;65:423-432.

CHAPTER 23

1. American College of Sports Medicine. *ACSM's Guidelines for Exercise Testing and Prescription*. 9th ed. Philadelphia: Lippincott Williams & Wilkins, 2013.

2. American College of Sports Medicine, Chodzko-Zajko WJ, Proctor DN, et al. American College of Sports Medicine position stand. Exercise and physical activity for older adults. *Med Sci Sports Exerc* 2009;41(7):1510-1530.

3. Arabi A, Baddoura R, Awada H, et al. Discriminative ability of dual-energy X-ray absorptiometry site selection in identifying patients with osteoporotic fractures. *Bone* 2007;40(4):1060-1065.

4. Arampatzis A, Peper A, Bierbaum S. Exercise of mechanisms for dynamic stability control increases stability performance in the elderly. *J Biomech* 2011;44(1):52-58.

5. Becker C. Pathophysiology and clinical manifestations of osteoporosis. *Clin Cornerstone* 2006;8(1):19-27.

6. Bemben DA, Bemben MG. Dose-response effect of 40 weeks of resistance training on bone mineral density in older adults. *Osteoporos Int* 2011;22(1):179-186.

7. Berger C, Goltzman D, Langsetmo L, et al. Peak bone mass from longitudinal data: Implications for the prevalence, pathophysiology, and diagnosis of osteoporosis. *J Bone Miner Res* 2010;25(9):1948-1957.

8. Bianchi ML, Baim S, Bishop NJ, et al. Official positions of the International Society for Clinical Densitometry (ISCD) on DXA evaluation in children and adolescents. *Pediatr Nephrol* 2010;25(1):37-47.

9. Blumsohn A, Marin F, Nickelsen T, et al. Early changes in biochemical markers of bone turnover and their relationship with bone mineral density changes after 24 months of treatment with teriparatide. *Osteoporos Int* 2011;22(6):1935-1946.

10. Bolland MJ, Avenell A, Baron JA, et al. Effect of calcium supplements on risk of myocardial infarction and cardiovascular events: Meta-analysis. *BMJ* 2010;341:c3691.

11. Bolland MJ, Grey A, Avenell A, Gamble GD, Reid IR. Calcium supplements with or without vitamin D and risk of cardiovascu-

lar events: Reanalysis of the Women's Health Initiative limited access dataset and meta-analysis. *BMJ* 2011;342:d2040.

12. Burge R, Dawson-Hughes B, Solomon DH, Wong JB, King A, Tosteson A. Incidence and economic burden of osteoporosis-related fractures in the United States, 2005-2025. *J Bone Miner Res* 2007;22(3):465-475.
13. Cauley JA, Ewing SK, Taylor BC, et al. Sex steroid hormones in older men: Longitudinal associations with 4.5-year change in hip bone mineral density—the osteoporotic fractures in men study. *J Clin Endocrinol Metab* 2010;95(9):4314-4323.
14. Cawthon PM. Gender differences in osteoporosis and fractures. *Clin Orthop Relat Res* 2011;469(7):1900-1905.
15. Clarke BL, Khosla S. Physiology of bone loss. *Radiol Clin North Am* 2010;48(3):483-495.
16. Cooper C, Cole ZA, Holroyd CR, et al. Secular trends in the incidence of hip and other osteoporotic fractures. *Osteoporos Int* 2011;22(5):1277-1288.
17. Gallacher SJ, Dixon T. Impact of treatments for postmenopausal osteoporosis (bisphosphonates, parathyroid hormone, strontium ranelate, and denosumab) on bone quality: A systematic review. *Calcif Tissue Int* 2010;87(6):469-484.
18. Gat-Yablonski G, Yackobovitch-Gavan M, Phillip M. Nutrition and bone growth in pediatrics. *Endocrinol Metab Clin North Am* 2009;38(3):565-586.
19. Granacher U, Muehlbauer T, Zahner L, Gollhofer A, Kressig RW. Comparison of traditional and recent approaches in the promotion of balance and strength in older adults. *Sports Med* 2011;41(5):377-400.
20. Griffith JF, Genant HK. New imaging modalities in bone. *Curr Rheumatol Rep* 2011;13(3):241-250
21. Guadalupe-Grau A, Fuentes T, Guerra B, Calbet JA. Exercise and bone mass in adults. *Sports Med* 2009;39(6):439-468.
22. Gusi N, Raimundo A, Leal A. Low-frequency vibratory exercise reduces the risk of bone fracture more than walking: A randomized controlled trial. *BMC Musculoskelet Disord* 2006;7:92.
23. Haentjens P, Magaziner J, Colon-Emeric CS, et al. Meta-analysis: Excess mortality after hip fracture among older women and men. *Ann Intern Med* 2010;152(6):380-390.
24. Heaney RP. Bone health. *Am J Clin Nutr* 2007;85(1):300S-303S.
25. Hess JA, Woollacott M. Effect of high-intensity strength-training on functional measures of balance ability in balance-impaired older adults. *J Manipulative Physiol Ther* 2005;28(8):582-590.
26. Hongsdusit N, von Muhlen D, Barrett-Connor E. A comparison between peripheral BMD and central BMD measurements in the prediction of spine fractures in men. *Osteoporos Int* 2006;17(6):872-877.
27. Hsia J, Heiss G, Ren H, et al. Calcium/vitamin D supplementation and cardiovascular events. *Circulation* 2007;115(7):846-854.
28. Jackowski SA, Erlandson MC, Mirwald RL, et al. Effect of maturational timing on bone mineral content accrual from childhood to adulthood: Evidence from 15 years of longitudinal data. *Bone* 2011;48(5):1178-1185.
29. Jacobs JW, Da Silva JA, Armbrecht G, Bijlsma JW, Verstappen SM. Prediction of vertebral fractures is specific for gender and site of bone mineral density measurement. *J Rheumatol* 2010;37(1):149-154.
30. Kanis JA, Bianchi G, Bilezikian JP, et al. Towards a diagnostic and therapeutic consensus in male osteoporosis. *Osteoporos Int* 2011; ePub ahead of print.
31. Kanis JA, Johansson H, Oden A, Dawson-Hughes B, Melton LJ 3rd, McCloskey EV. The effects of a FRAX revision for the USA. *Osteoporos Int* 2010;21(1):35-40.
32. Kanis JA, McCloskey EV, Johansson H, Oden A, Melton LJ 3rd, Khaltaev N. A reference standard for the description of osteoporosis. *Bone* 2008;42(3):467-475.
33. Kukuljan S, Nowson CA, Sanders KM, et al. Independent and combined effects of calcium-vitamin D3 and exercise on bone structure and strength in older men: An 18-month factorial design randomized controlled trial. *J Clin Endocrinol Metab* 2011;96(4):955-963.
34. Leboime A, Confavreux CB, Mehsen N, Paccou J, David C, Roux C. Osteoporosis and mortality. *Joint Bone Spine* 2010;77 Suppl. 2:S107-112.
35. Lewiecki EM, Gordon CM, Baim S, et al. International Society for Clinical Densitometry 2007 adult and pediatric official positions. *Bone* 2008;43(6):1115-1121.
36. Li WC, Chen YC, Yang RS, Tsauo JY. Effects of exercise programmes on quality of life in osteoporotic and osteopenic postmenopausal women: A systematic review and meta-analysis. *Clin Rehabil* 2009;23(10):888-896.
37. Marques EA, Wanderley F, Machado L, et al. Effects of resistance and aerobic exercise on physical function, bone mineral density, OPG and RANKL in older women. *Exp Gerontol* 2011;46(7):524-532.
38. Martyn-St James M, Carroll S. Meta-analysis of walking for preservation of bone mineral density in postmenopausal women. *Bone* 2008;43(3):521-531.
39. Menant JC, Close JC, Delbaere K, et al. Relationships between serum vitamin D levels, neuromuscular and neuropsychological function and falls in older men and women. *Osteoporos Int* 2012;23(3)981-989.
40. Nattiv A, Loucks AB, Manore MM, et al. American College of Sports Medicine position stand. The female athlete triad. *Med Sci Sports Exerc* 2007;39(10):1867-1882.
41. Nikander R, Sievanen H, Heinonen A, Daly RM, Uusi-Rasi K, Kannus P. Targeted exercise against osteoporosis: A systematic review and meta-analysis for optimising bone strength throughout life. *BMC Med* 2010;8:47.
42. Rachner TD, Khosla S, Hofbauer LC. Osteoporosis: Now and the future. *Lancet* 2011;377(9773):1276-1287.
43. Robling AG. Is bone's response to mechanical signals dominated by muscle forces? *Med Sci Sports Exerc* 2009;41(11):2044-2049.
44. Robling AG, Turner CH. Mechanical signaling for bone modeling and remodeling. *Crit Rev Eukaryot Gene Expr* 2009;19(4):319-338.
45. Silverman NE, Nicklas BJ, Ryan AS. Addition of aerobic exercise to a weight loss program increases BMD, with an associated reduction in inflammation in overweight postmenopausal women. *Calcif Tissue Int* 2009;84(4):257-265.
46. Slatkovska L, Alibhai SM, Beyene J, Cheung AM. Effect of whole-body vibration on BMD: A systematic review and meta-analysis. *Osteoporos Int* 2010;21(12):1969-1980.
47. Spangler M, Phillips BB, Ross MB, Moores KG. Calcium supplementation in postmenopausal women to reduce the risk of osteoporotic fractures. *Am J Health Syst Pharm* 2011;68(4):309-318.
48. Syed FA, Ng AC. The pathophysiology of the aging skeleton. *Curr Osteoporos Rep* 2010;8(4):235-240.

49. Tosteson AN, Burge RT, Marshall DA, Lindsay R. Therapies for treatment of osteoporosis in US women: Cost-effectiveness and budget impact considerations. *Am J Manag Care* 2008;14(9):605-615.

50. Turner CH, Warden SJ, Bellido T, et al. Mechanobiology of the skeleton. *Sci Signal* 2009;2(68):pt 3.

51. Vainionpaa A, Korpelainen R, Sievanen H, Vihriala E, Leppaluoto J, Jamsa T. Effect of impact exercise and its intensity on bone geometry at weight-bearing tibia and femur. *Bone* 2007;40(3):604-611.

52. Verschueren SM, Roelants M, Delecluse C, Swinnen S, Vanderschueren D, Boonen S. Effect of 6-month whole body vibration training on hip density, muscle strength, and postural control in postmenopausal women: A randomized controlled pilot study. *J Bone Miner Res* 2004;19(3):352-359.

53. von Stengel S, Kemmler W, Engelke K, Kalender WA. Effects of whole body vibration on bone mineral density and falls: Results of the randomized controlled ELVIS study with postmenopausal women. *Osteoporos Int* 2011;22(1):317-325.

54. Williams MA, Stewart KJ. Impact of strength and resistance training on cardiovascular disease risk factors and outcomes in older adults. *Clin Geriatr Med* 2009;25(4):703-714, ix.

CHAPTER 24

1. American College of Sports Medicine. ACSM position stand on the recommended quantity and quality of exercise for developing and maintaining cardiorespiratory and muscular fitness and flexibility in healthy adults. *Med Sci Sports Exerc* 1998;30:975-991.

2. American College of Sports Medicine. *ACSM's Guidelines for Exercise Testing and Prescription.* 9th ed. Philadelphia: Lippincott Williams & Wilkins, 2013.

3. American College of Sports Medicine. Progression models in resistance training for healthy adults. *Med Sci Sports Exerc* 2009;41:687-708.

4. Barbis JM. Prevention and management of low back pain. In: Rothman J, Levine RE, eds. *Prevention Practice: Strategies for Physical Therapy and Occupational Therapy* (pp. 63-72). Philadelphia: Saunders, 1992.

5. Biering-Sörensen F. Physical measurements as risk indicators for low-back trouble over a one year period. *Spine* 1984;9:106-119.

6. Bigos S, Bowyer O, Braen G, Brown K, Deyo R, Haldeman S, Hart J, Johnson E, Keller R, Kiddo D, Liang M, Nelson R, Nordin M, Owen B, Pope M, Schwartz R, Stewart D, Susman J, Triano J, Tripp L, Turk D, Watts C, Weinstein J. *Acute Low Back Problems in Adults.* Clinical Practice Guideline No. 14. AHCPR publication 95-0642. Rockville, MD: Agency for Health Care Policy and Research, Public Health Service, U.S. Department of Health and Human Services, 1994.

7. Bigos SJ, Davis GE. Scientific application of sports medicine principles for acute low back pain. *J Orthop Sports Phys Ther* 1996;24:192-207.

8. Bogduk N. *Clinical Anatomy of the Lumbar Spine and Sacrum.* 4th ed. New York: Churchill Livingstone, 2005.

9. Cady LD, Bischoff DP, O'Connell ER, Thomas PC, Allan JH. Strength and fitness and subsequent back injuries in firefighters. *J Occup Med* 1979;21:269-272.

10. Cassidy JD, Côté P, Carroll LJ, Kristman V. Incidence and course of low back pain episodes in the general population. *Spine* 2005;30:2817-2823.

11. Centers for Disease Control and Prevention (CDC). Adult Obesity Facts. www.cdc.gov/obesity/data/adult.html [accessed April 25, 2013].

12. Cherkin DC, Deyo RA, Wheeler K, Ciol MA. Physician variation in diagnostic testing for low back pain. Who you see is what you get. *Arthritis Rheum* 1994;37:15-22.

13. Chou R, Fu R, Carrino JA, Deyo RA. Imaging strategies for low-back pain: Systematic review and meta-analysis. *Lancet* 2009;373:463-472.

14. Compton WM, Conway KP, Stinson FS, Grant BF. Changes in the prevalence of major depression and comorbid substance use disorders in the United States between 1991-1992 and 2001-2002. *Am J Psychiatry* 2006;163:2141-2147.

15. Croft PR, Dunn KM, Raspe H. Course and prognosis of back pain in primary care: The epidemiological perspective. *Pain* 2006;122:1-3.

16. Croft PR, Macfarlane GJ, Papageorgiou AC, Silman AJ. Outcome of low back pain in general practice: A one year follow-up study. *BMJ* 1998;316:1356-1359.

17. Currie SR, Wang J. More data on major depression as an antecedent risk factor for first onset of chronic back pain. *Psychol Med* 2005;35:1275-1282.

18. Devereaux M. Low back pain. *Med Clin North Am* 2009;93:477-501, x.

19. Deyo RA. Diagnostic evaluation of LBP: Reaching a specific diagnosis is often impossible. *Arch Intern Med* 2002;162:1444-1448.

20. Deyo RA, Diehl AK, Rosenthal M. How many days of bed rest for acute low back pain? A randomized clinical trial. *New Engl J Med* 1986;315:1064-1070.

21. Deyo RA, Mirza SK, Turner JA, Martin BI. Overtreating chronic back pain: Time to back off? *J Am Board Fam Med* 2009;22:62-68.

22. Deyo RA, Weinstein JN. Low back pain. *New Engl J Med* 2001;344:363-370.

23. D'Orazio BP, ed. *Low Back Pain Handbook.* Boston: Butterworth-Heinemann, 1999.

24. D'Orazio BP, Tritsch C, Vath SA, Rennie MA. Postoperative lumbar rehabilitation. In: D'Orazio BP, ed. *Low Back Pain Handbook* (pp. 277-308). Boston: Butterworth-Heinemann, 1999.

25. Freburger JK, Carey TS, Holmes GM, Wallace AS, Castel LD, Darter JD, Jackman AM. Exercise prescription for chronic back or neck pain: Who prescribes it? who gets it? What is prescribed? *Arthritis Rheum* 2009;61:192-200.

26. Freburger JK, Holmes GM, Agans RP, Jackman AM, Darter JD, Wallace AS, Castel LD, Kalsbeek WD, Carey TS. The rising prevalence of chronic low back pain. *Arch Intern Med* 2009;169:251-258.

27. Friedly J, Standaert C, Chan L. Epidemiology of spine care: The back pain dilemma. *Phys Med Rehabil Clin N Am* 2010;21:659-677.

28. Fritz JM, Delitto A, Erhard RE. Comparison of classification-based physical therapy with therapy based on clinical practice guidelines for patients with acute low back pain. *Spine* 2003;28:1363-1372.

29. Frymoyer JW, Cats-Baril WL. An overview of the incidences and costs of low back pain. *Orthop Clin N Am* 1991;22:263-271.

30. Gibson JNA, Waddell G. Surgery for degenerative lumbar spondylosis. *Cochrane Database Syst Rev* 2005;4:CD001352.

31. Goldberg MS, Scott SC, Mayo NE. A review of the association between cigarette smoking and the development of nonspecific back pain and related outcomes. *Spine* 2000;25:995-1014.
32. Haldeman S, Dagenais S. A supermarket approach to the evidence-informed management of chronic low back pain. *Spine J* 2008;8:1-7.
33. Harreby M, Hesselsˉe G, Kjer J, Neergaard K. Low back pain and physical exercise in leisure time in 38-year-old men and women: A 25-year prospective cohort study of 640 school children. *Eur Spine J* 1997;6:181-186.
34. Harreby M, Kjer J, Hesselsˉe G, Neergaard K. Epidemiological aspects and risk factors for low back pain in 38-year-old men and women: A 25-year prospective cohort study of 640 school children. *Eur Spine J* 1996;5:312-318.
35. Hartvigsen J, Morso L, Bendix T, Manniche C. Supervised and non-supervised Nordic walking in the treatment of chronic low back pain: A single blind randomized clinical trial. *BMC Musculoskelet Disord* 2010;11:30.
36. Hayden JA, van Tulder MW, Malmivaara A, Koes BW. Exercise therapy for treatment of non-specific low back pain. *Cochrane Database Syst Rev* 2006;3:CD000335.
37. Hayden JA, van Tulder MW, Malmivaara A, Koes BW. Meta-analysis: Exercise therapy for treatment of non-specific low back pain. *Ann Intern Med* 2005;142:765-775.
38. Heitz CA, Hilfiker R, Bachmann LM, Joronen H, Lorenz T, Uebelhart D, Klipstein A, Brunner F. Comparison of risk factors predicting return to work between patients with subacute and chronic non-specific low back pain: Systematic review. *Eur Spine J* 2009;18:1829-1835.
39. Hestbaek L, Leboeuf-Yde C, Kyvik KO, Manniche C. The course of low back pain from adolescence to adulthood: Eight-year follow-up of 9600 twins. *Spine* 2006;31:468-472.
40. Hestbaek L, Leboeuf-Yde C, C Manniche. Low back pain: What is the long-term course? A review of studies of general patient populations. *Eur Spine J* 2005;12:149-165.
41. Heymans MW, van Tulder MW, Esmail R, Bombardier C, Koes BW. Back schools for non-specific low-back pain. *Cochrane Database Syst Rev* 2004;4:CD000261.
42. Hochschuler, SH, Cotler HB, Guyer RD. The tissue origin of mechanical low back pain and sciatica as identified by spinal microsurgery. In: *Rehabilitation of the Spine* (pp. 595-600). St. Louis: Mosby-Year Book, 1993.
43. Hochschuler S, Reznik B. *Treat Your Own Back Without Surgery.* 2nd ed. Alameda, CA: Hunter House, 2002.
44. Jackson R. Postural dynamics: Functional causes of low back pain. In: D'Orazio BP, ed. *Low Back Pain Handbook* (pp. 159-191). Boston: Butterworth-Heinemann, 1999.
45. Jarvik JG, Deyo RA. Diagnostic evaluation of low back pain with emphasis on imaging. *Ann Intern Med* 2002;37:586-597.
46. Jarvik JG, Hollingworth W, Heagerty PJ, Haynor DR, Boyko EJ, Deyo RA. Three-year incidence of low back pain in an initially asymptomatic cohort: Clinical and imaging risk factors. *Spine* 2005;30:1541-1548.
47. Jayson MI. ABC of work related disorders: Back pain. *BMJ* 1996;313:355-358.
48. Juker D, McGill SM, Kropf P, Steffen T. Quantitative intramuscular myoelectric activity of lumbar portions of psoas and the abdominal wall during a wide variety of tasks. *Med Sci Sports Exerc* 1998;30:301-310.
49. Kaminsky LA, ed. *ACSM's Health-Related Physical Fitness Assessment Manual.* 3rd ed. Baltimore: Lippincott Williams & Wilkins, 2010.
50. Karjalainen K, Malmivaara A, van Tulder M, Roine R, Juhiainen M, Hurri H, Koes B. Multidisciplinary biopsychosocial rehabilitation for subacute low-back pain among working age adults. *Cochrane Database Syst Rev* 2003;2:CD002193.
51. Kent PM, Keating J. The epidemiology of low back pain in primary care. *Chiropr Osteop* 2005;13:13 [Online]. Doi:10.1186/1746-1340-13-13.
52. Koes BW, van Tulder M, Lin CW, Macedo LG, McAuley J, Maher C. An updated overview of clinical guidelines for the management of non-specific low back pain in primary care. *Eur Spine J* 2010;19:2075-2094.
53. Koes BW, van Tulder MW, Thomas S. Diagnosis and treatment of low back pain. *BMJ* 2006;332:1430-1434.
54. Last AR, Hulbert K. Chronic low back pain: Evaluation and management. *Am Fam Physician* 2009;79:1067-1074.
55. Leboeuf-Yde C. Body weight and low back pain. A systematic literature review of 56 journal articles reporting on 65 epidemiological studies. *Spine* 2000;25:226-237.
56. Leboeuf-Yde C, Kyvik KO, Bruun NH. Low back pain and lifestyle. Part II: Obesity. Information from a population-based sample of 29,424 twin subjects. *Spine* 1999;24:779-784.
57. Linton SJ. Chronic back pain: Integrating psychological and physical therapy. *Behav Med* 1994;20:101-104.
58. Loney PL, Stratford PW. The prevalence of low back pain in adults: A methodological review of the literature. *Phys Ther* 1999;79:384-396.
59. Long A, Donelson R, Fung T. Does it matter which exercise? *Spine* 2004;29:2593-2602.
60. Long A, May S, Fung T. Specific directional exercises for patients with low back pain: A case series. *Physiother Can* 2008;60:307-317.
61. Luomajoki H, Kool J, de Bruin ED, Airaksinen O. Improvement in low back movement control, decreased pain and disability, resulting from specific exercise intervention. *Sports Med Arthrosc Rehabil Ther Technol* 2010;2:11.
62. Luoto S, Heliovaara M, Hurri H, Alaranta M. Static back endurance and the risk of low back pain. *Clin Biomech* 1995;10:323-324.
63. Maher CG. Effective physical treatment for chronic low back pain. *Orthop Clin N Am* 2004;35:57-64.
64. Manniche C. Clinical benefit of intensive dynamic exercises for low back pain. *Scand J Med Sci Sports* 1996;6:82-87.
65. Manniche C, Lundberg E, Christensen I, Bentzen L, Hesselsˉe G. Intensive dynamic back exercises for chronic low back pain: A clinical trial. *Pain* 1991;47:53-63.
66. Mannion AF, Helbling D, Pulkovski N, Sprott H. Spinal segmental stabilisation exercises for chronic low back pain: Programme adherence and its influence on clinical outcome. *Eur Spine J* 2009;18:1881-1891.
67. Matsui H, Kanamori M, Ishihara H, Yudoh K, Naruse Y, Tsuji H. Familial predisposition for lumbar degenerative disc disease. A case-control study. *Spine* 1998;23:1029-1034.
68. Maus T. Imaging the back pain patient. *Phys Med Rehabil Clin N Am* 2010;21:725-766.
69. McGill SM. *Low Back Disorders: Evidence-Based Prevention and Rehabilitation.* Champaign, IL: Human Kinetics, 2002.

70. McGill SM. Low back exercises: Evidence for improving exercise regimens. *Phys Ther* 1998;78:754-765.

71. McGill SM. Low back exercises: Prescription for the healthy back and when recovering from injury. In: Roitman JL, ed. *ACSM's Resource Manual for Guidelines for Exercise Testing and Prescription* (pp. 116-126). 3rd ed. Baltimore: Williams & Wilkins, 1998.

72. McKenzie RA. *The Lumbar Spine: Mechanical Diagnosis and Therapy*, vols. 1 and 2. 2nd ed. Waikanae, New Zealand: Spinal Publications New Zealand, 2003.

73. Moffroid MT. Endurance of trunk muscles in persons with chronic low back pain: Assessment, performance, training. *J Rehabil Res Dev* 1997;34:440-447.

74. Moore JE. Chronic low back pain and psychosocial issues. *Phys Med Rehabil Clin N Am* 2010;21:801-815.

75. National Research Council and Institute of Medicine. *Musculoskeletal Disorders and the Workplace: Low Back and Upper Extremities*. Washington, DC: National Academy Press, 2001.

76. Ng JK-F, Richardson CA, Kippers V, Parnianpour M. Relationship between muscle fiber composition and functional capacity of back muscles in healthy subjects and patients with back pain. *J Orthop Sports Phys Ther* 1998;26:389-402.

77. Nordin M, Welser S, Campello MA, Pietrek M. Self-care techniques for acute episodes of low back pain. *Best Pract Res Clin Rheumatol* 2002;16:89-104.

78. Oesch P, Kool J, Hagen KB, Bachmann S. Effectiveness of exercise on work disability in patients with non-acute non-specific low back pain: Systematic review and meta-analysis of randomised controlled trials. *J Rehabil Med* 2010;42:193-205.

79. Oldridge NB, Stoll JE. Spinal disorders and low back pain. In: Skinner JS, ed. *Exercise Testing and Exercise Prescription for Special Cases* (pp. 139-152). 2nd ed. Philadelphia: Lea & Febiger, 1997.

80. Ostelo RWJG, de Vet HCW, Waddell G, Kerckhoffs MR, Leffers P, van Tulder MW. Rehabilitation after lumbar disc surgery. *Cochrane Database Syst Rev* 2002:CD003007.

81. Pollock ML, Leggett SH, Graves JE, Jones A, Fulton M, Cirulli J. Effective resistance training of lumbar extensor strength. *Am J Sports Med* 1989;17:624-629.

82. Rainville J, Sobel J, Hartigan C, Monlux G, Bean J. Decreasing disability in chronic back pain through aggressive spine rehabilitation. *J Rehabil Res Dev* 1997;34:383-393.

83. Ritz S, Lorren T, Simpson S, Mondry T, Comer M. Rehabilitation of degenerative disease of the spine. In: Hochschuler SH, Cotler HB, Guyer RD, eds. *Rehabilitation of the Spine: Science and Practice* (pp. 457-477). St. Louis: Mosby, 1993.

84. Sato T, Ito T, Hirano T, Morita O, Kikuchi R, Endo N, Tanabe N. Low back pain in childhood and adolescence: Assessment of sports activities. *Eur Spine J* 2011;20:94-99.

85. Schonstein E, Kenny DT, Keating J, Koes BW. Work conditioning, work hardening and functional restoration for workers with back and neck pain. *Cochrane Database Syst Rev* 2003;3;CD001822.

86. Slade SC, Molloy E, Keating JL. "Listen to me, tell me": A qualitative study of partnership in care for people with non-specific chronic low back pain. *Clin Rehabil* 2009;23:270-280.

87. Slade SC, Molloy E, Keating JL. People with non-specific chronic low back pain who have participated in exercise programs have preferences about exercise: A qualitative study. *Aust J Physiother* 2009;55:115-121.

88. Smedley J, Inskip H, Cooper C, Coggon D. Natural history of low back pain. A longitudinal study in nurses. *Spine* 1998;23:2422-2426.

89. Smeets RJ, Severens JL, Beelen S, Vlaeyen JW, Knottnerus JA. More is not always better: Cost-effectiveness analysis of combined, single behavioral and single physical rehabilitation programs for chronic low back pain. *Eur J Pain* 2009;13:71-81.

90. Snook SH. Self-care guidelines for the management of nonspecific low back pain. *J Occup Rehabil* 2004;14:243-253.

91. Snook SH, Webster BS, McGorry RW. The reduction of chronic nonspecific low back pain through the control of early morning lumbar flexion: 3-year follow-up. *J Occup Rehabil* 2002;12:13-19.

92. Snook SH, Webster BS, McGorry RW, Fogleman MT, McCann KB. The reduction of chronic nonspecific low back pain through the control of early morning lumbar flexion: A randomized controlled trial. *Spine* 1998;23:2601-2607.

93. Spitzer WO, LeBlanc FE, Dupuis M, Abenhaim L, Belanger AY, Bloch R, Bombardier C, Cruess RL, Drouin G, Duval-Hesler N, Lafamme J, Lamoureux G, Nachemson A, Page JJ, Rossignol M, Salmi LR, Salois-Arsenault S, Suissa S, Wood-Dauphinee S. Scientific approach to the assessment and management of activity-related spinal disorders. A monograph for clinicians. Report of the Quebec Task Force on Spinal Disorders. *Spine* 1987;12(7S):S5-S59.

94. Thomas E, Silman AJ, Croft PR, Papageorgiou AC, Jayson MIV, Macfarlane GJ. Predicting who develops chronic low back pain in primary care: A prospective study. *BMJ* 1999;318:1662-1667.

95. Van den Hoogen HJM, Koes BW, van Eijk JTM, Bouter LM, DevillÈ W. On the course of low back pain in general practice: A one year follow up study. *Ann Rheum Dis* 1998;57:13-19.

96. Van Middelkoop M, Rubinstein SM, Verhagen AP, Ostelo RW, Koes BW, van Tulder MW. Exercise therapy for chronic nonspecific low-back pain. *Best Pract Res Clin Rheumatol* 2010;24:193-204.

97. Van Tulder M, Becker A, Bekkering T, Breen A, del Real MT, Hutchinson A, Koes B, Laerum E, Malmivaara A. European guidelines for the management of acute nonspecific low back pain in primary care. *Eur Spine J* 2006;15(Suppl. 2):S169-S191.

98. Videman T, BattiÈ MC. A critical review of the epidemiology of idiopathic back pain. In: Weinstein JN, Gordon SL, eds. *Low Back Pain: A Scientific and Clinical Overview* (pp. 317-332). Rosemont, IL: American Academy of Orthopaedic Surgeons, 1996.

99. VingÅrd E, Nachemson A. Work-related influences on neck and low back pain. In: Nachemson A, Jonsson E, eds. *Neck and Back Pain: The Scientific Evidence of Causes, Diagnosis, and Treatment* (pp. 97-126). Philadelphia: Lippincott Williams & Wilkins, 2000.

100. Waddell G. *The Back Pain Revolution*. 2nd ed. Edinburgh: Churchill Livingstone, 2004.

101. Waddell G. Subgroups within ìnonspecificî low back pain. *J Rheum* 2005;32:395-396.

102. Waddell G, Burton AK. Concepts of rehabilitation for the management of low back pain. *Best Pract Res Clin Rheumatol* 2005;19:655-670.

103. Weisel SW, Feffer HL, Rothman RH. A lumbar spine algorithm. In: Weinstein JN, Wiesel SW, eds. *The Lumbar Spine* (pp. 358-368). Philadelphia: Saunders, 1990.

104. White AA III, Panjabi MK. The clinical biomechanics of spine pain. In: *Clinical Biomechanics of the Spine* (pp. 379-474). 2nd ed. Philadelphia: Lippincott, 1990.

105. Williams PC. Lesions of the lumbosacral spine, part I. *J Bone Joint Surg* 1937;19:343-363.

106. Williams PC. Lesions of the lumbosacral spine, part II. *J Bone Joint Surg* 1937;19:690-703.

107. Wyke B. The neurological basis of thoracic spine pain. *Rheumatol Phys Med* 1970;10:356-367.

CHAPTER 25

1. Aksnes AK, Hjeltnes N, Wahlstrom EO, Katz A, Zierath JR, Wallberg-Henriksson H. Intact glucose transport in morphologically altered denervated skeletal muscle from quadriplegic patients. *Am J Physiol* 1996;271(3 Pt 1):E593-600.

2. American College of Sports Medicine. *ACSM's Guidelines for Exercise and Prescription.* 9th ed. Baltimore: Lippincott Williams & Wilkins, 2013.

3. American Spinal Injury Association. *International Standards for Neurological Classification of Spinal Cord Injury.* Chicago: American Spinal Injury Association, 2006.

4. Barbeau H, Ladouceur M, Mirbagheri MM, Kearney RE. The effect of locomotor training combined with functional electrical stimulation in chronic spinal cord injured subjects: Walking and reflex studies. *Brain Res Rev* 2002;40(1-3):274-291.

5. Barbeau H, Ladouceur M, Norman KE, Pepin A, Leroux A. Walking after spinal cord injury: Evaluation, treatment, and functional recovery. *Arch Phys Med Rehabil* 1999;80(2):225-235.

6. Barstow TJ, Scremin AM, Mutton DL, Kunkel CF, Cagle TG, Whipp BJ. Gas exchange kinetics during functional electrical stimulation in subjects with spinal cord injury. *Med Sci Sports Exerc* 1995;27(9):1284-1291.

7. Barstow TJ, Scremin AM, Mutton DL, Kunkel CF, Cagle TG, Whipp BJ. Peak and kinetic cardiorespiratory responses during arm and leg exercise in patients with spinal cord injury. *Spinal Cord* 2000;38(6):340-345.

8. Bauman WA, Alexander LR, Zhong Y-G, Spungen AM. Stimulated leg ergometry training improves body composition and HDL-cholesterol values. *J Am Paraplegia Soc* 1994;17(4):201. [Abstract]

9. Bauman WA, Raza M, Chayes Z, Machac J. Tomographic thallium-201 myocardial perfusion imaging after intravenous dipyridamole in asymptomatic subjects with quadriplegia. *Arch Phys Med Rehabil* 1993;74:740-744.

10. Bauman WA, Schwartz E, Song IS, et al. Dual-energy X-ray absorptiometry overestimates bone mineral density of the lumbar spine in persons with spinal cord injury. *Spinal Cord* 2009 Aug;47(8):628-633.

11. Bauman WA, Spungen AM. Disorders of carbohydrate and lipid metabolism in veterans with paraplegia or quadriplegia: A model of premature aging. *Metabolism* 1994;43(6):749-756.

12. Bauman WA, Spungen AM. Metabolic changes in persons after spinal cord injury. *Phys Med Rehabil Clin N Am* 2000;11(1):109-140.

13. Bauman WA, Spungen AM, Wang J, Pierson RN. The relationship between energy expenditure and lean tissue in monozygotic twins discordant for spinal cord injury. *J Rehabil Res Dev* 2004;41:1-8.

14. Behrman AL, Harkema SJ. Locomotor training after human spinal cord injury: A series of case studies. *Phys Ther* 2000;80(7):688-700.

15. Behrman AL, Lawless-Dixon AR, Davis SB, et al. Locomotor training progression and outcomes after incomplete spinal cord injury. *Phys Ther* 2005;85(12):1356-1371.

16. Bernard PL, Mercier J, Varray A, Prefaut C. Influence of lesion level on the cardioventilatory adaptations in paraplegic wheelchair athletes during muscular exercise. *Spinal Cord* 2000;38(1):16-25.

17. Bloomfield SA, Mysiw WJ, Jackson RD. Bone mass and endocrine adaptations to training in spinal cord injured individuals. *Bone* 1996;19(1):61-68.

18. Bose P, William W, Telford R, et al. Neuroplasticity following spinal cord injury (SCI) and locomotor training. Presented at the Society for Neuroscience annual meeting, 2005, Washington, DC.

19. Braddom RL, Rocco JF. Autonomic dysreflexia: A survey of current treatment. *Am J Phys Med Rehabil* 1991;70:234-241.

20. Bryce TN, Ragnarsson KT. Pain after spinal cord injury. *Phys Med Rehabil Clin N Am* 2000;11(1):157-168.

21. Buchholz AC, McGillivray CF, Pencharz PB. Differences in resting metabolic rate between paraplegic and able-bodied subjects are explained by differences in body composition. *Am J Clin Nutr* 2003;77(2):371-378.

22. Buchholz AC, Pencharz PB. Energy expenditure in chronic spinal cord injury. *Cur Opin Clin Nutrit Metab Care* 2004;7(6):635-639.

23. Bursell JP, Little JW, Stiens SA. Electrodiagnosis in spinal cord injured persons with new weakness or sensory loss: Central and peripheral etiologies. *Arch Phys Med Rehabil* 1999;80(8):904-909.

24. Chawla JC, Bar C, Creber I, Price J, Andrews B. Techniques for improving the strength and fitness of spinal cord injured patients. *Paraplegia* 1979-1980;17:185-189.

25. Collins EG, Gater D, Kiratli BJ, Butler J, Hanson K, Langbein W. Energy cost of physical activities in persons with spinal cord injury. *Med Sci Sports Exerc* 2010;42(4):691-700.

26. Cooper RA. A force/energy optimization model for wheelchair athletics. *IEEE Trans Syst Man Cybern* 1990;20(2):444-449.

27. Cooper RA, Baldini FD, Langbein WE, Robertson RN, Bennett P, Monical S. Prediction of pulmonary function in wheelchair users. *Paraplegia* 1993;31(9):560-570.

28. Coutts KD, Rhodes EC, McKenzie DC. Maximal exercise responses of tetraplegics and paraplegics. *J Appl Physiol* 1983;55(2):479-482.

29. Crane L, Klerk K, Ruhl A, Warner P, Ruhl C, Roach KE. The effect of exercise on pulmonary function in persons with quadriplegia. *Paraplegia* 1994;32:435-441.

30. Curt A, Nitsche B, Rodic B, Schurch B, Dietz V. Assessment of autonomic dysreflexia in patients with spinal cord injury. *J Neurol Neurosurg Psychiatry* 1997;62(5):473-477.

31. Curtis KA, Black K. Shoulder pain in female wheelchair basketball players. *J Orthop Sports Phys Ther* 1999;29(4):225-231.

32. Dallmeijer AJ, Hopman MTE, van As HHJ, van der Woude LHV. Physical capacity and physical strain in persons with tetraplegia: The role of sport activity. *Spinal Cord* 1996;34:729-735.

33. Davis GM. Exercise capacity of individuals with paraplegia. *Med Sci Sports Exerc* 1993;25(4):423-432.

34. Davis GM, Shephard RJ. Strength training for wheelchair users. *Br J Sports Med* 1990;24(1):25-30.

35. Davis GM, Tanhoffer RA, Tanhoffer AIP, Pithon KR, Estigoni EH, Raymond J. Energy expenditures during wheelchair propulsion derived from a body-worn sensor versus indirect calorimetry. *Med Sci Sports Exerc* 2010;42(5):335. [Abstract]

36. Denis F. The three column spine and its significance in the classification of acute thoracolumbar spinal injuries. *Spine* 1983;8(8):817-831.

37. DeVivo MJ, Krause JS, Lammertse DP. Recent trends in mortality and causes of death among persons with spinal cord injury. *Arch Phys Med Rehabil* 1999;80(11):1411-1419.

38. Dietz V, Harkema SJ. Locomotor activity in spinal cord-injured persons. *J Appl Physiol* 2004;96(5):1954-1960.

39. Dixit S. Bradycardia associated with high cervical spinal cord injury. *Surg Neurol* 1995;43:514.

40. Dobkin B, Barbeau H, Deforge D, et al. The evolution of walking-related outcomes over the first 12 weeks of rehabilitation for incomplete traumatic spinal cord injury: The multicenter randomized Spinal Cord Injury Locomotor Trial. *Neurorehabil Neural Repair* 2007;21(1):25-35.

41. Faghri PD, Glaser RM, Figoni SF. Functional electrical stimulation leg cycle ergometer exercise: Training effects on cardiorespiratory responses of spinal cord injured subjects at rest and during submaximal exercise. *Arch Phys Med Rehabil* 1992;73(11):1085-1093.

42. Ferguson-Pell MW. Technical considerations: Seat cushion selection. In: Todd SP, ed. Choosing a wheelchair system (pp. 47-73). *J Rehabil Res Dev Clin Suppl* 2. Baltimore: Department of Veterans Affairs, Veterans Health Administration, Rehabilitation Research and Development Service, Scientific and Technical Publications Section, 1992.

43. Figoni SF. Exercise responses and quadriplegia. *Med Sci Sports Exerc* 1993;25(4):433-441.

44. Figoni SF. Overuse shoulder problems after spinal cord injury: A conceptual model of risk and protective factors. *Clin Kinesiol* 2009;63(2):12-22.

45. Figoni SF, Glaser RM, Collins SR. Peak physiologic responses of trained quadriplegics during arm, leg and hybrid exercise in two postures. *Med Sci Sports Exerc* 1995;27(5):S83. [Abstract]

46. Figoni SF, Kiratli BJ, Myers JN. Spinal cord dysfunction. In: Myers J, Nieman D, eds. *ACSM's Resources for Clinical Exercise Physiology: Musculoskeletal, Neuromuscular, Neoplastic, Immunologic, and Hematologic Conditions* (pp. 58-78). 2nd ed. Baltimore: Lippincott Williams & Wilkins, 2010.

47. Figoni SF, McLain L, Bell AA, et al. Accessibility of physical fitness facilities in the Kansas City metropolitan area. *Top Spinal Cord Inj Rehabil* 1998;3(3):66-78.

48. Figoni SF, Rodgers MM, Glaser RM, Hooker SP, Feghri PD, Ezenwa BN, Mathews T, Suryaprasad AG, Gupta SC. Physiologic responses of paraplegics and quadriplegics to passive and active leg cycle ergometry. *J Am Paraplegia Soc* 1990;13(3):33-39.

49. Figoni SF, Vanlandewijck Y. Spinal cord injury. In: Riebe D, ed. Chapter 10. Exercise prescription for other clinical populations. In: Pescatello LS, senior ed. *ACSM's Guidelines for Exercise Testing and Prescription.* 9th ed. Baltimore: Lippincott Williams & Wilkins, 2013.

50. Finnie AK, Buchholz AC, Martin Ginis KA; SHAPE SCI Research Group. Current coronary heart disease risk assessment tools may underestimate risk in community-dwelling persons with chronic spinal cord injury. *Spinal Cord* 2008;46(9):608-615.

51. Franklin BA, Swantek KI, Grais SL, Johnstone KS, Gordon S, Timmis GC. Field test estimation of maximal oxygen consumption in wheelchair users. *Arch Phys Med Rehabil* 1990;71:574-578.

52. Garland DE, Stewart CA, Adkins RH, et al. Osteoporosis after spinal cord injury. *J Orthop Res* 1992;10(3):371-378.

53. Gass GC, Watson J, Camp EM, Court HJ, McPherson LM, Redhead P. The effects of physical training on high level spinal lesion patients. *Scand J Rehabil Med* 1980;12:61-65.

54. Gater DR Jr. Obesity after spinal cord injury. *Phys Med Rehabil Clin N Am* 2007;18(2):333-351, vii.

55. Gater DR. Pathophysiology of obesity after spinal cord injury. *Top Spinal Cord Inj Rehabil* 2007;12(4):20-34.

56. Gater DR, Yates JW, Clasey JL. Relationship between glucose intolerance and body composition in spinal cord injury. *Med Sci Sports Exerc* 2000;32(5):S148. [Abstract]

57. Ginis KA, Hicks AL, Latimer AE, et al. The development of evidence-informed physical activity guidelines for adults with spinal cord injury. *Spinal Cord* 2011, June 7; ePub ahead of print, doi:10.1038/sc.2011.63.

58. Glaser RM. Functional neuromuscular stimulation. Exercise conditioning of spinal cord injured patients. *Int J Sports Med* 1994;15(3):142-148.

59. Glaser RM, Janssen TWJ, Suryaprasad AG, Gupta SC, Mathews T. The physiology of exercise. In: Apple DF, ed. *Physical Fitness: A Guide for Individuals with Spinal Cord Injury* (pp. 1-24). Baltimore: Department of Veterans Affairs, Veterans Health Administration, Rehabilitation R&D Service, Scientific and Technical Publications Section, 1996.

60. Glaser RM, Sawka MN, Brune MF, Wilde SW. Physiological responses to maximal effort wheelchair ergometry and arm crank ergometry. *J Appl Physiol* 1980;48:1060-1064.

61. Glinsky J, Harvey L, Korten M, Drury C, Chee S, Gandevia SC. Short-term progressive resistance exercise may not be effective at increasing wrist strength in people with tetraplegia: A randomised controlled trial. *Aust J Physiother* 2008;54(2): 103-108.

62. Go BK, DeVivo MJ, Richards JS. The epidemiology of spinal cord injury. In: Stover SL, DeLisa JA, Whiteneck GG, eds. *Spinal Cord Injury: Clinical Outcomes from the Model Systems* (pp. 21-51). Rockville, MD: Aspen, 1995.

63. Hangartner TN, Rodgers MM, Glaser RM, Barre PS. Tibial bone density loss in spinal cord injured patients: Effects of FES exercise. *J Rehabil Res Dev* 1994;31(1):50-61.

64. Hartkopp A, Harridge SD, Mizuno M, et al. Effect of training on contractile and metabolic properties of wrist extensors in spinal cord-injured individuals. *Muscle Nerve* 2003;27(1):72-80.

65. Haskell WL, Lee IM, Pate RR, et al. Physical activity and public health: Updated recommendation for adults from the American College of Sports Medicine and the American Heart Association. *Med Sci Sports Exerc* 2007;39(8):1423-1434.

66. Hettinga DM, Andrews BJ. Oxygen consumption during functional electrical stimulation-assisted exercise in persons with spinal cord injury: Implications for fitness and health. *Sports Med* 2008;38(10):825-838.

67. Heymsfeld SB, Lichtman S, Baumgartner RN, et al. Body composition of humans: Comparison of two improved four-compartment models that differ in expense, technical complexity, and radiation exposure. *Am J Clin Nutr* 1990;52(1):52-58.

68. Hicks AL, Adams MM, Ginis KM, et al. Long-term body-weight-supported treadmill training and subsequent follow-up in persons with chronic SCI: Effects on functional walking ability and measures of subjective well-being. *Spinal Cord* 2005;43(5):291-298.

69. Hicks AL, Martin KA, Ditor DS, et al. Long-term exercise training in persons with spinal cord injury: Effects on strength, arm ergometry performance and psychological well-being. *Spinal Cord* 2003;41(1):34-43.

70. Hjeltnes N. Oxygen uptake and cardiac output in graded arm exercise in paraplegics with low level spinal lesions. *Scand J Rehabil Med* 1977;9:107-113.

71. Hjeltnes N, Jansen T. Physical endurance capacity, functional status and medical complications in spinal cord injured subjects with long-standing lesions. *Paraplegia* 1990;28:428-432.

72. Hooker SP, Figoni SF, Rodgers MM, Glaser RM, Mathews T, Suryaprasad AG, Gupta SC. Physiologic effects of electrical stimulation leg cycle exercise training in spinal cord injured persons. *Arch Phys Med Rehabil* 1992;73(5):470-476.

73. Hooker SP, Greenwood JD, Hatae DT, et al. Oxygen uptake and heart rate relationship in persons with spinal cord injury. *Med Sci Sports Exerc* 1993;25(10):1115-1119.

74. Jacobs PL. Effects of resistance and endurance training in persons with paraplegia. *Med Sci Sports Exerc* 2009;41:992-997.

75. Jacobs PL, Johnson B, Mahoney ET. Physiologic responses to electrically assisted and frame-supported standing in persons with paraplegia. *J Spinal Cord Med* 2003;26(4):384-389.

76. Jacobs PL, Mahoney ET, Robbins A, Nash M. Hypokinetic circulation in persons with paraplegia. *Med Sci Sports Exerc* 2002;34(9):1401-1407.

77. Jacobs PL, Nash MS. Exercise recommendations for individuals with spinal cord injury. *Sports Med* 2004;34(11):727-751.

78. Janssen TWJ, van Oers CAJM, Rozendaal EP, Willemsen EM, Hollander AP, van der Woude LHV. Changes in physical strain and physical capacity in men with spinal cord injuries. *Med Sci Sports Exerc* 1996;28(5):551-559.

79. Janssen TWJ, van Oers CAJM, van der Woude LHV, Hollander AP. Physical strain in daily life of wheelchair users with spinal cord injuries. *Med Sci Sports Exerc* 1994;26(6):661-670.

80. Janssen TW, van Oers CA, van Kamp GJ, TenVoorde BJ, van der Woude LH, Hollander AP. Coronary heart disease risk indicators, aerobic power, and physical activity in men with spinal cord injuries. *Arch Phys Med Rehabil* 1997;78(7):697-705.

81. Jones LM, Legge M, Goulding A. Healthy body mass index values often underestimate body fat in men with spinal cord injury. *Arch Phys Med Rehabil* 2003;84(7):1068-1071.

82. Katalinic OM, Harvey LA, Herbert RD, Moseley AM, Lannin NA, Schurr K. Stretch for the treatment and prevention of contractures. *Cochrane Database Syst Rev* 2010;8(9):CD007455.

83. Katz RT. Management of spasticity. *Am J Phys Med Rehabil* 1988;67:108-116.

84. Kavanagh T. Physical training in heart transplant recipients. *J Cardiovasc Risk* 1996;3(2):154-159.

85. Kawamoto M, Sakimura S, Takasaki M. Transient increase of parasympathetic tone in patients with cervical spinal cord trauma. *Anaesth Intensive Care* 1993;21:218-221.

86. Knutsson E, Lewenhaupt-Olsson E, Thorsen M. Physical work capacity and physical conditioning in paraplegic patients. *Paraplegia* 1973;11(3):205-216.

87. Langbein WE, Edwards SC, Louie EK, Hwang MH, Nemchausky BA. Wheelchair exercise and digital echocardiography for the detection of heart disease. *Rehabil Res Dev Rep* 1996;34:324-325.

88. Langbein WE, Maki KC, Edwards LC, Hwang MH, Sibley P, Fehr L. Initial clinical evaluation of a wheelchair ergometer for diagnostic exercise testing: A technical note. *J Rehabil Res Dev* 1994;31(4):317-325.

89. Laskin JJ, Ashley EA, Olenik LM, Burnham R, Cumming DC, Steadward RD, Wheeler GD. Electrical stimulation assisted rowing exercise in spinal cord injured people. A pilot study. *Paraplegia* 1993;31(8):534-541.

90. Lassau-Wray ER, Ward GR. Varying physiological response to arm-crank exercise in specific spinal injuries. *J Physiol Anthropol Appl Human Sci* 2000;19(1):5-12.

91. Laughton GE, Buchholz AC, Martin Ginis KA, Goy RE; SHAPE SCI Research Group. Lowering body mass index cutoffs better identifies obese persons with spinal cord injury. *Spinal Cord* 2009;47(10):757-762.

92. Lavis TD, Scelza WM. Cardiovascular health and fitness in persons with spinal cord injury. *Phys Med Rehabil Clin N Am* 2007;18:317-331.

93. Loudon JK, Cagle PE, Figoni SF, Nau KL, Klein RM. A submaximal all-extremity exercise test to predict maximal oxygen consumption. *Med Sci Sports Exerc* 1998;30(8):1299-1303.

94. Lundberg A. Wheelchair driving: Evaluation of a new training outfit. *Scand J Rehabil Med* 1987;12:67-72.

95. Maynard FM, Karunas RS, Adkins RH, Richards JS, Waring III WP. Management of the neuromusculoskeletal systems. In: Stover SL, DeLisa JA, Whiteneck GG, eds. *Spinal Cord Injury: Clinical Outcomes from the Model Systems* (pp. 145-169). Bethesda, MD: Aspen, 1995.

96. McLean KP, Skinner JS. Effect of body training position on outcomes of an aerobic training study on individuals with quadriplegia. *Arch Phys Med Rehabil* 1995;76:139-150.

97. Mohr T, Podenphant J, Biering-Sorensen F, Galbo H, Thamsborg G, Kjaer M. Increased bone mineral density after prolonged electrically induced cycle training of paralyzed limbs in spinal cord injured man. *Calcif Tissue Int* 1997;61(1):22-25.

98. Mulroy SJ, Thompson L, Kemp B, et al. Strengthening and optimal movements for painful shoulders (STOMPS) in chronic spinal cord injury: A randomized controlled trial. *Phys Ther* 2011;91(3):305-324.

99. Mutton DL, Scremin AME, Barstow TJ, Scott MD, Kunkel CF, Cagle TG. Physiologic responses during functional electrical stimulation leg cycling and hybrid exercise in spinal cord injured subjects. *Arch Phys Med Rehabil* 1997;78:712-718.

100. Naftchi NE. Mechanism of autonomic dysreflexia. Contributions of catecholamine and peptide neurotransmitters. *Ann N Y Acad Sci* 1990;579:133-148.

101. Nash MS, Bilsker MS, Kearney HM, Ramirez JN, Applegate B, Green BA. Effects of electrically-stimulated exercise and passive motion on echocardiographically derived wall motion and cardiodynamic function in tetraplegic persons. *Paraplegia* 1995;33(2):80-89.

102. Nash MS, Bilsker S, Marcillo AE, Isaac SM, Botelho LA, Klose J, Green BA, Rountree MT, Shea JD. Reversal of adaptive left ventricular atrophy following electrically stimulated exercise training in human tetraplegics. *Paraplegia* 1991;29:590-599.

103. Nash MS, Jacobs PL, Johnson BM, Field-Fote E. Metabolic and cardiac responses to robotic-assisted locomotion in motor-complete tetraplegia: A case report. *J Spinal Cord Med* 2004;27(1):78-82.

104. Needham-Shropshire BM, Broton JG, Cameron TL, Klose KJ. Improved motor function in tetraplegics following neuromuscular stimulation-assisted arm ergometry. *J Spinal Cord Med* 1997;20:49-55.

105. Nilsson S, Staff PH, Pruett ED. Physical work capacity and the effect of training on subjects with long-standing paraplegia. *Scand J Rehabil Med* 1975;7(2):51-56.

106. Petrofsky JS. Thermoregulatory stress during rest and exercise in heat in patients with a spinal cord injury. *Eur J Appl Physiol* 1992;64(6):503-507.

107. Phillips SM, Stewart BG, Mahoney DJ, et al. Body-weight-support treadmill training improves blood glucose regulation in persons with incomplete spinal cord injury. *J Appl Physiol* 2004;97(2):716-724.

108. Phillips W, Burkett LN, Munro R, Davis M, Pomeroy K. Relative changes in blood flow with functional electrical stimulation during exercise of the paralyzed lower limbs. *Paraplegia* 1995;33(2):90-93.

109. Physiotherapy Evidence Database. www.pedro.org.au/wp-content/uploads/PEDro_scale.pdf [accessed April 25, 2013].

110. Physiotherapy Evidence Database. www.scireproject.com/ tables /methods-of-systematic-review-table-1-five-levels-of-evidence [accessed April 16, 2011].

111. Ragnarsson KT, Pollack S, O'Daniel W, Edgar R, Petrofsky J, Nash MS. Clinical evaluation of computerized functional electrical stimulation after spinal cord injury: A multicenter pilot study. *Arch Phys Med Rehabil* 1988;69:672-677.

112. Reame NE. A prospective study of the menstrual cycle and spinal cord injury. *Am J Phys Med Rehabil* 1992;71:15-21.

113. Rimaud D, Calmels P, Devillard X. Training programs in spinal cord injury. Ann Réadapt Méd Phys 2005;48(5):259-269.

114. Rodgers MM, Glaser RM, Figoni SF, et al. Musculoskeletal responses of spinal cord injured individuals to functional neuromuscular stimulation-induced knee extension exercise training. *J Rehabil Res Dev* 1991;28(4):19-26.

115. Ropper AH. Trauma of the head and spine. In: Isselbacher KJ, Braunwald E, Wilson JD, Martin JB, Fauci AS, Kasper DL, eds. *Harrison's Principles of Internal Medicine.* 13th ed. St. Louis: McGraw-Hill, 1994.

116. Schmid A, Huonker M, Barturen JM, Stahl F, Schmidt-Trucksass A, Konig D, Grathwohl D, Lehmann M, Keul J. Catecholamines, heart rate, and oxygen uptake during exercise in persons with spinal cord injury. *J Appl Physiol* 1998;85(2):635-641.

117. Schmidt KD, Chan CW. Thermoregulation and fever in normal persons and in those with spinal cord injuries. *Mayo Clin Proc* 1992;67(5):469-475.

118. Schneider DA, Sedlock DA, Gass E, Gass G. VO_2peak and the gas-exchange anaerobic threshold during incremental arm cranking in able-bodied and paraplegic men. *Eur J Appl Physiol Occup Physiol* 1999;80(4):292-297.

119. SCI Action Canada. Physical Activity Guidelines for Adults with Spinal Cord Injury—Health Care Professional. www.sciaction-canada.ca/docs/guidelines/Physical-Activity-Guidelines-for-Adults-with-a-Spinal-Cord-Injury-Health-Care-Professional.pdf [accessed April 25, 2013].

120. SCI Info Pages. Spinal cord injury range of motion exercises. www.sci-info-pages.com/range.html [accessed April 25, 2013].

121. Sharkey BJ, Graetzer DG. Specificity of exercise, training, and testing. In: Durstine JL, King AC, Painter PL, Roitman JL, Zwiren LD, eds. *ACSM's Resource Manual for Guidelines for Exercise Testing and Prescription* (pp. 82-92). 2nd ed. Philadelphia: Lea & Febiger, 1993.

122. Sipski ML, Delisa JA, Schweer SA. Functional electrical stimulation bicycle ergometry: Patient perceptions. *Am J Phys Med Rehabil* 1989;68(3):147-149.

123. Spinal Cord Injury Network—Facts and Figures at a Glance, January 2008. http://images.main.uab.edu/spinalcord/pdffiles/Facts08.pdf [accessed April 25, 2013].

124. Spungen AM, Bauman WA, Wang J, Pierson RN. Measurement of body fat in individuals with tetraplegia: A comparison of eight clinical methods. *Paraplegia* 1995;33(7):402-408.

125. Subarrao JV, Garrison SJ. Heterotopic ossification: Diagnosis and management, current concepts and controversies. *J Spinal Cord Med* 1999;22(4):273-283.

126. Subarrao JV, Klopfstein J, Turpin R. Prevalence and impact of wrist and shoulder pain in patients with spinal cord injury. *J Spinal Cord Med* 1995;18(1):9-13.

127. Taylor AW, McDonell E, Brassard L. The effects of an arm ergometer training programme on wheelchair subjects. *Paraplegia* 1986;24:105-114.

128. Tsitouras PD, Zhong YG, Spungen AM, Bauman WA. Serum testosterone and growth hormone/insulin-like growth factor-I in adults with spinal cord injury. *Horm Metab Res* 1995;27(6):287-292.

129. Ugalde V, Litwiller SE, Gater DR. Physiatric anatomic principles, bladder and bowel anatomy for the physiatrist. *Phys Med Rehabil: State of the Art Reviews* 1996;10:547-568.

130. Uniform Data Systems for Medical Rehabilitation. www.udsmr.org/WebModules/FIM/Fim_About.aspx [accessed April 25, 2013].

131. University of Miami School of Medicine. Movement disorders in spinal cord injury: Range of motion (ROM) exercises: Passive stretching (lower extremity). http://calder.med.miami.edu/pointis/lower.html [accessed April 25, 2013].

132. University of Miami School of Medicine. Movement disorders in spinal cord injury: Range of motion (ROM) exercises: Passive stretching (upper extremity). http://calder.med.miami.edu/pointis/upper.html [accessed April 25, 2013].

133. Van Der Woude LHV, Veeger HEJ, Rozendal RH, Van Ingen Schenau GJ, Rooth F, Van Nierop P. Wheelchair racing: Effects of rim diameter and speed on physiology and technique. *Med Sci Sports Exerc* 1988;20(5):492-500.

134. Vinet A, Bernard PL, Poulain M, Varray A, Le Gallais D, Micallef JP. Validation of an incremental field test for the direct assessment of peak oxygen uptake in wheelchair-dependent athletes. *Spinal Cord* 1996;34(5):288-293.

135. Wang YH, Huang TS. Impaired adrenal reserve in men with spinal cord injury: Results of low- and high-dose adrenocorticotropin stimulation tests. *Arch Phys Med Rehabil* 1999;80(8):863-866.

136. Warburton DER, Sproule S, Krassioukov A, Eng JJ. Cardiovascular health and exercise following spinal cord injury. In: Eng JJ, Teasell RW, Miller WC, et al., eds. Spinal Cord Injury Rehabilitation Evidence. Version 3.0. Vancouver, BC. 2010:1-38. http://spinaltalk.com.au/wp-content/uploads/2012/07/SCIRE-Cardiovascular-Health-and-Exercise-Following-SCI.pdf [accessed April 25, 2013].

137. Wernig A, Nanassy A, Muller S. Maintenance of locomotor abilities following Laufband (treadmill) therapy in para- and tetraplegic persons: Follow-up studies. *Spinal Cord* 1998;36(11):744-749.

138. Wheelchair Track & Field, USA. 2011-2012 Competition Rules for Track, Field & Road Racing. www.wsusa.org/images/Wsusa/Rulebooks/2011/2011_wtfusa_rulebook.pdf [accessed April 25, 2013].

139. Wolfe DL, Martin Ginis KA, Latimer AE, et al. Physical activity and SCI. In: Eng JJ, Teasell RW, Miller WC, et al., eds. Spinal Cord Injury Rehabilitation Evidence. Version 3.0. Vancouver, BC, 2010. http://scireproject.com/rehabilitation-evidence/physical-activity [accessed April 25, 2013].

140. Wolff J. *Das gesetz der transformation der knochen*. Berlin: Ahirshwald, 1892.

141. Young RR. Spasticity: A review. *Neurology* 1994;44(Suppl. 9):S12-S20.

CHAPTER 26

1. Andreasen A, Stenager E, Dalgas U. The effect of exercise therapy on fatigue in multiple sclerosis. *Mult Scler* 2011;17(9):1041-1054.
2. Armstrong LE, Winant DM, Swasey PR, Seidle ME, Carter AL, Gehlsen G. Using isokinetic dynamometry to test ambulatory patients with multiple sclerosis. *Phys Ther* 1983;63(8):1274-1279.
3. Ascherio A, Munger KL. Environmental risk factors for multiple sclerosis. Part II: Noninfectious factors. *Ann Neurol* 2007;61(6):504-513.
4. Bakshi R. Fatigue associated with multiple sclerosis: Diagnosis, impact and management. *Mult Scler* 2003;9(3):219-227.
5. Benedetti MG, Piperno R, Simoncini L, Bonato P, Tonini A, Giannini S. Gait abnormalities in minimally impaired multiple sclerosis patients. *Mult Scler* 1999;5(5):363-368.
6. Borg GA. Psychophysical bases of perceived exertion. *Med Sci Sports Exerc* 1982;14(5):377-381.
7. Bourdette DN, Prochazka AV, Mitchell W, Licari P, Burks J. Health care costs of veterans with multiple sclerosis: Implications for the rehabilitation of MS. VA Multiple Sclerosis Rehabilitation Study Group. *Arch Phys Med Rehabil* 1993;74(1):26-31.
8. Burden of illness of multiple sclerosis: Part I: Cost of illness. The Canadian Burden of Illness Study Group. *Can J Neurol Sci* 1998;25(1):23-30.
9. Carroll CC, Gallagher PM, Seidle ME, Trappe SW. Skeletal muscle characteristics of people with multiple sclerosis. *Arch Phys Med Rehabil* 2005;86(2):224-229.
10. Cartlidge NE. Autonomic function in multiple sclerosis. *Brain* 1972;95(4):661-664.
11. Chan A., Heck CS. Mobility in multiple sclerosis: More than just a physical problem. *Int J MS Care* 2000;3:35-40.
12. Chiara T, Martin AD, Davenport PW, Bolser DC. Expiratory muscle strength training in persons with multiple sclerosis having mild to moderate disability: Effect on maximal expiratory pressure, pulmonary function, and maximal voluntary cough. *Arch Phys Med Rehabil* 2006;87(4):468-473.
13. Chung LH, Remelius JG, van Emmerik RE, Kent-Braun JA. Leg power asymmetry and postural control in women with multiple sclerosis. *Med Sci Sports Exerc* 2008;40(10):1717-1724.
14. Compston A. Treatment and management of multiple sclerosis. In Compston A, Ebers G, Lassmann H, McDonald I, Matthews B, Wekerle H, eds. *McAlpine's Multiple Sclerosis* (pp. 437-498). London: Churchill Livingstone, 1998.
15. Corcos DM, Chen CM, Quinn NP, McAuley J, Rothwell JC. Strength in Parkinson's disease: Relationship to rate of force generation and clinical status. *Ann Neurol* 1996;39:79-88.
16. Davis SL, Wilson TE, Vener JM, Crandall CG, Petajan JH, White AT. Pilocarpine-induced sweat gland function in individuals with multiple sclerosis. *J Appl Physiol* 2005;98(5):1740-1744.
17. DeBolt LS, McCubbin JA. The effects of home-based resistance exercise on balance, power, and mobility in adults with multiple sclerosis. *Arch Phys Med Rehabil* 2004;85(2):290-297.
18. de Haan A, de Ruiter CJ, Der Woude LH, Jongen PJ. Contractile properties and fatigue of quadriceps muscles in multiple sclerosis. *Muscle Nerve* 2000;23(10):1534-1541.
19. de Ruiter CJ, Jongen PJ, van der Woude LH, de Haan A. Contractile speed and fatigue of adductor pollicis muscle in multiple sclerosis. *Muscle Nerve* 2001;24(9):1173-1180.
20. Dorfman LJ, Howard JE, McGill KC. Motor unit firing rates and firing rate variability in the detection of neuromuscular disorders. *Electroencephalogr Clin Neurophysiol* 1989;73(3):215-224.
21. Duquette P, Pleines J, Girard M, Charest L, Senecal-Quevillon M, Masse C. The increased susceptibility of women to multiple sclerosis. *Can J Neurol Sci* 1992;19(4):466-471.
22. Ewing C, Bernard CC. Insights into the aetiology and pathogenesis of multiple sclerosis. *Immunol Cell Biol* 1998;76(1):47-54.
23. Fatigue and Multiple Sclerosis: Evidence-Based Management Strategies for Fatigue in Multiple Sclerosis. Multiple Sclerosis Council for Clinical Practice Guidelines, 1998. http://mssociety.ca/en/pdf/livingWell.pdf [accessed April 25, 2013].
24. Ferrante P, Mancuso R, Pagani E, et al. Molecular evidences for a role of HSV-1 in multiple sclerosis clinical acute attack. *J Neurovirol* 2000;6 Suppl. 2:S109-114.
25. Finlayson ML, Peterson EW, Cho CC. Risk factors for falling among people aged 45 to 90 years with multiple sclerosis. *Arch Phys Med Rehabil* 2006;87(9):1274-1279.
26. Fisk JD, Pontefract A, Ritvo PG, Archibald CJ, Murray TJ. The impact of fatigue on patients with multiple sclerosis. [See comments]. *Can J Neurol Sci* 1994;21:9-14.
27. Foglio K, Clini E, Facchetti D, et al. Respiratory muscle function and exercise capacity in multiple sclerosis. *Eur Respir J* 1994;7:23-28.
28. Franceschini M, Rampello A, Bovolenta F, Aiello M, Tzani P, Chetta A. Cost of walking, exertional dyspnoea and fatigue in individuals with multiple sclerosis not requiring assistive devices. *J Rehabil Med* 2010;42(8):719-723.
29. Freeman J, Allison R. Group exercise classes in people with multiple sclerosis: A pilot study. *Physiother Res Int* 2004;9(2):104-107.
30. Garner DJ, Widrick JJ. Cross-bridge mechanisms of muscle weakness in multiple sclerosis. *Muscle Nerve* 2003;27(4):456-464.
31. Gehlsen GM, Grigsby SA, Winant DM. Effects of an aquatic fitness program on the muscular strength and endurance of patients with multiple sclerosis. *Phys Ther* 1984;64(5):653-657.
32. Green AJ, Barcellos LF, Rimmler JB, et al. Sequence variation in the transforming growth factor-beta1 (TGFB1) gene and multiple sclerosis susceptibility. *J Neuroimmunol* 2001;116(1):116-124.
33. Grigorova V, Ivanov I, Stamblieva K. Effect of sensory inputs alteration and central sensory disinteraction on postural sway and optokinetic reflex maintaining simultaneously body balance. *Acta Physiol Pharmacol Bulg* 2001;26(3):177-180.
34. Gutierrez GM, Chow JW, Tillman MD, McCoy SC, Castellano V, White LJ. Resistance training improves gait kinematics in persons with multiple sclerosis. *Arch Phys Med Rehabil* 2005;86(9):1824-1829.
35. Hafler DA. Multiple sclerosis. *J Clin Invest* 2004;113(6):788-794.
36. Haskell WL, Lee IM, Pate RR, et al. Physical activity and public health: Updated recommendation for adults from the American College of Sports Medicine and the American Heart Association. *Med Sci Sports Exerc* 2007;39(8):1423-1434.
37. Hayes HA, Gappmaier E, LaStayo PC. Effects of high-intensity resistance training on strength, mobility, balance, and fatigue in individuals with multiple sclerosis: A randomized controlled trial. *J Neurol Phys Ther* 2011;35(1):2-10.
38. Hogancamp WE, Rodriguez M, Weinshenker BG. The epidemiology of multiple sclerosis. *Mayo Clin Proc* 1997;72(9):871-878.

39. Honarmand K, Akbar N, Kou N, Feinstein A. Predicting employment status in multiple sclerosis patients: The utility of the MS functional composite. *J Neurol* 2011;258(2):244-249.

40. Ingram DA, Thompson AJ, Swash M. Central motor conduction in multiple sclerosis: Evaluation of abnormalities revealed by transcutaneous magnetic stimulation of the brain. *J Neurol Neurosurg Psychiatry* 1988;51(4):487-494.

41. Inman RP. Disability indices, the economic costs of illness, and social insurance: The case of multiple sclerosis. *Acta Neurol Scand Suppl* 1984;101:46-55.

42. Jacobson DL, Gange SJ, Rose NR, Graham NM. Epidemiology and estimated population burden of selected autoimmune diseases in the United States. *Clin Immunol Immunopathol* 1997;84(3):223-243.

43. Johnson KP, Brooks BR, Cohen JA, et al. Extended use of glatiramer acetate (Copaxone) is well tolerated and maintains its clinical effect on multiple sclerosis relapse rate and degree of disability. 1998 [classical article]. *Neurology* 2001;57(12 Suppl. 5):S46-53.

44. Kantarci OH, de Andrade M, Weinshenker BG. Identifying disease modifying genes in multiple sclerosis. *J Neuroimmunol* 2002;123(1-2):144-159.

45. Karst GM, Venema DM, Roehrs TG, Tyler AE. Center of pressure measures during standing tasks in minimally impaired persons with multiple sclerosis. *J Neurol Phys Ther* 2005;29(4):170-180.

46. Kenney WL, Humphrey RH, Bryant CX. *ACSM's Guidelines for Exercise Testing and Prescription*. 5th ed. Baltimore: Williams & Wilkins, 1995.

47. Kent-Braun JA, Ng AV, Castro M, et al. Strength, skeletal muscle composition, and enzyme activity in multiple sclerosis. *J Appl Physiol* 1997;83:1998-2004.

48. Kent-Braun JA, Sharma KR, Miller RG, Weiner MW. Postexercise phosphocreatine resynthesis is slowed in multiple sclerosis. *Muscle Nerve* 1994;17:835-841.

49. Kent-Braun JA, Sharma KR, Weiner MW, Miller RG. Effects of exercise on muscle activation and metabolism in multiple sclerosis. *Muscle Nerve* 1994;17:1162-1169.

50. Kileff J, Ashburn A. A pilot study of the effect of aerobic exercise on people with moderate disability multiple sclerosis. *Clin Rehabil* 2005;19(2):165-169.

51. Krupp LB, Alvarez LA, LaRocca NG, Scheinberg LC. Fatigue in multiple sclerosis. *Arch Neurol* 1988;45(4):435-437.

52. Kurtzke JF. Rating neurologic impairment in multiple sclerosis: An expanded disability status scale (EDSS). *Neurology* 1983;33(11):1444-1452.

53. Lambert CP, Archer RL, Evans WJ. Muscle strength and fatigue during isokinetic exercise in individuals with multiple sclerosis. *Med Sci Sports Exerc* 2001;33(10):1613-1619.

54. Liu C, Blumhardt LD. Randomized, double-blind, placebo-controlled study of subcutaneous interferon beta-1a in relapsing-remitting multiple sclerosis: A categorical disability trend analysis. *Mult Scler* 2002;8(1):10-14.

55. Lublin FD, Reingold SC. Defining the clinical course of multiple sclerosis: Results of an international survey. National Multiple Sclerosis Society (USA) Advisory Committee on Clinical Trials of New Agents in Multiple Sclerosis. *Neurology* 1996;46(4):907-911.

56. Mancardi GL, Sardanelli F, Parodi RC, et al. Effect of copolymer-1 on serial gadolinium-enhanced MRI in relapsing remitting multiple sclerosis. *Neurology* 1998;50(4):1127-1133.

57. Martin CL, Phillips BA, Kilpatrick TJ, et al. Gait and balance impairment in early multiple sclerosis in the absence of clinical disability. *Mult Scler* 2006;12(5):620-628.

58. Matthews B, Compston A, Ebers G, et al. Symptoms and signs of multiple sclerosis. In Compston A, Ebers G, Lassmann H, McDonald I, Matthews B, Wekerle H, eds. *McAlpine's Multiple Sclerosis* (pp. 145-190). London: Churchill Livingstone, 1998.

59. McIntosh-Michaelis SA, Roberts MH, Wilkinson SM, et al. The prevalence of cognitive impairment in a community survey of multiple sclerosis. *Br J Clin Psychol* 1991;30(Pt 4):333-348.

60. Monassier L, Brandt CM, Bousquet P. Effects of centrally-acting glutamatergic modulators on cardiovascular responses to stress in humans. *J Cardiol* 2001;37 Suppl. 1:77-84.

61. Moritani T, deVries HA. Neural factors versus hypertrophy in the time course of muscle strength gain. *Am J Phys Med* 1979;58(3):115-130.

62. Motl RW, Goldman M. Physical inactivity, neurological disability, and cardiorespiratory fitness in multiple sclerosis. *Acta Neurol Scand* 2011;123(2):98-104.

63. Motl RW, McAuley E, Snook EM. Physical activity and multiple sclerosis: A meta-analysis. *Mult Scler* 2005;11(4):459-463.

64. National Institute of Neurological Disorders and Stroke. Multiple Sclerosis: Hope Through Research. www.ninds.nih.gov/disorders/multiple_sclerosis/detail_multiple_sclerosis.htm.

65. National Multiple Sclerosis Society. www.nmss.org [accessed April 25, 2013].

66. Nelson SR, Di Fabio RP, Anderson JH. Vestibular and sensory interaction deficits assessed by dynamic platform posturography in patients with multiple sclerosis. *Ann Otol Rhinol Laryngol* 1995;104(1):62-68.

67. Ng AV, Dao HT, Miller RG, Gelinas DF, Kent-Braun JA. Blunted pressor and intramuscular metabolic responses to voluntary isometric exercise in multiple sclerosis. *J Appl Physiol* 2000;88(3):871-880.

68. Ng AV, Kent-Braun JA. Quantitation of lower physical activity in persons with multiple sclerosis. *Med Sci Sports Exerc* 1997;29:517-523.

69. Ng AV, Miller RG, Gelinas D, Kent-Braun JA. Functional relationships of central and peripheral muscle alterations in multiple sclerosis. *Muscle Nerve* 2004;29(6):843-852.

70. Ng AV, Miller RG, Kent-Braun JA. Central motor drive is increased during voluntary muscle contractions in multiple sclerosis. *Muscle Nerve* 1997;20:1213-1218.

71. Noronha MJ, Vas CJ, Aziz H. Autonomic dysfunction (sweating responses) in multiple sclerosis. *J Neurol Neurosurg Psychiatry* 1968;31(1):19-22.

72. Noseworthy JH, Lucchinetti C, Rodriguez M, Weinshenker BG. Multiple sclerosis. *N Engl J Med* 2000;343(13):938-952.

73. Oken BS, Kishiyama S, Zajdel D, et al. Randomized controlled trial of yoga and exercise in multiple sclerosis. *Neurology* 2004;62(11):2058-2064.

74. Olgiati R, Burgunder JM, Mumenthaler M. Increased energy cost of walking in multiple sclerosis: Effect of spasticity, ataxia, and weakness. *Arch Phys Med Rehabil* 1988;69(10):846-849.

75. Olgiati R, Di Prampero PE. Effet de l'exercice physique sur l'adaptation a l'effort dans la sclerose en plaques [Effect of physical exercise on adaptation to energy expenditure in multiple sclerosis]. *Schweiz Med Wochenschr* 1986;116:374-377.

76. Paty DW, Li DK, UBC MS/MRI Study Group and IFNB Multiple Sclerosis Study Group. Interferon beta-lb is effective in relapsing-remitting multiple sclerosis. II. MRI analysis results of a multicenter, randomized, double-blind, placebo-controlled trial. 1993 [classical article]. *Neurology* 2001;57(12 Suppl. 5):S10-15.

77. Pepin EB, Hicks RW, Spencer MK, Tran ZV, Jackson CG. Pressor response to isometric exercise in patients with multiple sclerosis. *Med Sci Sports Exerc* 1996;28(6):656-660.

78. Petajan JH, Gappmaier E, White AT, Spencer MK, Mino L, Hicks RW. Impact of aerobic training on fitness and quality of life in multiple sclerosis. *Ann Neurol* 1996;39(4):432-441.

79. Pilutti LA, Lelli DA, Paulseth JE, et al. Effects of 12 weeks of supported treadmill training on functional ability and quality of life in progressive multiple sclerosis: A pilot study. *Arch Phys Med Rehabil* 2011;92(1):31-36.

80. Polman CH, Reingold SC, Banwell B, et al. Diagnostic criteria for multiple sclerosis: 2010 revisions to the McDonald criteria. *Ann Neurol* 2011;69(2):292-302.

81. Ponichtera JA. Concentric and eccentric isokinetic lower extremity strength in multiple sclerosis and able-bodied. *J Orthop Sports Phys Ther* 2006;16:114-122.

82. Rampello A, Franceschini M, Piepoli M, et al. Effect of aerobic training on walking capacity and maximal exercise tolerance in patients with multiple sclerosis: A randomized crossover controlled study. *Phys Ther* 2007;87(5):545-555.

83. Rao SM, Leo GJ, Bernardin L, Unverzagt F. Cognitive dysfunction in multiple sclerosis. I. Frequency, patterns, and prediction. *Neurology* 1991;41:685-691.

84. Remelius JG, Hamill J, Kent-Braun J, van Emmerik RE. Gait initiation in multiple sclerosis. *Motor Control* 2008;12(2):93-108.

85. Rice CL, Vollmer TL, Bigland-Ritchie B. Neuromuscular responses of patients with multiple sclerosis. *Muscle Nerve* 1992;15(10):1123-1132.

86. Rietberg MB, Brooks D, Uitdehaag BM, Kwakkel G. Exercise therapy for multiple sclerosis. *Cochrane Database Syst Rev* 2005;(1):CD003980.

87. Rodgers MM, Mulcare JA, King DL, Mathews T, Gupta SC, Glaser RM. Gait characteristics of individuals with multiple sclerosis before and after a 6-month aerobic training program. *J Rehabil Res Dev* 1999;36(3):183-188.

88. Rodriguez M. Multiple sclerosis: Basic concepts and hypotheses. *Mayo Clin Proc* 1989(64):570-576.

89. Romberg A, Virtanen A, Ruutiainen J, et al. Effects of a 6-month exercise program on patients with multiple sclerosis: A randomized study. *Neurology* 2004;63(11):2034-2038.

90. Rosenblum D, Saffir M. The natural history of multiple sclerosis and its diagnosis. *Phys Med Rehabil Clin N Am* 1998;9(3):537-549, v.

91. Rudick RA, Goodkin DE, Jacobs LD, et al. Impact of interferon beta-la on neurologic disability in relapsing multiple sclerosis. 1997. *Neurology* 2001;57(12 Suppl. 5):S25-30.

92. Sabapathy NM, Minahan CL, Turner GT, Broadley SA. Comparing endurance- and resistance-exercise training in people with multiple sclerosis: A randomized pilot study. *Clin Rehabil* 2011;25(1):14-24.

93. Sadovnick AD. Genetic epidemiology of multiple sclerosis: A survey. *Ann Neurol* 1994;36 Suppl. 2:S194-203.

94. Sadovnick AD, Baird PA, Ward RH. Multiple sclerosis: Updated risks for relatives. *Am J Med Genet* 1988;29(3):533-541.

95. Savci S, Inal-Ince D, Arikan H, et al. Six-minute walk distance as a measure of functional exercise capacity in multiple sclerosis. *Disabil Rehabil* 2005;27(22):1365-1371.

96. Schmidt S, Barcellos LF, DeSombre K, et al. Association of polymorphisms in the apolipoprotein E region with susceptibility to and progression of multiple sclerosis. *Am J Hum Genet* 2002;70(3):708-717.

97. Schulz KH, Gold SM, Witte J, et al. Impact of aerobic training on immune-endocrine parameters, neurotrophic factors, quality of life and coordinative function in multiple sclerosis. *J Neurol Sci* 2004;225(1-2):11-18.

98. Schwid SR, Thornton CA, Pandya S, et al. Quantitative assessment of motor fatigue and strength in MS. *Neurology* 1999;53(4):743-750.

99. Scott PJ, Huskisson EC. Measurement of functional capacity with visual analogue scales. *Rheumatol Rehabil* 1977;16:257-259.

100. Senaratne MP, Carroll D, Warren KG, Kappagoda T. Evidence for cardiovascular autonomic nerve dysfunction in multiple sclerosis. *J Neurol Neurosurg Psychiatr* 1984;47:947-952.

101. Sharma KR, Kent-Braun J, Mynhier MA, Weiner MW, Miller RG. Evidence of an abnormal intramuscular component of fatigue in multiple sclerosis. *Muscle Nerve* 1995;18:1403-1411.

102. Sheean GL, Murray NM, Rothwell JC, Miller DH, Thompson AJ. An electrophysiological study of the mechanism of fatigue in multiple sclerosis. *Brain* 1997;120:299-315.

103. Silber E, Sharief MK. Axonal degeneration in the pathogenesis of multiple sclerosis. *J Neurol Sci* 1999;170(1):11-18.

104. Silby WA. *Therapeutic Claims in Multiple Sclerosis: A Guide to Treatments.* New York: Demos, 1996.

105. Simon JH, Jacobs LD, Campion M, et al. Magnetic resonance studies of intramuscular interferon beta-1a for relapsing multiple sclerosis. The Multiple Sclerosis Collaborative Research Group. *Ann Neurol* 1998;43(1):79-87.

106. Simpson S Jr, Blizzard L, Otahal P, Van der Mei I, Taylor B. Latitude is significantly associated with the prevalence of multiple sclerosis: A meta-analysis. *J Neurol Neurosurg Psychiatry* 2011;82(10):1132-1141.

107. Skurvydas A. The effect of multiple sclerosis and gender on central and peripheral fatigue during 2-min MVC. *Clinical Neurophysiology* 2011;122:767-776.

108. Slawta JN, McCubbin JA, Wilcox AR, Fox SD, Nalle DJ, Anderson G. Coronary heart disease risk between active and inactive women with multiple sclerosis. *Med Sci Sports Exerc* 2002;34(6):905-912.

109. Snook EM, Motl RW. Effect of exercise training on walking mobility in multiple sclerosis: A meta-analysis. *Neurorehabil Neural Repair* 2009;23(2):108-116.

110. *Stedman's Medical Dictionary for the Health Professions and Nursing.* Baltimore: Lippincott Williams & Wilkins, 2005.

111. Sterman AB, Coyle PK, Panasci DJ, Grimson R. Disseminated abnormalities of cardiovascular autonomic functions in multiple sclerosis. *Neurology* 1985;35(11):1665-1668.

112. Surakka J, Romberg A, Ruutiainen J, et al. Effects of aerobic and strength exercise on motor fatigue in men and women with multiple sclerosis: A randomized controlled trial. *Clin Rehabil* 2004;18(7):737-746.

113. Svensson B, Gerdle B, Elert J. Endurance training in patients with multiple sclerosis: Five case studies. *Phys Ther* 1994;74:1017-1026.

114. Taylor NF, Dodd KJ, Prasad D, Denisenko S. Progressive resistance exercise for people with multiple sclerosis. *Disabil Rehabil* 2006;28(18):1119-1126.

115. Thoumie P, Lamotte D, Cantalloube S, Faucher M, Amarenco G. Motor determinants of gait in 100 ambulatory patients with multiple sclerosis. *Mult Scler* 2005;11(4):485-491.

116. van den Berg M, Dawes H, Wade DT, et al. Treadmill training for individuals with multiple sclerosis: A pilot randomised trial. *J Neurol Neurosurg Psychiatry* 2006;77(4):531-533.

117. van der Kamp W, Maertens de Noordhout A, Thompson PD, Rothwell JC, Day BL, Marsden CD. Correlation of phasic muscle strength and corticomotoneuron conduction time in multiple sclerosis. *Ann Neurol* 1991;29:6-12.

118. Van Emmerik RE, Remelius JG, Johnson MB, Chung LH, Kent-Braun JA. Postural control in women with multiple sclerosis: Effects of task, vision and symptomatic fatigue. *Gait Posture* 2010;32(4):608-614.

119. Waxman SG. Conduction in myelinated, unmyelinated, and demyelinated fibers. *Arch Neurol* 1977;34(10):585-589.

120. White AT, Petajan JH. Physiological measures of therapeutic response to interferon beta-1a treatment in remitting-relapsing MS. *Clin Neurophysiol* 2004;115(10):2364-2371.

121. White AT, Wilson TE, Davis SL, Petajan JH. Effect of precooling on physical performance in multiple sclerosis. *Mult Scler* 2000;6(3):176-180.

122. White LJ, Dressendorfer RH. Exercise and multiple sclerosis. *Sports Med* 2004;34(15):1077-1100.

123. White LJ, McCoy SC, Castellano V, et al. Resistance training improves strength and functional capacity in persons with multiple sclerosis. *Mult Scler* 2004;10(6):668-674.

124. Yates HA, Vardy TC, Kuchera ML, Ripley BD, Johnson JC. Effects of osteopathic manipulative treatment and concentric and eccentric maximal-effort exercise on women with multiple sclerosis: A pilot study. *J Am Osteopath Assoc* 2002;102(5):267-275.

125. Zwibel HL. Contribution of impaired mobility and general symptoms to the burden of multiple sclerosis. *Adv Ther* 2009;26(12):1043-1057.

CHAPTER 27

1. Ahlborg L, Andersson C, Julin P. Whole-body vibration training compared with resistance training: Effect on spasticity, muscle strength, and motor performance in adults with cerebral palsy. *J Rehabil Med* 2006;38:302-308.

2. Albright AL. Spasticity and movement disorders in cerebral palsy. *J Child Neurol* 1996;11(Suppl. 1):S1-4.

3. American College of Sports Medicine. *ACSM's Exercise Management for Persons with Chronic Diseases and Disabilities*. 2nd ed. Champaign, IL: Human Kinetics, 2003.

4. American College of Sports Medicine. *ACSM's Guidelines for Exercise Testing and Prescription*. 9th ed. Philadelphia: Lippincott Williams & Wilkins, 2013.

5. American College of Sports Medicine. *ACSM's Resources for Clinical Exercise Physiology: Musculoskeletal, Neuromuscular, Neoplastic, Immunologic and Hematologic Conditions* (p. 304). Philadelphia: Lippincott Williams & Wilkins, 2002.

6. Andersson C, Grooten W, Hellsten M, Kaping K, Mattsson E. Adults with cerebral palsy: Walking ability after progressive strength training. *Dev Med Child Neurol* 2003;45:220-228.

7. Andersson C, Mattsson E. Adults with cerebral palsy: A survey describing problems, needs, and resources, with special emphasis on locomotion. *Dev Med Child Neurol* 2001;43:76-82.

8. Ando N, Ueda S. Functional deterioration in adults with cerebral palsy. *Clin Rehabil* 2000;14:300-306.

9. Arias F, Romero R, Joist H, Kraus FT. Thrombophilia: A mechanism of disease in women with adverse pregnancy outcome and thrombotic lesions in the placenta. *J Matern Fetal Med* 1998;7:277-286.

10. Ayalon M, Ben-Sira D, Hutzler Y, Gilad T. Reliability of isokinetic strength measurements of the knee in children with cerebral palsy. *Dev Med Child Neurol* 2000;42:398-402.

11. Bachrach SJ, Walter RS, Trzcinski K. Use of glycopyrrolate and other anticholinergic medications for sialorrhea in children with cerebral palsy. *Clin Pediatr* 1998;37(8):485.

12. Barry MJ. Physical therapy interventions for patients with movement disorders due to cerebral palsy. *J Child Neurol* 1996;11(Suppl. 1):S51-60.

13. Bax DA, Siersema PD, Van Vliet AH, Kuipers EJ, Kusters JG. Molecular alterations during development of esophageal adenocarcinoma. *J Surg Oncol* 2005;92:89-98, discussion 99.

14. Bax M. Terminology and classification of cerebral palsy. *Dev Med Child Neurol* 1964;6:295-307.

15. Berg K. Effect of physical activation and of improved nutrition on the body composition of school children with cerebral palsy. *Acta Paediatr Scand Suppl* 1970;204:Suppl 204:53-69.

16. Berg KO, Wood-Dauphinee SL, Williams JI, Maki B. Measuring balance in the elderly: Validation of an instrument. *Can J Public Health* 1992;83(Suppl 2):S7-11.

17. Blundell SW, Shepherd RB, Dean CM, Adams RD, Cahill BM. Functional strength training in cerebral palsy: A pilot study of a group circuit training class for children aged 4-8 years. *Clin Rehabil* 2003;17:48-57.

18. Booth MY, Yates CC, Edgar TS, Bandy WD. Serial casting vs combined intervention with botulinum toxin A and serial casting in the treatment of spastic equinus in children. *Pediatr Phys Ther* 2003;15:216-220.

19. Brown SH, Lewis CA, McCarthy JM, Doyle ST, Hurvitz EA. The effects of internet-based home training on upper limb function in adults with cerebral palsy. *Neurorehabil Neural Repair* 2010;24(6):575-583.

20. Bryanton C, Bosse J, Brien M, McLean J, McCormick A, Sveistrup H. Feasibility, motivation, and selective motor control: Virtual reality compared to conventional home exercise in children with cerebral palsy. *Cyberpsychol Behav* 2006;9:123-128.

21. Butler JM, Scianni A, Ada L. Effect of cardiorespiratory training on aerobic fitness and carryover to activity in children with cerebral palsy: A systematic review. *Int J Rehabil Res* 2009;33(2):97-103.

22. Campbell SK, Vander Linden DW, Palisano RJ, eds. *Physical Therapy for Children*. 3rd ed. St. Louis: Saunders Elsevier, 2005.

23. Centers for Disease Control and Prevention. Economic costs associated with mental retardation, cerebral palsy, hearing loss, and vision impairment—United States, 2003. *MMWR* 2004;53(3):57-59.

24. Cerebral Palsy International Sports and Recreation Association (CPISRA). Classification and sports rule manual. 9th ed. Retrieved from www.gkef-fgda.org/descargas/cp-isra.pdf [accessed April 25, 2013).

25. Chad KE, Bailey DA, McKay HA, Zello GA, Snyder RE. The effect of a weight-bearing physical activity program on bone mineral content and estimated volumetric density in children with spastic cerebral palsy. *J Pediatr* 1999;135:115-117.

26. Cohen H, Blatchly CA, Gombash LL. A study of the clinical test of sensory interaction and balance. *Phys Ther* 1993;73:346-351, discussion 351-344.

27. Crawford A, Hollingsworth HH, Morgan K, Gray DB. People with mobility impairments: Physical activity and quality of participation. *Disabil Health J* 2008;1:7-13.

28. Damiano DL, Abel MF. Functional outcomes of strength training in spastic cerebral palsy. *Arch Phys Med Rehabil* 1998;79:119-125.

29. Damiano DL, DeJong SL. A systematic review of the effectiveness of treadmill training and body weight support in pediatric rehabilitation. *J Neurol Phys Ther* 2009;33:27-44.

30. Damiano DL, Kelly LE, Vaughn CL. Effects of quadriceps femoris muscle strengthening on crouch gait in children with spastic diplegia. *Phys Ther* 1995;75:658-667, discussion 668-671.

31. Darrah J, Wessel J, Nearingburg P, O'Connor M. Evaluation of a community fitness program for adolescents with cerebral palsy. *Pediatr Phys Ther* 1999;11:18-23.

32. Davis DW. Review of cerebral palsy, part II: Identification and intervention. *Neonatal Netw* 1997;16(4):19-25, quiz 26-9.

33. Delgado MR, et al. Practice parameter: Pharmacologic treatment of spasticity in children and adolescents with cerebral palsy (an evidence-based review). *Neurology* 2010;7(4):336.

34. DeLuca PA. The musculoskeletal management of children with cerebral palsy. *Pediatr Clin North Am* 1996;43:1135-1150.

35. Dodd KJ, Taylor NF, Graham HK. A randomized clinical trial of strength training in young people with cerebral palsy. *Dev Med Child Neurol* 2003;45:652-657.

36. Duncan PW, Weiner DK, Chandler J, Studenski S. Functional reach: A new clinical measure of balance. *J Gerontol* 1990;45:M192-197.

37. Eicher PS, Batshaw ML. Cerebral palsy. *Pediatr Clin North Am* 1993;40:537-551.

38. Fernandez JE, Pitetti KH, Betzen MT. Physiological capacities of individuals with cerebral palsy. *Hum Factors* 1990;32:457-466.

39. Flett P, Saunders B. Ophthalmic assessment of physically disabled children attending a rehabilitation centre. *J Paediatr Child Health* 1993;29:132-135.

40. Fowler EG, Kolobe THA, Damiano DL, Thorpe DE, Morgan DW, Brunstrom JE, Coster WJ, Henderson RC, Pitetti KH, Rimmer JH, Rose J, Stevenson RD. Promotion of physical fitness and prevention of secondary conditions for children with cerebral palsy: Section on pediatrics research summit proceedings. *Phys Ther* 2007;87(11):1-16.

41. Franklin DL, Luther F, Curzon ME. The prevalence of malocclusion in children with cerebral palsy. *Eur J Orthod* 1996;18:637-643.

42. Freud S. *Die infantile cerebrallahmung. Specielle Pathologie und Therapie,* vol. IX (p. 327). Vienna: Alfred Holder, 1897.

43. Gajdosik CG, Cicirello N. Secondary conditions of the musculoskeletal system in adolescents and adults with cerebral palsy. *Phys Occup Ther Pediatr* 2001;21:49-68.

44. Glanzman AM, Kim H, Swaminathan K, Beck T. Efficacy of botulinum toxin A, serial casting, and combined treatment for spastic equinus: A retrospective analysis. *Dev Med Child Neurol* 2004;46:807-811.

45. Gorter H, Holty L, Rameckers EEA, Elvers HJWH, Oostendorp RAB. Changes in endurance and walking ability through functional physical training in children with cerebral palsy. *Pediatr Phys Ther* 2009;21(1):31-37.

46. Grether JK, Nelson KB. Maternal infection and cerebral palsy in infants of normal birth weight. *JAMA* 1997;278:207-211.

47. Grether JK, Nelson KB, Cummins SK. Twinning and cerebral palsy: Experience in four northern California counties, births 1983 through 1985. *Pediatrics* 1993;92:854-858.

48. *International Classification of Functioning, Disability and Health.* Geneva: World Health Organization, 2001.

49. Jahnsen R, Villien L, Stanghelle JK, Holm I. Fatigue in adults with cerebral palsy in Norway compared with the general population. *Dev Med Child Neurol* 2003;45:296-303.

50. Keefer DJ, Wayland T, Caputo JL, Apperson K, McGreal S, Morgan DW. Within- and between-day stability of treadmill walking $\dot{V}O_2$ in children with hemiplegic cerebral palsy. Stability of walking $\dot{V}O_2$ in children with CP. *Gait Posture* 2005;22:177-181.

51. Kelly M, Darrah J. Aquatic exercise for children with cerebral palsy. *Dev Med Child Neurol* 2005;47:838-842.

52. Khalili MA. Quantitative sports and functional classification (QSFC) for disabled people with spasticity. *Br J Sports Med* 2004;38:310-313.

53. King W, Levin R, Schmidt R, Oestreich A, Heubi JE. Prevalence of reduced bone mass in children and adults with spastic quadriplegia. *Dev Med Child Neurol* 2003;45:12-16.

54. Klingbeil H, Baer HR, Wilson PE. Aging with a disability. *Arch Phys Med Rehabil* 2004;85:S68-73, quiz S74-75.

55. Koman LA, Smith BP, Shilt JS. Cerebral palsy. *Lancet* 2004;363:1619-1631.

56. Kraus FT. Cerebral palsy and thrombi in placental vessels of the fetus: Insights from litigation. *Hum Pathol* 1997;28:246-248.

57. Krigger KW. Cerebral palsy: An overview. *Am Fam Physician* 2006;73:91-100.

58. Laskin J. Cerebral palsy. In: Durstine LJ, Moore GE, Painter PL, Roberts SO, eds. *ACSM's Exercise Management for Persons with Chronic Diseases and Disabilities* (p. 345). 3rd ed. Champaign, IL: Human Kinetics, 2009.

59. Liptak GS. Health and well being of adults with cerebral palsy. *Curr Opin Neurol* 2008;21(2):136-142.

60. Liu JM, Li S, Lin Q, Li Z. Prevalence of cerebral palsy in China. *Int J Epidemiol* 1999;28:949-954.

61. Loudon JK, Cagle PE, Figoni SF, Nau KL, Klein RM. A submaximal all-extremity exercise test to predict maximal oxygen consumption. *Med Sci Sports Exerc* 1998;30:1299-1303.

62. Lundberg A. Maximal aerobic capacity of young people with spastic cerebral palsy. *Dev Med Child Neurol* 1978;20:205-210.

63. Lundberg A. Oxygen consumption in relation to work load in students with cerebral palsy. *J Appl Physiol* 1976;40:873-875.

64. MacKeith RC, MacKenzie ICK, Polani PE. Memorandum on terminology and classification of ìcerebral palsy.î *Cereb Palsy Bull* 1959;5:27-35.

65. Maeland S, Jahnsen R, Opheim A, Froslie KF, Moe-Nilssen R, Stanghelle JK. No effect on gait function of progressive resistance exercise in adults with cerebral palsy: A single-blind randomized controlled trial. *Adv Physiother* 2009;11:1-7.

66. Marge M. Health promotion for persons with disabilities: Moving beyond rehabilitation. *Am J Health Promot* 1988;2:29-35.

67. Mathias S, Nayak US, Isaacs B. Balance in elderly patients: The ìget up and goî test. *Arch Phys Med Rehabil* 1986;67:387-389.

68. Mattsson E, Andersson C. Oxygen cost, walking speed, and perceived exertion in children with cerebral palsy when walking with anterior and posterior walkers. *Dev Med Child* 1997;39:671-676.

69. Mergler S, Evenhuis HM, Boot AM, De Man SA, Bindels-De Heus KGCB, Huijbers WAR, Penning C. Epidemiology of low bone mineral density and fractures in children with severe cerebral palsy: A systematic review. *Dev Med Child Neurol* 2009;51:773-778.

70. Mockford M, Caulton JM. Systematic review of progressive strength training in children and adolescents with cerebral palsy who are ambulatory. *Pediatr Phys Ther* 2008;20(4):318-333.

71. Morton JF, Brownlee M, McFayden AK. The effects of progressive resistance training for children with cerebral palsy. *Clin Rehabil* 2005;19:283-289.

72. Morton R. New surgical interventions for cerebral palsy and the place of gait analysis. *Dev Med Child Neurol* 1999;41:424-428.

73. Murphy KP, Molnar GE, Lankasky K. Medical and functional status of adults with cerebral palsy. *Dev Med Child Neurol* 1995;37:1075-1084.

74. Murphy NA, Irwin MC, Hoff C. Intrathecal baclofen therapy in children with cerebral palsy: Efficacy and complications. *Arch Phys Med Rehabil* 2002;83:1721-1725.

75. Mutch L, Alberman E, Hagberg B, Kodama K, Perat MV. Cerebral palsy epidemiology: Where are we now and where are we going? *Dev Med Child Neurol* 1992;34:547-551.

76. Nelson KB, Ellenberg JH. Childhood neurological disorders in twins. *Paediatr Perinat Epidemiol* 1995;9:135-145.

77. O'Connell DG, Barnhart R. Improvement in wheelchair propulsion in pediatric wheelchair users through resistance training: A pilot study. *Arch Phys Med Rehabil* 1995;76:368-372.

78. Odding E, Roebroeck ME, Stam HJ. The epidemiology of cerebral palsy: Incidence, impairments and risk factors. *Disabil Rehabil* 2006;28:183-191.

79. Overeynder JC, Turk MA. Cerebral palsy and aging: A framework for promoting the health of older persons with cerebral palsy. *Top Geriatr Rehabil* 1998;13:19-24.

80. Palisano RJ, Snider LM, Orlin MN. Recent advances in physical and occupational therapy for children with cerebral palsy. *Semin Pediatr Neurol* 2004;11:66-77.

81. Parker DF, Carriere L, Hebestreit H, Bar-Or O. Anaerobic endurance and peak muscle power in children with spastic cerebral palsy. *Am J Dis Child* 1992;146:1069-1073.

82. Parkes J, McCullough N, Madden A, McCahey E. The health of children with cerebral palsy and stress in their parents. *J Adv Nurs* 2009;65(11):2311-2323.

83. Petersen MC, Palmer FB. Advances in prevention and treatment of cerebral palsy. *Ment Retard Dev Disabil Res Rev* 2001;7:30-37.

84. Pin T. The effectiveness of passive stretching in children with cerebral palsy. *Dev Med Child Neurol* 2006;48(10):855.

85. Pitetti KH, Fernandez J, Lanciault M. Feasibility of an exercise program for adults with cerebral palsy: A pilot study. *Adapt Phys Act Q* 1991;8:333-341.

86. Podsiadlo D, Richardson S. The timed ìUp & Goî: A test of basic functional mobility for frail elderly persons. *J Am Geriatr Soc* 1991;39:142-148.

87. Reddihough DS, Collins KJ. The epidemiology and causes of cerebral palsy. *Aust J Physiother* 2003;49:7-12.

88. Reuben DB, Siu AL. An objective measure of physical function of elderly outpatients. The Physical Performance Test. *J Am Geriatr Soc* 1990;38:1105-1112.

89. Rimmer JH. Physical fitness levels of persons with cerebral palsy. *Dev Med Child Neurol* 2001;43:208-212.

90. Rogers A, Furler BL, Brinks S, Darrah J. A systematic review of the effectiveness of aerobic exercise interventions for children with cerebral palsy: An AACPDM evidence report. *Dev Med Child Neurol* 2008;50:808-814.

91. Rose J, Haskell WL, Gamble JG. A comparison of oxygen pulse and respiratory exchange ratio in cerebral palsied and nondisabled children. *Arch Phys Med Rehabil* 1993;74:702-705.

92. Rose J, Medeiros JM, Parker R. Energy cost index as an estimate of energy expenditure of cerebral-palsied children during assisted ambulation. *Dev Med Child Neurol* 1985;27:485-490.

93. Rosenbaum P. Cerebral palsy: What parents and doctors want to know. *BMJ* 2003;26:970-974.

94. Rosenbaum P, Paneth N, Leviton A, et al. A report: the definition and classification of cerebral palsy. April 2006. *Dev Med Child Neurol* 2007;49(6):480.

95. Saito N, Ebara S, Ohotsuka K, Kumeta H, Takaoka K. Natural history of scoliosis in spastic cerebral palsy. *Lancet* 1998;351:1687-1692.

96. Samson JF, Sie LT, de Groot L. Muscle power development in preterm infants with periventricular flaring or leukomalacia in relation to outcome at 18 months. *Dev Med Child Neurol* 2002;44:734-740.

97. Sandstrom K, Alinder J, Oberg B. Descriptions of functioning and health and relations to a gross motor classification in adults with cerebral palsy. *Disabil Rehabil* 2004;26:1023-1031.

98. Sanger TD, Delgado MR, Gaebler-Spira D, Hallett M, Mink JW. Classification and definition of disorders causing hypertonia in childhood. *Pediatrics* 2003;111:e89-97.

99. Schlough K, Nawoczenski D, Case LE, Nolan K, Wigglesworth JK. The effects of aerobic exercise on endurance, strength, function and self-perception in adolescents with spastic cerebral palsy: A report of three case studies. *Pediatr Phys Ther* 2005;17:234-250.

100. Scholtes VA, Becher JG, Beelen A, Lankhorst GJ. Clinical assessment of spasticity in children with cerebral palsy: A critical review of available instruments. *Dev Med Child Neurol* 2006;48:64-73.

101. Schwartz L, Engel JM, Jensen MP. Pain in persons with cerebral palsy. *Arch Phys Med Rehabil* 1999;80:1243-1246.

102. Scianni A, Butler JM, Ada L, Teixeira-Salmela LF. Muscle strengthening is not effective in children and adolescents with cerebral palsy: A systematic review. *Aust J Physiother* 2009;55:81-87.

103. Shinohara TA, Suzuki N, Oba M, Kawasumi M, Kimizuka M, Mita K. Effect of exercise at the AT point for children with cerebral palsy. *Bull Hosp Jt Dis* 2002-2003;61:63-67.

104. Stanley F, Blair E, Alberman E. *Cerebral Palsies: Epidemiology and Causal Pathways.* London: Mac Keith Press, 1999.

105. *Stedman's Medical Dictionary.* 28th ed. Philadelphia: Lippincott Williams & Wilkins, 2006.

106. Strauss D, Shavelle R. Life expectancy of adults with cerebral palsy. *Dev Med Child Neurol* 1998;40:369-375.

107. Taylor NF, Dodd KJ, Graham HK. Test-retest reliability of hand-held dynamometric strength testing in young people with cerebral palsy. *Arch Phys Med Rehabil* 2004;85:77-80.

108. Taylor NF, Dodd KJ, Larkin H. Adults with cerebral palsy benefit from participating in a strength training programme at a community gymnasium. *Disabil Rehabil* 2004;26:1128-1134.

109. Thorarensen O, Ryan S, Hunter J, Younkin DP. Factor V Leiden mutation: An unrecognized cause of hemiplegic cerebral palsy, neonatal stroke, and placental thrombosis. *Ann Neurol* 1997;42:372-375.

110. Thorpe DE, Reilly MA. The effect of aquatic resistive exercise on lower extremity strength, energy expenditure, functional mobility, balance and self-perception in an adult with cerebral palsy: A retrospective case report. *J Aquatic Phys Ther* 2000;8: 18-24.

111. Tinetti ME. Performance-oriented assessment of mobility problems in elderly patients. *J Am Geriatr Soc* 1986;34:119-126.

112. Tobimatsu Y, Nakamura R, Kusano S, Iwasaki Y. Cardiorespiratory endurance in people with cerebral palsy measured using an arm ergometer. *Arch Phys Med Rehabil* 1998;79:991-993.

113. Turk MA. Secondary conditions and disability. In: Field MJ, Jette AM, Martin L, eds. *Workshop on Disability in America*. Washington, DC: National Academies Press, 2006.

114. Turk MA, Geremski CA, Rosenbaum PF, Weber RJ. The health status of women with cerebral palsy. *Arch Phys Med Rehabil* 1997;78:S10-17.

115. United Cerebral Palsy. Exercise principles and guidelines for persons with cerebral palsy and neuromuscular disorders. Retrieved from www.ucpaindy.org/wp-content/uploads/2010/03/Exercise-Principles-and-Guidelines-for-Persons-with-Cerebral-Palsy-and-Neuromuscular-Disorders.pdf [accessed April 25, 2013].

116. Unnithan VB, Dowling JJ, Frost G, Bar-Or O. Role of cocontraction in the O_2 cost of walking in children with cerebral palsy. *Med Sci Sports Exerc* 1996;28:1498-1504.

117. Van den Berg-Emons RJ, Van Baak MA, Speth L, Saris WH. Physical training of school children with spastic cerebral palsy: Effects on daily activity, fat mass and fitness. *Int J Rehabil Res* 1998;21:179-194.

118. Van der Dussen L, Nieuwstraten W, Roebroeck M, Stam HJ. Functional level of young adults with cerebral palsy. *Clin Rehabil* 2001;15:84-91.

119. Verrotti A, Greco R, Spalice A, Chiarelli F, Iannetti P. Pharmacotherapy of spasticity in children with cerebral palsy. *Pediatr Neurol* 2006;34(1):1.

120. Verschuren O, Ketelaar M, Takken T, Helders PJM, Gorter JW. Exercise programs for children with cerebral palsy: A systematic review of the literature. *Am J Phys Med Rehabil* 2008;87:404-417.

121. Verschuren O, Takken T. Aerobic capacity in children and adolescents with cerebral palsy. *Res Dev Disabil* 2010;31(6):1352-1357.

122. Vogtle LK. Pain in adults with cerebral palsy: Impact and solutions. *Dev Med Child Neurol* 2009;51(Suppl. 4):113-121.

123. Weightman A, Preston N, Levesley M, Holt R, Mon-Williams M, Clarke M, Cozens AJ, Bhakta B. Home based computer-assisted upper limb exercise for young children with cerebral palsy: A feasibility study investigating impact on motor control and functional outcome. *J Rehabil Med* 2011;43(4):359-363.

124. Wiart L, Darrah J, Kembhavi G. Stretching with children with cerebral palsy: What do we know and where are we going? *Pediatr Phys Ther* 2008;20(2):173.

125. Wiley ME, Damiano DL. Lower-extremity strength profiles in spastic cerebral palsy. *Dev Med Child Neurol* 1998;40:100-107.

126. Williams H, Pountney T. Effects of a static bicycling programme on the functional ability of young people with cerebral palsy who are non-ambulant. *Dev Med Child Neurol* 2007;49(7):522-527.

127. Williams K, Hennessy E, Alberman E. Cerebral palsy: Effects of twinning, birthweight, and gestational age. *Arch Dis Child Fetal Neonatal Ed* 1996;75:F178-182.

128. Willoughby KL, Dodd KJ, Shields N. A systematic review of the effectiveness of treadmill training for children with cerebral palsy. *Disabil Rehabil* 2009;31(24):1971-1979.

129. Yoon BH, Jun JK, Romero R, Park KH, Gomez R, Choi JH, Kim IO. Amniotic fluid inflammatory cytokines (interleukin-6, interleukin-1beta, and tumor necrosis factor-alpha), neonatal brain white matter lesions, and cerebral palsy. *Am J Obstet Gynecol* 1997;177:19-26.

130. Zhao H. Changes of calf muscle-tendon biomechanical properties induced by passive-stretching and active-movement training in children with cerebral palsy. *J Appl Physiol* 2011;111(2):435.

CHAPTER 28

1. Health disparities experienced by black or African Americans—United States. *MMWR* 2005;54:1-3.

2. Prevalence and most common causes of disability among adults—United States, 2005. *MMWR* 2009;58:421-426.

3. Ada L, Dean CM, Hall JM, Bampton J, Crompton S. A treadmill and overground walking program improves walking in persons residing in the community after stroke: A placebo-controlled, randomized trial. *Arch Phys Med Rehabil* 2003;84:1486-1491.

4. Anderson GL, Limacher M, Assaf AR, Bassford T, Beresford SA, Black H, Bonds D, Brunner R, Brzyski R, Caan B, Chlebowski R, Curb D, Gass M, Hays J, Heiss G, Hendrix S, Howard BV, Hsia J, Hubbell A, Jackson R, Johnson KC, Judd H, Kotchen JM, Kuller L, LaCroix AZ, Lane D, Langer RD, Lasser N, Lewis CE, Manson J, Margolis K, Ockene J, O'Sullivan MJ, Phillips L, Prentice RL, Ritenbaugh C, Robbins J, Rossouw JE, Sarto G, Stefanick ML, Van Horn L, Wactawski-Wende J, Wallace R, Wassertheil-Smoller S. Effects of conjugated equine estrogen in postmenopausal women with hysterectomy: The Women's Health Initiative randomized controlled trial. *JAMA* 2004;291:1701-1712.

5. Chu KS, Eng JJ, Dawson AS, Harris JE, Ozkaplan A, Gylfadottir S. Water-based exercise for cardiovascular fitness in people with chronic stroke: A randomized controlled trial. *Arch Phys Med Rehabil* 2004;85:870-874.

6. Cress ME, Meyer M. Maximal voluntary and functional performance levels needed for independence in adults aged 65 to 97 years. *Phys Ther* 2003;83:37-48.

7. Dean CM, Richards CL, Malouin F. Task-related circuit training improves performance of locomotor tasks in chronic stroke: A randomized, controlled pilot trial. *Arch Phys Med Rehabil* 2000;81:409-417.

8. Duncan PW. Stroke recovery and rehabilitation research. *J Rehabil Res Dev* 2002;39:ix-xi.

9. Eich HJ, Mach H, Werner C, Hesse S. Aerobic treadmill plus Bobath walking training improves walking in subacute stroke: A randomized controlled trial. *Clin Rehabil* 2004;18:640-651.

10. Engardt M, Knutsson E, Jonsson M, Sternhag M. Dynamic muscle strength training in stroke patients: Effects on knee extension torque, electromyographic activity, and motor function. *Arch Phys Med Rehabil* 1995;76:419-425.

11. Gurm HS, Yadav JS, Fayad P, Katzen BT, Mishkel GJ, Bajwa TK, Ansel G, Strickman NE, Wang H, Cohen SA, Massaro JM, Cutlip

DE. Long-term results of carotid stenting versus endarterectomy in high-risk patients. *N Engl J Med* 2008;358:1572-1579.

12. James AH, Bushnell CD, Jamison MG, Myers ER. Incidence and risk factors for stroke in pregnancy and the puerperium. *Obstet Gynecol* 2005;106:509-516.
13. Johnston SC, Gress DR, Browner WS, Sidney S. Short-term prognosis after emergency department diagnosis of TIA. *JAMA* 2000;284:2901-2906.
14. Katz-Leurer M, Shochina M, Carmeli E, Friedlander Y. The influence of early aerobic training on the functional capacity in patients with cerebrovascular accident at the subacute stage. *Arch Phys Med Rehabil* 2003;84:1609-1614.
15. Kelly-Hayes M, Beiser A, Kase CS, Scaramucci A, D'Agostino RB, Wolf PA. The influence of gender and age on disability following ischemic stroke: The Framingham Study. *J Stroke Cerebrovasc Dis* 2003;12:119-126.
16. Kittner SJ, Stern BJ, Feeser BR, Hebel R, Nagey DA, Buchholz DW, Earley CJ, Johnson CJ, Macko RF, Sloan MA, Wityk RJ, Wozniak MA. Pregnancy and the risk of stroke. *N Engl J Med* 1996;335:768-774.
17. Laufer Y, Dickstein R, Chefez Y, Marcovitz E. The effect of treadmill training on the ambulation of stroke survivors in the early stages of rehabilitation: A randomized study. *J Rehabil Res Dev* 2001;38:69-78.
18. Lee IM, Paffenbarger RS Jr. Physical activity and stroke incidence: The Harvard Alumni Health Study. *Stroke* 1998;29:2049-2054.
19. Liu M, Tsuji T, Hase K, Hara Y, Fujiwara T. Physical fitness in persons with hemiparetic stroke. *Keio J Med* 2003;52:211-219.
20. Lloyd-Jones D, Adams RJ, Brown TM, Carnethon M, Dai S, De Simone G, Ferguson TB, Ford E, Furie K, Gillespie C, Greenlund K, Haase N, Hailpern S, Ho M, Howard V, Kissela B, Kittner S, Lackland D, Lisabeth L, Marelli A, McDermott MM, Meigs J, Mozaffarian D, Mussolino M, Nichol G, Roger VL, Rosamond W, Sacco R, Sorlie P, Stafford R, Thom T, Wasserthiel-Smoller S, Wong ND, Wylie-Rosett J. Heart disease and stroke statistics—2010 update: A report from the American Heart Association. *Circulation* 2010;121:e46-215.
21. Luft A, Macko RF, Forrester LW, Villagra F, Ivey F, Sorkin JD, Whitall J, McCombe-Waller S, Katzel L, Goldberg AP, Hanley DF. Treadmill exercise activates subcortical neural networks and improves walking after stroke: A randomized controlled trial. *Stroke* 2008;39:3341-3350.
22. MacKay-Lyons MJ, Makrides L. Cardiovascular stress during a contemporary stroke rehabilitation program: Is the intensity adequate to induce a training effect? *Arch Phys Med Rehabil* 2002;83:1378-1383.
23. MacKay-Lyons MJ, Makrides L. Exercise capacity early after stroke. *Arch Phys Med Rehabil* 2002;83:1697-1702.
24. Macko RF, Katzel LI, Yataco A, Tretter LD, DeSouza CA, Dengel DR, Smith GV, Silver KH. Low-velocity graded treadmill stress testing in hemiparetic stroke patients. *Stroke* 1997;28:988-992.
25. Macko RF, Smith GV, Dobrovolny CL, Sorkin JD, Goldberg AP, Silver KH. Treadmill training improves fitness reserve in chronic stroke patients. *Arch Phys Med Rehabil* 2001;82:879-884.
26. Muntner P, Garrett E, Klag MJ, Coresh J. Trends in stroke prevalence between 1973 and 1991 in the US population 25 to 74 years of age. *Stroke* 2002;33:1209-1213.
27. Pang MY, Eng JJ, Dawson AS. Relationship between ambulatory capacity and cardiorespiratory fitness in chronic stroke: Influence of stroke-specific impairments. *Chest* 2005;127:495-501.
28. Pang MY, Eng JJ, Dawson AS, McKay HA, Harris JE. A community-based fitness and mobility exercise program for older adults with chronic stroke: A randomized, controlled trial. *J Am Geriatr Soc* 2005;53:1667-1674.
29. Pohl M, Mehrholz J, Ritschel C, Ruckriem S. Speed-dependent treadmill training in ambulatory hemiparetic stroke patients: A randomized controlled trial. *Stroke* 2002;33:553-558.
30. Potempa K, Lopez M, Braun LT, Szidon JP, Fogg L, Tincknell T. Physiological outcomes of aerobic exercise training in hemiparetic stroke patients. *Stroke* 1995;26:101-105.
31. Rimmer JH, Riley B, Creviston T, Nicola T. Exercise training in a predominantly African-American group of stroke survivors. *Med Sci Sports Exerc* 2000;32:1990-1996.
32. Rosser N, Floel A. Pharmacological enhancement of motor recovery in subacute and chronic stroke. *NeuroRehabilitation* 2008;23:95-103.
33. Salbach NM, Mayo NE, Wood-Dauphinee S, Hanley JA, Richards CL, Cote R. A task-orientated intervention enhances walking distance and speed in the first year post stroke: A randomized controlled trial. *Clin Rehabil* 2004;18:509-519.
34. Schellinger PD, Bryan RN, Caplan LR, Detre JA, Edelman RR, Jaigobin C, Kidwell CS, Mohr JP, Sloan M, Sorensen AG, Warach S. Evidence-based guideline: The role of diffusion and perfusion MRI for the diagnosis of acute ischemic stroke. Report of the Therapeutics and Technology Assessment Subcommittee of the American Academy of Neurology. *Neurology* 75:2010;177-185.
35. Stewart DG. Stroke rehabilitation. 1. Epidemiologic aspects and acute management. *Arch Phys Med Rehabil* 1999;80:S4-7.
36. Teixeira-Salmela LF, Olney SJ, Nadeau S, Brouwer B. Muscle strengthening and physical conditioning to reduce impairment and disability in chronic stroke survivors. *Arch Phys Med Rehabil* 80:1999;1211-1218.
37. Tsuji T, Liu M, Tsujiuchi K, Chino N. Bridging activity as a mode of stress testing for persons with hemiplegia. *Arch Phys Med Rehabil* 1999;80:1060-1064.
38. Wassertheil-Smoller S, Hendrix SL, Limacher M, Heiss G, Kooperberg C, Baird A, Kotchen T, Curb JD, Black H, Rossouw JE, Aragaki A, Safford M, Stein E, Laowattana S, Mysiw WJ. Effect of estrogen plus progestin on stroke in postmenopausal women: The Women's Health Initiative: A randomized trial. *JAMA* 2003;289:2673-2684.
39. Williams LS, Ghose SS, Swindle RW. Depression and other mental health diagnoses increase mortality risk after ischemic stroke. *Am J Psychiatry* 2004;161:1090-1095.

CHAPTER 29

1. American College of Sports Medicine. *ACSM's Certified News* 1997;8(1):2.
2. American College of Sports Medicine. *ACSM's Guidelines for Exercise Testing and Prescription.* 9th ed. Baltimore: Lippincott Williams & Wilkins, 2013.
3. Armstrong N, Williams L, Balding J, Gentle P, Kirby B. Cardiopulmonary fitness, physical activity patterns, and selected coronary risk factor variables in 11 to 16 year olds. *Pediatr Exerc Sci* 1991;3:219-228.
4. Bar-Or O, Rowland TW. *Pediatric Exercise Medicine.* Champaign, IL: Human Kinetics, 2004.

5. Cahill BR, ed. *Proceedings of the Conference on Strength Training and the Prepubescent.* Chicago: American Orthopedic Society for Sports Medicine, 1988.

6. Chang RKR, Gurvitz M, Rodriquez S, Hong E, Klitzner TS. Current practice of exercise stress testing among pediatric cardiology and pulmonology centers in the United States. *Pediatr Cardiol* 2006;27:110-116.

7. Chestnut C. Theoretical overview: Bone development, peak bone mass, bone loss, and fracture risk. *Am J Med* 1991;91(5B):25-45.

8. Clark HH. Joint and body range of movement. *Phys Fitness Res Dig* 1975;5:16-18.

9. Giardini A, Fenton M, Andrews RE, Derrick G, Burch M. Peak oxygen uptake correlates with survival without clinical deterioration in ambulatory children with dilated cardiomyopathy. *Circulation* 2011;124:1713-1718.

10. Grodjinovsky A, Inbar O, Dotan R, Bar-Or O. Training effect on the anaerobic performance of children as measured by the Wingate anaerobic test. In: Berg K, Eriksson BO, eds. *Children and Exercise IX* (pp. 139-145). Baltimore: University Park Press, 1979.

11. Hebestreit H. Exercise testing in children – what works, what doesn't, and where to go? *Paediatr Respir Rev* 2004;5(Suppl. A):S11-S14.

12. Hebestreit HU, Bar-Or O. Differences between children and adults for exercise testing and exercise prescription. In: Skinner JS, ed. *Exercise Testing and Prescription for Special Cases: Theoretical Basis and Clinical Application* (pp. 68-84). Philadelphia: Lippincott Williams & Wilkins, 2005.

13. Kimball TR. Pediatric stress echocardiography. *Pediatr Cardiol* 2002;23:347-357.

14. Kramer WJ, Fleck SJ. *Strength Training for Young Athletes.* Champaign, IL: Human Kinetics, 1993.

15. Lauer RM, Lee J, Clarke WR. Factors affecting the relationship between childhood and adult cholesterol levels: The Muscatine study. *Pediatrics* 1988;82:309-318.

16. Mahon AD. Exercise training. In: Armstrong N, Van Mechelen W, eds. *Pediatric Exercise Science and Medicine* (pp. 201-219). Oxford, UK: Oxford University Press, 2000.

17. McGill HC, McMahan A, Zieske AW, Walcott JV, Malcom GT, Tracy RE, Strong JP. Effects of nonlipid risk in youth with a favorable lipoprotein profile. *Circulation* 2001;103:1546-1550.

18. Milne C, Seefeldt V, Reuschlein P. Relationship between grade, sex, race, and motor development in young children. *Res Q* 1976;47:726.

19. National Association for Sport and Physical Education. *Physical Activity for Children: A Statement of Guidelines for Children Ages 5-12.* 2nd ed. Reston, VA: NASPE, 2004.

20. Owens S, Gutin B. Exercise testing of the child with obesity. *Pediatr Cardiol* 1999;20:79-83.

21. Paridon SM, et al. Clinical stress testing in the pediatric age group: A statement from the American Heart Association Council on Cardiovascular Disease in the Young, Committee on Atherosclerosis, Hypertension, and Obesity in Youth. *Circulation* 2006;113:1905-1920.

22. Pfeiffer RD, Francis RS. Effects of strength training on muscle development in pre-pubescent, pubescent, and postpubescent males. *Phys Sportsmed* 1986;14:134-143.

23. Robertson RJ. *Perceived Exertion for Practitioners: Rating Effort With the OMNI Pictures System* (pp. 141-150). Champaign, IL: Human Kinetics, 2004.

24. Rotstein AR, Dotan R, Bar-Or O, Tenenbaum G. Effects of training on anaerobic threshold, maximal aerobic power and anaerobic performance of preadolescent boys. *Int J Sports Med* 1986;7:281-286.

25. Rowland TW. *Developmental Exercise Physiology.* Champaign, IL: Human Kinetics, 1996.

26. Rowland TW. *Exercise and Children's Health* (pp. 27-83). Champaign, IL: Human Kinetics, 1990.

27. Rowland TW. *Pediatric Exercise Testing.* Champaign, IL: Human Kinetics, 1993.

28. Sale DG. Strength training in children. In: Gisolfi CV, Lamb DR, eds. *Perspectives in Exercise Science and Sports Medicine,* vol. 2 (pp. 165-222). Indianapolis: Benchmark Press, 1989.

29. Sallis JF, Patrick K. Physical activity guidelines for adolescents: Consensus statement. *Pediatr Exerc Sci* 1994;6:302-314.

30. Sargeant AJ, Dolan P, Thorne A. Effects of supplementary physical activity on body composition, aerobic, and anaerobic power in 13-year-old boys. In: Binkborst RA, Kemper HCG, Saris WH, eds. *Children and Exercise XI* (pp. 140-150). Champaign, IL: Human Kinetics, 1985.

31. Sewell L, Micheli LJ. Strength training for children. *J Pediatr Orthop* 1986;6:143-146.

32. Stephens P Jr, Paridon SM. Exercise testing in pediatrics. *Pediatr Clin North Am* 2004;51:1569-1587.

33. Tanaka, H., Monahan, K. D., and Seals, D. R. Age-Predicted Maximal Heart Rate Revisited. J Am Coll Cardiol 2001; 37:153-156.

34. Vrijens J. Muscle strength development in pre and postpubescent age. *Med Sport* 1978;11:152-158.

35. Weltman A, Janney C, Rians CB, Strand K, Berg B, Tippett S, Wise J, Cahill BR, Katch FI. The effects of hydraulic resistance strength training in pre-pubescent males. *Med Sci Sports Exerc* 1986;18:629-638.

CHAPTER 30

1. Adamo ML, Farrar RP. Resistance training, and IGF involvement in the maintenance of muscle mass during the aging process. *Ageing Res Rev* 2006;5:310-331.

2. Antunes-Correa LM, Kanamura BY, et al. Exercise training improves neurovascular control and functional capacity in heart failure patients regardless of age. *Eur J Cardiovasc Prev Rehabil* 2011; ePub ahead of print.

3. Aronow WS, Fleg JL, Pepine CJ, et al. ACCF/AHA 2011 expert consensus document on hypertension in the elderly: A report of the American College of Cardiology Foundation Task Force on Clinical Expert Consensus Documents. *Circulation* 2011;123:2434-2506.

4. Boden WE, O'Rourke RA, Teo KK, et al. Optimal medical therapy with or without PCI for stable coronary disease. *N Engl J Med* 2007;356:1503-1516.

5. Braunstein JB, Anderson GF, Gerstenblith G, et al. Noncardiac comorbidity increases preventable hospitalizations and mortality among Medicare beneficiaries with chronic heart failure. *J Am Coll Cardiol* 2003;42:1226-1233.

6. Buchner D. Exercise. In: *Merck Manual of Geriatrics* (pp. 295-305). Whitehouse Station, NJ: Merck Research Laboratories, 2000.

7. Busby MJ, Shefrin EA, Fleg JL. Prevalence and long-term significance of exercise-induced frequent or repetitive ventricular ectopic beats in apparently healthy volunteers. *J Am Coil Cardiol* 1989;14:1659-1665.

8. Carvalho EE, Crescêncio JC, Elias J Jr, Brito LB, Gallo L Jr, Simões MV. Improved endothelial function and reversal of myocardial perfusion defects after aerobic physical training in a patient with microvascular myocardial ischemia. *Am J Phys Med Rehabil* 2011;90:59-64.
9. Chodzko-Zajko WJ, Proctor DN, Fiatarone Singh MA, Minson CT, Nigg CR, Salem GJ, Skinner JS. American College of Sports Medicine position stand. Exercise and physical activity for older adults. *Med Sci Sports Exerc* 2009;41:1510-1530.
10. Chung HY, Lee EK, Choi YJ, et al. Molecular inflammation as an underlying mechanism of the aging process and age-related diseases. *J Dent Res* 2011;90:830-840.
11. Deschenes MR. Effects of aging on muscle fibre type and size. *Sports Med* 2004;34:809-824.
12. Downes TR, Nomeir AM, Smith KM, Stewart KP, Little WC. Mechanism of altered pattern of left ventricular filling with aging in subjects without cardiac disease. *Am J Cardiol* 1989;64:523-527.
13. Erbs S, Höllriegel R, Linke A, et al. Exercise training in patients with advanced chronic heart failure (NYHA IIIb) promotes restoration of peripheral vasomotor function, induction of endogenous regeneration, and improvement of left ventricular function. *Circ Heart Fail* 2010;3:486-494.
14. *Fact Book Fiscal Year 2010* (p. 47). National Institutes of Health, National Heart, Lung, and Blood Institute, 2011.
15. Fielding RA, LeBrasseur NK, Cuoco A, et al. High-velocity resistance training increases skeletal muscle peak power in older women. *J Am Geriatr Soc* 2002;50:655-662.
16. Fleg JL, Lakatta EG. Normal aging of the cardiovascular system. In: Aronow WS, Fleg JL, Rich MW, eds. *Cardiovascular Disease in the Elderly* (pp. 1-43). 4th ed. New York: Informa Healthcare USA, 2008.
17. Fleg JL, Lakatta EG. Role of muscle loss in the age-associated reduction in VO2max. *J Appl Physiol* 1988;65:1147-1151.
18. Fleg JL, Morrell CH, Bos AG, et al. Accelerated longitudinal decline of aerobic capacity in healthy older adults. *Circulation* 2005;112:674-682.
19. Fleg JL, O'Connor FC, Gerstenblith G, et al. Impact of age on the cardiovascular response to dynamic upright exercise in healthy men and women. *J Appl Physiol* 1995;78:890-900.
20. Fleg JL, Schulman S, O'Connor FC, et al. Effect of acute β-adrenergic receptor blockade on age-associated changes in cardiovascular performance during dynamic exercise. *Circulation* 1994;90:2333-2341.
21. Fleg JL, Tzankoff SP, Lakatta EG. Age-related augmentation of plasma catecholamines during dynamic exercise in healthy males. *J Appl Physiol* 1985;59:1033-1039.
22. Forman DE, Rich MW, Alexander KP, et al. Cardiac care for older adults: Time for a new paradigm. *J Am Coll Cardiol* 2011;57:1801-1810.
23. Frankel JE, Bean JF, Frontera WR. Exercise in the elderly: Research and clinical practice. *Clin Geriatr Med* 2006;22:239-256.
24. Fried LP, Ferrucci L, Darer J, Williamson JD, Anderson G. Untangling the concepts of disability, frailty, and comorbidity: Implications for improved targeting and care. *J Gerontol A Biol Sci Med Sci* 2004;59:255-263.
25. Gallagher D, Ruts E, Visser M, et al. Weight stability masks sarcopenia in elderly men and women. *Am J Physiol Endocrinol Metab* 2000;279:E366-E375.
26. Gerstenblith G, Frederiksen J, Yin FC, et al. Echocardiographic assessment of a normal adult aging population. *Circulation* 1977;56:273-278.
27. Goldman DP, McGlynn EA. U.S. Health Care. Facts About Cost, Access, and Quality. 2005. http://www.rand.org/content/dam/rand/pubs/corporate_pubs/2005/RAND_CP484.1.pdf [accessed April 25, 2013].
28. Gulley SP, Rasch EK, Chan L. The complex web of health: Relationships among chronic conditions, disability, and health services. *Public Health Rep* 2011;126(4):495-507.
29. Hammill BG, Curtis LH, Schulman KA, Shellan DJ. Relationship between cardiac rehabilitation and long-term risks of death and myocardial infarction among elderly Medicare beneficiaries. *Circulation* 2010;121:63-70.
30. Hees PS, Fleg JL, Lakatta EG, Shapiro EP. Left ventricular remodeling with age in normal men versus women: Novel insights using three-dimensional magnetic resonance imaging. *Am J Cardiol* 2002;90:1231-1236.
31. Hills AP, Shultz SP, Soares MJ, Byrne NM, Hunter GR, King NA, Misra A. Resistance training for obese, type 2 diabetic adults: A review of the evidence. *Obes Rev* 2010;11:740-749.
32. Hollmann W, Strüder HK, Tagarakis CV, King G. Physical activity and the elderly. *Eur J Cardiovasc Prev Rehabil* 2007;14(6):730-739.
33. Jackson AS, Sui X, Hebert JR, Church TS, Blair SN. Role of lifestyle and aging on the longitudinal change in cardiorespiratory fitness. *Arch Int Med* 2009;169:1781-1787.
34. Kallman DA, Plato CC, Tobin JD. The role of muscle loss in the age-related decline of grip strength: Cross-sectional and longitudinal perspectives. *J Gerontol* 1990;45:M82-88.
35. Kaufman G. Polypharmacy in older adults. *Nurs Stand* 2011;25(38):49-55.
36. Keogh JW, Kilding A, Pidgeon P, Ashley L, Gillis D. Physical benefits of dancing for healthy older adults: A review. *J Aging Phys Act* 2009;17(4):479-500.
37. Lakatta E, Wang M, Najjar SS. Arterial aging and subclinical arterial disease are fundamentally intertwined at macroscopic and molecular levels. *Med Clin North Am* 2009;93:583-604.
38. Liu H, Frank A. Tai chi as a balance improvement exercise for older adults: A systematic review. *J Geriatr Phys Ther* 2010;33(3):103-109.
39. Manolagas S. Aging and the musculoskeletal system. In: *Merck Manual of Geriatrics* (pp. 467-472). Whitehouse Station, NJ: Merck Research Laboratories, 2000.
40. Marques EA, Mota J, Machado L, Sousa F, Coelho M, Moreira P, Carvalho J. Multicomponent training program with weight-bearing exercises elicits favorable bone density, muscle strength, and balance adaptations in older women. *Calcif Tissue Int* 2011;88:117-129.
41. Maurer MS, Shefrin EA, Fleg JL. Prevalence and prognostic significance of exercise-induced supraventricular tachycardia in the elderly. *Am J Cardiol* 1995;75:788-792.
42. Miyamura M, Honda Y. Oxygen uptake and cardiac output during peak treadmill and bicycle exercise. *J Appl Physiol* 1972;32:185-188.
43. Mora S, Cook N, Buring JE, Ridker PM, Lee IM. Physical activity and reduced risk of cardiovascular events: Potential mediating mechanisms. *Circulation* 2007;116:2110-2118.
44. Nelson M, Layne J, Bernstein M, et al. The effects of multidimensional home-based exercise on functional performance in elderly people. *J Gerontol A Biol Sci Med Sci* 2004;59(2):154-160.

45. Nelson ME, Rejeski WJ, Blair SN, et al. Physical activity and public health in older adults: Recommendation from the American College of Sports Medicine and the American Heart Association. *Med Sci Sports Exerc* 2007;39:435-445.

46. Orr R. Contribution of muscle weakness to postural instability in the elderly. A systematic review. *Eur J Phys Rehabil Med* 2010;46:183-220.

47. Pate RR, Pratt M, Blair SN, et al. Physical activity and public health. A recommendation from the Centers for Disease Control and Prevention and the American College of Sports Medicine. *JAMA* 1995;273:402-407.

48. Paterson DH, Jones GR, Rice CL. Ageing and physical activity: Evidence to develop exercise recommendations for older adults. *Can J Public Health* 2007;98 Suppl. 2:S69-108.

49. Peeters P, Mets T. The 6-minute walk as an appropriate exercise test in elderly patients with chronic heart failure. *J Gerontol A Biol Sci Med Sci* 1996;51A:M147-M151.

50. Rolland Y, Dupuy C, Abellan van Kan G, Gillette S, Vellas B. Treatment strategies for sarcopenia and frailty. *Med Clin North Am* 2011;95:427-438.

51. Ryan AS. Exercise in aging: Its important role in mortality, obesity and insulin resistance. *Aging Health* 2010;6:551-563.

52. Sattelmair J, Pertman J, Ding EL, et al. Dose response between physical activity and risk of coronary artery disease. *Circulation* 2011;124(7):789-795.

53. Sattelmair JR, Pertman JH, Forman DE. Effects of physical activity on cardiovascular and noncardiovascular outcomes in older adults. *Clin Geriatr Med* 2009;25:677-702.

54. Schulman SP, Lakatta EG, Fleg JL, et al. Age-related decline in left ventricular filling at rest and exercise. *Am J Physiol* 1992;263:H1932-H1938.

55. Serra-Rexach JA, Bustamante-Ara N, Hierro Villarán M, et al. Short-term, light- to moderate-intensity exercise training improves leg muscle strength in the oldest old: A randomized controlled trial. *J Am Geriatr Soc* 2011;59:594-602.

56. Simonsick EM, Fan E, Fleg JL. Estimating cardiovascular fitness in well-functioning older adults: Treadmill validation of the long distance corridor walk. *J Am Geriatr Soc* 2006;54:127-132.

57. Skelton DA, Greig CA, Davies JM, Young A. Strength, power and related functional ability of healthy people aged 65–89 years. *Age Ageing* 1994;23:371-377.

58. Syed FA, Ng AC. The pathophysiology of the aging skeleton. *Curr Osteoporos Rep* 2010;8:235-240.

59. Talbot LA, Metter EJ, Fleg JL. Leisure-time physical activities and their relationship to cardiorespiratory fitness in healthy men and women 18–95 years old. *Med Sci Sports Exerc* 2000;32:417-425.

60. U.S. Census Bureau. National Population Projections. www.census.gov/population/projections [accessed April 25, 2013].

61. Vale RG, de Oliveira RD, Pernambuco CS, et al. Effects of muscle strength and aerobic training on basal serum levels of IGF-1 and cortisol in elderly women. *Arch Gerontol Geriatr* 2009;49(3):343-347.

62. Williams MA, Fleg JL, Ades PA, et al. Secondary prevention of coronary heart disease in the elderly (with emphasis on patients > or =75 years of age): An American Heart Association scientific statement from the Council on Clinical Cardiology Subcommittee on Exercise, Cardiac Rehabilitation, and Prevention. *Circulation* 2002;105:1735-1743.

63. Wisloff U, Stoylen A, Loennechen JP, et al. Superior cardiovascular effect of aerobic interval training versus moderate continuous training in heart failure patients: A randomized study. *Circulation* 2007;115(24):3086-3094.

64. Yanowitz FG, LaMonte MJ. Physical activity and health in the elderly. *Curr Sports Med Rep* 2002;1:354-361.

CHAPTER 31

1. American Psychiatric Association. *Diagnostic and Statistical Manual of Mental Disorders*. 4th ed., revised ed. Washington, DC: American Psychiatric Association, 2000.

2. Aneshensel C, Huba G. Depression, alcohol use, and smoking over one year: A four-wave longitudinal causal model. *J Abnorm Psychol* 1983;92:134-150.

3. Argyropoulos SV, Wilson SJ. Sleep disturbances in depression and the effects of antidepressants. *Int Rev Psychiatry* 2005;17:237-245.

4. Babyak M, Blumenthal J, Herman S, Khatri P, Doraiswamy P, Moore K, Craighead W, Baldewicz T, Krishnan K. Exercise training for major depression: Maintenance of therapeutic benefit at 10 months. *Psychosom Med* 2000;62:633-638.

5. Barkham M, Stiles WB, Connell J, Twigg E, Leach C, Lucock M, Mellor-Clark J, Bower P, King M, Shapiro DA, Hardy GE, Greenberg L, Angus L. Effects of psychological therapies in randomized trials and practice-based studies. *Br J Clin Psychol* 2008;47:397-415.

6. Barth J, Schumacher M, Herrmann-Lingen C. Depression as a risk factor for mortality in patients with coronary heart disease: A meta-analysis. *Psychosom Med* 2004;66:802-813.

7. Beck A, Rush A, Shaw B, Emery G. *Cognitive Therapy of Depression*. New York: Guilford Press, 1979.

8. Beck A, Steer R, Brown G. *Beck Depression Inventory*. 2nd ed. San Antonio: Psychological Corporation, 1996.

9. Beck J. *Cognitive Therapy: Basics and Beyond*. New York: Guilford Press, 1995.

10. Belmaker RH, Agam G. Major depressive disorder. *N Engl J Med* 2008;358:55-68.

11. Berkman LF, Blumenthal J, Burg M, Carney RM, Catellier D, Cowan MJ, Czajkowski SM, DeBusk R, Hosking J, Jaffe A, Kaufmann PG, Mitchell P, Norman J, Powell LH, Raczynski JM, Schneiderman N. Effects of treating depression and low perceived social support on clinical events after myocardial infarction: The Enhancing Recovery in Coronary Heart Disease Patients (ENRICHD) Randomized Trial. *JAMA* 2003;289:3106-3116.

12. Blumenthal JA, Babyak MA, Doraiswamy PM, Watkins L, Hoffman BM, Barbour KA, Herman S, Craighead WE, Brosse AL, Waugh R, Hinderliter A, Sherwood A. Exercise and pharmacotherapy in the treatment of major depressive disorder. *Psychosom Med* 2007;69:587-596.

13. Blumenthal J, Babyak M, Moore K, Craighead W, Herman S, Khatri P, Waugh R, Napolitano M, Forman L, Appelbaum M, Doraiswamy P, Krishnan K. Effects of exercise training on older adults with major depression. *Arch Intern Med* 1999;159:2349-2356.

14. Blumenthal JA, Sherwood A, Rogers SD, Babyak MA, Doraiswamy PM, Watkins L, Hoffman BM, O'Connell C, Johnson JJ, Patidar SM, Waugh R, Hinderliter A. Understanding prognostic benefits of exercise and antidepressant therapy for persons with depression and heart disease: The UPBEAT study—rationale, design, and methodological issues. *Clin Trials* 2007;4:548-559.

15. Bonnet F, Irving K, Terra J, Nony P, Berthezene F, Moulin P. Anxiety and depression are associated with unhealthy lifestyle

in patients at risk of cardiovascular disease. *Atherosclerosis* 2005;178:339-344.

16. Brosse A, Sheets E, Lett H, Blumenthal J. Exercise and the treatment of clinical depression in adults: Recent findings and future directions. *Sports Med* 2002;32:741-760.
17. Bruce EC, Musselman DL. Depression, alterations in platelet function, and ischemic heart disease. *Psychosom Med* 2005;67 Suppl. 1:S34-36.
18. Burns D. *Feeling Good. The New Mood Therapy.* New York: Avon Books, 1999.
19. Carney RM, Blumenthal JA, Freedland KE, Stein PK, Howells WB, Berkman LF, Watkins LL, Czajkowski SM, Hayano J, Domitrovich PP, Jaffe AS. Low heart rate variability and the effect of depression on post-myocardial infarction mortality. *Arch Intern Med* 2005;165:1486-1491.
20. Cassano P, Fava M. Depression and public health: An overview. *J Psychosom Res* 2002;53:849-857.
21. Chambless DL, Ollendick TH. Empirically supported psychological interventions: Controversies and evidence. *Annu Rev Psychol* 2001;52:685-716.
22. Cuijpers P, van Straten A, Andersson G, van Oppen P. Psychotherapy for depression in adults: A meta-analysis of comparative outcome studies. *J Consult Clin Psychol* 2008;76:909-922.
23. Davidson KW, Rieckmann N, Clemow L, Schwartz JE, Shimbo D, Medina V, Albanese G, Kronish I, Hegel M, Burg MM. Enhanced depression care for patients with acute coronary syndrome and persistent depressive symptoms: Coronary psychosocial evaluation studies randomized controlled trial. *Arch Intern Med* 2010;170:600-608.
24. De Groot M, Anderson R, Freedland K, Clouse R, Lustman P. Association of depression and diabetes complications: A meta-analysis. *Psychosom Med* 2001;63:619-630.
25. Demyttenaere K, Enzlin P, Dewe W, Boulanger B, De Bie J, De Troyer W. Compliance with antidepressants in a primary care setting: Beyond lack of efficacy and adverse events. *J Clin Psychiatry* 2001;62:30-33.
26. DiMatteo M, Lepper H, Croghan T. Depression is a risk factor for noncompliance with medical treatment: Meta-analysis of the effects of anxiety and depression on patient adherence. *Arch Intern Med* 2000;160:2101-2107.
27. Dunn A, Trivedi M, Kampert J, Clark C, Chambliss H. Exercise treatment for depression: Efficacy and dose response. *Am J Prev Med* 2005;28:1-8.
28. Eaton W, Anthony J, Gallo J, Cai G, Tien A, Romanoski A, Lyketsos C, Chen LS. Natural history of Diagnostic Interview Schedule/DSM-IV major depression. The Baltimore Epidemiologic Catchment Area follow-up. *Arch Gen Psychiatry* 1997;54:993-999.
29. Fournier JC, DeRubeis RJ, Hollon SD, Dimidjian S, Amsterdam JD, Shelton RC, Fawcett J. Antidepressant drug effects and depression severity: A patient-level meta-analysis. *JAMA* 2010;303:47-53.
30. Fraser R, Ingram MC, Anderson NH, Morrison C, Davies E, Connell JM. Cortisol effects on body mass, blood pressure, and cholesterol in the general population. *Hypertension* 1999;33:1364-1368.
31. Gehi A, Haas D, Pipkin S, Whooley M. Depression and medication adherence in outpatients with coronary heart disease: Findings from the heart and soul study. *Arch Intern Med* 2005;165:2508-2513.
32. Glassman AH, O'Connor CM, Califf RM, Swedberg K, Schwartz P, Bigger JT, Krishnan KRR, van Zyl LT, Swenson JR, Finkel MS, Landau C, Shapiro PA, Pepine CJ, Mardekian J, Harrison WM, Barton D, McLvor M. Sertraline treatment of major depression in patients with acute MI or unstable angina. *JAMA* 2002;288:701-709.
33. Glazer K, Emery C, Frid D, Banyasz R. Psychological predictors of adherence and outcomes among patients in cardiac rehabilitation. *J Cardiopulm Rehabil* 2002;22:40-46.
34. Greenberg PE, Kessler RC, Birnbaum HG, Leong SA, Lowe SW, Berglund PA, Corey-Lisle PK. The economic burden of depression in the United States: How did it change between 1990 and 2000? *J Clin Psychiatry* 2003;64:1465-1475.
35. Hansen R, Gaynes B, Thieda P, Gartlehner G, Deveaugh-Geiss A, Krebs E, Lohr K. Meta-analysis of major depressive disorder relapse and recurrence with second-generation antidepressants. *Psychiatr Serv* 2008;59:1121-1130.
36. Hansson GK. Inflammation, atherosclerosis, and coronary artery disease. *N Engl J Med* 2005;352:1685-1695.
37. Herman S, Blumenthal J, Babyak M, Khatri P, Craighead W, Krishnan K, Doraiswamy P. Exercise therapy for depression in middle-aged and older adults: Predictors of early dropout and treatment failure. *Health Psychol* 2002;21:553-563.
38. Hirschfeld R, Keller M, Panico S, Arons B, Barlow D, Davidoff F, Endicott J, Froom J, Goldstein M, Gorman J, Guthrie D, Marek R, Maurer T, Meyer R, Phillips K, Ross J, Schwenk T, Sharfstein S, Thase M, Wyatt R. The National Depressive and Manic-Depressive Association consensus statement on the undertreatment of depression. *JAMA* 1997;277:333-340.
39. Hoffman BM, Babyak MA, Craighead WE, Sherwood A, Doraiswamy PM, Coons MJ, Blumenthal JA. Exercise and pharmacotherapy in patients with major depression: One-year follow-up of the SMILE study. *Psychosom Med* 2011;73:127-133.
40. Hollon S, Thase M, Markowitz J. Treatment and prevention of depression. *Psychol Sci Public Interest* 2002;3:39-77.
41. Joffe R, Sokolov S, Streiner D. Antidepressant treatment of depression: A metanalysis. *Can J Psychiary* 1996;41:613-616.
42. Johnson J, Weissman M, Klerman G. Service utilization and social morbidity associated with depressive symptoms in the community. *JAMA* 1992;267:1478-1483.
43. Katon W, Rutter C, Simon G, Lin E, Ludman E, Ciechanowski P, Kinder L, Young B, Von Korff M. The association of comorbid depression with mortality in patients with type 2 diabetes. *Diabetes Care* 2005;28:2668-2672.
44. Kessler R, Berglund P, Demler O, Jin R, Walters E. Lifetime prevalence and age-of-onset distributions of DSM-IV disorders in the national comorbidity survey replication. *Arch Gen Psychiatry* 2005;62:593-602.
45. Kessler R, Chiu W, Demler O, Walters E. Prevalence, severity, and comorbidity of 12-month DSM-IV disorders in the national comorbidity survey replication. *Arch Gen Psychiatry* 2005;62:617-627.
46. King AC, Oman RF, Brassington GS, Bliwise DL, Haskell WL. Moderate-intensity exercise and self-rated quality of sleep in older adults: A randomized controlled trial. *JAMA* 1997;277:32-37.
47. Kirsch I, Deacon BJ, Huedo-Medina TB, Scoboria A, Moore TJ, Johnson BT. Initial severity and antidepressant benefits: A meta-analysis of data submitted to the Food and Drug Administration. *PLoS Med* 2008;5:e45.

48. Lavoie K, Fleet R, Lesperance F, Arsenault A, Laurin C, Frasure-Smith N, Bacon S. Are exercise stress tests appropriate for assessing myocardial ischemia in patients with major depressive disorder? *Am Heart J* 2004;148:621-627.

49. Lehmann H. Clinical evaluation and natural course of depression. *J Clin Psychiatry* 1983;44:5-10.

50. Lesperance F, Frasure-Smith N, Koszycki D, Laliberte MA, van Zyl LT, Baker B, Swenson JR, Ghatavi K, Abramson BL, Dorian P, Guertin MC. Effects of citalopram and interpersonal psychotherapy on depression in patients with coronary artery disease: The Canadian Cardiac Randomized Evaluation of Antidepressant and Psychotherapy Efficacy (CREATE) trial. *JAMA* 2007;297:367-379.

51. Lesperance F, Frasure-Smith N, Theroux P, Irwin M. The association between major depression and levels of soluble intercellular adhesion molecule 1, interleukin-6, and C-reactive protein in patients with recent acute coronary syndromes. *Am J Psychiatry* 2004;161:271-277.

52. Lett H, Blumenthal J, Babyak M, Sherwood A, Strauman T, Robins C, Newman M. Depression as a risk factor for coronary artery disease: Evidence, mechanisms, and treatment. *Psychosom Med* 2004;66:305-315.

53. Lichtman JH, Bigger JT, Blumenthal JA, Frasure-Smith N, Kaufmann PG, Lesperance F, Mark DB, Sheps DS, Taylor CB, Froelicher ES. Depression and coronary heart disease: Recommendations for screening, referral, and treatment: A science advisory from the American Heart Association Prevention Committee of the Council on Cardiovascular Nursing, Council on Clinical Cardiology, Council on Epidemiology and Prevention, and Interdisciplinary Council on Quality of Care and Outcomes Research: Endorsed by the American Psychiatric Association. *Circulation* 2008;118:1768-1775.

54. Lin E, Katon W, Von Korff M, Rutter C, Simon G, Oliver M, Ciechanowski P, Ludman E, Bush T, Young B. Relationship of depression and diabetes self-care, medication adherence, and preventive care. *Diabetes Care* 2004;27:2154-2160.

55. Manber R, Chambers AS. Insomnia and depression: A multifaceted interplay. *Curr Psychiatry Rep* 2009;11:437-442.

56. Mather A, Rodriguez C, Guthrie M, McHarg A, Reid I, McMurdo M. Effects of exercise on depressive symptoms in older adults with poorly responsive depressive disorder. *Br J Psychiatry* 2002;180:411-415.

57. McNeil J, LeBlanc E, Joyner M. The effect of exercise on depressive symptoms in the moderately depressed elderly. *Psychol Aging* 1991;6:487-488.

58. Mead G, Morley W, Campbell P, Greig C, McMurdo M, Lawlor D. Exercise for depression. *Cochrane Database Syst Rev* 2009;3:CD004366, doi 101002/14651858SCD004366.

59. Melfi C, Chawla A, Croghan T, Hanna M, Kennedy S, Sredl K. The effects of adherence to antidepressant treatment guidelines on relapse and recurrence of depression. *Arch Gen Psychiatry* 1998;55:1128-1132.

60. Montgomery P, Dennis J. Physical exercise for sleep problems in adults aged 60+. *Cochrane Database Syst Rev* 2009;4:CD003404, doi 10.1002/14651858.CD003404.

61. O'Connor CM, Jiang W, Kuchibhatla M, Silva SG, Cuffe MS, Callwood DD, Zakhary B, Stough WG, Arias RM, Rivelli SK, Krishnan R. Safety and efficacy of sertraline for depression in patients with heart failure: Results of the SADHART-CHF (Sertraline Against Depression and Heart Disease in Chronic Heart Failure) trial. *J Am Coll Cardiol* 2010;56:692-699.

62. Olfson M, Marcus S, Druss B, Elinson L, Tanielian T, Pincus H. National trends in the outpatient treatment of depression. *JAMA* 2002;287:203-209.

63. O'Neal H, Dunn A, Martinsen E. Depression and exercise. *Int J Sport Psychol* 2000;31:110-135.

64. Pirl W, Roth A. Diagnosis and treatment of depression in cancer patients. *Oncology* 1999;13:1293-1301.

65. Plant E, Sachs-Ericsson N. Racial and ethnic differences in depression: The roles of social support and meeting basic needs. *J Consult Clin Psych* 2004;72:41-52.

66. Radloff L. The CES-D Scale: A self-report depression scale for research in the general population. *Appl Psychol Meas* 1977;1:385-401.

67. Reid KJ, Baron KG, Lu B, Naylor E, Wolfe L, Zee PC. Aerobic exercise improves self-reported sleep and quality of life in older adults with insomnia. *Sleep Med* 2010;11:934-940.

68. Robinson W, Geske J, Prest L, Barnacle R. Depression treatment in primary care. *J Am Board Fam Pract* 2005;18:79-86.

69. Rosmond R, Bjorntorp P. The hypothalamic-pituitary-adrenal axis activity as a predictor of cardiovascular disease, type 2 diabetes and stroke. *J Intern Med* 2000;247:188-197.

70. Ruo B, Rumsfeld J, Pipkin S, Whooley M. Relation between depressive symptoms and treadmill exercise capacity in the heart and soul study. *Am J Cardiol* 2004;94:96-99.

71. Rush A, Trivendi M, Wisniewski S, Nierenberg A, Stewart J, Warden D, Niederehe G, Thase M, Lavori P, Lebowitz B, McGrath P, Rosenbaum J, Sackeim H, Kupfer D, Luther J, Fava M. Acute and longer-term outcomes in depressed out-patients requiring one or several treatment steps: A STAR*D report. *Am J Psychiatry* 2006;163:1905-1917.

72. Sherwood A, Hinderliter AL, Watkins LL, Waugh RA, Blumenthal JA. Impaired endothelial function in coronary heart disease patients with depressive symptomatology. *J Am Coll Cardiol* 2005;46:656-659.

73. Simon G. Evidence review: Efficacy and effectiveness of antidepressant treatment in primary care. *Gen Hosp Psychiatry* 2002;24:213-224.

74. Singh N, Clements K, Fiatarone M. A randomized controlled trial of the effect of exercise on sleep. *Sleep* 1997;20:95-101.

75. Singh N, Clements K, Singh M. The efficacy of exercise as a long-term antidepressant in elderly subjects: A randomized controlled trial. *J Gerontol* 2001;56A:M497-M504.

76. Singh N, Stavrinos T, Scarbek Y, Galambos G, Liber C, Singh M. A randomized controlled trial of high versus low intensity weight training versus general practitioner care for clinical depression in older adults. *J Gerontol A Biol Sci Med Sci* 2005;60A:768-776.

77. Spitzer R, Kroenke K, Williams J. Validation and utility of a self-report version of PRIME-MD: The PHQ Primary Care Study. *JAMA* 1999;282:1737-1744.

78. Trivedi M, Fava M, Wisniewski S, Thase M, Quitkin F, Warden D, Ritz L, Nierenberg A, Lebowitz B, Biggs M, Luther J, Shores-Wilson K, Rush A, Star*D Study Team. Medication augmentation after the failure of SSRIs for depression. *N Engl J Med* 2006;354:1243-1252.

79. Tworoger S, Yasui Y, Vitiello M, Schwartz R, Ulrich C, Aiello E, Irwin M, Bowen D, Potter J, McTiernan A. Effects of a yearlong moderate-intensity exercise and a stretching intervention on sleep quality in postmenopausal women. *Sleep* 2003;26:830-836.

80. Ustun T, Ayuso-Mateos J, Chatterji S, Mathers C, Murray C. Global burden of depressive disorders in the year 2000. *Br J Psychiatry* 2004;184:386-392.

81. van Melle JP, de Jonge P, Honig A, Schene AH, Kuyper AMG, Crijns HJGM, Schins A, Tulner D, van den Berg MP, Ormel J. Effects of antidepressant treatment following myocardial infarction. *Br J Psychiatry* 2007;190:460-466.

82. Waldman SV, Blumenthal JA, Babyak MA, Sherwood A, Sketch M, Davidson J, Watkins LL. Ethnic differences in the treatment of depression in patients with ischemic heart disease. *Am Heart J* 2009;157:77-83.

83. Wang JT, Hoffman B, Blumenthal JA. Management of depression in patients with coronary heart disease: Association, mechanisms, and treatment implications for depressed cardiac patients. *Expert Opin Pharmacother* 2011;12:85-98.

84. Wang P, Lane M, Olfson M, Pincus H, Wells K, Kessler R. Twelve-month use of mental health services in the United States. *Arch Gen Psychiatry* 2005;62:629-640.

85. Weissman M, Bland R, Canino G, Faravelli C, Greenwald S, Hwu H, Joyce P, Karam E, Lee C, Lellouch J, Lepine J, Newman S, Rubio-Stipec M, Wells J, Wickramaratne P, Wittchen H, Yeh E. Cross-national epidemiology of major depression and bipolar disorder. *JAMA* 1996;276:293-299.

86. Weissman M, Markowitz J, Klerman T. *Comprehensive Guide to Interpersonal Psychotherapy*. New York: Basic Books, 2000.

87. Wells K, Stewart A, Hays R, Burnam M, Rogers W, Daniels M, et al. The functioning and well-being of depressed patients. Results from the Medical Outcomes Study. *JAMA* 1989;262:914-919.

88. Wittchen H, Holsboer F, Jacobi F. Met and unmet needs in the management of depressive disorder in the community and primary care: The size and breadth of the problem. *J Clin Psychiatry* 2001;62:23-28.

89. Youngstedt S. Effects of exercise on sleep. *Clin Sports Med* 2005;24:355-365.

90. Youngstedt S, O'Connor P, Dishman R. The effects of actue exercise on sleep: A quantitative synthesis. *Sleep* 1997;20:203-214.

91. Zigmond A, Snaith R. The hospital anxiety and depression scale. *Acta Psychiatr Scand* 1983;67:361-370.

CHAPTER 32

1. ACSM. *ACSM's Guidelines for Exercise Testing and Prescription*. 9th ed. Baltimore: Lippincott Williams & Wilkins, 2013.

2. Agiovlasitis S, McCubbin JA, Yun J, Mpitsos G, Pavol MJ. Effects of Down syndrome on three-dimensional motion during walking at different speeds. *Gait Posture* 2009;30:345-350.

3. Agiovlasitis S, McCubbin JA, Yun J, Pavol MJ, Widrick JJ. Economy and preferred speed of walking in adults with and without Down syndrome. *Adapt Phys Activ Q* 2009;26:118-130.

4. Alexander D. Prevention of mental retardation: Four decades of research. *Ment Retard Dev Disabil Res Rev* 1998;4:50-58.

5. Auxter D, Pfyfer J, Huettig C. *Principles and Methods of Adapted Physical Education and Recreation* (pp. 428-460). New York: McGraw-Hill, 2001.

6. Baynard T, Pitetti KH, Guerra M, Unnithan VB, Fernhall B. Age-related changes in aerobic capacity in individuals with mental retardation: A 20-yr review. *Med Sci Sports Exerc* 2008;40:1984-1989.

7. Beasley CR. Effects of a jogging program on cardiovascular fitness and work performances of mentally retarded persons. *Am J Ment Defic* 1982;86:609-613.

8. Bundschuh E, Cureton K. Effect of bicycle ergometer conditioning on the physical work capacity of mentally retarded adolescents. *Am Correct Ther J* 1982;36:159-163.

9. Carmeli E, Kessel S, Coleman R, Ayalon M. Effects of a treadmill walking program on muscle strength and balance in elderly people with Down syndrome. *J Gerontol A Biol Sci Med Sci* 2002;57:M106-110.

10. Casey AF, Rasmussen R, Mackenzie SJ, Glenn J. Dual-energy x-ray absorptiometry to measure the influence of a 16-week community-based swim training program on body fat in children and adolescents with intellectual disabilities. *Arch Phys Med Rehabil* 2010;91:1064-1069.

11. Cowley PM, Ploutz-Snyder LL, Baynard T, Heffernan K, Jae SY, Hsu S, Lee M, Pitetti KH, Reiman MP, Fernhall B. Physical fitness predicts functional tasks in individuals with Down syndrome. *Med Sci Sports Exerc* 2010;42:388-393.

12. Cowley PM, Ploutz-Snyder LL, Baynard T, Heffernan KS, Young Jae S, Hsu S, Lee M, Pitetti KH, Reiman MP, Fernhall B. The effect of progressive resistance training on leg strength, aerobic capacity and functional tasks of daily living in persons with Down syndrome. *Disabil Rehabil* 2011;33(23-24):2229-2236.

13. Croce R, Horvat M. Effects of reinforcement based exercise on fitness and work productivity in adults with mental retardation. *Adapt Phys Activ Q* 1992;9:148-178.

14. Croce RV, Pitetti KH, Horvat M, Miller J. Peak torque, average power, and hamstring/quadriceps ratios in non-disabled adults and adults with mental retardation. *Arch Phys Med Rehabil* 1996;77:369-372.

15. Davis W, Sinning WE. Muscle stiffness in Down syndrome and other mentally handicapped subjects: A research note. *J Mot Behav* 1987;19:130-144.

16. Definition of Intellectual Disability. American Association on Intellectual and Developmental Disabilities. www.aaidd.org/content_100.cfm?navID=21[accessed July 25, 2011].

17. Elmahgoub SM, Lambers S, Stegen S, Van Laethem C, Cambier D, Calders P. The influence of combined exercise training on indices of obesity, physical fitness and lipid profile in overweight and obese adolescents with mental retardation. *Eur J Pediatr* 2009;168:1327-1333.

18. Eyman RK, Call TL. Life expectancy of persons with Down syndrome. *Am J Ment Retard* 1991;95:603-612.

19. Fernhall B. Mental retardation. In: *ACSM's Exercise Management for Persons with Chronic Disease and Disability* (pp. 221-226). Champaign, IL: Human Kinetics, 1997.

20. Fernhall B. Mental retardation. In: Durstine J, Moore G, eds. *ACSM's Exercise Management for Persons with Chronic Diseases and Disabilities* (pp. 304-310). Champaign, IL: Human Kinetics, 2003.

21. Fernhall B. Mental retardation. In: LeMura L, von Duvillard S, eds. *Clinical Exercise Physiology: Application and Physiological Principles* (pp. 617-627). Philadelphia: Lippincott Williams & Wilkins, 2004.

22. Fernhall B, Baynard T, Collier SR, Figueroa A, Goulopoulou S, Kamimori GH, Pitetti KH. Catecholamine response to maximal exercise in persons with Down syndrome. *Am J Cardiol* 2009;103:724-726.

23. Fernhall B, McCubbin J, Pitetti K, Rintala P, Rimmer J, Millar AL, de Silva A. Prediction of maximal heart rate in individuals with mental retardation. *Med Sci Sports Exerc* 2001;33:1655-1660.

24. Fernhall B, Otterstetter M. Attenuated responses to sympatho-excitation in individuals with Down syndrome. *J Appl Physiol* 2003;94:2158-2165.

25. Fernhall B, Pitetti K. Leg strength is related to endurance run performance in children and adolescents with mental retardation. *Pediatr Exerc Sci* 2000;12:324-333.

26. Fernhall B, Pitetti K. Limitations to work capacity in individuals with intellectual disabilities. *Clin Exerc Physiol* 2001;3:176-185.

27. Fernhall B, Pitetti K, Millar AL, Hensen T, Vukovich M. Cross validation of the 20 m shuttle run in children with mental retardation. *Adapt Phys Activ Q* 2000;17:402-412.

28. Fernhall B, Pitetti K, Rimmer JH, McCubbin JA, Rintala P, Millar AL, Kittredge J, Burkett LN. Cardiorespiratory capacity of individuals with mental retardation including Down syndrome. *Med Sci Sports Exerc* 1996;28:366-371.

29. Fernhall B, Pitetti K, Stubbs N, Stadler L. Validity and reliability of the 1/2 mile run-walk as an indicator of aerobic fitness in children with mental retardation. *Pediatr Exerc Sci* 1996;8:130-142.

30. Fernhall B, Pitetti KH, Vukovich MD, Stubbs N, Hensen T, Winnick JP, Short FX. Validation of cardiovascular fitness field tests in children with mental retardation. *Am J Ment Retard* 1998;102:602-612.

31. Fernhall B, Szymanski LM, Gorman P, Kamimoir G. Both atenolol and propanolol blunt and fibrinolytic response to exercise but not resting fibrinolytic potential. *Am J Cardiol* 2000;86:1398-1400.

32. Fernhall B, Tymeson G. Graded exercise testing of mentally retarded adults: A study of feasibility. *Arch Phys Med Rehabil* 1987;63:363-365.

33. Fernhall B, Tymeson G. Validation of cardiovascular fitness field tests for persons with mental retardation. *Adapt Phys Activ Q* 1988;5:49-59.

34. Fernhall B, Tymeson G, Millar AL, Burkett LN. Cardiovascular fitness testing and fitness levels of adolescents and adults with mental retardation including Down syndrome. *Educ Train Ment Retard* 1989;68:363-365.

35. Fernhall B, Tymeson G, Webster G. Cardiovascular fitness of mentally retarded adults. *Adapt Phys Activ Q* 1988;5:12-28.

36. Figueroa A, Collier SR, Baynard T, Giannopoulou I, Goulopoulou S, Fernhall B. Impaired vagal modulation of heart rate in individuals with Down syndrome. *Clin Auton Res* 2005;15:45-50.

37. Frey G, McCubbin JA, Hannigan-Downs S, Kasser S, Skaggs S. Physical fitness of trained runners with and without mild mental retardation. *Adapt Phys Activ Q* 1999;16:126-137.

38. Garber CE, Blissmer B, Deschenes MR, Franklin BA, Lamonte MJ, Lee IM, Nieman DC, Swain DP. Quantity and quality of exercise for developing and maintaining cardiorespiratory, musculoskeletal, and neuromotor fitness in apparently healthy adults: Guidance for prescribing exercise. *Med Sci Sports Exerc* 2011;43:1334-1359.

39. Grolla E, Andrighetto G, Parmigiani P, Hladnik U, Ferrari G, Bernardelle R, Dal Lago M, Albarello A, Baschirotto G, Filippi G, Lovato R, Dolcetta D. Specific treatment of Prader-Willi syndrome through cyclical rehabilitation programmes. *Disabil Rehabil* 2011;33(19-20):1837-1847.

40. Guerra M, Cuadrado F, Llorens N, Fernhall B. Comparison of the chronotropic exercise response in nondisabled individuals and individuals with Down syndrome. *Med Sci Sports Exerc* 2000;32:S235.

41. Guerra M, Llorens N, Fernhall B. Chronotropic incompetence in persons with Down syndrome. *Arch Phys Med Rehabil* 2003;84:1604-1608.

42. Guerra M, Roman B, Geronimo C, Cuadrado E, Fernhall B. Physical fitness levels of sedentary and active individuals with Down syndrome. *Adapt Phys Activ Q* 2000;17:310-321.

43. Hayden MF. Mortality among people with mental retardation living in the United States: Research review and policy application. *Ment Retard* 1998;36:345-359.

44. Heller T, McCubbin JA, Drum C, Peterson J. Physical activity and nutrition health promotion interventions: What is working for people with intellectual disabilities? *Intellect Dev Disabil* 2011;49:26-36.

45. Horvat M, Croce R, Pitetti KH, Fernhall B. Comparison of isokinetic peak force and work parameters in youth with and without mental retardation. *Med Sci Sports Exerc* 1999;31:1190-1195.

46. Horvat M, Pitetti KH, Corce R. Isokinetic torque, average power, and flexion/extension ratios in nondisabled adults and adults with mental retardation. *J Orthop Sports Phys Ther* 1997;6:395-399.

47. Hussey C, Maurer JF, Schofield LJ Jr. Physical education training for adult retardates in a sheltered workshop setting. *J Clin Psychol* 1976;32:701-705.

48. Kasch F, Zasueta S. Physical capacities of mentally retarded children. *Acta Peadiatr Scand* 1971;217:217-218.

49. Krebs P. Mental retardation. In: Winnick J. *Adapted Physical Education and Sport* (pp. 153-176). Champaign, IL: Human Kinetics, 1990.

50. Machek MA, Stopka CB, Tillman MD, Sneed SM, Naugle KE. The effects of a supervised resistance-training program on Special Olympics athletes. *J Sport Rehabil* 2008;17:372-379.

51. McCubbin J, Rintala P, Frey G. Correlational study of three cardiorespiratory fitness tests for men with mental retardation. *Adapt Phys Activ Q* 1997;14:43-50.

52. Mendonca GV, Pereira FD. Influence of long-term exercise training on submaximal and peak aerobic capacity and locomotor economy in adult males with Down's syndrome. *Med Sci Monit* 2009;15:CR33-39.

53. Mendonca GV, Pereira FD, Fernhall B. Effects of combined aerobic and resistance exercise training in adults with and without Down syndrome. *Arch Phys Med Rehabil* 2011;92:37-45.

54. Mendonca GV, Pereira FD, Fernhall B. Reduced exercise capacity in persons with Down syndrome: Cause, effect, and management. *Ther Clin Risk Manag* 2010;6:601-610.

55. Mendonca GV, Pereira FD, Morato PP, Fernhall B. Walking economy of adults with Down syndrome. *Int J Sports Med* 2010;31:10-15.

56. Millar AL, Fernhall B, Burkett LN, Tymeson G. Effect of aerobic training in adolescents with Down syndrome. *Med Sci Sports Exerc* 1993;25:260-264.

57. Miller DT, Adam MP, Aradhya S, Biesecker LG, Brothman AR, Carter NP, Church DM, Crolla JA, Eichler EE, Epstein CJ, Faucett WA, Feuk L, Friedman JM, Hamosh A, Jackson L, Kaminsky EB, Kok K, Krantz ID, Kuhn RM, Lee C, Ostell JM, Rosenberg C, Scherer SW, Spinner NB, Stavropoulos DJ, Tepperberg JH, Thorland EC, Vermeesch JR, Waggoner DJ, Watson MS, Martin

CL, Ledbetter DH. Consensus statement: Chromosomal microarray is a first-tier clinical diagnostic test for individuals with developmental disabilities or congenital anomalies. *Am J Hum Genet* 2010;86:749-764.

58. Montgomery DL, Reid G, Seidl C. The effects of two physical fitness programs designed for mentally retarded adults. *Can J Sport Sci* 1988;13:73-78.
59. Murdoch J, Rodger J, Rao S, Fletcher C, Dunnigan M. Down's syndrome: An atheroma-free model? *BMJ* 1977;2:226-228.
60. Ozmen T, Ryildirim NU, Yuktasir B, Beets MW. Effects of school-based cardiovascular-fitness training in children with mental retardation. *Pediatr Exerc Sci* 2007;19:171-178.
61. Peran S, Gil JL, Ruiz F, Fernandez-Pastor V. Development of physical response after athletics training in adolescents with Down's syndrome. *Scand J Med Sci Sports* 1997;7:283-288.
62. Perkins EA, Moran JA. Aging adults with intellectual disabilities. *JAMA* 2010;304:91-92.
63. Pitetti KH, Boneh S. Cardiovascular fitness as related to leg strength in adults with mental retardation. *Med Sci Sports Exerc* 1995;27:423-428.
64. Pitetti KH, Campbell KD. Mentally retarded individuals—a population at risk? *Med Sci Sports Exerc* 1991;23:586-593.
65. Pitetti KH, Climstein M, Mays MJ, Barret PJ. Isokinetic arm and leg strength of adults with Down syndrome: A comparative study. *Arch Phys Med Rehabil* 1992;73:847-850.
66. Pitetti K, Fernhall B. Aerobic capacity as related to leg strength in youths with mental retardation. *Pediatr Exerc Sci* 1997;9:223-236.
67. Pitetti KH, Rimmer JH, Fernhall B. Physical fitness and adults with mental retardation: An overview of current research and future directions. *Sports Med* 1993;16:23-56.
68. Pitetti K, Tan DM. Cardiorespiratory responses of mentally retarded adults to air brake ergometry and treadmill exercise. *Arch Phys Med Rehabil* 1990;17:318-321.
69. Pitetti KH, Tan DM. Effects of a minimally supervised exercise program for mentally retarded adults. *Med Sci Sports Exerc* 1991;23:594-601.
70. Pommering TL, Brose JA, Randolph E, Murray TF, Purdy RW, Cadamagnani PE, Foglesong JE. Effects of an aerobic exercise program on community-based adults with mental retardation. *Ment Retard* 1994;32:218-226.
71. Rimmer JH, Heller T, Wang E, Valerio I. Improvements in physical fitness in adults with Down syndrome. *Am J Ment Retard* 2004;109:165-174.
72. Rimmer J, Kelly LE. Effects of a resistance training program on adults with mental retardation. *Adapt Phys Activ Q* 1991;8:146-153.
73. Rimmer JH, Yamaki K, Lowry BM, Wang E, Vogel LC. Obesity and obesity-related secondary conditions in adolescents with intellectual/developmental disabilities. *J Intellect Disabil Res* 2010;54:787-794.
74. Roizen NJ, Patterson D. Down's syndrome. *Lancet* 2003;361:1281-1289.
75. Schurrer R, Weltman A, Brammel H. Effects of physical training on cardiovascular fitness and behavior patterns of mentally retarded adults. *Am J Ment Defic* 1985;90:167-170.
76. Shields N, Taylor NF, Dodd KJ. Effects of a community-based progressive resistance training program on muscle performance and physical function in adults with Down syndrome: A randomized controlled trial. *Arch Phys Med Rehabil* 2008;89:1215-1220.
77. Causes and Prevention of Intellectual Disabilities, The ARC of United States. www.thearc.org/page.aspx?pid=2453 [accessed July 25, 2011].
78. Suomi R. Self directed strength training: Its effect on leg strength in men with mental retardation. *Arch Phys Med Rehabil* 1998;79:323-328.
79. Suomi R, Surburg PR, Lecius P. Effects of hydraulic resistance strength training on isokinetic measures of leg strength in men with mental retardation. *Adapt Phys Activ Q* 1995;12:377-387.
80. Tamse TR, Tillman MD, Stopka CB, Weimer AC, Abrams GL, Issa IM. Supervised moderate intensity resistance exercise training improves strength in Special Olympic athletes. *J Strength Cond Res* 2010;24:695-700.
81. Teo-Koh SM, McCubbin JA. Relationship between peak VO_2 and 1-mile walk test performance of adolescents with mental retardation. *Pediatr Exerc Sci* 1999;11:144-157.
82. Tomporowski PD, Ellis NR. Effects of exercise on the physical fitness, intelligence, and adaptive behavior of institutionalized mentally retarded adults. *Appl Res Ment Retard* 1985;5:329-337.
83. Tomporowski PD, Jameson LD. Effects of a physical training program on the exercise behavior of institutionalized mentally retarded adults. *Adapt Phys Activ Q* 1985;2:197-205.
84. Tsimaras VK, Fotiadou EG. Effect of training on the muscle strength and dynamic balance ability of adults with down syndrome. *J Strength Cond Res* 2004;18:343-347.
85. Tsimaras V, Giagazoglou P, Fotiadou E, Christoulas K, Angelopoulou N. Jog-walk training in cardiorespiratory fitness of adults with Down syndrome. *Percept Mot Skills* 2003;96:1239-1251.
86. Varela AM, Sardinha L, Pitetti KH. Effects of aerobic rowing training in adults with Down syndrome. *Am J Ment Retard.* 2001;106:135-44.
87. Weber R, French R. Down's syndrome adolescents and strength training. *Clin Kinesiol* 1988;42:13-21.
88. Yla-Herttuala S, Luoma J, Nikkari T, Kivimaki T. Down's syndrome and atherosclerosis. *Atherosclerosis* 1989;76:269-272.

Index

Note: Page numbers followed by an italicized *f* or *t* refer to the figure or table on that page, respectively.

B

G

H

T

About the Editors

Jonathan K. Ehrman, PhD, FACSM, is the associate program director of preventive cardiology and director of the clinical weight management program at Henry Ford Hospital in Detroit. He has a 28-year background in clinical exercise physiology and is certified as an ACSM clinical exercise specialist and program director. He previously served as the chair of the exercise specialist credentialing committee for ACSM.

Dr. Ehrman is author of more than 100 manuscripts and abstracts as well as four books and chapters. He was senior editor of the sixth edition of *ACSM's Resource Manual for Guidelines for Exercise Testing and Prescription.* He is also the umbrella editor for the ACSM certification texts published in 2013. He is a fellow of the American College of Sports Medicine as well as a member of the American Association of Cardiovascular and Pulmonary Rehabilitation and the American Heart Association. Dr. Ehrman earned his PhD in clinical exercise physiology from The Ohio State University.

Paul M. Gordon, PhD, MPH, FACSM, is an ACSM clinical exercise specialist and professor and chair in the Department of Health, Human Performance, and Recreation at Baylor University. He has 15 years of teaching experience in clinical exercise physiology curricula and has directed several cardiopulmonary rehabilitation programs. Dr. Gordon has served as an examiner and coordinator for the ACSM exercise specialist certification and is a contributing author for the *ACSM's Guidelines for Exercise Testing and Prescription.*

Dr. Gordon is a fellow of the American College of Sports Medicine, a fellow of the Centers for Disease Control Physical Activity Research Program, and a member of the National Institutes of Health Study Section. He earned his PhD in exercise physiology and an MPH in epidemiology from the University of Pittsburgh.

Paul S. Visich, PhD, MPH, has more than 20 years of experience in clinical exercise physiology and is the chair of the Exercise and Sport Performance Department within the Westbrook College of Health Professions at the University of New England. He worked 12 years in a clinical setting that included cardiac and pulmonary rehabilitation and primary disease prevention. His research interests involve the assessment of cardiovascular disease risk factors in children, the influence of resistance training in elderly populations, and altitude physiology.

Dr. Visich was a member of the Registered Clinical Exercise Physiology Committee, previous chair for the Professional Education Committee, and author for Certified News for the American College of Sports Medicine. He earned a PhD in exercise physiology and an MPH in epidemiology from the University of Pittsburgh.

Steven J. Keteyian, PhD, FACSM, has more than 30 years of experience working as a clinical exercise physiologist. He is program director of preventive cardiology at the Henry Ford Hospital in Detroit. Over the course of his career, Dr. Keteyian has focused on exercise and physical activity in both healthy individuals and those with chronic diseases. He is the author of more than 100 scientific articles and chapters in books as well as four textbooks.

Dr. Keteyian is a fellow of the American Association of Cardiovascular and Pulmonary Rehabilitation and a member of the American Heart Association. He is a fellow of the American College of Sports Medicine. He earned his PhD from Wayne State University in Detroit.